Intermediate Algebra

Eighth Edition

Marvin L. Bittinger

Indiana University—Purdue University
at Indianapolis

Addison-
Wesley

Boston   San Francisco   New York
London   Toronto   Sydney   Tokyo   Singapore   Madrid
Mexico City   Munich   Paris   Capetown   Hong Kong   Montreal

Publisher	Jason A. Jordan
Assistant Editor	Susan Estey
Managing Editor	Ron Hampton
Production Supervisor	Kathleen A. Manley
Production Coordinator	Jane DePasquale
Design Direction	Susan Carsten
Text Designer	Rebecca Lloyd Lemna
Editorial and Production Services	Martha Morong/Quadrata, Inc.
Art Editor	Janet Theurer
Photo Researcher	Naomi Kornhauser
Marketing Managers	Craig Bleyer and Laura Rogers
Illustrators	Scientific Illustrators, Gayle Hayes, and Maria Sas
Compositor	The Beacon Group
Cover Designer	Jeannet Leendertse
Cover Photographs	© PhotoDisc, 1998 (background); © Comstock, Inc., 1998
Senior Prepress Buyer	Caroline Fell
Manufacturing Supervisor	Ralph Mattivello

PHOTO CREDITS

1, 38, © Hopfer/Stock Imagery **60,** Westlight **69,** Reuters/Wolfgang Rattay/Archive Photos **83,** Tom McCarthy/PhotoEdit **88,** Michael Newman/PhotoEdit **90,** Charles Krebs/Tony Stone Images **101,** AP/Wide World Photos **104,** Archive Photos **113,** Reuters/Wolfgang Rattay/Archive Photos **131,** Keith Wood/Tony Stone Images **147,** Archive Photos **162,** Lee Snider/The Image Works **171,** Archive Photos **227,** Robert Brenner/PhotoEdit **244,** © 1999 Michael Howell/Envision **259,** Robert Brenner/PhotoEdit **272,** Michael Newman/PhotoEdit **299,** Reuters/Corbis-Bettmann **304,** Zigy Kaluzny/Tony Stone Images **360,** Reuters/Corbis-Bettmann **373,** Bill Gallery/Stock Boston **419,** Spencer Grant/PhotoEdit **423,** Brian Spurlock **424, 428 (left),** AP/Wide World Photos **428 (right),** Reuters/Scott Olson/Archive Photos **440,** Bill Gallery/Stock Boston **451,** Everett Johnson/Tony Stone Images **459,** © 1999 D & I MacDonald/Envision **504,** Everett Johnson/Tony Stone Images **521,** Bruce Hands/The Image Works **532,** Reuters/Ira Strikstein/Archive **558,** Andrea Mohin/NYT Pictures **581,** Bruce Hands/The Image Works **603, 615,** Yva Momatiuk & John Eastcott/The Image Works **616,** AP/Wide World Photos **626,** Akos Szilvasi/Stock Boston **661,** © 1999 Bill Ellzey **667,** Mark E. Gibson/Mira **671,** AP/Wide World Photos **674 (left),** From *Classic Baseball Cards,* by Bert Randolph Sugar, copyright © 1977 by Dover Publishing, Inc. **674 (right),** AP/Wide World Photos **681,** Tony Arruza/The Image Works **683,** PhotoEdit **717,** Tony Arruza/The Image Works **725,** David Young-Wolff/PhotoEdit

LIBRARY OF CONGRESS CATALOGING-IN-PUBLICATION DATA

Bittinger, Marvin L.
    Intermediate algebra—8th ed./Marvin L. Bittinger.
       p. cm.
    Includes index.
    ISBN 0-201-95960-7
    1. Algebra. I. Title.
QA154.2.B54   1999
512.9—dc21   98-49224
                                     CIP

4 5 6 7 8 9 10—VH—0100

Contents

Contents

Final Examination   727

Appendixes

Preface

This text is in a series of texts that includes the following:

Bittinger: *Basic Mathematics,* Eighth Edition

Bittinger: *Fundamental Mathematics,* Second Edition

Bittinger: *Introductory Algebra,* Eighth Edition

Bittinger: *Intermediate Algebra,* Eighth Edition

Bittinger: *Intermediate Algebra,* Eighth Edition (Alternate Version)

Bittinger/Beecher: *Introductory and Intermediate Algebra: A Combined Approach*

Intermediate Algebra, Eighth Edition, is a significant revision of the Seventh Edition, particularly with respect to design, art program, pedagogy, features, and supplements package. Its unique approach, which has been developed and refined over eight editions, continues to blend the following elements in order to bring students success:

- **Writing style.** The author writes in a clear easy-to-read style that helps students progress from concepts through examples and margin exercises to section exercises.

- **Problem-solving approach.** The basis for solving problems and real-data applications is a five-step process (*Familiarize, Translate, Solve, Check,* and *State*) introduced early in the text and used consistently throughout. This problem-solving approach provides students with a consistent framework for solving applications. (See pages 91, 97, 255, and 360.)

- **Real data.** Real-data applications aid in motivating students by connecting the mathematics to their everyday lives. Extensive research was conducted to find new applications that relate mathematics to the real world.

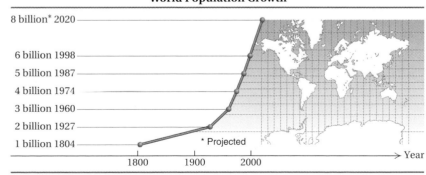

World Population Growth

8 billion* 2020
6 billion 1998
5 billion 1987
4 billion 1974
3 billion 1960
2 billion 1927
1 billion 1804
* Projected
1800 1900 2000 → Year

Source: U.S. Bureau of the Census

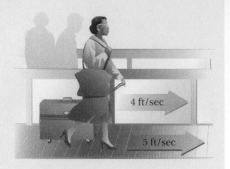

4 ft/sec

5 ft/sec

- **Art program.** The art program has been expanded to improve the visualization of mathematical concepts and to enhance the real-data applications.

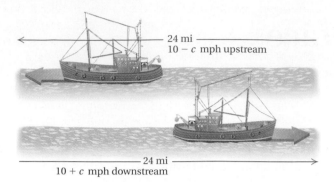

24 mi
$10 - c$ mph upstream

24 mi
$10 + c$ mph downstream

- **Reviewer feedback.** The author solicits feedback from reviewers and students to help fulfill student and instructor needs.

- **Accuracy.** The manuscript is subjected to an extensive accuracy-checking process to eliminate errors.

- **Supplements package.** All ancillary materials are closely tied with the text and created by members of the author team to provide a complete and consistent package for both students and instructors.

What's New in the Eighth Edition?

The style, format, and approach of the Seventh Edition have been strengthened in this new edition in a number of ways.

Updated Applications Extensive research has been done to make the applications in the Eighth Edition even more up to date and realistic. A large number of the applications are new to this edition, and many are drawn from the fields of business and economics, life and physical sciences, social sciences, and areas of general interest such as sports and daily life. To encourage students to understand the relevance of mathematics, many applications are enhanced by graphs and drawings similar to those found in today's newspapers and magazines. Many applications are also titled for quick and easy reference, and most real-data applications are credited with a source line. (See pages 209, 216, 310, 580, and 666.)

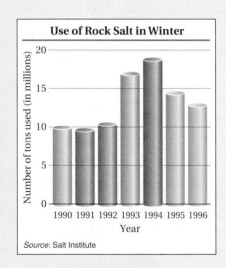

Use of Rock Salt in Winter

Number of tons used (in millions)

1990 1991 1992 1993 1994 1995 1996
Year

Source: Salt Institute

Improving Your Math Study Skills Occurring at least once in every chapter, and referenced in the table of contents, these mini-lessons provide students with concrete techniques to improve studying and test-taking. These features can be covered in their entirety at the beginning of the course, encouraging good study habits early on, or they can be used as they occur in the text, allowing students to learn them gradually. These features can also be used in conjunction with Marvin L. Bittinger's "Math Study Skills" Videotape, which is free to adopters. Please contact your Addison Wesley Longman sales consultant for details on how to obtain this videotape. (See pages 10, 128, 416, and 670.)

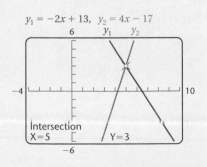

$y_1 = -2x + 13, \ y_2 = 4x - 17$

Intersection
X=5   Y=3

Calculator Spotlights Designed specifically for the beginning algebra student, these optional features include graphing-calculator instruction and practice exercises (see pages 182, 234, 305, and 458). Answers to all Calculator Spotlight exercises appear at the back of the text.

New Art and Design To enhance the greater emphasis on real data and applications, we have extensively increased the number of pieces of technical and situational art (see pages 101, 149, 218, and 605). Students gain a visual understanding of the algebra when they see a graphical interpretation of the algebraic solution. For this reason, numerous graphs have been added in the examples and exercises. The use of color has been carried out in a methodical and precise manner so that its use carries a consistent meaning, which enhances the readability of the text. For example, the use of both red and blue in mathematical art increases understanding of the concepts. When two lines are graphed using the same set of axes, one is usually red and the other blue. Note that equation labels are the same color as the corresponding line to aid in understanding. Answer lines have also been deleted from all section exercise sets to allow room for more exercises and additional art to better illustrate the exercises.

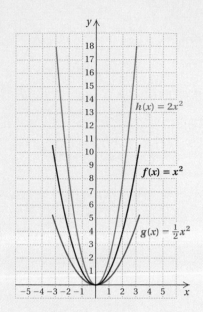

World Wide Web Integration The World Wide Web is a powerful resource available to more and more people every day. In an effort to get students more involved in using this source of information, we have added a World Wide Web address (www.mathmax.com) to every chapter opener (see pages 147, 299, and 603). Students can go to this page on the World Wide Web to further explore the subject matter of the chapter-opening application. Selected exercise sets, marked on the first page of the exercise set with an icon (see pages 199, 393, and 541), have additional practice-problem worksheets that can be downloaded from this site. Additional, more extensive, Summary and Review pages for each chapter, as well as other supplementary material, can also be downloaded for instructor and student use.

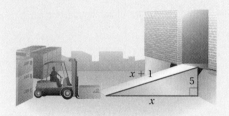

Algebraic–Graphical Connections To give students a better visual understanding of algebra, we have included algebraic–graphical connections in the Eighth Edition (see pages 415, 491, and 523). This feature gives the algebra more meaning by connecting the algebra to a graphical interpretation.

Collaborative Learning Features An icon located at the end of an exercise set signals the existence of a Collaborative Learning Activity correlating to that section in Irene Doo's *Collaborative Learning Activities Manual* (see pages 106, 202, 312, and 462). Please contact your Addison Wesley Longman sales consultant for details on ordering this supplement.

Exercises The deletion of answer lines in the exercise sets has allowed us to include more exercises in the Eighth Edition. Exercises are paired, meaning that each even-numbered exercise is very much like the odd-numbered one that precedes it. This gives the instructor several options: If an instructor wants the student to have answers available, the odd-numbered exercises are assigned; if an instructor wants the student to practice (perhaps for a test), with no answers available, then the even-numbered exercises are assigned. In this way, each exercise set actually serves as two exercise sets. Answers to all odd-numbered exercises, with the exception of the Thinking and Writing exercises, and *all* Skill Maintenance exercises are provided at the back of the text. If an instructor wants the student to have access to all the answers, a complete answer book is available.

Skill Maintenance Exercises The Skill Maintenance exercises have been enhanced by the inclusion of 60% more exercises in this edition (see pages 82, 332, and 444. These exercises focus on the four Objectives for

Retesting listed at the beginning of each chapter, but they also review concepts from other sections of the text in order to prepare students for the Final Examination. Section and objective codes appear next to each Skill Maintenance exercise for easy reference. Answers to all Skill Maintenance exercises appear at the back of the book.

Synthesis Exercises These exercises now appear in every exercise set, Summary and Review, Chapter Test, and Cumulative Review. Synthesis exercises help build critical thinking skills by requiring students to synthesize or combine learning objectives from the section being studied as well as preceding sections in the book. In addition, the Critical Thinking feature from the Seventh Edition has been incorporated into the Synthesis exercises for the Eighth Edition (see pages 164, 286, and 504).

Thinking and Writing Exercises ◈ Two Thinking and Writing exercises (denoted by the maze icon) have been added to the Synthesis section of every exercise set and Summary and Review. Designed to develop comprehension of critical concepts, these exercises encourage students to both think and write about key mathematical ideas in the chapter (see pages 38, 476, and 576).

Content We have made the following improvements to the content of *Intermediate Algebra.*

Through thorough research and an extensive review process, *Intermediate Algebra*, Eighth Edition, has experienced major changes in the development and reorganization of the table of contents. The following list includes only twelve of those changes. For a complete conversion guide, contact your Addison Wesley Longman sales representative.

- Review topics, formerly located in Chapter 1, have been streamlined and reorganized into a two-part review in Chapter R: real-number computation review in Sections R.1–R.3 and algebraic manipulation review in Sections R.4–R.7. This provides the instructor with flexible options when choosing the amount of review to include in this course.

- Section 1.4 ("Sets, Interval Notation, and Inequalities") has been expanded with the introduction of interval notation. Where appropriate, solution sets are now given in both set-builder notation and interval notation. Interval notation enhances the preparation for later courses.

- To provide a solid background for college algebra, functions are introduced early in the Eighth Edition (Section 2.2, "Functions and Graphs," and Section 2.3, "Finding Domain and Range") and are then integrated throughout. This early function approach allows for many uses of functions such as expansion of graphing applications in the polynomial, rational, radical, quadratic, exponential, and logarithmic topics.

- Modeling data with functions is now presented in three new sections: Section 2.7 ("Mathematical Modeling with Linear Functions"), Section 7.7 ("Mathematical Modeling with Quadratic Functions"), and Section 8.7 ("Mathematical Modeling with Exponential and Logarithmic Functions"). The study of applications becomes more interactive when students learn to model real data with functions and make predictions from graphs.

- Section 3.7 ("Business and Economics Applications") has been added to broaden the base of function applications. This section provides the background for business students who plan to take Business Calculus and Finite Mathematics.

- Objective 4.2e is new to Section 4.2 ("Multiplication of Polynomials"). Here students prepare for later courses by simplifying notation like $f(a + h)$ and $f(a + h) - f(a)$.

- Section 5.4 ("Complex Rational Expressions"), formerly located near the end of the chapter on rational expressions, is now placed earlier in this chapter, separating the rational expressions (Sections 5.1–5.4) and the rational equations (Sections 5.5–5.8).

- The early placement of rational numbers as exponents in Section 6.2 ("Rational Numbers as Exponents") allows for greater inclusion of rational exponents throughout the chapter. This especially smoothes the explanation of simplifying and calculating with rational exponents in Sections 6.3–6.5.

- The coverage of quadratic applications has been expanded with Section 7.3 ("Applications Involving Quadratic Equations"), which is now placed immediately after the introduction of the quadratic formula, and with Section 7.7 ("Mathematical Modeling with Quadratic Functions").

- Chapter 8 (*Exponential and Logarithmic Functions*) immediately follows the chapter on quadratics. Early presentation of function topics has allowed a smoother transition between types of functions throughout the text. Section 8.2 ("Inverse and Composite Functions") is now included immediately before the section in which it is needed (Section 8.3, "Logarithmic Functions").

- Placing the chapter on conic sections at the end of the text provides flexibility for those who consider this material optional at the intermediate level. The section on ellipses and hyperbolas in the Seventh Edition is now expanded into two sections: Section 9.2 ("Conic Sections: Ellipses") and Section 9.3 ("Conic Sections: Hyperbolas").

- Four new appendixes (*Handling Dimension Symbols, Determinants and Cramer's Rule, Elimination Using Matrices,* and *The Algebra of Functions*) have been added to the Eighth Edition. As reviewers have requested, these topics are now placed in a more optional format.

Learning Aids

Interactive Worktext Approach   The pedagogy of this text is designed to provide an interactive learning experience between the student and the exposition, annotated examples, art, margin exercises, and exercise sets. This approach provides students with a clear set of learning objectives, involves them with the development of the material, and provides immediate and continual reinforcement and assessment.

Section objectives are keyed by letter not only to section subheadings, but also to exercises in the exercise sets and Summary and Review, as well as answers to the Pretest, Chapter Test, and Cumulative Review questions. This enables students to easily find appropriate review material if they are unable to work a particular exercise.

Throughout the text, students are directed to numerous *margin exercises,* which provide immediate reinforcement of the concepts covered in each section.

Review Material   The Eighth Edition of *Intermediate Algebra* continues to provide many opportunities for students to prepare for final assessment.

Now in a two-column format, the *Summary and Review* appears at the end of each chapter and provides an extensive set of review exercises. Reference codes beside each exercise or direction line preceding it allow the student to easily return to the objective being reviewed (see pages 143, 515, and 675).

Also included at the end of every chapter but Chapters R and 1 is a *Cumulative Review*, which reviews material from all preceding chapters. At the back of the text are answers to all Cumulative Review exercises, together with section and objective references, so that students know exactly what material to study if they miss a review exercise (see pages 297, 371, and 519).

Objectives for Retesting are covered in each Summary and Review and Chapter Test, and are also included in the Skill Maintenance exercises and in the Printed Test Bank (see pages 228, 300, and 374).

For Extra Help Many valuable study aids accompany this text. Below the list of objectives found at the beginning of each section are references to appropriate videotape, audiotape, tutorial software, and CD-ROM programs to make it easy for the student to find the correct support materials.

Testing The following assessment opportunities exist in the text.

The *Diagnostic Pretest,* provided at the beginning of the text, can place students in the appropriate chapter for their skill level by identifying familiar material and specific trouble areas (see page xxi).

Chapter Pretests can then be used to place students in a specific section of the chapter, allowing them to concentrate on topics with which they have particular difficulty (see pages 70, 148, and 452).

Chapter Tests allow students to review and test comprehension of chapter skills, as well as the four Objectives for Retesting from earlier chapters (see pages 221, 369, and 599).

Answers to all Diagnostic Pretest, Chapter Pretest, and Chapter Test questions are found at the back of the book, along with appropriate section and objective references.

Supplements for the Instructor

Annotated Instructor's Edition
0-201-33876-9

The *Annotated Instructor's Edition* is a specially bound version of the student text with answers to all margin exercises, exercise sets, and chapter tests printed in a special color near the corresponding exercises.

Instructor's Solutions Manual
0-201-43845-3

The *Instructor's Solutions Manual* by Judith A. Penna contains brief worked-out solutions to all even-numbered exercises in the exercise sets and answers to all Thinking and Writing exercises.

Printed Test Bank/Instructor's Resource Guide
by Donna DeSpain
0-201-43846-1

The test-bank section of this supplement contains the following:

- Three alternate test forms for each chapter, with questions in the same topic order as the objectives presented in the chapter

- Five alternate test forms for each chapter, modeled after the Chapter Tests in the text
- Three alternate test forms for each chapter, designed for a 50-minute class period
- Two multiple-choice versions of each Chapter Test
- Two cumulative review tests for each chapter, with the exception of Chapters R and 1
- Eight final examinations: three with questions organized by chapter, three with questions scrambled as in the Cumulative Reviews, and two with multiple-choice questions
- Answers for the Chapter Tests and Final Examination

The resource-guide section contains the following:

- A conversion guide from the Seventh Edition to the Eighth Edition
- Extra practice exercises (with answers) for 40 of the most difficult topics in the text
- Critical Thinking exercises and answers
- Black-line masters of grids and number lines for transparency masters or test preparation
- Indexes to the videotapes and audiotapes that accompany the text
- Three-column chapter Summary and Review listing objectives, brief procedures, worked-out examples, multiple-choice problems similar to the example, and the answers to those problems
- Instructor support material for the CD-ROM

Collaborative Learning Activities Manual
0-201-35993-6

The *Collaborative Learning Activities Manual,* written by Irene Doo of Austin Community College, features group activities that are tied to sections of the text via an icon . Instructions for classroom setup are also included in the manual.

Answer Book
0-201-43847-X

The *Answer Book* contains answers to all exercises in the exercise sets in the text. Instructors can make quick reference to all answers or have quantities of these booklets made available for sale if they want students to have access to all the answers.

TestGen-EQ
0-201-38161-3 (Windows), 0-201-38168-0 (Macintosh)

This test generation software is available in Windows and Macintosh versions. TestGen-EQ's friendly graphical interface enables instructors to easily view, edit, and add questions, transfer questions to tests, and print tests in a variety of fonts and forms. Search and sort features help the instructor quickly locate questions and arrange them in a preferred order. Six question formats are available, including short-answer, true–false, multiple-choice, essay, matching, and bimodal formats. A built-in question editor gives the instructor the ability to create graphs, import graphics, insert mathematical symbols and templates, and insert variable numbers or text. Computerized testbanks include algorithmically defined problems organized according to each textbook. An "Export to HTML" feature lets instructors create practice tests for the World Wide Web.

QuizMaster-EQ
0-201-38161-3 (Windows), 0-201-38168-0 (Macintosh)

QuizMaster-EQ enables instructors to create and save tests and quizzes using TestGen-EQ so students can take them on a computer network. Instructors can set preferences for how and when tests are administered. QuizMaster-EQ automatically grades the exams and allows the instructor to view or print a variety of reports for individual students, classes, or courses. This software is available for both Windows and Macintosh and is fully networkable.

Supplements for the Student

Student's Solutions Manual
0-201-34025-9

The *Student's Solutions Manual* by Judith A. Penna contains fully worked-out solutions with step-by-step annotations for all the odd-numbered exercises in the exercise sets in the text, with the exception of the Thinking and Writing exercises. It may be purchased by your students from Addison Wesley Longman.

"Steps to Success" Videotapes
0-201-30361-2

Steps to Success is a complete revision of the existing series of videotapes, based on extensive input from both students and instructors. These videotapes feature an engaging team of mathematics teachers who present comprehensive coverage of each section of the text in a student-interactive format. The lecturers' presentations include examples and problems from the text and support an approach that emphasizes visualization and problem solving. A video icon [icon] at the beginning of each section references the appropriate videotape number.

"Math Study Skills for Students" Videotape
0-201-84521-0

Designed to help students make better use of their math study time, this videotape help students improve retention of concepts and procedures taught in classes from basic mathematics through intermediate algebra. Through carefully-crafted graphics and comprehensive on-camera explanation, Marvin L. Bittinger helps viewers focus on study skills that are commonly overlooked.

Audiotapes
0-201-43406-7

The audiotapes are designed to lead students through the material in each text section. Bill Saler explains solution steps to examples, cautions students about common errors, and instructs them at certain points to stop the tape and do exercises in the margin. He then reviews the margin-exercise solutions, pointing out potential errors. An audiotape icon ⌒ at the beginning of each section references the appropriate audiotape number.

InterAct Math Tutorial Software
0-201-38105-2 (Windows), 0-201-38112-5 (Macintosh)

InterAct Math Tutorial Software has been developed and designed by professional software engineers working closely with a team of experienced developmental-math teachers. This software includes exercises that are linked one-to-one with the odd-numbered exercises in the text and require the same computational and problem-solving skills as their companion exercises in the text. Each exercise has an example and an interactive guided solution that are designed to involve students in the solution process and to help them identify precisely where they are having trouble. In addition, the software recognizes common student errors and provides students with appropriate customized feedback. With its sophisticated answer recognition capabilities, *InterAct Math Tutorial Software* recognizes equivalent forms of the same answer for any kind of input. It also tracks for each section student activity and scores that can then be printed out. A disk icon ▦ at the beginning of each section identifies section coverage. *InterAct Math Tutorial Software* is available for Windows and Macintosh computers.

World Wide Web Supplement (www.mathmax.com)

This on-line supplement provides additional practice and learning resources for the student of intermediate algebra. For each book chapter, students can find additional practice exercises, Web links for further exploration, and expanded Summary and Review pages that review and reinforce the concepts and skills learned throughout the chapter. In addition, students can download a plug-in for Addison Wesley Longman's *InterAct Math Tutorial Software* that allows them to access additional tutorial problems directly through their Web browser. Students and instructors can also learn about the other supplements available for the MathMax series via sample audio clips and complete descriptions of other services provided by Addison Wesley Longman.

MathMax Multimedia CD-ROM
for Intermediate Algebra
0-201-39736-6 (Windows and Macintosh)

The Intermediate Algebra CD provides an active environment using graphics, animations, and audio narration to build on some of the unique and proven features of the MathMax series. Highlighting key concepts from the book, the content of the CD is tightly and consistently integrated with the *Intermediate Algebra* text and retains references to the *Intermediate Algebra* numbering scheme so that students can move smoothly between the CD and other *Intermediate Algebra* supplements. The CD includes Addison Wesley Longman's *InterAct Math Tutorial Software* so that students can practice additional tutorial problems. An interactive Summary and Review section allows students to review and practice what they have learned in each chapter; and multimedia presentations reiterate important study skills described throughout the book. A CD-ROM icon ◐ at the beginning of each section indicates section coverage. The Intermediate Algebra CD is available for both Windows and Macintosh computers. Contact your Addison Wesley Longman sales consultant for a demonstration.

Your author and his team have committed themselves to publishing an accessible, clear, accomplishable, error-free book and supplements package that will provide the student with a successful learning experience and will foster appreciation and enjoyment of mathematics. As part of our continual effort to accomplish this goal, we welcome your comments and suggestions at the following email address:

Marv Bittinger
exponent@aol.com

Acknowledgments

Many of you have helped to shape the Eighth Edition by reviewing, participating in telephone surveys and focus groups, filling out questionnaires, and spending time with us on your campuses. Our deepest appreciation to all of you and in particular to the following:

Jamie Ashby, *Texarkana College*
James Baglio, *Normandale Community College*
Sharon Balk, *Northeast Iowa Community College*
Shirley Beil, *Normandale Community College*
Boyd Benson, *Rio Hondo College*
Ben Cornelius, *Oregon Institute of Technology*
Cynthia Coulter, *Catawba Valley Community College*
Sherry Crabtree, *Northwest Shoals Community College*
Patrick Cross, *University of Oklahoma–Norman*
William Culbertson, *Northwest State Community College*
Laura Davis, *Garland City Community College*
Andres Delgado, *Orange County Community College*
Mimi Elwell, *Lake Michigan College*
Mike Flodin, *Tacoma Community College*
Timothy Foster, *Garrett Community College*
John Fourlis, *Harrisburg Area Community College*
Ormsin Gardner, *Winona State University*
Stephanie Gardner, *College of Eastern Utah*
Dennis Houk, *West Shore Community College*
Cheryl Ingrams, *Crowder College*
Charles Jordan, *Lawson State University*
Joseph Jordan, *John Tyler Community College*
Judy Kasabian, *El Camino College*
Joann Kelly, *Palm Beach Community College*
Joann Kendall, *College of the Mainland*
Linda Long, *Ricks College*
Sang Lee, *DeVry Institute of Technology*
Fred Lemmerhirt, *Waubonsee Community College*
Jann McInnis, *Florida Community College–Jacksonville*
Rudy Maglio, *Oakton Community College*
Greg Millican, *Northeast Alabama State Jr. College*
Steven Mondy, *Normandale Community College*
Michael Montano, *Riverside Community College City Campus*
Mika Moteki, *Cardinal Stritch University*
Frank Mulvaney, *Delaware County Community College*
Ellen O'Connell, *Triton College*
Mark Omadt, *Anoka-Ramsey Community College*
Helen Paszek, *Nash Community College*

Barbara Phillips, *Roane State Community College*
Marilyn Platt, *Gaston College*
John Pleasants, *Orange County Community College*
Margaret Rauch, *Texarkana College*
Robert Sampolski, *Oakton Community College*
Bud Sather, *Flathead Valley Community College*
Carole Shapero, *Oakton Community College*
Mimi Shuler, *Gulfcoast Community College*
Sam Sochis, *Barstow College*
Roman Sznajder, *Bowie State University*
Sharon Testone, Ph.D., *Onondaga Community College*
Rich Vanamerongen, *Portland Community College*
Sharon Walker, *Catawba Valley Community College*
Joyce Wellington, *Southeastern Community College*
Jaqueline Wing, *Angelina College*
Gerad Zimmerman, *Tacoma Community College*

We also wish to recognize the following people who wrote scripts, presented lessons on camera, and checked the accuracy of the videotapes:

Barbara Johnson, *Indiana University—Purdue University at Indianapolis*
Judith A. Penna, *Indiana University—Purdue University at Indianapolis*
Patricia Schwarzkopf, *University of Delaware*
Clen Vance, *Houston Community College*

I wish to thank Jason Jordan, my publisher and friend at Addison Wesley Longman, for his encouragement, for his marketing insight, and for providing me with the environment of creative freedom. The unwavering support of the Developmental Math group and the endless hours of hard work by Martha Morong and Janet Theurer have led to products of which I am immensely proud.

I also want to thank Judy Beecher, my co-author on many books and my developmental editor on this text. Her steadfast loyalty, vision, and encouragement have been invaluable. In addition to writing the Student's Solutions Manual, Judy Penna has continued to provide strong leadership in the preparation of the printed supplements, videotapes, and interactive CD-ROM. Other strong support has come from Donna DeSpain for the Printed Test Bank; Bill Saler for the audiotapes; Irene Doo for the Collaborative Learning Activities Manual; and Irene Doo, Shane Goodwin, Barbara Johnson, and Judy Penna for their accuracy checking.

M.L.B.

Diagnostic Pretest

Chapter R

1. Subtract: $1.45 - (-2.12)$.

2. Multiply: $-\frac{5}{6}\left(\frac{2}{15}\right)$.

3. Simplify: $2x - 2[x - (4 + 3x)]$.

4. Simplify: $\left[\dfrac{-3x^2y^{-3}}{2x^{-1}y^4}\right]^{-2}$.

Chapter 1

Solve.

5. $3(x + 1) = 2 - (x - 2)$

6. $2 \le 4 - x \le 7$

7. $|2x + 3| > 1$

8. Solve for P: $A = P + Prt$.

9. *Perimeter of NBA Court.* The perimeter of an NBA-sized basketball court is 288 ft. The length is 44 ft longer than the width. Find the dimensions of the court. (**Source:** National Basketball Association)

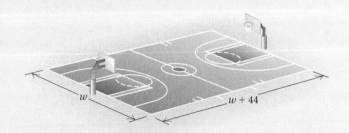

Chapter 2

Graph on a plane.

10. $3x - 4y = 12$

11. $x = -1$

12. $f(x) = 3 - x^2$

13. Find an equation of the line containing the pair of points $(1, 3)$ and $(-1, 5)$.

14. Find an equation of the line containing the point $(0, 3)$ and perpendicular to the line $3x - y = 7$.

15. For the function f given by $f(x) = 3 - x^2$, find $f(-1)$, $f(0)$, and $f(2)$.

16. Find the domain and the range of the function whose graph is shown below.

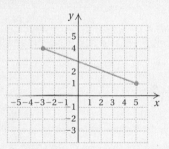

Chapter 3

17. Graph on a plane: $3x \le 6 - y$.

18. Solve: $\begin{aligned} 3x - y &= 5, \\ x + 2y &= 3. \end{aligned}$

19. Solve: $\begin{aligned} x - 4y + 2z &= -1, \\ 2x + y - z &= 8, \\ -x - 3y + z &= -5. \end{aligned}$

20. *Marine Travel.* A motorboat took 6 hr to make a downstream trip with a 3-mph current. The return trip against the same current took 8 hr. Find the speed of the boat in still water.

21. Graph. Find the coordinates of any vertices formed.
$$\begin{aligned} x &\ge 3, \\ x &\le 6 - 3y, \\ x - 2y &\le 6 \end{aligned}$$

Chapter 4

22. Multiply: $(3x - 5y)^2$.

23. Factor: $x^4 - 1$.

24. Solve: $x^2 - 18 = 7x$.

25. Three times the square of a number is 2 more than five times the number. Find the number.

26. *Sports Car Sales.* The sales of sports cars rise and fall due to the introduction of new or redesigned models such as the Corvette in 1997. Sales S, in thousands, for recent years can be approximated by the polynomial function

$$S(x) = 0.1x^4 - 1.6x^3 + 11.15x^2 - 38.04x + 112.65,$$

where x is the number of years since 1990.

a) Use the equation to approximate sports car sales in 2000.

b) Use the graph at right to approximate $S(5)$.

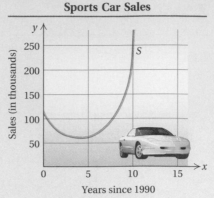

Sports Car Sales

Sales (in thousands)

Years since 1990

Source: Autodata

Chapter 5

27. Divide and simplify:

$$\frac{x^2 - 9}{x^2 + 3x + 2} \div \frac{x^2 - 6x + 9}{2x + 4}.$$

28. Simplify:

$$\frac{x - \dfrac{1}{x}}{1 + \dfrac{1}{x}}.$$

29. Find the domain of the function f given by

$$f(x) = \frac{x^2 - 3x + 2}{x^2 - 9}.$$

30. Solve:

$$\frac{3}{x} + \frac{2}{x - 2} = \frac{1}{x}.$$

31. *Print Work.* One folding machine in a print shop can fold an order of pamphlets in 4 hr. Another machine can do the same job in 3 hr. How long will it take if both folding machines are used?

Chapter 6

For Questions 32–34, assume that all expressions under radicals represent nonnegative numbers.

32. Multiply and simplify: $(\sqrt{6} + \sqrt{8x^3})(\sqrt{6} - 2\sqrt{2y^2})$.

33. Add and simplify: $\sqrt{75} + \sqrt{300} + 3\sqrt{27}$.

34. Rationalize the denominator: $\dfrac{2\sqrt{x} - \sqrt{y}}{\sqrt{x} - \sqrt{y}}$.

35. Solve: $\sqrt[3]{x - 5} = -2$.

36. Find the domain of the function f given by $f(x) = \sqrt{6 - 3x}$.

Chapter 7

Solve.

37. $x^2 + 2x + 5 = 0$

38. $(x - 3)^2 + (x - 3) - 12 = 0$

39. $\dfrac{x}{2} = \dfrac{x + 1}{x + 3}$

40. Graph $f(x) = 2x^2 + 5x + 3$. Label the vertex and the line of symmetry.

Chapter 8

41. Find the inverse of $f(x) = 2x - 5$.

42. Graph: $f(x) = 2^x$.

43. Graph: $f(x) = \log_2 x$.

44. Express as a single logarithm: $\frac{1}{2} \log_3 x - \log_3 y$.

45. Convert to an exponential equation: $y = \log_2 x$.

46. Solve: $\log_4 1 = x$.

47. Solve: $\log_2 (x - 1) + \log_2 (x + 1) = 1$.

Chapter 9

48. Graph: $\dfrac{x^2}{25} + \dfrac{y^2}{9} = 1$.

49. Graph: $x^2 + y^2 - 6x + 4y + 9 = 0$.

50. Solve:

$$x^2 - y^2 = 6,$$
$$xy = 4.$$

Intermediate Algebra

Eighth Edition

R

Review of Basic Algebra

Introduction

Why do we study mathematics? One reason is to enable us to apply mathematics to applications and problem solving. Before we can do so, however, there are certain basic skills of algebra to review.

Chapter R is a review of basic algebra. Because students come to this course with various backgrounds, we have organized Chapter R in two parts. Part 1 covers the operations on the real numbers. Part 2 covers the basic manipulations with expressions involving variables. The background of the students may allow either or both parts to be skipped. The Pretest on the next page can be used to determine how much review is needed.

An Application	The Mathematics
The area A of a circle with radius r is given by $A = \pi r^2$. The standard compact disc used for software and music has a radius of 6 cm. Find the area of such a CD (ignoring the hole in the middle). This problem appears as Exercise 37 in Exercise Set R.4.	We use the formula $A = \pi r^2$ with $r = 6$: $A = \underbrace{\pi r^2}_{} = \pi \cdot 6^2$ $= 36\pi \approx 113.1 \text{ cm}^2.$ This is an algebraic expression.

World Wide Web   For more information, visit us at www.mathmax.com

Pretest: Chapter R

Part 1

1. Which of the following numbers are rational?

$$-43, \sqrt{2}, -\frac{2}{3}, 2.3\overline{76}, \pi$$

2. Use $<$ or $>$ for ▨ to write a true sentence:

-8 ▨ -1.

3. Write another inequality with the same meaning as $x > 2$.

4. Is the following true or false? $-4 > 5$

5. Find the absolute value: $|-3.6|$.

Add, subtract, multiply, or divide, if possible.

6. $-8 + 24$

7. $-3.4 - 8.2$

8. $\left(-\frac{3}{5}\right)\left(\frac{10}{7}\right)$

9. $\frac{-200}{-25}$

Write exponential notation.

10. $q \cdot q \cdot q$

11. $64 \cdot 64 \cdot 64 \cdot 64 \cdot 64$

12. Rewrite using a positive exponent: x^{-3}.

Simplify.

13. $12 - 3 \cdot 2 + 10$

14. $\dfrac{7(5 - 2 \cdot 3) - 3^2}{4^2 + 3^2}$

Part 2

15. Translate to an algebraic expression: Forty-seven percent of some number.

16. Evaluate $x - 3y$ for $x = 5$ and $y = -4$.

17. Use multiplying by 1 to find an equivalent expression with the given denominator.

$$\frac{5}{4}; \quad 28x$$

18. Simplify: $\dfrac{54x}{-36x}$.

19. Multiply: $-4(x - 3y)$.

20. Factor: $6x - 18xy + 24$.

21. Collect like terms: $5x - 8y + 23 - 6x + 14y - 12$.

22. Find an equivalent expression: $-(-2x + 5y - 24)$.

Simplify.

23. $8(x + 2) - 6(2x + 12)$

24. $(5x^3)(-4x^{-6})$

25. $\dfrac{36x^3y^{-10}}{-9x^5y^{-8}}$

26. $(-5x^{-6}y^5)^{-2}$

Convert to scientific notation.

27. 0.0000000786

28. $457,890,000,000$

Convert to decimal notation.

29. 7.89×10^{13}

30. 7.89×10^{-6}

Part 1 Operations
R.1 The Set of Real Numbers

a | Set Notation and the Set of Real Numbers

A **set** is a collection of objects. In mathematics, we usually consider sets of numbers. The set we consider most in algebra is the **set of real numbers.** There is a real number for every point on the number line. Some commonly used sets of numbers are **subsets** of, or sets contained within, the set of real numbers. We begin by examining some subsets of the set of real numbers.

The set containing the numbers -5, 0, and 3 can be named $\{-5, 0, 3\}$. This method of describing sets is known as the **roster method**. In the following box, we use the roster method to describe three frequently used subsets of real numbers. Note that three dots are used to indicate that the pattern continues without end.

NATURAL NUMBERS (OR COUNTING NUMBERS)

Those numbers used for counting: $\{1, 2, 3, \ldots\}$

WHOLE NUMBERS

The set of natural numbers with 0 included: $\{0, 1, 2, 3, \ldots\}$

INTEGERS

The set of whole numbers and their opposites:

$\{\ldots, -4, -3, -2, -1, 0, 1, 2, 3, 4, \ldots\}$

The integers can be illustrated on the number line as follows:

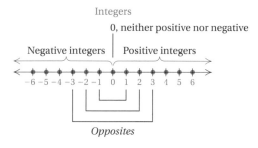

The set of integers extends infinitely to the left and to the right of 0. The **opposite** of a number is found by reflecting it across the number 0. Thus the opposite of 3 is -3. The opposite of -4 is 4. The opposite of 0 is 0. We read a symbol like -3 as either "the opposite of 3" or "negative 3."

The natural numbers are called **positive integers.** The opposites of the natural numbers (those to the left of 0) are called **negative integers.** Zero is neither positive nor negative.

Do Exercises 1–3 (in the margin at the right).

Find the opposite of the number.

1. 9

2. -6

3. 0

Answers on page A-2

4. Name the set consisting of the first seven odd whole numbers using both roster notation and set-builder notation.

Each point on the number line corresponds to a real number. In order to fill in the remaining numbers on our number line, we must describe two other subsets of the real numbers. And to do that, we need another type of set notation.

Set-builder notation is used to specify conditions under which a number is in a set. For example, the set of all odd natural numbers less than 9 can be described as follows:

$$\{x \mid x \text{ is an odd natural number less than 9}\}.$$

The set of ⟵
all numbers x ⟵
such that ⟵
x is an odd natural number less than 9. ⟵

We can easily write another name for this set using roster notation, as follows:

$$\{1, 3, 5, 7\}.$$

Example 1 Name the set consisting of the first six even whole numbers using both roster notation and set-builder notation.

Roster notation: $\{0, 2, 4, 6, 8, 10\}$

Set-builder notation: $\{x \mid x \text{ is one of the first six even whole numbers}\}$

Do Exercise 4.

Convert to decimal notation by long division and determine whether it is terminating or repeating.

5. $\dfrac{11}{16}$

The advantage of set-builder notation is that we can use it to describe very large sets that may be difficult to describe using roster notation. Such is the case when we try to name the set of **rational numbers.** Rational numbers can be named using fractional notation. The following are examples of rational numbers:

$$\frac{5}{8}, \frac{12}{-7}, \frac{-17}{15}, -\frac{9}{7}, \frac{39}{1}, \frac{0}{6}.$$

Note that

$$-\frac{9}{7} = \frac{-9}{7} = \frac{9}{-7}.$$

We can now describe the set of rational numbers.

6. $\dfrac{14}{13}$

> **Rational numbers** are numbers whose decimal representation either terminates or repeats. A rational number can be expressed as an integer divided by a nonzero integer:
>
> $$\left\{\frac{p}{q} \;\middle|\; p \text{ is an integer, } q \text{ is an integer, and } q \neq 0\right\}.$$

For example, each of the following is a rational number:

$$\frac{5}{8} = \underbrace{0.625}_{} \qquad \text{and} \qquad \frac{6}{11} = \underbrace{0.545454\ldots}_{} = 0.\overline{54}.$$

Terminating   Repeating

The bar in $0.\overline{54}$ indicates the repeating part of the decimal notation.

Note that $\frac{39}{1} = 39$. Thus the set of rational numbers contains the integers.

Do Exercises 5 and 6.

Answers on page A-2

The real-number line has a point for every rational number.

However, there are many points on the line for which there is no rational number. These points correspond to what are called **irrational numbers.**

Numbers like π, $\sqrt{2}$, $-\sqrt{10}$, $\sqrt{13}$, and $-1.898898889\ldots$ are examples of irrational numbers. Decimal notation for irrational numbers *neither* terminates *nor* repeats. Recall that decimal notation for rational numbers either terminates or repeats.

> **Irrational numbers** are numbers whose decimal representation neither terminates nor repeats. They cannot be represented as the quotient of two integers.

Nevertheless, we often approximate irrational numbers using rational numbers. For example, $\sqrt{2}$ is the number for which $\sqrt{2} \cdot \sqrt{2} = 2$. There is no rational number that can be multiplied by itself to get 2. But 1.4142 is a rational approximation of $\sqrt{2}$ because

$$(1.4142)^2 = 1.99996164 \approx 2.$$

The symbol $\approx$ means "is approximately equal to." We can find rational approximations for square roots and other irrational numbers using a calculator.

Calculator Spotlight

Approximating Square Roots and π. Eventually in this text, we will be using only graphing calculators in working the Calculator Spotlights. Because of this and because there is extensive variance in keystroke procedures among calculators, we will focus on the keystrokes for a graphing calculator such as a TI-82 or a TI-83. For the use of any other calculator, consult the manual for that particular model. In this feature, we will frequently refer to a graphing calculator as a grapher.

Square roots are found by pressing $\boxed{\text{2nd}}\ \boxed{\sqrt{\ }}$. ($\sqrt{\ }$ is the second operation associated with the $\boxed{x^2}$ key.)

To find an approximation for $\sqrt{48}$, we press $\boxed{\text{2nd}}\ \boxed{\sqrt{\ }}\ \boxed{4}\ \boxed{8}$ $\boxed{\text{ENTER}}$. The approximation 6.92820323 is displayed.

To find an approximation for π, we press $\boxed{\text{2nd}}\ \boxed{\pi}\ \boxed{\text{ENTER}}$. The approximation 3.141592654 is displayed.

Exercises

Approximate each of the following.

1. $\sqrt{87}$ **2.** $\sqrt{457}$ **3.** $23\sqrt{40}$ **4.** $31 + \sqrt{47}$

5. π **6.** $33 \cdot \pi$ **7.** $\pi \cdot 56.8$ **8.** $5\pi + 8\sqrt{237}$

7. Given the numbers

$20,\ -10,\ -5.34,\ 18.999,$

$\dfrac{11}{45},\ \sqrt{7},\ -\sqrt{2},\ 0,\ -\dfrac{2}{3},$

$9.34334333433334\ldots:$

a) Name the natural numbers.

b) Name the whole numbers.

c) Name the integers.

d) Name the irrational numbers.

e) Name the rational numbers.

f) Name the real numbers.

The set of all rational numbers, combined with the set of all irrational numbers, gives us the set of **real numbers.**

> **Real numbers** are either rational or irrational:
>
> $\{x \mid x \text{ is a rational number or } x \text{ is an irrational number}\}.$

Every point on the number line represents some real number and every real number is represented by some point on a number line.

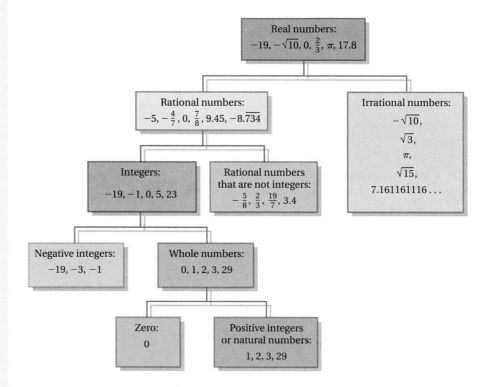

The following figure shows the relationships among various kinds of real numbers.

Do Exercise 7.

Answers on page A-2

b Order for the Real Numbers

Real numbers are named in order on the number line, with larger numbers named further to the right. For any two numbers on the line, the one to the left is less than the one to the right.

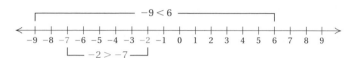

We use the symbol **<** to mean "**is less than.**" The sentence $-9 < 6$ means "-9 is less than 6." The symbol **>** means "**is greater than.**" The sentence $-2 > -7$ means "-2 is greater than -7." A handy mental device is to think of > or < as an arrowhead that points to the smaller number.

Examples Use either < or > for ▨ to write a true sentence.

2. 4 ▨ 9 Since 4 is to the left of 9, 4 is less than 9, so $4 < 9$.

3. -8 ▨ 3 Since -8 is to the left of 3, we have $-8 < 3$.

4. 7 ▨ -12 Since 7 is to the right of -12, then $7 > -12$.

5. -21 ▨ -5 Since -21 is to the left of -5, we have $-21 < -5$.

6. -2.7 ▨ $-\dfrac{3}{2}$ Since $-\dfrac{3}{2} = -1.5$ and -2.7 is to the left of -1.5, we have $-2.7 < -\dfrac{3}{2}$.

7. $1\dfrac{1}{4}$ ▨ -2.7 The answer is $1\dfrac{1}{4} > -2.7$.

8. 4.79 ▨ 4.97 The answer is $4.79 < 4.97$.

9. -8.45 ▨ 1.32 The answer is $-8.45 < 1.32$.

10. $\dfrac{5}{8}$ ▨ $\dfrac{7}{11}$ We convert to decimal notation $\left(\dfrac{5}{8} = 0.625 \text{ and } \dfrac{7}{11} = 0.6363\ldots\right)$ and compare. Thus, $\dfrac{5}{8} < \dfrac{7}{11}$.

Do Exercises 8–16.

All positive real numbers are greater than zero and all negative real numbers are less than zero.

> If x is a positive real number, then $x > 0$.
> If x is a negative real number, then $x < 0$.

Note that $-8 < 5$ and $5 > -8$ are both true. These are **inequalities**. Every true inequality yields another true inequality if we interchange the numbers or variables and reverse the direction of the inequality sign.

> $a < b$ also has the meaning $b > a$.

Insert < or > for ▨ to write a true sentence.

8. -5 ▨ -4

9. $-\dfrac{1}{4}$ ▨ $-\dfrac{1}{2}$

10. 87 ▨ 67

11. -9.8 ▨ $-4\dfrac{2}{3}$

12. 6.78 ▨ -6.77

13. $-\dfrac{4}{5}$ ▨ -0.86

14. $\dfrac{14}{29}$ ▨ $\dfrac{17}{32}$

15. $-\dfrac{12}{13}$ ▨ $-\dfrac{14}{15}$

16. 1.8 ▨ 1.08

Answers on page A-2

Write another inequality with the same meaning.

17. $x > 6$

18. $-4 < 7$

Write true or false.

19. $6 \geq -9.4$

20. $-18 \leq -18$

21. $-7.6 \leq -10\frac{4}{5}$

Graph the inequality.

22. $x > -1$

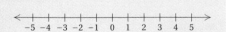

23. $x \leq 5$

Answers on page A-2

Examples   Write another inequality with the same meaning.

11. $a < -5$ The inequality $-5 > a$ has the same meaning.

12. $-3 > -8$ The inequality $-8 < -3$ has the same meaning.

Do Exercises 17 and 18.

Expressions like $a \leq b$ and $b \geq a$ are also **inequalities**. We read $a \leq b$ as "**a is less than or equal to b.**" We read $a \geq b$ as "**a is greater than or equal to b.**"

Examples   Write true or false.

13. $-8 \leq 5.7$ True since $-8 < 5.7$ is true.

14. $-8 \leq -8$ True since $-8 = -8$ is true.

15. $-7 \geq 4\frac{1}{3}$ False since neither $-7 > 4\frac{1}{3}$ nor $-7 = 4\frac{1}{3}$ is true.

Do Exercises 19–21.

c | Graphing Inequalities on a Number Line

Some replacements for the variable in an inequality make it true and some make it false. A replacement that makes it true is called a **solution**. The set of all solutions is called the **solution set**. A **graph** of an inequality is a drawing that represents its solution set.

Example 16   Graph the inequality $x > -3$ on a number line.

The solutions consist of all real numbers greater than -3, so we shade all numbers greater than -3. Note that -3 is not a solution. We indicate this by using a parenthesis at -3.

The graph represents the solution set $\{x \mid x > -3\}$. Numbers in this set include -2.6, -1, 0, π, $\sqrt{2}$, $3\frac{7}{8}$, 5, and 123.

Example 17   Graph the inequality $x \leq 2$ on a number line.

We make a drawing that represents the solution set $\{x \mid x \leq 2\}$. This time the graph consists of 2 as well as the numbers less than 2. We shade all numbers to the left of 2 and use a bracket at 2 to indicate that it is also a solution.

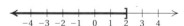

Do Exercises 22 and 23.

d Absolute Value

From the number line, we see that numbers like 6 and −6 are the same distance from zero. We call the distance from zero the **absolute value** of the number.

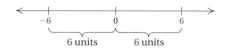

> The **absolute value** of a number is its distance from zero on the number line. We use the symbol $|x|$ to represent the absolute value of a number x.

To find absolute value:

1. If a number is negative, make it positive.
2. If a number is positive or zero, leave it alone.

Examples   Find the absolute value.

18. $|-7|$ The distance of −7 from 0 is 7, so $|-7|$ is 7.

19. $|12|$ The distance of 12 from 0 is 12, so $|12|$ is 12.

20. $|0|$ The distance of 0 from 0 is 0, so $|0|$ is 0.

21. $\left|\dfrac{4}{5}\right| = \dfrac{4}{5}$

22. $|-3.86| = 3.86$

Do Exercises 24–27.

Find the absolute value.

24. $\left|-\dfrac{1}{4}\right|$

25. $|2|$

26. $\left|\dfrac{3}{2}\right|$

27. $|-2.3|$

Answers on page A-2

Improving Your Math Study Skills

Tips for Using This Textbook

Throughout this textbook, you will find a feature called "Improving Your Math Study Skills." At least one such topic is included in each chapter. Each topic title is listed in the table of contents beginning on p. iii.

One of the most important ways to improve your math study skills is to learn the proper use of the textbook. Here we highlight a few points that we consider most helpful.

- **Be sure to note the special symbols a , b , c , and so on, that correspond to the objectives you are to be able to perform.** They appear in many places throughout the text. The first time you see them is in the margin at the beginning of each section. The second time is in the subheadings of each section, and the third time is in the exercise set. You will also find them next to the skill maintenance exercises in each exercise set and in the review exercises at the end of the chapter, as well as in the answers to the pretests, the chapter tests, and the cumulative reviews. These objective symbols allow you to refer back whenever you need to review a topic.

- **Note the symbols in the margin under the list of objectives at the beginning of each section.** These refer to the many distinctive study aids that accompany the book.

- **Read and study each step of each example.** The examples include important side comments that explain each step. These carefully chosen examples and notes prepare you for success in the exercise set.

- **Stop and do the margin exercises as you study a section.** When our students come to us troubled about how they are doing in the course, the first question we ask is "Are you doing the margin exercises when directed to do so?" This is one of the most effective ways to enhance your ability to learn mathematics from this text. Don't deprive yourself of its benefits!

- **When you study the book, don't mark the points that you think are important, but mark the points you do not understand!** This book includes many design features that highlight important points. Use your efforts to mark where you are having trouble. Then when you go to class, a math lab, or a tutoring session, you will be prepared to ask questions that home in on your difficulties rather than spending time going over what you already understand.

- **If you are having trouble, consider using the *Student's Solutions Manual*, which contains worked-out solutions to the odd-numbered exercises in the exercise sets.**

- **Try to keep one section ahead of your syllabus.** If you study ahead of your lectures, you can concentrate on what is being explained in them, rather than trying to write everything down. You can then take notes only of special points or of questions related to what is happening in class.

Exercise Set R.1

a Given the numbers -6, 0, 1, $-\frac{1}{2}$, -4, $\frac{7}{9}$, 12, $-\frac{6}{5}$, 3.45, $5\frac{1}{2}$, $\sqrt{3}$, $\sqrt{25}$, $-\frac{12}{3}$, $0.131331333133331\ldots$:

1. Name the natural numbers.

2. Name the whole numbers.

3. Name the rational numbers.

4. Name the integers.

5. Name the real numbers.

6. Name the irrational numbers.

Given the numbers $-\sqrt{5}$, -3.43, -11, 12, 0, $\frac{11}{34}$, $-\frac{7}{13}$, π, $-3.565665666566665\ldots$:

7. Name the whole numbers.

8. Name the natural numbers.

9. Name the integers.

10. Name the rational numbers.

11. Name the irrational numbers.

12. Name the real numbers.

Use roster notation to name the set.

13. The set of all letters in the word "math"

14. The set of all letters in the word "solve"

15. The set of all positive integers less than 13

16. The set of all odd whole numbers less than 13

17. The set of all even natural numbers

18. The set of all negative integers greater than -4

Use set-builder notation to name the set.

19. $\{0, 1, 2, 3, 4, 5\}$

20. $\{4, 5, 6, 7, 8, 9, 10\}$

21. The set of all rational numbers

22. The set of all real numbers

23. The set of real numbers greater than -3

24. The set of all real numbers less than or equal to 21

b Use either $<$ or $>$ for ▨ to write a true sentence.

25. 13 ▨ 0

26. 18 ▨ 0

27. -8 ▨ 2

28. 7 ▨ -7

29. -8 ▨ 8

30. 0 ▨ -11

31. -8 ▨ -3

32. -6 ▨ -3

33. -2 ▨ -12

34. -7 ▨ -10

35. -9.9 ▨ -2.2

36. $-13\frac{1}{5}$ ▨ $\frac{11}{250}$

37. $37\frac{1}{5}$ ▨ $-1\frac{67}{100}$

38. -13.99 ▨ -8.45

39. $\frac{6}{13}$ ▨ $\frac{13}{25}$

40. $-\frac{14}{15}$ ▨ $-\frac{27}{53}$

Write an inequality with the same meaning.

41. $-8 > x$

42. $x < 7$

43. $-12.7 \leq y$

44. $10\frac{2}{3} \geq t$

Write true or false.

45. $6 \leq -6$

46. $-7 \leq -7$

47. $5 \geq -8.4$

48. $-11 \geq -13\frac{1}{2}$

c Graph the inequality.

49. $x < -2$

50. $x < -1$

51. $x \leq -2$

52. $x \geq -1$

53. $x > -3.3$

54. $x < 0$

55. $x \geq 2$

56. $x \leq 0$

d Find the absolute value.

57. $|-6|$

58. $|-3|$

59. $|28|$

60. $|16|$

61. $|-35|$

62. $|465|$

63. $\left|-\frac{2}{3}\right|$

64. $\left|-\frac{13}{8}\right|$

65. $\left|\frac{0}{7}\right|$

66. $|16.4|$

Synthesis Exercises

Exercises designated as *Synthesis Exercises* differ from those found in the main body of the exercise set. The icon ◈ denotes synthesis exercises that are writing exercises. Writing exercises are meant to be answered in one or more complete sentences. Because answers to writing exercises often vary, they are not listed at the back of the book. These and the other synthesis exercises will often challenge you to put together two or more objectives at once.

67. ◈ Explain the difference between a rational number and an irrational number in as many ways as you can.

68. ◈ List five examples of rational numbers that are not integers and explain why they are not.

Use either $\leq$ or $\geq$ for ▓ to write a true sentence.

69. $|-3|$ ▓ 5

70. $|-5|$ ▓ $|-2|$

71. $|4|$ ▓ $|-7|$

72. $|-8|$ ▓ $|8|$

73. List the following numbers in order from least to greatest.

$$\frac{1}{11}, \quad 1.1\%, \quad \frac{2}{7}, \quad 0.3\%, \quad 0.11, \quad \frac{1}{8}\%, \quad 0.009, \quad \frac{99}{1000}, \quad 0.286, \quad \frac{1}{8}, \quad 1\%, \quad \frac{9}{100}$$

R.2 Operations with Real Numbers

We now review the addition, subtraction, multiplication, and division of real numbers.

a | Addition

To explain addition of real numbers, we can use the number line. To perform the addition $a + b$, we start at a and then move according to b. If b is positive, we move to the right. If b is negative, we move to the left.

Examples

1. $6 + (-8) = -2$

Start at 6. Move 8 units to the left. The answer is -2.

2. $-3 + 7 = 4$

Start at -3. Move 7 units to the right. The answer is 4.

3. $-1 + (-5) = -6$

Start at -1. Move 5 units to the left. The answer is -6.

Do Exercises 1–4.

You may have noticed some patterns in the preceding examples. These lead us to rules for adding without using the number line.

> **RULES FOR ADDITION OF REAL NUMBERS**
>
> **1.** *Positive numbers*: Add the numbers. The result is positive.
> **2.** *Negative numbers*: Add absolute values. Make the answer negative.
> **3.** *A positive and a negative number*: Subtract the smaller absolute value from the larger. Then:
>   **a)** If the positive number has the greater absolute value, make the answer positive.
>   **b)** If the negative number has the greater absolute value, make the answer negative.
>   **c)** If the numbers have the same absolute value, make the answer 0.
> **4.** *One number is zero*: The sum is the other number.

Rule 4 is known as the **Identity Property of 0.** It says that for any real number a, $a + 0 = a$.

Objectives

a   Add real numbers.

b   Find the opposite, or additive inverse, of a number.

c   Subtract real numbers.

d   Multiply real numbers.

e   Divide real numbers.

For Extra Help

TAPE 1   TAPE 1A   MAC WIN   CD-ROM

Add using a number line.

1. $-3 + 5$

2. $9 + (-5)$

3. $6 + (-10)$

4. $-5 + 5$

Answers on page A-2

Add.

5. $-7 + (-11)$

6. $-8.9 + (-9.7)$

7. $-\frac{6}{5} + \left(-\frac{23}{5}\right)$

8. $-\frac{3}{10} + \left(-\frac{2}{5}\right)$

9. $-7 + 7$

10. $-7.4 + 0$

11. $4 + (-7)$

12. $-7.8 + 4.5$

13. $\frac{3}{8} + \left(-\frac{5}{8}\right)$

14. $-\frac{3}{5} + \frac{7}{10}$

Find the opposite, or additive inverse, of the number.

15. -14

16. $\frac{2}{3}$

17. 0

Answers on page A-2

Examples   Add without using a number line.

4. $-13 + (-8) = -21$ Two negatives. *Think*: Add the absolute values, 13 and 8, getting 21. Make the answer *negative*, -21.

5. $-2.1 + 8.5 = 6.4$ The absolute values are 2.1 and 8.5. Subtract 2.1 from 8.5. The positive number has the larger absolute value, so the answer is *positive*, 6.4.

6. $-48 + 31 = -17$ The absolute values are 48 and 31. Subtract 31 from 48. The negative number has the larger absolute value, so the answer is *negative*, -17.

7. $2.6 + (-2.6) = 0$ The numbers have the same absolute value. The sum is 0.

8. $-\frac{5}{9} + 0 = -\frac{5}{9}$ One number is zero. The sum is $-\frac{5}{9}$.

9. $-\frac{3}{4} + \frac{9}{4} = \frac{6}{4} = \frac{3}{2}$

10. $-\frac{2}{3} + \frac{5}{8} = -\frac{16}{24} + \frac{15}{24} = -\frac{1}{24}$

Do Exercises 5–14.

b   Opposites, or Additive Inverses

Suppose we add two numbers that are **opposites,** such as 4 and -4. The result is 0. When opposites are added, the result is always 0. Such numbers are also called **additive inverses.** Every real number has an opposite, or additive inverse.

> Two numbers whose sum is 0 are called **opposites,** or **additive inverses,** of each other.

Examples   Find the opposite, or additive inverse, of the number.

11. 8.6   The opposite of 8.6 is -8.6 because $8.6 + (-8.6) = 0$.

12. 0   The opposite of 0 is 0 because $0 + 0 = 0$.

13. $-\frac{7}{9}$ The opposite of $-\frac{7}{9}$ is $\frac{7}{9}$ because $-\frac{7}{9} + \frac{7}{9} = 0$.

To name the opposite, or additive inverse, we use the symbol $-$, and read the symbolism $-a$ as "the opposite of a" or "the additive inverse of a."

Do Exercises 15–17.

A symbol such as -8 is usually read "negative 8." It could be read "the opposite of 8," because the opposite of 8 is -8. It could also be read "the additive inverse of 8," because the additive inverse of 8 is -8. When a variable is involved, as in a symbol like $-x$, it can be read "the opposite of x" or "the additive inverse of x" but *not* "negative x," because we do not know whether the symbol represents a positive number, a negative number, or 0.

Example 14 Evaluate $-x$ and $-(-x)$ for (a) $x = 23$ and (b) $x = -5$.

a) If $x = 23$, then $-x = -23 = -23$.   The opposite, or additive inverse, of 23 is −23.

   If $x = 23$, then $-(-x) = -(-23) = 23$.   The opposite of the opposite of 23 is 23.

b) If $x = -5$, then $-x = -(-5) = 5$.
   If $x = -5$, then $-(-x) = -(-(-5)) = -(5) = -5$.

Note in Example 14(b) that an extra set of parentheses is used to show that we are substituting the negative number -5 for x. Symbolism like $--x$ is not considered meaningful.

Do Exercises 18–21.

We can use the symbolism $-a$ for the opposite of a to restate the definition of opposite.

> For any real number a, the **opposite**, or **additive inverse**, of a, which is $-a$, is such that
> $$a + (-a) = (-a) + a = 0.$$

Signs of Numbers

A negative number is sometimes said to have a "negative sign." A positive number is said to have a "positive sign." When we replace a number with its opposite, or additive inverse, we can say that we have "changed its sign."

Examples   Change the sign. (Find the opposite, or additive inverse.)

15. -3 $-(-3) = 3$ **16.** $-\frac{3}{8}$ $-\left(-\frac{3}{8}\right) = \frac{3}{8}$

17. 0 $-0 = 0$ **18.** 14 $-(14) = -14$

Do Exercise 22.

We can now use the concept of opposite to give a more formal definition of absolute value.

> For any real number a, the **absolute value** of a, denoted $|a|$, is given by
> $$|a| = \begin{cases} a, & \text{if } a \geq 0 \\ -a, & \text{if } a < 0. \end{cases}$$
> For example, $|8| = 8$ and $|0| = 0$.
> For example, $|-5| = -(-5) = 5$.
>
> (The absolute value of a is a if a is nonnegative. The absolute value of a is the opposite of a if a is negative.)

c | Subtraction

Subtraction is defined as follows.

> The **difference** $a - b$ is the number that when added to b gives a. That is, $a - b = c$, if c is a number such that $a = c + b$.

---

18. Evaluate $-a$ for $a = 9$.

19. Evaluate $-a$ for $a = -\frac{3}{5}$.

20. Evaluate $-(-a)$ for $a = -5.9$.

21. Evaluate $-(-a)$ for $a = \frac{2}{3}$.

22. Change the sign.
a) 11

b) -17

c) 0

d) x

e) $-x$

Answers on page A-2

Subtract.

23. $8 - (-9)$

24. $-10 - 6$

25. $5 - 8$

26. $-23.7 - 5.9$

27. $-2 - (-5)$

28. $-\dfrac{11}{12} - \left(-\dfrac{23}{12}\right)$

29. $\dfrac{2}{3} - \left(-\dfrac{5}{6}\right)$

30. a) $17 - 23$

    b) $-17 - 23$

    c) $-17 - (-23)$

31. Look for a pattern and complete.

$$4 \cdot 5 = 20 \qquad -2 \cdot 5 =$$
$$3 \cdot 5 = 15 \qquad -3 \cdot 5 =$$
$$2 \cdot 5 = \qquad -4 \cdot 5 =$$
$$1 \cdot 5 = \qquad -5 \cdot 5 =$$
$$0 \cdot 5 = \qquad -6 \cdot 5 =$$
$$-1 \cdot 5 =$$

Multiply.

32. $-4 \cdot 6$

33. $(3.5)(-8.1)$

34. $-\dfrac{4}{5} \cdot 10$

35. Look for a pattern and complete.

$$4(-5) = -20 \qquad -1(-5) =$$
$$3(-5) = -15 \qquad -2(-5) =$$
$$2(-5) = \qquad -3(-5) =$$
$$1(-5) = \qquad -4(-5) =$$
$$0(-5) = \qquad -5(-5) =$$

Multiply.

36. $-8(-9)$

37. $\left(-\dfrac{4}{5}\right) \cdot \left(-\dfrac{2}{3}\right)$

38. $(-4.7)(-9.1)$

Answers on page A-2

For example, $3 - 5 = -2$ because $3 = -2 + 5$. That is, -2 is the number that when added to 5 gives 3. Although this illustrates the formal definition of subtraction, we generally use the following when we subtract.

> For any real numbers a and b,
> $$a - b = a + (-b).$$
> (We can subtract by adding the opposite (additive inverse) of the number being subtracted.)

Examples   Subtract.

19. $3 - 5 = 3 + (-5) = -2$ **Changing the sign of 5 and adding**

20. $7 - (-3) = 7 + (3) = 10$ **Changing the sign of −3 and adding**

21. $-19.4 - 5.6 = -19.4 + (-5.6) = -25$

22. $-\dfrac{4}{3} - \left(-\dfrac{2}{5}\right) = -\dfrac{4}{3} + \dfrac{2}{5} = -\dfrac{20}{15} + \dfrac{6}{15} = -\dfrac{14}{15}$

Do Exercises 23–30.

d | Multiplication

We know how to multiply positive numbers. What happens when we multiply a positive number and a negative number?

Do Exercise 31.

> To multiply a positive number and a negative number, multiply their absolute values. Then make the answer negative.

Examples   Multiply.

23. $-3 \cdot 5 = -15$ **24.** $6 \cdot (-7) = -42$

25. $(-1.2)(4.5) = -5.4$ **26.** $3 \cdot \left(-\dfrac{1}{2}\right) = \dfrac{3}{1} \cdot \left(-\dfrac{1}{2}\right) = -\dfrac{3}{2}$

Note in Example 25 that the parentheses indicate multiplication.

Do Exercises 32–34.

What happens when we multiply two negative numbers?

Do Exercise 35.

> To multiply two negative numbers, multiply their absolute values. The answer is positive.

Examples   Multiply.

27. $-3 \cdot (-5) = 15$ **28.** $-5.2(-10) = 52$

29. $(-8.8)(-3.5) = 30.8$ **30.** $\left(-\dfrac{3}{4}\right) \cdot \left(-\dfrac{5}{2}\right) = \dfrac{15}{8}$

Do Exercises 36–38.

e Division

> The **quotient** a/b is defined to be the number c (if it exists) that when multiplied by b gives a. That is, $a/b = c$, if c is a number such that $a = cb$.

The definition of division parallels the one for subtraction. Using this definition and the rules for multiplying, we can see how to handle signs when dividing.

Examples Divide.

31. $\dfrac{10}{-2} = -5$, because $-5 \cdot (-2) = 10$

32. $\dfrac{-32}{4} = -8$, because $-8 \cdot (4) = -32$

33. $\dfrac{-25}{-5} = 5$, because $5 \cdot (-5) = -25$

34. $\dfrac{40}{-4} = -10$

35. $-10 \div 5 = -2$

36. $\dfrac{-10}{-40} = \dfrac{1}{4}$, or 0.25

37. $\dfrac{-10}{-3} = \dfrac{10}{3}$, or $3.\overline{3}$

The rules for division and multiplication are the same.

> To multiply or divide two real numbers:
>
> 1. Multiply or divide the absolute values.
> 2. If the signs are the same, then the answer is positive.
> 3. If the signs are different, then the answer is negative.

Do Exercises 39–42.

Division by Zero

We cannot divide a nonzero number n by zero. Let's see why. By the definition of division, $n/0$ would be some number that when multiplied by 0 gives n. But when any number is multiplied by 0, the result is 0. The only possibility for n would be 0.

Consider $0/0$. We might say that it is 5 because $5 \cdot 0 = 0$. We might also say that it is -8 because $-8 \cdot 0 = 0$. In fact, $0/0$ could be any number at all. So, division by 0 does not make sense. Division by 0 is undefined and not possible.

Examples Divide, if possible.

38. $\dfrac{7}{0}$ Undefined: Division by 0.

39. $\dfrac{0}{7} = 0$ The quotient is 0 because $0 \cdot 7 = 0$.

40. $\dfrac{4}{x - x}$ Undefined: $x - x = 0$ for any x.

Do Exercises 43–46.

Divide.

39. $\dfrac{-28}{-14}$

40. $125 \div (-5)$

41. $\dfrac{-75}{25}$

42. $-4.2 \div (-21)$

Divide, if possible.

43. $\dfrac{8}{0}$

44. $\dfrac{0}{9}$

45. $\dfrac{17}{2x - 2x}$

46. $\dfrac{3x - 3x}{x - x}$

Answers on page A-2

Find the reciprocal of the number.

47. $\dfrac{3}{8}$

48. $-\dfrac{4}{5}$

49. 18

50. -4.3

51. 0.5

52. Complete the following table.

	Opposite (Additive Inverse)	Reciprocal (Multiplicative Inverse)
$\dfrac{2}{3}$	$-\dfrac{2}{3}$	$\dfrac{3}{2}$
$\dfrac{4}{5}$		
$-\dfrac{3}{4}$		
0.25		
8		
-5		
0		

Divide by multiplying by the reciprocal of the divisor.

53. $-\dfrac{3}{4} \div \dfrac{7}{8}$

54. $-\dfrac{12}{5} \div \left(-\dfrac{7}{15}\right)$

Answers on page A-2

Division and Reciprocals

Two numbers whose product is 1 are called **reciprocals** (or **multiplicative inverses**) of each other.

> Every nonzero real number a has a **reciprocal** (or **multiplicative inverse**) $1/a$. The reciprocal of a positive number is positive. The reciprocal of a negative number is negative.

Examples   Find the reciprocal of the number.

41. $\dfrac{4}{5}$ The reciprocal is $\dfrac{5}{4}$, because $\dfrac{4}{5} \cdot \dfrac{5}{4} = 1$.

42. 8   The reciprocal is $\dfrac{1}{8}$, because $8 \cdot \dfrac{1}{8} = 1$.

43. $-\dfrac{2}{3}$ The reciprocal is $-\dfrac{3}{2}$, because $-\dfrac{2}{3} \cdot \left(-\dfrac{3}{2}\right) = 1$.

44. 0.25   The reciprocal is $\dfrac{1}{0.25}$ or 4, because $0.25 \cdot 4 = 1$.

Remember that a number and its reciprocal (multiplicative inverse) have the same sign. Do *not* change the sign when taking the reciprocal of a number. When finding an opposite (additive inverse), change the sign.

Do Exercises 47–52.

We know that we can subtract by adding an opposite, or additive inverse. Similarly, we can divide by multiplying by a reciprocal.

> For any real numbers a and b, $b \neq 0$,
> $$a \div b = \frac{a}{b} = a \cdot \frac{1}{b}.$$
> (To divide, we can multiply by the reciprocal of the divisor.)

We sometimes say that we "invert the divisor and multiply."

Examples   Divide by multiplying by the reciprocal of the divisor.

45. $\dfrac{1}{4} \div \dfrac{3}{5} = \dfrac{1}{4} \cdot \dfrac{5}{3} = \dfrac{5}{12}$ "Inverting" the divisor, $\frac{3}{5}$, and multiplying

46. $\dfrac{2}{3} \div \left(-\dfrac{4}{9}\right) = \dfrac{2}{3} \cdot \left(-\dfrac{9}{4}\right) = -\dfrac{18}{12}$, or $-\dfrac{3}{2}$

Do Exercises 53 and 54.

The following properties can be used to make sign changes in fractional notation.

> For any numbers a and b, $b \neq 0$,
> $$\frac{-a}{b} = \frac{a}{-b} = \frac{a}{b} \quad \text{and} \quad \frac{-a}{-b} = \frac{a}{b}.$$

Exercise Set R.2

a Add.

1. $-10 + (-18)$

2. $-13 + (-12)$

3. $7 + (-2)$

4. $7 + (-5)$

5. $-8 + (-8)$

6. $-6 + (-6)$

7. $7 + (-11)$

8. $9 + (-12)$

9. $-16 + 6$

10. $-21 + 11$

11. $-26 + 0$

12. $0 + (-32)$

13. $-8.4 + 9.6$

14. $-6.3 + 8.2$

15. $-2.62 + (-6.24)$

16. $-2.73 + (-8.46)$

17. $-\dfrac{5}{9} + \dfrac{2}{9}$

18. $-\dfrac{3}{7} + \dfrac{1}{7}$

19. $-\dfrac{11}{12} + \left(-\dfrac{5}{12}\right)$

20. $-\dfrac{3}{8} + \left(-\dfrac{7}{8}\right)$

21. $\dfrac{2}{5} + \left(-\dfrac{3}{10}\right)$

22. $-\dfrac{3}{4} + \dfrac{1}{8}$

23. $-\dfrac{2}{5} + \dfrac{3}{4}$

24. $-\dfrac{5}{6} + \left(-\dfrac{7}{8}\right)$

b Evaluate $-a$ for each of the following.

25. $a = -4$

26. $a = -9$

27. $a = 3.7$

28. $a = 0$

Find the opposite (additive inverse).

29. 10

30. $-\dfrac{2}{3}$

31. 0

32. $-2x$

c Subtract.

33. $3 - 7$

34. $8 - 13$

35. $-5 - 9$

36. $-6 - 14$

37. $23 - 23$

38. $23 - (-23)$

39. $-23 - 23$

40. $-23 - (-23)$

41. $-6 - (-11)$

42. $-7 - (-12)$

43. $10 - (-5)$

44. $28 - (-16)$

45. $15.8 - 27.4$

46. $17.2 - 34.9$

47. $-18.01 - 11.24$

48. $-19.04 - 15.76$

49. $-\dfrac{21}{4} - \left(-\dfrac{7}{4}\right)$

50. $-\dfrac{16}{5} - \left(-\dfrac{3}{5}\right)$

51. $-\dfrac{1}{3} - \left(-\dfrac{1}{12}\right)$

52. $-\dfrac{7}{8} - \left(-\dfrac{5}{2}\right)$

53. $-\dfrac{3}{4} - \dfrac{5}{6}$

54. $-\dfrac{2}{3} - \dfrac{4}{5}$

55. $\dfrac{1}{3} - \dfrac{4}{5}$

56. $-\dfrac{4}{7} - \left(-\dfrac{5}{9}\right)$

$\boxed{\text{d}}$ Multiply.

57. $3(-7)$

58. $5(-8)$

59. $-2 \cdot 4$

60. $-5 \cdot 9$

61. $-8(-3)$

62. $-5(-7)$

63. $-7 \cdot 16$

64. $-8 \cdot 19$

65. $-6(-5.7)$

66. $-7(-6.1)$

67. $-\dfrac{3}{5} \cdot \dfrac{4}{7}$

68. $-\dfrac{5}{4} \cdot \dfrac{11}{3}$

69. $-3\left(-\dfrac{2}{3}\right)$

70. $-5\left(-\dfrac{3}{5}\right)$

71. $-3(-4)(5)$

72. $-6(-8)(9)$

73. $(-4.2)(-6.3)$

74. $(-7.4)(-9.6)$

75. $-\dfrac{9}{11} \cdot \left(-\dfrac{11}{9}\right)$

76. $-\dfrac{13}{7} \cdot \left(-\dfrac{5}{2}\right)$

77. $-\dfrac{2}{3} \cdot \left(-\dfrac{2}{3}\right) \cdot \left(-\dfrac{2}{3}\right)$

78. $-\dfrac{4}{5} \cdot \left(-\dfrac{4}{5}\right) \cdot \left(-\dfrac{4}{5}\right)$

e Divide, if possible.

79. $\dfrac{-8}{4}$

80. $\dfrac{-16}{2}$

81. $\dfrac{56}{-8}$

82. $\dfrac{63}{-7}$

83. $-77 \div (-11)$

84. $-48 \div (-6)$

85. $\dfrac{-5.4}{-18}$

86. $\dfrac{-8.4}{-12}$

87. $\dfrac{5}{0}$

88. $\dfrac{92}{0}$

89. $\dfrac{0}{32}$

90. $\dfrac{0}{17}$

91. $\dfrac{9}{y - y}$

92. $\dfrac{2x - 2x}{2x - 2x}$

Find the reciprocal of the number.

93. $\dfrac{3}{4}$

94. $\dfrac{9}{10}$

95. $-\dfrac{7}{8}$

96. $-\dfrac{5}{6}$

97. 25

98. -65

99. 0.2

100. 0.8

101. $-\dfrac{a}{b}$

102. $\dfrac{1}{8x}$

Divide.

103. $\dfrac{2}{7} \div \left(-\dfrac{11}{3}\right)$

104. $\dfrac{3}{5} \div \left(-\dfrac{6}{7}\right)$

105. $-\dfrac{10}{3} \div \left(-\dfrac{2}{15}\right)$

106. $-\dfrac{12}{5} \div \left(-\dfrac{3}{10}\right)$

107. $18.6 \div (-3.1)$

108. $39.9 \div (-13.3)$

109. $(-75.5) \div (-15.1)$

110. $(-12.1) \div (-0.11)$

111. $-48 \div 0.4$

112. $520 \div (-0.13)$

113. $\dfrac{3}{4} \div \left(-\dfrac{2}{3}\right)$

114. $\dfrac{5}{8} \div \left(-\dfrac{1}{2}\right)$

115. $-\dfrac{5}{4} \div \left(-\dfrac{3}{4}\right)$

116. $-\dfrac{5}{9} \div \left(-\dfrac{5}{6}\right)$

117. $-\dfrac{2}{3} \div \left(-\dfrac{4}{9}\right)$

118. $-\dfrac{3}{5} \div \left(-\dfrac{5}{8}\right)$

119. $-\dfrac{3}{8} \div \left(-\dfrac{8}{3}\right)$

120. $-\dfrac{5}{8} \div \left(-\dfrac{5}{6}\right)$

121. $-6.6 \div 3.3$

122. $-44.1 \div (-6.3)$

123. $\dfrac{-12}{-13}$

124. $\dfrac{-1.9}{20}$

125. $\dfrac{48.6}{-30}$

126. $\dfrac{-17.8}{3.2}$

127. $\dfrac{-9}{17 - 17}$

128. $\dfrac{-8}{-6 + 6}$

129. Complete the following table.

Number	Opposite (Additive Inverse)	Reciprocal (Multiplicative Inverse)
$\dfrac{2}{3}$		
$-\dfrac{5}{4}$		
0		
1		
-4.5		
$x, x \neq 0$		

Skill Maintenance

The exercises that follow begin an important feature called *skill maintenance exercises*. These exercises provide an ongoing review of any preceding objective in the book. You will see them in virtually every exercise set. It has been found that this kind of extensive review can significantly improve your performance on a final examination.

Given the numbers $\sqrt{3}$, -12.47, -13, 26, π, 0, $-\dfrac{23}{32}$, $\dfrac{7}{11}$, $4.57557555755557\ldots$:   [R.1a]

130. Name the whole numbers.

131. Name the natural numbers.

132. Name the integers.

133. Name the irrational numbers.

134. Name the rational numbers.

135. Name the real numbers.

Use either $<$ or $>$ for $\blacksquare$ to write a true sentence.   [R.1b]

136. $-7 \ \blacksquare \ 8$

137. $5 \ \blacksquare \ \dfrac{3}{8}$

138. $-45.6 \ \blacksquare \ -23.8$

139. $123 \ \blacksquare \ -10$

Synthesis

140. ◆ Explain in your own words why $\dfrac{7}{0}$ is undefined.

141. ◆ Explain in your own words why a positive number divided by a negative number is negative.

142. What number can be added to 11.7 to obtain $-7\dfrac{3}{4}$?

143. The reciprocal of an electric resistance is called **conductance**. When two resistors are connected in parallel, the conductance is the sum of the conductances,

$$\dfrac{1}{r_1} + \dfrac{1}{r_2}.$$

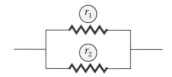

Find the conductance of two resistors of 12 ohms and 6 ohms when connected in parallel.

R.3 Exponential Notation and Order of Operations

a | Exponential Notation

Exponential notation is a shorthand device. For $3 \cdot 3 \cdot 3 \cdot 3$, we write 3^4. In the **exponential notation** 3^4, the number 3 is called the **base** and the number 4 is called the **exponent**.

> Exponential notation a^n, where n is an integer greater than 1, means
> $$\underbrace{a \cdot a \cdot a \cdots a \cdot a.}_{n \text{ factors}}$$
> We read "a^n" as "a to the nth power," or simply "a to the nth."
> We can read "a^2" as "a-squared" and "a^3" as "a-cubed."

CAUTION! a^n does *not* mean to multiply n times a. For example, 3^2 means $3 \cdot 3$, or 9, not $3 \cdot 2$, or 6.

Examples Write exponential notation.

1. $7 \cdot 7 \cdot 7 = 7^3$

2. $xxxxx = x^5$

3. $\frac{2}{3} \cdot \frac{2}{3} \cdot \frac{2}{3} \cdot \frac{2}{3} = \left(\frac{2}{3}\right)^4$

Do Exercises 1–3.

Examples Evaluate.

4. $5^2 = 5 \cdot 5 = 25$

5. $\left(\frac{1}{2}\right)^3 = \frac{1}{2} \cdot \frac{1}{2} \cdot \frac{1}{2} = \frac{1}{8}$

6. $\left(\frac{7}{8}\right)^2 = \frac{7}{8} \cdot \frac{7}{8} = \frac{49}{64}$

7. $(0.1)^4 - (0.1)(0.1)(0.1)(0.1)$
$\qquad - 0.0001$

8. $(-5)^3 = (-5)(-5)(-5)$
$\qquad = -125$

9. $(-10)^4 = (-10)(-10)(-10)(-10)$
$\qquad = 10,000$

Do Exercises 4–8.

When an exponent is an integer greater than 1, it tells how many times the base occurs as a factor. What happens when the exponent is 1 or 0? We cannot have the base occurring as a factor 1 time or 0 times because there are no products. Look for a pattern below. Think of dividing by 10 on the right.

On this side, the exponents decrease by 1 at each step.

$10^4 = 10 \cdot 10 \cdot 10 \cdot 10 = 10,000$
$10^3 = 10 \cdot 10 \cdot 10 = 1000$
$10^2 = 10 \cdot 10 = 100$
$10^1 = ?$
$10^0 = ?$

On this side, we divide by 10 at each step.

In order for the pattern to continue, 10^1 would have to be 10 and 10^0 would have to be 1. We will *agree* that exponents of 1 and 0 have that meaning.

Objectives

a Rewrite expressions with whole-number exponents, and evaluate exponential expressions.

b Rewrite expressions with or without negative integers as exponents.

c Simplify expressions using the rules for order of operations.

For Extra Help

TAPE 1   TAPE 1B   MAC WIN   CD-ROM

Write exponential notation.

1. $8 \cdot 8 \cdot 8 \cdot 8$

2. $mmmmmm$

3. $\dfrac{7}{8} \cdot \dfrac{7}{8} \cdot \dfrac{7}{8}$

Evaluate.

4. 3^4

5. $\left(\dfrac{1}{4}\right)^2$

6. $(-10)^6$

7. $(0.2)^3$

8. $(5.8)^4$

Answers on page A-3

Rewrite without exponents.

9. 8^1

10. $(-31)^1$

11. 3^0

12. $(-7)^0$

13. y^0, where $y \neq 0$

Rewrite using a positive exponent.

14. m^{-4}

15. $(-4)^{-3}$

16. $\dfrac{1}{x^{-3}}$

Answers on page A-3

> For any number a, we agree that a^1 means a.
> For any nonzero number a, we agree that a^0 means 1.

Examples   Rewrite without an exponent.

10. $4^1 = 4$ **11.** $(-97)^1 = -97$

12. $6^0 = 1$ **13.** $(-37.4)^0 = 1$

Let's consider a justification for not defining 0^0. By examining the pattern $3^0 = 1$, $2^0 = 1$, and $1^0 = 1$, we might think that 0^0 should be 1. However, by examining the pattern $0^3 = 0$, $0^2 = 0$, and $0^1 = 0$, we might think that 0^0 should be 0. To avoid this confusion, mathematicians agree *not* to define 0^0.

Do Exercises 9–13.

b   Negative Integers as Exponents

How shall we define negative integers as exponents? Look for a pattern below. Again, think of dividing by 10 on the right.

$$
\begin{array}{c|c|c}
\text{On this side,} & 10^2 = 100 & \text{On this side,} \\
\text{the exponents} & 10^1 = 10 & \text{we divide by 10} \\
\text{decrease by 1} & 10^0 = 1 & \text{at each step.} \\
\text{at each step.} & 10^{-1} = ? & \\
\downarrow & 10^{-2} = ? & \downarrow
\end{array}
$$

In order for the pattern to continue, 10^{-1} would have to be $\frac{1}{10}$ and 10^{-2} would have to be $\frac{1}{100}$. This leads to the following agreement.

> For any real number a that is nonzero and any integer n,
> $$\frac{1}{a^n} = a^{-n}.$$
> (To take the reciprocal of a^n, change the sign of the exponent.)

The numbers a^n and a^{-n} are reciprocals because

$$a^n \cdot a^{-n} = a^n \cdot \frac{1}{a^n} = \frac{a^n}{a^n} = 1.$$

Examples   Rewrite using a positive exponent. Evaluate, if possible.

14. $y^{-5} = \dfrac{1}{y^5}$ **15.** $\dfrac{1}{t^{-4}} = t^4$ (t^{-4} and t^4 are reciprocals)

16. $(-2)^{-3} = \dfrac{1}{(-2)^3} = \dfrac{1}{(-2)(-2)(-2)}$

$$= \frac{1}{-8} = -\frac{1}{8}$$

CAUTION!   A negative exponent does *not* necessarily indicate that an answer is negative! For example, 3^{-2} means $1/3^2$, or $1/9$, not -9.

Do Exercises 14–16.

Examples Rewrite using a negative exponent.

17. $\dfrac{1}{x^2} = x^{-2}$

18. $\dfrac{1}{(-7)^4} = (-7)^{-4}$

Do Exercises 17 and 18.

c Order of Operations

What does $8 + 2 \cdot 5^3$ mean? If we add 8 and 2 and multiply by 5^3, or 125, we get 1250. If we multiply 2 times 125 and add 8, we get 258. Both results cannot be correct. To avoid such difficulties, we make agreements about which operations should be done first.

RULES FOR ORDER OF OPERATIONS

1. Do all the calculations within grouping symbols, like parentheses, before operations outside.

2. Evaluate all exponential expressions.

3. Do all multiplications and divisions in order from left to right.

4. Do all additions and subtractions in order from left to right.

Most computers and calculators are programmed using these rules.

Example 19 Simplify: $-43 \cdot 56 - 17$.

There are no parentheses or powers so we start with the third step.

$$-43 \cdot 56 - 17 = -2408 - 17 \quad \text{Carrying out all multiplications and divisions in order from left to right}$$

$$= -2425 \quad \text{Carrying out all additions and subtractions in order from left to right}$$

Example 20 Simplify and compare: $(8 - 10)^2$ and $8^2 - 10^2$.

$$(8 - 10)^2 = (-2)^2 = 4;$$
$$8^2 - 10^2 = 64 - 100 = -36$$

We see that $(8 - 10)^2$ and $8^2 - 10^2$ are *not* the same.

Example 21 Simplify: $3^4 + 62 \cdot 8 - 2(29 + 33 \cdot 4)$.

$$3^4 + 62 \cdot 8 - 2(29 + 33 \cdot 4)$$

$$= 3^4 + 62 \cdot 8 - 2(29 + 132) \quad \text{Carrying out operations inside parentheses first; doing the multiplication}$$

$$= 3^4 + 62 \cdot 8 - 2(161) \quad \text{Completing the addition inside parentheses}$$

$$= 81 + 62 \cdot 8 - 2(161) \quad \text{Evaluating exponential expressions}$$

$$= 81 + 496 - 322 \quad \text{Doing all multiplications}$$

$$= 577 - 322 \quad \text{Doing all additions and subtractions in order from left to right}$$

$$= 255$$

Do Exercises 19–21.

Rewrite using a negative exponent.

17. $\dfrac{1}{q^3}$

18. $\dfrac{1}{(-5)^4}$

Simplify.

19. $43 - 52 \cdot 80$

20. $62 \cdot 8 + 4^3 - (5^2 - 64 \div 4)$

21. Simplify and compare:

$$(7 - 4)^2 \quad \text{and} \quad 7^2 - 4^2.$$

Answers on page A-3

Simplify.

22. $6 - \{5 - [2 - (8 + 20)]\}$

23. $5 + \{6 - [2 + (5 - 2)]\}$

Simplify.

24. $\dfrac{8 \cdot 7 - |6 - 8|}{5^2 + 6^3}$

25. $\dfrac{(8 - 3)^2 + (7 - 10)^2}{3^2 - 2^3}$

When parentheses occur within parentheses, we can make them different shapes, such as [ ] (also called "brackets") and { } (usually called "braces"). All of these have the same meaning. When parentheses occur within parentheses, computations in the *innermost* ones are to be done first.

Example 22 Simplify: $5 - \{6 - [3 - (7 + 3)]\}$.

$$5 - \{6 - [3 - (7 + 3)]\} = 5 - \{6 - [3 - 10]\} \quad \text{Adding } 7 + 3$$
$$= 5 - \{6 - [-7]\} \quad \text{Subtracting } 3 - 10$$
$$= 5 - 13 \quad \text{Subtracting } 6 - (-7)$$
$$= -8$$

Example 23 Simplify: $7 - [3(2 - 5) - 4(2 + 3)]$.

$$7 - [3(2 - 5) - 4(2 + 3)] = 7 - [3(-3) - 4(5)] \quad \begin{array}{l}\text{Doing the calculations in}\\ \text{the innermost parentheses}\\ \text{first}\end{array}$$
$$= 7 - [-9 - 20]$$
$$= 7 - [-29]$$
$$= 36$$

Do Exercises 23 and 24.

In addition to the usual grouping symbols—parentheses, brackets, and braces—a fraction bar and absolute-value signs can act as grouping symbols.

Example 24 Calculate: $\dfrac{12|7 - 9| + 8 \cdot 5}{3^2 + 2^3}$.

An equivalent expression with brackets as grouping symbols is

$$[12|7 - 9| + 8 \cdot 5] \div [3^2 + 2^3].$$

What this shows, in effect, is that we do the calculations in the numerator and in the denominator, and then divide the results:

$$\frac{12|7 - 9| + 8 \cdot 5}{3^2 + 2^3} = \frac{12|-2| + 8 \cdot 5}{9 + 8}$$
$$= \frac{12(2) + 8 \cdot 5}{17}$$
$$= \frac{24 + 40}{17} = \frac{64}{17}.$$

Do Exercises 24 and 25.

Answers on page A-3

Calculator Spotlight

Negative Numbers. To enter a negative number on a graphing calculator, we use the opposite key $(-)$, found in the lower row of a TI-82 or TI-83, rather than the subtraction key, located in the far right column. To enter -5, we press $(-)$ 5 ENTER. The display then reads -5.

To enter $-\frac{7}{8}$, we press $(-)$ 7 $/$ 8 ENTER. The display then gives the answer in decimal notation: $-.875$.

Exponents and Powers. To find 3^5, we press the following keystrokes: 3 $\wedge$ 5 ENTER. The answer 243 is displayed.

Since raising a number to the second power is so common, the calculator has a special "x-squared" key, x^2. To find 1.5^2, we press 1 $.$ 5 x^2 ENTER. The display will show 2.25.

To enter a power like $(-39)^4$, we press $($ $(-)$ 3 9 $)$ $\wedge$ 4 ENTER. The answer is 2,313,441.

To find -39^4, think of it first as -1×39^4. We then press $(-)$ 3 9 $\wedge$ 4 ENTER. The answer is $-2,313,441$.

Order of Operations and Grouping Symbols. Let's consider simplifying the expression $3 + 4 \cdot 2$. We know that by following the rules for order of operations, we have $3 + 4 \cdot 2 = 3 + 8 = 11$. That is, we multiply first and then add. We can evaluate this expression by pressing the following keys: 3 $+$ 4 $\times$ 2 ENTER. We obtain the answer 11.

Sometimes we do need grouping symbols such as $($ and $)$. To do a calculation like $-7(2 - 9) - 20$, we press the following keys: $(-)$ 7 $($ 2 $-$ 9 $)$ $-$ 2 0 ENTER. The multiplication key $\times$ could be used but this is generally not necessary. The answer is 29.

To do a calculation like $-8 - (-2.3)$, we press these keys: $(-)$ 8 $-$ $(-)$ 2 $.$ 3 ENTER. The answer is -5.7. Note that we did not need grouping symbols because of the way in which the calculator is programmed.

To calculate the value of an expression like

$$\frac{38 + 142}{2 - 47},$$

we also use grouping symbols:

$$(38 + 142) \div (2 - 47).$$

We then press $($ 3 8 $+$ 1 4 2 $)$ $\div$ $($ 2 $-$ 4 7 $)$ ENTER. The answer is -4.

Exercises

Press the appropriate keys so that your calculator displays each of the following numbers.

1. -3 2. -508 3. -0.17 4. $-\frac{5}{8}$

Evaluate.

5. 4^5 6. 7^9

7. 19^2 8. 5.718^2

9. 1.8^4 10. 23.04^3

11. $\left(\frac{17}{32}\right)^5$ 12. $\left(\frac{17}{32}\right)^9$

13. $(-7)^6$ 14. $(-17)^5$

15. $(-104)^3$ 16. -7^6

17. -17^5 18. -104^3

Calculate.

19. $36 \div 2 \cdot 3 - 4 \cdot 4$ 20. $68 - 8 \div 4 + 3 \cdot 5$

21. $36 \div (2 \cdot 3 - 4) \cdot 4$ 22. $-8 + 4(7 - 9) + 5$

23. $-3[2 + (-5)]$

24. $7[4 - (-3)] + 5[3^2 - (-4)]$

25. $(15 + 3)^3 + 4(12 - 7)^2$

26. $50.6 - 8.9 \times 3.01 + 4(5^2 - 224.7)$

27. $3.2 + 4.7[159.3 - 2.1(60.3 - 59.4)]$

28. $\{(150 \cdot 5) \div [(3 \cdot 16) \div (8 \cdot 3)]\} + 25(12 \div 4)$

29. $\left(\frac{28}{89} + 42.8 \times 17.01\right)^3 \div \left(\frac{678}{119} - \frac{23.2}{46.08}\right)^2$

30. $\dfrac{178 - 38}{5 + 30}$

31. $785 - \dfrac{5^4 - 285}{17 + 3 \cdot 51}$

32. $12^5 - 12^4 + 11^5 \div 11^3 - 10.2^2$

33. $\dfrac{311 - 17^2}{2 - 13}$

34. $\dfrac{32.1^3 - (54/67)^2}{78.6^5} - \dfrac{285 - 5^4}{17 + 3 \cdot 51}$

Improving Your Math Study Skills

Forming Math Study Groups, by James R. Norton

Dr. James Norton has taught at the University of Phoenix and Scottsdale Community College. He has extensive experience with the use of study groups to learn mathematics.

The use of math study groups for learning has become increasingly common in recent years. Some instructors regard them as a primary source of learning, while others let students form groups on their own.

A study group generally consists of study partners who help each other learn the material and do the homework. You will probably meet outside of class at least once or twice a week. Here are some do's and don'ts to make your study group more valuable.

- DO make the group up of no more than four or five people. Research has shown clearly that this size works best.

- DO trade phone numbers so that you can get in touch with each other for help between team meetings.

- DO make sure that everyone in the group has a chance to contribute.

- DON'T let a group member copy from others without contributing. If this should happen, one member should speak with that student privately; if the situation continues, that student should be asked to leave the group.

- DON'T let the "A" students be passive. The group needs them! The benefits to even the best students

are twofold: (1) Other students will benefit from their expertise and (2) the bright students will learn the material better by teaching it to someone else.

- DON'T let the slower students be passive either. *Everyone* can contribute something, and being in a group will actually improve their self-esteem as well as their performance.

How do you form study groups if the instructor has not already done so? A good place to begin is to get together with three or four friends and arrange a study time. If you don't know anyone, start getting acquainted with other people in the class during the first week of the semester.

What should you look for in a study partner?

- Do you live near each other to make it easy to get together?

- What are your class schedules like? Are you both on campus? Do you have free time?

- What about work schedules, athletic practice, and other out-of-school commitments that you might have to work around?

Making use of a study group is not a form of "cheating." You are merely helping each other learn. So long as everyone in the group is both contributing and doing the work, this method will bring you great success!

Exercise Set R.3

a Write exponential notation.

1. $4 \cdot 4 \cdot 4 \cdot 4 \cdot 4$

2. $6 \cdot 6 \cdot 6$

3. $5 \cdot 5 \cdot 5 \cdot 5 \cdot 5 \cdot 5$

4. $x \cdot x \cdot x \cdot x$

5. mmm

6. $ttttt$

7. $\dfrac{7}{12} \cdot \dfrac{7}{12} \cdot \dfrac{7}{12} \cdot \dfrac{7}{12}$

8. $(3.8)(3.8)(3.8)(3.8)(3.8)$

9. $(123.7)(123.7)$

10. $\left(-\dfrac{4}{5}\right)\left(-\dfrac{4}{5}\right)\left(-\dfrac{4}{5}\right)$

Evaluate.

11. 2^7

12. 9^3

13. $(-2)^5$

14. $(-7)^2$

15. $\left(\dfrac{1}{3}\right)^4$

16. $(0.1)^6$

17. $(-4)^3$

18. $(-3)^4$

19. $(-5.6)^2$

20. $\left(\dfrac{2}{3}\right)^4$

21. 5^1

22. $(\sqrt{6})^1$

23. 34^0

24. $\left(\dfrac{5}{2}\right)^1$

25. $(\sqrt{6})^0$

26. $(-4)^0$

27. $\left(\dfrac{7}{8}\right)^1$

28. $(-87)^0$

b Rewrite using a positive exponent. Evaluate, if possible.

29. y^{-5}

30. x^{-6}

31. $\dfrac{1}{a^{-2}}$

32. $\dfrac{1}{y^{-7}}$

33. $(-11)^{-1}$

34. $(-4)^{-3}$

Rewrite using a negative exponent.

35. $\dfrac{1}{3^4}$

36. $\dfrac{1}{9^2}$

37. $\dfrac{1}{b^3}$

38. $\dfrac{1}{n^5}$

39. $\dfrac{1}{(-16)^2}$

40. $\dfrac{1}{(-8)^6}$

c Simplify.

41. $12 - 4(5 - 1)$

42. $6 - 4(8 - 5)$

43. $9[8 - 7(5 - 2)]$

44. $10[7 - 4(8 - 5)]$

45. $[5(8 - 6) + 12] - [24 - (8 - 4)]$

46. $[9(7 - 4) + 19] - [25 - (7 + 3)]$

47. $[64 \div (-4)] \div (-2)$

48. $[48 \div (-3)] \div \left(-\dfrac{1}{4}\right)$

49. $19(-22) + 60$

50. $30 \cdot 10 - 18 \cdot 25$

51. $(5 + 7)^2; \quad 5^2 + 7^2$

52. $(9 - 12)^2; \quad 9^2 - 12^2$

53. $2^3 + 2^4 - 20 \cdot 30$

54. $7 \cdot 8 - 3^2 - 2^3$

55. $5^3 + 36 \cdot 72 - (18 + 25 \cdot 4)$

56. $4^3 + 20 \cdot 10 + 7^2 - 23$

57. $(13 \cdot 2 - 8 \cdot 4)^2$

58. $(9 \cdot 8 + 3 \cdot 3)^2$

59. $4000 \cdot (1 + 0.12)^3$

60. $5000 \cdot (4 + 1.16)^2$

61. $(20 \cdot 4 + 13 \cdot 8)^2 - (39 \cdot 15)^3$

62. $(43 \cdot 6 - 14 \cdot 7)^3 + (33 \cdot 34)^2$

63. $18 - 2 \cdot 3 - 9$

64. $18 - (2 \cdot 3 - 9)$

65. $(18 - 2 \cdot 3) - 9$

66. $(18 - 2)(3 - 9)$

67. $[24 \div (-3)] \div \left(-\dfrac{1}{2}\right)$

68. $[(-32) \div (-2)] \div (-2)$

69. $15 \cdot (-24) + 50$

70. $30 \cdot 20 - 15 \cdot 24$

71. $4 \div (8 - 10)^2 + 1$

72. $16 \div (19 - 15)^2 - 7$

73. $6^3 + 25 \cdot 71 - (16 + 25 \cdot 4)$

74. $5^3 + 20 \cdot 40 + 8^2 - 29$

75. $5000 \cdot (1 + 0.16)^3$

76. $4000 \cdot (3 + 1.14)^2$

77. $4 \cdot 5 - 2 \cdot 6 + 4$

78. $8(7 - 3)/4$

79. $4 \cdot (6 + 8)/(4 + 3)$

80. $4^3/8$

81. $[2 \cdot (5 - 3)]^2$

82. $5^3 - 7^2$

83. $8(-7) + 6(-5)$

84. $10(-5) + 1(-1)$

85. $19 - 5(-3) + 3$

86. $14 - 2(-6) + 7$

87. $9 \div (-3) + 16 \div 8$

88. $-32 - 8 \div 4 - (-2)$

89. $7 + 10 - (-10 \div 2)$

90. $(3 - 8)^2$

91. $5^2 - 8^2$

92. $28 - 10^3$

93. $20 + 4^3 \div (-8)$

94. $2 \times 10^3 - 5000$

95. $-7(3^4) + 18$

96. $6[9 - (3 - 4)]$

97. $9[(8 - 11) - 13]$

98. $1000 \div (-100) \div 10$

99. $256 \div (-32) \div (-4)$

100. $\dfrac{20 - 6^2}{9^2 + 3^2}$

101. $\dfrac{5^2 - |4^3 - 8|}{9^2 - 2^2 - 1^5}$

102. $\dfrac{4|6 - 7| - 5 \cdot 4}{6 \cdot 7 - 8|4 - 1|}$

103. $\dfrac{30(8 - 3) - 4(10 - 3)}{10|2 - 6| - 2(5 + 2)}$

104. $\dfrac{5^3 - 3^2 + 12 \cdot 5}{-32 \div (-16) \div (-4)}$

Skill Maintenance

Find the absolute value. [R.1d]

105. $\left| -\dfrac{9}{7} \right|$

106. $|2.3|$

107. $|0|$

108. $|-900|$

Compute. [R.2a, c, d]

109. $23 - 56$

110. $-23 - 56$

111. $-23 - (-56)$

112. $-23 + (-56)$

113. $(-10)(2.3)$

114. $(-10)(-2.3)$

115. $10(-2.3)$

116. $\left(-\dfrac{2}{3} \right)\left(-\dfrac{15}{16} \right)$

Synthesis

117. ◈ Explain the meaning of a negative exponent in as many ways as you can.

118. ◈ Students often use the memory device PEMDAS, or "Please Excuse My Dear Aunt Sally," to remember the rules for order of operations. Explain how this works.

Simplify.

119. $(-2)^0 - (-2)^3 - (-2)^{-1} + (-2)^4 - (-2)^{-2}$

120. $2(6^1 \cdot 6^{-1} - 6^{-1} \cdot 6^0)$

121. Place parentheses in this statement to make it true: $9 \cdot 5 + 2 - 8 \cdot 3 + 1 = 22$.

The symbol [icon] means to use your calculator to work a particular exercise.

122. [icon] Find each of the following.

$12345679 \cdot 9 = ?$

$12345679 \cdot 18 = ?$

$12345679 \cdot 27 = ?$

Then look for a pattern and find $12345679 \cdot 36$ without the use of a calculator.

123. [icon] Find $(0.2)^{(-0.2)^{-1}}$.

124. [icon] Compare $(\pi)^{\sqrt{2}}$ and $(\sqrt{2})^{\pi}$.

Chapter R   Review of Basic Algebra

[Collaborative Learning Manual]

Use the order of operations to modify an expression.

Part 2 Manipulations

R.4 Introduction to Algebraic Expressions

The study of algebra involves the use of equations to solve problems. Equations are constructed from algebraic expressions. The purpose of Part 2 of this chapter is to introduce you to the types of expressions encountered in algebra and ways in which we can manipulate them.

Algebraic Expressions and Their Use

In arithmetic, you worked with expressions such as

$$91 + 76, \qquad 26 - 17, \qquad 14 \cdot 35, \qquad \frac{7}{8}, \quad \text{and} \quad 5^2 - 3^2.$$

In algebra, we use these as well as expressions like

$$x + 76, \qquad 26 - q, \qquad 14 \cdot x, \qquad \frac{d}{t}, \quad \text{and} \quad x^2 - y^2.$$

When a letter is used to stand for various numbers, it is called a **variable**. If a letter represents one particular number, it is called a **constant**. Let d = the number of hours in a day. Then d is a constant. Let t = the number of hours that a passenger jet has been flying. Then t is a variable, because t changes as the flight continues.

An **algebraic expression** consists of variables, numbers, and operation signs. All the expressions above are examples of algebraic expressions. When an equals sign is placed between two expressions, an **equation** is formed.

Algebraic expressions and equations occur frequently in applied and problem-solving situations. For example, consider the following bar graph, which you might see in a newspaper or a magazine.

Top Participatory Sports

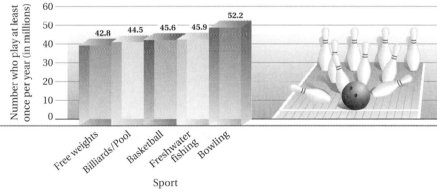

Source: American Sports Data, Inc.

Suppose we want to determine how many more people participate in bowling than in billiards/pool. Using algebra, we can translate the problem to an equation. It might be done as follows.

1. Translate to an equation and solve: How many more people participate in freshwater fishing than in free weights?

Number of participants in billiards/pool	plus	How many more people	is	Number of participants in bowling
↓	↓	↓	↓	↓
44.5	+	x	=	52.2

Note that we have an algebraic expression on the left. To find the number x, we can subtract 44.5 on both sides of the equation:

$$44.5 + x = 52.2$$
$$44.5 + x - 44.5 = 52.2 - 44.5$$
$$x = 7.7.$$

We obtain the answer, 7.7 million.

Do Exercise 1.

a | Translating to Algebraic Expressions

To translate problems to equations, we need to know that certain words correspond to certain symbols, as shown in the following tables.

KEY WORDS			
Addition	**Subtraction**	**Multiplication**	**Division**
add	subtract	multiply	divide
sum	difference	product	divided by
plus	minus	times	quotient
increased by	decreased by	twice	ratio
more than	less than	of	per

2. Translate to an algebraic expression: Sixteen less than some number.

Phrase	Algebraic Expression
Five *more than* some number	$n + 5$
Half *of* a number	$\frac{1}{2}t$ or $\frac{t}{2}$
Five *more than* three *times* some number	$3p + 5$
The *difference* of two numbers	$x - y$
Six *less than* the *product* of two numbers	$rs - 6$
Seventy-six percent *of* some number	$0.76z$ or $\frac{76}{100}z$

Note that expressions like rs represent products and can also be written as $r \cdot s$, $r \times s$, $(r)(s)$, or $r(s)$. The multipliers r and s are also called **factors**.

Example 1 Translate to an algebraic expression: Eight less than some number.

We can use any variable we wish, such as x, y, t, m, n, and so on. Here we let t represent the number. If we knew the number to be 23, then the translation of "eight less than 23" would be $23 - 8$. If we knew the number to be 345, then the translation of "eight less than 345" would be $345 - 8$. Since we are using a variable for the number, the translation is

$$t - 8. \quad \text{Caution! } 8 - t \text{ would be incorrect.}$$

Do Exercise 2.

Answers on page A-3

Example 2 Translate to an algebraic expression: Twenty-two more than some number.

This time we let y represent the number. If we knew the number to be 47, then the translation would be $47 + 22$, or $22 + 47$. If we knew the number to be 17.95, then the translation would be $17.95 + 22$, or $22 + 17.95$. Since we are using a variable, the translation is

$$y + 22, \quad \text{or} \quad 22 + y.$$

Example 3 Translate to an algebraic expression: Five less than forty-three percent of the quotient of two numbers.

We let r and s represent the two numbers.

$$(0.43) \cdot \frac{r}{s} - 5 \qquad 43\% = 0.43$$

Five less than   forty-three percent   of   the quotient of two numbers

Do Exercises 3–7.

b Evaluating Algebraic Expressions

When we replace a variable with a number, we say that we are **substituting** for the variable. This process is called **evaluating the expression.**

Example 4 Evaluate $x - y$ for $x = 83$ and $y = 49$.

We substitute 83 for x and 49 for y and carry out the subtraction:

$$x - y = 83 - 49 = 34.$$

The number 34 is called the **value** of the expression.

Example 5 Evaluate a/b for $a = -63$ and $b = 7$.

We substitute -63 for a and 7 for b and carry out the division:

$$\frac{a}{b} = \frac{-63}{7} = -9.$$

Do Exercises 8–11.

Example 6 Evaluate the expression $3xy + z$ for $x = 2$, $y = -5$, and $z = 7$.

We substitute and carry out the calculations according to the rules for order of operations:

$$3xy + z = 3(2)(-5) + 7$$
$$= -30 + 7$$
$$= -23.$$

Do Exercises 12 and 13 on the following page.

Translate to an algebraic expression.

3. Forty-seven more than some number

4. Sixteen minus some number

5. One-fourth of some number

6. Six more than eight times some number

7. Eight less than ninety-nine percent of the quotient of two numbers

8. Evaluate $a + b$ for $a = 48$ and $b = 36$.

9. Evaluate $x - y$ for $x = -97$ and $y = 29$.

10. Evaluate a/b for $a = 400$ and $b = -8$.

11. Evaluate $8t$ for $t = 15$.

Answers on page A-3

12. Evaluate $4x + 5y$ for $x = -2$ and $y = 10$.

13. Evaluate $7ab - c$ for $a = -3$, $b = 4$, and $c = 62$.

14. Find the area of a triangle when h is 24 ft and b is 8 ft.

15. Evaluate $(x - 3)^2$ for $x = 11$.

16. Evaluate $x^2 - 6x + 9$ for $x = 11$.

17. Evaluate $8 - x^3 + 10 \div 5y^2$ for $x = 4$ and $y = 6$.

Geometric formulas must often be evaluated in applied problems. In the next example, we use the formula for the area A of a triangle with a base of length b and a height of length h:

$$A = \tfrac{1}{2}bh.$$

Example 7   The base of a triangular sail is 8 m and the height is 6.4 m. Find the area of the sail.

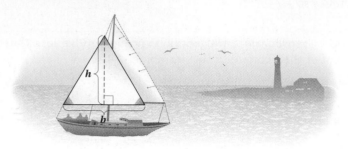

We substitute 8 for b and 6.4 for h and multiply:

$$A = \tfrac{1}{2}bh = \tfrac{1}{2} \cdot 8 \cdot 6.4$$
$$= 25.6 \text{ m}^2.$$

Do Exercise 14.

Example 8   Evaluate $5 + 2(a - 1)^2$ for $a = 4$.

$$
\begin{aligned}
5 + 2(a - 1)^2 &= 5 + 2(4 - 1)^2 && \text{Substituting}\\
&= 5 + 2(3)^2 && \text{Working within parentheses first}\\
&= 5 + 2(9) && \text{Simplifying } 3^2\\
&= 5 + 18 && \text{Multiplying}\\
&= 23 && \text{Adding}
\end{aligned}
$$

The rules for order of operations tell us to divide before we multiply when division appears first, reading left to right. Similarly, if subtraction appears before addition, we subtract before we add.

Example 9   Evaluate $9 - x^3 + 6 \div 2y^2$ for $x = 2$ and $y = 5$.

$$
\begin{aligned}
9 - x^3 + 6 \div 2y^2 &= 9 - 2^3 + 6 \div 2(5)^2 && \text{Substituting}\\
&= 9 - 8 + 6 \div 2 \cdot 25 && \text{Simplifying } 2^3 \text{ and } 5^2\\
&= 9 - 8 + 3 \cdot 25 && \text{Dividing}\\
&= 9 - 8 + 75 && \text{Multiplying}\\
&= 1 + 75 && \text{Subtracting}\\
&= 76 && \text{Adding}
\end{aligned}
$$

Do Exercises 15–17.

Answers on page A-3

Exercise Set R.4

a Translate to an algebraic expression.

1. Seven more than some number

2. Two less than some number

3. Twelve times some number

4. Twice a number

5. Sixty-five percent of some number

6. Thirty-nine percent of some number

7. Nine less than twice a number

8. Four more than half a number

9. Eight more than ten percent of a number

10. Five less than ten percent of some number

11. One less than the difference of two numbers

12. Two more than the product of two numbers

13. Ninety miles for every g gallons of gasoline

14. One hundred twenty words for every t seconds

15. The sum of a and b

16. The square of the sum of a and b

17. The square of x minus y

18. The square of x minus the square of y

19. A driver drove at a speed of 75 mph on an interstate highway in Arizona for t hours. How far did the driver travel?

20. Sarah drove on her sales route at a speed of r mph for 4 hours. How far did she travel?

b Evaluate.

21. $23z$, for $z = -4$

22. $57y$, for $y = -8$

23. $\dfrac{a}{b}$, for $a = -24$ and $b = -8$

24. $\dfrac{x}{y}$, for $x = 30$ and $y = -6$

25. $\dfrac{m - n}{8}$, for $m = 36$ and $n = 4$

26. $\dfrac{5}{p + q}$, for $p = 20$ and $q = 30$

27. $\dfrac{5z}{y}$, for $z = 9$ and $y = 2$

28. $\dfrac{18m}{n}$, for $m = 7$ and $n = 18$

29. $2c \div 3b$, for $b = 4$ and $c = 6$

30. $4x - y$, for $x = 3$ and $y = -2$

31. $25 - r^2 + s \div r^2$, for $r = 3$ and $s = 27$

32. $n^3 - 2 + p \div n^2$, for $n = 2$ and $p = 12$

33. $m + n(5 + n^2)$, for $m = 15$ and $n = 3$

34. $a^2 - 3(a - b)$, for $a = 10$ and $b = -8$

Simple Interest. The **simple interest** on a principal of P dollars at interest rate r for t years is given by $I = Prt$.

35. Find the simple interest on a principal of $7345 at 6% for 1 yr.

36. Find the simple interest on a principal of $18,000 at 4.6% for 2 yr. (*Hint:* 4.6% = 0.046.)

37. *Area of a Compact Disc.* The area A of a circle with radius r is given by $A = \pi r^2$. The circumference C of the circle is given by $C = 2\pi r$. The standard compact disc used for software and music has a radius of 6 cm. Find the area and the circumference of such a CD (ignoring the hole in the middle). Use 3.14 for π.

38. *Area of a Parallelogram.* The area A of a parallelogram with base b and height h is given by $A = bh$. Find the area of a parallelogram-shaped horse-feeding area with a height of 24.3 ft and a base of 67.8 ft.

Skill Maintenance

Evaluate.   [R.3a]

39. 3^5

40. $(-3)^5$

41. $(-10)^2$

42. $(-10)^4$

43. $(-5.3)^2$

44. $\left(\dfrac{3}{5}\right)^2$

45. $(4.5)^0$

46. $(4.5)^1$

47. $(3x)^1$

48. $(3x)^0$

Synthesis

49. ◈ If the base and the height of a triangle are doubled, does its area double? Explain.

50. ◈ If the base and the height of a parallelogram are doubled, does its area double? Explain.

Translate to an equation.

51. The distance d that a rapid transit train in the Denver airport travels in time t at a speed r is given by speed times time. Write an equation for d.

52. You invest P dollars at 6.7% simple interest. Write an equation for the number of dollars N in the account 1 yr from now.

Evaluate.

53. $\dfrac{x + y}{2} + \dfrac{3y}{2}$, for $x = 2$ and $y = 4$

54. $\dfrac{2.56y}{3.2x}$, for $y = 3$ and $x = 4$

R.5 Equivalent Algebraic Expressions

a ┃ Equivalent Expressions

When solving equations and performing other operations in algebra, we manipulate expressions in various ways. For example, rather than

$$x + 2x,$$

we might write

$$3x,$$

knowing that the two expressions represent the same number for any allowable replacement of x. In that sense, the expressions $x + 2x$ and $3x$ are **equivalent**.

> Two expressions that have the same value for all *allowable* replacements are called **equivalent expressions.**

Example 1 Complete the table by evaluating each of the expressions $x + 2x$ and $3x$ for the given values. Then determine whether the expressions are equivalent.

	$x + 2x$	$3x$
$x = -2$		
$x = 5$		
$x = 0$		

We substitute and find the value of each expression. For example, for $x = -2$, $x + 2x = -2 + 2(-2) = -2 - 4 = -6$.

	$x + 2x$	$3x$
$x = -2$	-6	-6
$x = 5$	15	15
$x = 0$	0	0

Note that the values of $x + 2x$ and $3x$ are the same for the given values of x. Indeed, they are the same for any allowable real-number replacement of x, though we cannot substitute them all to find out. The expressions $x + 2x$ and $3x$ are **equivalent**.

Do Exercises 1 and 2.

b ┃ Equivalent Fractional Expressions

For the remainder of this section, we will consider several laws of real numbers that will allow us to find equivalent expressions.

> **THE IDENTITY PROPERTY OF 1**
>
> For any real number a,
>
> $$a \cdot 1 = 1 \cdot a = a.$$
>
> (The number 1 is the **multiplicative identity**.)

Objectives

a ┃ Determine whether two expressions are equivalent by completing a table of values.

b ┃ Find equivalent fractional expressions by multiplying by 1, and simplify fractional expressions.

c ┃ Use the commutative and the associative laws to find equivalent expressions.

d ┃ Use the distributive laws to find equivalent expressions by multiplying and factoring.

For Extra Help

TAPE 2   TAPE 2A   MAC WIN   CD-ROM

Complete the table by evaluating each expression for the given values. Then determine whether the expressions are equivalent.

1.

	$6x - x$	$5x$
$x = -2$		
$x = 8$		
$x = 0$		

2.

	$(x + 3)^2$	$x^2 + 9$
$x = -2$		
$x = 5$		
$x = 4.8$		

Answers on page A-3

3. Use multiplying by 1 to find an expression equivalent to $\frac{2}{7}$ with a denominator of $7y$.

4. Use multiplying by 1 to find an expression equivalent to $\frac{3}{5}$ with a denominator of $10x$.

Simplify.

5. $\dfrac{2y}{3y}$

6. $-\dfrac{20m}{12m}$

We will often refer to the use of the identity property of 1 as "multiplying by 1." We can use multiplying by 1 to change from one fractional expression to an equivalent one with a different denominator.

Example 2 Use multiplying by 1 to find an expression equivalent to $\frac{3}{5}$ with a denominator of $5x$.

We multiply by 1, using x/x as a name for 1:

$$\frac{3}{5} = \frac{3}{5} \cdot 1 = \frac{3}{5} \cdot \frac{x}{x} = \frac{3x}{5x}.$$

Note that the expressions 3/5 and $3x/5x$ are equivalent. They have the same value for any allowable replacement. Note too that 0 is not an allowable replacement in $3x/5x$, but for all nonzero real numbers, the expressions 3/5 and $3x/5x$ have the same value.

Do Exercises 3 and 4.

In algebra, we consider an expression like 3/5 to be a "simplified" form of $3x/5x$. To find such simplified expressions, we reverse the identity property of 1 in order to "remove a factor of 1."

Example 3 Simplify: $\dfrac{3x}{5x}$.

We do the reverse of what we did in Example 2:

$$\frac{3x}{5x} = \frac{3 \cdot x}{5 \cdot x} \qquad \text{We look for the largest common factor of the numerator and the denominator and factor each.}$$

$$= \frac{3}{5} \cdot \frac{x}{x} \qquad \text{Factoring the expression}$$

$$= \frac{3}{5} \cdot 1 \qquad \frac{x}{x} = 1$$

$$= \frac{3}{5}. \qquad \text{Removing a factor of 1 using the identity property of 1 in reverse}$$

Example 4 Simplify: $-\dfrac{24y}{16y}$.

$$-\frac{24y}{16y} = -\frac{3 \cdot 8y}{2 \cdot 8y} \qquad \text{We look for the largest common factor of the numerator and the denominator and factor each.}$$

$$= -\frac{3}{2} \cdot \frac{8y}{8y} \qquad \text{Factoring the expression}$$

$$= -\frac{3}{2} \cdot 1 \qquad \frac{8y}{8y} = 1$$

$$= -\frac{3}{2} \qquad \text{Removing a factor of 1 using the identity property of 1 in reverse}$$

Do Exercises 5 and 6.

Answers on page A-3

c | The Commutative and the Associative Laws

Let's examine the expressions $x + y$ and $y + x$, as well as xy and yx.

Example 5   Evaluate $x + y$ and $y + x$ for $x = 5$ and $y = 8$.

We substitute 5 for x and 8 for y in both expressions:

$$x + y = 5 + 8 = 13; \qquad y + x = 8 + 5 = 13.$$

Example 6   Evaluate xy and yx for $x = 26$ and $y = 13$.

We substitute 26 for x and 13 for y in both expressions:

$$xy = 26 \cdot 13 = 338; \qquad yx = 13 \cdot 26 = 338.$$

Do Exercises 7 and 8.

Note that the expressions $x + y$ and $y + x$ have the same values no matter what the variables stand for. Thus they are equivalent. They illustrate that when we add two numbers, the order in which we add does not matter. Similarly, when we multiply two numbers, the order in which we multiply does not matter. Thus the expressions xy and yx are equivalent. They have the same values no matter what the variables stand for. These are examples of general patterns or laws.

▶ **THE COMMUTATIVE LAWS**

Addition.   For any numbers a and b,

$$a + b = b + a.$$

(We can change the order when adding without affecting the answer.)

Multiplication.   For any numbers a and b,

$$ab = ba.$$

(We can change the order when multiplying without affecting the answer.)

Using a commutative law, we know that $x + 4$ and $4 + x$ are equivalent. Similarly, $5x$ and $x(5)$ are equivalent. Thus, in an algebraic expression, we can replace one with the other and the result will be equivalent to the original expression.

Now let's examine the expressions $a + (b + c)$ and $(a + b) + c$. Note that these expressions use parentheses as grouping symbols, and they also involve three numbers. Calculations within parentheses are to be done first.

Example 7   Evaluate $a + (b + c)$ and $(a + b) + c$ for $a = 4$, $b = 8$, and $c = 5$.

$$
\begin{aligned}
a + (b + c) &= 4 + (8 + 5) &&\text{Substituting}\\
&= 4 + 13 &&\text{Calculating within parentheses first:}\\
& &&\text{adding 8 and 5}\\
&= 17;\\
(a + b) + c &= (4 + 8) + 5 &&\text{Substituting}\\
&= 12 + 5 &&\text{Calculating within parentheses first:}\\
& &&\text{adding 4 and 8}\\
&= 17
\end{aligned}
$$

9. Evaluate

$a + (b + c)$ and $(a + b) + c$.

for $a = 10$, $b = 9$, and $c = 2$.

10. Evaluate

$a \cdot (b \cdot c)$ and $(a \cdot b) \cdot c$

for $a = 11$, $b = 5$, and $c = 8$.

11. Use the commutative laws to write an expression equivalent to each of $y + 5$, ab, and $8 + mn$.

12. Use the commutative and the associative laws to write at least three expressions equivalent to $(2 \cdot x) \cdot y$.

Answers on page A-3

Example 8 Evaluate $a \cdot (b \cdot c)$ and $(a \cdot b) \cdot c$ for $a = 7$, $b = 4$, and $c = 2$.

$$a \cdot (b \cdot c) = 7 \cdot (4 \cdot 2) \qquad (a \cdot b) \cdot c = (7 \cdot 4) \cdot 2$$
$$= 7 \cdot 8 \qquad\qquad\qquad = 28 \cdot 2$$
$$= 56; \qquad\qquad\qquad = 56$$

Do Exercises 9 and 10.

When only addition is involved, parentheses can be placed any way we please without affecting the answer. Likewise, when only multiplication is involved, parentheses can be placed any way we please without affecting the answer.

► **THE ASSOCIATIVE LAWS**

Addition.   For any numbers a, b, and c,
$$a + (b + c) = (a + b) + c.$$
(Numbers can be grouped in any manner for addition.)

Multiplication.   For any numbers a, b, and c,
$$a \cdot (b \cdot c) = (a \cdot b) \cdot c.$$
(Numbers can be grouped in any manner for multiplication.)

When only additions or only multiplications are involved, parentheses may be placed any way we please. Thus we often omit them. For example,

$$x + (y + 3) \text{ means } x + y + 3, \quad \text{and} \quad l(wh) \text{ means } lwh.$$

Example 9   Use the commutative and the associative laws to write at least three expressions equivalent to $(x + 8) + y$.

a) $(x + 8) + y = x + (8 + y)$ Using the associative law first and then
    $\qquad\qquad\quad = x + (y + 8)$ the commutative law

b) $(x + 8) + y = y + (x + 8)$ Using the commutative law and then
    $\qquad\qquad\quad = y + (8 + x)$ the commutative law again

c) $(x + 8) + y = (8 + x) + y$ Using the commutative law first and then
    $\qquad\qquad\quad = 8 + (x + y)$ the associative law

Do Exercises 11 and 12.

d | The Distributive Laws

Let's now examine two laws, each of which involves two operations. The first involves multiplication and addition.

Example 10   Evaluate $8(x + y)$ and $8x + 8y$ for $x = 4$ and $y = 5$.

$$8(x + y) = 8(4 + 5) \qquad \text{Substituting}$$
$$= 8(9) \qquad\qquad \text{Adding}$$
$$= 72; \qquad\qquad \text{Multiplying}$$
$$8x + 8y = 8 \cdot 4 + 8 \cdot 5 \qquad \text{Substituting}$$
$$= 32 + 40 \qquad\qquad \text{Multiplying}$$
$$= 72 \qquad\qquad\quad \text{Adding}$$

The expressions $8(x + y)$ and $8x + 8y$ in Example 10 are equivalent. This fact is the result of a law called *the distributive law of multiplication over addition*.

▶ **THE DISTRIBUTIVE LAW OF MULTIPLICATION OVER ADDITION**

For any numbers a, b, and c,
$$a(b + c) = ab + ac.$$
(We can add and then multiply, or we can multiply and then add.)

Do Exercises 13 and 14.

The other distributive law involves multiplication and subtraction.

Example 11 Evaluate $\frac{1}{2}(a - b)$ and $\frac{1}{2}a - \frac{1}{2}b$ for $a = 42$ and $b = 78$.

$$\frac{1}{2}(a - b) = \frac{1}{2}(42 - 78) \qquad \frac{1}{2}a - \frac{1}{2}b = \frac{1}{2} \cdot 42 - \frac{1}{2} \cdot 78$$
$$= \frac{1}{2}(-36) \qquad\qquad\qquad = 21 - 39$$
$$= -18; \qquad\qquad\qquad\qquad = -18$$

The expressions $\frac{1}{2}(a - b)$ and $\frac{1}{2}a - \frac{1}{2}b$ in Example 11 are equivalent. This fact is the result of a law called *the distributive law of multiplication over subtraction*.

▶ **THE DISTRIBUTIVE LAW OF MULTIPLICATION OVER SUBTRACTION**

For any real numbers a, b, and c,
$$a(b - c) = ab - ac.$$
(We can subtract and then multiply, or we can multiply and then subtract.)

We often refer to "*the* distributive law" when we mean *either* or *both* of these laws.

Do Exercises 15 and 16.

Multiplying Expressions with Variables

The distributive laws are the basis of multiplication in algebra as well as in arithmetic. In the following examples, note that we multiply each number or letter inside the parentheses by the factor outside.

Examples Multiply.

12. $4(x - 2) = 4 \cdot x - 4 \cdot 2 = 4x - 8$

13. $b(s - t + f) = bs - bt + bf$

14. $-3(y + 4) = -3 \cdot y + (-3) \cdot 4 = -3y - 12$

15. $-2x(y - 1) = -2x \cdot y - (-2x) \cdot 1 = -2xy + 2x$

Do Exercises 17–19.

---

13. Evaluate $10(x + y)$ and $10x + 10y$ for $x = 7$ and $y = 11$.

14. Evaluate $9(a + b)$ and $9a + 9b$ for $a = 5$ and $b = -2$.

15. Evaluate $5(a - b)$ and $5a - 5b$ for $a = 10$ and $b = 8$.

16. Evaluate $\frac{2}{3}(p - q)$ and $\frac{2}{3}p - \frac{2}{3}q$ for $p = 60$ and $q = 24$.

Multiply.

17. $8(y - 10)$

18. $a(x + y - z)$

19. $10\left(4x - 6y + \frac{1}{2}z\right)$

Answers on page A-3

20. List the terms of

$$-5x - 7y + 67t - \frac{4}{5}.$$

Factor.

21. $9x + 9y$

22. $ac - ay$

23. $6x - 12$

24. $35x - 25y + 15w + 5$

25. $bs + bt - bw$

Factoring Expressions with Variables

The reverse of multiplying is called **factoring**. Factoring an expression involves factoring its *terms*. **Terms** of algebraic expressions are the parts separated by plus signs.

Example 16 List the terms of $3x - 4y - 2z$.

We first find an equivalent expression that uses addition signs:

$$3x - 4y - 2z = 3x + (-4y) + (-2z). \quad \text{Using the property } a - b = a + (-b)$$

Thus the terms are $3x$, $-4y$, and $-2z$.

Do Exercise 20.

Now we can consider the reverse of multiplying, *factoring*.

> ▶ To **factor** an expression is to find an equivalent expression that is a product. If $N = ab$, then a and b are **factors** of N.

Examples Factor.

17. $8x + 8y = 8(x + y)$ 8 and $x + y$ are factors.

18. $cx - cy = c(x - y)$ c and $x - y$ are factors.

Whenever the terms of an expression have a factor in common, we can "remove" that factor, or "factor it out," using the distributive laws. We proceed as in Examples 17 and 18, but we may have to factor some of the terms first in order to display the common factor.

Generally, we try to factor out the largest factor common to all the terms. In the following example, we might factor out 3, but there is a larger factor common to the terms, 9. So we factor out the 9.

Example 19 Factor: $9x + 27y$.

$$9x + 27y = 9 \cdot x + 9 \cdot (3y) = 9(x + 3y)$$

We often have to supply a factor of 1 when factoring out a common factor, as in the next example, which is a formula about simple interest.

Example 20 Factor: $P + Prt$.

$$P + Prt = P \cdot 1 + P \cdot rt \quad \text{Writing } P \text{ as a product of } P \text{ and } 1$$
$$= P(1 + rt) \quad \text{Using the distributive law}$$

> *CAUTION!* It is a common error to omit this 1. If you do so, when you factor out P, you will leave out an entire term.

Do Exercises 21–25.

Answers on page A-3

Exercise Set R.5

a Complete the table by evaluating each expression for the given values. Then determine whether the expressions are equivalent.

1.

	$2x + 3x$	$5x$
$x = -2$		
$x = 5$		
$x = 0$		

2.

	$7x - 2x$	$5x$
$x = -2$		
$x = 5$		
$x = 0$		

3.

	$4x + 8x$	$4(x + 3x)$
$x = -1$		
$x = 3.2$		
$x = 0$		

4.

	$5(x - 2)$	$5x - 2$
$x = -1$		
$x = 4.6$		
$x = 0$		

b Use multiplying by 1 to find an equivalent expression with the given denominator.

5. $\dfrac{7}{8}$; $8x$

6. $\dfrac{4}{3}$; $3a$

7. $\dfrac{3}{4}$; $8a$

8. $\dfrac{3}{10}$; $50y$

Simplify.

9. $\dfrac{25x}{15x}$

10. $\dfrac{36y}{18y}$

11. $-\dfrac{100a}{25a}$

12. $\dfrac{-625t}{15t}$

c Use a commutative law to find an equivalent expression.

13. $w + 3$

14. $y + 5$

15. rt

16. cd

17. $4 + cd$

18. $pq + 14$

19. $yz + x$

20. $s + qt$

Use an associative law to find an equivalent expression.

21. $m + (n + 2)$

22. $5 \cdot (p \cdot q)$

23. $(7 \cdot x) \cdot y$

24. $(7 + p) + q$

Use the commutative and the associative laws to find three equivalent expressions.

25. $(a + b) + 8$

26. $(4 + x) + y$

27. $7 \cdot (a \cdot b)$

28. $(8 \cdot m) \cdot n$

d Multiply.

29. $4(a + 1)$

30. $3(c + 1)$

31. $8(x - y)$

32. $7(b - c)$

33. $-5(2a + 3b)$ **34.** $-2(3c + 5d)$ **35.** $2a(b - c + d)$ **36.** $5x(y - z + w)$

37. $2\pi r(h + 1)$ **38.** $P(1 + rt)$ **39.** $\dfrac{1}{2}h(a + b)$ **40.** $\dfrac{1}{4}\pi r(1 + s)$

List the terms of each of the following.

41. $4a - 5b + 6$ **42.** $5x - 9y + 12$ **43.** $2x - 3y - 2z$ **44.** $5a - 7b - 9c$

Factor.

45. $24x + 24y$ **46.** $9a + 9b$ **47.** $7p - 7$ **48.** $22x - 22$

49. $7x - 21$ **50.** $6y - 36$ **51.** $xy + x$ **52.** $ab + a$

53. $2x - 2y + 2z$ **54.** $3x + 3y - 3z$ **55.** $3x + 6y - 3$ **56.** $4a + 8b - 4$

57. $ab + ac - ad$ **58.** $xy - xz + xw$ **59.** $\dfrac{1}{4}\pi rr + \dfrac{1}{4}\pi rs$ **60.** $\dfrac{1}{2}ah + \dfrac{1}{2}bh$

Skill Maintenance

Translate to an algebraic expression. [R.4a]

61. The square of the sum of two numbers

62. The sum of the squares of two numbers

Subtract. [R.2c]

63. $-34.2 - 67.8$ **64.** $-\dfrac{11}{5} - \left(-\dfrac{17}{10}\right)$

Multiply. [R.2d]

65. $-\dfrac{1}{4}\left(-\dfrac{1}{2}\right)$ **66.** $0.23(-200)$

Synthesis

67. ◈ If we omit the parentheses in the expression $a(b + c)$, will the result be equivalent to $ab + c$? Explain how you would decide.

68. ◈ Look at the answers to Exercises 1 and 2 and at the expressions. What conclusion might you reach about the expressions $2x + 3x$ and $7x - 2x$? How would you verify it?

Make substitutions to determine whether the pair of expressions is equivalent.

69. $x^2 + y^2$; $(x + y)^2$ **70.** $(a - b)(a + b)$; $a^2 - b^2$ **71.** $x^2 \cdot x^3$; x^5 **72.** $\dfrac{x^8}{x^4}$; x^y

R.6 Simplifying Algebraic Expressions

There are many situations in algebra in which we want to find either an alternative or a simpler expression equivalent to a given one. We consider some of these possibilities in this section.

a Collecting Like Terms

If two terms have the same letter, or letters, we say that they are **like terms**, or **similar terms**. If two terms have no letters at all but are just numbers, they are also similar terms. We can simplify by **collecting** or **combining like terms**, using the distributive laws, which we can apply on the right, because of the commutative law of multiplication.

Examples Collect like terms.

1. $3x + 5x = (3 + 5)x = 8x$ Factoring out the x using the distributive law

2. $x - 3x = 1 \cdot x - 3 \cdot x = (1 - 3)x = -2x$

3. $2x + 3y - 5x - 2y = 2x + 3y + (-5x) + (-2y)$ Subtracting by adding an opposite

$$= 2x + (-5x) + 3y + (-2y)$$ Using a commutative law

$$= (2 - 5)x + (3 - 2)y$$ Using the distributive law

$$= -3x + y$$ Adding

4. $3x + 2x + 5 + 7 = (3 + 2)x + (5 + 7) = 5x + 12$

5. $4.2x - 6.7y - 5.8x + 23y = (4.2 - 5.8)x + (-6.7 + 23)y$

$$= -1.6x + 16.3y$$

6. $-\dfrac{1}{4}a + \dfrac{1}{2}b - \dfrac{3}{5}a - \dfrac{2}{5}b = \left(-\dfrac{1}{4} - \dfrac{3}{5}\right)a + \left(\dfrac{1}{2} - \dfrac{2}{5}\right)b$

$$= \left(-\dfrac{5}{20} - \dfrac{12}{20}\right)a + \left(\dfrac{5}{10} - \dfrac{4}{10}\right)b$$

$$= -\dfrac{17}{20}a + \dfrac{1}{10}b$$

You need not write the intervening steps when you can do the computations mentally.

Do Exercises 1–6.

b Multiplying by −1 and Removing Parentheses

What happens when we multiply a number by −1?

Examples

7. $-1 \cdot 9 = -9$ **8.** $-1 \cdot \left(-\dfrac{3}{5}\right) = \dfrac{3}{5}$ **9.** $-1 \cdot 0 = 0$

Do Exercises 7–9.

Objectives

a Simplify an expression by collecting like terms.

b Simplify an expression by removing parentheses and collecting like terms.

For Extra Help

TAPE 2   TAPE 2A   MAC   CD-ROM
                       WIN

Collect like terms.

1. $9x + 11x$

2. $5x - 12x$

3. $5x + x$

4. $x - 7x$

5. $22x - 2.5y + 1.4x + 6.4y$

6. $\dfrac{2}{3}x - \dfrac{3}{4}y + \dfrac{4}{5}x - \dfrac{5}{6}y + 23$

Multiply.

7. $-1 \cdot 24$

8. $-1 \cdot 0$

9. $-1 \cdot (-10)$

Answers on page A-4

Find an equivalent expression
without parentheses.

10. $-(9x)$

11. $-(-24t)$

For each opposite, find an
equivalent expression without
parentheses.

12. $-(7 - y)$

13. $-(x - y)$

14. $-(9x + 6y + 11)$

15. $-(23x - 7y - 2)$

16. $-(-3x - 2y - 1)$

Answers on page A-4

> ▶ **THE PROPERTY OF −1**
>
> For any number a,
>
> $$-1 \cdot a = -a.$$
>
> (Negative 1 times a is the opposite of a; in other words, changing the sign
> is the same as multiplying by −1.)

From the property of −1, we know that we can replace − with −1 or the
reverse, in any expression. In that way, we can find an equivalent expression for an opposite.

Examples   Find an equivalent expression without parentheses.

10. $-(3x) = -1(3x)$ Replacing − with −1 using the property of −1
$= (-1 \cdot 3)x$ Using an associative law
$= -3x$ Multiplying

11. $-(-9y) = -1(-9y)$ Replacing − with −1
$= [-1(-9)]y$ Using an associative law
$= 9y$ Multiplying

Do Exercises 10 and 11.

Examples   For each opposite, find an equivalent expression without
parentheses.

12. $-(4 + x) = -1(4 + x)$ Replacing − with −1
$= -1 \cdot 4 + (-1) \cdot x$ Multiplying using the distributive law
$= -4 + (-x)$ Replacing $-1 \cdot x$ with $-x$
$= -4 - x$ Adding an opposite is the same as subtracting.

13. $-(3x - 2y + 4) = -1(3x - 2y + 4)$
$= -1 \cdot 3x - (-1)2y + (-1)4$ Using the distributive law
$= -3x - (-2y) + (-4)$ Multiplying
$= -3x + [-(-2y)] + (-4)$ Adding an opposite
$= -3x + 2y - 4$

14. $-(a - b) = -1(a - b)$
$= -1 \cdot a - (-1) \cdot b$
$= -a + [-(-1)b]$
$= -a + b = b - a$

Example 14 illustrates something that you should remember, because it
is a convenient shortcut.

> ▶ For any real numbers a and b,
>
> $$-(a - b) = b - a.$$
>
> (The opposite of $a - b$ is $b - a$.)

Do Exercises 12–16.

The examples above show that we can find an equivalent expression for
an opposite by multiplying every term by −1. We could also say that we
change the sign of every term inside the parentheses. Thus we can skip
some steps.

Example 15 Find an equivalent expression without parentheses:

$$-\left(-9t + 7z - \tfrac{1}{4}w\right).$$

We have

$$-\left(-9t + 7z - \tfrac{1}{4}w\right) = 9t - 7z + \tfrac{1}{4}w. \quad \begin{array}{l}\textbf{Changing the sign}\\\textbf{of every term}\end{array}$$

Do Exercises 17–19.

In some expressions commonly encountered in algebra, there are parentheses preceded by subtraction signs. These parentheses can be removed by changing the sign of *every* term inside. In this way, we simplify by finding a less complicated equivalent expression.

Examples Remove parentheses and simplify.

16. $6x - (4x + 2) = 6x + [-(4x + 2)]$ **Subtracting by adding the opposite**
$$= 6x - 4x - 2 \quad \textbf{Changing the sign of every term inside}$$
$$= 2x - 2 \quad \textbf{Collecting like terms}$$

17. $3y - 4 - (9y - 7) = 3y - 4 - 9y + 7$
$$= -6y + 3, \text{ or } 3 - 6y$$

In Example 16, we see the reason for the word "simplify." The expression $2x - 2$ is equivalent to $6x - (4x + 2)$ but it is shorter.

If parentheses are preceded by an addition sign, *no* signs are changed when they are removed.

Examples Remove parentheses and simplify.

18. $3y + (3x - 8) - (5 - 12y) = 3y + 3x - 8 - 5 + 12y$
$$= 15y + 3x - 13$$

19. $\tfrac{1}{3}(15x - 4) - (5x + 2y) + 1 = \tfrac{1}{3} \cdot 15x - \tfrac{1}{3} \cdot 4 - 5x - 2y + 1$
$$= 5x - \tfrac{4}{3} - 5x - 2y + 1$$
$$= -2y - \tfrac{1}{3}$$

Do Exercises 20–23.

We now consider subtracting an expression consisting of several terms preceded by a number other than -1.

Examples Remove parentheses and simplify.

20. $x - 3(x + y) = x + [-3(x + y)]$ **Subtracting by adding the opposite**
$$= x - 3x - 3y \quad \begin{array}{l}\textbf{Removing parentheses by multiplying}\\\textbf{\textit{x} + \textit{y} by }-3\end{array}$$
$$= -2x - 3y \quad \textbf{Collecting like terms}$$

> **Caution!** A common error is to forget to change this sign. *Remember*: When multiplying like this, change the sign of *every* term inside the parentheses.

21. $3y - 2(4y - 5) = 3y - 8y + 10$ $\begin{array}{l}\textbf{Removing parentheses by multiplying}\\\textbf{4\textit{y} − 5 by }-2\end{array}$
$$= -5y + 10 \quad \textbf{Collecting like terms}$$

Do Exercises 24–26.

Find an equivalent expression without parentheses.

17. $-(-2x - 5z + 24)$

18. $-(3x - 2y)$

19. $-\left(\tfrac{1}{4}t + 41w - 5d - 23\right)$

Remove parentheses and simplify.

20. $6x - (3x + 8)$

21. $6y - 4 - (2y - 5)$

22. $6x - (9y - 4) - (8x + 10)$

23. $7x - (-9y - 4) + (8x - 10)$

Remove parentheses and simplify.

24. $x - 2(y + x)$

25. $3x - 5(2y - 4x)$

26. $(4a - 3b) - \tfrac{1}{4}(4a - 3) + 5$

Answers on page A-4

Simplify.

27. $15x - \{2[2(x - 5) - 6(x + 3)] + 4\}$

28. $9a + \{3a - 2[(a - 4) - (a + 2)]\}$

Answers on page A-4

When expressions with parentheses contain variables, we still work from the inside out when simplifying, using the rules for order of operations.

Example 22 Simplify: $6y - \{4[3(y - 2) - 4(y + 2)] - 3\}$.

$6y - \{4[3(y - 2) - 4(y + 2)] - 3\}$

$= 6y - \{4[3y - 6 - 4y - 8] - 3\}$ **Multiplying to remove the innermost parentheses using the distributive law**

$= 6y - \{4[-y - 14] - 3\}$ **Collecting like terms in the inner parentheses**

$= 6y - \{-4y - 56 - 3\}$ **Multiplying to remove the inner parentheses using the distributive law**

$= 6y - \{-4y - 59\}$ **Collecting like terms**

$= 6y + 4y + 59$ **Removing parentheses**

$= 10y + 59$ **Collecting like terms**

Do Exercises 27 and 28.

Improving Your Math Study Skills

Getting Started in a Math Class: The First-Day Handout or Syllabus

There are many ways in which to improve your math study skills. We have already considered some tips on using this book (see Section R.1). We now consider some more general tips.

- **Textbook.** On the first day of class, most instructors distribute a handout that lists the textbook and other materials needed in the course. If possible, call the instructor or the department office before the term begins to find out which textbook you will be using and visit the bookstore to pick it up. This way, you can purchase the book before class starts and be ready to begin studying.

- **Attendance.** The handout may also describe the attendance policy for your class. Some instructors take attendance at every class, while others use different methods to track students' attendance. Regardless of the policy, you should plan to attend every class. Missing even one class can cause you to fall behind. If attendance counts toward your course grade, find out if there is a way to make up for missed days.

    If you do miss a class, call the instructor as soon as possible to find out what material was covered and what was assigned for the next class. If you have a study partner, call this person; ask if you can make a copy of his or her notes and find out what the homework assignment was.

- **Homework.** The first-day handout may also detail how homework is handled. Find out when, and how often, homework will be assigned, whether homework is collected or graded, and whether there will be quizzes over the homework material.

If the homework will be graded, find out what part of the final grade it will determine. Ask what the policy is for late homework. If you do miss a homework deadline, be sure to do the assigned homework anyway, as this is the best way to learn the material.

- **Grading.** The handout may also provide information on how your grade will be calculated at the end of the term. Typically, there will be tests during the term and a final exam at the end of the term. Frequently, homework is counted as part of the grade calculation, as are the quizzes. Find out how many tests will be given, if there is an option for make-up tests, or if any test grades will be dropped at the end of the term.

    Some instructors keep the class grades on a computer. If this is the case, find out if you can receive current grade reports throughout the term.

- **Get to know your classmates.** It can be a big help in a math class to get to know your fellow students. You might consider forming a study group. If you do so, find out their phone numbers and schedules so that you can coordinate study time for homework or tests.

- **Get to know your instructor.** It can, of course, help immensely to get to know your instructor. Trivial though it may seem, get basic information like his or her name, how he or she can be contacted outside of class, and where the office is. Learn about your instructor's teaching style and try to adapt your learning to it.

Exercise Set R.6

a Collect like terms.

1. $7x + 5x$

2. $6a + 9a$

3. $8b - 11b$

4. $9c - 12c$

5. $14y + y$

6. $13x + x$

7. $12a - a$

8. $15x - x$

9. $t - 9t$

10. $x - 6x$

11. $5x - 3x + 8x$

12. $3x - 11x + 2x$

13. $3x - 5y + 8x$

14. $4a - 9b + 10a$

15. $3c + 8d - 7c + 4d$

16. $12a + 3b - 5a + 6b$

17. $4x - 7 + 18x + 25$

18. $13p + 5 - 4p + 7$

19. $1.3x + 1.4y - 0.11x - 0.47y$

20. $0.17a + 1.7b - 12a - 38b$

21. $\dfrac{2}{3}a + \dfrac{5}{6}b - 27 - \dfrac{4}{5}a - \dfrac{7}{6}b$

22. $-\dfrac{1}{4}x - \dfrac{1}{2}x + \dfrac{1}{4}y + \dfrac{1}{2}y - 34$

The **perimeter** of a rectangle is the distance around it. The perimeter P is given by $P = 2l + 2w$.

23. Find an equivalent expression for the perimeter formula $P = 2l + 2w$ by factoring.

24. The standard football field has $l = 360$ ft and $w = 160$ ft. Evaluate both expressions to find the perimeter. (See Exercise 23.)

b For each opposite, find an equivalent expression without parentheses.

25. $-(-2c)$

26. $-(-5y)$

27. $-(b + 4)$

28. $-(a + 9)$

29. $-(b - 3)$

30. $-(x - 8)$

31. $-(t - y)$

32. $-(r - s)$

33. $-(x + y + z)$

34. $-(r + s + t)$

35. $-(8x - 6y + 13)$

36. $-(9a - 7b + 24)$

37. $-(-2c + 5d - 3e + 4f)$

38. $-(-4x + 8y - 5w + 9z)$

39. $-\left(-1.2x + 56.7y - 34z - \dfrac{1}{4}\right)$

40. $-\left(-x + 2y - \dfrac{2}{3}z - 56.3w\right)$

Simplify by removing parentheses and collecting like terms.

41. $a + (2a + 5)$

42. $x + (5x + 9)$

43. $4m - (3m - 1)$

44. $5a - (4a - 3)$

45. $5d - 9 - (7 - 4d)$

46. $6x - 7 - (9 - 3x)$

47. $-2(x + 3) - 5(x - 4)$

48. $-9(y + 7) - 6(y - 3)$

49. $5x - 7(2x - 3) - 4$

50. $8y - 4(5y - 6) + 9$

51. $8x - (-3y + 7) + (9x - 11)$

52. $-5t + (4t - 12) - 2(3t + 7)$

53. $\dfrac{1}{4}(24x - 8) - \dfrac{1}{2}(-8x + 6) - 14$

54. $-\dfrac{1}{2}(10t - w) + \dfrac{1}{4}(-28t + 4) + 1$

Simplify.

55. $7a - [9 - 3(5a - 2)]$

56. $14b - [7 - 3(9b - 4)]$

57. $5\{-2 + 3[4 - 2(3 + 5)]\}$

58. $7\{-7 + 8[5 - 3(4 + 6)]\}$

59. $[10(x + 3) - 4] + [2(x - 1) + 6]$

60. $[9(x + 5) - 7] + [4(x - 12) + 9]$

61. $[7(x + 5) - 19] - [4(x - 6) + 10]$

62. $[6(x + 4) - 12] - [5(x - 8) + 11]$

63. $3\{[7(x - 2) + 4] - [2(2x - 5) + 6]\}$

64. $4\{[8(x - 3) + 9] - [4(3x - 7) + 2]\}$

65. $4\{[5(x - 3) + 2^2] - 3[2(x + 5) - 9^2]\}$

66. $3\{[6(x - 4) + 5^2] - 2[5(x + 8) - 10^2]\}$

67. $2y + \{8[3(2y - 5) - (8y + 9)] + 6\}$

68. $7b - \{5[4(3b - 8) - (9b + 10)] + 14\}$

Skill Maintenance

Add. [R.2a]

69. $17 + (-54)$

70. $-17 + (-54)$

71. $-13.78 + (-9.32)$

72. $-\dfrac{2}{3} + \dfrac{7}{8}$

Divide. [R.2e]

73. $-256 \div 16$

74. $-256 \div (-16)$

75. $256 \div (-16)$

76. $-\dfrac{3}{8} \div \dfrac{9}{4}$

Multiply. [R.5d]

77. $8(a - b)$

78. $-8(2a - 3b + 4)$

79. $6x(a - b + 2c)$

80. $\dfrac{2}{3}(24x - 12y + 15)$

Factor. [R.5d]

81. $24a - 24$

82. $24a - 16b$

83. $ab - ac + a$

84. $15p + 45q - 10$

Synthesis

85. ◈ Explain in your own words the meaning of the statement $(-x)^2 = x^2$. Determine whether it is true or not. Explain why or why not.

86. ◈ Explain in your own words the meaning of the statement $ab = (-a)(-b)$. Determine whether it is true or not. Explain why or why not.

Insert one pair of parentheses to convert the false statement into a true statement.

87. $3 - 8^2 + 9 = 34$

88. $2 \cdot 7 + 3^2 \cdot 5 = 104$

89. $5 \cdot 2^3 \div 3 - 4^4 = 40$

90. $2 - 7 \cdot 2^2 + 9 = -11$

Simplify.

91. $[11(a - 3) + 12a] - \{6[4(3b - 7) - (9b + 10)] + 11\}$

92. $-3[9(x - 4) + 5x] - 8\{3[5(3y + 4)] - 12\}$

93. $z - \{2z + [3z - (4z + 5x) - 6z] + 7z\} - 8z$

94. $\{x + [f - (f + x)] + [x - f]\} + 3x$

95. $x - \{x + 1 - [x + 2 - (x - 3 - \{x + 4 - [x - 5 + (x - 6)]\})]\}$

Simplify algebraic expressions as a group.

Collaborative
Learning Manual

R.7 Properties of Exponents and Scientific Notation

We often need to find ways to determine *equivalent* exponential expressions. We do this with several rules or properties regarding exponents.

a | Multiplication and Division

To see how to multiply, or simplify, in an expression such as $a^3 \cdot a^2$, we use the definition of exponential notation:

$$a^3 \cdot a^2 = \underbrace{a \cdot a \cdot a}_{3 \text{ factors}} \cdot \underbrace{a \cdot a}_{2 \text{ factors}} = a^5$$

The exponent in a^5 is the *sum* of those in $a^3 \cdot a^2$. In general, the exponents are added when we multiply, but note that the base must be the same in all factors. This is true for any integer exponents, even those that may be negative or zero.

> **THE PRODUCT RULE**
>
> For any number a and any integers m and n,
>
> $$a^m \cdot a^n = a^{m+n}.$$
>
> (When multiplying with exponential notation, add the exponents if the bases are the same.)

Examples Multiply and simplify.

1. $x^4 \cdot x^3 = x^{4+3} = x^7$

2. $4^5 \cdot 4^{-3} = 4^{5+(-3)} = 4^2 = 16$

3. $(-2)^{-3}(-2)^7 = (-2)^{-3+7}$
$= (-2)^4 = 16$

4. $(8x^n)(6x^{2n}) = 8 \cdot 6x^{n+2n}$
$= 48x^{3n}$

5. $(8x^4y^{-2})(-3x^{-3}y) = 8 \cdot (-3) \cdot x^4 \cdot x^{-3} \cdot y^{-2} \cdot y^1$ Using the associative and the commutative laws

$= -24x^{4-3}y^{-2+1}$ Using the product rule

$= -24xy^{-1} = -\dfrac{24x}{y}$
↑

Note that we give answers using positive exponents. In some situations, this may not be appropriate, but we do so here.

Do Exercises 1–7.

Consider this division:

$$\frac{8^5}{8^3} = \frac{8 \cdot 8 \cdot 8 \cdot 8 \cdot 8}{8 \cdot 8 \cdot 8} = \frac{8 \cdot 8 \cdot 8}{8 \cdot 8 \cdot 8} \cdot 8 \cdot 8 = 8 \cdot 8 = 8^2.$$

We can obtain the result by subtracting exponents. This is always the case, even if exponents are negative or zero.

Multiply and simplify.

1. $8^{-3}8^7$

2. y^7y^{-2}

3. $(9x^{-4})(-2x^7)$

4. $(-3x^{-4})(25x^{-10})$

5. $(-7x^{3n})(6x^{5n})$

6. $(5x^{-3}y^4)(-2x^{-9}y^{-2})$

7. $(4x^{-2}y^4)(15x^2y^{-3})$

Answers on page A-4

Divide and simplify.

8. $\dfrac{4^8}{4^5}$

9. $\dfrac{5^4}{5^{-2}}$

10. $\dfrac{10^{-8}}{10^{-2}}$

11. $\dfrac{45x^{5n}}{-9x^{3n}}$

12. $\dfrac{42y^7x^6}{-21y^{-3}x^{10}}$

13. $\dfrac{33a^5b^{-2}}{22a^2b^{-4}}$

> **THE QUOTIENT RULE**
>
> For any nonzero number a and any integers m and n,
>
> $$\dfrac{a^m}{a^n} = a^{m-n}.$$
>
> (When dividing with exponential notation, subtract the exponent of the denominator from the exponent of the numerator, if the bases are the same.)

Examples Divide and simplify.

6. $\dfrac{5^7}{5^3} = 5^{7-3} = 5^4$ **Subtracting exponents using the quotient rule**

7. $\dfrac{5^7}{5^{-3}} = 5^{7-(-3)} = 5^{7+3} = 5^{10}$ **Subtracting exponents (adding an opposite)**

8. $\dfrac{9^{-2}}{9^5} = 9^{-2-5} = 9^{-7} = \dfrac{1}{9^7}$

9. $\dfrac{7^{-4}}{7^{-5}} = 7^{-4-(-5)} = 7^{-4+5} = 7^1 = 7$

10. $\dfrac{16x^4y^7}{-8x^3y^9} = \dfrac{16}{-8} \cdot \dfrac{x^4}{x^3} \cdot \dfrac{y^7}{y^9} = -2xy^{-2} = -\dfrac{2x}{y^2}$

> The answers $\dfrac{-2x}{y^2}$ or $\dfrac{2x}{-y^2}$ would also be correct here.

11. $\dfrac{40x^{-2n}}{4x^{5n}} = \dfrac{40}{4}x^{-2n-5n} = 10x^{-7n} = \dfrac{10}{x^{7n}}$

12. $\dfrac{14x^7y^{-3}}{4x^5y^{-5}} = \dfrac{14}{4} \cdot \dfrac{x^7}{x^5} \cdot \dfrac{y^{-3}}{y^{-5}} = \dfrac{7}{2}x^2y^2$

In exercises such as Examples 6–12 above, it may help to think as follows: After writing the base, write the top exponent. Then write a subtraction sign. Then write the bottom exponent. Then do the subtraction. For example,

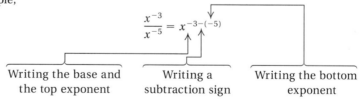

Writing the base and the top exponent   Writing a subtraction sign   Writing the bottom exponent

Do Exercises 8–13.

Answers on page A-4

b Raising Powers to Powers and Products and Quotients to Powers

When an expression inside parentheses is raised to a power, the inside expression is the base. Consider an expression like $(5^2)^4$. In this case, we are raising 5^2 to the fourth power:

$$(5^2)^4 = (5^2)(5^2)(5^2)(5^2)$$
$$= (5 \cdot 5)(5 \cdot 5)(5 \cdot 5)(5 \cdot 5)$$
$$= 5 \cdot 5 \cdot 5 \cdot 5 \cdot 5 \cdot 5 \cdot 5 \cdot 5 \qquad \textbf{Using an associative law}$$
$$= 5^8.$$

Note that here we could have multiplied the exponents:

$$(5^2)^4 = 5^{2 \cdot 4} = 5^8.$$

Likewise, $(y^8)^3 = (y^8)(y^8)(y^8) = y^{24}$. Once again, we get the same result if we multiply the exponents:

$$(y^8)^3 = y^{8 \cdot 3} = y^{24}.$$

> **THE POWER RULE**
>
> For any real number a and any integers m and n,
> $$(a^m)^n = a^{mn}.$$
> (To raise a power to a power, multiply the exponents.)

Examples Simplify.

13. $(x^5)^7 = x^{5 \cdot 7}$ **Multiply exponents.**
$= x^{35}$

14. $(y^{-2})^{-2} = y^{(-2)(-2)}$
$= y^4$

15. $(x^{-5})^4 = x^{-5 \cdot 4}$
$= x^{-20} = \dfrac{1}{x^{20}}$

16. $(x^4)^{-2t} = x^{4(-2t)}$
$= x^{-8t} = \dfrac{1}{x^{8t}}$

Do Exercises 14–16.

Let's compare $2a^3$ and $(2a)^3$:

$$2a^3 = 2 \cdot a \cdot a \cdot a \qquad \text{The base is } a.$$

and

$$(2a)^3 = (2a)(2a)(2a) \qquad \text{The base is } 2a.$$
$$= (2 \cdot 2 \cdot 2)(a \cdot a \cdot a) \qquad \text{Using the associative law of multiplication}$$
$$= 2^3 a^3 = 8a^3.$$

We see that $2a^3$ and $(2a)^3$ are *not* equivalent. We also see that we can evaluate the power $(2a)^3$ by raising each factor to the power 3. This leads us to the following rule for raising a product to a power.

> **RAISING A PRODUCT TO A POWER**
>
> For any real numbers a and b and any integer n,
> $$(ab)^n = a^n b^n.$$
> (To raise a product to the nth power, raise each factor to the nth power.)

Simplify.

14. $(3^7)^6$

15. $(z^{-4})^{-5}$

16. $(t^2)^{-7m}$

Answers on page A-4

Simplify.

17. $(2xy)^3$

18. $(4x^{-2}y^7)^2$

19. $(-2x^4y^2)^5$

20. $(10x^{-4}y^7z^{-2})^3$

Simplify.

21. $\left(\dfrac{x^{-3}}{y^4}\right)^{-3}$

22. $\left(\dfrac{3x^2y^{-3}}{y^5}\right)^2$

23. $\left[\dfrac{-3a^{-5}b^3}{2a^{-2}b^{-4}}\right]^{-3}$

Answers on page A-4

Examples Simplify.

17. $(3x^2y^{-2})^3 = 3^3(x^2)^3(y^{-2})^3 = 3^3x^6y^{-6} = 27x^6y^{-6} = \dfrac{27x^6}{y^6}$

18. $(5x^3y^{-5}z^2)^4 = 5^4(x^3)^4(y^{-5})^4(z^2)^4 = 625x^{12}y^{-20}z^8 = \dfrac{625x^{12}z^8}{y^{20}}$

Do Exercises 17–20.

There is a similar rule for raising a quotient to a power.

> **RAISING A QUOTIENT TO A POWER**
>
> For any real numbers a and b, $b \neq 0$, and any integer n,
>
> $$\left(\frac{a}{b}\right)^n = \frac{a^n}{b^n}.$$
>
> (To raise a quotient to the nth power, raise the numerator to the nth power and divide by the denominator to the nth power.)

Examples Simplify. Write the answer using positive exponents.

19. $\left(\dfrac{x^2}{y^{-3}}\right)^{-5} = \dfrac{x^{2\cdot(-5)}}{y^{-3\cdot(-5)}} = \dfrac{x^{-10}}{y^{15}} = \dfrac{1}{x^{10}y^{15}}$

20. $\left(\dfrac{2x^3y^{-2}}{3y^4}\right)^5 = \dfrac{(2x^3y^{-2})^5}{(3y^4)^5} = \dfrac{2^5(x^3)^5(y^{-2})^5}{3^5(y^4)^5} = \dfrac{32x^{15}y^{-10}}{243y^{20}}$

$$= \dfrac{32x^{15}y^{-10-20}}{243} = \dfrac{32x^{15}y^{-30}}{243} = \dfrac{32x^{15}}{243y^{30}}$$

21. $\left[\dfrac{-3a^{-5}b^3}{2a^{-2}b^{-4}}\right]^{-2} = \dfrac{(-3a^{-5}b^3)^{-2}}{(2a^{-2}b^{-4})^{-2}} = \dfrac{(-3)^{-2}(a^{-5})^{-2}(b^3)^{-2}}{2^{-2}(a^{-2})^{-2}(b^{-4})^{-2}}$

$$= \dfrac{\dfrac{1}{(-3)^2}a^{10}b^{-6}}{\dfrac{1}{2^2}a^4b^8} = \dfrac{2^2}{(-3)^2}a^{10-4}b^{-6-8} = \dfrac{4}{9}a^6b^{-14} = \dfrac{4a^6}{9b^{14}}$$

An alternative way to carry out Example 21 is to first write the expression with a positive exponent, as follows:

$$\left[\dfrac{-3a^{-5}b^3}{2a^{-2}b^{-4}}\right]^{-2} = \left[\dfrac{2a^{-2}b^{-4}}{-3a^{-5}b^3}\right]^2 = \dfrac{(2a^{-2}b^{-4})^2}{(-3a^{-5}b^3)^2} = \dfrac{2^2(a^{-2})^2(b^{-4})^2}{(-3)^2(a^{-5})^2(b^3)^2}$$

$$= \dfrac{4a^{-4}b^{-8}}{9a^{-10}b^6} = \dfrac{4}{9}a^{-4-(-10)}b^{-8-6} = \dfrac{4}{9}a^6b^{-14} = \dfrac{4a^6}{9b^{14}}.$$

Do Exercises 21–23.

c Scientific Notation

There are many kinds of symbolism, or *notation,* for numbers. You are already familiar with fractional notation, decimal notation, and percent notation. Now we study another, **scientific notation,** which is especially useful when calculations involve very large or very small numbers and when estimating.

The following are examples of scientific notation:

The distance from the sun to the planet Pluto:

3.664×10^9 mi = 3,664,000,000 mi

The diameter of a helium atom:

2.2×10^{-8} cm = 0.000000022 cm

> ► **Scientific notation** for a number is an expression of the type
>
> $M \times 10^n,$
>
> where n is an integer, M is greater than or equal to 1 and less than 10 ($1 \le M < 10$), and M is expressed in decimal notation. 10^n is also considered to be scientific notation when $M = 1$.

You should try to make conversions to scientific notation mentally as much as possible. Here is a handy mental device.

> A positive exponent in scientific notation indicates a large number (greater than one) and a negative exponent indicates a small number (less than one).

Examples Convert mentally to scientific notation.

22. Light travels 9,460,000,000,000 km in one year.

$9,460,000,000,000 = 9.46 \times 10^{12}$ 9.460,000,000,000.

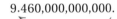
                                                                    12 places

Large number, so the exponent is positive.

23. The mass of a grain of sand is 0.0648 g (grams).

$0.0648 = 6.48 \times 10^{-2}$ 0.06.48

                                        2 places

Small number, so the exponent is negative.

Examples Convert mentally to decimal notation.

24. $4.893 \times 10^5 = 489,300$ 4.89300.

                                            5 places

Positive exponent, so the answer is a large number.

25. $8.7 \times 10^{-8} = 0.000000087$ 0.00000008.7

                                            8 places

Negative exponent, so the answer is a small number.

Each of the following is *not* scientific notation.

$\underline{13.95} \times 10^{13},$ $\underline{0.468} \times 10^{-8}$

This number is greater than 10.   This number is less than 1.

Do Exercises 24–27.

Convert to scientific notation.

24. Light travels 5,880,000,000,000 mi in one year.

25. 0.000000000257

Convert to decimal notation.

26. 4.567×10^{-13}

27. The distance from the earth to the sun is 9.3×10^7 mi.

Calculator Spotlight

 Scientific Notation. To enter a number in scientific notation, we first enter the decimal portion of the number; then we press [2nd] [EE] followed by the exponent. For example, to enter 1.849×10^{-11}, we press

| 1 | . | 8 | 4 | 9 |

| 2nd | EE | (−) | 1 | 1 | .

The display will read,

| 1.849 E $^-$11 | .

Exercises

Enter in scientific notation.

1. 260,000,000

2. 0.00000000006709

Answers on page A-4

Multiply and write scientific notation for the answer.

28. $(9.1 \times 10^{-17})(8.2 \times 10^3)$

29. $(1.12 \times 10^{-8})(5 \times 10^{-7})$

Divide and write scientific notation for the answer.

30. $\dfrac{4.2 \times 10^5}{2.1 \times 10^2}$

31. $\dfrac{1.1 \times 10^{-4}}{2.0 \times 10^{-7}}$

32. *Light from the Sun to Earth.* The distance from Earth to the sun is about 93,000,000 mi. Light travels 1.86×10^5 mi in 1 sec. About how many seconds does it take light from the sun to reach Earth? Write scientific notation for the answer.

33. *Mass of Jupiter.* The mass of the planet Jupiter is about 318 times the mass of Earth. Write scientific notation for the mass of Jupiter. See Example 29.

Answers on page A-4

Multiplying and dividing in scientific notation is easy because we can use the properties of exponents.

Example 26 Multiply and write scientific notation for the answer: $(3.1 \times 10^5)(4.5 \times 10^{-3})$.

We apply the commutative and the associative laws to get

$$(3.1 \times 10^5)(4.5 \times 10^{-3}) = (3.1 \times 4.5)(10^5 \times 10^{-3}) = 13.95 \times 10^2.$$

To find scientific notation for the result, we convert 13.95 to scientific notation and then simplify:

$$13.95 \times 10^2 = (1.395 \times 10^1) \times 10^2 = 1.395 \times 10^3.$$

Do Exercises 28 and 29.

Example 27 Divide and write scientific notation for the answer:

$$\frac{6.4 \times 10^{-7}}{8.0 \times 10^6}.$$

$$\frac{6.4 \times 10^{-7}}{8.0 \times 10^6} = \frac{6.4}{8.0} \times \frac{10^{-7}}{10^6} \qquad \boxed{\text{Factoring shows two divisions.}}$$

$$= 0.8 \times 10^{-13} \qquad \textbf{Doing the divisions separately}$$

$$= (8.0 \times 10^{-1}) \times 10^{-13} \qquad \textbf{Converting 0.8 to scientific notation}$$

$$= 8.0 \times 10^{-14} \qquad \boxed{\text{The answer } 0.8 \times 10^{-13} \text{ is not scientific notation.}}$$

Do Exercises 30 and 31.

Example 28 *Light from the Sun to Neptune.* The planet Neptune is about 2,790,000,000 mi from the sun. Light travels 1.86×10^5 mi in 1 sec. About how many seconds does it take light from the sun to reach Neptune? Write scientific notation for the answer.

The time it takes light to travel from the sun to Neptune is

$$\frac{2{,}790{,}000{,}000}{1.86 \times 10^5} = \frac{2.79 \times 10^9}{1.86 \times 10^5} = \frac{2.79}{1.86} \times \frac{10^9}{10^5} = 1.5 \times 10^4 \text{ sec.}$$

Example 29 *Mass of the Sun.* The mass of Earth is about 5.98×10^{24} kg. The mass of the sun is about 333,000 times the mass of Earth. Write scientific notation for the mass of the sun.

The mass of the sun is 333,000 times the mass of Earth. We convert to scientific notation and multiply:

$$(333{,}000)(5.98 \times 10^{24}) = (3.33 \times 10^5)(5.98 \times 10^{24})$$
$$= (3.33 \times 5.98)(10^5 \times 10^{24})$$
$$= 19.9134 \times 10^{29}$$
$$= (1.99134 \times 10^1) \times 10^{29}$$
$$= 1.99134 \times 10^{30} \text{ kg.}$$

Do Exercises 32 and 33.

Exercise Set R.7

a Multiply and simplify.

1. $3^6 \cdot 3^3$

2. $8^2 \cdot 8^6$

3. $6^{-6} \cdot 6^2$

4. $9^{-5} \cdot 9^3$

5. $8^{-2} \cdot 8^{-4}$

6. $9^{-1} \cdot 9^{-6}$

7. $b^2 \cdot b^{-5}$

8. $a^4 \cdot a^{-3}$

9. $a^{-3} \cdot a^4 \cdot a^2$

10. $x^{-8} \cdot x^5 \cdot x^3$

11. $(2x)^3 \cdot (3x)^2$

12. $(9y)^2 \cdot (2y)^3$

13. $(14m^2n^3)(-2m^3n^2)$

14. $(6x^5y^{-2})(-3x^2y^3)$

15. $(-2x^{-3})(7x^{-8})$

16. $(6x^{-4}y^3)(-4x^{-8}y^{-2})$

17. $(15x^{4t})(7x^{-6t})$

18. $(9x^{-4n})(-4x^{-8n})$

Divide and simplify.

19. $\dfrac{8^9}{8^2}$

20. $\dfrac{7^8}{7^2}$

21. $\dfrac{6^3}{6^{-2}}$

22. $\dfrac{5^{10}}{5^{-3}}$

23. $\dfrac{10^{-3}}{10^6}$

24. $\dfrac{12^{-4}}{12^8}$

25. $\dfrac{9^{-4}}{9^{-6}}$

26. $\dfrac{2^{-7}}{2^{-5}}$

27. $\dfrac{x^{-4n}}{x^{6n}}$

28. $\dfrac{y^{-3t}}{y^{8t}}$

29. $\dfrac{w^{-11q}}{w^{-6q}}$

30. $\dfrac{m^{-7t}}{m^{-5t}}$

31. $\dfrac{a^3}{a^{-2}}$

32. $\dfrac{y^4}{y^{-5}}$

33. $\dfrac{27x^7z^5}{-9x^2z}$

34. $\dfrac{24a^5b^3}{-8a^4b}$

35. $\dfrac{-24x^6y^7}{18x^{-3}y^9}$

36. $\dfrac{14a^4b^{-3}}{-8a^8b^{-5}}$

37. $\dfrac{-18x^{-2}y^3}{-12x^{-5}y^5}$

38. $\dfrac{-14a^{14}b^{-5}}{-18a^{-2}b^{-10}}$

b Simplify.

39. $(4^3)^2$

40. $(5^4)^5$

41. $(8^4)^{-3}$

42. $(9^3)^{-4}$

43. $(6^{-4})^{-3}$

44. $(7^{-8})^{-5}$

45. $(5a^2b^2)^3$

46. $(2x^3y^4)^5$

47. $(-3x^3y^{-6})^{-2}$

48. $(-3a^2b^{-5})^{-3}$

49. $(-6a^{-2}b^3c)^{-2}$

50. $(-8x^{-4}y^5z^2)^{-4}$

51. $\left(\dfrac{4^{-3}}{3^4}\right)^3$

52. $\left(\dfrac{5^2}{4^{-3}}\right)^{-3}$

53. $\left(\dfrac{2x^3y^{-2}}{3y^{-3}}\right)^3$

54. $\left(\dfrac{-4x^4y^{-2}}{5x^{-1}y^4}\right)^{-4}$

55. $\left(\dfrac{125a^2b^{-3}}{5a^4b^{-2}}\right)^{-5}$

56. $\left(\dfrac{-200x^3y^{-5}}{8x^5y^{-7}}\right)^{-4}$

57. $\left(\dfrac{-6^5y^4z^{-5}}{2^{-2}y^{-2}z^3}\right)^6$

58. $\left(\dfrac{9^{-2}x^{-4}y}{3^{-3}x^{-3}y^2}\right)^8$

59. $[(-2x^{-4}y^{-2})^{-3}]^{-2}$

60. $[(-4a^{-4}b^{-5})^{-3}]^4$

61. $\left(\dfrac{3a^{-2}b}{5a^{-7}b^5}\right)^{-7}$

62. $\left(\dfrac{2x^2y^{-2}}{3x^8y^7}\right)^9$

63. $\dfrac{10^{2a+1}}{10^{a+1}}$

64. $\dfrac{11^{b+2}}{11^{3b-3}}$

65. $\dfrac{9a^{x-2}}{3a^{2x+2}}$

66. $\dfrac{-12x^{a+1}}{4x^{2-a}}$

67. $\dfrac{45x^{2a+4}y^{b+1}}{-9x^{a+3}y^{2+b}}$

68. $\dfrac{-28x^{b+5}y^{4+c}}{7x^{b-5}y^{c-4}}$

69. $(8^x)^{4y}$

70. $(7^{2p})^{3q}$

71. $(12^{3-a})^{2b}$

72. $(x^{a-1})^{3b}$

73. $(5x^{a-1}y^{b+1})^{2c}$

74. $(4x^{3a}y^{2b})^{5c}$

75. $\dfrac{4x^{2a+3}y^{2b-1}}{2x^{a+1}y^{b+1}}$

76. $\dfrac{25x^{a+b}y^{b-a}}{-5x^{a-b}y^{b+a}}$

c Convert the number to scientific notation.

77. 47,000,000,000

78. 2,600,000,000,000

79. A fast-food chain uses 250,000,000 lb of potatoes each year.

80. Each year there are 1,095,000 new single-family homes built in the United States.

81. 0.000000016

82. 0.000000263

Convert the number to decimal notation.

83. 6.73×10^8

84. 9.24×10^7

85. The wavelength of a certain red light is 6.6×10^{-5} cm.

86. The mass of an electron is 9.11×10^{-28} g.

87. The population of Mexico eats 1×10^9 tortillas a day.

88. An electron has a charge of 4.8×10^{-11} electrostatic units.

Multiply and write the answer in scientific notation.

89. $(2.3 \times 10^6)(4.2 \times 10^{-11})$

90. $(6.5 \times 10^3)(5.2 \times 10^{-8})$

91. $(2.34 \times 10^{-8})(5.7 \times 10^{-4})$

92. $(3.26 \times 10^{-6})(8.2 \times 10^9)$

Divide and write the answer in scientific notation.

93. $\dfrac{8.5 \times 10^8}{3.4 \times 10^5}$

94. $\dfrac{5.1 \times 10^6}{3.4 \times 10^3}$

95. $\dfrac{4.0 \times 10^{-6}}{8.0 \times 10^{-3}}$

96. $\dfrac{7.5 \times 10^{-9}}{2.5 \times 10^{-4}}$

Write the answers to Exercises 97–106 in scientific notation.

97. *Paper Weight.* A ream of copier paper weighs about 4.961 lb. A ream contains 500 sheets of paper. How much does a sheet of copier paper weigh?

98. *Orbit of Venus.* Venus has a nearly circular orbit of the sun. The average distance from the sun to Venus is about 6.71×10^7 mi. How far does Venus travel in one orbit?

99. *Alpha Centauri.* Other than the sun, the star closest to Earth is Alpha Centauri. Its distance from Earth is about 2.4×10^{13} mi. One light-year = the distance that light travels in one year = 5.88×10^{12} mi. How many light-years is it from Earth to Alpha Centauri?

100. *World Insect Population.* World population is expected to be 6.1 billion by the year 2000. It is estimated that there are 1 million insects for every human being on Earth. How many insects will there be in the year 2000?

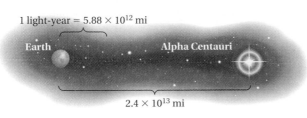

1 light-year = 5.88×10^{12} mi

Earth

Alpha Centauri

2.4×10^{13} mi

101. About how many seconds are there in 2000 yr? Assume that there are 365 days in one year.

102. *Manufacturing of Pennies.* Each day the U.S. Mint manufactures 39 million pennies. If manufacturing operates approximately 250 days a year, how many pennies does it manufacture in one year?

103. The distance that light travels in 100 yr is approximately 5.87×10^{14} mi. How far does light travel in 13 weeks?

104. *Popcorn Consumption.* On average, Americans eat 6.5 million gal of popcorn each day. How much popcorn do they eat in one year?

105. There are 300,000 words in the English language. The average person knows about 10,000 of them. What part of the total number of words does the average person know?

106. The average discharge at the mouth of the Amazon River is 4,200,000 cubic feet per second. How much water is discharged from the Amazon River in one hour? one year?

Skill Maintenance

Simplify.   [R.3c], [R.6b]

107. $9x - (-4y + 8) + (10x - 12)$

108. $-6t - (5t - 13) + 2(4 - 6t)$

109. $4^2 + 30 \cdot 10 - 7^3 + 16$

110. $5^4 - 38 \cdot 24 - (16 - 4 \cdot 18)$

111. $20 - 5 \cdot 4 - 8$

112. $20 - (5 \cdot 4 - 8)$

Synthesis

113. ◈ A $20 bill weighs about 2.2×10^{-3} lb. A criminal claims to be carrying $5 million in $20 bills in his suitcase. Is this possible? Why or why not?

114. ◈ 〰 When a calculator indicates that $5^{17} = 7.629394531 \times 10^{11}$, you know that an approximation is being made. How can you tell? (*Hint*: What should the ones digit be?)

Simplify.

115. $\dfrac{(2^{-2})^{-4} \cdot (2^3)^{-2}}{(2^{-2})^2 \cdot (2^5)^{-3}}$

116. $\left[\dfrac{(-3x^{-2}y^5)^{-3}}{(2x^4y^{-8})^{-2}} \right]^2$

117. $\left[\left(\dfrac{a^{-2}}{b^7} \right)^{-3} \cdot \left(\dfrac{a^4}{b^{-3}} \right)^2 \right]^{-1}$

Simplify. Assume that variables in exponents represent integers.

118. $(m^{x-b}n^{x+b})^x(m^b n^{-b})^x$

119. $\left[\dfrac{(2x^a y^b)^3}{(-2x^a y^b)^2} \right]^2$

120. $(x^b y^a \cdot x^a y^b)^c$

Summary and Review Exercises: Chapter R

Important Properties and Formulas

Properties of Real Numbers

Commutative Laws: $a + b = b + a,\quad ab = ba$

Associative Laws: $a + (b + c) = (a + b) + c,\quad a(bc) = (ab)c$

Distributive Laws: $a(b + c) = ab + ac,\quad a(b - c) = ab - ac$

Inverses: $a + (-a) = 0,\quad a \cdot \dfrac{1}{a} = 1$

Identity Property of 0: $a + 0 = a$

Identity Property of 1: $1 \cdot a = a$

Property of -1: $-1 \cdot a = -a$

Properties of Exponents: $a^1 = a,\qquad a^0 = 1,\qquad a^{-n} = \dfrac{1}{a^n}$

Product Rule: $a^m \cdot a^n = a^{m+n}$

Power Rule: $(a^m)^n = a^{mn}$

Quotient Rule: $\dfrac{a^m}{a^n} = a^{m-n}$

Raising a Product to a Power: $(ab)^n = a^n b^n$

Raising a Quotient to a Power: $\left(\dfrac{a}{b}\right)^n = \dfrac{a^n}{b^n}$

Scientific Notation: $M \times 10^n$, or 10^n, where M is such that $1 \le M < 10$.

The review exercises that follow are for practice. Answers are given at the back of the book. If you miss an exercise, restudy the objective indicated in blue next to the exercise or the direction line that precedes it.

Part 1

1. Which of the following numbers are rational?
   [R.1a]

   $2,\ \sqrt{3},\ -\dfrac{2}{3},\ 0.45\overline{45},\ -23.788$

2. Use set-builder notation to name the set of all real numbers less than or equal to 46.   [R.1a]

3. Use $<$ or $>$ for ▮ to write a true sentence: [R.1b]
   -3.9 ▮ 2.9.

4. Write another inequality with the same meaning as $19 > x$.   [R.1b]

Is each of the following true or false?   [R.1b]

5. $-13 \ge 5$

6. $7.01 \le 7.01$

Graph the inequality on a number line.   [R.1c]

7. $x > -4$

8. $x \le 1$

Find the absolute value.   [R.1d]

9. $|-7.23|$

10. $|9 - 9|$

Add, subtract, multiply, or divide, if possible.
[R.2a, c, d, e]

11. $6 + (-8)$

12. $-3.8 + (-4.1)$

13. $\dfrac{3}{4} + \left(-\dfrac{13}{7}\right)$

14. $-8 - (-3)$

15. $-17.3 - 9.4$

16. $\dfrac{3}{2} - \left(-\dfrac{13}{4}\right)$

17. $(-3.8)(-2.7)$

18. $-\dfrac{2}{3}\left(\dfrac{9}{14}\right)$

19. $-6(-7)(4)$

20. $-12 \div 3$

21. $\dfrac{-84}{-4}$

22. $\dfrac{49}{-7}$

23. $\dfrac{5}{6} \div \left(-\dfrac{10}{7}\right)$

24. $-\dfrac{5}{2} \div \left(-\dfrac{15}{16}\right)$

25. $\dfrac{21}{0}$

26. $-108 \div 4.5$

Evaluate $-a$ for each of the following.   [R.2b]

27. $a = -7$

28. $a = 2.3$

29. $a = 0$

Write using exponential notation.   [R.3a]

30. $a \cdot a \cdot a \cdot a \cdot a$

31. $\left(-\dfrac{7}{8}\right)\left(-\dfrac{7}{8}\right)\left(-\dfrac{7}{8}\right)$

32. Rewrite using a positive exponent: a^{-4}.   [R.3b]

33. Rewrite using a negative exponent: $\dfrac{1}{x^8}$.   [R.3b]

Simplify.   [R.3c]

34. $2^3 - 3^4 + (13 \cdot 5 + 67)$

35. $64 \div (-4) + (-5)(20)$

Part 2

Translate to an algebraic expression.   [R.4a]

36. Five times some number

37. Twenty-eight percent of some number

38. Nine less than t

39. Eight less than the quotient of two numbers

Evaluate. [R.4b]

40. $5x - 7$, for $x = -2$

41. $\dfrac{x - y}{2}$, for $x = 4$ and $y = 20$

42. The area A of a rectangle is given by the length l times the width w: $A = lw$. Find the area of a rectangular rug that measures 7 ft by 12 ft.

w

l

Complete the table by evaluating each expression for the given values. Then determine whether the expressions are equivalent. [R.5a]

43.

	$x^2 - 5$	$(x - 5)^2$
$x = -1$		
$x = 10$		
$x = 0$		

44.

	$2x - 14$	$2(x - 7)$
$x = -1$		
$x = 10$		
$x = 0$		

45. Use multiplying by 1 to find an equivalent expression with the given denominator: [R.5b]

$$\frac{7}{3}; \quad 9x.$$

46. Simplify: $\dfrac{-84x}{7x}$. [R.5b]

Use a commutative law to find an equivalent expression. [R.5c]

47. $11 + a$

48. $8y$

Use an associative law to find an equivalent expression. [R.5c]

49. $(9 + a) + b$

50. $8(xy)$

Multiply. [R.5d]

51. $-3(2x - y)$

52. $1ab(2c + 1)$

Factor. [R.5d]

53. $5x + 10y - 5z$

54. $ptr + pts$

Collect like terms. [R.6a]

55. $2x + 6y - 5x - y$

56. $7c - 6 + 9c + 2 - 4c$

57. Find an equivalent expression: $-(-9c + 4d - 3)$. [R.6b]

Simplify. [R.6b]

58. $4(x - 3) - 3(x - 5)$

59. $12x - 3(2x - 5)$

60. $7x - [4 - 5(3x - 2)]$

61. $4m - 3[3(4m - 2) - (5m + 2) + 12]$

Multiply or divide, and simplify. [R.7a]

62. $(2x^4 y^{-3})(-5x^3 y^{-2})$

63. $\dfrac{-15x^2 y^{-5}}{10x^6 y^{-8}}$

Simplify. [R.7b]

64. $(-3a^{-4}bc^3)^{-2}$

65. $\left[\dfrac{-2x^4 y^{-4}}{3x^{-2} y^6}\right]^{-4}$

Multiply or divide, and write scientific notation for the answer. [R.7c]

66. $\dfrac{2.2 \times 10^7}{3.2 \times 10^{-3}}$

67. $(3.2 \times 10^4)(4.1 \times 10^{-6})$

68. A sheet of plastic has a thickness of 0.00015 m. The sheet is 1.2 m by 79 m. Find the volume of the sheet and express the answer in scientific notation.

Synthesis

69. ◈ Explain and compare the commutative, associative, and distributive laws.

70. ◈ Give two expressions that are equivalent. Give two expressions that are not equivalent and explain why they are not.

71. Simplify: $(x^y \cdot x^{3y})^3$. [R.7b]

72. If $a = 2^x$ and $b = 2^{x+5}$, find $a^{-1}b$. [R.7a]

73. Which of the following pairs are equivalent?
   a) $3x - 3y$ b) $3x - y$
   c) $x^{-2} x^5$ d) x^{-10}
   e) x^{-3} f) $(x^{-2})^5$
   g) $x(yz)$ h) $x(y + z)$
   i) $3(x - y)$ j) $xy + xz$

Test: Chapter R

Part 1

1. Which of the following numbers are irrational?

$$-43, \quad \sqrt{7}, \quad -\frac{2}{3}, \quad 2.3\overline{76}, \quad \pi$$

2. Use set-builder notation to name the set of real numbers greater than 20.

3. Use $<$ or $>$ for ▨ to write a true sentence:

-4.5 ▨ -8.7.

4. Write another inequality with the same meaning as $a \leq 5$.

Is each of the following true or false?

5. $-6 \geq -6$ **6.** $-8 \leq -6$

7. Graph $x > -2$ on a number line.

Find the absolute value.

8. $|0|$

9. $\left| -\frac{7}{8} \right|$

Add, subtract, multiply, or divide, if possible.

10. $7 + (-9)$

11. $-5.3 + (-7.8)$

12. $-\frac{5}{2} + \left(-\frac{7}{2} \right)$

13. $-6 - (-5)$

14. $-18.2 - 11.5$

15. $\frac{19}{4} - \left(-\frac{3}{2} \right)$

16. $(-4.1)(8.2)$

17. $-\frac{4}{5}\left(-\frac{15}{16} \right)$

18. $-6(-4)(-11)2$

19. $-75 \div (-5)$

20. $\frac{-10}{2}$

21. $-\frac{5}{2} \div \left(-\frac{15}{16} \right)$

22. $-459.2 \div 5.6$

23. $\frac{-3}{0}$

Evaluate $-a$ for each of the following.

24. $a = -13$

25. $a = 0$

26. Write exponential notation: $q \cdot q \cdot q \cdot q$.

27. Rewrite using a negative exponent: $\frac{1}{a^9}$.

Simplify.

28. $1 - (2 - 5)^2 + 5 \div 10 \cdot 4^2$

29. $\dfrac{7(5 - 2 \cdot 3) - 3^2}{4^2 - 3^2}$

Part 2

Translate to an algebraic expression.

30. Nine more than t

31. Twelve less than the quotient of two numbers

32. Evaluate $3x - 3y$ for $x = 2$ and $y = -4$.

33. The area A of a triangle is given by $A = \frac{1}{2}bh$. Find the area of a stamp whose base measures 3 cm and whose height measures 2.5 cm.

2.5 cm

3 cm

Answers

1. _____
2. _____
3. _____
4. _____
5. _____
6. _____
7. _____
8. _____
9. _____
10. _____
11. _____
12. _____
13. _____
14. _____
15. _____
16. _____
17. _____
18. _____
19. _____
20. _____
21. _____
22. _____
23. _____
24. _____
25. _____
26. _____
27. _____
28. _____
29. _____
30. _____
31. _____
32. _____
33. _____

Complete a table by evaluating each expression for $x = -1$, 10, and 0. Then determine whether the expressions are equivalent. Answer yes or no.

34. $x(x - 3)$; $x^2 - 3x$

35. $3x + 5x^2$; $8x^2$

36. Use multiplying by 1 to find an equivalent expression with the given denominator.

$$\frac{3}{4}; \quad 36x$$

37. Simplify:

$$\frac{-54x}{-36x}.$$

Use a commutative law to find an equivalent expression.

38. pq

39. $t + 4$

Use an associative law to find an equivalent expression.

40. $3 + (t + w)$

41. $(4a)b$

Multiply.

42. $-2(3a - 4b)$

43. $3\pi r(s + 1)$

Factor.

44. $ab - ac + 2ad$

45. $2ah + h$

Collect like terms.

46. $6y - 8x + 4y + 3x$

47. $4a - 7 + 17a + 21$

48. Find an equivalent expression: $-(-9x + 7y - 22)$.

Simplify.

49. $-3(x + 2) - 4(x - 5)$

50. $4x - [6 - 3(2x - 5)]$

51. $3a - 2[5(2a - 5) - (8a + 4) + 10]$

Multiply or divide, and simplify.

52. $\dfrac{-12x^3y^{-4}}{8x^7y^{-6}}$

53. $(3a^4b^{-2})(-2a^5b^{-3})$

54. $(5a^{4n})(-10a^{5n})$

55. $\dfrac{-60x^{3t}}{12x^{7t}}$

Simplify.

56. $(-3a^{-3}b^2c)^{-4}$

57. $\left[\dfrac{-5a^{-2}b^8}{10a^{10}b^{-4}}\right]^{-4}$

58. Convert to scientific notation: 0.0000437.

Multiply or divide, and write scientific notation for the answer.

59. $(8.7 \times 10^{-9})(4.3 \times 10^{15})$

60. $\dfrac{1.2 \times 10^{-12}}{6.4 \times 10^{-7}}$

61. *Mass of Pluto.* The mass of Earth is 5.98×10^{24} kg. The mass of the planet Pluto is about 0.002 times the mass of Earth. Find the mass of Pluto and express the answer in scientific notation.

Synthesis

62. The music on a compact disc is recorded using small pits, or pieces of information, in the disc surface. There can be up to 600 megabytes on a single disc. A megabyte is 1 million bytes and a byte is 8 pieces of information. How many pits, or pieces of information, can be on the surface of a compact disc? Write scientific notation for your answer.

63. Which of the following pairs are equivalent?

a) $x^{-3}x^{-4}$ b) x^{12}

c) x^{-12} d) $5x + 5$

e) $(x^{-3})^{-4}$ f) $5(x + 1)$

g) $5x$ h) $5 + 5x$

i) $5(xy)$ j) $(5x)y$

1

Solving Linear Equations and Inequalities

An Application

The Mathematics

Michael Johnson set a world record of 19.32 sec in the men's 200-m dash in the 1996 Olympics. The equation

$$R = -0.0513t + 19.32$$

can be used to predict the world record in the men's 200-m dash t years after 1996. Determine those years for which the world record will be less than 19.0 sec.

We can translate the problem to the following sentence:

$$\underbrace{-0.0513t + 19.32} < 19.0.$$

This is an inequality.

This problem appears as Example 15 in Section 1.4.

 For more information, visit us at www.mathmax.com

Pretest: Chapter 1

Solve.

1. $-6x = 42$

2. $-3.2 + y = 18.1$

3. $9 - 2x = 11$

4. $4 - 3(y + 1) = 2(y - 3) - 7$

5. $-7x < 21$

6. $2x - 1 \leq 5x + 8$

7. $-1 \leq 3x + 2 \leq 4$

8. $|2y - 7| = 9$

9. $4a + 5 < -3 \ or \ 4a + 5 > 3$

10. $|3a + 5| > 1$

11. $|3x + 4| = |x - 7|$

12. $-8(x - 7) \geq 10(2x + 3) - 12$

13. Solve for r: $P = 3rq$.

14. Find the distance between 8.2 and 10.6.

Solve.

15. *Perimeter of a College Volleyball Court.* The perimeter of a college volleyball court is 180 ft and the length is twice the width. What are the dimensions?

16. *Records in the Women's 100-m Dash.* Florence Griffith Joyner set a world record of 10.49 sec in the women's 100-m dash in 1988. The equation

$$R = -0.0433t + 10.49$$

can be used to predict the world record in the women's 100-m dash t years after 1988. Determine those years for which the world record will be less than 10.35 sec.

17. *Real Estate Commission.* The following is a typical real estate commission on the selling price of a house:

7% for the first $100,000, and

4% for the amount that exceeds $100,000.

A realtor receives a commission of $10,120 for selling a house. What was the selling price?

18. Find the intersection:

$$\{-4, -3, 1, 4, 9\} \cap \{-3, -2, 2, 4, 5\}.$$

19. Find the union: $\{-1, 1\} \cup \{0\}$.

20. Simplify: $\left| \dfrac{-5y}{2x} \right|$.

Objectives for Retesting

The objectives to be tested in addition to the material in this chapter are as follows.

[R.2a, c]   Add and subtract real numbers.

[R.2d, e]   Multiply and divide real numbers.

[R.5d]   Use the distributive laws to find equivalent expressions by multiplying and factoring.

[R.6b]   Simplify an expression by removing parentheses and collecting like terms.

1.1 Solving Equations

a | Equations and Solutions

In order to solve problems, we must be able to solve *equations*.

> An **equation** is a number sentence that says that the expressions on either side of the equals sign, =, represent the same number.

Here are some examples:

$$3 + 5 = 8, \quad 15 - 10 = 2 + 3, \quad x + 8 = 23, \quad 5x - 2 = 9 - x.$$

Equations have expressions on each side of the equals sign. The sentence "$15 - 10 = 2 + 3$" asserts that the expressions $15 - 10$ and $2 + 3$ name the same number. It is a *true* equation.

Some equations are true. Some are false. Some are neither true nor false.

Examples Determine whether the equation is true, false, or neither.

1. $1 + 10 = 11$ The equation is *true*.

2. $7 - 8 = 9 - 13$ The equation is *false*.

3. $x - 9 = 3$ The equation is *neither* true nor false, because we do not know what number x represents.

Do Exercises 1–3.

If an equation contains a variable, then the equation may be neither true nor false. Some replacements may make it true and some may make it false.

> The replacements making an equation true are called its **solutions**. The set of all solutions is called the **solution set.** When we find all solutions, no matter how, we say that we have **solved** the equation.

To determine whether a number is a solution of an equation, we evaluate the algebraic expression on each side of the equation by substitution. If the values are the same, then the number is a solution.

Example 4 Determine whether -21 is a solution of $x + 6 = -15$.

We have

$$\frac{x + 6 = -15}{-21 + 6 \ ? \ -15}$$
$$-15 \ | \qquad \text{TRUE}$$

Writing the equation

Substituting -21 for x

Since the left-hand and the right-hand sides are the same, we have a solution. No other number makes the equation true, so the only solution is the number -21.

Consider the following equations.

a) $3 + 4 = 7$
b) $5 - 1 = 2$
c) $21 + 2 = 24$
d) $x - 5 = 12$
e) $9 - x = x$
f) $13 + 2 = 15$

1. Which equations are true?

2. Which equations are false?

3. Which equations are neither true nor false?

Answers on page A-5

Determine whether the given number is a solution of the given equation.

4. 8; $x + 5 = 13$

5. -4; $7x = 16$

6. 5; $2x + 3 = 13$

7. Determine whether

$3x + 2 = 11$ and $x = 3$

are equivalent.

8. Determine whether

$4 - 5x = -11$ and $x = -3$

are equivalent.

Answers on page A-5

Example 5   Determine whether 18 is a solution of $2x - 3 = 5$.

We have

$$2x - 3 = 5 \qquad \text{Writing the equation}$$
$$2 \cdot 18 - 3 \;?\; 5 \qquad \text{Substituting 18 for } x$$
$$36 - 3$$
$$33 \qquad \text{FALSE}$$

Since the left-hand and the right-hand sides are not the same, we do not have a solution.

Do Exercises 4–6.

Equivalent Equations

Consider the equation

$$x = 5.$$

The solution of this equation is easily "seen" to be 5. If we replace x with 5, we get

$$5 = 5, \quad \text{which is true.}$$

Now consider the equation $2x + 3 = 13$. In Margin Exercise 6, we see that the solution of this equation is also 5, but the fact that 5 is the solution is not so obvious. We now consider principles that allow us to start with an equation like $2x + 3 = 13$, and end up with an *equivalent equation* like $x = 5$, in which the variable is alone on one side, and for which the solution is easy to find.

   Equations with the same solutions are called **equivalent equations**.

Do Exercises 7 and 8.

b   The Addition Principle

One of the principles we use in solving equations involves adding. Consider the equation $a = b$. It says that a and b represent the same number. Suppose that $a = b$ is true and then add a number c to a. We will get the same answer if we add c to b, because a and b are the same number.

> **THE ADDITION PRINCIPLE**
> For any real numbers a, b, and c,
> $$a = b \quad \text{is equivalent to} \quad a + c = b + c.$$

When we use the addition principle, we sometimes say that we "add the same number on both sides of an equation." This is also true for subtraction, because we can express every subtraction as an addition; that is,

$$a - c = b - c \quad \text{is equivalent to} \quad a + (-c) = b + (-c).$$

The addition principle also tells us that we can "subtract the same number on both sides of an equation."

Example 6 Solve: $x + 6 = -15$.

$$x + 6 = -15$$

$$x + 6 - 6 = -15 - 6 \qquad \text{Using the addition principle: adding } -6 \text{ on both sides or subtracting 6 on both sides. Note that 6 and } -6 \text{ are opposites.}$$

$$x + 0 = -21 \qquad \text{Simplifying}$$

$$x = -21 \qquad \text{Using the identity property of 0: } x + 0 = x$$

CHECK: $$\frac{x + 6 = -15}{-21 + 6 \; ? \; -15}$$
$$-15 \; | \qquad \text{TRUE}$$

Substituting -21 for x

The solution is -21.

In Example 6, we wanted to get x alone so that we could easily see the solution, so we added the opposite of 6. This eliminated the 6 on the left, giving us the *additive identity* 0, which when added to x is x. We started with $x + 6 = -15$; using the addition principle, we derived a simpler equation, $x = -21$, from which it was easy to *see* the solution. The equations $x + 6 = -15$ and $x = -21$ are *equivalent*.

Example 7 Solve: $y - 4.7 = 13.9$.

$$y - 4.7 = 13.9$$

$$y - 4.7 + 4.7 = 13.9 + 4.7 \qquad \text{Using the addition principle: adding 4.7 on both sides. Note that } -4.7 \text{ and 4.7 are opposites.}$$

$$y + 0 = 18.6 \qquad \text{Simplifying}$$

$$y = 18.6 \qquad \text{Using the identity property of 0: } y + 0 = y$$

CHECK: $$\frac{y - 4.7 = 13.9}{18.6 - 4.7 \; ? \; 13.9}$$
$$13.9 \; | \qquad \text{TRUE}$$

Substituting 18.6 for y

The solution is 18.6.

Example 8 Solve: $-\frac{3}{8} + x = -\frac{5}{7}$.

$$-\frac{3}{8} + x = -\frac{5}{7}$$

$$\frac{3}{8} + \left(-\frac{3}{8}\right) + x = \frac{3}{8} + \left(-\frac{5}{7}\right) \qquad \text{Using the addition principle: adding } \frac{3}{8}$$

$$0 + x = \frac{3}{8} - \frac{5}{7}$$

$$x = \frac{3}{8} \cdot \frac{7}{7} - \frac{5}{7} \cdot \frac{8}{8} \qquad \text{Multiplying by 1 to obtain the least common denominator}$$

$$x = \frac{21}{56} - \frac{40}{56}$$

$$x = -\frac{19}{56}$$

CHECK: $$\frac{-\frac{3}{8} + x = -\frac{5}{7}}{-\frac{3}{8} + \left(-\frac{19}{56}\right) \; ? \; -\frac{5}{7}}$$
$$-\frac{3}{8} \cdot \frac{7}{7} + \left(-\frac{19}{56}\right) \Big|$$
$$-\frac{21}{56} + \left(-\frac{19}{56}\right) \Big|$$
$$-\frac{40}{56} \Big|$$
$$-\frac{5}{7} \Big| \qquad \text{TRUE}$$

The solution is $-\frac{19}{56}$.

Solve using the addition principle.

9. $x + 9 = 2$

10. $x + \dfrac{1}{4} = -\dfrac{3}{5}$

11. $13 = -25 + y$

12. $y - 61.4 = 78.9$

Do Exercises 9–12.

c | The Multiplication Principle

Suppose that $a = b$ is true and we multiply a by a nonzero number c. We get the same answer if we multiply b by c, because a and b are the same number.

> ▶ **THE MULTIPLICATION PRINCIPLE**
>
> For any real numbers a, b, and c, $c \neq 0$,
> $$a = b \quad \text{is equivalent to} \quad a \cdot c = b \cdot c.$$

Example 9   Solve: $\frac{4}{5}x = 22$.

$$\frac{4}{5}x = 22$$

$$\frac{5}{4} \cdot \frac{4}{5}x = \frac{5}{4} \cdot 22 \qquad \text{Multiplying by } \tfrac{5}{4}, \text{ the reciprocal of } \tfrac{4}{5}$$

$$1 \cdot x = \frac{55}{2} \qquad \text{Multiplying and simplifying}$$

$$x = \frac{55}{2} \qquad \text{Using the identity property of 1: } 1 \cdot x = x$$

CHECK:
$$\frac{4}{5}x = 22$$
$$\frac{4}{5} \cdot \frac{55}{2} \;?\; 22$$
$$22 \;\big|\; \qquad \text{TRUE}$$

The solution is $\frac{55}{2}$.

In Example 9, in order to get x alone, we multiplied by the *multiplicative inverse*, or *reciprocal*, of $\frac{4}{5}$. When we multiplied, we got the *multiplicative identity* 1 times x, or $1 \cdot x$, which simplified to x. This enabled us to eliminate the $\frac{4}{5}$ on the left.

This multiplication principle also tells us that we can "divide on both sides by a nonzero number." This is because division is the same as multiplying by a reciprocal. That is,

$$\frac{a}{c} = \frac{b}{c} \quad \text{is equivalent to} \quad a \cdot \frac{1}{c} = b \cdot \frac{1}{c}, \quad \text{when } c \neq 0.$$

In practice, if the number in front of the variable, the **coefficient**, is in fractional notation, it is more convenient to "multiply" on both sides by a reciprocal. If the coefficient is in decimal notation or is an integer, it is more convenient to "divide" by the coefficient.

Example 10 Solve: $4x = 9$.

$$4x = 9$$

$$\frac{4x}{4} = \frac{9}{4} \qquad \text{Using the multiplication principle: multiplying by } \tfrac{1}{4} \text{ on both sides or dividing by the coefficient, 4, on both sides}$$

$$1 \cdot x = \frac{9}{4} \qquad \text{Simplifying}$$

$$x = \frac{9}{4} \qquad \text{Using the identity property of 1: } 1 \cdot x = x$$

CHECK:
$$\begin{array}{c}4x = 9 \\ \hline 4 \cdot \tfrac{9}{4} \ ? \ 9 \\ 9 \ | \qquad \text{TRUE} \end{array}$$

The solution is $\frac{9}{4}$.

Do Exercises 13–15.

Example 11 Solve: $5.5 = -0.05y$.

$$5.5 = -0.05y$$

$$\frac{5.5}{-0.05} = \frac{-0.05y}{-0.05} \qquad \text{Dividing by } -0.05 \text{ on both sides}$$

$$\frac{5.5}{-0.05} = 1 \cdot y$$

$$-110 = y$$

The check is left to the student. The solution is -110.

Note that equations are reversible. That is, $a = b$ is equivalent to $b = a$. Thus, when we solve $5.5 = -0.05y$, we can reverse it and solve $-0.05y = 5.5$ if we wish.

Do Exercise 16.

Example 12 Solve: $-\dfrac{x}{4} = 10$.

$$-\frac{x}{4} = 10$$

$$-\frac{1}{4}x = 10$$

$$-4 \cdot \left(-\frac{1}{4}\right)x = -4 \cdot 10 \qquad \text{Multiplying by } -4 \text{ on both sides}$$

$$1 \cdot x = -40 \qquad \text{Simplifying}$$

$$x = -40$$

The check is left to the student. The solution is -40.

Do Exercises 17 and 18.

Solve using the multiplication principle.

13. $8x = 10$

14. $-\dfrac{3}{7}y = 21$

15. $-4x = -\dfrac{6}{7}$

16. Solve: $-12.6 = 4.2y$.

Solve.

17. $-\dfrac{x}{8} = 17$

18. $-x = -5$

Answers on page A-5

19. Solve: $-4 + 9x = 8$.

20. Clear the fractions and solve:

$$\frac{2}{3} - \frac{5}{6}y = \frac{1}{3}.$$

Answers on page A-5

---

d | **Using the Principles Together**

Let's see how we can use the addition and multiplication principles together.

Example 13   Solve: $3x - 4 = 13$.

$$3x - 4 = 13$$
$$3x - 4 + 4 = 13 + 4 \qquad \text{Using the addition principle: adding 4}$$
$$3x = 17 \qquad \text{Simplifying}$$
$$\frac{3x}{3} = \frac{17}{3} \qquad \text{Dividing by 3}$$
$$x = \frac{17}{3}$$

CHECK:

$$\begin{array}{c|c} \hline 3x - 4 = 13 \\ \hline 3 \cdot \frac{17}{3} - 4 \ ? \ 13 \\ 17 - 4 \ \Big| \\ 13 \ \Big| \qquad \text{TRUE} \end{array}$$

The solution is $\frac{17}{3}$, or $5\frac{2}{3}$.

> In algebra, "improper" fractional notation, such as $\frac{17}{3}$, is quite "proper." We will generally use such notation rather than $5\frac{2}{3}$.

Do Exercise 19.

In a situation such as Example 13, it is easier to first use the addition principle. In a situation in which fractions or decimals are involved, it may be easier to use the multiplication principle first to clear them, but it is not mandatory.

Example 14   Clear the fractions and solve: $\frac{3}{16}x + \frac{1}{8} = \frac{11}{8}$.

We multiply on both sides by the least common denominator—in this case, 16:

$$\frac{3}{16}x + \frac{1}{8} = \frac{11}{8} \qquad \text{The LCM is 16.}$$
$$16\left(\frac{3}{16}x + \frac{1}{8}\right) = 16\left(\frac{11}{8}\right) \qquad \text{Multiplying by 16}$$
$$16 \cdot \frac{3}{16}x + 16 \cdot \frac{1}{8} = 22 \qquad \begin{array}{l}\text{Carrying out the multiplication.}\\ \text{We use the distributive law on the left, being}\\ \text{careful to multiply } \textit{both} \text{ terms by 16.}\end{array}$$
$$3x + 2 = 22 \qquad \text{Simplifying}$$
$$3x + 2 - 2 = 22 - 2 \qquad \text{Subtracting 2}$$
$$3x = 20$$
$$\frac{3x}{3} = \frac{20}{3} \qquad \text{Dividing by 3}$$
$$x = \frac{20}{3}.$$

The number $\frac{20}{3}$ checks and is the solution.

Do Exercise 20.

Example 15 Clear the decimals and solve: $12.4 - 5.12x = 3.14x$.

We multiply on both sides by a power of ten—in this case, 10^2 or 100—to clear the equation of decimals:

$$12.4 - 5.12x = 3.14x$$

$$100(12.4 - 5.12x) = 100(3.14x) \qquad \text{Multiplying by 100}$$

$$100(12.4) - 100(5.12x) = 314x \qquad \text{Carrying out the multiplication. We use the distributive law on the left.}$$

$$1240 - 512x = 314x \qquad \text{Simplifying}$$

$$1240 - 512x + 512x = 314x + 512x \qquad \text{Adding } 512x$$

$$1240 = 826x \qquad \text{Simplifying}$$

$$\frac{1240}{826} = \frac{826x}{826} \qquad \text{Dividing by 826}$$

$$x = \frac{1240}{826}, \text{ or } \frac{620}{413}$$

The solution is $\frac{620}{413}$.

Do Exercise 21.

When there are like terms in an equation, we collect them on each side. Then if there are still like terms on opposite sides, we get them on the same side using the addition principle.

Example 16 Solve: $8x + 6 - 2x = -4x - 14$.

$$8x + 6 - 2x = -4x - 14$$

$$6x + 6 = -4x - 14 \qquad \text{Collecting like terms on the left}$$

$$4x + 6x + 6 = 4x - 4x - 14 \qquad \text{Adding } 4x$$

$$10x + 6 = -14 \qquad \text{Collecting like terms}$$

$$10x + 6 - 6 = -14 - 6 \qquad \text{Subtracting 6}$$

$$10x = -20$$

$$\frac{10x}{10} = \frac{-20}{10} \qquad \text{Dividing by 10}$$

$$x = -2 \qquad \text{Simplifying}$$

CHECK:

$$\begin{array}{c|c} \multicolumn{2}{c}{8x + 6 - 2x = -4x - 14} \\ \hline 8(-2) + 6 - 2(-2) \ ? & -4(-2) - 14 \\ -16 + 6 + 4 & 8 - 14 \\ -6 & -6 \qquad \text{TRUE} \end{array}$$

The solution is -2.

Do Exercises 22–24.

Special Cases

There are equations with no solution.

Example 17 Solve: $-8x + 5 = 14 - 8x$.

$$-8x + 5 = 14 - 8x$$

$$5 = 14 \qquad \text{Adding } 8x, \text{ we get a false equation.}$$

No matter what number we try for x, we get a false sentence. Thus the equation has no solution.

---

21. Clear the decimals and solve:

$$6.3x - 9.1 = 3x.$$

Solve.

22. $\dfrac{5}{2}x + \dfrac{9}{2}x = 21$

23. $1.4x - 0.9x + 0.7 = -2.2$

24. $-4x + 2 + 5x = 3x - 15$

Answers on page A-5

Solve.

25. $4 + 7x = 7x + 9$

26. $3 + 9x = 9x + 3$

Solve.

27. $30 + 7(x - 1) = 3(2x + 7)$

28. $3(y - 1) - 1 = 2 - 5(y + 5)$

Answers on page A-5

There are equations for which any real number is a solution.

Example 18   Solve: $-8x + 5 = 5 - 8x$.

$$-8x + 5 = 5 - 8x$$
$$5 = 5 \qquad \text{Adding } 8x, \text{ we get a true equation.}$$

Replacing x with any real number gives a true sentence. Thus any real number is a solution.

Do Exercises 25 and 26.

Equations Containing Parentheses

Equations containing parentheses can often be solved by first multiplying to remove parentheses and then proceeding as before.

Example 19   Solve: $3(7 - 2x) = 14 - 8(x - 1)$.

$$3(7 - 2x) = 14 - 8(x - 1)$$
$$21 - 6x = 14 - 8x + 8 \qquad \text{Multiplying, using the distributive law, to remove parentheses}$$
$$21 - 6x = 22 - 8x \qquad \text{Collecting like terms}$$
$$21 - 6x + 8x = 22 - 8x + 8x \qquad \text{Adding } 8x$$
$$21 + 2x = 22 \qquad \text{Collecting like terms}$$
$$21 + 2x - 21 = 22 - 21 \qquad \text{Subtracting 21}$$
$$2x = 1 \qquad \text{Simplifying}$$
$$\frac{2x}{2} = \frac{1}{2} \qquad \text{Dividing by 2}$$
$$x = \frac{1}{2}$$

CHECK:

$$\begin{array}{c|c}
\multicolumn{2}{c}{3(7 - 2x) = 14 - 8(x - 1)} \\
\hline
3\left(7 - 2 \cdot \tfrac{1}{2}\right) \;?\; & 14 - 8\left(\tfrac{1}{2} - 1\right) \\
3(7 - 1) & 14 - 8\left(-\tfrac{1}{2}\right) \\
3 \cdot 6 & 14 + 4 \\
18 & 18 \qquad \text{TRUE}
\end{array}$$

The solution is $\frac{1}{2}$.

Do Exercises 27 and 28.

Here is a summary of the method of solving equations, as used in the preceding examples.

AN EQUATION-SOLVING PROCEDURE

1. Clear the equation of fractions or decimals if that is needed.
2. If parentheses occur, multiply to remove them using the distributive law.
3. Collect like terms on each side of the equation, if necessary.
4. Use the addition principle to get all like terms with letters on one side and all other terms on the other side.
5. Collect like terms on each side again, if necessary.
6. Use the multiplication principle to solve for the variable.

Exercise Set 1.1

Remember to review the objectives before doing the exercises.

a Determine whether the given number is a solution of the given equation.

1. 17; $x + 23 = 40$

2. 24; $47 - x = 23$

3. -8; $2x - 3 = -18$

4. -10; $3x + 14 = -27$

5. 45; $\dfrac{-x}{9} = -2$

6. 32; $\dfrac{-x}{8} = -3$

7. 10; $2 - 3x = 21$

8. -11; $4 - 5x = 59$

9. 19; $5x + 7 = 102$

10. 9; $9y + 5 = 86$

11. -11; $7(y - 1) = 84$

12. -13; $x + 5 = 5 + x$

b Solve using the addition principle. Don't forget to check.

13. $y + 6 = 13$

14. $x + 7 = 14$

15. $-20 = x - 12$

16. $-27 = y - 17$

17. $-8 + x = 19$

18. $-8 + r = 17$

19. $-12 + z = -51$

20. $-37 + x = -89$

21. $p - 2.96 = 83.9$

22. $z - 14.9 = -5.73$

23. $-\dfrac{3}{8} + x = -\dfrac{5}{24}$

24. $x + \dfrac{1}{12} = -\dfrac{5}{6}$

c Solve using the multiplication principle. Don't forget to check.

25. $3x = 18$

26. $5x = 30$

27. $-11y = 44$

28. $-4x = 124$

29. $-\dfrac{x}{7} = 21$

30. $-\dfrac{x}{3} = -25$

31. $-96 = -3z$

32. $-120 = -8y$

33. $4.8y = -28.8$ **34.** $0.39t = -2.73$ **35.** $\dfrac{3}{2}t = -\dfrac{1}{4}$ **36.** $-\dfrac{7}{6}y = -\dfrac{7}{8}$

d Solve using the principles together. Don't forget to check.

37. $6x - 15 = 45$ **38.** $4x - 7 = 81$ **39.** $5x - 10 = 45$

40. $6z - 7 = 11$ **41.** $9t + 4 = -104$ **42.** $5x + 7 = -108$

43. $-\dfrac{7}{3}x + \dfrac{2}{3} = -18$ **44.** $-\dfrac{9}{2}y + 4 = -\dfrac{91}{2}$ **45.** $\dfrac{6}{5}x + \dfrac{4}{10}x = \dfrac{32}{10}$

46. $\dfrac{9}{5}y + \dfrac{4}{10}y = \dfrac{66}{10}$ **47.** $0.9y - 0.7y = 4.2$ **48.** $0.8t - 0.3t = 6.5$

49. $8x + 48 = 3x - 12$ **50.** $15x + 40 = 8x - 9$ **51.** $7y - 1 = 23 - 5y$

52. $3x - 15 = 15 - 3x$ **53.** $3x - 4 = 5 + 12x$ **54.** $9t - 4 = 14 + 15t$

55. $5 - 4a = a - 13$ **56.** $6 - 7x = x - 14$ **57.** $3m - 7 = -7 - 4m - m$

58. $5x - 8 = -8 + 3x - x$

59. $5x + 3 = 11 - 4x + x$

60. $6y + 20 = 10 + 3y + y$

61. $-7 + 9x = 9x - 7$

62. $-3t + 4 = 5 - 3t$

63. $6y - 8 = 9 + 6y$

64. $5 - 2y = -2y + 5$

65. $2(x + 7) = 4x$

66. $3(y + 6) = 9y$

67. $80 = 10(3t + 2)$

68. $27 = 9(5y - 2)$

69. $180(n - 2) = 900$

70. $210(x - 3) = 840$

71. $5y - (2y - 10) = 25$

72. $8x - (3x - 5) = 40$

73. $7(3x + 6) = 11 - (x + 2)$

74. $9(2x + 8) = 20 - (x + 5)$

75. $5(4x - 3) - 2(6 - 8x) + 10(-2x + 7) = -4(9 - 12x)$

76. $9(4x + 7) - 3(5x - 8) = 6\left(\frac{2}{3} - x\right) - 5\left(\frac{3}{5} + 2x\right)$

77. $2[9 - 3(-2x - 4)] = 12x + 42$

78. $-40x + 45 = 3[7 - 2(7x - 4)]$

79. $\frac{1}{8}(16y + 8) - 17 = -\frac{1}{4}(8y - 16)$

80. $\frac{1}{6}(12t + 48) - 20 = -\frac{1}{8}(24t - 144)$

81. $3[5 - 3(4 - t)] - 2 = 5[3(5t - 4) + 8] - 26$

82. $6[4(8 - y) - 5(9 + 3y)] - 21 = -7[3(7 + 4y) - 4]$

83. $\dfrac{2}{3}\left(\dfrac{7}{8} + 4x\right) - \dfrac{5}{8} = \dfrac{3}{8}$

84. $\dfrac{3}{4}\left(3x - \dfrac{1}{2}\right) + \dfrac{2}{3} = \dfrac{1}{3}$

Skill Maintenance

The exercises that follow begin an important feature called *skill maintenance exercises*. These exercises provide an ongoing review of any preceding objective in the book. You will see them in virtually every exercise set. It has been found that this kind of extensive review can significantly improve your performance on a final examination.

Multiply or divide, and simplify.   [R.7a]

85. $a^{-9} \cdot a^{23}$

86. $\dfrac{a^{-9}}{a^{23}}$

87. $(6x^5y^{-4})(-3x^{-3}y^{-7})$

88. $\dfrac{6x^5y^{-4}}{-3x^{-3}y^{-7}}$

Multiply.   [R.5d]

89. $2(6 - 10x)$

90. $-1(5 - 6x)$

91. $-4(3x - 2y + z)$

92. $5(-2x + 7y - 4)$

Factor.   [R.5d]

93. $2x - 6y$

94. $-4x - 24y$

95. $4x - 10y + 2$

96. $-10x + 35y - 20$

97. Name the set consisting of the positive integers less than 10, using both roster notation and set-builder notation.   [R.1a]

98. Name the set consisting of the negative integers greater than −9 using both roster notation and set-builder notation.   [R.1a]

Synthesis

Exercises designed as *Synthesis Exercises* differ from those found in the main body of the exercise set. The icon ◈ denotes synthesis exercises that are writing exercises. Writing exercises are meant to be answered in one or more complete sentences. Because answers to writing exercises often vary, they are not listed at the back of the book. These and other synthesis exercises will often challenge you to put together two or more objectives at once.

99. ◈ Explain the difference between equivalent expressions and equivalent equations.

100. ◈ Explain how the distributive and commutative laws can be used to express $3x + 6y + 4x + 2y$ as $7x + 8y$.

Solve. (The symbol ⎍ indicates an exercise designed to be solved with any type of calculator.)

101. ⎍ $4.23x - 17.898 = -1.65x - 42.454$

102. ⎍ $-0.00458y + 1.7787 = 13.002y - 1.005$

103. $\dfrac{3x}{2} + \dfrac{5x}{3} - \dfrac{13x}{6} - \dfrac{2}{3} = \dfrac{5}{6}$

104. $\dfrac{2x - 5}{6} + \dfrac{4 - 7x}{8} = \dfrac{10 + 6x}{3}$

105. $x - \{3x - [2x - (5x - (7x - 1))]\} = x + 7$

106. $23 - 2\{4 + 3(x - 1)\} + 5\{x - 2(x + 3)\} = 7\{x - 2[5 - (2x + 3)]\}$

Collaborative Learning Manual

Solve linear equations as a group.

1.2 Formulas and Applications

a | Evaluating and Solving Formulas

A **formula** is a "recipe" for doing a certain type of calculation. Formulas are often given by equations.

Example 1 *Time Interval of Appointments.* In an effort to minimize waiting time for patients in a doctor's office without increasing the physician's idle time, the following formula was developed to compute the time interval of appointments (**Source**: Michael Goiten, Massachusetts General Hospital):

$$I = 1.08 \frac{T}{N},$$

where I is the interval time, in minutes, between scheduled appointments, T is the total number of minutes that a physician spends with patients per day, and N is the total number of scheduled appointments.

Dr. Peggy Carey determines that she has a total of 8 hr a day to see patients and wants to keep 35 appointments. What is the interval time I of each appointment?

We substitute $8 \cdot 60$, or 480, min for the time T and 35 for the number of appointments N. Then

$$I = 1.08 \frac{T}{N} = 1.08 \frac{480}{35} \approx 14.8.$$

Thus the interval time for her appointments is about 15 min.

Do Exercise 1.

Suppose that the doctor in Example 1 knows that she wants the interval time of her appointments to be 20 min and that she wants to keep 35 appointments. How can she compute the number of minutes (or hours) that she needs to spend in her office? To find out, we can substitute 20 for I and 35 for N in the formula and solve for T:

$$I = 1.08 \frac{T}{N}$$

$$20 = 1.08 \frac{T}{35} \qquad \text{Substituting}$$

$$35 \cdot 20 = 35 \cdot 1.08 \frac{T}{35} \qquad \text{Multiplying by 35 on both sides}$$

$$700 = 1.08T \qquad \text{Simplifying}$$

$$\frac{700}{1.08} = \frac{1.08T}{1.08} \qquad \text{Dividing by 1.08}$$

$$648.15 \approx T. \qquad \text{Simplifying}$$

Thus the doctor would spend about 648.15 min, or 648.15/60 or 10.8 hr, in her office per day. If the doctor wanted to make a calculation like this repeatedly, it might be easier to first solve for T, getting it alone on one side. We "solve" the formula for T using the same equation-solving principles that we did in Section 1.1.

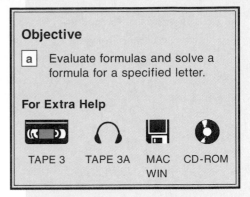
1. Suppose that the doctor in Example 1 determines that she wants to spend more time with her patients, so she decides to keep 30 appointments in the same 8-hr day. What is the interval time I of each appointment?

Answer on page A-6

2. Solve for m: $F = \dfrac{mv^2}{r}$.

(This is a physics formula.)

3. Solve for c: $P = \dfrac{3}{5}(c + 10)$.

Example 2 Solve for T: $I = 1.08\dfrac{T}{N}$.

$$I = 1.08\dfrac{T}{N} \qquad \text{We want this letter alone.}$$

$$N \cdot I = N \cdot 1.08\dfrac{T}{N} \qquad \text{Multiplying by } N \text{ on both sides}$$

$$NI = 1.08T \qquad \text{Simplifying}$$

$$\dfrac{NI}{1.08} = \dfrac{1.08T}{1.08} \qquad \text{Dividing by 1.08}$$

$$\dfrac{NI}{1.08} = T \qquad \text{Simplifying}$$

When we solve a formula for a given letter, we use the same principles that we would use to solve any equation.

> To solve a formula for a given letter, identify the letter, and:
>
> **1.** Multiply on both sides to clear the fractions or decimals, if necessary.
>
> **2.** If parentheses occur, multiply to remove them using the distributive law.
>
> **3.** Collect like terms on each side, if necessary. This may require factoring if a variable is in more than one term.
>
> **4.** Using the addition principle, get all terms with the letter to be solved for on one side of the equation and all other terms on the other side.
>
> **5.** Collect like terms again, if necessary.
>
> **6.** Solve for the letter in question using the multiplication principle.

Do Exercise 2.

Example 3 Solve for b: $A = \frac{5}{2}(b - 20)$.

$$A = \tfrac{5}{2}(b - 20) \qquad \text{We want this letter alone.}$$

$$2A = 5(b - 20) \qquad \text{Multiplying by 2 to clear the fraction}$$

$$2A = 5b - 100 \qquad \text{Removing parentheses}$$

$$2A + 100 = 5b \qquad \text{Adding 100}$$

$$\dfrac{2A + 100}{5} = b, \quad \text{or} \quad b = \dfrac{2A}{5} + 20 \qquad \text{Dividing by 5}$$

Do Exercise 3.

Example 4 Solve for r: $H = 2r + 3m$.

$$H = 2r + 3m \qquad \text{We want this letter alone.}$$

$$H - 3m = 2r \qquad \text{Subtracting } 3m$$

$$\dfrac{H - 3m}{2} = r \qquad \text{Dividing by 2}$$

Answers on page A-6

To see how the addition and multiplication principles apply to formulas, compare the following.

A. Solve.

$7x + 2 = 16$

$7x = 16 - 2$

$7x = 14$

$x = \dfrac{14}{7} = 2$

B. Solve.

$7x + 2 = 16$

$7x = 16 - 2$

$x = \dfrac{16 - 2}{7}$

C. Solve for x.

$ax + b = c$

$ax = c - b$

$x = \dfrac{c - b}{a}$

In (A) we solved as we did before. In (B) we did not carry out the calculations. In (C) we could not carry out the calculations because we had unknown numbers.

Do Exercise 4.

Example 5 *Area of a Trapezoid.* Solve for a: $A = \frac{1}{2}h(a + b)$. (To find the area of a trapezoid, take half the product of the height, h, and the sum of the lengths of the parallel sides, a and b.)

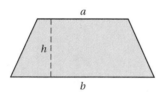

$A = \frac{1}{2}h(a + b)$ We want this letter alone.

$2A = h(a + b)$ **Multiplying by 2 to clear the fraction**

$2A = ha + hb$ **Using the distributive law**

$2A - hb = ha$ **Subtracting hb**

$\dfrac{2A - hb}{h} = a,$ or $a = \dfrac{2A}{h} - \dfrac{hb}{h} = \dfrac{2A}{h} - b$

Note that there is more than one correct form of the answer to Example 5. Such is a common occurrence in formula solving.

Do Exercise 5.

Example 6 *Simple Interest.* Solve for P: $A = P + Prt$. (To find the amount A to which principal P, in dollars, will grow at simple interest rate R, in t years, add the principal P to the interest, Prt.)

$A = P + Prt$ We want this letter alone.

$A = P(1 + rt)$ **Factoring (or collecting like terms)**

$\dfrac{A}{1 + rt} = P$ **Dividing by $1 + rt$ on both sides**

Do Exercise 6.

4. Solve for m: $H = 2r + 3m$.

5. Solve for b: $A = \dfrac{1}{2}h(a + b)$.

6. Solve for Q: $T - Q + Qiy$.

Answers on page A-6

7. *Chess Ratings.* Use the formula given in Example 7.

  a) Find the chess rating of a player who plays 6 games in a tournament, winning 2 games and losing 4. The average rating of his opponents is 1384.

  b) Solve the formula for W.

Example 7   *Chess Ratings.* The formula

$$R = r + \frac{400(W - L)}{N}$$

is used to establish a chess player's rating R, after he or she has played N games, where W is the number of wins, L is the number of losses, and r is the average rating of the opponents (**Source:** U.S. Chess Federation).

a) Find the chess rating of a player who plays 8 games in a tournament, winning 5 games and losing 3. The average rating of her opponents is 1205.

b) Solve the formula for L.

a) We substitute 8 for N, 5 for W, 3 for L, and 1205 for r in the formula. Then we calculate R:

$$R = r + \frac{400(W - L)}{N}$$

$$= 1205 + \frac{400(5 - 3)}{8} = 1305.$$

b) We solve as follows:

$$R = r + \frac{400(W - L)}{N} \qquad \text{We want this letter alone.}$$

$$NR = N\left[r + \frac{400(W - L)}{N}\right] \qquad \text{Multiplying by } N \text{ to clear the fraction}$$

$$NR = N \cdot r + N \cdot \frac{400(W - L)}{N} \qquad \text{Multiplying using the distributive law}$$

$$NR = Nr + 400(W - L) \qquad \text{Simplifying}$$

$$NR - Nr = 400(W - L) \qquad \text{Subtracting } Nr$$

$$NR - Nr = 400W - 400L \qquad \text{Using the distributive law}$$

$$NR - Nr - 400W = -400L \qquad \text{Subtracting } 400W$$

$$\frac{NR - Nr - 400W}{-400} = L. \qquad \text{Dividing by } -400$$

Other correct forms of the answer are

$$L = \frac{Nr + 400W - NR}{400} \quad \text{and} \quad L = W - \frac{NR - Nr}{400}.$$

Do Exercise 7.

Answers on page A-6

Exercise Set 1.2

a Solve for the given letter.

1. *Distance Formula*:
$$d = rt, \quad \text{for } r$$
(Distance *d*, speed *r*, time *t*)

Speed *r*   Time *t*

Distance *d*

2. $d = rt$,   for *t*

3. *Area of a Parallelogram*:
$$A = bh, \quad \text{for } h$$
(Area *A*, base *b*, height *h*)

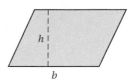

4. *Volume of a Sphere*:
$$V = \frac{4}{3}\pi r^3, \quad \text{for } r^3$$
(Volume *V*, radius *r*)

5. *Perimeter of a Rectangle*:
$$P = 2l + 2w, \quad \text{for } w$$
(Perimeter *P*, length *l*, and width *w*)

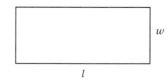

6. $P = 2l + 2w$,   for *l*

7. *Area of a Triangle*:
$$A = \frac{1}{2}bh, \quad \text{for } b$$

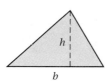

8. $A = \frac{1}{2}bh$,   for *h*

9. *Average of Two Numbers*:
$$A = \frac{a + b}{2}, \quad \text{for } a$$

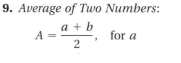

a $A = \dfrac{a+b}{2}$ b

10. $A = \dfrac{a + b}{2}$,   for *b*

11. *Force*:
$$F = ma, \quad \text{for } m$$
(Force *F*, mass *m*, acceleration *a*)

12. $F = ma$,   for *a*

13. *Simple Interest*:
$$I = Prt, \quad \text{for } t$$
(Interest *I*, principal *P*, interest rate *r*, time *t*)

14. $I = Prt$,   for *P*

15. *Relativity*:
$$E = mc^2, \quad \text{for } c^2$$
(Energy *E*, mass *m*, speed of light *c*)

16. $E = mc^2$, for m

17. $Q = \dfrac{p - q}{2}$, for p

18. $Q = \dfrac{p - q}{2}$, for q

19. $Ax + By = c$, for y

20. $Ax + By = c$, for x

21. $I = 1.08\dfrac{T}{N}$, for N

22. $F = \dfrac{mv^2}{r}$, for v^2

Projecting Birth Weight. Ultrasonic images of 29-week-old fetuses can be used to predict weight. One model, or formula, developed by Thurnau,* is $P = 9.337da - 299$; a second model, developed by Weiner,[†] is $P = 94.593c + 34.227a - 2134.616$. For both formulas, P is the estimated fetal weight, in grams, d is the diameter of the fetal head, in centimeters, c is the circumference of the fetal head, in centimeters, and a is the circumference of the fetal abdomen, in centimeters.

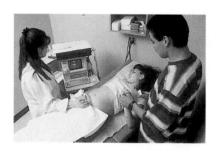

23. a) Use Thurnau's model to estimate the birth weight of a 29-week-old fetus when the diameter of the fetal head is 8.5 cm and the circumference of the fetal abdomen is 24.1 cm.
   b) Solve the formula for a.

24. a) Use Weiner's model to estimate the birth weight of a 29-week-old fetus when the circumference of the fetal head is 26.7 cm and the circumference of the fetal abdomen is 24.1 cm.
   b) Solve the formula for c.

*Thurnau, G. R., R. K. Tamura, R. E. Sabbagha, et al. *Am. J. Obstet Gynecol* 1983; **145**:557.
[†]Weiner, C. P., R. E. Sabbagha, N. Vaisrub, et al. *Obstet Gynecol* 1985; **65**:812.

25. *Young's Rule in Medicine.* Young's rule for determining the amount of a medicine dosage for a child is given by

$$c = \frac{ad}{a + 12},$$

where a is the child's age and d is the usual adult dosage, in milligrams. (*Warning!* Do not apply this formula without checking with a physician!) (**Source:** Olsen, June Looby, et al., *Medical Dosage Calculations*, 6th ed. Reading, MA: Addison Wesley Longman, p. A-31.)

a) The usual adult dosage of a particular medication is 250 mg. Find the dosage for a child of age 3.

b) Solve the formula for d.

26. *Full-Time-Equivalent Students.* Colleges accommodate students who need to take different total-credit-hour loads. They determine the number of "full-time-equivalent" students, F, using the formula

$$F = \frac{n}{15},$$

where n is the total number of credits students enroll in for a given semester.

a) Determine the number of full-time-equivalent students on a campus in which students register for 42,690 credits.

b) Solve the formula for n.

27. *Male Caloric Needs.* The number of calories K needed each day by a moderately active man who weighs w kilograms, is h centimeters tall, and is a years old can be estimated by the formula

$$K = 19.18w + 7h - 9.52a + 92.4.$$

(**Source:** Parker, M., *She Does Math.* Mathematical Association of America, p. 96.)

a) Marv is moderately active, weighs 97 kg, is 185 cm tall, and is 56 yr old. What are his caloric needs?

b) Solve the formula for a, for h, and for w.

28. *Female Caloric Needs.* The number of calories K needed each day by a moderately active woman who weighs w pounds, is h inches tall, and is a years old can be estimated by the formula

$$K = 917 + 6(w + h - a).$$

(**Source:** Parker, M., *She Does Math.* Mathematical Association of America, p. 96.)

a) Elaine is moderately active, weighs 120 lb, is 67 in. tall, and is 23 yr old. What are her caloric needs?

b) Solve the formula for a, for h, and for w.

Skill Maintenance

Divide. [R.2e]

29. $\dfrac{80}{-16}$

30. $-2000 \div (-8)$

31. $-\dfrac{1}{2} \div \dfrac{1}{4}$

32. $120 \div (-4.8)$

33. $-\dfrac{2}{3} \div \left(-\dfrac{5}{6}\right)$

34. $\dfrac{-90}{-15}$

35. $\dfrac{-90}{15}$

36. $\dfrac{-80}{16}$

37. ◈ The equations

$$P = 2l + 2w \quad \text{and} \quad w = \frac{P}{2} - l$$

are equivalent formulas involving the perimeter P, length l, and width w of a rectangle. Devise a problem for which the second of the two formulas would be more useful.

38. ◈ Devise an application in which it would be useful to solve the equation $d = rt$ for r. (See Exercise 1.)

Solve.

39. $A = \pi rs + \pi r^2$, for s

40. $s = v_1 t + \frac{1}{2} at^2$, for a; for v_1

41. $\dfrac{P_1 V_1}{T_1} = \dfrac{P_2 V_2}{T_2}$, for V_1; for P_2

42. $\dfrac{P_1 V_1}{T_1} = \dfrac{P_2 V_2}{T_2}$, for T_2; for P_1

43. In Exercise 13, you solved the formula $I = Prt$ for t. Now use it to find how long it will take a deposit of $75 to earn $3 interest when invested at 5% simple interest.

44. The area of the shaded triangle ABE is 20 cm². Find the area of the trapezoid.

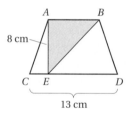

45. *Horsepower of an Engine.* The horsepower of an engine can be calculated by the formula

$$H = W\left(\frac{v}{234}\right)^3,$$

where W is the weight, in pounds, of the car, including the driver, fluids, and fuel, and v is the maximum velocity, or speed, in miles per hour, of the car attained a quarter mile after beginning acceleration.

a) Find the horsepower of a V-6, 2.8-liter engine if $W = 2700$ lb and $v = 83$ mph.

b) Find the horsepower of a 4-cylinder, 2.0-liter engine if $W = 3100$ lb and $v = 73$ mph.

1.3 Applications and Problem Solving

a Five Steps for Problem Solving

Since you have already studied some algebra, you have probably had some experience with problem solving. The following five-step strategy can be very helpful.

Of the five steps, probably the most important is the first one: becoming familiar with the problem situation. Here are some hints for familiarization.

To *familiarize* yourself with the problem situation:

- If a problem is given in words, read it carefully.
- Reread the problem, perhaps aloud. Try to verbalize the problem to yourself.
- List the information given and the question to be answered. Choose a variable (or variables) to represent the unknown(s) and clearly state what the variable represents. Be descriptive! For example, let L = length (in meters), d = distance (in miles), and so on.
- Make a drawing and label it with known information. Also, indicate unknown information, using specific units if given.
- Find further information if necessary. Look up a formula at the back of this book or in a reference book. Talk to a reference librarian or an expert in the field.
- Make a table that lists all the information you have collected. Look for patterns that may help in the translation to an equation.
- Guess or estimate the answer.

Example 1 *Installing Seamless Guttering.* The Beechers are installing seamless guttering on their house. This type of guttering is delivered on a continuous roll. Sections are then cut from the roll as needed. The Beechers know that they need 127 ft of guttering for six separate sections and that the four shortest sections will be the same size. The longest piece of guttering is 13 ft more than twice the length of the mid-size piece. The shortest piece of guttering is 10 ft less than the mid-size piece. How long is each piece of guttering?

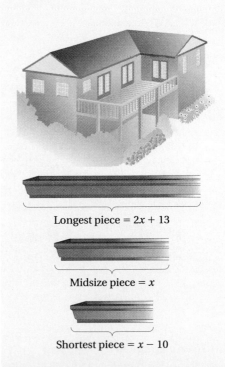

Longest piece = $2x + 13$

Midsize piece = x

Shortest piece = $x - 10$

1. *Board Cutting.* A 106-in. board is cut into three pieces. The shortest length is two-thirds of the mid-size length and the longest length is 15 in. less than two times the mid-size length. Find the length of each board.

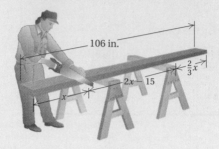

1. **Familiarize.** We let x = the length of the mid-size piece, $2x + 13$ = the length of the longest piece, and $x - 10$ = the length of the shortest piece. We might also make a drawing.

2. **Translate.** The sum of the length of the longest piece, plus the length of the mid-size piece, plus four times the length of the shortest piece is 127 ft. This gives us the following translation:

Longest piece	plus	Mid-size piece	plus	Four times the shortest piece	is	Total length
$(2x + 13)$	$+$	x	$+$	$4(x - 10)$	$=$	$127.$

3. **Solve.** We solve the equation, as follows:

$$(2x + 13) + x + 4(x - 10) = 127$$
$$2x + 13 + x + 4x - 40 = 127 \qquad \text{Using the distributive law}$$
$$7x - 27 = 127 \qquad \text{Collecting like terms}$$
$$7x - 27 + 27 = 127 + 27 \qquad \text{Adding 27}$$
$$7x = 154 \qquad \text{Collecting like terms}$$
$$\frac{7x}{7} = \frac{154}{7} \qquad \text{Dividing by 7}$$
$$x = 22. \qquad \text{Simplifying}$$

4. **Check.** Do we have an answer to the *problem*? If the length of the mid-size piece is 22 ft, then the length of the longest piece is

$$2 \cdot 22 + 13, \text{ or } 57 \text{ ft,}$$

and the length of the shortest piece is

$$22 - 10, \text{ or } 12 \text{ ft.}$$

The sum of the lengths of the longest piece, the mid-size piece, and four times the shortest piece must be 127 ft:

$$57 \text{ ft} + 22 \text{ ft} + 4(12 \text{ ft}) = 127 \text{ ft.}$$

These lengths check.

5. **State.** The length of the longest piece is 57 ft, the length of the mid-size piece is 22 ft, and the length of the shortest piece is 12 ft.

Do Exercise 1.

Answer on page A-6

Example 2 Three numbers are such that the second is 6 less than three times the first and the third is 2 more than two-thirds of the first. The sum of the three numbers is 150. Find the largest of the three numbers.

1. **Familiarize.** Three numbers are involved, and we want to find the largest. We list the information in a table, letting x = the first number.

First number	x
Second number	6 less than three times the first
Third number	2 more than two-thirds the first

2. **Translate.** We want to write an equation in just one variable. To do so, we need to express the second and third numbers using x (that is, "in terms of x"). To do so, we expand the table, as follows.

First number	x	x
Second number	6 less than three times the first	$3x - 6$
Third number	2 more than two-thirds the first	$\frac{2}{3}x + 2$

We know that the sum of the numbers is 150. Substituting each of the expressions into the equation for the sum, we obtain a new equation:

First plus Second plus Third is 150

$$x + (3x - 6) + \left(\tfrac{2}{3}x + 2\right) = 150.$$

3. **Solve.** We solve the equation:

$$x + (3x - 6) + \left(\tfrac{2}{3}x + 2\right) = 150$$
$$\left(1 + 3 + \tfrac{2}{3}\right)x - 4 = 150 \qquad \text{Collecting like terms}$$
$$\tfrac{14}{3}x - 4 = 150$$
$$\tfrac{14}{3}x - 4 + 4 = 150 + 4 \qquad \text{Adding 4}$$
$$\tfrac{14}{3}x = 154 \qquad \text{Simplifying}$$
$$\tfrac{3}{14} \cdot \tfrac{14}{3}x = \tfrac{3}{14} \cdot 154 \qquad \text{Multiplying by } \tfrac{3}{14}$$
$$x = 33.$$

This equation could also have been solved by first multiplying by 3 on both sides to clear fractions.

Next, we substitute to find the other two numbers:

Second: $3x - 6 = 3 \cdot 33 - 6 = 93$;
Third: $\frac{2}{3}x + 2 = \frac{2}{3} \cdot 33 + 2 = 24$.

4. **Check.** We consider the original problem. There are three numbers: 33, 93, and 24. We note that 93 is 6 less than three times 33, and that two-thirds of 33 plus 2 is 24. Then we check the sum:

$$33 + 93 + 24 = 150.$$

These numbers check.

5. **State.** The problem asks for the largest number, so the answer is "The largest of the three numbers is 93."

Do Exercise 2.

2. Three numbers are such that the second is 6 less than three times the first, and the third is 2 more than two-thirds of the second. The sum of the three numbers is 172. Find the smallest number.

Answer on page A-6

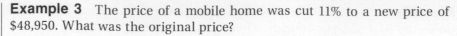

3. A clothing store drops the price of running suits 25% to a sale price of $93. What was the former price?

Example 3   The price of a mobile home was cut 11% to a new price of $48,950. What was the original price?

1. Familiarize.   We let x = the original price.

2. Translate.

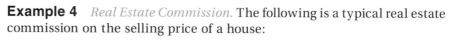

Original price	minus	11%	of	Original price	is	New price	
↓	↓	↓	↓	↓	↓	↓	
x	−	11%	·	x	=	48,950	**Translation**

3. Solve.   We solve the equation:

$$x - 11\% \cdot x = 48{,}950$$
$$1x - 0.11x = 48{,}950 \qquad \text{Replacing 11\% with 0.11}$$
$$\left.\begin{array}{r}(1 - 0.11)x = 48{,}950 \\ 0.89x = 48{,}950\end{array}\right\} \quad \text{Collecting like terms}$$
$$x = 55{,}000. \qquad \text{Dividing by 0.89}$$

4. Check.   Note that 11% of $55,000 is $6050. Subtracting this decrease from $55,000 (the original price), we get $48,950 (the new price). This checks.

5. State.   The original price was $55,000.

Do Exercise 3.

Example 4   *Real Estate Commission.* The following is a typical real estate commission on the selling price of a house:

> 7% for the first $100,000,   and
>
> 4% for the amount that exceeds $100,000.

A realtor receives a commission of $10,320 for selling a house. What was the selling price?

1. Familiarize.   Let's make a guess or estimate to become familiar with the problem. Suppose that a house sells for $225,000. The realtor's commission is then

> 7% of $100,000 = 0.07($100,000) = $7000

plus

> 4% times ($225,000 − $100,000) = 0.04($125,000) = $5000.

The total commission is $7000 + $5000, or $12,000. Our sample calculation not only makes us familiar with the problem, it also tells us that the house in the problem sold for less than this one since $10,320 is less than $12,000. We let S = the selling price of the house.

2. Translate.   We translate as follows:

Commission on the first $100,000	plus	Commission on the amount that exceeds $100,000	is	Total commission
↓	↓	↓	↓	↓
7% · 100,000	+	4%(S − 100,000)	=	10,320.

Answer on page A-6

3. Solve. We solve the equation:

$$7\% \cdot 100{,}000 + 4\%(S - 100{,}000) = 10{,}320$$

$$0.07 \cdot 100{,}000 + 0.04(S - 100{,}000) = 10{,}320 \qquad \text{Converting to decimal notation}$$

$$7000 + 0.04S - 0.04 \cdot 100{,}000 = 10{,}320 \qquad \text{Using the distributive law}$$

$$7000 + 0.04S - 4000 = 10{,}320 \qquad \text{Simplifying}$$

$$0.04S + 3000 = 10{,}320 \qquad \text{Collecting like terms}$$

$$0.04S + 3000 - 3000 = 10{,}320 - 3000 \qquad \text{Subtracting 3000}$$

$$0.04S = 7320 \qquad \text{Simplifying}$$

$$\frac{0.04S}{0.04} = \frac{7320}{0.04} \qquad \text{Dividing by 0.04}$$

$$S = \$183{,}000. \qquad \text{Simplifying}$$

4. Check. Performing the check is similar to the sample calculation in the *Familiarize* step. We leave the check to the student.

5. State. The selling price of the house was $183,000.

Do Exercise 4.

Example 5 *Coca-Cola Co.* The equation

$$y = 0.0662x + 0.4602$$

can be used to approximate the total revenue *y*, in billions of dollars, of the Coca-Cola Co. *x* years after 1990—that is,

$\quad x = 0$ corresponds to 1990,

$\quad x = 1$ corresponds to 1991,

$\quad x = 8$ corresponds to 1998, and so on

(**Source**: Coca-Cola Bottling Consolidated). (This equation was derived using a procedure called *regression*. Its development belongs to a later course, although we will consider how to use it in a Calculator Spotlight on pp. 212–213.)

a) Predict the total revenue in the years 2000 and 2004.

b) In what year will the total revenue be about $1.2546 billion?

Since a formula is given, we will not use the five-step problem-solving strategy.

a) To predict the total revenue for 2000, we note that $2000 - 1990 = 10$. Thus we substitute 10 for *x*:

$$y = 0.0662x + 0.4602 = 0.0662(10) + 0.4602 = 1.1222.$$

To find the total revenue for 2004, we note that $2004 - 1990 = 14$. Thus we substitute 14 for *x*:

$$y = 0.0662x + 0.4602 = 0.0662(14) + 0.4602 = 1.387.$$

Thus we predict total revenue to be $1.1222 billion in 2000 and $1.387 billion in 2004.

4. *Real Estate Commission.* The following is a typical real estate commission on the selling price of a house:

7% for the first $100,000

and

4% for the amount that exceeds $100,000.

A realtor receives a commission of $13,400 for selling a house. What was the selling price?

5. *Coca-Cola Co.* Refer to Example 5.

a) Predict the total revenue in 2005 and 2010.

b) In what year will the total revenue be about $2.1152 billion?

Answers on page A-6

6. *Picture Framing.* Repeat Example 6, given that the piece of wood is 120 in. long.

b) To determine the year in which total revenue will be about $1.2546 billion, we first substitute 1.2546 for y. Then we solve for x. Note that we choose not to clear decimals because each number has the same number of decimal places. However, you can do so if you wish.

$$y = 0.0662x + 0.4602$$

$$1.2546 = 0.0662x + 0.4602 \quad \text{Substituting 1.2546 for } y$$

$$1.2546 - 0.4602 = 0.0662x + 0.4602 - 0.4602 \quad \text{Subtracting 0.4602}$$

$$0.7944 = 0.0662x$$

$$\frac{0.7944}{0.0662} = \frac{0.0662x}{0.0662} \quad \text{Dividing by 0.0662}$$

$$12 = x$$

The answer 12 gives us the number of years after 1990. To find that year, we add 12 to 1990:

$$1990 + 12 = 2002.$$

Thus, if the equation continues to be valid, the total revenue of Coca-Cola Co. will be approximately $1.2546 billion in 2002.

Do Exercise 5 on the preceding page.

Example 6 *Picture Framing.* A piece of wood trim that is 100 in. long is to be cut into two pieces, and those pieces are each to be cut into four pieces to form a square frame. The length of a side of one square is to be $1\frac{1}{2}$ times the length of a side of the other. How should the wood be cut?

1. **Familiarize.** We first make a drawing of the situation, letting $s =$ the length of a side of the small square and $\frac{3}{2}s =$ the length of a side of the larger square. Note that the *perimeter* (distance around) of each square is four times the length of a side and that the sum of the two perimeters must be 100 in.

2. **Translate.** We reword the problem and translate.

	Perimeter of small square	plus	Perimeter of larger square	is	100 in.
Rewording:					
Translating:	$4s$	$+$	$4\left(\frac{3}{2}s\right)$	$=$	100

3. **Solve.** We solve the equation:

$$4s + 4\left(\tfrac{3}{2}s\right) = 100$$

$$4s + 6s = 100 \quad \text{Multiplying}$$

$$10s = 100 \quad \text{Collecting like terms}$$

$$\frac{10s}{10} = \frac{100}{10} \quad \text{Dividing by 10}$$

$$s = 10. \quad \text{Simplifying}$$

4. **Check.** If 10 in. is the length of the smaller side, then $\frac{3}{2}(10)$, or 15 in., is the length of the larger side. The sum of the two perimeters is $4 \cdot 10 + 4 \cdot 15 = 40 + 60 = 100$, so the lengths check.

5. **State.** The wood should be cut into two pieces, one 40 in. long and the other 60 in. long. Each piece should then be cut into four pieces of the same length in order to form the frames.

Do Exercise 6.

100 in.

s

$1\frac{1}{2}s$, or $\frac{3}{2}s$

Answer on page A-6

Example 7 *Triangular Peace Monument.* In Indianapolis, Indiana, children have turned a vacant lot into a monument for peace. This community project brought together neighborhood volunteers, businesses, and government in hopes of showing children positive nonviolent ways of dealing with conflict. A landscape architect used the children's drawings and ideas to design a triangular-shaped peace garden as shown in the figure below. Two sides of the triangle are formed by Indiana Avenue and Senate Avenue. The third side is formed by an apartment building. The second angle of the triangle is 7° more than the first. The third angle is 7° less than twice the first. How large are the angles?

7. The second angle of a triangle is three times as large as the first. The measure of the third angle is 25° greater than that of the first angle. How large are the angles?

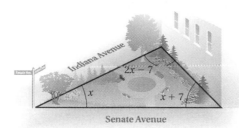

Source: Mentoring in the City Program, Marian College, Indianapolis, Indiana

1. Familiarize. We make a drawing as above. We let

the measure of the first angle = x.

Then the measure of the second angle = $x + 7$

and the measure of the third angle = $2x - 7$.

2. Translate. To translate, we need to recall a geometric fact. (You might, as part of step 1, look it up in a geometry book or in the list of formulas at the back of this book.) Remember, the measures of the angles of a triangle add up to 180°.

Measure of first angle	plus	Measure of second angle	plus	Measure of third angle	is	180°
x	$+$	$(x + 7)$		$(2x - 7)$	$-$	180

3. Solve. We solve the equation:

$$x + (x + 7) + (2x - 7) = 180$$
$$4x = 180 \qquad \text{Collecting like terms}$$
$$\frac{4x}{4} = \frac{180}{4} \qquad \text{Dividing by 4}$$
$$x = 45.$$

Possible measures for the angles are as follows:

First angle: $x = 45°$;

Second angle: $x + 7 = 45 + 7 = 52°$;

Third angle: $2x - 7 = 2(45) - 7 = 83°$.

4. Check. Consider 45°, 52°, and 83°. The second is 7° more than the first and the third is 7° less than twice the first. The sum is 180°. The answer checks.

5. State. The measures of the angles are 45°, 52°, and 83°.

Do Exercise 7.

Answer on page A-6

8. *Artist's Prints.* The artist in Example 8 saves three other prints for his own archives. These are also numbered consecutively. The sum of the numbers is 1266. Find the numbers of each of those prints.

The following are examples of **consecutive integers:** 19, 20, 21, 22; as well as $-34, -33, -32, -31$. Note that consecutive integers can be represented in the form $x, x + 1, x + 2$, and so on.

The following are examples of **consecutive even integers:** 20, 22, 24, 26; as well as $-34, -32, -30, -28$. Note that consecutive even integers can be represented in the form $x, x + 2, x + 4, x + 6$, and so on.

The following are examples of **consecutive odd integers:** 17, 19, 21, 23; as well as $-77, -75, -73, -71$. Note that consecutive odd integers can be represented in the form $x, x + 2, x + 4, x + 6$, and so on.

Example 8 *Artist's Prints.* Often an artist will number in sequence a limited number of prints in order to enhance their value. An artist creates 500 prints and saves three for his children. The numbers of those prints are consecutive integers whose sum is 189. Find the numbers of each of those prints.

1. **Familiarize.** The numbers of the prints are consecutive integers. Thus we let $x =$ the first integer, $x + 1 =$ the second, and $x + 2 =$ the third.

2. **Translate.** We translate as follows:

$$\underbrace{\text{First integer}}_{x} + \underbrace{\text{Second integer}}_{(x + 1)} + \underbrace{\text{Third integer}}_{(x + 2)} = \underbrace{189}_{189.}$$

3. **Solve.** We solve the equation:

$$x + (x + 1) + (x + 2) = 189$$
$$3x + 3 = 189 \qquad \text{Collecting like terms}$$
$$3x + 3 - 3 = 189 - 3 \qquad \text{Subtracting 3}$$
$$3x = 186 \qquad \text{Simplifying}$$
$$\frac{3x}{3} = \frac{186}{3} \qquad \text{Dividing by 3}$$
$$x = 62. \qquad \text{Simplifying}$$

Then $x + 1 = 62 + 1 = 63$ and $x + 2 = 62 + 2 = 64$.

4. **Check.** The numbers are 62, 63, and 64. These are consecutive integers and their sum is 189. The numbers check.

5. **State.** The numbers of the prints are 62, 63, and 64.

Do Exercise 8.

b | **Basic Motion Problems**

When a problem deals with speed, distance, and time, we can expect to use the following **motion formula.**

> **THE MOTION FORMULA**
> Distance = Rate (or speed) · Time
> $d = rt$

Answer on page A-6

We will consider motion problems several times throughout the text. Here we consider a basic problem to provide a foundation for those that occur later.

Example 9 *Moving Sidewalks.* A moving sidewalk in O'Hare Airport is 300 ft long and moves at a speed of 5 ft/sec. If Kate walks at a speed of 4 ft/sec, how long will it take her to travel the 300 ft using the moving sidewalk?

1. **Familiarize.** First read the problem very carefully. This may even involve speaking aloud to a fellow student or to yourself. In this case, organizing the information in a table can be very helpful.

Distance to be traveled	300 ft
Speed of Kate	4 ft/sec
Speed of the moving sidewalk	5 ft/sec
Total speed of Kate on the sidewalk	?
Time required	?

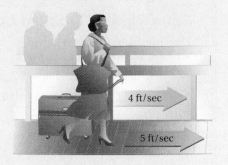

4 ft/sec

5 ft/sec

We might now try to determine, possibly with the aid of outside references, what relationships exist among the various quantities in the problem. Since Kate is walking on the sidewalk in the same direction in which it is moving, the two speeds can be added to determine the total speed of Kate on the sidewalk. We can then complete the table, letting t = the time, in seconds, required to travel 300 ft on the moving sidewalk.

Distance to be traveled	300 ft
Speed of Kate	4 ft/sec
Speed of the moving sidewalk	5 ft/sec
Total speed of Kate on the sidewalk	9 ft/sec
Time required	t

2. **Translate.** To translate, we use the motion formula $d = rt$ and substitute 300 ft for d and 9 ft/sec for r:

$$d = rt$$
$$300 = 9 \cdot t.$$

3. **Solve.** We solve the equation:

$$300 = 9t$$
$$\frac{300}{9} = \frac{9t}{9} \qquad \text{Dividing by 9}$$
$$\frac{100}{3} = t. \qquad \text{Simplifying}$$

4. **Check.** At a speed of 9 ft/sec and in a time of 100/3, or $33\frac{1}{3}$ sec, Kate would travel $d = 9 \cdot \frac{100}{3} = 300$ ft. This answer checks.

5. **State.** Kate will travel the 300 ft in 100/3, or $33\frac{1}{3}$ sec.

Do Exercise 9.

9. *Marine Travel.* Tim's fishing boat travels 12 km/h in still water. How long will it take him to travel 25 km upstream if the river's current is 3 km/h? 25 km downstream if the river's current is 3 km/h?

Answer on page A-6

Improving Your Math Study Skills

Extra Tips on Problem Solving

The following tips, which are focused on problem solving, summarize some points already considered and include some new ones.

- Get in the habit of using all five steps for problem solving.

1. **Familiarize yourself with the problem situation.** Some suggestions for this are given on p. 91.
2. **Translate the problem to an equation.** As you study more mathematics, you will find that the translation may be to some other kind of mathematical language, such as an inequality.
3. **Solve the equation.** If the translation is to some other kind of mathematical language, you would carry out some kind of mathematical manipulation—in the case of an inequality, you would solve it.
4. **Check the answer in the original problem.** This does not mean to check in the translated equation. It means to go back to the original worded problem.
5. **State the answer to the problem clearly.**

For Step 4, some further comment on checking is appropriate. *You may be able to translate to an equation and to solve the equation, but you may find that none of the solutions of the equation is a solution of the original problem.* To see how this can happen, consider this example.

Example

The sum of two consecutive even integers is 537. Find the integers.

1. **Familiarize.** Suppose we let $x =$ the first number. Then $x + 2 =$ the second number.

2. **Translate.** The problem can be translated to the following equation: $x + (x + 2) = 537$.
3. **Solve.** We solve the equation as follows:

$$2x + 2 = 537$$
$$2x = 535$$
$$x = \frac{535}{2}, \text{ or } 267.5.$$

Then $x + 2 = 269.5$.

4. **Check.** The numbers are not only not even, but they are not integers.
5. **State.** The problem has no solution.

The following are some other tips.

- **To be good at problem solving, do lots of problems.** Learning to solve problems is similar to learning other skills such as golf. At first you may not be successful, but the more you practice and work at improving your skills, the more successful you will become. For problem solving, do more than just two or three odd-numbered assigned problems. Do them all, and if you have time, do the even-numbered problems as well. Then find another book on the same subject and do problems in that book.

- **Look for patterns when solving problems.** You will eventually see patterns in similar kinds of problems. For example, there is a pattern in the way that you solve problems involving consecutive integers.

- **When translating to an equation, or some other mathematical language, consider the dimensions of the variables and the constants in the equation.** The variables that represent length should all be in the same unit, those that represent money should all be in dollars or in cents, and so on.

Exercise Set 1.3

a Solve.

Tour de France. The Tour de France is a famous cycling race held each year, mostly in France. The 1997 race, which was won by Jan Ullrich of Germany, included 21 parts, or stages, of various lengths for a total of 3870 km. (**Source:** Tour de France Web Site) Use this information for Exercises 1 and 2.

Colmar=164
Ulrich= *1.64Km*
x
y

1. *Stage 18.* This stage ran from Colmar to Montbeliard at a length of 164 km. If at one point the winner, Ullrich, was twice as far from Colmar as from Montbeliard, how many kilometers of the stage had he completed?

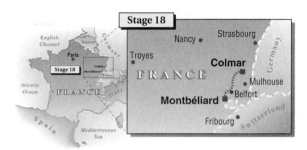

2. *Stage 21.* This stage ran from EuroDisney to Paris at a length of 150 km. If at one point Ullrich was six times as far from EuroDisney as from Paris, how many kilometers of the stage had he completed?

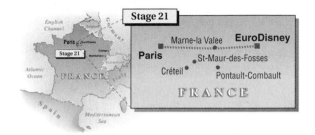

3. Sixteen plus three times a number is eleven times the number. What is the number?

16+ 3x = 11x.
-3x -3x
16 = 8x   2
8   16

4. Seven times a number is 18 plus five times the number. Find the number.

5. Six less than ten times a number is 60 more than seven times the number. Find the number.

10x-6

6. Fifteen more than three times a number is the same as 10 less than six times the number. What is the number?

7. Three numbers are such that the second is 10 more than five times the first, and the third is 4 less than four-fifths of the second. The sum of the three numbers is 244. Find the smallest number.

5x+10   = 244

8. Three numbers are such that the second is 14 less than six times the first and the third is 2 more than three-fourths of the first. The sum of the three numbers is 112. Find the largest of the three numbers.

9. *Purchasing.* Elka pays $1187.20 for a computer. The price includes a 6% sales tax. What is the price of the computer itself?

$$1187.20 \times .06$$

10. *Pricing.* The Sound Connection prices its blank audiotape cassettes by raising the wholesale price 50% and adding 25¢. What is the tape's wholesale price if the tape sells for $1.99?

11. *Solar Panel.* The cross section of a support for a solar energy panel is triangular. One angle of the triangle is five times as large as the first angle. The third angle is 2° less than the first angle. How large are the angles?

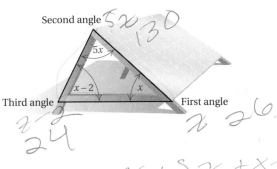

Second angle $5x$ 30

5x

$x - 2$ x

Third angle   First angle

2
24   x 26.

$$x + 5x + x - 2 = 180$$
$$6x + x - 2 = 180$$
$$ +2 \quad +2$$
$$6x + x = 182$$

12. *Cross Section of a Roof.* In a triangular cross section of a roof, the second angle is twice as large as the first. The third angle is 20° greater than the first angle. How large are the angles?

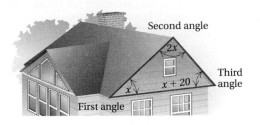

Second angle

$2x$

x $x + 20$ Third angle

First angle

$$6x + x = 182$$
$$\frac{2x}{7} \qquad \frac{x = 182}{7}$$

13. *Perimeter of NBA Court.* The perimeter of an NBA-sized basketball court is 288 ft. The length is 44 ft longer than the width. (**Source:** National Basketball Association) Find the dimensions of the court.

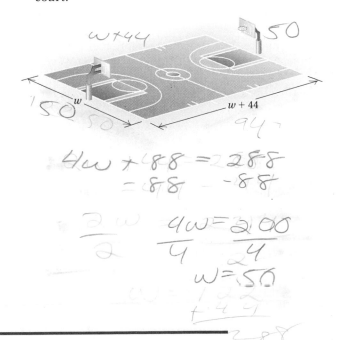

$w + 44$ 50

50   w $w + 44$
50   94

$$4w + 88 = 288$$
$$ -88 \quad -88$$
$$\frac{2w}{2} \quad \frac{4w = 200}{4} \quad \frac{4}{4}$$
$$w = 50$$
$$+44$$
$$288$$

14. *Tennis Court.* The width of a standard tennis court used for playing doubles is 42 ft less than the length. The perimeter of the court is 228 ft. Find the dimensions of the court.

l

$l - 42$

15. A piece of wire that is 100 cm long is to be cut into two pieces, each to be bent to make a square. The length of a side of one square is to be twice the length of a side of the other. How should the wire be cut?

16. A rope that is 168 ft long is to be cut into three pieces such that the second piece is 6 ft less than three times the first, and the third is 2 ft more than two-thirds of the second. Find the length of the longest piece.

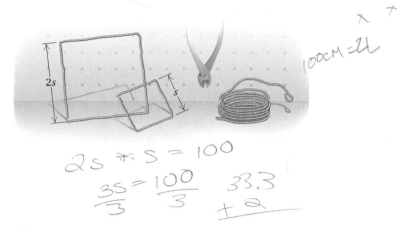

$100CM = 2u$ x x

$2s * S = 100$

$\dfrac{3S}{3} = \dfrac{100}{3}$ 33.3

$+2$

17. *Real Estate Commission.* The following is a real estate commission on the selling price of a house:

    7% for the first $100,000, and
    5% for the amount that exceeds $100,000.

A realtor receives a commission of $15,250 for selling a house. What was the selling price?

18. *Real Estate Commission.* The following is a real estate commission on the selling price of a house:

    8% for the first $100,000, and
    3% for the amount that exceeds $100,000.

A realtor receives a commission of $9200 for selling a house. What was the selling price?

19. Find three consecutive odd integers such that the sum of the first, two times the second, and three times the third is 70.

20. Find three consecutive even integers such that the sum of the first, five times the second, and four times the third is 1226.

$1 + 2 \cdot 3 + 3 \cdot 3,$

$7 + 18 + 33$ 9 22

18 $9, 11, 13$

21. *Consecutive Post-Office Box Numbers.* The sum of the numbers on two adjacent post-office boxes is 697. What are the numbers?

22. *Consecutive Page Numbers.* The sum of the page numbers on the facing pages of the book *True Success: A New Philosophy of Excellence,* by Tom Morris, is 373. What are the page numbers?

$9 + (2 \times 11) + (3 \times 13) = 70$

23. *Records in the Women's 100-m Dash.* In 1988, Florence Griffith Joyner set a world record of 10.49 sec in the women's 100-m dash. The equation

$$R = -0.0433t + 10.49$$

can be used to predict the world record in the women's 100-m dash t years after 1988 (**Source**: International Amateur Athletic Foundation).

a) Predict the world record in the year 2000.
b) In what year will the record be 9.624 sec?

24. *Nike, Inc.* The equation

$$y = 0.69606x + 1.68722$$

can be used to approximate the total revenue y, in billions of dollars, of Nike, Inc., x years after 1990—that is,

$x = 0$ corresponds to 1990,

$x = 1$ corresponds to 1991,

$x = 10$ corresponds to 2000, and so on.

(**Source**: Nike, Inc.)

a) Predict the total revenue in 1999, 2002, and 2008.
b) In what year will the total revenue be about $12.82418 billion?

25. *Carpet Cleaning.* A carpet company charges $75 to clean the first 200 sq ft of carpet. There is an additional charge of 25¢ per square foot for any footage that exceeds 200 sq ft and $1.40 per step for any carpeting on a staircase. A customer's cleaning bill was $253.95. This included the cleaning of a staircase with 13 steps. In addition to the staircase, how many square feet of carpet did the customer have cleaned?

26. *Picture Purchases.* Latent Images sells school photographs. Its prices are as follows:

1 8 × 10 2 5 × 7 12 Wallet-size	$14.95
plus $1.35 per sheet of 6 extra wallet-sizes	

The Hoepners have three children in school. How many wallet-size photos did they buy if their total bill for the photos is $57?

27. An editorial assistant receives an 8% raise, bringing her salary to $20,466. What was her salary before the raise?

28. After a salesman receives a 5% raise, his new salary is $19,530. What was his old salary?

29. *Swimming.* Fran swims at a speed of 5 mph in still water. The Lazy River flows at a speed of 2.3 mph. How long will it take Fran to swim 1.8 mi upstream? 1.8 mi downstream?

[handwritten: 1.8 + 2.3   2/3   18/73]

30. *Boating.* A paddleboat moves at a rate of 14 km/h in still water. If the river's current moves at a rate of 7 km/h, how long will it take the boat to travel 56 km downstream? 56 km upstream?

31. *Cruising Altitude.* A commercial airliner has been instructed to climb from its present altitude of 8000 ft to a cruising altitude of 29,000 ft. The plane ascends at a rate of 3500 ft/min. How long will it take to reach the cruising altitude?

[handwritten: 8000 + 3500 z = 29 000   − 8000]

32. *Flight into a Headwind.* An airplane traveling 390 mph in still air encounters a 65-mph headwind. How long will it take the plane to travel 725 mi into the wind?

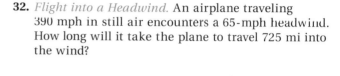
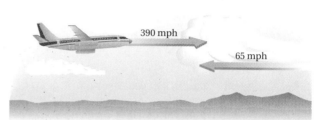

390 mph
65 mph

[handwritten: x − 8% · x = 20 466   x − .08 · x = 20466   1x − .08x = 20466   (1 − .08)x = 20466   .92x = 20466   over .92 .92   x =]

Skill Maintenance

Simplify. [R.3c]

33. $-44 \cdot 55 - 22$

34. $16 \cdot 8 + 200 \div 25 \cdot 10$

35. $(5 - 12)^2$

36. $5^2 - 12^2$

37. $5^2 - 2 \cdot 5 \cdot 12 + 12^2$

38. $(5 - 12)(5 + 12)$

39. $\dfrac{12|8 - 10| + 9 \cdot 6}{5^4 + 4^5}$

40. $\dfrac{(9 - 4)^2 + (8 - 11)^2}{4^2 + 2^2}$

41. $[-64 \div (-4)] \div (-16)$

42. $-64 \div [-4 \div (-16)]$

43. $2^{13} \div 2^5 \div 2^3$

44. $2^{13} \cdot 2^5 \cdot 2^3$

Synthesis

45. ◆ Write a problem for a classmate to solve. Devise the problem so that the solution is, "The material should be cut into two pieces, one 30 cm long and the other 45 cm long."

46. ◆ How can a guess or estimate help prepare you for the *Translate* step when solving problems?

47. ◆ Solve: "The sum of three consecutive integers is 55." Find the integers. Why is it important to check the solution from the *Solve* step in the original problem?

48. ◆ Why is it important to clearly label what a variable represents in an applied problem?

49. A literature professor owns 400 novels. The number of horror novels is 46% of the number of science fiction novels; the number of science fiction novels is 65% of the number of romance novels; and the number of mystery novels is 17% of the number of horror novels. How many science fiction novels does the professor have? Round to the nearest one.

50. *NBA Shot Clock.* The National Basketball Association operates what is called a 24-second clock. This means that the offensive team has 24 sec in which to attempt a field goal. If it does not make such an attempt, it loses the ball. The NBA arrived at 24 sec by dividing the total number of seconds in a game by the average number of points scored per game. (***Source***: National Basketball Association) There are 48 min in a game. What is the average number of points scored in a game?

51. The yearly changes in the population census of a city for three consecutive years are, respectively, a 20% increase, a 30% increase, and a 20% decrease. What is the total percent change from the beginning to the end of the third year, to the nearest percent?

52. The height and sides of a triangle are represented by four consecutive integers. The height is the first integer and the base is the third integer. The perimeter of the triangle is 42 in. Find the area of the triangle.

53. 🖩 *Home Prices.* Panduski's real estate prices increased 6% from 1996 to 1997 and 2% from 1997 to 1998. From 1998 to 1999, prices dropped 1%. If a house sold for $117,743 in 1999, what was its worth in 1996? Round to the nearest dollar.

54. *Test Scores.* Tico's scores on four tests are 83, 91, 78, and 81. How many points above the average must Tico score on the next test in order to raise his average 2 points?

55. Your watch loses $1\frac{1}{2}$ sec every hour. You have a friend whose watch gains 1 sec every hour. The watches show the same time now. After how many more seconds will they show the same time again?

56. The salary of an employee was reduced n% during a recession. By what percent would the company then have to raise the salary in order to bring it back to where it was before?

57. Consider the geometric figure below. Suppose that $L \parallel M$, $m\angle 8 = 5x + 25$, and $m\angle 4 = 8x + 4$. Find $m\angle 2$ and $m\angle 1$.

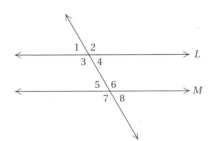

58. Suppose the figure *ABCD* below is a square. Point *A* is folded onto the midpoint of $\overline{AB}$ and point *D* is folded onto the midpoint of $\overline{DC}$. The perimeter of the smaller figure formed is 25 in. Find the area of the square *ABCD*.

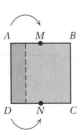

Use the five-step problem-solving strategy.

Collaborative
Learning Manual

1.4 Sets, Interval Notation, and Inequalities

a Inequalities

We can extend our equation-solving skills to the solving of inequalities (see Section R.1).

> An **inequality** is any sentence containing $<$, $>$, $\leq$, $\geq$, or $\neq$.

Examples of inequalities are the following:

$$-2 < a, \quad x > 4, \quad x + 3 \leq 6, \quad 6 - 7y \geq 10y - 4, \quad \text{and} \quad 5x \neq 10.$$

Any replacement for the variables that makes an inequality true is called a **solution**. The set of all solutions is called the **solution set.** When all the solutions of an inequality have been found, we say that we have **solved** the inequality.

Examples Determine whether the given number is a solution of the inequality.

1. $x + 3 < 6$; 5

We substitute and get $5 + 3 < 6$, or $8 < 6$, a false sentence. Therefore, 5 is not a solution.

2. $2x - 3 > -3$; 1

We substitute and get $2(1) - 3 > -3$, or $-1 > -3$, a true sentence. Therefore, 1 is a solution.

3. $4x - 1 \leq 3x + 2$; 3

We substitute and get $4(3) - 1 \leq 3(3) + 2$, or $11 \leq 11$, a true sentence. Therefore, 3 is a solution.

Do Exercises 1–3.

b Inequalities and Interval Notation

The **graph** of an inequality is a drawing that represents its solutions. An inequality in one variable can be graphed on the number line.

Example 4 Graph $x < 4$ on the number line.

The solutions are all real numbers less than 4, so we shade all numbers less than 4 on the number line. To indicate that 4 is not a solution, we use a right parenthesis ")" at 4.

We can write the solution set for $x < 4$ using **set-builder notation** (see Section R.1):

$$\{x \mid x < 4\}.$$

This is read

"The set of all x such that x is less than 4."

Determine whether the given number is a solution of the inequality.

1. $3 - x < 2$; 8

2. $3x + 2 > -1$; -2

3. $3x + 2 \leq 4x - 3$; 5

Answers on page A-6

Another way to write solutions of an inequality in one variable is to use **interval notation.** Interval notation uses parentheses ( ) and brackets [ ].

If a and b are real numbers such that $a < b$, we define the interval (a, b) as the set of all numbers between but not including a and b—that is, the set of all x for which $a < x < b$. Thus,

$$(a, b) = \{x \mid a < x < b\}.$$

The points a and b are the **endpoints**. The parentheses indicate that the endpoints are *not* included in the graph.

The interval $[a, b]$ is defined as the set of all numbers x for which $a \le x \le b$. Thus,

$$[a, b] = \{x \mid a \le x \le b\}.$$

The brackets indicate that the endpoints *are* included in the graph.*

CAUTION! Do not confuse the *interval* (a, b) with the *ordered pair* (a, b) used in connection with an equation in two variables in the plane, as in Chapter 3. The context in which the notation appears usually makes the meaning clear.

The following intervals include one endpoint and exclude the other:

$$(a, b] = \{x \mid a < x \le b\}. \quad \text{The graph excludes } a \text{ and includes } b.$$

$$[a, b) = \{x \mid a \le x < b\}. \quad \text{The graph includes } a \text{ and excludes } b.$$

Some intervals extend without bound in one or both directions. We use the symbols ∞, read "infinity," and $-\infty$, read "negative infinity," to name these intervals. The notation (a, ∞) represents the set of all numbers greater than a—that is,

$$(a, \infty) = \{x \mid x > a\}.$$

Similarly, the notation $(-\infty, a)$ represents the set of all numbers less than a—that is,

$$(-\infty, a) = \{x \mid x < a\}.$$

The notations $[a, \infty)$ and $(-\infty, a]$ are used when we want to include the endpoints. The interval $(-\infty, \infty)$ names the set of all real numbers.

$$(-\infty, \infty) = \{x \mid x \text{ is a real number}\}$$

Interval notation is summarized in the following table.

---

*Some books use the representations ⊷⊶ and ⊷⊶ instead of, respectively, ⟨———⟩ and [———] .

Chapter 1 Solving Linear Equations and Inequalities

INTERVALS: NOTATION AND GRAPHS

Interval Notation	Set Notation	Graph
(a, b)	$\{x \mid a < x < b\}$	
$[a, b]$	$\{x \mid a \le x \le b\}$	
$[a, b)$	$\{x \mid a \le x < b\}$	
$(a, b]$	$\{x \mid a < x \le b\}$	
(a, ∞)	$\{x \mid x > a\}$	
$[a, \infty)$	$\{x \mid x \ge a\}$	
$(-\infty, b)$	$\{x \mid x < b\}$	
$(-\infty, b]$	$\{x \mid x \le b\}$	
$(-\infty, \infty)$	$\{x \mid x \text{ is a real number}\}$	

Examples   Write interval notation for the given set or graph.

5. $\{x \mid -4 < x < 5\} = (-4, 5)$

6. $\{x \mid x \ge -2\} = [-2, \infty)$

7.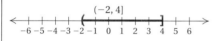

8.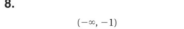

Do Exercises 4–7.

c   Solving Inequalities

Two inequalities are **equivalent** if they have the same solution set. For example, the inequalities $x > 4$ and $4 < x$ are equivalent. Just as the addition principle for equations gives us equivalent equations, the addition principle for inequalities gives us equivalent inequalities.

> **THE ADDITION PRINCIPLE FOR INEQUALITIES**
>
> For any real numbers a, b, and c:
>
> $$a < b \text{ is equivalent to } a + c < b + c;$$
> $$a > b \text{ is equivalent to } a + c > b + c.$$
>
> Similar statements hold for $\le$ and $\ge$.

Since subtracting c is the same as adding $-c$, there is no need for a separate subtraction principle.

Write interval notation for the given set or graph.

4. $\{x \mid -4 \le x < 5\}$

5. $\{x \mid x \le -2\}$

6.

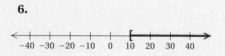

7.

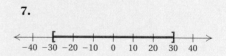

Answers on page A-6

Solve and graph.

8. $x + 6 > 9$

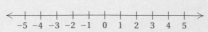

9. $x + 4 \leq 7$

10. Solve and graph:

$$2x - 3 \geq 3x - 1.$$

Answers on page A-6

Example 9 Solve and graph: $x + 5 > 1$.

We have

$$x + 5 > 1$$
$$x + 5 - 5 > 1 - 5 \qquad \text{Using the addition principle:} \\ \text{adding } -5 \text{ or subtracting } 5$$
$$x > -4.$$

We used the addition principle to show that the inequalities $x + 5 > 1$ and $x > -4$ are equivalent. The solution set is $\{x \mid x > -4\}$ and consists of an infinite number of solutions. We cannot possibly check them all. Instead, we can perform a partial check by substituting one member of the solution set (here we use -1) into the original inequality:

$$\frac{x + 5 > 1}{-1 + 5 \;?\; 1}$$
$$\qquad 4 \;\mid\; \qquad \text{TRUE}$$

Since $4 > 1$ is true, we have our check. The solution set is $\{x \mid x > -4\}$, or $(-4, \infty)$. The graph is as follows:

Do Exercises 8 and 9.

Example 10 Solve and graph: $4x - 1 \geq 5x - 2$.

We have

$$4x - 1 \geq 5x - 2$$
$$4x - 1 + 2 \geq 5x - 2 + 2 \qquad \text{Adding 2}$$
$$4x + 1 \geq 5x \qquad \text{Simplifying}$$
$$4x + 1 - 4x \geq 5x - 4x \qquad \text{Subtracting } 4x$$
$$1 \geq x. \qquad \text{Simplifying}$$

We know that $1 \geq x$ has the same meaning as $x \leq 1$. You can check that any number less than or equal to 1 is a solution. The solution set is $\{x \mid 1 \geq x\}$ or, more commonly, $\{x \mid x \leq 1\}$. Using interval notation, we write that the solution set is $(-\infty, 1]$. The graph is as follows:

Do Exercise 10.

The multiplication principle for inequalities is somewhat different from the multiplication principle for equations. Consider this true inequality:

$$-4 < 9. \qquad \text{True}$$

If we multiply both numbers by 2, we get another true inequality:

$$-4(2) < 9(2), \quad \text{or} \quad -8 < 18. \qquad \text{True}$$

If we multiply both numbers by -3, we get a false inequality:

$$-4(-3) < 9(-3), \quad \text{or} \quad 12 < -27. \qquad \text{False}$$

However, if we now *reverse* the inequality symbol above, we get a true inequality:

$$12 > -27. \qquad \text{True}$$

> **THE MULTIPLICATION PRINCIPLE FOR INEQUALITIES**
>
> For any real numbers a and b, and any *positive* number c:
>
> $a < b$ is equivalent to $ac < bc$;
>
> $a > b$ is equivalent to $ac > bc$.
>
> For any real numbers a and b, and any *negative* number c:
>
> $a < b$ is equivalent to $ac > bc$;
>
> $a > b$ is equivalent to $ac < bc$.
>
> Similar statements hold for $\leq$ and $\geq$.

Since division by c is the same as multiplication by $1/c$, there is no need for a separate division principle.

Example 11 Solve and graph: $3y < \frac{3}{4}$.

We have

$$3y < \frac{3}{4}$$

$$\frac{1}{3} \cdot 3y < \frac{1}{3} \cdot \frac{3}{4} \quad \text{Multiplying by } \frac{1}{3}.$$
$$\text{The symbol stays the same.}$$

$$y < \frac{1}{4}.$$

Any number less than $\frac{1}{4}$ is a solution. The solution set is $\left\{y \mid y < \frac{1}{4}\right\}$, or $\left(-\infty, \frac{1}{4}\right)$. The graph is as follows:

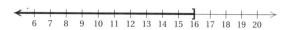

Example 12 Solve and graph: $-5x \geq -80$.

We have

$$-5x \geq -80$$

$$\frac{-5x}{-5} \leq \frac{-80}{-5} \quad \text{Dividing by } -5. \text{ The}$$
$$\text{symbol must be reversed.}$$

$$x \leq 16.$$

The solution set is $\{x \mid x \leq 16\}$, or $(-\infty, 16]$. The graph is as follows:

Do Exercises 11–13.

Solve and graph.

11. $5y \leq \dfrac{3}{2}$

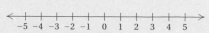

12. $-2y > 10$

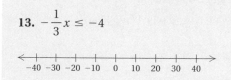

13. $-\dfrac{1}{3}x \leq -4$

Answers on page A-6

Solve.

14. $6 - 5y \geq 7$

15. $3x + 5x < 4$

16. $17 - 5(y - 2) \leq$
$45y + 8(2y - 3) - 39y$

We use the addition and multiplication principles together in solving inequalities in much the same way as in solving equations.

Example 13 Solve: $16 - 7y \geq 10y - 4$.

We have

$$16 - 7y \geq 10y - 4$$
$$-16 + 16 - 7y \geq -16 + 10y - 4 \qquad \text{Adding } -16$$
$$-7y \geq 10y - 20 \qquad \text{Collecting like terms}$$
$$-10y + (-7y) \geq -10y + 10y - 20 \qquad \text{Adding } -10y$$
$$-17y \geq -20 \qquad \text{Collecting like terms}$$

$$\frac{-17y}{-17} \leq \frac{-20}{-17} \qquad \text{Dividing by } -17. \text{ The symbol must be reversed.}$$

$$y \leq \frac{20}{17}. \qquad \text{Simplifying}$$

The solution set is $\left\{y \mid y \leq \frac{20}{17}\right\}$, or $\left(-\infty, \frac{20}{17}\right]$.

In some cases, you can avoid the concern about multiplying or dividing by a negative number by using the addition principle in a different way. Let's rework Example 13 by adding $7y$ instead of $-10y$:

$$16 - 7y \geq 10y - 4$$
$$16 - 7y + 7y \geq 10y - 4 + 7y \qquad \text{Adding } 7y$$
$$16 \geq 17y - 4 \qquad \text{Collecting like terms}$$
$$16 + 4 \geq 17y - 4 + 4 \qquad \text{Adding } 4$$
$$20 \geq 17y \qquad \text{Collecting like terms}$$
$$\frac{20}{17} \geq \frac{17y}{17} \qquad \text{Dividing by } 17$$
$$\frac{20}{17} \geq y.$$

Note that $\frac{20}{17} \geq y$ is equivalent to $y \leq \frac{20}{17}$.

Example 14 Solve: $-3(x + 8) - 5x > 4x - 9$.

We have

$$-3(x + 8) - 5x > 4x - 9$$
$$-3x - 24 - 5x > 4x - 9 \qquad \text{Using the distributive law}$$
$$-24 - 8x > 4x - 9 \qquad \text{Collecting like terms}$$
$$-24 - 8x + 8x > 4x - 9 + 8x \qquad \text{Adding } 8x$$
$$-24 > 12x - 9 \qquad \text{Collecting like terms}$$
$$-24 + 9 > 12x - 9 + 9 \qquad \text{Adding } 9$$
$$-15 > 12x \qquad \text{Simplifying}$$

$$\frac{-15}{12} > \frac{12x}{12} \qquad \text{Dividing by } 12. \text{ The symbol stays the same.}$$

$$-\frac{5}{4} > x.$$

The solution set is $\left\{x \mid -\frac{5}{4} > x\right\}$, or $\left\{x \mid x < -\frac{5}{4}\right\}$, or $\left(-\infty, -\frac{5}{4}\right)$.

Do Exercises 14–16.

d Applications and Problem Solving

Many problem-solving and applied situations translate to inequalities. In addition to "is less than" and "is more than," other phrases are commonly used.

Phrase	Translation
a "is at most" 17	$a \leq 17$
a "is at least" 5	$a \geq 5$
a "can't exceed" 12	$a \leq 12$
a "is a better price than" *b*	$a < b$

Example 15 *Records in the Men's 200-m Dash.* Michael Johnson set a world record of 19.32 sec in the men's 200-m dash in the 1996 Olympics (**Source:** International Amateur Athletic Foundation). The equation

$$R = -0.0513t + 19.32$$

can be used to predict the world record in the men's 200-m dash *t* years after 1996. Determine (in terms of an inequality) those years for which the world record will be less than 19.0 sec.

1. **Familiarize.** We already have a formula. To become more familiar with it, we might make a substitution for *t*. Suppose we want to know the record after 20 yr, in the year 2016. We substitute 20 for *t*:

   $$R = -0.0513(20) + 19.32 = 18.294 \text{ sec}.$$

   We see that by 2016, the record will be less than 19.0 sec. To predict the exact year in which the 19.0-sec mark will be broken, we could make other guesses that are less than 20. Instead, we proceed to the next step.

2. **Translate.** The record *R* is to be *less than* 19.0 sec. Thus we have

   $$R < 19.0.$$

   We replace *R* with $-0.0513t + 19.32$ to find the times *t* that solve the inequality:

   $$-0.0513t + 19.32 < 19.0.$$

3. **Solve.** We solve the inequality:

   $$-0.0513t + 19.32 < 19.0$$
   $$-0.0513t < -0.32 \qquad \text{Subtracting 19.32}$$
   $$t > 6.24. \qquad \text{Dividing by } -0.0513 \text{ and rounding}$$

4. **Check.** A partial check is to substitute a value for *t* greater than 6.24. We did that in the *Familiarize* step.

5. **State.** The record will be less than 19.0 sec for races occurring more than 6.24 yr after 1996, or approximately after 2002.

Do Exercise 17.

17. *Records in the Men's 200-m Dash.* Refer to Example 15. Determine (in terms of an inequality) those years for which the world record will be less than 19.1 sec.

Answer on page A-6

18. *Salary Plans.* A painter can be paid in one of two ways:

Plan A: $500 plus $4 per hour;

Plan B: Straight $9 per hour.

Suppose that the job takes *n* hours. For what values of *n* is plan A better for the painter?

Example 16 *Salary Plans.* On a new job, Rose can be paid in one of two ways: *Plan A* is a salary of $600 per month, plus a commission of 4% of sales; and *Plan B* is a salary of $800 per month, plus a commission of 6% of sales in excess of $10,000. For what amount of monthly sales is plan A better than plan B, if we assume that sales are always more than $10,000?

1. Familiarize. Listing the given information in a table will be helpful.

Plan A: Monthly Income	Plan B: Monthly Income
$600 salary 4% of sales *Total*: $600 + 4% of sales	$800 salary 6% of sales over $10,000 *Total*: $800 + 6% of sales over $10,000

Next, suppose that Rose sold $12,000 in one month. Which plan would be better? Under plan A, she would earn $600 plus 4% of $12,000, or

$$600 + 0.04(12,000) = \$1080.$$

Since with plan B commissions are paid only on sales in excess of $10,000, Rose would earn $800 plus 6% of ($12,000 − $10,000), or

$$800 + 0.06(12,000 - 10,000) = \$920.$$

This shows that for monthly sales of $12,000, plan A is better. Similar calculations will show that for sales of $30,000 a month, plan B is better. To determine *all* values for which plan A pays more money, we must solve an inequality that is based on the calculations above.

2. Translate. We let *S* represent the amount of monthly sales. If we examine the calculations in the *Familiarize* step, we see that the monthly income from plan A is $600 + 0.04S$ and from plan B is $800 + 0.06(S - 10,000)$. Thus we want to find all values of *S* for which

Income from plan A	is greater than	Income from plan B
$600 + 0.04S$	$>$	$800 + 0.06(S - 10,000).$

3. Solve. We solve the inequality:

$$600 + 0.04S > 800 + 0.06(S - 10,000)$$

$600 + 0.04S > 800 + 0.06S \quad 600$ **Using the distributive law**

$600 + 0.04S > 200 + 0.06S$ **Collecting like terms**

$\quad\quad\quad 400 > 0.02S$ **Subtracting both 200 and 0.04S**

$\quad\quad 20,000 > S$, or $S < 20,000.$ **Dividing by 0.02**

4. Check. For $S = 20,000$, the income from plan A is

$$600 + 4\% \cdot 20,000, \text{ or } \$1400.$$

The income from plan B is

$$800 + 6\% \cdot (20,000 - 10,000), \text{ or } \$1400.$$

This confirms that for sales of $20,000, Rose's pay is the same under either plan.

In the *Familiarize* step, we saw that for sales of $12,000, plan A pays more. Since $12,000 < 20,000$, this is a partial check. Since we cannot check all possible values of *S*, we will stop here.

5. State. For monthly sales of less than $20,000, plan A is better.

Do Exercise 18.

Answer on page A-6

Exercise Set 1.4

a Determine whether the given numbers are solutions of the inequality.

1. $x - 2 \geq 6$; $-4, 0, 4, 8$

2. $3x + 5 \leq -10$; $-5, -10, 0, 27$

3. $t - 8 > 2t - 3$; $0, -8, -9, -3, -\frac{7}{8}$

4. $5y - 7 < 8 - y$; $2, -3, 0, 3, \frac{2}{3}$

b Write interval notation for the given set or graph.

5. $\{x \mid x < 5\}$

6. $\{t \mid t \geq -5\}$

7. $\{x \mid -3 \leq x \leq 3\}$

8. $\{t \mid -10 < t \leq 10\}$

9.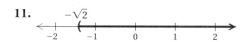

10.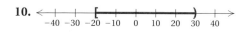

11. $-\sqrt{2}$
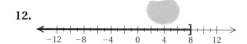

12.

c Solve and graph.

13. $x + 2 > 1$

14. $x + 8 > 4$

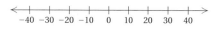

15. $y + 3 < 9$

16. $y + 4 < 10$

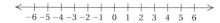

17. $a - 9 \leq -31$

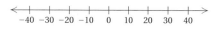

18. $a + 6 \leq -14$

19. $t + 13 \geq 9$

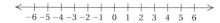

20. $x - 8 \leq 17$

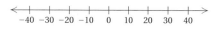

21. $y - 8 > -14$

22. $y - 9 > -18$

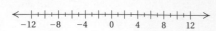

23. $x - 11 \leq -2$

24. $y - 18 \leq -4$

25. $8x \geq 24$

26. $8t < -56$

27. $0.3x < -18$

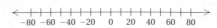

28. $0.6x < 30$

Solve.

29. $-9x \geq -8.1$

30. $-5y \leq 3.5$

31. $-\frac{3}{4}x \geq -\frac{5}{8}$

32. $-\frac{1}{8}y \leq -\frac{9}{8}$

33. $2x + 7 < 19$

34. $5y + 13 > 28$

35. $5y + 2y \leq -21$

36. $-9x + 3x \geq -24$

37. $2y - 7 < 5y - 9$

38. $8x - 9 < 3x - 11$

39. $0.4x + 5 \leq 1.2x - 4$

40. $0.2y + 1 > 2.4y - 10$

41. $5x - \frac{1}{12} \leq \frac{5}{12} + 4x$

42. $2x - 3 < \frac{13}{4}x + 10 - 1.25x$

43. $4(4y - 3) \geq 9(2y + 7)$

44. $2m + 5 \geq 16(m - 4)$

45. $3(2 - 5x) + 2x < 2(4 + 2x)$

46. $2(0.5 - 3y) + y > (4y - 0.2)8$

47. $5[3m - (m + 4)] > -2(m - 4)$

48. $[8x - 3(3x + 2)] - 5 \geq 3(x + 4) - 2x$

49. $3(r - 6) + 2 > 4(r + 2) - 21$

50. $5(t + 3) + 9 < 3(t - 2) + 6$

51. $19 - (2x + 3) \leq 2(x + 3) + x$

52. $13 - (2c + 2) \geq 2(c + 2) + 3c$

53. $\frac{1}{4}(8y + 4) - 17 < -\frac{1}{2}(4y - 8)$

54. $\frac{1}{3}(6x + 24) - 20 > -\frac{1}{4}(12x - 72)$

55. $2[4 - 2(3 - x)] - 1 \geq 4[2(4x - 3) + 7] - 25$

56. $5[3(7 - t) - 4(8 + 2t)] - 20 \leq -6[2(6 + 3t) - 4]$

57. $\frac{4}{5}(7x - 6) < 40$

58. $\frac{2}{3}(4x - 3) > 30$

59. $\frac{3}{4}(3 + 2x) + 1 \geq 13$

60. $\frac{7}{8}(5 - 4x) - 17 \geq 38$

61. $\frac{3}{4}\left(3x - \frac{1}{2}\right) - \frac{2}{3} < \frac{1}{3}$

62. $\frac{2}{3}\left(\frac{7}{8} - 4x\right) - \frac{5}{8} < \frac{3}{8}$

63. $0.7(3x + 6) \geq 1.1 - (x + 2)$

64. $0.9(2x + 8) < 20 - (x + 5)$

65. $a + (a - 3) \leq (a + 2) - (a + 1)$

66. $0.8 - 4(b - 1) > 0.2 + 3(4 - b)$

d Solve.

Body Mass Index. The *body mass index I* can be used to determine an individual's risk of heart disease. An index less than 25 indicates a low risk. The body mass index is given by the formula, or model,

$$I = \frac{700W}{H^2},$$

where W is the weight, in pounds, and H is the height, in inches.

67. a) Marv weighs 214 lb and his height is 73 in. What is his body mass index?
  b) Determine (in terms of an inequality) those weights W for Marv that will keep him in the low-risk category.

68. a) Elaine weighs 110 lb and her height is 67 in. What is her body mass index?
  b) Determine (in terms of an inequality) those weights W for Elaine that will keep her in the low-risk category.

69. *Grades.* You are taking a European history course in which there will be 4 tests, each worth 100 points. You have scores of 89, 92, and 95 on the first three tests. You must make a total of at least 360 in order to get an A. What scores on the last test will give you an A?

70. *Grades.* You are taking a literature course in which there will be 5 tests, each worth 100 points. You have scores of 94, 90, and 89 on the first three tests. You must make a total of at least 450 in order to get an A. What scores on the fourth test will keep you eligible for an A?

71. *Insurance Claims.* After a serious automobile accident, most insurance companies will replace the damaged car with a new one if repair costs exceed 80% of the N.A.D.A., or "blue-book," value of the car. Miguel's car recently sustained $9200 worth of damage but was not replaced. What was the blue-book value of his car?

72. *Phone Rates.* A long-distance telephone call using Down East Calling costs 20 cents for the first minute and 16 cents for each additional minute. The same call, placed on Long Call Systems, costs 19 cents for the first minute and 18 cents for each additional minute. For what length phone calls is Down East Calling less expensive?

73. *Salary Plans.* Toni can be paid in one of two ways:

Plan A: A salary of $400 per month plus a commission of 8% of gross sales;

Plan B: A salary of $610 per month, plus a commission of 5% of gross sales.

For what amount of gross sales should Toni select plan A?

74. *Salary Plans.* Branford can be paid for his masonry work in one of two ways:

Plan A: $300 plus $9.00 per hour;

Plan B: Straight $12.50 per hour.

Suppose that the job takes n hours. For what values of n is plan B better for Branford?

75. *Checking-Account Rates.* The Hudson Bank offers two checking-account plans. Their Anywhere plan charges 20¢ per check whereas their Acu-checking plan costs $2 per month plus 12¢ per check. For what numbers of checks per month will the Acu-checking plan cost less?

76. *Insurance Benefits.* Bayside Insurance offers two plans. Under plan A, Giselle would pay the first $50 of her medical bills and 20% of all bills after that. Under plan B, Giselle would pay the first $250 of bills, but only 10% of the rest. For what amount of medical bills will plan B save Giselle money? (Assume that her bills will exceed $250.)

77. *Wedding Costs.* The Arnold Inn offers two plans for wedding parties. Under plan A, the inn charges $30 for each person in attendance. Under plan B, the inn charges $1300 plus $20 for each person in excess of the first 25 who attend. For what size parties will plan B cost less? (Assume that more than 25 guests will attend.)

78. *Investing.* Lillian is about to invest $20,000, part at 6% and the rest at 8%. What is the most that she can invest at 6% and still be guaranteed at least $1500 in interest per year?

79. *Converting Dress Sizes.* The formula

$$I = 2(s + 10)$$

can be used to convert dress sizes s in the United States to dress sizes I in Italy. For what dress sizes in the United States will dress sizes in Italy be larger than 36?

80. *Temperatures of Solids.* The formula

$$C = \frac{5}{9}(F - 32)$$

can be used to convert Fahrenheit temperatures F to Celsius temperatures C.

a) Gold is a solid at Celsius temperatures less than 1063°C. Find the Fahrenheit temperatures for which gold is a solid.

b) Silver is a solid at Celsius temperatures less than 960.8°C. Find the Fahrenheit temperatures for which silver is a solid.

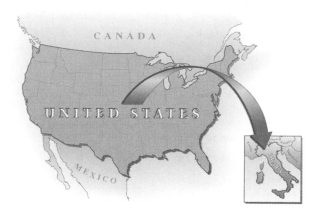

81. *Consumption of Bottled Water.* People are consuming more bottled water each year. The number N of gallons per year that each person drinks t years after 1990 is approximated by

$$N = 0.733t + 8.398.$$

a) How many gallons of bottled water did each person drink in 1990 ($t = 0$)? in 1995 ($t = 5$)? in 2000?

b) For what years will the amount of bottled water drunk be at least 15 gal?

82. *Dewpoint Spread.* Pilots use the **dewpoint spread,** or the difference between the current temperature and the dewpoint (the temperature at which dew occurs), to estimate the height of the cloud cover. Each 3° of dewpoint spread corresponds to an increased height of cloud cover of 1000 ft. A plane, flying with limited instruments, must have a cloud cover greater than 3500 ft. What dewpoint spreads will allow the plane to fly?

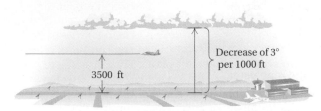

Decrease of 3° per 1000 ft

3500 ft

Skill Maintenance

Simplify. [R.6b]

83. $3a - 6(2a - 5b)$

84. $2(x - y) + 10(3x - 7y)$

85. $4(a - 2b) - 6(2a - 5b)$

86. $-3(2a - 3b) + 8b$

Factor. [R.5d]

87. $30x - 70y - 40$

88. $-12a + 30ab$

89. $-8x + 24y - 4$

90. $10n - 45mn + 100m$

Add or subtract. [R.2a, c]

91. $-2.3 - 8.9$

92. $-2.3 + 8.9$

93. $-2.3 + (-8.9)$

94. $-2.3 - (-8.9)$

Synthesis

95. ◈ Explain in your own words why the inequality symbol must be reversed when both sides of an inequality are multiplied or divided by a negative number.

96. ◈ Graph the solution set of each of the following on a number line and compare:

$$x < 2, \quad x > 2, \quad x = 2, \quad x \le 2, \quad \text{and} \quad x \ge 2.$$

97. *Supply and Demand.* The supply S and demand D for a certain product are given by

$$S = 460 + 94p \quad \text{and} \quad D = 2000 - 60p.$$

a) Find those values of p for which supply exceeds demand.

b) Find those values of p for which supply is less than demand.

Determine whether the statement is true or false. If false, give a counterexample.

98. For any real numbers x and y, if $x < y$, then $x^2 < y^2$.

99. For any real numbers a, b, c, and d, if $a < b$ and $c < d$, then $a + c < b + d$.

100. Determine whether the inequalities

$$x < 3 \quad \text{and} \quad 0 \cdot x < 0 \cdot 3$$

are equivalent. Give reasons to support your answer.

Solve.

101. $x + 5 \le 5 + x$

102. $x + 8 < 3 + x$

103. $x^2 + 1 > 0$

1.5 Intersections, Unions, and Compound Inequalities

The steroid cholesterol is a fat-related compound in tissues of cells. One commonly used measure of cholesterol is *total cholesterol*. The following table shows the health-risk levels of this measure of cholesterol.

Type of Cholesterol	Normal Risk	Borderline Risk	High Risk
Total	Total ≤ 200	200 < Total < 240	Total ≥ 240

A total-cholesterol level T between 200 and 240 is considered border-line risk. We can express this by the sentence

$$200 < T \quad and \quad T < 240$$

or more simply by

$$200 < T < 240.$$

This is an example of a *compound inequality*. **Compound inequalities** consist of two or more inequalities joined by the word *and* or the word *or*. We now "solve" such sentences—that is, we find the set of all solutions.

a Intersections of Sets and Conjunctions of Inequalities

The **intersection** of two sets A and B is the set of all members that are common to A and B. We denote the intersection of sets A and B as

$$A \cap B.$$

The intersection of two sets is often illustrated as shown at right.

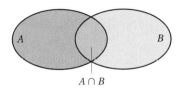

$A \cap B$

Example 1   Find the intersection:

$$\{1, 2, 3, 4, 5\} \cap \{-2, -1, 0, 1, 2, 3\}.$$

The numbers 1, 2, and 3 are common to the two sets, so the intersection is $\{1, 2, 3\}$.

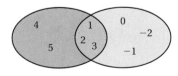

Do Exercises 1 and 2.

When two or more sentences are joined by the word *and* to make a compound sentence, the new sentence is called a **conjunction** of the sentences. The following is a conjunction of inequalities:

$$-2 < x \quad and \quad x < 1.$$

Objectives

a Find the intersection of two sets. Solve and graph conjunctions of inequalities.

b Find the union of two sets. Solve and graph disjunctions of inequalities.

c Solve applied problems involving conjunctions and disjunctions of inequalities.

For Extra Help

TAPE 4   TAPE 3B   MAC WIN   CD-ROM

1. Find the intersection:

$$\{0, 3, 5, 7\} \cap \{0, 1, 3, 11\}.$$

2. Shade the intersection of sets A and B.

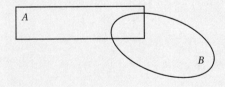

Answers on page A-7

3. Graph and write interval notation:

$$-1 < x \text{ and } x < 4.$$

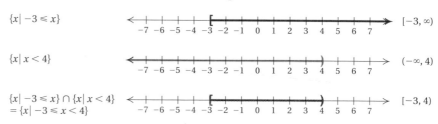

←——+——+——+——+——+——+——+——+——+——+——→
  −5 −4 −3 −2 −1 0 1 2 3 4 5

In order for a conjunction to be true, each individual sentence must be true. *The solution set of a conjunction is the intersection of the solution sets of the individual sentences.* Consider the conjunction

$$-2 < x \quad \text{and} \quad x < 1.$$

The graphs of each separate sentence are shown below, and the intersection is the last graph. We use both set-builder and interval notations.

$\{x \mid -2 < x\}$ ←——+——+——+——+——(——+——+——+——+——+——+——+——+——→ $(-2, \infty)$
                     −7 −6 −5 −4 −3 −2 −1 0 1 2 3 4 5 6 7

$\{x \mid x < 1\}$ ←——+——+——+——+——+——+——+——)——+——+——+——+——+——→ $(-\infty, 1)$
                     −7 −6 −5 −4 −3 −2 −1 0 1 2 3 4 5 6 7

$\{x \mid -2 < x\} \cap \{x \mid x < 1\}$ ←——+——+——+——+——(——+——)——+——+——+——+——+——→ $(-2, 1)$
$= \{x \mid -2 < x \text{ and } x < 1\}$ −7 −6 −5 −4 −3 −2 −1 0 1 2 3 4 5 6 7

Because there are numbers that are both greater than −2 and less than 1, the conjunction $-2 < x$ and $x < 1$ can be abbreviated by $-2 < x < 1$. Thus the interval $(-2, 1)$ can be represented as $\{x \mid -2 < x < 1\}$, the set of all numbers that are *simultaneously* greater than −2 *and* less than 1. Note that, in general, for $a < b$,

$$a < x \quad \text{and} \quad x < b \quad \text{can be abbreviated} \quad a < x < b;$$

and $\quad b > x \quad$ and $\quad x > a \quad$ can be abbreviated $\quad b > x > a.$

> ▶ The word "and" corresponds to "intersection" and to the symbol "∩". In order for a number to be a solution of a conjunction, it must be in *both* solution sets.

Do Exercise 3.

Example 2   Solve and graph: $-1 \leq 2x + 5 < 13$.

This inequality is an abbreviation for the conjunction

$$-1 \leq 2x + 5 \quad \text{and} \quad 2x + 5 < 13.$$

The word *and* corresponds to set *intersection*, ∩. To solve the conjunction, we solve each of the two inequalities separately and then find the intersection of the solution sets:

$$-1 \leq 2x + 5 \quad \text{and} \quad 2x + 5 < 13$$
$$-6 \leq 2x \qquad \text{and} \qquad 2x < 8 \qquad \text{Subtracting 5}$$
$$-3 \leq x \qquad \text{and} \qquad x < 4. \qquad \text{Dividing by 2}$$

We now abbreviate the answer:

$$-3 \leq x < 4.$$

The solution set is $\{x \mid -3 \leq x < 4\}$, or, in interval notation, $[-3, 4)$. The graph is the intersection of the two separate solution sets.

$\{x \mid -3 \leq x\}$ ←——+——+——+——[——+——+——+——+——+——+——+——+——+——→ $[-3, \infty)$
                 −7 −6 −5 −4 −3 −2 −1 0 1 2 3 4 5 6 7

$\{x \mid x < 4\}$ ←——+——+——+——+——+——+——+——+——+——+——)——+——+——→ $(-\infty, 4)$
                 −7 −6 −5 −4 −3 −2 −1 0 1 2 3 4 5 6 7

$\{x \mid -3 \leq x\} \cap \{x \mid x < 4\}$ ←——+——+——+——[——+——+——+——+——+——+——)——+——+——→ $[-3, 4)$
$= \{x \mid -3 \leq x < 4\}$ −7 −6 −5 −4 −3 −2 −1 0 1 2 3 4 5 6 7

Answer on page A-7

The steps above are generally combined as follows:

$-1 \le 2x + 5 < 13$ **2x + 5 appears in both inequalities.**

$-6 \le 2x < 8$ **Subtracting 5**

$-3 \le x < 4.$ **Dividing by 2**

Such an approach saves some writing and will prove useful in Section 1.6.

Do Exercise 4.

Sometimes there is no way to solve both parts of a conjunction at the same time.

Example 3 Solve and graph: $2x - 5 \ge -3$ *and* $5x + 2 \ge 17$.

We first solve each inequality separately:

$$2x - 5 \ge -3 \quad and \quad 5x + 2 \ge 17$$
$$2x \ge 2 \quad and \quad 5x \ge 15$$
$$x \ge 1 \quad and \quad x \ge 3.$$

Next, we find the intersection of the two separate solution sets:

$\{x \mid x \ge 1\}$

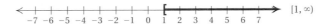

$[1, \infty)$

$\{x \mid x \ge 3\}$

$[3, \infty)$

$\{x \mid x \ge 1\} \cap \{x \mid x \ge 3\}$
$= \{x \mid x \ge 3\}$

$[3, \infty)$

The numbers common to both sets are those that are greater than or equal to 3. Thus the solution set is $\{x \mid x \ge 3\}$, or, in interval notation, $[3, \infty)$. You should check that any number in $[3, \infty)$ satisfies the conjunction whereas numbers outside $[3, \infty)$ do not.

Do Exercise 5.

Sometimes two sets have no elements in common. In such a case, we say that the intersection of the two sets is the empty set, denoted { } or ∅. Two sets with an empty intersection are said to be **disjoint**.

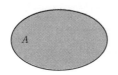

$A \cap B = \varnothing.$

Example 4 Solve and graph: $2x - 3 > 1$ *and* $3x - 1 < 2$.

We solve each inequality separately:

$$2x - 3 > 1 \quad and \quad 3x - 1 < 2$$
$$2x > 4 \quad and \quad 3x < 3$$
$$x > 2 \quad and \quad x < 1.$$

The solution set is the intersection of the solution sets of the individual inequalities.

4. Solve and graph:

$$-22 < 3x - 7 \le 23.$$

$$\overset{\longleftarrow}{\underset{-12 \quad -8 \quad -4 \quad 0 \quad 4 \quad 8 \quad 12}{\rule{0pt}{0pt}}}\longrightarrow$$

5. Solve and graph:

$$3x + 4 < 10 \ and \ 2x - 7 < -13.$$

$$\overset{\longleftarrow}{\underset{-5 \ -4 \ -3 \ -2 \ -1 \ 0 \ 1 \ 2 \ 3 \ 4 \ 5}{\rule{0pt}{0pt}}}\longrightarrow$$

Answers on page A-7

6. Solve and graph.

$3x - 7 \leq -13$ *and* $4x + 3 > 8$.

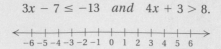

7. Solve: $-4 \leq 8 - 2x \leq 4$.

8. Find the union:

$\{0, 1, 3, 4\} \cup \{0, 1, 7, 9\}$.

9. Shade the union of sets A and B.

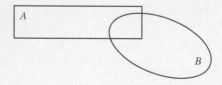

$\{x \mid x > 2\}$ ⟵━━━━━━━━(━━━━⟶ $(2, \infty)$
  −7 −6 −5 −4 −3 −2 −1 0 1 2 3 4 5 6 7

$\{x \mid x < 1\}$ ⟵━━━━━)━━━━━━━━━⟶ $(-\infty, 1)$
  −7 −6 −5 −4 −3 −2 −1 0 1 2 3 4 5 6 7

$\{x \mid x > 2\} \cap \{x \mid x < 1\}$ ⟵━━━━━━━━━━━━━⟶ $\varnothing$
$= \{x \mid x > 2 \ and \ x < 1\}$ −7 −6 −5 −4 −3 −2 −1 0 1 2 3 4 5 6 7
$= \varnothing$

Since no number is both greater than 2 and less than 1, the solution set is the empty set, $\varnothing$.

Do Exercise 6.

Example 5 Solve: $3 \leq 5 - 2x < 7$.

We have

$$3 \leq 5 - 2x < 7$$
$$3 - 5 \leq 5 - 2x - 5 < 7 - 5 \qquad \text{Subtracting 5}$$
$$-2 \leq \quad -2x \quad < 2 \qquad \text{Simplifying}$$

$$\frac{-2}{-2} \geq \frac{-2x}{-2} > \frac{2}{-2} \qquad \begin{array}{l} \text{Dividing by } -2. \text{ The symbols} \\ \text{must be reversed.} \end{array}$$

$$1 \geq x > -1. \qquad \text{Simplifying}$$

The solution set is $\{x \mid 1 \geq x > -1\}$, or $\{x \mid -1 < x \leq 1\}$, since the inequalities $1 \geq x > -1$ and $-1 < x \leq 1$ are equivalent. The solution, in interval notation, is $(-1, 1]$.

Do Exercise 7.

b Unions of Sets and Disjunctions of Inequalities

The **union** of two sets A and B is the collection of elements belonging to A and/or B. We denote the union of A and B by

$$A \cup B.$$

The union of two sets is often pictured as shown below.

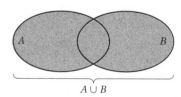

$$A \cup B$$

Example 6 Find the union: $\{2, 3, 4\} \cup \{3, 5, 7\}$.

The numbers in either or both sets are 2, 3, 4, 5, and 7, so the union is $\{2, 3, 4, 5, 7\}$.

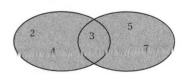

Do Exercises 8 and 9.

When two or more sentences are joined by the word *or* to make a compound sentence, the new sentence is called a **disjunction** of the sentences. Here are three examples:

$$x < -3 \quad or \quad x > 3;$$
$$y \text{ is an odd number} \quad or \quad y \text{ is a prime number;}$$
$$x < 0 \quad or \quad x = 0 \quad or \quad x > 0.$$

In order for a disjunction to be true, at least one of the individual sentences must be true. *The solution set of a disjunction is the union of the individual solution sets.* Consider the disjunction

$$x < -3 \quad or \quad x > 3.$$

The graphs of each separate sentence are shown below, and the union is the last graph. Again, we use both set-builder and interval notations.

$\{x \mid x < -3\}$ $(-\infty, -3)$

$\{x \mid x > 3\}$ $(3, \infty)$

$\{x \mid x < -3\} \cup \{x \mid x > 3\}$ $(-\infty, -3)$
$= \{x \mid x < -3 \text{ or } x > 3\}$ $\cup (3, \infty)$

Answers to disjunctions can rarely be written in a shorter manner. The solution set of $x < -3$ *or* $x > 3$ is simply written $\{x \mid x < -3 \text{ or } x > 3\}$, or $(-\infty, -3) \cup (3, \infty)$.

> ▶ The word "or" corresponds to "union" and the symbol "∪". In order for a number to be in the solution set of a disjunction, it must be in *at least one* of the solution sets.

Do Exercise 10.

Example 7 Solve and graph: $7 + 2x < -1$ *or* $13 - 5x \le 3$.

We solve each inequality separately, retaining the word *or*:

$$7 + 2x < -1 \quad or \quad 13 - 5x \le 3$$
$$2x < -8 \quad or \quad -5x \le -10$$

Dividing by −5. The symbol must be reversed.

$$x < -4 \quad or \quad x \ge 2.$$

To find the solution set of the disjunction, we consider the individual graphs. We graph $x < -4$ and then $x \ge 2$. Then we take the union of the graphs.

$\{x \mid x < -4\}$ $(-\infty, -4)$

$\{x \mid x \ge 2\}$ $[2, \infty)$

$\{x \mid x < -4 \text{ or } x \ge 2\}$ $(-\infty, -4)$ $\cup [2, \infty)$

The solution set is $\{x \mid x < -4 \text{ or } x \ge 2\}$, or, in interval notation, $(-\infty, -4) \cup [2, \infty)$.

Answer on page A-7

10. Graph and write interval notation:

$$x \le -2 \text{ or } x > 4.$$

Solve and graph.

11. $x - 4 < -3$ *or* $x - 3 \geq 3$

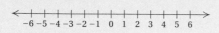

12. $-2x + 4 \leq -3$ *or* $x + 5 < 3$

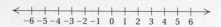

13. Solve:

$-3x - 7 < -1$ *or* $x + 4 < -1$.

14. Solve and graph:

$5x - 7 \leq 13$ *or* $2x - 1 \geq -7$.

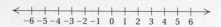

CAUTION! A compound inequality like

$$x < -4 \quad or \quad x \geq 2,$$

as in Example 7, *cannot* be expressed as $2 \leq x < -4$ because to do so would be to say that x is *simultaneously* less than -4 and greater than or equal to 2. No number is both less than -4 *and* greater than or equal to 2, but many are less than -4 *or* greater than or equal to 2.

Do Exercises 11 and 12.

Example 8 Solve: $-2x - 5 < -2$ *or* $x - 3 < -10$.

We solve the individual inequalities separately, retaining the word *or*:

$$-2x - 5 < -2 \quad or \quad x - 3 < -10$$
$$-2x < 3 \quad or \quad x < -7$$

Reversing the symbol

$$x > -\tfrac{3}{2} \quad or \quad x < -7.$$

Keep the word "or."

The solution set is $\left\{x \mid x < -7 \ or \ x > -\tfrac{3}{2}\right\}$, or, in interval notation, $(-\infty, -7) \cup \left(-\tfrac{3}{2}, \infty\right)$.

Do Exercise 13.

Example 9 Solve: $3x - 11 < 4$ *or* $4x + 9 \geq 1$.

We solve the individual inequalities separately, retaining the word *or*:

$$3x - 11 < 4 \quad or \quad 4x + 9 \geq 1$$
$$3x < 15 \quad or \quad 4x \geq -8$$
$$x < 5 \quad or \quad x \geq -2.$$

To find the solution set, we first look at the individual graphs.

$\{x \mid x < 5\}$ $(-\infty, 5)$

$\{x \mid x \geq -2\}$ $[-2, \infty)$

$\{x \mid x < 5\} \cup \{x \mid x \geq -2\}$
$= \{x \mid x < 5 \ or \ x \geq -2\}$ $(-\infty, \infty)$
= The set of all real numbers

Since any number is either less than 5 or greater than or equal to -2, the two sets fill the entire number line. Thus the solution set is the set of all real numbers.

Do Exercise 14.

c | Applications and Problem Solving

Example 10 *Converting Dress Sizes.* The equation

$$I = 2(s + 10)$$

can be used to convert dress sizes s in the United States to dress sizes I in Italy. Which dress sizes in the United States correspond to dress sizes between 32 and 46 in Italy?

1. **Familiarize.** We have a formula for converting the dress sizes. Thus we can substitute a value into the formula. For a dress of size 6 in the United States, we get the corresponding dress size in Italy as follows:

$$I = 2(6 + 10) = 2 \cdot 16 = 32.$$

This familiarizes us with the formula and also tells us that the United States sizes that we are looking for must be larger than size 6.

2. **Translate.** We want the Italian sizes *between* 32 and 46, so we want to find those values of s for which

$$32 < I < 46 \qquad \text{\textit{I} is between 32 and 46}$$

or

$$32 < 2(s + 10) < 46. \qquad \text{Substituting } 2(s + 10) \text{ for } I$$

Thus we have translated the problem to an inequality.

3. **Solve.** We solve the inequality:

$$32 < 2(s + 10) < 46$$
$$\frac{32}{2} < \frac{2(s + 10)}{2} < \frac{46}{2} \qquad \text{Dividing by 2}$$
$$16 < s + 10 < 23 \qquad \text{Simplifying}$$
$$6 < s < 13. \qquad \text{Subtracting 10}$$

4. **Check.** We substitute some values as we did in the *Familiarize* step.

5. **State.** Dress sizes between 6 and 13 in the United States correspond to dress sizes between 32 and 46 in Italy.

Do Exercise 15.

15. *Converting Dress Sizes.* Refer to Example 10. Which dress sizes in the United States correspond to dress sizes between 36 and 58 in Italy?

Answer on page A-7

Improving Your Math Study Skills

Learning Resources and Time Management

Two other topics to consider in enhancing your math study skills are learning resources and time management.

Learning Resources

- **Textbook supplements.** Are you aware of all the supplements that exist for this textbook? Many details are given in the preface. Now that you are more familiar with the book, let's discuss them.

  1. The *Student's Solutions Manual* contains worked-out solutions to the odd-numbered exercises in the exercise sets. Consider obtaining a copy if you are having trouble. It should be your first choice if you can make an additional purchase.

  2. An extensive set of *videotapes* supplement this text. These may be available to you on your campus at a learning center or math lab. Check with your instructor.

  3. *Tutorial software* also accompanies the text. If not available in the campus learning center, you can order it by calling the number 1-800-322-1377.

- **The Internet.** Our on-line World Wide Web supplement provides additional practice resources. If you have internet access, you can reach this site through the address:

  http://www.mathmax.com

  It contains many helpful ideas as well as links to other resources for learning mathematics.

- **Your college or university.** Your own college or university probably has resources to enhance your math learning.

  1. For example, is there a learning lab or tutoring center for drop-in tutoring?

  2. Are there special lab classes or formal tutoring sessions tailored for the specific course you are taking?

  3. Perhaps there is a bulletin board or network where you can locate the names of experienced private tutors.

- **Your instructor.** Although it may seem obvious, students neglect to consider the most underused resource available to them: their instructor. Find out your instructor's office hours and make it a point to visit when you need additional help.

Time Management

- **Juggling time.** Have reasonable expectations about the time you need to study math. Unreasonable expectations may lead to lower grades and frustrations. Working 40 hours per week and taking 12 hours of credit is equivalent to working two full-time jobs. Can you handle such a load? As a rule of thumb, your ratio of work hours to credit load should be about 40/3, 30/6, 20/9, 10/12, and 5/14. Budget about 2–3 hours of homework and studying per hour of class.

- **Daily schedule.** Make an hour-by-hour schedule of your typical week. Include work, college, home, personal, sleep, study, and leisure times. Be realistic about the amount of time needed for sleep and home duties. If possible, try to schedule time for study when you are most alert.

Exercise Set 1.5

a, **b** Find the intersection or union.

1. {9, 10, 11} ∩ {9, 11, 13}

2. {1, 5, 10, 15} ∩ {5, 15, 20}

3. {a, b, c, d} ∩ {b, f, g}

4. {m, n, o, p} ∩ {m, o, p}

5. {9, 10, 11} ∪ {9, 11, 13}

6. {1, 5, 10, 15} ∪ {5, 15, 20}

7. {a, b, c, d} ∪ {b, f, g}

8. {m, n, o, p} ∪ {m, o, p}

9. {2, 5, 7, 9} ∩ {1, 3, 4}

10. {a, e, i, o, u} ∩ {m, q, w, s, t}

11. {3, 5, 7} ∪ ∅

12. {3, 5, 7} ∩ ∅

a Graph and write interval notation.

13. $-4 < a$ *and* $a \leq 1$

<───┼──┼──┼──┼──┼──┼──┼──┼──┼──┼──┼──┼──┼───>
   −6 −5 −4 −3 −2 −1 0 1 2 3 4 5 6

14. $-\frac{5}{2} \leq m$ *and* $m < \frac{3}{2}$

   −6 −5 −4 −3 −2 −1 0 1 2 3 4 5 6

15. $1 < x < 6$

<───┼──┼──┼──┼──┼──┼──┼──┼──┼──┼──┼──┼──┼───>
   −6 −5 −4 −3 −2 −1 0 1 2 3 4 5 6

16. $-3 \leq y \leq 4$

<───┼──┼──┼──┼──┼──┼──┼──┼──┼──┼──┼──┼──┼───>
   −6 −5 −4 −3 −2 −1 0 1 2 3 4 5 6

Solve and graph.

17. $-10 \leq 3x + 2$ *and* $3x + 2 < 17$

<───┼──┼──┼──┼──┼──┼──┼──┼──┼──┼──┼──┼──┼───>
    6 −5 −4 −3 −2 −1 0 1 2 3 4 5 6

18. $-11 < 4x - 3$ *and* $4x - 3 \leq 13$

<───┼──┼──┼──┼──┼──┼──┼──┼──┼──┼──┼──┼──┼───>
   −6 −5 −4 −3 −2 −1 0 1 2 3 4 5 6

19. $3x + 7 \geq 4$ *and* $2x - 5 \geq -1$

<───┼──┼──┼──┼──┼──┼──┼──┼──┼──┼──┼──┼──┼───>
   −6 −5 −4 −3 −2 −1 0 1 2 3 4 5 6

20. $4x - 7 < 1$ *and* $7 - 3x > -8$

<───┼──┼──┼──┼──┼──┼──┼──┼──┼──┼──┼──┼──┼───>
   −6 −5 −4 −3 −2 −1 0 1 2 3 4 5 6

21. $4 - 3x \geq 10$ *and* $5x - 2 > 13$

<───┼──┼──┼──┼──┼──┼──┼──┼──┼──┼──┼──┼──┼───>
   −6 −5 −4 −3 −2 −1 0 1 2 3 4 5 6

22. $5 - 7x > 19$ *and* $2 - 3x < -4$

<───┼──┼──┼──┼──┼──┼──┼──┼──┼──┼──┼──┼──┼───>
   −6 −5 −4 −3 −2 −1 0 1 2 3 4 5 6

Solve.

23. $-4 < x + 4 < 10$

24. $-6 < x + 6 \leq 8$

25. $6 > -x \geq -2$

26. $3 > -x \geq -5$

27. $1 < 3y + 4 \leq 19$

28. $5 \leq 8x + 5 \leq 21$

29. $-10 \leq 3x - 5 \leq -1$

30. $-18 \leq -2x - 7 < 0$

31. $2 < x + 3 \leq 9$

32. $-6 \leq x + 1 < 9$

33. $-6 \leq 2x - 3 < 6$

34. $4 > -3m - 7 \geq 2$

35. $-\frac{1}{2} < \frac{1}{4}x - 3 \leq \frac{1}{2}$

36. $-\frac{2}{3} \leq 4 - \frac{1}{4}x < \frac{2}{3}$

37. $-3 < \frac{2x - 5}{4} < 8$

38. $-4 \leq \frac{7 - 3x}{5} \leq 4$

b Graph and write interval notation.

39. $x < -2$ *or* $x > 1$

40. $x < -4$ *or* $x > 0$

41. $x \leq -3$ *or* $x > 1$

42. $x \leq -1$ *or* $x > 3$

Solve and graph.

43. $x + 3 < -2$ *or* $x + 3 > 2$

44. $x - 2 < -1$ *or* $x - 2 > 3$

45. $2x - 8 \leq -3$ *or* $x - 1 \geq 3$

46. $x - 5 \leq -4$ *or* $2x - 7 \geq 3$

47. $7x + 4 \geq -17$ *or* $6x + 5 \geq -7$

48. $4x - 4 < -8$ *or* $4x - 4 < 12$

Solve.

49. $7 > -4x + 5$ *or* $10 \leq -4x + 5$

50. $6 > 2x - 1$ *or* $-4 \leq 2x - 1$

51. $3x - 7 > -10$ *or* $5x + 2 \leq 22$

52. $3x + 2 < 2$ *or* $4 - 2x < 14$

53. $-2x - 2 < -6$ *or* $-2x - 2 > 6$

54. $-3m - 7 < -5$ *or* $-3m - 7 > 5$

55. $\dfrac{2}{3}x - 14 < -\dfrac{5}{6}$ *or* $\dfrac{2}{3}x - 14 > \dfrac{5}{6}$

56. $\dfrac{1}{4} - 3x \leq -3.7$ *or* $\dfrac{1}{4} - 5x \geq 4.8$

57. $\dfrac{2x - 5}{6} \leq -3$ *or* $\dfrac{2x - 5}{6} \geq 4$

58. $\dfrac{7 - 3x}{5} < -4$ *or* $\dfrac{7 - 3x}{5} > 4$

---

c Solve.

59. *Pressure at Sea Depth.* The equation

$$P = 1 + \frac{d}{33}$$

gives the pressure P, in atmospheres (atm), at a depth of d feet in the sea. For what depths d is the pressure at least 1 atm and at most 7 atm?

60. *Temperatures of Liquids.* The formula

$$C = \tfrac{5}{9}(F - 32)$$

can be used to convert Fahrenheit temperatures F to Celsius temperatures C.

a) Gold is a liquid for Celsius temperatures C such that $1063° \leq C < 2660°$. Find such an inequality for the corresponding Fahrenheit temperatures.

b) Silver is a liquid for Celsius temperatures C such that $960.8° \leq C < 2180°$. Find such an inequality for the corresponding Fahrenheit temperatures.

61. *Solid-Waste Generation.* The equation

$$W = 0.05t + 4.3$$

can be used to estimate the average number of pounds W of solid waste produced daily by each person in the United States, t years after 1991. For what years will waste production range from 5.0 to 5.25 lb per person per day?

62. *Aerobic Exercise.* In order to achieve maximum results from aerobic exercise, one should maintain one's heart rate at a certain level. A 30-yr-old woman with a resting heart rate of 60 beats per minute should keep her heart rate between 138 and 162 beats per minute while exercising. She checks her pulse for 10 sec while exercising. What should the number of beats be?

63. *Body Mass Index.* See Exercises 67 and 68 in Exercise Set 1.4. Marv's height is 73 in. What weights W will allow Marv to keep his body mass index I between 20 and 30?

64. *Young's Rule in Medicine.* See Exercise 25 in Exercise Set 1.2. An 8-yr-old child needs medication. What adult dosage can be used if a child's dosage must stay between 100 mg and 200 mg?

Skill Maintenance

Find the absolute value.   [R.1d]

65. $|-3.2|$

66. $|-5| + |7|$

67. $|-5 + 7|$

68. $|7 - 7|$

Simplify.   [R.7a, b]

69. $(-2x^{-4}y^6)^5$

70. $(-4a^5b^{-7})(5a^{-12}b^8)$

71. $\dfrac{-4a^5b^{-7}}{5a^{-12}b^8}$

72. $(5p^6q^{11})^2$

73. $\left(\dfrac{56a^5b^{-6}}{28a^7b^{-8}}\right)^{-3}$

74. $\left(\dfrac{125p^{11}q^{12}}{25p^6q^8}\right)^2$

Synthesis

75. ◆ Explain why the conjunction $3 < x$ *and* $x < 5$ is equivalent to $3 < x < 5$, but the disjunction $3 < x$ *or* $x < 5$ is not.

76. ◆ Describe the circumstances under which $[a, b] \cup [c, d] = [a, d]$.

77. What is the union of the set of all rational numbers with the set of all irrational numbers? the intersection?

78. *Minimizing Tolls.* A $3.00 toll is charged to cross the bridge from Sanibel Island to mainland Florida. A six-month pass, costing $15.00, reduces the toll to $0.50 per crossing. A one-year pass, costing $60, allows for free crossings. How many crossings per month does it take, on average, for the six-month pass to be the more economical choice?

Solve.

79. $x - 10 < 5x + 6 \le x + 10$

80. $4m - 8 > 6m + 5$ *or* $5m - 8 < -2$

81. $-\frac{2}{15} \le \frac{2}{3}x - \frac{2}{5} \le \frac{2}{15}$

82. $2[5(3 - y) - 2(y - 2)] > y + 4$

83. $3x < 4 - 5x < 5 + 3x$

84. $2x - \frac{3}{4} < -\frac{1}{10}$ *or* $2x - \frac{3}{4} > \frac{1}{10}$

85. $x + 4 < 2x - 6 \le x + 12$

86. $2x + 3 \le x - 6$ *or* $3x - 2 \le 4x + 5$

1.6 Absolute-Value Equations and Inequalities

a | Properties of Absolute Value

We can think of the **absolute value** of a number as its distance from zero on the number line. Recall the formal definition from Section R.2.

> The absolute value of x, denoted $|x|$, is defined as follows:
>
> $$x \geq 0 \implies |x| = x; \qquad x < 0 \implies |x| = -x.$$

Some simple properties of absolute value allow us to manipulate or simplify algebraic expressions.

> **PROPERTIES OF ABSOLUTE VALUE**
>
> a) For any real numbers a and b, $|ab| = |a| \cdot |b|$.
>
>   (The absolute value of a product is the product of the absolute values.)
>
> b) $\left| \dfrac{a}{b} \right| = \dfrac{|a|}{|b|}$, provided that $b \neq 0$.
>
>   (The absolute value of a quotient is the quotient of the absolute values.)
>
> c) $|-a| = |a|$
>
>   (The absolute value of the opposite of a number is the same as the absolute value of the number.)

Examples Simplify, leaving as little as possible inside the absolute-value signs.

1. $|5x| = |5| \cdot |x| = 5|x|$
2. $|-3y| = |-3| \cdot |y| = 3|y|$
3. $|7x^2| = |7| \cdot |x^2| = 7|x^2| = 7x^2$ Since x^2 is never negative for any number x
4. $\left| \dfrac{6x}{-3x^2} \right| = \left| \dfrac{2}{-x} \right| = \dfrac{|2|}{|-x|} = \dfrac{2}{|x|}$

Do Exercises 1–5.

b | Distance on a Number Line

The number line below shows that the distance between -3 and 2 is 5.

5 units

Another way to find the distance between two numbers on a number line is to take the absolute value of the difference, as follows:

$$|-3 - 2| = |-5| = 5, \quad \text{or} \quad |2 - (-3)| = |5| = 5.$$

Note that the order in which we subtract does not matter because we are taking the absolute value after we have subtracted.

Objectives

a   Simplify expressions containing absolute-value symbols.

b   Find the distance between two points on a number line.

c   Solve equations with absolute-value expressions.

d   Solve equations with two absolute-value expressions.

e   Solve inequalities with absolute-value expressions.

For Extra Help

TAPE 4   TAPE 4A   MAC WIN   CD-ROM

Simplify, leaving as little as possible inside the absolute-value signs.

1. $|7x|$

2. $|x^8|$

3. $|5a^2b|$

4. $\left| \dfrac{7a}{b^2} \right|$

5. $|-9x|$

Answers on page A-7

Find the distance between the points.

6. −6, −35

7. 19, 14

8. 0, p

9. Solve: $|x| = 6$. Then graph using a number line.

10. Solve: $|x| = -6$.

11. Solve: $|p| = 0$.

Answers on page A-7

> For any real numbers a and b, the **distance** between them is $|a - b|$.

We should note that the distance is also $|b - a|$, because $a - b$ and $b - a$ are opposites and hence have the same absolute value.

Example 5   Find the distance between −8 and −92 on a number line.

$$|-8 - (-92)| = |84| = 84, \quad \text{or} \quad |-92 - (-8)| = |-84| = 84$$

Example 6   Find the distance between x and 0 on the number line.

$$|x - 0| = |x|$$

Do Exercises 6–8.

c   Equations with Absolute Value

Example 7   Solve: $|x| = 4$. Then graph using the number line.

Note that $|x| = |x - 0|$, so that $|x - 0|$ is the distance from x to 0. Thus solutions of the equation $|x| = 4$, or $|x - 0| = 4$, are those numbers x whose distance from 0 is 4. Those numbers are −4 and 4. The solution set is $\{-4, 4\}$. The graph consists of just two points, as shown.

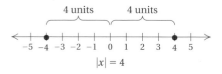

$|x| = 4$

Example 8   Solve: $|x| = 0$.

The only number whose absolute value is 0 is 0 itself. Thus the solution is 0. The solution set is $\{0\}$.

Example 9   Solve: $|x| - -7$.

The absolute value of a number is always nonnegative. There is no number whose absolute value is −7. Thus there is no solution. The solution set is $\varnothing$.

Examples 7–9 lead us to the following principle for solving linear equations with absolute value.

> **THE ABSOLUTE-VALUE PRINCIPLE**
>
> For any positive number p and any algebraic expression X:
>
> **a)** The solutions of $|X| = p$ are those numbers that satisfy $X = -p$ or $X = p$.
>
> **b)** The equation $|X| = 0$ is equivalent to the equation $X = 0$.
>
> **c)** The equation $|X| = -p$ has no solution.

Do Exercises 9–11.

We can use the absolute-value principle with the addition and multiplication principles to solve equations with absolute value.

Example 10 Solve: $2|x| + 5 = 9$.

We first use the addition and multiplication principles to get $|x|$ by itself. Then we use the absolute-value principle.

$$2|x| + 5 = 9$$
$$2|x| = 4 \qquad \text{Subtracting 5}$$
$$|x| = 2 \qquad \text{Dividing by 2}$$
$$x = -2 \quad or \quad x = 2 \qquad \text{Using the absolute-value principle}$$

The solutions are -2 and 2. The solution set is $\{-2, 2\}$.

Do Exercises 12–14.

Example 11 Solve: $|x - 2| = 3$.

We can consider solving this equation in two different ways.

METHOD 1 This method allows us to see the meaning of the solutions graphically. The solution set consists of those numbers that are 3 units from 2 on the number line.

The solutions of $|x - 2| = 3$ are -1 and 5. The solution set is $\{-1, 5\}$.

METHOD 2 This method is more efficient. We use the absolute-value principle, replacing X with $x - 2$ and p with 3. Then we solve each equation separately.

$$|X| = p$$
$$|x - 2| = 3$$
$$x - 2 = -3 \quad or \quad x - 2 = 3 \qquad \text{Absolute-value principle}$$
$$x = -1 \quad or \quad x = 5$$

The solutions are -1 and 5. The solution set is $\{-1, 5\}$.

Do Exercise 15.

Example 12 Solve: $|2x + 5| = 13$.

We use the absolute-value principle, replacing X with $2x + 5$ and p with 13:

$$|X| = p$$
$$|2x + 5| = 13$$
$$2x + 5 = -13 \quad or \quad 2x + 5 = 13 \qquad \text{Absolute-value principle}$$
$$2x = -18 \quad or \quad 2x = 8$$
$$x = -9 \quad or \quad x = 4.$$

The solutions are -9 and 4. The solution set is $\{-9, 4\}$.

Do Exercise 16.

Solve.

12. $|3x| = 6$

13. $4|x| + 10 = 27$

14. $3|x| - 2 = 10$

15. Solve: $|x - 4| = 1$. Use two methods as in Example 11.

16. Solve: $|3x - 4| = 17$.

Answers on page A-8

17. Solve: $|6 + 2x| = -3$.

Solve.

18. $|5x - 3| = |x + 4|$

19. $|x - 3| = |x + 10|$

20. Solve: $|x| = 5$. Then graph using a number line.

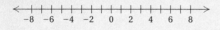

Example 13   Solve: $|4 - 7x| = -8$.

Since absolute value is always nonnegative, this equation has no solution. The solution set is $\varnothing$.

Do Exercise 17.

d   Equations with Two Absolute-Value Expressions

Sometimes equations have two absolute-value expressions. Consider $|a| = |b|$. This means that a and b are the same distance from 0. If a and b are the same distance from 0, then either they are the same number or they are opposites of each other.

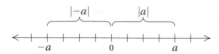

Example 14   Solve: $|2x - 3| = |x + 5|$.

Either $2x - 3 = x + 5$ or $2x - 3 = -(x + 5)$. We solve each equation:

$$2x - 3 = x + 5 \quad or \quad 2x - 3 = -(x + 5)$$
$$x - 3 = 5 \quad or \quad 2x - 3 = -x - 5$$
$$x = 8 \quad or \quad 3x - 3 = -5$$
$$x = 8 \quad or \quad 3x = -2$$
$$x = 8 \quad or \quad x = -\tfrac{2}{3}.$$

The solutions are 8 and $-\tfrac{2}{3}$. The solution set is $\left\{8, -\tfrac{2}{3}\right\}$.

Example 15   Solve: $|x + 8| = |x - 5|$.

$$x + 8 = x - 5 \quad or \quad x + 8 = -(x - 5)$$
$$8 = -5 \quad or \quad x + 8 = -x + 5$$
$$8 = -5 \quad or \quad 2x = -3$$
$$8 = -5 \quad or \quad x = -\tfrac{3}{2}$$

The first equation has no solution. The second equation has $-\tfrac{3}{2}$ as a solution. The solution set is $\left\{-\tfrac{3}{2}\right\}$.

Do Exercises 18 and 19.

e   Inequalities with Absolute Value

We can extend our methods for solving equations with absolute value to those for solving inequalities with absolute value.

Example 16   Solve: $|x| = 4$. Then graph using a number line.

From Example 7, we know that the solutions are -4 and 4. The solution set is $\{-4, 4\}$. The graph consists of just two points, as shown here.

Do Exercise 20.

Example 17 Solve: $|x| < 4$. Then graph.

The solutions of $|x| < 4$ are the solutions of $|x - 0| < 4$ and are those numbers x whose distance from 0 is less than 4. We can check by substituting or by looking at the number line that numbers like $-3, -2, -1, -\frac{1}{2}$, $-\frac{1}{4}, 0, \frac{1}{4}, \frac{1}{2}, 1, 2$, and 3 are all solutions. In fact, the solutions are all the real numbers x between -4 and 4, such that $-4 < x < 4$. The solution set is $\{x \mid -4 < x < 4\}$ or, in interval notation, $(-4, 4)$. The graph is as follows.

$(-4, 4)$

$|x| < 4$

Do Exercise 21.

Example 18 Solve: $|x| \geq 4$. Then graph.

The solutions of $|x| \geq 4$ are solutions of $|x - 0| \geq 4$ and are those numbers whose distance from 0 is greater than or equal to 4—in other words, those numbers x such that $x \leq -4$ or $x \geq 4$. The solution set is $\{x \mid x \leq -4$ or $x \geq 4\}$, or $(-\infty, -4] \cup [4, \infty)$. The graph is as follows.

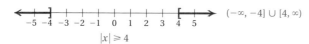

$(-\infty, -4] \cup [4, \infty)$

$|x| \geq 4$

Do Exercise 22.

Examples 16–18 illustrate three cases of solving equations and inequalities with absolute value. The expression inside the absolute-value signs can be something besides a single variable. The following is a general principle for solving.

> For any positive number p and any algebraic expression X:
>
> **a)** The solutions of $|X| = p$ are those numbers that satisfy $X = -p$ or $X = p$.
>
> As an example, replacing X with $5x - 1$ and p with 8, we see that the solutions of $|5x - 1| = 8$ are those numbers x for which
>
> $5x - 1 = -8$ *or* $5x - 1 = 8$
>
> $5x = -7$ *or* $5x = 9$
>
> $x = -\frac{7}{5}$ *or* $x = \frac{9}{5}$.
>
> The solution set is $\left\{-\frac{7}{5}, \frac{9}{5}\right\}$.
>
>
>
> **b)** The solutions of $|X| < p$ are those numbers that satisfy $-p < X < p$.
>
> As an example, replacing X with $6x + 7$ and p with 5, we see that the solutions of $|6x + 7| < 5$ are those numbers x for which
>
> $-5 < 6x + 7 < 5$
>
> $-12 < 6x < -2$
>
> $-2 < x < -\frac{1}{3}$.
>
> The solution set is $\left\{x \mid -2 < x < -\frac{1}{3}\right\}$, or $\left(-2, -\frac{1}{3}\right)$.
>
>

(continued)

22. Solve: $|x| \geq 5$. Then graph.

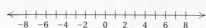

Answers on page A-8

23. Solve: $|2x - 3| < 7$. Then graph.

24. Solve: $|7 - 3x| \leq 4$.

25. Solve: $|3x + 2| \geq 5$. Then graph.

c) The solutions of $|X| > p$ are those numbers that satisfy $X < -p$ or $X > p$.

As an example, replacing X with $2x - 9$ and p with 4, we see that the solutions of $|2x - 9| > 4$ are those numbers x for which

$$2x - 9 < -4 \quad or \quad 2x - 9 > 4$$
$$2x < 5 \quad or \quad 2x > 13$$
$$x < \tfrac{5}{2} \quad or \quad x > \tfrac{13}{2}.$$

The solution set is $\{x \mid x < \tfrac{5}{2} \text{ or } x > \tfrac{13}{2}\}$, or $\left(-\infty, \tfrac{5}{2}\right) \cup \left(\tfrac{13}{2}, \infty\right)$.

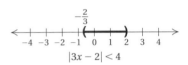

Example 19   Solve: $|3x - 2| < 4$. Then graph.

We use part (b). In this case, X is $3x - 2$ and p is 4:

$$|X| < p$$
$$|3x - 2| < 4 \qquad \text{Replacing } X \text{ with } 3x - 2 \text{ and } p \text{ with } 4$$
$$-4 < 3x - 2 < 4$$
$$-2 < 3x < 6$$
$$-\tfrac{2}{3} < x < 2.$$

The solution set is $\{x \mid -\tfrac{2}{3} < x < 2\}$, or $\left(-\tfrac{2}{3}, 2\right)$. The graph is as follows.

$$-\tfrac{2}{3}$$

```
←—+—+—+—+(+—+—+)—+—+→
  -4 -3 -2 -1  0  1  2  3  4
        |3x - 2| < 4
```

Example 20   Solve: $|8 - 4x| \leq 5$.

We use part (b). In this case, X is $8 - 4x$ and p is 5:

$$|X| \leq p$$
$$|8 - 4x| \leq 5 \qquad \text{Replacing } X \text{ with } 8 - 4x \text{ and } p \text{ with } 5$$
$$-5 \leq 8 - 4x \leq 5$$
$$-13 \leq -4x \leq -3$$
$$\tfrac{13}{4} \geq x \geq \tfrac{3}{4}. \qquad \begin{array}{l}\text{Dividing by } -4 \text{ and reversing} \\ \text{the inequality symbols}\end{array}$$

The solution set is $\{x \mid \tfrac{13}{4} \geq x \geq \tfrac{3}{4}\}$, or $\{x \mid \tfrac{3}{4} \leq x \leq \tfrac{13}{4}\}$, or $\left[\tfrac{3}{4}, \tfrac{13}{4}\right]$.

Example 21   Solve: $|4x + 2| \geq 6$.

We use part (c). In this case, X is $4x + 2$ and p is 6:

$$|X| \geq p$$
$$|4x + 2| \geq 6 \qquad \text{Replacing } X \text{ with } 4x + 2 \text{ and } p \text{ with } 6$$
$$4x + 2 \leq -6 \quad or \quad 4x + 2 \geq 6$$
$$4x \leq -8 \quad or \quad 4x \geq 4$$
$$x \leq -2 \quad or \quad x \geq 1.$$

The solution set is $\{x \mid x \leq -2 \text{ or } x \geq 1\}$, or $(-\infty, -2] \cup [1, \infty)$.

Do Exercises 23–25.

Exercise Set 1.6

a Simplify, leaving as little as possible inside absolute-value signs.

1. $|9x|$

2. $|26x|$

3. $|2x^2|$

4. $|8x^2|$

5. $|-2x^2|$

6. $|-20x^2|$

7. $|-6y|$

8. $|-17y|$

9. $\left|\dfrac{-2}{x}\right|$

10. $\left|\dfrac{y}{3}\right|$

11. $\left|\dfrac{x^2}{-y}\right|$

12. $\left|\dfrac{x^4}{-y}\right|$

13. $\left|\dfrac{-8x^2}{2x}\right|$

14. $\left|\dfrac{9y}{3y^2}\right|$

b Find the distance between the points on a number line.

15. $-8,\ -46$

16. $-7,\ -32$

17. $36,\ 17$

18. $52,\ 18$

19. $-3.9,\ 2.4$

20. $-1.8,\ -3.7$

21. $-5,\ 0$

22. $\frac{2}{3},\ -\frac{5}{6}$

c Solve.

23. $|x| = 3$

24. $|x| = 5$

25. $|x| = -3$

26. $|x| = -9$

27. $|q| = 0$

28. $|y| = 7.4$

29. $|x - 3| = 12$

30. $|3x - 2| = 6$

31. $|2x - 3| = 4$

32. $|5x + 2| = 3$

33. $|4x - 9| = 14$

34. $|9y - 2| = 17$

35. $|x| + 7 = 18$

36. $|x| - 2 = 6.3$

37. $574 = 283 + |t|$

38. $-562 = -2000 + |x|$

39. $|5x| = 40$ **40.** $|2y| = 18$ **41.** $|3x| - 4 = 17$ **42.** $|6x| + 8 = 32$

43. $7|w| - 3 = 11$ **44.** $5|x| + 10 = 26$ **45.** $\left|\dfrac{2x - 1}{3}\right| = 5$ **46.** $\left|\dfrac{4 - 5x}{6}\right| = 7$

47. $|m + 5| + 9 = 16$ **48.** $|t - 7| - 5 = 4$ **49.** $10 - |2x - 1| = 4$ **50.** $2|2x - 7| + 11 = 25$

$$3x - 4 = -2 \qquad 3x - 4 = 2 \quad (2/3, 2)$$
$$+4 \quad +4 \qquad +4 \quad +4$$
$$-3x + 4 = -12 \qquad \frac{3x}{3} = \frac{2}{3} \quad x = 2/3 \qquad \frac{3}{3}x = \frac{6}{3} \quad x = \frac{2}{1}$$

51. $|3x - 4| = -2$ **52.** $|x - 6| = -8$ **53.** $\left|\dfrac{5}{9} + 3x\right| = \dfrac{1}{6}$ **54.** $\left|\dfrac{2}{3} - 4x\right| = \dfrac{4}{5}$

$$3x - 4 = -2 \qquad 3x - 4 = -2$$
$$+2 \quad +4$$
$$\frac{3x}{3} = \frac{2}{3} \qquad 2/3$$

d Solve.

55. $|3x + 4| = |x - 7|$ **56.** $|2x - 8| = |x + 3|$ **57.** $|x + 3| = |x - 6|$

58. $|x - 15| = |x + 8|$ **59.** $|2a + 4| = |3a - 1|$ **60.** $|5p + 7| = |4p + 3|$

61. $|y - 3| = |3 - y|$ **62.** $|m - 7| = |7 - m|$ **63.** $|5 - p| = |p + 8|$

64. $|8 - q| = |q + 19|$ **65.** $\left|\dfrac{2x - 3}{6}\right| = \left|\dfrac{4 - 5x}{8}\right|$ **66.** $\left|\dfrac{6 - 8x}{5}\right| = \left|\dfrac{7 + 3x}{2}\right|$

67. $\left|\dfrac{1}{2}x - 5\right| = \left|\dfrac{1}{4}x + 3\right|$ **68.** $\left|2 - \dfrac{2}{3}x\right| = \left|4 + \dfrac{7}{8}x\right|$

e Solve.

69. $|x| < 3$

70. $|x| \leq 5$

71. $|x| \geq 2$

72. $|y| > 12$

73. $|x - 1| < 1$

74. $|x + 4| \leq 9$

75. $|x + 4| \leq 1$

76. $|x - 2| > 6$

77. $|2x - 3| \leq 4$

70. $|5x + 2| \leq 3$

79. $|2y - 7| > 10$

80. $|3y - 4| > 8$

81. $|4x - 9| \geq 14$

82. $|9y - 2| \geq 17$

83. $|y - 3| < 12$

84. $|p - 2| < 6$

85. $|2x + 3| \leq 4$

86. $|5x + 2| \leq 13$

87. $|4 - 3y| > 8$

88. $|7 - 2y| > 5$

89. $|9 - 4x| \geq 14$

90. $|2 - 9p| \geq 17$

91. $|3 - 4x| < 21$

92. $|-5 - 7x| \leq 30$

93. $\left| \dfrac{1}{2} + 3x \right| \geq 12$

94. $\left| \dfrac{1}{4}y - 6 \right| > 24$

95. $\left| \dfrac{x - 7}{3} \right| < 4$

96. $\left| \dfrac{x + 5}{4} \right| \leq 2$

97. $\left| \dfrac{2 - 5x}{4} \right| \geq \dfrac{2}{3}$

98. $\left| \dfrac{1 + 3x}{5} \right| > \dfrac{7}{8}$

99. $|m + 5| + 9 \leq 16$

100. $|t - 7| + 3 \geq 4$

101. $7 - |3 - 2x| \geq 5$

102. $16 \leq |2x - 3| + 9$

103. $\left| \dfrac{2x - 1}{0.0059} \right| \leq 1$

104. $\left| \dfrac{3x - 2}{5} \right| \geq 1$

Skill Maintenance

Translate to an algebraic expression. [R.4a]

105. Forty-nine percent of some number

106. Six more than four times a number

107. Fifty less than the quotient of two numbers

108. The sum of one number and three times another

Compute. [R.2a, c, d, e]

109. $-43.5 + (-5.8)$

110. $-43.5 - (-5.8)$

111. $-43.5(-5.8)$

112. $-43.5 \div (-5.8)$

113. $-\dfrac{7}{8} \div \dfrac{3}{4}$

114. $-\dfrac{4}{5} \cdot \dfrac{5}{4}$

115. $-\dfrac{7}{8} - \dfrac{3}{4}$

116. $-\dfrac{7}{8} + \dfrac{3}{4}$

Synthesis

117. ◆ Explain in your own words why the solutions of the inequality $|x + 5| \le 2$ can be interpreted as "all those numbers x whose distance from -5 is at most 2 units."

118. ◆ Explain in your own words why the interval $[6, \infty)$ is only part of the solution set of $|x| \ge 6$.

119. *Motion of a Spring.* A weighted spring is bouncing up and down so that its distance d above the ground satisfies the inequality $|d - 6\text{ ft}| \le \frac{1}{2}$ ft. Find all possible distances d.

120. *Container Sizes.* A container company is manufacturing rectangular boxes of various sizes. The length of any box must exceed the width by at least 3 in., but the perimeter cannot exceed 24 in. What widths are possible?

$$l \ge w + 3,$$
$$2l + 2w \le 24$$

Solve.

121. $|x + 5| = x + 5$

122. $1 - \left|\frac{1}{4}x + 8\right| = \frac{3}{4}$

123. $|7x - 2| = x + 4$

124. $|x - 1| = x - 1$

125. $|x - 6| \le -8$

126. $|3x - 4| > -2$

127. $|x + 5| > x$

128. $\left|\frac{5}{9} + 3x\right| < -\frac{1}{6}$

129. $|x| \ge 0$

130. $2 \le |x - 1| \le 5$

Find an equivalent inequality with absolute value.

131. $-3 < x < 3$

132. $-5 \le y \le 5$

133. $x \le -6 \text{ or } x \ge 6$

134. $-5 < x < 1$

135. $x < -8 \text{ or } x > 2$

Collaborative
Learning Manual

Create and solve equations with absolute value
as a group.

Summary and Review Exercises: Chapter 1

The review exercises that follow are for practice. Answers are given at the back of the book. If you miss an exercise, restudy the objective indicated in blue after the exercise or the direction line that precedes it. Beginning with this chapter, certain objectives, from four particular sections of preceding chapters, will be retested on the chapter test. The objectives to be tested in addition to the material in this chapter are [R.2a, c], [R.2d, e], [R.5d], and [R.6b].

Solve.   [1.1b, c, d]

1. $-11 + y = -3$

2. $-7x = -3$

3. $-\frac{5}{3}x + \frac{7}{3} = -5$

4. $6(2x - 1) = 3 - (x + 10)$

5. $2.4x + 1.5 = 1.02$

6. $2(3 - x) - 4(x + 1) = 7(1 - x)$

Solve for the indicated letter.   [1.2a]

7. $C = \frac{4}{11}d + 3$,   for d

8. $A = 2a - 3b$,   for b

9. *Interstate Mile Markers.* If you are traveling on a U.S. interstate highway, you will notice numbered markers every mile to tell your location in case of an accident or other emergency. In many states, the numbers on the markers increase from west to east. (*Source:* Federal Highway Administration, Ed Rotalewski) The sum of two mile markers on I-70 in Utah is 371. Find the numbers on the markers. [1.3a]

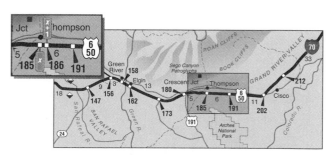

10. *Rope Cutting.* A piece of rope 27 m long is cut into two pieces so that one piece is four-fifths as long as the other. Find the length of each piece.   [1.3a]

11. *Population Growth.* The population of Newcastle grew 12% from one year to the next to a total of 179,200. What was the former population? [1.3a]

12. *Moving Sidewalk.* A moving sidewalk in an airport is 360 ft long and moves at a speed of 6 ft/sec. If Arnie walks at a speed of 3 ft/sec, how long will it take him to walk the length of the moving sidewalk? [1.3b]

Write interval notation for the given set or graph. [1.4b]

13. $\{x \mid -8 \le x < 9\}$

14.
```
   <—+++++++++++++++|++++—>
  -80 -60 -40 -20  0  20  40  60  80
```

Solve and graph. Write interval notation for the solution set. [1.4c]

15. $x - 2 \le -4$

16. $x + 5 > 6$

Solve. [1.4c]

17. $a + 7 \le -14$

18. $y - 5 \ge -12$

19. $4y > -16$

20. $-0.3y < 9$

21. $-6x - 5 < 13$

22. $4y + 3 \le -6y - 9$

23. $-\frac{1}{2}x - \frac{1}{4} > \frac{1}{2} - \frac{1}{4}x$

24. $0.3y - 8 < 2.6y + 15$

25. $-2(x - 5) \ge 6(x + 7) - 12$

26. *Moving Costs.* Musclebound Movers charges $85 plus $40 an hour to move households across town. Champion Moving charges $60 an hour for cross-town moves. For what lengths of time is Champion more expensive? [1.4d]

27. *Investments.* You are going to invest $30,000, part at 13% and part at 15%, for one year. What is the most that can be invested at 13% in order to make at least $4300 interest in one year? [1.4d]

Graph and write interval notation. [1.5a, b]

28. $-2 \le x < 5$

29. $x \le -2$ or $x > 5$

Solve. [1.5a, b]

30. $2x - 5 < -7$ and $3x + 8 \ge 14$

31. $-4 < x + 3 \le 5$

32. $-15 < -4x - 5 < 0$

33. $3x < -9$ or $-5x < -5$

34. $2x + 5 < -17$ or $-4x + 10 \le 34$

35. $2x + 7 \le -5$ or $x + 7 \ge 15$

36. *Records in the Women's 100-m Dash.* Florence Griffith Joyner set a world record of 10.49 sec in the women's 100-m dash in 1988. The equation

$$R = -0.0433t + 10.49$$

can be used to predict the world record in the women's 100-m dash t years after 1988. For what years was the record between 10.15 and 10.35 sec? [1.5c]

Simplify. [1.6a]

37. $\left| -\dfrac{3}{x} \right|$

38. $\left| \dfrac{2x}{y^2} \right|$

39. $\left| \dfrac{12y}{-3y^2} \right|$

40. Find the distance between -23 and 39. [1.6b]

Solve. [1.6c, d]

41. $|x| = 6$

42. $|x - 2| = 7$

43. $|2x + 5| = |x - 9|$

44. $|5x + 6| = -8$

Solve. [1.6e]

45. $|2x + 5| < 12$

46. $|x| \ge 3.5$

47. $|3x - 4| \ge 15$

48. $|x| < 0$

49. Find the intersection: [1.5a]
$\{1, 2, 5, 6, 9\} \cap \{1, 3, 5, 9\}$.

50. Find the union: [1.5b]
$\{1, 2, 5, 6, 9\} \cup \{1, 3, 5, 9\}$.

Skill Maintenance

Calculate. [R.2a, c, d, e]

51. $-23 + 56$

52. $-\frac{2}{3} - \left(-\frac{5}{6}\right)$

53. $-\frac{2}{3} \div \left(-\frac{5}{6}\right)$

54. $-45(-52.2)$

55. Multiply: $10(2x - 3y + 7)$. [R.5d]

56. Factor: $40x - 8y + 16$. [R.5d]

57. Simplify: $-8 + 2(2x - y) - 6(2x + 5y) + 20$. [R.6b]

Synthesis

58. ◆ Find the error or errors in each of the following steps: [1.4c]

$$7 - 9x + 6x < -9(x + 2) + 10x$$
$$7 - 9x + 6x < -9x + 2 + 10x \quad (1)$$
$$7 + 6x > 2 + 10x \quad (2)$$
$$-4x > 8 \quad (3)$$
$$x > -2. \quad (4)$$

59. ◆ How does the word "solve" vary in meaning in this chapter?

60. Solve: $|2x + 5| \le |x + 3|$. [1.6d, e]

Test: Chapter 1

Solve.

1. $-12x = -8$

2. $x - \frac{3}{5} = \frac{2}{3}$

3. $0.7x - 0.1 = 2.1 - 0.3x$

4. $5(3x + 6) = 6 - (x + 8)$

5. $|x - 3| = 9$

6. Solve $A = \dfrac{B - C}{3}$ for B.

Solve.

7. *Room Dimensions.* A room has a perimeter of 48 ft. The width is two-thirds of the length. What are the dimensions of the room?

8. *IKON Copiers.* IKON Office Solutions rents a Canon GP30F copier for $240 per month plus 1.8¢ per copy (**Source**: IKON Office Solutions, Keith Palmer). A law firm needs to lease a copy machine for use during a special case that they anticipate will take 3 months. They allot a budget of $1500. How many copies can they make to stay within budget?

9. *Population Decrease.* The population of Oldcastle dropped 12% from one year to the next to a total of 158,400. What was the former population?

10. *Angles in a Triangle.* The degree measures of the angles of a triangle are three consecutive integers. Find the measures of the angles.

11. *Boating.* A paddleboat moves at a rate of 12 mph in still water. If the river's current moves at a rate of 3 mph, how long will it take the boat to travel 36 mi downstream? 36 mi upstream?

Write interval notation for the given set or graph.

12. $\{x \mid -3 < x \le 2\}$

13.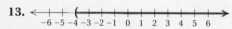

Solve and graph. Write interval notation for the solution set.

14. $x - 2 \le 4$

15. $-4y - 3 \ge 5$

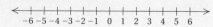

Solve.

16. $-0.6y < 30$

17. $3a - 5 \le -2a + 6$

18. $-5y - 1 > -9y + 3$

19. $4(5 - x) < 2x + 5$

20. $-8(2x + 3) + 6(4 - 5x) \ge 2(1 - 7x) - 4(4 + 6x)$

Answers

1. _____

2. _____

3. _____

4. _____

5. _____

6. _____

7. _____

8. _____

9. _____

10. _____

11. _____

12. _____

13. _____

14. _____

15. _____

16. _____

17. _____

18. _____

19. _____

20. _____

Answers

Solve.

21. _Moving Costs._ Motivated Movers charges $105 plus $30 an hour to move households across town. Quick-Pak Moving charges $80 an hour for cross-town moves. For what lengths of time is Quick-Pak more expensive?

22. _Pressure at Sea Depth._ The equation

$$P = 1 + \frac{d}{33}$$

gives the pressure P, in atmospheres (atm), at a depth of d feet in the sea. For what depths d is the pressure at least 2 atm and at most 8 atm?

Graph and write interval notation.

23. $-3 \le x \le 4$

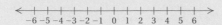

24. $x < -3 \ or \ x > 4$

Solve.

25. $5 - 2x \le 1 \ and \ 3x + 2 \ge 14$

26. $-3 < x - 2 < 4$

27. $-11 \le -5x - 2 < 0$

28. $-3x > 12 \ or \ 4x > -10$

29. $x - 7 \le -5 \ or \ x - 7 \ge -10$

30. $3x - 2 < 7 \ or \ x - 2 > 4$

Simplify.

31. $\left| \dfrac{7}{x} \right|$

32. $\left| \dfrac{-6x^2}{3x} \right|$

33. Find the distance between 4.8 and -3.6.

Solve.

34. $|x| = 9$

35. $|x| > 3$

36. $|4x - 1| < 4.5$

37. $|-5x - 3| \ge 10$

38. $|x + 10| = |x - 12|$

39. $|2 - 5x| = -10$

40. $\left| \dfrac{6 - x}{7} \right| \le 15$

41. Find the intersection:
$\{1, 3, 5, 7, 9\} \cap \{3, 5, 11, 13\}.$

42. Find the union:
$\{1, 3, 5, 7, 9\} \cup \{3, 5, 11, 13\}.$

Skill Maintenance

Calculate.

43. $\frac{3}{4} + \left(-\frac{5}{8} \right)$

44. $\frac{3}{4} - \left(-\frac{5}{8} \right)$

45. $-\frac{2}{3} \div \left(-\frac{5}{6} \right)$

46. $-45(-52.2)$

47. Multiply: $-8(2a - 3b)$.

48. Factor: $6a - 10b + 12$.

49. Simplify: $9 - 3(2x - y) - 6(2x + 5y) + 20$.

Synthesis

Solve.

50. $|3x - 4| \le -3$

51. $7x < 8 - 3x < 6 + 7x$

Graphs, Functions, and Applications

2

An Application	**The Mathematics**

The graph at right approximates the weekly U.S. revenue, in millions of dollars, from the recent movie *Air Force One*. The revenue is a function *f* of the number of weeks since the movie was released. No equation is given for the function. What was the movie revenue for week 5? In other words, we need to find $f(5)$.

This problem appears as Example 9 in Section 2.2.

We locate 5 on the horizontal axis and move directly up until we reach the graph. Then we move across to the vertical axis. We estimate that value to be about $13 million—that is, $f(5) = 13$.

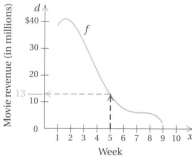

Source: Exhibitor Relations Co., Inc.

For more information, visit us at www.mathmax.com

Pretest: Chapter 2

Graph on a plane.

1. $2x - 5y = 20$

2. $x = 4$

3. $y = x - 2$

4. $f(x) = -2$

5. $f(x) = 3 - x^2$

6. Find the slope and the y-intercept: $y = 5x - 3$.

7. Find the slope, if it exists, of the line containing the points $(7, 4)$ and $(-5, 4)$.

8. Find an equation of the line containing the points $(-3, 7)$ and $(-8, 2)$.

9. Find an equation of the line having the given slope and containing the given point.
$$m = 2; \quad (-2, 3)$$

10. Find an equation of the line containing the given point and parallel to the given line.
$$(2, 5); \quad 2x - 7y = 10$$

11. Find an equation of the line containing the given point and perpendicular to the given line.
$$(2, 5); \quad 2x - 7y = 10$$

Determine whether the graphs of the pair of lines are parallel or perpendicular.

14. Find the intercepts of $2x - 3y = 12$.

12. $3y - 2x = 21$,
$3x + 2y = 8$

13. $y = 3x + 7$,
$y = 3x - 4$

15. For the function f given by $f(x) = |x| - 3$, find $f(0)$, $f(-2)$, and $f(4)$.

16. Find the domain:
$$f(x) = \frac{3}{2x - 5}.$$

17. Determine whether each of the following is the graph of a function.

a)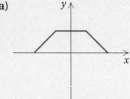

b)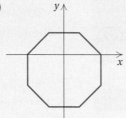

18. *Spending on Recorded Music.* Consider the following graph showing the average amount of spending per person per year on recorded music.

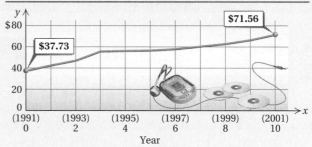

Average Amount Spent per Person
(Data for 1997–2001 are projections.)

Year

Source: Veronis, Suhler & Associates, Industry Sources

a) Use the two points $(0, \$37.73)$ and $(10, \$71.56)$ to find a linear function that fits the data.

b) Use the function to predict the average amount of spending on recorded music in 2011.

Objectives for Retesting

The objectives to be tested in addition to the material in this chapter are as follows.

[1.3a]   Solve applied problems by translating to equations.

[1.4c]   Solve an inequality using the addition and multiplication principles and then graph the inequality.

[1.5a, b]   Solve conjunctions and disjunctions of inequalities.

[1.6c, e]   Solve equations and inequalities with absolute-value expressions.

2.1 Graphs of Equations

Today's print and electronic media make extensive use of graphs. This is due in part to the ease with which some graphs can be prepared by computer and in part to the large quantity of information that a graph can display.

The following are examples of circle graphs, or pie charts, bar graphs, and line graphs.

Circle Graph

Circle graphs, or *pie charts,* are often used to show what percent of the whole each particular item in a group represents. The following circle graph allows us to see where large companies spend their advertising dollars.

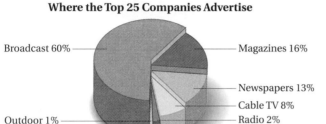

Where the Top 25 Companies Advertise

Broadcast 60%
Magazines 16%
Newspapers 13%
Cable TV 8%
Radio 2%
Outdoor 1%

Source: Interep Research

Bar Graph

Bar graphs are convenient for showing comparisons. In the following bar graph, each group is paired with the number of deaths caused by the disease(s) of that group.

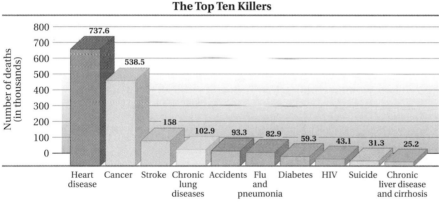

The Top Ten Killers

Number of deaths (in thousands)

Heart disease 737.6
Cancer 538.5
Stroke 158
Chronic lung diseases 102.9
Accidents 93.3
Flu and pneumonia 82.9
Diabetes 59.3
HIV 43.1
Suicide 31.3
Chronic liver disease and cirrhosis 25.2

Source: National Center for Health Statistics

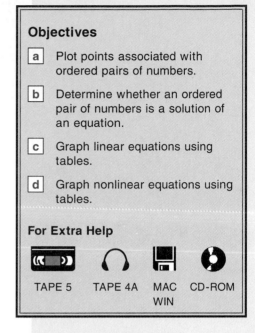

Line Graph

Line graphs are often used to show change over time. Certain points are first plotted to represent given information. When segments are drawn to connect the points, a line graph is formed. In the following line graph, *years* are shown on the horizontal axis. With each year, there is associated a number on the vertical axis representing the number of videotape rentals, in billions, in that year.

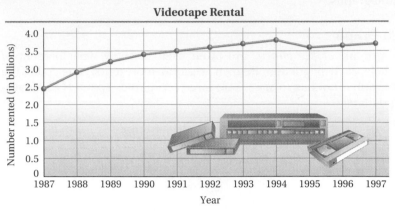

Videotape Rental

Source: Adams Media Research, Carmel Valley, California

a | Plotting Ordered Pairs

We have already learned to graph numbers and inequalities in one variable on a line. To enable us to graph an equation that contains two variables, we now learn to graph pairs of numbers on a plane.

On a number line, each point is the graph of a number. On a plane, each point is the graph of a number pair. (See the graph below.) We use two perpendicular number lines called **axes**. They cross at a point called the **origin**. The arrows show the positive directions on the axes. Consider the **ordered pair** (2, 3). The numbers in an ordered pair are called **coordinates**. In (2, 3), the **first coordinate** is 2 and the **second coordinate** is 3. (The first coordinate is sometimes called the **abscissa** and the second the **ordinate**.) To plot (2, 3), we start at the origin and move horizontally to the 2. Then we move up vertically 3 units and make a "dot."

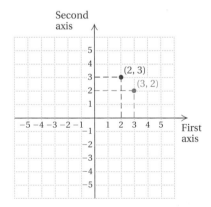

The point (3, 2) is also plotted in the figure. Note that (3, 2) and (2, 3) are different points. The order of the numbers in the pair is indeed important. They are called *ordered pairs* because it makes a difference which number comes first. The coordinates of the origin are (0, 0) even though it is usually labeled with the number 0, or not labeled at all. In general, the first axis is the *x*-axis and the second axis is the *y*-axis. We call this the **Cartesian coordinate system** in honor of the great French mathematician and philosopher René Descartes (1596–1650).

Example 1   Plot the points $(-4, 3)$, $(-5, -3)$, $(0, 4)$, and $(2.5, -2)$.

To plot $(-4, 3)$, we note that the first number, -4, tells us the distance in the first, or horizontal, direction. We move 4 units *left*. The second number tells us the distance in the second, or vertical, direction. We move 3 units *up*. The point $(-4, 3)$ is then marked, or plotted.

The points $(-5, -3)$, $(0, 4)$, and $(2.5, -2)$ are plotted in the same manner.

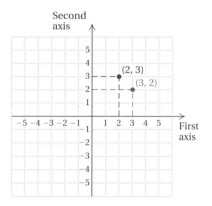

Do Exercises 1–10.

Quadrants

The axes divide the plane into four regions called **quadrants**, as shown here. In region I (the *first* quadrant), both coordinates of a point are positive. In region II (the *second* quadrant), the first coordinate is negative and the second coordinate is positive. In the *third* quadrant, both coordinates are negative, and in the *fourth* quadrant, the first coordinate is positive and the second is negative.

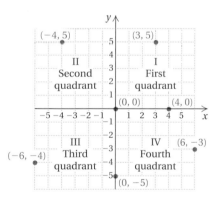

Points with one or more 0's as coordinates, such as $(0, -5)$, $(4, 0)$, and $(0, 0)$, are on axes and *not* in quadrants.

Do Exercises 11 and 12.

b | Solutions of Equations

If an equation has two variables, its solutions are pairs of numbers. When such a solution is written as an ordered pair, the first number listed in the pair generally replaces the variable that occurs first alphabetically.

Plot the points on the graph below.

1. $(6, 4)$ **2.** $(4, 6)$

3. $(-3, 5)$ **4.** $(5, -3)$

5. $(-4, -3)$ **6.** $(4, -2)$

7. $(0, 3)$ **8.** $(3, 0)$

9. $(0, -4)$ **10.** $(-4, 0)$

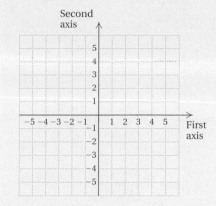

11. What can you say about the coordinates of a point in the third quadrant?

12. What can you say about the coordinates of a point in the fourth quadrant?

Answers on page A-9

13. Determine whether $(2, -4)$ is a solution of $5b - 3a = 34$.

Example 2 Determine whether each of the following pairs is a solution of $5b - 3a = 34$: $(2, 8)$ and $(-1, 6)$.

For $(2, 8)$, we substitute 2 for a and 8 for b (alphabetical order of variables):

$$\begin{array}{c|c} \multicolumn{2}{c}{5b - 3a = 34} \\ \hline 5 \cdot 8 - 3 \cdot 2 \; ? \; 34 \\ 40 - 6 \\ 34 & \text{TRUE} \end{array}$$

Thus, $(2, 8)$ is a solution of the equation.
For $(-1, 6)$, we substitute -1 for a and 6 for b:

$$\begin{array}{c|c} \multicolumn{2}{c}{5b - 3a = 34} \\ \hline 5 \cdot 6 - 3 \cdot (-1) \; ? \; 34 \\ 30 + 3 \\ 33 & \text{FALSE} \end{array}$$

Thus, $(-1, 6)$ is *not* a solution of the equation.

14. Determine whether $(2, -4)$ is a solution of $7p + 5q = -6$.

Do Exercises 13 and 14.

Example 3 Show that the pairs $(-4, 3)$, $(0, 1)$, and $(4, -1)$ are solutions of $y = 1 - \frac{1}{2}x$. Then graph the line containing the three points and use the graph to help determine another pair that is a solution.

We replace x with the first coordinate and y with the second coordinate of each pair:

$$\begin{array}{c|c} \multicolumn{2}{c}{y = 1 - \frac{1}{2}x} \\ \hline 3 \; ? \; 1 - \frac{1}{2} \cdot (-4) \\ 1 + 2 \\ 3 & \text{TRUE} \end{array} \qquad \begin{array}{c|c} \multicolumn{2}{c}{y = 1 - \frac{1}{2}x} \\ \hline 1 \; ? \; 1 - \frac{1}{2} \cdot 0 \\ 1 - 0 \\ 1 & \text{TRUE} \end{array} \qquad \begin{array}{c|c} \multicolumn{2}{c}{y = 1 - \frac{1}{2}x} \\ \hline -1 \; ? \; 1 - \frac{1}{2} \cdot 4 \\ 1 - 2 \\ -1 & \text{TRUE} \end{array}$$

In each of the three cases, the substitution results in a true equation. Thus all the pairs are solutions.

15. Use the graph in Example 3 to find at least two more points that are solutions.

We plot the points as shown at right. Note that the three points appear to "line up." That is, they appear to be on a straight line. We use a ruler and draw a line passing through $(-4, 3)$, $(0, 1)$, and $(4, -1)$.

The line appears to pass through $(2, 0)$ as well. Let's see if this pair is a solution of $y = 1 - \frac{1}{2}x$:

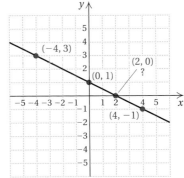

$$\begin{array}{c|c} \multicolumn{2}{c}{y = 1 - \frac{1}{2}x} \\ \hline 0 \; ? \; 1 - \frac{1}{2} \cdot 2 \\ 1 - 1 \\ 0 & \text{TRUE} \end{array}$$

We see that $(2, 0)$ is a solution.

Do Exercise 15.

Answers on page A-10

Example 3 leads us to believe that any point on the line that passes through $(-4, 3)$, $(0, 1)$, and $(4, -1)$ represents a solution of $y = 1 - \frac{1}{2}x$. In fact, every solution of $y = 1 - \frac{1}{2}x$ is represented by a point on that line and every point on that line represents a solution. The line is said to be the *graph* of the equation.

> The **graph** of an equation is a drawing that represents all its solutions.

c | Graphs of Linear Equations

Equations like $5b - 3a = 34$ and $y = 1 - \frac{1}{2}x$ are said to be **linear** because the graph of their solutions is a line. In general, any equation equivalent to one of the form $y = mx + b$ or $Ax + By = C$, where m, b, A, B, and C are constants (that is, they stay the same) and A and B are not both 0, is linear.

To graph a linear equation:

1. Select a value for one variable and calculate the corresponding value of the other variable. Form an ordered pair using alphabetical order as indicated by the variables.

2. Repeat step (1) to obtain at least two other ordered pairs. Two ordered pairs are essential. A third serves as a check.

3. Plot the ordered pairs and draw a straight line passing through the points.

Example 4   Graph: $y = 2x$.

We find some ordered pairs that are solutions. This time we list the pairs in a table. To find an ordered pair, we can choose *any* number for x and then determine y. For example, if we choose 3 for x, then $y = 2 \cdot 3 = 6$ (substituting into the equation $y = 2x$). We choose some negative values for x, as well as some positive ones. If a number takes us off the graph paper, we generally do not use it. Next, we plot these points. If we plotted *many* such points, they would appear to make a solid line. We draw the line with a ruler and label it $y = 2x$.

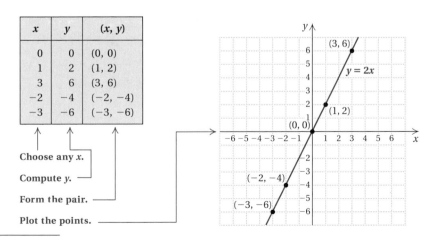

x	y	(x, y)
0	0	$(0, 0)$
1	2	$(1, 2)$
3	6	$(3, 6)$
-2	-4	$(-2, -4)$
-3	-6	$(-3, -6)$

Choose any x.

Compute y.

Form the pair.

Plot the points.

Do Exercises 16 and 17.

Graph.

16. $y = -2x$

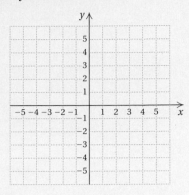

17. $y = \frac{1}{2}x$

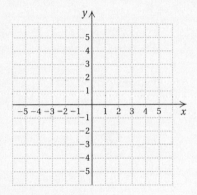

Answers on page A-10

Graph.

18. $y = 2x + 3$

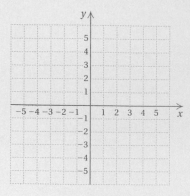

19. $y = -\dfrac{1}{2}x - 3$

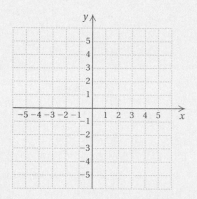

Answers on page A-10

Example 5 Graph: $y = -\frac{1}{2}x + 3$.

By choosing even integers for x, we can avoid fractional values when calculating y. For example, if we choose 4 for x, we get

$$y = -\tfrac{1}{2}x + 3 = -\tfrac{1}{2}(4) + 3 = -2 + 3 = 1.$$

When x is -6, we get

$$y = -\tfrac{1}{2}x + 3 = -\tfrac{1}{2}(-6) + 3 = 3 + 3 = 6$$

and when x is 0, we get

$$y = -\tfrac{1}{2}x + 3 = -\tfrac{1}{2}(0) + 3 = 0 + 3 = 3.$$

Results are often listed in a table, as shown below. The points corresponding to each pair are then plotted.

x	y	(x, y)
4	1	(4, 1)
-6	6	(-6, 6)
0	3	(0, 3)

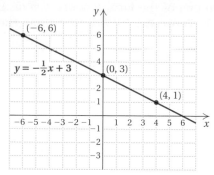

Note that the three points line up. If they did not, we would know that we had made a mistake. When only two points are plotted, an error is harder to detect. We use a ruler or other straightedge to draw a line through the points and then label the graph. Every point on the line represents a solution of $y = -\frac{1}{2}x + 3$.

Do Exercises 18 and 19.

Calculating ordered pairs is usually easiest when y is isolated on one side of the equation, as in $y = 2x$ and $y = -\frac{1}{2}x + 3$. To graph an equation in which y is not isolated, we can use the addition and multiplication principles (see Sections 1.1 and 1.2) to first solve for y.

Example 6 Graph: $3x + 5y = 10$.

We first solve for y:

$$3x + 5y = 10$$
$$3x + 5y - 3x = 10 - 3x \qquad \text{Subtracting } 3x$$
$$5y = 10 - 3x \qquad \text{Simplifying}$$
$$\tfrac{1}{5} \cdot 5y = \tfrac{1}{5} \cdot (10 - 3x) \qquad \text{Multiplying by } \tfrac{1}{5}, \text{ or dividing by 5}$$
$$y = \tfrac{1}{5} \cdot (10) - \tfrac{1}{5} \cdot (3x) \qquad \text{Using the distributive law}$$
$$= 2 - \tfrac{3}{5}x$$
$$= -\tfrac{3}{5}x + 2.$$

Thus the equation $3x + 5y = 10$ is equivalent to $y = -\frac{3}{5}x + 2$. We now find three ordered pairs, using multiples of 5 for x to avoid fractions.

x	y	(x, y)
0	2	(0, 2)
5	−1	(5, −1)
−5	5	(−5, 5)

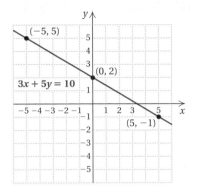

We plot the points, draw the line, and label the graph as shown.

Do Exercises 20 and 21.

d Graphing Nonlinear Equations

There are many equations whose graphs are not straight lines. Let's graph some of these **nonlinear equations.**

Example 7 Graph: $y = x^2 - 5$.

We select numbers for x and find the corresponding values for y. For example, if we choose -2 for x, we get $y = (-2)^2 - 5 = 4 - 5 = -1$. The table lists several ordered pairs.

x	y
0	−5
−1	−4
1	−4
−2	−1
2	−1
−3	4
3	4

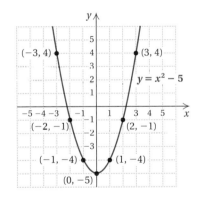

Next, we plot the points. The more points plotted, the clearer the shape of the graph becomes. Since the value of $x^2 - 5$ grows rapidly as x moves away from the origin, the graph rises steeply on either side of the y-axis.

Do Exercise 22.

Graph.

20. $4y - 3x = -8$

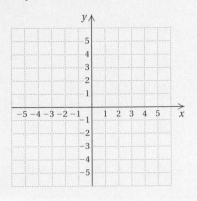

21. $5x + 2y = 4$

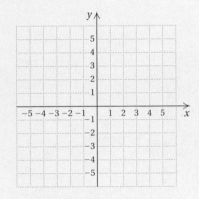

22. Graph: $y = 4 - x^2$.

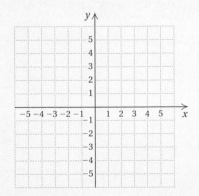

Answers on page A-10

23. Graph: $y = \dfrac{2}{x}$.

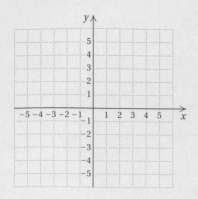

24. Graph: $y = 4 - |x|$.

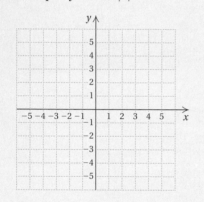

Example 8 Graph: $y = 1/x$.

We select x-values and find the corresponding y-values. The table lists the ordered pairs $\left(3, \frac{1}{3}\right)$, $\left(2, \frac{1}{2}\right)$, $(1, 1)$, and so on.

x	y
3	$\frac{1}{3}$
2	$\frac{1}{2}$
1	1
$\frac{1}{2}$	2
$-\frac{1}{2}$	-2
-1	-1
-2	$-\frac{1}{2}$
-3	$-\frac{1}{3}$

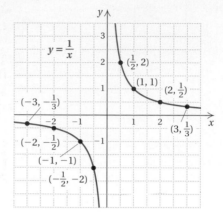

We plot these points, noting that each first coordinate is paired with its reciprocal. Since $1/0$ is undefined, we cannot use 0 as a first coordinate. Thus there are two "branches" to this graph—one on each side of the y-axis. Note that for x-values far to the right or far to the left of 0, the graph approaches, but does not touch, the x-axis; and for x-values close to 0, the graph approaches, but does not touch, the y-axis.

Do Exercise 23.

Example 9 Graph: $y = |x|$.

We select numbers for x and find the corresponding values for y. For example, if we choose -1 for x, we get $y = |-1| = 1$. Several ordered pairs are listed in the table below.

x	y
-3	3
-2	2
-1	1
0	0
1	1
2	2
3	3

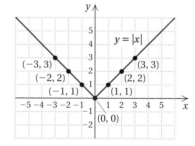

We plot these points, noting that the absolute value of a positive number is the same as the absolute value of its opposite. Thus the x-values 3 and -3 both are paired with the y-value 3. Note that the graph is V-shaped and centered at the origin.

Do Exercise 24.

With equations like $y = -\frac{1}{2}x + 3$, $y = x^2 - 5$, and $y = |x|$, which we have graphed in this section, it is understood that y is the **dependent variable** and x is the **independent variable,** since y is calculated after first choosing x and y is expressed in terms of x.

Answers on page A-10

Calculator Spotlight

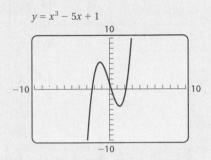

Introduction to the Use of a Graphing Calculator: Windows and Graphs

Viewing Windows. In this chapter, we begin to create graphs using a graphing calculator and computer graphing software, referred to simply as **graphers**. Most of the coverage will refer to a TI-83 graphing calculator (and with adaptation to a TI-82) but in a somewhat generic manner, discussing features common to virtually all graphers. Although some reference to keystrokes will be mentioned, in general, exact details on keystrokes will be covered in the manual for your particular grapher.

One feature common to all graphers is the **viewing window.** This refers to the rectangular screen in which a graph appears. Windows are described by four numbers, [**L**, **R**, **B**, **T**], which represent the **L**eft and **R**ight endpoints of the *x*-axis and the **B**ottom and **T**op endpoints of the *y*-axis. A WINDOW feature can be used to set these dimensions. Below is a window setting of $[-20, 20, -5, 5]$ with axis scaling denoted as Xscl = 5 and Yscl = 1, which means that there are 5 units between tick marks on the *x*-axis and 1 unit between tick marks on the *y*-axis. Graphs are made up of black rectangular dots called **pixels.** Roughly speaking, the notation Xres = 1 is an indicator of the number of pixels used in making a graph.* We will usually leave it at 1 and not refer to it unless needed.

Axis scaling must be chosen with care, because tick marks become blurred and indistinguishable when too many appear. On some graphers, a setting of $[-10, 10, -10, 10]$, Xscl = 1, Yscl = 1, Xres = 1 is considered **standard**. On a TI-83 grapher, there is a ZStandard feature for automatic standard window.

The primary use for a grapher is to graph equations. For example, let's graph the equation $y = x^3 - 5x + 1$. The equation can be entered using the notation $y = x \wedge 3 - 5x + 1$. Some software uses Basic notation, in which case the equation might be entered as $y = x \wedge 3 - 5 * x + 1$. We obtain the following graph in the standard viewing window.

$$y = x^3 - 5x + 1$$

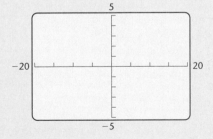

In general, choosing a window that best reveals a graph's characteristics involves some trial and error and, in some cases, some knowledge about the shape of that graph. We will consider this in more detail in Section 4.1.

```
WINDOW
Xmin = -20
Xmax = 20
Xscl = 5
Ymin = -5
Ymax = 5
Yscl = 1
Xres = 1
```

---

*Xres sets pixel resolution at 1 through 8 for graphs of equations. At Xres = 1, equations are evaluated and graphed at each pixel on the *x*-axis. At Xres = 8, equations are evaluated and graphed at every eighth pixel on the *x*-axis. The resolution is better for smaller Xres values than for larger values.

(continued)

To graph an equation like $3x + 5y = 10$, most graphers require that the equation be solved for y, that is, "$y = \ldots$." Thus we must rewrite and enter the equation as

$$y = \frac{-3x + 10}{5}, \quad \text{or} \quad y = -\frac{3}{5}x + 2.$$

(See Example 6.) Its graph is shown below in the standard window.

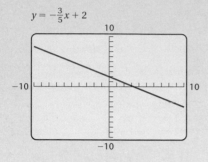

$y = -\frac{3}{5}x + 2$

Exercises

Use a grapher to graph each of the following equations. Select the standard window $[-10, 10, -10, 10]$ and axis scaling $\text{Xscl} = 1$, $\text{Yscl} = 1$.

1. $y = 2x - 2$ **2.** $y = -3x + 1$

3. $y = \frac{2}{5}x + 4$ **4.** $y = -\frac{3}{5}x - 1$

5. $y = 2.085x + 15.08$ **6.** $y = -\frac{4}{5}x + \frac{13}{7}$

7. $2x - 3y = 18$ **8.** $5y + 3x = 4$

9. $y = x^2$ **10.** $y = (x + 4)^2$

11. $y = 8 - x^2$ **12.** $y = 4 - 3x - x^2$

13. $y + 10 = 5x^2 - 3x$ **14.** $y - 2 = x^3$

15. $y = x^3 - 7x - 2$ **16.** $y = x^4 - 3x^2 + x$

17. $y = |x|$ (On most graphers, this is entered as $y - \text{abs}(x)$.)

18. $y = |x - 5|$ **19.** $y = |x| - 5$

20. $y = 9 - |x|$

Exercise Set 2.1

a Plot the following points.

1. $A(4, 1)$, $B(2, 5)$, $C(0, 3)$, $D(0, -5)$, $E(6, 0)$, $F(-3, 0)$, $G(-2, -4)$, $H(-5, 1)$, $J(-6, 6)$

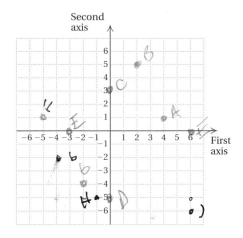

2. $A(-3, -5)$, $B(1, 3)$, $C(0, 7)$, $D(0, -2)$, $E(5, 0)$, $F(-4, 0)$, $G(1, -7)$, $H(-6, 4)$, $J(-3, 3)$

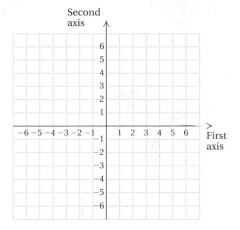

3. Plot the points $M(2, 3)$, $N(5, -3)$, and $P(-2, -3)$. Draw $\overline{MN}$, $\overline{NP}$, and $\overline{MP}$. ($\overline{MN}$ means the line segment from M to N.) What kind of geometric figure is formed? What is its area?

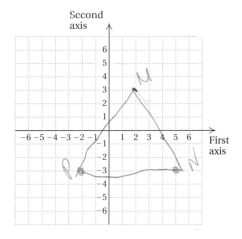

4. Plot the points $Q(-4, 3)$, $R(5, 3)$, $S(2, -1)$, and $T(-7, -1)$. Draw $\overline{QR}$, $\overline{RS}$, $\overline{ST}$, and $\overline{TQ}$. What kind of figure is formed? What is its area?

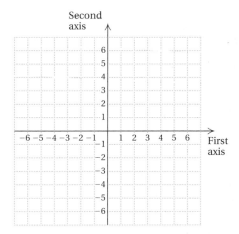

b Determine whether the given point is a solution of the equation.

5. $(1, -1)$; $y = 2x - 3$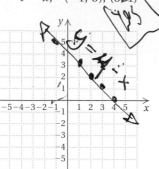

6. $(3, 4)$; $3s + t = 4$

7. $(3, 5)$; $4x - y = 7$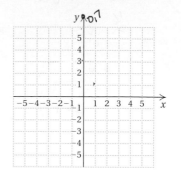

8. $(2, -1)$; $4r + 3s = 5$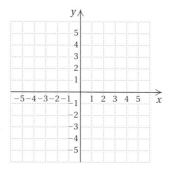

9. $\left(0, \dfrac{3}{5}\right)$; $2a + 5b = 7$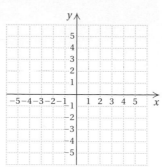

10. $(-5, 1)$; $2p - 3q = -13$

In Exercises 11–16, an equation and two ordered pairs are given. Show that each pair is a solution of the equation. Then graph the two points and use the line containing the graph to determine another solution. Answers may vary.

11. $y = 4 - x$; $(-1, 5), (3, 1)$

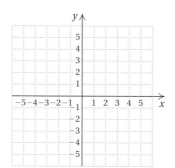

12. $y = x - 3$; $(5, 2), (-1, -4)$

13. $3x + y = 7$; $(2, 1), (4, -5)$

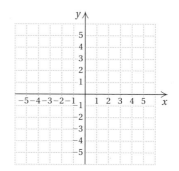

14. $y = \dfrac{1}{2}x + 3$; $(4, 5), (-2, 2)$

15. $6x - 3y = 3$; $(1, 1), (-1, -3)$

16. $4x - 2y = 10$; $(0, -5), (4, 3)$

c Graph.

17. $y = x - 1$

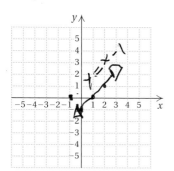

18. $y = x + 1$

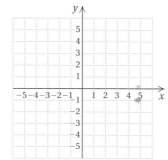

19. $y = x$

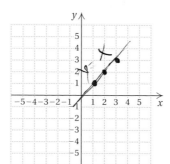

20. $y = -3x$

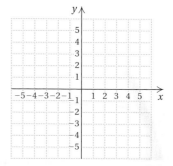

21. $y = \dfrac{1}{4}x$

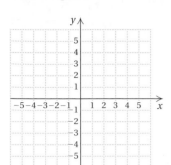

22. $y = \dfrac{1}{3}x$

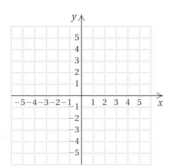

23. $y = 3 - x$

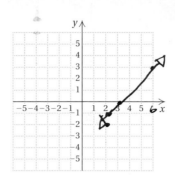

24. $y = x + 3$

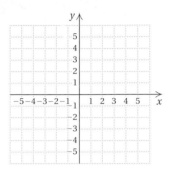

25. $y = 5x - 2$

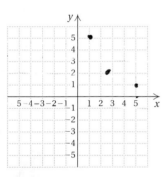

26. $y = \dfrac{1}{4}x + 2$

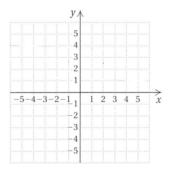

27. $y = \dfrac{1}{2}x + 1$

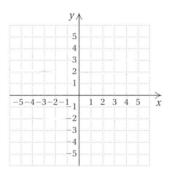

28. $y = \dfrac{1}{3}x - 4$

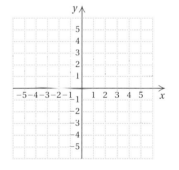

29. $x + y = 5$

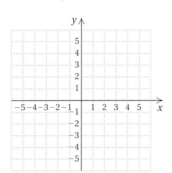

30. $x + y = -4$

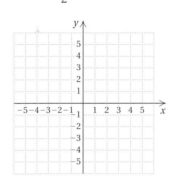

31. $y = -\dfrac{5}{3}x - 2$

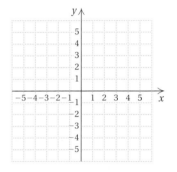

32. $y = -\dfrac{5}{2}x + 3$

33. $x + 2y = 8$

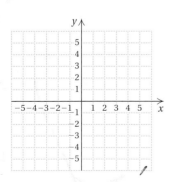

34. $x + 2y = -6$

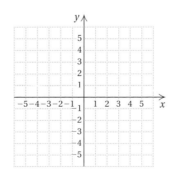

35. $y = \dfrac{3}{2}x + 1$

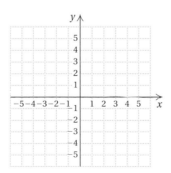

36. $y = -\dfrac{1}{2}x - 3$

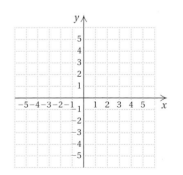

37. $8y + 2x = 4$

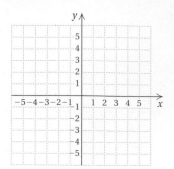

38. $6x - 3y = -9$

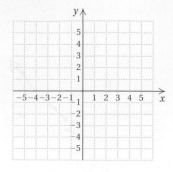

39. $8y + 2x = -4$

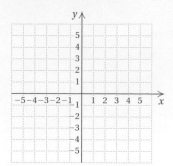

40. $6y + 2x = 8$

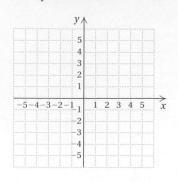

41. *FedEx Mailing Costs.* The cost y, in dollars, of mailing a FedEx Priority Overnight package weighing 1 lb or more is given by the equation

$$y = 2.085x + 15.08,$$

where x is the number of pounds (***Source:*** Federal Express Corporation).

a) Find the cost of mailing packages weighing 5 lb, 7 lb, and 10 lb.

b) Graph the equation and then use the graph to estimate the cost of mailing a $6\frac{1}{2}$-lb package.

c) If a package costs $177.71 to mail, how much does it weigh?

42. *Value of an Office Machine.* The value of Dupliographic's color copier is given by the equation

$$v = -0.68t + 3.4,$$

where v is the value, in thousands of dollars, t years from the date of purchase.

a) Find the value after 1 yr, 2 yr, 4 yr, and 5 yr.

b) Graph the equation and then use the graph to estimate the value of the copier after $2\frac{1}{2}$ yr.

c) After what amount of time is the value of the copier $1500?

d Graph.

43. $y = x^2$

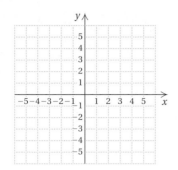

44. $y = -x^2$ (*Hint:* $-x^2 = -1 \cdot x^2$.)

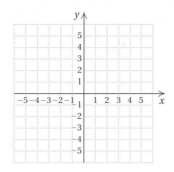

45. $y = x^2 + 2$

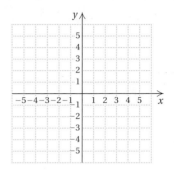

46. $y = 3 - x^2$

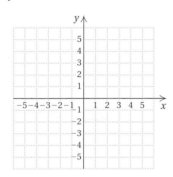

47. $y = x^2 - 3$

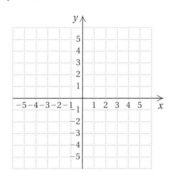

48. $y = x^2 - 3x$

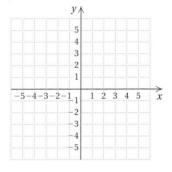

49. $y = -\dfrac{1}{x}$

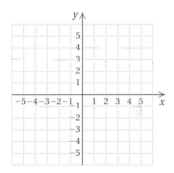

50. $y - \dfrac{3}{x}$

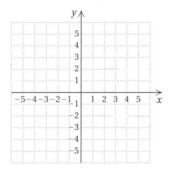

51. $y = |x - 2|$

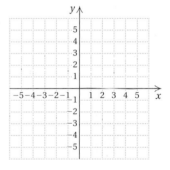

52. $y = |x| + 2$

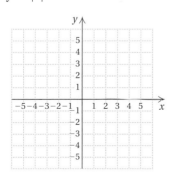

53. $y = x^3$

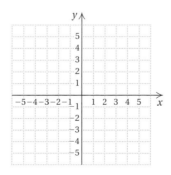

54. $y = x^3 - 2$

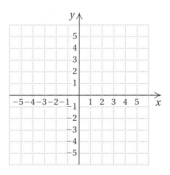

Solve. [1.5a, b]

55. $-3 < 2x - 5 \leq 10$

56. $2x - 5 \geq -10$ *or*
$-4x - 2 < 10$

57. $3x - 5 \leq -12$ *or*
$3x - 5 \geq 12$

58. $-13 < 3x + 5 < 23$

Solve. [1.3a]

59. *Landscaping.* Grass seed is being spread on a triangular traffic island. If the grass seed can cover an area of 200 ft² and the island's base is 16 ft long, how tall a triangle can the seed fill?

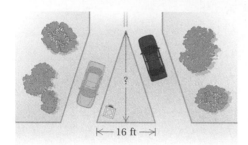

← 16 ft →

60. Three numbers are such that the second is 6 less than three times the first and the third is 2 more than two-thirds of the first. The sum of the three numbers is 150. Find the smallest of the three numbers.

61. *Taxi Fare.* The fare for a taxi ride from Johnson Street to Elm Street is $5.20. The driver charges $1.00 for the first $\frac{1}{2}$ mi and 30¢ for each additional $\frac{1}{4}$ mi. How far is it from Johnson Street to Elm Street?

62. *Real Estate Commission.* The following is a typical real estate commission on the selling price of a house:

　　7% for the first $100,000,   and

　　4% for the amount that exceeds $100,000.

A realtor receives a commission of $16,200 for selling a house. What was the selling price?

Synthesis

63. ◈ Using the equation $y = |x|$, explain why it is "risky" to draw a graph after plotting only two points.

64. ◈ Without making a drawing, how can you tell that the graph of $y = x - 30$ passes through three quadrants?

Use a grapher to graph each of the equations in Exercises 65–68. Use a standard viewing window of $[-10, 10, -10, 10]$, Xscl = 1, and Yscl = 1.

65. $y = x^3 - 3x + 2$

66. $y = x - |x|$

67. $y = \dfrac{1}{x - 2}$

68. $y = \dfrac{1}{x^2}$

In Exercises 69–72, try to find an equation for the given graph.

69.

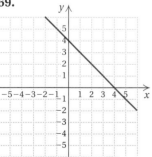

70.

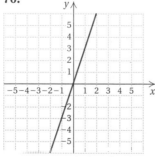

71.

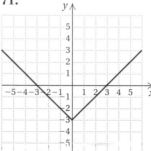

72.

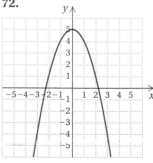

2.2 Functions and Graphs

a | Identifying Functions

We now develop one of the most important concepts in mathematics, **functions**. We have actually been studying functions all through this text; we just haven't identified them as such. In much the same way that ordered pairs form correspondences between first and second coordinates, a *function* is a correspondence from one set to another. For example:

> To each student in a college, there corresponds his or her student ID.
>
> To each item in a store, there corresponds its price.
>
> To each real number, there corresponds the cube of that number.

In each case, the first set is called the **domain** and the second set is called the **range**. This kind of correspondence is called a **function**. Given a member of the domain, there is *just one* member of the range to which it corresponds.

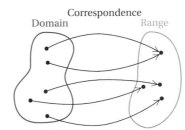

Correspondence

Domain   Range

Objectives

a Determine whether a correspondence is a function.

b Given a function described by an equation, find function values (outputs) for specified values (inputs).

c Draw the graph of a function.

d Determine whether a graph is that of a function using the vertical-line test.

e Solve applied problems involving functions and their graphs.

For Extra Help

TAPE 5   TAPE 4B   MAC WIN   CD-ROM

Example 1   Determine whether the correspondence is a function.

f:

Domain	Range
1 ⟶	$107.4
2 ⟶	$ 34.1
3 ⟶	$ 29.6
4 ⟶	$ 19.6

g:

Domain	Range
3 ⟶	5
4 ⟶	9
5 ⟶	−7
6 ⟶	

h:

Domain	Range
Chicago ⟶	Cubs / White Sox
Baltimore ⟶	Orioles
San Diego ⟶	Padres

p:

Domain	Range
Cubs ⟶	Chicago
White Sox ⟶	Chicago
Orioles ⟶	Baltimore
Padres ⟶	San Diego

The correspondence f is a function because each member of the domain is matched to *only one* member of the range.

The correspondence g is also a function because each member of the domain is matched to *only one* member of the range.

The correspondence h *is not* a function because one member of the domain, Chicago, is matched to *more than one* member of the range.

The correspondence p is a function because each member of the domain is matched to *only one* member of the range.

> A **function** is a correspondence between a first set, called the **domain**, and a second set, called the **range**, such that each member of the domain corresponds to *exactly one* member of the range.

Do Exercises 1–4.

Determine whether the correspondence is a function.

1.

Domain	Range
Cheetah ⟶	70 mph
Human ⟶	28 mph
Lion ⟶	50 mph
Chicken ⟶	9 mph

2.

Domain	Range
A ⟶	a
B	b
C	c
D	d
	e

3.

Domain	Range
−2	4
2	
−3	9
3	
0 ⟶	0

4.

Domain	Range
4	−2
	2
9	−3
	3
0 ⟶	0

Answers on page A-12

Determine whether each of the following is a function.

5. *Domain*
A set of numbers

Correspondence
Square each number and subtract 10.

Range
A set of numbers

6. *Domain*
A set of polygons

Correspondence
Find the area of each polygon.

Range
A set of numbers

Answers on page A-12

Example 2   Determine whether the correspondence is a function.

Domain	*Correspondence*	*Range*
a) A family	Each person's weight	A set of positive numbers
b) The integers	Each number's square	A set of nonnegative integers
c) The set of all states	Each state's members of the U.S. Senate	A set of U.S. Senators

a) The correspondence *is* a function because each person has *only one* weight.

b) The correspondence *is* a function because each integer has *only one* square.

c) The correspondence *is not* a function because each state has two U.S. Senators.

Do Exercises 5 and 6.

When a correspondence between two sets is not a function, it is still an example of a **relation**.

> A **relation** is a correspondence between a first set, called the **domain**, and a second set, called the **range**, such that each member of the domain corresponds to *at least one* member of the range.

Thus, although the correspondences of Examples 1 and 2 are not all functions, they *are* all relations. A function is a special type of relation— one in which each member of the domain is paired with *exactly one* member of the range.

b   Finding Function Values

Most functions considered in mathematics are described by equations like $y = 2x + 3$ or $y = 4 - x^2$. We graph the function $y = 2x + 3$ by first performing calculations like the following:

for $x = 4$, $y = 2x + 3 = 2 \cdot 4 + 3 = 11$;

for $x = -5$, $y = 2x + 3 = 2 \cdot (-5) + 3 = -7$;

for $x = 0$, $y = 2x + 3 = 2 \cdot 0 + 3 = 3$; and so on.

For $y = 2x + 3$, the **inputs** (members of the domain) are values of x substituted into the equation. The **outputs** (members of the range) are the resulting values of y. If we call the function f, we can use x to represent an arbitrary *input* and $f(x)$—read "f of x," or "f at x," or "the value of f at x"— to represent the corresponding *output*. In this notation, the function given by $y = 2x + 3$ is written as $f(x) = 2x + 3$ and the calculations above can be written more concisely as follows:

$f(4) = 2 \cdot 4 + 3 = 11$;

$f(-5) = 2 \cdot (-5) + 3 = -7$;

$f(0) = 2 \cdot 0 + 3 = 3$; and so on.

Thus instead of writing "when $x = 4$, the value of y is 11," we can simply write "$f(4) = 11$," which can also be read as "f of 4 is 11" or "for the input 4, the output of f is 11."

It helps to think of a function as a machine. Think of $f(4) = 11$ as putting a member of the domain (an input), 4, into the machine. The machine knows the correspondence $f(x) = 2x + 3$, multiplies 4 by 2 and adds 3, and gives out a member of the range (the output), 11.

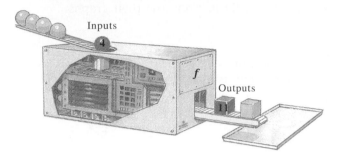

Inputs

4

f

Outputs

11

CAUTION! The notation $f(x)$ *does not mean "f times x"* and should not be read that way.

Example 3 A function f is given by $f(x) = 3x^2 - 2x + 8$. Find each of the indicated function values.

a) $f(0)$ b) $f(1)$

c) $f(-5)$ d) $f(7a)$

One way to find function values when a formula is given is to think of the formula with blanks, or placeholders, as follows:

$$f(\ \ \ \ ) = 3\ \ \ \ ^2 - 2\ \ \ \ + 8.$$

To find an output for a given input, we think: "Whatever goes in the blank on the left goes in the blank(s) on the right." With this in mind, let's complete the example.

a) $f(0) = 3 \cdot 0^2 - 2 \cdot 0 + 8 = 8$

b) $f(1) = 3 \cdot 1^2 - 2 \cdot 1 + 8 = 3 \cdot 1 - 2 + 8 = 3 - 2 + 8 = 9$

c) $f(-5) = 3(-5)^2 - 2 \cdot (-5) + 8 = 3 \cdot 25 + 10 + 8 = 75 + 10 + 8 = 93$

d) $f(7a) = 3(7a)^2 - 2(7a) + 8 = 3 \cdot 49a^2 - 14a + 8 = 147a^2 - 14a + 8$

Do Exercise 7.

Example 4 Find the indicated function value.

a) $f(5)$, for $f(x) = 3x + 2$ b) $g(-2)$, for $g(r) = 5r^2 + 3r$

c) $h(4)$, for $h(x) = 7$ d) $F(a + 1)$, for $F(x) = 5x - 8$

a) $f(5) = 3 \cdot 5 + 2 = 15 + 2 = 17$

b) $g(-2) = 5(-2)^2 + 3(-2) = 5(4) - 6 = 20 - 6 = 14$

c) For the function given by $h(x) = 7$, all inputs share the same output, 7. Thus, $h(4) = 7$. The function h is an example of a **constant function.**

d) $F(a + 1) = 5(a + 1) - 8 = 5a + 5 - 8 = 5a - 3$

Do Exercise 8.

7. Find the indicated function values for the following function:
$$f(x) = 2x^2 + 3x - 4.$$
a) $f(0)$
b) $f(8)$
c) $f(-5)$
d) $f(7a)$

8. Find the indicated function value.
a) $f(-6)$, for $f(x) = 5x - 3$
b) $g(-1)$, for $g(r) = 5r^2 + 3r$
c) $h(55)$, for $h(x) = 7$
d) $F(a + 2)$, for $F(x) = 5x - 8$

Answers on page A-12

Graph.

9. $f(x) = x - 4$

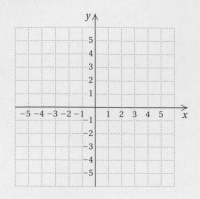

10. $g(x) = 5 - x^2$

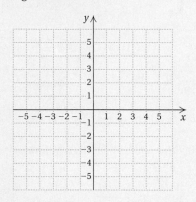

11. $t(x) = 3 - |x|$

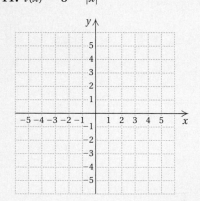

Answers on page A-12

| c | **Graphs of Functions** |

To graph a function, we find ordered pairs (x, y) or $(x, f(x))$, plot them, and connect the points. Note that y and $f(x)$ are used interchangeably—that is, $y = f(x)$—when working with functions and their graphs.

Example 5   Graph: $f(x) = x + 2$.

A list of some function values is shown in this table. We plot the points and connect them. The graph is a straight line.

x	$f(x)$
-4	-2
-3	-1
-2	0
-1	1
0	2
1	3
2	4
3	5
4	6

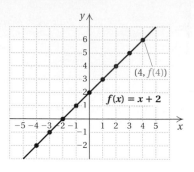

Example 6   Graph: $g(x) = 4 - x^2$.

We calculate some function values and draw the curve.

$$g(0) = 4 - 0^2 = 4 - 0 = 4,$$
$$g(-1) = 4 - (-1)^2 = 4 - 1 = 3,$$
$$g(2) = 4 - 2^2 = 4 - 4 = 0,$$
$$g(-3) = 4 - (-3)^2 = 4 - 9 = -5$$

x	$g(x)$
-3	-5
-2	0
-1	3
0	4
1	3
2	0
3	-5

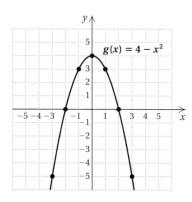

Example 7   Graph: $h(x) = |x|$.

A list of some function values is shown in the following table. We plot the points and connect them. The graph is a V-shaped "curve" that rises on either side of the vertical axis.

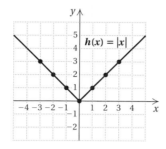

x	$h(x)$
-3	3
-2	2
-1	1
0	0
1	1
2	2
3	3

Do Exercises 9–11 on the preceding page.

d | The Vertical-Line Test

Consider the graph of the function f described by $f(x) = x^2 - 5$. Its graph is shown at right. It is also the graph of the equation $y = x^2 - 5$.

To find a function value, like $f(3)$, from a graph, we locate the input on the horizontal axis, move directly up or down to the graph of the function, and then move left or right to find the output on the vertical axis. Thus, $f(3) = 4$. Keep in mind that members of the domain are found on the horizontal axis and members of the range are found on the vertical axis.

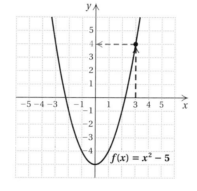

When one member of the domain is paired with two or more different members of the range, the correspondence is not a function. Thus, when a graph contains two or more different points with the same first coordinate, the graph cannot represent a function. Points sharing a common first coordinate are vertically above or below each other (see the following graph).

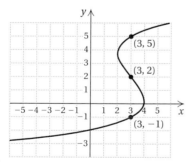

Since 3 is paired with more than one member of the range, the graph does not represent a function.

This observation leads to the *vertical-line test*.

> ▶ **THE VERTICAL-LINE TEST**
>
> A graph represents a function if it is impossible to draw a vertical line that intersects the graph more than once.

Calculator Spotlight

〰 Graphs and Function Values.   To graph the function $f(x) = 2x^2 + x$, we enter it into the grapher as $y_1 = 2x^2 + x$ and use the GRAPH feature.

To find function values, we can use the CALC menu and choose Value. To evaluate the function $f(x) = 2x^2 + x$ at $x = -2$, that is, to find $f(-2)$, we press ⎡2nd⎤ ⎡CALC⎤ ⎡1⎤. We enter $x = -2$. The function value, or y-value, $y = -6$ appears together with a trace indicator showing the point $(-2, 6)$.

Functions can also be evaluated using the TABLE feature or the Y-VARS feature. Consult the manual for your particular grapher for specific keystrokes.

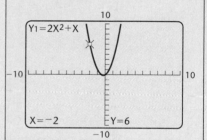

Exercises

1. Graph $f(x) = x^2 + 3x - 4$.
   Then find $f(-5)$, $f(-4.7)$, $f(11)$, and $f(2/3)$.
2. Graph $f(x) = 3.7 - x^2$.
   Then find $f(-5)$, $f(-4.7)$, $f(11)$, and $f(2/3)$.
3. Graph
   $f(x) = 4 - 1.2x - 3.4x^2$.
   Then find $f(-5)$, $f(-4.7)$, $f(11)$, and $f(2/3)$.

Determine whether each of the following is a graph of a function.

12.

13.

14.

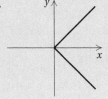

15.

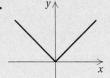

Answers on page A-12

Example 8 Determine whether each of the following is the graph of a function.

a)

b)

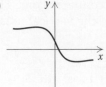

c)

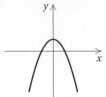

d)

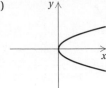

a) The graph *is not* that of a function because a vertical line can cross the graph at more than one point.

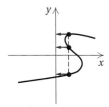

b) The graph *is* that of a function because no vertical line can cross the graph at more than one point. This can be confirmed with a ruler or straightedge.

c) The graph *is* that of a function.

d) The graph *is not* that of a function. There is a vertical line that can cross the graph more than once.

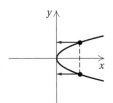

Do Exercises 12–15.

e | Applications of Functions and Their Graphs

Functions are often described by graphs, whether or not an equation is given. To use a graph in an application, we note that each point on the graph represents a pair of values.

Example 9 *Movie Revenue.* The following graph approximates the weekly U.S. revenue, in millions of dollars, from the recent movie *Air Force One*. The revenue is a function f of the number of weeks x since the movie was released. No equation is given for the function.

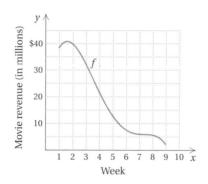

Use the graph to answer the following.

a) What was the movie revenue for week 1? That is, find $f(1)$.

b) What was the movie revenue for week 5? That is, find $f(5)$.

a) To estimate the revenue for week 1, we locate 1 on the horizontal axis and move directly up until we reach the graph. Then we move across to the vertical axis. We estimate that value to be about $38 million—that is, $f(1) = 38$.

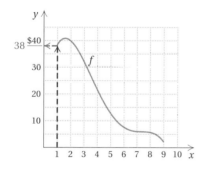
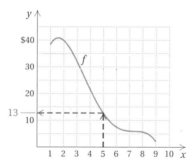

b) To estimate the revenue for week 5, we locate 5 on the horizontal axis and move directly up until we reach the graph. Then we move across to the vertical axis. We estimate that value to be about $13 million—that is, $f(5) = 13$.

Do Exercises 16 and 17.

Referring to the graph in Example 9:

16. What was the movie revenue for week 2?

17. What was the movie revenue for week 6?

Answers on page A-12

Calculator Spotlight

 TRACE and TABLE Features

TRACE Feature. There are two ways in which we can determine the coordinates of points on a graph drawn by a grapher. One approach is to use the TRACE key. When the TRACE feature is activated, the cursor appears on the line (or curve) that has been graphed and the coordinates at that point are displayed.

Let's consider the equation or function $y = x^3 - 5x + 1$ considered in the Calculator Spotlight in Section 2.1. Here we graph the equation in the window $[-5, 5, -10, 10]$ and press TRACE:

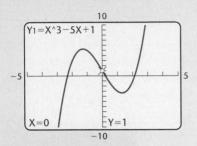

The coordinates at the bottom indicate that the cursor is at the point with coordinates $(0, 1)$. By using the arrow keys, we can obtain coordinates of other points. For example, if we press the left arrow key ◁ seven times, we move the cursor to the location shown below, obtaining a point on the graph with coordinates $(-0.7446809, 4.3104418)$.

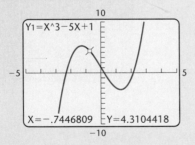

TABLE Feature. Another way to find the coordinates of solutions of equations makes use of the TABLE feature. We first press 2nd TBLSET . For the equation above, let's set TblStart = 0.3 and △Tbl = 1.

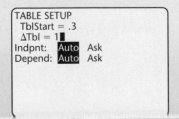

This means that the table's x-values will start at 0.3 and increase by 1. (We could choose other values for TblStart and △Tbl.) By setting Indpnt and Depend to Auto, we obtain the following when we press 2nd TABLE :

X	Y₁	
.3	−.473	
1.3	−3.303	
2.3	1.667	
3.3	20.437	
4.3	59.007	
5.3	123.38	
6.3	219.55	
X = .3		

The arrow keys allow us to scroll up and down the table and extend it to other values not initially shown.

X	Y₁	
12.3	1800.4	
13.3	2287.1	
14.3	2853.7	
15.3	3506.1	
16.3	4250.2	
17.3	5092.2	
18.3	6038	
X = 18.3		

Exercises

1. Use the TRACE feature to find five different ordered-pair solutions of the equation $y = x^3 - 5x + 1$.

2. For $y = x^3 - 5x + 1$, use the TABLE feature to construct a table starting with $x = 10$ and △Tbl = 5. Find the value of y when x is 10. Then find the value of y when x is 35.

3. Adjust the table settings to Indpnt: Ask. How does the table change? Enter a number of your choice and see what happens. Use this setting to find the value of y when $x = 28$.

Exercise Set 2.2

a Determine whether the correspondence is a function.

1. Domain   Range

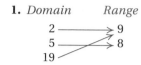

2. Domain   Range

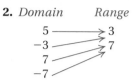

3. Domain   Range

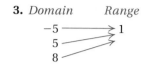

4. Domain   Range

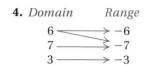

5. Domain   Range

(Girl's Age,   (Average Daily Weight
in months)   Gain, in grams)

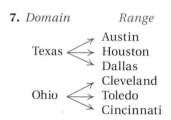

Source: *American Family Physician,*
December 1993, p. 1435.

6. Domain   Range

(Year)   (Consumption of Diet Cola,
in gallons per person)

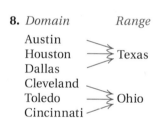

Source: U.S. Department of Agriculture
Economic Research Service

7. Domain   Range

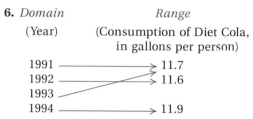

8. Domain   Range

Austin
Houston   Texas
Dallas
Cleveland
Toledo   Ohio
Cincinnati

	Domain	Correspondence	Range
9.	A family	Each person's eye color	A set of colors
10.	A textbook	An even-numbered page in the book	A set of pages
11.	A set of avenues	An intersecting road	A set of cross streets
12.	A math class	Each person's seat number	A set of numbers
13.	A set of numbers	Square each number and then add 4.	A set of positive numbers
14.	A set of shapes	The perimeter of each shape	A set of positive numbers

Find the function values.

15. $f(x) = x + 5$
  a) $f(4)$ b) $f(7)$
  c) $f(-3)$ d) $f(0)$
  e) $f(2.4)$ f) $f\left(\frac{2}{3}\right)$

16. $g(t) = t - 6$
  a) $g(0)$ b) $g(6)$
  c) $g(13)$ d) $g(-1)$
  e) $g(-1.08)$ f) $g\left(\frac{7}{8}\right)$

17. $h(p) = 3p$
  a) $h(-7)$ b) $h(5)$
  c) $h(14)$ d) $h(0)$
  e) $h\left(\frac{2}{3}\right)$ f) $h(a + 1)$

18. $f(x) = -4x$
  a) $f(6)$ b) $f\left(-\frac{1}{2}\right)$
  c) $f(a - 1)$ d) $f(11.8)$
  e) $f(0)$ f) $f(-1)$

19. $g(s) = 3s + 4$
  a) $g(1)$ b) $g(-7)$
  c) $g(6.7)$ d) $g(0)$
  e) $g(-10)$ f) $g\left(\frac{2}{3}\right)$

20. $h(x) = 19$, a constant function
  a) $h(4)$ b) $h(-6)$
  c) $h(12.5)$ d) $h(0)$
  e) $h\left(\frac{2}{3}\right)$ f) $h(1234)$

21. $f(x) = 2x^2 - 3x$
  a) $f(0)$ b) $f(-1)$
  c) $f(2)$ d) $f(10)$
  e) $f(-5)$ f) $f(4a)$

22. $f(x) = 3x^2 - 2x + 1$
  a) $f(0)$ b) $f(1)$
  c) $f(-1)$ d) $f(10)$
  e) $f(2a)$ f) $f(-3)$

23. $f(x) = |x| + 1$
  a) $f(0)$ b) $f(-2)$
  c) $f(2)$ d) $f(-3)$
  e) $f(-10)$ f) $f(a - 1)$

24. $g(t) = |t - 1|$
  a) $g(4)$ b) $g(-2)$
  c) $g(-1)$ d) $g(100)$
  e) $g(-50)$ f) $g(a + 1)$

25. $f(x) = x^3$
  a) $f(0)$ b) $f(-1)$
  c) $f(2)$ d) $f(10)$
  e) $f(-5)$ f) $f(-10)$

26. $f(x) = x^4 - 3$
  a) $f(1)$ b) $f(-1)$
  c) $f(0)$ d) $f(2)$
  e) $f(-2)$ f) $f(10)$

27. *Archaeology.* The function H described by

$$H(x) = 2.75x + 71.48$$

can be used to predict the height, in centimeters, of a woman whose *humerus* (the bone from the elbow to the shoulder) is x cm long. If a humerus is known to be from a female, how tall was she if the bone is **(a)** 32 cm long? **(b)** 35 cm long?

28. Refer to Exercise 27. When a humerus is from a male, the function

$$M(x) = 2.89x + 70.64$$

can be used to find the male's height, in centimeters. If a humerus is known to be from a male, how tall was the male if the bone is **(a)** 30 cm long? **(b)** 35 cm long?

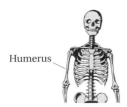

Humerus

29. *Pressure at Sea Depth.* The function $P(d) = 1 + (d/33)$ gives the pressure, in *atmospheres* (atm), at a depth of d feet in the sea. Note that $P(0) = 1$ atm, $P(33) = 2$ atm, and so on. Find the pressure at 20 ft, 30 ft, and 100 ft.

30. *Temperature as a Function of Depth.* The function $T(d) = 10d + 20$ gives the temperature, in degrees Celsius, inside the earth as a function of the depth d, in kilometers. Find the temperature at 5 km, 20 km, and 1000 km.

31. *Melting Snow.* The function $W(d) = 0.112d$ approximates the amount, in centimeters, of water that results from d centimeters of snow melting. Find the amount of water that results from snow melting from depths of 16 cm, 25 cm, and 100 cm.

32. *Temperature Conversions.* The function $C(F) = \frac{5}{9}(F - 32)$ determines the Celsius temperature that corresponds to F degrees Fahrenheit. Find the Celsius temperature that corresponds to 62°F, 77°F, and 23°F.

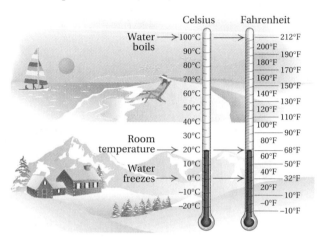

	Celsius	Fahrenheit
Water boils	100°C	212°F
	90°C	200°F / 190°F
	80°C	180°F / 170°F
	70°C	160°F
	60°C	150°F / 140°F
	50°C	130°F / 120°F
	40°C	110°F / 100°F
	30°C	90°F
Room temperature	20°C	80°F / 68°F
	10°C	60°F / 50°F
Water freezes	0°C	40°F / 32°F
	−10°C	20°F / 10°F
	−20°C	−0°F / −10°F

c Graph the function.

33. $f(x) = 3x - 1$

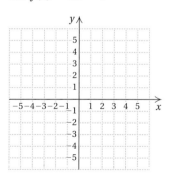

34. $g(x) = 2x + 5$

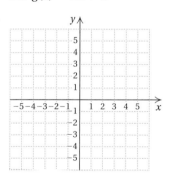

35. $g(x) = -2x + 3$

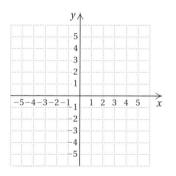

36. $f(x) = -\frac{1}{2}x + 2$

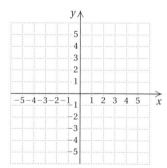

37. $f(x) = \frac{1}{2}x + 1$

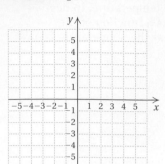

38. $f(x) = -\frac{3}{4}x - 2$

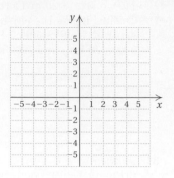

39. $f(x) = 2 - |x|$

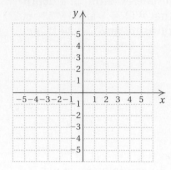

40. $f(x) = |x| - 4$

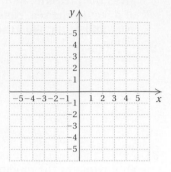

41. $f(x) = x^2$

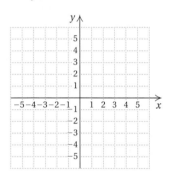

42. $f(x) = x^2 - 1$

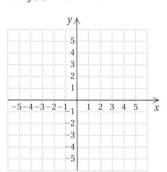

43. $f(x) = x^2 - x - 2$

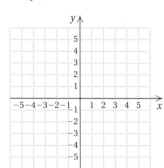

44. $f(x) = x^2 + 6x + 5$

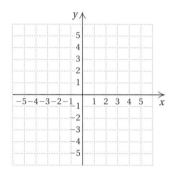

d Determine whether each of the following is the graph of a function.

45.

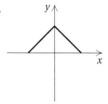

46.

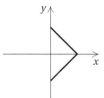

47.

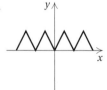

48.

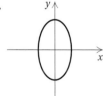

49.

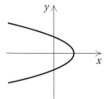

50.

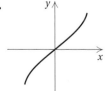

51.

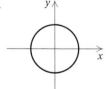

52.

e

Cholesterol Level and Risk of a Heart Attack. The following graph shows the annual heart attack rate per 10,000 men as a function of blood cholesterol level.

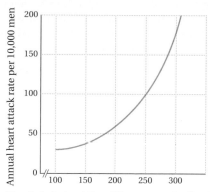

Blood cholesterol (in milligrams per deciliter)

Source: Copyright 1989, CSPI. Adapted from *Nutrition Action Healthletter* (1875 Connecticut Avenue, N.W., Suite 300, Washington, DC 20009-5728)

Videotape Rentals. For Exercises 55–58, use the following graph, which shows the number of videotape rentals as a function of time.

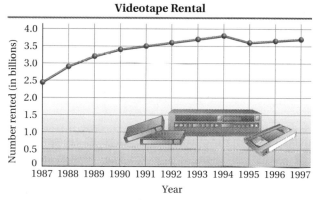

Source: Adams Media Research, Carmel Valley, California

53. Approximate the annual heart attack rate per 10,000 men for those whose blood cholesterol level is 225 mg/dl.

54. Approximate the annual heart attack rate per 10,000 men for those whose blood cholesterol level is 275 mg/dl.

55. Approximate the number of videotape rentals in 1997.

56. Approximate the number of videotape rentals in 1994.

57. In what year did the greatest number of rentals occur?

58. In what year did the least number of rentals occur?

Skill Maintenance

Solve.

59. Find three consecutive even integers such that the sum of the first, twice the second, and three times the third is 124. [1.3a]

60. A piece of wire 32.8 ft long is to be cut into two pieces and each of those pieces is to be bent into a square. The length of a side of one square is to be 2.2 ft longer than the length of a side of the other square. How should the wire be cut? [1.3a]

61. The surface area of a rectangular solid of length l, width w, and height h is given by

$$S = 2lh + 2lw + 2wh.$$

Solve for l. [1.2a]

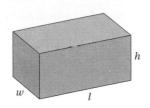

62. Solve the formula in Exercise 61 for w. [1.2a]

Solve. [1.4c]

63. $4x - 2 \leq 5x + 7$

64. $4x - 2 > 8$

65. $-6x + 2x - 32 < 64$

66. $-5x \geq 120$

Solve. [1.6c, e]

67. $|5x + 7| = 10$

68. $|4x - 2| \leq 10$

69. $|3x - 7| \geq 24$

70. $\left| \dfrac{3x + 8}{10} \right| > 12$

Synthesis

Researchers at Yale University have suggested that the following graphs may represent three different aspects of love.

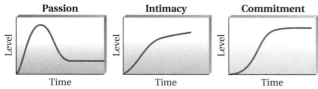

Source: "A Triangular Theory of Love," by R. J. Sternberg, 1986, *Psychological Review*, **93**(2), 119–135. Copyright 1986 by the American Psychological Association, Inc. Reprinted by permission.

71. ◈ In what unit would you measure time if the horizontal length of each graph were 10 units? Why?

72. ◈ Do you agree with the researchers that these graphs should be shaped as they are? Why or why not?

73. ◈ Is it possible for a function to have more numbers as outputs than as inputs? Why or why not?

74. ◈ Look up the word "function" in a dictionary. Explain how that definition might be related to the mathematical one given in this section.

For Exercises 75 and 76, let $f(x) = 3x^2 - 1$ and $g(x) = 2x + 5$.

75. Find $f(g(-4))$ and $g(f(-4))$.

76. Find $f(g(-1))$ and $g(f(-1))$.

77. Suppose that a function g is such that $g(-1) = -7$ and $g(3) = 8$. Find a formula for g if $g(x)$ is of the form $g(x) = mx + b$, where m and b are constants.

2.3 Finding Domain and Range

a Finding Domain and Range

The solutions of an equation in two variables consist of a set of ordered pairs. An arbitrary set of ordered pairs is called a **relation**. When a set of ordered pairs is such that no two different pairs share a common first coordinate, we have a function. The **domain** is the set of all first coordinates and the **range** is the set of all second coordinates.

Example 1 Find the domain and the range of the function f whose graph is shown below.

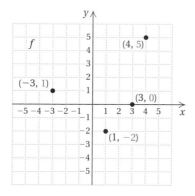

This is a rather simple function. Its graph contains just four ordered pairs and it can be written as

$$\{(-3, 1), (1, -2), (3, 0), (4, 5)\}.$$

We can determine the domain and the range by reading the x- and y-values directly from the graph.

The domain is the set of all first coordinates, $\{-3, 1, 3, 4\}$. The range is the set of all second coordinates, $\{1, -2, 0, 5\}$.

Do Exercise 1.

Example 2 For the function f whose graph is shown below, determine each of the following.

a) The number in the range that is paired with 1 (from the domain). That is, find $f(1)$.

b) The domain of f

c) The numbers in the domain that are paired with 1 (from the range). That is, find all x such that $f(x) = 1$.

d) The range of f

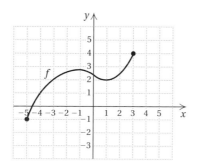

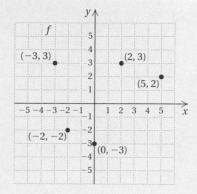
1. Find the domain and the range of the function f whose graph is shown below.

Answer on page A-12

2. For the function f whose graph is shown below, determine each of the following.

a) The number in the range that is paired with the input 1. That is, find $f(1)$.

b) The domain of f

c) The number in the domain that is paired with 4

d) The range of f

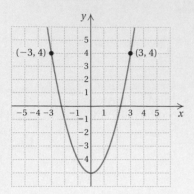

a) To determine which number in the range is paired with 1 in the domain, we locate 1 on the horizontal axis. Next, we find the point on the graph of f for which 1 is the first coordinate. From that point, we can look to the vertical axis to find the corresponding y-coordinate, 2. The input 1 has the output 2—that is, $f(1) = 2$.

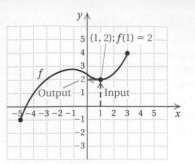

b) The domain of the function is the set of all x-values, or inputs, of the points on the graph. These extend from −5 to 3 and can be viewed as the curve's shadow, or projection, onto the x-axis. Thus the domain is the set $\{x \mid -5 \le x \le 3\}$, or, in interval notation, $[-5, 3]$.

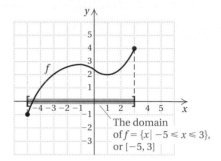

c) To determine which numbers in the domain are paired with 1 in the range, we locate 1 on the vertical axis. From there, we look left and right to the graph of f to find any points (inputs) for which 1 is the second coordinate (output). One such point exists, $(-4, 1)$. For this function, we note that $x = -4$ is the only member of the domain paired with 1. For other functions, there might be more than one member of the domain paired with a member of the range.

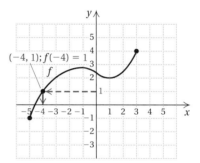

d) The range of the function is the set of all y-values, or outputs, of the points on the graph. These extend from −1 to 4 and can be viewed as the curve's shadow, or projection, onto the y-axis. Thus the range is the set $\{y \mid -1 \le y \le 4\}$, or, in interval notation, $[-1, 4]$.

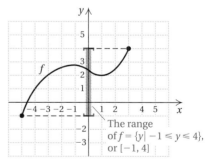

Do Exercise 2.

Answers on page A-12

When a function is given by an equation or formula, the domain is understood to be the largest set of real numbers (inputs) for which function values (outputs) can be calculated. That is, the domain is the set of all possible allowable inputs into the formula. To find the domain, think, "What can we substitute?"

Example 3   Find the domain: $f(x) = |x|$.

We ask, "What can we substitute?" Is there any number x for which we cannot calculate $|x|$? The answer is no. Thus the domain of f is the set of all real numbers.

Example 4   Find the domain: $f(x) = \dfrac{3}{2x - 5}$.

We ask, "What can we substitute?" Is there any number x for which we cannot calculate $3/(2x - 5)$? Since $3/(2x - 5)$ cannot be calculated when the denominator $2x - 5$ is 0, we solve the following equation to find those real numbers that must be excluded from the domain of f:

$$2x - 5 = 0 \qquad \text{Setting the denominator equal to 0}$$
$$2x = 5 \qquad \text{Adding 5}$$
$$x = \tfrac{5}{2}. \qquad \text{Dividing by 2}$$

Thus $\frac{5}{2}$ is not in the domain, whereas all other real numbers are.

The domain of f is $\left\{x \mid x \text{ is a real number and } x \neq \frac{5}{2}\right\}$. In interval notation, the domain is $\left(-\infty, \frac{5}{2}\right) \cup \left(\frac{5}{2}, \infty\right)$.

Do Exercises 3 and 4.

The task of determining the domain and the range of a function is one that we will return to several times as we consider other types of functions in this book.

The following is a review of the function concepts considered in Sections 2.2 and 2.3.

Function Concepts

- Formula for f: $f(x) = x^2 - 7$
- For every input of f, there is exactly one output.
- 1 is an input; -6 is an output.
- $f(1) = -6$
- $(1, -6)$ is on the graph.
- Domain = The set of all inputs
  = The set of all real numbers
- Range = The set of all outputs
  = $[-7, \infty)$

Graph

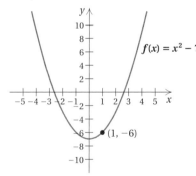

$f(x) = x^2 - 7$

$(1, -6)$

3. Find the domain:
$$f(x) = x^3 - |x|.$$

4. Find the domain:
$$f(x) = \dfrac{4}{3x + 2}.$$

Answers on page A-12

Calculator Spotlight

 Determining Domain and Range. Use a grapher to graph the function in the given viewing window. Then determine the domain and the range.

a) $f(x) = 3 - |x|$, $[-10, 10, -10, 10]$

b) $f(x) = x^3 - x$, $[-3, 3, -4, 4]$

c) $f(x) = \dfrac{12}{x}$, or $12x^{-1}$, $[-14, 14, -14, 14]$

d) $f(x) = x^4 - 2x^2 - 3$, $[-4, 4, -6, 6]$

We have the following.

a) $y = 3 - |x|$

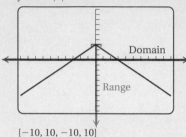

$[-10, 10, -10, 10]$

Domain = all real numbers;
range = $(-\infty, 3]$

b) $y = x^3 - x$

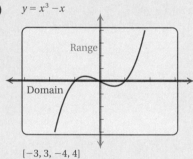

$[-3, 3, -4, 4]$

Domain = all real numbers;
range = all real numbers

c) $y = \dfrac{12}{x}$, or $12x^{-1}$

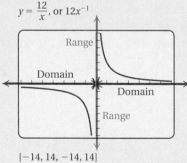

$[-14, 14, -14, 14]$

The number 0 is excluded as an input.
Domain = $(-\infty, 0) \cup (0, \infty)$;
range = $(-\infty, 0) \cup (0, \infty)$

d) $y = x^4 - 2x^2 - 3$

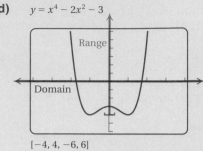

$[-4, 4, -6, 6]$

Domain = all real numbers;
range = $[-4, \infty)$

We can confirm our results using the TRACE feature, moving the cursor from left to right along the curve. We can also use the TABLE feature. In Example (d), it might not appear as though the domain is all real numbers because the graph seems "thin," but reexamining the formula shows that we can indeed substitute any real number.

Exercises

Use a grapher to graph the function in the given viewing window. Then determine the domain and the range.

1. $f(x) = |x| - 7$, $[-10, 10, -10, 10]$

2. $f(x) = 2 + 3x - x^3$, $[-5, 5, -5, 5]$

3. $f(x) = \dfrac{-16}{x}$, or $-16x^{-1}$, $[-20, 20, -20, 20]$

4. $f(x) = x^4 - 2x^2 - 7$, $[-4, 4, -9, 9]$

Exercise Set 2.3

a In Exercises 1–12, the graph is that of a function. Determine for each one **(a)** $f(1)$; **(b)** the domain; **(c)** all x-values such that $f(x) = 2$; and **(d)** the range. An open dot indicates that the point does not belong to the graph.

1.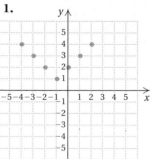

2.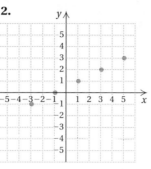

3.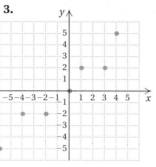

4.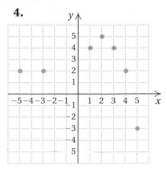

5.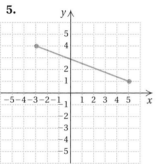

6.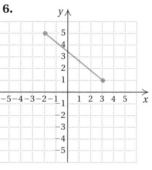

7.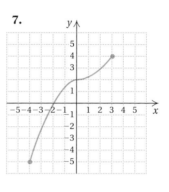

8.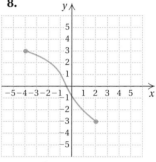

9.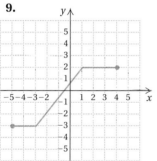

10.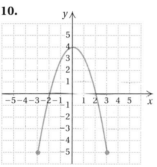

11.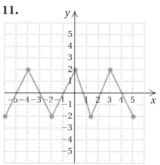

12.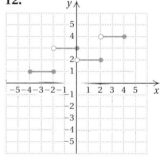

Find the domain.

13. $f(x) = \dfrac{2}{x + 3}$

14. $f(x) = \dfrac{7}{5 - x}$

15. $f(x) = 2x + 1$

16. $f(x) = 4 - 5x$

17. $f(x) = x^2 + 3$

18. $f(x) = x^2 - 2x + 3$

19. $f(x) = \dfrac{8}{5x - 14}$

20. $f(x) = \dfrac{x - 2}{3x + 4}$

21. $f(x) = |x| - 4$

22. $f(x) = |x - 4|$

23. $f(x) = \dfrac{4}{|2x - 3|}$

24. $f(x) = \dfrac{x^2 - 3x}{|4x - 7|}$

25. $g(x) = \dfrac{1}{x - 1}$

26. $g(x) = \dfrac{-11}{4 + x}$

27. $g(x) = x^2 - 2x + 1$

28. $g(x) = 8 - x^2$

29. $g(x) = x^3 - 1$

30. $g(x) = 4x^3 + 5x^2 - 2x$

31. $g(x) = \dfrac{7}{20 - 8x}$

32. $g(x) = \dfrac{2x - 3}{6x - 12}$

33. $g(x) = |x + 7|$

34. $g(x) = |x| + 1$

35. $g(x) = \dfrac{-2}{|4x + 5|}$

36. $g(x) = \dfrac{x^2 + 2x}{|10x - 20|}$

37. For the function f whose graph is shown below, find $f(-1)$, $f(0)$, and $f(1)$.

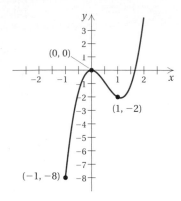

38. For the function g whose graph is shown below, find all the x-values for which $g(x) = 1$.

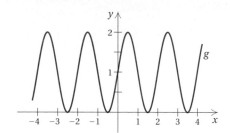

Skill Maintenance

Solve. [1.4d]

39. On a new job, Anthony can be paid in one of two ways:

  Plan A: A salary of $800 per month, plus a commission of 5% of sales;

  Plan B: A salary of $1000 per month, plus a commission of 7% of sales in excess of $15,000.

For what amount of monthly sales is plan B better than plan A, if we assume that sales are always more than $15,000?

40. You are taking a business course for which there will be 4 tests, each worth 100 points. You have scores of 92, 90, and 88 on the first three tests. You must make a total of at least 360 points in order to get an A. What scores on the fourth test will keep you eligible for an A?

Solve. [1.6c]

41. $|x| = 8$

42. $|x| = -8$

43. $|x - 7| = 11$

44. $|2x + 3| = 13$

45. $|3x - 4| = |x + 2|$

46. $|5x - 6| = |3 - 8x|$

47. $|3x - 8| = -11$

48. $|3x - 8| = 0$

Synthesis

49. ◆ Explain the difference between the domain and the range of a function.

50. ◆ For a given function f, it is known that $f(2) = -3$. Give as many interpretations of this fact as you can.

51. 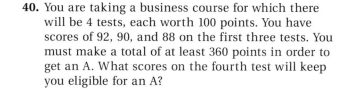 Determine the range of each of the functions in Exercises 13, 18, 21, and 22.

52. Determine the range of each of the functions in Exercises 26, 27, 28, and 34.

2.4 Linear Functions: Graphs and Slope

We now turn our attention to functions whose graphs are straight lines. Such functions are called **linear** and can be written in the form $f(x) = mx + b$.

> **LINEAR FUNCTION**
>
> A **linear function** f is any function that can be described by $f(x) = mx + b$.

In this section, we consider the effects of the constants m and b on the graphs of linear functions.

a The Constant *b*: The *y*-Intercept

Let's first explore the effect of the constant b.

Example 1 Graph $y = 2x$ and $y = 2x + 3$ using the same set of axes. Compare the graphs.

We first make a table of solutions of both equations.

	y	y
x	$y = 2x$	$y = 2x + 3$
0	0	3
1	2	5
-1	-2	1
2	4	7
-2	-4	-1

Next, we plot these points. Drawing a red line for $y = 2x$ and a blue line for $y = 2x + 3$, we note that the graph of $y = 2x + 3$ is simply the graph of $y = 2x$ shifted, or *translated,* up 3 units. The lines are parallel.

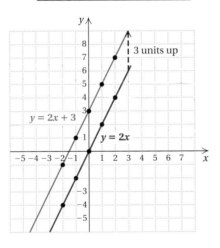

Do Exercises 1 and 2.

Example 2 Graph $f(x) = \frac{1}{3}x$ and $g(x) = \frac{1}{3}x - 2$ using the same set of axes. Compare the graphs.

We first make a table of solutions of both equations. By choosing multiples of 3, we can avoid fractions.

	$f(x)$	$g(x)$
x	$f(x) = \frac{1}{3}x$	$g(x) = \frac{1}{3}x - 2$
0	0	-2
3	1	-1
-3	-1	-3
6	2	0

Objectives

a Find the y-intercept of a line from the equation $y = mx + b$ or $f(x) = mx + b$.

b Given two points on a line, find the slope; given a linear equation, derive the equivalent slope–intercept equation and determine the slope and the y-intercept.

c Solve applied problems involving slope.

For Extra Help

TAPE 6   TAPE 5A   MAC WIN   CD-ROM

1. Graph $y = 3x$ and $y = 3x - 6$ using the same set of axes. Compare the graphs.

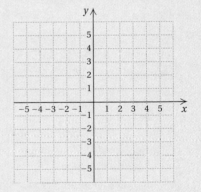

2. Graph $y = -2x$ and $y = -2x + 3$ using the same set of axes. Compare the graphs.

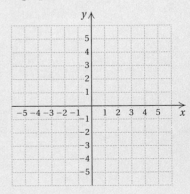

Answers on page A-13

3. Graph $f(x) = \frac{1}{3}x$ and $g(x) = \frac{1}{3}x + 2$ using the same set of axes. Compare the graphs.

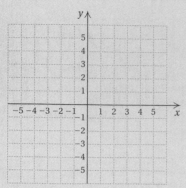

We then plot these points. Drawing a line for $f(x) = \frac{1}{3}x$ and a line for $g(x) = \frac{1}{3}x - 2$, we see that the graph of $g(x) = \frac{1}{3}x - 2$ is simply the graph of $f(x) = \frac{1}{3}x$ shifted, or translated, down 2 units.

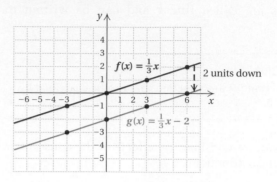

Note that in Example 1, the graph of $y = 2x + 3$ passed through the point (0, 3) and in Example 2, the graph of $g(x) = \frac{1}{3}x - 2$ passed through the point (0, −2). In general, the graph of $y = mx + b$ is a line parallel to $y = mx$, passing through the point (0, b). The point (0, b) is called the **y-intercept** because it is the point at which the graph crosses the y-axis. Often it is convenient to refer to the number b as the y-intercept. The constant b has the effect of moving the graph of $y = mx$ up or down $|b|$ units to obtain the graph of $y = mx + b$.

Do Exercise 3.

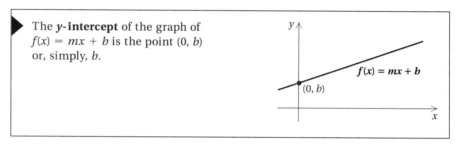

The **y-intercept** of the graph of $f(x) = mx + b$ is the point (0, b) or, simply, b.

Example 3 Find the y-intercept: $y = -5x + 4$.

$$y = -5x + 4 \qquad (0, 4), \text{ or simply } 4, \text{ is the } y\text{-intercept.}$$

Example 4 Find the y-intercept: $f(x) = 6.3x - 7.8$.

$$f(x) = 6.3x - 7.8 \qquad (0, -7.8), \text{ or simply } -7.8, \text{ is the } y\text{-intercept.}$$

Do Exercises 4 and 5.

b The Constant *m*: Slope

Look again at the graphs in Examples 1 and 2. Note that the slant of each red line seems to match the slant of each corresponding blue line. This leads us to believe that the number *m* in the equation $y = mx + b$ is related to the slant of the line. The following definition enables us to visualize this slant and attach a number, a geometric ratio, or *slope*, to the line.

Find the y-intercept.

4. $y = 7x + 8$

5. $f(x) = -6x - \frac{2}{3}$

Answers on page A-13

The **slope** of a line containing points (x_1, y_1) and (x_2, y_2) is given by

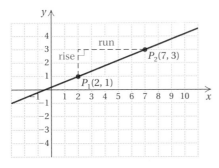

$$m = \frac{\text{rise}}{\text{run}}$$

$$= \frac{\text{change in } y}{\text{change in } x} = \frac{y_2 - y_1}{x_2 - x_1} = \frac{y_1 - y_2}{x_1 - x_2}.$$

Consider a line with two points marked P_1 and P_2. As we move from P_1 to P_2, the y-coordinate changes from 1 to 3 and the x-coordinate changes from 2 to 7. The change in y is $3 - 1$, or 2. The change in x is $7 - 2$, or 5.

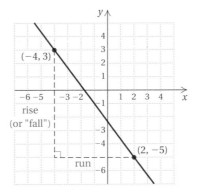

We call the change in y the **rise** and the change in x the **run**. The ratio rise/run is the same for any two points on a line. We call this ratio the **slope**. Slope describes the slant of a line. The slope of the line in the graph above is given by

$$\frac{\text{rise}}{\text{run}}, \quad \text{or} \quad \frac{\text{change in } y}{\text{change in } x}, \quad \text{or} \quad \frac{2}{5}.$$

Whenever x increases by 5 units, y increases by 2 units. Equivalently, whenever x increases by 1 unit, y increases by $\frac{2}{5}$ unit.

Example 5 Graph the line containing the points $(-4, 3)$ and $(2, -5)$ and find the slope.

The graph is shown at right. Going from $(-4, 3)$ to $(2, -5)$, we see that the change in y, or the rise, is $-5 - 3$, or -8. The change in x, or the run, is $2 - (-4)$, or 6.

$$\text{Slope} = \frac{\text{rise}}{\text{run}} = \frac{\text{change in } y}{\text{change in } x}$$

$$= \frac{-5 - 3}{2 - (-4)}$$

$$= \frac{-8}{6} = -\frac{8}{6}, \text{ or } -\frac{4}{3}$$

The formula

$$m = \frac{y_2 - y_1}{x_2 - x_1} = \frac{y_1 - y_2}{x_1 - x_2}$$

tells us that we can subtract in two ways. We must remember, however, to subtract the x-coordinates in the same order that we subtract the y-coordinates.

Exercises

Use the window settings $[-6, 6, -4, 4]$, Xscl = 1, Yscl = 1 for Exercises 1–4.

1. Graph

$$y = x + 1, \quad y = 2x + 1,$$
$$y = 3x + 1, \quad y = 10x + 1.$$

What do you think the graph of $y = 247x + 1$ will look like?

2. Graph

$$y = x, \quad y = \frac{7}{8}x,$$
$$y = 0.47x, \quad y = \frac{2}{31}x.$$

What do you think the graph of $y = 0.000018x$ will look like?

3. Graph

$$y = -x, \quad y = -2x,$$
$$y = -5x, \quad y = -10x.$$

What do you think the graph of $y = -247x$ will look like?

4. Graph

$$y = -x - 1,$$
$$y = -\frac{3}{4}x - 1,$$
$$y = -0.38x - 1,$$
$$y = -\frac{5}{32}x - 1.$$

What do you think the graph of $y = -0.000043x - 1$ will look like?

Graph the line through the given points and find its slope.

6. $(-1, -1)$ and $(2, -4)$

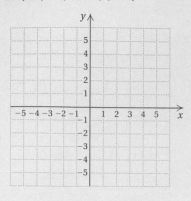

7. $(0, 2)$ and $(3, 1)$

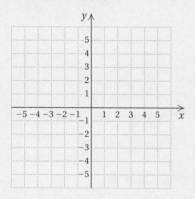

8. Find the slope of the line $f(x) = -\frac{2}{3}x + 1$. Use the points $(9, -5)$ and $(3, -1)$.

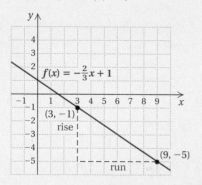

Answers on page A-13

Let's do Example 5 again:

$$\text{Slope} = \frac{\text{change in } y}{\text{change in } x} = \frac{3 - (-5)}{-4 - 2} = \frac{8}{-6} = -\frac{8}{6} = -\frac{4}{3}.$$

We see that both ways give the same slope value.

The slope of a line tells how it slants. A line with positive slope slants up from left to right. The larger the positive number, the steeper the slant. A line with negative slope slants downward from left to right. The smaller the negative number, the steeper the line.

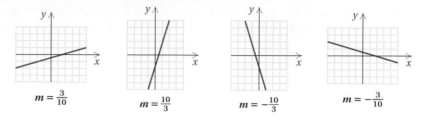

$$m = \frac{3}{10} \qquad m = \frac{10}{3} \qquad m = -\frac{10}{3} \qquad m = -\frac{3}{10}$$

Do Exercises 6 and 7.

How can we find the slope from a given equation? Let's consider the equation $y = 2x + 3$, which is in the form $y = mx + b$. We can find two points by choosing convenient values for x, say 0 and 1, and substituting to find the corresponding y-values. We find two points on the line to be $(0, 3)$ and $(1, 5)$. The slope of the line is found as follows, using the definition of slope:

$$m = \frac{\text{change in } y}{\text{change in } x}$$

$$= \frac{5 - 3}{1 - 0} = \frac{2}{1} = 2.$$

The slope is 2. Note that this is also the coefficient of the x-term in the equation $y = 2x + 3$.

Do Exercise 8.

We see that the slope of the line $y = mx + b$ is indeed the constant m, the coefficient of x.

> The **slope** of the line $y = mx + b$ is m.

From a linear equation in the form $y = mx + b$, we can read directly the slope and the y-intercept of the graph.

> The equation $y = mx + b$ is called the **slope-intercept equation.**
> The slope is m and the y-intercept is $(0, b)$.

Note that any graph of an equation $y = mx + b$ passes the vertical-line test and thus represents a function.

Example 6 Find the slope and the y-intercept of $y = 5x - 4$.

Since the equation is already in the form $y = mx + b$, we simply read the slope and the y-intercept from the equation:

$$y = 5x - 4.$$

The slope is 5.   The y-intercept is $(0, -4)$.

Example 7 Find the slope and the y-intercept of $2x + 3y = 8$.

We first solve for y so we can easily read the slope and the y-intercept:

$2x + 3y = 8$

$\quad 3y = -2x + 8$ Subtracting $2x$

$\quad \dfrac{3y}{3} = \dfrac{-2x + 8}{3}$ Dividing by 3

$\quad\quad y = -\dfrac{2}{3}x + \dfrac{8}{3}.$ Finding the form $y = mx + b$

The slope is $-\frac{2}{3}$.   The y-intercept is $\left(0, \frac{8}{3}\right)$.

Do Exercises 9 and 10.

c | Applications

Slope has many real-world applications. For example, numbers like 2%, 3%, and 6% are often used to represent the *grade* of a road, a measure of how steep a road on a hill or mountain is. A 3% grade $\left(3\% = \frac{3}{100}\right)$ means that for every horizontal distance of 100 ft, the road rises 3 ft, and a -3% grade means that for every horizontal distance of 100 ft, the road drops 3 ft. An athlete might change the grade of a treadmill during a workout. An escape ramp on an airliner might have a slope of about -0.6.

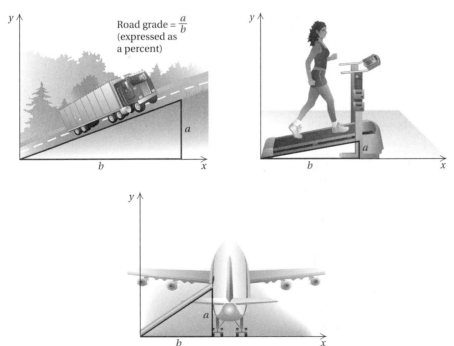

Find the slope and the y-intercept.

9. $f(x) = -8x + 23$

10. $5x - 10y = 25$

Answers on page A-13

11. *Capital Outlay for U.S. Defense.* Find the rate of change of capital outlay for U.S. defense.

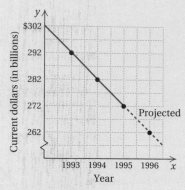

Source: Statistical Abstract of the United States

Architects and carpenters use slope when designing and building stairs, ramps, or roof pitches. Another application occurs in hydrology. When a river flows, the strength or force of the river depends on how far the river falls vertically compared to how far it flows horizontally. Slope can also be considered as a **rate of change.**

Example 8 *Amount Spent on Cancer Research.* The amount spent on cancer research has increased steadily over the years, as shown in the following graph. Find the rate of change of that amount.

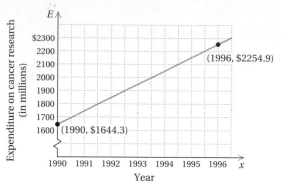

Source: The New England Journal of Medicine

First, we determine the coordinates of two points on the graph. In this case, they are given as (1990, $1644.3) and (1996, $2254.9). Then we compute the slope, or rate of change, as follows:

$$\text{Slope} = \text{Rate of change} = \frac{\text{change in } y}{\text{change in } x}$$

$$= \frac{\$2254.9 - \$1644.3}{1996 - 1990} = \frac{\$610.6}{6} \approx 101.8 \, \frac{\$}{\text{yr}}.$$

This result tells us that each year the amount spent on cancer research has increased by about $101.8 million.

Do Exercise 11.

Answer on page A-13

Exercise Set 2.4

a, **b** Find the slope and the *y*-intercept.

1. $y = 4x + 5$

2. $y = -5x + 10$

3. $f(x) = -2x - 6$

4. $g(x) = -5x + 7$

5. $y = -\frac{3}{8}x - \frac{1}{5}$

6. $y = \frac{15}{7}x + \frac{16}{5}$

7. $g(x) = 0.5x - 9$

8. $f(x) = -3.1x + 5$

9. $2x - 3y = 8$

10. $-8x - 7y = 24$

11. $9x = 3y + 6$

12. $9y + 36 - 4x = 0$

13. $3 - \frac{1}{4}y = 2x$

14. $5x = \frac{2}{3}y - 10$

15. $17y + 4x + 3 = 7 + 4x$

16. $3y - 2x = 5 + 9y - 2x$

b Find the slope of the line.

17.

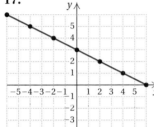

18.

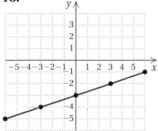

19.

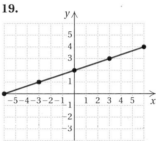

20.

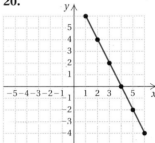

Find the slope of the line containing the given pair of points.

21. (6, 9) and (4, 5)

22. (8, 7) and (2, −1)

23. (9, −4) and (3, −8)

24. (17, −12) and (−9, −15)

25. (−16.3, 12.4) and (−5.2, 8.7)

26. (14.4, −7.8) and (−12.5, −17.6)

c Find the slope (or rate of change).

27. Find the slope (or grade) of the treadmill.

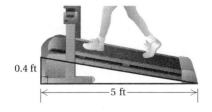

0.4 ft

5 ft

28. Find the slope (or pitch) of the roof.

2.6 ft

8.2 ft

29. Find the slope (or head) of the river.

43.33 ft

1238 ft

30. Public buildings regularly include steps with 7-in. risers and 11-in. treads. Find the grade of such a stairway.

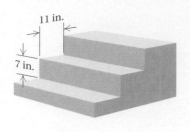

31. Find the rate of change of the tuition and fees at public two-year colleges.

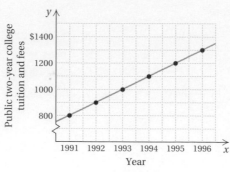

Source: *Statistical Abstract of the United States*

32. Find the rate of change of the cost of a formal wedding.

Source: *Modern Bride Magazine*

Skill Maintenance

Simplify.   [R.3c], [R.6b]

33. $3^2 - 24 \cdot 56 + 144 \div 12$

34. $9\{2x - 3[5x + 2(-3x + y^0 - 2)]\}$

35. $10\{2x + 3[5x - 2(-3x + y^1 - 2)]\}$

36. $5^4 \div 625 \div 5^2 \cdot 5^7 \div 5^3$

Solve.   [1.3a]

37. One side of a square is 5 yd less than a side of an equilateral triangle. If the perimeter of the square is the same as the perimeter of the triangle, what is the length of a side of the square? of the triangle?

Solve.   [1.6c, e]

38. $|5x - 8| \geq 32$

39. $|5x - 8| < 32$

40. $|5x - 8| = 32$

41. $|5x - 8| = -32$

Synthesis

42. ◈ 〰 A student makes a mistake when using a grapher to draw $4x + 5y = 12$ and the following screen appears. Use algebra to show that a mistake has been made. What do you think the mistake was?

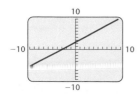

43. ◈ 〰 A student makes a mistake when using a grapher to draw $5x - 2y = 3$ and the following screen appears. Use algebra to show that a mistake has been made. What do you think the mistake was?

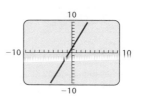

2.5 More on Graphing Linear Equations

a Graphing Using Intercepts

The **x-intercept** of the graph of a linear equation or function is the point at which the graph crosses the x-axis. The **y-intercept** is the point at which the graph crosses the y-axis. We know from geometry that only one line can be drawn through two given points. Thus, if we know the intercepts, we can graph the line. To ensure that a computation error has not been made, it is a good idea to calculate a third point as a check.

Many equations of the type $Ax + By = C$ can be graphed conveniently using intercepts.

> A **y-intercept** is a point $(0, b)$. To find b, let $x = 0$ and solve for y.
> An **x-intercept** is a point $(a, 0)$. To find a, let $y = 0$ and solve for x.

Example 1 Find the intercepts of $3x + 2y = 12$ and then graph the line.

To find the y-intercept, we let $x = 0$ and see what y must be. Covering up the $3x$ or ignoring it amounts to letting x be 0. So we cover up $3x$ and see that $2y = 12$. Then y must be 6. The y-intercept is $(0, 6)$.

To find the x-intercept, we can cover up the y-term and look at the rest of the equation. We have $3x = 12$, or $x = 4$. The x-intercept is $(4, 0)$.

We plot these points and draw the line, using a third point as a check. We choose $x = 6$ and solve for y:

$$3(6) + 2y = 12$$
$$18 + 2y = 12$$
$$2y = -6$$
$$y = -3.$$

We plot $(6, -3)$ and note that it is on the line.

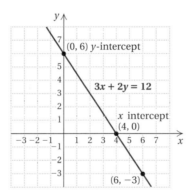

Do Exercise 1.

b Graphing Using the Slope and the y-Intercept

We can also graph a line using its slope and y-intercept.

Example 2 Graph: $y = -\frac{2}{3}x + 1$.

This equation is in slope–intercept form, $y = mx + b$. The y-intercept is $(0, 1)$. We plot $(0, 1)$. We can think of the slope as $\frac{-2}{3}$. Starting at the y-intercept and using the slope, we find another point by moving 2 units down (since the numerator is *negative* and corresponds to the change in y) and 3 units to the right (since the denominator is *positive* and corresponds to the change in x). We get to a new point, $(3, -1)$. In a similar manner, we can move from the point $(3, -1)$ to find another point, $(6, -3)$.

Objectives

a Graph linear equations using intercepts.

b Given a linear equation in slope–intercept form, use the slope and the y-intercept to graph the line.

c Graph linear equations of the form $x = a$ or $y = b$.

d Given the equations of two lines, determine whether their graphs are parallel or whether they are perpendicular.

For Extra Help

TAPE 6   TAPE 5A   MAC WIN   CD-ROM

1. Find the intercepts of $4y - 12 = -6x$ and then graph the line.

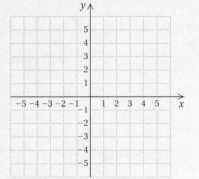

Answer on page A-14

Graph using the slope and the *y*-intercept.

2. $y = \dfrac{3}{2}x + 1$

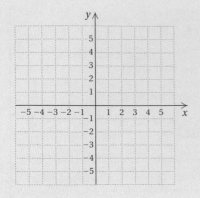

3. $f(x) = \dfrac{3}{4}x - 2$

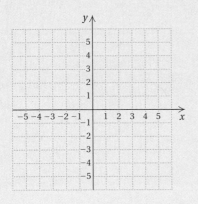

4. $g(x) = -\dfrac{3}{5}x + 5$

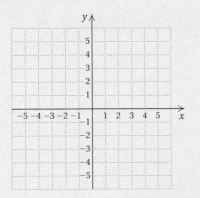

Suppose we think of the slope $-\dfrac{2}{3}$ as $\dfrac{2}{-3}$. Then we can start again at (0, 1), but this time we move 2 units up (since the numerator is *positive* and corresponds to the change in *y*) and 3 units to the left (since the denominator is *negative* and corresponds to the change in *x*). We get another point on the graph, $(-3, 3)$, and from it we can obtain $(-6, 5)$ and others in a similar manner. We plot the points and draw the line.

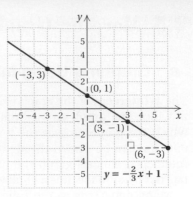

Example 3   Graph: $f(x) = \frac{2}{5}x + 4$.

First, we plot the *y*-intercept, (0, 4). We then consider the slope $\frac{2}{5}$. Starting at the *y*-intercept and using the slope, we find another point by moving 2 units up (since the numerator is *positive* and corresponds to the change in *y*) and 5 units to the right (since the denominator is *positive* and corresponds to the change in *x*). We get to a new point, (5, 6).

We can also think of the slope $\frac{2}{5}$ as $\frac{-2}{-5}$. We again start at the *y*-intercept, (0, 4). We move 2 units down (since the numerator is *negative* and corresponds to the change in *y*) and 5 units to the left (since the denominator is *negative* and corresponds to the change in *x*). We get to another new point, $(-5, 2)$. We plot the points and draw the line.

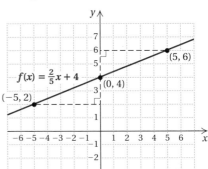

Do Exercises 2–5. (Exercise 5 is on the following page.)

c   Horizontal and Vertical Lines

Some equations have graphs that are parallel to one of the axes. This happens when either *A* or *B* is 0 in $Ax + By = C$. These equations have a missing variable. In the following example, *x* is missing.

Example 4   Graph: $y = 3$.

Since *x* is missing, any number for *x* will do. Thus all ordered pairs (*x*, 3) are solutions. The graph is a **horizontal line** parallel to the *x*-axis.

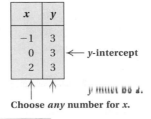

x	*y*	
-1	3	
0	3	← *y*-intercept
2	3	

↑ Choose *any* number for *x*.

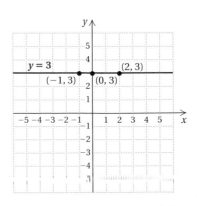

What about the slope of a horizontal line? In Example 4, consider the points $(-1, 3)$ and $(2, 3)$, which are on the line $y = 3$. The change in y is $3 - 3$, or 0. The change in x is $-1 - 2$, or -3. Thus,

$$m = \frac{3 - 3}{-1 - 2} = \frac{0}{-3} = 0.$$

Any two points on a horizontal line have the same y-coordinate. Thus the change in y is always 0, so the slope is 0.

We can also determine the slope by noting that $y = 3$ can be written in slope–intercept form as $y = 0x + 3$, or $f(x) = 0x + 3$. From this equation, we read that the slope is 0. A function of this type is called a **constant function.** We can express it in the form $y = b$, or $f(x) = b$. Its graph is a horizontal line that crosses the y-axis at $(0, b)$.

Do Exercises 6 and 7.

In the following example, y is missing and the graph is parallel to the y-axis.

Example 5 Graph: $x = -2$.

Since y is missing, any number for y will do. Thus all ordered pairs $(-2, y)$ are solutions. The graph is a **vertical line** parallel to the y-axis.

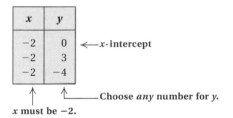

x	y	
-2	0	←—x-intercept
-2	3	
-2	-4	

↑ x must be -2. ↑ Choose *any* number for y.

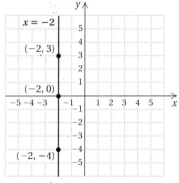

This graph is not the graph of a function because it fails the vertical-line test. There is a vertical line (itself) that crosses the graph more than once.

What about the slope of a vertical line? In Example 5, consider the points $(-2, 3)$ and $(-2, -4)$, which are on the line $x = -2$. The change in y is $3 - (-4)$, or 7. The change in x is $-2 - (-2)$, or 0. Thus,

$$m = \frac{3 - (-4)}{-2 - (-2)} = \frac{7}{0}. \qquad \textbf{Undefined}$$

Since division by 0 is not defined, the slope of this line is not defined. Any two points on a vertical line have the same x-coordinate. Thus the change in x is always 0, so the slope of any vertical line is undefined.

The following summarizes horizontal and vertical lines and their equations.

> ▶ **HORIZONTAL LINE**
>
> The graph of $y = b$, or $f(x) = b$, is a horizontal line with y-intercept $(0, b)$. It is the graph of a constant function with slope 0.
>
> **VERTICAL LINE**
>
> The graph of $x = a$ is a vertical line through the point $(a, 0)$. The slope is undefined. It is not the graph of a function.

Do Exercises 8–10 on the following page.

---

5. $y = -\dfrac{5}{3}x - 4$

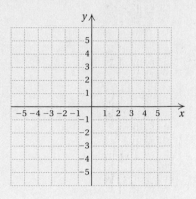

Graph and determine the slope.

6. $f(x) = -4$

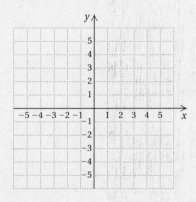

7. $y = 3.6$

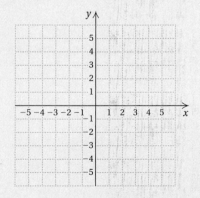

Answers on page A-14

Graph.

8. $x = -5$

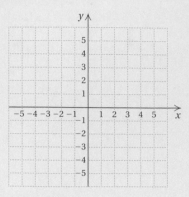

9. $8x - 5 = 19$
(*Hint*: Solve for *x*.)

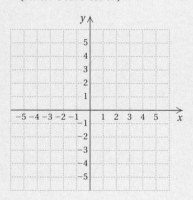

10. Determine, if possible, the slope of each line.

a) $x = -12$
b) $y = 6$
c) $2y + 7 = 11$
d) $x = 0$
e) $y = -\frac{3}{4}$
f) $10 - 5x = 15$

We have graphed linear equations in several ways in this chapter. Although, in general, you can use any method that works best for you, the following are some guidelines.

> To graph a linear equation:
>
> 1. Is the equation of the type $x = a$ or $y = b$? If so, the graph will be a line parallel to an axis; $x = a$ is vertical and $y = b$ is horizontal.
> 2. If the line is of the type $y = mx$, both intercepts are the origin $(0, 0)$. Plot $(0, 0)$ and one other point.
> 3. If the line is of the type $y = mx + b$, plot the *y*-intercept and one other point.
> 4. If the equation is of the form $Ax + By = C$, but not of the form $x = a$, $y = b$, $y = mx$, or $y = mx + b$, graph using intercepts. If the intercepts are too close together, choose another point farther from the origin.
> 5. In all cases, use a third point as a check.

d | Parallel and Perpendicular Lines

Parallel Lines

If two lines are vertical, they are parallel. How can we tell whether nonvertical lines are parallel? We examine their slopes and *y*-intercepts.

> Two nonvertical lines are **parallel** if they have the same slope and different *y*-intercepts.

Example 6 Determine whether the graphs of

$$y - 3x = 1 \quad \text{and} \quad 3x + 2y = -2$$

are parallel.

To determine whether lines are parallel, we first find their slopes. Thus we first find the slope–intercept form of each equation by solving for *y*:

$$
\begin{aligned}
y - 3x &= 1 & 3x + 2y &= -2 \\
y &= 3x + 1; & 2y &= -3x - 2 \\
& & y &= \tfrac{1}{2}(-3x - 2) \\
& & &= \tfrac{1}{2}(-3x) - \tfrac{1}{2}(2) \\
& & &= -\tfrac{3}{2}x - 1.
\end{aligned}
$$

The slopes, 3 and $-\frac{3}{2}$, are different. Thus the lines are not parallel, as the graphs shown at right confirm.

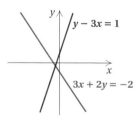

Answers on page A-14

Example 7   Determine whether the graphs of

$$3x - y = -5 \quad \text{and} \quad y - 3x = -2$$

are parallel.

We first find the slope–intercept form of each equation by solving for y:

$$3x - y = -5 \qquad\qquad y - 3x = -2$$
$$-y = -3x - 5 \qquad\qquad y = 3x - 2.$$
$$y = 3x + 5;$$

The slopes, 3, are the same. The
y-intercepts are different. Thus the lines
are parallel, as the graphs seem to confirm.

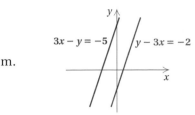

Do Exercises 11–13.

If one line is vertical and another is horizontal, they are perpendicular.
Otherwise, how can we tell whether two lines are perpendicular?

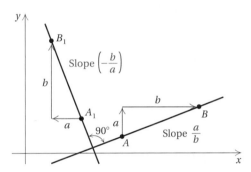

Consider a line $\overleftrightarrow{AB}$, as shown in the figure above, with slope a/b. Then
think of rotating the line 90° to get a line $\overleftrightarrow{A_1 B_1}$ perpendicular to $\overleftrightarrow{AB}$. For
the new line, the rise and the run are interchanged, but the run is now
negative. Thus the slope of the new line is $-b/a$, which is the opposite of
the reciprocal of the slope of the first line. Also note that when we multiply the slopes, we get

$$\frac{a}{b}\left(-\frac{b}{a}\right) = -1.$$

This is the condition under which lines will be perpendicular.

> ▶ Two nonvertical lines are **perpendicular** if the product of their slopes is
> -1. (If one line has slope m, the slope of a line perpendicular to it is $-1/m$.
> That is, to find the slope of a line perpendicular to a given line, we take
> the reciprocal of the given slope and change the sign.)
>
> Lines are also perpendicular if one of them is vertical ($x = a$) and one of
> them is horizontal ($y = b$).

Example 8   Determine whether the graphs of $5y = 4x + 10$ and $4y = -5x + 4$ are perpendicular.

To determine whether the lines are perpendicular, we determine
whether the product of their slopes is -1. We first find the slope–intercept
form of each equation by solving for y.

Determine whether the graphs of
the given pair of lines are parallel.

11. $x + 4 = y,$
$\quad y - x = -3$

12. $y + 4 = 3x,$
$\quad 4x - y = -7$

13. $y = 4x + 5,$
$\quad 2y = 8x + 10$

Answers on page A-14

Determine whether the graphs of the given pair of lines are perpendicular.

14. $2y - x = 2$,
$y + 2x = 4$

15. $3y = 2x + 15$,
$2y = 3x + 10$

Answers on page A-14

We have

$$5y = 4x + 10 \qquad\qquad 4y = -5x + 4$$
$$y = \tfrac{1}{5}(4x + 10) \qquad\qquad y = \tfrac{1}{4}(-5x + 4)$$
$$= \tfrac{1}{5}(4x) + \tfrac{1}{5}(10) \qquad\qquad = \tfrac{1}{4}(-5x) + \tfrac{1}{4}(4)$$
$$= \tfrac{4}{5}x + 2; \qquad\qquad = -\tfrac{5}{4}x + 1.$$

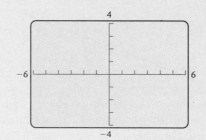

The slope of the first line is $\frac{4}{5}$, and the slope of the second line is $-\frac{5}{4}$. The product of the slopes is -1; that is, $\frac{4}{5} \cdot \left(-\frac{5}{4}\right) = -1$. Thus the lines are perpendicular.

Do Exercises 14 and 15.

Calculator Spotlight

 Squaring Viewing Windows; Visualizing Parallel and Perpendicular Lines

Squaring a Viewing Window. Consider the $[-10, 10, -10, 10]$ viewing window on the left below. Note that the distance between units is not visually the same on both axes. In this case, the length of the interval shown on the y-axis is about two-thirds of the length of the interval on the x-axis. If we change the dimensions of the window to $[-6, 6, -4, 4]$, we get a graph for which the units are visually about the same on both axes. Creating such a window is called **squaring**. On a TI-83 grapher, there is a ZSQUARE feature for automatic window squaring. This feature alters the standard window dimensions to $[-15.1613, 15.1613, -10, 10]$.

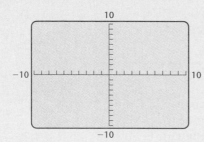

Each of the following is a graph of the line $y = 2x - 3$, but the viewing windows are different. When the window is square, as shown on the right, we get an accurate representation of the *slope* of the line. This is important when we need to visualize perpendicular lines.

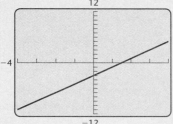

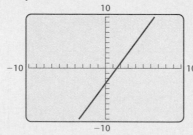

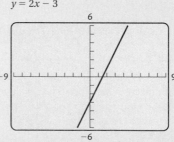

Squared window

Exercises

1. Graph each pair of equations in Margin Exercises 11–13. Check visually whether the lines appear to be parallel.
2. Graph each pair of equations in Margin Exercises 14 and 15 using a squared viewing window. Check visually whether the lines appear to be perpendicular.

Exercise Set 2.5

a Find the intercepts and then graph the line.

1. $x - 2 = y$

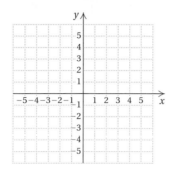

2. $x + 3 = y$

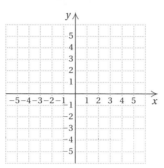

3. $x + 3y = 6$

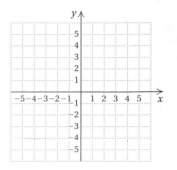

4. $x - 2y = 4$

5. $2x + 3y = 6$

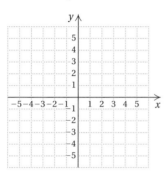

6. $5x - 2y = 10$

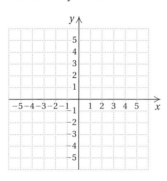

7. $f(x) = -2 - 2x$

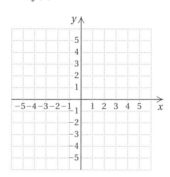

8. $g(x) = 5x - 5$

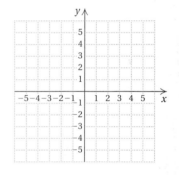

9. $5y = -15 + 3x$

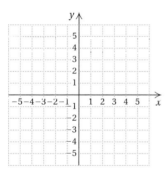

10. $5x - 10 = 5y$

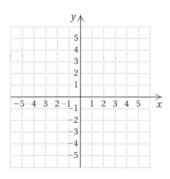

11. $2x - 3y = 6$

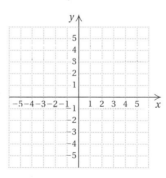

12. $4x + 5y = 20$

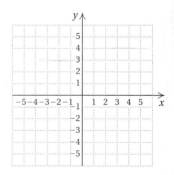

13. $2.8y - 3.5x = -9.8$

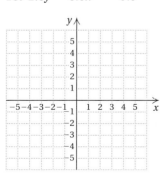

14. $10.8x - 22.68 = 4.2y$

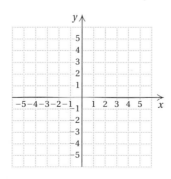

15. $5x + 2y = 7$

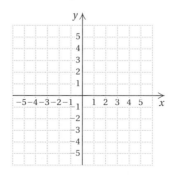

16. $3x - 4y = 10$

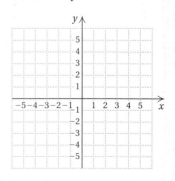

b Graph using the slope and the y-intercept.

17. $y = \dfrac{5}{2}x + 1$

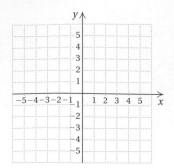

18. $y = \dfrac{2}{5}x - 4$

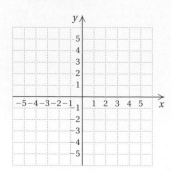

19. $f(x) = -\dfrac{5}{2}x - 4$

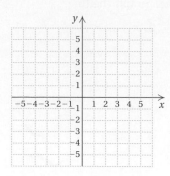

20. $f(x) = \dfrac{2}{5}x + 3$

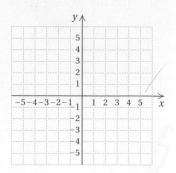

21. $x + 2y = 4$

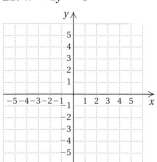

22. $x - 3y = 6$

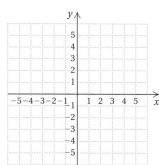

23. $4x - 3y = 12$

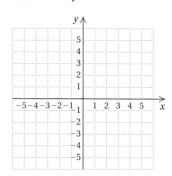

24. $2x + 6y = 12$

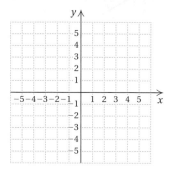

25. $f(x) = \dfrac{1}{3}x - 4$

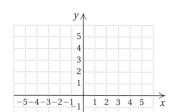

26. $g(x) = -0.25x + 2$

27. $5x + 4 \cdot f(x) = 4$
(*Hint*: Solve for $f(x)$.)

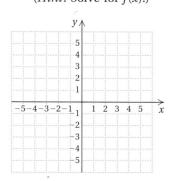

28. $3 \cdot f(x) = 4x + 6$

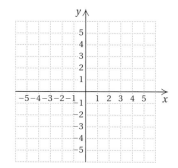

c Graph and, if possible, determine the slope.

29. $x = 1$

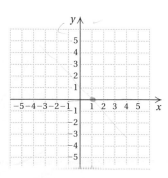

30. $x = -4$

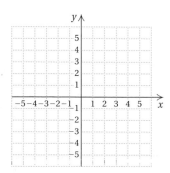

31. $y = -1$

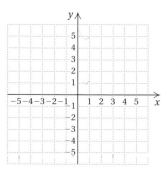

32. $y = \dfrac{3}{2}$

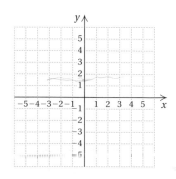

33. $f(x) = -6$

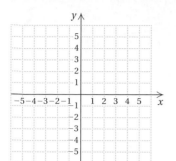

34. $f(x) = 2$

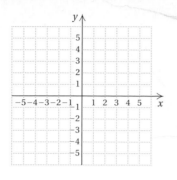

35. $y = 0$

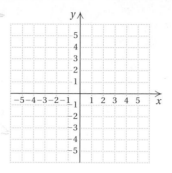

36. $x = 0$

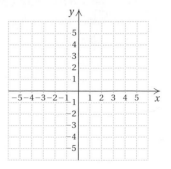

37. $2 \cdot f(x) + 5 = 0$

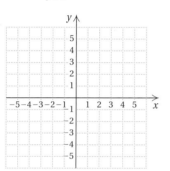

38. $4 \cdot g(x) + 3x = 12 + 3x$

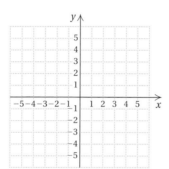

39. $7 - 3x = 4 + 2x$

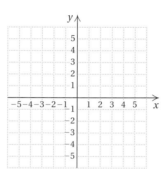

40. $3 - f(x) = 2$

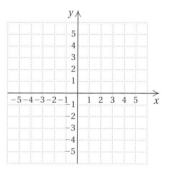

d Determine whether the graphs of the given pair of lines are parallel.

41. $x + 6 = y$,
$y - x = -2$

42. $2x - 7 = y$,
$y - 2x = 8$

43. $y + 3 = 5x$,
$3x - y = -2$

44. $y + 8 = -6x$,
$-2x + y = 5$

45. $y = 3x + 9$,
$2y = 6x - 2$

46. $y + 7x = -9$,
$-3y = 21x + 7$

47. $12x = 3$,
$-7x = 10$

48. $5y = -2$,
$\frac{3}{4}x = 16$

Determine whether the graphs of the given pair of lines are perpendicular.

49. $y = 4x - 5$,
$4y = 8 - x$

50. $2x - 5y = -3$,
$2x + 5y = 4$

51. $x + 2y = 5$,
$2x + 4y = 8$

52. $y = -x + 7$,
$y = x + 3$

53. $2x - 3y = 7$,
$2y - 3x = 10$

54. $x = y$,
$y = -x$

55. $2x = 3$,
$-3y = 6$

56. $-5y = 10$,
$y = -\frac{4}{9}$

Write in scientific notation. [R.7c]

57. 53,000,000,000

58. 0.000047

59. 0.018

60. 99,902,000

Write in decimal notation. [R.7c]

61. 2.13×10^{-5}

62. 9.01×10^{8}

63. 2×10^{4}

64. 8.5677×10^{-2}

Factor. [R.5d]

65. $9x - 15y$

66. $12a + 21ab$

67. $21p - 7pq + 14p$

68. $64x - 128y + 256$

Synthesis

69. ◈ Under what conditions will the x- and the y-intercepts of a line be the same? What would the equation for such a line look like?

70. ◈ Explain why the slope of a vertical line is undefined but the slope of a horizontal line is 0.

71. Find an equation of a horizontal line that passes through the point $(-2, 3)$.

72. Find an equation of a vertical line that passes through the point $(-2, 3)$.

73. Find the value of a such that the graphs of $5y = ax + 5$ and $\frac{1}{4}y = \frac{1}{10}x - 1$ are parallel.

74. Find the value of k such that the graphs of $x + 7y = 70$ and $y + 3 = kx$ are perpendicular.

75. Write an equation of the line that has x-intercept $(-3, 0)$ and y-intercept $\left(0, \frac{2}{5}\right)$.

76. Find the coordinates of the point of intersection of the graphs of the equations $x = -4$ and $y = 5$.

77. Write an equation for the x-axis. Is this equation a function?

78. Write an equation for the y-axis. Is this equation a function?

79. Find the value of m in $y = mx + 3$ so that the x-intercept of its graph will be $(4, 0)$.

80. Find the value of b in $2y = -7x + 3b$ so that the y-intercept of its graph will be $(0, -13)$.

81. Match each sentence with the most appropriate graph.

a) The rate at which fluids were given intravenously was doubled after 3 hr.

b) The rate at which fluids were given intravenously was gradually reduced to 0.

c) The rate at which fluids were given intravenously remained constant for 5 hr.

d) The rate at which fluids were given intravenously was gradually increased.

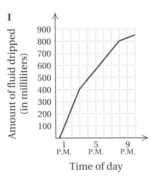

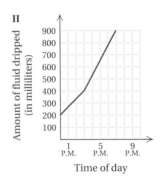

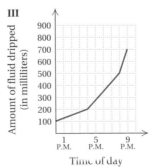

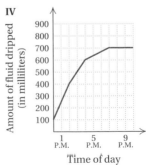

Practice graphing and identifying the graphs of linear equations.

Collaborative Learning Manual

2.6 Finding Equations of Lines

In this section, we will learn to find a linear equation for a line for which we have been given two pieces of information.

a | Finding an Equation of a Line When the Slope and the *y*-Intercept Are Given

If we know the slope and the *y*-intercept of a line, we can find an equation of the line using the slope–intercept equation $y = mx + b$.

Example 1 A line has slope -0.7 and *y*-intercept $(0, 13)$. Find an equation of the line.

We use the slope–intercept equation and substitute -0.7 for m and 13 for b:

$$y = mx + b$$
$$y = -0.7x + 13.$$

Do Exercise 1.

b | Finding an Equation of a Line When the Slope and a Point Are Given

If we know the slope of a line and a certain point on that line, we can find an equation of the line using the slope–intercept equation

$$y = mx + b.$$

Example 2 Find an equation of the line with slope 5 and containing the point $\left(\frac{1}{2}, -1\right)$.

The point $\left(\frac{1}{2}, -1\right)$ is on the line, so it is a solution. Thus we can substitute $\frac{1}{2}$ for x and -1 for y in $y = mx + b$. We also substitute 5 for m, the slope. Then we solve for b:

$$y = mx + b$$
$$-1 = 5 \cdot \left(\tfrac{1}{2}\right) + b \quad \text{Substituting}$$
$$-1 = \tfrac{5}{2} + b$$
$$-1 - \tfrac{5}{2} = b$$
$$-\tfrac{2}{2} - \tfrac{5}{2} = b$$
$$-\tfrac{7}{2} = b. \quad \text{Solving for } b$$

We then use the equation $y = mx + b$ and substitute 5 for m and $-\frac{7}{2}$ for b:

$$y = 5x - \tfrac{7}{2}.$$

Do Exercises 2–5.

Answers on page A-15

Objectives

a | Find an equation of a line when the slope and the *y*-intercept are given.

b | Find an equation of a line when the slope and a point are given.

c | Find an equation of a line when two points are given.

d | Given a line and a point not on the given line, find an equation of the line parallel to the line and containing the point, and find an equation of the line perpendicular to the line and containing the point.

e | Solve applied problems involving linear functions.

For Extra Help

TAPE 6   TAPE 5B   MAC WIN   CD-ROM

1. A line has slope 3.4 and *y*-intercept $(0, -8)$. Find an equation of the line.

Find an equation of the line with the given slope and containing the given point.

2. $m = -5$, $(-4, 2)$

3. $m = 3$, $(1, -2)$

4. $m = 8$, $(3, 5)$

5. $m = -\dfrac{2}{3}$, $(1, 4)$

6. Find an equation of the line containing the points $(4, -3)$ and $(1, 2)$.

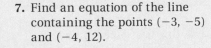

c | Finding an Equation of a Line When Two Points Are Given

We can also use the slope–intercept equation to find an equation of a line when two points are given.

Example 3 Find an equation of the line containing the points $(2, 3)$ and $(-6, 1)$.

First, we find the slope:

$$m = \frac{3 - 1}{2 - (-6)} = \frac{2}{8}, \text{ or } \frac{1}{4}.$$

Now we have the slope and two points. We then proceed as we did in Example 2, using either point. We choose $(2, 3)$ and substitute 2 for x, 3 for y, and $\frac{1}{4}$ for m:

$$y = mx + b$$
$$3 = \frac{1}{4} \cdot 2 + b \qquad \text{Substituting}$$
$$3 = \frac{1}{2} + b$$
$$3 - \frac{1}{2} = \frac{1}{2} + b - \frac{1}{2}$$
$$\frac{6}{2} - \frac{1}{2} = b$$
$$\frac{5}{2} = b. \qquad \text{Solving for } b$$

Finally, we use the equation $y = mx + b$ and substitute $\frac{1}{4}$ for m and $\frac{5}{2}$ for b:

$$y = \frac{1}{4}x + \frac{5}{2}.$$

7. Find an equation of the line containing the points $(-3, -5)$ and $(-4, 12)$.

Do Exercises 6 and 7.

d | Finding an Equation of a Line Parallel or Perpendicular to a Given Line Through a Point Off the Line

We can also use the method of Example 2 to find equations of lines through a point off the line parallel and perpendicular to a given line.

Example 4 Find an equation of the line containing the point $(-1, 3)$ and parallel to the line $2x + y = 10$.

An equation parallel to the given line $2x + y = 10$ must have the same slope. To find that slope, we first find the slope–intercept equation by solving for y:

$$2x + y = 10$$
$$y = -2x + 10.$$

Thus the new line through $(-1, 3)$ must also have slope -2.

We then proceed as in Example 2, substituting -1 for x and 3 for y in $y = mx + b$. We also substitute -2 for m, the slope. Then we solve for b:

$$y = mx + b$$
$$3 = -2(-1) + b \qquad \text{Substituting}$$
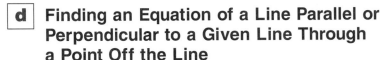
$$1 = b. \qquad \text{Solving for } b$$

Answers on page A-15

We then use the equation $y = mx + b$ and substitute -2 for m and 1 for b:

$$y = -2x + 1.$$

The given line $y = -2x + 10$ and the new line $y = -2x + 1$ have the same slope but different y-intercepts. Thus their graphs are parallel.

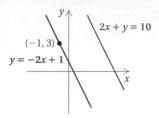

Do Exercise 8.

Example 5 Find an equation of the line containing the point $(2, -3)$ and perpendicular to the line $4y - x = 20$.

To find the slope of the given line, we first find its slope–intercept form by solving for y:

$$4y - x = 20$$
$$4y = x + 20$$
$$\frac{4y}{4} = \frac{x + 20}{4}$$
$$y = \tfrac{1}{4}x + 5.$$

We know that the slope of the perpendicular line must be the opposite of the reciprocal of $\tfrac{1}{4}$. Thus the new line through $(2, -3)$ must have slope -4.

We now substitute 2 for x and -3 for y in $y = mx + b$. We also substitute -4 for m, the slope. Then we solve for b:

$$y = mx + b$$
$$-3 = -4(2) + b \qquad \text{Substituting}$$
$$-3 = -8 + b$$
$$5 = b. \qquad \text{Solving for } b$$

Finally, we use the equation $y = mx + b$ and substitute -4 for m and 5 for b:

$$y = -4x + 5.$$

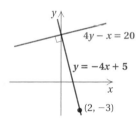

Do Exercise 9.

e | Applications of Linear Functions

When the essential parts of a problem are described in mathematical language, we say that we have a **mathematical model.** We have already studied many kinds of mathematical models in this text—for example, the formulas in Section 1.2 and the functions in Section 2.2. Here we study linear functions as models.

8. Find an equation of the line containing the point $(2, -1)$ and parallel to the line $8x = 7y - 24$.

9. Find an equation of the line containing the point $(5, 4)$ and perpendicular to the line $2x - 4y = 12$.

Answers on page A-15

10. *Cable TV Service.* Clear County Cable TV Service charges a $25 installation fee and $20 per month for basic service.

a) Formulate a linear function for cost $C(t)$ for t months of cable TV service.

b) Graph the model.

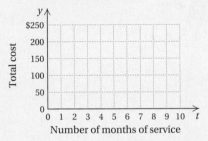

Number of months of service

c) Use the model to determine the cost of $8\frac{1}{2}$ months of service.

Example 6 *Cost Projections.* Cleartone Communications charges $50 for a cellular telephone and $40 per month for phone calls under its economy plan.

a) Formulate a linear function for total cost $C(t)$, where t is the number of months of telephone usage.

b) Graph the model.

c) Use the model to determine the cost of $3\frac{1}{2}$ months of service.

a) The problem describes a situation in which a monthly fee is charged after an initial purchase has been made. After 1 month of service, the total cost is

$$\$50 + \$40 \cdot 1 = \$90.$$

After 2 months of service, the total cost is

$$\$50 + \$40 \cdot 2 = \$130.$$

We can generalize that after t months of service, the total cost $C(t)$ is $C(t) = 50 + 40t$, where $t \geq 0$ (since there cannot be a negative number of months). Note that $C(t)$ is a way of saying that the cost C of the phone is a function of time t.

b) Before graphing, we rewrite the model in slope–intercept form:

$$C(t) = 40t + 50.$$

We note that the y-intercept is $(0, 50)$ and the slope, or rate of change, is $40 per month. We plot $(0, 50)$, and from there we count $40 *up* and 1 month *to the right*. This takes us to $(1, 90)$. We then draw a line through the points, calculating a third value as a check:

$$C(4) = 40 \cdot 4 + 50 = 210.$$

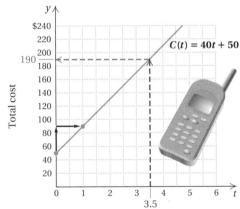

Number of months of service

c) To find the cost for $3\frac{1}{2}$ months, we determine $C(3.5)$:

$$C(3.5) = 40(3.5) + 50 = 190.$$

Thus it would cost $190 for $3\frac{1}{2}$ months of service.

Do Exercise 10.

Answers on page A-15

Exercise Set 2.6

a Find an equation of the line having the given slope and y-intercept.

1. Slope: -8; y-intercept: $(0, 4)$

2. Slope: 5; y-intercept: $(0, -3)$

3. Slope: 2.3; y-intercept: $(0, -1)$

4. Slope: -9.1; y-intercept: $(0, 2)$

Find a linear function $f(x) = mx + b$ whose graph has the given slope and y-intercept.

5. Slope: $-\frac{7}{3}$; y-intercept: $(0, -5)$

6. Slope: $\frac{4}{5}$; y-intercept: $(0, 28)$

7. Slope: $\frac{2}{3}$; y-intercept: $\left(0, \frac{5}{8}\right)$

8. Slope: $-\frac{7}{8}$; y-intercept: $\left(0, -\frac{7}{11}\right)$

b Find an equation of the line having the given slope and containing the given point.

9. $m = 5$, $(4, 3)$

10. $m = 4$, $(5, 2)$

11. $m = -3$, $(9, 6)$

12. $m = -2$, $(2, 8)$

13. $m = 1$, $(-1, -7)$

14. $m = 3$, $(-2, -2)$

15. $m = -2$, $(8, 0)$

16. $m = -3$, $(-2, 0)$

17. $m = 0$, $(0, -7)$

18. $m = 0$, $(0, 4)$

19. $m = \frac{2}{3}$, $(1, -2)$

20. $m = -\frac{4}{5}$, $(2, 3)$

c Find an equation of the line containing the given pair of points.

21. $(1, 4)$ and $(5, 6)$

22. $(2, 5)$ and $(4, 7)$

23. $(-3, -3)$ and $(2, 2)$

24. $(-1, -1)$ and $(9, 9)$

25. $(-4, 0)$ and $(0, 7)$

26. $(0, -5)$ and $(3, 0)$

27. $(-2, -3)$ and $(-4, -6)$

28. $(-4, -7)$ and $(-2, -1)$

29. $(0, 0)$ and $(6, 1)$

30. $(0, 0)$ and $(-4, 7)$

31. $\left(\frac{1}{4}, -\frac{1}{2}\right)$ and $\left(\frac{3}{4}, 6\right)$

32. $\left(\frac{2}{3}, \frac{3}{2}\right)$ and $\left(-3, \frac{5}{6}\right)$

d Write an equation of the line containing the given point and parallel to the given line.

33. $(3, 7)$; $x + 2y = 6$

34. $(0, 3)$; $2x - y = 7$

35. $(2, -1)$; $5x - 7y = 8$

36. $(-4, -5)$; $2x + y = -3$

37. $(-6, 2)$; $3x = 9y + 2$

38. $(-7, 0)$; $2y + 5x = 6$

Write an equation of the line containing the given point and perpendicular to the given line.

39. $(2, 5)$; $2x + y = 3$

40. $(4, 1)$; $x - 3y = 9$

41. $(3, -2)$; $3x + 4y = 5$

42. $(-3, -5)$; $5x - 2y = 4$

43. $(0, 9)$; $2x + 5y = 7$

44. $(-3, -4)$; $6y - 3x = 2$

e   Solve.

45. *Moving Costs.* Musclebound Movers charges $85 plus $40 an hour to move households across town.

   a) Formulate a linear function for total cost $C(t)$ for t hours of moving.
   b) Graph the model.
   c) Use the model to determine the cost of $6\frac{1}{2}$ hr of moving service.

46. *Deluxe Cable TV Service.* Twin Cities Cable TV Service charges a $35 installation fee and $20 per month for basic service.

   a) Formulate a linear function for total cost $C(t)$ for t months of cable TV service.
   b) Graph the model.
   c) Use the model to determine the cost of 9 months of service.

47. *Value of a Fax Machine.* FaxMax bought a multifunction fax machine for $750. The value $V(t)$ of the machine depreciates (declines) at a rate of $25 per month.

   a) Formulate a linear function for the value $V(t)$ of the machine after t months.
   b) Graph the model.
   c) Use the model to determine the value of the machine after 13 months.

48. *Value of a Computer.* SendUp Graphics bought a computer for $3800. The value $V(t)$ of the computer depreciates at a rate of $50 per month.

   a) Formulate a linear function for the value $V(t)$ of the computer after t months.
   b) Graph the model.
   c) Use the model to determine the value of the computer after $10\frac{1}{2}$ months.

Skill Maintenance

Solve.   [1.4c], [1.5a], [1.6c, d, e]

49. $2x + 3 > 51$

50. $|2x + 3| = 51$

51. $2x + 3 \leq 51$

52. $2x + 3 \leq 5x - 4$

53. $|2x + 3| \leq 13$

54. $|2x + 3| = |x - 4|$

55. $|5x - 4| = -8$

56. $-12 \leq 2x + 3 < 51$

Synthesis

Determine m and b in each application and explain their meaning.

57. ◆ *Cost of a Movie Ticket.* The average price $P(t)$, in dollars, of a movie ticket can be estimated by the function

   $P(t) = 0.1522t + 4.29$,

where t is the number of years since 1990 (*Source:* Motion Picture Association of America).

58. ◆ *Cost of a Taxi Ride.* The cost $C(d)$, in dollars, of a taxi ride in Pelham is given by

   $C(d) = 0.75d + 2$,

where d is the number of miles traveled.

2.7 Mathematical Modeling with Linear Functions

We have considered many linear functions or models in this chapter. Many of them were derived from data gathered in real-world situations. How can we find a linear model from the data? In this section, we consider two methods in the examples and another, called **regression**, in a Calculator Spotlight.

a Data Analysis: Drawing a Linear Model

Often we will encounter data in an application. To determine whether a linear function fits the data, we graph ordered pairs of data, forming a **scatterplot**. Then we decide whether we can draw a straight line that fits the data well.

Let's look at two sets of data and their scatterplots. Does it appear that a linear function could fit either set of data?

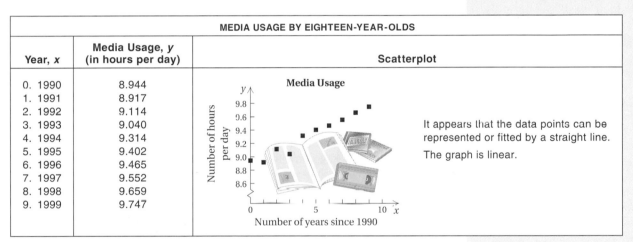

MEDIA USAGE BY EIGHTEEN-YEAR-OLDS		
Year, x	Media Usage, y (in hours per day)	Scatterplot
0. 1990	8.944	
1. 1991	8.917	
2. 1992	9.114	
3. 1993	9.040	
4. 1994	9.314	It appears that the data points can be represented or fitted by a straight line.
5. 1995	9.402	
6. 1996	9.465	The graph is linear.
7. 1997	9.552	
8. 1998	9.659	
9. 1999	9.747	

Source: Veronis, Suhler & Associates, Inc., New York

GROWTH OF WORLD WIDE WEB SITES		
Year, x	Number of Web Sites, y (in millions)	Scatterplot
0. 1995	8	
1. 1996	11	
2. 1997	27	It appears that the data points cannot be represented by a straight line.
3. 1998	50	
4. 1999	78	The graph is nonlinear.
5. 2000	140	

Source: International Data Corporation, 1996

1. *Study Time and Test Scores.* A professor gathered the following data comparing study time and test scores.

a) Make a scatterplot of the data (graph the ordered pairs).

b) Use extrapolation to estimate the test score received if one has studied for 23 hr.

Study Time (in hours)	Test Score (in percent)
19	83
20	85
21	88
22	91
23	?

Looking at the scatterplots, we see that the media usage data seem to be rising in a manner to suggest that a linear function might fit, although a "perfect" straight line cannot be drawn through the data points. However, a linear function does not seem to fit the web site data.

Let's try to use the data in the first table to make a prediction.

Example 1 *Media Usage by Eighteen-Year-Olds.* Consider the preceding data and scatterplot. Use extrapolation to estimate media usage in the year 2000 ($x = 10$).

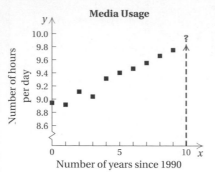
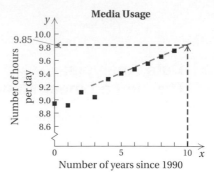

We analyze the data and note that they tend to follow a straight line past 1994 ($x = 4$). Keeping this in mind, we draw a "representative" line through the data and beyond. To estimate a value for the year 2000 ($x = 10$), we draw a vertical line up from 10 to the representative line. We then move to the left and read a value from the vertical axis. That value is about 9.85.

When we estimate to find a "go-beyond value," as we did in Example 1, we are using a process called **extrapolation**. Note that the process of drawing a representative line is arbitrary. Thus the answer we provide might differ from the one you find. This is not important in our work here. The idea is to use a line to go beyond the data to make an estimated prediction.

Do Exercise 1.

b Data Analysis: Using Two Points to Find a Linear Model

In the following example, we refine the process of finding a linear model. First, we choose two points and find an equation for the linear function through these points. Then we use the equation to make a prediction.

Answers on page A-16

Example 2 *Media Usage by Eighteen-Year-Olds.* Choose two points from the data (this can vary) regarding media usage.

a) Use the two points to find a linear function that fits the data.

b) Use the function to predict media usage in the year 2000 ($x = 10$).

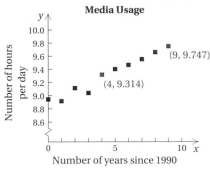

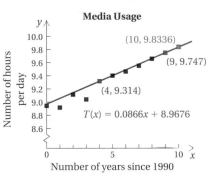

2. *Media Usage.* Repeat Example 2 using the data points (5, 9.402) and (9, 9.747). Compare the results with those of Example 2.

a) We note in the data chart on p. 209 that media usage in 1994 ($x = 4$) was about 9.314 hr per day, while in 1999 it is projected to be about 9.747. We let $T(x)$ = the number of hours per day. We determine a slope–intercept equation $T(x) = mx + b$ using the procedure of Section 2.6(c) with the data points (4, 9.314) and (9, 9.747).

We first find the slope, or rate of change:

$$m = \frac{9.747 \text{ hr/day} - 9.314 \text{ hr/day}}{9 \text{ yr} - 4 \text{ yr}}$$

$$= \frac{0.433 \text{ hr/day}}{5 \text{ yr}}$$

$$= 0.0866 \text{ hr/day per year.}$$

Now we know the slope and two points. We proceed to find b using the slope and either one of the points. We choose to use (4, 9.314) and substitute 4 for x, 9.314 for y, and 0.0866 for m:

$$y = mx + b$$
$$9.314 = 0.0866(4) + b \qquad \text{Substituting}$$
$$9.314 = 0.3464 + b$$
$$8.9676 = b. \qquad\qquad \text{Solving for } b$$

Next, we use the equation $y = mx + b$, substituting $T(x)$ for y, 0.0866 for m, and 8.9676 for b. This gives us a linear function:

$$T(x) = 0.0866x + 8.9676.$$

b) To predict media usage in the year 2000 ($x = 10$), we find $T(10)$:

$$T(10) = 0.0866(10) + 8.9676 = 9.8336.$$

Thus media usage in 2000 will be about 9.83 hr per day.

Note that this prediction differs slightly from the number 9.85 found in Example 1. Such is the nature of the process of making predictions. How much you rely on such information depends on the type of application. Nevertheless, modeling applications have extensive use in areas such as life science, physical science, business, and social science. To see this, look through an economics book or a medical journal.

Do Exercise 2.

Answers on page A-16

Linear Regression

Another procedure, called **linear regression,** can be used to fit a linear function to data. The strength of this procedure is that it makes use of *all* the data, not just two points. Most graphing calculators have a REGRESSION feature, which we discuss in the following Calculator Spotlight. The mathematical basis for regression belongs to a course in statistics and/or calculus.

Calculator Spotlight

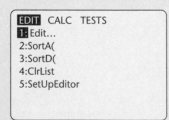 Linear Regression: Fitting a Linear Function to Data. We now consider **linear regression,** a procedure that can be used to fit a linear function to a set of data. Although the complete basis for this method belongs to a statistics and/or calculus course, we consider it here because we can carry out the procedure easily using technology. The grapher gives us the powerful capability to find linear models and make predictions.

Example Consider the data (on p. 209) on media usage by eighteen-year-olds. If we look at the scatterplot, it appears that the data points can be modeled by a linear function.

a) Fit a regression line to the data using the REGRESSION feature on a grapher.

b) Make a scatterplot of the data. Then graph the regression line with the scatterplot.

c) Use the linear model to predict media usage in the year 2000 ($x = 10$).

We proceed as follows.

a) We can fit a linear function to the data using linear regression. We show this using a TI-83 grapher.

```
EDIT  CALC  TESTS
1: Edit...
2: SortA(
3: SortD(
4: ClrList
5: SetUpEditor
```

1. Press [STAT] and then [1] or [ENTER] to enter the data.

2. If there are any data in column one, clear the numbers by pressing the arrow keys until L_1 is highlighted. Then press [CLEAR] and [▽]. Repeat for each column in which data appear, using [▷] or [◁] to move between columns.

3. Go back to the first column and enter the years by typing the number and then the down arrow key. Then move to the second column and enter the hours of media usage in a similar manner. You will not be able to see all the data points at once on the display.

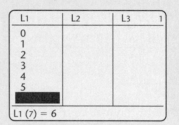

4. Equations are calculated using the STAT-CALC menu. Press [STAT] [▷] [4] to choose LinReg($ax + b$). Then press [VARS] [▷] [1] [1] to copy Y_1 on the screen, as shown below. The regression equation will now automatically be copied as Y_1 on the Y= screen. If an equation is currently entered as Y_1, it must be cleared *before* the regression equation can be found.

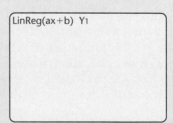

(continued)

5. Next, press ENTER . We now have a linear regression equation that fits the data.

```
LinReg
y=ax+b
a=.0968121212
b=8.879745455
```

Rounding the coefficients to four decimal places, we find that the regression line is $y = 0.0968x + 8.8797$.

b) To make a scatterplot, or xy-graph, of the data, we can use the STAT PLOT feature. (If the scatterplot had not been shown on p. 209, we would begin this example by creating a scatterplot to determine whether it appears as though a linear function could fit the data points.)

1. Press 2nd STAT PLOT . Turn on Plot 1 by pressing ENTER twice. Note that the highlighting indicates that the data in the first list L1 are related to the independent variable, x, and that the data in the second list L2 are related to the dependent variable, y_1.

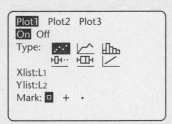

```
Plot1  Plot2  Plot3
On  Off
Type:  ⠿  ⟋  ▥
       ⊞··  ⊞   ⟋
Xlist:L1
Ylist:L2
Mark: ▫  +  ·
```

2. Press the Y= screen. You will see that the regression function has already been entered. Using the ZOOMSTAT feature—that is, pressing ZOOM 9 , we get the graph of the data points and the regression function.

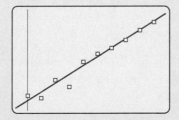

c) To predict media usage in the year 2000, we evaluate Y_1 for $x = 10$. Press VARS ▷ 1 1 . We get Y_1 on the screen. Then press (1 0) ENTER . This gives $Y_1(10) = 9.847866667$, as shown below.

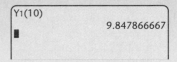

```
Y1(10)
                9.847866667
■
```

Thus we predict media usage in the year 2000 to be about 9.85 hr/day. Note this agrees exactly with the estimate found by "eyeballing" the data in Example 1 and is close to that found with a linear function formed by using only two data points in Example 2. This may not always be the case.

Exercises

1. Use the data on study time and test scores in Margin Exercise 1.

a) Fit a regression line to the data using the REGRESSION feature on a grapher.

b) Make a scatterplot of the data. Then graph the regression line with the scatterplot.

c) Use the linear model to predict the test score received if one has studied for 23 hr.

d) Compare your answers with those found in Margin Exercise 1.

Improving Your Math Study Skills

Better Test Taking

How often do you make the following statement after taking a test: "I was able to do the homework, but I froze during the test"? Instructors have heard this comment for years, and in most cases, it is merely a coverup for a lack of proper study habits. Here are two related tips, however, to help you with this difficulty. Both are intended to make test taking less stressful by getting you to practice good test-taking habits on a daily basis.

- **Treat *every* homework exercise as if it were a test question.** If you had to work a problem at your job with no backup answer provided, what would you do? You would probably work it very deliberately, checking and rechecking every step. You might work it more than one time, or you might try to work it another way to check the result. Try to use this approach when doing your homework. Treat every exercise as though it were a test question and no answers were provided at the back of the book.

- **Be sure that you do questions without answers as part of every homework assignment whether or not the instructor has assigned them!** One reason a test may seem such a different task from homework is that questions on a test lack answers. That is the reason for taking a test: to see if you can answer the questions without assistance. As part of your test preparation, be sure you do some exercises for which you do not have the answers. Thus when you take a test, you are doing a more familiar task.

The purpose of doing your homework using these approaches is to give you more test-taking practice beforehand. Let's make a sports analogy here. At a basketball game, the players take lots of practice shots before the game. They play the first half, go to the locker room, and come out for the second half. What do they do before the second half, even though they have just played 20 minutes of basketball? They shoot baskets again! We suggest the same approach here. Create more and more situations in which you practice taking test questions by treating each homework exercise like a test question and by doing exercises for which you have no answers. Good luck! Please send me an e-mail (exponent@aol.com) and let me know how it works for you.

Exercise Set 2.7

a

1. *Net Sales of the Gap.*

Years, x (since 1990)		Net Sales, S (in billions)
0.	1990	$1.9
1.	1991	2.5
2.	1992	3.0
3.	1993	3.3
4.	1994	3.7
5.	1995	3.4
6.	1996	5.3

Source: The Gap, Inc.

a) Make a scatterplot of the data.
b) Draw a representative graph of a linear function.
c) Predict net sales for The Gap in 1999 and 2001.

2. *Earnings Per Share of Toys "R" Us.*

Years, x (since 1992)		Earnings per Share, E
0.	1992	$1.15
1.	1993	1.47
2.	1994	1.63
3.	1995	1.85
4.	1996	1.73

Source: Toys Я Us

a) Make a scatterplot of the data.
b) Draw a representative graph of a linear function.
c) Predict earnings per share for Toys "R" Us in 1999 and 2001.

For each set of data, **(a)** make a scatterplot of the data, **(b)** draw a representative graph of a linear function, and **(c)** predict the number of home runs in 1998 and 2000.

3. *Home Runs Per Game in the National League.*

Years, x (since 1992)		Average Number of Home Runs per Game, H
0.	1992	1.30
1.	1993	1.72
2.	1994	1.91
3.	1995	1.90
4.	1996	2.17
5.	1997	1.83

Source: Major League Baseball

4. *Home Runs Per Game in the American League.*

Years, x (since 1992)		Average Number of Home Runs per Game, H
0.	1992	1.57
1.	1993	1.83
2.	1994	2.23
3.	1995	2.14
4.	1996	2.48
5.	1997	2.09

Source: Major League Baseball

5. *Bird Watching.* Bird watching has been on the increase in recent years, as shown by the following graph.

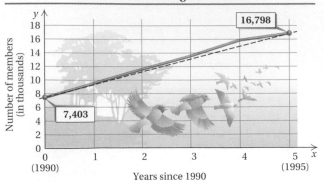

The American Birding Association

Number of members (in thousands)

16,798

7,403

0 (1990) 1 2 3 4 5 (1995) *x*

Years since 1990

Source: American Birding Association

a) Use the two points (0, 7403) and (5, 16,798) to find a linear function that fits the data.
b) Graph the function.
c) Use the function to predict the number of members of the American Birding Association in 2000 and 2010.

6. *The National Debt.* The national debt has been increasing for several years, as shown by the following graph.

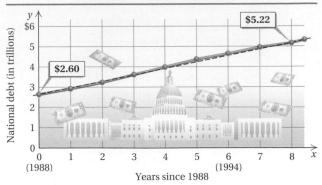

Growth of the National Debt

National debt (in trillions)

$5.22

$2.60

0 (1988) 1 2 3 4 5 6 (1994) 7 8 *x*

Years since 1988

Source: U.S. Department of the Treasury

a) Use the two points (0, $2.6) and (8, $5.22) to find a linear function that fits the data.
b) Graph the function.
c) Use the function to predict the national debt in 2000 and 2010.

7. *Home Runs Per Game in the National League.* Use the data from Exercise 3.

a) Choose two points and find a linear function that fits the data. Answers may vary depending on the two points chosen.
b) Graph the function on the scatterplot.
c) Use the function to predict the average number of home runs in 1999 and 2002.

8. *Home Runs Per Game in the American League.* Use the data from Exercise 4.

a) Choose two points and find a linear function that fits the data. Answers may vary.
b) Graph the function on the scatterplot.
c) Use the function to predict the average number of home runs in 1999 and 2002.

9. *Net Sales of The Gap.* Use the data from Exercise 1.

a) Choose two points and find a linear function that fits the data. Answers may vary.
b) Graph the function on the scatterplot.
c) Use the function to predict net sales of The Gap in 2000 and 2005.

10. *Earnings Per Share of Toys "R" Us.* Use the data from Exercise 2.

a) Choose two points and find a linear function that fits the data. Answers may vary.
b) Graph the function on the scatterplot.
c) Use the function to predict earnings per share of Toys "R" Us in 2000 and 2010.

11. *Study Time vs. Grades.* A mathematics instructor asked her students to keep track of how much time each spent studying the chapter on percent notation in her basic mathematics course. She collected the information, together with test scores from that chapter's test, in the table below.

Study Time, x (in hours)	Test Grade, y (in percent)
9	74
11	94
13	81
15	86
16	87
17	81
21	87
23	92

a) Choose two points from the data (this can vary) and find a linear function that fits the data.
b) Use the function to predict the test scores of someone who has studied for 18 hr; for 25 hr.

12. *Maximum Heart Rate.* A person exercising should not exceed a maximum heart rate, which depends on his or her gender, age, and resting heart rate. The following table shows data relating resting heart rate and maximum heart rate for a 20-yr-old woman.

Resting Heart Rate, r (in beats per minute)	Maximum Heart Rate, M (in beats per minute)
50	170
60	172
70	174
80	176

Source: American Heart Association

a) Choose two points from the data and find a linear function that fits the data.
b) Use the function to predict the maximum heart rate of a woman whose resting heart rate is 62; whose resting heart rate is 75.

Skill Maintenance

Solve. [1.3a]

13. The price of a radio, including 5% sales tax, is $36.75. Find the price of the radio before the tax was added.

14. A basketball team increases its final score by 7 points in each of three games. The total of the three scores was 228. What was the score in the last game?

Solve for the indicated letter. [1.2a]

15. $Ax + By = C$, for y

16. $3x - 7y = 12$, for y

17. $P = \frac{2}{3}q - y$, for q

18. $A = \frac{1}{2}bh$, for b

Write interval notation for the set. [1.4b]

19. $\{x \mid -9 \le x \le 2\}$

20. $\{x \mid -9 \le x < 2\}$

21. $\{x \mid x \ge -9\}$

22. $\{x \mid x < -9\}$

Determine whether the graph might be modeled by a linear function. Give reasons why or why not.

23. ◆

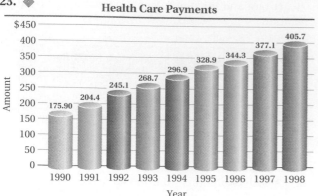

Health Care Payments

Source: Healthcare Financing Administration, U.S. Department of Health and Human Resources

24. ◆

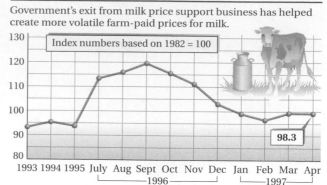

Milk Prices Shaken Up

Government's exit from milk price support business has helped create more volatile farm-paid prices for milk.

Source: U.S. Department of Agriculture

 Use the REGRESSION feature on a grapher for Exercises 25–28.

25. *Home Runs Per Game in the National League.*

a) Fit a regression line to the data in Exercise 3.
b) Make a scatterplot of the data. Then graph the regression line with the scatterplot.
c) Use the linear model to predict the average number of home runs per game in 2002 ($x = 10$).

26. *Net Sales of The Gap.*

a) Fit a regression line to the data in Exercise 1.
b) Make a scatterplot of the data. Then graph the regression line with the scatterplot.
c) Use the linear model to predict the net sales of The Gap in 2001 ($x = 11$).

The following table contains data relating infant mortality rate, life expectancy, and average daily caloric intake. Use it for Exercises 27 and 28.

Country	Average Daily Caloric Intake, c, in 1992	Life Expectancy, L, Projected in 2000	Infant Mortality Rate, M (per 1000 births)
Argentina	2880	72.3	26.1
Bolivia	2100	62.0	60.2
Canada	3482	80.0	5.5
Dominican Republic	2359	70.4	40.8
Germany	3443	76.7	22.2
Haiti	1707	50.2	98.4
Mexico	3181	75.0	20.7
United States	3671	76.3	6.2

Source: *Universal Almanac*; *Statistical Abstract of the United States*

27. *Infant Mortality Rate as a Function of Daily Caloric Intake.*

a) Fit a regression line, $M(c) = mc + b$, to the data using the REGRESSION feature on a grapher.
b) Make a scatterplot of the data. Then graph the regression line with the scatterplot.
c) Use the linear model to estimate the infant mortality rate for Australia, which has an average daily caloric intake of 3216.
d) Use the linear model to estimate the infant mortality rate for Venezuela, which has an average daily caloric intake of 2622.

28. *Life Expectancy as a Function of Daily Caloric Intake.*

a) Fit a regression line, $L(c) = mc + b$, to the data using the REGRESSION feature on a grapher.
b) Make a scatterplot of the data. Then graph the regression line with the scatterplot.
c) Use the linear model to estimate life expectancy for Australia, which has an average daily caloric intake of 3216.
d) Use the linear model to estimate life expectancy for Venezuela, which has an average daily caloric intake of 2622.

Make predictions from a set of data.

Summary and Review Exercises: Chapter 2

Important Properties and Formulas

Slope $= m = \dfrac{y_2 - y_1}{x_2 - x_1}$, or $\dfrac{y_1 - y_2}{x_1 - x_2}$

Equations of Lines and Linear Functions

Horizontal Line: $f(x) = b$, or $y = b$; slope $= 0$ Vertical Line: $x = a$, slope is undefined.

Slope–Intercept Equation: $f(x) = mx + b$, or $y = mx + b$

Parallel Lines: $m_1 = m_2, b_1 \neq b_2$ Perpendicular Lines: $m_1 = -\dfrac{1}{m_2}$

The objectives to be tested in addition to the material in this chapter are [1.3a], [1.4c], [1.5a, b], and [1.6c, e].

1. Show that the ordered pairs $(0, -2)$ and $(-1, -5)$ are solutions of the equation $3x - y = 2$. Then use the graph of the two points to determine another solution. Answers may vary. Show your work. [2.1a, b]

Graph. [2.1c, d]

2. $y = -3x + 2$ **3.** $y = \frac{5}{2}x - 3$

4. $y = |x - 3|$ **5.** $y = 3 - x^2$

Determine whether the correspondence is a function. [2.2a]

6.

7.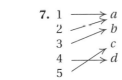

Find the function values. [2.2b]

8. $g(x) = -2x + 5$; $g(0)$ and $g(-1)$

9. $f(x) = 3x^2 - 2x + 7$; $f(0)$ and $f(-1)$

10. The function described by $C(t) = 645t + 9800$ can be used to estimate the average cost of tuition at a state university t years after 1997. Estimate the average cost of tuition at a state university in 2010. [2.2b]

Determine whether each of the following is the graph of a function. [2.2d]

11. 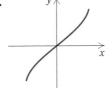   **12.**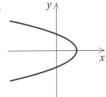

13. For the following graph of a function f, determine (a) $f(2)$; (b) the domain; (c) all x-values such that $f(x) = 2$; and (d) the range. [2.3a]

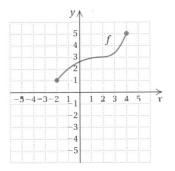

Find the domain. [2.3a]

14. $f(x) = \dfrac{5}{x - 4}$ **15.** $g(x) = x - x^2$

Find the slope and the y-intercept. [2.4a, b]

16. $y = -3x + 2$ **17.** $4y + 2x = 8$

18. Find the slope, if it exists, of the line containing the points $(13, 7)$ and $(10, -4)$. [2.4b]

Find the intercepts. Then graph the equation. [2.5a]

19. $2y + x = 4$ **20.** $2y = 6 - 3x$

Graph using the slope and the y-intercept. [2.5b]

21. $g(x) = -\frac{2}{3}x - 4$ **22.** $f(x) = \frac{5}{2}x + 3$

Graph. [2.5c]

23. $x = -3$ **24.** $f(x) = 4$

Determine whether the graphs of the given pair of lines are parallel or perpendicular. [2.5d]

25. $y + 5 = -x,$
    $x - y = 2$

26. $3x - 5 = 7y,$
    $7y - 3x = 7$

27. $4y + x = 3,$
    $2x + 8y = 5$

28. $x = 4,$
    $y = -3$

29. Find a linear function $f(x) = mx + b$ whose graph has the given slope and y-intercept: [2.6a]

   slope: 4.7; y-intercept: $(0, -23)$.

30. Find an equation of the line having the given slope and containing the given point: [2.6b]

   $m = -3;$ $(3, -5)$.

31. Find an equation of the line containing the given pair of points: [2.6c]

   $(-2, 3),$ and $(-4, 6)$.

32. Find an equation of the line containing the given point and parallel to the given line: [2.6d]

   $(14, -1);$ $5x + 7y = 8$.

33. Find an equation of the line containing the given point and perpendicular to the given line: [2.6d]

   $(5, 2);$ $3x + y = 5$.

Use the following table of data for Exercise 34.

Year, x, since 1990	Health Care Payments, H (in billions)
0. 1990	$175.9
1. 1991	204.4
2. 1992	245.1
3. 1993	268.7
4. 1994	328.9
5. 1995	344.3
6. 1996	377.1
7. 1997	405.7

Source: Adams Media Research, Carmel, California

34. a) Use the two points $(0, 175.9)$ and $(7, 405.7)$ to find a linear function that fits the data.
   b) Use the function to predict health care payments in 2000 and 2005. [2.7b]

Skill Maintenance

35. The Sound Connection prices its blank audiotape cassettes by raising the wholesale price 40% and adding 35¢. What is the wholesale price of one tape if the tapes are to sell for $1.61 each? [1.3a]

36. Solve: $2x - 4(x - 12) < -3(2x + 5) - 14$. [1.4c]

Solve. [1.5a, b]

37. $-2x + 7 < -14$ *or* $-2x + 7 > 14$

38. $-29 \le -2x + 7 \le 29$

Solve. [1.6c, e]

39. $|4x - 7| = 19$ **40.** $|4x - 7| \le 19$

41. $|4x - 7| \ge 19$

Synthesis

42. ◈ Explain the usefulness of the slope concept when describing a line. [2.4b, c], [2.5b], [2.6a, b, c, d]

43. ◈ Explain the meaning of the notation $f(x)$ when considering a function in as many ways as you can. [2.2b]

Use the REGRESSION feature on a grapher for Exercise 44. [2.7b]

44. a) Fit a regression line, $H(x) = mx + b$, to the data in Exercise 34.
   b) Make a scatterplot of the data. Then graph the regression line with the scatterplot.
   c) Use the linear model to predict health care payments in 2005.

Test: Chapter 2

Determine whether the given points are solutions of the equation.

1. $(2, -3);$ $y = 5 - 4x$

2. $(2, -3);$ $5b - 7a = 10$

Graph.

3. $y = -2x - 5$

4. $f(x) = -\dfrac{3}{5}x$

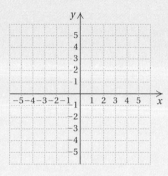

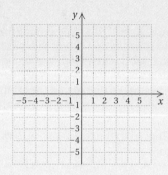

5. $g(x) = 2 - |x|$

6. $y = \dfrac{4}{x}$

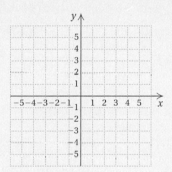

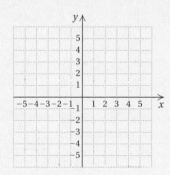

7. *Median Age of Cars.* The function

$$A(t) = 0.233t + 5.87$$

can be used to estimate the median age of cars in the United States t years after 1990 (**Source**: The Polk Co.). (In this context, we mean that if the median age of cars is 3 yr, then half the cars are older than 3 yr and half are younger.)

a) Find the median age of cars in 2002.
b) In what year will the median age of cars be 7.734 yr?

Determine whether the correspondence is a function.

8.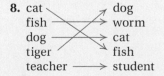

cat → dog
fish → worm
dog → cat
tiger → fish
teacher → student

9.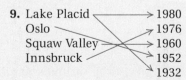

Lake Placid → 1980
Oslo → 1976
Squaw Valley → 1960
Innsbruck → 1952
→ 1932

Find the function values.

10. $f(x) = -3x - 4;$ $f(0)$ and $f(-2)$

11. $g(x) = x^2 + 7;$ $g(0)$ and $g(-1)$

12. _____

13. _____

14. a) _____

b) _____

15. a) _____

b) _____

c) _____

d) _____

16. _____

17. _____

18. _____

19. _____

20. _____

21. _____

Determine whether each of the following is the graph of a function.

12.

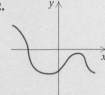

13.

14. _Movie Revenue._ The following graph approximates the weekly revenue, in millions of dollars, from the recent movie _Jurassic Park—The Lost World_. The revenue is given as a function of the week. Use the graph to answer the following.

a) What was the movie revenue for week 1?

b) What was the movie revenue for week 5?

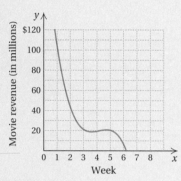

Source: Exhibitor Relations Co., Inc.

15. For the following graph of function _f_, determine **(a)** $f(2)$; **(b)** the domain; **(c)** all _x_-values such that $f(x) = 2$; and **(d)** the range.

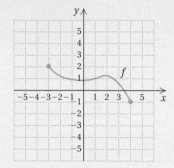

Find the domain.

16. $g(x) = 5 - x^2$

17. $f(x) = \dfrac{8}{2x + 3}$

Find the slope and the _y_-intercept.

18. $f(x) = -\dfrac{3}{5}x + 12$

19. $-5y - 2x = 7$

Find the slope, if it exists, of the line containing the following points.

20. $(-2, -2)$ and $(6, 3)$

21. $(-3.1, 5.2)$ and $(-4.4, 5.2)$

22. Find the slope, or rate of change, of the graph at right.

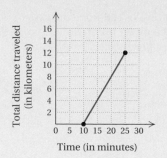
Time (in minutes)

23. Find the intercepts. Then graph the equation.

$$2x + 3y = 6$$

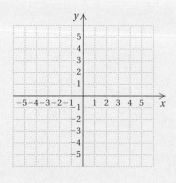

24. Graph using the slope and the y-intercept:

$$f(x) = -\frac{2}{3}x - 1.$$

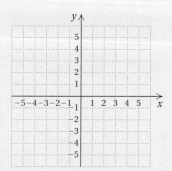

Graph.

25. $y = f(x) = -3$

26. $2x = -4$

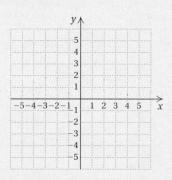

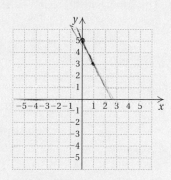

Determine whether the graphs of the given pair of lines are parallel or perpendicular.

27. $4y + 2 = 3x,$
$-3x + 4y = -12$

28. $y = -2x + 5,$
$2y - x = 6$

29. Find an equation of the line that has the given characteristics:

slope: -3; y-intercept: $(0, 4.8)$.

30. Find a linear function $f(x) = mx + b$ whose graph has the given slope and y-intercept:

slope: 5.2; y-intercept: $\left(0, -\frac{5}{8}\right)$.

31. Find an equation of the line having the given slope and containing the given point:

$m = -4$; $(1, -2)$.

32. Find an equation of the line containing the given pair of points:

$(4, -6)$ and $(-10, 15)$.

22. _____

23. _____

24. _____

25. _____

26. _____

27. _____

28. _____

29. _____

30. _____

31. _____

32. _____

33. Find an equation of the line containing the given point and parallel to the given line:

$(4, -1); \quad x - 2y = 5.$

34. Find an equation of the line containing the given point and perpendicular to the given line:

$(2, 5); \quad x + 3y = 2.$

Sales of Books on Tape. Sales of books on audiotape have increased in recent years. Use the following table of data for Exercise 35.

Year, x, since 1990	Books on Tape Sales, S (in billions)
0. 1990	$0.3
1. 1991	0.4
2. 1992	0.7
3. 1993	0.9
4. 1994	1.2
5. 1995	1.4
6. 1996	1.6

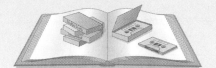

Source: Audio Book Club

35. **a)** Use the two points (0, 0.3) and (6, 1.6) to find a linear function that fits the data.
b) Use the function to predict sales of books on audiotape in 1998 and 2000.

Skill Maintenance

Solve.

36. A rope that is 344 ft long is to be cut into three pieces such that the second piece is 6 ft shorter than three times the first and the third is 2 ft longer than two-thirds of the second. Find the length of the longest piece.

Solve.

37. $2x - 4(9 - 7x) \geq 3(4x - 6) + 8$

38. $9 - 7x < -23 \text{ or } 9 - 7x > 23$

39. $|9 - 7x| < 23$

40. $|9 - 7x| = 23$

Synthesis

41. Find k such that the line $3x + ky = 17$ is perpendicular to the line $8x - 5y = 26.$

42. Find a formula for a function f for which $f(-2) = 3.$

Cumulative Review: Chapters R–2

1. Evaluate $a^3 + b^0 - c$ for $a = 3$, $b = 7$, and $c = -3$.

Simplify.

2. $|4.3 - 2.1|$

3. $\left| -\dfrac{2}{3} \right|$

4. $-\dfrac{1}{3} - \left(-\dfrac{5}{6} \right)$

5. $-3.2(-11.4)$

6. $2x - 4(3x - 8)$

7. $(-16x^3y^{-4})(-2x^5y^3)$

8. $\dfrac{27x^0y^3}{-3x^2y^5}$

9. $3(x - 7) - 4[2 - 5(x + 3)]$

10. $-128 \div 16 + 32 \cdot (-10)$

11. $2^3 - (4 \cdot 2 - 3)^2 + 23^0 \cdot 16^1$

Solve.

12. $x + 9.4 = -12.6$

13. $\dfrac{2}{3}x - \dfrac{1}{4} = -\dfrac{4}{5}x$

14. $-2.4t = -48$

15. $4x + 7 = -14$

16. $3n - (4n - 2) = 7$

17. Solve $W = Ax + By$ for x.

18. Solve $M = A + 4AB$ for A.

Solve.

19. $y - 12 \leq -5$

20. $6x - 7 < 2x - 13$

21. $5(1 - 2x) + x < 2(3 + x)$

22. $x + 3 < -1 \ or \ x + 9 \geq 1$

23. $-3 < x + 4 \leq 8$

24. $-8 \leq 2x - 4 \leq -1$

25. $|x| = 8$

26. $|y| > 4$

27. $|4x - 1| \leq 7$

Graph on a plane.

28. $y = -2x + 3$

29. $3x = 2y + 6$

30. $4x + 16 = 0$

31. $-2y = -6$

32. $f(x) = \dfrac{2}{3}x + 1$

33. $g(x) = 5 - |x|$

34. Find an equation of the line containing the point $(-4, -6)$ and perpendicular to the line whose equation is $4y - x = 3$.

35. Find an equation of the line containing the point $(-4, -6)$ and parallel to the line whose equation is $4y - x = 3$.

36. For the following graph of function f, determine (a) $f(15)$; (b) the domain; (c) all x-values such that $f(x) = 14$; and (d) the range.

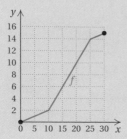

37. *Walks per Game in the National League.*

Years, x, since 1992	Average Number of Walks Per Game, W
0. 1992	6.15
1. 1993	6.26
2. 1994	6.47
3. 1995	6.62
4. 1996	7.00
5. 1997	6.50

Source: Major League Baseball

a) Use the two points $(0, 6.15)$ and $(4, 7)$ to find a linear function $W(x) = mx + b$ that fits the data.
b) Make a scatterplot of the data. Graph the function found in part (a) on the scatterplot.
c) Use the function found in part (a) to estimate the average number of walks in 1999 and 2000.

38. Find the slope and the y-intercept of $-4y + 9x = 12$.

39. Find the slope, if it exists, of the line containing the points $(2, 7)$ and $(-1, 3)$.

40. Find an equation of the line with slope -3 and containing the point $(2, -11)$.

41. Find an equation of the line containing the points $(-6, 3)$ and $(4, 2)$.

Solve.

42. Nine plus five times a number is 173.4. Find the number.

43. The perimeter of a lot is 80 m. The length exceeds the width by 6 m. Find the dimensions.

44. Seventeen more than seven times a number is 3 less than twelve times the number. What is the number?

45. After David receives a 20% raise in salary, his new salary is $10,800. What was the old salary?

Synthesis

46. Wayside Auto Sales discovers that when $1000 is spent on radio advertising, weekly sales increase by $101,000. When $1250 is spent on radio advertising, weekly sales increase by $126,000. Assuming that sales increase according to a linear equation, by what would sales increase when $1500 is spent on radio advertising?

47. Simplify: $(6x^{a+2}y^{b+2})(-2x^{a-2}y^{y+1})$.

48. Solve: $x + 5 < 3x - 7 \le x + 13$.

49. Which pairs of the following four equations represent perpendicular lines?

(1) $7y - 3x = 21$
(2) $3x - 7y = 12$
(3) $7y + 3x = 21$
(4) $3y + 7x = 12$

Systems of Equations

3

An Application	**The Mathematics**
A nontoxic floor wax can be made by combining lemon juice and food-grade linseed oil. The amount of oil should be twice the amount of lemon juice. How much of each ingredient is needed in order to make 32 oz of floor wax?	We let x = the amount of lemon juice, in ounces, and y = the amount of linseed oil, in ounces. We can then translate the problem as:

$$\left.\begin{array}{l} x + y = 32, \\ y = 2x. \end{array}\right\}$$

This is a *system of equations.*

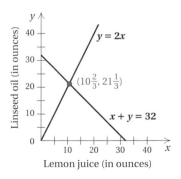

This problem appears as Exercise 6 in Exercise Set 3.3.

World Wide Web For more information, visit us at www.mathmax.com

Pretest: Chapter 3

1. Solve this system by graphing:
   $$y = x + 1,$$
   $$y + x = 3.$$

2. Solve the system of equations in Exercise 1 by the substitution method.

3. Solve this system by the elimination method:
   $$3x + 5y = 1,$$
   $$4x + 3y = -6.$$

4. Classify the system of equations in Exercise 1 as consistent or inconsistent.

5. Classify the system of equations in Exercise 1 as dependent or independent.

6. Solve:
   $$3x + 5y - 2z = 7,$$
   $$2x + y - 3z = -5,$$
   $$4x - 2y + z = 3.$$

Solve.

7. *Mixed Nuts.* The Nutty Professor sells cashews for $6.75 per pound and Brazil nuts for $5.00 per pound. How much of each type should be used to make a 50-lb mixture that sells for $5.70 per pound?

8. *Investments.* Two investments are made totaling $7500. For a certain year, the investments yielded $516 in simple interest. Part of the $7500 is invested at 8% and part at 6%. Find the amount invested at each rate.

9. *Salt-Water Mixtures.* Mixture A is 32% salt and the rest water. Mixture B is 58% salt and the rest water. How many pounds of each mixture should be combined in order to obtain 120 lb of a mixture that is 44% salt?

10. *Marine Travel.* A motorboat took 4 hr to make a trip downstream with a 6-mph current. The return trip against the same current took 5 hr. Find the speed of the boat in still water.

11. Graph on a plane: $6x - 2y < 12$.

12. Graph the following system of inequalities. Find the coordinates of any vertices formed.
    $$x + y \leq 16,$$
    $$3x + 6y \leq 60,$$
    $$x \geq 0,$$
    $$y \geq 0$$

13. For the following total-cost and total-revenue functions, find (a) the total-profit function and (b) the break-even point.
    $$C(x) = 90{,}000 + 15x,$$
    $$R(x) = 26x.$$

Objectives for Retesting

The objectives to be tested in addition to the material in this chapter are as follows.

[1.1d] Solve equations using the addition and multiplication principles together, removing parentheses where appropriate.

[1.2a] Evaluate formulas and solve a formula for a specified letter.

[2.2b] Given a function described by an equation, find function values (outputs) for specified values (inputs).

[2.4b] Given a linear equation, derive the equivalent slope–intercept equation and determine the slope and the y intercept.

3.1 Systems of Equations in Two Variables

We can solve many applied problems more easily by translating to two or more equations in two or more variables than by translating to a single equation. Let's look at such a problem.

Real Estate Merger

In 1996, the Simon Property Group and the DeBartolo Realty Corporation merged to form the largest real estate company in the United States, owning 183 shopping centers in 32 states (***Source:*** Simon Property Group; DeBartolo Realty Corporation). Before the merger, Simon owned twice as many properties as DeBartolo. How many properties did each own originally?

The Indianapolis Star
REAL ESTATE GIANTS MERGE

DeBartolo shareholders would receive 0.68 share of Simon common stock for each share of DeBartolo common stock. Simon also would agree to repay $1.5 billion in DeBartolo debt. At Tuesday's closing price of $23.625 a share for common stock, the transaction is valued at roughly $3 billion.

Executives say the proposed company, Simon DeBartolo Group, would be the largest real estate company in the United States, worth $7.5 billion.

Not included in the deal: DeBartolo's ownership stake in the San Francisco 49ers, or the Indiana Pacers, owned separately by the Simon

To solve, we let $x =$ the number of properties originally owned by Simon and $y =$ the number of properties originally owned by DeBartolo. There are two statements to translate.

First we look at the total number of properties involved:

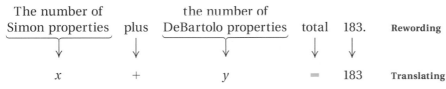

The number of Simon properties | plus | the number of DeBartolo properties | total | 183. **Rewording**

x $+$ y $=$ 183 **Translating**

The second statement compares the number of properties that each company held before merging:

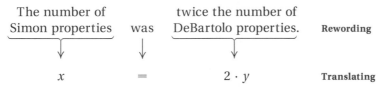

The number of Simon properties | was | twice the number of DeBartolo properties. **Rewording**

x $=$ $2 \cdot y$ **Translating**

We have now translated the problem to a **pair**, or **system, of equations**:

$$x + y = 183,$$
$$x = 2y.$$

We can also write this system in function notation by solving each equation for y:

$$x + y = 183 \longrightarrow y = 183 - x \longrightarrow f(x) = 183 - x,$$
$$x = 2y \longrightarrow y = \tfrac{1}{2}x \longrightarrow g(x) = \tfrac{1}{2}x.$$

Solve the system graphically.

1. $-2x + y = 1,$
    $3x + y = 1$

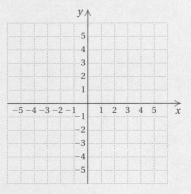

2. $f(x) = \tfrac{1}{2}x,$
    $g(x) = -\tfrac{1}{4}x + \tfrac{3}{2}$

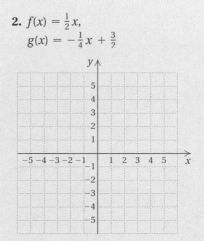

Answers on page A-19

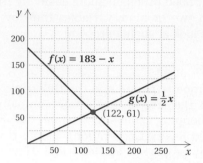
To solve this system, whether it is written in function notation or not, we graph each equation and look for the point of intersection of the graphs. (We may need to use detailed graphing paper to determine the point of intersection exactly.)

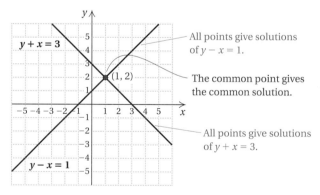

As we see in the graph above, the ordered pair (122, 61) is the intersection and thus the solution—that is, $x = 122$ and $y = 61$. This tells us that Simon originally owned 122 properties and DeBartolo owned 61.

a Solving Systems of Equations Graphically

One Solution

A **solution** of a system of two equations in two variables is an ordered pair that makes *both* equations true. If we graph a system of equations or functions, the point at which the graphs intersect will be a solution of *both* equations or functions.

Example 1 Solve this system graphically:

$$y - x = 1,$$
$$y + x = 3.$$

We draw the graph of each equation using any method studied in Chapter 2.

The point of intersection has coordinates that make *both* equations true. The solution seems to be the point (1, 2). However, since graphing alone is not perfectly accurate, solving by graphing may give only approximate answers. Thus we check the pair (1, 2) as follows.

CHECK:

$$\begin{array}{c}y - x = 1 \\ \hline 2 - 1 \ ? \ 1 \\ 1 \ | \qquad \text{TRUE}\end{array} \qquad \begin{array}{c}y + x = 3 \\ \hline 2 + 1 \ ? \ 3 \\ 3 \ | \qquad \text{TRUE}\end{array}$$

The solution is (1, 2).

Do Exercises 1 and 2 on the preceding page.

No Solution

Sometimes the equations in a system have graphs that are parallel lines.

Example 2   Solve graphically:

$$f(x) = -3x + 5,$$
$$g(x) = -3x - 2.$$

We graph the functions. The graphs have the same slope, -3, and different y-intercepts, so they are parallel. There is no point at which they cross, so the system has no solution. No matter what point we try, it will *not* check in *both* equations. The solution set is thus the empty set, denoted $\varnothing$ or { }.

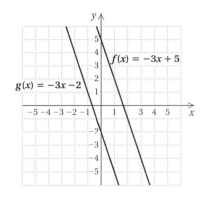

3. Solve graphically:

$$y + 2x = 3,$$
$$y + 2x = -4.$$

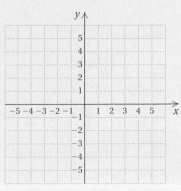

> **CONSISTENT AND INCONSISTENT SYSTEMS**
>
> A system of equations has a solution. ⟹ It is **consistent**.
>
> A system of equations has no solution. ⟹ It is **inconsistent**.

The system in Example 1 is consistent. The system in Example 2 is inconsistent.

Do Exercises 3 and 4.

Infinitely Many Solutions

Sometimes the equations in a system have the same graph.

Example 3   Solve graphically:

$$3y - 2x = 6,$$
$$-12y + 8x = -24.$$

We graph the equations and see that the graphs are the same. Thus any solution of one of the equations is a solution of the other. Each equation has an infinite number of solutions, two of which are shown on the graph.

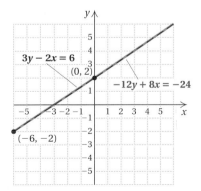

We check one such solution, $(0, 2)$, which is the y-intercept of each equation.

4. Classify each of the systems in Margin Exercises 1–3 as consistent or inconsistent.

Calculator Spotlight

Use the TABLE feature to check the results of Examples 2 and 3.

Answers on page A-19

5. Solve graphically:

$$2x - 5y = 10,$$
$$-6x + 15y = -30.$$

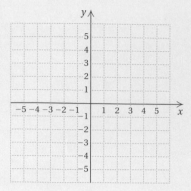

6. Classify each system in Margin Exercises 1, 2, 3, and 5 as dependent or independent.

7. a) Solve $x + 1 = \frac{2}{3}x$ algebraically.

b) Solve $x + 1 = \frac{2}{3}x$ graphically using Method 1.

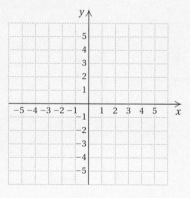

c) Compare your answers to parts (a) and (b).

Answers on page A-19

CHECK:

$$\begin{array}{c} 3y - 2x = 6 \\ \hline 3(2) - 2(0) \ ? \ 6 \\ 6 - 0 \\ 6 \end{array} \quad \text{TRUE}$$

$$\begin{array}{c} -12y + 8x = -24 \\ \hline -12(2) + 8(0) \ ? \ -24 \\ -24 + 0 \\ -24 \end{array} \quad \text{TRUE}$$

On your own, check that $(-6, -2)$ is a solution of both equations. If $(0, 2)$ and $(-6, -2)$ are solutions, then all points on the line containing them will be solutions. The system has an infinite number of solutions.

> **DEPENDENT AND INDEPENDENT SYSTEMS**
>
> A system of two equations in two variables:
>
>   has infinitely many solutions. ⟹ It is **dependent**.
>
>   has one solution or no solutions. ⟹ It is **independent**.

The system in Example 3 is dependent. The systems in Examples 1 and 2 are independent.

When we graph a system of two equations, one of the following three things can happen.

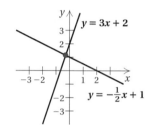

One solution.
Graphs intersect.
Equations are *consistent* and *independent*.

No solution.
Graphs are parallel.
Equations are *inconsistent* and *independent*.

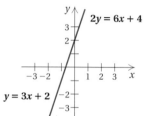

Infinitely many solutions.
Equations have the same graph. Equations are *consistent* and *dependent*.

Do Exercises 5 and 6.

▲G Algebraic–Graphical Connection

To bring together the concepts of Chapters 1–3, let's look at equation solving from both algebraic and graphical viewpoints.

Consider the equation $-2x + 13 = 4x - 17$. Let's solve it algebraically as we did in Chapter 1:

$$\begin{aligned} -2x + 13 &= 4x - 17 \\ 13 &= 6x - 17 \qquad \text{Adding } 2x \\ 30 &= 6x \qquad \text{Adding } 17 \\ 5 &= x. \qquad \text{Dividing by } 6 \end{aligned}$$

Could we also solve the equation graphically? The answer is yes, as we see in the following two methods.

METHOD 1 Solve $-2x + 13 = 4x - 17$ graphically.

We let

$$f(x) = -2x + 13 \quad \text{and} \quad g(x) = 4x - 17.$$

Graphing the system of equations, we get the graph shown at right.

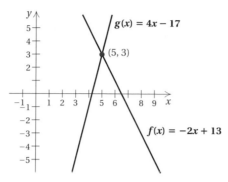

The point of intersection of the two graphs is (5, 3). Note that the x-coordinate of the intersection is 5. This value for x is the solution of the equation $-2x + 13 = 4x - 17$.

Do Exercises 7 and 8. (Exercise 7 is on the preceding page.)

METHOD 2 Solve $-2x + 13 = 4x - 17$ graphically.

Adding $-4x$ and 17 on both sides, we obtain the form

$$-6x + 30 = 0.$$

This time we let

$$f(x) = -6x + 30 \quad \text{and} \quad g(x) = 0.$$

Since $g(x) = 0$, or $y = 0$, is the x-axis, we need only graph $f(x) = -6x + 30$ and see where it crosses the x-axis.

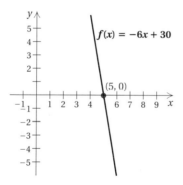

Note that the x-intercept of $f(x) = -6x + 30$ is (5, 0), or just 5. This x-value is the solution of the equation $-2x + 13 = 4x - 17$.

Do Exercise 9.

Let's compare the two methods. Using Method 1, we graph two functions. The solution of the original equation is the x-coordinate of the point of intersection. Using Method 2, we find that the solution of the original equation is the x-intercept of the graph.

Do Exercise 10.

Methods 1 and 2 are helpful when we are using a graphing calculator or computer graphing software. See the Calculator Spotlight that follows.

8. Solve $\frac{1}{2}x + 3 = 2$ graphically using Method 1.

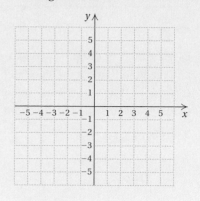

9. a) Solve $x + 1 = \frac{2}{3}x$ graphically using Method 2.

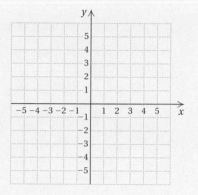

b) Compare your answers to Margin Exercises 7(a), 7(b), and 9(a).

10. Solve $\frac{1}{2}x + 3 = 2$ graphically using Method 2.

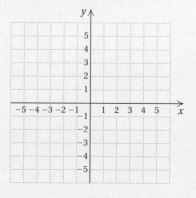

Answers on page A-19

Calculator Spotlight

 The INTERSECT and ZERO Features

INTERSECT Feature. Most graphers have an INTERSECT feature that can be used to find the intersection of the graphs of two functions. Let's use it to find the point of intersection of the two functions

$$f(x) = -2x + 13,$$
$$g(x) = 4x - 17.$$

We enter the functions as

$$y_1 = -2x + 13,$$
$$y_2 = 4x - 17.$$

Then we graph the functions, adjusting the viewing window in order to see the point of intersection.

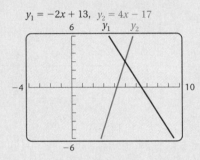

Next, we use the INTERSECT feature (see the CALC menu) to obtain the point of intersection (5, 3).

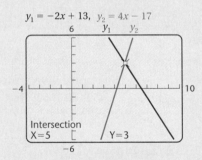

The first coordinate of the point of intersection, 5, is the solution of the equation $-2x + 13 = 4x - 17$.

ZERO Feature. Most graphers have a ZERO, or ROOT, feature that allows us to solve an equation quickly. In this context, the word "zero" refers to an input for which the output of a function is 0.

To use this feature, we must first have a 0 on one side of the equation. So to solve

$$-2x + 13 = 4x - 17,$$

we subtract $4x$ and add 17 on both sides to get $-6x + 30 = 0$. Graphing $y = -6x + 30$ and using the ZERO feature, we obtain a screen like the following.

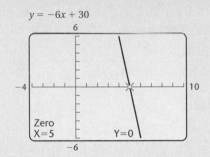

We see that $-6x + 30 = 0$ when $x = 5$, so 5 is the solution of the equation $-2x + 13 = 4x - 17$.

Exercises

Use the INTERSECT feature on your grapher to solve the equation.

1. $x + 1 = \frac{1}{2}x$

2. $\frac{1}{2}x + 3 = 2$

3. $-\frac{3}{4}x + 6 = 2x - 1$

4. $-3x + 4 = 3x - 4$

5. $2.4 - 1.8x = 6.8 - x^2$

6. $x^3 - 3x - 5 = 0$

7.–12. Use the ZERO, or ROOT, feature on your grapher to solve each of the equations in Exercises 1–6.

Exercise Set 3.1

a Solve the system of equations graphically. Then classify the system as consistent or inconsistent and as dependent or independent.

1. $x + y = 4$,
$x - y = 2$

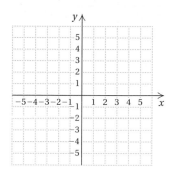

CHECK: $x + y = 4$
$$\frac{}{?}$$

$x - y = 2$
$$\frac{}{?}$$

2. $x - y = 3$,
$x + y = 5$

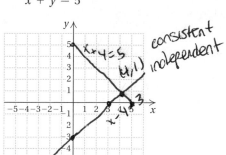

consistent
independent

CHECK: $x - y = 3$
$$\frac{}{?}$$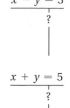

$x + y = 5$
$$\frac{}{?}$$

3. $2x - y = 4$,
$2x + 3y = -4$

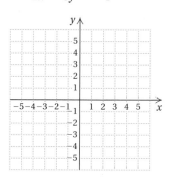

CHECK: $2x - y = 4$
$$\frac{}{?}$$

$2x + 3y = -4$
$$\frac{}{?}$$

4. $3x + y = 5$,
$x - 2y = 4$

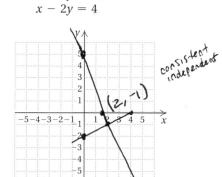

consistent
independent

$(2, -1)$

CHECK: $3x + y = 5$
$$\frac{}{?}$$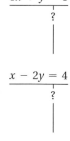

$x - 2y = 4$
$$\frac{}{?}$$

5. $2x + y = 6$,
$3x + 4y = 4$

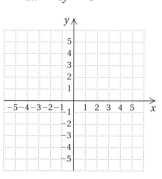

6. $2y = 6 - x$,
$3x - 2y = 6$

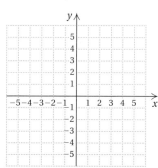

7. $f(x) = x - 1$,
$g(x) = -2x + 5$

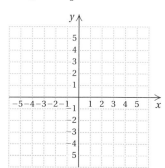

8. $f(x) = x + 1$,
$g(x) = \frac{2}{3}x$

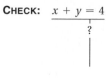

9. $2u + v = 3,$
$2u = v + 7$

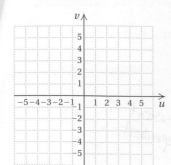

10. $2b + a = 11,$
$a - b = 5$

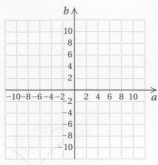

11. $f(x) = -\frac{1}{3}x - 1,$
$g(x) = \frac{4}{3}x - 6$

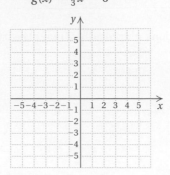

12. $f(x) = -\frac{1}{4}x + 1,$
$g(x) = \frac{1}{2}x - 2$

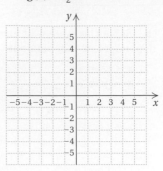

13. $6x - 2y = 2,$
$9x - 3y = 1$

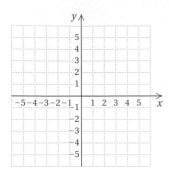

14. $y - x = 5,$
$2x - 2y = 10$

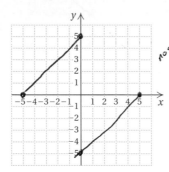

no solution

15. $2x - 3y = 6,$
$3y - 2x = -6$

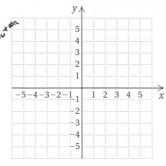

16. $y = 3 - x,$
$2x + 2y = 6$

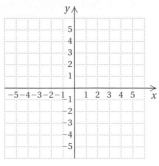

17. $x = 4,$
$y = -5$

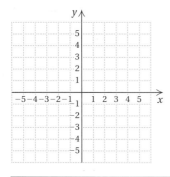

18. $x = -3,$
$y = 2$

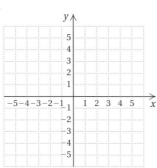

19. $y = -x - 1,$
$4x - 3y = 17$

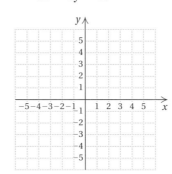

20. $a + 2b = -3,$
$b - a = 6$

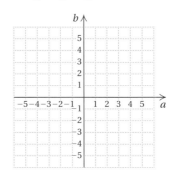

Skill Maintenance

Solve. [1.1d]

21. $3x + 4 = x - 2$

22. $\frac{3}{4}x + 2 = \frac{2}{5}x - 5$

23. $4x - 5x = 8x - 9 + 11x$

24. $5(10 - 4x) = -3(7x - 4)$

Synthesis

25. ◆ Explain how to find the solution of $\frac{3}{4}x + 2 = \frac{2}{5}x - 5$ in two ways graphically and in two ways algebraically.

26. ◆ Write a system of equations with the given solution. Answers may vary.

a) $(4, -3)$ b) No solution
c) Infinitely many solutions

Use a grapher to find the point of intersection of the pair of equations. Round all answers to the nearest hundredth. You may need to solve for y first.

27. $2.18x + 7.81y = 13.78,$
$5.79x - 3.45y = 8.94$

28. $f(x) = 123.52x + 89.32,$
$g(x) = -89.22x + 33.76$

Chapter 3 Systems of Equations

Copyright © 1999 Addison Wesley Longman

3.2 Solving by Substitution or Elimination

Consider this system of equations:

$$5x + 9y = 2,$$
$$4x - 9y = 10.$$

What is the solution? It is rather difficult to tell exactly by graphing. It would appear that fractions are involved. It turns out that the solution is

$$\left(\frac{4}{3}, -\frac{14}{27}\right).$$

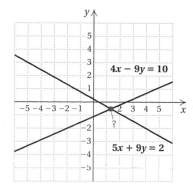

Solving by graphing, though useful in many applied situations, is not always fast or accurate in cases where solutions are not integers. We need techniques involving algebra to determine the solution exactly. Because they use algebra, they are called **algebraic methods.**

Objectives

a Solve systems of equations in two variables by the substitution method.

b Solve systems of equations in two variables by the elimination method.

c Solve applied problems by solving systems of two equations using substitution or elimination.

For Extra Help

TAPE 7   TAPE 6A   MAC WIN   CD-ROM

a The Substitution Method

One nongraphical method for solving systems is known as the **substitution method.**

Example 1 Solve this system:

$$x + y = 4, \quad (1)$$
$$x = y + 1. \quad (2)$$

Equation (2) says that x and $y + 1$ name the same number. Thus we can substitute $y + 1$ for x in equation (1):

$$x + y = 4 \qquad \text{Equation (1)}$$
$$(y + 1) + y = 4. \qquad \text{Substituting } y + 1 \text{ for } x$$

Since this equation has only one variable, we can solve for y using methods learned earlier:

$$(y + 1) + y = 4$$
$$2y + 1 = 4 \qquad \text{Removing parentheses and collecting like terms}$$
$$2y = 3 \qquad \text{Subtracting 1}$$
$$y = \frac{3}{2}. \qquad \text{Dividing by 2}$$

We return to the original pair of equations and substitute $\frac{3}{2}$ for y in *either* equation so that we can solve for x. Calculation will be easier if we choose equation (2) since it is already solved for x:

$$x = y + 1 \qquad \text{Equation (2)}$$
$$= \frac{3}{2} + 1 \qquad \text{Substituting } \frac{3}{2} \text{ for } y$$
$$= \frac{3}{2} + \frac{2}{2} = \frac{5}{2}.$$

We obtain the ordered pair $\left(\frac{5}{2}, \frac{3}{2}\right)$. Even though we solved for y *first*, it is still the *second* coordinate. We check to be sure that the ordered pair is a solution.

Solve by the substitution method.

1. $x + y = 6$,
   $y = x + 2$

2. $y = 7 - x$,
   $2x - y = 8$

(*Caution*: Use parentheses when you substitute, being careful about removing them. Remember to solve for both variables.)

Solve by the substitution method.

3. $2y + x = 1$,
   $y - 2x = 8$

4. $8x + 5y = 184$,
   $x - y = -3$

Calculator Spotlight

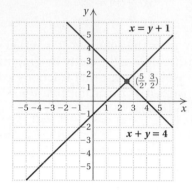 Use the TABLE feature to check the solutions of the systems in Examples 1 and 2 and Margin Exercises 1–4.

Answers on page A-19

CHECK:

$$\begin{array}{c|c} x + y = 4 \\ \hline \frac{5}{2} + \frac{3}{2} \;?\; 4 \\ \frac{8}{2} \\ 4 & \text{TRUE} \end{array} \qquad \begin{array}{c|c} x = y + 1 \\ \hline \frac{5}{2} \;?\; \frac{3}{2} + 1 \\ \frac{3}{2} + \frac{2}{2} \\ \frac{5}{2} & \text{TRUE} \end{array}$$

Since $\left(\frac{5}{2}, \frac{3}{2}\right)$ checks, it is the solution. Even though exact fractional solutions are difficult to determine graphically, a graph can help us to visualize whether the solution is reasonable.

Do Exercises 1 and 2.

Suppose neither equation of a pair has a variable alone on one side. We then solve one equation for one of the variables.

Example 2   Solve this system:

$$2x + y = 6, \quad \textbf{(1)}$$
$$3x + 4y = 4. \quad \textbf{(2)}$$

First, we solve one equation for one variable. Since the coefficient of y is 1 in equation (1), it is the easier one to solve for y:

$$y = 6 - 2x. \quad \textbf{(3)}$$

Next, we substitute $6 - 2x$ for y in equation (2) and solve for x:

$$3x + 4(6 - 2x) = 4 \qquad \text{Substituting } 6 - 2x \text{ for } y$$

Remember to use parentheses when you substitute. Then remove them carefully.

$$3x + 24 - 8x = 4 \qquad \text{Multiplying to remove parentheses}$$
$$24 - 5x = 4 \qquad \text{Collecting like terms}$$
$$-5x = -20 \qquad \text{Subtracting 24}$$
$$x = 4. \qquad \text{Dividing by } -5$$

We return to either of the original equations, (1) or (2), or equation (3), which we solved for y. It is generally easier to use an equation like (3), where we have solved for the specific variable. We substitute 4 for x in equation (3) and solve for y:

$$y = 6 - 2x$$
$$= 6 - 2(4) = 6 - 8 = -2.$$

We obtain the ordered pair $(4, -2)$.

CHECK:

$$\begin{array}{c|c} 2x + y = 6 \\ \hline 2(4) + (-2) \;?\; 6 \\ 8 - 2 \\ 6 & \text{TRUE} \end{array} \qquad \begin{array}{c|c} 3x + 4y = 4 \\ \hline 3(4) + 4(-2) \;?\; 4 \\ 12 - 8 \\ 4 & \text{TRUE} \end{array}$$

Since $(4, -2)$ checks, it is the solution.

Do Exercises 3 and 4.

b The Elimination Method

The **elimination method** for solving systems of equations makes use of the *addition principle* for equations. Some systems are much easier to solve using the elimination method rather than the substitution method.

Example 3 Solve this system:

$$2x - 3y = 0, \quad \textbf{(1)}$$
$$-4x + 3y = -1. \quad \textbf{(2)}$$

The key to the advantage of the elimination method for solving this system involves the $-3y$ in one equation and the $3y$ in the other. These terms are opposites. If we add them, these terms will add to 0, and in effect, the variable y will have been "eliminated."

We will use the addition principle for equations, adding the same number on both sides of the equation. According to equation (2), $-4x + 3y$ and -1 are the same number. Thus we can use a vertical form and add $-4x + 3y$ to the left side of equation (1) and -1 to the right side:

$$
\begin{array}{ll}
2x - 3y = 0 & \textbf{(1)} \\
\underline{-4x + 3y = -1} & \textbf{(2)} \\
-2x + 0y = -1 & \text{Adding} \\
-2x + 0 = -1 & \\
-2x = -1. &
\end{array}
$$

We have eliminated the variable y, which is why we call this the *elimination method*. We now have an equation with just one variable, which we solve for x:

$$-2x = -1$$
$$x = \tfrac{1}{2}.$$

Next, we substitute $\tfrac{1}{2}$ for x in either equation and solve for y:

$$
\begin{array}{ll}
2 \cdot \tfrac{1}{2} - 3y = 0 & \text{Substituting in equation (1)} \\
1 - 3y = 0 & \\
-3y = -1 & \text{Subtracting 1} \\
y = \tfrac{1}{3}. & \text{Dividing by } -3
\end{array}
$$

CHECK:

$$
\begin{array}{c|c}
2x - 3y = 0 & -4x + 3y = -1 \\
\hline
2\left(\tfrac{1}{2}\right) - 3\left(\tfrac{1}{3}\right) \; ? \; 0 & -4\left(\tfrac{1}{2}\right) + 3\left(\tfrac{1}{3}\right) \; ? \; -1 \\
1 - 1 \;\bigm| & -2 + 1 \;\bigm| \\
0 \;\bigm|\quad \text{TRUE} & -1 \;\bigm|\quad \text{TRUE}
\end{array}
$$

Since $\left(\tfrac{1}{2}, \tfrac{1}{3}\right)$ checks, it is the solution. We can also see this in the graph shown at right.

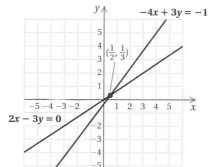

Do Exercises 5 and 6.

Solve by the elimination method.

5. $5x + 3y = 17,$
$\quad -5x + 2y = 3$

6. $-3a + 2b = 0,$
$\quad 3a - 4b = -1$

Answers on page A-19

7. Solve by the elimination method:

$$2y + 3x = 12,$$
$$-4y + 5x = -2.$$

In order to eliminate a variable, we sometimes use the multiplication principle to multiply one or both of the equations by a particular number before adding.

Example 4   Solve this system:

$$3x + 3y = 15,   \textbf{(1)}$$
$$2x + 6y = 22.   \textbf{(2)}$$

If we add directly, we get $5x + 9y = 37$, but we have not eliminated a variable. However, note that if the $3y$ in equation (1) were $-6y$, we could eliminate y. Thus we multiply by -2 on both sides of equation (1) and add:

$$
\begin{array}{ll}
-6x - 6y = -30 & \text{Multiplying equation (1) by } -2 \text{ on both sides} \\
\underline{2x + 6y = 22} & \text{Equation (2)} \\
-4x + 0 = -8 & \text{Adding} \\
{-4x} = -8 & \\
{x} = 2. & \text{Solving for } x
\end{array}
$$

Then

$$
\begin{array}{ll}
2 \cdot 2 + 6y = 22 & \text{Substituting 2 for } x \text{ in Equation (2)} \\
4 + 6y = 22 & \\
6y = 18 & \text{Solving for } y \\
y = 3. &
\end{array}
$$

We obtain $(2, 3)$, or $x = 2$, $y = 3$. This checks, so it is the solution.

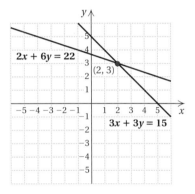

Do Exercise 7.

Sometimes we must multiply twice in order to make two terms opposites.

Example 5   Solve this system:

$$2x + 3y = 17,   \textbf{(1)}$$
$$5x + 7y = 29.   \textbf{(2)}$$

We must first multiply in order to make one pair of terms with the same variable opposites. We decide to do this with the x-terms in each equation. We multiply equation (1) by 5 and equation (2) by -2. Then we get $10x$ and $-10x$, which are opposites.

$$
\begin{array}{lll}
\textit{From Equation (1):} & 10x + 15y = 85 & \text{Multiplying by 5} \\
\textit{From Equation (2):} & \underline{-10x - 14y = -50} & \text{Multiplying by } -2 \\
& 0 + y = 27 & \text{Adding} \\
& y = 27. & \text{Solving for } y
\end{array}
$$

Answer on page A-19

Then

$$2x + 3 \cdot 27 = 17 \qquad \text{Substituting 27 for } y \text{ in equation (1)}$$

$$\left. \begin{aligned} 2x + 81 &= 17 \\ 2x &= -64 \\ x &= -32. \end{aligned} \right\} \quad \text{Solving for } x$$

CHECK:

$2x + 3y = 17$	
$2(-32) + 3(27)$? 17	
$-64 + 81$	
17	TRUE

$5x + 7y = 29$	
$5(-32) + 7(27)$? 29	
$-160 + 189$	
29	TRUE

We obtain $(-32, 27)$, or $x = -32$, $y = 27$, as the solution.

Do Exercises 8 and 9.

Some systems have no solution, as we saw graphically in Section 3.1. How do we recognize such systems if we are solving using an algebraic method?

Example 6 Solve this system:

$$y + 3x = 5, \qquad \textbf{(1)}$$
$$y + 3x = -2. \qquad \textbf{(2)}$$

If we find the slope–intercept equations for this system, we get

$$y = -3x + 5,$$
$$y = -3x - 2.$$

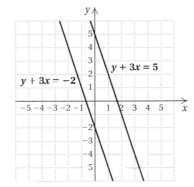

The graphs are parallel lines. The system has no solution.

Let's see what happens if we attempt to solve the system by the elimination method. We multiply by -1 on both sides of equation (2) and add:

$$\begin{array}{ll} y + 3x = 5 & \text{Equation (1)} \\ \underline{-y - 3x = 2} & \text{Multiplying equation (2) by } -1 \\ 0 = 7. & \text{Adding, we obtain a false equation.} \end{array}$$

The x-terms and the y-terms are eliminated and we end up with a *false* equation. Thus, if we obtain a false equation when solving algebraically, we know that the system has no solution. The system is inconsistent and independent.

Do Exercise 10.

Solve by the elimination method.

8. $4x + 5y = -8,$
$7x + 9y = 11$

9. $4x - 5y = 38,$
$7x - 8y = -22$

10. Solve by the elimination method:

$$y + 2x = 3,$$
$$y + 2x = -1.$$

Answers on page A-19

11. Solve by the elimination method:

$$2x - 5y = 10,$$
$$-6x + 15y = -30.$$

12. Clear the decimals. Then solve.

$$0.02x + 0.03y = 0.01,$$
$$0.3x - 0.1y = 0.7$$

(*Hint*: Multiply the first equation by 100 and the second one by 10.)

13. Clear the fractions. Then solve.

$$\frac{3}{5}x + \frac{2}{3}y = \frac{1}{3},$$
$$\frac{3}{4}x - \frac{1}{3}y = \frac{1}{4}$$

Some systems have infinitely many solutions. How can we recognize such a situation when we are solving systems using an algebraic method?

Example 7 Solve this system:

$$3y - 2x = 6, \qquad \textbf{(1)}$$
$$-12y + 8x = -24. \qquad \textbf{(2)}$$

The graphs are the same line. The system has an infinite number of solutions.

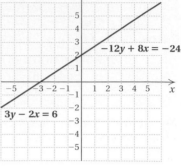

Suppose we try to solve this system by the elimination method:

$$12y - 8x = 24 \qquad \text{**Multiplying equation (1) by 4**}$$
$$\underline{-12y + 8x = -24} \qquad \text{**Equation (2)**}$$
$$\ 0 = 0. \qquad \text{**Adding, we obtain a true equation.**}$$

We have eliminated both variables, and what remains is a true equation. It can be expressed as $0 \cdot x + 0 \cdot y = 0$, and is true for all numbers x and y. If a pair is a solution of one of the original equations, then it will be a solution of the other. The system has an infinite number of solutions. The system is consistent and dependent.

> When solving a system of two linear equations in two variables:
>
> 1. If a false equation is obtained, such as $0 = 7$, then the system has no solution. The system is inconsistent and independent.
>
> 2. If a true equation is obtained, such as $0 = 0$, then the system has an infinite number of solutions. The system is consistent and dependent.

Do Exercise 11.

In carrying out the elimination method, it helps to first write the equations in the form $Ax + By = C$. When decimals or fractions occur, it also helps to *clear* before solving.

Example 8 Solve this system:

$$0.2x + 0.3y = 1.7,$$
$$\tfrac{1}{7}x + \tfrac{1}{5}y = \tfrac{29}{35}.$$

We have

$$0.2x + 0.3y = 1.7, \xrightarrow{\text{Multiplying by 10}} 2x + 3y = 17,$$
$$\tfrac{1}{7}x + \tfrac{1}{5}y = \tfrac{29}{35} \xrightarrow{\text{Multiplying by 35}} 5x + 7y = 29.$$

We multiplied by 10 to clear the decimals. Multiplication by 35, the least common denominator, clears the fractions. The problem is now identical to Example 5. The solution is $(-32, 27)$, or $x = -32$, $y = 27$.

Do Exercises 12 and 13.

Answers on page A-19

To use the elimination method to solve systems of two equations:

1. Write both equations in the form $Ax + By = C$.
2. Clear any decimals or fractions.
3. Choose a variable to eliminate.
4. Make the chosen variable's terms opposites by multiplying one or both equations by appropriate numbers if necessary.
5. Eliminate a variable by adding the sides of the equations and then solve for the remaining variable.
6. Substitute in either of the original equations to find the value of the other variable.

Comparing Methods

The following table is a summary that compares the graphical, substitution, and elimination methods for solving systems of equations.

When deciding which method to use, consider this table and directions from your instructor. The situation is analogous to having a piece of wood to cut and three different types of saws available. Although all three saws can cut the wood, the "best" choice depends on the particular piece of wood, the type of cut being made, and your level of skill with each saw.

Method	Strengths	Weaknesses
Graphical	Can "see" solutions.	Inexact when solutions involve numbers that are not integers. Solution may not appear on the part of the graph drawn.
Substitution	Yields exact solutions. Convenient to use when a variable has a coefficient of 1.	Can introduce extensive computations with fractions. Cannot "see" solutions quickly.
Elimination	Yields exact solutions. Convenient to use when no variable has a coefficient of 1. The preferred method for systems of 3 or more equations in 3 or more variables (see Section 3.4).	Cannot "see" solutions quickly.

c Solving Applied Problems Involving Two Equations

Many applied problems are easier to solve if we first translate to a system of two equations rather than to a single equation. Using substitution or elimination, we will solve some fairly easy problems here; then in Section 3.3 we will consider more complicated problems. Let's begin with the real-world application involving the Simon–DeBartolo merger that we solved graphically in Section 3.1.

14. *Basketball Scoring.* The Central College Cougars made 40 field goals in a recent basketball game, some 2-pointers and the rest 3-pointers. Altogether the 40 baskets counted for 89 points. How many of each type of field goal was made?

Answer on page A-19

Example 9   *Real Estate Merger.* In 1996, the Simon Property Group and the DeBartolo Realty Corporation merged to form the largest real estate company in the United States, owning 183 shopping centers in 32 states. Before the merger, Simon owned twice as many properties as DeBartolo. How many properties did each own originally?

1., 2. Familiarize and **Translate.**   These steps were actually completed at the beginning of Section 3.1. The resulting system of equations is

$$x + y = 183, \qquad \textbf{(1)}$$
$$x = 2y, \qquad \textbf{(2)}$$

where x is the number of properties originally owned by Simon and y is the number of properties originally owned by DeBartolo.

3. Solve.   We solve the system of equations. We note that one equation already has a variable by itself on one side, so substitution can be used. (Keep in mind that unless you are directed otherwise, you can use the method you prefer.)

$$x + y = 183$$
$$2y + y = 183 \qquad \text{Substituting } 2y \text{ for } x \text{ in equation (1)}$$
$$3y = 183 \qquad \text{Collecting like terms}$$
$$y = 61 \qquad \text{Solving for } y$$

Then we substitute 61 for y in equation (2) and solve for x:

$$x = 2y = 2(61) = 122.$$

4. Check.   The sum of 122 and 61 is 183, so the total number of properties is correct. Since 122 is twice 61, the numbers check.

5. State.   Before the merger, Simon owned 122 properties and DeBartolo owned 61.

Do Exercise 14.

Example 10   *Architecture.* The architects who designed the John Hancock Building in Chicago created a visually appealing building that slants on the sides. Thus the ground floor is a rectangle that is larger than the rectangle formed by the top floor. The ground floor has a perimeter of 860 ft. The length is 100 ft more than the width. Find the length and the width.

1. Familiarize.   We first make a drawing and label it. We recall, or look up, the formula for perimeter: $P = 2l + 2w$. This formula can be found at the back of the book.

2. Translate.   We translate as follows:

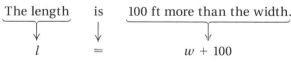

The perimeter   is   860 ft.

$$2l + 2w \quad = \quad 860$$

We then translate the second statement:

The length   is   100 ft more than the width.

$$l \quad = \quad w + 100$$

$l = w + 100$ w

We now have a system of equations:

$$2l + 2w = 860, \qquad \textbf{(1)}$$
$$l = w + 100. \qquad \textbf{(2)}$$

3. Solve. For comparison, we use both the substitution and elimination methods. Let's first use *substitution*, substituting $w + 100$ for l in equation (1):

$2(w + 100) + 2w = 860$	Substituting in equation (1)
$2w + 200 + 2w = 860$	Multiplying to remove parentheses on the left
$4w + 200 = 860$	Collecting like terms
$4w = 660$	
$w = 165.$	Solving for w

Next, we substitute 165 for w in equation (2) and solve for l:

$$l = 165 + 100 = 265.$$

To use *elimination*, we first rewrite both equations in the form $Ax + By = C$. Since equation (1) is already in this form, we need only rewrite equation (2):

$l = w + 100$	Equation (2)
$l - w = 100.$	Subtracting w

Now we solve the system

$$2l + 2w = 860, \qquad \textbf{(1)}$$
$$l - w = 100. \qquad \textbf{(2)}$$

We multiply by 2 on both sides of equation (2) and add:

$2l + 2w = 860$	Equation (1)
$2l - 2w = 200$	Multiplying by 2 on both sides of equation (2)
$4l \qquad = 1060$	Adding
$l = 265.$	Solving for l

Next, we substitute 265 for l in the equation $l = w + 100$ and solve for w:

$$265 = w + 100$$
$$165 = w.$$

Which method do you prefer to use?

4. Check. Consider the dimensions 265 ft and 165 ft. The length is 100 ft more than the width. The perimeter is 2(265 ft) + 2(165 ft), or 860 ft. The dimensions 265 ft and 165 ft check in the original problem.

5. State. The length is 265 ft and the width is 165 ft.

Do Exercise 15.

15. *Architecture.* The top floor of the John Hancock Building is also a rectangle, but its perimeter is 520 ft. The width is 60 ft less than the length. Find the length and the width.

a) First, use substitution to solve the resulting system.

b) Second, use elimination to solve the resulting system.

Answers on page A-19

Improving Your Math Study Skills

Studying for Tests and Making the Most of Tutoring Sessions

This math study skill feature focuses on the very important task of test preparation.

Test-Taking Tips

- **Make up your own test questions as you study.** You have probably become accustomed by now to the section and objective codes that appear throughout the book. After you have done your homework over a particular objective, write one or two questions on your own that you think might be on a test. You will be amazed at the insight this will provide. You are actually carrying out a task similar to what a teacher does in preparing an exam.

- **Do an overall review of the chapter focusing on the objectives and the examples.** This should be accompanied by a study of any class notes you may have taken.

- **Do the review exercises at the end of the chapter.** Check your answers at the back of the book. If you have trouble with an exercise, use the objective symbol as a guide to go back and do further study of that objective. These review exercises are very much like a sample test.

- **Do the chapter test at the end of the chapter.** This is like taking a second sample test. Check the answers and objective symbols at the back of the book.

- **Ask former students for old exams.** Working such exams can be very helpful and allows you to see what various professors think is important.

- **When taking a test, read each question carefully and try to do all the questions the first time through, but pace yourself.** Answer all the questions, and mark those to recheck if you have time at the end. Very often, your first hunch will be correct.

- **Try to write your test in a neat and orderly manner.** Very often, your instructor tries to give you partial credit when grading an exam. If your test paper is sloppy and disorderly, it is difficult to verify the partial credit. Doing your work neatly can ease such a task for the instructor. Try using an erasable pen to make your writing darker and therefore more readable.

- **What about the student who says, "I could do the work at home, but on the test I made silly mistakes"?** Yes, all of us, including instructors, make silly computational mistakes in class, on homework, and on tests. But your instructor, if he or she has taught for some time, is probably aware that 90% of students who make such comments in truth do not have the required depth of knowledge of the subject matter, and such silly mistakes often are a sign that the student has not mastered the material. There is no way we can make that analysis for you. It will have to be unraveled by some careful soul searching on your part or by a conference with your instructor.

Making the Most of Tutoring and Help Sessions

Often you will determine that a tutoring session would be helpful. The following comments may help you to make the most of such sessions.

- **Work on the topics before you go to the help or tutoring session. Do not go to such sessions viewing yourself as an empty cup and the tutor as a magician who will pour in the learning.** The primary source of your ability to learn is within you. We have seen so many students over the years go to help or tutoring sessions with no advanced preparation. You are often wasting your time and perhaps your money if you are paying for such sessions. Go to class, study the textbook, and mark trouble spots. Then use the help and tutoring sessions to deal with these difficulties most efficiently.

- **Do not be afraid to ask questions in these sessions!** The more you talk to your tutor, the more the tutor can help you with your difficulties.

- **Try being a "tutor" yourself.** Explaining a topic to someone else—a classmate, your instructor—is often the best way to learn it.

Exercise Set 3.2

---

a Solve the system by the substitution method.

1. $2x + 5y = 4,$
$\quad x = 3 - 3y$

2. $5x - 2y = 23,$
$\quad y = 8 - 4x$

3. $9x - 2y = -6,$
$\quad 7x + 8 = y$

4. $x = 3y - 3,$
$\quad x + 2y = 9$

5. $5m + n = 8,$
$\quad 3m - 4n = 14$

6. $4x + y = -1,$
$\quad x - 2y = 11$

7. $4x + 13y = 5,$
$\quad -6x + y = 13$

8. $-5a + b = -23,$
$\quad 6a + 7b = 3$

b Solve the system by the elimination method.

9. $\ x + 3y = 7,$
$\quad -x + 4y = 7$

10. $\ x + y = 9,$
$\quad 2x - y = -3$

11. $9x + 5y = 6,$
$\quad 2x - 5y = -17$

12. $8x - 3y = 16,$
$\quad 8x + 3y = -8$

13. $5x + 3y = 19,$
$\quad 2x - 5y = 11$

14. $3x + 2y = 3,$
$\quad 9x - 8y = -2$

15. $5r - 3s = 24,$
$\quad 3r + 5s = 28$

16. $5x - 7y = -16,$
$\quad 2x + 8y = 26$

17. $0.3x - 0.2y = 4,$
$\quad 0.2x + 0.3y = 1$

18. $\ \ 0.7x - 0.3y = 0.5,$
$\quad -0.4x + 0.7y = 1.3$

19. $\frac{1}{2}x + \frac{1}{3}y = 4,$
$\quad \frac{1}{4}x + \frac{1}{3}y = 3$

20. $\frac{2}{3}x + \frac{1}{7}y = -11,$
$\quad \frac{1}{7}x - \frac{1}{3}y = -10$

21. $\frac{2}{5}x + \frac{1}{2}y = 2,$
$\frac{1}{2}x - \frac{1}{6}y = 3$

22. $\frac{1}{3}x + \frac{1}{5}y = 7,$
$\frac{1}{6}x - \frac{2}{5}y = -4$

23. $2x + 3y = 1,$
$4x + 6y = 2$

24. $3x - 2y = 1,$
$-6x + 4y = -2$

25. $2x - 4y = 5,$
$2x - 4y = 6$

26. $3x - 5y = -2,$
$5y - 3x = 7$

27. $5x - 9y = 7,$
$7y - 3x = -5$

28. $a - 2b = 16,$
$b + 3 = 3a$

29. $3(a - b) = 15,$
$4a = b + 1$

30. $10x + y = 306,$
$10y + x = 90$

31. $x - \frac{1}{10}y = 100,$
$y - \frac{1}{10}x = -100$

32. $\frac{1}{8}x + \frac{3}{5}y = \frac{19}{2},$
$-\frac{3}{10}x - \frac{7}{20}y = -1$

33. $0.05x + 0.25y = 22,$
$0.15x + 0.05y = 24$

34. $1.3x - 0.2y = 12,$
$0.4x + 17y = 89$

$\boxed{c}$ Solve.

35. *Racquetball Court.* A regulation racquetball court has a perimeter of 120 ft, with a length that is twice the width. Find the length and the width of such a court.

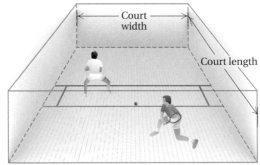

36. *Soccer Field.* The perimeter of a soccer field is 340 m. The length exceeds the width by 50 m. Find the length and the width.

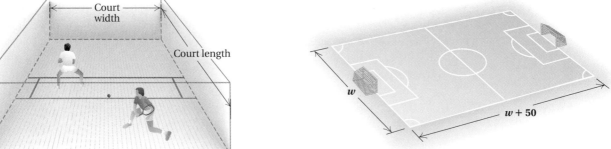

---

37. *Supplementary Angles.* **Supplementary angles** are angles whose sum is 180°. Two supplementary angles are such that one angle is 12° less than three times the other. Find the measures of the angles.

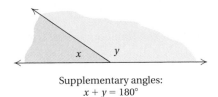

Supplementary angles:
$x + y = 180°$

38. *Complementary Angles.* **Complementary angles** are angles whose sum is 90°. Two complementary angles are such that one angle is 6° more than five times the other. Find the measures of the angles.

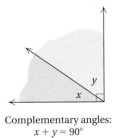

Complementary angles:
$x + y = 90°$

39. The sum of two numbers is −42. The first number minus the second is 52. What are the numbers?

40. The difference between two numbers is 11. Twice the smaller plus three times the larger is 123. What are the numbers?

41. *Hockey Points.* Hockey teams receive two points when they win a game and one point when they tie. One season, a team won a championship with 60 points. They won 9 more games than they tied. How many wins and how many ties did the team have?

42. *Airplane Seating.* An airplane has a total of 152 seats. The number of coach-class seats is 5 more than six times the number of first-class seats. How many of each type of seat are there on the plane?

43. Find the slope of the line $y = 1.3x - 7$. [2.4b]

44. Simplify: $-9(y + 7) - 6(y - 4)$. [R.6b]

45. Solve $A = \dfrac{pq}{7}$ for p. [1.2a]

46. Find the slope of the line containing the points $(-2, 3)$ and $(-5, -4)$. [2.4b]

Solve. [1.1d]

47. $-4x + 5(x - 7) = 8x - 6(x + 2)$

48. $-12(2x - 3) = 16(4x - 5)$

Given the function $f(x) = 3x^2 - x + 1$, find each of the following function values. [2.2b]

49. $f(0)$

50. $f(-1)$

51. $f(1)$

52. $f(10)$

53. $f(-2)$

54. $f(2a)$

55. $f(-4)$

56. $f(1.8)$

Synthesis

57. ◆ Describe a method that could be used to create inconsistent systems of equations.

58. ◆ Describe a method that could be used to create dependent systems of equations.

59. ◢◣ Use the INTERSECT feature to solve the following system of equations. You may need to first solve for y. Round answers to the nearest hundredth.

$$3.5x - 2.1y = 106.2,$$
$$4.1x + 16.7y = -106.28$$

60. Solve:

$$\frac{x + y}{2} - \frac{x - y}{5} = 1,$$
$$\frac{x - y}{2} + \frac{x + y}{6} = -2.$$

61. Solve for x and y in terms of a and b:

$$5x + 2y = a,$$
$$x - y = b.$$

62. Determine a and b for which $(-4, -3)$ will be a solution of the system

$$ax + by = -26,$$
$$bx - ay = 7.$$

63. The points $(0, -3)$ and $\left(-\frac{3}{2}, 6\right)$ are two of the solutions of the equation $px - qy = -1$. Find p and q.

64. For $y = mx + b$, two solutions are $(1, 2)$ and $(-3, 4)$. Find m and b.

65. The solution of this system is $(-5, -1)$. Find A and B.

$$Ax - 7y = -3,$$
$$x - By = -1$$

66. Find an equation to pair with $6x + 7y = -4$ such that $(-3, 2)$ is a solution of the system.

Collaborative Learning Manual

Compare the three methods for solving systems of equations in two variables.

3.3 Solving Applied Problems: Systems of Two Equations

a | Total-Value and Mixture Problems

Systems of equations can be a useful tool in solving applied problems. Using systems often makes the *Translate* step easier than using a single equation. The first kind of problem we consider involves quantities of items sold and the total value of the items. We refer to this type of problem as a **total-value problem.**

Example 1   *Retail Sales of Gloves.* In one day, Glovers, Inc., sold 20 pairs of gloves. Fleece gloves sold for $24.95 a pair and Gore-Tex gloves for $37.50. Receipts totaled $687.25. How many of each kind of glove were sold?

1. **Familiarize.**   To familiarize ourselves with the problem situation, let's guess that Glovers sold 12 pairs of fleece gloves and 8 pairs of Gore-Tex gloves. The total is 20. How much money was taken in? Since fleece gloves sold for $24.95 a pair and Gore-Tex for $37.50, the total received would then be

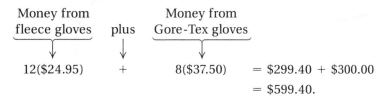

$$12(\$24.95) + 8(\$37.50) = \$299.40 + \$300.00$$
$$= \$599.40.$$

Although the total number of pairs is correct, our guess is incorrect because the problem states that the total amount received was $687.25. Since $599.40 is less than $687.25, more of the expensive gloves were sold than we had guessed. We could now adjust our guess accordingly. Instead, let's use an algebraic approach that avoids guessing.

We let p = the number of pairs of fleece gloves sold and q = the number of pairs of Gore-Tex gloves sold. It helps to organize the information in a table as follows:

Kind of Glove	Fleece	Gore-Tex	Total
Number Sold	p	q	20	$\rightarrow p + q = 20$
Price	$24.95	$37.50	
Amount Taken In	24.95p	37.50q	687.25	$\rightarrow 24.95p + 37.50q = 687.25$

2. **Translate.**   The first row of the table and the first sentence of the problem tell us that a total of 20 pairs of gloves was sold. Thus we have one equation:

$$p + q = 20.$$

Since each pair of fleece gloves costs $24.95 and p pairs were sold, 24.95p is the amount taken in from the sale of fleece gloves. Similarly, 37.50q is the amount taken in from the sale of q pairs of the Gore-Tex gloves. From the third row of the table and the third sentence of the problem, we get the second equation:

$$24.95p + 37.50q = 687.25.$$

1. *Retail Sales of Sweatshirts.* Sandy's Sweatshirt Shop sells college sweatshirts. White sweatshirts sell for $18.95 each and red ones sell for $19.50 each. If receipts for 30 shirts total $572.90, how many of each color did the shop sell? Complete the following table, letting w = the number of white sweatshirts and r = the number of red sweatshirts.

Kind of Sweatshirts	White	Red	Totals	
Number Sold	w	r	30	$\rightarrow$ ( ) + r = 30
Price	$18.95	$19.50		
Amount Taken In	18.95w			$\rightarrow$ 18.95w + ( ) = ( )

Answer on page A-20

We can multiply by 100 on both sides of this equation in order to clear the decimals. This gives us the following system of equations as a translation:

$$p + q = 20, \qquad \textbf{(1)}$$
$$2495p + 3750q = 68{,}725. \qquad \textbf{(2)}$$

3. Solve. We choose to use the elimination method to solve the system. We eliminate p by multiplying equation (1) by -2495 and adding it to equation (2):

$$
\begin{array}{rll}
-2495p - 2495q = & -49{,}900 & \text{Multiplying equation (1) by } -2495 \\
\underline{2495p + 3750q = } & \underline{68{,}725} & \\
1255q = & 18{,}825 & \text{Adding} \\
q = & 15. & \text{Solving for } q
\end{array}
$$

To find p, we substitute 15 for q in equation (1) and solve for p:

$$
\begin{array}{ll}
p + q = 20 & \text{Equation (1)} \\
p + 15 = 20 & \text{Substituting 15 for } q \\
p = 5. & \text{Solving for } p
\end{array}
$$

We obtain (5, 15), or $p = 5$, $q = 15$.

4. Check. We check in the original problem. Remember that p is the number of pairs of fleece gloves and q the number of pairs of Gore-Tex gloves:

Number of gloves: $p + q = 5 + 15 = 20$
Money from fleece gloves: $\$24.95p = 24.95 \times 5 = \$124.75$
Money from Gore-Tex gloves: $\$37.50q = 37.50 \times 15 = \underline{\$562.50}$
                                              Total = $687.25

The numbers check.

5. State. The store sold 5 pairs of fleece gloves and 15 pairs of Gore-Tex gloves.

Do Exercise 1.

The following problem, similar to Example 1, is called a **mixture problem.**

Example 2 *Blending Teas.* Tara's Tea Terrace sells loose Black tea for 95¢ per ounce and Lapsang Souchong tea for $1.43 per ounce. Tara wants to mix the two teas to get a 1-lb mixture, called Imperial Blend, that sells for $1.10 per ounce. How many ounces of each type of tea should Tara use?

1. Familiarize. To familiarize ourselves with the problem situation, we make a guess and do some calculations. The total amount of tea is to be 1 lb, or 16 oz. Let's try 12 oz of Black tea and 4 oz of Lapsang Souchong.

The sum of the amounts of tea is $12 + 4$, or 16.

The values of these amounts of tea are found by multiplying the cost per ounce, in dollars, by the number of ounces and adding:

$0.95(12) + $1.43(4)$, or $17.12.

The desired cost is $1.10 per ounce. If we multiply $1.10 by 16, we get $16($1.10)$, or $17.60. This does not agree with $17.12, but these calculations help us to translate.

We let a = the number of ounces of Black tea and b = the number of ounces of Lapsang Souchong tea. Next, we organize the information in a table, as follows.

	Black Tea	Lapsang Souchong	Imperial Blend	
Number of Ounces	a	b	16	→ $a + b = 16$
Price per Ounce	$0.95	$1.43	$1.10	
Value of Tea	0.95a	1.43b	16 · 1.10, or 17.60	→ $0.95a + 1.43b = 17.60$

2. Translate. The total amount of tea is 16 oz, so we have

$$a + b = 16.$$

The value of the Black tea is $0.95a$ and the value of the Lapsang Souchong tea is $1.43b$. These amounts are in dollars. Since the total is to be $16($1.10)$, or $17.60, we have

$$0.95a + 1.43b = 17.60.$$

We can multiply by 100 on both sides of this equation to clear the decimals. Thus we have the translation, a system of equations:

$$a + b = 16, \qquad \textbf{(1)}$$
$$95a + 143b = 1760. \qquad \textbf{(2)}$$

3. Solve. We decide to use substitution, although elimination could be used as we did in Example 1. When equation (1) is solved for b, we get $b = 16 - a$. Substituting $16 - a$ for b in equation (2) and solving gives us

$95a + 143(16 - a) = 1760$	Substituting
$95a + 2288 - 143a = 1760$	Using the distributive law
$-48a = -528$	Subtracting 2288 and collecting like terms
$a = 11.$	

We have $a = 11$. Substituting this value in the equation $b = 16 - a$, we obtain $b = 16 - 11$, or 5.

2. *Blending Coffees.* The Coffee Counter charges $9.00 per pound for Kenyan French Roast coffee and $8.00 per pound for Sumatran coffee. How much of each type should be used to make a 20-lb blend that sells for $8.40 per pound?

Answer on page A-20

3. *Client Investments.* Kaufman Financial Corporation makes investments for corporate clients. It makes an investment of $3700 for one year at simple interest, yielding $297. Part of the money is invested at 7% and the rest at 9%. How much was invested at each rate?

Do the *Familiarize* and *Translate* steps by completing the following table. Let $x =$ the number of dollars invested at 7% and $y =$ the number of dollars invested at 9%.

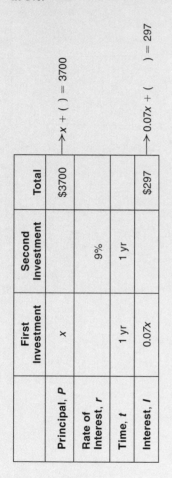

4. Check. We check in a manner similar to our guess in the *Familiarize* step. The total amount of tea is $11 + 5$, or 16 oz. The value of the tea is

$0.95(11) + $1.43(5)$, or $17.60.

Thus the amounts of tea check.

5. State. The Imperial Blend can be made by mixing 11 oz of Black tea with 5 oz of Lapsang Souchong tea.

Do Exercise 2 on the preceding page.

Example 3   *Student Loans.* Enid's student loans totaled $9600. Part was a Perkins loan made at 5% interest and the rest was a Federal Education Loan made at 8% interest. After one year, Enid's loans accumulated $633 in interest. What was the amount of each loan?

1. Familiarize. Listing the given information in a table will help. The columns in the table come from the formula for simple interest: $I = Prt$. We let $x =$ the number of dollars in the Perkins loan and $y =$ the number of dollars in the Federal Education loan.

	Perkins Loan	Federal Loan	Total	
Principal	x	y	$9600	→ $x + y = 9600$
Rate of Interest	5%	8%		
Time	1 yr	1 yr		
Interest	$0.05x$	$0.08y$	$633	→ $0.05x + 0.08y = 633$

2. Translate. The total of the amounts of the loans is found in the first row of the table. This gives us one equation:

$x + y = 9600.$

Look at the last row of the table. The interest, or **yield**, totals $633. This gives us a second equation:

$5\%x + 8\%y = 633,$ or $0.05x + 0.08y = 633.$

After we multiply on both sides to clear the decimals, we have

$5x + 8y = 63,300.$

3. Solve. Using either elimination or substitution, we solve the resulting system:

$x + y = 9600,$
$5x + 8y = 63,300.$

We find that $x = 4500$ and $y = 5100$.

4. Check. The sum is $4500 + $5100, or $9600. The interest from $4500 at 5% for one year is 5%($4500), or $225. The interest from $5100 at 8% for one year is 8%($5100), or $408. The total interest is $225 + $408, or $633. The numbers check in the problem.

5. State. The Perkins loan was for $4500 and the Federal Education loan was for $5100.

Do Exercise 3.

Example 4 *Mixing Fertilizers.* Yardbird Gardening carries two kinds of fertilizer containing nitrogen and water. "Gently Green" is 5% nitrogen and "Sun Saver" is 15% nitrogen. Yardbird Gardening needs to combine the two types of solution to make 90 L of a solution that is 12% nitrogen. How much of each brand should be used?

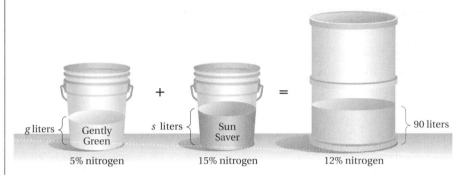

| g liters { Gently Green | s liters { Sun Saver | 90 liters |
| 5% nitrogen | 15% nitrogen | 12% nitrogen |

1. **Familiarize.** We first make a drawing and a guess to become familiar with the problem.

   We choose two numbers that total 90 L—say, 40 L of Gently Green and 50 L of Sun Saver—for the amounts of each fertilizer. Will the resulting mixture have the correct percentage of fertilizer? To find out, we multiply as follows:

   5%(40 L) = 2 L of nitrogen   and   15%(50 L) = 7.5 L of nitrogen.

   Thus the total amount of nitrogen in the mixture is 2 L + 7.5 L, or 9.5 L. The final mixture of 90 L is supposed to have 12% nitrogen. Now

   12%(90 L) = 10.8 L.

   Since 9.5 L and 10.8 L are not the same, our guess is incorrect. But these calculations help us to become familiar with the problem and to make the translation.

   We let g = the number of liters of Gently Green and s = the number of liters of Sun Saver.

   The information can be organized in a table, as follows.

	Gently Green	Sun Saver	Mixture	
Number of Liters	g	s	90	→ $g + s = 90$
Percent of Nitrogen	5%	15%	12%	
Amount of Nitrogen	$0.05g$	$0.15s$	0.12×90, or 10.8 liters	→ $0.05g + 0.15s = 10.8$

2. **Translate.** If we add g and s in the first row, we get 90, and this gives us one equation:

   $g + s = 90.$

   If we add the amounts of nitrogen listed in the third row, we get 10.8, and this gives us another equation:

   5%g + 15%s = 10.8,   or $0.05g + 0.15s = 10.8.$

4. *Mixing Cleaning Solutions.* King's Service Station uses two kinds of cleaning solution containing acid and water. "Attack" is 2% acid and "Blast" is 6% acid. They want to mix the two to get 60 qt of a solution that is 5% acid. How many quarts of each should they use?

Do the *Familiarize* and *Translate* steps by completing the following table. Let a = the number of quarts of Attack and b = the number of quarts of Blast.

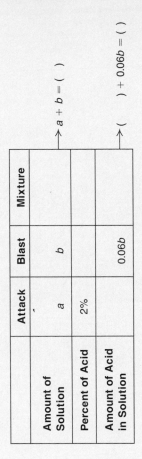

	Attack	Blast	Mixture	
Amount of Solution	a	b		$\rightarrow a + b = (\quad)$
Percent of Acid	2%			
Amount of Acid in Solution		$0.06b$		$\rightarrow (\quad) + 0.06b = (\quad)$

Answer on page A-20

After clearing the decimals, we have the following system:

$$g + s = 90, \qquad \textbf{(1)}$$
$$5g + 15s = 1080. \qquad \textbf{(2)}$$

3. Solve. We solve the system using elimination. We multiply equation (1) by -5 and add the result to equation (2):

$$
\begin{array}{ll}
-5g - 5s = -450 & \textbf{Multiplying equation (1) by -5} \\
\underline{5g + 15s = 1080} & \\
\qquad 10s = 630 & \textbf{Adding} \\
\qquad\quad s = 63; & \textbf{Dividing by 10}
\end{array}
$$

$$
\begin{array}{ll}
g + 63 = 90 & \textbf{Substituting in equation (1) of the system} \\
\quad\ g = 27. & \textbf{Solving for g}
\end{array}
$$

4. Check. Remember that g is the number of liters of Gently Green, with 5% nitrogen, and s is the number of liters of Sun Saver, with 15% nitrogen.

Total number of liters of mixture: $g + s = 27 + 63 = 90$

Amount of nitrogen: $5\%(27) + 15\%(63) = 1.35 + 9.45 = 10.8 \text{ L}$

Percentage of nitrogen in mixture: $\dfrac{10.8}{90} = 0.12 = 12\%$

The numbers check in the original problem.

5. State. Yardbird Gardening should mix 27 L of Gently Green with 63 L of Sun Saver.

Do Exercise 4.

b   Motion Problems

When a problem deals with speed, distance, and time, we can expect to use the following *motion formula*.

> **THE MOTION FORMULA**
> Distance = Rate (or speed) · Time
> $$d = rt$$

We have five steps for problem solving. The following tips are also helpful when solving motion problems.

> **TIPS FOR SOLVING MOTION PROBLEMS**
>
> 1. Draw a diagram using an arrow or arrows to represent distance and the direction of each object in motion.
> 2. Organize the information in a table or chart.
> 3. Look for as many things as you can that are the same, so you can write equations.

Example 5 *Auto Travel.* Your brother leaves on a trip, forgetting his suit-case. You know that he normally drives at a speed of 55 mph. You do not discover the suitcase until 1 hr after he has left. If you follow him at a speed of 65 mph, how long will it take you to catch up with him?

1. Familiarize. We first make a drawing.

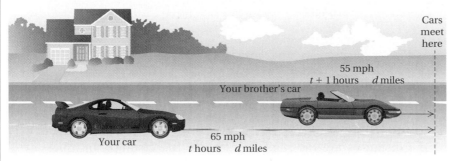

From the drawing, we see that when you catch up with your brother, the distances from home are the same. Let's call the distance *d*. If we let *t* = the time for you to catch your brother, then *t* + 1 = the time traveled by your brother at a slower speed. We organize the information in a table.

$$d \quad = \quad r \quad \cdot \quad t$$

	Distance	Rate	Time	
Brother	*d*	55	*t* + 1	→ $d = 55(t + 1)$
You	*d*	65	*t*	→ $d = 65t$

2. Translate. Using $d = rt$ in each row of the table, we get an equation. Thus we have a system of equations:

$$d = 55(t + 1), \qquad \textbf{(1)}$$
$$d = 65t. \qquad\qquad \textbf{(2)}$$

3. Solve. We solve the system using the substitution method:

$65t = 55(t + 1)$ **Substituting 65*t* for *d* in equation (1)**

$65t = 55t + 55$ **Multiplying to remove parentheses on the right**

$\left.\begin{array}{l} 10t = 55 \\ t = 5.5. \end{array}\right\}$ **Solving for *t***

Your time is 5.5 hr, which means that your brother's time is 5.5 + 1, or 6.5 hr.

4. Check. At 65 mph, you will travel 65 · 5.5, or 357.5 mi, in 5.5 hr. At 55 mph, your brother will travel 55 · 6.5, or the same 357.5 mi, in 6.5 hr. The numbers check.

5. State. You will overtake your brother in 5.5 hr.

Do Exercise 5.

5. *Train Travel.* A train leaves Barstow traveling east at 35 km/h. One hour later, a faster train leaves Barstow, also traveling east on a parallel track at 40 km/h. How far from Barstow will the faster train catch up with the slower one?

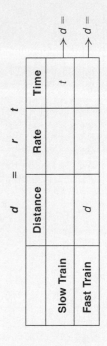

Answer on page A-20

6. *Air Travel.* An airplane flew for 4 hr with a 20-mph tailwind. The return flight against the same wind took 5 hr. Find the speed of the plane in still air.

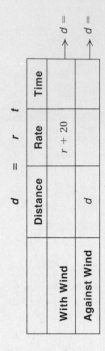

	Distance	Rate	Time	
		= r · t		
With Wind	d	$r + 20$		$= d$
Against Wind	d			$= d$

Example 6 *Marine Travel.* A Coast-Guard patrol boat travels 4 hr on a trip downstream with a 6-mph current. The return trip against the same current takes 5 hr. Find the speed of the boat in still water.

Downstream, $r + 6$
6-mph current, 4 hours, d miles

Upstream, $r - 6$
6-mph current, 5 hours, d miles

1. **Familiarize.** We first make a drawing. From the drawing, we see that the distances are the same. We let d = the distance, in miles, and r = the speed of the boat in still water, in miles per hour. Then, when the boat is traveling downstream, its speed is $r + 6$ (the current helps the boat along). When it is traveling upstream, its speed is $r - 6$ (the current holds the boat back). We can organize the information in a table. We use the formula $d = rt$.

	Distance	Rate	Time	
	d	= r · t		
Downstream	d	$r + 6$	4	→ $d = (r + 6)4$
Upstream	d	$r - 6$	5	→ $d = (r - 6)5$

2. **Translate.** From each row of the table, we get an equation, $d = rt$:

$$d = 4r + 24, \quad \textbf{(1)}$$
$$d = 5r - 30. \quad \textbf{(2)}$$

3. **Solve.** We solve the system using the substitution method:

$4r + 24 = 5r - 30$ **Substituting 4r + 24 for d in equation (2)**

$\left.\begin{array}{l} 24 = r - 30 \\ 54 = r. \end{array}\right\}$ **Solving for r**

4. **Check.** If $r = 54$, then $r + 6 = 60$; and $60 \cdot 4 = 240$, the distance traveled downstream. If $r = 54$, then $r - 6 = 48$; and $48 \cdot 5 = 240$, the distance traveled upstream. The distances are the same. In this type of problem, a problem-solving tip to keep in mind is "Have I found what the problem asked for?" We could solve for a certain variable but still have not answered the question of the original problem. For example, we might have found speed when the problem wanted distance. In this problem, we want the speed of the boat in still water, and that is r.

5. **State.** The speed in still water is 54 mph.

Do Exercise 6.

Answer on page A-20

Exercise Set 3.3

a Solve.

1. *Retail Sales.* Paint Town sold 45 paintbrushes, one kind at $8.50 each and another at $9.75 each. In all, $398.75 was taken in for the brushes. How many of each kind were sold?

2. *Retail Sales.* Mountainside Fleece sold 40 neckwarmers. Solid-color neckwarmers sold for $9.90 each and print ones sold for $12.75 each. In all, $421.65 was taken in for the neckwarmers. How many of each type were sold?

3. *Sales of Pharmaceuticals.* The Diabetic Express recently charged $15.75 for a vial of Humulin insulin and $12.95 for a vial of Novolin insulin. If a total of $959.35 was collected for 65 vials of insulin, how many vials of each type were sold?

4. *Fundraising.* The St. Mark's Community Barbecue served 250 dinners. A child's plate cost $3.50 and an adult's plate cost $7.00. A total of $1347.50 was collected. How many of each type of plate was served?

5. *Radio Airplay.* Omar must play 12 commercials during his 1-hr radio show. Each commercial is either 30 sec or 60 sec long. If the total commercial time during that hour is 10 min, how many commercials of each type does Omar play?

6. *Nontoxic Floor Wax.* A nontoxic floor wax can be made by combining lemon juice and food-grade linseed oil. The amount of oil should be twice the amount of lemon juice. How much of each ingredient is needed in order to make 32 oz of floor wax? (The mix should be spread with a rag and buffed when dry.)

7. *Catering.* Casella's Catering is planning a wedding reception. The bride and groom would like to serve a nut mixture containing 25% peanuts. Casella has available mixtures that are either 40% or 10% peanuts. How much of each type should be mixed to get a 10-lb mixture that is 25% peanuts?

8. *Blending Granola.* Deep Thought Granola is 25% nuts and dried fruit. Oat Dream Granola is 10% nuts and dried fruit. How much of Deep Thought and how much of Oat Dream should be mixed to form a 20-lb batch of granola that is 19% nuts and dried fruit?

9. *Ink Remover.* Etch Clean Graphics uses one cleanser that is 25% acid and a second that is 50% acid. How many liters of each should be mixed to get 10 L of a solution that is 40% acid?

10. *Livestock Feed.* Soybean meal is 16% protein and corn meal is 9% protein. How many pounds of each should be mixed to get a 350-lb mixture that is 12% protein?

11. *Student Loans.* Lomasi's two student loans totaled $12,000. One of her loans was at 6% simple interest and the other at 9%. After one year, Lomasi owed $855 in interest. What was the amount of each loan?

12. *Investments.* An executive nearing retirement made two investments totaling $15,000. In one year, these investments yielded $1432 in simple interest. Part of the money was invested at 9% and the rest at 10%. How much was invested at each rate?

13. *Food Science.* The following bar graph shows the milk fat percentages in three dairy products. How many pounds each of whole milk and cream should be mixed to form 200 lb of milk for cream cheese?

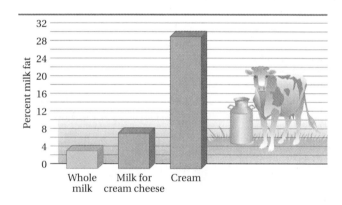

14. *Automotive Maintenance.* "Arctic Antifreeze" is 18% alcohol and "Frost No-More" is 10% alcohol. How many liters of each should be mixed to get 20 L of a mixture that is 15% alcohol?

15. *Teller Work.* Ashford goes to a bank and gets change for a $50 bill consisting of all $5 bills and $1 bills. There are 22 bills in all. How many of each kind are there?

16. *Making Change.* Cecilia makes a $9.25 purchase at the bookstore with a $20 bill. The store has no bills and gives her the change in quarters and fifty-cent pieces. There are 30 coins in all. How many of each kind are there?

b Solve.

17. *Train Travel.* A train leaves Danville Junction and travels north at a speed of 75 mph. Two hours later, a second train leaves on a parallel track and travels north at 125 mph. How far from the station will they meet?

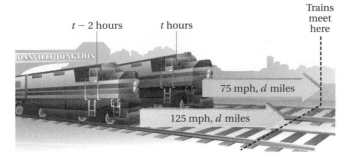

18. *Car Travel.* Two cars leave Denver traveling in opposite directions. One car travels at a speed of 80 km/h and the other at 96 km/h. In how many hours will they be 528 km part?

19. *Canoeing.* Alvin paddled for 4 hr with a 6-km/h current to reach a campsite. The return trip against the same current took 10 hr. Find the speed of Alvin's canoe in still water.

20. *Boating.* Mia's motorboat took 3 hr to make a trip downstream with a 6-mph current. The return trip against the same current took 5 hr. Find the speed of the boat in still water.

21. *Car Travel.* Donna is late for a sales meeting after traveling from one town to another at a speed of 32 mph. If she had traveled 4 mph faster, she could have made the trip in $\frac{1}{2}$ hr less time. How far apart are the towns?

22. *Air Travel.* Rod is a pilot for Crossland Airways. He computes his flight time against a headwind for a trip of 2900 mi at 5 hr. The flight would take 4 hr and 50 min if the headwind were half as great. Find the headwind and the plane's air speed.

23. *Air Travel.* Two planes travel toward each other from cities that are 780 km apart at rates of 190 km/h and 200 km/h. They started at the same time. In how many hours will they meet?

24. *Motorcycle Travel.* Sally and Rocky travel on motorcycles toward each other from Chicago and Indianapolis, which are about 350 km apart, and they are biking at rates of 110 km/h and 90 km/h. They started at the same time. In how many hours will they meet?

25. *Air Travel.* Two airplanes start at the same time and fly toward each other from points 1000 km apart at rates of 420 km/h and 330 km/h. After how many hours will they meet?

26. *Truck and Car Travel.* A truck and a car leave a service station at the same time and travel in the same direction. The truck travels at 55 mph and the car at 40 mph. They can maintain CB radio contact within a range of 10 mi. When will they lose contact?

27. 🖩 *Point of No Return.* A plane flying the 3458-mi trip from New York City to London has a 50-mph tailwind. The flight's *point of no return* is the point at which the flight time required to return to New York is the same as the time required to continue to London. If the speed of the plane in still air is 360 mph, how far is New York from the point of no return?

28. 🖩 *Point of No Return.* A plane is flying the 2553-mi trip from Los Angeles to Honolulu into a 60-mph headwind. If the speed of the plane in still air is 310 mph, how far from Los Angeles is the plane's point of no return? (See Exercise 27.)

Skill Maintenance

Given the function $f(x) = 4x - 7$, find each of the following function values.   [2.2b]

29. $f(0)$

30. $f(-1)$

31. $f(1)$

32. $f(10)$

33. $f(-2)$

34. $f(2a)$

35. $f(-4)$

36. $f(1.8)$

37. $f\left(\frac{3}{4}\right)$

38. $f(-2.5)$

39. $f(-3h)$

40. $f(1000)$

Synthesis

41. ◆ List three or four study tips for someone beginning this exercise set.

42. ◆ Write a problem similar to Example 1 for a classmate to solve. Design the problem so the answer is "The florist sold 14 hanging plants and 9 flats of petunias."

43. *Automotive Maintenance.* The radiator in Michelle's car contains 16 L of antifreeze and water. This mixture is 30% antifreeze. How much of this mixture should she drain and replace with pure antifreeze so that there will be a mixture of 50% antifreeze?

44. *Physical Exercise.* Natalie jogs and walks to school each day. She averages 4 km/h walking and 8 km/h jogging. The distance from home to school is 6 km and Natalie makes the trip in 1 hr. How far does she jog in a trip?

45. *Fuel Economy.* Ellen Jordan's station wagon gets 18 miles per gallon (mpg) in city driving and 24 mpg in highway driving. The car is driven 465 mi on 23 gal of gasoline. How many miles were driven in the city and how many were driven on the highway?

46. *Gender.* Phil and Phyllis are siblings. Phyllis has twice as many brothers as she has sisters. Phil has the same number of brothers as sisters. How many girls and how many boys are in the family?

47. *Wood Stains.* Bennett Custom Flooring has 0.5 gal of stain that is 20% brown and 80% neutral. A customer orders 1.5 gal of a stain that is 60% brown and 40% neutral. How much pure brown stain and how much neutral stain should be added to the original 0.5 gal in order to make up the order?

48. 〽 See Exercise 47. Let x = the amount of pure brown stain added to the original 0.5 gal. Find a function $P(x)$ that can be used to determine the percentage of brown stain in the 1.5 gal mixture. On a grapher, draw the graph of P and use ZOOM and TRACE or the TABLE feature to confirm the answer to Exercise 47.

3.4 Systems of Equations in Three Variables

a | Solving Systems in Three Variables

Objective

a   Solve systems of three equations in three variables.

For Extra Help

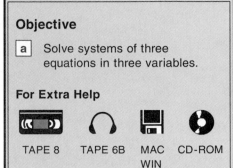

TAPE 8   TAPE 6B   MAC   CD-ROM
                         WIN

A **linear equation in three variables** is an equation equivalent to one of the type $Ax + By + Cz = D$. A solution of a system of three equations in three variables is an ordered triple (x, y, z) that makes *all three* equations true.

The substitution method can be used to solve systems of three equations, but it is not efficient unless a variable has already been eliminated from one or more of the equations. Therefore, we will use only the elimination method—essentially the same procedure for systems of three equations as for systems of two equations. The first step is to eliminate a variable and obtain a system of two equations in two variables.

Example 1   Solve the following system of equations:

$$x + y + z = 4, \quad \textbf{(1)}$$
$$x - 2y - z = 1, \quad \textbf{(2)}$$
$$2x - y - 2z = -1. \quad \textbf{(3)}$$

a) We first use *any* two of the three equations to get an equation in two variables. In this case, let's use equations (1) and (2) and add to eliminate z:

$$
\begin{array}{ll}
x + y + z = 4 & \textbf{(1)} \\
\underline{x - 2y - z = 1} & \textbf{(2)} \\
2x - y \phantom{{}+z} = 5. & \textbf{(4)} \quad \text{Adding}
\end{array}
$$

b) We use a different pair of equations and eliminate the *same variable* that we did in part (a). Let's use equations (1) and (3) and again eliminate z. Be careful! A common error is to eliminate a different variable the second time.

$$
\begin{array}{ll}
x + y + z = 4 & \textbf{(1)} \\
2x - y - 2z = -1 & \textbf{(3)}
\end{array}
$$

$$
\begin{array}{ll}
2x + 2y + 2z = 8 & \text{Multiplying equation (1) by 2} \\
\underline{2x - y - 2z = -1} & \textbf{(3)} \\
4x + y \phantom{{}- 2z} = 7 & \textbf{(5)} \quad \text{Adding}
\end{array}
$$

c) Now we solve the resulting system of equations, (4) and (5). That solution will give us two of the numbers. Note that we now have two equations in two variables. Had we eliminated two different variables in parts (a) and (b), this would not be the case.

$$
\begin{array}{ll}
2x - y = 5 & \textbf{(4)} \\
\underline{4x + y = 7} & \textbf{(5)} \\
6x \phantom{{}- y} = 12 & \text{Adding} \\
x = 2
\end{array}
$$

We can use either equation (4) or (5) to find y. We choose equation (5):

$$
\begin{array}{ll}
4x + y = 7 & \textbf{(5)} \\
4(2) + y = 7 & \text{Substituting 2 for } x \\
8 + y = 7 & \\
y = -1. &
\end{array}
$$

1. Solve. Don't forget to check.

$$4x - y + z = 6,$$
$$-3x + 2y - z = -3,$$
$$2x + y + 2z = 3$$

d) We now have $x = 2$ and $y = -1$. To find the value for z, we use any of the original three equations and substitute to find the third number, z. Let's use equation (1) and substitute our two numbers in it:

$$x + y + z = 4 \qquad \textbf{(1)}$$
$$2 + (-1) + z = 4 \qquad \text{Substituting 2 for } x \text{ and } -1 \text{ for } y$$
$$\left. \begin{array}{r} 1 + z = 4 \\ z = 3. \end{array} \right\} \quad \text{Solving for } z$$

We have obtained the ordered triple $(2, -1, 3)$. We check as follows, substituting $(2, -1, 3)$ into each of the three equations using alphabetical order.

CHECK:

$$\begin{array}{r} x + y + z = 4 \\ \hline 2 + (-1) + 3 \ ? \ 4 \\ 4 \ \end{array} \quad \text{TRUE}$$

$$\begin{array}{r} x - 2y - z = 1 \\ \hline 2 - 2(-1) - 3 \ ? \ 1 \\ 2 + 2 - 3 \\ 1 \ \end{array} \quad \text{TRUE}$$

$$\begin{array}{r} 2x - y - 2z = -1 \\ \hline 2(2) - (-1) - 2 \cdot 3 \ ? \ -1 \\ 4 + 1 - 6 \\ -1 \ \end{array} \quad \text{TRUE}$$

The triple $(2, -1, 3)$ checks and is the solution.

To use the elimination method to solve systems of three equations:

1. Write all equations in the standard form $Ax + By + Cz = D$.

2. Clear any decimals or fractions.

3. Choose a variable to eliminate. Then use *any* two of the three equations to eliminate that variable, getting an equation in two variables.

4. Next, use a different pair of equations and get another equation in *the same two variables*. That is, eliminate the same variable that you did in step (3).

5. Solve the resulting system (pair) of equations. That will give two of the numbers.

6. Then use any of the original three equations to find the third number.

Do Exercise 1.

Example 2 Solve this system:

$$4x - 2y - 3z = 5, \qquad \textbf{(1)}$$
$$-8x - y + z = -5, \qquad \textbf{(2)}$$
$$2x + y + 2z = 5. \qquad \textbf{(3)}$$

a) The equations are in standard form and do not contain decimals or fractions.

b) We decide to eliminate the variable y since the y-terms are opposites in equations (2) and (3). We add:

$$\begin{array}{rr} -8x - y + z = -5 & \textbf{(2)} \\ 2x + y + 2z = 5 & \textbf{(3)} \\ \hline -6x \qquad + 3z = 0. & \textbf{(4)} \quad \text{Adding} \end{array}$$

Answer on page A-20

c) We use another pair of equations to get an equation in the same two variables, x and z. That is, we eliminate the same variable y that we did in step (b). We use equations (1) and (3) and eliminate y:

$$4x - 2y - 3z = 5 \quad \textbf{(1)}$$
$$2x + y + 2z = 5 \quad \textbf{(3)}$$

$$
\begin{array}{ll}
4x - 2y - 3z = 5 & \textbf{(1)} \\
\underline{4x + 2y + 4z = 10} & \text{Multiplying equation (3) by 2} \\
8x \quad\quad + z = 15. & \textbf{(5)} \quad \text{Adding}
\end{array}
$$

d) Now we solve the resulting system of equations (4) and (5). That will give us two of the numbers:

$$-6x + 3z = 0, \quad \textbf{(4)}$$
$$8x + z = 15. \quad \textbf{(5)}$$

We multiply equation (5) by -3. $\left(\text{We could also have multiplied equation (4) by } -\frac{1}{3}.\right)$

$$
\begin{array}{ll}
-6x + 3z = \quad 0 & \textbf{(4)} \\
\underline{-24x - 3z = -45} & \text{Multiplying equation (5) by } -3 \\
-30x \quad\quad = -45 & \text{Adding} \\
x = \frac{-45}{-30} = \frac{3}{2}
\end{array}
$$

We now use equation (5) to find z:

$$
\begin{array}{ll}
8x + z = 15 & \textbf{(5)} \\
8\left(\frac{3}{2}\right) + z = 15 & \text{Substituting } \frac{3}{2} \text{ for } x \\
12 + z = 15 & \\
z = 3. & \text{Solving for } z
\end{array}
$$

e) Next, we use any of the original equations and substitute to find the third number, y. We choose equation (3) since the coefficient of y there is 1:

$$
\begin{array}{ll}
2x + y + 2z = 5 & \textbf{(3)} \\
2\left(\frac{3}{2}\right) + y + 2(3) = 5 & \text{Substituting } \frac{3}{2} \text{ for } x \text{ and } 3 \text{ for } z \\
3 + y + 6 = 5 & \\
y + 9 = 5 & \text{Solving for } y \\
y = -4. &
\end{array}
$$

The solution is $\left(\frac{3}{2}, -4, 3\right)$. The check is as follows.

CHECK:

$$
\begin{array}{c|c}
4x - 2y - 3z = 5 \\ \hline
4 \cdot \frac{3}{2} - 2(-4) - 3(3) \; ? \; 5 \\
6 + 8 - 9 \\
5 \quad\quad \text{TRUE}
\end{array}
\qquad
\begin{array}{c|c}
-8x - y + z = -5 \\ \hline
-8 \cdot \frac{3}{2} - (-4) + 3 \; ? \; -5 \\
-12 + 4 + 3 \\
-5 \quad\quad \text{TRUE}
\end{array}
$$

$$
\begin{array}{c|c}
2x + y + 2z = 5 \\ \hline
2 \cdot \frac{3}{2} + (-4) + 2(3) \; ? \; 5 \\
3 - 4 + 6 \\
5 \quad\quad \text{TRUE}
\end{array}
$$

Do Exercise 2.

2. Solve. Don't forget to check.

$$2x + y - 4z = 0,$$
$$x - y + 2z = 5,$$
$$3x + 2y + 2z = 3$$

Answer on page A-20

3.4 **Systems of Equations in Three Variables**

265

3. Solve. Don't forget to check.

$$x + y + z = 100,$$
$$x - y = -10,$$
$$x - z = -30$$

In Example 3, two of the equations have a missing variable.

Example 3 Solve this system:

$$x + y + z = 180, \qquad \textbf{(1)}$$
$$x - z = -70, \qquad \textbf{(2)}$$
$$2y - z = 0. \qquad \textbf{(3)}$$

We note that there is no y in equation (2). In order to have a system of two equations in the variables x and z, we need to find another equation without a y. We use equations (1) and (3) to eliminate y:

$$x + y + z = 180 \qquad \textbf{(1)}$$
$$2y - z = 0 \qquad \textbf{(3)}$$

$$
\begin{array}{ll}
-2x - 2y - 2z = -360 & \text{Multiplying equation (1) by } -2 \\
 2y - z = 0 \quad \textbf{(3)} & \\
\hline
-2x - 3z = -360. \quad \textbf{(4)} & \text{Adding}
\end{array}
$$

Now we solve the resulting system of equations (2) and (4):

$$x - z = -70 \qquad \textbf{(2)}$$
$$-2x - 3z = -360 \qquad \textbf{(4)}$$

$$
\begin{array}{ll}
2x - 2z = -140 & \text{Multiplying equation (2) by 2} \\
-2x - 3z = -360 \quad \textbf{(4)} & \\
\hline
 -5z = -500 & \text{Adding} \\
z = 100. &
\end{array}
$$

To find x, we substitute 100 for z in equation (2) and solve for x:

$$x - z = -70$$
$$x - 100 = -70$$
$$x = 30.$$

To find y, we substitute 100 for z in equation (3) and solve for y:

$$2y - z = 0$$
$$2y - 100 = 0$$
$$2y = 100$$
$$y = 50.$$

The triple (30, 50, 100) is the solution. The check is left to the student.

Do Exercise 3.

It is possible for a system of three equations to have no solution, that is, to be inconsistent. An example is the system

$$x + y + z = 14,$$
$$x + y + z = 11,$$
$$2x - 3y + 4z = -3.$$

Note the first two equations. It is not possible for a sum of three numbers to be both 14 and 11. Thus the system has no solution. We will not consider such systems here, nor will we consider systems with infinitely many solutions, which also exist.

Answer on page A-20

Exercise Set 3.4

a | Solve.

1. $x + y + z = 2$,
$2x - y + 5z = -5$,
$-x + 2y + 2z = 1$

2. $2x - y - 4z = -12$,
$2x + y + z = 1$,
$x + 2y + 4z = 10$

3. $2x - y + z = 5$,
$6x + 3y - 2z = 10$,
$x - 2y + 3z = 5$

4. $x - y + z = 4$,
$3x + 2y + 3z = 7$,
$2x + 9y + 6z = 5$

5. $2x - 3y + z = 5$,
$x + 3y + 8z = 22$,
$3x - y + 2z = 12$

6. $6x - 4y + 5z = 31$,
$5x + 2y + 2z = 13$,
$x + y + z = 2$

7. $3a - 2b + 7c = 13$,
$a + 8b - 6c = -47$,
$7a - 9b - 9c = -3$

8. $x + y + z = 0$,
$2x + 3y + 2z = -3$,
$-x + 2y - 3z = -1$

9. $2x + 3y + z = 17$,
$x - 3y + 2z = -8$,
$5x - 2y + 3z = 5$

10. $2x + y - 3z = -4$,
$4x - 2y + z = 9$,
$3x + 5y - 2z = 5$

11. $2x + y + z = -2$,
$2x - y + 3z = 6$,
$3x - 5y + 4z = 7$

12. $2x + y + 2z = 11$,
$3x + 2y + 2z = 8$,
$x + 4y + 3z = 0$

13. $x - y + z = 4$,
$5x + 2y - 3z = 2$,
$3x - 7y + 4z = 8$

14. $2x + y + 2z = 3$,
$x + 6y + 3z = 4$,
$3x - 2y + z = 0$

15. $4x - y - z = 4$,
$2x + y + z = -1$,
$6x - 3y - 2z = 3$

16. $a + 2b + c = 1,$
$7a + 3b - c = -2,$
$a + 5b + 3c = 2$

17. $2r + 3s + 12t = 4,$
$4r - 6s + 6t = 1,$
$r + s + t = 1$

18. $10x + 6y + z = 7,$
$5x - 9y - 2z = 3,$
$15x - 12y + 2z = -5$

19. $4a + 9b = 8,$
$8a + 6c = -1,$
$6b + 6c = -1$

20. $3p + 2r = 11,$
$q - 7r = 4,$
$p - 6q = 1$

21. $x + y + z = 57,$
$-2x + y = 3,$
$x - z = 6$

22. $x + y + z = 105,$
$10y - z = 11,$
$2x - 3y = 7$

23. $r + s = 5,$
$3s + 2t = -1,$
$4r + t = 14$

24. $a - 5c = 17,$
$b + 2c = -1,$
$4a - b - 3c = 12$

Skill Maintenance

Solve for the indicated letter. [1.2a]

25. $F = 3ab$, for a

26. $Q = 4(a + b)$, for a

27. $F = \frac{1}{2}t(c - d)$, for c

28. $F = \frac{1}{2}t(c - d)$, for d

29. $Ax - By = c$, for y

30. $Ax + By = c$, for y

Find the slope and the y-intercept. [2.4b]

31. $y = -\frac{2}{3}x - \frac{5}{4}$

32. $y = 5 - 4x$

33. $2x - 5y = 10$

34. $7x - 6.4y = 20$

Synthesis

35. ◈ Explain a procedure that could be used to solve a system of four equations in four variables.

36. ◈ Is it possible for a system of three equations to have exactly two ordered triples in its solution set? Why or why not?

Solve.

37. $w + x + y + z = 2,$
$w + 2x + 2y + 4z = 1,$
$w - x + y + z = 6,$
$w - 3x - y + z = 2$

38. $w + x - y + z = 0,$
$w - 2x - 2y + z = 5,$
$w - 3x - y + z = 4,$
$2w - x - y + 3z = 7$

3.5 Solving Applied Problems: Systems of Three Equations

a | Using Systems of Three Equations

Solving systems of three or more equations is important in many applications occurring in the natural and social sciences, business, and engineering.

Example 1 *Architecture.* In a triangular cross-section of a roof, the largest angle is 70° greater than the smallest angle. The largest angle is twice as large as the remaining angle. Find the measure of each angle.

1. **Familiarize.** We first make a drawing. Since we do not know the size of any angle, we use x, y, and z for the measures of the angles. We let x = the smallest angle, z = the largest angle, and y = the remaining angle.

2. **Translate.** In order to translate the problem, we need to make use of a geometric fact—that is, the sum of the measures of the angles of a triangle is 180°. This fact about triangles gives us one equation:

$$x + y + z = 180.$$

There are two statements in the problem that we can translate directly.

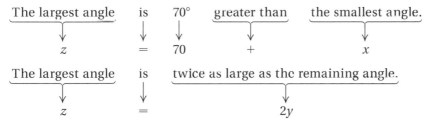

The largest angle	is	70°	greater than	the smallest angle.
z	$=$	70	$+$	x

The largest angle	is	twice as large as the remaining angle.
z	$=$	$2y$

We now have a system of three equations:

$$
\begin{aligned}
x + y + z &= 180, \\
x + 70 &= z, \quad\text{or} \\
2y &= z;
\end{aligned}
\qquad
\begin{aligned}
x + y + z &= 180, \\
x \quad\;\; - z &= -70, \\
2y - z &= 0.
\end{aligned}
$$

3. **Solve.** The system was solved in Example 3 of Section 3.4. The solution is (30, 50, 100).

4. **Check.** The sum of the numbers is 180. The largest angle measures 100° and the smallest measures 30°. The largest angle is 70° greater than the smallest. The remaining angle measures 50°. The largest angle is twice as large as the remaining angle. We do have an answer to the problem.

5. **State.** The measures of the angles of the triangle are 30°, 50°, and 100°.

Do Exercise 1.

Objective

a | Solve applied problems using systems of three equations.

For Extra Help

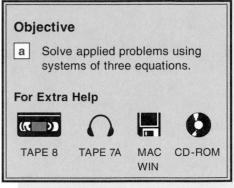

TAPE 8   TAPE 7A   MAC   CD-ROM
                     WIN

1. *Triangle Measures.* One angle of a triangle is twice as large as a second angle. The remaining angle is 20° greater than the first angle. Find the measure of each angle.

Answer on page A-20

Example 2 *Cholesterol Levels.* Americans have become very conscious of their cholesterol levels. Recent studies indicate that a child's intake of cholesterol should be no more than 300 mg per day. By eating 1 egg, 1 cupcake, and 1 slice of pizza, a child consumes 302 mg of cholesterol. If the child eats 2 cupcakes and 3 slices of pizza, he or she takes in 65 mg of cholesterol. By eating 2 eggs and 1 cupcake, a child consumes 567 mg of cholesterol. How much cholesterol is in each item?

1. **Familiarize.** After we have read the problem a few times, it becomes clear that an egg contains considerably more cholesterol than the other foods. Let's guess that one egg contains 200 mg of cholesterol and one cupcake contains 50 mg. Because of the third sentence in the problem, it would follow that a slice of pizza contains 52 mg of cholesterol since $200 + 50 + 52 = 302$.

   To see if our guess satisfies the other statements in the problem, we find the amount of cholesterol that 2 cupcakes and 3 slices of pizza would contain: $2 \cdot 50 + 3 \cdot 52 = 256$. Since this does not match the 65 mg listed in the fourth sentence of the problem, our guess was incorrect. Rather than guess again, we examine how we checked our guess and let e, c, and s = the number of milligrams of cholesterol in an egg, a cupcake, and a slice of pizza, respectively.

2. **Translate.** By rewording some of the sentences in the problem, we can translate it into three equations.

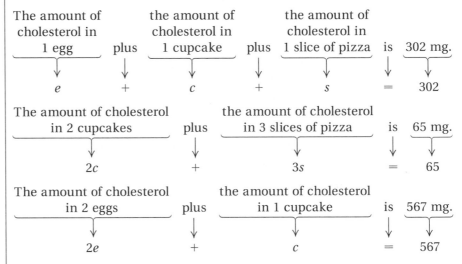

We now have a system of three equations:

$$e + c + s = 302,$$
$$2c + 3s = 65,$$
$$2e + c = 567.$$

3. **Solve.** We solve and get $e = 274$, $c = 19$, $s = 9$, or $(274, 19, 9)$.

4. **Check.** The sum of 274, 19, and 9 is 302 so the total cholesterol in 1 egg, 1 cupcake, and 1 slice of pizza checks. Two cupcakes and three slices of pizza would contain $2 \cdot 19 + 3 \cdot 9$, or 65 mg, while two eggs and one cupcake would contain $2 \cdot 274 + 19$, or 567 mg of cholesterol. The answer checks.

5. **State.** An egg contains 274 mg of cholesterol, a cupcake contains 19 mg of cholesterol, and a slice of pizza contains 9 mg of cholesterol.

Do Exercise 2.

2. *Client Investments.* Kaufman Financial Corporation makes investments for corporate clients. One year, a client receives $1620 in simple interest from three investments that total $25,000. Part is invested at 5%, part at 6%, and part at 7%. There is $11,000 more invested at 7% than at 6%. How much was invested at each rate?

Answer on page A-20

Exercise Set 3.5

a Solve.

1. *Restaurant Management.* Kyle works at Dunkin® Donuts, where a 10-oz cup of coffee costs 95¢, a 14-oz cup costs $1.15, and a 20-oz cup costs $1.50. During one busy period, Kyle served 34 cups of coffee, emptying five 96-oz pots while collecting a total of $39.60. How many cups of each size did Kyle fill?

2. *Restaurant Management.* McDonald's® recently sold small soft drinks for 89¢, medium soft drinks for 99¢, and large soft drinks for $1.19. During a lunch-time rush, Chris sold 55 soft drinks for a total of $54.95. The number of small and large drinks, combined, was 5 fewer than the number of medium drinks. How many drinks of each size were sold?

10 oz   14 oz   20 oz
$0.95 $1.15 $1.50

small   medium   large
$0.89 $0.99 $1.19

3. *Triangle Measures.* In triangle *ABC*, the measure of angle *B* is three times that of angle *A*. The measure of angle *C* is 20° more than that of angle *A*. Find the measure of each angle.

4. *Triangle Measures.* In triangle *ABC*, the measure of angle *B* is twice the measure of angle *A*. The measure of angle *C* is 80° more than that of angle *A*. Find the measure of each angle.

5. *Automobile Pricing.* A recent basic model of a particular automobile had a price of $12,685. The basic model with the added features of automatic transmission and power door locks was $14,070. The basic model with air conditioning (AC) and power door locks was $13,580. The basic model with AC and automatic transmission was $13,925. What was the individual cost of each of the three options?

6. *Telemarketing.* Sven, Tillie, and Isaiah can process 740 telephone orders per day. Sven and Tillie together can process 470 orders, while Tillie and Isaiah together can process 520 orders per day. How many orders can each person process alone?

7. *Lens Production.* When Sight-Rite's three polishing machines, A, B, and C, are all working, 5700 lenses can be polished in one week. When only A and B are working, 3400 lenses can be polished in one week. When only B and C are working, 4200 lenses can be polished in one week. How many lenses can be polished in a week by each machine alone?

8. *Welding Rates.* Elrod, Dot, and Wendy can weld 74 linear feet per hour when working together. Elrod and Dot together can weld 44 linear feet per hour, while Elrod and Wendy can weld 50 linear feet per hour. How many linear feet per hour can each weld alone?

9. *Investments.* A business class divided an imaginary investment of $80,000 among three mutual funds. The first fund grew by 10%, the second by 6%, and the third by 15%. Total earnings were $8850. The earnings from the first fund were $750 more than the earnings from the third. How much was invested in each fund?

10. *Advertising.* In a recent year, companies spent a total of $84.8 billion on newspaper, television, and radio ads. The total amount spent on television and radio ads was only $2.6 billion more than the amount spent on newspaper ads alone. The amount spent on newspaper ads was $5.1 billion more than what was spent on television ads. How much was spent on each form of advertising? (*Hint*: Let the variables represent numbers of billions of dollars.)

11. *Twin Births.* In the United States, the highest incidence of fraternal twin births occurs among Asian-Americans, then African-Americans, and then Caucasians. Of every 15,400 births, the total number of fraternal twin births for all three is 739, where there are 185 more for Asian-Americans than African-Americans and 231 more for Asian-Americans than Caucasians. How many births of fraternal twins are there for each group out of every 15,400 births?

12. *Crying Rate.* The sum of the average number of times a man, a woman, and a one-year-old child cry each month is 71.7. A one-year-old cries 46.4 more times than a man. The average number of times a one-year-old cries per month is 28.3 more than the average number of times combined that a man and a woman cry. What is the average number of times per month that each cries?

13. *Nutrition.* A dietician in a hospital prepares meals under the guidance of a physician. Suppose that for a particular patient a physician prescribes a meal to have 800 calories, 55 g of protein, and 220 mg of vitamin C. The dietician prepares a meal of roast beef, baked potatoes, and broccoli according to the data in the following table.

	Calories	Protein (in grams)	Vitamin C (in milligrams)
Roast Beef, 3 oz	300	20	0
Baked Potato	100	5	20
Broccoli, 156 g	50	5	100

How many servings of each food are needed in order to satisfy the doctor's orders?

14. *Nutrition.* Repeat Exercise 13 but replace the broccoli with asparagus, for which one 180-g serving contains 50 calories, 5 g of protein, and 44 mg of vitamin C. Which meal would you prefer eating?

15. *Golf.* On an 18-hole golf course, there are par-3 holes, par-4 holes, and par-5 holes. A golfer who shoots par on every hole has a total of 70. There are twice as many par-4 holes as there are par-5 holes. How many of each type of hole are there on the golf course?

16. *Golf.* On an 18-hole golf course, there are par-3 holes, par-4 holes, and par-5 holes. A golfer who shoots par on every hole has a total of 72. The sum of the number of par-3 holes and the number of par-5 holes is 8. How many of each type of hole are there on the golf course?

17. The sum of three numbers is 5. The first number minus the second plus the third is 1. The first minus the third is 3 more than the second. Find the numbers.

18. The sum of three numbers is 26. Twice the first minus the second is 2 less than the third. The third is the second minus three times the first. Find the numbers.

19. *Basketball Scoring.* The New York Knicks recently scored a total of 92 points on a combination of 2-point field goals, 3-point field goals, and 1-point foul shots. Altogether, the Knicks made 50 baskets and 19 more 2-pointers than foul shots. How many shots of each kind were made?

20. *History.* Find the year in which the first U.S. transcontinental railroad was completed. The following are some facts about the number. The sum of the digits in the year is 24. The ones digit is 1 more than the hundreds digit. Both the tens and the ones digits are multiples of 3.

Skill Maintenance

Determine whether the correspondence is a function.   [2.2a]

21.

Domain	*Range*
(State)	(City)

California → Los Angeles
→ San Francisco
↗ San Diego
Kansas → Kansas City
→ Topeka
↗ Wichita

22.

Domain	*Range*
(Boy's Age, in months)	(Average Daily Weight Gain, in grams)

2 ———————→ 24.3
9 ———————→ 11.7
16 ———————→ 8.2
23 ———————→ 7.0

Source: *American Family Physician*, December 1993, p. 1435

23. Find the domain of the function: [2.3a]

$$f(x) = \frac{x - 5}{x + 7}.$$

24. Find the domain and the range of the function: [2.3a]

$$g(x) = 5 - x^2.$$

25. Find an equation of the line with slope $-\frac{3}{5}$ and y-intercept $(0, -7)$.   [2.6a]

26. Simplify: $\dfrac{(a^2 b^3)^5}{a^7 b^{16}}$.   [R.7b]

Synthesis

27. ◈ Exercise 8 can be solved mentally after a careful reading of the problem. How is this possible?

28. ◈ A theater audience of 100 people consists of adults, senior citizens, and children. The ticket prices are $10 each for adults, $3 each for senior citizens, and $0.50 each for children. The total amount of money taken in is $100. How many adults, senior citizens, and children are in attendance? Does there seem to be some information missing? Do some careful reasoning and explain.

29. Find the sum of the angle measures at the tips of the star in this figure.

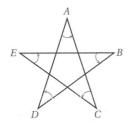

30. *Sharing Raffle Tickets.* Hal gives Tom as many raffle tickets as Tom has and Gary as many as Gary has. In like manner, Tom then gives Hal and Gary as many tickets as each then has. Similarly, Gary gives Hal and Tom as many tickets as each then has. If each finally has 40 tickets, with how many tickets does Tom begin?

Solve a system of equations in three variables by choosing a different variable to eliminate.

Collaborative Learning Manual

3.6 Systems of Linear Inequalities in Two Variables

A **graph** of an inequality is a drawing that represents its solutions. An inequality in one variable can be graphed on the number line. An inequality in two variables can be graphed on a coordinate plane.

A **linear inequality** is one that we can get from a related linear equation by changing the equals symbol to an inequality symbol. The graph of a linear inequality is a region on one side of a line. This region is called a **half-plane**. The graph sometimes includes the graph of the related line at the boundary of the half-plane.

a | Solutions of Inequalities in Two Variables

The solutions of an inequality in two variables are ordered pairs.

Examples Determine whether the ordered pair is a solution of the inequality $5x - 4y > 13$.

1. $(-3, 2)$

We have

$$5x - 4y > 13$$

$$\begin{array}{c|c} 5(-3) - 4 \cdot 2 \ ? \ 13 & \text{We use alphabetical order to} \\ -15 - 8 & \text{replace } x \text{ with } -3 \text{ and } y \text{ with 2.} \\ -23 & \text{FALSE} \end{array}$$

Since $-23 > 13$ is false, $(-3, 2)$ is not a solution.

2. $(6, -7)$

We have

$$5x - 4y > 13$$

$$\begin{array}{c|c} 5(6) - 4(-7) \ ? \ 13 & \text{Replacing } x \text{ with 6 and } y \text{ with } -7 \\ 30 + 28 & \\ 58 & \text{TRUE} \end{array}$$

Since $58 > 13$ is true, $(6, -7)$ is a solution.

Do Exercises 1 and 2.

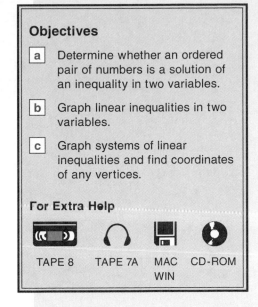

Objectives

a Determine whether an ordered pair of numbers is a solution of an inequality in two variables.

b Graph linear inequalities in two variables.

c Graph systems of linear inequalities and find coordinates of any vertices.

For Extra Help

TAPE 8   TAPE 7A   MAC   CD-ROM
                        WIN

1. Determine whether $(1, -4)$ is a solution of $4x - 5y < 12$.

2. Determine whether $(4, -3)$ is a solution of $3y - 2x \leq 6$.

Answers on page A-20

b | Graphing Inequalities in Two Variables

Example 3 Graph: $y < x$.

We first graph the line $y = x$ for comparison. Every solution of $y = x$ is an ordered pair like (3, 3), where the first and second coordinates are the same. The graph of $y = x$ is shown on the left below. We draw it dashed because these points are *not* solutions of $y < x$.

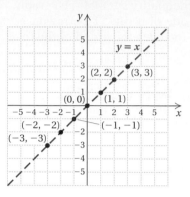 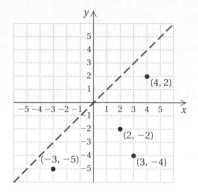

Now look at the graph on the right above. Several ordered pairs are plotted on the half-plane below $y = x$. Each is a solution of $y < x$. We can check a pair (4, 2) as follows:

$$\frac{y < x}{2 \; ? \; 4} \quad \text{TRUE}$$

It turns out that any point on the same side of $y = x$ as (4, 2) is also a solution. Thus, if you know that one point in a half-plane is a solution, then all points in that half-plane are solutions. In this text, we will usually indicate this by color shading. We shade the half-plane below $y = x$.

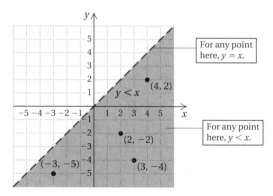

For any point here, $y = x$.

For any point here, $y < x$.

Example 4   Graph: $8x + 3y \geq 24$.

First, we sketch the line $8x + 3y = 24$. Points on the line $8x + 3y = 24$ are also in the graph of $8x + 3y \geq 24$, so we draw the line solid. This indicates that all points on the line are solutions. The rest of the solutions are in the half-plane either to the left or to the right of the line. To determine which, we select a point that is not on the line and determine whether it is a solution of $8x + 3y \geq 24$. We try $(-3, 4)$ as a test point:

$$\begin{array}{c|c} 8x + 3y \geq 24 & \\ \hline 8(-3) + 3(4) \;?\; 24 & \\ -24 + 12 & \\ -12 & \text{FALSE} \end{array}$$

We see that $-12 \geq 24$ is *false*. Since $(-3, 4)$ is not a solution, none of the points in the half-plane containing $(-3, 4)$ is a solution. Thus the points in the opposite half-plane are solutions. We shade that half-plane and obtain the graph shown below.

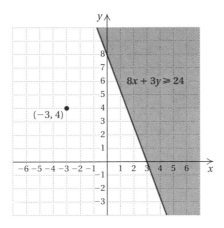

To graph an inequality in two variables:

1. Replace the inequality symbol with an equals sign and graph this related equation.

2. If the inequality symbol is $<$ or $>$, draw the line dashed. If the inequality symbol is $\leq$ or $\geq$, draw the line solid.

3. The graph consists of a half-plane that is either above or below or to the left or right of the line and, if the line is solid, the line as well. To determine which half-plane to shade, choose a point not on the line as a test point. Substitute to determine whether that point is a solution. If so, shade the half-plane containing that point. If not, shade the opposite half-plane.

Example 5   Graph: $6x - 2y < 12$.

1. We first graph the related equation $6x - 2y = 12$.

2. Since the inequality uses the symbol $<$, points on the line are not solutions of the inequality, so we draw a dashed line.

3. To determine which half-plane to shade, we consider a test point *not* on the line. We try $(0, 0)$ and substitute:

$$\begin{array}{c|c} 6x - 2y < 12 & \\ \hline 6(0) - 2(0) \;?\; 12 & \\ 0 - 0 & \\ 0 & \text{TRUE} \end{array}$$

Graph.

3. $6x - 3y < 18$

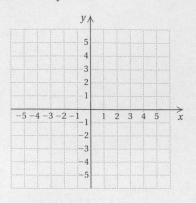

4. $4x + 3y \geq 12$

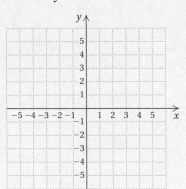

Since the inequality $0 < 12$ is *true*, the point $(0, 0)$ is a solution; each point in the half-plane containing $(0, 0)$ is a solution. Thus each point in the opposite half-plane is *not* a solution. The graph is shown below.

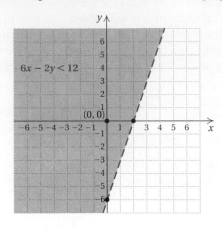

Do Exercises 3 and 4.

Example 6 Graph $x > -3$ on a plane.

There is a missing variable in this inequality. If we graph the inequality on the number line, its graph is as follows:

However, we can also write this inequality as $x + 0y > -3$ and consider graphing it in the plane. We use the same technique that we have used with the other examples. We first graph the related equation $x = -3$ in the plane. We draw the boundary with a dashed line. The rest of the graph is a half-plane to the right or left of the line $x = -3$. To determine which, we consider a test point, $(2, 5)$:

$$\begin{array}{c} x + 0y > -3 \\ \hline 2 + 0(5) \;?\; -3 \\ 2 \;\big|\; \qquad \text{TRUE} \end{array}$$

Since $(2, 5)$ is a solution, all the pairs in the half-plane containing $(2, 5)$ are solutions. We shade that half-plane.

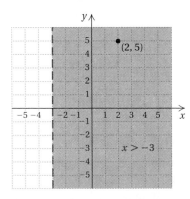

We see that the solutions of $x > -3$ are all those ordered pairs whose first coordinates are greater than -3.

Answers on page A-20

Example 7 Graph $y \leq 4$ on a plane.

We first graph $y = 4$ using a solid line. We then use $(2, -3)$ as a test point and substitute:

$$\frac{0x + y \leq 4}{\begin{array}{c} 0(2) + (-3) \ ? \ 4 \\ -3 \ | \qquad \text{TRUE} \end{array}}$$

We see that $(2, -3)$ is a solution, so all points in the half-plane containing $(2, -3)$ are solutions. Note that this half-plane consists of all ordered pairs whose second coordinate is less than or equal to 4.

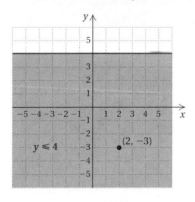

Do Exercises 5 and 6.

c | Systems of Linear Inequalities

The following is an example of a system of two linear inequalities in two variables:

$$x + y \leq 4,$$
$$x - y < 4.$$

A **solution** of a system of linear inequalities is an ordered pair that is a solution of *both* inequalities. We now graph solutions of systems of linear inequalities. To do so, we graph each inequality and determine where the graphs overlap, or intersect. That will be a region in which the ordered pairs are solutions of both inequalities.

Example 8 Graph the solutions of the system

$$x + y \leq 4,$$
$$x - y < 4.$$

We graph the inequality $x + y \leq 4$ by first graphing the equation $x + y = 4$ using a solid red line. We consider $(0, 0)$ as a test point and find that it is a solution, so we shade all points on that side of the line using red shading. The arrows at the ends of the line also indicate the half-plane, or region, that contains the solutions.

Graph on a plane.

5. $x < 3$

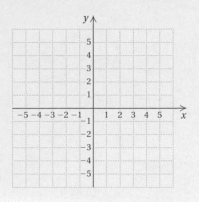

6. $y \geq -4$

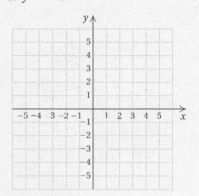

Answers on page A-21

7. Graph:

$$x + y \geq 1,$$
$$y - x \geq 2.$$

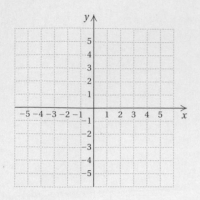

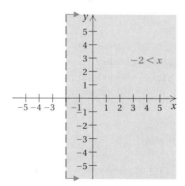

Next, we graph $x - y < 4$. We begin by graphing the equation $x - y = 4$ using a dashed blue line and consider $(0, 0)$ as a test point. Again, $(0, 0)$ is a solution so we shade that side of the line using blue shading. The solution set of the system is the region that is shaded both red and blue and part of the line $x + y = 4$.

Do Exercise 7.

Example 9 Graph: $-2 < x \leq 5$.

This is actually a system of inequalities:

$$-2 < x,$$
$$x \leq 5.$$

We graph the equation $-2 = x$ and see that the graph of the first inequality is the half-plane to the right of the line $-2 = x$ (see the graph on the left below).

Next, we graph the second inequality, starting with the line $x = 5$, and find that its graph is the line and also the half-plane to the left of it (see the graph on the right below).

8. Graph: $-3 \leq y < 4$.

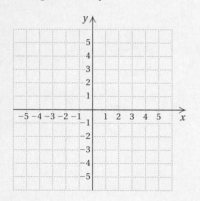

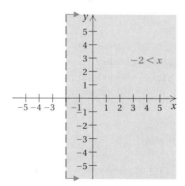

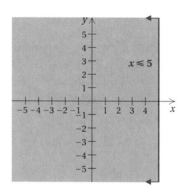

We shade the intersection of these graphs.

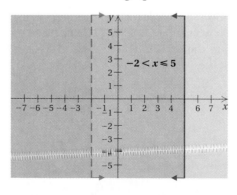

Do Exercise 8.

Answers on page A-21

A system of inequalities may have a graph that consists of a polygon and its interior. In *linear programming,* which is a topic rich in application that you may study in a later course, it is important to be able to find the vertices of such a polygon.

Example 10 Graph the following system of inequalities. Find the coordinates of any vertices formed.

$$6x - 2y \leq 12, \quad \textbf{(1)}$$
$$y - 3 \leq 0, \quad \textbf{(2)}$$
$$x + y \geq 0 \quad \textbf{(3)}$$

We graph the lines $6x - 2y = 12$, $y - 3 = 0$, and $x + y = 0$ using solid lines. The regions for each inequality are indicated by the arrows at the ends of the lines. We then note where the regions overlap and shade the region of solutions using one color.

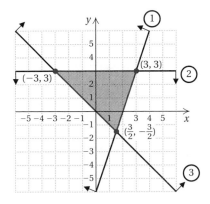

To find the vertices, we solve three different systems of equations. The system of equations from inequalities (1) and (2) is

$$6x - 2y = 12,$$
$$y - 3 = 0.$$

Solving, we obtain the vertex (3, 3).

The system of equations from inequalities (1) and (3) is

$$6x - 2y = 12,$$
$$x + y = 0.$$

Solving, we obtain the vertex $\left(\frac{3}{2}, -\frac{3}{2}\right)$.

The system of equations from inequalities (2) and (3) is

$$y - 3 = 0,$$
$$x + y = 0.$$

Solving, we obtain the vertex $(-3, 3)$.

Do Exercise 9.

9. Graph the system of inequalities. Find the coordinates of any vertices formed.

$$5x + 6y \leq 30,$$
$$0 \leq y \leq 3,$$
$$0 \leq x \leq 4$$

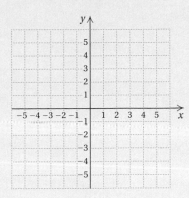

Answer on page A-21

10. Graph the system of inequalities. Find the coordinates of any vertices formed.

$$2x + 4y \leq 8,$$
$$x + y \leq 3,$$
$$x \geq 0,$$
$$y \geq 0$$

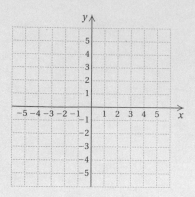

Example 11 Graph the following system of inequalities. Find the coordinates of any vertices formed.

$$x + y \leq 16, \quad \textbf{(1)}$$
$$3x + 6y \leq 60, \quad \textbf{(2)}$$
$$x \geq 0, \quad \textbf{(3)}$$
$$y \geq 0 \quad \textbf{(4)}$$

We graph each inequality using solid lines. The regions for each inequality are indicated by the arrows at the ends of the lines. We then note where the regions overlap and shade the region of solutions using one color.

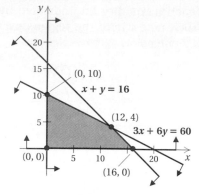

To find the vertices, we solve four different systems of equations. The system of equations from inequalities (1) and (2) is

$$x + y = 16,$$
$$3x + 6y = 60.$$

Solving, we obtain the vertex (12, 4).
   The system of equations from inequalities (1) and (4) is

$$x + y = 16,$$
$$y = 0.$$

Solving, we obtain the vertex (16, 0).
   The system of equations from inequalities (3) and (4) is

$$x = 0,$$
$$y = 0.$$

The vertex is (0, 0).
   The system of equations from inequalities (2) and (3) is

$$3x + 6y = 60,$$
$$x = 0.$$

Solving, we obtain the vertex (0, 10).

Do Exercise 10.

Answer on page A-21

Exercise Set 3.6

a Determine whether the given ordered pair is a solution of the given inequality.

1. $(-3, 3)$; $3x + y < -5$

2. $(6, -8)$; $4x + 3y \geq 0$

3. $(5, 9)$; $2x - y > -1$

4. $(5, -2)$; $6y - x > 2$

b Graph the inequality on a plane.

5. $y > 2x$

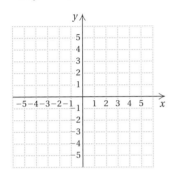

6. $y < 3x$

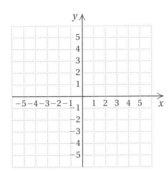

7. $y < x + 1$

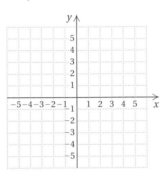

8. $y \leq x - 3$

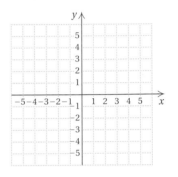

9. $y > x - 2$

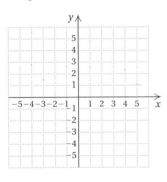

10. $y \geq x + 4$

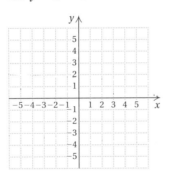

11. $x + y < 4$

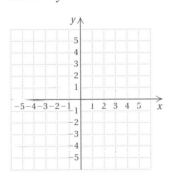

12. $x - y \geq 3$

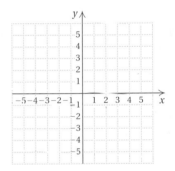

13. $3x + 4y \leq 12$

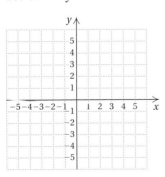

14. $2x + 3y < 6$

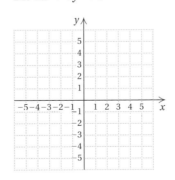

15. $2y - 3x > 6$

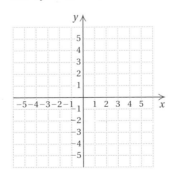

16. $2y - x \leq 4$

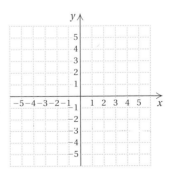

17. $3x - 2 \leq 5x + y$

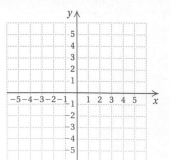

18. $2x - 2y \geq 8 + 2y$

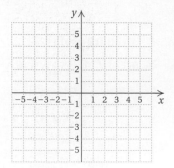

19. $x < 5$

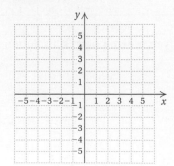

20. $y \geq -2$

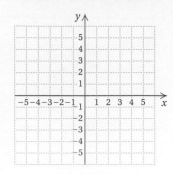

21. $y > 2$

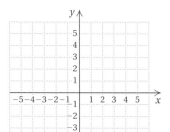

22. $x \leq -4$

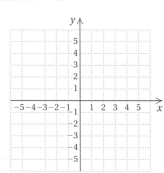

23. $2x + 3y \leq 6$

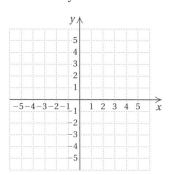

24. $7x + 2y \geq 21$

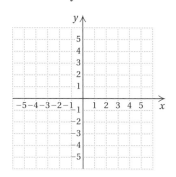

c Graph the system of inequalities. Find the coordinates of any vertices formed.

25. $y \geq x$,
$\quad y \leq -x + 2$

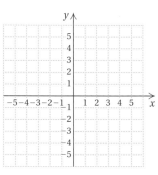

26. $y \geq x$,
$\quad y \leq -x + 4$

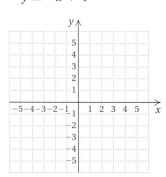

27. $y > x$,
$\quad y < -x + 1$

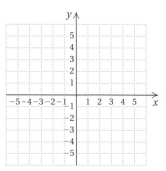

28. $y < x$,
$\quad y > -x + 3$

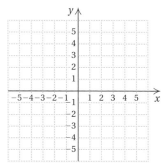

29. $y \geq -2$,
  $x \geq 1$

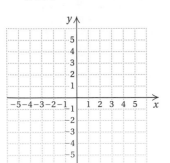

30. $y \leq -2$,
  $x \geq 2$

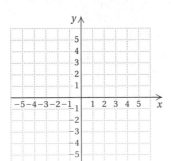

31. $x \leq 3$,
  $y \geq -3x + 2$

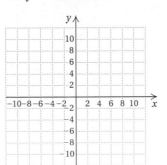

32. $x \geq -2$,
  $y \leq -2x + 3$

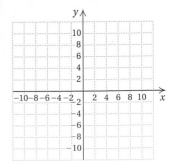

33. $y \geq -2$,
  $y \geq x + 3$

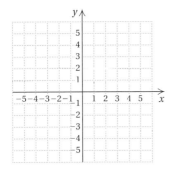

34. $y \leq 4$,
  $y \geq -x + 2$

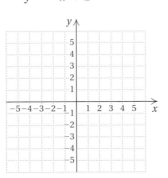

35. $x + y \leq 1$,
  $x - y \leq 2$

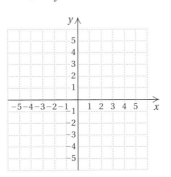

36. $x + y \leq 3$,
  $x - y \leq 4$

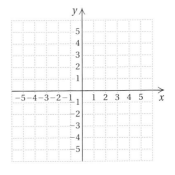

37. $y - 2x \geq 1$,
  $y - 2x \leq 3$

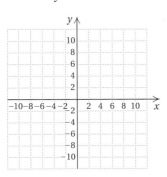

38. $y + 3x \geq 0$,
  $y + 3x \leq 2$

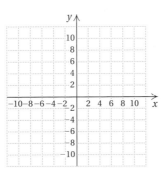

39. $y \leq 2x + 1$,
  $y \geq -2x + 1$,
  $x \leq 2$

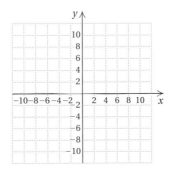

40. $x - y \leq 2$,
  $x + 2y \geq 8$,
  $y \leq 4$

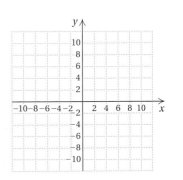

41. $x + 2y \leq 12$,
  $2x + y \leq 12$,
  $x \geq 0$,
  $y \geq 0$

42. $4y - 3x \geq -12$,
  $4y + 3x \geq -36$,
  $y \leq 0$,
  $x \leq 0$

43. $8x + 5y \leq 40$,
  $x + 2y \leq 8$,
  $x \geq 0$,
  $y \geq 0$

44. $y - x \geq 1$,
  $y - x \leq 3$,
  $2 \leq x \leq 5$

Solve. [1.1d]

45. $5(3x - 4) = -2(x + 5)$

46. $4(3x + 4) = 2 - x$

47. $2(x - 1) + 3(x - 2) - 4(x - 5) = 10$

48. $10x - 8(3x - 7) = 2(4x - 1)$

49. $5x + 7x = -144$

50. $0.5x - 2.34 + 2.4x = 7.8x - 9$

Given the function $f(x) = |2 - x|$, find each of the following function values.   [2.2b]

51. $f(0)$

52. $f(-1)$

53. $f(1)$

54. $f(10)$

55. $f(-2)$

56. $f(2a)$

57. $f(-4)$

58. $f(1.8)$

Synthesis

59. ◆ Do all systems of linear inequalities have solutions? Why or why not?

60. ◆ When graphing linear inequalities, Ron always shades above the line when he sees a $\geq$ symbol. Is this wise? Why or why not?

61. *Luggage Size.* Unless an additional fee is paid, most major airlines will not check any luggage that is more than 62 in. long. The U.S. Postal Service will ship a package only if the sum of the package's length and girth (distance around its midsection) does not exceed 108 in. Concert Productions is ordering several 62-in. long trunks that will be both mailed and checked as luggage. Using w and h for width and height (in inches), respectively, write and graph an inequality that represents all acceptable combinations of width and height.

62. *Exercise Danger Zone.* It is dangerous to exercise when the weather is hot and humid. The solutions of the following system of inequalities give a "danger zone" for which it is dangerous to exercise intensely:

$$4H - 3F < 70,$$
$$F + H > 160,$$
$$2F + 3H > 390,$$

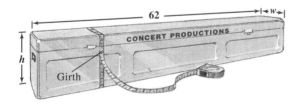

where F is the temperature, in degrees Fahrenheit, and H is the humidity.

a) Draw the danger zone by graphing the system of inequalities.

b) Is it dangerous to exercise when $F = 80°$ and $H = 80\%$?

In Exercises 63–66, use a grapher with a SHADE feature to graph the inequality.

63. $3x + 6y > 2$

64. $x - 5y \leq 10$

65. $13x - 25y + 10 \leq 0$

66. $2x + 5y > 0$

67. Use a grapher with a SHADE feature to check your answers to Exercises 25–38. Then use the INTERSECT feature to determine any point(s) of intersection.

3.7 Business and Economics Applications

a | Break-Even Analysis

When a company manufactures x units of a product, it invests money. This is **total cost** and can be thought of as a function C, where $C(x)$ is the total cost of producing x units. When the company sells x units of the product, it takes in money. This is **total revenue** and can be thought of as a function R, where $R(x)$ is the total revenue from the sale of x units. **Total profit** is the money taken in less the money spent, or total revenue minus total cost. Total profit from the production and sale of x units is a function P given by

$$\text{Profit} = \text{Revenue} - \text{Cost}, \quad \text{or} \quad P(x) = R(x) - C(x).$$

If $R(x)$ is greater than $C(x)$, the company has a profit. If $C(x)$ is greater than $R(x)$, the company has a loss. When $R(x) = C(x)$, the company breaks even.

There are two kinds of costs. First, there are costs like rent, insurance, machinery, and so on. These costs, which must be paid whether a product is produced or not, are called **fixed costs.** When a product is being produced, there are costs for labor, materials, marketing, and so on. These are called **variable costs,** because they vary according to the amount of the product being produced. The sum of the fixed costs and the variable costs gives the **total cost** of producing a product.

Example 1 *Manufacturing Radios.* Ergs, Inc., is planning to make a new kind of radio. Fixed costs will be $90,000, and it will cost $15 to produce each radio (variable costs). Each radio sells for $26.

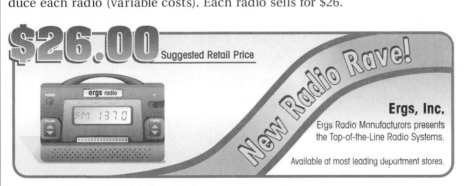

a) Find the total cost $C(x)$ of producing x radios.

b) Find the total revenue $R(x)$ from the sale of x radios.

c) Find the total profit $P(x)$ from the production and sale of x radios.

d) What profit or loss will the company realize from the production and sale of 3000 radios? of 14,000 radios?

e) Graph the total-cost, total-revenue, and total-profit functions using the same set of axes. Determine the break-even point.

a) Total cost is given by

$$C(x) = (\text{Fixed costs}) \text{ plus } (\text{Variable costs}),$$

$$\text{or} \quad C(x) = \quad 90,000 \quad + \quad 15x,$$

where x is the number of radios produced.

Objectives

a Given total-cost and total-revenue functions, find the total-profit function and the break-even point.

b Given supply and demand functions, find the equilibrium point.

For Extra Help

TAPE 8   TAPE 7B   MAC WIN   CD-ROM

1. *Manufacturing Radios.* Refer to Example 1. Suppose that fixed costs are $80,000, and it costs $20 to produce each radio. Each radio sells for $36.

a) Find the total cost $C(x)$ of producing x radios.

b) Find the total revenue $R(x)$ from the sale of x radios.

c) Find the total profit $P(x)$ from the production and sale of x radios.

d) What profit or loss will the company realize from the production and sale of 4000 radios? of 16,000 radios?

e) Graph the total-cost, total-revenue, and total-profit functions using the same set of axes. Determine the break-even point.

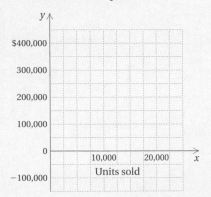

Answers on page A-22

b) Total revenue is given by

$$R(x) = 26x.$$ $26 times the number of radios sold. We assume that every radio produced is sold.

c) Total profit is given by

$$P(x) = R(x) - C(x)$$
$$= 26x - (90,000 + 15x)$$
$$= 11x - 90,000.$$

d) Profits will be

$$P(3000) = 11 \cdot 3000 - 90,000 = -\$57,000$$

when 3000 radios are produced and sold, and

$$P(14,000) = 11 \cdot 14,000 - 90,000 = \$64,000$$

when 14,000 radios are produced and sold. Thus the company loses $57,000 if only 3000 radios are sold, but makes $64,000 if 14,000 are sold.

e) The graphs of each of the three functions are shown below:

$$C(x) = 90,000 + 15x, \quad \textbf{(1)}$$
$$R(x) = 26x, \quad \textbf{(2)}$$
$$P(x) = 11x - 90,000. \quad \textbf{(3)}$$

$R(x)$, $C(x)$, and $P(x)$ are all in dollars.

Equation (2) has a graph that goes through the origin and has a slope of 26. Equation (1) has an intercept on the y-axis of 90,000 and has a slope of 15. Equation (3) has an intercept on the $-axis of $-90,000$ and has a slope of 11. It is shown by the dashed line. The red dashed line shows a "negative" profit, which is a loss. (That is what is known as "being in the red.") The black dashed line shows a "positive" profit, or gain. (That is what is known as "being in the black.")

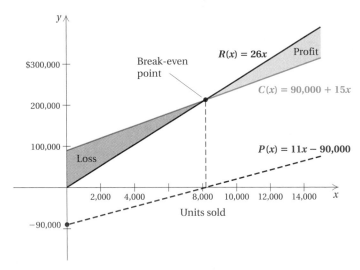

Profits occur when the revenue is greater than the cost. Losses occur when the revenue is less than the cost. The **break-even point** occurs where the graphs of R and C cross. Thus to find the break-even point, we solve a system:

$$C(x) = 90,000 + 15x,$$
$$R(x) = 26x.$$

Since both revenue and cost are in *dollars* and they are equal at the break-even point, the system can be rewritten as

$$d = 90,000 + 15x, \quad \textbf{(1)}$$
$$d = 26x \quad \textbf{(2)}$$

and solved using substitution:

$26x = 90,000 + 15x$ **Substituting 26x for d in equation (1)**

$11x = 90,000$

$x \approx 8181.8.$

    The firm will break even if it produces and sells about 8182 radios (8181 will yield a tiny loss and 8182 a tiny gain), and takes in a total of $R(8182) = 26 \cdot 8182 = \$212,732$ in revenue. Note that the x-coordinate of the break-even point is also the x-intercept of the profit function. It can also be found by solving $P(x) = 0$.

Do Exercise 1 on the preceding page.

b | Supply and Demand

As the price of coffee varies, the amount sold varies. The table and graph below both show that consumer demand goes down as the price goes up and the demand goes up as the price goes down.

DEMAND FUNCTION, D

Price, p, per Kilogram	Quantity, $D(p)$ (in millions of kilograms)
$ 8.00	25
9.00	20
10.00	15
11.00	10
12.00	5

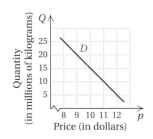

    As the price of coffee varies, the amount available varies. The table and graph below both show that sellers will supply less as the price goes down, but will supply more as the price goes up.

SUPPLY FUNCTION, S

Price, p, per Kilogram	Quantity, $S(p)$ (in millions of kilograms)
$ 9.00	5
9.50	10
10.00	15
10.50	20
11.00	25

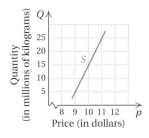

Calculator Spotlight

Linear regression can be used to find equations for supply and demand. See the Calculator Spotlight in Section 2.7.

2. Find the equilibrium point for the following supply and demand functions:

$$D(p) = 1000 - 46p,$$
$$S(p) = 300 + 4p.$$

Let's look at the above graphs together. We see that as price increases, demand decreases. As price increases, supply increases. The point of intersection of the demand and supply functions is called the **equilibrium point.** At the equilibrium point, the amount that the seller will supply is the same amount that the consumer will buy. The situation is analogous to a buyer and a seller negotiating the price of an item. The equilibrium point is the price and quantity on which they finally agree.

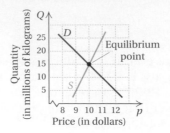

Any ordered pair of coordinates from the graph is (price, quantity), because the horizontal axis is the price axis and the vertical axis is the quantity axis. If D is a demand function and S is a supply function, then the equilibrium point is where demand equals supply:

$$D(p) = S(p).$$

Example 2   Find the equilibrium point for the following demand and supply functions:

$$D(p) = 1000 - 60p, \quad \textbf{(1)}$$
$$S(p) = 200 + 4p. \quad \textbf{(2)}$$

Since both demand and supply are *quantities* and they are equal at the equilibrium point, we rewrite the system as

$$q = 1000 - 60p, \quad \textbf{(1)}$$
$$q = 200 + 4p. \quad \textbf{(2)}$$

We substitute $200 + 4p$ for q in equation (1) and solve:

$$200 + 4p = 1000 - 60p$$
$$200 + 64p = 1000 \qquad \text{Adding } 60p \text{ on both sides}$$
$$64p = 800 \qquad \text{Subtracting 200 on both sides}$$
$$p = \tfrac{800}{64} = 12.5.$$

Thus the equilibrium price is $12.50 per unit.

To find the equilibrium quantity, we substitute $12.50 into either $D(p)$ or $S(p)$. We use $S(p)$:

$$S(12.5) = 200 + 4(12.5)$$
$$= 200 + 50 = 250.$$

Thus the equilibrium quantity is 250 units, and the equilibrium point is ($12.50, 250).

Do Exercise 2.

Calculator Spotlight

Use the INTERSECT feature on your grapher to solve Example 2 and Margin Exercise 2.

Answer on page A-22

Exercise Set 3.7

a For each of the following pairs of total-cost and total-revenue functions, find **(a)** the total-profit function and **(b)** the break-even point.

1. $C(x) = 25x + 270{,}000$;
$R(x) = 70x$

2. $C(x) = 45x + 300{,}000$;
$R(x) = 65x$

3. $C(x) = 10x + 120{,}000$;
$R(x) = 60x$

4. $C(x) = 30x + 49{,}500$;
$R(x) = 85x$

5. $C(x) = 20x + 10{,}000$;
$R(x) = 100x$

6. $C(x) = 40x + 22{,}500$;
$R(x) = 85x$

7. $C(x) = 22x + 16{,}000$;
$R(x) = 40x$

8. $C(x) = 15x + 75{,}000$;
$R(x) = 55x$

9. $C(x) = 50x + 195{,}000$;
$R(x) = 125x$

10. $C(x) = 34x + 928{,}000$;
$R(x) = 128x$

Solve.

11. *Manufacturing Lamps.* City Lights is planning to manufacture a new type of lamp. For the first year, the fixed costs for setting up production are $22,500. The variable costs for producing each lamp are $40. The revenue from each lamp is $85. Find the following.

a) The total cost $C(x)$ of producing x lamps
b) The total revenue $R(x)$ from the sale of x lamps
c) The total profit $P(x)$ from the production and sale of x lamps
d) The profit or loss from the production and sale of 3000 lamps; of 400 lamps
e) The break-even point

12. *Computer Manufacturing.* Sky View Electronics is planning to introduce a new line of computers. For the first year, the fixed costs for setting up production are $125,100. The variable costs for producing each computer are $750. The revenue from each computer is $1050. Find the following.

a) The total cost $C(x)$ of producing x computers
b) The total revenue $R(x)$ from the sale of x computers
c) The total profit $P(x)$ from the production and sale of x computers
d) The profit or loss from the production and sale of 400 computers; of 700 computers
e) The break-even point

13. *Manufacturing Caps.* Martina's Custom Printing is planning on adding painter's caps to its product line. For the first year, the fixed costs for setting up production are $16,404. The variable costs for producing a dozen caps are $6.00. The revenue on each dozen caps is $18.00. Find the following.

a) The total cost $C(x)$ of producing x dozen caps
b) The total revenue $R(x)$ from the sale of x dozen caps
c) The total profit $P(x)$ from the production and sale of x dozen caps
d) The profit or loss from the production and sale of 3000 dozen caps; of 1000 dozen caps
e) The break-even point

14. *Sport Coat Production.* Sarducci's is planning a new line of sport coats. For the first year, the fixed costs for setting up production are $10,000. The variable costs for producing each coat are $20. The revenue from each coat is $100. Find the following.

a) The total cost $C(x)$ of producing x coats
b) The total revenue $R(x)$ from the sale of x coats
c) The total profit $P(x)$ from the production and sale of x coats
d) The profit or loss from the production and sale of 2000 coats; of 50 coats
e) The break-even point

b) Find the equilibrium point for each of the following pairs of demand and supply functions.

15. $D(p) = 1000 - 10p;$
$S(p) = 230 + p$

16. $D(p) = 2000 - 60p;$
$S(p) = 460 + 94p$

17. $D(p) = 760 - 13p;$
$S(p) = 430 + 2p$

18. $D(p) = 800 - 43p;$
$S(p) = 210 + 16p$

19. $D(p) = 7500 - 25p;$
$S(p) = 6000 + 5p$

20. $D(p) = 8800 - 30p;$
$S(p) = 7000 + 15p$

21. $D(p) = 1600 - 53p;$
$S(p) = 320 + 75p$

22. $D(p) = 5500 - 40p;$
$S(p) = 1000 + 85p$

Skill Maintenance

Find the slope and the y-intercept. [2.4b]

23. $5y - 3x = 8$

24. $6x + 7y - 9 = 4$

25. $2y = 3.4x + 98$

26. $\dfrac{x}{3} + \dfrac{y}{4} = 1$

Synthesis

27. ◆ Variable costs and fixed costs are often compared to the slope and the y-intercept, respectively, of an equation of a line. Explain why this analogy is valid.

28. ◆ In this section, we examined supply and demand functions for coffee. Does it seem realistic to you for the graph of D to have a constant slope? Why or why not?

Summary and Review Exercises: Chapter 3

The objectives to be tested in addition to the material in this chapter are [1.1d], [1.2a], [2.2b], and [2.4b].

Solve graphically. Then classify the system as consistent or inconsistent and as dependent or independent.   [3.1a]

1. $4x - y = -9,$
$x - y = -3$

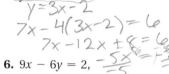

2. $15x + 10y = -20,$
$3x + 2y = -4$

3. $y - 2x = 4,$
$y - 2x = 5$

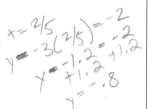

Solve by the substitution method.   [3.2a]

4. $7x - 4y = 6,$
$y - 3x = -2$

5. $y = x + 2,$
$y - x = 8$

6. $9x - 6y = 2,$
$x = 4y + 5$

Solve by the elimination method.   [3.2b]

7. $8x - 2y = 10,$
$-4y - 3x = -17$

8. $4x - 7y = 18,$
$9x + 14y = 40$

9. $3x - 5y = -4,$
$5x - 3y = 4$

10. $1.5x - 3 = -2y,$
$3x + 4y = 6$

11. *Music Spending.* Sean has $37 to spend. He can spend all of it on two compact discs and a cassette, or he can buy one CD and two cassettes and have $5.00 left over. What is the price of a CD? of a cassette?   [3.3a]

12. *Orange Drink Mixtures.* "Orange Thirst" is 15% orange juice and "Quencho" is 5% orange juice. How many liters of each should be combined in order to get 10 L of a mixture that is 10% orange juice?   [3.3a]

13. *Train Travel.* A train leaves Watsonville at noon traveling north at 44 mph. One hour later, another train, going 52 mph, travels north on a parallel track. How many hours will the second train travel before it overtakes the first train?   [3.3b]

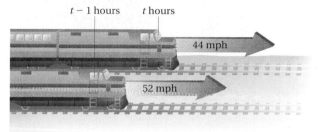

Solve.   [3.4a]

14. $x + 2y + z = 10,$
$2x - y + z = 8,$
$3x + y + 4z = 2$

15. $3x + 2y + z = 3,$
$6x - 4y - 2z = -34,$
$-x + 3y - 3z = 14$

16. $2x - 5y - 2z = -4,$
$7x + 2y - 5z = -6,$
$-2x + 3y + 2z = 4$

17. $x + y + 2z = 1,$
$x - y + z = 1,$
$x + 2y + z = 2$

18. *Triangle Measures.* In triangle ABC, the measure of angle A is four times the measure of angle C, and the measure of angle B is 45° more than the measure of angle C. What are the measures of the angles of the triangle?   [3.5a]

19. *Money Mixtures.* Elaine has $194, consisting of $20, $5, and $1 bills. The number of $1 bills is 1 less than the total number of $20 and $5 bills. If she has 39 bills in her purse, how many of each denomination does she have?   [3.5a]

Graph.   [3.6b]

20. $2x + 3y < 12$

21. $y \le 0$

22. $x + y \ge 1$

Graph. Find the coordinates of any vertices formed. [3.6c]

23. $y \ge -3,$
$x \ge 2$

24. $x + 3y \ge -1,$
$x + 3y \le 4$

25. $x - 3y \le 3,$
$x + 3y \ge 9,$
$y \le 6$

26. *Bed Manufacturing.* Kregel Furniture is planning to produce a new type of bed. For the first year, the fixed costs for setting up production are $35,000. The variable costs for producing each bed are $175. The revenue from each bed is $300. Find the following.   [3.7a]

   a) The total cost $C(x)$ of producing x beds
   b) The total revenue $R(x)$ from the sale of x beds
   c) The total profit from the production and sale of x beds
   d) The profit or loss from the production and sale of 1200 beds; of 200 beds
   e) The break-even point

27. Find the equilibrium point for the following demand and supply functions: [3.7b]

$$D(p) = 120 - 13p,$$
$$S(p) = 60 + 7p.$$

Skill Maintenance

28. Solve: $4x - 5(x + 8) = -9(1 - 3x)$.   [1.1d]

29. Solve $Q = at - 4t$ for t.   [1.2a]

30. Given the function $f(x) = 8 - 3x$, find $f(0)$ and $f(-2)$.   [2.2b]

31. For $5x - 8y = 40$, find the slope and the y-intercept.   [2.4b]

Synthesis

32. ◈ Briefly compare the strengths and the weaknesses of the graphical, substitution, and elimination methods as applied to the solution of two equations in two variables.   [3.1a], [3.2a, b]

33. ◈ Explain the advantages of using a system of equations to solve an applied problem.   [3.2a], [3.3a, b], [3.5a]

34. Solve graphically: [3.1a]
$$y = x + 2,$$
$$y = x^2 + 2.$$

35. *Height Estimation in Anthropology.* An anthropologist can use linear functions to estimate the height of a male or female, given the length of certain bones. The *femur* is the large bone from the hip to the knee, as shown below. Let $x =$ the length of the femur, in centimeters. Then the height, in centimeters, of a male with a femur of length x is given by the function

$$M(x) = 1.88x + 81.31.$$

The height, in centimeters, of a female with a femur of length x is given by the function

$$F(x) = 1.95x + 72.85.$$

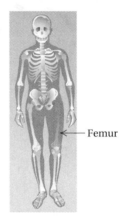
Femur

   A 45-cm femur was uncovered at an anthropological dig.   [3.1a], [3.2a]

   a) If we assume that it was from a male, how tall was he?
   b) If we assume that it was from a female, how tall was she?
   c) Graph each equation and find the point of intersection of the graphs of the equations.
   d) For what length of a male femur and a female femur, if any, would the height be the same?

Test: Chapter 3

Solve graphically. Then classify the system as consistent or inconsistent and as dependent or independent.

1. $y = 3x + 7,$
$3x + 2y = -4$

2. $y = 3x + 4,$
$y = 3x - 2$

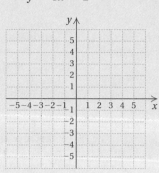

3. $y - 3x = 6,$
$6x - 2y = -12$

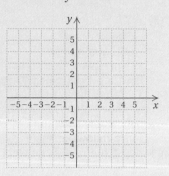

Solve by the substitution method.

4. $x + 3y = -8,$
$4x - 3y = 23$

5. $2x + 4y = -6,$
$y = 3x - 9$

Solve by the elimination method.

6. $4x - 6y = 3,$
$6x - 4y = -3$

7. $4y + 2x = 18,$
$3x + 6y = 26$

Solve.

8. *Saline Solutions.* Saline (saltwater) solutions are often used for sore throats. A nurse has a saline solution that is 34% salt and the rest water. She also has a 61% saline solution. She wants to use a 50% solution with a particular patient without wasting the existing solutions. How many milliliters (mL) of each solution would be needed in order to obtain 120 mL of a mixture that is 50% salt?

9. *Chicken Dinners.* High Flyin' Wings charges $12 for a bucket of chicken wings and $7 for a chicken dinner. After filling 28 orders for buckets and dinners during a football game, the waiters had collected $281. How many buckets and how many dinners did they sell?

10. *Air Travel.* An airplane flew for 5 hr with a 20-km/h tailwind and returned in 7 hr against the same wind. Find the speed of the plane in still air.

11. *Tennis Court.* The perimeter of a standard tennis court used for playing doubles is 288 ft. The width of the court is 42 ft less than the length. Find the length and the width.

Answers

1. _____

2. _____

3. _____

4. _____

5. _____

6. _____

7. _____

8. _____

9. _____

10. _____

11. _____

12. _____

13. _____

14. _____

15. a) _____

b) _____

c) _____

d) _____

e) _____

16. _____

17. _____

18. _____

19. _____

20. _____

21. _____

22. _____

23. _____

12. Solve:

$$6x + 2y - 4z = 15,$$
$$-3x - 4y + 2z = -6,$$
$$4x - 6y + 3z = 8.$$

13. Find the equilibrium point for the following demand and supply functions:

$$D(p) = 79 - 8p,$$
$$S(p) = 37 + 6p.$$

Solve.

14. _Repair Rates._ An electrician, a carpenter, and a plumber are hired to work on a house. The electrician earns $21 per hour, the carpenter $19.50 per hour, and the plumber $24 per hour. The first day on the job, they worked a total of 21.5 hr and earned a total of $469.50. If the plumber worked 2 hr more than the carpenter did, how many hours did the electrician work?

15. _Manufacturing Tennis Rackets._ Sweet Spot Manufacturing is planning to produce a new type of tennis racket. For the first year, the fixed costs for setting up production are $40,000. The variable costs for producing each racket are $30. The sales department predicts that 1500 rackets can be sold during the first year. The revenue from each racket is $80. Find the following.

a) The total cost $C(x)$ of producing x rackets
b) The total revenue $R(x)$ from the sale of x rackets
c) The total profit from the production and sale of x rackets
d) The profit or loss from the production and sale of 1200 rackets; of 200 rackets
e) The break-even point

Graph. Find the coordinates of any vertices formed.

16. $x - 6y < -6$

17. $x + y \geq 3,$
    $x - y \geq 5$

18. $2y - x \geq -7,$
    $2y + 3x \leq 15,$
    $y \leq 0,$
    $x \leq 0$

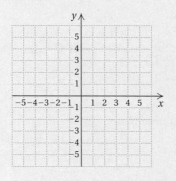

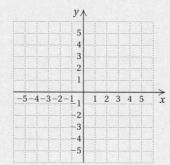

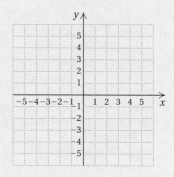

Skill Maintenance

19. Solve: $-8(t + 8) + 3(4 - t) = -15.$

20. Solve $P = 4a - 3b$ for a.

21. For $7x = 14 - 2y$, find the slope and the y-intercept.

22. Given the function $f(x) = x^2 - 8$, find $f(0)$ and $f(-3)$.

Synthesis

23. The graph of the function $f(x) = mx + b$ contains the points $(-1, 3)$ and $(-2, -4)$. Find m and b.

Cumulative Review: Chapters R–3

Evaluate for $a = 2$ and $b = -5$.

1. $\dfrac{a + b}{3}$

2. $ab - 2a$

Simplify.

3. $|-13|$

4. $\left|\dfrac{0}{2}\right|$

5. $(-2.1)(3.8)(-11.0)$

6. $\left(\dfrac{5}{9}\right) \div \left(-\dfrac{7}{3}\right)$

7. $9x - 3(2x - 11)$

8. $[5(6 - 3) + 2] - [4(5 - 6) + 11]$

9. $\dfrac{-10a^7b^{-11}}{25a^{-4}b^{22}}$

10. $4b + 2 - [7 - 6(2b + 1)]$

11. $\dfrac{y^4}{y^{-6}}$

Solve.

12. $6y - 5(3y - 4) = 10$

13. $-3 + 5x = 2x + 15$

14. $A = \pi r^2 h$, for h

15. $L = \dfrac{1}{3}m(k + p)$, for p

16. $5x + 8 > 2x + 5$

17. $2x - 10 \le -4 \ or \ x - 4 \ge 3$

18. $-12 \le -3x + 1 < 0$

19. $|8y - 3| \ge 15$

20. $|x + 1| = 4$

Graph on a plane.

21. $3y = 9$

22. $y = -\dfrac{1}{2}x - 3$

23. $3x - 1 = y$

24. $y > 3x - 4$

25. $3x + 5y = 15$

26. $2x - y \le 6$

27. Solve graphically:
$$2x - y = 7,$$
$$x + 3y = 0.$$

Solve.

28. $3x + 4y = 4,$
$\quad x - 2y = 2$

29. $3x + y = 4,$
$\quad 6x - y = 5$

30. $4x + 3y = -2,$
$\quad 2x - 5y = -12$

31. $\quad 2x + 5y - 3z = -11,$
$\quad -5x + 3y - 2z = -7,$
$\quad 3x - 2y + 5z = 12$

32. $\quad x - y + z = 1,$
$\quad 2x + y + z = 3,$
$\quad x + y - 2z = 4$

Graph. Find the coordinates of any vertices formed.

33. $x + y \le -3,$
$\quad x - y \le 1$

34. $x \ge 0,$
$\quad y \le x - 2$

35. The following is the graph of a function f. No equation is given for the function. Use the graph to approximate $f(-1)$, $f(0)$, and $f(2)$.

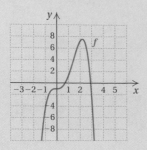

36. For the function f whose graph is shown below, determine **(a)** the domain, **(b)** the range, **(c)** $f(-3)$, and **(d)** any input for which $f(x) = 5$.

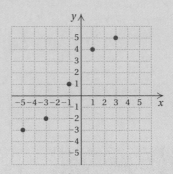

37. Find the domain of the function given by

$$f(x) = \frac{7}{2x - 1}.$$

38. Given $g(x) = 1 - 2x^2$, find $g(-1)$, $g(0)$, and $g(3)$.

39. Find the slope and the y-intercept of $5y - 4x = 20$.

40. Find the slope and the y-intercept of $2x + 4y = 7$.

41. Find an equation of the line with slope -3 and containing the point $(5, 2)$.

42. Find an equation of the line parallel to $3x - 9y = 2$ and containing the point $(-6, 2)$.

Solve.

43. *Wire Cutting.* Rolly's Electric wants to cut a piece of copper wire 10 m long into two pieces, one of them two-thirds as long as the other. How should the wire be cut?

44. *Soap Mixtures.* "Soakem" is 34% salt and the rest water. "Rinsem" is 61% salt and the rest water. How many ounces of each would be needed to obtain 120 oz of a mixture that is 50% salt?

45. *Inventory.* The Everton College store paid $1728 for an order of 45 calculators. The store paid $9 for each scientific calculator. The others, all graphing calculators, cost the store $58 each. How many of each type of calculator was ordered?

46. *Utility Costs.* One month Ladi and Bo spent $680 for electricity, rent, and telephone. The electric bill was one-fourth of the rent and the rent was $400 more than the phone bill. How much was the electric bill?

47. *Growth of Bicycling.* In the United States, the number N of bicyclists, in millions, t years since 1997 can be approximated by the function

$$N(t) = 3t + 114.$$

a) How many bicyclists will there be in the year 2000?

b) What is the slope, or rate of change, of the number of bicyclists?

c) When will there be more than 150,000,000 bicyclists?

48. *Test Scores.* Linda has a total of 225 on three tests. The sum of the scores on the first and second tests exceeds her third score by 61. Her first score exceeds her second by 6. Find the three scores.

Synthesis

49. Simplify: $(6x^{a+2}y^{b+2})(-2x^{a-2}y^{y+1})$.

50. *Radio Advertising.* An automotive dealer discovers that when $1000 is spent on radio advertising, weekly sales increase by $101,000. When $1250 is spent on radio advertising, weekly sales increase by $126,000. Assuming that sales increase according to a linear function, by what amount would sales increase when $1500 is spent on radio advertising?

51. Given that $f(x) = mx + b$ and that $f(5) = -3$ when $f(-4) = 2$, find m and b.

4

Polynomials and Polynomial Functions

Introduction

A polynomial is a type of algebraic expression that contains one or more terms. In this chapter, we learn to add, subtract, and multiply polynomials. We also learn the important skill of reversing multiplication called factoring. This process can be used to solve equations and applied problems.

Functions defined in terms of polynomials are called polynomial functions. We have already studied two kinds of polynomial functions: linear and constant functions. In this chapter, we extend our study of polynomial functions.

An Application	The Mathematics

An Application

Fireworks are typically launched from a mortar with an upward velocity (initial speed) of about 64 ft/sec. The height h, in feet, of a "weeping willow" display, t seconds after having been launched from an 80-ft-high rooftop, is given by the function

$$h(t) = -16t^2 + 64t + 80.$$

After how long will the cardboard shell from the fireworks reach the ground?

This problem appears as Example 8 in Section 4.7.

The Mathematics

We find the length of time by solving the equation

$$-16t^2 + 64t + 80 = 0.$$

This is a **quadratic** polynomial **equation**.

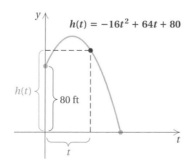

$h(t) = -16t^2 + 64t + 80$

80 ft

Pretest: Chapter 4

1. Given $P(x) = 2x^2 - 3x + 1$, find $P(-1)$ and $P(0)$.

2. Add: $(13x^2y - 4xy^2 + 3xy) + (4xy^2 - 7x^2y - 2xy)$.

3. Subtract:
$(5m^3 - 3m^2 + 6m + 3) - (6m - 9 - m^2 + 4m^2)$.

4. Arrange the polynomial $4xy^5 - 3x^6y^2 + x^2y^3 - 2y$ in descending powers of y.

Multiply.

5. $(x^2 - 1)(x^2 - 2x + 1)$

6. $(2y + 5z)(4y - z)$

7. $(a + 3b)(a - 3b)$

8. $(5t - 3m^2)^2$

Factor.

9. $4x^2 + 4x - 3$

10. $50m^2 + 40m + 8$

11. $4t^6 + 4t^3$

12. $a^2 + 6a + 8$

13. $x^2 - 49y^2$

14. $y^3 + 3y^2 + 4y + 12$

15. Solve: $6x + 8 = 9x^2$.

16. *Garden Design.* Ignacio is planning a garden to be 25 m longer than it is wide. The garden will have an area of 7500 m². What will its dimensions be?

17. *Height of a Baseball.* Suppose that a baseball is thrown upward with an initial velocity of 80 ft/sec from a height of 224 ft. Its height h after t seconds is given by the function
$$h(t) = -16t^2 + 80t + 224.$$
What is the height of the ball after 0 sec? 2 sec?

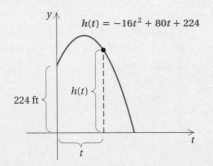

Objectives for Retesting

The objectives to be tested in addition to the material in this chapter are as follows.

[2.2d]   Determine whether a graph is that of a function using the vertical-line test.
[2.5b]   Given a linear equation in slope–intercept form, use the slope and the y-intercept to graph the line.
[3.2a, b]   Solve systems of equations in two variables.
[3.2c], [3.3a, b] Solve applied problems by solving systems of two equations.

4.1 Introduction to Polynomials and Polynomial Functions

A **polynomial** is a particular kind of algebraic expression. Let's examine an application before we consider definitions and manipulations involving polynomials.

Movie Revenue. The polynomial

$$-0.1672x^4 + 3.2956x^3 - 21.7343x^2 + 47.9053x + 8.5393,$$
$$1 \le x \le 9.5,$$

can be used to approximate the weekly revenue, in millions of dollars, from the recent movie *Air Force One*. It can also be used to express the revenue as a function of time as

$$f(x) = -0.1672x^4 + 3.2956x^3 - 21.7343x^2 + 47.9053x + 8.5393,$$
$$1 \le x \le 9.5,$$

where x is the number of weeks since the movie was released. The graph of the function is shown below.

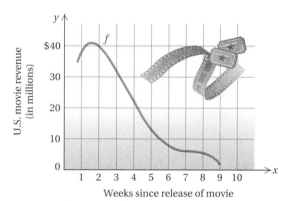

Source: Exhibitor Relations Co, Inc.

Although we will not be considering graphs of polynomial functions in detail in this chapter (other than in Calculator Spotlights), this situation gives us an idea of how polynomial functions can occur in applied problems.

a   Polynomial Expressions

The following are examples of *monomials*:

$$0, \quad -3, \quad z, \quad 8x, \quad -7y^2, \quad 4a^2b^3, \quad 1.3p^4q^5r^7.$$

> A **monomial** is an expression like $ax^ny^mz^q$, where a is a real number and n, m, and q are nonnegative integers. More specifically, a monomial is a constant or a constant times some variable or variables raised to powers that are nonnegative integers.

Expressions like these are called **polynomials in one variable:**

$$5x^2, \quad 8a, \quad 2, \quad 2x + 3, \quad -7x + 5, \quad 2y^2 + 5y - 3,$$
$$5a^4 - 3a^2 + \tfrac{1}{4}a - 8, \quad b^6 + 3b^5 - 8b + 7b^4 + \tfrac{1}{2}.$$

Expressions like these are called **polynomials in several variables:**

$$5a - ab^2 + 7b + 2, \quad 9xy^2z - 4x^3z + (-14x^4y^2) + 9, \quad 15x^3y^2.$$

Objectives

a Identify the degree of each term and the degree of a polynomial; identify terms, coefficients, monomials, binomials, and trinomials; arrange polynomials in ascending or descending order; and identify the leading coefficient.

b Evaluate a polynomial function for given inputs.

c Collect like terms in a polynomial and add polynomials.

d Find the opposite of a polynomial and subtract polynomials.

For Extra Help

TAPE 9   TAPE 7B   MAC WIN   CD-ROM

A **polynomial** is a monomial or a combination of sums and/or differences of monomials.

The following are algebraic expressions that are not polynomials:

(1) $\dfrac{y^2 - 3}{y^2 + 4}$,   (2) $8x^4 - 2x^3 + \dfrac{1}{x}$,   (3) $\dfrac{2xy}{x^3 - y^3}$.

Expressions (1) and (3) are not polynomials because they represent quotients. Expression (2) is not a polynomial because

$$\frac{1}{x} = x^{-1};$$

this is not a monomial because the exponent is negative.

The polynomial $5x^3y - 7xy^2 - y^3 + 2$ has four **terms:**

$$5x^3y, \qquad -7xy^2, \qquad -y^3, \quad \text{and} \quad 2.$$

The **coefficients** of the terms are 5, -7, -1, and 2.

The **degree of a term** is the sum of the exponents of the variables, if there are variables. The degree of a constant term is 0, except when the constant term is 0. Mathematicians agree that the polynomial 0 has no degree. This is because we can express 0 as $0 = 0x^5 = 0x^8$, and so on, using any exponent we wish. The **degree of a polynomial** is the same as the degree of its term of highest degree.

The **leading term** of a polynomial is the term of highest degree. Its coefficient is called the **leading coefficient.**

Example 1 Identify the terms, the degree of each term, and the degree of the polynomial. Then identify the leading term and the leading coefficient.

$$2x^3 + 8x^2 - 17x - 3$$

Term	$2x^3$	$8x^2$	$-17x$	-3
Degree	3	2	1	0
Degree of Polynomial	3			
Leading Term	$2x^3$			
Leading Coefficient	2			

Example 2 Identify the terms, the degree of each term, and the degree of the polynomial. Then identify the leading term and the leading coefficient.

$$6x^2 + 8x^2y^3 - 17xy - 24xy^2z^4 + 2y + 3$$

Term	$6x^2$	$8x^2y^3$	$-17xy$	$-24xy^2z^4$	$2y$	3
Degree	2	5	2	7	1	0
Degree of Polynomial	7					
Leading Term	$-24xy^2z^4$					
Leading Coefficient	-24					

Do Exercises 1–3.

The following are some names for certain kinds of polynomials.

Type	Definition	Examples
Monomial	A polynomial of one term	4, $-3p$, $5x^2$, $-7a^2b^3$, 0, xyz
Binomial	A polynomial of two terms	$2x + 7$, $a - 3b$, $5x^2 + 7y^3$
Trinomial	A polynomial of three terms	$x^2 - 7x + 12$, $4a^2 + 2ab + b^2$

Do Exercise 4.

We generally arrange polynomials in one variable so that the exponents *decrease* from left to right, which is **descending order.** Sometimes they may be written so that the exponents *increase* from left to right, which is **ascending order.** In general, if an exercise is written in a particular order, we write the answer in that same order.

Example 3 Consider $12 + x^2 - 7x$. Arrange in descending order and then in ascending order.

$$\text{Descending order:} \quad 12 + x^2 - 7x = x^2 - 7x + 12$$
$$\text{Ascending order:} \quad 12 + x^2 - 7x = 12 - 7x + x^2$$

Do Exercise 5.

Example 4 Consider $x^4 + 2 - 5x^2 + 3x^3y + 7xy^2$. Arrange in descending powers of x and then in ascending powers of x.

$$\text{Descending powers of } x: \quad x^4 + 3x^3y - 5x^2 + 7xy^2 + 2$$
$$\text{Ascending powers of } x: \quad 2 + 7xy^2 - 5x^2 + 3x^3y + x^4$$

Do Exercise 6.

b   Evaluating Polynomial Functions

A polynomial function is one like

$$P(x) = 5x^7 + 3x^5 - 4x^2 - 5,$$

where the algebraic expression used to describe the function is a polynomial. To find the outputs of a polynomial function for a given input, we substitute the input for each occurrence of the variable as we did in Section 2.2.

Example 5 For the polynomial function

$$P(x) = -x^2 + 4x - 1,$$

find $P(2)$, $P(10)$, and $P(-10)$.

$$P(2) = -2^2 + 4(2) - 1 \qquad \text{We square the input before taking its opposite.}$$
$$= -4 + 8 - 1 = 3;$$

$$P(10) = -10^2 + 4(10) - 1$$
$$= -100 + 40 - 1 = -61;$$

$$P(-10) = -(-10)^2 + 4(-10) - 1$$
$$= -100 - 40 - 1 = -141$$

Do Exercise 7.

4. Consider the following polynomials.

a) $3x^2 - 2$
b) $5x^3 + 9x - 3$
c) $4x^2$
d) $-7y$
e) -3
f) $8x^3 - 2x^2$
g) $-4y^2 - 5 - 5y$
h) $5 - 3x$

Identify the monomials, the binomials, and the trinomials.

5. a) Arrange in ascending order:

$$5 - 6x^2 + 7x^3 - x^4 + 10x.$$

b) Arrange in descending order:

$$5 - 6x^2 + 7x^3 - x^4 + 10x.$$

6. a) Arrange in ascending powers of y:

$$5x^4y - 3y^2 + 3x^2y^3 + x^3 - 5.$$

b) Arrange in descending powers of y:

$$5x^4y - 3y^2 + 3x^2y^3 + x^3 - 5.$$

7. For the polynomial function

$$P(x) = x^2 - 2x + 5,$$

find $P(0)$, $P(4)$, and $P(-2)$.

Answers on page A-24

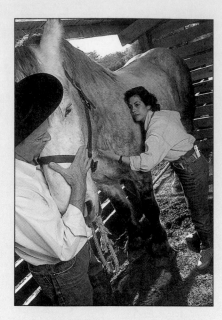

8. *Veterinary Medicine.* Refer to the function and the graph of Example 6.

   a) Evaluate $C(3)$ to find the concentration 3 hr after injection.

   b) Use only the graph at right to estimate $C(9)$.

Example 6 *Veterinary Medicine.* Gentamicin is an antibiotic frequently used by veterinarians. The concentration, in micrograms per milliliter (mcg/mL), of Gentamicin in a horse's bloodstream t hours after injection can be approximated by the polynomial function

$$C(t) = -0.005t^4 + 0.003t^3 + 0.35t^2 + 0.5t.$$

a) Evaluate $C(2)$ to find the concentration 2 hr after injection.

b) Use only the graph below to estimate $C(4)$.

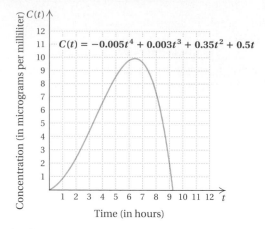

a) We evaluate the function for $t = 2$:

$$C(2) = -0.005(2)^4 + 0.003(2)^3 + 0.35(2)^2 + 0.5(2)$$
$$= -0.005(16) + 0.003(8) + 0.35(4) + 0.5(2)$$
$$= -0.08 + 0.024 + 1.4 + 1$$
$$= 2.344.$$

<small>We carry out the calculation using the rules for order of operations.</small>

   The concentration after 2 hr is about 2.344 mcg/mL.

b) To estimate $C(4)$, the concentration after 4 hr, we locate 4 on the horizontal axis. From there we move vertically to the graph of the function and then horizontally to the $C(t)$-axis. This locates a value of about 6.5. Thus,

$$C(4) \approx 6.5.$$

   The concentration after 4 hr is about 6.5 mcg/mL.

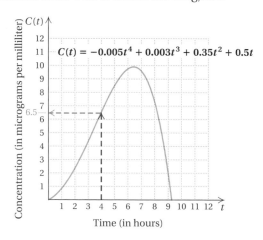

Do Exercise 8.

Answers on page A-24

Calculator Spotlight

Viewing Windows; Graphing and Evaluating Polynomial Functions. We can graph and evaluate polynomial functions using a grapher. Consider the function

$$f(x) = x^3 - x^2 - 8x + 5.$$

We enter the function as $y_1 = x^3 - x^2 - 8x + 5$. Suppose we first graph it using the viewing window $[-2, 2, -5, 5]$, as shown in the first window below.

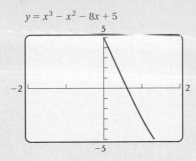

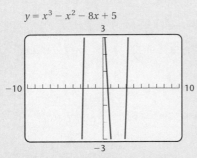

This window does not reveal much about the curvature of the graph. Nor does the graph in the second window above, which uses the window $[-10, 10, -3, 3]$. Finding a "good" window that reveals a graph's characteristics and curvature involves some trial and error and mathematical experience. The more you know about the shape of a graph, the more quickly you will be able to create a "good" graph. For now, be willing to experiment. For example, if we change to the window $[-4, 4, -14, 14]$, we can see more of the graph and get a better look at its curvature.

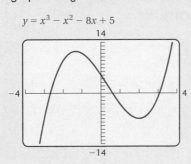

To evaluate the function at $x = -3$, that is, to find $f(-3)$, we use the CALC menu and choose VALUE.

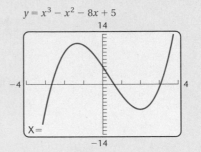

We enter $x = -3$. The function value, or y-value, $y = -7$, appears, together with a trace indicator showing the point $(-3, -7)$.

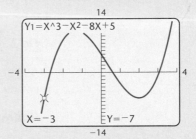

Polynomial functions can also be evaluated using the TABLE feature (see Section 2.2) or the VARS or Y-VARS feature. To use VARS or Y-VARS, we select the variable y_1 and then find a function value by enclosing the desired x-value in parentheses.

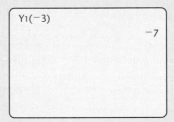

Exercises

1. Graph the function in Example 5 and find the given function values.
2. Graph the function in Example 6 and find the given function values.

Graph the given function using an appropriate window that best reveals the curvature of the graph and find the given function values.

3. $P(x) = x^3 - 3x - 1$; $P(0)$, $P(-4)$, $P(6.23)$
4. $f(x) = x^4 - 2x^2$; $f(-2)$, $f(0)$, $f(1)$, $f(-12.7)$
5. $f(x) = -1 + 2x + \frac{1}{2}x^2 - \frac{1}{3}x^3$; $f(-2)$, $f(0)$, $f(5.4)$
6. $g(x) = 280x - 0.4x^2$; $g(0)$, $g(350)$, $g(700)$
7. $f(x) = x^4 - x^3 - 11x^2 + 9x + 18$; $f(-3)$, $f(3)$, $f(-2)$, $f(20)$, $f(0)$

Collect like terms.

9. $3y - 4x + 6xy^2 - 2xy^2$

10. $3xy^3 + 2x^3y + 5xy^3 - 8x + 15 - 3x^2y - 6x^2y + 11x - 8$

Add.

11. $(3x^3 + 4x^2 - 7x - 2) + (-7x^3 - 2x^2 + 3x + 4)$

12. $(7y^5 - 5) + (3y^5 - 4y^2 + 10)$

13. $(5p^2q^4 - 2p^2q^2 - 3q) + (-6p^2q^2 + 3q + 5)$

Answers on page A-24

c | Adding Polynomials

When two terms have the same variable(s) raised to the same power(s), they are called **like terms,** or **similar terms,** and they can be "collected," or "combined," using the distributive laws, adding or subtracting the coefficients as follows.

Examples   Collect like terms.

7. $3x^2 - 4y + 2x^2 = 3x^2 + 2x^2 - 4y$ **Rearranging using the commutative law for addition**

$$= (3 + 2)x^2 - 4y$$ **Using the distributive law**
$$= 5x^2 - 4y$$

8. $9x^3 + 5x - 4x^2 - 2x^3 + 5x^2 = 7x^3 + x^2 + 5x$ **We usually perform the middle steps mentally and write just the answer.**

9. $3x^2y + 5xy^2 - 3x^2y - xy^2 = 4xy^2$

Do Exercises 9 and 10.

The sum of two polynomials can be found by writing a plus sign between them and then combining like terms.

Example 10   Add: $(-3x^3 + 2x - 4) + (4x^3 + 3x^2 + 2)$.

$$(-3x^3 + 2x - 4) + (4x^3 + 3x^2 + 2) = x^3 + 3x^2 + 2x - 2$$

Example 11   Add: $13x^3y + 3x^2y - 5y$ and $x^3y + 4x^2y - 3xy$.

$$(13x^3y + 3x^2y - 5y) + (x^3y + 4x^2y - 3xy) = 14x^3y + 7x^2y - 3xy - 5y$$

Do Exercises 11–13.

Using columns to add is sometimes helpful. To do so, we write the polynomials one under the other, listing like terms under one another and leaving spaces for missing terms.

Example 12   Add: $4ax^2 + 4bx - 5$ and $-6ax^2 + 8$.

$$\begin{array}{r} 4ax^2 + 4bx - 5 \\ -6ax^2 \qquad + 8 \\ \hline -2ax^2 + 4bx + 3 \end{array}$$

d | Subtracting Polynomials

If the sum of two polynomials is 0, they are called **opposites,** or **additive inverses,** of each other. For example,

$$(3x^2 - 5x + 2) + (-3x^2 + 5x - 2) = 0,$$

so the opposite of $3x^2 - 5x + 2$ is $-3x^2 + 5x - 2$. We can say the same thing using algebraic symbolism, as follows:

The opposite of $(3x^2 - 5x + 2)$ is $(-3x^2 + 5x - 2)$.

$-$ $(3x^2 - 5x + 2)$ $=$ $-3x^2 + 5x - 2$

Thus, $-(3x^2 - 5x + 2)$ and $-3x^2 + 5x - 2$ are equivalent.

The *opposite* of a polynomial P can be symbolized by $-P$ or by replacing each term with its opposite. The two expressions for the opposite are equivalent.

Example 13   Write two equivalent expressions for the opposite of

$7xy^2 - 6xy - 4y + 3$.

First expression: $-(7xy^2 - 6xy - 4y + 3)$ Writing an inverse
                                                          sign in front

Second expression: $-7xy^2 + 6xy + 4y - 3$ Writing the opposite of
                                                          each term
                                                          (see also Section R.6)

Do Exercises 14–16.

To subtract a polynomial, we add its opposite.

Example 14   Subtract: $(-5x^2 + 4) - (2x^2 + 3x - 1)$.

$(-5x^2 + 4) - (2x^2 + 3x - 1)$
$= (-5x^2 + 4) + (-2x^2 - 3x + 1)$ Adding the opposite of the
                                            polynomial being subtracted

$= -7x^2 - 3x + 5$

With practice, you will find that you can skip some steps, by mentally taking the opposite of each term and then combining like terms. Eventually, all you will write is the answer.

To use columns for subtraction, we mentally change the signs of the terms being subtracted.

Do Exercises 17–19.

Example 15   Subtract:

$(4x^2y - 6x^3y^2 + x^2y^2) - (4x^2y + x^3y^2 + 3x^2y^3 - 8x^2y^2)$.

Write: (Subtract)

$\begin{array}{l} 4x^2y - 6x^3y^2 \qquad\qquad + x^2y^2 \\ -(4x^2y + \quad x^3y^2 + 3x^2y^3 - 8x^2y^2) \end{array}$

Think: (Add)

$\begin{array}{l} 4x^2y - 6x^3y^2 \qquad\qquad + x^2y^2 \\ -4x^2y - \quad x^3y^2 - 3x^2y^3 + 8x^2y^2 \\ \hline \qquad\quad - 7x^3y^2 - 3x^2y^3 + 9x^2y^2 \end{array}$

Take the opposite of each term mentally and add.

Do Exercises 20–22.

Write two equivalent expressions for the opposite, or additive inverse.

14. $4x^3 - 5x^2 + \dfrac{1}{4}x - 10$

15. $8xy^2 - 4x^3y^2 - 9x - \dfrac{1}{5}$

16. $-9y^5 - 8y^4 + \dfrac{1}{2}y^3 - y^2 + y - 1$

Subtract.

17. $(6x^2 + 4) - (3x^2 - 1)$

18. $(9y^3 - 2y - 4) - (-5y^3 - 8)$

19. $(-3p^2 + 5p - 4) - (-4p^2 + 11p - 2)$

Subtract.

20. $(2y^5 - y^4 + 3y^3 - y^2 - y - 7) - (-y^5 + 2y^4 - 2y^3 + y^2 - y - 4)$

21. $(4p^4q - 5p^3q^2 + p^2q^3 + 2q^4) - (-5p^4q + 5p^3q^2 - 3p^2q^3 - 7q^4)$

22. $\left(\dfrac{3}{2}y^3 - \dfrac{1}{2}y^2 + 0.3\right) - \left(\dfrac{1}{2}y^3 + \dfrac{1}{2}y^2 - \dfrac{4}{3}y + 0.2\right)$

Answers on page A-24

Calculator Spotlight

 Checking Operations with a Table or a Graph

TABLE Feature. The TABLE feature can be used as a partial check that polynomials have been added or subtracted correctly. To check whether

$$(-5x^2 + 4) - (2x^2 + 3x - 1) = 7x^2 - 3x + 5$$

is correct, we enter

$$y_1 = (-5x^2 + 4) - (2x^2 + 3x - 1)$$

and

$$y_2 = 7x^2 - 3x + 5$$

and look at a table of values.

X	Y₁	Y₂
−1	1	15
0	5	5
1	−5	9
2	−29	27
3	−67	59
4	−119	105
5	−185	165

X = −1

If our subtraction is correct, the y_1- and y_2-values will be the same, regardless of the table settings used. We see that y_1 and y_2 are not the same, so the subtraction is not correct.

GRAPH Feature. We can also check the subtraction above using the GRAPH feature.

$$y_1 = (-5x^2 + 4) - (2x^2 + 3x - 1),$$
$$y_2 = 7x^2 - 3x + 5$$

We see that the graphs differ, so the subtraction is not correct.

Exercises

Use the TABLE or GRAPH feature to check whether each of the following is correct.

1. $(6x^2 + 4) - (3x^2 - 1) = 3x^2 + 5$ (Margin Exercise 17)

2. $(-3x^3 + 2x - 4) + (4x^3 + 3x^2 + 2) = x^3 - 4x^2 + 2x - 6$ (Example 10)

Use the TABLE or GRAPH feature to check whether your answers to the following are correct.

3. Margin Exercise 11

4. Margin Exercise 18 (Use x for the variable.)

5. Margin Exercise 19

6. Why can't the grapher be used to check Example 12?

Exercise Set 4.1

a Identify the terms, the degree of each term, and the degree of the polynomial. Then identify the leading term and the leading coefficient.

1. $-9x^4 - x^3 + 7x^2 + 6x - 8$

2. $y^3 - 5y^2 + y + 1$

3. $t^3 + 4t^7 + s^2t^4 - 2$

4. $a^2 + 9b^5 - a^4b^3 - 11$

5. $u^7 + 8u^2v^6 + 3uv + 4u - 1$

6. $2p^6 + 5p^4w^4 - 13p^3w + 7p^2 - 10$

Arrange in descending powers of y.

7. $23 - 4y^3 + 7y - 6y^2$

8. $5 - 8y + 6y^2 + 11y^3 - 18y^4$

9. $x^2y^2 + x^3y - xy^3 + 1$

10. $x^3y - x^2y^2 + xy^3 + 6$

11. $2by - 9b^5y^5 - 8b^2y^3$

12. $dy^6 - 2d^7y^2 + 3cy^5 - 7y - 2d$

Arrange in ascending powers of x.

13. $12x + 5 + 8x^5 - 4x^3$

14. $-3x^2 + 8x + 2$

15. $-9x^3y + 3xy^3 + x^2y^2 + 2x^4$

16. $5x^2y^2 - 9xy + 8x^3y^2 - 5x^4$

17. $4ax - 7ab + 4x^6 - 7ax^2$

18. $5xy^8 - 3ax^5 + 4ax^3 - 12a + 5x^5$

b Evaluate the polynomial function for the given values of the variable.

19. $P(x) = 3x^2 - 2x + 5$; $P(4)$, $P(-2)$, $P(0)$

20. $f(x) = -7x^3 + 10x^2 - 13$; $f(4)$, $f(-1)$, $f(0)$

21. $p(x) = 9x^3 + 8x^2 - 4x - 9$; $p(-3)$, $p(0)$, $p(1)$, $p\left(\frac{1}{2}\right)$

22. $Q(x) = 6x^3 - 11x - 4$; $Q(-2)$, $Q\left(\frac{1}{3}\right)$, $Q(0)$, $Q(10)$

23. *Falling Distance.* The distance $s(t)$, in feet, traveled by a body falling freely from rest in t seconds is approximated by the function given by
$$s(t) = 16t^2.$$
  a) A paintbrush falls from a scaffold and takes 3 sec to hit the ground. How high is the scaffold?
  b) A stone is dropped from a cliff and takes 8 sec to hit the ground. How high is the cliff?

24. *Electing Officers.* For a club consisting of n people, the number of ways in which a president, vice president, and treasurer can be elected can be found using the polynomial function given by
$$p(n) = n^3 - 3n^2 + 2n.$$
  a) The Stage Right drama club has 12 members. In how many ways can a president, vice president, and treasurer be elected?
  b) The Southside Rugby Club has 20 members. In how many ways can they elect a president, vice president, and treasurer?

$s(t) = 16t^2$

25. *Books on Audiotape.* More people are listening to books on audiotape. Sales S, in billions of dollars, can be approximated by the polynomial function
$$S(x) = 0.0095x^4 - 0.1305x^3 + 0.5636x^2 - 0.5175x + 0.2718,$$
where x is the number of years since 1989. The graph is shown below.

Sales (in billions)

$2.0
1.5
1.0
0.5

1 2 3 4 5 6 7 8 9 10 x
Years since 1989

Source: Audio Book Club

a) Evaluate $S(7)$ to find the sales in 1996 ($x = 7$).
b) Use only the graph to estimate $S(4)$.

26. *Median Income by Age.* The polynomial function
$$I(x) = -0.0560x^4 + 7.9980x^3 - 436.1840x^2 + 11,627.8376x - 90,625.0001,$$
$$13 \le x \le 65,$$
can be used to approximate the median income I by age x of a person living in the United States. The graph is shown below.

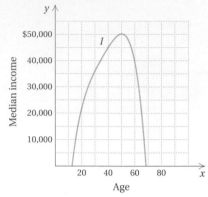

Median income

$50,000
40,000
30,000
20,000
10,000

20 40 60 80 x
Age

Source: U.S. Bureau of the Census;
The Conference Board: Simmons Bureau of Labor Statistics

a) Evaluate $I(22)$ to estimate the median income of a 22-year-old.
b) Use only the graph to estimate $I(40)$.

27. *Total Revenue.* An electronics firm is marketing a new kind of cellular telephone. The firm determines that when it sells x phones, its total revenue is
$$R(x) = 280x - 0.4x^2 \text{ dollars.}$$
a) What is the total revenue from the sale of 75 phones?
b) What is the total revenue from the sale of 100 phones?

28. *Total Cost.* The electronics firm determines that the total cost, in dollars, of producing x phones is given by
$$C(x) = 5000 + 0.6x^2.$$
a) What is the total cost of producing 75 phones?
b) What is the total cost of producing 100 phones?

Total Profit. **Total profit P** is defined as total revenue R minus total cost C, and is given by the function
$$P(x) = R(x) - C(x).$$
For each of the following, find the total profit $P(x)$.

29. $R(x) = 280x - 0.4x^2,$ $C(x) = 7000 + 0.6x^2$

30. $R(x) = 280x - 0.7x^2,$ $C(x) = 8000 + 0.5x^2$

c Collect like terms.

31. $6x^2 - 7x^2 + 3x^2$

32. $-2y^2 - 7y^2 + 5y^2$

33. $7x - 2y - 4x + 6y$

34. $a - 8b - 5a + 7b$

35. $3a + 9 - 2 + 8a - 4a + 7$

36. $13x + 14 - 6 - 7x + 3x + 5$

37. $3a^2b + 4b^2 - 9a^2b - 6b^2$

38. $5x^2y^2 + 4x^3 - 8x^2y^2 - 12x^3$

39. $8x^2 - 3xy + 12y^2 + x^2 - y^2 + 5xy + 4y^2$

40. $a^2 - 2ab + b^2 + 9a^2 + 5ab - 4b^2 + a^2$

41. $4x^2y - 3y + 2xy^2 - 5x^2y + 7y + 7xy^2$

42. $3xy^2 + 4xy - 7xy^2 + 7xy + x^2y$

Add.

43. $(3x^2 + 5y^2 + 6) + (2x^2 - 3y^2 - 1)$

44. $(11y^2 + 6y - 3) + (9y^2 - 2y + 9)$

45. $(2a - c + 3b) + (4a - 2b + 2c)$

46. $(8x + z - 7y) + (5x + 10y - 4z)$

47. $(a^2 - 3b^2 + 4c^2) + (-5a^2 + 2b^2 - c^2)$

48. $(x^2 - 5y^2 - 9z^2) + (-6x^2 + 9y^2 - 2z^2)$

49. $(x^2 + 3x - 2xy - 3) + (-4x^2 - x + 3xy + 2)$

50. $(5a^2 - 3b + ab + 6) + (-a^2 + 8b - 8ab - 4)$

51. $(7x^2y - 3xy^2 + 4xy) + (-2x^2y - xy^2 + xy)$

52. $(7ab - 3ac + 5bc) + (13ab - 15ac - 8bc)$

53. $(2r^2 + 12r - 11) + (6r^2 - 2r + 4) + (r^2 - r - 2)$

54. $(5x^2 + 19x - 23) + (-7x^2 - 11x + 12) + (-x^2 - 9x + 8)$

55. $\left(\frac{2}{3}xy + \frac{5}{6}xy^2 + 5.1x^2y\right) + \left(-\frac{4}{5}xy + \frac{3}{4}xy^2 - 3.4x^2y\right)$

56. $\left(\frac{1}{8}xy - \frac{3}{5}x^3y^2 + 4.3y^3\right) + \left(-\frac{1}{3}xy - \frac{3}{4}x^3y^2 - 2.9y^3\right)$

---

d Write two equivalent expressions for the opposite of the polynomial.

57. $5x^3 - 7x^2 + 3x - 6$

58. $-8y^4 - 18y^3 + 4y - 9$

59. $-13y^2 + 6ay^4 - 5by^2$

60. $9ax^5y^3 - 8by^5 - abx - 16ay$

Subtract.

61. $(7x - 2) - (-4x + 5)$

62. $(8y + 1) - (-5y - 2)$

63. $(-3x^2 + 2x + 9) - (x^2 + 5x - 4)$

64. $(-9y^2 + 4y + 8) - (4y^2 + 2y - 3)$

65. $(5a + c - 2b) - (3a + 2b - 2c)$

66. $(z + 8x - 4y) - (4x + 6y - 3z)$

67. $(3x^2 - 2x - x^3) - (5x^2 - x^3 - 8x)$

68. $(8y^2 - 4y^3 - 3y) - (3y^2 - 9y - 7y^3)$

69. $(5a^2 + 4ab - 3b^2) - (9a^2 - 4ab + 2b^2)$

70. $(9y^2 - 14yz - 8z^2) - (12y^2 - 8yz + 4z^2)$

71. $(6ab - 4a^2b + 6ab^2) - (3ab^2 - 10ab - 12a^2b)$

72. $(10xy - 4x^2y^2 - 3y^3) - (-9x^2y^2 + 4y^3 - 7xy)$

73. $(0.09y^4 - 0.052y^3 + 0.93) - (0.03y^4 - 0.084y^3 + 0.94y^2)$

74. $(1.23x^4 - 3.122x^3 + 1.11x) - (0.79x^4 - 8.734x^3 + 0.04x^2 + 6.71x)$

75. $\left(\frac{5}{8}x^4 - \frac{1}{4}x^2 - \frac{1}{2}\right) - \left(-\frac{3}{8}x^4 + \frac{3}{4}x^2 + \frac{1}{2}\right)$

76. $\left(\frac{5}{6}y^4 - \frac{1}{2}y^2 - 7.8y + \frac{1}{3}\right) - \left(-\frac{3}{8}y^4 + \frac{3}{4}y^2 + 3.4y - \frac{1}{5}\right)$

Skill Maintenance

Graph.   [2.1c, d]

77. $f(x) = \frac{2}{3}x - 1$ **78.** $g(x) = |x| - 1$ **79.** $g(x) = \dfrac{4}{x - 3}$ **80.** $f(x) = 1 - x^2$

Multiply.   [R.5d]

81. $3(y - 2)$ **82.** $-10(x + 2y - 7)$ **83.** $-14(3p - 2q - 10)$ **84.** $\frac{2}{3}(12w - 9t + 30)$

Graph using the slope and the y-intercept.   [2.5b]

85. $y = \frac{4}{3}x + 2$ **86.** $y = -0.4x + 1$ **87.** $y = 0.4x - 3$ **88.** $y = -\frac{2}{3}x - 4$

Synthesis

89. ◈ Is the sum of two binomials always a binomial? Why or why not?

90. ◈ Annie claims that she can add any two polynomials but finds subtraction difficult. What advice would you offer her?

91. The number of spheres in a triangular pyramid with x layers is given by the function
$$N(x) = \frac{1}{6}x^3 + \frac{1}{2}x^2 + \frac{1}{3}x.$$
The volume of a sphere of radius r is given by the function
$$V(r) = \frac{4}{3}\pi r^3,$$
where π can be approximated as 3.14.

Chocolate Heaven has a window display of truffles piled in a triangular pyramid formation 5 layers deep. If the diameter of each truffle is 3 cm, find the volume of chocolate in the display.

92. *Surface Area.* Find a polynomial function that gives the outside surface area of a box like this one, with an open top and dimensions as shown.

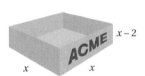

Perform the indicated operations. Assume that the exponents are natural numbers.

93. $(47x^{4a} + 3x^{3a} + 22x^{2a} + x^a + 1) + (37x^{3a} + 8x^{2a} + 3)$

94. $(3x^{6a} - 5x^{5a} + 4x^{3a} + 8) - (2x^{6a} + 4x^{4a} + 3x^{3a} + 2x^{2a})$

95. 〰 Use the TABLE and GRAPH features of a grapher to check your answers to Exercises 43, 61, and 63.

96. 〰 A student who is trying to graph $f(x) = 0.05x^4 - x^2 + 5$ gets the following screen. How can the student tell at a glance that a mistake has been made?

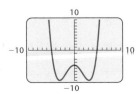

Collaborative Learning Manual

Determine the polynomial function for the number of handshakes possible in a group.

4.2 Multiplication of Polynomials

a | Multiplication of Any Two Polynomials

Multiplying Monomials

To multiply monomials, we first multiply their coefficients. Then we multiply the variables using the commutative and associative laws and the rules for exponents that we studied in Chapter R.

Examples Multiply and simplify.

1. $(10x^2)(8x^5) = (10 \cdot 8)(x^2 \cdot x^5)$
$\qquad\qquad\quad = 80x^{2+5}$ **Adding exponents**
$\qquad\qquad\quad = 80x^7$

2. $(-8x^4y^7)(5x^3y^2) = (-8 \cdot 5)(x^4 \cdot x^3)(y^7 \cdot y^2)$
$\qquad\qquad\qquad\quad = -40x^{4+3}y^{7+2}$ **Adding exponents**
$\qquad\qquad\qquad\quad = -40x^7y^9$

Do Exercises 1–3.

Multiplying Monomials and Binomials

The distributive law is the basis for multiplying polynomials other than monomials. We first multiply a monomial and a binomial.

Example 3 Multiply: $2x(3x - 5)$.

$2x \cdot (3x - 5) = 2x \cdot 3x - 2x \cdot 5$ **Using the distributive law**
$\qquad\qquad\quad = 6x^2 - 10x$ **Multiplying monomials**

Example 4 Multiply: $3a^2b(a^2 - b^2)$.

$3a^2b \cdot (a^2 - b^2) = 3a^2b \cdot a^2 - 3a^2b \cdot b^2$ **Using the distributive law**
$\qquad\qquad\qquad = 3a^4b - 3a^2b^3$

Do Exercises 4 and 5.

Multiplying Binomials

Next, we multiply two binomials. To do so, we use the distributive law twice, first considering one of the binomials as a single expression and multiplying it by each term of the other binomial.

Example 5 Multiply: $(3y^2 + 4)(y^2 - 2)$.

$(\boxed{3y^2 + 4})(y^2 - 2) = (\boxed{3y^2 + 4}) \cdot y^2 - (\boxed{3y^2 + 4}) \cdot 2$ **Using the distributive law**

$\qquad\qquad\qquad = [3y^2 \cdot y^2 + 4 \cdot y^2] - [3y^2 \cdot 2 + 4 \cdot 2]$ **Using the distributive law**

$\qquad\qquad\qquad = 3y^2 \cdot y^2 + 4 \cdot y^2 - 3y^2 \cdot 2 - 4 \cdot 2$ **Removing parentheses**

$\qquad\qquad\qquad = 3y^4 + 4y^2 - 6y^2 - 8$ **Multiplying the monomials**

$\qquad\qquad\qquad = 3y^4 - 2y^2 - 8$ **Collecting like terms**

Multiply.

1. $(9y^2)(-2y)$

2. $(4x^3y)(6x^5y^2)$

3. $(-5xy^7z^4)(18x^3y^2z^8)$

Multiply.

4. $(-3y)(2y + 6)$

5. $(2xy)(4y^2 - 5)$

Multiply.

6. $(5x^2 - 4)(x + 3)$

7. $(2y + 3)(3y - 4)$

Answers on page A-25

Multiply.

8. $(p - 3)(p^3 + 4p^2 - 5)$

9. $(2x^3 + 4x - 5)(x - 4)$

Multiply.

10. $(-4x^3 - 2x + 1) \times$
$(-2x^2 - 3x + 6)$

11. $(a^2 - 2ab + b^2) \times$
$(a^3 + 3ab - b^2)$

A visualization of
$(x + 7)(x + 4)$ using areas

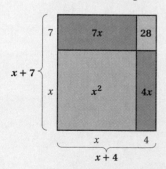

Answers on page A-25

Do Exercises 6 and 7 on the preceding page.

Multiplying Any Two Polynomials

To find a quick way to multiply any two polynomials, let's consider another example.

Example 6   Multiply: $(p + 2)(p^4 - 2p^3 + 3)$.

By the distributive law, we have

$$(\;p + 2\;)(p^4 - 2p^3 + 3)$$
$$= (\;p + 2\;)(p^4) - (\;p + 2\;)(2p^3) + (\;p + 2\;)(3)$$
$$= p(p^4) + 2(p^4) - p(2p^3) - 2(2p^3) + p(3) + 2(3)$$
$$= p^5 + 2p^4 - 2p^4 - 4p^3 + 3p + 6$$
$$= p^5 - 4p^3 + 3p + 6. \qquad \text{Collecting like terms}$$

Do Exercises 8 and 9.

From the preceding examples, we can see how to multiply any two polynomials.

> To multiply two polynomials P and Q, select one of the polynomials, say P. Then multiply each term of P by every term of Q and collect like terms.

We can use columns to save time doing long multiplications. We multiply each term at the top by every term at the bottom, keeping like terms in columns and adding spaces for missing terms. Then we add.

Example 7   Multiply: $(5x^3 + x - 4)(-2x^2 + 3x + 6)$.

$$
\begin{array}{r}
5x^3 + \quad x - \quad 4 \\
-2x^2 + \quad 3x + \quad 6 \\
\hline
30x^3 \qquad\qquad + 6x - 24 \\
15x^4 \qquad\qquad + 3x^2 - 12x \\
-10x^5 \qquad - 2x^3 + 8x^2 \\
\hline
-10x^5 + 15x^4 + 28x^3 + 11x^2 - 6x - 24
\end{array}
$$

Multiplying by 6
Multiplying by 3x
Multiplying by $-2x^2$

Do Exercises 10 and 11.

b   Product of Two Binomials Using the FOIL Method

We now consider some **special products.** There are rules for faster multiplication in certain situations.

Let's find a faster special-product rule for the product of two binomials. Consider $(x + 7)(x + 4)$. We multiply each term of $(x + 7)$ by each term of $(x + 4)$:

$$(x + 7)(x + 4) = x \cdot x + x \cdot 4 + 7 \cdot x + 7 \cdot 4.$$

This multiplication illustrates a pattern that occurs whenever two binomials are multiplied:

$$
\underbrace{\text{First}}_{\text{terms}} \quad \underbrace{\text{Outside}}_{\text{terms}} \quad \underbrace{\text{Inside}}_{\text{terms}} \quad \underbrace{\text{Last}}_{\text{terms}}
$$

$$(x + 7)(x + 4) = x \cdot x \;+\; 4x \;+\; 7x \;+\; 7(4) \;=\; x^2 + 11x + 28.$$

This special method of multiplying is called the **FOIL method.**

THE FOIL METHOD

To multiply two binomials, $A + B$ and $C + D$, multiply the **F**irst terms AC, the **O**utside terms AD, the **I**nside terms BC, and then the **L**ast terms BD. Then collect like terms, if possible.

$$(A + B)(C + D) = AC + AD + BC + BD$$

1. Multiply First terms: AC.
2. Multiply Outside terms: AD.
3. Multiply Inside terms: BC.
4. Multiply Last terms: BD.

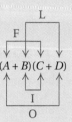

FOIL

Examples Multiply.

$$
\begin{array}{cccc}
\text{F} & \text{O} & \text{I} & \text{L}
\end{array}
$$

8. $(x + 5)(x - 8) = x^2 - 8x + 5x - 40$
$\qquad\qquad\qquad = x^2 - 3x - 40 \qquad$ **Collecting like terms**

We write the result in descending order since the original binomials are in descending order.

$$
\begin{array}{cccc}
\text{F} & \text{O} & \text{I} & \text{L}
\end{array}
$$

9. $(3xy + 2x)(x^2 + 2xy^2) = 3x^3y + 6x^2y^3 + 2x^3 + 4x^2y^2$

10. $(2x - 3)(y + 2) = 2xy + 4x - 3y - 6$

11. $(2x + 3y)(x - 4y) = 2x^2 - 8xy + 3xy - 12y^2$
$\qquad\qquad\qquad\quad\; = 2x^2 - 5xy - 12y^2 \qquad$ **Collecting like terms**

Do Exercises 12–14.

 Squares of Binomials

We can use the FOIL method to develop special products for the square of a binomial:

$$
\begin{aligned}
(A + B)^2 &= (A + B)(A + B) \\
&= A^2 + AB + AB + B^2 \\
&= A^2 + 2AB + B^2;
\end{aligned}
\qquad
\begin{aligned}
(A - B)^2 &= (A - B)(A - B) \\
&= A^2 - AB - AB + B^2 \\
&= A^2 - 2AB + B^2.
\end{aligned}
$$

▶ The **square of a binomial** is the square of the first term, plus twice the product of the two terms, plus the square of the last term.

$$(A + B)^2 = A^2 + 2AB + B^2;$$
$$(A - B)^2 = A^2 - 2AB + B^2$$

Multiply.

12. $(y - 4)(y + 10)$

13. $(p + 5q)(2p - 3q)$

14. $(x^2y + 2x)(xy^2 + y^2)$

A visualization of $(A + B)^2$ using areas

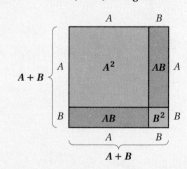

Calculator Spotlight

Use the TABLE and GRAPH features to check the results of Examples 8 and 12.

Answers on page A-25

Multiply.

15. $(a - b)^2$

16. $(x + 8)^2$

17. $(3x - 7)^2$

18. $\left(m^3 + \dfrac{1}{4}n\right)^2$

Calculator Spotlight

Use the TABLE and GRAPH features to check the results of Example 16 and Margin Exercise 17.

Multiply.

19. $(x + 8)(x - 8)$

20. $(4y - 7)(4y + 7)$

21. $(2.8a + 4.1b)(2.8a - 4.1b)$

22. $\left(3w - \dfrac{3}{5}q^2\right)\left(3w + \dfrac{3}{5}q^2\right)$

Answers on page A-25

Examples   Multiply.

$$(A - B)^2 = A^2 - 2\ A\ B + B^2$$

12. $(y - 5)^2 = y^2 - 2(y)(5) + 5^2 = y^2 - 10y + 25$

$$(A + B)^2 = A^2 + 2\ A\ B + B^2$$

13. $(2x + 3y)^2 = (2x)^2 + 2(2x)(3y) + (3y)^2 = 4x^2 + 12xy + 9y^2$

14. $(3x^2 + 5xy^2)^2 = (3x^2)^2 + 2(3x^2)(5xy^2) + (5xy^2)^2$
$$= 9x^4 + 30x^3y^2 + 25x^2y^4$$

15. $\left(\tfrac{1}{2}a^2 - b^3\right)^2 = \left(\tfrac{1}{2}a^2\right)^2 - 2\left(\tfrac{1}{2}a^2\right)(b^3) + (b^3)^2 = \tfrac{1}{4}a^4 - a^2b^3 + b^6$

Do Exercises 15–18.

d   Products of Sums and Differences

Another special case of a product of two binomials is the product of a sum and a difference. Note the following:

$$
\begin{array}{cccc}
F & O & I & L \\
\downarrow & \downarrow & \downarrow & \downarrow
\end{array}
$$
$$(A + B)(A - B) = A^2 - AB + AB - B^2 = A^2 - B^2$$

> The product of the sum and the difference of the same two terms is the square of the first term minus the square of the second term (the difference of their squares).
>
> $$(A + B)(A - B) = A^2 - B^2$$ This is called a **difference of squares.**

Examples   Multiply. (Say the rule as you work.)

$$(A + B)(A - B) = A^2 - B^2$$

16. $(y + 5)(y - 5) = y^2 - 5^2 = y^2 - 25$

17. $(2xy^2 + 3x)(2xy^2 - 3x) = (2xy^2)^2 - (3x)^2 = 4x^2y^4 - 9x^2$

18. $(0.2t - 1.4m)(0.2t + 1.4m) = (0.2t)^2 - (1.4m)^2 = 0.04t^2 - 1.96m^2$

19. $\left(\tfrac{2}{3}n - m^2\right)\left(\tfrac{2}{3}n + m^2\right) = \left(\tfrac{2}{3}n\right)^2 - (m^2)^2 = \tfrac{4}{9}n^2 - m^4$

Do Exercises 19–22.

Examples   Multiply.

20. $(\ 5y + 4\ + 3x)(\ 5y + 4\ - 3x) = (\ 5y + 4\ )^2 - (3x)^2$
$$= 25y^2 + 40y + 16 - 9x^2$$

Here we treat the binomial $5y + 4$ as the first expression, A, and $3x$ as the second, B.

21. $(3xy^2 + 4y)(-3xy^2 + 4y) = -(3xy^2)^2 + (4y)^2 = 16y^2 - 9x^2y^4$

Do Exercises 23 and 24 on the following page.

Try to multiply polynomials mentally, even when several types are mixed. First check to see what types of polynomials are to be multiplied. Then use the quickest method. Sometimes we might use more than one method. Remember that FOIL *always* works for multiplying binomials!

Example 22 Multiply: $(a - 5b)(a + 5b)(a^2 - 25b^2)$.

We first note that $a - 5b$ and $a + 5b$ can be multiplied using the rule $(A - B)(A + B) = A^2 - B^2$. Then we square, using $(A - B)^2 = A^2 - 2AB + B^2$.

$$
\begin{aligned}
(a - 5b)(a + 5b)(a^2 - 25b^2) &= (a^2 - 25b^2)(a^2 - 25b^2) \\
&= (a^2 - 25b^2)^2 \\
&= (a^2)^2 - 2(a^2)(25b^2) + (25b^2)^2 \\
&= a^4 - 50a^2b^2 + 625b^4
\end{aligned}
$$

Do Exercise 25.

e | Using Function Notation

AG Algebraic–Graphical Connection

Let's stop for a moment and look back at what we have done in this section. We have shown, for example, that

$$(x - 2)(x + 2) = x^2 - 4,$$

that is, $x^2 - 4$ and $(x - 2)(x + 2)$ are equivalent expressions.
From the viewpoint of functions, if

$$f(x) = x^2 - 4$$

and

$$g(x) = (x - 2)(x + 2),$$

then for any given input x, the outputs $f(x)$ and $g(x)$ are identical. Thus the graphs of these functions are identical and we say that f and g represent the same function. Functions like these are graphed in detail in Chapter 7.

x	$f(x)$	$g(x)$
3	5	5
2	0	0
1	-3	-3
0	-4	-4
-1	-3	-3
-2	0	0
-3	5	5

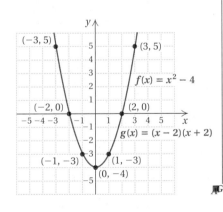

Multiply.

23. $(7x^2y + 2y)(-2y + 7x^2y)$

24. $(2x + 3 - 5y)(2x + 3 + 5y)$

25. Multiply:

$(3x + 2y)(3x - 2y)(9x^2 + 4y^2)$.

Answers on page A-25

26. Given $f(x) = x^2 + 2x - 7$, find and simplify $f(a + 1)$ and $f(a + h) - f(a)$.

Our work with multiplying can be used when evaluating functions.

Example 23 Given $f(x) = x^2 - 4x + 5$, find and simplify $f(a + 3)$ and $f(a + h) - f(a)$.

To find $f(a + 3)$, we replace x with $a + 3$. Then we simplify:

$$f(a + 3) = (a + 3)^2 - 4(a + 3) + 5$$
$$= a^2 + 6a + 9 - 4a - 12 + 5 = a^2 + 2a + 2.$$

To find $f(a + h) - f(a)$, we replace x with $a + h$ for $f(a + h)$ and x with a for $f(a)$. Then we simplify:

$$f(a + h) - f(a) = [(a + h)^2 - 4(a + h) + 5] - [a^2 - 4a + 5]$$
$$= a^2 + 2ah + h^2 - 4a - 4h + 5 - a^2 + 4a - 5$$
$$= 2ah + h^2 - 4h.$$

Answer on page A-25

Do Exercise 26.

Improving Your Math Study Skills

A Checklist of Your Study Skills

You are now about halfway through this textbook as well as the course. How are you doing? If you are struggling, we might ask if you are making full use of the study skills that we have suggested in these inserts. To determine this, review the following list of all study skill suggestions made so far and answer the questions "yes" or "no."

Study Skill Questions	Yes	No
1. Are you doing a thorough job of reading the book?		
2. Are you stopping and working the margin exercises when directed to do so?		
3. Are you doing your homework as soon as possible after class?		
4. Are you doing your homework at a specified time and in a quiet setting?		
5. Have you found a study group in which to work?		
6. Are you consistently trying to apply the five-step problem-solving strategy when working applied problems?		
7. Are you asking questions in class and in tutoring sessions?		

Study Skill Questions	Yes	No
8. Are you doing lots of even-numbered exercises for which answers are not available?		
9. Are you keeping one section ahead of your syllabus?		
10. Are you using the book supplements, such as the *Student's Solutions Manual* and the *InterAct Math Tutorial Software*?		
11. When you study the book, are you marking the points that you do not understand as a source for in-class questions?		
12. Are you reading and studying each step of each example?		
13. Are you using the objective code symbols (a , b , c , etc.) that appear at the beginning of each section, throughout the section, and in the exercise sets, the summary–reviews, and the answers for the chapter tests?		

If you have answered "no" seven or more times and are struggling in the course, you need to improve your study skills by following more of these suggestions.

A consultation with your instructor regarding your situation is strongly advised.

Exercise Set 4.2

a Multiply.

1. $8y^2 \cdot 3y$

2. $-5x^2 \cdot 6xy$

3. $2x(-10x^2y)$

4. $-7ab^2(4a^2b^2)$

5. $(5x^5y^4)(-2xy^3)$

6. $(2a^2bc^2)(-3ab^5c^4)$

7. $2z(7 - x)$

8. $4a(a^2 - 3a)$

9. $6ab(a + b)$

10. $2xy(2x - 3y)$

11. $5cd(3c^2d - 5cd^2)$

12. $a^2(2a^2 - 5a^3)$

13. $(5x + 2)(3x - 1)$

14. $(2a - 3b)(4a - b)$

15. $(s + 3t)(s - 3t)$

16. $(y + 4)(y - 4)$

17. $(x - y)(x - y)$

18. $(a + 2b)(a + 2b)$

19. $(x^3 + 8)(x^3 - 5)$

20. $(2x^4 - 7)(3x^3 + 5)$

21. $(a^2 - 2b^2)(a^2 - 3b^2)$

22. $(2m^2 - n^2)(3m^2 - 5n^2)$

23. $(x - 4)(x^2 + 4x + 16)$

24. $(y + 3)(y^2 \quad 3y + 9)$

25. $(x + y)(x^2 - xy + y^2)$

26. $(a - b)(a^2 + ab + b^2)$

27. $(a^2 + a - 1)(a^2 + 4a - 5)$

28. $(x^2 - 2x + 1)(x^2 + x + 2)$

29. $(4a^2b - 2ab + 3b^2)(ab - 2b + a)$

30. $(2x^2 + y^2 - 2xy)(x^2 - 2y^2 - xy)$

31. $\left(x + \frac{1}{4}\right)\left(x + \frac{1}{4}\right)$

32. $\left(b - \frac{1}{3}\right)\left(b - \frac{1}{3}\right)$

33. $\left(\frac{1}{2}x - \frac{2}{3}\right)\left(\frac{1}{4}x + \frac{1}{3}\right)$

34. $\left(\frac{2}{3}a + \frac{1}{6}b\right)\left(\frac{1}{3}a - \frac{5}{6}b\right)$

35. $(1.3x - 4y)(2.5x + 7y)$

36. $(40a - 0.24b)(0.3a + 10b)$

Multiply.

37. $(a + 8)(a + 5)$

38. $(x + 2)(x + 3)$

39. $(y + 7)(y - 4)$

40. $(y - 2)(y + 3)$

41. $\left(3a + \frac{1}{2}\right)^2$

42. $\left(2x - \frac{1}{3}\right)^2$

43. $(x - 2y)^2$

44. $(2s + 3t)^2$

45. $\left(b - \frac{1}{3}\right)\left(b - \frac{1}{2}\right)$

46. $\left(x - \frac{1}{2}\right)\left(x - \frac{1}{4}\right)$

47. $(2x + 9)(x + 2)$

48. $(3b + 2)(2b - 5)$

49. $(20a - 0.16b)^2$

50. $(10p^2 + 2.3q)^2$

51. $(2x - 3y)(2x + y)$

52. $(2a - 3b)(2a - b)$

53. $(x^3 + 2)^2$

54. $(y^4 - 7)^2$

55. $(2x^2 - 3y^2)^2$

56. $(3s^2 + 4t^2)^2$

57. $(a^3b^2 + 1)^2$

58. $(x^2y - xy^3)^2$

59. $(0.1a^2 - 5b)^2$

60. $(6p + 0.45q^2)^2$

61. *Compound Interest.* Suppose that P dollars is invested in a savings account at interest rate i, compounded annually, for 2 yr. The amount A in the account after 2 yr is given by

$$A = P(1 + i)^2.$$

Find an equivalent expression for A.

62. *Compound Interest.* Suppose that P dollars is invested in a savings account at interest rate i, compounded semiannually, for 1 yr. The amount A in the account after 1 yr is given by

$$A = P\left(1 + \frac{i}{2}\right)^2.$$

Find an equivalent expression for A.

d Multiply.

63. $(d + 8)(d - 8)$

64. $(y - 3)(y + 3)$

65. $(2c + 3)(2c - 3)$

66. $(1 - 2x)(1 + 2x)$

67. $(6m - 5n)(6m + 5n)$

68. $(3x + 7y)(3x - 7y)$

69. $(x^2 + yz)(x^2 - yz)$

70. $(2a^2 + 5ab)(2a^2 - 5ab)$

71. $(-mn + m^2)(mn + m^2)$

72. $(1.6 + pq)(-1.6 + pq)$

73. $(-3pq + 4p^2)(4p^2 + 3pq)$

74. $(-10xy + 5x^2)(5x^2 + 10xy)$

75. $\left(\frac{1}{2}p - \frac{2}{3}q\right)\left(\frac{1}{2}p + \frac{2}{3}q\right)$

76. $\left(\frac{3}{5}ab + 4c\right)\left(\frac{3}{5}ab - 4c\right)$

77. $(x + 1)(x - 1)(x^2 + 1)$

78. $(y - 2)(y + 2)(y^2 + 4)$

79. $(a - b)(a + b)(a^2 - b^2)$

80. $(2x - y)(2x + y)(4x^2 - y^2)$

81. $(a + b + 1)(a + b - 1)$

82. $(m + n + 2)(m + n - 2)$

83. $(2x + 3y + 4)(2x + 3y - 4)$

84. $(3a - 2b + c)(3a - 2b - c)$

85. Given $f(x) = 5x + x^2$, find and simplify.

  a) $f(t - 1)$

  b) $f(a + h) - f(a)$

86. Given $f(x) = 4 + 3x - x^2$, find and simplify.

  a) $f(p + 1)$

  b) $f(a + h) - f(a)$

87. Given $f(x) = 2 - 4x - 3x^2$, find and simplify.

  a) $f(a + 2)$

  b) $f(a + h) - f(a)$

88. Given $f(x) = 4 - x^2$, find and simplify.

  a) $f(t - 3)$

  b) $f(a + h) - f(a)$

Skill Maintenance

Solve. [3.3b]

89. *Auto Travel.* Rachel leaves on a business trip, forgetting her briefcase. Her sister discovers Rachel's briefcase 2 hr later, and knowing that Rachel needs its contents for her sales presentation and that Rachel normally travels at a speed of 55 mph, she decides to follow her at a speed of 75 mph. After how long will Rachel's sister catch up with her?

90. *Air Travel.* An airplane flew for 5 hr against a 20-mph headwind. The return trip with the wind took 4 hr. Find the speed of the plane in still air.

Solve. [3.2a, b]

91. $5x + 9y = 2,$
  $4x - 9y = 10$

92. $x + 4y = 13,$
  $5x - 7y = -16$

93. $2x - 3y = 1,$
  $4x - 6y = 2$

94. $9x - 8y = -2,$
  $3x + 2y = 3$

Synthesis

95. ◆ Find two binomials whose product is $x^2 - 9$ and explain how you decided on those two binomials.

96. ◆ Find two binomials whose product is $x^2 - 6x + 9$ and explain how you decided on those two binomials.

97. ⟋⟍ Use the TABLE and GRAPH features of a grapher to check your answers to Exercises 28, 40, and 77.

98. ⟋⟍ Use the TABLE and GRAPH features of a grapher to determine whether each of the following is correct.

  a) $(x - 1)^2 = x^2 - 1$

  b) $(x - 2)(x + 3) = x^2 + x - 6$

  c) $(x - 1)^3 = x^3 - 3x^2 + 3x - 1$

  d) $(x + 1)^4 = x^4 + 1$

Multiply. Assume that variables in exponents represent natural numbers.

99. $(z^{n^2})^{n^3}(z^{4n^3})^{n^2}$

100. $y^3 z^n (y^{3n} z^3 - 4yz^{2n})$

101. $(r^2 + s^2)^2 (r^2 + 2rs + s^2)(r^2 - 2rs + s^2)$

102. $(y - 1)^6 (y + 1)^6$

103. $\left(3x^5 - \frac{5}{11}\right)^2$

104. $(4x^2 + 2xy + y^2)(4x^2 - 2xy + y^2)$

105. $(x^a + y^b)(x^a - y^b)(x^{2a} + y^{2b})$

106. $\left(x - \frac{1}{7}\right)\left(x^2 + \frac{1}{7}x + \frac{1}{49}\right)$

107. $(x - 1)(x^2 + x + 1)(x^3 + 1)$

108. $(x^{a-b})^{a+b}$

Derive the formulas for the squares of binomials.

4.3 Introduction to Factoring

Factoring is the reverse of multiplication. To **factor** an expression is to find an equivalent expression that is a product. For example, reversing a type of multiplication we have considered, we know that

$$x^2 - 9 = (x + 3)(x - 3).$$

We say that $x + 3$ and $x - 3$ are **factors** of $x^2 - 9$ and that $(x + 3)(x - 3)$ is a **factorization**.

> To **factor** a polynomial is to express it as a product.
>
> A **factor** of a polynomial P is a polynomial that can be used to express P as a product.
>
> A **factorization** of a polynomial P is an expression that names P as a product.

CAUTION! Be careful not to confuse terms with factors! The terms of $x^2 - 9$ are x^2 and -9. Terms are used to form sums. Factors of $x^2 - 9$ are $x - 3$ and $x + 3$. Factors are used to form products.

Do Exercise 1.

a Terms with Common Factors

To factor polynomials quickly, we consider the special-product rules, but we first factor out the largest common factor.

To multiply a monomial and a polynomial with more than one term, we multiply each term by the monomial using the distributive laws. To factor, we do the reverse. We express a polynomial as a product using the distributive laws in reverse. Compare.

Multiply

$5x(x^2 - 3x + 1)$
$= 5x \cdot x^2 - 5x \cdot 3x + 5x \cdot 1$
$= 5x^3 - 15x^2 + 5x$

Factor

$5x^3 - 15x^2 + 5x$
$= 5x \cdot x^2 - 5x \cdot 3x + 5x \cdot 1$
$= 5x(x^2 - 3x + 1)$

Example 1 Factor: $4y^2 - 8$.

$4y^2 - 8 = 4 \cdot y^2 - 4 \cdot 2$ **4 is the largest common factor.**

$ = 4(y^2 - 2)$ **Factoring out the common factor 4**

In some cases, there is more than one common factor. In Example 2 below, for instance, 5 is a common factor, x^3 is a common factor, and $5x^3$ is a common factor. If there is more than one common factor, we generally choose the one with the largest coefficient and the largest exponent.

Examples Factor.

2. $5x^4 - 20x^3 = 5x^3(x - 4)$ **Try to write your answer directly. Multiply mentally to check your answer.**

3. $12x^2y - 20x^3y = 4x^2y(3 - 5x)$

4. $10p^6q^2 - 4p^5q^3 + 2p^4q^4 = 2p^4q^2(5p^2 - 2pq + q^2)$

---

1. Consider

$$x^2 - 4x - 5 = (x - 5)(x + 1).$$

a) What are the factors of $x^2 - 4x - 5$?

b) What are the terms of $x^2 - 4x - 5$?

Factor.

2. $3x^2 - 6$

3. $4x^5 - 8x^3$

4. $9y^4 - 15y^3 + 3y^2$

5. $6x^2y - 21x^3y^2 + 3x^2y^3$

Answers on page A-25

Factor out a common factor with a negative coefficient.

6. $-8x + 32$

7. $-3x^2 - 15x + 9$

Answers on page A-25

The polynomials in Examples 1–4 have been **factored completely.** They cannot be factored further. The factors in the resulting factorization are said to be **prime polynomials.**

Do Exercises 2–5 on the preceding page.

When a factor contains more than one term, it is usually desirable for the leading coefficient to be positive. To achieve this may require factoring out a common factor with a negative coefficient.

Examples Factor out a common factor with a negative coefficient.

5. $-4x - 24 = -4(x + 6)$

6. $-2x^2 + 6x - 10 = -2(x^2 - 3x + 5)$

Do Exercises 6 and 7.

Example 7 *Height of a Thrown Object.* Suppose that a softball is thrown upward with an initial velocity of 64 ft/sec. Its height h, in feet, after t seconds is given by the function

$$h(t) = -16t^2 + 64t.$$

a) Find an equivalent expression for $h(t)$ by factoring out a common factor with a negative coefficient.

b) Check your factoring by evaluating both expressions for $h(t)$ at $t = 1$.

a) We factor out $-16t$ as follows:

$$h(t) = -16t^2 + 64t = -16t(t - 4).$$

b) We check as follows:

$$h(1) = -16 \cdot 1^2 + 64 \cdot 1 = 48;$$

$$h(1) = -16 \cdot 1(1 - 4) = 48. \qquad \text{Using the factorization}$$

Do Exercise 8 on the following page.

b | Factoring by Grouping

In expressions of four or more terms, there may be a *common binomial factor.* We proceed as in the following examples.

Example 8 Factor: $(a - b)(x + 5) + (a - b)(x - y^2)$.

$$(a - b)(x + 5) + (a - b)(x - y^2) = (a - b)[(x + 5) + (x - y^2)]$$
$$= (a - b)(2x + 5 - y^2)$$

Do Exercises 9 and 10 on the following page.

In Example 9, we factor two parts of the expression. Then we factor as in Example 8.

Example 9 Factor: $y^3 + 3y^2 + 4y + 12$.

$$
\begin{aligned}
y^3 + 3y^2 + 4y + 12 &= (y^3 + 3y^2) + (4y + 12) \quad &\text{Grouping}\\
&= y^2(y + 3) + 4(y + 3) \quad &\text{Factoring each binomial}\\
&= (y^2 + 4)(y + 3) \quad &\text{Factoring out the common factor } y + 3
\end{aligned}
$$

Example 10 Factor: $3x^3 - 6x^2 - x + 2$.

$$
\begin{aligned}
3x^3 - 6x^2 - x + 2 &= (3x^3 - 6x^2) + (-x + 2)\\
&= 3x^2(x - 2) - 1(x - 2) \quad &\text{Check: } -1(x - 2) = -x + 2\\
&= (3x^2 - 1)(x - 2) \quad &\text{Factoring out the common factor } x - 2
\end{aligned}
$$

Example 11 Factor: $4x^3 - 15 + 20x^2 - 3x$.

$$
\begin{aligned}
4x^3 - 15 + 20x^2 - 3x &= 4x^3 + 20x^2 - 3x - 15 \quad &\text{Rearranging}\\
&= 4x^2(x + 5) - 3(x + 5) \quad &\text{Check: } -3(x + 5) = -3x - 15\\
&= (4x^2 - 3)(x + 5) \quad &\text{Factoring out the common factor } x + 5
\end{aligned}
$$

Not all polynomials with four terms can be factored by grouping. An example is

$$x^3 + x^2 + 3x - 3.$$

Note that neither $x^2(x + 1) + 3(x - 1)$ nor $x(x^2 + 3) + (x^2 - 3)$ nor any other groupings allow us to factor out a common binomial.

Do Exercises 11 and 12 on the following page.

c Factoring Trinomials: $x^2 + bx + c$

We now consider factoring trinomials of the type $x^2 + bx + c$. We use a refined trial-and-error process that is based on the FOIL method.

Constant Term Positive

Recall the FOIL method of multiplying two binomials:

$$
\begin{array}{cccc}
\text{F} & \text{O} & \text{I} & \text{L}\\
(x + 3)(x + 5) = x^2 & + 5x & + 3x & + 15\\
= x^2 & & + 8x & + 15.
\end{array}
$$

The product is a trinomial. In this example, the leading term has a coefficient of 1. The constant term is positive. To factor $x^2 + 8x + 15$, we think of FOIL in reverse. We multiplied x times x to get the first term of the trinomial. Thus the first term of each binomial factor is x. We want to find numbers p and q such that

$$x^2 + 8x + 15 = (x + p)(x + q).$$

---

8. *Height of a Rocket.* A model rocket is launched upward with an initial velocity of 96 ft/sec. Its height h, in feet, after t seconds is given by the function

$$h(t) = -16t^2 + 96t.$$

a) Find an equivalent expression for $h(t)$ by factoring out a common factor with a negative coefficient.

b) Check your factoring by evaluating both expressions for $h(t)$ at $t = 2$.

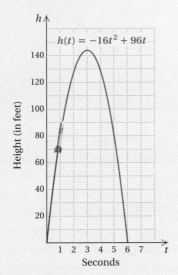

Factor.

9. $(p + q)(x + 2) + (p + q)(x + y)$

10. $(y + 3)(y - 21) + (y + 3)(y + 10)$

Answers on page A-25

Factor by grouping.

11. $5y^3 + 2y^2 - 10y - 4$

12. $x^3 + 5x^2 + 4x - 20$

Factor. Check by multiplying.

13. $x^2 + 5x + 6$

14. $y^2 + 7y + 10$

Calculator Spotlight

Use the TABLE and GRAPH features to check the results of Example 12 and Margin Exercises 13 and 14.

Factor.

15. $m^2 - 8m + 12$

16. $24 - 11t + t^2$

Answers on page A-25

To get the middle term and the last term of the trinomial, we look for two numbers whose product is 15 and whose sum is 8. Those numbers are 3 and 5. Thus the factorization is

$$(x + 3)(x + 5), \quad \text{or} \quad (x + 5)(x + 3)$$

by the commutative law of multiplication. In general,

$$(x + p)(x + q) = x^2 + (p + q)x + pq.$$

To factor, we can use this equation in reverse.

Example 12 Factor: $x^2 + 9x + 8$.

Think of FOIL in reverse. The first term of each factor is x. We are looking for numbers p and q such that

$$x^2 + 9x + 8 = (x + p)(x + q) = x^2 + (p + q)x + pq$$

We look for two numbers p and q whose product is 8 and whose sum is 9. Since both 8 and 9 are positive, we need consider only positive factors.

Pairs of Factors	Sums of Factors	
2, 4	6	
1, 8	9 ←	The numbers we need are 1 and 8.

The factorization is $(x + 1)(x + 8)$. We can check by multiplying.

Do Exercises 13 and 14.

> When the constant term of a trinomial is positive, we look for two factors with the same sign (both positive or both negative). The sign is that of the middle term.

Example 13 Factor: $y^2 - 9y + 20$.

Since the constant term, 20, is positive and the coefficient of the middle term, -9, is negative, we look for a factorization of 20 in which both factors are negative. Their sum must be -9.

Pairs of Factors	Sums of Factors	
$-1, -20$	-21	
$-2, -10$	-12	
$-4, -5$	-9 ←	The numbers we need are -4 and -5.

The factorization is $(y - 4)(y - 5)$.

Do Exercises 15 and 16.

Constant Term Negative

> When the constant term of a trinomial is negative, we look for two factors whose product is negative. One of them must be positive and the other negative. Their sum must be the coefficient of the middle term.

Example 14 Factor: $x^3 - x^2 - 30x$.

Always look first for the largest common factor. This time x is the common factor. We first factor it out:

$$x^3 - x^2 - 30x = x(x^2 - x - 30).$$

Now consider $x^2 - x - 30$. Since the constant term, -30, is negative, we look for a factorization of -30 in which one factor is positive and one factor is negative. The sum of the factors must be -1, the coefficient of the middle term, so the negative factor must have the larger absolute value. Thus we consider only pairs of factors in which the negative factor has the larger absolute value.

Pairs of Factors	Sums of Factors
1, −30	−29
2, −15	−13
3, −10	−7
5, −6	−1 ← The numbers we want are 5 and −6.

The factorization of $x^2 - x - 30$ is $(x + 5)(x - 6)$. But do not forget the common factor! The factorization of the original trinomial is $x(x + 5)(x - 6)$.

Do Exercises 17–19.

Example 15 Factor: $x^2 + 17x - 110$.

Since the constant term, -110, is negative, we look for a factorization of -110 in which one factor is positive and one factor is negative. Their sum must be 17, so the positive factor must have the larger absolute value.

Pairs of Factors	Sums of Factors	
−1, 110	109	We consider only pairs of factors in which the positive term has the larger absolute value.
−2, 55	53	
−5, 22	17 ←	The numbers we need are −5 and 22.
−10, 11	1	

The factorization is $(x - 5)(x + 22)$.

Do Exercises 20–22.

Some trinomials are not factorable.

Example 16 Factor: $x^2 - x - 7$.

There are no factors of -7 whose sum is -1. This trinomial is *not* factorable into binomials.

Do Exercise 23.

---

17. a) Factor $x^2 - x - 20$.

b) Explain why you would not consider these pairs of factors in factoring $x^2 - x - 20$.

Pairs of Factors	Products of Factors
1, 20	
2, 10	
4, 5	
−1, −20	
−2, −10	
−4, −5	

Factor.

18. $x^3 - 3x^2 - 54x$

19. $2x^3 - 2x^2 - 84x$

Factor.

20. $x^3 + 4x^2 - 12x$

21. $y^2 - 4y - 12$

22. $x^2 - 110 - x$

23. Factor: $x^2 + x - 5$.

Answers on page A-26

Factor.

24. $x^2 - 5xy + 6y^2$

25. $p^2 - 6pq - 16q^2$

Factor.

26. $x^4 - 9x^2 + 14$

27. $p^6 + p^3 - 6$

Answers on page A-26

To factor $x^2 + bx + c$:

1. First arrange in descending order.
2. Use a trial-and-error procedure that looks for factors of c whose sum is b.
   - If c is positive, then the signs of the factors are the same as the sign of b.
   - If c is negative, then one factor is positive and the other is negative. (If the sum of the two factors is the opposite of b, changing the signs of each factor will give the desired factors whose sum is b.)
3. Check your result by multiplying.

The procedure considered here can also be applied to a trinomial with more than one variable.

Example 17   Factor: $x^2 - 2xy - 48y^2$.

We look for numbers p and q such that

$$x^2 - 2xy - 48y^2 = (x + py)(x + qy).$$

Our thinking is much the same as if we were factoring $x^2 - 2x - 48$. We look for factors of -48 whose sum is -2. Those factors are 6 and -8. Then

$$x^2 - 2xy - 48y^2 = (x + 6y)(x - 8y).$$

We can check by multiplying.

Do Exercises 24 and 25.

Sometimes a trinomial like $x^4 + 2x^2 - 15$ can be factored using the following method. We can first think of the trinomial as $(x^2)^2 + 2x^2 - 15$, or we can make a substitution (perhaps, just mentally), letting $u = x^2$. Then the trinomial becomes

$$u^2 + 2u - 15.$$

We factor this trinomial and if a factorization is found, we replace each occurrence of u with x^2.

Example 18   Factor: $x^4 + 2x^2 - 15$.

We let $u = x^2$. Then consider $u^2 + 2u - 15$. The constant term is negative and the middle term is positive. Thus we look for pairs of factors of -15, one positive and one negative, such that the positive factor has the larger absolute value and the sum of the factors is 2.

Pairs of Factors	Sums of Factors
-1, 15	14
-3, 5	2 ←

The numbers we need are -3 and 5.

The desired factorization of $u^2 + 2u - 15$ is

$$(u - 3)(u + 5).$$

Replacing u with x^2, we obtain the following factorization of the original trinomial:

$$(x^2 - 3)(x^2 + 5).$$

Do Exercises 26 and 27.

Exercise Set 4.3

a Factor.

1. $6a^2 + 3a$

2. $4x^2 + 2x$

3. $x^3 + 9x^2$

4. $y^3 + 8y^2$

5. $8x^2 - 4x^4$

6. $6x^2 + 3x^4$

7. $4x^2y - 12xy^2$

8. $5x^2y^3 + 15x^3y^2$

9. $3y^2 - 3y - 9$

10. $5x^2 - 5x + 15$

11. $4ab - 6ac + 12ad$

12. $8xy + 10xz - 14xw$

13. $10a^4 + 15a^2 - 25a - 30$

14. $12t^5 - 20t^4 + 8t^2 - 16$

Factor out a common factor with a negative coefficient.

15. $-5x - 45$

16. $-3t + 18$

17. $-6a - 84$

18. $-8t + 40$

19. $-2x^2 + 2x - 24$

20. $-2x^2 + 16x - 20$

21. $-3y^2 + 24y$

22. $-7x^2 - 56y$

23. $-3y^3 + 12y^2 - 15y + 24$

24. $-4m^4 - 32m^3 + 64m - 12$

25. *Counting Spheres in a Pile.* The number N of spheres in a triangular pile like the one shown here is given by the polynomial function
$$N(x) = \frac{1}{6}x^3 + \frac{1}{2}x^2 + \frac{1}{3}x,$$
where x is the number of layers and $N(x)$ is the number of spheres. Find an equivalent expression for $N(x)$ by factoring out a common factor.

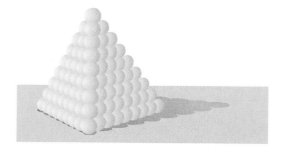

26. *Surface Area of a Silo.* A silo is a structure that is shaped like a right circular cylinder with a half sphere on top. The surface area of a silo of height h and radius r (including the area of the base) is given by the polynomial $2\pi rh + \pi r^2$. Find an equivalent expression by factoring out a common factor.

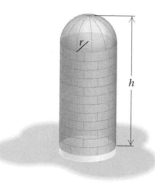

27. *Height of a Baseball.* A baseball is popped up with an upward velocity of 72 ft/sec. Its height h, in feet, after t seconds is given by

$$h(t) = -16t^2 + 72t.$$

a) Find an equivalent expression for $h(t)$ by factoring out a common factor with a negative coefficient.

b) Perform a partial check of part (a) by evaluating both expressions for $h(t)$ at $t = 2$.

28. *Number of Diagonals.* The number of diagonals of a polygon having n sides is given by the polynomial function

$$P(n) = \tfrac{1}{2}n^2 - \tfrac{3}{2}n.$$

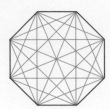

Find an equivalent expression for $P(n)$ by factoring out a common factor.

29. *Total Revenue.* Household Sound is marketing a new kind of stereo. The firm determines that when it sells x stereos, the total revenue R is given by the polynomial function

$$R(x) = 280x - 0.4x^2 \text{ dollars.}$$

Find an equivalent expression for $R(x)$ by factoring out $0.4x$.

30. *Total Cost.* Household Sound determines that the total cost C of producing x stereos is given by the polynomial function

$$C(x) = 0.18x + 0.6x^2.$$

Find an equivalent expression for $C(x)$ by factoring out $0.6x$.

b Factor.

31. $a(b - 2) + c(b - 2)$

32. $a(x^2 - 3) - 2(x^2 - 3)$

33. $(x - 2)(x + 5) + (x - 2)(x + 8)$

34. $(m - 4)(m + 3) + (m - 4)(m - 3)$

35. $a^2(x - y) + a^2(x - y)$

36. $3x^2(x - 6) + 3x^2(x - 6)$

37. $ac + ad + bc + bd$

38. $xy + xz + wy + wz$

39. $b^3 - b^2 + 2b - 2$

40. $y^3 - y^2 + 3y - 3$

41. $y^3 - 8y^2 + y - 8$

42. $t^3 + 6t^2 - 2t - 12$

43. $24x^3 - 36x^2 + 72x - 108$

44. $10a^4 + 15a^2 - 25a - 30$

45. $a^4 - a^3 + a^2 + a$

46. $p^6 + p^5 - p^3 + p^2$

47. $2y^4 + 6y^2 + 5y^2 + 15$

48. $2xy + x^2y - 6 - 3x$

c Factor.		
49. $x^2 + 13x + 36$	**50.** $x^2 + 9x + 18$	**51.** $t^2 - 8t + 15$
52. $y^2 - 10y + 21$	**53.** $x^2 - 8x - 33$	**54.** $t^2 - 15 - 2t$
55. $2y^2 - 16y + 32$	**56.** $2a^2 - 20a + 50$	**57.** $p^2 + 3p - 54$
58. $m^2 + m - 72$	**59.** $12x + x^2 + 27$	**60.** $10y + y^2 + 24$
61. $y^2 - \dfrac{2}{3}y + \dfrac{1}{9}$	**62.** $p^2 + \dfrac{2}{5}p + \dfrac{1}{25}$	**63.** $t^2 - 4t + 3$
64. $y^2 - 14y + 45$	**65.** $5x + x^2 - 14$	**66.** $x + x^2 - 90$
67. $x^2 + 5x + 6$	**68.** $y^2 + 8y + 7$	**69.** $56 + x - x^2$
70. $32 + 4y - y^2$	**71.** $32y + 4y^2 - y^3$	**72.** $56x + x^2 - x^3$
73. $x^4 + 11x^2 - 80$	**74.** $y^4 + 5y^2 - 84$	**75.** $x^2 - 3x + 7$
76. $x^2 + 12x + 13$	**77.** $x^2 + 12xy + 27y^2$	**78.** $p^2 - 5pq - 24q^2$

79. $x^4 + 50x^2 + 49$

80. $p^4 + 80p^2 + 79$

81. $x^6 + 11x^3 + 18$

82. $x^6 - x^3 - 42$

83. $x^8 - 11x^4 + 24$

84. $x^8 - 7x^4 + 10$

Skill Maintenance

Solve. [3.3a]

85. *Mixing Rice.* Countryside Rice is 90% white rice and 10% wild rice. Mystic Rice is 50% wild rice. How much of each type should be used to create a 25-lb batch of rice that is 35% wild rice?

86. *Wages.* Takako worked a total of 17 days last month at her father's restaurant. She earned $50 a day during the week and $60 a day during the weekend. Last month Takako earned $940. How many weekdays did she work?

Determine whether each of the following is the graph of a function. [2.2d]

87.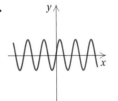

88.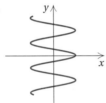

89.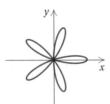

90.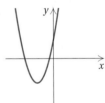

Find the domain of *f*. [2.3a]

91. $f(x) = x^2 - 2$

92. $f(x) = 3 - 2x$

93. $f(x) = \dfrac{3}{4x - 7}$

94. $f(x) = 3 - |x|$

Synthesis

95. ◆ Under what conditions would it be easier to evaluate a polynomial function after it has been factored?

96. ◆ Checking the factorization of a second-degree polynomial by making a single replacement is only a *partial* check. Write an *incorrect* factorization and explain how evaluating both the polynomial and the factorization might not catch a possible error.

97. Find all integers *m* for which $x^2 + mx + 75$ can be factored.

98. Find all integers *q* for which $x^2 + qx - 32$ can be factored.

99. One of the factors of $x^2 - 345x - 7300$ is $x + 20$. Find the other factor.

100. Use a grapher to show that
$$(x^2 - 3x + 2)^4 = x^8 + 81x^4 + 16$$
is *not* correct.

101. Use the TABLE and GRAPH features of a grapher to check your answers to Exercises 42, 49, and 54.

102. Factor: $7y^{2a+b} - 5y^{a+b} + 3y^{a+2b}$.

4.4 Factoring Trinomials: $ax^2 + bx + c$, $a \neq 1$

Now we learn to factor trinomials of the type $ax^2 + bx + c$, $a \neq 1$. We use two methods: the grouping method and the FOIL method. Although one is discussed before the other, this should not be taken as a recommendation of one form over the other.

Objectives

a Factor trinomials of the type $ax^2 + bx + c$, $a \neq 1$, by the grouping method.

b Factor trinomials of the type $ax^2 + bx + c$, $a \neq 1$, by the FOIL method.

For Extra Help

TAPE 9   TAPE 8B   MAC   CD-ROM
                   WIN

a The Grouping Method

The first method of factoring trinomials of the type $ax^2 + bx + c$, $a \neq 1$, is known as the **grouping method.** It involves not only trial and error and FOIL, but also factoring by grouping.

We know that

$$x^2 + 7x + 10 = x^2 + 2x + 5x + 10$$
$$= x(x + 2) + 5(x + 2)$$
$$= (x + 5)(x + 2),$$

but what if the leading coefficient is not 1? Consider $6x^2 + 23x + 20$. The method for factoring this trinomial is similar to what we just did with $x^2 + 7x + 10$.

To factor $ax^2 + bx + c$, $a \neq 1$, using the **grouping method**:

1. Factor out the largest common factor.
2. Multiply the leading coefficient a and the constant c.
3. Try to factor the product ac so that the sum of the factors is b. That is, find integers p and q such that $pq = ac$ and $p + q = b$.
4. Split the middle term. That is, write it as a sum using the factors found in step (3).
5. Factor by grouping.

Example 1   Factor: $6x^2 + 23x + 20$.

1. First, factor out a common factor, if any. There is none (other than 1 or -1).
2. Multiply the leading coefficient, 6, and the constant, 20: $6 \cdot 20 = 120$.
3. Then look for a factorization of 120 in which the sum of the factors is the coefficient of the middle term, 23.

Pairs of Factors	Sums of Factors		Pairs of Factors	Sums of Factors
1, 120	121		5, 24	29
2, 60	62		6, 20	26
3, 40	43		8, 15	23
4, 30	34		10, 12	22

4. Next, split the middle term as a sum or a difference using the factors found in step (3):

$$6x^2 + 23x + 20 = 6x^2 + 8x + 15x + 20.$$

Factor by the grouping method.

1. $4x^2 + 4x - 3$

2. $4x^2 + 37x + 9$

Factor by the grouping method.

3. $10y^4 - 7y^3 - 12y^2$

4. $36a^3 + 21a^2 + a$

Answers on page A-26

5. Factor by grouping as follows:

$$6x^2 + 23x + 20 = 6x^2 + 8x + 15x + 20$$
$$= 2x(3x + 4) + 5(3x + 4) \qquad \text{Factoring by grouping; see Section 4.3}$$
$$= (2x + 5)(3x + 4).$$

We could also split the middle term as $15x + 8x$. We still get the same factorization, although the factors are in a different order. Note the following:

$$6x^2 + 23x + 20 = 6x^2 + 15x + 8x + 20$$
$$= 3x(2x + 5) + 4(2x + 5)$$
$$= (3x + 4)(2x + 5).$$

Check by multiplying: $(3x + 4)(2x + 5) = 6x^2 + 23x + 20$.

Do Exercises 1 and 2.

Example 2   Factor: $6x^4 - 116x^3 - 80x^2$.

1. First, factor out the largest common factor, if any. The expression $2x^2$ is common to all three terms: $2x^2(3x^2 - 58x - 40)$.

2. Now, factor the trinomial $3x^2 - 58x - 40$. Multiply the leading coefficient, 3, and the constant, -40: $3(-40) = -120$.

3. Next, try to factor -120 so that the sum of the factors is -58.

Pairs of Factors	Sums of Factors
1, −120	−119
2, −60	−58
8, −15	−7
10, −12	−2

4. Split the middle term, $-58x$, as follows: $-58x = 2x - 60x$.

5. Factor by grouping:

$$3x^2 - 58x - 40 = 3x^2 + 2x - 60x - 40 \qquad \text{Substituting } 2x - 60x \text{ for } -58x$$
$$= x(3x + 2) - 20(3x + 2) \qquad \text{Factoring by grouping}$$
$$= (x - 20)(3x + 2).$$

The factorization of $3x^2 - 58x - 40$ is $(x - 20)(3x + 2)$. But don't forget the common factor! We must include it to get a factorization of the original trinomial:

$$6x^4 - 116x^3 - 80x^2 = 2x^2(x - 20)(3x + 2).$$

Do Exercises 3 and 4.

b | The FOIL Method

Now we consider the **FOIL method** for factoring trinomials of the type $ax^2 + bx + c, a \neq 1$. Consider the following multiplication.

$$
\begin{array}{ccccc}
& \text{F} & \text{O} & \text{I} & \text{L} \\
& \downarrow & \downarrow & \downarrow & \downarrow \\
(3x + 2)(4x + 5) = & 12x^2 & +\ 15x & +\ 8x & +\ 10 \\
& \downarrow & & \downarrow & \downarrow \\
= & 12x^2 & + & 23x & +\ 10
\end{array}
$$

To factor $12x^2 + 23x + 10$, we must reverse what we just did. We look for two binomials whose product is this trinomial. The product of the First terms must be $12x^2$. The product of the Outside terms plus the product of the Inside terms must be $23x$. The product of the Last terms must be 10. We know from the preceding discussion that the answer is

$$(3x + 2)(4x + 5).$$

In general, however, finding such an answer involves trial and error. We use the following method.

To factor trinomials of the type $ax^2 + bx + c, a \neq 1$, using the **FOIL method:**

1. Factor out the largest common factor.
2. Find two **First** terms whose product is ax^2:

$$(\blacksquare x + \quad)(\blacksquare x + \quad) = ax^2 + bx + c.$$

$$\underline{\hspace{4cm}} \text{FOIL}$$

3. Find two **Last** terms whose product is c:

$$(\ x + \blacksquare)(\ x + \blacksquare) = ax^2 + bx + c.$$

$$\underline{\hspace{4cm}} \text{FOIL}$$

4. Repeat steps (2) and (3) until a combination is found for which the sum of the **Outside** and **Inside** products is bx:

$$(\blacksquare x + \blacksquare)(\blacksquare x + \blacksquare) = ax^2 + bx + c.$$

$$\lfloor \text{I} \rfloor \qquad\qquad \text{FOIL}$$

$$\underline{\hspace{2cm}} \text{O} \underline{\hspace{2cm}}$$

Example 3   Factor: $3x^2 + 10x - 8$.

1. First, we factor out the largest common factor, if any. There is none (other than 1 or -1).
2. Next, we factor the first term, $3x^2$. The only possibility is $3x \cdot x$. The desired factorization is then of the form $(3x + \ )(x + \ )$.
3. We then factor the last term, -8, which is negative. The possibilities are $(-8)(1)$, $8(-1)$, $2(-4)$, and $(-2)(4)$.

Calculator Spotlight

Use the TABLE and GRAPH features to check the results of Example 3 and Margin Exercise 5.

Factor by the FOIL method.

5. $3x^2 - 13x - 56$

4. We look for combinations of factors from steps (2) and (3) such that the sum of the outside and the inside products is the middle term, $10x$:

$$\overset{3x}{\overbrace{(3x - 8)(x + 1)}} = 3x^2 - 5x - 8; \qquad \overset{-3x}{\overbrace{(3x + 8)(x - 1)}} = 3x^2 + 5x - 8;$$

$\underset{-8x}{\underbrace{}}$ Wrong middle term $\underset{8x}{\underbrace{}}$ Wrong middle term

$$\overset{-12x}{\overbrace{(3x + 2)(x - 4)}} = 3x^2 - 10x - 8; \qquad \overset{12x}{\overbrace{(3x - 2)(x + 4)}} = 3x^2 + 10x - 8$$

$\underset{2x}{\underbrace{}}$ Wrong middle term $\underset{-2x}{\underbrace{}}$ Correct middle term!

There are four other possibilities that we could try, but we have a factorization: $(3x - 2)(x + 4)$.

Do Exercises 5 and 6.

Example 4   Factor: $6x^6 - 19x^5 + 10x^4$.

1. First, we factor out the largest common factor, if any. The expression x^4 is common to all terms, so we factor it out: $x^4(6x^2 - 19x + 10)$.

2. Next, we factor the trinomial $6x^2 - 19x + 10$. We factor the first term, $6x^2$, and get $6x, x$ or $3x, 2x$. We then have these as possibilities for factorizations: $(3x + \;\;)(2x + \;\;)$ or $(6x + \;\;)(x + \;\;)$.

3. We then factor the last term, 10, which is positive. The possibilities are 10, 1 and $-10, -1$ and 5, 2 and $-5, -2$.

6. $3x^2 + 5x + 2$

4. We look for combinations of factors from steps (2) and (3) such that the sum of the outside and the inside products is the middle term, $-19x$. The sign of the middle term is negative, but the sign of the last term, 10, is positive. Thus the signs of both factors of the last term, 10, must be negative. From our list of factors in step (3), we can use only $-10, -1$ and $-5, -2$ as possibilities. This reduces the possibilities for factorizations by half. We begin by using these factors with $(3x + \;\;)(2x + \;\;)$. Should we not find the correct factorization, we will consider $(6x + \;\;)(x + \;\;)$.

$$\overset{-3x}{\overbrace{(3x - 10)(2x - 1)}} = 6x^2 - 23x + 10; \qquad \overset{-30x}{\overbrace{(3x - 1)(2x - 10)}} = 6x^2 - 32x + 10;$$

$\underset{-20x}{\underbrace{}}$ Wrong middle term $\underset{-2x}{\underbrace{}}$ Wrong middle term

$$\overset{-6x}{\overbrace{(3x - 5)(2x - 2)}} = 6x^2 - 16x + 10; \qquad \overset{-15x}{\overbrace{(3x - 2)(2x - 5)}} = 6x^2 - 19x + 10$$

$\underset{-10x}{\underbrace{}}$ Wrong middle term $\underset{-4x}{\underbrace{}}$ Correct middle term!

We have a correct answer. We need not consider $(6x + \;\;)(x + \;\;)$.

Answers on page A-26

Look again at the possibility $(3x - 1)(2x - 10)$. Without multiplying, we can reject such a possibility, noting that

$$(3x - 1)(2x - 10) = 2(3x - 1)(x - 5).$$

The expression $2x - 10$ has a common factor, 2. But we removed the largest common factor before we began. If this expression were a factorization, then 2 would have to be a common factor along with x^4. Thus, as we saw when we multiplied, $(3x - 1)(2x - 10)$ cannot be part of the factorization of the original trinomial. Given that we factored out the largest common factor at the outset, we can now eliminate factorizations that have a common factor.

The factorization of $6x^2 - 19x + 10$ is $(3x - 2)(2x - 5)$. But do not forget the common factor! We must include it in order to get a complete factorization of the original trinomial:

$$6x^6 - 19x^5 + 10x^4 = x^4(3x - 2)(2x - 5).$$

Here is another tip that might speed up your factoring. Suppose in Example 4 that we considered the possibility

$$(3x + 2)(2x + 5) = 6x^2 + 19x + 10.$$

We might have tried this before noting that using all plus signs would give us a plus sign for the middle term. If we change *both* signs, however, we get the correct answer before including the common factor:

$$(3x - 2)(2x - 5) = 6x^2 - 19x + 10.$$

Do Exercises 7 and 8.

TIPS FOR FACTORING $ax^2 + bx + c$, $a \neq 1$, USING THE FOIL METHOD

1. If the largest common factor has been factored out of the original trinomial, then no binomial factor can have a common factor (other than 1 or -1).

2. a) If the signs of all the terms are positive, then the signs of all the terms of the binomial factors are positive.

   b) If a and c are positive and b is negative, then the signs of the factors of c are negative.

   c) If a is positive and c is negative, then the factors of c will have opposite signs.

3. Be systematic about your trials. Keep track of those you have tried and those you have not.

4. Changing the signs of the factors of c will change the sign of the middle term.

Keep in mind that this method of factoring trinomials of the type $ax^2 + bx + c$ involves trial and error. As you practice, you will find that you can make better and better guesses.

Do Exercises 9 and 10.

Factor.

7. $24y^2 - 46y + 10$

8. $20x^5 - 46x^4 + 24x^3$

Factor.

9. $3x^2 + 19x + 20$

10. $16x^2 - 12 + 16x$

Answers on page A-26

Factor.

11. $21x^2 - 5xy - 4y^2$

The procedure considered here can also be applied to a trinomial with more than one variable.

Example 5   Factor: $60m^2 + 44mn - 22n^2$.

1. First, we factor out the largest common factor, if any. In this polynomial, 2 is common to all terms, so we factor it out:

$$2(30m^2 + 22mn - 11n^2).$$

2. Next, we factor the trinomial $30m^2 + 22mn - 11n^2$. We factor the first term, $30m^2$, and get the following possibilities:

$$30m \cdot m, \qquad 15m \cdot 2m, \qquad 10m \cdot 3m, \quad \text{and} \quad 6m \cdot 5m.$$

We then have these as possibilities for factorizations:

$$(30m + \blacksquare)(m + \blacksquare), \qquad (15m + \blacksquare)(2m + \blacksquare),$$
$$(10m + \blacksquare)(3m + \blacksquare), \qquad (6m + \blacksquare)(5m + \blacksquare).$$

3. We then factor the last term, $-11n^2$, which is negative. The possibilities are $-11n \cdot n$ and $11n \cdot (-n)$.

4. We look for combinations of factors from steps (2) and (3) such that the sum of the outside and the inside products is the middle term, $22mn$. Since the coefficient of the middle term is positive, let's begin our search using $11n \cdot (-n)$. Should we not find the correct factorization, we will consider $-11n \cdot n$.

$$(30m + 11n)(m - n) = 30m^2 - 19mn - 11n^2;$$
$$(30m - n)(m + 11n) = 30m^2 + 329mn - 11n^2;$$
$$(15m + 11n)(2m - n) = 30m^2 + 7mn - 11n^2;$$
$$(15m - n)(2m + 11n) = 30m^2 + 163mn - 11n^2;$$
$$(10m + 11n)(3m - n) = 30m^2 + 22mn - 11n^2 \longleftarrow \text{Correct middle term}$$

We have a correct answer: $30m^2 + 22mn - 11n^2$. The factorization of $30m^2 + 22mn - 11n^2$ is $(10m + 11n)(3m - n)$. But do not forget the common factor! The complete factorization of the original trinomial is

$$60m^2 + 44mn - 22n^2 = 2(10m + 11n)(3m - n).$$

Do Exercises 11 and 12.

12. $60a^2 + 123ab - 27b^2$

Answers on page A-26

Exercise Set 4.4

a , b Factor.

1. $3x^2 - 14x - 5$

2. $8x^2 - 6x - 9$

3. $10y^3 + y^2 - 21y$

4. $6x^3 + x^2 - 12x$

5. $3c^2 - 20c + 32$

6. $12b^2 - 8b + 1$

7. $35y^2 + 34y + 8$

8. $9a^2 + 18a + 8$

9. $4t + 10t^2 - 6$

10. $8x + 30x^2 - 6$

11. $8x^2 - 16 - 28x$

12. $18x^2 - 24 - 6x$

13. $12x^3 - 31x^2 + 20x$

14. $15x^3 - 19x^2 - 10x$

15. $14x^4 - 19x^3 - 3x^2$

16. $70x^4 - 68x^3 + 16x^2$

17. $3a^2 - a - 4$

18. $6a^2 - 7a - 10$

19. $9x^2 + 15x + 4$

20. $6y^2 - y - 2$

21. $3 + 35z - 12z^2$

22. $8 - 6a - 9a^2$

23. $-4t^2 - 4t + 15$

24. $-12a^2 + 7a - 1$

25. $3x^3 - 5x^2 - 2x$

26. $18y^3 - 3y^2 - 10y$

27. $24x^2 - 2 - 47x$

28. $15y^2 - 10 - 15y$

29. $21x^2 + 37x + 12$

30. $10y^2 + 23y + 12$

31. $40x^4 + 16x^2 - 12$

32. $24y^4 + 2y^2 - 15$

33. $12a^2 - 17ab + 6b^2$

34. $20p^2 - 23pq + 6q^2$

35. $2x^2 + xy - 6y^2$

36. $8m^2 - 6mn - 9n^2$

37. $12x^2 - 58xy + 56y^2$

38. $30p^2 + 21pq - 36q^2$

39. $9x^2 - 30xy + 25y^2$

40. $4p^2 + 12pq + 9q^2$

41. $3x^6 + 4x^3 - 4$

42. $2p^8 + 11p^4 + 15$

43. *Height of a Thrown Baseball.* Suppose that a baseball is thrown upward with an initial velocity of 80 ft/sec from a height of 224 ft. Its height h after t seconds is given by the function
$$h(t) = -16t^2 + 80t + 224.$$
a) What is the height of the ball after 0 sec? 1 sec? 3 sec? 4 sec? 6 sec?
b) Find an equivalent expression for $h(t)$ by factoring.

44. *Height of a Thrown Rock.* Suppose that a rock is thrown upward with an initial velocity of 96 ft/sec from a height of 880 ft. Its height h after t seconds is given by the function
$$h(t) = -16t^2 + 96t + 880.$$
a) What is the height of the rock after 0 sec? 1 sec? 3 sec? 8 sec? 10 sec?
b) Find an equivalent expression for $h(t)$ by factoring.

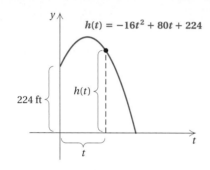

Skill Maintenance

Solve. [3.4a]

45. $x + 2y - z = 0,$
$4x + 2y + 5z = 6,$
$2x - y + z = 5$

46. $2x + y + 2z = 5,$
$4x - 2y - 3z = 5,$
$-8x - y + z = -5$

47. $2x + 9y + 6z = 5,$
$x - y + z = 4,$
$3x + 2y + 3z = 7$

48. $x - 3y + 2z = -8,$
$2x + 3y + z = 17,$
$5x - 2y + 3z = 5$

Determine whether the graphs of the given pair of lines are parallel or perpendicular. [2.5d]

49. $y - 2x = 18,$
$2x - 7 = y$

50. $21x + 7 = -3y,$
$y + 7x = -9$

51. $2x + 5y = 4,$
$2x - 5y = -3$

52. $y + x = 7,$
$y - x = 3$

Find an equation of the line containing the given pair of points. [2.6c]

53. $(-2, -3)$ and $(5, -4)$

54. $(2, -3)$ and $(5, -4)$

55. $(-10, 3)$ and $(7, -4)$

56. $\left(-\frac{2}{3}, 1\right)$ and $\left(\frac{4}{3}, -4\right)$

Synthesis

57. ◆ Explain how to use the grouping method to factor trinomials of the type $ax^2 + bx + c, a \neq 1$.

58. ◆ Explain how to use the FOIL method to factor trinomials of the type $ax^2 + bx + c, a \neq 1$.

59. ☂ Use the TABLE and GRAPH features of a grapher to check your answers to Exercises 2, 13, and 24.

60. ☂ Use the TABLE and GRAPH features of a grapher to check your answers to Exercises 4, 11, and 28.

Factor. Assume that variables in exponents represent positive integers.

61. $p^2q^2 + 7pq + 12$

62. $2x^4y^6 - 3x^2y^3 - 20$

63. $x^2 - \frac{4}{25} + \frac{3}{5}x$

64. $y^2 - \frac{8}{49} + \frac{2}{7}y$

65. $y^2 + 0.4y - 0.05$

66. $t^2 + 0.6t - 0.27$

67. $7a^2b^2 + 6 + 13ab$

68. $9x^2y^2 - 4 + 5xy$

69. $3x^2 + 12x - 495$

70. $15t^3 - 60t^2 - 315t$

71. $216x + 78x^2 + 6x^3$

72. $\frac{1}{4}p^2 - \frac{2}{5}p + \frac{4}{25}$

73. $x^{2a} + 5x^a - 24$

74. $4x^{2a} - 4x^a - 3$

4.5 Special Factoring

In this section, we consider some special factoring methods. When we recognize certain types of polynomials, we can factor more quickly using these special methods. Most of them are the reverse of the methods of special multiplication.

a ┃ Trinomial Squares

Consider the trinomial $x^2 + 6x + 9$. To factor it, we can use the method considered in Section 4.3. We look for factors of 9 whose sum is 6. We see that these factors are 3 and 3 and the factorization is

$$x^2 + 6x + 9 = (x + 3)(x + 3) = (x + 3)^2.$$

Note that the result is the square of a binomial. We also call $x^2 + 6x + 9$ a **trinomial square,** or **perfect-square trinomial.** We can certainly use the procedures of Sections 4.3 and 4.4 to factor trinomial squares, but we want to develop an even faster procedure.

In order to do so, we must first be able to recognize when a trinomial is a square.

> How to recognize a **trinomial square:**
> a) Two of the terms must be squares, such as A^2 and B^2.
> b) There must be no minus sign before either A^2 or B^2.
> c) Multiplying A and B (which are the square roots of A^2 and B^2) and doubling the result should give either the remaining term, $2AB$, or its opposite, $-2AB$.

Examples Determine whether the polynomial is a trinomial square.

1. $x^2 + 10x + 25$

a) Two terms are squares: x^2 and 25.

b) There is no minus sign before either x^2 or 25.

c) If we multiply the square roots of x^2 and 25, x and 5, and double the product, we get $10x$, the remaining term.

Thus this is a trinomial square.

2. $4x + 16 + 3x^2$

a) Only one term, 16, is a square ($3x^2$ is not a square because 3 is not a perfect square and $4x$ is not a square because x is not a square).

Thus this is not a trinomial square.

3. $100y^2 + 81 - 180y$
(It can help to first write this in descending order: $100y^2 - 180y + 81$.)

a) Two of the terms, $100y^2$ and 81, are squares.

b) There is no minus sign before either $100y^2$ or 81.

c) If we multiply the square roots of $100y^2$ and 81, $10y$ and 9, and double the product, we get the opposite of the remaining term: $2(10y)(9) = 180y$, which is the opposite of $-180y$.

Thus this is a trinomial square.

Do Exercise 1.

Objectives

a	Factor trinomial squares.
b	Factor differences of squares.
c	Factor certain polynomials with four terms by grouping and possibly using the factoring of a trinomial square or the difference of squares.
d	Factor sums and differences of cubes.

For Extra Help

TAPE 10   TAPE 8B   MAC WIN   CD-ROM

---

1. Which of the following are trinomial squares?

a) $x^2 + 6x + 9$

b) $x^2 - 8x + 16$

c) $x^2 + 6x + 11$

d) $4x^2 + 25 - 20x$

e) $16x^2 - 20x + 25$

f) $16 + 14x + 5x^2$

g) $x^2 + 8x - 16$

h) $x^2 - 8x - 16$

Answer on page A-26

Factor.

2. $x^2 + 14x + 49$

3. $9y^2 - 30y + 25$

4. $16x^2 + 72xy + 81y^2$

5. $16x^4 - 40x^2y^3 + 25y^6$

Calculator Spotlight

Use the TABLE and GRAPH features to check the results of Example 4 and Margin Exercise 2.

Factor.

6. $-8a^2 + 24ab - 18b^2$

7. $3a^2 - 30ab + 75b^2$

Answers on page A-26

The factors of a trinomial square are two identical binomials. We use the following equations.

$$A^2 + 2AB + B^2 = (A + B)^2;$$
$$A^2 - 2AB + B^2 = (A - B)^2$$

Example 4 Factor: $x^2 - 10x + 25$.

$$x^2 - 10x + 25 = (x - 5)^2 \qquad$$ We find the square terms and write their square roots with a minus sign between them.

Note the sign!

Example 5 Factor: $16y^2 + 49 + 56y$.

$$16y^2 + 49 + 56y = 16y^2 + 56y + 49 \qquad$$ Writing in descending order

$$= (4y + 7)^2 \qquad$$ We find the square terms and write their square roots with a plus sign between them.

Example 6 Factor: $-20xy + 4y^2 + 25x^2$.

We have

$$-20xy + 4y^2 + 25x^2 = 4y^2 - 20xy + 25x^2 \qquad$$ Writing descending order in y

$$= (2y - 5x)^2.$$

This square can also be expressed as

$$25x^2 - 20xy + 4y^2 = (5x - 2y)^2.$$

Do Exercises 2–5.

In factoring, we must always remember to look *first* for the largest factor common to all the terms.

Example 7 Factor: $2x^2 - 12xy + 18y^2$.

Always remember to look first for a common factor. This time the largest common factor is 2.

$$2x^2 - 12xy + 18y^2 = 2(x^2 - 6xy + 9y^2) \qquad$$ Removing the common factor 2

$$= 2(x - 3y)^2 \qquad$$ Factoring the trinomial square

Example 8 Factor: $-4y^2 - 144y^8 + 48y^5$.

$$-4y^2 - 144y^8 + 48y^5$$
$$= -4y^2(1 + 36y^6 - 12y^3) \qquad$$ Removing the common factor $-4y^2$
$$= -4y^2(1 - 12y^3 + 36y^6) \qquad$$ Changing order
$$= -4y^2(1 - 6y^3)^2 \qquad$$ Factoring the trinomial square

Do Exercises 6 and 7.

b | Differences of Squares

The following are differences of squares:

$$x^2 - 9, \qquad 49 - 4y^2, \qquad a^2 - 49b^2.$$

To factor a difference of two expressions that are squares, we can use a pattern for multiplying a sum and a difference that we used earlier.

> ▶ To factor a difference of two squares, write the square root of the first expression *plus* the square root of the second *times* the square root of the first *minus* the square root of the second.
> $$A^2 - B^2 = (A + B)(A - B)$$

You should memorize this rule, and then say it as you work.

Example 9   Factor: $x^2 - 9$.

$$x^2 - 9 = x^2 - 3^2 = (x + 3)(x - 3)$$

Example 10   Factor: $25y^6 - 49x^2$.

$$
\begin{array}{cccccccccc}
A^2 & - & B^2 & = & (A & + & B)(A & - & B) \\
\downarrow & & \downarrow & & \downarrow & & \downarrow \ \ \downarrow & & \downarrow
\end{array}
$$
$$25y^6 - 49x^2 = (5y^3)^2 - (7x)^2 = (5y^3 + 7x)(5y^3 - 7x)$$

Example 11   Factor: $x^2 - \frac{1}{16}$.

$$x^2 - \frac{1}{16} = x^2 - \left(\tfrac{1}{4}\right)^2 = \left(x + \tfrac{1}{4}\right)\left(x - \tfrac{1}{4}\right)$$

Do Exercises 8–10.

Common factors should always be removed. Removing common factors actually eases the factoring process because the type of factoring to be done becomes clearer.

Example 12   Factor: $5 - 5x^2y^6$.

There is a common factor, 5.

$$
\begin{aligned}
5 - 5x^2y^6 &= 5(1 - x^2y^6) && \text{Removing the common factor 5}\\
&= 5[1^2 - (xy^3)^2] && \text{Recognizing the difference of squares}\\
&= 5(1 + xy^3)(1 - xy^3) && \text{Factoring the difference of squares}
\end{aligned}
$$

Example 13   Factor: $2x^4 - 8y^4$.

There is a common factor, 2.

$$
\begin{aligned}
2x^4 - 8y^4 &= 2(x^4 - 4y^4) && \text{Removing the common factor 2}\\
&= 2[(x^2)^2 - (2y^2)^2] && \text{Recognizing the difference of squares}\\
&= 2(x^2 + 2y^2)(x^2 - 2y^2) && \text{Factoring the difference of squares}
\end{aligned}
$$

Factor.

8. $y^2 - 4$

9. $49x^4 - 25y^{10}$

10. $m^2 - \dfrac{1}{9}$

Calculator Spotlight

〰 Use the TABLE and GRAPH features to check the results of Example 11 and Margin Exercise 8.

Answers on page A-26

Factor.

11. $25x^2y^2 - 4a^2$

12. $9x^2 - 16y^2$

13. $20x^2 - 5y^2$

14. $81x^4y^2 - 16y^2$

15. Factor: $a^3 + a^2 - 16a - 16$.

Factor completely.

16. $x^2 + 2x + 1 - p^2$

17. $y^2 - 8y + 16 - 9m^2$

18. $x^2 + 8x + 16 - 100t^2$

19. $64p^2 - (x^2 + 8x + 16)$

Calculator Spotlight

Use the TABLE and GRAPH features to check the results of Example 15.

Answers on page A-26

Example 14   Factor: $16x^4y - 81y$.

There is a common factor, y.

$$16x^4y - 81y = y(16x^4 - 81) \qquad \text{Removing the common factor } y$$
$$= y[(4x^2)^2 - 9^2]$$
$$= y(4x^2 + 9)(4x^2 - 9) \qquad \text{Factoring the difference of squares}$$
$$= y(4x^2 + 9)(2x + 3)(2x - 3) \qquad \begin{array}{l}\text{Factoring } 4x^2 - 9, \\ \text{which is also a} \\ \text{difference of squares}\end{array}$$

In Example 14, it may be tempting to try to factor $4x^2 + 9$. Note that it is a sum of two expressions that are squares, but it cannot be factored further.

CAUTION!   If the greatest common factor has been removed, then you cannot factor a sum of squares further. In particular,

$$A^2 + B^2 \neq (A + B)^2.$$

Consider $25x^2 + 225$. This is a case in which we have a sum of squares, but there is a common factor, 25. Factoring, we get $25(x^2 + 9)$. Now $x^2 + 9$ cannot be factored further.

Note too in Example 14 that one of the factors, $4x^2 - 9$, could be factored further. Whenever that is possible, you should do so. That way you will be factoring *completely*.

Do Exercises 11–14.

c | More Factoring by Grouping

Sometimes when factoring a polynomial with four terms completely, we might get a factor that can be factored further using other methods we have learned.

Example 15   Factor completely: $x^3 + 3x^2 - 4x - 12$.

$$x^3 + 3x^2 - 4x - 12 = x^2(x + 3) - 4(x + 3)$$
$$= (x^2 - 4)(x + 3)$$
$$= (x + 2)(x - 2)(x + 3)$$

Do Exercise 15.

A difference of squares can have more than two terms. For example, one of the squares may be a trinomial. We can factor by a type of grouping.

Example 16   Factor completely: $x^2 + 6x + 9 - y^2$.

$$x^2 + 6x + 9 - y^2 = (x^2 + 6x + 9) - y^2 \qquad \begin{array}{l}\text{Grouping as a trinomial} \\ \text{minus } y^2 \text{ to show a} \\ \text{difference of squares}\end{array}$$
$$= (x + 3)^2 - y^2$$
$$= (x + 3 + y)(x + 3 - y)$$

Do Exercises 16–19.

d | Sums or Differences of Cubes

We can factor the sum or the difference of two expressions that are cubes. Consider the following products:

$$(A + B)(A^2 - AB + B^2) = A(A^2 - AB + B^2) + B(A^2 - AB + B^2)$$
$$= A^3 - A^2B + AB^2 + A^2B - AB^2 + B^3$$
$$= A^3 + B^3$$

and

$$(A - B)(A^2 + AB + B^2) = A(A^2 + AB + B^2) - B(A^2 + AB + B^2)$$
$$= A^3 + A^2B + AB^2 - A^2B - AB^2 - B^3$$
$$= A^3 - B^3.$$

The above equations (reversed) show how we can factor a sum or a difference of two cubes.

> $A^3 + B^3 = (A + B)(A^2 - AB + B^2)$;
> $A^3 - B^3 = (A - B)(A^2 + AB + B^2)$

Note that what we are considering here is a sum or a difference of cubes. We are not cubing a binomial. For example, $(A + B)^3$ is *not* the same as $A^3 + B^3$. The table of cubes in the margin is helpful.

Example 17 Factor: $x^3 - 27$.

We have

$$\begin{array}{cc} A^3 & - B^3 \\ \downarrow & \downarrow \\ x^3 - 27 = x^3 & - 3^3. \end{array}$$

In one set of parentheses, we write the cube root of the first term, x. Then we write the cube root of the second term, -3. This gives us the expression $x - 3$:

$$(x - 3)(\qquad).$$

To get the next factor, we think of $x - 3$ and do the following:

- Square the first term: x^2.
- Multiply the terms and then change the sign: $3x$.
- Square the second term: 9.

$$(x - 3)(x^2 + 3x + 9).$$
$$(A - B)(A^2 + AB + B^2)$$

Note that we cannot factor $x^2 + 3x + 9$. It is not a trinomial square nor can it be factored by trial and error. Check this on your own.

Do Exercises 20 and 21.

Example 18 Factor: $125x^3 + y^3$.

We have

$$125x^3 + y^3 = (5x)^3 + y^3.$$

In one set of parentheses, we write the cube root of the first term, $5x$. Then we write a plus sign, and then the cube root of the second term, y:

$$(5x + y)(\qquad).$$

N	N^3
0.2	0.008
0.1	0.001
0	0
1	1
2	8
3	27
4	64
5	125
6	216
7	343
8	512
9	729
10	1000

Factor.

20. $x^3 - 8$

21. $64 - y^3$

Calculator Spotlight

Use the TABLE and GRAPH features to check the results of Example 17 and Margin Exercises 20 and 21 and to show that

$$x^3 + 8 = (x + 2)^3$$

is not correct.

Answers on page A-26

Factor.

22. $27x^3 + y^3$

23. $8y^3 + z^3$

Factor.

24. $m^6 - n^6$

25. $16x^7y + 54xy^7$

26. $729x^6 - 64y^6$

27. $x^3 - 0.027$

Answers on page A-26

To get the next factor, we think of $5x + y$ and do the following:

Square the first term: $25x^2$.
Multiply the terms and then change the sign: $-5xy$.
Square the second term: y^2.

$(5x + y)(25x^2 - 5xy + y^2).$
$(A + B)(\ A^2\ -\ AB + B^2)$

Do Exercises 22 and 23.

Example 19   Factor: $128y^7 - 250x^6y$.

We first look for the largest common factor:

$$128y^7 - 250x^6y = 2y(64y^6 - 125x^6) = 2y[(4y^2)^3 - (5x^2)^3]$$
$$= 2y(4y^2 - 5x^2)(16y^4 + 20x^2y^2 + 25x^4).$$

Example 20   Factor: $a^6 - b^6$.

We can express this polynomial as a difference of squares:

$$(a^3)^2 - (b^3)^2.$$

We factor as follows:

$$a^6 - b^6 = (a^3 + b^3)(a^3 - b^3).$$

One factor is a sum of two cubes, and the other factor is a difference of two cubes. We factor them:

$$(a + b)(a^2 - ab + b^2)(a - b)(a^2 + ab + b^2).$$

We have now factored completely.

In Example 20, had we thought of factoring first as a difference of two cubes, we would have had

$$(a^2)^3 - (b^2)^3 = (a^2 - b^2)(a^4 + a^2b^2 + b^4)$$
$$= (a + b)(a - b)(a^4 + a^2b^2 + b^4).$$

In this case, we might have missed some factors; $a^4 + a^2b^2 + b^4$ can be factored as $(a^2 - ab + b^2)(a^2 + ab + b^2)$, but we probably would not have known to do such factoring.

Example 21   Factor: $64a^6 - 729b^6$.

We have

$64a^6 - 729b^6 = (8a^3 - 27b^3)(8a^3 + 27b^3)$ **Factoring a difference of squares**
$\qquad\qquad = [(2a)^3 - (3b)^3][(2a)^3 + (3b)^3].$

Each factor is a sum or a difference of cubes. We factor each:

$$= (2a - 3b)(4a^2 + 6ab + 9b^2)(2a + 3b)(4a^2 - 6ab + 9b^2).$$

▶ | | |
|---|---|
| Sum of cubes: | $A^3 + B^3 = (A + B)(A^2 - AB + B^2)$; |
| Difference of cubes: | $A^3 - B^3 = (A - B)(A^2 + AB + B^2)$; |
| Difference of squares: | $A^2 - B^2 = (A + B)(A - B)$; |
| Sum of squares: | $A^2 + B^2$ cannot be factored using real numbers if the largest common factor has been removed. |

Do Exercises 24–27.

Exercise Set 4.5

a Factor.

1. $x^2 - 4x + 4$

2. $y^2 - 16y + 64$

3. $y^2 + 18y + 81$

4. $x^2 + 8x + 16$

5. $x^2 + 1 + 2x$

6. $x^2 + 1 - 2x$

7. $9y^2 + 12y + 4$

8. $25x^2 - 60x + 36$

9. $-18y^2 + y^3 + 81y$

10. $24a^2 + a^3 + 144a$

11. $12a^2 + 36a + 27$

12. $20y^2 + 100y + 125$

13. $2x^2 - 40x + 200$

14. $32x^2 + 48x + 18$

15. $1 - 8d + 16d^2$

16. $64 + 25y^2 - 80y$

17. $y^4 - 8y^2 + 16$

18. $y^4 - 18y^2 + 81$

19. $0.25x^2 + 0.30x + 0.09$

20. $0.04x^2 - 0.28x + 0.49$

21. $p^2 - 2pq + q^2$

22. $m^2 + 2mn + n^2$

23. $a^2 + 4ab + 4b^2$

24. $49p^2 - 14pq + q^2$

25. $25a^2 - 30ab + 9b^2$

26. $49p^2 - 84pq + 36q^2$

27. $y^6 + 26y^3 + 169$

28. $p^6 - 10p^3 + 25$

29. $16x^{10} - 8x^5 + 1$

30. $9x^{10} + 12x^5 + 4$

31. $x^4 + 2x^2y^2 + y^4$

32. $p^6 - 2p^3q^4 + q^8$

b Factor.

33. $x^2 - 16$

34. $y^2 - 9$

35. $p^2 - 49$

36. $m^2 - 64$

37. $p^2q^2 - 25$

38. $a^2b^2 - 81$

39. $6x^2 - 6y^2$

40. $8x^2 - 8y^2$

41. $4xy^4 - 4xz^4$

42. $25ab^4 - 25az^4$

43. $4a^3 - 49a$

44. $9x^3 - 25x$

45. $3x^8 - 3y^8$ **46.** $2a^9 - 32a$ **47.** $9a^4 - 25a^2b^4$ **48.** $16x^6 - 121x^2y^4$

49. $\frac{1}{36} - z^2$ **50.** $\frac{1}{100} - y^2$ **51.** $0.04x^2 - 0.09y^2$ **52.** $0.01x^2 - 0.04y^2$

c Factor.

53. $m^3 - 7m^2 - 4m + 28$ **54.** $x^3 + 8x^2 - x - 8$ **55.** $a^3 - ab^2 - 2a^2 + 2b^2$

56. $p^2q - 25q + 3p^2 - 75$ **57.** $(a + b)^2 - 100$ **58.** $(p - 7)^2 - 144$

59. $a^2 + 2ab + b^2 - 9$ **60.** $x^2 - 2xy + y^2 - 25$ **61.** $r^2 - 2r + 1 - 4s^2$

62. $c^2 + 4cd + 4d^2 - 9p^2$ **63.** $2m^2 + 4mn + 2n^2 - 50b^2$ **64.** $12x^2 + 12x + 3 - 3y^2$

65. $9 - (a^2 + 2ab + b^2)$ **66.** $16 - (x^2 - 2xy + y^2)$

d Factor.

67. $z^3 + 27$ **68.** $a^3 + 8$ **69.** $x^3 - 1$ **70.** $c^3 - 64$

71. $y^3 + 125$ **72.** $x^3 + 1$ **73.** $8a^3 + 1$ **74.** $27x^3 + 1$

75. $y^3 - 8$

76. $p^3 - 27$

77. $8 - 27b^3$

78. $64 - 125x^3$

79. $64y^3 + 1$

80. $125x^3 + 1$

81. $8x^3 + 27$

82. $27y^3 + 64$

83. $a^3 - b^3$

84. $x^3 - y^3$

85. $a^3 + \frac{1}{8}$

86. $b^3 + \frac{1}{27}$

87. $2y^3 - 128$

88. $3z^3 - 3$

89. $24a^3 + 3$

90. $54x^3 + 2$

91. $rs^3 + 64r$

92. $ab^3 + 125a$

93. $5x^3 - 40z^3$

94. $2y^3 - 54z^3$

95. $x^3 + 0.001$

96. $y^3 + 0.125$

97. $64x^6 - 8t^6$

98. $125c^6 - 8d^6$

99. $2y^4 - 128y$

100. $3z^5 - 3z^2$

101. $z^6 - 1$

102. $t^6 + 1$

103. $t^6 + 64y^6$

104. $p^6 - q^6$

Solve. [3.2a, b]

105. $7x - 2y = -11,$
$2x + 7y = 18$

106. $y = 3x - 8,$
$4x - 6y = 100$

107. $x - y = -12,$
$x + y = 14$

108. $7x - 2y = -11,$
$2y - 7x = -18$

Graph the given system of inequalities and determine coordinates of any vertices found. [3.6c]

109. $x - y \le 5,$
$x + y \ge 3$

110. $x - y \le 5,$
$x + y \ge 3,$
$x \le 6$

111. $x - y \ge 5,$
$x + y \le 3,$
$x \ge 1$

112. $x - y \ge 5,$
$x + y \le 3$

Given the line and a point not on the line, find an equation through the point parallel to the given line, and find an equation through the point perpendicular to the given line. [2.6d]

113. $x - y = 5;\ (-2, -4)$

114. $2x - 3y = 6;\ (1, -7)$

115. $y = -\frac{1}{2}x + 3;\ (4, 5)$

116. $x - 4y = -10;\ (6, 0)$

Synthesis

117. ◈ Under what conditions, if any, can the sum of two squares be factored? Explain.

118. ◈ 〜 Explain how you could use factoring or graphing to explain why $x^3 - 8 \ne (x - 2)^3$.

119. Given that $P(x) = x^2$, use factoring to simplify $P(a + h) - P(a)$.

120. Given that $P(x) = x^4$, use factoring to simplify $P(a + h) - P(a)$.

121. Given that $P(x) = x^3$, use factoring to simplify $P(a + h) - P(a)$.

122. 〜 Use the TABLE and GRAPH features of a grapher to check your answers to Exercises 6, 36, and 69.

123. ◈ Explain how the geometric model below can be used to verify the formula for factoring $a^3 - b^3$.

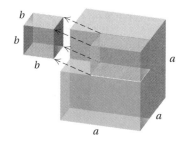

124. *Volume of Carpeting.* The volume of a carpet that is rolled up can be estimated by the polynomial $\pi R^2 h - \pi r^2 h$.

a) Factor the polynomial.
b) Use both the original and the factored forms to find the volume of a roll for which $R = 50$ cm, $r = 10$ cm, and $h = 4$ m. Use 3.14 for π.

Factor. Assume that variables in exponents represent positive integers.

125. $x^4y^4 - 8x^2y^2 + 16$

126. $\frac{1}{49}p^2 - \frac{8}{63}p + \frac{16}{81}$

127. $5c^{100} - 80d^{100}$

128. $9x^{2n} - 6x^n + 1$

129. $x^{4a} - y^{2b}$

130. $y^8 - 256$

131. $x^{6a} + y^{3b}$

132. $a^3x^3 - b^3y^3$

133. $3x^{3a} + 24y^{3b}$

134. $\frac{8}{27}x^3 + \frac{1}{64}y^3$

135. $\frac{1}{24}x^3y^3 + \frac{1}{3}z^3$

136. $7x^3 - \frac{7}{8}$

137. $(x + y)^3 - x^3$

138. $(1 - x)^3 + (x - 1)^6$

139. $(a + 2)^3 - (a - 2)^3$

140. $y^4 - 8y^3 - y + 8$

4.6 Factoring: A General Strategy

a | A General Factoring Strategy

Factoring is an important algebraic skill, used for solving equations and many other manipulations of algebraic symbolism. We now consider polynomials of many types and learn to use a general strategy for factoring. The key is to recognize the type of polynomial to be factored.

A STRATEGY FOR FACTORING

a) Always look for a common factor (other than 1 or −1). If there are any, factor out the largest one.

b) Then look at the number of terms.

Two terms: Try factoring as a difference of squares first. Next, try factoring as a sum or a difference of cubes. Do *not* try to factor a *sum* of squares: $A^2 + B^2$.

Three terms: Determine whether the expression is a trinomial square. If it is, you know how to factor. If not, use trial and error.

Four or more terms: Try factoring by grouping and removing a common binomial factor. Next, try grouping into a difference of squares, one of which is a trinomial.

c) Always *factor completely*. If a factor with more than one term can be factored, you should factor it.

Example 1 Factor: $10a^2x - 40b^2x$.

a) We look first for a common factor:

$$10x(a^2 - 4b^2).$$ **Factoring out the largest common factor**

b) The factor $a^2 - 4b^2$ has only two terms. It is a difference of squares. We factor it, keeping the common factor: $10x(a + 2b)(a - 2b)$.

c) Have we factored completely? Yes, because none of the factors with more than one term can be factored further into a polynomial of smaller degree.

Example 2 Factor: $x^6 - y^6$.

a) We look for a common factor. There isn't one (other than 1 or −1).

b) There are only two terms. It is a difference of squares: $(x^3)^2 - (y^3)^2$. We factor it: $(x^3 + y^3)(x^3 - y^3)$. One factor is a sum of two cubes, and the other factor is a difference of two cubes. We factor them:

$$(x + y)(x^2 - xy + y^2)(x - y)(x^2 + xy + y^2).$$

c) We have factored completely because none of the factors can be factored further into a polynomial of smaller degree.

Do Exercises 1–3.

Example 3 Factor: $10x^6 + 40y^2$.

a) We remove the largest common factor: $10(x^6 + 4y^2)$.

b) In the parentheses, there are two terms, a sum of squares, which cannot be factored.

c) We have factored $10x^6 + 40y^2$ completely as $10(x^6 + 4y^2)$.

Factor completely.

1. $3y^3 - 12x^2y$

2. $7a^3 - 7$

3. $64x^6 - 729y^6$

Answers on page A-27

Factor.

4. $3x - 6 - bx^2 + 2bx$

5. $5y^4 + 20x^6$

6. $6x^2 - 3x - 18$

7. $a^3 - ab^2 - a^2b + b^3$

8. $3x^2 + 18ax + 27a^2$

9. $2x^2 - 20x + 50 - 18b^2$

Calculator Spotlight

Use the TABLE and GRAPH features to check the results of Example 5 and Margin Exercises 2 and 6.

Answers on page A-27

Example 4 Factor: $2x^2 + 50a^2 - 20ax$.

a) We remove the largest common factor: $2(x^2 + 25a^2 - 10ax)$.

b) In the parentheses, there are three terms. The trinomial is a square. We factor it: $2(x - 5a)^2$.

c) None of the factors with more than one term can be factored further.

Example 5 Factor: $6x^2 - 20x - 16$.

a) We remove the largest common factor: $2(3x^2 - 10x - 8)$.

b) In the parentheses, there are three terms. The trinomial is not a square. We factor: $2(x - 4)(3x + 2)$.

c) We cannot factor further.

Example 6 Factor: $3x + 12 + ax^2 + 4ax$.

a) There is no common factor (other than 1 or −1).

b) There are four terms. We try grouping to remove a common binomial factor:

$$3(x + 4) + ax(x + 4) \qquad \text{Factoring two grouped binomials}$$
$$= (3 + ax)(x + 4). \qquad \text{Removing the common binomial factor}$$

c) None of the factors with more than one term can be factored further.

Example 7 Factor: $y^2 - 9a^2 + 12y + 36$.

a) There is no common factor (other than 1 or −1).

b) There are four terms. We try grouping to remove a common binomial factor, but that is not possible. We try grouping as a difference of squares:

$$(y^2 + 12y + 36) - 9a^2 = (y + 6)^2 - (3a)^2$$
$$= (y + 6 + 3a)(y + 6 - 3a). \qquad \text{Factoring the difference of squares}$$

c) No factor with more than one term can be factored further.

Example 8 Factor: $x^3 - xy^2 + x^2y - y^3$.

a) There is no common factor (other than 1 or −1).

b) There are four terms. We try grouping to remove a common binomial factor:

$$x(x^2 - y^2) + y(x^2 - y^2) \qquad \text{Factoring two grouped binomials}$$
$$= (x + y)(x^2 - y^2). \qquad \text{Removing the common binomial factor}$$

c) The factor $x^2 - y^2$ can be factored further, giving

$$(x + y)(x + y)(x - y). \qquad \text{Factoring a difference of squares}$$

None of the factors with more than one term can be factored further, so we have factored completely.

Do Exercises 4–9.

Exercise Set 4.6

a Factor completely.

1. $y^2 - 225$

2. $x^2 - 400$

3. $2x^2 + 11x + 12$

4. $8a^2 + 18a - 5$

5. $5x^4 - 20$

6. $3xy^2 - 75x$

7. $p^2 + 36 + 12p$

8. $a^2 + 49 + 14a$

9. $2x^2 - 10x - 132$

10. $3y^2 - 15y - 252$

11. $9x^2 - 25y^2$

12. $16a^2 - 81b^2$

13. $m^6 - 1$

14. $64t^6 - 1$

15. $x^2 + 6x - y^2 + 9$

16. $t^2 + 10t - p^2 + 25$

17. $250x^3 - 128y^3$

18. $27a^3 - 343b^3$

19. $8m^3 + m^6 - 20$

20. $-37x^2 + x^4 + 36$

21. $ac + cd - ab - bd$

22. $xw - yw + xz - yz$

23. $50b^2 - 5ab - a^2$

24. $9c^2 + 12cd - 5d^2$

25. $-7x^2 + 2x^3 + 4x - 14$

26. $9m^2 + 3m^3 + 8m + 24$

27. $2x^3 + 6x^2 - 8x - 24$

28. $3x^3 + 6x^2 - 27x - 54$

29. $16x^3 + 54y^3$

30. $250a^3 + 54b^3$

31. $36y^2 - 35 + 12y$

32. $2b - 28a^2b + 10ab$

33. $a^8 - b^8$

34. $2x^4 - 32$

35. $a^3b - 16ab^3$

36. $x^3y - 25xy^3$

37. $\frac{1}{16}x^2 - \frac{1}{6}xy^2 + \frac{1}{9}y^4$ **38.** $36x^2 + 15x + \frac{25}{16}$ **39.** $5x^3 - 5x^2y - 5xy^2 + 5y^3$ **40.** $a^3 - ab^2 + a^2b - b^3$

41. $42ab + 27a^2b^2 + 8$ **42.** $-23xy + 20x^2y^2 + 6$ **43.** $8y^4 - 125y$ **44.** $64p^4 - p$

45. $a^2 - b^2 - 6b - 9$ **46.** $m^2 - n^2 - 8n - 16$

Skill Maintenance

Solve.   [3.2c]

47. *Exam Scores.* There are 75 questions on a college entrance examination. Two points are awarded for each correct answer, and one half point is deducted for each incorrect answer. A score of 100 indicates how many correct and how many incorrect answers, assuming that all questions are answered?

48. *Perimeter.* A pentagon with all five sides the same length has the same perimeter as an octagon in which all eight sides are the same length. One side of the pentagon is 2 less than three times the length of one side of the octagon. Find the perimeters.

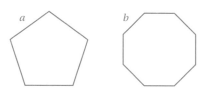

Synthesis

49. ◈ In your own words, outline a procedure that can be used to try to factor a polynomial.

50. ◈ Emily has factored a particular polynomial as $(a - b)(x - y)$. George factors the same polynomial and gets $(b - a)(y - x)$. Who is correct and why?

Factor. Assume that variables in exponents represent natural numbers.

51. $30y^4 - 97xy^2 + 60x^2$

52. $3x^2y^2z + 25xyz^2 + 28z^3$

53. $5x^3 - \frac{5}{27}$

54. $x^4 - 50x^2 + 49$

55. $(x - p)^2 - p^2$

56. $s^6 - 729t^6$

57. $(y - 1)^4 - (y - 1)^2$

58. $27x^{6s} + 64y^{3t}$

59. $4x^2 + 4xy + y^2 - r^2 + 6rs - 9s^2$

60. $c^4d^4 - a^{16}$

61. $c^{2w+1} + 2c^{w+1} + c$

62. $24x^{2a} - 6$

63. $3(x + 1)^2 + 9(x + 1) - 12$

64. $8(a - 3)^2 - 64(a - 3) + 128$

65. $x^6 - 2x^5 + x^4 - x^2 + 2x - 1$

66. $1 - \frac{x^{27}}{1000}$

67. $y^9 - y$

68. $(m - 1)^3 - (m + 1)^3$

4.7 Applications of Polynomial Equations and Functions

Whenever two polynomials are set equal to each other, we have a **polynomial equation.** Some examples of polynomial equations are

$$4x^3 + x^2 + 5x = 6x - 3,$$
$$x^2 - x = 6,$$

and

$$3y^4 + 2y^2 + 2 = 0.$$

A second-degree polynomial equation in one variable is often called a **quadratic equation.** Of the equations listed above, only $x^2 - x = 6$ is a quadratic equation.

Polynomial equations, and quadratic equations in particular, occur frequently in applications, so the ability to solve them is an important skill. One way of solving certain polynomial equations involves factoring.

a The Principle of Zero Products

When we multiply two or more numbers, the product is 0 if one of the factors is 0. Conversely, if a product is 0, then at least one of the factors must be 0. This property of 0 gives us a new principle for solving equations.

> **THE PRINCIPLE OF ZERO PRODUCTS**
>
> For any real numbers a and b:
>
> If $ab = 0$, then $a = 0$ or $b = 0$ (or both).
> If $a = 0$ or $b = 0$, then $ab = 0$.

To solve an equation using the principle of zero products, we first write it in *standard form*: with 0 on one side of the equation and the leading coefficient positive.

Example 1   Solve: $x^2 - x = 6$.

In order to use the principle of zero products, we must have 0 on one side of the equation, so we subtract 6 on both sides:

$$x^2 - x - 6 = 0. \qquad \text{Getting 0 on one side}$$

We need a factorization on the other side, so we factor the polynomial:

$$(x - 3)(x + 2) = 0. \qquad \text{Factoring}$$

We now have two expressions, $x - 3$ and $x + 2$, whose product is 0. Using the principle of zero products, we set each expression or factor equal to 0:

$$x - 3 = 0 \quad or \quad x + 2 = 0. \qquad \text{Using the principle of zero products}$$

This gives us two simple linear equations. We solve them separately,

$$x = 3 \quad or \quad x = -2,$$

and check in the original equation as follows.

1. Solve: $x^2 + 8 = 6x$.

CHECK:

$$\begin{array}{c} x^2 - x = 6 \\ \hline 3^2 - 3 \ ? \ 6 \\ 9 - 3 \\ 6 \end{array} \qquad \text{TRUE}$$

$$\begin{array}{c} x^2 - x = 6 \\ \hline (-2)^2 - (-2) \ ? \ 6 \\ 4 + 2 \\ 6 \end{array} \qquad \text{TRUE}$$

The numbers 3 and -2 are both solutions.

> To solve an equation using the principle of zero products:
>
> **1.** Obtain a 0 on one side of the equation.
>
> **2.** Factor the other side.
>
> **3.** Set each factor equal to 0.
>
> **4.** Solve the resulting equations.

Do Exercise 1.

When you solve an equation using the principle of zero products, you may wish to check by substitution as we did in Example 1. Such a check will detect errors in solving.

CAUTION! When using the principle of zero products, it is important that there is a 0 on one side of the equation. If neither side of the equation is 0, the procedure will not work.

2. Solve: $5y + 2y^2 = 3$.

For example, consider $x^2 - x = 6$ in Example 1 as

$$x(x - 1) = 6.$$

Suppose we reasoned as follows, setting factors equal to 6:

$$x = 6 \quad or \quad x - 1 = 6 \qquad \text{This step is incorrect!}$$
$$x = 7.$$

Neither 6 nor 7 checks, as shown below:

$$\begin{array}{c} x(x - 1) = 6 \\ \hline 6(6 - 1) \ \big| \ 6 \\ 6(5) \\ 30 \end{array} \qquad \text{FALSE}$$

$$\begin{array}{c} x(x - 1) = 6 \\ \hline 7(7 - 1) \ \big| \ 6 \\ 7(6) \\ 42 \end{array} \qquad \text{FALSE}$$

Example 2   Solve: $7y + 3y^2 = -2$.

Since there must be a 0 on one side of the equation, we add 2 to get 0 on the right-hand side and arrange in descending order. Then we factor and use the principle of zero products.

$$7y + 3y^2 = -2$$
$$3y^2 + 7y + 2 = 0 \qquad \text{Getting 0 on one side}$$
$$(3y + 1)(y + 2) = 0 \qquad \text{Factoring}$$
$$3y + 1 = 0 \quad or \quad y + 2 = 0 \qquad \text{Using the principle of zero products}$$
$$y = -\tfrac{1}{3} \quad or \qquad y = -2$$

The solutions are $-\tfrac{1}{3}$ and -2.

Do Exercise 2.

Answers on page A-27

Example 3 Solve: $5b^2 = 10b$.

$$5b^2 = 10b$$

$5b^2 - 10b = 0$ Getting 0 on one side

$5b(b - 2) = 0$ Factoring

$5b = 0$ or $b - 2 = 0$ Using the principle of zero products

$b = 0$ or $b = 2$

The solutions are 0 and 2.

Do Exercise 3.

Example 4 Solve: $x^2 - 6x + 9 = 0$.

$x^2 - 6x + 9 = 0$ Getting 0 on one side

$(x - 3)(x - 3) = 0$ Factoring

$x - 3 = 0$ or $x - 3 = 0$ Using the principle of zero products

$x = 3$ or $x = 3$

There is only one solution, 3.

Do Exercise 4.

Example 5 Solve: $3x^3 - 9x^2 = 30x$.

$$3x^3 - 9x^2 = 30x$$

$3x^3 - 9x^2 - 30x = 0$ Getting 0 on one side

$3x(x^2 - 3x - 10) = 0$ Factoring out a common factor

$3x(x + 2)(x - 5) = 0$ Factoring the trinomial

$3x = 0$ or $x + 2 = 0$ or $x - 5 = 0$ Using the principle of zero products

$x = 0$ or $x = -2$ or $x = 5$

The solutions are 0, -2, and 5.

Do Exercise 5.

Example 6 Given that $f(x) = 3x^2 - 4x$, find all values of x for which $f(x) = 4$.

We want all numbers x for which $f(x) = 4$. Since $f(x) = 3x^2 - 4x$, we must have

$3x^2 - 4x = 4$ Setting $f(x)$ equal to 4

$3x^2 - 4x - 4 = 0$ Getting 0 on one side

$(3x + 2)(x - 2) = 0$ Factoring

$3x + 2 = 0$ or $x - 2 = 0$

$x = -\frac{2}{3}$ or $x = 2$.

We can check as follows.

$$f\left(-\tfrac{2}{3}\right) = 3\left(-\tfrac{2}{3}\right)^2 - 4\left(-\tfrac{2}{3}\right) = 3 \cdot \tfrac{4}{9} + \tfrac{8}{3} = \tfrac{4}{3} + \tfrac{8}{3} = \tfrac{12}{3} = 4;$$

$$f(2) = 3(2)^2 - 4(2) = 3 \cdot 4 - 8 = 12 - 8 = 4.$$

To have $f(x) = 4$, we must have $x = -\frac{2}{3}$ or $x = 2$.

Do Exercise 6.

3. Solve: $8b^2 = 16b$.

4. Solve: $25 + x^2 = -10x$.

5. Solve: $x^3 + x^2 = 6x$.

6. Given that $f(x) = 10x^2 + 13x$, find all values of x for which $f(x) = 3$.

Answers on page A-27

7. Find the domain of the function G if

$$G(x) = \frac{2x - 9}{x^2 - 3x - 28}.$$

8. Consider solving the equation

$$x^2 - 6x + 8 = 0$$

graphically.

a) Below is the graph of

$$f(x) = x^2 - 6x + 8.$$

Use *only* the graph to find the x-intercepts of the graph.

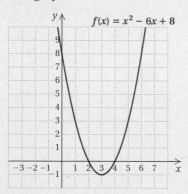

b) Use *only* the graph to find the solutions of
$x^2 - 6x + 8 = 0$.

c) Compare your answers to parts (a) and (b).

Answers on page A-27

Example 7   Find the domain of F if $F(x) = \dfrac{x - 2}{x^2 + 2x - 15}$.

The domain of F is the set of all values for which

$$\frac{x - 2}{x^2 + 2x - 15}$$

is a real number. Since division by 0 is undefined, $F(x)$ cannot be calculated for any x-value for which the denominator, $x^2 + 2x - 15$, is 0. To make sure these values are *excluded,* we solve:

$$
\begin{aligned}
x^2 + 2x - 15 &= 0 && \text{Setting the denominator equal to 0}\\
(x - 3)(x + 5) &= 0 && \text{Factoring}\\
x - 3 = 0 \quad &or \quad x + 5 = 0\\
x = 3 \quad &or \quad\quad x = -5. && \text{These are the values to \textit{exclude.}}
\end{aligned}
$$

The domain of F is $\{x \mid x$ is a real number *and* $x \neq -5$ *and* $x \neq 3\}$.

Do Exercise 7.

🄰🄶 Algebraic–Graphical Connection

We now consider graphical connections with the algebraic equation-solving concepts.

In Chapter 2, we briefly considered the graph of a quadratic function

$$f(x) = ax^2 + bx + c, \quad a \neq 0.$$

For example, the graph of the function $f(x) = x^2 + 6x + 8$ and its x-intercepts are shown below.

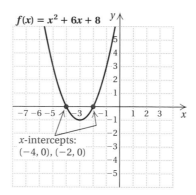

The x-intercepts are $(-4, 0)$ and $(-2, 0)$. These pairs are also the points of intersection of the graphs of $f(x) = x^2 + 6x + 8$ and $g(x) = 0$ (the x-axis).

In this section, we began studying how to solve quadratic equations like $x^2 + 6x + 8 = 0$ using factoring.

$$
\begin{aligned}
x^2 + 6x + 8 &= 0\\
(x + 4)(x + 2) &= 0 && \text{Factoring}\\
x + 4 = 0 \quad &or \quad x + 2 = 0 && \text{Principle of zero products}\\
x = -4 \quad &or \quad\quad x = -2
\end{aligned}
$$

We see that the solutions of $0 = x^2 + 6x + 8$, -4 and -2, are the first coordinates of the x-intercepts, $(-4, 0)$ and $(-2, 0)$, of the graph of $f(x) = x^2 + 6x + 8$,

Do Exercise 8.

Calculator Spotlight

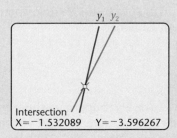 Solving Polynomial Equations. See the Calculator Spotlight on the INTERSECT and ZERO features in Section 3.1.

INTERSECT Feature. Consider solving the equation

$$x^3 = 3x + 1.$$

Doing so amounts to finding the x-coordinates of the point(s) of intersection of the two functions

$$f(x) = x^3 \quad \text{and} \quad g(x) = 3x + 1.$$

We enter the functions as

$$y_1 = x^3 \quad \text{and} \quad y_2 = 3x + 1$$

and then graph. We use a $[-3, 3, -5, 8]$ window to see the curvature and possible points of intersection.

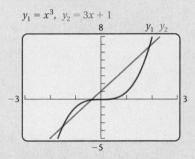

There appear to be at least three points of intersection. Let's adjust the viewing window to isolate the point of intersection on the left.

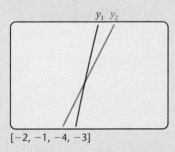

$[-2, -1, -4, -3]$

Using the INTERSECT feature (see the CALC menu), we see that the point of intersection is about $(-1.53, -3.60)$.

y₁ y₂

Intersection
X=-1.532089   Y=-3.596267

In a similar manner, we find the other points of intersection to be about $(-0.35, -0.04)$ and $(1.88, 6.64)$. The solutions of $x^3 = 3x + 1$ are the x-coordinates of these points:

$$-1.53, \quad -0.35, \quad \text{and} \quad 1.88.$$

ZERO Feature. A ZERO, or ROOT, feature can be used to solve an equation. The word "zero" in this context refers to an input, or x-value, for which the output of a function is 0.

To use such a feature, we must first get a 0 on one side of the equation. Thus, to solve $x^3 = 3x + 1$, we consider $x^3 - 3x - 1 = 0$ by subtracting $3x$ and then 1. Graphing $y = x^3 - 3x - 1$ and using the ZERO feature, we obtain a screen like the following.

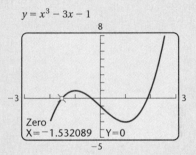

We move the tracer near the left-hand solution. We see that $x^3 - 3x - 1 = 0$ when $x = -1.53$, so -1.53 is an approximate solution of the equation $x^3 = 3x + 1$. Proceeding in a similar manner, we can approximate the other solutions as -0.35 and 1.88.

Exercises

Using the INTERSECT feature, solve the equation.

1. $x^2 = 10 - 3x$ (Check by factoring.)
2. $2x + 24 = x^2$ (Check by factoring.)
3. $x^3 = 3x - 2$
4. $x^4 - 2x^2 = 0$

Using the ZERO feature, solve the equation.

5. $0.4x^2 = 280x$
6. $\frac{1}{3}x^3 - \frac{1}{2}x^2 = 2x - 1$
7. $x^2 = 0.1x^4 + 0.4$
8. $0 = 2x^4 - 4x^2 + 2$
9. $x^4 + x^3 = 4x^2 + 2x - 4$
10. $11x^2 - 9x - 18 = x^4 - x^3$

9. *Fireworks.* Suppose that a bottle rocket is launched upward with an initial velocity of 96 ft/sec from a height of 880 ft. Its height h, in feet, after t seconds is given by

$$h(t) = -16t^2 + 96t + 880.$$

After how long will the rocket reach the ground?

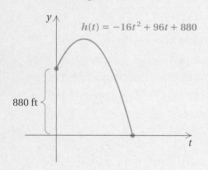

Answer on page A-27

b | **Applications and Problem Solving**

Some problems can be translated to quadratic equations. The problem-solving process is the same one we use for other kinds of applied problems.

Example 8   *Fireworks Displays.* Fireworks are typically launched from a mortar with an upward velocity (initial speed) of about 64 ft/sec. The height h, in feet, of a "weeping willow" display, t seconds after having been launched from an 80-ft-high rooftop, is given by the function

$$h(t) = -16t^2 + 64t + 80.$$

After how long will the cardboard shell from the fireworks reach the ground?

1. **Familiarize.**   We first make a drawing and label it, using the information provided. If we wanted to, we could evaluate $h(t)$ for a few values of t. Note that t cannot be negative, since it represents time from launch.

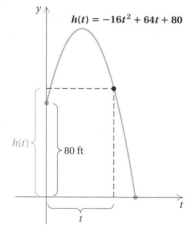

2. **Translate.**   Since we are asked to determine how long it will take for the shell to reach the ground, we are interested in the value of t for which $h(t) = 0$:

$$-16t^2 + 64t + 80 = 0.$$

3. **Solve.**   We solve by factoring:

$$-16t^2 + 64t + 80 = 0$$
$$\left.\begin{array}{c} -16(t^2 - 4t - 5) = 0 \\ -16(t - 5)(t + 1) = 0 \end{array}\right\} \quad \text{Factoring}$$
$$t - 5 = 0 \quad or \quad t + 1 = 0$$
$$t = 5 \quad or \qquad t = -1.$$

The solutions appear to be 5 and −1.

4. **Check.**   Remember: A solution of the equation may not be a solution of the original problem. Since t cannot be negative, we need check only 5: $h(5) = -16 \cdot 5^2 + 64 \cdot 5 + 80 = 0$. The number 5 checks.

5. **State.**   The cardboard shell will reach the ground about 5 sec after the fireworks display has been launched.

Do Exercise 9.

The following example involves the **Pythagorean theorem,** which relates the lengths of the sides of a right triangle. A **right triangle** has a 90°, or right, angle, which is denoted by a symbol like ⌐, The longest side, opposite the 90° angle, is called the **hypotenuse.** The other sides, called **legs,** form the two sides of the right angle.

> **THE PYTHAGOREAN THEOREM**
>
> The sum of the squares of the lengths of the legs of a right triangle is equal to the square of the length of the hypotenuse:
>
> $$a^2 + b^2 = c^2.$$

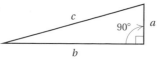

Example 9 *Skateboard Ramp.* The lengths of the sides of a right triangle formed by a skateboard ramp form three consecutive integers. Find all three lengths.

1. **Familiarize.** We first make a drawing. We let

   x = the length of the first side.

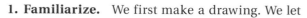

Since the lengths are consecutive integers, we know that

   $x + 1$ = the length of the second side

and

   $x + 2$ = the length of the third side.

The hypotenuse is always the longest side of a right triangle, so we know that it is the side with length $x + 2$.

2. **Translate.** Applying the Pythagorean theorem, we obtain the following translation:

   $$a^2 + b^2 = c^2$$
   $$x^2 + (x + 1)^2 = (x + 2)^2.$$

3. **Solve.** We solve the equation:

   $$x^2 + (x^2 + 2x + 1) = x^2 + 4x + 4 \qquad \text{Squaring the binomials}$$
   $$2x^2 + 2x + 1 = x^2 + 4x + 4 \qquad \text{Collecting like terms}$$
   $$x^2 - 2x - 3 = 0 \qquad \text{Subtracting to get 0 on one side}$$
   $$(x - 3)(x + 1) = 0 \qquad \text{Factoring}$$
   $$x - 3 = 0 \quad or \quad x + 1 = 0$$
   $$x = 3 \quad or \qquad x = -1.$$

4. **Check.** The integer -1 cannot be a length of a side because it is negative. When $x = 3$, $x + 1 = 4$, and $x + 2 = 5$; and $3^2 + 4^2 = 5^2$. So 3, 4, and 5 check.

5. **State.** The consecutive integers 3, 4, and 5 are the lengths of the sides.

Do Exercise 10.

10. *Loading-Dock Ramp.* The lengths of the sides of a right triangle formed by a ramp on a loading dock are such that one leg has length 5 m. The lengths of the other sides are consecutive integers. Find the lengths of the other sides of the triangle.

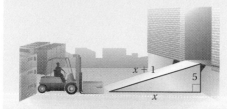

Answer on page A-27

11. *Games in a Sports League.* Use

$$N(n) = n^2 - n$$

for the following.

a) A men's basketball league has 16 teams. If we assume that each team plays every other team twice, what is the total number of games played?

b) Another basketball league plays a total of 42 games. If we assume that each team plays every other team twice, how many teams are in the league?

Answers on page A-27

Example 10 *Games in a Sports League.* In a sports league of n teams in which each team plays every other team twice, the total number N of games to be played is given by the function

$$N(n) = n^2 - n.$$

a) A women's college volleyball league has 6 teams. If we assume that each team plays every other team twice, what is the total number of games to be played?

Team	Wins	Losses
Senators	6	1
Tigers	5	2
Phillies	4	3
Giants	3	4
Friends	2	5
Phenatics	1	6

b) Another volleyball league plays a total of 72 games. If we assume that each team plays every other team twice, how many teams are in the league?

We solve as follows.

a) We evaluate the polynomial function for $n = 6$:

$$N(6) = n^2 - n = 6^2 - 6 = 36 - 6 = 30.$$

The league plays a total of 30 games.

b) We solve using the five-step problem-solving procedure.

1. **Familiarize.** We are given that n is the number of teams in the league and $N(n)$ is the total number of games played. In (a), we have familiarized ourselves with the problem by making the substitution.

2. **Translate.** To find the number n of teams in a league in which 72 games are played, we solve for n such that $N(n) = 72$:

$$n^2 - n = 72. \qquad \text{Substituting 72 for } N \text{ and } n^2 - n \text{ for } N(n)$$

3. **Solve.** We solve the equation:

$$n^2 - n = 72$$
$$n^2 - n - 72 = 0 \qquad \text{Subtracting 72 to get 0 on one side}$$
$$(n - 9)(n + 8) = 0 \qquad \text{Factoring}$$
$$n - 9 = 0 \quad or \quad n + 8 = 0 \qquad \text{Using the principle of zero products}$$
$$n = 9 \quad or \qquad n = -8.$$

4. **Check.** The solutions of the equation are 9 and -8. Since the number of teams cannot be negative, -8 cannot be a solution. But 9 checks, since $9^2 - 9 = 72$.

5. **State.** There are 9 teams in the league.

Do Exercise 11.

Exercise Set 4.7

a Solve.

1. $x^2 + 3x = 28$

2. $y^2 - 4y = 45$

3. $y^2 + 9 = 6y$

4. $r^2 + 4 = 4r$

5. $x^2 + 20x + 100 = 0$

6. $y^2 + 10y + 25 = 0$

7. $9x + x^2 + 20 = 0$

8. $8y + y^2 + 15 = 0$

9. $x^2 + 8x = 0$

10. $t^2 + 9t = 0$

11. $x^2 - 25 = 0$

12. $p^2 - 49 = 0$

13. $z^2 = 144$

14. $y^2 = 64$

15. $y^2 + 2y = 63$

16. $a^2 + 3a = 40$

17. $32 + 4x - x^2 = 0$

18. $27 + 12t + t^2 = 0$

19. $3b^2 + 8b + 4 = 0$

20. $9y^2 + 15y + 4 = 0$

21. $8y^2 - 10y + 3 = 0$

22. $4x^2 + 11x + 6 = 0$

23. $6z - z^2 = 0$

24. $8y - y^2 = 0$

25. $12z^2 + z = 6$

26. $6x^2 - 7x = 10$

27. $7x^2 - 7 = 0$

28. $4y^2 - 36 = 0$

29. $21r^2 + r - 10 = 0$

30. $12a^2 - 5a - 28 = 0$

31. $15y^2 = 3y$

32. $18x^2 = 9x$

33. $14 = x(x - 5)$

34. $x(x - 5) = 24$

35. $2x^3 - 2x^2 = 12x$

36. $50y + 5y^3 = 35y^2$

37. $2x^3 = 128x$

38. $147y = 3y^3$

39. $t^4 - 26t^2 + 25 = 0$

40. $x^4 - 13x^2 + 36 = 0$

41. Given that $f(x) = x^2 + 12x + 40$, find all values of x such that $f(x) = 8$.

42. Given that $f(x) = x^2 + 14x + 50$, find all values of x such that $f(x) = 5$.

43. Given that $g(x) = 2x^2 + 5x$, find all values of x such that $g(x) = 12$.

44. Given that $g(x) = 2x^2 - 15x$, find all values of x such that $g(x) = -7$.

45. Given that $h(x) = 12x + x^2$, find all values of x such that $h(x) = -27$.

46. Given that $h(x) = 4x - x^2$, find all values of x such that $h(x) = -32$.

Find the domain of the function f given by each of the following.

47. $f(x) = \dfrac{3}{x^2 - 4x - 5}$

48. $f(x) = \dfrac{2}{x^2 - 7x + 6}$

49. $f(x) = \dfrac{x}{6x^2 - 54}$

50. $f(x) = \dfrac{2x}{5x^2 - 20}$

51. $f(x) = \dfrac{x - 5}{9x - 18x^2}$

52. $f(x) = \dfrac{1 + x}{3x - 15x^2}$

53. $f(x) = \dfrac{7}{5x^3 - 35x^2 + 50x}$

54. $f(x) = \dfrac{3}{2x^3 - 2x^2 - 12x}$

b Solve.

55. The square of a number plus the number is 132. What is the number?

56. If four times the square of a number is 45 more than eight times the number, what is the number?

57. *Area of an Envelope.* An envelope is 4 cm longer than it is wide. The area is 96 cm². Find the length and the width.

58. *Book Area.* A book is 5 cm longer than it is wide. Find the length and the width if the area is 84 cm².

59. *Tent Design.* The triangular entrance to a tent is 2 ft taller than it is wide. The area of the entrance is 12 ft². Find the height and the base.

60. *Sailing.* A triangular sail is 9 m taller than it is wide. The area is 56 m². Find the height and the base of the sail.

Games in a Sports League. Use $N(n) = n^2 - n$ of Example 10 for Exercises 61 and 62.

61. A men's soccer league plays a total of 56 games. If we assume that each team plays every other team twice, how many teams are in the league?

62. A women's basketball league plays a total of 90 games. If we assume that each team plays every other team twice, how many teams are in the league?

63. *Flower Bed Design.* A flower bed is to be 3 m longer than it is wide. The flower bed will have an area of 108 m^2. What will its dimensions be?

64. *Workbench Design.* The length of the top of a workbench is 4 ft greater than the width. The area is 96 ft^2. Find the length and the width.

65. *Antenna Wires.* A wire is stretched from the ground to the top of an antenna tower, as shown. The wire is 20 ft long. The height of the tower is 4 ft greater than the distance d from the tower's base to the end of the wire. Find the distance d and the height of the tower.

66. *Ladder Location.* The foot of an extension ladder is 9 ft from a wall. The height that the ladder reaches on the wall and the length of the ladder are consecutive integers. How long is the ladder?

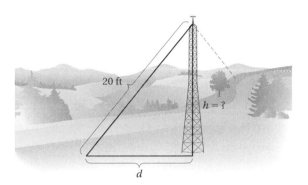

67. Three consecutive even integers are such that the square of the third is 76 more than the square of the second. Find the three integers.

68. Three consecutive even integers are such that the square of the first plus the square of the third is 136. Find the three integers.

69. *Geometry.* If each of the sides of a square is lengthened by 6 cm, the area becomes 144 cm^2. Find the length of a side of the original square.

70. *Geometry.* If each of the sides of a square is lengthened by 4 m, the area becomes 49 m^2. Find the length of a side of the original square.

71. *Sandbox Dimensions.* The area of the square base of a sandbox is 12 more than its perimeter. Find the length of a side.

72. *Framing a Picture.* A picture frame measures 12 cm by 20 cm, and 84 cm^2 of picture shows. Find the width of the frame.

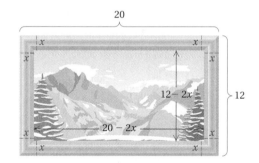

73. *Parking Lot Design.* A rectangular parking lot is 50 ft longer than it is wide. Determine the dimensions of the parking lot if it measures 250 ft diagonally.

74. *Framing a Picture.* A picture frame measures 14 cm by 20 cm, and 160 cm^2 of picture shows. Find the width of the frame.

75. *Triangle Dimensions.* One leg of a right triangle has length 9 m. The other sides have lengths that are consecutive integers. Find these lengths.

76. *Triangle Dimensions.* One leg of a right triangle is 10 cm. The other sides have lengths that are consecutive even integers. Find these lengths.

77. *Safety Flares.* Suppose that a flare is launched upward with an initial velocity of 80 ft/sec from a height of 224 ft. Its height h, in feet, after t seconds is given by

$$h(t) = -16t^2 + 80t + 224.$$

After how long will the flare reach the ground?

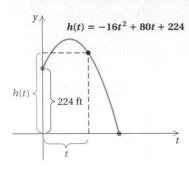

78. *Camcorder Production.* Suppose that the cost of making x video cameras is $C(x) = \frac{1}{9}x^2 + 2x + 1$, where $C(x)$ is in thousands of dollars. If the revenue from the sale of x video cameras is given by $R(x) = \frac{5}{36}x^2 + 2x$, where $R(x)$ is in thousands of dollars, how many cameras must be sold in order for the firm to break even? (See Section 3.7.)

Skill Maintenance

Find the distance between the given pair of points on the number line. [1.6b]

79. $-3, 4$

80. $-3, -4$

81. $3, -4$

82. $-7.8, -10.3$

83. $3.6, 4.9$

84. $-\frac{3}{5}, \frac{2}{3}$

85. $-123, 568$

86. $0, -1023$

Find an equation of the line containing the given pair of points. [2.6c]

87. $(-2, 7)$ and $(-8, -4)$

88. $(-2, 7)$ and $(8, -4)$

89. $(-2, 7)$ and $(8, 4)$

90. $(-24, 10)$ and $(-86, -42)$

Synthesis

91. ◆ Explain how one could write a quadratic equation that has 5 and -3 as solutions. Can the number of solutions of a quadratic equation exceed two? Why or why not?

92. ◆ Suppose that you are given a detailed graph of $y = P(x)$, where $P(x)$ is a polynomial. How could you use the graph to solve the equation $P(x) = 0$? $P(x) = 4$?

93. Following is the graph of $f(x) = -x^2 - 2x + 3$. Use *only* the graph to solve $-x^2 - 2x + 3 = 0$ and $-x^2 - 2x + 3 \geq -5$.

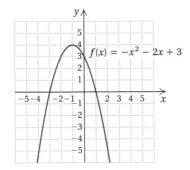

94. Following is the graph of $f(x) = x^4 - 3x^3$. Use *only* the graph to solve $x^4 - 3x^3 = 0$, $x^4 - 3x^3 \leq 0$, and $x^4 - 3x^3 > 0$.

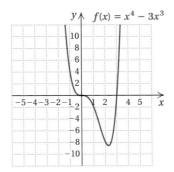

95. Use the TABLE feature to check that -5 and 3 are not in the domain of F, as shown in Example 7.

96. Use the TABLE feature to check your answers to Exercises 47, 50, and 53.

97. Use a grapher to solve each equation.

  a) $x^4 - 3x^3 - x^2 + 5 = 0$
  b) $x^4 - 3x^3 - x^2 + 5 = 5$
  c) $x^4 - 3x^3 - x^2 + 5 = -8$
  d) $x^4 = 1 + 3x^3 + x^2$

98. Solve each of the following equations.

  a) $(8x + 11)(12x^2 - 5x - 2) = 0$
  b) $(3x^2 - 7x - 20)(x - 5) = 0$
  c) $3x^3 + 6x^2 - 27x - 54 = 0$
     (*Hint*: Factor by grouping.)
  d) $2x^3 + 6x^2 = 8x + 24$

Create and solve quadratic equations as a group.

Collaborative Learning Manual

Summary and Review Exercises: Chapter 4

Important Properties and Formulas

Factoring Formulas: $A^2 - B^2 = (A + B)(A - B)$, $A^2 + 2AB + B^2 = (A + B)^2$,
$A^2 - 2AB + B^2 = (A - B)^2$, $A^3 + B^3 = (A + B)(A^2 - AB + B^2)$,
$A^3 - B^3 = (A - B)(A^2 + AB + B^2)$

The Principle of Zero Products: **For any real numbers a and b:**
If $ab = 0$, then $a = 0$ or $b = 0$.
If $a = 0$ or $b = 0$, then $ab = 0$.

The objectives to be tested in addition to the material in this chapter are [2.2d], [2.5b], [3.2a, b, c], and [3.3a, b].

1. Given the polynomial   [4.1a]
$$3x^6y - 7x^8y^3 + 2x^3 - 3x^2:$$
   a) Identify the degree of each term and the degree of the polynomial.
   b) Identify the leading term and the leading coefficient.
   c) Arrange in ascending powers of x.
   d) Arrange in descending powers of y.

Evaluate the polynomial function for the given values. [4.1b]

2. $P(x) = x^3 - x^2 + 4x;$ $P(0)$ and $P(-1)$

3. $P(x) = 4 - 2x - x^2;$ $P(-2)$ and $P(5)$

4. *NASCAR Attendance.* Attendance at NASCAR auto races has grown rapidly over the past 10 years. Attendance A, in millions, can be approximated by the polynomial function
$$A(x) = 0.00107x^3 + 0.01399x^2 + 0.12387x + 2.71762,$$
where x is the number of years since 1987.   [4.1b]
   a) Use the equation to approximate the attendance in 1999.
   b) Use the graph below to approximate $A(8)$.

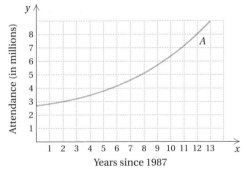

Source: NASCAR

Collect like terms.   [4.1c]

5. $4x^2y - 3xy^2 - 5x^2y + xy^2$

6. $3ab - 10 + 5ab^2 - 2ab + 7ab^2 + 14$

Add, subtract, or multiply.   [4.1c, d], [4.2a, b, c, d]

7. $(-6x^3 - 4x^2 + 3x + 1) + (5x^3 + 2x + 6x^2 + 1)$

8. $(4x^3 - 2x^2 - 7x + 5) + (8x^2 - 3x^3 - 9 + 6x)$

9. $(-9xy^2 - xy + 6x^2y) + (-5x^2y - xy + 4xy^2) + (12x^2y - 3xy^2 + 6xy)$

10. $(3x - 5) - (-6x + 2)$

11. $(4a - b + 3c) - (6a - 7b - 4c)$

12. $(9p^2 - 4p + 4) - (-7p^2 + 4p + 4)$

13. $(6x^2 - 4xy + y^2) - (2y^2 + 3xy - 2y^2)$

14. $(3x^2y)(-6xy^3)$

15. $(x^4 - 2x^2 + 3)(x^4 + x^2 - 1)$

16. $(4ab + 3c)(2ab - c)$

17. $(2x + 5y)(2x - 5y)$

18. $(2x - 5y)^2$

19. $(5x^2 - 7x + 3)(4x^2 + 2x - 9)$

20. $(x^2 + 4y^3)^2$

21. $(x - 5)(x^2 + 5x + 25)$

22. $\left(x - \tfrac{1}{3}\right)\left(x - \tfrac{1}{6}\right)$

23. Given that $f(x) = x^2 - 2x - 7$, find and simplify $f(a - 1)$ and $f(a + h) - f(a)$.   [4.2e]

Factor.   [4.3a, b, c], [4.4a, b], [4,5a, b, c, d], [4.6a]

24. $9y^4 - 3y^2$

25. $15x^4 - 18x^3 + 21x^2 - 9x$

26. $a^2 - 12a + 27$

27. $3m^2 + 14m + 8$

28. $25x^2 + 20x + 4$

29. $4y^2 - 16$

30. $ax + 2bx - ay - 2by$

31. $4x^4 + 4x^2 + 20$

32. $27x^3 - 8$

33. $0.064b^3 - 0.125c^3$

34. $y^5 - y$

35. $2z^8 - 16z^6$

36. $54x^6y - 2y$

37. $1 + a^3$

38. $36x^2 - 120x + 100$

39. $6t^2 + 17pt + 5p^2$

40. $x^3 + 2x^2 - 9x - 18$

41. $a^2 - 2ab + b^2 - 4t^2$

Solve.   [4.7a]

42. $x^2 - 20x = -100$

43. $6b^2 - 13b + 6 = 0$

44. $8y^2 = 14y$

45. $r^2 = 16$

46. Given that $f(x) = x^2 - 7x - 40$, find all values of x such that $f(x) = 4$.

47. Find the domain of the function f given by

$$f(x) = \frac{x - 3}{3x^2 + 19x - 14}.$$

Solve.   [4.7b]

48. *Photograph Dimensions.* A photograph is 3 in. longer than it is wide. When a 2-in. matte border is placed around the photograph, the total area of the photograph and the border is 108 in². Find the dimensions of the photograph.

49. The sum of the squares of three consecutive odd integers is 83. Find the integers.

50. *Area.* The area of a square is 7 more than six times the length of a side. What is the length of a side of the square?

Skill Maintenance

51. *Television Sales.* At the beginning of the month, J.C.'s Appliances had 150 TVs in stock. During the month, they sold 45% of their conventional TVs and 60% of their surround-sound TVs. They sold 78 TVs altogether. How many of each type did they sell? [3.2c]

Solve.   [3.2b]

52. $4x - 5y = 11,$
$3x + 7y = 2$

53. $2x - 3y = 7,$
$3y - 2x = 4$

Determine whether each of the following is the graph of a function.   [2.2d]

54.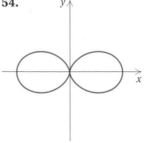

55.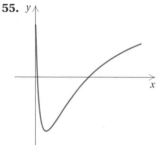

Use the slope and the y-intercept to graph the equation.   [2.5b]

56. $y = -\frac{1}{2}x - 3$

57. $y = \frac{2}{3}x + 2$

Synthesis

58. ◆ In this chapter, we learned to solve equations that we could not have solved before. Describe these new equations and the way we go about solving them. How is the procedure different from those we have used before now?   [4.7a]

59. ◆ Explain the error in each of the following.
a) $(a + 3)^2 - a^2 + 9$ [4.2c]
b) $a^3 + b^3 = (a + b)(a^2 - 2ab + b^2)$ [4.5d]
c) $(a - b)(a - b) = a^2 - b^2$ [4.2c]
d) $(x + 3)(x - 4) = x^2 - 12$ [4.2b]
e) $(p + 7)(p - 7) = p^2 + 49$ [4.2d]
f) $(t - 3)^2 = t^2 - 9$ [4.2c]

Factor.   [4.5d]

60. $128x^6 - 2y^6$

61. $(x + 1)^3 - (x - 1)^3$

62. Multiply: $[a - (b - 1)][(b - 1)^2 + a(b - 1) + a^2]$. [4.5d]

00. Solve. $64x^7 - x$.   [4.7a]

Test: Chapter 4

1. Given the polynomial
$$3xy^3 - 4x^2y + 5x^5y^4 - 2x^4y:$$
   a) Identify the degree of each term and the degree of the polynomial.
   b) Identify the leading term and the leading coefficient.
   c) Arrange in ascending powers of x.
   d) Arrange in descending powers of y.

2. Given that $P(x) = 2x^3 + 3x^2 - x + 4$, find $P(0)$ and $P(-2)$.

3. *Sports Car Sales.* Sales of sports cars experience rises and falls due generally to new or redesigned models such as the Corvette in 1997. Sales S, in thousands, for recent years can be approximated by the polynomial function
$$S(x) = 0.1x^4 - 1.6x^3 + 11.15x^2 - 38.04x + 112.65,$$
   where x is the number of years since 1990.
   a) Use the equation to approximate sports car sales in 2000.
   b) Use the graph at right to approximate $S(5)$.

Sports Car Sales

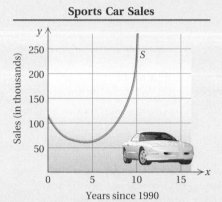

Source: Autodata

4. Collect like terms: $5xy - 2xy^2 - 2xy + 5xy^2$.

Add, subtract, or multiply.

5. $(-6x^3 + 3x^2 - 4y) + (3x^3 - 2y - 7y^2)$

6. $(4a^3 - 2a^2 + 6a - 5) + (3a^3 - 3a + 2 - 4a^2)$

7. $(5m^3 - 4m^2n - 6mn^2 - 3n^3) + (9mn^2 - 4n^3 + 2m^3 + 6m^2n)$

8. $(9a - 4b) - (3a + 4b)$

9. $(4x^2 - 3x + 7) - (-3x^2 + 4x - 6)$

10. $(6y^2 - 2y - 5y^3) - (4y^2 - 7y - 6y^3)$

11. $(-4x^2y)(-16xy^2)$

12. $(6a - 5b)(2a + b)$

13. $(x - y)(x^2 - xy - y^2)$

14. $(3m^2 + 4m - 2)(-m^2 - 3m + 5)$

15. $(4y - 9)^2$

16. $(x - 2y)(x + 2y)$

17. Given that $f(x) = x^2 - 5x$, find and simplify $f(a + 10)$ and $f(a + h) - f(a)$.

Answers

1. a) _____
   b) _____
   c) _____
   d) _____
2. _____
3. a) _____
   b) _____
4. _____
5. _____
6. _____
7. _____
8. _____
9. _____
10. _____
11. _____
12. _____
13. _____
14. _____
15. _____
16. _____
17. _____

18. _____

19. _____

20. _____

21. _____

22. _____

23. _____

24. _____

25. _____

26. _____

27. _____

28. _____

29. _____

30. _____

31. _____

32. _____

33. _____

34. _____

35. _____

36. _____

37. _____

38. _____

39. _____

40. _____

41. _____

42. _____

43. _____

44. _____

45. _____

46. _____

47. _____

Factor.

18. $9x^2 + 7x$

19. $24y^3 + 16y^2$

20. $y^3 + 5y^2 - 4y - 20$

21. $p^2 - 12p - 28$

22. $12m^2 + 20m + 3$

23. $9y^2 - 25$

24. $3r^3 - 3$

25. $9x^2 + 25 - 30x$

26. $(z + 1)^2 - b^2$

27. $x^8 - y^8$

28. $y^2 + 8y + 16 - 100t^2$

29. $20a^2 - 5b^2$

30. $24x^2 - 46x + 10$

31. $16a^7b + 54ab^7$

Solve.

32. $x^2 - 18 = 3x$

33. $5y^2 - 125 = 0$

34. $2x^2 + 21 = -17x$

35. Given that $f(x) = 3x^2 - 15x + 11$, find all values of x such that $f(x) = 11$.

36. Find the domain of the function f given by

$$f(x) = \frac{3 - x}{x^2 + 2x + 1}.$$

Solve.

37. The square of a number plus the number is 156. What is the number?

38. *Photograph Dimensions.* A photograph is 3 cm longer than it is wide. Its area is 40 cm². Find its length and its width.

39. *Ladder Location.* The foot of an extension ladder is 10 ft from a wall. The ladder is 2 ft longer than the distance that it reaches up the wall. How far up the wall does the ladder reach?

40. *Area.* The area of a square is 5 more than four times the length of a side. What is the length of a side of the square?

41. *Number of Games in a League.* If there are n teams in a league and each team plays every other team once, the total number of games played is given by the polynomial function $f(n) = \frac{1}{2}n^2 - \frac{1}{2}n$. Find an equivalent expression for $f(n)$ by factoring out the largest common factor.

Skill Maintenance

42. *Auto Travel.* At noon, two cars leave from the same location traveling in opposite directions at different speeds. After 7 hr, they are 651 mi apart. One car is traveling 15 mph slower than the other. What are their speeds?

43. Solve:
$$5x - 6y = 12,$$
$$4x + 9y = 2.$$

44. Determine whether the following graph is that of a function.

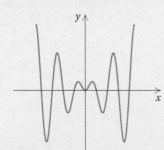

45. Use the slope and the y-intercept to graph $y = -\frac{2}{3}x - 4$.

Synthesis

46. Factor: $6x^{2n} - 7x^n - 20$.

47. If $pq = 5$ and $(p + q)^2 = 29$, find the value of $p^2 + q^2$.

Cumulative Review: Chapters R–4

1. Evaluate $\dfrac{2m - n}{4}$ for $m = 3$ and $n = 2$.

Simplify.

2. $|0|$

3. $-8 - (-4)$

4. $-3a + (6a - 1) - 4(a + 1)$

5. $2[5(3x + 4) - 2x] - [7(3x + 4) - 8]$

6. $[(-3a^{-6}b^2)^5]^{-2}$

7. $(x^2 + 4x - xy - 9) + (-3x^2 - 3x + 8)$

8. $(6x^2 - 3x + 2x^3) - (8x^2 - 9x + 2x^3)$

9. $(a^2 - a - 3) \cdot (a^2 + 2a - 3)$

10. $(x + 4)(x + 9)$

Solve.

11. $8 - 3x = 6x - 10$

12. $\frac{1}{2}x - 3 = \frac{7}{2}$

13. $A = \frac{1}{2}h(a + b)$, for b

14. $6x - 1 \le 3(5x + 2)$

15. $4x - 3 < 2$ or $x - 3 > 1$

16. $|2x - 3| < 7$

17. $x + y + z = -5,$
$x - z = 10,$
$y - z = 12$

18. $2x + 5y = -2,$
$5x + 3y = 14$

19. $3x - y = 7,$
$2x + 2y = 5$

20. $x + 2y - z = 0,$
$3x + y - 2z = -1,$
$x - 4y + z = -2$

21. $2x + 3y = 2,$
$-4x + 3y = -7$

22. $-3a + 2b = 0,$
$3a - 4b = -1$

23. $2x + 2y - 4z = 1,$
$-2x - 4y + 8z = -1,$
$4x + 4y + 4z = 5$

24. $11x + x^2 + 24 = 0$

25. $2x^2 - 15x = -7$

26. Given that $f(x) = 3x^2 + 4x$, find all values of x such that $f(x) = 4$.

27. Find the domain of the function F given by
$$F(x) = \frac{x + 7}{x^2 - 2x - 15}.$$

Factor.

28. $3x^3 - 12x^2$

29. $2x^4 + x^3 + 2x + 1$

30. $x^2 + 5x - 14$

31. $20a^2 - 23a + 6$

32. $4x^2 - 25$

33. $2x^2 - 28x + 98$

34. $a^3 + 64$

35. $8x^3 - 1$

36. $4a^3 + a^6 - 12$

37. $4x^4y^2 - x^2y^4$

Graph.

38. $x < 1$ or $x \ge 2$

39. $y = -2x$

40. $y = \frac{1}{2}x$

41. $4y + 3x = 12 + 3x$

42. $6y + 24 = 0$

43. $y > x + 6$

44. $2x + y \le 2$

45. $f(x) = x^2 - 3$

46. $g(x) = 4 - |x|$

Graph. List the vertices.

47. $y \geq -x,$
$y \leq 2x + 1$

48. $2x + 3y \leq 6,$
$5x - 5y \leq 15,$
$x \geq 0$

49. Find an equation of the line containing the point $(3, 7)$ and parallel to the line $x + 2y = 6$.

50. Find an equation of the line containing the point $(3, -2)$ and perpendicular to the line $3x + 4y = 5$.

51. Find an equation of the line containing the points $(-1, 4)$ and $(-1, 0)$.

52. Find an equation of the line with slope -3 and through the point $(2, 1)$.

53. *Diaper Changes.* A family with twins did an experiment on the cost of diapering their two children. Disposables were used on one child and cloth diapers from a diaper service were used on the other child. After the twins were out of diapers, calculations showed that for both children, $2886 was spent on diapering. Using disposables cost $546 more than using the diaper service. How much did each cost?

54. *Casualty Loss.* When filing your income tax, you may claim a casualty loss only if it exceeds 10% of your adjusted gross income plus $100. If the loss was $1500, can you claim it if your adjusted gross income is $13,500?

55. *Gasoline Prices.* Caracas, Venezuela, has the most inexpensive gas in the world. The total cost of 10 gal of gas in Caracas and 5 gal of gas in Lagos, Nigeria, is $2.90. The difference in the price per gallon in each location is $0.16. What is the price per gallon in each city?

56. *Lens Polishing.* In a factory there are three polishing machines, A, B, and C. When all three of them are working, 5700 lenses can be polished in one week. When only A and B are working, 3400 lenses can be polished in one week. When only B and C are working, 4200 lenses can be polished in one week. How many lenses can be polished in a week by each machine?

57. *Solid-Waste Generation.* The function

$$W(t) = 0.05t + 4.3$$

can be used to estimate the number of pounds W of solid waste produced daily, on average, by each person in the United States for t years since 1991.

a) Estimate the number of pounds of solid waste that will be produced in 2001 by finding $W(10)$.
b) In what year will solid-waste production be 6.25 lb per day?
c) For what years will waste production range from 6.0 to 6.25 lb per person per day?

Synthesis

58. Solve: $|x + 1| \leq |x - 3|$.

59. Bert and Sally Ng are mathematics professors at a state university. Together they have 46 years of service. Two years ago, Bert had taught 2.5 times as long as Sally. How long has each taught at the university?

5

Rational Expressions, Equations, and Functions

Introduction

A rational expression is the ratio of two polynomials. In this chapter, we learn to manipulate rational expressions and solve rational equations. This will allow us to solve some new types of applied problems.

A rational function is a function described by a simplified rational expression. We will connect rational functions to many of the concepts of this chapter.

5.1 Rational Expressions and Functions: Multiplying, Dividing, and Simplifying

5.2 LCMs, LCDs, Addition, and Subtraction

5.3 Division of Polynomials

5.4 Complex Rational Expressions

5.5 Solving Rational Equations

5.6 Applications and Problem Solving

5.7 Formulas and Applications

5.8 Variation and Applications

An Application

The intensity *I* of a television signal varies inversely as the square of the distance *d* from the transmitter. If the intensity is 23 watts per square meter (W/m²) at a distance of 2 km, what is the intensity at a distance of 6 km?

This problem appears as Example 10 in Section 5.8.

The Mathematics

The equation of variation is

$$I = \frac{92}{d^2}.$$

This is a rational equation.

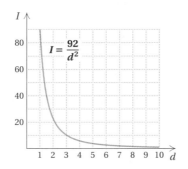

Pretest: Chapter 5

1. Find the LCM of $x^3 - 2x^2$ and $2x^2 - 6x + 4$.

Perform the indicated operations and simplify.

2. $\dfrac{5}{y^2 + 8y + 16} - \dfrac{4}{y^2 + 9y + 20}$

3. $\dfrac{a - 1}{1 - a} + \dfrac{3a + 3}{a^2 - 1}$

4. $\dfrac{2}{y + 8} + \dfrac{24}{y^2 - 64} - \dfrac{3}{2y - 16}$

5. $\dfrac{y^2 - 8y - 33}{y^2 - 6y + 8} \div \dfrac{y^2 + y - 6}{y^2 - 15y + 44}$

6. $\dfrac{y^2 + 6y + 9}{3y^2 - 27} \cdot \dfrac{2y - 6}{y^2 + 4y + 3}$

7. Simplify: $\dfrac{4 - \dfrac{4}{y}}{4 + \dfrac{4}{y}}$.

8. Divide: $(x^2 + 8x - 3) \div (x - 4)$.

Solve.

9. $\dfrac{x}{11} + \dfrac{x}{6} = 2$

10. $\dfrac{3}{y + 7} - \dfrac{6}{y^2 + 10y + 21} = \dfrac{4}{y + 3}$

11. Solve for b: $M = \dfrac{ab + c}{ab}$.

12. Find the domain of f if
$$f(x) = \dfrac{2x + 5}{x^2 + 7x + 10}.$$

13. Find an equation of variation in which y varies jointly as x and the square of z and inversely as w, and $y = 1$ when $x = 5$, $z = 4$, and $w = 20$.

14. *Patio Concrete.* The amount of concrete necessary to construct a patio varies directly as the area of the patio. You note on the package that it takes 4.5 yd^3 of concrete to construct a 200-ft^2 patio. How many cubic yards of concrete will you need to construct a 250-ft^2 patio?

15. *Air Travel.* Airplane A travels 50 km/h faster than airplane B. Airplane A travels 875 km in the same time that it takes airplane B to travel 700 km. Find the speed of each plane.

16. *Keyboarding Time.* One administrative assistant can keyboard a report in 4 hr. Another assistant can keyboard the same report in 3 hr. How long would it take them to keyboard the same report working together?

Objectives for Retesting

The objectives to be tested in addition to the material in this chapter are as follows.

[2.3a]   Find the domain and the range of a function.
[3.6b]   Graph linear inequalities in two variables.
[4.4a, b]   Factor trinomials of the type $ax^2 + bx + c$, $a \neq 1$, by the grouping method or the FOIL method.
[4.5d]   Factor sums and differences of cubes.

5.1 Rational Expressions and Functions: Multiplying, Dividing, and Simplifying

a | Rational Expressions and Functions

An expression that consists of the quotient of two polynomials, where the polynomial in the denominator is nonzero, is called a **rational expression.** The following are examples of rational expressions:

$$\frac{7}{8}, \quad \frac{z}{-6}, \quad \frac{a}{b}, \quad \frac{8}{y + 5}, \quad \frac{t^4 - 5t}{t^2 - 3t - 28}, \quad \frac{x^2 + 7xy - 4}{x^3 - y^3}.$$

Note that every rational number is a rational expression.

Rational expressions indicate division. Thus we cannot make a replacement of the variable that allows a denominator to be 0.

Example 1 Find all numbers for which the rational expression

$$\frac{2x + 1}{x - 3}$$

is undefined.

When x is replaced with 3, the denominator is 0, and the expression is undefined:

$$\frac{2x + 1}{x - 3} = \frac{2 \cdot 3 + 1}{3 - 3} = \boxed{\frac{7}{0}}. \leftarrow \text{Undefined}$$

You can check some replacements other than 3 to see that it appears that 3 is the only replacement that is not allowable. Thus the rational expression is undefined for the number 3.

You may have noticed that the procedure in Example 1 is similar to one we have performed when finding the domain of a function.

Example 2 Find the domain of f if $f(x) = \dfrac{2x + 1}{x - 3}$.

The domain is the set of all allowable replacements (see Section 2.3). We begin by determining the replacements that are *not* allowable. We can do this by setting the denominator equal to 0 and solving for x:

$$x - 3 = 0$$
$$x = 3.$$

The domain of f is $\{x \mid x$ is a real number *and* $x \neq 3\}$, or, in interval notation, $(-\infty, 3) \cup (3, \infty)$.

Do Exercises 1 and 2.

Objectives

a Find all numbers for which a rational expression is undefined or that are not in the domain of a rational function.

b Multiply a rational expression by 1, using an expression like A/A.

c Simplify rational expressions.

d Multiply rational expressions and simplify.

e Divide rational expressions and simplify.

For Extra Help

TAPE 11   TAPE 9B   MAC WIN   CD-ROM

1. Find all numbers for which the rational expression

$$\frac{x^2 - 4x + 9}{2x + 5}$$

is undefined.

2. Find the domain of f if

$$f(x) = \frac{x^2 - 4x + 9}{2x + 5}.$$

Write interval notation for the answer.

Answers on page A-30

AG Algebraic–Graphical Connection

Although graphing rational functions is not an objective of this text, other than in an optional Calculator Spotlight, it is worthwhile to look at a graph of the function in Example 2.

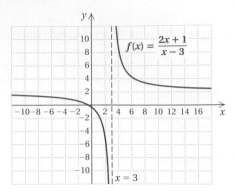

Note that the graph consists of two unconnected "branches." If a vertical line were drawn at $x = 3$, shown dashed here, it would not touch the graph of f. Thus 3 is not in the domain of f.

AG

Calculator Spotlight

Graphs of Rational Functions. Consider the rational function given by

$$f(x) = \frac{2x + 1}{x - 3}.$$

Let's consider its graph and compare it with the preceding hand-drawn graph. Two versions of the graph are shown below.

CONNECTED Mode

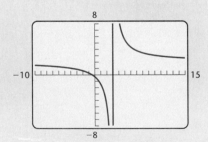

DOT Mode

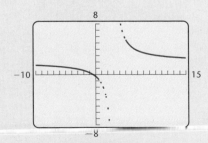

Here we see a disadvantage of the grapher. CONNECTED mode can lead to an *incorrect* graph. Because, in CONNECTED mode, a grapher connects plotted points with line segments, it connects branches of the graph, making it appear as though the vertical line $x = 3$ is part of the graph.

On the other hand, in DOT mode, a grapher simply plots dots representing coordinates of points.

If you have a choice in plotting functions given by rational expressions, use DOT mode.

Exercises

Graph each of the following functions using DOT mode.

1. $f(x) = \dfrac{4}{x - 2}$ **2.** $f(x) = \dfrac{x}{x + 2}$

3. $f(x) = \dfrac{x^2 - 1}{x^2 + x - 6}$ **4.** $f(x) = \dfrac{x^2 - 4}{x - 1}$

5. $f(x) = \dfrac{10}{x^2 + 4}$ **6.** $f(x) = \dfrac{8}{x^2 - 4}$

7. $f(x) = \dfrac{2x + 3}{3x^2 + 7x - 6}$ **8.** $f(x) = \dfrac{2x^3}{x^2 + 1}$

Example 3 Find all numbers for which the rational expression

$$\frac{t^4 - 5t}{t^2 - 3t - 28}$$

is undefined.

The rational expression will be undefined for a replacement that makes the denominator 0. To determine those replacements to exclude, we set the denominator equal to 0 and solve:

$t^2 - 3t - 28 = 0$	Setting the denominator equal to 0
$(t - 7)(t + 4) = 0$	Factoring
$t - 7 = 0$ *or* $t + 4 = 0$	Using the principle of zero products
$t = 7$ *or* $t = -4$.	

Thus the expression is undefined for the replacements 7 and -4.

Example 4 Find the domain of g if

$$g(t) = \frac{t^4 - 5t}{t^2 - 3t - 28}.$$

We proceed as we did in Example 3. The expression is undefined for the replacements 7 and -4. Thus the domain is $\{t \mid t$ is a real number *and* $t \neq 7$ *and* $t \neq -4\}$, or, in interval notation, $(-\infty, -4) \cup (-4, 7) \cup (7, \infty)$.

Do Exercises 3 and 4.

b | Finding Equivalent Rational Expressions

Calculations with rational expressions are similar to those with rational numbers.

> To multiply two rational expressions, multiply numerators and multiply denominators.

For example, we have the following:

$$\frac{3}{5} \cdot \frac{2}{7} = \frac{3 \cdot 2}{5 \cdot 7} = \frac{6}{35}$$

and

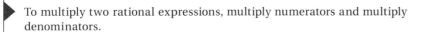

$$\frac{x + 3}{y - 4} \cdot \frac{x^3}{y + 5} = \frac{(x + 3)x^3}{(y - 4)(y + 5)}. \qquad \text{Multiplying numerators and multiplying denominators}$$

For purposes of our work in this chapter, it is better in the second example above to leave the numerator $(x + 3)x^3$ and the denominator $(y - 4)(y + 5)$ in factored form because it is easier to simplify if we do not multiply.

To simplify, we first consider multiplying by 1. Any rational expression with the same numerator and denominator is a symbol for 1:

$$\frac{73}{73} = 1, \qquad \frac{x - y}{x - y} = 1, \qquad \frac{4x^2 - 5}{4x^2 - 5} = 1, \qquad \frac{-1}{-1} = 1, \qquad \frac{x + 5}{x + 5} = 1.$$

We can multiply by 1 to get equivalent expressions. For example,

$$\frac{7}{9} \cdot \frac{4}{4} = \frac{7 \cdot 4}{9 \cdot 4} = \frac{28}{36}.$$

3. Find all numbers for which the rational expression

$$\frac{t^2 - 9}{t^2 - 7t + 10}$$

is undefined.

4. Find the domain of g if

$$g(t) = \frac{t^2 - 9}{t^2 - 7t + 10}.$$

Write interval notation for the answer.

Answers on page A-30

Multiply.

5. $\dfrac{3x + 2y}{5x + 4y} \cdot \dfrac{x}{x}$

6. $\dfrac{2x^2 - y}{3x + 4} \cdot \dfrac{3x + 2}{3x + 2}$

7. $\dfrac{-1}{-1} \cdot \dfrac{2a - 5}{a - b}$

8. Simplify by removing a factor of 1:

$\dfrac{128}{160}.$

Answers on page A-30

As another example, let's multiply $(x + y)/5$ by 1, using the symbol $(x - y)/(x - y)$:

$$\frac{x + y}{5} \cdot \frac{x - y}{x - y} = \frac{(x + y)(x - y)}{5(x - y)}. \qquad \text{Multiplying by } \frac{x - y}{x - y}, \text{ which is 1}$$

We know that the expressions

$$\frac{x + y}{5} \quad \text{and} \quad \frac{(x + y)(x - y)}{5(x - y)}$$

are equivalent. This means that they will name the same number for all replacements that do not make a denominator 0.

Examples Multiply to obtain an equivalent expression.

5. $\dfrac{x^2 + 3}{x - 1} \cdot 1 = \dfrac{x^2 + 3}{x - 1} \cdot \dfrac{x + 1}{x + 1} = \dfrac{(x^2 + 3)(x + 1)}{(x - 1)(x + 1)}$ Using $\dfrac{x + 1}{x + 1}$ for 1

6. $\dfrac{-1}{-1} \cdot \dfrac{x - 4}{x - y} = \dfrac{-1 \cdot (x - 4)}{-1 \cdot (x - y)} = \dfrac{-x + 4}{-x + y}$

Do Exercises 5–7.

c | Simplifying Rational Expressions

We simplify rational expressions using the identity property of 1 in reverse. That is, we "remove" factors that are equal to 1. We first factor the numerator and the denominator and then factor the rational expression, so that a factor is equal to 1. We also say, accordingly, that we "remove a factor of 1."

Example 7 Simplify by removing a factor of 1: $\dfrac{120}{320}.$

$$\frac{120}{320} = \frac{40 \cdot 3}{40 \cdot 8} \qquad \text{Factoring the numerator and the denominator, looking for common factors}$$

$$= \frac{40}{40} \cdot \frac{3}{8} \qquad \text{Factoring the rational expression}$$

$$= 1 \cdot \frac{3}{8} \qquad \frac{40}{40} = 1$$

$$= \frac{3}{8} \qquad \text{Removing a factor of 1}$$

Do Exercise 8.

Examples Simplify by removing a factor of 1.

8. $\dfrac{5x^2}{x} = \dfrac{5x \cdot x}{1 \cdot x} \qquad \text{Factoring the numerator and the denominator}$

$\qquad = \dfrac{5x}{1} \cdot \dfrac{x}{x} \qquad \text{Factoring the rational expression}$

$\qquad = 5x \cdot 1 \qquad \dfrac{x}{x} = 1$

$\qquad = 5x \qquad \text{Removing a factor of 1}$

In this example, we supplied a 1 in the denominator. This can always be done, but it is not necessary.

9. $\dfrac{4a + 8}{2} = \dfrac{2(2a + 4)}{2 \cdot 1}$ Factoring the numerator and the denominator

$\qquad = \dfrac{2}{2} \cdot \dfrac{2a + 4}{1}$ Factoring the rational expression

$\qquad = \dfrac{2a + 4}{1}$ Removing a factor of 1

$\qquad = 2a + 4$

Do Exercises 9 and 10.

Examples   Simplify by removing a factor of 1.

10. $\dfrac{2x^2 + 4x}{6x^2 + 2x} = \dfrac{2x(x + 2)}{2x(3x + 1)}$ Factoring the numerator and the denominator

$\qquad = \dfrac{2x}{2x} \cdot \dfrac{x + 2}{3x + 1}$ Factoring the rational expression

$\qquad = \dfrac{x + 2}{3x + 1}$ Removing a factor of 1

11. $\dfrac{x^2 - 1}{2x^2 - x - 1} = \dfrac{(x - 1)(x + 1)}{(2x + 1)(x - 1)}$ Factoring the numerator and the denominator

$\qquad = \dfrac{x - 1}{x - 1} \cdot \dfrac{x + 1}{2x + 1}$ Factoring the rational expression

$\qquad = \dfrac{x + 1}{2x + 1}$ Removing a factor of 1

12. $\dfrac{9x^2 + 6xy - 3y^2}{12x^2 - 12y^2} = \dfrac{3(x + y)(3x - y)}{3(4)(x + y)(x - y)}$ Factoring the numerator and the denominator

$\qquad = \dfrac{3(x + y)}{3(x + y)} \cdot \dfrac{3x - y}{4(x - y)}$ Factoring the rational expression

$\qquad = \dfrac{3x - y}{4(x - y)}$ Removing a factor of 1

For purposes of later work, we generally do not multiply out the numerator and the denominator when simplifying rational expressions.

CANCELING.   Canceling is a shortcut that you may have used for removing a factor of 1 when working with fractional notation or rational expressions. With great concern, we mention it here as a possible way to speed up your work. **Canceling may be done for removing factors of 1 only in products.** It *cannot* be done in sums or when adding expressions together. Our concern is that canceling be done with care and understanding. Example 12 might have been done faster as follows:

$$\frac{9x^2 + 6xy - 3y^2}{12x^2 - 12y^2} = \frac{\cancel{3}\,(x + y)(3x - y)}{\cancel{3}(4)(x + y)(x - y)}$$ When a factor of 1 is noted, it is "canceled" as shown.

$$= \frac{3x - y}{4(x - y)}.$$ Removing a factor of 1: $\dfrac{3(x + y)}{3(x + y)} = 1$

Simplify by removing a factor of 1.

9. $\dfrac{7x^2}{x}$

10. $\dfrac{6a + 9}{3}$

Answers on page A-30

Simplify by removing a factor of 1.

11. $\dfrac{6x^2 + 4x}{4x^2 + 8x}$

12. $\dfrac{2y^2 + 6y + 4}{y^2 - 1}$

13. $\dfrac{20a^2 - 80b^2}{16a^2 - 64ab + 64b^2}$

Multiply. Simplify by removing a factor of 1, if possible.

14. $\dfrac{(x - y)^3}{x + y} \cdot \dfrac{3x + 3y}{x^2 - y^2}$

15. $\dfrac{a^3 + b^3}{a^2 - b^2} \cdot \dfrac{a^2 - 2ab + b^2}{a^2 - ab + b^2}$

CAUTION! The difficulty with canceling is that it can be applied incorrectly in situations such as the following:

$$\dfrac{\cancel{2} + 3}{\cancel{2}} = 3, \qquad \dfrac{\cancel{4} + 1}{\cancel{4} + 2} = \dfrac{1}{2}, \qquad \dfrac{1\cancel{5}}{\cancel{5}4} = \dfrac{1}{4}.$$

Wrong!   Wrong!   Wrong!

In each of these situations, the expressions canceled were *not* factors of 1. Factors are parts of products. For example, in $2 \cdot 3$, 2 and 3 are factors, but in $2 + 3$, 2 and 3 are *not* factors. **If you can't factor, you can't cancel! If in doubt, don't cancel!**

Do Exercises 11–13.

d | Multiplying and Simplifying

After multiplying, we generally simplify, if possible. That is one reason why we leave the numerator and the denominator in factored form. Even so, we might need to factor them further in order to simplify.

Examples   Multiply. Then simplify by removing a factor of 1.

13. $\dfrac{x + 2}{x - 3} \cdot \dfrac{x^2 - 4}{x^2 + x - 2} = \dfrac{(x + 2)(x^2 - 4)}{(x - 3)(x^2 + x - 2)}$ Multiplying the numerators and the denominators

$$= \dfrac{(x + 2)(x + 2)(x - 2)}{(x - 3)(x + 2)(x - 1)}$$ Factoring the numerator and the denominator

$$= \dfrac{(x + 2)\cancel{(x + 2)}(x - 2)}{(x - 3)\cancel{(x + 2)}(x - 1)}$$ Removing a factor of 1: $\dfrac{x + 2}{x + 2} = 1$

$$= \dfrac{(x + 2)(x - 2)}{(x - 3)(x - 1)}$$ Simplifying

14. $\dfrac{a^3 - b^3}{a^2 - b^2} \cdot \dfrac{a^2 + 2ab + b^2}{a^2 + ab + b^2}$

$$= \dfrac{(a^3 - b^3)(a^2 + 2ab + b^2)}{(a^2 - b^2)(a^2 + ab + b^2)}$$

$$= \dfrac{(a - b)(a^2 + ab + b^2)(a + b)(a + b)}{(a - b)(a + b)(a^2 + ab + b^2) \cdot 1}$$ Factoring the numerator and the denominator

$$= \dfrac{\cancel{(a - b)}\cancel{(a^2 + ab + b^2)}\cancel{(a + b)}(a + b)}{\cancel{(a - b)}\cancel{(a + b)}\cancel{(a^2 + ab + b^2)} \cdot 1}$$ Removing a factor of 1: $\dfrac{(a - b)(a^2 + ab + b^2)(a + b)}{(a - b)(a^2 + ab + b^2)(a + b)} = 1$

$$= \dfrac{a + b}{1}$$ Simplifying

$$= a + b$$

Do Exercises 14 and 15.

Answers on page A-30

Chapter 5   Rational Expressions, Equations, and Functions

380

e | Dividing and Simplifying

Two expressions are reciprocals (or multiplicative inverses) of each other if their product is 1. To find the reciprocal of a rational expression, we interchange the numerator and the denominator.

The reciprocal of $\dfrac{x + 2y}{x + y - 1}$ is $\dfrac{x + y - 1}{x + 2y}$.

The reciprocal of $y - 8$ is $\dfrac{1}{y - 8}$.

Do Exercises 16–18.

> We divide with rational expressions as we do with real numbers. We multiply by the reciprocal of the divisor.

We sometimes say, "invert the divisor and multiply."

For example,

$$\frac{2}{3} \div \frac{4}{5} = \frac{2}{3} \cdot \frac{5}{4} = \frac{10}{12} = \frac{5 \cdot 2}{6 \cdot 2} = \frac{5}{6} \cdot \frac{2}{2} = \frac{5}{6} \cdot 1 = \frac{5}{6}.$$

Find the reciprocal.

16. $\dfrac{x + 3}{x - 5}$

17. $x + 7$

18. $\dfrac{1}{y^3 - 9}$

Answers on page A-30

Divide. Simplify by removing a factor of 1, if possible.

19. $\dfrac{x^2 + 7x + 10}{2x - 4} \div \dfrac{x^2 - 3x - 10}{x - 2}$

20. $\dfrac{a^2 - b^2}{ab} \div \dfrac{a^2 - 2ab + b^2}{2a^2b^2}$

21. Perform the indicated operations and simplify:

$\dfrac{a^3 + 8}{a - 2} \div (a^2 - 2a + 4) \cdot (a - 2)^2.$

Examples Divide. Simplify by removing a factor of 1, if possible.

15. $\dfrac{x - 2}{x + 1} \div \dfrac{x + 5}{x - 3} = \dfrac{x - 2}{x + 1} \cdot \dfrac{x - 3}{x + 5}$ Multiplying by the reciprocal of the divisor

$\qquad = \dfrac{(x - 2)(x - 3)}{(x + 1)(x + 5)}$

16. $\dfrac{a^2 - 1}{a - 1} \div \dfrac{a^2 - 2a + 1}{a + 1}$

$\qquad = \dfrac{a^2 - 1}{a - 1} \cdot \dfrac{a + 1}{a^2 - 2a + 1}$ Multiplying by the reciprocal of the divisor

$\qquad = \dfrac{(a^2 - 1)(a + 1)}{(a - 1)(a^2 - 2a + 1)}$ Multiplying the numerators and the denominators

$\qquad = \dfrac{(a + 1)(a - 1)(a + 1)}{(a - 1)(a - 1)(a - 1)}$ Factoring the numerator and the denominator

$\qquad = \dfrac{(a + 1)\cancel{(a - 1)}(a + 1)}{(a - 1)\cancel{(a - 1)}(a - 1)}$ Removing a factor of 1: $\dfrac{a - 1}{a - 1} = 1$

$\qquad = \dfrac{(a + 1)(a + 1)}{(a - 1)(a - 1)}$ Simplifying

Do Exercises 19 and 20.

Example 17 Perform the indicated operations and simplify:

$$\dfrac{c^3 - d^3}{(c + d)^2} \div (c - d) \cdot (c + d).$$

Using the rules for order of operations, we do the division first:

$\dfrac{c^3 - d^3}{(c + d)^2} \div (c - d) \cdot (c + d)$

$\qquad = \dfrac{c^3 - d^3}{(c + d)^2} \cdot \dfrac{1}{c - d} \cdot (c + d)$

$\qquad = \dfrac{(c - d)(c^2 + cd + d^2)(c + d)}{(c + d)(c + d)(c - d)}$

$\qquad = \dfrac{\cancel{(c - d)}(c^2 + cd + d^2)\cancel{(c + d)}}{(c + d)\cancel{(c + d)}\cancel{(c - d)}}$ Removing a factor of 1: $\dfrac{(c - d)(c + d)}{(c - d)(c + d)} = 1$

$\qquad = \dfrac{c^2 + cd + d^2}{c + d}.$

Do Exercise 21.

> **TIP**
>
> The procedures covered in this chapter are by their nature rather long. It may help to write out lots of steps as you do the problems. If you have difficulty, consider taking a clean sheet of paper and starting over. Don't squeeze your work into a small amount of space. When using lined paper, consider using two spaces at a time, with the paper's line representing the fraction bar.

Answers on page A-30

Exercise Set 5.1

a Find all numbers for which the rational expression is undefined.

1. $\dfrac{5t^2 - 64}{3t + 17}$

2. $\dfrac{x^2 + x + 105}{5x - 45}$

3. $\dfrac{x^3 - x^2 + x + 2}{x^2 + 12x + 35}$

4. $\dfrac{x^2 - 3x - 4}{x^2 - 18x + 77}$

Find the domain. Write interval notation for the answer.

5. $f(t) = \dfrac{5t^2 - 64}{3t + 17}$

6 $f(x) = \dfrac{x^2 + x + 105}{5x - 45}$

7. $f(x) = \dfrac{x^3 - x^2 + x + 2}{x^2 + 12x + 35}$

8. $f(x) = \dfrac{x^2 - 3x - 4}{x^2 - 18x + 77}$

b Multiply to obtain an equivalent expression. Do not simplify.

9. $\dfrac{7x}{7x} \cdot \dfrac{x + 2}{x + 8}$

10. $\dfrac{2 - y^2}{8 - y} \cdot \dfrac{-1}{-1}$

11. $\dfrac{q - 5}{q + 3} \cdot \dfrac{q + 5}{q + 5}$

12. $\dfrac{p + 1}{p + 4} \cdot \dfrac{p - 4}{p - 4}$

c Simplify by removing a factor of 1.

13. $\dfrac{15y^5}{5y^4}$

14. $\dfrac{7w^3}{28w^2}$

15. $\dfrac{16p^3}{24p^7}$

16. $\dfrac{48t^5}{56t^{11}}$

17. $\dfrac{9a - 27}{9}$

18. $\dfrac{6a - 30}{6}$

19. $\dfrac{12x - 15}{21}$

20. $\dfrac{18a - 2}{22}$

21. $\dfrac{4y - 12}{4y + 12}$

22. $\dfrac{8x + 16}{8x - 16}$

23. $\dfrac{t^2 - 16}{t^2 - 8t + 16}$

24. $\dfrac{p^2 - 25}{p^2 + 10p + 25}$

25. $\dfrac{x^2 - 9x + 8}{x^2 + 3x - 4}$

26. $\dfrac{y^2 + 8y - 9}{y^2 - 5y + 4}$

27. $\dfrac{w^3 - z^3}{w^2 - z^2}$

28. $\dfrac{a^2 - b^2}{a^3 + b^3}$

---

d Multiply and simplify.

29. $\dfrac{x^4}{3x + 6} \cdot \dfrac{5x + 10}{5x^7}$

30. $\dfrac{10t}{6t - 12} \cdot \dfrac{20t - 40}{30t^3}$

31. $\dfrac{x^2 - 16}{x^2} \cdot \dfrac{x^2 - 4x}{x^2 - x - 12}$

32. $\dfrac{y^2 + 10y + 25}{y^2 - 9} \cdot \dfrac{y^2 - 3y}{y + 5}$

33. $\dfrac{y^2 - 16}{2y + 6} \cdot \dfrac{y + 3}{y - 4}$

34. $\dfrac{m^2 - n^2}{4m + 4n} \cdot \dfrac{m + n}{m - n}$

35. $\dfrac{x^2 - 2x - 35}{2x^3 - 3x^2} \cdot \dfrac{4x^3 - 9x}{7x - 49}$

36. $\dfrac{y^2 - 10y + 9}{y^2 - 1} \cdot \dfrac{y + 4}{y^2 - 5y - 36}$

37. $\dfrac{c^3 + 8}{c^2 - 4} \cdot \dfrac{c^2 - 4c + 4}{c^2 - 2c + 4}$

38. $\dfrac{x^3 - 27}{x^2 - 9} \cdot \dfrac{x^2 - 6x + 9}{x^2 + 3x + 9}$

39. $\dfrac{x^2 - y^2}{x^3 - y^3} \cdot \dfrac{x^2 + xy + y^2}{x^2 + 2xy + y^2}$

40. $\dfrac{4x^2 - 9y^2}{8x^3 - 27y^3} \cdot \dfrac{4x^2 + 6xy + 9y^2}{4x^2 + 12xy + 9y^2}$

e Divide and simplify.

41. $\dfrac{12x^8}{3y^4} \div \dfrac{16x^3}{6y}$

42. $\dfrac{9a^7}{8b^2} \div \dfrac{12a^2}{24b^7}$

43. $\dfrac{3y + 15}{y} \div \dfrac{y + 5}{y}$

44. $\dfrac{6x + 12}{x} \div \dfrac{x + 2}{x^3}$

45. $\dfrac{y^2 - 9}{y} \div \dfrac{y + 3}{y + 2}$

46. $\dfrac{x^2 - 4}{x} \div \dfrac{x - 2}{x + 4}$

47. $\dfrac{4a^2 - 1}{a^2 - 4} \div \dfrac{2a - 1}{a - 2}$

48. $\dfrac{25x^2 - 4}{x^2 - 9} \div \dfrac{5x - 2}{x + 3}$

49. $\dfrac{x^2 - 16}{x^2 - 10x + 25} \div \dfrac{3x - 12}{x^2 - 3x - 10}$

50. $\dfrac{y^2 - 36}{y^2 - 8y + 16} \div \dfrac{3y - 18}{y^2 - y - 12}$

51. $\dfrac{y^3 + 3y}{y^2 - 9} \div \dfrac{y^2 + 5y - 14}{y^2 + 4y - 21}$

52. $\dfrac{a^3 + 4a}{a^2 - 16} \div \dfrac{a^2 + 8a + 15}{a^2 + a - 20}$

53. $\dfrac{x^3 - 64}{x^3 + 64} \div \dfrac{x^2 - 16}{x^2 - 4x + 16}$

54. $\dfrac{8y^3 + 27}{64y^3 - 1} \div \dfrac{4y^2 - 9}{16y^2 + 4y + 1}$

Perform the indicated operations and simplify.

55. $\dfrac{r^2 - 4s^2}{r + 2s} \div (r + 2s) \cdot \dfrac{2s}{r - 2s}$

56. $\dfrac{d^2 - d}{d^2 - 6d + 8} \cdot \dfrac{d - 2}{d^2 + 5d} \div \dfrac{5d}{d^2 - 9d + 20}$

Skill Maintenance

In Exercises 57–60, the graph is that of a function. Determine the domain and the range. [2.3a]

57.

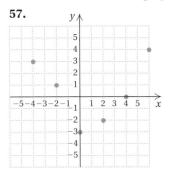

58.

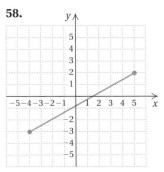

59.

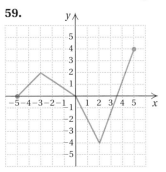

60.

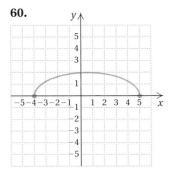

Factor. [4.4a, b]

61. $6a^2 + 5ab - 25b^2$

62. $9a^2 - 30ab + 25b^2$

63. $10x^2 - 80x + 70$

64. $10x^2 - 13x + 4$

65. $21p^2 + p - 10$

66. $12m^2 - 26m - 10$

67. $2x^3 - 16x^2 - 66x$

68. $10y^2 + 80y - 650$

Synthesis

69. ◈ Is it possible to understand how to simplify rational expressions without first understanding how to multiply? Why or why not?

70. ◈ Nancy *incorrectly* simplifies $(x + 2)/x$ as follows:

$$\frac{x + 2}{x} = \frac{\cancel{x} + 2}{\cancel{x}} = 1 + 2 = 3.$$

She insists that this is correct because when x is replaced with 1, her answer checks. Explain her error.

71. Let
$$g(x) = \frac{2x + 3}{4x - 1}.$$
Find $g(5)$, $g(0)$, $g\left(\frac{1}{4}\right)$, and $g(a + h)$.

72. 📈 Use the TABLE feature on a grapher to determine whether each of the following is correct.

a) $\dfrac{x^2 - 16}{x - 4} = x + 4$ b) $\dfrac{x^2 - 16}{x + 2} = x - 8$

Simplify.

73. $\dfrac{x(x + 1) - 2(x + 3)}{(x + 1)(x + 2)(x + 3)}$

74. $\dfrac{2x - 5(x + 2) - (x - 2)}{x^2 - 4}$

75. $\dfrac{m^2 - t^2}{m^2 + t^2 + m + t + 2mt}$

76. $\dfrac{a^3 - 2a^2 + 2a - 4}{a^3 - 2a^2 - 3a + 6}$

5.2 LCMs, LCDs, Addition, and Subtraction

a Finding LCMs by Factoring

To add rational expressions when denominators are different, we first find a common denominator. Let's review the procedure used in arithmetic first. To do the addition

$$\frac{5}{42} + \frac{7}{12},$$

we find a common denominator. We look for the least common multiple (LCM) of 42 and 12. That number becomes the least common denominator (LCD).

To find the LCM, we factor both numbers completely (into primes).

$$42 = 2 \cdot 3 \cdot 7 \longleftarrow \boxed{\text{Any multiple of 42 has these factors.}}$$

$$12 = 2 \cdot 2 \cdot 3 \longleftarrow \boxed{\text{Any multiple of 12 has these factors.}}$$

The LCM is the number that has 2 as a factor twice, 3 as a factor once, and 7 as a factor once: LCM = $2 \cdot 2 \cdot 3 \cdot 7$, or 84.

> To obtain the LCM, use each factor the greatest number of times that it occurs in any one prime factorization.

Example 1 Find the LCM of 18 and 24.

$$18 = \boxed{3 \cdot 3} \cdot 2$$
$$24 = \boxed{2 \cdot 2 \cdot 2} \cdot 3$$

The LCM is $\boxed{3 \cdot 3} \cdot \boxed{2 \cdot 2 \cdot 2}$, or 72.

Do Exercises 1 and 2.

Now let's return to adding $\frac{5}{42}$ and $\frac{7}{12}$:

$$\frac{5}{42} + \frac{7}{12} = \frac{5}{2 \cdot 3 \cdot 7} + \frac{7}{2 \cdot 2 \cdot 3}. \quad \text{Factoring the denominators}$$

The LCD is the LCM of the denominators, $2 \cdot 2 \cdot 3 \cdot 7$. To get this LCD in the first denominator, we need a factor of 2. In the second denominator, we need a factor of 7. We multiply by 1, as follows:

$$\frac{5}{2 \cdot 3 \cdot 7} \cdot \frac{2}{2} + \frac{7}{2 \cdot 2 \cdot 3} \cdot \frac{7}{7} = \frac{10}{2 \cdot 2 \cdot 3 \cdot 7} + \frac{49}{2 \cdot 2 \cdot 3 \cdot 7}$$
$$= \frac{59}{2 \cdot 2 \cdot 3 \cdot 7} = \frac{59}{84}.$$

Multiplying the first fraction by $\frac{2}{2}$ gave us an equivalent fraction with a denominator that is the LCD. Multiplying the second fraction by $\frac{7}{7}$ also gave us an equivalent fraction with a denominator that is the LCD. Once we had a common denominator, we added the numerators.

Do Exercises 3 and 4.

Objectives

a Find the LCM of several algebraic expressions by factoring.

b Add and subtract rational expressions.

c Simplify combined additions and subtractions of rational expressions.

For Extra Help

TAPE 11   TAPE 9B   MAC WIN   CD-ROM

Find the LCM by factoring.

1. 18, 30

2. 12, 18, 24

Add, first finding the LCD of the denominators.

3. $\dfrac{5}{12} + \dfrac{11}{30}$

4. $\dfrac{7}{12} + \dfrac{13}{18} + \dfrac{1}{24}$

Answers on page A-30

Find the LCM.

5. a^2b^2, $5a^3b$

We find the LCM of algebraic expressions in the same way that we find the LCM of natural numbers.

Examples

2. Find the LCM of $12xy^2$ and $15x^3y$.

We factor each expression completely. To find the LCM, we use each factor the greatest number of times that it occurs in any one prime factorization.

$$12xy^2 = \boxed{2 \cdot 2 \cdot 3} \cdot x \cdot \boxed{y \cdot y} \;;$$
$$15x^3y = 3 \cdot \boxed{5 \cdot x \cdot x \cdot x} \cdot y$$

Factoring

$12xy^2$ is a factor.

$$\text{LCM} = 2 \cdot 2 \cdot 3 \cdot 5 \cdot x \cdot x \cdot x \cdot y \cdot y = 60x^3y^2$$

$15x^3y$ is a factor.

6. $y^2 + 7y + 12$, $y^2 + 8y + 16$, $y + 4$

3. Find the LCM of $x^2 + 2x + 1$, $5x^2 - 5x$, and $x^2 - 1$.

$$x^2 + 2x + 1 = \boxed{(x + 1)(x + 1)} \;;$$
$$5x^2 - 5x = \boxed{5x(x - 1)} \;;$$
$$x^2 - 1 = (x + 1)(x - 1)$$

Factoring

$$\text{LCM} = 5x(x + 1)(x + 1)(x - 1)$$

4. Find the LCM of $x^2 - y^2$, $x^3 + y^3$, and $x^2 + 2xy + y^2$.

$$x^2 - y^2 = \boxed{(x - y)}\,(x + y);$$
$$x^3 + y^3 = (x + y)\,\boxed{(x^2 - xy + y^2)} \;;$$
$$x^2 + 2xy + y^2 = \boxed{(x + y)(x + y)}$$

Factoring

7. $x^2 - 9$, $x^3 - x^2 - 6x$, $2x^2$

$$\text{LCM} = (x - y)(x + y)(x + y)(x^2 - xy + y^2)$$

The opposite, or additive inverse, of an LCM is also an LCM. For example, if $(x + 2)(x - 3)$ is an LCM, then $-(x + 2)(x - 3)$ is also an LCM. We can name the latter $(x + 2)(-1)(x - 3)$, or $(x + 2)(3 - x)$. If, when we are finding LCMs, factors that are opposites occur, we do not use both of them. For example, if $a - b$ occurs in one factorization and $b - a$ occurs in another, we do not use both, since they are opposites.

Example 5 Find the LCM of $x^2 - y^2$ and $3y - 3x$.

8. $a^2 - b^2$, $2b - 2a$

$$x^2 - y^2 = \boxed{(x + y)(x - y)} \longleftarrow$$

We can use $(x - y)$ or $(y - x)$, but we do not use both.

$$3y - 3x = \boxed{3}\,(y - x), \text{ or } -3(x - y)$$

$$\text{LCM} = 3(x + y)(x - y), \text{ or } 3(x + y)(y - x), \text{ or } -3(x + y)(x - y)$$

In most cases, we would use the form $\text{LCM} = 3(x + y)(x - y)$.

Do Exercises 5–8.

Answers on page A-30

b Adding and Subtracting Rational Expressions

> When denominators are the same, add or subtract the numerators and keep the same denominator.

Example 6 Add: $\dfrac{3 + x}{x} + \dfrac{4}{x}$.

$$\frac{3 + x}{x} + \frac{4}{x} = \frac{3 + x + 4}{x} = \frac{7 + x}{x} \longleftarrow$$

> *CAUTION!* This expression does *not* simplify to 7: $\dfrac{7 + x}{x} \neq 7$.

Example 6 shows that

$$\frac{3 + x}{x} + \frac{4}{x} \quad \text{and} \quad \frac{7 + x}{x}$$

are equivalent expressions. They name the same number for all allowable replacements.

Example 7 Add: $\dfrac{4x^2 - 5xy}{x^2 - y^2} + \dfrac{2xy - y^2}{x^2 - y^2}$.

$$\frac{4x^2 - 5xy}{x^2 - y^2} + \frac{2xy - y^2}{x^2 - y^2} = \frac{4x^2 - 3xy - y^2}{x^2 - y^2} \qquad \text{Adding the numerators}$$

$$= \frac{(4x + y)(x - y)}{(x + y)(x - y)} \qquad \text{Factoring the numerator and the denominator}$$

$$= \frac{(4x + y)\cancel{(x - y)}}{(x + y)\cancel{(x - y)}} \qquad \text{Removing a factor of 1: } \frac{x - y}{x - y} = 1$$

$$= \frac{4x + y}{x + y}$$

Do Exercises 9 and 10.

Example 8 Subtract: $\dfrac{4x + 5}{x + 3} - \dfrac{x - 2}{x + 3}$.

$$\frac{4x + 5}{x + 3} - \frac{x - 2}{x + 3} = \frac{4x + 5 - (x - 2)}{x + 3} \longleftarrow \quad \text{Subtracting numerators}$$

$$= \frac{4x + 5 - x + 2}{x + 3}$$

> A common error: Forgetting these parentheses. If you forget them, you will be subtracting only *part* of the numerator, $x - 2$.

$$= \frac{3x + 7}{x + 3}$$

Do Exercises 11 and 12.

Add.

9. $\dfrac{5 + y}{y} + \dfrac{7}{y}$

10. $\dfrac{2x^2 + 5x - 9}{x - 5} + \dfrac{x^2 - 19x + 4}{x - 5}$

Subtract.

11. $\dfrac{a}{b + 2} - \dfrac{b}{b + 2}$

12. $\dfrac{4y + 7}{x^2 + y^2} - \dfrac{3y - 5}{x^2 + y^2}$

Calculator Spotlight

Use the TABLE feature to determine whether the addition or subtraction in Examples 6 and 8 is correct. Then check your answers to Margin Exercises 9 and 10.

Answers on page A-30

Add.

13. $\dfrac{b}{3b} + \dfrac{b^3}{-3b}$

14. $\dfrac{3x^2 + 4}{x - 5} + \dfrac{x^2 - 7}{5 - x}$

Subtract.

15. $\dfrac{3}{4y} - \dfrac{7x}{-4y}$

16. $\dfrac{4x^2}{2x - y} - \dfrac{7x^2}{y - 2x}$

Calculator Spotlight

Use the TABLE feature to determine whether the addition in Example 9 is correct. Then check your answer to Margin Exercise 14.

Answers on page A-30

> When one denominator is the opposite, or additive inverse, of the other, multiply one expression by 1, using $-1/-1$. This gives a common denominator.

Example 9   Add: $\dfrac{a}{2a} + \dfrac{a^3}{-2a}$.

$$\dfrac{a}{2a} + \dfrac{a^3}{-2a} = \dfrac{a}{2a} + \dfrac{a^3}{-2a} \cdot \dfrac{-1}{-1} \qquad \text{Multiplying by 1, using } \dfrac{-1}{-1}$$

This is equal to 1 (not -1).

$$= \dfrac{a}{2a} + \dfrac{-a^3}{2a}$$

$$= \dfrac{a - a^3}{2a} \qquad \text{Adding numerators}$$

$$= \dfrac{a(1 - a^2)}{2a} \qquad \text{Factoring}$$

$$= \dfrac{\cancel{a}(1 - a^2)}{2\cancel{a}} \qquad \text{Removing a factor of 1: } \dfrac{a}{a} = 1$$

$$= \dfrac{1 - a^2}{2}$$

Example 10   Subtract: $\dfrac{x^2}{5y} - \dfrac{x^3}{-5y}$.

$$\dfrac{x^2}{5y} - \dfrac{x^3}{-5y} = \dfrac{x^2}{5y} - \dfrac{x^3}{-5y} \cdot \dfrac{-1}{-1} \qquad \text{Multiplying by } \dfrac{-1}{-1}$$

$$= \dfrac{x^2}{5y} - \dfrac{-x^3}{5y}$$

$$= \dfrac{x^2 - (-x^3)}{5y} \qquad \boxed{\text{Don't forget these parentheses!}}$$

$$= \dfrac{x^2 + x^3}{5y}$$

Example 11   Subtract: $\dfrac{5x}{x - 2y} - \dfrac{3y - 7}{2y - x}$.

$$\dfrac{5x}{x - 2y} - \dfrac{3y - 7}{2y - x} = \dfrac{5x}{x - 2y} - \dfrac{3y - 7}{2y - x} \cdot \dfrac{-1}{-1}$$

$$= \dfrac{5x}{x - 2y} - \dfrac{-3y + 7}{x - 2y} \qquad \boxed{\begin{array}{l}\text{Remember: } (2y - x)(-1) = \\ -2y + x = x - 2y.\end{array}}$$

$$= \dfrac{5x - (-3y + 7)}{x - 2y} \qquad \text{Subtracting numerators}$$

$$= \dfrac{5x + 3y - 7}{x - 2y}$$

Do Exercises 13–16.

> When denominators are different, but not additive inverses of each other, first find equivalent rational expressions with the least common denominator (LCD) and then add or subtract the numerators.

Example 12 Add: $\dfrac{2a}{5} + \dfrac{3b}{2a}$.

We first find the LCD:

$$\left.\begin{array}{c} 5 \\ 2a \end{array}\right\} \quad \text{LCD} = 5 \cdot 2a, \text{ or } 10a.$$

Now we multiply each expression by 1. We choose symbols for 1 that will give us the LCD in each denominator. In this case, we use $2a/2a$ and $5/5$:

$$\frac{2a}{5} \cdot \frac{2a}{2a} + \frac{3b}{2a} \cdot \frac{5}{5} = \frac{4a^2}{10a} + \frac{15b}{10a} = \frac{4a^2 + 15b}{10a}.$$

Multiplying the first term by $2a/2a$ gave us a denominator of $10a$. Multiplying the second term by $\frac{5}{5}$ also gave us a denominator of $10a$.

Example 13 Add: $\dfrac{3x^2 + 3xy}{x^2 - y^2} + \dfrac{2 - 3x}{x - y}$.

We first find the LCD of the denominators:

$$\left.\begin{array}{l} x^2 - y^2 = (x + y)(x - y) \\ x - y = x - y \end{array}\right\} \quad \text{LCD} = (x + y)(x - y).$$

We now multiply by 1 to get the LCD in the second expression. Then we add and simplify if possible.

$$\frac{3x^2 + 3xy}{(x + y)(x - y)} + \frac{2 - 3x}{x - y} \cdot \frac{x + y}{x + y} \qquad \text{Multiplying by 1 to get the LCD}$$

$$= \frac{3x^2 + 3xy}{(x + y)(x - y)} + \frac{(2 - 3x)(x + y)}{(x - y)(x + y)}$$

$$= \frac{3x^2 + 3xy}{(x + y)(x - y)} + \frac{2x + 2y - 3x^2 - 3xy}{(x - y)(x + y)} \qquad \text{Multiplying in the numerator}$$

$$= \frac{3x^2 + 3xy + 2x + 2y - 3x^2 - 3xy}{(x + y)(x - y)} \qquad \text{Adding the numerators}$$

$$= \frac{2x + 2y}{(x + y)(x - y)} \qquad \text{Combining like terms}$$

$$= \frac{2(x + y)}{(x + y)(x - y)} \qquad \text{Factoring the numerator}$$

$$= \frac{2\cancel{(x + y)}}{\cancel{(x + y)}(x - y)} \qquad \text{Removing a factor of 1: } \frac{x + y}{x + y} = 1$$

$$= \frac{2}{x - y}$$

Do Exercises 17 and 18.

Add.

17. $\dfrac{3x}{7} + \dfrac{4y}{3x}$

18. $\dfrac{2xy - 2x^2}{x^2 - y^2} + \dfrac{2x + 3}{x + y}$

Answers on page A-30

Subtract.

19. $\dfrac{a}{a+3} - \dfrac{a-4}{a}$

20. $\dfrac{4y-5}{y^2-7y+12} - \dfrac{y+7}{y^2+2y-15}$

21. Perform the indicated operations and simplify:

$$\frac{8x}{x^2-1} + \frac{2}{1-x} - \frac{4}{x+1}.$$

Answers on page A-30

Example 14 Subtract: $\dfrac{2y+1}{y^2-7y+6} - \dfrac{y+3}{y^2-5y-6}.$

$\dfrac{2y+1}{y^2-7y+6} - \dfrac{y+3}{y^2-5y-6}$

$= \dfrac{2y+1}{(y-6)(y-1)} - \dfrac{y+3}{(y-6)(y+1)}$ $\qquad$ LCD $=(y-6)(y-1)(y+1)$

$= \dfrac{2y+1}{(y-6)(y-1)} \cdot \dfrac{y+1}{y+1} - \dfrac{y+3}{(y-6)(y+1)} \cdot \dfrac{y-1}{y-1}$ $\qquad$ Multiplying by 1 to get the LCD

$= \dfrac{(2y+1)(y+1) - (y+3)(y-1)}{(y-6)(y-1)(y+1)}$ $\qquad$ Subtracting the numerators

$= \dfrac{(2y^2+3y+1) - (y^2+2y-3)}{(y-6)(y-1)(y+1)}$ $\qquad$ Multiplying. Note the use of parentheses.

$= \dfrac{2y^2+3y+1-y^2-2y+3}{(y-6)(y-1)(y+1)}$

$= \dfrac{y^2+y+4}{(y-6)(y-1)(y+1)}$ $\qquad$ The numerator cannot be factored. The rational expression is simplified.

We generally do not multiply out a numerator or a denominator if it has three or more factors (other than monomials). This will be helpful when we solve equations.

Do Exercises 19 and 20.

c Combined Additions and Subtractions

Example 15 Perform the indicated operations and simplify.

$\dfrac{2x}{x^2-4} + \dfrac{5}{2-x} - \dfrac{1}{2+x}$

$= \dfrac{2x}{(x-2)(x+2)} + \dfrac{5}{2-x} - \dfrac{1}{2+x}$

$= \dfrac{2x}{(x-2)(x+2)} + \dfrac{5}{2-x} \cdot \dfrac{-1}{-1} - \dfrac{1}{x+2}$ $\qquad$ Multiplying by $\dfrac{-1}{-1}$

$= \dfrac{2x}{(x-2)(x+2)} + \dfrac{-5}{x-2} - \dfrac{1}{x+2}$ $\qquad$ LCD $=(x-2)(x+2)$

$= \dfrac{2x}{(x-2)(x+2)} + \dfrac{-5}{x-2} \cdot \dfrac{x+2}{x+2} - \dfrac{1}{x+2} \cdot \dfrac{x-2}{x-2}$ $\qquad$ Multiplying by 1 to get the LCD

$= \dfrac{2x - 5(x+2) - (x-2)}{(x-2)(x+2)}$ $\qquad$ Adding and subtracting the numerators

$= \dfrac{2x - 5x - 10 - x + 2}{(x-2)(x+2)}$ $\qquad$ Removing parentheses

$= \dfrac{-4x-8}{(x-2)(x+2)}$

$= \dfrac{-4(x+2)}{(x-2)(x+2)}$ $\qquad$ Removing a factor of 1: $\dfrac{x+2}{x+2}=1$

$= \dfrac{-4}{x-2}$, or $-\dfrac{4}{x-2}$

Another correct form of the answer is $4/(2-x)$. It is found by multiplying by $-1/-1$.

Do Exercise 21.

Exercise Set 5.2

a Find the LCM by factoring.

1. 15, 40

2. 12, 32

3. 18, 48

4. 45, 54

5. 30, 105

6. 24, 60

7. 9, 15, 5

8. 27, 35, 63

Add, first finding the LCD.

9. $\dfrac{5}{6} + \dfrac{4}{15}$

10. $\dfrac{5}{12} + \dfrac{13}{18}$

11. $\dfrac{7}{36} + \dfrac{1}{24}$

12. $\dfrac{11}{30} + \dfrac{19}{75}$

13. $\dfrac{3}{4} + \dfrac{7}{30} + \dfrac{1}{16}$

14. $\dfrac{5}{8} + \dfrac{7}{12} + \dfrac{11}{40}$

Find the LCM.

15. $21x^2y, \quad 7xy$

16. $18a^2b, \quad 50ab^3$

17. $y^2 - 100, \quad 10y + 100$

18. $r^2 - s^2, \quad rs + s^2$

19. $15ab^2, \quad 3ab, \quad 10a^3b$

20. $6x^2y^2, \quad 9x^3y, \quad 15y^3$

21. $5y - 15, \quad y^2 - 6y + 9$

22. $x^2 + 10x + 25, \quad x^2 + 2x - 15$

23. $y^2 - 25, \quad 5 - y$

24. $x^2 - 36, \quad 6 - x$

25. $2r^2 - 5r - 12, \quad 3r^2 - 13r + 4, \quad r^2 - 16$

26. $2x^2 - 5x - 3, \quad 2x^2 - x - 1, \quad x^2 - 6x + 9$

27. $x^5 + 4x^3, \quad x^3 - 4x^2 + 4x$

28. $9x^3 + 9x^2 - 18x, \quad 6x^5 + 24x^4 + 24x^3$

29. $x^5 - 2x^4 + x^3, \quad 2x^3 + 2x, \quad 5x + 5$

30. $x^5 - 4x^4 + 4x^3, \quad 3x^2 - 12, \quad 2x + 4$

b Add or subtract. Simplify by removing a factor of 1, if possible.

31. $\dfrac{x - 2y}{x + y} + \dfrac{x + 9y}{x + y}$

32. $\dfrac{a - 8b}{a + b} + \dfrac{a + 13b}{a + b}$

33. $\dfrac{4y + 3}{y - 2} - \dfrac{y - 2}{y - 2}$

34. $\dfrac{3t + 2}{t - 4} - \dfrac{t - 4}{t - 4}$

35. $\dfrac{a^2}{a - b} + \dfrac{b^2}{b - a}$

36. $\dfrac{r^2}{r - s} + \dfrac{s^2}{s - r}$

37. $\dfrac{6}{y} - \dfrac{7}{-y}$

38. $\dfrac{4}{x} - \dfrac{9}{-x}$

39. $\dfrac{4a - 2}{a^2 - 49} + \dfrac{5 + 3a}{49 - a^2}$

40. $\dfrac{2y - 3}{y^2 - 1} - \dfrac{4 - y}{1 - y^2}$

41. $\dfrac{a^3}{a - b} + \dfrac{b^3}{b - a}$

42. $\dfrac{x^3}{x^2 - y^2} + \dfrac{y^3}{y^2 - x^2}$

Add or subtract. Simplify by removing a factor of 1, if possible. If a denominator has three or more factors (other than monomials), leave it factored.

43. $\dfrac{y - 2}{y + 4} + \dfrac{y + 3}{y - 5}$

44. $\dfrac{x - 2}{x + 3} + \dfrac{x + 2}{x - 4}$

45. $\dfrac{4xy}{x^2 - y^2} + \dfrac{x - y}{x + y}$

46. $\dfrac{5ab}{a^2 - b^2} + \dfrac{a + b}{a - b}$

47. $\dfrac{9x + 2}{3x^2 - 2x - 8} + \dfrac{7}{3x^2 + x - 4}$

48. $\dfrac{3y + 2}{2y^2 - y - 10} + \dfrac{8}{2y^2 - 7y + 5}$

49. $\dfrac{4}{x + 1} + \dfrac{x + 2}{x^2 - 1} + \dfrac{3}{x - 1}$

50. $\dfrac{-2}{y + 2} + \dfrac{5}{y - 2} + \dfrac{y + 3}{y^2 - 4}$

51. $\dfrac{x-1}{3x+15} - \dfrac{x+3}{5x+25}$

52. $\dfrac{y-2}{4y+8} - \dfrac{y+6}{5y+10}$

53. $\dfrac{5ab}{a^2-b^2} - \dfrac{a-b}{a+b}$

54. $\dfrac{6xy}{x^2-y^2} - \dfrac{x+y}{x-y}$

55. $\dfrac{3y}{y^2-7y+10} - \dfrac{2y}{y^2-8y+15}$

56. $\dfrac{5x}{x^2-6x+8} - \dfrac{3x}{x^2-x-12}$

57. $\dfrac{y}{y^2-y-20} + \dfrac{2}{y+4}$

58. $\dfrac{6}{y^2+6y+9} + \dfrac{5}{y^2-9}$

59. $\dfrac{3y+2}{y^2+5y-24} + \dfrac{7}{y^2+4y-32}$

60. $\dfrac{3y+2}{y^2-7y+10} + \dfrac{2y}{y^2-8y+15}$

61. $\dfrac{3x-1}{x^2+2x-3} - \dfrac{x+4}{x^2-9}$

62. $\dfrac{3p-2}{p^2+2p-24} - \dfrac{p-3}{p^2-16}$

c Perform the indicated operations and simplify.

63. $\dfrac{1}{x+1} - \dfrac{x}{x-2} + \dfrac{x^2+2}{x^2-x-2}$

64. $\dfrac{2}{y+3} - \dfrac{y}{y-1} + \dfrac{y^2+2}{y^2+2y-3}$

65. $\dfrac{x-1}{x-2} - \dfrac{x+1}{x+2} + \dfrac{x-6}{x^2-4}$

66. $\dfrac{y-3}{y-4} - \dfrac{y+2}{y+4} + \dfrac{y-7}{y^2-16}$

67. $\dfrac{4x}{x^2 - 1} + \dfrac{3x}{1 - x} - \dfrac{4}{x - 1}$

68. $\dfrac{5y}{1 - 2y} - \dfrac{2y}{2y + 1} + \dfrac{3}{4y^2 - 1}$

69. $\dfrac{1}{x + y} + \dfrac{1}{y - x} - \dfrac{2x}{x^2 - y^2}$

70. $\dfrac{1}{b - a} + \dfrac{1}{a + b} - \dfrac{2b}{a^2 - b^2}$

Skill Maintenance

Graph.　[3.6b]

71. $2x - 3y > 6$　　　　**72.** $y - x > 3$　　　　**73.** $5x + 3y \le 15$　　　　**74.** $5x - 3y \le 15$

Factor.　[4.5d]

75. $t^3 - 8$　　　　**76.** $q^3 + 125$　　　　**77.** $23x^4 + 23x$　　　　**78.** $64a^3 - 27b^3$

79. Simplify: $\dfrac{15x^{-7}y^{12}z^4}{35x^{-2}y^6z^{-3}}$.　[R.7a]

80. Find an equation of the line that passes through the point $(-2, 3)$ and is perpendicular to the line $5y + 4x = 7$.　[2.6d]

Synthesis

81. ◈ Janine found that the sum of two rational expressions was $(3 - x)/(x - 5)$, but the answer at the back of the book was $(x - 3)/(5 - x)$. Was Janine's answer correct? Why or why not?

82. ◈ Many students make the mistake of always multiplying the denominators when looking for a least common denominator. Use Example 14 to explain why this approach can yield results that are more difficult to simplify.

83. Determine the domain and the range of the function graphed below.

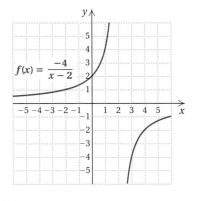

$f(x) = \dfrac{-4}{x - 2}$

84. 〰 Use the TABLE feature on a grapher to check your answers to Exercises 44, 49, 65, and 66.

Find the LCM.

85. 18, 42, 82, 120, 300, 700

86. $x^8 - x^4,\ x^5 - x^2,\ x^5 - x^3,\ x^5 + x^2$

87. The LCM of two expressions is $8a^4b^7$. One of the expressions is $2a^3b^7$. List all possibilities for the other expression.

Perform the indicated operations and simplify.

88. $\dfrac{b - c}{a - (b - c)} - \dfrac{b - a}{(b - a) - c}$

89. $\dfrac{x + y + 1}{y - (x + 1)} + \dfrac{x + y - 1}{x - (y - 1)} - \dfrac{x - y - 1}{1 - (y - x)}$

90. $\dfrac{x^2}{3x^2 - 5x - 2} - \dfrac{2x}{3x + 1} \cdot \dfrac{1}{x - 2}$

91. $\dfrac{x}{x^4 - y^4} - \dfrac{1}{x^2 + 2xy + y^2}$

5.3 Division of Polynomials

A rational expression represents division. "Long" division of polynomials, like division of real numbers, relies on our multiplication and subtraction skills.

a Divisor a Monomial

We first consider division by a monomial. When we are dividing a monomial by a monomial, we can use the rules of exponents and subtract exponents when the bases are the same. (We studied this in Section R.7.) For example,

$$\frac{45x^{10}}{3x^4} = \frac{45}{3}x^{10-4} = 15x^6, \qquad \frac{48a^2b^5}{-3ab^2} = \frac{48}{-3}a^{2-1}b^{5-2} = -16ab^3.$$

When we divide a polynomial by a monomial, we break up the division into a sum of quotients of monomials. To do so, we use the rule for addition using fractional notation in reverse. That is, since

$$\frac{A}{C} + \frac{B}{C} = \frac{A+B}{C}, \quad \text{we know that} \quad \frac{A+B}{C} = \frac{A}{C} + \frac{B}{C}.$$

Example 1 Divide $12x^3 + 8x^2 + x + 4$ by $4x$.

$$\frac{12x^3 + 8x^2 + x + 4}{4x}$$
Writing a fractional expression

$$= \frac{12x^3}{4x} + \frac{8x^2}{4x} + \frac{x}{4x} + \frac{4}{4x}$$
Dividing each term of the numerator by the monomial: This is the reverse of addition.

$$= 3x^2 + 2x + \frac{1}{4} + \frac{1}{x}$$
Doing the four indicated divisions

Do Exercise 1.

Example 2 Divide: $(8x^4y^5 - 3x^3y^4 + 5x^2y^3) \div x^2y^3$.

$$\frac{8x^4y^5 - 3x^3y^4 + 5x^2y^3}{x^2y^3} = \frac{8x^4y^5}{x^2y^3} - \frac{3x^3y^4}{x^2y^3} + \frac{5x^2y^3}{x^2y^3}$$
$$= 8x^2y^2 - 3xy + 5$$

You should try to write only the answer.

> To divide a polynomial by a monomial, divide each term by the monomial.

Do Exercises 2 and 3.

b Divisor Not a Monomial

When the divisor is not a monomial, we use a procedure very much like long division in arithmetic.

1. Divide: $\dfrac{x^3 + 16x^2 + 6x}{2x}$.

Divide.

2. $(15y^5 - 6y^4 + 18y^3) \div 3y^2$

3. $(x^4y^3 + 10x^3y^2 + 16x^2y) \div 2x^2y$

Answers on page A-31

4. Divide and check:

$$x - 2\overline{)x^2 + 3x - 10}.$$

5. Divide and check:

$(2x^4 + 3x^3 - x^2 - 7x + 9) \div (x + 4).$

Example 3   Divide $x^2 + 5x + 8$ by $x + 3$.

We have

$$
\begin{array}{r}
x \\
x + 3\overline{)x^2 + 5x + 8} \\
x^2 + 3x \\
2x
\end{array}
$$

— Divide the first term by the first term: $x^2/x = x$.

— Multiply x above by the divisor, $x + 3$.

— Subtract: $(x^2 + 5x) - (x^2 + 3x) = x^2 + 5x - x^2 - 3x = 2x.$

We now "bring down" the other terms of the dividend—in this case, 8.

$$
\begin{array}{r}
x + 2 \\
x + 3\overline{)x^2 + 5x + 8} \\
x^2 + 3x \\
2x + 8 \\
2x + 6 \\
2
\end{array}
$$

— Divide the first term by the first term: $2x/x = 2.$

— The 8 has been "brought down."

— Multiply 2 above by the divisor, $x + 3$.

— Subtract: $(2x + 8) - (2x + 6) = 2x + 8 - 2x - 6 = 2.$

The answer is $x + 2$, R 2, or

$$x + 2 + \frac{2}{x + 3}.$$

This expression is the remainder over the divisor.

Note that the answer is not a polynomial unless the remainder is 0.

To check, we multiply the quotient by the divisor and add the remainder to see if we get the dividend:

Divisor		Quotient		Remainder		Dividend
$(x + 3)$	$\cdot$	$(x + 2)$	$+$	2	$=$	$(x^2 + 5x + 6) + 2$
					$=$	$x^2 + 5x + 8$

The answer checks.

Example 4   Divide: $(5x^4 + x^3 - 3x^2 - 6x - 8) \div (x - 1)$.

$$
\begin{array}{r}
5x^3 + 6x^2 + 3x - 3 \\
x - 1\overline{)5x^4 + x^3 - 3x^2 - 6x - 8} \\
5x^4 - 5x^3 \\
6x^3 - 3x^2 \\
6x^3 - 6x^2 \\
3x^2 - 6x \\
3x^2 - 3x \\
-3x - 8 \\
-3x + 3 \\
-11
\end{array}
$$

— Subtract: $(5x^4 + x^3) - (5x^4 - 5x^3) = 6x^3.$

— Subtract: $(6x^3 - 3x^2) - (6x^3 - 6x^2) = 3x^2.$

— Subtract: $(3x^2 - 6x) - (3x^2 - 3x) = -3x.$

— Subtract: $(-3x - 8) - (-3x + 3) = -11.$

The answer is $5x^3 + 6x^2 + 3x - 3$, R -11; or

$$5x^3 + 6x^2 + 3x - 3 + \frac{-11}{x - 1}.$$

Do Exercises 4 and 5.

Answers on page A-31

Always remember when dividing polynomials to arrange the polynomials in descending order. In a polynomial division, if there are *missing* terms in the dividend, either write them with 0 coefficients or leave space for them. For example, in $125y^3 - 8$, we say that "the y^2- and y-terms are **missing**." We could write them in as follows: $125y^3 + 0y^2 + 0y - 8$.

Example 5 Divide: $(125y^3 - 8) \div (5y - 2)$.

$$
\begin{array}{r}
25y^2 + 10y + 4 \\
5y - 2 \overline{\smash{)}125y^3 + 0y^2 + 0y - 8} \\
\underline{125y^3 - 50y^2} \\
50y^2 + 0y \\
\underline{50y^2 - 20y} \\
20y - 8 \\
\underline{20y - 8} \\
0
\end{array}
$$

$\longleftarrow$ When there are missing terms, we can write them in.

$\longleftarrow$ Subtract: $125y^3 - (125y^3 - 50y^2) = 50y^2$.

The answer is $25y^2 + 10y + 4$.

Do Exercise 6.

Another way to deal with missing terms is to leave space for them, as we see in Example 6.

Example 6 Divide: $(x^4 - 9x^2 - 5) \div (x - 2)$.

Note that the x^3- and x-terms are missing in the dividend.

$$
\begin{array}{r}
x^3 + 2x^2 - 5x - 10 \\
x - 2 \overline{\smash{)}x^4 \quad\quad - 9x^2 \quad\quad - 5} \\
\underline{x^4 - 2x^3} \\
2x^3 - 9x^2 \\
\underline{2x^3 - 4x^2} \\
-5x^2 \\
\underline{-5x^2 + 10x} \\
-10x - 5 \\
\underline{-10x + 20} \\
-25
\end{array}
$$

$\longleftarrow$ We leave spaces for missing terms.

$\longleftarrow$ Subtract: $x^4 - (x^4 - 2x^3) = 2x^3$.

$\longleftarrow$ Subtract: $(2x^3 - 9x^2) - (2x^3 - 4x^2) = -5x^2$.

The answer is $x^3 + 2x^2 - 5x - 10$, R -25, or

$$x^3 + 2x^2 - 5x - 10 + \frac{-25}{x - 2}.$$

Do Exercises 7 and 8.

---

6. Divide and check:

$$(9y^4 + 14y^2 - 8) \div (3y + 2).$$

Divide and check.

7. $(y^3 - 11y^2 + 6) \div (y - 3)$

8. $(x^3 + 9x^2 - 5) \div (x - 1)$

Answers on page A-31

9. Divide and check:

$(y^3 - 11y^2 + 6) \div (y^2 - 3).$

When dividing, we may "come out even" (have a remainder of 0) or we may not. If not, how long should we keep working? We continue until the degree of the remainder is less than the degree of the divisor, as in the next example.

Example 7   Divide: $(6x^3 + 9x^2 - 5) \div (x^2 - 2x).$

$$
\begin{array}{r}
6x + 21 \\
x^2 - 2x \overline{\smash{)}6x^3 + 9x^2 + 0x - 5} \\
\underline{6x^3 - 12x^2} \\
21x^2 + 0x \\
\underline{21x^2 - 42x} \\
42x - 5
\end{array}
$$

Again we have a missing term, so we can write it in.

The degree of the remainder is less than the degree of the divisor, so we are finished.

The answer is $6x + 21$, R $42x - 5$, or

$$6x + 21 + \frac{42x - 5}{x^2 - 2x}.$$

Do Exercise 9.

c | Synthetic Division

To divide a polynomial by a binomial of the type $x - a$, we can streamline the general procedure by a process called **synthetic division.**

Compare the following. In **A** we perform a division. In **B** we also divide but we do not write the variables.

A.
$$
\begin{array}{r}
4x^2 + 5x + 11 \\
x - 2 \overline{\smash{)}4x^3 - 3x^2 + x + 7} \\
\underline{4x^3 - 8x^2} \\
5x^2 + x \\
\underline{5x^2 - 10x} \\
11x + 7 \\
\underline{11x - 22} \\
29
\end{array}
$$

B.
$$
\begin{array}{r}
4 + 5 + 11 \\
1 - 2 \overline{\smash{)}4 - 3 + 1 + 7} \\
\underline{4 - 8} \\
5 + 1 \\
\underline{5 - 10} \\
11 + 7 \\
\underline{11 - 22} \\
29
\end{array}
$$

In **B** there is still some duplication of writing. Also, since we can subtract by adding the opposite, we can use 2 instead of -2 and then add instead of subtracting.

Answer on page A-31

C. *Synthetic Division*

a) $2\underline{)}\,4 - 3 + 1 + 7$

$\overline{}$

$\quad\ 4$

Write the 2, the opposite of −2 in the divisor $x - 2$, and the coefficients of the dividend.

Bring down the first coefficient.

b) $2\underline{)}\,4 - 3 + 1 + 7$

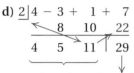

$\quad 4 \quad 5$

Multiply 4 by 2 to get 8. Add 8 and −3.

c) $2\underline{)}\,4 - 3 + \ 1 + 7$

$\qquad\quad 8 \quad 10$

$\overline{\quad 4 \quad 5 \quad 11}$

Multiply 5 by 2 to get 10. Add 10 and 1.

d) $2\underline{)}\,4 - 3 + \ 1 + \ 7$

$\qquad\quad 8 \quad 10 \quad 22$

$\overline{\quad 4 \quad 5 \quad 11\ |\ 29}$

Multiply 11 by 2 to get 22. Add 22 and 7.

Quotient Remainder

The last number, 29, is the remainder. The other numbers are the coefficients of the quotient, with that of the term of highest degree first, as follows.

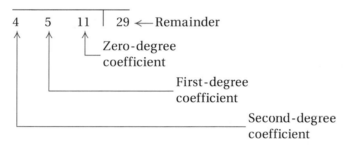

The answer is $4x^2 + 5x + 11$, R 29, or $4x^2 + 5x + 11 + \dfrac{29}{x - 2}$.

> ▶ It is important to remember that in order for this method to work, the divisor must be of the form $x - a$, that is, a variable minus a constant. The coefficient of the variable must be 1.

Example 8 Use synthetic division to divide:

$$(x^3 + 6x^2 - x - 30) \div (x - 2).$$

We have

$2\underline{)}\,1 \quad 6 \quad -1 \quad -30$

$\qquad\quad 2 \quad 16 \quad 30$

$\overline{\quad 1 \quad 8 \quad 15\ |\ \ 0}$

The answer is $x^2 + 8x + 15$, R 0, or just $x^2 + 8x + 15$.

Do Exercise 10.

10. Use synthetic division to divide:

$$(2x^3 - 4x^2 + 8x - 8) \div (x - 3).$$

Answer on page A-31

Use synthetic division to divide.

11. $(x^3 - 2x^2 + 5x - 4) \div (x + 2)$

When there are missing terms, be sure to write 0's for their coefficients.

Examples Use synthetic division to divide.

9. $(2x^3 + 7x^2 - 5) \div (x + 3)$

There is no x-term, so we must write a 0 for its coefficient. Note that $x + 3 = x - (-3)$, so we write -3 in the left corner.

$$
\begin{array}{r|rrrr}
-3 & 2 & 7 & 0 & -5 \\
   &   & -6 & -3 & 9 \\
\hline
   & 2 & 1 & -3 & 4 \\
\end{array}
$$

The answer is $2x^2 + x - 3$, R 4, or $2x^2 + x - 3 + \dfrac{4}{x + 3}$.

10. $(x^3 + 4x^2 - x - 4) \div (x + 4)$

$$
\begin{array}{r|rrrr}
-4 & 1 & 4 & -1 & -4 \\
   &   & -4 & 0 & 4 \\
\hline
   & 1 & 0 & -1 & 0 \\
\end{array}
$$

The answer is $x^2 - 1$.

11. $(x^4 - 1) \div (x - 1)$

$$
\begin{array}{r|rrrrr}
1 & 1 & 0 & 0 & 0 & -1 \\
  &   & 1 & 1 & 1 & 1 \\
\hline
  & 1 & 1 & 1 & 1 & 0 \\
\end{array}
$$

The answer is $x^3 + x^2 + x + 1$.

12. $(8x^5 - 6x^3 + x - 8) \div (x + 2)$

$$
\begin{array}{r|rrrrrr}
-2 & 8 & 0 & -6 & 0 & 1 & -8 \\
   &   & -16 & 32 & -52 & 104 & -210 \\
\hline
   & 8 & -16 & 26 & -52 & 105 & -218 \\
\end{array}
$$

The answer is $8x^4 - 16x^3 + 26x^2 - 52x + 105$, R -218, or

$$8x^4 - 16x^3 + 26x^2 - 52x + 105 + \dfrac{-218}{x + 2}.$$

12. $(y^3 + 1) \div (y + 1)$

Do Exercises 11 and 12.

Answers on page A-31

Exercise Set 5.3

a Divide and check.

1. $\dfrac{24x^6 + 18x^5 - 36x^2}{6x^2}$

2. $\dfrac{30y^8 - 15y^6 + 40y^4}{5y^4}$

3. $\dfrac{45y^7 - 20y^4 + 15y^2}{5y^2}$

4. $\dfrac{60x^8 + 44x^5 - 28x^3}{4x^3}$

5. $(32a^4b^3 + 14a^3b^2 - 22a^2b) \div 2a^2b$

6. $(7x^3y^4 - 21x^2y^3 + 28xy^2) \div 7xy$

b Divide.

7. $(x^2 + 10x + 21) \div (x + 3)$

8. $(y^2 - 8y + 16) \div (y - 4)$

9. $(a^2 - 8a - 16) \div (a + 4)$

10. $(y^2 - 10y - 25) \div (y - 5)$

11. $(x^2 + 7x + 14) \div (x + 5)$

12. $(t^2 - 7t - 9) \div (t - 3)$

13. $(4y^3 + 6y^2 + 14) \div (2y + 4)$

14. $(6x^3 - x^2 - 10) \div (3x + 4)$

15. $(10y^3 + 6y^2 - 9y + 10) \div (5y - 2)$

16. $(6x^3 - 11x^2 + 11x - 2) \div (2x - 3)$

17. $(2x^4 - x^3 - 5x^2 + x - 6) \div (x^2 + 2)$

18. $(3x^4 + 2x^3 - 11x^2 - 2x + 5) \div (x^2 - 2)$

19. $(2x^5 - x^4 + 2x^3 - x) \div (x^2 - 3x)$

20. $(2x^5 + 3x^3 + x^2 - 4) \div (x^2 + x)$

c Use synthetic division to divide.

21. $(x^3 - 2x^2 + 2x - 5) \div (x - 1)$

22. $(x^3 - 2x^2 + 2x - 5) \div (x + 1)$

23. $(a^2 + 11a - 19) \div (a + 4)$

24. $(a^2 + 11a - 19) \div (a - 4)$

25. $(x^3 - 7x^2 - 13x + 3) \div (x - 2)$

26. $(x^3 - 7x^2 - 13x + 3) \div (x + 2)$

27. $(3x^3 + 7x^2 - 4x + 3) \div (x + 3)$

28. $(3x^3 + 7x^2 - 4x + 3) \div (x - 3)$

29. $(y^3 - 3y + 10) \div (y - 2)$

30. $(x^3 - 2x^2 + 8) \div (x + 2)$

31. $(3x^4 - 25x^2 - 18) \div (x - 3)$

32. $(6y^4 + 15y^3 + 28y + 6) \div (y + 3)$

33. $(x^3 - 8) \div (x - 2)$

34. $(y^3 + 125) \div (y + 5)$

35. $(y^4 - 16) \div (y - 2)$

36. $(x^5 - 32) \div (x - 2)$

Skill Maintenance

Graph.　[3.6b]

37. $2x - 3y < 6$

38. $5x + 3y \le 15$

39. $y > 4$

40. $x \le -2$

Graph.　[2.2c]

41. $f(x) = x^2$

42. $g(x) = x^2 - 3$

43. $f(x) = 3 - x^2$

44. $f(x) = x^2 + 6x + 6$

Solve.　[4.7a]

45. $x^2 - 5x = 0$

46. $25y^2 = 64$

47. $12x^2 = 17x + 5$

48. $12x^2 + 11x + 2 = 0$

Synthesis

49. ◈ Do addition, subtraction, multiplication, and division of polynomials always result in a polynomial? Why or why not?

50. ◈ Explain how synthetic division can be useful when factoring a polynomial.

51. Let $f(x) = 4x^3 + 16x^2 - 3x - 45$. Find $f(-3)$ and then solve $f(x) = 0$.

52. ⌁ Use the TRACE feature on a grapher to check your answer to Exercise 51.

53. Let $f(x) = 6x^3 - 13x^2 - 79x + 140$. Find $f(4)$ and then solve $f(x) = 0$.

54. ⌁ Use the TRACE feature on a grapher to check your answer to Exercise 53.

5.4 Complex Rational Expressions

a   A **complex rational expression** is a rational expression that contains rational expressions within its numerator and/or its denominator. Here are some examples:

$$\frac{1 + \dfrac{5}{x}}{4x}, \quad \frac{\dfrac{x-y}{x+y}}{\dfrac{2x-y}{3x+y}}, \quad \frac{\dfrac{2}{3}}{\dfrac{4}{5}}, \quad \frac{\dfrac{3x}{5} - \dfrac{2}{x}}{\dfrac{4x}{3} + \dfrac{7}{6x}}.$$

The rational expressions within each complex rational expression are red.

There are two methods that can be used to simplify complex rational expressions. We will consider both of them. Use the one that works best for you or the one that your instructor directs you to use.

Method 1: Multiplying by the LCM of All the Denominators

> *Method 1.* To simplify a complex rational expression:
>
> **1.** First, find the LCM of all the denominators of all the rational expressions occurring within both the numerator and the denominator of the complex rational expression.
>
> **2.** Multiply by 1 using LCM/LCM.
>
> **3.** If possible, simplify by removing a factor of 1.

Example 1   Simplify: $\dfrac{x + \dfrac{1}{5}}{x - \dfrac{1}{3}}$.

We first find the LCM of all the denominators of all the rational expressions occurring in both the numerator and the denominator of the complex rational expression. The denominators are 3 and 5. The LCM of these denominators is $3 \cdot 5$, or 15. We multiply by 15/15.

$$\frac{x + \dfrac{1}{5}}{x - \dfrac{1}{3}} = \left(\frac{x + \dfrac{1}{5}}{x - \dfrac{1}{3}}\right) \cdot \frac{15}{15} \qquad \text{Multiplying by 1}$$

$$= \frac{\left(x + \dfrac{1}{5}\right) \cdot 15}{\left(x - \dfrac{1}{3}\right) \cdot 15} \qquad \text{Multiplying the numerators and the denominators}$$

$$= \frac{15x + \dfrac{1}{5} \cdot 15}{15x - \dfrac{1}{3} \cdot 15} \qquad \text{Carrying out the multiplications using the distributive laws}$$

$$= \frac{15x + 3}{15x - 5} \qquad \text{No further simplification is possible.}$$

Do Exercise 1.

Objective

a   Simplify complex rational expressions.

For Extra Help

TAPE 11   TAPE 10A   MAC WIN   CD-ROM

1. Simplify. Use Method 1.

$$\frac{y + \dfrac{1}{2}}{y - \dfrac{1}{7}}$$

Answer on page A-32

2. Simplify. Use Method 1.

$$\frac{1 - \dfrac{1}{x}}{1 - \dfrac{1}{x^2}}$$

Answer on page A-32

Example 2   Simplify: $\dfrac{1 + \dfrac{1}{x}}{1 - \dfrac{1}{x^2}}$.

We first find the LCM of all the denominators of all the rational expressions occurring in both the numerator and the denominator of the complex rational expression. The denominators are x and x^2. The LCM of these denominators is x^2. We multiply by x^2/x^2.

$$\frac{1 + \dfrac{1}{x}}{1 - \dfrac{1}{x^2}} = \left(\frac{1 + \dfrac{1}{x}}{1 - \dfrac{1}{x^2}}\right) \cdot \frac{x^2}{x^2} \qquad \text{Multiplying by 1}$$

$$= \frac{\left(1 + \dfrac{1}{x}\right) \cdot x^2}{\left(1 - \dfrac{1}{x^2}\right) \cdot x^2} \qquad \begin{array}{l}\text{Multiplying the numerators}\\ \text{and the denominators}\end{array}$$

$$= \frac{x^2 + \dfrac{1}{x} \cdot x^2}{x^2 - \dfrac{1}{x^2} \cdot x^2} \qquad \begin{array}{l}\text{Carrying out the multiplications}\\ \text{using the distributive laws}\end{array}$$

$$= \frac{x^2 + x}{x^2 - 1}$$

$$= \frac{x(x + 1)}{(x + 1)(x - 1)} \qquad \text{Factoring}$$

$$= \frac{x\cancel{(x + 1)}}{\cancel{(x + 1)}(x - 1)} \qquad \text{Removing a factor of 1: } \frac{x + 1}{x + 1} = 1$$

$$= \frac{x}{x - 1}$$

Do Exercise 2.

Example 3   Simplify: $\dfrac{\dfrac{1}{a} + \dfrac{1}{b}}{\dfrac{1}{a^3} + \dfrac{1}{b^3}}$.

We first find the LCM of all the denominators of all the rational expressions occurring in both the numerator and the denominator of the complex rational expression. The denominators are a, b, a^3, and b^3. The LCM of these denominators is $a^3 b^3$. We multiply by $a^3 b^3 / a^3 b^3$.

$$\frac{\dfrac{1}{a} + \dfrac{1}{b}}{\dfrac{1}{a^3} + \dfrac{1}{b^3}} = \left(\frac{\dfrac{1}{a} + \dfrac{1}{b}}{\dfrac{1}{a^3} + \dfrac{1}{b^3}}\right) \cdot \frac{a^3 b^3}{a^3 b^3} \qquad \text{Multiplying by 1}$$

$$= \frac{\left(\dfrac{1}{a} + \dfrac{1}{b}\right) \cdot a^3 b^3}{\left(\dfrac{1}{a^3} + \dfrac{1}{b^3}\right) \cdot a^3 b^3} \qquad \begin{array}{l}\text{Multiplying the numerators and}\\ \text{the denominators}\end{array}$$

Then

$$= \frac{\dfrac{1}{a} \cdot a^3 b^3 + \dfrac{1}{b} \cdot a^3 b^3}{\dfrac{1}{a^3} \cdot a^3 b^3 + \dfrac{1}{b^3} \cdot a^3 b^3}$$ **Carrying out the multiplications using a distributive law**

$$= \frac{a^2 b^3 + a^3 b^2}{b^3 + a^3} = \frac{a^2 b^2 (b + a)}{(b + a)(b^2 - ba + a^2)}$$ **Factoring**

$$= \frac{a^2 b^2 \cancel{(b + a)}}{\cancel{(b + a)}(b^2 - ba + a^2)}$$ **Removing a factor of 1:** $\dfrac{b + a}{b + a} = 1$

$$= \frac{a^2 b^2}{b^2 - ba + a^2}.$$

Do Exercises 3 and 4.

Method 2: Adding or Subtracting in the Numerator and the Denominator

Method 2. To simplify a complex rational expression:

1. Add or subtract, as necessary, to get a single rational expression in the numerator.

2. Add or subtract, as necessary, to get a single rational expression in the denominator.

3. Divide the numerator by the denominator.

4. If possible, simplify by removing a factor of 1.

We will redo Examples 1–3 using this method.

Example 4 Simplify: $\dfrac{x + \dfrac{1}{5}}{x - \dfrac{1}{3}}$.

$$\frac{x + \dfrac{1}{5}}{x - \dfrac{1}{3}} = \frac{x \cdot \dfrac{5}{5} + \dfrac{1}{5}}{x - \dfrac{1}{3}} = \frac{\dfrac{5x + 1}{5}}{x - \dfrac{1}{3}}$$ **To get a single rational expression in the numerator, we note that the LCM in the numerator is 5. We multiply by 1 and add.**

$$= \frac{\dfrac{5x + 1}{5}}{x \cdot \dfrac{3}{3} - \dfrac{1}{3}} = \frac{\dfrac{5x + 1}{5}}{\dfrac{3x - 1}{3}}$$ **To get a single rational expression in the denominator, we note that the LCM in the denominator is 3. We multiply by 1 and subtract.**

$$= \frac{5x + 1}{5} \cdot \frac{3}{3x - 1}$$ **Multiplying by the reciprocal of the denominator**

$$= \frac{15x + 3}{15x - 5}$$ **No further simplification is possible.**

If you feel more comfortable doing so, you can always write denominators of 1 where there are no denominators. In this case, you could start out by writing

$$\frac{\dfrac{x}{1} + \dfrac{1}{5}}{\dfrac{x}{1} - \dfrac{1}{3}}.$$

Simplify. Use Method 1.

3. $\dfrac{\dfrac{1}{a} + \dfrac{1}{b}}{\dfrac{1}{a} - \dfrac{1}{b}}$

4. $\dfrac{\dfrac{1}{a} - \dfrac{1}{b}}{\dfrac{1}{a^3} - \dfrac{1}{b^3}}$

Simplify. Use Method 2.

5. $\dfrac{y + \dfrac{1}{2}}{y - \dfrac{1}{7}}$

6. $\dfrac{1 - \dfrac{1}{x}}{1 - \dfrac{1}{x^2}}$

Answers on page A-32

Simplify. Use Method 2.

7. $\dfrac{\dfrac{1}{a} + \dfrac{1}{b}}{\dfrac{1}{a} - \dfrac{1}{b}}$

8. $\dfrac{\dfrac{1}{a} - \dfrac{1}{b}}{\dfrac{1}{a^3} - \dfrac{1}{b^3}}$

COMPARING METHODS. It is difficult to say which method is preferable. For expressions like

$$\dfrac{\dfrac{3x + 1}{x - 5}}{\dfrac{2 - x}{x + 3}} \quad \text{or} \quad \dfrac{\dfrac{3}{x} - \dfrac{2}{x}}{\dfrac{1}{x + 1} + \dfrac{5}{x + 1}},$$

Method 2 is probably easier to use since little or no work is required to write the expression as a quotient of two rational expressions.

On the other hand, using Method 1 for expressions like

$$\dfrac{\dfrac{3}{a^2b} - \dfrac{4}{bc^3}}{\dfrac{1}{b^3c} + \dfrac{2}{ac^4}}$$

or

$$\dfrac{\dfrac{5}{a^2 - b^2} + \dfrac{2}{a^2 + 2ab + b^2}}{\dfrac{1}{a - b} + \dfrac{4}{a + b}}$$

will require fewer steps. Either method, however, will give a correct answer.

Answers on page A-32

Example 5 Simplify: $\dfrac{1 + \dfrac{1}{x}}{1 - \dfrac{1}{x^2}}$.

$$\dfrac{1 + \dfrac{1}{x}}{1 - \dfrac{1}{x^2}} = \dfrac{1 \cdot \dfrac{x}{x} + \dfrac{1}{x}}{1 \cdot \dfrac{x^2}{x^2} - \dfrac{1}{x^2}} \qquad \begin{array}{l}\text{Finding the LCM in the numerator}\\ \text{and multiplying by 1}\\[1ex] \text{Finding the LCM in the denominator}\\ \text{and multiplying by 1}\end{array}$$

$$= \dfrac{\dfrac{x}{x} + \dfrac{1}{x}}{\dfrac{x^2}{x^2} - \dfrac{1}{x^2}}$$

$$= \dfrac{\dfrac{x + 1}{x}}{\dfrac{x^2 - 1}{x^2}} \qquad \begin{array}{l}\text{Adding in the numerator and}\\ \text{subtracting in the denominator}\end{array}$$

$$= \dfrac{x + 1}{x} \cdot \dfrac{x^2}{x^2 - 1} \qquad \begin{array}{l}\text{Multiplying by the reciprocal}\\ \text{of the denominator}\end{array}$$

$$= \dfrac{(x + 1) \cdot x \cdot x}{x(x - 1)(x + 1)} \qquad \text{Removing a factor of 1: } \dfrac{x(x + 1)}{x(x + 1)} = 1$$

$$= \dfrac{x}{x - 1}$$

Do Exercises 5 and 6 on the preceding page.

Example 6 Simplify: $\dfrac{\dfrac{1}{a} + \dfrac{1}{b}}{\dfrac{1}{a^3} + \dfrac{1}{b^3}}$.

The LCM in the numerator is ab, and the LCM in the denominator is a^3b^3.

$$\dfrac{\dfrac{1}{a} + \dfrac{1}{b}}{\dfrac{1}{a^3} + \dfrac{1}{b^3}} = \dfrac{\dfrac{1}{a} \cdot \dfrac{b}{b} + \dfrac{1}{b} \cdot \dfrac{a}{a}}{\dfrac{1}{a^3} \cdot \dfrac{b^3}{b^3} + \dfrac{1}{b^3} \cdot \dfrac{a^3}{a^3}} = \dfrac{\dfrac{b}{ab} + \dfrac{a}{ab}}{\dfrac{b^3}{a^3b^3} + \dfrac{a^3}{a^3b^3}}$$

$$= \dfrac{\dfrac{b + a}{ab}}{\dfrac{b^3 + a^3}{a^3b^3}} \qquad \begin{array}{l}\text{Adding in the numerator}\\ \text{and the denominator}\end{array}$$

$$= \dfrac{b + a}{ab} \cdot \dfrac{a^3b^3}{b^3 + a^3} \qquad \begin{array}{l}\text{Multiplying by the reciprocal}\\ \text{of the denominator}\end{array}$$

$$= \dfrac{(b + a)a^3b^3}{ab(b^3 + a^3)}$$

$$= \dfrac{(b + a) \cdot ab \cdot a^2b^2}{ab(b + a)(b^2 - ab + a^2)} \qquad \text{Removing a factor of 1: } \dfrac{ab(b + a)}{ab(b + a)} = 1$$

$$= \dfrac{a^2b^2}{b^2 - ab + a^2}$$

Do Exercises 7 and 8.

Exercise Set 5.4

a Simplify. Use either Method 1 or Method 2 as you wish or as directed by your instructor.

1. $\dfrac{\dfrac{1}{a} + 2}{\dfrac{1}{a} - 1}$

2. $\dfrac{\dfrac{1}{t} + 6}{\dfrac{1}{t} - 5}$

3. $\dfrac{x - \dfrac{1}{x}}{x + \dfrac{1}{x}}$

4. $\dfrac{y + \dfrac{1}{y}}{y - \dfrac{1}{y}}$

5. $\dfrac{\dfrac{3}{x} + \dfrac{4}{y}}{\dfrac{4}{x} - \dfrac{3}{y}}$

6. $\dfrac{\dfrac{2}{y} + \dfrac{5}{z}}{\dfrac{1}{y} - \dfrac{4}{z}}$

7. $\dfrac{\dfrac{9x^2 - y^2}{xy}}{\dfrac{3x - y}{y}}$

8. $\dfrac{\dfrac{a^2 - 16b^2}{ab}}{\dfrac{a + 4b}{b}}$

9. $\dfrac{a - \dfrac{3a}{b}}{b - \dfrac{b}{a}}$

10. $\dfrac{1 - \dfrac{2}{3x}}{x - \dfrac{4}{9x}}$

11. $\dfrac{\dfrac{1}{a} + \dfrac{1}{b}}{\dfrac{a^2 - b^2}{ab}}$

12. $\dfrac{\dfrac{1}{x} - \dfrac{1}{y}}{\dfrac{x^2 - y^2}{xy}}$

13. $\dfrac{\dfrac{1}{x + h} - \dfrac{1}{x}}{h}$

14. $\dfrac{\dfrac{1}{a - h} - \dfrac{1}{a}}{h}$

It may help you to write h as $\dfrac{h}{1}$.

15. $\dfrac{\dfrac{x^2 - x - 12}{x^2 - 2x - 15}}{\dfrac{x^2 + 8x + 12}{x^2 - 5x - 14}}$

16. $\dfrac{\dfrac{y^2 - y - 6}{y^2 - 5y - 14}}{\dfrac{y^2 + 6y + 5}{y^2 - 6y - 7}}$

17. $\dfrac{\dfrac{1}{x+2}+\dfrac{4}{x-3}}{\dfrac{2}{x-3}-\dfrac{7}{x+2}}$

18. $\dfrac{\dfrac{1}{y-4}+\dfrac{1}{y+5}}{\dfrac{6}{y+5}+\dfrac{2}{y-4}}$

19. $\dfrac{\dfrac{6}{x^2-4}-\dfrac{5}{x+2}}{\dfrac{7}{x^2-4}-\dfrac{4}{x-2}}$

20. $\dfrac{\dfrac{1}{x^2-1}+\dfrac{5}{x^2-5x+4}}{\dfrac{1}{x^2-1}+\dfrac{2}{x^2+3x+2}}$

21. $\dfrac{\dfrac{1}{z^2}-\dfrac{1}{w^2}}{\dfrac{1}{z^3}+\dfrac{1}{w^3}}$

22. $\dfrac{\dfrac{1}{b^2}-\dfrac{1}{c^2}}{\dfrac{1}{b^3}-\dfrac{1}{c^3}}$

Skill Maintenance

Factor.   [4.3a, c], [4.5d]

23. $4x^3 + 20x^2 + 6x$

24. $y^3 + 8$

25. $y^3 - 8$

26. $2x^3 - 32x^2 + 126x$

27. $1000x^3 + 1$

28. $1 - 1000a^3$

29. $y^3 - 64x^3$

30. $\frac{1}{8}a^3 - 343$

31. Solve for s: $T = \dfrac{r+s}{3}$.   [1.2a]

32. Graph: $f(x) = -3x + 2$.   [2.2c]

33. Given that $f(x) = x^2 - 3$, find $f(-5)$.   [2.2b]

34. Solve: $|2x - 5| = 7$.   [1.6c]

Synthesis

35. ◆ In arithmetic, we are taught that

$$\frac{a}{b} \div \frac{c}{d} = \frac{a}{b} \cdot \frac{d}{c}.$$

(To divide using fractional notation, we invert and multiply.) Use Method 1 to explain why we can do this.

36. ◆ Explain why it is easier to use Method 1 than Method 2 to simplify the following expression.

$$\dfrac{\dfrac{a}{b}+\dfrac{c}{d}}{\dfrac{a}{b}-\dfrac{c}{d}}$$

For each function in Exercises 37–40, find and simplify $\dfrac{f(a+h)-f(a)}{h}$.

37. $f(x) = \dfrac{3}{x^2}$

38. $f(x) = \dfrac{5}{x}$

39. $f(x) = \dfrac{1}{1-x}$

40. $f(x) = \dfrac{x}{1+x}$

41. Use the TABLE feature on a grapher to check your answers to Exercises 1, 3, 10, 19, and 20.

Simplify.

42. $\dfrac{5x^{-1} - 5y^{-1} + 10x^{-1}y^{-1}}{6x^{-1} - 6y^{-1} + 12x^{-1}y^{-1}}$

43. $\left[\dfrac{\dfrac{x+3}{x-3}+1}{\dfrac{x+3}{x-3}-1}\right]^8$

Find the reciprocal and simplify.

44. $x^2 - \dfrac{1}{x}$

45. $\dfrac{1-\dfrac{1}{a}}{a-1}$

46. $\dfrac{a^3+b^3}{a+b}$

47. $x^2 + x + 1 + \dfrac{1}{x} + \dfrac{1}{x^2}$

5.5 Solving Rational Equations

a | Rational Equations

In Sections 5.1–5.4, we studied operations with *rational expressions*. These expressions do not have equals signs. Although we can perform the operations and simplify, we cannot solve them. Note the following examples:

$$\frac{x^2 - 6x + 9}{x^2 - 4} \cdot \frac{x - 2}{x - 3}, \qquad \frac{x + y}{x - y} \div \frac{x^2 + y}{x^2 - y^2}, \quad \text{and} \quad \frac{a + 7}{a^2 - 16} + \frac{5}{5a - 15}.$$

Operation signs occur. There are no equals signs!

Most often, the result of our calculation is another rational expression that is not cleared of fractions.

    Equations *do have* equals signs, and we can clear them of fractions as we did in Section 1.1. A **rational**, or **fractional**, **equation** is an equation containing one or more rational expressions. Here are some examples:

$$\frac{2}{3} - \frac{5}{6} = \frac{1}{x}, \qquad x + \frac{6}{x} = 5, \quad \text{and} \quad \frac{2x}{x - 3} - \frac{6}{x} = \frac{18}{x^2 - 3x}.$$

There are equals signs as well as operation signs.

> To solve a rational equation, the first step is to clear the equation of fractions. To do this, multiply on both sides of the equation by the LCM of all the denominators. Then carry out the equation-solving process as discussed in Chapter 1.

Example 1   Solve: $\dfrac{2}{3} - \dfrac{5}{6} = \dfrac{1}{x}$.

The LCM of all the denominators is $6x$, or $2 \cdot 3 \cdot x$. Using the multiplication principle of Chapter 1, we multiply by the LCM on both sides of the equation.

$$(2 \cdot 3 \cdot x) \cdot \left(\frac{2}{3} - \frac{5}{6}\right) = (2 \cdot 3 \cdot x) \cdot \frac{1}{x} \qquad \text{\small Multiplying by the LCM on both sides}$$

$$2 \cdot 3 \cdot x \cdot \frac{2}{3} - 2 \cdot 3 \cdot x \cdot \frac{5}{6} = 2 \cdot 3 \cdot x \cdot \frac{1}{x} \qquad \text{\small Multiplying to remove parentheses}$$

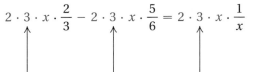

> When clearing fractions, be sure to multiply *every* term in the equation by the LCM. We are *not* multiplying by 1.

$$2 \cdot x \cdot 2 - x \cdot 5 = 2 \cdot 3$$
$$4x - 5x = 6$$
$$-x = 6$$
$$-1 \cdot x = 6$$
$$x = -6$$

CHECK:

$$\frac{\dfrac{2}{3} - \dfrac{5}{6} = \dfrac{1}{x}}{\dfrac{2}{3} - \dfrac{5}{6} \;?\; \dfrac{1}{-6}}$$
$$\frac{4}{6} - \frac{5}{6} \quad \Big| \quad -\frac{1}{6}$$
$$-\frac{1}{6} \quad \Big| \qquad \text{TRUE}$$

The solution is -6.

Do Exercise 1.

Objective

a   Solve rational equations.

For Extra Help

TAPE 12   TAPE 10B   MAC WIN   CD-ROM

1. Solve: $\dfrac{2}{3} + \dfrac{5}{6} = \dfrac{1}{x}$.

Answer on page A-32

2. Solve: $\dfrac{y - 4}{5} - \dfrac{y + 7}{2} = 5$.

Calculator Spotlight

Using a grapher, we can do a partial check of Example 3. We enter

$$y_1 = \frac{2x}{x - 3} - \frac{6}{x}$$

and

$$y_2 = \frac{18}{x^2 - 3x}.$$

We then use the TABLE feature and see that no values of y_1 and y_2 agree. Graphs can be drawn in DOT mode, but viewing windows will need to be changed and/or the ZOOM feature will be needed to see that the graphs do not intersect.

X	Y₁	Y₂
−2	3.8	1.8
−1.5	4.6667	2.6667
−1	6.5	4.5
−.5	12.286	10.286
0	ERROR	ERROR
.5	−12.4	−14.4
1	−7	−9

X = −2

Use the TABLE feature to check the results of Examples 2 and 4 and Margin Exercises 2 and 3.

Answers on page A-32

Example 2 Solve: $\dfrac{x + 1}{2} - \dfrac{x - 3}{3} = 3$.

The LCM of all the denominators is $2 \cdot 3$, or 6. We multiply by the LCM on both sides of the equation.

$$2 \cdot 3 \cdot \left(\frac{x + 1}{2} - \frac{x - 3}{3} \right) = 2 \cdot 3 \cdot 3 \qquad \text{Multiplying by the LCM on both sides}$$

$$2 \cdot 3 \cdot \frac{x + 1}{2} - 2 \cdot 3 \cdot \frac{x - 3}{3} = 2 \cdot 3 \cdot 3 \qquad \text{Multiplying to remove parentheses}$$

$$3(x + 1) - 2(x - 3) = 18 \qquad \text{Simplifying}$$

$$\left. \begin{array}{r} 3x + 3 - 2x + 6 = 18 \\ x + 9 = 18 \end{array} \right\} \quad \text{Multiplying and collecting like terms}$$

$$x = 9$$

CHECK:
$$\frac{x + 1}{2} - \frac{x - 3}{3} = 3$$

$$\frac{9 + 1}{2} - \frac{9 - 3}{3} \stackrel{?}{} 3$$

$$5 - 2 $$

$$3 \quad \text{TRUE}$$

> **CAUTION!** *Clearing fractions* is a valid procedure only when solving equations, *not* when adding, subtracting, multiplying, or dividing rational expressions.

The solution is 9.

Do Exercise 2.

> When we multiply on both sides of an equation by the LCM, the resulting equation might yield numbers that are *not* solutions of the original equation. Thus we must *always* check possible solutions in the original equation.
>
> 1. If you have carried out all algebraic procedures correctly, you need only check to see if a number makes a denominator 0 in the original equation.
> 2. To be sure that no computational errors have been made and that you indeed have a solution, a complete check is necessary, as we did in Chapter 1.

The next example illustrates the importance of checking all possible solutions.

Example 3 Solve: $\dfrac{2x}{x - 3} - \dfrac{6}{x} = \dfrac{18}{x^2 - 3x}$.

The LCM of the denominators is $x(x - 3)$. We multiply by $x(x - 3)$.

$$x(x - 3)\left(\frac{2x}{x - 3} - \frac{6}{x} \right) = x(x - 3)\left(\frac{18}{x^2 - 3x} \right) \qquad \begin{array}{l} \text{Multiplying by the} \\ \text{LCM on both sides} \end{array}$$

$$x(x - 3) \cdot \frac{2x}{x - 3} - x(x - 3) \cdot \frac{6}{x} = x(x - 3)\left(\frac{18}{x^2 - 3x} \right) \qquad \begin{array}{l} \text{Multiplying to} \\ \text{remove parentheses} \end{array}$$

$$2x^2 - 6(x - 3) = 18 \qquad \text{Simplifying}$$

$$2x^2 - 6x + 18 = 18$$

$$2x^2 - 6x = 0$$

$$2x(x - 3) = 0 \qquad \text{Factoring}$$

$$2x = 0 \quad \text{or} \quad x - 3 = 0 \qquad \begin{array}{l} \text{Using the principle} \\ \text{of zero products} \end{array}$$

$$x = 0 \quad \text{or} \qquad x = 3$$

The numbers 0 and 3 are possible solutions. We look at the original equation and see that each makes a denominator 0. We can also carry out a check, as follows.

CHECK:

For 0:

$$\frac{2x}{x-3} - \frac{6}{x} = \frac{18}{x^2 - 3x}$$

$$\begin{array}{c|c} \dfrac{2(0)}{0-3} - \dfrac{6}{0} & \dfrac{18}{0^2 - 3(0)} \\[2ex] 0 - \dfrac{6}{0} & \dfrac{18}{0} \end{array} \quad \text{UNDEFINED}$$

For 3:

$$\frac{2x}{x-3} - \frac{6}{x} = \frac{18}{x^2 - 3x}$$

$$\begin{array}{c|c} \dfrac{2(3)}{3-3} - \dfrac{6}{3} & \dfrac{18}{3^2 - 3(3)} \\[2ex] \dfrac{6}{0} - 2 & \dfrac{18}{0} \end{array} \quad \text{UNDEFINED}$$

The equation has *no solution*.

Do Exercise 3.

Example 4 Solve: $\dfrac{x^2}{x-2} = \dfrac{4}{x-2}$.

The LCM of the denominators is $x - 2$. We multiply by $x - 2$.

$$(x-2) \cdot \frac{x^2}{x-2} = (x-2) \cdot \frac{4}{x-2}$$

$$x^2 = 4 \qquad \text{Simplifying}$$

$$x^2 - 4 = 0$$

$$(x+2)(x-2) = 0$$

$$x = -2 \quad \text{or} \quad x = 2 \qquad \text{Using the principle of zero products}$$

CHECK:

For 2:

$$\frac{x^2}{x-2} = \frac{4}{x-2}$$

$$\begin{array}{c|c} \dfrac{2^2}{2-2} & \dfrac{4}{2-2} \\[2ex] \dfrac{4}{0} & \dfrac{4}{0} \end{array} \quad \text{UNDEFINED}$$

For -2:

$$\frac{x^2}{x-2} = \frac{4}{x-2}$$

$$\begin{array}{c|c} \dfrac{(-2)^2}{-2-2} & \dfrac{4}{-2-2} \\[2ex] \dfrac{4}{-4} & \dfrac{4}{-4} \\[2ex] -1 & -1 \end{array} \quad \text{TRUE}$$

The number -2 is a solution, but 2 is not (it results in division by 0).

Do Exercise 4.

Example 5 Solve: $\dfrac{2}{x-1} = \dfrac{3}{x+1}$.

The LCM of the denominators is $(x - 1)(x + 1)$. We multiply by $(x - 1)(x + 1)$.

$$(x-1)(x+1) \cdot \frac{2}{x-1} = (x-1)(x+1) \cdot \frac{3}{x+1} \qquad \text{Multiplying}$$

$$2(x+1) = 3(x-1) \qquad \text{Simplifying}$$

$$2x + 2 = 3x - 3$$

$$5 = x$$

We leave the check to the student. The number 5 checks and is the solution.

3. Solve:

$$\frac{4x}{x+5} + \frac{20}{x} = \frac{100}{x^2 + 5x}.$$

4. Solve: $\dfrac{x^2}{x-3} = \dfrac{9}{x-3}$.

Answers on page A-32

Solve.

5. $\dfrac{2}{x-1} = \dfrac{3}{x+2}$

6. $\dfrac{2}{x^2-9} + \dfrac{5}{x-3} = \dfrac{3}{x+3}$

Example 6   Solve: $\dfrac{2}{x+5} + \dfrac{1}{x-5} = \dfrac{16}{x^2-25}$.

The LCM of the denominators is $(x+5)(x-5)$. We multiply by $(x+5)(x-5)$.

$$(x+5)(x-5) \cdot \left[\dfrac{2}{x+5} + \dfrac{1}{x-5} \right] = (x+5)(x-5) \cdot \dfrac{16}{x^2-25}$$

$$(x+5)(x-5) \cdot \dfrac{2}{x+5} + (x+5)(x-5) \cdot \dfrac{1}{x-5} = (x+5)(x-5) \cdot \dfrac{16}{x^2-25}$$

$$2(x-5) + (x+5) = 16$$
$$2x - 10 + x + 5 = 16$$
$$3x - 5 = 16$$
$$3x = 21$$
$$x = 7$$

CHECK:

$$\dfrac{2}{x+5} + \dfrac{1}{x-5} = \dfrac{16}{x^2-25}$$

$$\begin{array}{c|c} \dfrac{2}{7+5} + \dfrac{1}{7-5} & \dfrac{16}{7^2-25} \\[2mm] \dfrac{2}{12} + \dfrac{1}{2} & \dfrac{16}{49-25} \\[2mm] \dfrac{8}{12} & \dfrac{16}{24} \\[2mm] \dfrac{2}{3} & \dfrac{2}{3} \end{array}$$ TRUE

The solution is 7.

Do Exercises 5 and 6.

Example 7   Given that $f(x) = x + 6/x$, find all values of x for which $f(x) = 5$.

Since $f(x) = x + 6/x$, we want to find all values of x for which

$$x + \dfrac{6}{x} = 5.$$

The LCM of the denominators is x. We multiply by x on both sides:

$$x\left(x + \dfrac{6}{x} \right) = x \cdot 5 \qquad \text{Multiplying by } x \text{ on both sides}$$

$$x \cdot x + x \cdot \dfrac{6}{x} = 5x$$

$$x^2 + 6 = 5x \qquad \text{Simplifying}$$
$$x^2 - 5x + 6 = 0 \qquad \text{Getting 0 on one side}$$
$$(x-3)(x-2) = 0 \qquad \text{Factoring}$$
$$x = 3 \quad or \quad x = 2. \qquad \text{Using the principle of zero products}$$

Answers on page A-32

CHECK:

For 3:

$$x + \frac{6}{x} = 5$$

$$\begin{array}{c|c} 3 + \dfrac{6}{3} & 5 \\ \hline 3 + 2 & \\ 5 & \text{TRUE} \end{array}$$

For 2:

$$x + \frac{6}{x} = 5$$

$$\begin{array}{c|c} 2 + \dfrac{6}{2} & 5 \\ \hline 2 + 3 & \\ 5 & \text{TRUE} \end{array}$$

The solutions are 2 and 3.

Do Exercise 7.

AG Algebraic–Graphical Connection

Let's make a visual check of Example 7 by looking at a graph. We can think of the equation

$$x + \frac{6}{x} = 5$$

as the intersection of the graphs of

$$f(x) = x + \frac{6}{x} \quad \text{and} \quad g(x) = 5.$$

We see in the following graph that there are two points of intersection, at $x = 2$ and at $x = 3$.

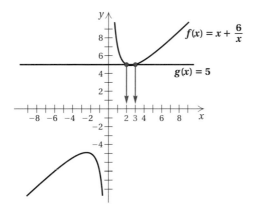

CAUTION! In this section, we have introduced a new use of the LCM. Before, you used the LCM in adding or subtracting rational expressions. Now we are working with equations. There are equals signs. We clear the fractions by multiplying on both sides of the equation by the LCM. This eliminates the denominators. *Do not* make the mistake of trying to "clear the fractions" when you do not have an equation!

7. Given that $f(x) = x - 12/x$, find all values of x for which $f(x) = 1$.

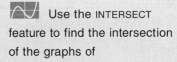
Answer on page A-32

Improving Your Math Study Skills

Are You Calculating or Solving?

At the beginning of this section, we noted that one of the common difficulties with this chapter is knowing for sure the task at hand. Are you combining expressions using operations to get another *rational expression,* or are you solving equations for which the results are numbers that are *solutions* of an equation? To learn to make these decisions, complete the following list by writing in the blank the type of answer you should get: "Rational expression" or "Solutions." You do not need to complete the mathematical operations.

Task	Answer (Just write "Rational expression" or "Solutions.")
1. Add: $\dfrac{4}{x-2} + \dfrac{1}{x+2}$.	
2. Solve: $\dfrac{4}{x-2} = \dfrac{1}{x+2}$.	
3. Subtract: $\dfrac{4}{x-2} - \dfrac{1}{x+2}$.	
4. Multiply: $\dfrac{4}{x-2} \cdot \dfrac{1}{x+2}$.	
5. Divide: $\dfrac{4}{x-2} \div \dfrac{1}{x+2}$.	
6. Solve: $\dfrac{4}{x-2} + \dfrac{1}{x+2} = \dfrac{26}{x^2-4}$.	
7. Perform the indicated operations and simplify: $\dfrac{4}{x-2} + \dfrac{1}{x+2} - \dfrac{26}{x^2-4}$.	
8. Solve: $\dfrac{x^2}{x-1} = \dfrac{1}{x-1}$.	
9. Solve: $\dfrac{2}{y^2-25} = \dfrac{3}{y-5} + \dfrac{1}{y-5}$.	
10. Solve: $\dfrac{x}{x+4} - \dfrac{4}{x-4} = \dfrac{x^2+16}{x^2-16}$.	
11. Perform the indicated operations and simplify: $\dfrac{x}{x+4} - \dfrac{4}{x-4} - \dfrac{x^2+16}{x^2-16}$.	
12. Solve: $\dfrac{5}{y-3} - \dfrac{30}{y^2-9} = 1$.	
13. Add: $\dfrac{5}{y-3} + \dfrac{30}{y^2-9} + 1$.	

Exercise Set 5.5

a Solve.

1. $\dfrac{y}{10} = \dfrac{2}{5} + \dfrac{3}{8}$

2. $\dfrac{3}{8} + \dfrac{1}{3} = \dfrac{t}{12}$

3. $\dfrac{1}{4} - \dfrac{5}{6} = \dfrac{1}{a}$

4. $\dfrac{5}{8} - \dfrac{2}{5} = \dfrac{1}{y}$

5. $\dfrac{x}{3} - \dfrac{x}{4} = 12$

6. $\dfrac{y}{5} - \dfrac{y}{3} = 15$

7. $x + \dfrac{8}{x} = -9$

8. $y + \dfrac{22}{y} = -13$

9. $\dfrac{3}{y} + \dfrac{7}{y} = 5$

10. $\dfrac{4}{3y} - \dfrac{3}{y} = \dfrac{10}{3}$

11. $\dfrac{1}{2} = \dfrac{z-5}{z+1}$

12. $\dfrac{x-6}{x+9} = \dfrac{2}{7}$

13. $\dfrac{3}{y+1} = \dfrac{2}{y-3}$

14. $\dfrac{4}{x-1} = \dfrac{3}{x+2}$

15. $\dfrac{y-1}{y-3} = \dfrac{2}{y-3}$

16. $\dfrac{x-2}{x-4} = \dfrac{2}{x-4}$

17. $\dfrac{x+1}{x} = \dfrac{3}{2}$

18. $\dfrac{y+2}{y} = \dfrac{5}{3}$

19. $\dfrac{1}{2} - \dfrac{4}{9x} = \dfrac{4}{9} - \dfrac{1}{6x}$

20. $-\dfrac{1}{3} - \dfrac{5}{4y} = \dfrac{3}{4} - \dfrac{1}{6y}$

21. $\dfrac{60}{x} - \dfrac{60}{x-5} = \dfrac{2}{x}$

22. $\dfrac{50}{y} - \dfrac{50}{y-2} = \dfrac{4}{y}$

23. $\dfrac{7}{5x-2} = \dfrac{5}{4x}$

24. $\dfrac{5}{y+4} = \dfrac{3}{y-2}$

25. $\dfrac{x}{x-2} + \dfrac{x}{x^2-4} = \dfrac{x+3}{x+2}$

26. $\dfrac{3}{y-2} + \dfrac{2y}{4-y^2} = \dfrac{5}{y+2}$

27. $\dfrac{6}{x^2-4x+3} - \dfrac{1}{x-3} = \dfrac{1}{4x-4}$

28. $\dfrac{1}{2x+10} = \dfrac{8}{x^2-25} - \dfrac{2}{x-5}$

29. $\dfrac{5}{y+3} = \dfrac{1}{4y^2-36} + \dfrac{2}{y-3}$

30. $\dfrac{7}{x-2} - \dfrac{8}{x+5} = \dfrac{1}{2x^2+6x-20}$

31. $\dfrac{a}{2a-6} - \dfrac{3}{a^2-6a+9} = \dfrac{a-2}{3a-9}$

32. $\dfrac{1}{x-2} = \dfrac{2}{x+4} + \dfrac{2x-1}{x^2+2x-8}$

33. $\dfrac{2x+3}{x-1} = \dfrac{10}{x^2-1} + \dfrac{2x-3}{x+1}$

34. $\dfrac{y}{y+1} + \dfrac{3y+5}{y^2+4y+3} = \dfrac{2}{y+3}$

35. $\dfrac{4}{x+3} + \dfrac{7}{x^2-3x+9} = \dfrac{108}{x^3+27}$

36. $\dfrac{3x}{x+2} + \dfrac{72}{x^3+8} = \dfrac{24}{x^2-2x+4}$

37. $\dfrac{5x}{x-7} - \dfrac{35}{x+7} = \dfrac{490}{x^2-49}$

38. $\dfrac{3x}{x+2} + \dfrac{6}{x} + 4 = \dfrac{12}{x^2+2x}$

39. $\dfrac{x-1}{3} + \dfrac{6x+1}{15} + \dfrac{2(x-2)}{13-7x} = \dfrac{2(x+2)}{5}$

For the given rational function f, find all values of x for which $f(x)$ has the indicated value.

40. $f(x) = 2x - \dfrac{15}{x}; \quad f(x) = 1$

41. $f(x) = 2x - \dfrac{6}{x}; \quad f(x) = 1$

42. $f(x) = \dfrac{x-5}{x+1}; \quad f(x) = \dfrac{3}{5}$

43. $f(x) = \dfrac{x-3}{x+2}; \quad f(x) = \dfrac{1}{5}$

44. $f(x) = \dfrac{12}{x} - \dfrac{12}{2x}; \quad f(x) = 8$

45. $f(x) = \dfrac{6}{x} - \dfrac{6}{2x}; \quad f(x) = 5$

Skill Maintenance

Factor.   [4.5d]

46. $4t^3 + 500$

47. $1 - t^6$

48. $a^3 + 8b^3$

49. $a^3 - 8b^3$

Synthesis

50. ◈ Explain how you might easily create rational equations for which there is no solution. (See Example 4 for a hint.)

51. ◈ Explain why it is sufficient, when checking a possible solution of a rational equation, to verify that the number in question does not make a denominator 0.

52. 〰 Use the TABLE feature of a grapher to check your answers to Exercises 7, 8, and 34.

53. 〰 Use the TABLE feature of a grapher to check your answers to Exercises 10, 25, and 38.

54. ◈ 〰
a) Use the INTERSECT feature of a grapher to find the points of intersection of the graphs of

$$f(x) = \dfrac{1}{1+x} + \dfrac{x}{1-x} \quad \text{and} \quad g(x) = \dfrac{1}{1-x} - \dfrac{x}{1+x}.$$

b) Use the algebraic methods of this section to check your answers to part (a).
c) Explain which procedure you prefer.

55. ◈ 〰
a) Use the INTERSECT feature of a grapher to find the points of intersection of the graphs of

$$f(x) = \dfrac{x+3}{x+2} - \dfrac{x+4}{x+3} \quad \text{and} \quad g(x) = \dfrac{x+5}{x+4} - \dfrac{x+6}{x+5}.$$

b) Use the algebraic methods of this section to check your answers to part (a).
c) Explain which procedure you prefer.

5.6 Applications and Problem Solving

a Work Problems

Example 1 *Cake Decorating.* Sara and Jeff work at Gourmet Perfection, Inc. Sara can decorate a six-tier wedding cake in 4 hr. Jeff can decorate the same cake in 5 hr. How long would it take them, working together, to decorate the cake?

Objectives

a Solve work problems and certain basic problems using rational equations.

b Solve applied problems involving proportions.

c Solve motion problems using rational equations.

For Extra Help

TAPE 12   TAPE 10B   MAC WIN   CD-ROM

1. **Familiarize.** We familiarize ourselves with the problem by considering two *incorrect* ways of translating the problem to mathematical language.

   a) A common *incorrect* way to translate the problem is to add the two times: 4 hr + 5 hr = 9 hr. Let's think about this. Sara can do the job alone in 4 hr. If Sara and Jeff work together, the time it takes them should be *less* than 4 hr. Thus we reject 9 hr as a solution, but we do have a partial check on any answer we get. The answer should be less than 4 hr.

   b) Another *incorrect* way to translate the problem is as follows. Suppose the two people split up the decorating job in such a way that Sara does half the decorating and Jeff does the other half. Then

   Sara decorates $\frac{1}{2}$ of the cake in $\frac{1}{2}(4$ hr$)$, or 2 hr,

   and

   Jeff decorates $\frac{1}{2}$ of the cake in $\frac{1}{2}(5$ hr$)$, or $2\frac{1}{2}$ hr.

   But time is wasted since Sara would finish her part $\frac{1}{2}$ hr earlier than Jeff. In effect, they have not worked together to complete the job as fast as possible. If Sara helps Jeff after completing her half, the entire job could be finished in a time somewhere between 2 hr and $2\frac{1}{2}$ hr.

   We proceed to a translation by considering how much of the job is finished in 1 hr, 2 hr, 3 hr, and so on. It takes Sara 4 hr to do the decorating job alone. Then in 1 hr, she can do $\frac{1}{4}$ of the job. It takes Jeff 5 hr to do the job alone. Then in 1 hr, he can do $\frac{1}{5}$ of the job. Working together, they can do

   $$\frac{1}{4} + \frac{1}{5}, \text{ or } \frac{9}{20} \text{ of the job in 1 hr.}$$

   In 2 hr, Sara can do $2\left(\frac{1}{4}\right)$ of the job and Jeff can do $2\left(\frac{1}{5}\right)$ of the job. Working together, they can do

   $$2\left(\frac{1}{4}\right) + 2\left(\frac{1}{5}\right), \text{ or } \frac{9}{10} \text{ of the job in 2 hr.}$$

   Continuing this reasoning, we can form a table like the following one.

1. *Trimming Shrubbery.* Alex can trim the shrubbery at Beecher Community College in 6 hr. Tanya can do the same job in 4 hr. How long would it take them, working together, to do the same trimming job?

	Fraction of the Job Completed		
Time	**Sarah**	**Jeff**	**Together**
1 hr	$\frac{1}{4}$	$\frac{1}{5}$	$\frac{1}{4} + \frac{1}{5}$, or $\frac{9}{20}$
2 hr	$2\left(\frac{1}{4}\right)$	$2\left(\frac{1}{5}\right)$	$2\left(\frac{1}{4}\right) + 2\left(\frac{1}{5}\right)$, or $\frac{9}{10}$
3 hr	$3\left(\frac{1}{4}\right)$	$3\left(\frac{1}{5}\right)$	$3\left(\frac{1}{4}\right) + 3\left(\frac{1}{5}\right)$, or $1\frac{7}{20}$
t hr	$t\left(\frac{1}{4}\right)$	$t\left(\frac{1}{5}\right)$	$t\left(\frac{1}{4}\right) + t\left(\frac{1}{5}\right)$

From the table, we see that if they work 3 hr, the fraction of the job that they will finish is $1\frac{7}{20}$, which is more of the job than needs to be done. We also see that the answer is somewhere between 2 hr and 3 hr. What we want is a number t such that the fraction of the job that is completed is 1; that is, the job is just completed—not more $\left(1\frac{7}{20}\right)$ and not less $\left(\frac{9}{10}\right)$.

2. Translate. From the table, we see that the time we want is some number t for which

$$t\left(\frac{1}{4}\right) + t\left(\frac{1}{5}\right) = 1, \quad \text{or} \quad \frac{t}{4} + \frac{t}{5} = 1,$$

where 1 represents the idea that the entire job is completed in time t.

3. Solve. We solve the equation:

$$\frac{t}{4} + \frac{t}{5} = 1$$

$$20\left(\frac{t}{4} + \frac{t}{5}\right) = 20 \cdot 1 \qquad \text{The LCM is } 2 \cdot 2 \cdot 5, \text{ or } 20.$$
$$\text{We multiply by 20.}$$

$$20 \cdot \frac{t}{4} + 20 \cdot \frac{t}{5} = 20 \qquad \text{Using the distributive law}$$

$$5t + 4t = 20 \qquad \text{Simplifying}$$

$$9t = 20$$

$$t = \frac{20}{9}, \text{ or } 2\frac{2}{9} \text{ hr.}$$

4. Check. The check can be done by using $\frac{20}{9}$ for t and substituting into the original equation:

$$\frac{20}{9}\left(\frac{1}{4}\right) + \frac{20}{9}\left(\frac{1}{5}\right) = \frac{5}{9} + \frac{4}{9} = \frac{9}{9} = 1.$$

We also have a partial check in what we learned from the *Familiarize* step. The answer, $2\frac{2}{9}$ hr, is between 2 hr and 3 hr (see the table), and it is less than 4 hr, the time it takes Sara working alone.

5. State. It takes $2\frac{2}{9}$ hr for them to complete the job working together.

▶ THE WORK PRINCIPLE

Suppose that a is the time it takes A to do a job, b is the time it takes B to do the same job, and t is the time it takes them to do the job working together. Then

$$\frac{t}{a} + \frac{t}{b} = 1.$$

Answer on page A-32

Do Exercise 1.

Example 2 *Building a Wall.* It takes Red 9 hr longer than Hannah to build a stone wall. Working together, they can build the wall in 20 hr. How long would it take each, working alone, to build the wall?

1. **Familiarize.** Comparing this problem to Example 1, we note that we do not know the times required by each person to build the wall had each worked alone. We let

   h = the amount of time, in hours, that it would take Hannah working alone.

   Then

   $h + 9$ = the amount of time, in hours, that it would take Red working alone.

   We also know that $t = 20$ hr = total time. Thus,

   $$\frac{20}{h} = \text{the fraction of the wall that Hannah could finish in 20 hr}$$

   and

   $$\frac{20}{h + 9} = \text{the fraction of the wall that Red could finish in 20 hr.}$$

2. **Translate.** Using the work principle, we know that

   $$\frac{t}{a} + \frac{t}{b} = 1 \qquad \text{Using the work principle}$$

   $$\frac{20}{h} + \frac{20}{h + 9} = 1. \qquad \text{Substituting } \frac{20}{h} \text{ for } \frac{t}{a} \text{ and } \frac{20}{h+9} \text{ for } \frac{t}{b}$$

3. **Solve.** We solve the equation:

   $$\frac{20}{h} + \frac{20}{h + 9} = 1$$

   $$h(h + 9)\left(\frac{20}{h} + \frac{20}{h + 9}\right) = h(h + 9) \cdot 1 \qquad \begin{array}{l}\text{We multiply by the LCM,} \\ \text{which is } h(h+9).\end{array}$$

   $$h(h + 9) \cdot \frac{20}{h} + h(h + 9) \cdot \frac{20}{h + 9} = h(h + 9) \qquad \text{Using the distributive law}$$

   $$(h + 9) \cdot 20 + h \cdot 20 = h^2 + 9h \qquad \text{Simplifying}$$

   $$20h + 180 + 20h = h^2 + 9h$$

   $$0 = h^2 - 31h - 180 \qquad \text{Getting 0 on one side}$$

   $$0 = (h - 36)(h + 5) \qquad \text{Factoring}$$

   $$h - 36 = 0 \quad \text{or} \quad h + 5 = 0 \qquad \begin{array}{l}\text{Using the principle of} \\ \text{zero products}\end{array}$$

   $$h = 36 \quad \text{or} \qquad h = -5.$$

4. **Check.** Since negative time has no meaning in the problem, we reject −5 as a solution to the original problem. The number 36 checks since if Hannah takes 36 hr alone and Red takes 36 + 9, or 45 hr alone, in 20 hr, working together, they would have completed

   $$\frac{20}{36} + \frac{20}{45} = \frac{5}{9} + \frac{4}{9}, \text{ or 1 wall.}$$

5. **State.** It would take Hannah 36 hr and Red 45 hr to build the wall alone.

Do Exercise 2.

2. *Filling a Water Tank.* Two pipes carry water to the same tank. Pipe A, working alone, can fill the tank three times as fast as pipe B. Together, the pipes can fill the tank in 24 hr. Find the time it would take each pipe to fill the tank alone.

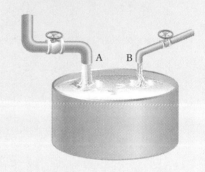

Answer on page A-32

b | Applications Involving Proportions

Any rational expression a/b represents a **ratio**. Percent can be considered a ratio. For example, 67% is the ratio of 67 to 100, or 67/100. The ratio of two different kinds of measure is called a **rate**. Speed is an example of a rate. Florence Griffith Joyner set a world record in a recent Olympics with a time of 10.49 sec in the 100-m dash. Her speed, or rate, was

$$\frac{100 \text{ m}}{10.49 \text{ sec}}, \quad \text{or} \quad 9.5\frac{\text{m}}{\text{sec}}. \qquad \text{Rounded to the nearest tenth}$$

> An equality of ratios, $A/B = C/D$, is called a **proportion**. The numbers named in a true proportion are said to be **proportional**.

We can use proportions to solve applied problems by expressing a ratio in two ways, as shown below. For example, suppose that it takes 8 gal of gas to drive for 120 mi, and we want to determine how much will be required to drive for 550 mi. If we assume that the car uses gas at the same rate throughout the trip, the ratios are the same, and we can write a proportion.

$$\begin{array}{l} \text{Miles} \longrightarrow \\ \text{Gallons} \longrightarrow \end{array} \frac{120}{8} = \frac{550}{x} \begin{array}{l} \longleftarrow \text{Miles} \\ \longleftarrow \text{Gallons} \end{array}$$

To solve this proportion, we note that the LCM is $8x$. Thus we multiply by $8x$.

$$8x \cdot \frac{120}{8} = 8x \cdot \frac{550}{x} \qquad \text{Multiplying by } 8x$$

$$x \cdot 120 = 8 \cdot 550 \qquad \text{Simplifying}$$

$$120x = 8 \cdot 550$$

$$x = \frac{8 \cdot 550}{120} \qquad \text{Dividing by 120}$$

$$x \approx 36.67$$

Thus, 36.67 gal will be required to drive for 550 mi.

It is common to use **cross products** to solve proportions, as follows:

$$\frac{120}{8} \bowtie \frac{550}{x} \qquad \text{If } \frac{A}{B} = \frac{C}{D}, \text{ then } AD = BC.$$

$$120 \cdot x = 8 \cdot 550 \qquad 120 \cdot x \text{ and } 8 \cdot 550 \text{ are called } \textit{cross products.}$$

$$x = \frac{8 \cdot 550}{120}$$

$$x \approx 36.67.$$

Example 3 *Calories Burned.* The author of this text generally exercises three times per week. The readout on a stairmaster machine tells him that if he exercises for 24 min, he will burn 356 calories. How many calories will he burn if he exercises for 30 min?

Source: Stairmaster® Free Climber 4400 PT; Jordan YMCA, Marv Bittinger

1. **Familiarize.** We let c = the number of calories burned in 30 min.

2. **Translate.** Next, we translate to a proportion. We make each side the ratio of minutes to number of calories, with minutes in the numerator and number of calories in the denominator.

$$\text{Minutes} \longrightarrow \frac{24}{356} = \frac{30}{c} \longleftarrow \text{Minutes}$$
$$\text{Calories} \longrightarrow \qquad\qquad \longleftarrow \text{Calories}$$

3. **Solve.** We solve the proportion:

$$\frac{24}{356} = \frac{30}{c}$$

$$24c = 356 \cdot 30 \qquad \text{Equating cross products}$$

$$c = \frac{356 \cdot 30}{24} \qquad \text{Dividing by 24}$$

$$= 445. \qquad \text{Multiplying and dividing}$$

4. **Check.** We substitute into the proportion and check cross products:

$$\frac{24}{356} = \frac{30}{445};$$

$$24 \cdot 445 = 10,680; \qquad 356 \cdot 30 = 10,680.$$

Since the cross products are the same, the answer checks.

5. **State.** In 30 min, he will burn 445 calories.

Do Exercise 3.

3. *Determining Medication Dosage.* To control a fever, a doctor suggests that a child who weighs 28 kg be given 420 mg of Tylenol. If dosage is proportional to the child's weight, how much Tylenol would be recommended for a child who weighs 35 kg?

Answer on page A-32

4. *Estimating Fish Populations.*
To determine the number of
fish in a lake, a conservationist catches 225 fish, tags
them, and releases them back
into the lake. Later, 108 fish
are caught, and it is found that
15 of them are tagged.
Estimate how many fish are in
the lake.

Example 4 *Estimating Wildlife Populations.* To determine the number
of deer in a forest, a conservationist catches 612 deer, tags them, and releases them. Later, 244 deer are caught, and it is found that 72 of them are
tagged. Estimate how many deer are in the forest.

1. **Familiarize.** We let $D =$ the number of deer in the forest. We assume
   that the tagged deer mix freely with the entire population, and that
   later when some deer have been captured, the ratio of the number of
   tagged deer to the number of deer captured is the same as the ratio
   of the number of deer originally tagged to the total number of deer in
   the forest. For example, given that 1 of every 3 deer captured later is
   tagged, we assume that 1 of every 3 deer in the forest was originally
   tagged.

2. **Translate.** We translate to a proportion, as follows:

   Deer tagged originally $\longrightarrow \dfrac{612}{D} = \dfrac{72}{244}$ $\longleftarrow$ Tagged deer caught
   Deer in forest $\longrightarrow$ $\phantom{\dfrac{612}{D}}$ $\longleftarrow$ Deer caught

3. **Solve.** We solve the proportion:

   $612 \cdot 244 = D \cdot 72$ Equating cross products

   $\dfrac{612 \cdot 244}{72} = D$ Dividing by 72

   $2074 = D.$ Multiplying and dividing

4. **Check.** We substitute into the proportion and check cross products:

   $\dfrac{612}{2074} = \dfrac{72}{244};$

   $612 \cdot 244 = 149{,}328;$ $2074 \cdot 72 = 149{,}328.$

5. **State.** We estimate that there are 2074 deer in the forest.

Do Exercise 4.

c | Motion Problems

We considered motion problems earlier in Sections 1.3 and 3.3. To translate
them, we know that we can use either the basic motion formula, $d = rt$,
or either of two formulas $r = d/t$, or $t = d/r$, which can be derived from
$d = rt$.

Example 5 *Bicycling.* A racer is bicycling 15 km/h faster than a person
on a mountain bike. In the time it takes the racer to travel 80 km, the person on the mountain bike has gone 50 km. Find the speed of each bicyclist.

Answer on page A-32

1. Familiarize. Let's guess that the person on the mountain bike is going 10 km/h. The racer would then be traveling 10 + 15, or 25 km/h. At 25 km/h, the racer will travel 80 km in $\frac{80}{25} = 3.2$ hr. Going 10 km/h, the mountain bike will cover 50 km in $\frac{50}{10} = 5$ hr. Since $3.2 \neq 5$, our guess was wrong, but we can see that if r is the rate, in kilometers per hour, of the slower bike, then the rate of the racer, who is traveling 15 km/h faster, is $r + 15$.

Making a drawing and organizing the facts in a chart can be helpful.

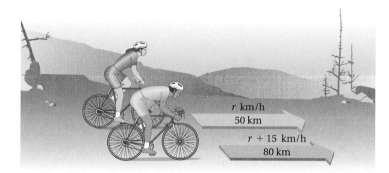

r km/h
50 km
r + 15 km/h
80 km

	Distance	Speed	Time	
Mountain Bike	50	r	t	$\rightarrow t = \dfrac{50}{r}$
Racing Bike	80	$r + 15$	t	$\rightarrow t = \dfrac{80}{r + 15}$

2. Translate. The time is the same for both bikes. Using the formula $t = d/r$ across both rows of the table, we find two expressions for time and can equate them as

$$\frac{50}{r} = \frac{80}{r + 15}.$$

3. Solve. We solve the equation:

$$\frac{50}{r} = \frac{80}{r + 15}$$

$$r(r + 15) \cdot \frac{50}{r} = r(r + 15) \cdot \frac{80}{r + 15} \qquad \text{The LCM is } r(r + 15).$$
$$\text{We multiply by } r(r + 15).$$

$$(r + 15) \cdot 50 = r \cdot 80 \qquad \text{Simplifying. We can also obtain this by equating cross products.}$$

$$50r + 750 = 80r \qquad \text{Using the distributive law}$$

$$750 = 30r \qquad \text{Subtracting } 50r$$

$$\frac{750}{30} = r \qquad \text{Dividing by 30}$$

$$25 = r.$$

4. Check. If our answer checks, the mountain bike is going 25 km/h and the racing bike is going 25 + 15, or 40 km/h.

Traveling 80 km at 40 km/h, the racer is riding for $\frac{80}{40} = 2$ hr. Traveling 50 km at 25 km/h, the person on the mountain bike is riding for $\frac{50}{25} = 2$ hr. Our answer checks since the two times are the same.

5. State. The speed of the racer is 40 km/h, and the speed of the person on the mountain bike is 25 km/h.

Do Exercise 5.

5. *Four-wheeler Travel.* Jaime's four-wheeler travels 8 km/h faster than Mara's. Jaime travels 69 km in the same time it takes Mara to travel 45 km. Find the speed of each person's four-wheeler.

Answer on page A-32

6. *River Travel.* The Delta Queen is a riverboat that travels the Ohio River. Suppose that it travels 246 mi downstream in the same time that it takes to travel 180 mi upstream. The speed of the current in the river is 5.5 mph. Find the speed of the boat in still water.

Example 6 *Air Travel.* An airplane flies 1062 mi with the wind. In the same amount of time, it can fly 738 mi against the wind. The speed of the plane in still air is 200 mph. Find the speed of the wind.

1. Familiarize. We first make a drawing. We let w = the speed of the wind and t = the time, and then organize the facts in a chart.

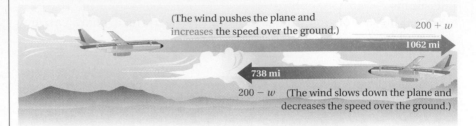

(The wind pushes the plane and increases the speed over the ground.) $200 + w$

1062 mi

738 mi

$200 - w$ (The wind slows down the plane and decreases the speed over the ground.)

	Distance	Speed	Time
With Wind	1062	200 + w	t
Against Wind	738	200 − w	t

$$t = \frac{1062}{200 + w}$$

$$t = \frac{738}{200 - w}$$

2. Translate. Using the formula $t = d/r$ across both rows of the table, we find two expressions for time and can equate them as

$$\frac{1062}{200 + w} = \frac{738}{200 - w}.$$

3. Solve. We solve the equation:

$$\frac{1062}{200 + w} = \frac{738}{200 - w}$$

$$(200 + w)(200 - w)\left(\frac{1062}{200 + w}\right) = (200 + w)(200 - w)\left(\frac{738}{200 - w}\right)$$

$$(200 - w)1062 = (200 + w)738 \qquad \text{We can also obtain this by equating cross products.}$$

$$212{,}400 - 1062w = 147{,}600 + 738w$$

$$64{,}800 = 1800w$$

$$w = 36.$$

4. Check. With the wind, the speed of the plane is $200 + 36$, or 236 mph. Dividing the distance, 1062 mi, by the speed, 236 mph, we get 4.5 hr. Against the wind, the speed of the plane is $200 - 36$, or 164 mph. Dividing the distance, 738 mi, by the speed, 164 mph, we get 4.5 hr. Since the times are the same, the answer checks.

5. State. The speed of the wind is 36 mph.

Do Exercise 6.

Answer on page A-32

Exercise Set 5.6

a Solve.

1. *Mail Order.* Jason, an experienced shipping clerk, can fill a certain order in 5 hr. Jessica, a new clerk, needs 9 hr to do the same job. Working together, how long will it take them to fill the order?

2. *Painting.* Alexander can paint a room in 4 hr. Alexandra can paint the same room in 3 hr. Working together, how long will it take them to paint the room?

3. *Filling a Pool.* An in-ground backyard pool can be filled in 12 hr if water enters through a pipe alone, or in 30 hr if water enters through a hose alone. If water is entering through both the pipe and the hose, how long will it take to fill the pool?

4. *Filling a Tank.* A tank can be filled in 18 hr by pipe A alone and in 22 hr by pipe B alone. How long will it take to fill the tank if both pipes are working?

5. *Printing.* One printing press can print an order of advertising brochures in 4.5 hr. Another press can do the same job in 5.5 hr. How long will it take if both presses are used?

> *Hint*: You may find that multiplying by $\frac{1}{10}$ on both sides of the equation will clear the decimals.

6. *Wood Cutting.* Hannah can clear a lot in 5.5 hr. Damon can do the same job in 7.5 hr. How long will it take them to clear the lot working together?

7. The sum of a number and five times its reciprocal is −6. Find the number.

8. The sum of a number and four times its reciprocal is −5. Find the number.

9. The reciprocal of 8 plus the reciprocal of 9 is the reciprocal of what number?

10. The reciprocal of 2 plus the reciprocal of 6 is the reciprocal of what number?

11. *Painting.* Juan can paint the neighbor's house four times as fast as Ariel. The year they worked together it took them 8 days. How long would it take each to paint the house alone?

12. *Newspaper Delivery.* Matt can deliver papers three times as fast as his father. If they work together, it takes them 1 hr. How long would it take each to deliver the papers alone?

13. *Cutting Firewood.* Jake can cut and split a cord of firewood in 6 fewer hr than Skyler can. When they work together, it takes them 4 hr. How long would it take each of them to do the job alone?

14. *Painting.* Sara takes 3 hr longer to paint a floor than Kate does. When they work together, it takes them 2 hr. How long would each take to do the job alone?

b Solve.

15. *Women's Hip Measurements.* For improved health, it is recommended that a woman's waist-to-hip ratio be 0.85 (or lower). (**Source:** David Schmidt, "Lifting Weight Myths," *Nutrition Action Newsletter,* 20, no. 4, October 1993) Marta's hip measurement is 40 in. To meet the recommendation, what should Marta's waist measurement be?

16. *Men's Hip Measurements.* It is recommended that a man's waist-to-hip ratio be 0.95 (or lower). Malcom's hip measurement is 40 in. To meet the recommendation, what should Malcom's waist measurement be?

Home Run Attack. In 1998, Mark McGwire, "Big Mac" of the St. Louis Cardinals, and Sammy Sosa, of the Chicago Cubs, were setting home-run records with astounding hitting. After 54 games, McGwire had hit 27 home runs and Sosa had hit 14. Could we predict, using proportions, that they would break all-time home-run records?

17. *McGwire's Record.* Assume that McGwire continued to hit home runs at the same rate.
   a) How many home runs would he hit in the normal 162-game season?
   b) Would he break Roger Maris's all-time record of 61 home runs?
   c) McGwire's actual final count was 70 home runs. How close is your prediction to the actual count? [f]

18. *Sosa's Record.* Assume that Sosa continued to hit home runs at the same rate.
   a) How many home runs would he hit in the normal 162-game season?
   b) Would he break Roger Maris's all-time record of 61 home runs?
   c) Sosa's actual final count was 66. How close is your prediction to the actual count?

19. *Coffee Consumption.* Coffee beans from 14 trees are required to produce 7.7 kg of coffee. (This is the average amount that each person in the United States drinks each year.) The beans from how many trees are required to produce 638 kg of coffee?

20. *Human Blood.* 10 cm^3 of a normal specimen of human blood contains 1.2 g of hemoglobin. How many grams does 32 cm^3 of the same blood contain?

21. *Estimating Wildlife Populations.* To determine the number of trout in a lake, a conservationist catches 112 trout, tags them, and releases them back into the lake. Later, 82 trout are caught; 32 of them are tagged. How many trout are in the lake?

22. *Estimating Wildlife Populations.* To determine the number of deer in a game preserve, a conservationist catches 318 deer, tags them, and lets them loose. Later, 168 deer are caught; 56 of them are tagged. How many deer are in the preserve?

23. *Weight on Mars.* The ratio of the weight of an object on Mars to the weight of an object on Earth is 0.4 to 1.

   a) How much will a 12-T rocket weigh on Mars?
   b) How much will a 120-lb astronaut weigh on Mars?

24. *Weight on Moon.* The ratio of the weight of an object on the moon to the weight of an object on Earth is 0.16 to 1.

   a) How much will a 12-T rocket weigh on the moon?
   b) How much will a 180-lb astronaut weigh on the moon?

25. Consider the numbers 1, 2, 3, and 5. If the same number is added to each of the numbers, it is found that the ratio of the first new number to the second is the same as the ratio of the third new number to the fourth. Find the number.

26. A rope is 28 ft long. How can the rope be cut in such a way that the ratio of the resulting two segments is 3 to 5?

c Solve.

27. *Boating.* The current in the Lazy River moves at a rate of 4 mph. Ken's dinghy motors 6 mi upstream in the same time that it takes to motor 12 mi downstream. What is the speed of the dinghy in still water?

28. *Kayaking.* The speed of the current in Catamount Creek is 3 mph. Ahmad's kayak can travel 4 mi upstream in the same time that it takes to travel 10 mi downstream. What is the speed of Ahmad's kayak in still water?

29. *Marine Travel.* Sandy's tugboat moves at a rate of 10 mph in still water. It travels 24 mi upstream and 24 mi downstream in a total time of 5 hr. What is the speed of the current?

30. *Shipping.* A barge moves at a rate of 7 km/h in still water. It travels 45 km upriver and 45 km downriver in a total time of 14 hr. What is the speed of the current?

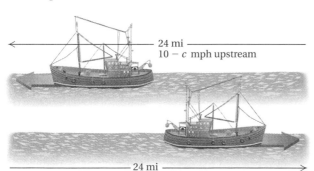

24 mi
$10 - c$ mph upstream

24 mi
$10 + c$ mph downstream

31. *Moving Sidewalks.* The moving sidewalk at O'Hare Airport in Chicago moves at a rate of 1.8 ft/sec. Walking on the moving sidewalk, Camille travels 105 ft forward in the time it takes to travel 51 ft in the opposite direction. How fast would Camille be walking on a nonmoving sidewalk?

32. *Moving Sidewalks.* Newark Airport's moving sidewalk moves at a rate of 1.7 ft/sec. Walking on the moving sidewalk, Benny can travel 120 ft forward in the same time it takes to travel 52 ft in the opposite direction. How fast would Benny be walking on a nonmoving sidewalk?

33. *Walking.* Rosanna walks 2 mph slower than Simone. In the time it takes Simone to walk 8 mi, Rosanna walks 5 mi. Find the speed of each person.

34. *Bus Travel.* A local bus travels 7 mph slower than the express. The express travels 90 mi in the time it takes the local to travel 75 mi. Find the speed of each bus.

Skill Maintenance

In Exercises 35–38, the graph is that of a function. Determine the domain and the range.   [2.3a]

35.

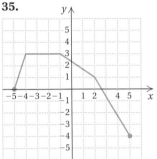

36.

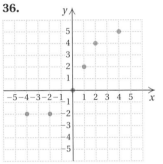

37.

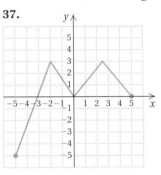

38.

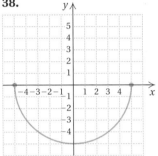

Graph.   [3.6b]

39. $x - 4y \geq 4$

40. $x \geq 3$

Graph.   [2.2c]

41. $f(x) = |x + 3|$

42. $f(x) = 5 - |x|$

Synthesis

43. ◈ Two steamrollers are being used to pave a parking lot. Will using the two steamrollers together take less than half as long as using the slower steamroller alone? Why or why not?

44. ◈ Write a work problem for a classmate to solve. Devise the problem so that the solution is "Jen takes 5 hr and Pablo takes 6 hr to complete the job alone."

45. *Travel by Car.* Melissa drives to work at 50 mph and arrives 1 min late. She drives to work at 60 mph and arrives 5 min early. How far does Melissa live from work?

46. *Escalators.* Together, a 100-cm-wide escalator and a 60-cm-wide escalator can empty a 1575-person auditorium in 14 min (*Source*: McGraw-Hill *Encyclopedia of Science and Technology*). The wider escalator moves twice as many people as the narrower one does. How many people per hour does the 60-cm-wide escalator move?

47. *Gas Mileage.* An automobile gets 22.5 miles per gallon (mpg) in city driving and 30 mpg in highway driving. The car is driven 465 mi on a full tank of 18.4 gal of gasoline. How many miles were driven in the city and how many were driven on the highway?

48. *Filling a Tank.* A tank can be filled in 9 hr and drained in 11 hr. How long will it take to fill the tank if the drain is left open?

49. *Clock Hands.* At what time after 4:00 will the minute hand and the hour hand of a clock first be in the same position?

50. Three trucks, A, B, and C, working together, can move a load of sand in t hours. When working alone, it takes A 1 extra hour to move the sand, B 6 extra hours, and C t extra hours. Find t.

5.7 Formulas and Applications

a | Formulas

Formulas occur frequently as mathematical models. Here we consider rational formulas. The procedure is as follows.

> To solve a rational formula for a given letter, identify the letter, and:
>
> 1. Multiply on both sides to clear fractions or decimals, if that is needed.
> 2. Multiply if necessary to remove parentheses.
> 3. Get all terms with the letter to be solved for on one side of the equation and all other terms on the other side, using the addition principle.
> 4. Factor out the unknown.
> 5. Solve for the letter in question, using the multiplication principle.

Example 1 *Optics.* The formula $f = L/d$ defines a camera's "f-stop," where L is the **focal length** (the distance from the lens to the film) and d is the **aperture** (the diameter of the lens). Solve the formula for d.

We solve this equation as we did the rational equations in Section 5.5:

$$f = \frac{L}{d} \quad \text{We want this letter alone.}$$

$$d \cdot f = d \cdot \frac{L}{d} \quad \begin{array}{l}\text{The LCM is } d.\\ \text{We multiply by } d.\end{array}$$

$$df = L \quad \text{Simplifying}$$

$$d = \frac{L}{f}. \quad \text{Dividing by } f$$

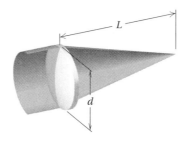

The formula $d = L/f$ can now be used to find the aperture if we know the focal length and the f-stop.

Do Exercise 1.

Example 2 *Astronomy.* The formula

$$L = \frac{dR}{D - d},$$

where D is the diameter of the sun, d is the diameter of the earth, R is the earth's distance from the sun, and L is some fixed distance, is used to calculate when lunar eclipses occur. Solve the formula for D.

$$L = \frac{dR}{D - d} \quad \text{We want this letter alone.}$$

$$(D - d) \cdot L = (D - d) \cdot \frac{dR}{D - d} \quad \begin{array}{l}\text{The LCM is } D - d.\\ \text{We multiply by } D - d.\end{array}$$

$$(D - d)L = dR \quad \text{Simplifying}$$

$$DL - dL = dR$$

$$DL = dR + dL \quad \text{Adding } dL$$

$$D = \frac{dR + dL}{L}. \quad \text{Dividing by } L$$

Objective

a Solve a formula for a letter.

For Extra Help

TAPE 12   TAPE 11A   MAC WIN   CD-ROM

1. *Boyle's Law.* The formula

$$\frac{PV}{T} = k$$

relates the pressure P, the volume V, and the temperature T of a gas. Solve the formula for T. (*Hint*: Begin by clearing the fraction.)

2. Solve the formula

$$I = \frac{pT}{M + pn}$$

for n.

Answers on page A-33

3. *The Doppler Effect.* The formula

$$F = \frac{sg}{s + v}$$

is used to determine the frequency F of a sound that is moving at velocity v toward a listener who hears the sound as frequency g. Here s is the speed of sound in a particular medium. Solve the formula for s.

4. *Work Formula.* The formula

$$\frac{t}{a} + \frac{t}{b} = 1$$

involves the total time t for some work to be done by two workers whose individual times are a and b. Solve the formula for t.

Answers on page A-33

Since D appears by itself on one side and not on the other, we have solved for D.

Do Exercise 2 on the preceding page.

Example 3   Solve the formula $L = \dfrac{dR}{D - d}$ for d.

We proceed as we did in Example 2 until we reach the equation

$$DL - dL = dR. \qquad \text{We want } d \text{ alone.}$$

We must now get all terms containing d alone on one side:

$$DL - dL = dR$$
$$DL = dR + dL \qquad \text{Adding } dL$$
$$DL = d(R + L) \qquad \text{Factoring out the letter } d$$
$$\frac{DL}{R + L} = d. \qquad \text{Dividing by } R + L$$

We now have d alone on one side, so we have solved the formula for d.

CAUTION!   If, when you are solving an equation for a letter, the letter appears on both sides of the equation, you know the answer is wrong. The letter must be alone on one side and *not* occur on the other.

Do Exercise 3.

Example 4   *Resistance.* The formula

$$\frac{1}{R} = \frac{1}{r_1} + \frac{1}{r_2}$$

involves the resistance R of two resistors r_1 and r_2 connected in parallel.* Solve the formula for r_1.

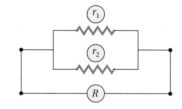

We multiply by the LCM, which is Rr_1r_2:

$$Rr_1r_2 \cdot \frac{1}{R} = Rr_1r_2 \cdot \left(\frac{1}{r_1} + \frac{1}{r_2} \right) \qquad \text{Multiplying by the LCM}$$

$$Rr_1r_2 \cdot \frac{1}{R} = Rr_1r_2 \cdot \frac{1}{r_1} + Rr_1r_2 \cdot \frac{1}{r_2} \qquad \begin{array}{l}\text{Multiplying to remove}\\ \text{parentheses}\end{array}$$

$$r_1r_2 = Rr_2 + Rr_1. \qquad \begin{array}{l}\text{Simplifying by removing}\\ \text{factors of 1}\end{array}$$

We might be tempted at this point to multiply by $1/r_2$ to get r_1 alone on the left, *but* note that there is an r_1 on the right. We must get all the terms involving r_1 on the *same side* of the equation.

$$r_1r_2 - Rr_1 = Rr_2 \qquad \text{Subtracting } Rr_1$$
$$r_1(r_2 - R) = Rr_2 \qquad \text{Factoring out } r_1$$
$$r_1 = \frac{Rr_2}{r_2 - R} \qquad \text{Dividing by } r_2 - R \text{ to get } r_1 \text{ alone}$$

Do Exercise 4.

---

*Note that R, r_1, and r_2 are all different variables. It is common to use subscripts, as in r_1 (read "r sub 1") and r_2, to distinguish variables.

Exercise Set 5.7

Solve.

1. $\dfrac{W_1}{W_2} = \dfrac{d_1}{d_2}$, for W_2

2. $\dfrac{W_1}{W_2} = \dfrac{d_1}{d_2}$, for d_1

3. $\dfrac{1}{R} = \dfrac{1}{r_1} + \dfrac{1}{r_2}$, for r_2

4. $\dfrac{1}{R} = \dfrac{1}{r_1} + \dfrac{1}{r_2}$, for R

5. $s = \dfrac{(v_1 + v_2)t}{2}$, for t

6. $s = \dfrac{(v_1 + v_2)t}{2}$, for v_1

7. $R = \dfrac{gs}{g + s}$, for s

8. $I = \dfrac{2V}{V + 2r}$, for V

9. $\dfrac{1}{p} + \dfrac{1}{q} = \dfrac{1}{f}$, for p

10. $\dfrac{1}{p} + \dfrac{1}{q} = \dfrac{1}{f}$, for f

11. $\dfrac{t}{a} + \dfrac{t}{b} = 1$, for a

12. $\dfrac{t}{a} + \dfrac{t}{b} = 1$, for b

13. $I = \dfrac{nE}{E + nr}$, for E

14. $I = \dfrac{nE}{E + nr}$, for n

15. $S = \dfrac{H}{m(t_1 - t_2)}$, for H

16. $S = \dfrac{H}{m(t_1 - t_2)}$, for t_1

17. $\dfrac{E}{e} = \dfrac{R + r}{r}$, for r

18. $\dfrac{E}{e} = \dfrac{R + r}{r}$, for e

19. $V = \dfrac{1}{3} \pi h^2 (3R - h)$, for R

20. $A = P(1 + rt)$, for r

21. *Interest.* The formula

$$P = \frac{A}{1 + r}$$

is used to determine what amount of principal P should be invested for one year at simple interest r in order to have A dollars after a year. Solve the formula for r.

22. *Average Speed.* The formula

$$v = \frac{d_2 - d_1}{t_2 - t_1}$$

gives an object's average speed v when that object has traveled d_1 miles in t_1 hours and d_2 miles in t_2 hours. Solve the formula for t_2.

23. *Escape Velocity.* The formula

$$\frac{V^2}{R^2} = \frac{2g}{R + h}$$

is used to find a satellite's *escape velocity* V, where R is a planet's radius, h is the satellite's height above the planet, and g is the planet's acceleration due to gravity. Solve the formula for h.

24. *Earned Run Average.* The formula

$$A = 9 \cdot \frac{R}{I}$$

gives a pitcher's *earned run average*, where A is the earned run average, R is the number of earned runs, and I is the number of innings pitched. How many earned runs were given up if a pitcher's earned run average is 2.4 after 45 innings? Solve the formula for I.

25. *Semester Average.* The formula

$$A = \frac{2Tt + Qq}{2T + Q}$$

gives a student's average A after T tests and Q quizzes, where each test counts as 2 quizzes, t is the test average, and q is the quiz average. Solve the formula for Q.

Skill Maintenance

Solve. [3.3a], [3.5a]

26. *Coin Value.* There are 50 dimes in a roll of dimes, 40 nickels in a roll of nickels, and 40 quarters in a roll of quarters. Rob has 12 rolls of coins with a total value of $70.00. He has 3 more rolls of nickels than dimes. How many of each roll of coin does he have?

27. *Audiotapes.* Esther wants to buy tapes for her work at the campus radio station. She needs some 30-min tapes and some 60-min tapes. She buys 12 tapes with a total recording time of 10 hr. How many of each length did she buy?

Given that $f(x) = x^3 - x$, find each of the following. [2.2b]

28. $f(-2)$

29. $f(2)$

30. $f(0)$

31. $f(2a)$

32. Find the slope of the line containing the points $(-2, 5)$ and $(8, -3)$. [2.4b]

33. Find an equation of the line containing the points $(-2, 5)$ and $(8, -3)$. [2.6c]

Synthesis

34. ◆ Which is easier to solve for x? Explain why.

$$\frac{1}{38} + \frac{1}{47} + \frac{1}{x} \quad \text{or} \quad \frac{1}{a} + \frac{1}{b} = \frac{1}{x}$$

35. ◆ Describe a situation in which the result of Margin Exercise 4,

$$t = \frac{ab}{a + b},$$

would be especially useful.

36. *Escape Velocity.* (Refer to Exercise 23.) A satellite's escape velocity is 6.5 mi/sec, the radius of the earth is 3960 mi, and the acceleration due to gravity is 32.2 ft/sec^2. How far is the satellite from the surface of the earth?

Collaborative Learning Manual

Develop a formula for calculating the amount of time required to complete a task when two or more people are working together.

5.8 Variation and Applications

We now extend our study of formulas and functions by considering applications involving variation.

a | Equations of Direct Variation

An electrician earns $21 per hour. In 1 hr, $21 is earned; in 2 hr, $42 is earned; in 3 hr, $63 is earned; and so on. We plot this information on a graph, using the number of hours as the first coordinate and the amount earned as the second coordinate to form a set of ordered pairs:

(1, 21),   (2, 42),

(3, 63),   (4, 84),

and so on.

Note that the ratio of the second coordinate to the first is the same number for each point:

$$\frac{21}{1} = 21, \qquad \frac{42}{2} = 21,$$

$$\frac{63}{3} = 21, \qquad \frac{84}{4} = 21,$$

and so on.

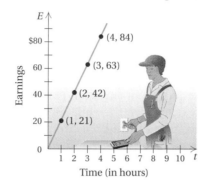

Whenever a situation produces pairs of numbers in which the *ratio is constant*, we say that there is **direct variation.** Here the amount earned varies directly as the time:

$$\frac{E}{t} = 21 \text{ (a constant)}, \quad \text{or} \quad E = 21t,$$

or, using function notation, $E(t) = 21t$. The equation is an equation of **direct variation.** The coefficient, 21 in the situation above, is called the **variation constant.** In this case, it is the rate of change of earnings with respect to time.

> **DIRECT VARIATION**
>
> If a situation gives rise to a linear function $f(x) = kx$, or $y = kx$, where k is a positive constant, we say that we have **direct variation,** or that **y varies directly as x,** or that **y is directly proportional to x.** The number k is called the **variation constant,** or **constant of proportionality.**

Example 1 Find the variation constant and an equation of variation in which y varies directly as x, and $y = 32$ when $x = 2$.

We know that (2, 32) is a solution of $y = kx$. Thus,

$y = kx$

$32 = k \cdot 2$ **Substituting**

$\frac{32}{2} = k$, or $k = 16$.   **Solving for k**

The variation constant, 16, is the rate of change of y with respect to x. The equation of variation is $y = 16x$.

Do Exercises 1 and 2.

1. Find the variation constant and an equation of variation in which y varies directly as x, and $y = 8$ when $x = 20$.

2. Find the variation constant and an equation of variation in which y varies directly as x, and $y = 5.6$ when $x = 8$.

The graph of $y = kx$, $k > 0$, always goes through the origin and rises from left to right. Note that as x increases, y increases. The constant k is also the slope of the line.

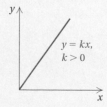

Answers on page A-33

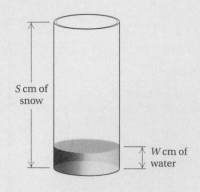

S cm of snow

W cm of water

3. *Ohm's Law.* Ohm's Law states that the voltage V in an electric circuit varies directly as the number of amperes I of electric current in the circuit. If the voltage is 10 volts when the current is 3 amperes, what is the voltage when the current is 15 amperes?

4. *An Ecology Problem.* The amount of garbage G produced in the United States varies directly as the number of people N who produce the garbage. It is known that 50 tons of garbage is produced by 200 people in 1 year. The population of the San Francisco–Oakland–San Jose area is 6,300,000. How much garbage is produced by this area in 1 year?

Answers on page A-33

b Applications of Direct Variation

Example 2 *Water from Melting Snow.* The number of centimeters W of water produced from melting snow varies directly as S, the number of centimeters of snow. Meteorologists have found that 150 cm of snow will melt to 16.8 cm of water. To how many centimeters of water will 200 cm of snow melt?

We first find the variation constant using the data and then find an equation of variation:

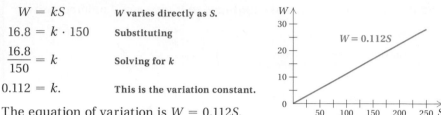

$W = kS$	W varies directly as S.
$16.8 = k \cdot 150$	Substituting
$\dfrac{16.8}{150} = k$	Solving for k
$0.112 = k.$	This is the variation constant.

The equation of variation is $W = 0.112S$.

Next, we use the equation to find how many centimeters of water will result from melting 200 cm of snow:

$$W = 0.112S$$
$$= 0.112(200) \quad \text{Substituting}$$
$$= 22.4.$$

Thus, 200 cm of snow will melt to 22.4 cm of water.

Do Exercises 3 and 4.

c Equations of Inverse Variation

A bus is traveling a distance of 20 mi. At a speed of 5 mph, the trip will take 4 hr; at 20 mph, it will take 1 hr; at 40 mph, it will take $\frac{1}{2}$ hr; and so on. We plot this information on a graph, using speed as the first coordinate and time as the second coordinate to determine a set of ordered pairs:

(5, 4),   (20, 1),
$\left(40, \frac{1}{2}\right)$,   and so on.

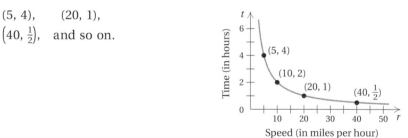

Note that the products of the coordinates are all the same number:

$5 \cdot 4 = 20$, $20 \cdot 1 = 20$, $40 \cdot \frac{1}{2} = 20$,   and so on.

Whenever a situation produces pairs of numbers in which the *product is constant*, we say that there is **inverse variation.** Here the time varies inversely as the speed:

$$rt = 20 \text{ (a constant)}, \quad \text{or} \quad t = \frac{20}{r}.$$

The equation is an equation of **inverse variation.** The coefficient, 20 in the situation above, is called the **variation constant.** Note that as the first number increases, the second number decreases.

> **INVERSE VARIATION**
>
> If a situation gives rise to a function $f(x) = k/x$, or $y = k/x$, where k is a positive constant, we say that we have **inverse variation**, or that **y varies inversely as x,** or that **y is inversely proportional to x.** The number k is called the **variation constant,** or **constant of proportionality.**

Example 3 Find the variation constant and an equation of variation in which y varies inversely as x, and $y = 32$ when $x = 0.2$.

We know that $(0.2, 32)$ is a solution of $y = k/x$. We substitute:

$$y = \frac{k}{x}$$

$$32 = \frac{k}{0.2} \qquad \text{Substituting}$$

$$(0.2)32 = k \qquad \text{Solving for } k$$

$$6.4 = k.$$

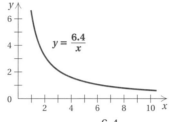

The variation constant is 6.4. The equation of variation is $y = \dfrac{6.4}{x}$.

Do Exercise 5.

d Applications of Inverse Variation

Example 4 *Building a Shed.* The time t required to do a job varies inversely as the number of people P who work on the job (assuming that all work at the same rate). It takes 4 hr for 12 people to build a woodshed. How long would it take 3 people to complete the same job?

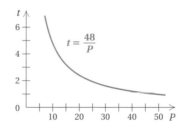

We first find the variation constant using the data and then find an equation of variation:

$$t = \frac{k}{P} \qquad t \text{ varies inversely as } P.$$

$$4 = \frac{k}{12} \qquad \text{Substituting}$$

$$48 = k. \qquad \text{Solving for } k \text{, the variation constant}$$

The equation of variation is $t = 48/P$. In this case, 48 is the total number of hours required to build the shed.

It is helpful to look at the graph of $y = k/x$, $k > 0$. The graph is like the one shown below for positive values of x. Note that as x increases, y decreases.

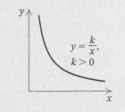

5. Find the variation constant and an equation of variation in which y varies inversely as x, and $y = 0.012$ when $x = 50$.

Answer on page A-33

6. Time and Speed. The time t required to drive a fixed distance varies inversely as the speed r. It takes 5 hr at 60 km/h to drive a fixed distance. How long would it take to drive that same distance at 40 km/h?

Next, we use the equation to find the time that it would take 3 people to do the job:

$$t = \frac{48}{P}$$

$$= \frac{48}{3} \qquad \text{Substituting}$$

$$= 16.$$

It would take 16 hr for 3 people to do the job.

Do Exercise 6.

e | Other Kinds of Variation

We now look at other kinds of variation. Consider the equation for the area of a circle, in which A and r are variables and π is a constant:

$$A = \pi r^2, \quad \text{or, as a function,} \quad A(r) = \pi r^2.$$

We say that the area *varies directly* as the square of the radius.

> y varies directly as the nth power of x if there is some positive constant k such that $y = kx^n$.

Example 5 Find an equation of variation in which y varies directly as the square of x, and $y = 12$ when $x = 2$.

7. Find an equation of variation in which y varies directly as the square of x, and $y = 175$ when $x = 5$.

We write an equation of variation and find k:

$$y = kx^2$$

$$12 = k \cdot 2^2 \qquad \text{Substituting}$$

$$12 = k \cdot 4$$

$$3 = k.$$

Thus, $y = 3x^2$.

Do Exercise 7.

From the law of gravity, we know that the weight W of an object *varies inversely* as the square of its distance d from the center of the earth:

$$W = \frac{k}{d^2}.$$

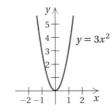

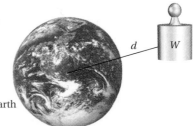

Earth

> y varies inversely as the nth power of x if there is some positive constant k such that
>
> $$y = \frac{k}{x^n}.$$

Answers on page A-33

Example 6 Find an equation of variation in which W varies inversely as the square of d, and $W = 3$ when $d = 5$.

$$W = \frac{k}{d^2}$$

$$3 = \frac{k}{5^2} \qquad \text{Substituting}$$

$$3 = \frac{k}{25}$$

$$75 = k$$

Thus, $W = \dfrac{75}{d^2}$.

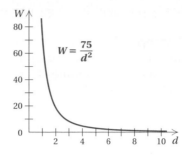

Do Exercise 8.

Consider the equation for the area A of a triangle with height h and base b: $A = \frac{1}{2}bh$. We say that the area *varies jointly* as the height and the base.

> y varies jointly as x and z if there is some positive constant k such that
>
> $$y = kxz.$$

Example 7 Find an equation of variation in which y varies jointly as x and z, and $y = 42$ when $x = 2$ and $z = 3$.

$$y = kxz$$

$$42 = k \cdot 2 \cdot 3 \qquad \text{Substituting}$$

$$42 = k \cdot 6$$

$$7 = k$$

Thus, $y = 7xz$.

Do Exercise 9.

The equation

$$y = k \cdot \frac{xz^2}{w}$$

asserts that y varies jointly as x and the square of z, and inversely as w.

Example 8 Find an equation of variation in which y varies jointly as x and z and inversely as the square of w, and $y = 105$ when $x = 3$, $z = 20$, and $w = 2$.

$$y = k \cdot \frac{xz}{w^2}$$

$$105 = k \cdot \frac{3 \cdot 20}{2^2} \qquad \text{Substituting}$$

$$105 = k \cdot 15$$

$$7 = k$$

Thus, $y = 7 \cdot \dfrac{xz}{w^2}$.

Do Exercise 10.

8. Find an equation of variation in which y varies inversely as the square of x, and $y = \frac{1}{4}$ when $x = 6$.

9. Find an equation of variation in which y varies jointly as x and z, and $y = 65$ when $x = 10$ and $z = 13$.

10. Find an equation of variation in which y varies jointly as x and the square of z and inversely as w, and $y = 80$ when $x = 4$, $z = 10$, and $w = 25$.

Answers on page A-33

11. *Distance of a Dropped Object.*
The distance s that an object falls when dropped from some point above the ground varies directly as the square of the time t that it falls. If the object falls 19.6 m in 2 sec, how far will the object fall in 10 sec?

12. *Electrical Resistance.* At a fixed temperature, the resistance R of a wire varies directly as the length l and inversely as the square of its diameter d. If the resistance is 0.1 ohm when the diameter is 1 mm and the length is 50 cm, what is the resistance when the length is 2000 cm and the diameter is 2 mm?

Answers on page A-33

---

f | Other Applications of Variation

Many problem situations can be described with equations of variation.

Example 9 *Volume of a Tree.* The volume of wood V in a tree varies jointly as the height h and the square of the girth g (girth is distance around). If the volume of a redwood tree is 216 m³ when the height is 30 m and the girth is 1.5 m, what is the height of a tree whose volume is 960 m³ and girth is 2 m?

We first find k using the first set of data. Then we solve for h using the second set of data.

$$V = khg^2$$
$$216 = k \cdot 30 \cdot 1.5^2$$
$$3.2 = k$$

Then the equation of variation is $V = 3.2hg^2$. We substitute the second set of data into the equation:

$$960 = 3.2 \cdot h \cdot 2^2$$
$$75 = h.$$

Therefore, the height of the tree is 75 m.

Example 10 *TV Signal.* The intensity I of a TV signal varies inversely as the square of the distance d from the transmitter. If the intensity is 23 watts per square meter (W/m²) at a distance of 2 km, what is the intensity at a distance of 6 km?

We first find k using the first set of data. Then we solve for I using the second set of data.

$$I = \frac{k}{d^2}$$
$$23 = \frac{k}{2^2}$$
$$92 = k$$

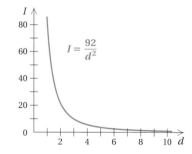

Then the equation of variation is $I = 92/d^2$. We substitute the second distance into the equation:

$$I = \frac{92}{d^2} = \frac{92}{6^2} \approx 2.56. \qquad \text{Rounded to the nearest hundredth}$$

Therefore, at 6 km, the intensity is about 2.56 W/m².

Do Exercises 11 and 12.

Exercise Set 5.8

a Find the variation constant and an equation of variation in which y varies directly as x and the following are true.

1. $y = 40$ when $x = 8$

2. $y = 54$ when $x = 12$

3. $y = 4$ when $x = 30$

4. $y = 3$ when $x = 33$

5. $y = 0.9$ when $x = 0.4$

6. $y = 0.8$ when $x = 0.2$

b Solve.

7. *Aluminum Usage.* The number N of aluminum cans used each year varies directly as the number of people using the cans. If 250 people use 60,000 cans in one year, how many cans are used each year in Dallas, which has a population of 1,008,000?

8. *Weekly Allowance.* According to Fidelity Investments *Investment Vision Magazine*, the average weekly allowance A of children varies directly as their grade level G. It is known that the average allowance of a 9th-grade student is $9.66 per week. What then is the average weekly allowance of a 4th-grade student?

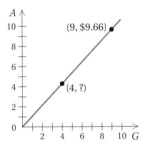

9. *Hooke's Law.* Hooke's law states that the distance d that a spring is stretched by a hanging object varies directly as the weight w of the object. If a spring is stretched 40 cm by a 3-kg barbell, what is the distance stretched by a 5-kg barbell?

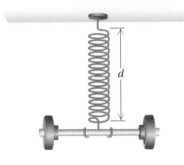

10. *Lead Pollution.* The average U.S. community of population 12,500 released about 385 tons of lead into the environment in a recent year (**Source:** *Conservation Matters,* Autumn 1995 issue. Boston: Conservation Law Foundation, p. 30). How many tons were released nationally? Use 250,000,000 as the U.S. population.

11. *Fat Intake.* The maximum number of grams of fat that should be in a diet varies directly as a person's weight. A person weighing 120 lb should have no more than 60 g of fat per day. What is the maximum daily fat intake for a person weighing 180 lb?

12. *Relative Aperture.* The relative aperture, or f-stop, of a 23.5-mm diameter lens is directly proportional to the focal length F of the lens. If a 150-mm focal length has an f-stop of 6.3, find the f-stop of a 23.5-mm diameter lens with a focal length of 80 mm.

13. *Mass of Water in Body.* The number of kilograms W of water in a human body varies directly as the mass of the body. A 96-kg person contains 64 kg of water. How many kilograms of water are in a 60-kg person?

14. *Weight on Mars.* The weight M of an object on Mars varies directly as its weight E on Earth. A person who weighs 95 lb on Earth weighs 38 lb on Mars. How much would a 100-lb person weigh on Mars?

c Find the variation constant and an equation of variation in which y varies inversely as x and the following are true.

15. $y = 14$ when $x = 7$

16. $y = 1$ when $x = 8$

17. $y = 3$ when $x = 12$

18. $y = 12$ when $x = 5$

19. $y = 0.1$ when $x = 0.5$

20. $y = 1.8$ when $x = 0.3$

d Solve.

21. *Work Rate.* The time T required to do a job varies inversely as the number of people P working. It takes 5 hr for 7 bricklayers to build a park wall. How long will it take 10 bricklayers to complete the job?

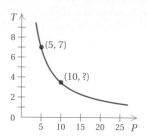

22. *Pumping Rate.* The time t required to empty a tank varies inversely as the rate r of pumping. If a pump can empty a tank in 45 min at the rate of 600 kL/min, how long will it take the pump to empty the same tank at the rate of 1000 kL/min?

23. *Current and Resistance.* The current I in an electrical conductor varies inversely as the resistance R of the conductor. If the current is $\frac{1}{2}$ ampere when the resistance is 240 ohms, what is the current when the resistance is 540 ohms?

24. *Wavelength and Frequency.* The wavelength W of a radio wave varies inversely as its frequency F. A wave with a frequency of 1200 kilohertz has a length of 300 meters. What is the length of a wave with a frequency of 800 kilohertz?

25. *Musical Pitch.* The pitch P of a musical tone varies inversely as its wavelength W. One tone has a pitch of 330 vibrations per second and a wavelength of 3.2 ft. Find the wavelength of another tone that has a pitch of 550 vibrations per second.

26. *Beam Weight.* The weight W that a horizontal beam can support varies inversely as the length L of the beam. Suppose that an 8-m beam can support 1200 kg. How many kilograms can a 14-m beam support?

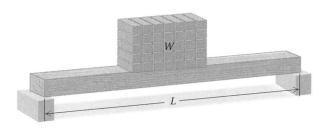

27. *Volume and Pressure.* The volume V of a gas varies inversely as the pressure P upon it. The volume of a gas is 200 cm^3 under a pressure of 32 kg/cm^2. What will be its volume under a pressure of 40 kg/cm^2?

28. *Rate of Travel.* The time t required to drive a fixed distance varies inversely as the speed r. It takes 5 hr at a speed of 80 km/h to drive a fixed distance. How long will it take to drive the same distance at a speed of 70 km/h?

e Find an equation of variation in which the following are true.

29. y varies directly as the square of x, and $y = 0.15$ when $x = 0.1$

30. y varies directly as the square of x, and $y = 6$ when $x = 3$

31. y varies inversely as the square of x, and $y = 0.15$ when $x = 0.1$

32. y varies inversely as the square of x, and $y = 6$ when $x = 3$

33. y varies jointly as x and z, and $y = 56$ when $x = 7$ and $z = 8$

34. y varies directly as x and inversely as z, and $y = 4$ when $x = 12$ and $z = 15$

35. y varies jointly as x and the square of z, and $y = 105$ when $x = 14$ and $z = 5$

36. y varies jointly as x and z and inversely as w, and $y = \frac{3}{2}$ when $x = 2$, $z = 3$, and $w = 4$

37. y varies jointly as x and z and inversely as the product of w and p, and $y = \frac{3}{28}$ when $x = 3$, $z = 10$, $w = 7$, and $p = 8$

38. y varies jointly as x and z and inversely as the square of w, and $y = \frac{12}{5}$ when $x = 16$, $z = 3$, and $w = 5$

---

f | Solve.

39. *Stopping Distance of a Car.* The stopping distance d of a car after the brakes have been applied varies directly as the square of the speed r. If a car traveling 60 mph can stop in 200 ft, how fast can a car travel and still stop in 72 ft?

40. *Boyle's Law.* The volume V of a given mass of a gas varies directly as the temperature T and inversely as the pressure P. If $V = 231$ cm^3 when $T = 42°$ and $P = 20$ kg/cm^2, what is the volume when $T = 30°$ and $P = 15$ kg/cm^2?

41. *Intensity of Light.* The intensity I of light from a light bulb varies inversely as the square of the distance d from the bulb. Suppose that I is 90 W/m^2 (watts per square meter) when the distance is 5 m. How much *further* would it be to a point where the intensity is 40 W/m^2?

42. *Weight of an Astronaut.* The weight W of an object varies inversely as the square of the distance d from the center of the earth. At sea level (3978 mi from the center of the earth), an astronaut weighs 220 lb. Find his weight when he is 200 mi above the surface of the earth and the spacecraft is not in motion.

43. *Earned-Run Average.* A pitcher's earned-run average E varies directly as the number R of earned runs allowed and inversely as the number I of innings pitched. In a recent year, Shawn Estes of the San Francisco Giants had an earned-run average of 3.18. He gave up 71 earned runs in 201 innings. How many earned runs would he have given up had he pitched 300 innings with the same average? Round to the nearest whole number.

44. *Atmospheric Drag.* Wind resistance, or atmospheric drag, tends to slow down moving objects. Atmospheric drag varies jointly as an object's surface area A and velocity v. If a car traveling at a speed of 40 mph with a surface area of 37.8 ft^2 experiences a drag of 222 N (Newtons), how fast must a car with 51 ft^2 of surface area travel in order to experience a drag force of 430 N?

45. *Water Flow.* The amount Q of water emptied by a pipe varies directly as the square of the diameter d. A pipe 5 in. in diameter will empty 225 gal of water over a fixed time period. If we assume the same kind of flow, how many gallons of water are emptied in the same amount of time by a pipe that is 9 in. in diameter?

46. *Weight of a Sphere.* The weight W of a sphere of a given material varies directly as its volume V, and its volume V varies directly as the cube of its diameter.

a) Find an equation of variation relating the weight W to the diameter d.

b) An iron ball that is 5 in. in diameter is known to weigh 25 lb. Find the weight of an iron ball that is 8 in. in diameter.

Skill Maintenance

Solve. [4.7a]

47. $(x - 3)(x + 4) = 0$

48. $x^2 - 6x + 9 = 0$

49. $12x^2 - 11x + 2 = 0$

50. $x^2 - 49 = 0$

51. Find an equation of the line with slope $-\frac{2}{3}$ and y-intercept $(0, -5)$. [2.6a]

52. Find an equation of the line having slope $-\frac{2}{7}$ and containing the point $(-4, 8)$. [2.6b]

53. *NASDAQ Stock Index.* Consider the graph (at right) of recent NASDAQ stock indices at the beginning of various months.

a) Use the two points $(1, 1213)$ and $(6, 1702)$ to find a linear function that fits the data.

b) Use the function to predict the NASDAQ stock index at the beginning of November and at the beginning of December. [2.7b]

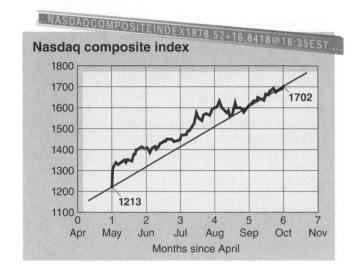

Synthesis

54. ◆ Write a variation problem for a classmate to solve. Design the problem so that the answer is "When Simone studies for 8 hr a week, her quiz score is 92."

55. ◆ If y varies directly as x and x varies inversely as z, how does y vary with regard to z? Why?

56. In each of the following equations, state whether y varies directly as x, inversely as x, or neither directly nor inversely as x.

a) $7xy = 14$

b) $x - 2y = 12$

c) $-2x + 3y = 0$

d) $x = \dfrac{3}{4}y$

e) $\dfrac{x}{y} = 2$

57. *Area of a Circle.* The area of a circle varies directly as the square of the length of a diameter. What is the variation constant?

58. *Volume and Cost.* A peanut butter jar in the shape of a right circular cylinder is 4 in. high and 3 in. in diameter and sells for $1.20. If we assume that cost is proportional to volume, how much should a jar 6 in. high and 6 in. in diameter cost?

Describe, in words, the variation given by the equation.

59. $Q = \dfrac{kp^2}{q^3}$

60. $W = \dfrac{km_1 M_1}{d^2}$

Collaborative Learning Manual

Model the height of a bouncing ball with an equation of direct variation.

Summary and Review Exercises: Chapter 5

Important Properties and Formulas

Direct Variation: $y = kx$

Inverse Variation: $y = \dfrac{k}{x}$

Joint Variation: $y = kxz$

Work Principle: $\dfrac{t}{a} + \dfrac{t}{b} = 1$, where a is the time needed for A to complete the job alone, b is the time needed for B to complete the job alone, and t is the time needed for A and B to complete the job working together.

The objectives to be tested in addition to the material in this chapter are [2.3a], [3.6b], [4.4a, b], and [4.5d].

1. Find all numbers for which the rational expression
$$\frac{x^2 - 3x + 2}{x^2 - 9}$$
is undefined. [5.1a]

2. Find the domain of f where [5.1a]
$$f(x) = \frac{x^2 - 3x + 2}{x^2 - 9}.$$

Simplify. [5.1c]

3. $\dfrac{4x^2 - 7x - 2}{12x^2 + 11x + 2}$

4. $\dfrac{a^2 + 2a + 4}{a^3 - 8}$

Find the LCM. [5.2a]

5. $6x^3, \quad 16x^2$

6. $x^2 - 49, \quad 3x + 1$

7. $x^2 + x - 20, \quad x^2 + 3x - 10$

Perform the indicated operations and simplify. [5.1d, e], [5.2b, c]

8. $\dfrac{y^2 - 64}{2y + 10} \cdot \dfrac{y + 5}{y + 8}$

9. $\dfrac{x^3 - 8}{x^2 - 25} \cdot \dfrac{x^2 + 10x + 25}{x^2 + 2x + 4}$

10. $\dfrac{9a^2 - 1}{a^2 - 9} \div \dfrac{3a + 1}{a + 3}$

11. $\dfrac{x^3 - 64}{x^2 - 16} \div \dfrac{x^2 + 5x + 6}{x^2 - 3x - 18}$

12. $\dfrac{x}{x^2 + 5x + 6} - \dfrac{2}{x^2 + 3x + 2}$

13. $\dfrac{2x^2}{x - y} + \dfrac{2y^2}{x + y}$

14. $\dfrac{3}{y + 4} - \dfrac{y}{y - 1} + \dfrac{y^2 + 3}{y^2 + 3y - 4}$

Divide.

15. $(16ab^3c - 10ab^2c^2 + 12a^2b^2c) \div (4ab)$ [5.3a]

16. $(y^2 - 20y + 64) \div (y - 6)$ [5.3b]

17. $(6x^4 + 3x^2 + 5x + 4) \div (x^2 + 2)$ [5.3b]

Divide using synthetic division. Show your work. [5.3c]

18. $(x^3 + 5x^2 + 4x - 7) \div (x - 4)$

19. $(3x^4 - 5x^3 + 2x - 7) \div (x + 1)$

Simplify. [5.4a]

20. $\dfrac{3 + \dfrac{3}{y}}{4 + \dfrac{4}{y}}$

21. $\dfrac{\dfrac{2}{a} + \dfrac{2}{b}}{\dfrac{4}{a^3} + \dfrac{4}{b^3}}$

22. $\dfrac{\dfrac{x^2 - 5x - 36}{x^2 - 36}}{\dfrac{x^2 + x - 12}{x^2 - 12x + 36}}$

23. $\dfrac{\dfrac{4}{x + 3} - \dfrac{2}{x^2 - 3x + 2}}{\dfrac{3}{x - 2} + \dfrac{1}{x^2 + 2x - 3}}$

Solve. [5.5a]

24. $\dfrac{x}{4} + \dfrac{x}{7} = 1$

25. $\dfrac{5}{3x + 2} = \dfrac{3}{2x}$

26. $\dfrac{4x}{x + 1} + \dfrac{4}{x} + 9 = \dfrac{4}{x^2 + x}$

27. $\dfrac{90}{x^2 - 3x + 9} - \dfrac{5x}{x + 3} = \dfrac{405}{x^3 + 27}$

28. $\dfrac{2}{x - 3} + \dfrac{1}{4x + 20} = \dfrac{1}{x^2 + 2x - 15}$

29. Given that
$$f(x) = \frac{6}{x} + \frac{4}{x},$$
find all x for which $f(x) = 5$.

30. *House Painting.* Danielle can paint the outside of a house in 12 hr. Bill can paint the same house in 9 hr. How long would it take them working together to paint the house? [5.6a]

31. *Boat Travel.* The current of the Gold River is 6 mph. A boat travels 50 mi downstream in the same time that it takes to travel 30 mi upstream. What is the speed of the boat in still water? [5.6c]

32. *Travel Distance.* Fred operates a potato-chip delivery route. He drives 800 mi in 3 days. How far will he travel in 15 days? [5.6b]

Solve for the indicated letter. [5.7a]

33. $R = \dfrac{gs}{g + s}$, for s; for g

34. $S = \dfrac{p}{a} + \dfrac{t}{b}$, for b; for t

35. Find an equation of variation in which y varies directly as x, and $y = 100$ when $x = 25$. [5.8a]

36. Find an equation of variation in which y varies inversely as x, and $y = 100$ when $x = 25$. [5.8c]

37. *Pumping Time.* The time t required to empty a tank varies inversely as the rate r of pumping. If a pump can empty a tank in 35 min at the rate of 800 kL per minute, how long will it take the pump to empty the same tank at the rate of 1400 kL per minute? [5.8d]

38. *Test Score.* The score N on a test varies directly as the number of correct responses a. Ellen answers 28 questions correctly and earns a score of 87. What would Ellen's score have been if she had answered 25 questions correctly? [5.8b]

39. *Power of Electric Current.* The power P expended by heat in an electric circuit of fixed resistance varies directly as the square of the current C in the circuit. A circuit expends 180 watts when a current of 6 amperes is flowing. What is the amount of heat expended when the current is 10 amperes? [5.8f]

Skill Maintenance

40. Find the domain and the range of the function graphed below. [2.3a]

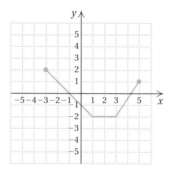

Graph on a plane. [3.6b]

41. $y - 2x \geq 4$ **42.** $x > -3$

Factor. [4.4a, b], [4.5d]

43. $125x^3 - 8y^3$ **44.** $6x^2 + 29x - 42$

Synthesis

45. ◆ Discuss at least three different uses of the LCM studied in this chapter. [5.2b], [5.4a], [5.5a]

46. ◆ You have learned to solve a new kind of equation in this chapter. Explain how this type differs from those you have studied previously and how the equation-solving process differs. [5.5a]

47. Solve: $\dfrac{5}{x - 13} - \dfrac{5}{x} = \dfrac{65}{x^2 - 13x}$. [5.5a]

48. Find the reciprocal and simplify: $\dfrac{a - b}{a^3 - b^3}$. [5.1c]

Test: Chapter 5

1. Find all numbers for which the rational expression
$$\frac{x^2 - 16}{x^2 - 3x + 2}$$
is undefined.

2. Find the domain of f where
$$f(x) = \frac{x^2 - 16}{x^2 - 3x + 2}.$$

Simplify.

3. $\dfrac{12x^2 + 11x + 2}{4x^2 - 7x - 2}$

4. $\dfrac{p^3 + 1}{p^2 - p - 2}$

5. Find the LCM of $x^2 + x - 6$ and $x^2 + 8x + 15$.

Perform the indicated operations and simplify.

6. $\dfrac{2x^2 + 20x + 50}{x^2 - 4} \cdot \dfrac{x + 2}{x + 5}$

7. $\dfrac{x}{x^2 + 11x + 30} - \dfrac{5}{x^2 + 9x + 20}$

8. $\dfrac{y^2 - 16}{2y + 6} \div \dfrac{y - 4}{y + 3}$

9. $\dfrac{x^2}{x - y} + \dfrac{y^2}{y - x}$

10. $\dfrac{1}{x + 1} - \dfrac{x + 2}{x^2 - 1} + \dfrac{3}{x - 1}$

11. $\dfrac{a}{a - b} + \dfrac{b}{a^2 + ab + b^2} - \dfrac{2}{a^3 - b^3}$

Divide.

12. $(20r^2s^3 + 15r^2s^2 - 10r^3s^3) \div (5r^2s)$

13. $(y^3 + 125) \div (y + 5)$

14. $(4x^4 + 3x^3 - 5x - 2) \div (x^2 + 1)$

Divide using synthetic division. Show your work.

15. $(x^3 + 3x^2 + 2x - 6) \div (x - 3)$

16. $(4x^3 - 6x^2 - 9) \div (x + 5)$

Simplify.

17. $\dfrac{1 - \dfrac{1}{x^2}}{1 - \dfrac{1}{x}}$

18. $\dfrac{\dfrac{1}{a^3} + \dfrac{1}{b^3}}{\dfrac{1}{a} + \dfrac{1}{b}}$

Solve.

19. $\dfrac{2}{x - 1} = \dfrac{3}{x + 3}$

20. $\dfrac{7x}{x + 3} + \dfrac{21}{x - 3} = \dfrac{126}{x^2 - 9}$

21. $\dfrac{2x}{x - 7} - \dfrac{14}{x} = \dfrac{98}{x^2 - 7x}$

22. $\dfrac{1}{3x - 6} - \dfrac{1}{x^2 - 4} = \dfrac{3}{x + 2}$

23. Given that
$$f(x) = \frac{2}{x - 1} + \frac{2}{x + 2},$$
find all x for which $f(x) = 1$.

24. *Completing a Puzzle.* Working together, Hans and Gretel can complete a jigsaw puzzle in 1.5 hr. Hans takes 4 hr longer than Gretel does when working alone. How long would it take Gretel to complete the puzzle?

25. *Bicycle Travel.* Dodi can bicycle at a rate of 12 mph when there is no wind. Against the wind, Dodi bikes 8 mi in the same time that it takes to bike 14 mi with the wind. What is the speed of the wind?

Answers

1. _____

2. _____

3. _____

4. _____

5. _____

6. _____

7. _____

8. _____

9. _____

10. _____

11. _____

12. _____

13. _____

14. _____

15. _____

16. _____

17. _____

18. _____

19. _____

20. _____

21. _____

22. _____

23. _____

24. _____

25. _____

26. *Predicting Paint Needs.* Lowell and Chris run a summer painting company to defray their college expenses. They need 4 gal of paint to paint 1700 ft² of clapboard. How much paint would they need for a building with 6000 ft² of clapboard?

26. _____

Solve for the indicated letter.

27. $T = \dfrac{ab}{a - b}$, for a; for b

27. _____

28. $Q = \dfrac{2}{a} - \dfrac{t}{b}$, for a

28. _____

29. Find an equation of variation in which Q varies jointly as x and y, and $Q = 25$ when $x = 2$ and $y = 5$.

30. Find an equation of variation in which y varies inversely as x, and $y = 10$ when $x = 25$.

29. _____

31. *Income vs Time.* Dean's income I varies directly as the time t worked. He gets a job that pays $275 for 40 hr of work. What is he paid for working 72 hr, assuming that there is no change in pay scale for overtime?

32. *Area of a Balloon.* The surface area of a balloon varies directly as the square of its radius. The area is 314 cm² when the radius is 5 cm. What is the area when the radius is 7 cm?

30. _____

31. _____

32. _____

Skill Maintenance

33. Find the domain and the range of the function graphed below.

34. Graph: $2x + 5y > 10$.

33. _____

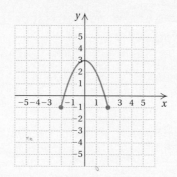

34. _____

35. _____

Factor.

35. $16t^2 - 24t - 72$

36. $64a^3 + 1$

36. _____

Synthesis

37. _____

37. Solve: $\dfrac{6}{x - 15} - \dfrac{6}{x} = \dfrac{90}{x^2 - 15x}$.

38. Find the LCM of $1 - t^6$ and $1 + t^6$.

39. Find the x- and y-intercepts of the function f given by

$$f(x) = \dfrac{\dfrac{5}{x + 4} - \dfrac{3}{x - 2}}{\dfrac{2}{x - 3} + \dfrac{1}{x + 4}}.$$

38. _____

39. _____

Cumulative Review: Chapters R–5

1. Evaluate $\dfrac{4a}{7b}$ for $a = 14$ and $b = 6$.

Simplify.

2. $|-12|$

3. $-\dfrac{1}{3} \div \dfrac{3}{8}$

4. $12b - [9 - 7(5b - 6)]$

5. $5^3 \div \dfrac{1}{2} - 2(3^2 + 2^2)$

6. $\left(\dfrac{2x^3 y^{-6}}{-4y^{-2}}\right)^2$

7. $(6p^2 - 2p + 5) - (-10p^2 + 6p + 5)$

8. $(6m - n)^2$

9. $(3a - 4b)(5a + 2b)$

10. $\dfrac{y^2 - 4}{3y + 33} \cdot \dfrac{y + 11}{y + 2}$

11. $\dfrac{9x^2 - 25}{x^2 - 16} \div \dfrac{3x + 5}{x - 4}$

12. $\dfrac{2x + 1}{4x - 12} - \dfrac{x - 2}{5x - 15}$

13. $\dfrac{2}{x + 2} + \dfrac{3}{x - 2} - \dfrac{x + 1}{x^2 - 4}$

14. $\dfrac{\dfrac{1}{x} + \dfrac{2}{y}}{\dfrac{3}{x} - \dfrac{1}{y}}$

15. $\dfrac{1 - \dfrac{2}{y^2}}{1 - \dfrac{1}{y^3}}$

16. $(2x^3 - 7x^2 + x - 3) \div (x + 2)$

Solve.

17. $9y - (5y - 3) = 33$

18. $F = \dfrac{9}{5}C + 32$, for C

19. $-3 < -2x - 6 < 0$

20. $|x| \geq 2.1$

21. $4x - 2y = 6,$
$6x - 3y = 9$

22. $x + 2y - 2z = 9,$
$2x - 3y + 4z = -4,$
$5x - 4y + 2z = 5$

23. $8x = 1 + 16x^2$

24. $14 + 3x = 2x^2$

25. $\dfrac{3x}{x - 2} - \dfrac{6}{x + 2} = \dfrac{24}{x^2 - 4}$

26. $\dfrac{6}{x - 5} = \dfrac{2}{2x}$

27. $P = \dfrac{3a}{a + b}$, for a

Solve.

28. $4x + 5y = -3,$
$x = 1 - 3y$

29. $x + 6y + 4z = -2,$
$4x + 4y + \ z = 2,$
$3x + 2y - 4z = 5$

Factor.

30. $4x^3 + 18x^2$

31. $8a^3 - 4a^2 - 6a + 3$

32. $x^2 + 8x - 84$

33. $6x^2 + 11x - 10$

34. $16y^2 - 81$

35. $t^2 - 16t + 64$

36. $64x^3 + 8$

37. $0.027b^3 - 0.008c^3$

38. $x^6 - x^2$

39. $20x^2 + 7x - 3$

Graph.

40. $y = -5x + 4$

41. $3x - 18 = 0$

42. $x + 3y < 4$

43. $x + y \geq 4$,
$x - y > 1$

44. Find an equation of the line with slope $-\frac{1}{2}$ passing through the point $(2, -2)$.

45. Find an equation of the line that is perpendicular to the line $2x + y = 5$ and passes through the point $(3, -1)$.

46. *Medical Expenses.* Medical expenses can be deducted on your tax return only if they exceed $7\frac{1}{2}\%$ of your adjusted gross income. If your expenses were $3400, what is the largest adjusted gross income for which no medical expenses can be deducted?

47. *Housing.* According to Census Bureau statistics, 20.4% of housing is in the West, 24.0% in the North Central, and 35.3% in the South. What percent is in the Northeast?

48. *Residential Taxes.* The residential taxes per $100 of property value are $2.43 more in Detroit than in Chicago. The taxes for $10,000 of property in Detroit and $20,000 of property in Chicago total $741. What are the residential property taxes per $100 of property value in each city?

49. *Hockey Results.* A hockey team played 81 games in a season. They won 1 fewer game than three times the number of ties and lost 8 fewer games than they won. How many games did they win? lose? tie?

50. *Painting Time.* Dave can paint the outside of his house in 15 hr. Bill can paint the same house in 12 hr. How long would it take them to paint the house together?

51. *Insurance Costs.* The cost c of an insurance policy varies directly as the age a of the insured. A 32-year-old person pays an annual premium of $152. What is the age of a person who pays $285?

52. Given that

$$f(x) = \frac{x - 2}{x^2 - 25},$$

find the domain.

53. Find the domain and the range of the function graphed below.

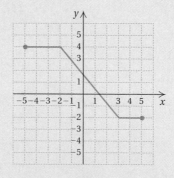

Synthesis

54. The graph of $y = ax^2 + bx + c$ contains the three points $(4, 2)$, $(2, 0)$, and $(1, 2)$. Find $a, b,$ and c.

55. Solve:

$$\frac{18}{x - 9} + \frac{10}{x + 5} = \frac{28x}{x^2 - 4x - 45}.$$

56. Solve: $16x^3 = x$.

6

Radical Expressions, Equations, and Functions

Introduction

In this chapter, we introduce square roots, cube roots, fourth roots, fifth roots, and so on. We study them in relation to radical expressions and functions and use them to solve applied problems like the one below. We also consider expressions with rational exponents and their related functions.

An Application	The Mathematics

The *wind chill temperature* T_w is what the temperature would have to be with no wind in order to give the same chilling effect as with wind. A formula for wind chill temperature is given by

$$T_w = 91.4 - \frac{(10.45 + 6.68\sqrt{v} - 0.447v)(457 - 5T)}{110},$$

where T is the actual temperature, in degrees Fahrenheit, and v is the wind speed, in miles per hour. Find the wind chill when $T = 10°F$ and $v = 20$ mph.

We substitute 10 for T and 20 for v in the formula:

$$T_w = 91.4 - \frac{(10.45 + 6.68\sqrt{20} - 0.447 \cdot 20)(457 - 5 \cdot 10)}{110}$$

$$\approx -24.7°.$$

This is a *radical equation*.

This problem appears as Exercise 44 in Exercise Set 6.7.

World Wide Web For more information, visit us at www.mathmax.com

Pretest: Chapter 6

Simplify. Assume that letters can represent *any* real number.

1. $\sqrt{t^2}$

2. $\sqrt[3]{27x^3}$

3. $\sqrt[12]{y^{12}}$

In Questions 4–20, assume that all expressions under the radical represent positive numbers.

Simplify.

4. $(\sqrt[3]{9a^2b})^2$

5. $\sqrt{45} - 3\sqrt{125} + 4\sqrt{80}$

Multiply and simplify.

6. $\sqrt{18x^2}\,\sqrt{8x^3}$

7. $(2\sqrt{6} - 1)^2$

8. Divide and simplify:

$\dfrac{\sqrt{52a^4}}{\sqrt{13a^3}}$.

9. Rationalize the denominator:

$\dfrac{3}{2 + \sqrt{5}}$.

Solve.

10. $\sqrt{2x - 1} = 5$

11. $\sqrt[3]{1 - 6x} = 2$

12. $\sqrt{3x + 1} - \sqrt{2x} = 1$

In Questions 13 and 14, give an exact answer and an approximation to three decimal places.

13. In a right triangle with leg $b = 5$ and hypotenuse $c = 8$, find the length of leg a.

14. The diagonal of a square has length 8 ft. Find the length of a side.

15. Use rational exponents to write a single radical expression:

$\sqrt{x}\,\sqrt[3]{x}$.

16. Determine whether $-1 + i$ is a solution of $x^2 + 2x + 4 = 0$.

17. Subtract: $(-2 + 7i) - (3 - 6i)$.

18. Multiply: $(4 + 3i)(3 - 4i)$.

19. Divide: $\dfrac{-4 + i}{3 - 2i}$.

20. Simplify: i^{87}.

Objectives for Retesting

The objectives to be tested in addition to the material in this chapter are as follows.

[4.7a]   Solve quadratic and other polynomial equations by first factoring and then using the principle of zero products.

[5.1d, e]   Multiply and divide rational expressions and simplify.

[5.5a]   Solve rational equations

[5.6a, b, c]   Solve applied problems involving work, proportions, and motion.

6.1 Radical Expressions and Functions

In this section, we consider roots, such as square roots and cube roots. We define the symbolism and consider methods of manipulating symbols to get equivalent expressions.

a Square Roots and Square-Root Functions

When we raise a number to the second power, we say that we have **squared** the number. Sometimes we may need to find the number that was squared. We call this process **finding a square root** of a number.

> The number c is a **square root** of a if $c^2 = a$.

For example:

5 is a square root of 25 because $5^2 = 5 \cdot 5 = 25$;

-5 is a square root of 25 because $(-5)^2 = (-5)(-5) = 25$.

The number -4 does not have a real-number square root because there is no real number b such that $b^2 = -4$.

> **SQUARE ROOTS**
> Every positive real number has two real-number square roots.
> The number 0 has just one square root, 0 itself.
> Negative numbers do not have real-number square roots.*

Example 1 Find the two square roots of 64.

The square roots of 64 are 8 and -8 because $8^2 = 64$ and $(-8)^2 = 64$.

Do Exercises 1–3.

> The **principal square root** of a nonnegative number is its nonnegative square root. The symbol $\sqrt{a}$ represents the principal square root of a. To name the negative square root of a, we can write $-\sqrt{a}$.

Examples Simplify.

2. $\sqrt{25} = 5$ *Remember:* $\sqrt{}$ indicates the principal (nonnegative) square root.

3. $\sqrt{\dfrac{81}{64}} = \dfrac{9}{8}$ **4.** $\sqrt{0.0049} = 0.07$

5. $\sqrt{0} = 0$ **6.** $-\sqrt{25} = -5$

7. $\sqrt{-25}$ Does not exist as a real number. Negative numbers do not have real-number square roots.

Do Exercises 4–10.

---

*In Section 6.8, we will consider an expansion of the real-number system, in which negative numbers do have square roots.

Objectives

a Find principal square roots and their opposites, approximate square roots, find outputs of square-root functions, graph square-root functions, and find the domains of square-root functions.

b Simplify radical expressions with perfect-square radicands.

c Find cube roots, simplifying certain expressions, and find outputs of cube-root functions.

d Simplify expressions involving odd and even roots.

For Extra Help

TAPE 13   TAPE 11B   MAC WIN   CD-ROM

Find the square roots.

1. 9   **2.** 36

3. 121

Simplify.

4. $\sqrt{1}$ **5.** $\sqrt{36}$

6. $\sqrt{\dfrac{81}{100}}$ **7.** $\sqrt{0.0064}$

Find the following.

8. a) $\sqrt{16}$

   b) $-\sqrt{16}$

   c) $\sqrt{-16}$

9. a) $\sqrt{49}$

   b) $-\sqrt{49}$

   c) $\sqrt{-49}$

10. a) $\sqrt{144}$

   b) $-\sqrt{144}$

   c) $\sqrt{-144}$

Answers on page A-35

It would be helpful to memorize the following table of exact square roots.

Table of Common Square Roots	
$\sqrt{1} = 1$	$\sqrt{196} = 14$
$\sqrt{4} = 2$	$\sqrt{225} = 15$
$\sqrt{9} = 3$	$\sqrt{256} = 16$
$\sqrt{16} = 4$	$\sqrt{289} = 17$
$\sqrt{25} = 5$	$\sqrt{324} = 18$
$\sqrt{36} = 6$	$\sqrt{361} = 19$
$\sqrt{49} = 7$	$\sqrt{400} = 20$
$\sqrt{64} = 8$	$\sqrt{441} = 21$
$\sqrt{81} = 9$	$\sqrt{484} = 22$
$\sqrt{100} = 10$	$\sqrt{529} = 23$
$\sqrt{121} = 11$	$\sqrt{576} = 24$
$\sqrt{144} = 12$	$\sqrt{625} = 25$
$\sqrt{169} = 13$	

Use a calculator to approximate each of the following square roots to three decimal places.

11. $\sqrt{17}$ **12.** $\sqrt{40}$

13. $\sqrt{1138}$ **14.** $-\sqrt{867.6}$

15. $\sqrt{\dfrac{22}{35}}$

16. $-\sqrt{\dfrac{2103.4}{67.82}}$

Identify the radicand.

17. $\sqrt{28 + x}$

18. $\sqrt{\dfrac{y}{y + 3}}$

Answers on page A-35

We found exact square roots in Examples 1–6. We often need to use rational numbers to *approximate* square roots that are irrational (see Section R.1). Such expressions can be found using a calculator with a square-root key.

Examples   Use a calculator to approximate each of the following.

		Using a calculator with a 10-digit readout	Rounded to three decimal places
8.	$\sqrt{11}$	3.316624790	3.317
9.	$\sqrt{487}$	22.06807649	22.068
10.	$-\sqrt{7297.8}$	-85.42716196	-85.427
11.	$\sqrt{\dfrac{463}{557}}$	.9117229728	0.912

Do Exercises 11–16.

> The symbol $\sqrt{}$ is called a **radical**.
> An expression written with a radical is called a **radical expression.**
> The expression written under the radical is called the **radicand**.

These are radical expressions:

$$\sqrt{5}, \qquad \sqrt{a}, \qquad -\sqrt{5x}, \qquad \sqrt{y^2 + 7}.$$

The radicands in these expressions are 5, a, $5x$, and $y^2 + 7$, respectively.

Example 12   Identify the radicand in $\sqrt{x^2 - 9}$.

The radicand in $\sqrt{x^2 - 9}$ is $x^2 - 9$.

Do Exercises 17 and 18.

Since each nonnegative real number x has exactly one principal square root, the symbol $\sqrt{x}$ represents exactly one real number and thus can be used to define a square-root function:

$$f(x) = \sqrt{x}.$$

The domain of this function is the set of nonnegative real numbers. In interval notation, the domain is $[0, \infty)$.

Example 13   For the given function, find the indicated function values:

$$f(x) = \sqrt{3x - 2}; \quad f(1), f(5), \text{ and } f(0).$$

We have

$f(1) = \sqrt{3 \cdot 1 - 2}$ Substituting

      $= \sqrt{3 - 2} = \sqrt{1} = 1;$ Simplifying and taking the square root

$f(5) = \sqrt{3 \cdot 5 - 2}$ Substituting

      $= \sqrt{13} \approx 3.606;$ Simplifying and approximating

$f(0) = \sqrt{3 \cdot 0 - 2}$ Substituting

      $= \sqrt{-2}.$ Negative radicand. No real-number function value exists.

Do Exercises 19 and 20 on the following page.

Example 14 Find the domain of $f(x) = \sqrt{x + 2}$.

The expression $\sqrt{x + 2}$ is a real number only when $x + 2$ is nonnegative. Thus the domain of $f(x) = \sqrt{x + 2}$ is the set of all x-values for which $x + 2 \geq 0$. We solve as follows:

$$x + 2 \geq 0$$
$$x \geq -2. \quad \text{Adding } -2$$

The domain of $f = \{x \mid x \geq -2\} = [-2, \infty)$.

Example 15 Graph: $f(x) = \sqrt{x}$.

We first find outputs as we did in Example 13. We can either select inputs that have exact outputs or use a calculator to make approximations. Once ordered pairs have been calculated, a smooth curve can be drawn.

x	$f(x) = \sqrt{x}$	$(x, f(x))$
0	0	$(0, 0)$
1	1	$(1, 1)$
3	1.7	$(3, 1.7)$
4	2	$(4, 2)$
7	2.6	$(7, 2.6)$
9	3	$(9, 3)$

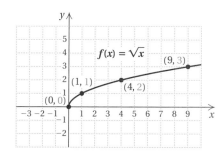

We can see from the table and the graph that the domain is $[0, \infty)$. The range is also the set of nonnegative real numbers $[0, \infty)$.

Do Exercises 21–24.

b | Finding $\sqrt{a^2}$

In the expression $\sqrt{a^2}$, the radicand is a perfect square. It is tempting to think that $\sqrt{a^2} = a$, but we see below that this is not the case.

Suppose $a = 5$. Then we have $\sqrt{5^2}$, which is $\sqrt{25}$, or 5.

Suppose $a = -5$. Then we have $\sqrt{(-5)^2}$, which is $\sqrt{25}$, or 5.

Suppose $a = 0$. Then we have $\sqrt{0^2}$, which is $\sqrt{0}$, or 0.

The symbol $\sqrt{a^2}$ never represents a negative number. It represents the principal square root of a^2. Note that if a represents a positive number or 0, then $\sqrt{a^2}$ represents a. If a is negative, then $\sqrt{a^2}$ represents the opposite of a. In all cases, the radical expression represents the absolute value of a.

> ▶ For any real number a, $\sqrt{a^2} = |a|$. The principal (nonnegative) square root of a^2 is the absolute value of a.

The absolute value is used to ensure that the principal square root is nonnegative, which is as it is defined.

Examples Find the following. Assume that letters can represent any real number.

16. $\sqrt{(-16)^2} = |-16|$, or 16

For the given function, find the indicated function values.

19. $g(x) = \sqrt{6x + 4}$; $g(0)$, $g(3)$, and $g(-5)$

20. $f(x) = -\sqrt{x}$; $f(4)$, $f(7)$, and $f(-3)$

Find the domain of the function.

21. $f(x) = \sqrt{x - 5}$

22. $g(x) = \sqrt{2x + 3}$

Graph.

23. $g(x) = -\sqrt{x}$

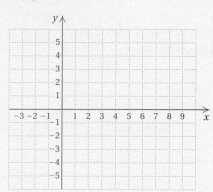

24. $f(x) = 2\sqrt{x + 3}$

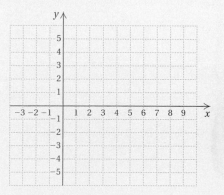

Answers on page A-35

Find the following. Assume that letters can represent *any* real number.

25. $\sqrt{y^2}$

26. $\sqrt{(-24)^2}$

27. $\sqrt{(5y)^2}$

28. $\sqrt{16y^2}$

29. $\sqrt{(x + 7)^2}$

30. $\sqrt{4(x - 2)^2}$

31. $\sqrt{49(y + 5)^2}$

32. $\sqrt{x^2 - 6x + 9}$

Find the following.

33. $\sqrt[3]{-64}$

34. $\sqrt[3]{27y^3}$

35. $\sqrt[3]{8(x + 2)^3}$

36. $\sqrt[3]{-\dfrac{343}{64}}$

Answers on page A-35

Chapter 6   Radical Expressions, Equations, and Functions

456

17. $\sqrt{(3b)^2} = |3b| = |3| \cdot |b| = 3|b|$

$|3b|$ can be simplified to $3|b|$ because the absolute value of any product is the product of the absolute values. That is, $|a \cdot b| = |a| \cdot |b|$.

18. $\sqrt{(x - 1)^2} = |x - 1|$

19. $\sqrt{x^2 + 8x + 16} = \sqrt{(x + 4)^2}$
$$= |x + 4| \leftarrow$$

CAUTION! $|x + 4|$ is *not* the same as $|x| + 4$.

Do Exercises 25–32.

c   Cube Roots

> The number c is the **cube root** of a if its third power is a—that is, if $c^3 = a$.

For example:

2 is the cube root of 8 because $2^3 = 2 \cdot 2 \cdot 2 = 8$;

-4 is the cube root of -64 because $(-4)^3 = (-4)(-4)(-4) = -64$.

We talk about *the* cube root of a number because of the following.

> Every real number has exactly one cube root in the system of real numbers. The symbol $\sqrt[3]{a}$ represents the cube root of a.

Examples   Find the following.

20. $\sqrt[3]{8} = 2$

21. $\sqrt[3]{-27} = -3$

22. $\sqrt[3]{-\dfrac{216}{125}} = -\dfrac{6}{5}$

23. $\sqrt[3]{0.001} = 0.1$

24. $\sqrt[3]{x^3} = x$

25. $\sqrt[3]{-8} = -2$

26. $\sqrt[3]{0} = 0$

27. $\sqrt[3]{-8y^3} = -2y$

When we are determining a cube root, no absolute-value signs are needed because a real number has just one cube root. The real-number cube root of a positive number is positive. The real-number cube root of a negative number is negative. The cube root of 0 is 0. That is, $\sqrt[3]{a^3} = a$ whether $a > 0$, $a < 0$, or $a = 0$.

Do Exercises 33–36.

Since the symbol $\sqrt[3]{x}$ represents exactly one real number, it can be used to define a function.

Example 28   For the given function, find the indicated function values.

$$f(x) = \sqrt[3]{x}; \quad f(125), f(-8), f(0), \text{ and } f(-10).$$

We have

$$f(125) = \sqrt[3]{125} = 5; \qquad f(-8) = \sqrt[3]{-8} = -2;$$
$$f(0) = \sqrt[3]{0} = 0; \qquad f(-10) = \sqrt[3]{-10} \approx -2.1544.$$

Do Exercise 37 on the following page.

The graph of $f(x) = \sqrt[3]{x}$ is shown below for reference. Note that the domain and the range *each* consists of the entire set of real numbers, $(-\infty, \infty)$.

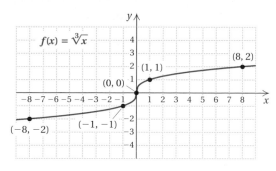

d | Odd and Even *k*th Roots

In the expression $\sqrt[k]{a}$, we call k the **index** and assume $k \geq 2$.

Odd Roots

The 5th root of a number a is the number c for which $c^5 = a$. There are also 7th roots, 9th roots, and so on. Whenever the number k in $\sqrt[k]{}$ is an odd number, we say that we are taking an **odd root.**

Every number has just one real-number odd root. If the number is positive, then the root is positive. If the number is negative, then the root is negative. If the number is 0, then the root is 0.

> If k is an *odd* natural number, then for any real number a,
> $$\sqrt[k]{a^k} = a.$$

Absolute-value signs are not needed when we are finding odd roots.

Examples   Find the following.

29. $\sqrt[5]{32} = 2$ 30. $\sqrt[5]{-32} = -2$

31. $-\sqrt[5]{32} = -2$ 32. $-\sqrt[5]{-32} = -(-2) = 2$

33. $\sqrt[7]{x^7} = x$ 34. $\sqrt[7]{128} = 2$

35. $\sqrt[7]{-128} = -2$ 36. $\sqrt[7]{0} = 0$

37. $\sqrt[5]{a^5} = a$ 38. $\sqrt[9]{(x-1)^9} = x - 1$

Do Exercises 38–44.

Even Roots

When the index k in $\sqrt[k]{}$ is an even number, we say that we are taking an **even root.** Every positive real number has two real-number kth roots when k is even. One of those roots is positive and one is negative. Negative real numbers do not have real-number kth roots when k is even. When we are finding even kth roots, absolute-value signs are sometimes necessary, as they are with square roots. When the index is 2, we do not write it. For example,

$$\sqrt{64} = 8, \qquad \sqrt[6]{64} = 2, \qquad -\sqrt[6]{64} = -2, \qquad \sqrt[6]{64x^6} = |2x| = 2|x|.$$

Note that in $\sqrt[6]{64x^6}$, we need absolute-value signs because a variable is involved.

---

37. For the given function, find the indicated function values.
$$g(x) = \sqrt[3]{x - 4}; \quad g(-23),$$
$$g(4), \ g(-1), \text{ and } g(11)$$

Find the following.

38. $\sqrt[5]{243}$ 39. $\sqrt[5]{-243}$

40. $\sqrt[5]{x^5}$ 41. $\sqrt[7]{y^7}$

42. $\sqrt[5]{0}$

43. $\sqrt[5]{-32x^5}$

44. $\sqrt[7]{(3x + 2)^7}$

Answers on page A-35

Find the following. Assume that letters can represent any real number.

45. $\sqrt[4]{81}$ **46.** $-\sqrt[4]{81}$

47. $\sqrt[4]{-81}$ **48.** $\sqrt[4]{0}$

49. $\sqrt[4]{16(x-2)^4}$

50. $\sqrt[6]{x^6}$

51. $\sqrt[8]{(x+3)^8}$

Examples Find the following. Assume that letters can represent any real number.

39. $\sqrt[4]{16} = 2$

40. $-\sqrt[4]{16} = -2$

41. $\sqrt[4]{-16}$ Does not exist as a real number.

42. $\sqrt[4]{81x^4} = 3|x|$

43. $\sqrt[6]{(y+7)^6} = |y+7|$

44. $\sqrt{81y^2} = 9|y|$

The following is a summary of how absolute value is used when we are taking even or odd roots.

> For any real number a:
> a) $\sqrt[k]{a^k} = |a|$ when k is an *even* natural number. We use absolute value when k is even unless a is nonnegative.
> b) $\sqrt[k]{a^k} = a$ when k is an *odd* natural number greater than 1. We do not use absolute value when k is odd.

Do Exercises 45–51.

Calculator Spotlight

Graphing Radical Functions. Graphing functions defined by radical expressions involves approximating roots. Since the square root of a negative number is not a real number, y-values may not exist for some x-values. For example, y-values for the graph of $f(x) = \sqrt{x-1}$ do not exist for x-values that are less than 1 because square roots of negative numbers would result.

We must enter $y = \sqrt{x-1}$ using parentheses around the radicand as $y_1 = \sqrt{(x-1)}$. Some graphers supply the left parenthesis automatically.

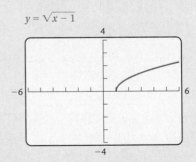

$y = \sqrt{x-1}$

Similarly, y-values for the graph of $f(x) = \sqrt{2-x}$ do not exist for x-values that are greater than 2.

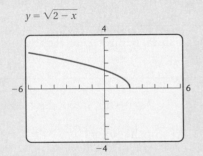

$y = \sqrt{2-x}$

Exercises

Graph each of the following functions. Then use the TABLE and TRACE features to determine the domain and the range of each function. (See also the Calculator Spotlight in Section 2.3.) The MATH feature contains $\boxed{\sqrt[3]{}}$ and $\boxed{\sqrt[x]{}}$ keys to enter kth roots.

1. $f(x) = \sqrt{x}$

2. $g(x) = \sqrt{x+2}$

3. $f(x) = \sqrt[3]{x}$

4. $f(x) = \sqrt[3]{x-2}$

5. $f(x) = \sqrt[4]{x-1}$

6. $F(x) = \sqrt[5]{6-x}$

7. $g(x) = 5 - \sqrt{x+3}$

8. $f(x) = 4 - \sqrt[3]{x}$

Use the GRAPH and TABLE features to determine whether each of the following is correct.

9. $\sqrt{x+4} = \sqrt{x} + 2$

10. $\sqrt{25x} = 5\sqrt{x}$

Answers on page A-35

Exercise Set 6.1

a Find the square roots.

1. 16   **2.** 225   **3.** 144   **4.** 9   **5.** 400   **6.** 81

Simplify.

7. $-\sqrt{\dfrac{49}{36}}$ **8.** $-\sqrt{\dfrac{361}{9}}$ **9.** $\sqrt{196}$

10. $\sqrt{441}$ **11.** $\sqrt{0.0036}$ **12.** $\sqrt{0.04}$

Use a calculator to approximate to three decimal places.

13. $\sqrt{347}$ **14.** $-\sqrt{1839.2}$ **15.** $\sqrt{\dfrac{285}{74}}$ **16.** $\sqrt{\dfrac{839.4}{19.7}}$

Identify the radicand.

17. $9\sqrt{y^2 + 16}$ **18.** $-3\sqrt{p^2 - 10}$ **19.** $x^4 y^5 \sqrt{\dfrac{x}{y-1}}$ **20.** $a^2 b^2 \sqrt{\dfrac{a^2 - b}{b}}$

For the given function, find the indicated function values.

21. $f(x) = \sqrt{5x - 10}$; $f(6), f(2), f(1)$, and $f(-1)$

22. $t(x) = -\sqrt{2x + 1}$; $t(4), t(0), t(-1)$, and $t\left(-\dfrac{1}{2}\right)$

23. $g(x) = \sqrt{x^2 - 25}$; $g(-6), g(3), g(6)$, and $g(13)$

24. $F(x) = \sqrt{x^2 + 1}$; $F(0), F(-1)$, and $F(-10)$

25. Find the domain of the function f in Exercise 21.

26. Find the domain of the function t in Exercise 22.

27. *Speed of a Skidding Car.* How do police determine how fast a car had been traveling after an accident has occurred? The function

$$S(x) = 2\sqrt{5x}$$

can be used to approximate the speed S, in miles per hour, of a car that has left a skid mark of length x, in feet. What was the speed of a car that left skid marks of length 30 ft? 150 ft?

28. *Parking-Lot Arrival Spaces.* The attendants at a parking lot park cars in temporary spaces before the cars are taken to permanent parking stalls. The number N of such spaces needed is approximated by the function

$$N(a) = 2.5\sqrt{a},$$

where a is the average number of arrivals in peak hours. What is the number of spaces needed when the average number of arrivals is 66? 100?

Graph.

29. $f(x) = 2\sqrt{x}$

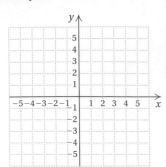

30. $g(x) = 3 - \sqrt{x}$

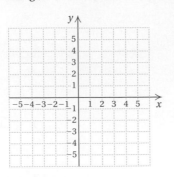

31. $F(x) = -3\sqrt{x}$

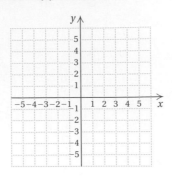

32. $f(x) = 2 + \sqrt{x - 1}$

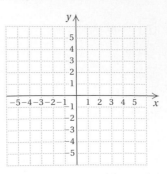

33. $f(x) = \sqrt{x}$

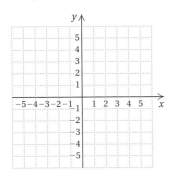

34. $g(x) = -\sqrt{x}$

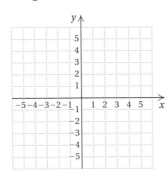

35. $f(x) = \sqrt{x - 2}$

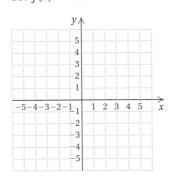

36. $g(x) = \sqrt{x + 3}$

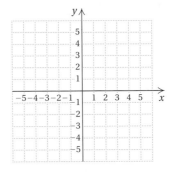

37. $f(x) = \sqrt{12 - 3x}$

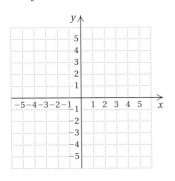

38. $g(x) = \sqrt{8 - 4x}$

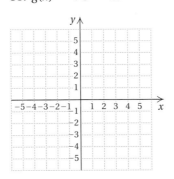

39. $g(x) = \sqrt{3x + 9}$

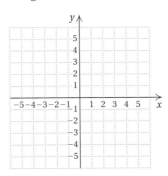

40. $f(x) = \sqrt{3x - 6}$

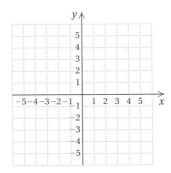

b Find the following. Assume that letters can represent *any* real number.

41. $\sqrt{16x^2}$

42. $\sqrt{25t^2}$

43. $\sqrt{(-12c)^2}$

44. $\sqrt{(-9d)^2}$

45. $\sqrt{(p + 3)^2}$

46. $\sqrt{(2 - x)^2}$

47. $\sqrt{x^2 - 4x + 4}$

48. $\sqrt{9t^2 - 30t + 25}$

49. $\sqrt[3]{27}$

50. $-\sqrt[3]{64}$

51. $\sqrt[3]{-64x^3}$

52. $\sqrt[3]{-125y^3}$

53. $\sqrt[3]{-216}$

54. $-\sqrt[3]{-1000}$

55. $\sqrt[3]{0.343(x+1)^3}$

56. $\sqrt[3]{0.000008(y-2)^3}$

For the given function, find the indicated function values.

57. $f(x) = \sqrt[3]{x+1}$; $f(7), f(26), f(-9)$, and $f(-65)$

58. $g(x) = -\sqrt[3]{2x-1}$; $g(-62), g(0), g(-13)$, and $g(63)$

59. $f(x) = -\sqrt[3]{3x+1}$; $f(0), f(-7), f(21)$, and $f(333)$

60. $g(t) = \sqrt[3]{t-3}$; $g(30), g(-5), g(1)$, and $g(67)$

d Find the following. Assume that letters can represent *any* real number.

61. $\sqrt[4]{625}$

62. $-\sqrt[4]{256}$

63. $\sqrt[5]{-1}$

64. $\sqrt[5]{-32}$

65. $\sqrt[5]{-\dfrac{32}{243}}$

66. $\sqrt[5]{-\dfrac{1}{32}}$

67. $\sqrt[6]{x^6}$

68. $\sqrt[8]{y^8}$

69. $\sqrt[4]{(5a)^4}$

70. $\sqrt[4]{(7b)^4}$

71. $\sqrt[10]{(-6)^{10}}$

72. $\sqrt[12]{(-10)^{12}}$

73. $\sqrt[414]{(a+b)^{414}}$

74. $\sqrt[1999]{(2a+b)^{1999}}$

75. $\sqrt[7]{y^7}$

76. $\sqrt[3]{(-6)^3}$

77. $\sqrt[5]{(x-2)^5}$

78. $\sqrt[9]{(2xy)^9}$

Skill Maintenance

Solve.   [4.7a]

79. $x^2 + x - 2 = 0$

80. $x^2 + x = 0$

81. $4x^2 - 49 = 0$

82. $2x^2 - 26x + 72 = 0$

83. $3x^2 + x = 10$

84. $4x^2 - 20x + 25 = 0$

85. $4x^3 - 20x^2 + 25x = 0$

86. $x^3 - x^2 = 0$

Simplify.   [R.7a, b]

87. $(a^3 b^2 c^5)^3$

88. $(5a^7 b^8)(2a^3 b)$

Synthesis

89. ◈ Does the nth root of x^2 always exist? Why or why not?

90. ◈ Explain how to formulate a radical expression that can be used to define a function f with a domain of $\{x \mid x \le 5\}$.

91. Find the domain of
$$f(x) = \frac{\sqrt{x+3}}{\sqrt{2-x}}.$$

92. 〰 Use a grapher to check your answers to Exercises 33, 37, and 39.

93. Use only the graph of $f(x) = \sqrt{x}$, shown below, to approximate $\sqrt{3}$, $\sqrt{5}$, and $\sqrt{10}$. Answers may vary.

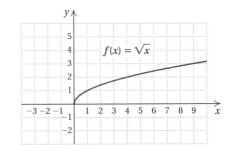

94. Use only the graph of $f(x) = \sqrt[3]{x}$, shown below, to approximate $\sqrt[3]{4}$, $\sqrt[3]{6}$, and $\sqrt[3]{-5}$. Answers may vary.

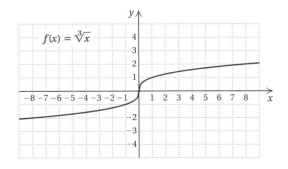

95. 〰 Use the TABLE, TRACE, and GRAPH features of a grapher to find the domain and the range of each of the following functions.
a) $f(x) = \sqrt[3]{x}$
b) $g(x) = \sqrt[3]{4x-5}$
c) $q(x) = 2 - \sqrt{x+3}$
d) $h(x) = \sqrt[4]{x}$
e) $t(x) = \sqrt[4]{x-3}$

Collaborative Learning Manual

Use the function concerning skid length to determine a safe distance from which to follow another vehicle.

6.2 Rathonal Numbers as Exponents

In this section, we give meaning to expressions such as $a^{1/3}$, $7^{-1/2}$, and $(3x)^{0.84}$, which have rational numbers as exponents. We will see that using such notation can help simplify certain radical expressions.

a ▮ Rational Exponents

Expressions like $a^{1/2}$, $5^{-1/4}$, and $(2y)^{4/5}$ have not yet been defined. We will define such expressions so that the general properties of exponents hold.

Consider $a^{1/2} \cdot a^{1/2}$. If we want to multiply by adding exponents, it must follow that $a^{1/2} \cdot a^{1/2} = a^{1/2+1/2}$, or a^1. Thus we should define $a^{1/2}$ to be a square root of a. Similarly, $a^{1/3} \cdot a^{1/3} \cdot a^{1/3} = a^{1/3+1/3+1/3}$, or a^1, so $a^{1/3}$ should be defined to mean $\sqrt[3]{a}$.

> For any nonnegative real number a and any natural number index n
> ($n \neq 1$), $a^{1/n}$ means $\sqrt[n]{a}$ (the nonnegative nth root of a).

Whenever we use rational exponents, we assume that the bases are nonnegative.

Examples Rewrite without rational exponents, and simplify, if possible.

1. $x^{1/2} = \sqrt{x}$

2. $27^{1/3} = \sqrt[3]{27} = 3$

3. $(abc)^{1/5} = \sqrt[5]{abc}$

Do Exercises 1–5.

Examples Rewrite with rational exponents.

4. $\sqrt[5]{7xy} = (7xy)^{1/5}$ We need parentheses around the radicand here.

5. $8\sqrt[3]{xy} = 8(xy)^{1/3}$

6. $\sqrt[7]{\dfrac{x^3 y}{9}} = \left(\dfrac{x^3 y}{9}\right)^{1/7}$

Do Exercises 6–9.

How should we define $a^{2/3}$? If the general properties of exponents are to hold, we have $a^{2/3} = (a^{1/3})^2$, or $(\sqrt[3]{a})^2$, or $\sqrt[3]{a^2}$. We define this accordingly.

> For any natural numbers m and n ($n \neq 1$) and any nonnegative real number a,
> $$a^{m/n} \text{ means } \sqrt[n]{a^m}, \text{ or } (\sqrt[n]{a})^m.$$

Examples Rewrite without rational exponents, and simplify, if possible.

7. $(27)^{2/3} = \sqrt[3]{27^2}$
        $= (\sqrt[3]{27})^2$
        $= 3^2 = 9$

8. $4^{3/2} = \sqrt[2]{4^3}$
       $= (\sqrt[2]{4})^3$
       $= 2^3 = 8$

Do Exercises 10–12.

Rewrite without rational exponents, and simplify, if possible.

1. $y^{1/4}$ **2.** $(3a)^{1/2}$

3. $16^{1/4}$ **4.** $(125)^{1/3}$

5. $(a^3 b^2 c)^{1/5}$

Rewrite with rational exponents.

6. $\sqrt[3]{19ab}$ **7.** $19\sqrt[3]{ab}$

8. $\sqrt[5]{\dfrac{x^2 y}{16}}$ **9.** $7\sqrt[4]{2ab}$

Rewrite without rational exponents, and simplify, if possible.

10. $x^{3/2}$ **11.** $8^{2/3}$

12. $4^{5/2}$

Rewrite with rational exponents.

13. $(\sqrt[3]{7abc})^4$ **14.** $\sqrt[5]{6^7}$

Answers on page A-36

Rewrite with positive exponents, and simplify, if possible.

15. $16^{-1/4}$ 16. $(3xy)^{-7/8}$

17. $81^{-3/4}$ 18. $7p^{3/4}q^{-6/5}$

19. $\left(\dfrac{11m}{7n}\right)^{-2/3}$

Answers on page A-36

Examples   Rewrite with rational exponents.

The index becomes the denominator of the rational exponent.

9. $\sqrt[3]{9^4} = 9^{4/3}$

10. $(\sqrt[4]{7xy})^5 = (7xy)^{5/4}$

Do Exercises 13 and 14 on the preceding page.

b Negative Rational Exponents

Negative rational exponents have a meaning similar to that of negative integer exponents.

> For any rational number m/n and any positive real number a,
> $$a^{-m/n} \quad \text{means} \quad \frac{1}{a^{m/n}},$$
> that is, $a^{m/n}$ and $a^{-m/n}$ are reciprocals.

Examples   Rewrite with positive exponents, and simplify, if possible.

11. $9^{-1/2} = \dfrac{1}{9^{1/2}} = \dfrac{1}{\sqrt{9}} = \dfrac{1}{3}$

12. $(5xy)^{-4/5} = \dfrac{1}{(5xy)^{4/5}}$ $(5xy)^{-4/5}$ is the reciprocal of $(5xy)^{4/5}$.

13. $64^{-2/3} = \dfrac{1}{64^{2/3}} = \dfrac{1}{(\sqrt[3]{64})^2} = \dfrac{1}{4^2} = \dfrac{1}{16}$

14. $4x^{-2/3}y^{1/5} = 4 \cdot \dfrac{1}{x^{2/3}} \cdot y^{1/5} = \dfrac{4y^{1/5}}{x^{2/3}}$

15. $\left(\dfrac{3r}{7s}\right)^{-5/2} = \left(\dfrac{7s}{3r}\right)^{5/2}$ Since $\left(\dfrac{a}{b}\right)^{-n} = \left(\dfrac{b}{a}\right)^{n}$

Do Exercises 15–19.

c Laws of Exponents

The same laws hold for rational-number exponents as for integer exponents. We list them for review.

> For any real number a and any rational exponents m and n:
>
> 1. $a^m \cdot a^n = a^{m+n}$ In multiplying, we can add exponents if the bases are the same.
>
> 2. $\dfrac{a^m}{a^n} = a^{m-n}$ In dividing, we can subtract exponents if the bases are the same.
>
> 3. $(a^m)^n = a^{m \cdot n}$ To raise a power to a power, we can multiply the exponents.
>
> 4. $(ab)^m = a^m b^m$ To raise a product to a power, we can raise each factor to the power.
>
> 5. $\left(\dfrac{a}{b}\right)^n = \dfrac{a^n}{b^n}$ To raise a quotient to a power, we can raise both the numerator and the denominator to the power.

Examples Use the laws of exponents to simplify.

16. $3^{1/5} \cdot 3^{3/5} = 3^{1/5+3/5} = 3^{4/5}$ **Adding exponents**

17. $\dfrac{7^{1/4}}{7^{1/2}} = 7^{1/4-1/2} = 7^{1/4-2/4} = 7^{-1/4} = \dfrac{1}{7^{1/4}}$ **Subtracting exponents**

18. $(7.2^{2/3})^{3/4} = 7.2^{2/3 \cdot 3/4} = 7.2^{6/12}$ **Multiplying exponents**

$= 7.2^{1/2}$

19. $(a^{-1/3}b^{2/5})^{1/2} = a^{-1/3 \cdot 1/2} \cdot b^{2/5 \cdot 1/2}$ **Raising a product to a power and multiplying exponents**

$= a^{-1/6}b^{1/5} = \dfrac{b^{1/5}}{a^{1/6}}$

Do Exercises 20–23.

Use the laws of exponents to simplify.
20. $7^{1/3} \cdot 7^{3/5}$

21. $\dfrac{5^{7/6}}{5^{5/6}}$

22. $(9^{3/5})^{2/3}$

23. $(p^{-2/3}q^{1/4})^{1/2}$

d   Simplifying Radical Expressions

Rational exponents can be used to simplify some radical expressions. The procedure is as follows.

> 1. Convert radical expressions to exponential expressions.
> 2. Use arithmetic and the laws of exponents to simplify.
> 3. Convert back to radical notation when appropriate.
>
> *Important*: This procedure works only when all expressions under radicals are nonnegative since rational exponents are not defined otherwise. No absolute-value signs will be needed.

Use rational exponents to simplify.
24. $\sqrt[4]{a^2}$

25. $\sqrt[4]{x^4}$

26. $\sqrt[6]{8}$

Examples Use rational exponents to simplify.

20. $\sqrt[6]{x^3} = x^{3/6}$ **Converting to an exponential expression**

$= x^{1/2}$ **Simplifying the exponent**

$= \sqrt{x}$ **Converting back to radical notation**

21. $\sqrt[6]{4} = 4^{1/6}$ **Converting to exponential notation**

$= (2^2)^{1/6}$ **Renaming 4 as 2^2**

$= 2^{2/6}$ **Using $(a^m)^n = a^{mn}$; multiplying exponents**

$= 2^{1/3}$ **Simplifying the exponent**

$= \sqrt[3]{2}$ **Converting back to radical notation**

Do Exercises 24–26.

Use rational exponents to simplify.
27. $\sqrt[12]{x^3 y^6}$

Example 22 Use rational exponents to simplify: $\sqrt[8]{a^2b^4}$.

$\sqrt[8]{a^2b^4} = (a^2b^4)^{1/8}$ **Converting to exponential notation**

$= a^{2/8} \cdot b^{4/8}$ **Using $(ab)^n = a^n b^n$**

$= a^{1/4} \cdot b^{1/2}$ **Simplifying the exponents**

$= a^{1/4} \cdot b^{2/4}$ **Rewriting $\frac{1}{2}$ with a denominator of 4**

$= (ab^2)^{1/4}$ **Using $a^n b^n = (ab)^n$**

$= \sqrt[4]{ab^2}$ **Converting back to radical notation**

28. $\sqrt[6]{a^{12}b^3}$

29. $\sqrt[5]{a^5 b^{10}}$

Do Exercises 27–29.

Answers on page A-36

30. Use rational exponents to write a single radical expression:

$$\sqrt[4]{7} \cdot \sqrt{3}.$$

Write a single radical expression.

31. $x^{2/3}y^{1/2}z^{5/6}$

32. $\dfrac{a^{1/2}b^{3/8}}{a^{1/4}b^{1/8}}$

Use rational exponents to simplify.

33. $\sqrt[14]{(5m)^2}$

34. $\sqrt[18]{m^3}$

35. $(\sqrt[6]{a^5b^3c})^{24}$

36. $\sqrt[5]{\sqrt{x}}$

Answers on page A-36

We can use properties of rational exponents to write a single radical expression for a product or a quotient.

Example 23   Use rational exponents to write a single radical expression for $\sqrt[3]{5} \cdot \sqrt{2}$.

$$
\begin{aligned}
\sqrt[3]{5} \cdot \sqrt{2} &= 5^{1/3} \cdot 2^{1/2} && \text{Converting to exponential notation} \\
&= 5^{2/6} \cdot 2^{3/6} && \text{Rewriting so that exponents have} \\
& && \text{a common denominator} \\
&= (5^2 \cdot 2^3)^{1/6} && \text{Using } a^n b^n = (ab)^n \\
&= \sqrt[6]{5^2 \cdot 2^3} && \text{Converting back to radical notation} \\
&= \sqrt[6]{200} && \text{Multiplying under the radical}
\end{aligned}
$$

Do Exercise 30.

Example 24   Write a single radical expression for $a^{1/2}b^{-1/2}c^{5/6}$.

$$
\begin{aligned}
a^{1/2}b^{-1/2}c^{5/6} &= a^{3/6}b^{-3/6}c^{5/6} && \text{Rewriting so that exponents have} \\
& && \text{a common denominator} \\
&= (a^3 b^{-3} c^5)^{1/6} && \text{Using } a^n b^n = (ab)^n \\
&= \sqrt[6]{a^3 b^{-3} c^5} && \text{Converting to radical notation}
\end{aligned}
$$

Do Exercises 31 and 32.

Examples   Use rational exponents to simplify.

25.
$$
\begin{aligned}
\sqrt[6]{(5x)^3} &= (5x)^{3/6} && \text{Converting to exponential notation} \\
&= (5x)^{1/2} && \text{Simplifying the exponent} \\
&= \sqrt{5x} && \text{Converting back to radical notation}
\end{aligned}
$$

26.
$$
\begin{aligned}
\sqrt[5]{t^{20}} &= t^{20/5} && \text{Converting to exponential notation} \\
&= t^4 && \text{Simplifying the exponent}
\end{aligned}
$$

27.
$$
\begin{aligned}
(\sqrt[3]{pq^2c})^{12} &= (pq^2c)^{12/3} && \text{Converting to exponential notation} \\
&= (pq^2c)^4 && \text{Simplifying the exponent} \\
&= p^4 q^8 c^4 && \text{Using } (ab)^n = a^n b^n
\end{aligned}
$$

28.
$$
\begin{aligned}
\sqrt{\sqrt[3]{x}} &= \sqrt{x^{1/3}} && \text{Converting the radicand to exponential notation} \\
&= (x^{1/3})^{1/2} && \text{Try to go directly to this step.} \\
&= x^{1/6} && \text{Multiplying exponents} \\
&= \sqrt[6]{x} && \text{Converting back to radical notation}
\end{aligned}
$$

Do Exercises 33–36.

Exercise Set 6.2

a Rewrite without rational exponents, and simplify, if possible.

1. $y^{1/7}$ **2.** $x^{1/6}$ **3.** $8^{1/3}$ **4.** $16^{1/2}$ **5.** $(a^3b^3)^{1/5}$

6. $(x^2y^2)^{1/3}$ **7.** $16^{3/4}$ **8.** $4^{7/2}$ **9.** $49^{3/2}$ **10.** $27^{4/3}$

Rewrite with rational exponents.

11. $\sqrt{17}$ **12.** $\sqrt{x^3}$ **13.** $\sqrt[3]{18}$ **14.** $\sqrt[3]{23}$ **15.** $\sqrt[5]{xy^2z}$

16. $\sqrt[7]{x^3y^2z^2}$ **17.** $(\sqrt{3mn})^3$ **18.** $(\sqrt[3]{7xy})^4$ **19.** $(\sqrt[7]{8x^2y})^5$ **20.** $(\sqrt[6]{2a^5b})^7$

b Rewrite with positive exponents, and simplify, if possible.

21. $27^{-1/3}$ **22.** $100^{-1/2}$ **23.** $100^{-3/2}$ **24.** $16^{-3/4}$ **25.** $x^{-1/4}$

26. $y^{-1/7}$ **27.** $(2rs)^{-3/4}$ **28.** $(5xy)^{-5/6}$ **29.** $2a^{3/4}b^{-1/2}c^{2/3}$ **30.** $5x^{-2/3}y^{4/5}$

31. $\left(\dfrac{7x}{8yz}\right)^{-3/5}$ **32.** $\left(\dfrac{2ab}{3c}\right)^{-5/6}$ **33.** $\dfrac{1}{x^{-2/3}}$ **34.** $\dfrac{1}{a^{-7/8}}$ **35.** $2^{-1/3}x^4y^{-2/7}$

36. $3^{-5/2}a^3b^{-7/3}$ **37.** $\dfrac{7x}{\sqrt[3]{z}}$ **38.** $\dfrac{6a}{\sqrt[4]{b}}$ **39.** $\dfrac{5a}{3c^{-1/2}}$ **40.** $\dfrac{2z}{5x^{-1/3}}$

c Use the laws of exponents to simplify. Write the answers with positive exponents.

41. $5^{3/4} \cdot 5^{1/8}$ **42.** $11^{2/3} \cdot 11^{1/2}$ **43.** $\dfrac{7^{5/8}}{7^{3/8}}$ **44.** $\dfrac{3^{5/8}}{3^{-1/8}}$ **45.** $\dfrac{4.9^{-1/6}}{4.9^{-2/3}}$

46. $\dfrac{2.3^{-3/10}}{2.3^{-1/5}}$ **47.** $(6^{3/8})^{2/7}$ **48.** $(3^{2/9})^{3/5}$ **49.** $a^{2/3} \cdot a^{5/4}$ **50.** $x^{3/4} \cdot x^{2/3}$

51. $(a^{2/3} \cdot b^{5/8})^4$ **52.** $(x^{-1/3} \cdot y^{-2/5})^{-15}$ **53.** $(x^{2/3})^{-3/7}$ **54.** $(a^{-3/2})^{2/9}$

d Use rational exponents to simplify. Write the answer in radical notation if appropriate.

55. $\sqrt[6]{a^2}$ **56.** $\sqrt[6]{t^4}$ **57.** $\sqrt[3]{x^{15}}$ **58.** $\sqrt[4]{a^{12}}$ **59.** $\sqrt[6]{x^{-18}}$

60. $\sqrt[5]{a^{-10}}$ **61.** $(\sqrt[3]{ab})^{15}$ **62.** $(\sqrt[7]{cd})^{14}$ **63.** $\sqrt[4]{32}$ **64.** $\sqrt[6]{81}$

65. $\sqrt[6]{4x^2}$ **66.** $\sqrt[3]{8y^6}$ **67.** $\sqrt{x^4y^6}$ **68.** $\sqrt[4]{16x^4y^2}$ **69.** $\sqrt[5]{32c^{10}d^{15}}$

Use rational exponents to write a single radical expression.

70. $\sqrt[3]{3}\sqrt{3}$ **71.** $\sqrt[3]{7} \cdot \sqrt[4]{5}$ **72.** $\sqrt[7]{11} \cdot \sqrt[6]{13}$ **73.** $\sqrt[4]{5} \cdot \sqrt[5]{7}$ **74.** $\sqrt[3]{y}\sqrt[5]{3y}$

75. $\sqrt{x}\sqrt[3]{2x}$ **76.** $(\sqrt[3]{x^2y^5})^{12}$ **77.** $(\sqrt[5]{a^2b^4})^{15}$ **78.** $\sqrt[4]{\sqrt{x}}$ **79.** $\sqrt[3]{\sqrt[6]{m}}$

80. $a^{2/3} \cdot b^{3/4}$ **81.** $x^{1/3} \cdot y^{1/4} \cdot z^{1/6}$ **82.** $\dfrac{x^{8/15} \cdot y^{7/5}}{x^{1/3} \cdot y^{-1/5}}$ **83.** $\left(\dfrac{c^{-4/5}d^{5/9}}{c^{3/10}d^{1/6}}\right)^3$ **84.** $\sqrt[3]{\sqrt{xy}}$

Skill Maintenance

Solve. [5.7a]

85. $A = \dfrac{ab}{a+b}$, for a **86.** $Q = \dfrac{st}{s-t}$, for s **87.** $Q = \dfrac{st}{s-t}$, for t **88.** $\dfrac{1}{t} = \dfrac{1}{a} - \dfrac{1}{b}$, for b

---

Synthesis

89. ◈ Find the domain of
$$f(x) = (x+5)^{1/2}(x+7)^{-1/2}$$
and explain how you found your answer.

90. ◈ Explain why $\sqrt[3]{x^6} = x^2$ for any value of x, but $\sqrt{x^6} = x^3$ only when $x \geq 0$.

91. 〰 Use the SIMULTANEOUS mode to graph
$$y_1 = x^{1/2}, \quad y_2 = 3x^{2/5} \quad y_3 = x^{4/7} \quad y_4 = \tfrac{1}{3}x^{3/4}$$
Then, looking only at coordinates, match each graph with its equation.

92. Simplify:
$$\left(\sqrt[10]{\sqrt[6]{\sqrt[5]{x^{15}}}}\right)^5 \left(\sqrt[5]{\sqrt[10]{\sqrt[6]{x^{15}}}}\right)^5.$$

Investigate the effect of the order of rational exponents on exponential functions.

6.3 Simplifying Radical Expressions

a   Multiplying and Simplifying Radical Expressions

Note that $\sqrt{4}\sqrt{25} = 2 \cdot 5 = 10$. Also $\sqrt{4 \cdot 25} = \sqrt{100} = 10$. Likewise,

$$\sqrt[3]{27}\sqrt[3]{8} = 3 \cdot 2 = 6 \quad \text{and} \quad \sqrt[3]{27 \cdot 8} = \sqrt[3]{216} = 6.$$

These examples suggest the following.

> **THE PRODUCT RULE FOR RADICALS**
>
> For any nonnegative real numbers a and b and any index k,
> $$\sqrt[k]{a} \cdot \sqrt[k]{b} = \sqrt[k]{a \cdot b}.$$
> (To multiply, multiply the radicands.)

Note that the index k must be the same throughout.

Examples   Multiply.

1. $\sqrt{3} \cdot \sqrt{5} = \sqrt{3 \cdot 5} = \sqrt{15}$

2. $\sqrt{5a}\sqrt{2b} = \sqrt{5a \cdot 2b} = \sqrt{10ab}$

3. $\sqrt[3]{4}\sqrt[3]{5} = \sqrt[3]{4 \cdot 5} = \sqrt[3]{20}$

4. $\sqrt[4]{\dfrac{y}{5}}\sqrt[4]{\dfrac{7}{x}} = \sqrt[4]{\dfrac{y}{5} \cdot \dfrac{7}{x}} = \sqrt[4]{\dfrac{7y}{5x}}$

A common error is to omit the index in the answer.

Do Exercises 1–4.

Keep in mind that the product rule can be used only when the indexes are the same. When indexes differ, we can use rational exponents as we did in Section 6.2.

Example 5   Multiply: $\sqrt{5x} \cdot \sqrt[4]{3y}$.

$$\sqrt{5x} \cdot \sqrt[4]{3y} = (5x)^{1/2}(3y)^{1/4} \qquad \text{Converting to exponential notation}$$

$$= (5x)^{2/4}(3y)^{1/4} \qquad \begin{array}{l}\text{Rewriting so that exponents}\\ \text{have a common denominator}\end{array}$$

$$= [(5x)^2(3y)]^{1/4} \qquad \text{Using } a^n b^n = (ab)^n$$

$$= \sqrt[4]{(25x^2)(3y)} \qquad \begin{array}{l}\text{Squaring } 5x \text{ and converting back}\\ \text{to radical notation}\end{array}$$

$$= \sqrt[4]{75x^2 y} \qquad \text{Multiplying under the radical}$$

Do Exercises 5 and 6.

We can reverse the product rule to simplify a product. We simplify the root of a product by taking the root of each factor separately.

> For any nonnegative real numbers a and b and any index k,
> $$\sqrt[k]{ab} = \sqrt[k]{a} \cdot \sqrt[k]{b}.$$
> (Take the kth root of each factor separately.)

Multiply.

1. $\sqrt{19}\sqrt{7}$

2. $\sqrt{3p}\sqrt{7q}$

3. $\sqrt[4]{403}\sqrt[4]{7}$

4. $\sqrt[3]{\dfrac{5}{p}} \cdot \sqrt[3]{\dfrac{2}{q}}$

Multiply.

5. $\sqrt{5}\sqrt[3]{2}$

6. $\sqrt{x}\sqrt[3]{5y}$

Answers on page A-36

Simplify by factoring.

7. $\sqrt{32}$

This process shows a way to factor and thus simplify radical expressions. Consider $\sqrt{20}$. The number 20 has the factor 4, which is a perfect square. Therefore,

$$\sqrt{20} = \sqrt{4 \cdot 5} \qquad \text{Factoring the radicand (4 is a perfect square)}$$
$$= \sqrt{4} \cdot \sqrt{5} \qquad \text{Factoring into two radicals}$$
$$= 2\sqrt{5}. \qquad \text{Taking the square root of 4}$$

> To simplify a radical expression by factoring:
>
> 1. Look for the largest factors of the radicand that are perfect kth powers (where k is the index).
> 2. Then take the kth root of the resulting factors.
> 3. A radical expression, with index k, is *simplified* when its radicand has no factors that are perfect kth powers.

Examples Simplify by factoring.

6. $\sqrt{50} = \sqrt{25 \cdot 2} = \sqrt{25} \cdot \sqrt{2} = 5\sqrt{2}$

This factor is a perfect square.

7. $\sqrt[3]{32} = \sqrt[3]{8 \cdot 4} = \sqrt[3]{8} \cdot \sqrt[3]{4} = 2\sqrt[3]{4}$

This factor is a perfect cube (third power).

8. $\sqrt[4]{48} = \sqrt[4]{16 \cdot 3} = \sqrt[4]{16} \cdot \sqrt[4]{3} = 2\sqrt[4]{3}$

Do Exercises 7 and 8.

8. $\sqrt[3]{80}$

In many situations, expressions under radicals never represent negative numbers. In such cases, absolute-value notation is not necessary. For this reason, we will henceforth assume that *all expressions under radicals are nonnegative.*

Examples Simplify by factoring. Assume that all expressions under radicals represent nonnegative numbers.

9. $\sqrt{5x^2} = \sqrt{x^2 \cdot 5} \qquad \text{Factoring the radicand}$
$\phantom{9. \sqrt{5x^2}} = \sqrt{x^2} \cdot \sqrt{5} \qquad \text{Factoring into two radicals}$
$\phantom{9. \sqrt{5x^2}} = x \cdot \sqrt{5} \qquad \text{Taking the square root of } x^2$

10. $\sqrt{18x^2y} = \sqrt{9x^2 \cdot 2y}$
$\phantom{10. \sqrt{18x^2y}} = \sqrt{9x^2} \cdot \sqrt{2y}$
$\phantom{10. \sqrt{18x^2y}} = 3x\sqrt{2y}$

Absolute-value notation is not needed because expressions under radicals are not negative.

11. $\sqrt{216x^5y^3} = \sqrt{36 \cdot 6 \cdot x^4 \cdot x \cdot y^2 \cdot y} \qquad \text{Factoring the radicand}$
$\phantom{11. \sqrt{216x^5y^3}} = \sqrt{36 \cdot x^4 \cdot y^2 \cdot 6 \cdot x \cdot y}$
$\phantom{11. \sqrt{216x^5y^3}} = \sqrt{36}\sqrt{x^4}\sqrt{y^2}\sqrt{6xy} \qquad \text{Factoring into several radicals}$
$\phantom{11. \sqrt{216x^5y^3}} = 6x^2y\sqrt{6xy} \qquad \text{Taking square roots}$

Answers on page A-36

Note: Had we not seen in Example 11 that $216 = 36 \cdot 6$, where 36 is the largest square factor of 216, we could have found the prime factorization

$$2 \cdot 2 \cdot 2 \cdot 3 \cdot 3 \cdot 3.$$

Each pair of factors makes a square, so

$$\sqrt{2 \cdot 2 \cdot 2 \cdot 3 \cdot 3 \cdot 3} = \sqrt{2^2 \cdot 3^2 \cdot 2 \cdot 3} = 2 \cdot 3\sqrt{2 \cdot 3} = 6\sqrt{6}.$$

12. $\sqrt[3]{16a^7b^{11}} = \sqrt[3]{8 \cdot 2 \cdot a^6 \cdot a \cdot b^9 \cdot b^2}$ Factoring the radicand. The index is 3, so we look for the largest powers that are multiples of 3 because these are perfect cubes.

$$= \sqrt[3]{8} \cdot \sqrt[3]{a^6} \cdot \sqrt[3]{b^9} \cdot \sqrt[3]{2ab^2} \qquad \text{Factoring into radicals}$$

$$= 2a^2b^3\sqrt[3]{2ab^2} \qquad \text{Taking cube roots}$$

Do Exercises 9–14.

Sometimes after we have multiplied, we can then simplify by factoring.

Examples   Multiply and simplify. Assume that all expressions under radicals represent nonnegative numbers.

13. $\sqrt{20}\,\sqrt{8} = \sqrt{20 \cdot 8} = \sqrt{4 \cdot 5 \cdot 4 \cdot 2} = 4\sqrt{10}$

14. $3\sqrt[3]{25} \cdot 2\sqrt[3]{5} = 6 \cdot \sqrt[3]{25 \cdot 5} = 6 \cdot \sqrt[3]{125}$

$$= 6 \cdot 5 = 30$$

15. $\sqrt[3]{18y^3}\,\sqrt[3]{4x^2} = \sqrt[3]{18y^3 \cdot 4x^2}$ Multiplying radicands

$$= \sqrt[3]{72y^3x^2}$$

$$= \sqrt[3]{8y^3 \cdot 9x^2} \qquad \text{Factoring the radicand}$$

$$= \sqrt[3]{8y^3}\,\sqrt[3]{9x^2} \qquad \text{Factoring into two radicals}$$

$$= 2y\sqrt[3]{9x^2} \qquad \text{Taking the cube root}$$

Do Exercises 15–18.

b   Dividing and Simplifying Radical Expressions

Note that $\dfrac{\sqrt[3]{27}}{\sqrt[3]{8}} = \dfrac{3}{2}$ and that $\sqrt[3]{\dfrac{27}{8}} = \dfrac{3}{2}$. This example suggests the following.

> **THE QUOTIENT RULE FOR RADICALS**
>
> For any nonnegative number a, any positive number b, and any index k,
>
> $$\frac{\sqrt[k]{a}}{\sqrt[k]{b}} = \sqrt[k]{\frac{a}{b}}.$$
>
> (To divide, divide the radicands. After doing this, you can sometimes simplify by taking roots.)

Examples   Divide and simplify. Assume that all expressions under radicals represent positive numbers.

16. $\dfrac{\sqrt{80}}{\sqrt{5}} = \sqrt{\dfrac{80}{5}} = \sqrt{16} = 4$ | We divide the radicands. |

Simplify by factoring. Assume that all expressions under radicals represent nonnegative numbers.

9. $\sqrt{300}$

10. $\sqrt{36y^2}$

11. $\sqrt{12a^2b}$

12. $\sqrt{12ab^3c^2}$

13. $\sqrt[3]{16}$

14. $\sqrt[3]{81x^4y^8}$

Multiply and simplify. Assume that all expressions under radicals represent nonnegative numbers.

15. $\sqrt{3}\,\sqrt{6}$

16. $\sqrt{18y}\,\sqrt{14y}$

17. $\sqrt[3]{3x^2y}\,\sqrt[3]{36x}$

18. $\sqrt{7a}\,\sqrt{21b}$

Answers on page A-36

Divide and simplify. Assume that all expressions under radicals represent positive numbers.

19. $\dfrac{\sqrt{75}}{\sqrt{3}}$

20. $\dfrac{14\sqrt{128xy}}{2\sqrt{2}}$

21. $\dfrac{\sqrt{50a^3}}{\sqrt{2a}}$

22. $\dfrac{4\sqrt[3]{250}}{7\sqrt[3]{2}}$

Simplify by taking the roots of the numerator and the denominator. Assume that all expressions under radicals represent positive numbers.

23. $\sqrt{\dfrac{25}{36}}$

24. $\sqrt{\dfrac{x^2}{100}}$

25. $\sqrt[3]{\dfrac{54x^5}{125}}$

26. Divide and simplify:

$$\dfrac{\sqrt[4]{x^3y^2}}{\sqrt[3]{x^2y}}.$$

Answers on page A-36

17. $\dfrac{5\sqrt[3]{32}}{\sqrt[3]{2}} = 5\sqrt[3]{\dfrac{32}{2}} = 5\sqrt[3]{16} = 5\sqrt[3]{8 \cdot 2} = 5\sqrt[3]{8}\sqrt[3]{2} = 5 \cdot 2\sqrt[3]{2} = 10\sqrt[3]{2}$

18. $\dfrac{\sqrt{72xy}}{2\sqrt{2}} = \dfrac{1}{2}\dfrac{\sqrt{72xy}}{\sqrt{2}} = \dfrac{1}{2}\sqrt{\dfrac{72xy}{2}} = \dfrac{1}{2}\sqrt{36xy} = \dfrac{1}{2}\sqrt{36}\sqrt{xy}$

$= \dfrac{1}{2} \cdot 6\sqrt{xy} = 3\sqrt{xy}$

Do Exercises 19–22.

We can reverse the quotient rule to simplify a quotient. We simplify the root of a quotient by taking the roots of the numerator and of the denominator separately.

> For any nonnegative number a, any positive number b, and any index k,
>
> $$\sqrt[k]{\dfrac{a}{b}} = \dfrac{\sqrt[k]{a}}{\sqrt[k]{b}}.$$
>
> (Take the kth roots of the numerator and of the denominator separately.)

Examples   Simplify by taking the roots of the numerator and the denominator. Assume that all expressions under radicals represent positive numbers.

19. $\sqrt[3]{\dfrac{27}{125}} = \dfrac{\sqrt[3]{27}}{\sqrt[3]{125}} = \dfrac{3}{5}$ We take the cube root of the numerator and of the denominator.

20. $\sqrt{\dfrac{25}{y^2}} = \dfrac{\sqrt{25}}{\sqrt{y^2}} = \dfrac{5}{y}$ We take the square root of the numerator and of the denominator.

21. $\sqrt{\dfrac{16x^3}{y^4}} = \dfrac{\sqrt{16x^3}}{\sqrt{y^4}} = \dfrac{\sqrt{16x^2 \cdot x}}{\sqrt{y^4}} = \dfrac{\sqrt{16x^2} \cdot \sqrt{x}}{\sqrt{y^4}} = \dfrac{4x\sqrt{x}}{y^2}$

22. $\sqrt[3]{\dfrac{27y^5}{343x^3}} = \dfrac{\sqrt[3]{27y^5}}{\sqrt[3]{343x^3}} = \dfrac{\sqrt[3]{27y^3 \cdot y^2}}{\sqrt[3]{343x^3}} = \dfrac{\sqrt[3]{27y^3} \cdot \sqrt[3]{y^2}}{\sqrt[3]{343x^3}} = \dfrac{3y\sqrt[3]{y^2}}{7x}$

We are assuming that no expression represents 0 or a negative number. Thus we need not be concerned about zero denominators.

Do Exercises 23–25.

When indexes differ, we can use rational exponents.

Example 23   Divide and simplify: $\dfrac{\sqrt[3]{a^2b^4}}{\sqrt{ab}}$.

$$\dfrac{\sqrt[3]{a^2b^4}}{\sqrt{ab}} = \dfrac{(a^2b^4)^{1/3}}{(ab)^{1/2}}$$ **Converting to exponential notation**

$$= \dfrac{a^{2/3}b^{4/3}}{a^{1/2}b^{1/2}}$$ **Using the product and power rules**

$$= a^{2/3-1/2}b^{4/3-1/2}$$ **Subtracting exponents**

$$= a^{4/6-3/6}b^{8/6-3/6} = a^{1/6}b^{5/6}$$

$$= (ab^5)^{1/6} = \sqrt[6]{ab^5}$$

Do Exercise 26.

Exercise Set 6.3

a Simplify by factoring. Assume that all expressions under radicals represent nonnegative numbers.

1. $\sqrt{24}$ **2.** $\sqrt{20}$ **3.** $\sqrt{90}$ **4.** $\sqrt{18}$

5. $\sqrt[3]{250}$ **6.** $\sqrt[3]{108}$ **7.** $\sqrt{180x^4}$ **8.** $\sqrt{175y^6}$

9. $\sqrt[3]{54x^8}$ **10.** $\sqrt[3]{40y^3}$ **11.** $\sqrt[3]{80t^8}$ **12.** $\sqrt[3]{108x^5}$

13. $\sqrt[4]{80}$ **14.** $\sqrt[4]{32}$ **15.** $\sqrt{32a^2b}$ **16.** $\sqrt{75p^3q^4}$

17. $\sqrt[4]{243x^8y^{10}}$ **18.** $\sqrt[4]{162c^4d^6}$ **19.** $\sqrt[5]{96x^7y^{15}}$ **20.** $\sqrt[5]{p^{14}q^9r^{23}}$

Multiply and simplify. Assume that all expressions under radicals represent nonnegative numbers.

21. $\sqrt{10}\,\sqrt{5}$ **22.** $\sqrt{6}\,\sqrt{3}$ **23.** $\sqrt{15}\,\sqrt{6}$ **24.** $\sqrt{2}\,\sqrt{32}$

25. $\sqrt[3]{2}\,\sqrt[3]{4}$ **26.** $\sqrt[3]{9}\,\sqrt[3]{3}$ **27.** $\sqrt{45}\,\sqrt{60}$ **28.** $\sqrt{24}\,\sqrt{75}$

29. $\sqrt{3x^3}\,\sqrt{6x^5}$ **30.** $\sqrt{5a^7}\,\sqrt{15a^3}$ **31.** $\sqrt{5b^3}\,\sqrt{10c^4}$ **32.** $\sqrt{2x^3y}\,\sqrt{12xy}$

33. $\sqrt[3]{5a^2}\,\sqrt[3]{2a}$

34. $\sqrt[3]{7x}\,\sqrt[3]{3x^2}$

35. $\sqrt[3]{y^4}\,\sqrt[3]{16y^5}$

36. $\sqrt[3]{s^2t^4}\,\sqrt[3]{s^4t^6}$

37. $\sqrt[4]{16}\,\sqrt[4]{64}$

38. $\sqrt[5]{64}\,\sqrt[5]{16}$

39. $\sqrt{12a^3b}\,\sqrt{8a^4b^2}$

40. $\sqrt{30x^3y^4}\,\sqrt{18x^2y^5}$

41. $\sqrt{2}\,\sqrt[3]{5}$

42. $\sqrt{6}\,\sqrt[3]{5}$

43. $\sqrt[4]{3}\,\sqrt{2}$

44. $\sqrt[3]{5}\,\sqrt[4]{2}$

45. $\sqrt{a}\,\sqrt[4]{a^3}$

46. $\sqrt[3]{x^2}\,\sqrt[6]{x^5}$

47. $\sqrt[5]{b^2}\,\sqrt{b^3}$

48. $\sqrt[4]{a^3}\,\sqrt[3]{a^2}$

49. $\sqrt{xy^3}\,\sqrt[3]{x^2y}$

50. $\sqrt[4]{9ab^3}\,\sqrt{3a^4b}$

---

b Divide and simplify. Assume that all expressions under radicals represent positive numbers.

51. $\dfrac{\sqrt{90}}{\sqrt{5}}$

52. $\dfrac{\sqrt{98}}{\sqrt{2}}$

53. $\dfrac{\sqrt{35q}}{\sqrt{7q}}$

54. $\dfrac{\sqrt{30x}}{\sqrt{10x}}$

55. $\dfrac{\sqrt[3]{54}}{\sqrt[3]{2}}$

56. $\dfrac{\sqrt[3]{40}}{\sqrt[3]{5}}$

57. $\dfrac{\sqrt{56xy^3}}{\sqrt{8x}}$

58. $\dfrac{\sqrt{52ab^3}}{\sqrt{13a}}$

59. $\dfrac{\sqrt[3]{96a^4b^2}}{\sqrt[3]{12a^2b}}$

60. $\dfrac{\sqrt[3]{189x^5y^7}}{\sqrt[3]{7x^2y^2}}$

61. $\dfrac{\sqrt{128xy}}{2\sqrt{2}}$

62. $\dfrac{\sqrt{48ab}}{2\sqrt{3}}$

63. $\dfrac{\sqrt[4]{48x^9y^{13}}}{\sqrt[4]{3xy^5}}$

64. $\dfrac{\sqrt[5]{64a^{11}b^{28}}}{\sqrt[5]{2ab^2}}$

65. $\dfrac{\sqrt[3]{a}}{\sqrt{a}}$

66. $\dfrac{\sqrt{x}}{\sqrt[4]{x}}$

67. $\dfrac{\sqrt[3]{a^2}}{\sqrt[4]{a}}$

68. $\dfrac{\sqrt[3]{x^2}}{\sqrt[5]{x}}$

69. $\dfrac{\sqrt[4]{x^2y^3}}{\sqrt[3]{xy}}$

70. $\dfrac{\sqrt[5]{a^4b^2}}{\sqrt[3]{ab^2}}$

Simplify.

71. $\sqrt{\dfrac{25}{36}}$

72. $\sqrt{\dfrac{49}{64}}$

73. $\sqrt{\dfrac{16}{49}}$

74. $\sqrt{\dfrac{100}{81}}$

75. $\sqrt[3]{\dfrac{125}{27}}$

76. $\sqrt[3]{\dfrac{343}{1000}}$

77. $\sqrt{\dfrac{49}{y^2}}$

78. $\sqrt{\dfrac{121}{x^2}}$

79. $\sqrt{\dfrac{25y^3}{x^4}}$

80. $\sqrt{\dfrac{36a^5}{b^6}}$

81. $\sqrt[3]{\dfrac{27a^4}{8b^3}}$

82. $\sqrt[3]{\dfrac{64x^7}{216y^6}}$

83. $\sqrt[4]{\dfrac{81x^4}{16}}$

84. $\sqrt[4]{\dfrac{81x^4}{y^8z^4}}$

85. $\sqrt[5]{\dfrac{32x^8}{y^{10}}}$

86. $\sqrt[5]{\dfrac{32b^{10}}{243a^{20}}}$

87. $\sqrt[6]{\dfrac{x^{13}}{y^6z^{12}}}$

88. $\sqrt[6]{\dfrac{p^9q^{24}}{r^{18}}}$

Skill Maintenance

Solve. [4.7b]

89. The sum of a number and its square is 90. Find the number.

90. The base of a triangle is 2 in. longer than the height. The area is 12 in^2. Find the height and the base.

Solve. [5.5a]

91. $\dfrac{12x}{x-4} - \dfrac{3x^2}{x+4} = \dfrac{384}{x^2-16}$

92. $\dfrac{2}{3} + \dfrac{1}{t} = \dfrac{4}{5}$

93. $\dfrac{18}{x^2-3x} = \dfrac{2x}{x-3} - \dfrac{6}{x}$

94. $\dfrac{4x}{x+5} + \dfrac{20}{x} = \dfrac{100}{x^2+5x}$

Synthesis

95. ◆ Is the quotient of two irrational numbers always an irrational number? Why or why not?

96. ◆ 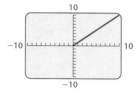 Ron is puzzled. When he uses a grapher to graph $y = \sqrt{x} \cdot \sqrt{x}$, he gets the following screen. Explain why Ron did not get the complete line $y_1 = x$.

97. *Pendulums.* The *period* of a pendulum is the time it takes to complete one cycle, swinging to and fro. For a pendulum that is L centimeters long, the period T is given by the function

$$T(L) = 2\pi\sqrt{\dfrac{L}{980}},$$

where T is in seconds. Find, to the nearest hundredth of a second, the period of a pendulum of length **(a)** 65 cm; **(b)** 98 cm; **(c)** 120 cm. Use a calculator's $\boxed{\pi}$ key if possible.

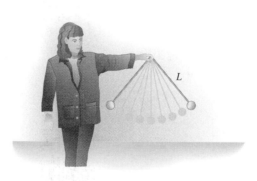

Simplify.

98. $\dfrac{\sqrt[3]{x^3-y^3}}{\sqrt[3]{x-y}}$

99. $\dfrac{\sqrt{44x^2y^9z}\,\sqrt{22y^9z^6}}{(\sqrt{11xy^{\prime 8}z^2})^2}$

100. Use a grapher to check your answers to Exercises 7, 12, 30, and 54.

6.4 Addition, Subtraction, and More Multiplication

a Addition and Subtraction

Any two real numbers can be added. For example, the sum of 7 and $\sqrt{3}$ can be expressed as $7 + \sqrt{3}$. We cannot simplify this sum. However, when we have **like radicals** (radicals having the same index and radicand), we can use the distributive laws to simplify by collecting like radical terms. For example,

$$7\sqrt{3} + \sqrt{3} = 7\sqrt{3} + 1 \cdot \sqrt{3} = (7 + 1)\sqrt{3} = 8\sqrt{3}.$$

Examples Add or subtract. Simplify by collecting like radical terms, if possible.

1. $6\sqrt{7} + 4\sqrt{7} = (6 + 4)\sqrt{7}$ **Using a distributive law (factoring out $\sqrt{7}$)**

$\qquad\qquad\quad = 10\sqrt{7}$

2. $8\sqrt[3]{2} - 7x\sqrt[3]{2} + 5\sqrt[3]{2} = (8 - 7x + 5)\sqrt[3]{2}$ **Factoring out $\sqrt[3]{2}$**

$\qquad\qquad\qquad\qquad\quad = (13 - 7x)\sqrt[3]{2}$

> These parentheses *are* necessary!

3. $6\sqrt[5]{4x} + 4\sqrt[5]{4x} - \sqrt[3]{4x} = (6 + 4)\sqrt[5]{4x} - \sqrt[3]{4x}$

$\qquad\qquad\qquad\qquad\quad = 10\sqrt[5]{4x} - \sqrt[3]{4x}$

> Note that these expressions have the *same* radicand, but they are *not* like radicals because they do not have the same index.

Do Exercises 1 and 2.

Sometimes we need to simplify radicals by factoring in order to obtain terms with like radicals.

Examples Add or subtract. Simplify by collecting like radical terms, if possible.

4. $3\sqrt{8} - 5\sqrt{2} = 3(\sqrt{4 \cdot 2}) - 5\sqrt{2}$ **Factoring 8**

$\qquad\qquad\quad = 3\sqrt{4} \cdot \sqrt{2} - 5\sqrt{2}$ **Factoring $\sqrt{4 \cdot 2}$ into two radicals**

$\qquad\qquad\quad = 3 \cdot 2\sqrt{2} - 5\sqrt{2}$ **Taking the square root of 4**

$\qquad\qquad\quad = 6\sqrt{2} - 5\sqrt{2}$

$\qquad\qquad\quad = (6 - 5)\sqrt{2}$ **Collecting like radical terms**

$\qquad\qquad\quad = \sqrt{2}$

5. $5\sqrt{2} - 4\sqrt{3}$ No simplification possible

6. $5\sqrt[3]{16y^4} + 7\sqrt[3]{2y} = 5\sqrt[3]{8y^3 \cdot 2y} + 7\sqrt[3]{2y}$ ⎫

$\qquad\qquad\qquad\quad = 5\sqrt[3]{8y^3} \cdot \sqrt[3]{2y} + 7\sqrt[3]{2y}$ ⎬ **Factoring the first radical**

$\qquad\qquad\qquad\quad = 5 \cdot 2y \cdot \sqrt[3]{2y} + 7\sqrt[3]{2y}$ **Taking the cube root of $8y^3$**

$\qquad\qquad\qquad\quad = 10y\sqrt[3]{2y} + 7\sqrt[3]{2y}$

$\qquad\qquad\qquad\quad = (10y + 7)\sqrt[3]{2y}$ **Collecting like radical terms**

Do Exercises 3–5.

Add or subtract. Simplify by collecting like radical terms, if possible.

1. $5\sqrt{2} + 8\sqrt{2}$

2. $7\sqrt[4]{5x} + 3\sqrt[4]{5x} - \sqrt{7}$

Add or subtract. Simplify by collecting like radical terms, if possible.

3. $7\sqrt{45} - 2\sqrt{5}$

4. $3\sqrt[3]{y^5} + 4\sqrt[3]{y^2} + \sqrt[3]{8y^6}$

5. $\sqrt{25x - 25} - \sqrt{9x - 9}$

Answers on page A-36

Multiply. Assume that all expressions under radicals represent nonnegative numbers.

6. $\sqrt{2}\,(5\sqrt{3} + 3\sqrt{7})$

7. $\sqrt[3]{a^2}\,(\sqrt[3]{3a} - \sqrt[3]{2})$

Multiply. Assume that all expressions under radicals represent nonnegative numbers.

8. $(\sqrt{3} - 5\sqrt{2})(2\sqrt{3} + \sqrt{2})$

9. $(\sqrt{a} + 2\sqrt{3})(3\sqrt{b} - 4\sqrt{3})$

Multiply. Assume that all expressions under radicals represent nonnegative numbers.

10. $(\sqrt{2} + \sqrt{5})(\sqrt{2} - \sqrt{5})$

11. $(\sqrt{p} - \sqrt{q})(\sqrt{p} + \sqrt{q})$

Multiply.

12. $(2\sqrt{5} - y)^2$

13. $(3\sqrt{6} + 2)^2$

Answers on page A-36

---

b More Multiplication

To multiply expressions in which some factors contain more than one term, we use the procedures for multiplying polynomials.

Examples Multiply.

7. $\sqrt{3}\,(x - \sqrt{5}) = \sqrt{3} \cdot x - \sqrt{3} \cdot \sqrt{5}$ Using a distributive law
$$= x\sqrt{3} - \sqrt{15}$$ Multiplying radicals

8. $\sqrt[3]{y}\,(\sqrt[3]{y^2} + \sqrt[3]{2}) = \sqrt[3]{y} \cdot \sqrt[3]{y^2} + \sqrt[3]{y} \cdot \sqrt[3]{2}$ Using a distributive law
$$= \sqrt[3]{y^3} + \sqrt[3]{2y}$$ Multiplying radicals
$$= y + \sqrt[3]{2y}$$ Simplifying $\sqrt[3]{y^3}$

Do Exercises 6 and 7.

Example 9 Multiply: $(4\sqrt{3} + \sqrt{2})(\sqrt{3} - 5\sqrt{2})$.

$$
\begin{aligned}
(4\sqrt{3} + \sqrt{2})(\sqrt{3} - 5\sqrt{2}) &= \overset{\text{F}}{4(\sqrt{3})^2} - \overset{\text{O}}{20\sqrt{3} \cdot \sqrt{2}} + \overset{\text{I}}{\sqrt{2} \cdot \sqrt{3}} - \overset{\text{L}}{5(\sqrt{2})^2} \\
&= 4 \cdot 3 - 20\sqrt{6} + \sqrt{6} - 5 \cdot 2 \\
&= 12 - 20\sqrt{6} + \sqrt{6} - 10 \\
&= 2 - 19\sqrt{6} \quad \text{Collecting like terms}
\end{aligned}
$$

Example 10 Multiply: $(\sqrt{a} + \sqrt{3})(\sqrt{b} + \sqrt{3})$. Assume that all expressions under radicals represent nonnegative numbers.

$$
\begin{aligned}
(\sqrt{a} + \sqrt{3})(\sqrt{b} + \sqrt{3}) &= \sqrt{a}\sqrt{b} + \sqrt{a}\sqrt{3} + \sqrt{3}\sqrt{b} + \sqrt{3}\sqrt{3} \\
&= \sqrt{ab} + \sqrt{3a} + \sqrt{3b} + 3
\end{aligned}
$$

Do Exercises 8 and 9.

Example 11 Multiply: $(\sqrt{5} + \sqrt{7})(\sqrt{5} - \sqrt{7})$.

$$(\sqrt{5} + \sqrt{7})(\sqrt{5} - \sqrt{7}) = (\sqrt{5})^2 - (\sqrt{7})^2$$ This is now a difference of two squares.
$$= 5 - 7 = -2$$

Example 12 Multiply: $(\sqrt{a} + \sqrt{b})(\sqrt{a} - \sqrt{b})$. Assume that all expressions under radicals represent nonnegative numbers.

$$(\sqrt{a} + \sqrt{b})(\sqrt{a} - \sqrt{b}) = (\sqrt{a})^2 - (\sqrt{b})^2$$
$$= a - b \longleftarrow \boxed{\text{No radicals}}$$

Expressions of the form $\sqrt{a} + \sqrt{b}$ and $\sqrt{a} - \sqrt{b}$ are called **conjugates**. Their product is always an expression that has no radicals.

Do Exercises 10 and 11.

Example 13 Multiply: $(\sqrt{3} + x)^2$.

$$(\sqrt{3} + x)^2 = (\sqrt{3})^2 + 2x\sqrt{3} + x^2$$ Squaring a binomial
$$= 3 + 2x\sqrt{3} + x^2$$

Do Exercises 12 and 13.

Exercise Set 6.4

a Add or subtract. Then simplify by collecting like radical terms, if possible. Assume that all expressions under radicals represent nonnegative numbers.

1. $7\sqrt{5} + 4\sqrt{5}$

2. $2\sqrt{3} + 9\sqrt{3}$

3. $6\sqrt[3]{7} - 5\sqrt[3]{7}$

4. $13\sqrt[5]{3} - 8\sqrt[5]{3}$

5. $4\sqrt[3]{y} + 9\sqrt[3]{y}$

6. $6\sqrt[4]{t} - 3\sqrt[4]{t}$

7. $5\sqrt{6} - 9\sqrt{6} - 4\sqrt{6}$

8. $3\sqrt{10} - 8\sqrt{10} + 7\sqrt{10}$

9. $4\sqrt[3]{3} - \sqrt{5} + 2\sqrt[3]{3} + \sqrt{5}$

10. $5\sqrt{7} - 8\sqrt[4]{11} + \sqrt{7} + 9\sqrt[4]{11}$

11. $8\sqrt{27} - 3\sqrt{3}$

12. $9\sqrt{50} - 4\sqrt{2}$

13. $8\sqrt{45} + 7\sqrt{20}$

14. $9\sqrt{12} + 16\sqrt{27}$

15. $18\sqrt{72} + 2\sqrt{98}$

16. $12\sqrt{45} - 8\sqrt{80}$

17. $3\sqrt[3]{16} + \sqrt[3]{54}$

18. $\sqrt[3]{27} - 5\sqrt[3]{8}$

19. $2\sqrt{128} - \sqrt{18} + 4\sqrt{32}$

20. $5\sqrt{50} - 2\sqrt{18} + 9\sqrt{32}$

21. $\sqrt{5a} + 2\sqrt{45a^3}$

22. $4\sqrt{3x^3} - \sqrt{12x}$

23. $\sqrt[3]{24x} - \sqrt[3]{3x^4}$

24. $\sqrt[3]{54x} - \sqrt[3]{2x^4}$

25. $5\sqrt[3]{32} - \sqrt[3]{108} + 2\sqrt[3]{256}$

26. $3\sqrt[3]{8x} - 4\sqrt[3]{27x} + 2\sqrt[3]{64x}$

b Multiply.

27. $\sqrt{5}(4 - 2\sqrt{5})$

28. $\sqrt{6}(2 + \sqrt{6})$

29. $\sqrt{3}(\sqrt{2} - \sqrt{7})$

30. $\sqrt{2}(\sqrt{5} - \sqrt{2})$

31. $\sqrt{3}(2\sqrt{5} - 3\sqrt{4})$

32. $\sqrt{2}(3\sqrt{10} - 2\sqrt{2})$

33. $\sqrt[3]{2}(\sqrt[3]{4} - 2\sqrt[3]{32})$

34. $\sqrt[3]{3}(\sqrt[3]{9} - 4\sqrt[3]{21})$

35. $\sqrt[3]{a}(\sqrt[3]{2a^2} + \sqrt[3]{16a^2})$

36. $\sqrt[3]{x}(\sqrt[3]{3x^2} - \sqrt[3]{81x^2})$

37. $(\sqrt{3} - \sqrt{2})(\sqrt{3} + \sqrt{2})$

38. $(\sqrt{5} + \sqrt{6})(\sqrt{5} - \sqrt{6})$

39. $(\sqrt{8} + 2\sqrt{5})(\sqrt{8} - 2\sqrt{5})$

40. $(\sqrt{18} + 3\sqrt{7})(\sqrt{18} - 3\sqrt{7})$

41. $(7 + \sqrt{5})(7 - \sqrt{5})$

42. $(4 - \sqrt{3})(4 + \sqrt{3})$

43. $(2 - \sqrt{3})(2 + \sqrt{3})$

44. $(11 - \sqrt{2})(11 + \sqrt{2})$

45. $(\sqrt{8} + \sqrt{5})(\sqrt{8} - \sqrt{5})$

46. $(\sqrt{6} - \sqrt{7})(\sqrt{6} + \sqrt{7})$

47. $(3 + 2\sqrt{7})(3 - 2\sqrt{7})$

48. $(6 - 3\sqrt{2})(6 + 3\sqrt{2})$

For the following exercises, assume that all expressions under radicals represent nonnegative numbers.

49. $(\sqrt{a} + \sqrt{b})(\sqrt{a} - \sqrt{b})$

50. $(\sqrt{x} - \sqrt{y})(\sqrt{x} + \sqrt{y})$

51. $(3 - \sqrt{5})(2 + \sqrt{5})$

52. $(2 + \sqrt{6})(4 - \sqrt{6})$

53. $(\sqrt{3} + 1)(2\sqrt{3} + 1)$

54. $(4\sqrt{3} + 5)(\sqrt{3} - 2)$

55. $(2\sqrt{7} - 4\sqrt{2})(3\sqrt{7} + 6\sqrt{2})$

56. $(4\sqrt{5} + 3\sqrt{3})(3\sqrt{5} - 4\sqrt{3})$

57. $(\sqrt{a} + \sqrt{2})(\sqrt{a} + \sqrt{3})$

58. $(2 - \sqrt{x})(1 - \sqrt{x})$

59. $(2\sqrt[3]{3} + \sqrt[3]{2})(\sqrt[3]{3} - 2\sqrt[3]{2})$

60. $(3\sqrt[4]{7} + \sqrt[4]{6})(2\sqrt[4]{9} - 3\sqrt[4]{6})$

61. $(2 + \sqrt{3})^2$

62. $(\sqrt{5} + 1)^2$

63. $(\sqrt[5]{9} - \sqrt[5]{3})(\sqrt[5]{8} + \sqrt[5]{27})$

64. $(\sqrt[3]{8x} - \sqrt[3]{5y})^2$

Skill Maintenance

Multiply or divide and simplify.   [5.1d, e]

65. $\dfrac{x^3 + 4x}{x^2 - 16} \div \dfrac{x^2 + 8x + 15}{x^2 + x - 20}$

66. $\dfrac{a^2 - 4}{a} \div \dfrac{a - 2}{a + 4}$

67. $\dfrac{a^3 + 8}{a^2 - 4} \cdot \dfrac{a^2 - 4a + 4}{a^2 - 2a + 4}$

68. $\dfrac{y^3 - 27}{y^2 - 9} \cdot \dfrac{y^2 - 6y + 9}{y^2 + 3y + 9}$

Simplify.   [5.4a]

69. $\dfrac{x - \dfrac{1}{3}}{x + \dfrac{1}{4}}$

70. $\dfrac{1 - \dfrac{1}{x}}{1 - \dfrac{1}{x^2}}$

71. $\dfrac{\dfrac{1}{p} - \dfrac{1}{q}}{\dfrac{1}{p^2} - \dfrac{1}{q^2}}$

Synthesis

72. ◈ Why do we need to know how to simplify radical expressions before we learn to add them?

73. ◈ In what way(s) is collecting like radical terms the same as collecting like monomial terms?

74. 〰 Graph the function $f(x) = \sqrt{(x - 2)^2}$. What is the domain?

75. 〰 Use a grapher to check your answers to Exercises 5, 22, and 58.

Multiply and simplify.

76. $\sqrt[4]{9 + 3\sqrt{5}} \cdot \sqrt[4]{9 - 3\sqrt{5}}$

77. $(\sqrt{x + 2} - \sqrt{x - 2})^2$

78. $(\sqrt{3} + \sqrt{5} - \sqrt{6})^2$

79. $\sqrt[3]{y}(1 - \sqrt[3]{y})(1 + \sqrt[3]{y})$

80. $(\sqrt[3]{9} - 2)(\sqrt[3]{9} + 4)$

81. $[\sqrt{3} + \sqrt{2 + \sqrt{1}}]^4$

6.5 More on Division of Radical Expressions

a Rationalizing Denominators

Sometimes in mathematics it is useful to find an equivalent expression without a radical in the denominator. This provides a standard notation for expressing results. The procedure for finding such an expression is called **rationalizing the denominator.** We carry this out by multiplying by 1.

Example 1 Rationalize the denominator: $\sqrt{\dfrac{7}{3}}$.

We multiply by 1, using $\sqrt{3}/\sqrt{3}$. We do this so that the denominator of the radicand will be a perfect square.

$$\sqrt{\frac{7}{3}} = \frac{\sqrt{7}}{\sqrt{3}} \cdot \frac{\sqrt{3}}{\sqrt{3}} = \frac{\sqrt{7} \cdot \sqrt{3}}{\sqrt{3} \cdot \sqrt{3}} = \frac{\sqrt{21}}{\sqrt{3^2}} = \frac{\sqrt{21}}{3}$$

└─ The radicand is a perfect square.

Do Exercise 1.

Example 2 Rationalize the denominator: $\sqrt[3]{\dfrac{7}{25}}$.

We first factor the denominator:

$$\sqrt[3]{\frac{7}{25}} = \sqrt[3]{\frac{7}{5 \cdot 5}}.$$

To get a perfect cube in the denominator, we consider the index 3 and the factors. We have 2 factors of 5, and we need 3 factors of 5. We achieve this by multiplying by 1, using $\sqrt[3]{5}/\sqrt[3]{5}$.

$$\sqrt[3]{\frac{7}{25}} = \sqrt[3]{\frac{7}{5 \cdot 5}} \cdot \frac{\sqrt[3]{5}}{\sqrt[3]{5}} \quad \text{Multiplying by } \frac{\sqrt[3]{5}}{\sqrt[3]{5}} \text{ to make the denominator}$$
$$\text{of the radicand a perfect cube}$$
$$= \frac{\sqrt[3]{7} \cdot \sqrt[3]{5}}{\sqrt[3]{5 \cdot 5} \cdot \sqrt[3]{5}}$$
$$= \frac{\sqrt[3]{35}}{\sqrt[3]{5^3}} \longleftarrow \text{The radicand is a perfect cube.}$$
$$= \frac{\sqrt[3]{35}}{5}.$$

Do Exercise 2.

Objectives

a Rationalize the denominator of a radical expression having one term in the denominator.

b Rationalize the denominator of a radical expression having two terms in the denominator.

For Extra Help

TAPE 14   TAPE 12B   MAC WIN   CD-ROM

1. Rationalize the denominator:

$$\sqrt{\frac{2}{5}}.$$

2. Rationalize the denominator:

$$\sqrt[3]{\frac{5}{4}}.$$

Answers on page A-37

3. Rationalize the denominator:

$$\sqrt{\frac{4a}{3b}}.$$

Rationalize the denominator.

4. $\dfrac{\sqrt[4]{7}}{\sqrt[4]{2}}$

5. $\sqrt[3]{\dfrac{3x^5}{2y}}$

6. Rationalize the denominator:

$$\dfrac{7x}{\sqrt[3]{4xy^5}}.$$

Answers on page A-37

Example 3   Rationalize the denominator: $\sqrt{\dfrac{2a}{5b}}$. Assume that all expressions under radicals represent positive numbers.

$$\sqrt{\frac{2a}{5b}} = \frac{\sqrt{2a}}{\sqrt{5b}} \qquad \text{Converting to a quotient of radicals}$$

$$= \frac{\sqrt{2a}}{\sqrt{5b}} \cdot \frac{\sqrt{5b}}{\sqrt{5b}} \qquad \text{Multiplying by 1}$$

$$= \frac{\sqrt{10ab}}{\sqrt{(5b)^2}} \qquad \text{The denominator is a perfect square.}$$

$$= \frac{\sqrt{10ab}}{5b}$$

Do Exercise 3.

Example 4   Rationalize the denominator: $\dfrac{\sqrt[3]{a}}{\sqrt[3]{9x}}$.

We factor the denominator:

$$\frac{\sqrt[3]{a}}{\sqrt[3]{9x}} = \frac{\sqrt[3]{a}}{\sqrt[3]{3 \cdot 3 \cdot x}}.$$

To choose the symbol for 1, we look at $3 \cdot 3 \cdot x$. To make it a cube, we need another 3 and two more x's. Thus we multiply by 1, using $\sqrt[3]{3x^2}/\sqrt[3]{3x^2}$:

$$\frac{\sqrt[3]{a}}{\sqrt[3]{9x}} = \frac{\sqrt[3]{a}}{\sqrt[3]{9x}} \cdot \frac{\sqrt[3]{3x^2}}{\sqrt[3]{3x^2}} \qquad \text{Multiplying by 1}$$

$$= \frac{\sqrt[3]{3ax^2}}{\sqrt[3]{(3x)^3}} \qquad \text{The denominator is a perfect cube.}$$

$$= \frac{\sqrt[3]{3ax^2}}{3x}.$$

Do Exercises 4 and 5.

Example 5   Rationalize the denominator: $\dfrac{3x}{\sqrt[5]{2x^2y^3}}$.

$$\frac{3x}{\sqrt[5]{2x^2y^3}} = \frac{3x}{\sqrt[5]{2 \cdot x \cdot x \cdot y \cdot y \cdot y}}$$

$$= \frac{3x}{\sqrt[5]{2x^2y^3}} \cdot \frac{\sqrt[5]{2^4x^3y^2}}{\sqrt[5]{2^4x^3y^2}}$$

$$= \frac{3x\sqrt[5]{16x^3y^2}}{\sqrt[5]{32x^5y^5}} \qquad \text{The denominator is a perfect fifth power.}$$

$$= \frac{3x\sqrt[5]{16x^3y^2}}{2xy}$$

$$= \frac{3\sqrt[5]{16x^3y^2}}{2y}$$

Do Exercise 6.

b Rationalizing When There Are Two Terms

Do Exercises 7 and 8.

Certain pairs of expressions containing square roots, such as $c - \sqrt{b}$, $c + \sqrt{b}$ and $\sqrt{a} - \sqrt{b}$, $\sqrt{a} + \sqrt{b}$, are called **conjugates**. The product of such a pair of conjugates has no radicals in it. (See Example 12 of Section 6.4). Thus when we wish to rationalize a denominator that has two terms and one or more of them involves a square-root radical, we multiply by 1 using the conjugate of the denominator to write a symbol for 1.

Examples What symbol for 1 would you use to rationalize the denominator?

	Expression	*Symbol for 1*	
6.	$\dfrac{3}{x + \sqrt{7}}$	$\dfrac{x - \sqrt{7}}{x - \sqrt{7}}$	Change the operation sign to obtain the conjugate. Use the conjugate for the numerator and denominator of the symbol for 1.
7.	$\dfrac{\sqrt{7} + 4}{3 - 2\sqrt{5}}$	$\dfrac{3 + 2\sqrt{5}}{3 + 2\sqrt{5}}$	

Do Exercises 9 and 10.

Example 8 Rationalize the denominator: $\dfrac{4 + \sqrt{2}}{\sqrt{5} - \sqrt{2}}$.

$$\frac{4 + \sqrt{2}}{\sqrt{5} - \sqrt{2}} = \frac{4 + \sqrt{2}}{\sqrt{5} - \sqrt{2}} \cdot \frac{\sqrt{5} + \sqrt{2}}{\sqrt{5} + \sqrt{2}} \quad \text{Multiplying by 1, using the conjugate of } \sqrt{5} - \sqrt{2}\text{, which is } \sqrt{5} + \sqrt{2}$$

$$= \frac{(4 + \sqrt{2})(\sqrt{5} + \sqrt{2})}{(\sqrt{5} - \sqrt{2})(\sqrt{5} + \sqrt{2})} \quad \text{Multiplying numerators and denominators}$$

$$= \frac{4\sqrt{5} + 4\sqrt{2} + \sqrt{2}\sqrt{5} + (\sqrt{2})^2}{(\sqrt{5})^2 - (\sqrt{2})^2}$$

$$= \frac{4\sqrt{5} + 4\sqrt{2} + \sqrt{10} + 2}{5 - 2}$$

$$= \frac{4\sqrt{5} + 4\sqrt{2} + \sqrt{10} + 2}{3}$$

Example 9 Rationalize the denominator: $\dfrac{4}{\sqrt{3} + x}$.

$$\frac{4}{\sqrt{3} + x} = \frac{4}{\sqrt{3} + x} \cdot \frac{\sqrt{3} - x}{\sqrt{3} - x}$$

$$= \frac{4(\sqrt{3} - x)}{(\sqrt{3} + x)(\sqrt{3} - x)}$$

$$= \frac{4\sqrt{3} - 4x}{3 - x^2}$$

Do Exercises 11 and 12.

Multiply.

7. $(c - \sqrt{b})(c + \sqrt{b})$

8. $(\sqrt{a} + \sqrt{b})(\sqrt{a} - \sqrt{b})$

What symbol for 1 would you use to rationalize the denominator?

9. $\dfrac{\sqrt{5} + 1}{\sqrt{3} - y}$

10. $\dfrac{1}{\sqrt{2} + \sqrt{3}}$

Rationalize the denominator.

11. $\dfrac{5 + \sqrt{2}}{1 - \sqrt{2}}$

12. $\dfrac{14}{3 + \sqrt{2}}$

Answers on page A-37

Improving Your Math Study Skills

Classwork: During and After Class

During Class

Asking Questions

Many students are afraid to ask questions in class. You will find that most instructors are not only willing to answer questions during class, but often encourage students to ask questions. In fact, some instructors would like more questions than are offered. Probably your question is one that other students in the class might have been afraid to ask!

Wait for an appropriate time to ask questions. Some instructors will pause to ask the class if they have questions. Use this opportunity to get clarification on any concept you do not understand.

After Class

Restudy Examples and Class Notes

As soon as possible after class, find some time to go over your notes. Read the appropriate sections from the textbook and try to correlate the text with your class notes. You may also want to restudy the examples in the textbook for added comprehension.

Often students make the mistake of doing the homework exercises without reading their notes or textbook. This is not a good idea, since you may lose the opportunity for a complete understanding of the concepts. Simply being able to work the exercises does not ensure that you know the material well enough to work problems on a test.

Videotapes

If you can find the time, visit the library, math lab, or media center to view the videotapes on the textbook. Look on the first page of each section in the textbook for the appropriate tape reference.

The videotapes provide detailed explanations of each objective and they may give you a different presentation than the one offered by your instructor. Being able to pause the tape while you take notes or work the examples and replaying the tape as many times as you need to are additional advantages to using the videos.

Also, consider studying the special tapes *Math Problem Solving in the Real World* and *Math Study Skills* prepared by the author. If these are not available in the media center, contact your instructor.

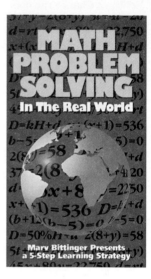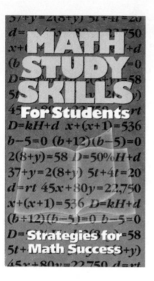

Software

If you would like additional practice on any section of the textbook, you can use the accompanying Interact Math Tutorial Software. This software can generate many different versions of the basic odd-numbered exercises for added practice. You can also ask the software to work out each problem step by step.

Ask your instructor about the availability of this software.

Exercise Set 6.5

a Rationalize the denominator. Assume that all expressions under radicals represent positive numbers.

1. $\sqrt{\dfrac{5}{3}}$

2. $\sqrt{\dfrac{8}{7}}$

3. $\sqrt{\dfrac{11}{2}}$

4. $\sqrt{\dfrac{17}{6}}$

5. $\dfrac{2\sqrt{3}}{7\sqrt{5}}$

6. $\dfrac{3\sqrt{5}}{8\sqrt{2}}$

7. $\sqrt[3]{\dfrac{16}{9}}$

8. $\sqrt[3]{\dfrac{3}{9}}$

9. $\dfrac{\sqrt[3]{3a}}{\sqrt[3]{5c}}$

10. $\dfrac{\sqrt[3]{7x}}{\sqrt[3]{3y}}$

11. $\dfrac{\sqrt[3]{2y^4}}{\sqrt[3]{6x^4}}$

12. $\dfrac{\sqrt[3]{3a^4}}{\sqrt[3]{7b^2}}$

13. $\dfrac{1}{\sqrt[4]{st}}$

14. $\dfrac{1}{\sqrt[3]{yz}}$

15. $\sqrt{\dfrac{3x}{20}}$

16. $\sqrt{\dfrac{7a}{32}}$

17. $\sqrt[3]{\dfrac{4}{5x^5y^2}}$

18. $\sqrt[3]{\dfrac{7c}{100ab^5}}$

19. $\sqrt[4]{\dfrac{1}{8x^7y^3}}$

20. $\dfrac{2x}{\sqrt[5]{18x^8y^6}}$

Rationalize the denominator. Assume that all expressions under radicals represent positive numbers.

21. $\dfrac{9}{6 - \sqrt{10}}$

22. $\dfrac{3}{8 + \sqrt{5}}$

23. $\dfrac{-4\sqrt{7}}{\sqrt{5} - \sqrt{3}}$

24. $\dfrac{34\sqrt{5}}{2\sqrt{5} - \sqrt{3}}$

25. $\dfrac{\sqrt{5} - 2\sqrt{6}}{\sqrt{3} - 4\sqrt{5}}$

26. $\dfrac{\sqrt{6} - 3\sqrt{5}}{\sqrt{3} - 2\sqrt{7}}$

27. $\dfrac{2 - \sqrt{a}}{3 + \sqrt{a}}$

28. $\dfrac{5 + \sqrt{x}}{8 - \sqrt{x}}$

29. $\dfrac{5\sqrt{3} - 3\sqrt{2}}{3\sqrt{2} - 2\sqrt{3}}$

30. $\dfrac{7\sqrt{2} + 4\sqrt{3}}{4\sqrt{3} - 3\sqrt{2}}$

31. $\dfrac{\sqrt{x} - \sqrt{y}}{\sqrt{x} + \sqrt{y}}$

32. $\dfrac{\sqrt{a} + \sqrt{b}}{\sqrt{a} - \sqrt{b}}$

Skill Maintenance

Solve. [5.5a]

33. $\dfrac{1}{2} - \dfrac{1}{3} = \dfrac{5}{t}$

34. $\dfrac{5}{x - 1} + \dfrac{9}{x^2 + x + 1} = \dfrac{15}{x^3 - 1}$

Divide and simplify. [5.1e]

35. $\dfrac{1}{x^3 - y^3} \div \dfrac{1}{(x - y)(x^2 + xy + y^2)}$

36. $\dfrac{2x^2 - x - 6}{x^2 + 4x + 3} \div \dfrac{2x^2 + x - 3}{x^2 - 1}$

Synthesis

37. ◈ A student *incorrectly* claims that

$$\frac{5 + \sqrt{2}}{\sqrt{18}} = \frac{5 + \sqrt{1}}{\sqrt{9}} = \frac{5 + 1}{3} = 2.$$

How could you convince the student that a mistake has been made? How would you explain the correct way of rationalizing the denominator?

38. ◈ A student considers the radical expression

$$\frac{11}{\sqrt[3]{4} - \sqrt[3]{5}}$$

and tries to rationalize the denominator by multiplying by

$$\frac{\sqrt[3]{4} + \sqrt[3]{5}}{\sqrt[3]{4} + \sqrt[3]{5}}.$$

Discuss the difficulties of such a plan.

39. 🖩 Use a grapher to check your answers to Exercises 15, 16, and 28.

40. Express each of the following as the product of two radical expressions.

a) $x - 5$ b) $x - a$

Simplify. (*Hint*: Rationalize the denominator.)

41. $\sqrt{a^2 - 3} - \dfrac{a^2}{\sqrt{a^2 - 3}}$

42. $\dfrac{1}{4 + \sqrt{3}} + \dfrac{1}{\sqrt{3}} + \dfrac{1}{\sqrt{3} - 4}$

6.6 Solving Radical Equations

a | The Principle of Powers

A **radical equation** has variables in one or more radicands—for example,

$$\sqrt[3]{2x} + 1 = 5, \qquad \sqrt{x} + \sqrt{4x - 2} = 7.$$

To solve such an equation, we need a new principle. Suppose that an equation $a = b$ is true. If we square both sides, we get another true equation: $a^2 = b^2$. This can be generalized.

> **THE PRINCIPLE OF POWERS**
>
> For any natural number n, if an equation $a = b$ is true, then $a^n = b^n$ is true.

However, if an equation $a^n = b^n$ is true, it *may not* be true that $a = b$. For example, $3^2 = (-3)^2$ is true, but $3 = -3$ is not true. Thus we must make a check when we solve an equation using the principle of powers.

Example 1 Solve: $\sqrt{x} - 3 = 4$.

$$\sqrt{x} - 3 = 4$$
$$\sqrt{x} = 7 \qquad \text{Adding to isolate the radical}$$
$$(\sqrt{x})^2 = 7^2 \qquad \text{Using the principle of powers (squaring)}$$
$$x = 49. \qquad \sqrt{x} \cdot \sqrt{x} = x$$

The number 49 is a possible solution. But we *must* make a check in order to be sure!

CHECK:
$$\frac{\sqrt{x} - 3 = 4}{\sqrt{49} - 3 \;?\; 4}$$
$$\begin{array}{c|c} 7 - 3 & \\ 4 & \text{TRUE} \end{array}$$

The solution is 49.

> The principle of powers does not always give equivalent equations. For this reason, a check is a must!

Example 2 Solve: $\sqrt{x} = -3$.

We might observe at the outset that this equation has no solution because the principal square root of a number is never negative. Let's continue as above for comparison.

$$\sqrt{x} = -3$$
$$(\sqrt{x})^2 = (-3)^2$$
$$x = 9$$

CHECK:
$$\frac{\sqrt{x} = -3}{\sqrt{9} \;?\; -3}$$
$$\begin{array}{c|c} 3 & \text{FALSE} \end{array}$$

The number 9 does *not* check. Thus the equation $\sqrt{x} = -3$ has no real-number solution. Note that the equation $x = 9$ has solution 9, but that $\sqrt{x} = -3$ has *no* solution. Thus the equations $x = 9$ and $\sqrt{x} = -3$ are *not* equivalent. That is, $\sqrt{9} \neq -3$.

Solve.

1. $\sqrt{x} - 7 = 3$

2. $\sqrt{x} = -2$

Solve.

3. $x + 2 = \sqrt{2x + 7}$

4. $x + 1 = 3\sqrt{x - 1}$

Do Exercises 1 and 2.

To solve an equation with a radical term, we first isolate the radical term on one side of the equation. Then we use the principle of powers.

Example 3   Solve: $x - 7 = 2\sqrt{x + 1}$.

The radical term is already isolated. We proceed with the principle of powers:

$$x - 7 = 2\sqrt{x + 1}$$
$$(x - 7)^2 = (2\sqrt{x + 1})^2 \qquad \text{Using the principle of powers (squaring)}$$
$$x^2 - 14x + 49 = 2^2(\sqrt{x + 1})^2 \qquad \text{Squaring the binomial on the left;} \\ \text{raising a product to a power on the right}$$
$$x^2 - 14x + 49 = 4(x + 1)$$
$$x^2 - 14x + 49 = 4x + 4$$
$$x^2 - 18x + 45 = 0$$
$$(x - 3)(x - 15) = 0 \qquad \text{Factoring}$$
$$x - 3 = 0 \quad or \quad x - 15 = 0 \qquad \text{Using the principle of zero products}$$
$$x = 3 \quad or \qquad x = 15.$$

The possible solutions are 3 and 15. We check.

For 3:

$$\frac{x - 7 = 2\sqrt{x + 1}}{3 - 7 \ ? \ 2\sqrt{3 + 1}}$$
$$\begin{array}{c|c} -4 & 2\sqrt{4} \\ & 2(2) \\ & 4 \qquad \text{FALSE} \end{array}$$

For 15:

$$\frac{x - 7 = 2\sqrt{x + 1}}{15 - 7 \ ? \ 2\sqrt{15 + 1}}$$
$$\begin{array}{c|c} 8 & 2\sqrt{16} \\ & 2(4) \\ & 8 \qquad \text{TRUE} \end{array}$$

The number 3 does *not* check, but the number 15 does check. The solution is 15.

The number 3 in Example 3 is sometimes called an *extraneous solution,* but such terminology is risky at best because the number 3 is in *no way* a solution of the original equation.

Do Exercises 3 and 4.

Example 4   Solve: $x = \sqrt{x + 7} + 5$.

$$x = \sqrt{x + 7} + 5$$
$$x - 5 = \sqrt{x + 7} \qquad \text{Subtracting 5 to isolate the radical term}$$
$$(x - 5)^2 = (\sqrt{x + 7})^2 \qquad \text{Using the principle of powers} \\ \text{(squaring both sides)}$$
$$x^2 - 10x + 25 = x + 7$$
$$x^2 - 11x + 18 = 0$$
$$(x - 9)(x - 2) = 0 \qquad \text{Factoring}$$
$$x = 9 \quad or \quad x = 2 \qquad \text{Using the principle of zero products}$$

The possible solutions are 9 and 2. Let's check.

For 9:

$$\frac{x = \sqrt{x + 7} + 5}{9 \ ? \ \sqrt{9 + 7} + 5}$$
$$\begin{array}{c|c} & 9 \qquad \text{TRUE} \end{array}$$

For 2:

$$\frac{x = \sqrt{x + 7} + 5}{2 \ ? \ \sqrt{2 + 7} + 5}$$
$$\begin{array}{c|c} & 8 \qquad \text{FALSE} \end{array}$$

Since 9 checks but 2 does not, the solution is 9.

AG Algebraic–Graphical Connection

We can visualize or check the solutions of a radical equation graphically. Consider the equation of Example 3:

$$x - 7 = 2\sqrt{x + 1}.$$

We can examine the solutions by graphing the equations

$$y = x - 7 \quad \text{and} \quad y = 2\sqrt{x + 1}$$

using the same set of axes. A hand-drawn graph of $y = 2\sqrt{x + 1}$ would involve approximating square roots on a calculator.

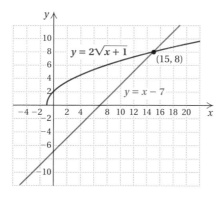

It appears from the graph that when $x = 15$, the values of $y = x - 7$ and $y = 2\sqrt{x + 1}$ are the same, 8. We can check this as we did in Example 3. Note too that the graphs *do not* intersect at $x = 3$, the "extraneous" solution.

AG

Calculator Spotlight

Solving Radical Equations. Consider the equation in Example 3. From each side of the equation, we form two new equations in the "$y=$" form:

$$y_1 = x - 7, \qquad (1)$$
$$y_2 = 2\sqrt{x + 1}. \qquad (2)$$

The graphs of y_1 and y_2 are as shown below.

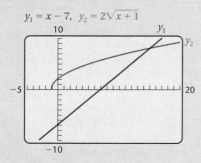

In some cases, the viewing window may need to be adjusted to view the intersection.

To find the solution, we use the CALC-INTERSECT features. The solution of the equation $x - 7 = 2\sqrt{x + 1}$ is the first coordinate at the bottom of the screen. Note that there is no point of intersection at $x = 3$, the "extra-

neous" solution. The number 3 is *not* a solution of the original equation.

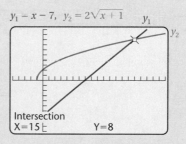

Exercises

Solve.

1. $x = \sqrt{x + 7} + 5$ (Example 4)
2. $x + 2 = \sqrt{2x + 7}$ (Margin Exercise 3)
   What happens to the equations at $x = -3$? Do the graphs intersect?
3. $x + 1 = 3\sqrt{x - 1}$ (Margin Exercise 4)

Approximate the solutions.

4. $x - 7.4 = 2\sqrt{x + 1.3}$
5. $x - 5.3 = \sqrt{x + 8.1}$
6. $x + 1.92 = 3.6\sqrt{x - 0.8}$

5. $x = \sqrt{x + 5} + 1$

Example 5 Solve: $\sqrt[3]{2x + 1} + 5 = 0$.

$$\sqrt[3]{2x + 1} + 5 = 0$$

$$\sqrt[3]{2x + 1} = -5 \qquad \text{Subtracting 5; this isolates the radical term}$$

$$(\sqrt[3]{2x + 1})^3 = (-5)^3 \qquad \text{Using the principle of powers (raising to the third power)}$$

$$2x + 1 = -125$$

$$2x = -126 \qquad \text{Subtracting 1}$$

$$x = -63$$

CHECK: $\dfrac{\sqrt[3]{2x + 1} + 5 = 0}{\sqrt[3]{2 \cdot (-63) + 1} + 5 \,?\, 0}$

$$\sqrt[3]{-125} + 5$$
$$-5 + 5$$
$$0 \quad | \qquad \text{TRUE}$$

The solution is -63.

Do Exercises 5 and 6.

b | Equations with Two Radical Terms

A general strategy for solving equations with two radical terms is as follows.

> To solve an equation with two radical terms:
>
> **1.** Isolate one of the radical terms.
>
> **2.** Use the principle of powers.
>
> **3.** If a radical remains, perform steps (1) and (2) again.
>
> **4.** Check possible solutions.

6. $\sqrt[4]{x - 1} - 2 = 0$

Example 6 Solve: $\sqrt{x - 3} + \sqrt{x + 5} = 4$.

$$\sqrt{x - 3} + \sqrt{x + 5} = 4$$

$$\sqrt{x - 3} = 4 - \sqrt{x + 5} \qquad \text{Subtracting } \sqrt{x + 5}\text{; this isolates one of the radical terms}$$

$$(\sqrt{x - 3})^2 = (4 - \sqrt{x + 5})^2 \qquad \text{Using the principle of powers (squaring both sides)}$$

> Here we are squaring a binomial. We square 4, then subtract twice the product of 4 and $\sqrt{x + 5}$, and then add the square of $\sqrt{x + 5}$. (See the rule in Section 4.2.)

$$x - 3 = 16 - 8\sqrt{x + 5} + (x + 5)$$

$$-3 = 21 - 8\sqrt{x + 5} \qquad \text{Subtracting } x \text{ and collecting like terms}$$

$$-24 = -8\sqrt{x + 5} \qquad \text{Isolating the remaining radical term}$$

$$3 = \sqrt{x + 5} \qquad \text{Dividing by } -8$$

$$3^2 = (\sqrt{x + 5})^2 \qquad \text{Squaring}$$

$$9 = x + 5$$

$$4 = x$$

The number 4 checks and is the solution.

Answers on page A-37

CAUTION! A common error in solving equations like $\sqrt{x-3} + \sqrt{x+5} = 4$ is to square the left side, obtaining $(x-3) + (x+5)$. That is wrong because the square of a sum is *not* the sum of the squares.

Example: $\sqrt{9} + \sqrt{16} = 7$, but $9 + 16 = 25$, not 7^2.

Solve.

7. $\sqrt{x} - \sqrt{x-5} = 1$

Example 7 Solve: $\sqrt{2x-5} = 1 + \sqrt{x-3}$.

$$\sqrt{2x-5} = 1 + \sqrt{x-3}$$

$(\sqrt{2x-5})^2 = (1 + \sqrt{x-3})^2$ One radical is already isolated; we square both sides.

$$2x - 5 = 1 + 2\sqrt{x-3} + (\sqrt{x-3})^2$$

$$2x - 5 = 1 + 2\sqrt{x-3} + (x-3)$$

$x - 3 = 2\sqrt{x-3}$ Isolating the remaining radical term

$(x-3)^2 = (2\sqrt{x-3})^2$ Squaring both sides

$$x^2 - 6x + 9 = 4(x-3)$$

$$x^2 - 6x + 9 = 4x - 12$$

$$x^2 - 10x + 21 = 0$$

$(x-7)(x-3) = 0$ Factoring

$x = 7 \quad or \quad x = 3$ Using the principle of zero products

The numbers 7 and 3 check and are the solutions.

8. $\sqrt{2x-5} - 2 = \sqrt{x-2}$

Do Exercises 7 and 8.

Example 8 Solve: $\sqrt{x+2} - \sqrt{2x+2} + 1 = 0$.

We first isolate one radical.

$$\sqrt{x+2} - \sqrt{2x+2} + 1 = 0$$

$\sqrt{x+2} = \sqrt{2x+2} - 1$ Adding $\sqrt{2x+2}$ and subtracting 1 to isolate a radical

$(\sqrt{x+2})^2 = (\sqrt{2x+2} - 1)^2$ Squaring both sides

$$x + 2 = (\sqrt{2x+2})^2 - 2\sqrt{2x+2} + 1$$

$$x + 2 = 2x + 2 - 2\sqrt{2x+2} + 1$$

$-x - 1 = -2\sqrt{2x+2}$ Isolating the remaining radical

$x + 1 = 2\sqrt{2x+2}$ Multiplying by -1

$(x+1)^2 = (2\sqrt{2x+2})^2$ Squaring both sides

$$x^2 + 2x + 1 = 4(2x+2)$$

$$x^2 + 2x + 1 = 8x + 8$$

$$x^2 - 6x - 7 = 0$$

$(x-7)(x+1) = 0$ Factoring

$x - 7 = 0 \quad or \quad x + 1 = 0$ Using the principle of zero products

$x = 7 \quad or \qquad x = -1$

9. Solve:

$$\sqrt{3x+1} - 1 - \sqrt{x+4} = 0.$$

The check is left to the student. The number 7 checks, but -1 does not. The solution is 7.

Do Exercise 9.

Answers on page A-37

10. How far to the horizon can you see through an airplane window at a height, or altitude, of 38,000 ft?

c Applications

Sighting to the Horizon. How far can you see from a given height? The function

$$D(h) = \sqrt{2h} \qquad \textbf{(1)}$$

can be used to approximate the distance D, in miles, that a person can see to the horizon from a height h, in feet.

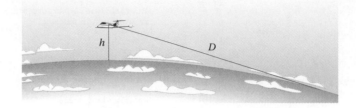

Example 9 How far to the horizon can you see through an airplane window at a height, or altitude, of 30,000 ft?

We substitute 30,000 for h in equation (1) and find an approximation using a calculator:

$$D(30,000) = \sqrt{2 \cdot 30,000}$$
$$\approx 245 \text{ mi}.$$

You can see for about 245 mi to the horizon.

11. A sailor climbs 40 ft up the mast of a ship to a crow's nest. How far can he see to the horizon?

Do Exercises 10 and 11.

Example 10 How far above sea level must a sailor climb on the mast of a ship in order to see 10.2 mi out to an iceberg?

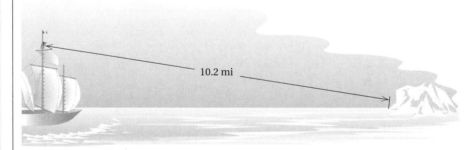

12. How far above sea level must a sailor climb on the mast of a ship in order to see 18.4 mi out to sea?

We substitute 10.2 for $D(h)$ in equation (1) and solve:

$$10.2 = \sqrt{2h}$$
$$(10.2)^2 = (\sqrt{2h})^2$$
$$104.04 = 2h$$
$$\frac{104.04}{2} = h$$
$$52.02 = h.$$

The sailor must climb to a height of about 52 ft in order to see 10.2 mi out to the iceberg.

Answers on page A-37

Do Exercise 12.

Exercise Set 6.6

a Solve.

1. $\sqrt{2x - 3} = 4$

2. $\sqrt{5x + 2} = 7$

3. $\sqrt{6x} + 1 = 8$

4. $\sqrt{3x} - 4 = 6$

5. $\sqrt{y + 7} - 4 = 4$

6. $\sqrt{x - 1} - 3 = 9$

7. $\sqrt{5y + 8} = 10$

8. $\sqrt{2y + 9} = 5$

9. $\sqrt[3]{x} = -1$

10. $\sqrt[3]{y} = -2$

11. $\sqrt{x + 2} = -4$

12. $\sqrt{y - 3} = -2$

13. $\sqrt[3]{x + 5} = 2$

14. $\sqrt[3]{x - 2} = 3$

15. $\sqrt[4]{y - 3} = 2$

16. $\sqrt[4]{x + 3} = 3$

17. $\sqrt[3]{6x + 9} + 8 = 5$

18. $\sqrt[3]{3y + 6} + 2 = 3$

19. $8 = \dfrac{1}{\sqrt{x}}$

20. $\dfrac{1}{\sqrt{y}} = 3$

b Solve.

21. $\sqrt{3y + 1} = \sqrt{2y + 6}$

22. $\sqrt{5x - 3} = \sqrt{2x + 3}$

23. $\sqrt{y - 5} + \sqrt{y} = 5$

24. $\sqrt{x - 9} + \sqrt{x} = 1$

25. $3 + \sqrt{z - 6} = \sqrt{z + 9}$

26. $\sqrt{4x - 3} = 2 + \sqrt{2x - 5}$

27. $\sqrt{20 - x} + 8 = \sqrt{9 - x} + 11$

28. $4 + \sqrt{10 - x} = 6 + \sqrt{4 - x}$

29. $\sqrt{4y + 1} - \sqrt{y - 2} = 3$

30. $\sqrt{y + 15} - \sqrt{2y + 7} = 1$

31. $\sqrt{x + 2} + \sqrt{3x + 4} = 2$

32. $\sqrt{6x + 7} - \sqrt{3x + 3} = 1$

33. $\sqrt{3x - 5} + \sqrt{2x + 3} + 1 = 0$

34. $\sqrt{2m - 3} + 2 - \sqrt{m + 7} = 0$

35. $2\sqrt{t - 1} - \sqrt{3t - 1} = 0$

36. $3\sqrt{2y + 3} - \sqrt{y + 10} = 0$

Use the formula $D(h) = \sqrt{2h}$ for Exercises 37–40.

37. How far to the horizon can you see through an airplane window at a height, or altitude, of 27,000 ft?

38. How far to the horizon can you see through an airplane window at a height, or altitude, of 32,000 ft?

39. How far above sea level must a pilot fly in order to see to a horizon that is 180 mi away?

40. A person can see 220 mi to the horizon through an airplane window. How high above sea level is the airplane?

Speed of a Skidding Car. How do police determine how fast a car had been traveling after an accident has occurred? The formula

$$S(x) = 2\sqrt{5x}$$

can be used to approximate the speed S, in miles per hour, of a car that has left a skid mark of length x, in feet. (See Exercise 27 in Section 6.1.) Use this formula for Exercises 41 and 42.

211 ft

41. How far will a car skid at 65 mph? at 75 mph?

42. How far will a car skid at 55 mph? at 90 mph?

43. Find the number such that twice its square root is 14.

44. Find the number such that the square root of 4 more than five times the number is 8.

45. Find the number such that the square root of twice the number minus 1, all added to 1, is the square root of the number plus 11.

46. Find the number such that the square root of 4 less than the number plus the square root of 1 more than the number is 5.

Solve. [5.6a]

47. *Painting a Room.* Julia can paint a room in 8 hr. George can paint the same room in 10 hr. How long will it take them, working together, to paint the same room?

48. *Delivering Leaflets.* Jeff can drop leaflets in mailboxes three times as fast as Grace can. If they work together, it takes them 1 hr to complete the job. How long would it take each to deliver the leaflets alone?

Solve. [5.6b]

49. *Bicycle Travel.* A cyclist traveled 702 mi in 14 days. At this same ratio, how far would the cyclists have traveled in 56 days?

50. *Earnings.* Dharma earned $696.64 working for 56 hr at a fruit stand. How many hours must she work in order to earn $1044.96?

Solve. [4.7a]

51. $x^2 + 2.8x = 0$

52. $3x^2 - 5x = 0$

53. $x^2 - 64 = 0$

54. $2x^2 = x + 21$

For each of the following functions, find and simplify $f(a + h) - f(a)$. [4.2e]

55. $f(x) = x^2$

56. $f(x) = x^2 - x$

57. $f(x) = 2x^2 - 3x$

58. $f(x) = 2x^2 + 3x - 7$

Synthesis

59. ◆ The principle of powers contains an "if–then" statement that becomes false when the parts are interchanged. Find another mathematical example of such an "if–then" statement.

60. ◆ Is checking necessary when the principle of powers is used with an odd power n? Why or why not?

61. 〰 Use a grapher to check your answers to Exercises 4, 9, 25, and 30.

62. ◆ 〰 Consider the equation
$$\sqrt{2x + 1} + \sqrt{5x - 4} = \sqrt{10x + 9}.$$
a) Use a grapher to solve the equation.
b) Solve the equation algebraically.
c) Explain the advantages and disadvantages of using each method. Which do you prefer?

Solve.

63. $\sqrt[3]{\dfrac{z}{4}} - 10 = 2$

64. $\sqrt[4]{z^2 + 17} = 3$

65. $\sqrt{\sqrt{y + 49} - \sqrt{y}} = \sqrt{7}$

66. $\sqrt[3]{x^2 + x + 15} - 3 = 0$

67. $\sqrt{\sqrt{x^2 + 9x + 34}} = 2$

68. $\sqrt{8 - b} = b\sqrt{8 - b}$

69. $\sqrt{x - 2} - \sqrt{x + 2} + 2 = 0$

70. $6\sqrt{y} + 6y^{-1/2} = 37$

71. $\sqrt{a^2 + 30a} = a + \sqrt{5a}$

72. $\sqrt{\sqrt{x + 4}} = \sqrt{x} - 2$

73. $\dfrac{x - 1}{\sqrt{x^2 + 3x + 6}} = \dfrac{1}{4}$

74. $\sqrt{x + 1} - \dfrac{2}{\sqrt{x + 1}} = 1$

75. $\sqrt{y^2 + 6} + y - 3 = 0$

76. $2\sqrt{x - 1} - \sqrt{3x - 5} = \sqrt{x - 9}$

77. $\sqrt{y + 1} - \sqrt{2y - 5} = \sqrt{y - 2}$

78. Evaluate: $\sqrt{7 + 4\sqrt{3}} - \sqrt{7 - 4\sqrt{3}}$.

6.7 Applications Involving Powers and Roots

a | Applications

There are many kinds of applied problems that involve powers and roots. Many also make use of right triangles and the Pythagorean theorem: $a^2 + b^2 = c^2$.

Example 1 *Vegetable Garden.* Benito and Dominique are planting a vegetable garden in the backyard. They decide that it will be a 30-ft by 40-ft rectangle and begin to lay it out using string. They soon realize that it is difficult to form the right angles and that it would be helpful to know the length of a diagonal. Find the length of a diagonal.

Using the Pythagorean theorem, $a^2 + b^2 = c^2$, we substitute 30 for a and 40 for b and then solve for c:

$$a^2 + b^2 = c^2$$
$$30^2 + 40^2 = c^2 \quad \text{Substituting}$$
$$900 + 1600 = c^2$$
$$2500 = c^2$$
$$\sqrt{2500} = c$$
$$50 = c.$$

$a = 30$ ft
$c = ?$
$b = 40$ ft

The length of the hypotenuse, or the diagonal, is 50 ft. Knowing this measurement would help in laying out the garden. Construction workers often use a procedure like this to lay out a right angle.

Example 2 Find the length of the hypotenuse of this right triangle. Give an exact answer and an approximation to three decimal places.

$$7^2 + 4^2 = c^2 \qquad \text{Substituting in the Pythagorean theorem}$$
$$49 + 16 = c^2$$
$$65 = c^2$$
Exact answer: $\quad c = \sqrt{65}$
Approximation: $\quad c \approx 8.062 \qquad \text{Using a calculator}$

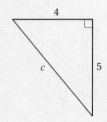

c
4
7

Example 3 Find the missing length b in this right triangle. Give an exact answer and an approximation to three decimal places.

$$1^2 + b^2 = (\sqrt{11})^2 \qquad \text{Substituting in the Pythagorean theorem}$$
$$1 + b^2 = 11$$
$$b^2 = 10$$
Exact answer: $\quad b = \sqrt{10}$
Approximation: $\quad b \approx 3.162$

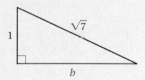

$\sqrt{11}$
1
b

Do Exercises 1–3.

Objective

a | Solve applied problems involving the Pythagorean theorem and powers and roots.

For Extra Help

TAPE 14   TAPE 13A   MAC WIN   CD-ROM

1. Find the length of the hypotenuse of this right triangle. Give an exact answer and an approximation to three decimal places.

4
c
5

2. Find the length of the leg of this right triangle. Give an exact answer and an approximation to three decimal places.

1
$\sqrt{7}$
b

3. Find the length of the hypotenuse of this right triangle. Give an exact answer and an approximation to three decimal places.

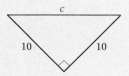

c
10
10

Answers on page A-37

4. *Baseball Diamond.* A baseball diamond is actually a square 90 ft on a side. Suppose a catcher fields a bunt along the third-base line 10 ft from home plate. How far would the catcher have to throw the ball to first base? Give an exact answer and an approximation to three decimal places.

Example 4 *Ramps for the Handicapped.* Laws regarding access ramps for the handicapped state that a ramp must be in the form of a right triangle, where every vertical length (leg) of 1 ft has a horizontal length (leg) of 12 ft. What is the length of a ramp with a 12-ft horizontal leg and a 1-ft vertical leg? Give an exact answer and an approximation to three decimal places.

We make a drawing and let h = the length of the ramp. It is the length of the hypotenuse of a right triangle whose legs are 12 ft and 1 ft. We substitute these values into the Pythagorean theorem to find h.

$$h^2 = 12^2 + 1^2$$
$$h^2 = 144 + 1$$
$$h^2 = 145$$
$$h = \sqrt{145}$$

Exact answer: $h = \sqrt{145}$ ft

Approximation: $h \approx 12.042$ ft

Do Exercise 4.

Example 5 *Road-Pavement Messages.* In a psychological study, it was determined that the proper length L of the letters of a word painted on pavement is given by

$$L = \frac{0.000169d^{2.27}}{h},$$

where d is the distance of a car from the lettering and h is the height of the eye above the road. All units are in feet. In other words, for a person h feet above the road, a message d feet away will be the most readable if the length of the letters is L.

Find L, given that $h = 4$ ft and $d = 180$ ft.

We substitute 4 for h and 180 for d and calculate L using a calculator with an exponential key $\boxed{y^x}$, or $\boxed{\wedge}$:

$$L = \frac{0.000169(180)^{2.27}}{4}$$

$$\approx 5.6 \text{ ft.}$$

Do Exercise 5.

5. Referring to Example 5, find L given that $h = 3$ ft and $d = 180$ ft. You will need a calculator with an exponential key $\boxed{y^x}$, or $\boxed{\wedge}$.

Answers on page A-37

Exercise Set 6.7

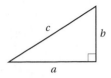

a In a right triangle, find the length of the side not given. Give an exact answer and an approximation to three decimal places.

1. $a = 3$, $b = 5$

2. $a = 8$, $b = 10$

3. $a = 15$, $b = 15$

4. $a = 8$, $b = 8$

5. $b = 12$, $c = 13$

6. $a = 5$, $c = 12$

7. $c = 7$, $a = \sqrt{6}$

8. $c = 10$, $a = 4\sqrt{5}$

9. $b = 1$, $c = \sqrt{13}$

10. $a = 1$, $c = \sqrt{12}$

11. $a = 1$, $c = \sqrt{n}$

12. $c = 2$, $a = \sqrt{n}$

In the following problems, give an exact answer and, where appropriate, an approximation to three decimal places.

13. *Bridge Expansion.* During the summer heat, a 2-mi bridge expands 2 ft in length. If we assume that the bulge occurs straight up the middle, how high is the bulge? (The answer may surprise you. In reality, bridges are built with expansion spaces to avoid such buckling.)

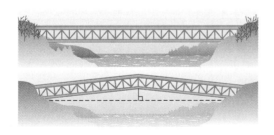

14. Triangle *ABC* has sides of lengths 25 ft, 25 ft, and 30 ft. Triangle *PQR* has sides of lengths 25 ft, 25 ft, and 40 ft. Which triangle has the greater area and by how much?

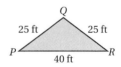

15. *Guy Wire.* How long is a guy wire reaching from the top of a 10-ft pole to a point on the ground 4 ft from the pole?

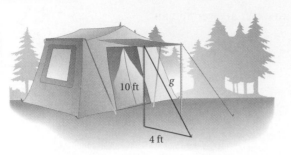

10 ft   g

4 ft

16. *Softball Diamond.* A slow-pitch softball diamond is actually a square 65 ft on a side. How far is it from home to second base?

17. Each side of a regular octagon has length *s*. Find a formula for the distance *d* between the parallel sides of the octagon.

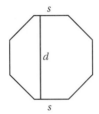

s

d

s

18. The two equal sides of an isosceles right triangle are of length *s*. Find a formula for the length of the hypotenuse.

s $h = ?$

s

19. *Road-Pavement Messages.* Using the formula of Example 5, find the length *L* of a road-pavement message when $h = 4$ ft and $d = 200$ ft.

20. *Road-Pavement Messages.* Using the formula of Example 5, find the length *L* of a road-pavement message when $h = 8$ ft and $d = 300$ ft.

21. The length and the width of a rectangle are given by consecutive integers. The area of the rectangle is 90 cm². Find the length of a diagonal of the rectangle.

22. The diagonal of a square has length $8\sqrt{2}$ ft. Find the length of a side of the square.

23. *Television Sets.* What does it mean to refer to a 20-in. TV set or a 25-in. TV set? Such units refer to the diagonal of the screen. A 20-in. TV set also has a width of 16 in. What is its height?

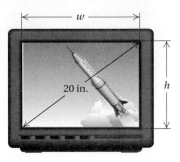

24. *Television Sets.* A 25-in. TV set has a screen with a height of 15 in. What is its width?

25. Find all points on the *x*-axis of a Cartesian coordinate system that are 5 units from the point (0, 4).

26. Find all points on the *y*-axis of a Cartesian coordinate system that are 5 units from the point (3, 0).

27. *Speaker Placement.* A stereo receiver is in a corner of a 12-ft by 14-ft room. Speaker wire will run under a rug, diagonally, to a speaker in the far corner. If 4 ft of slack is required on each end, how long a piece of wire should be purchased?

28. *Distance Over Water.* To determine the width of a pond, a surveyor locates two stakes at either end of the pond and uses instrumentation to place a third stake so that the distance across the pond is the length of a hypotenuse. If the third stake is 90 m from one stake and 70 m from the other, how wide is the pond?

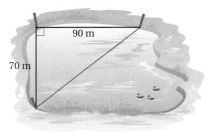

29. *Plumbing.* Plumbers use the Pythagorean theorem to calculate pipe length. If a pipe is to be offset, as shown in the figure, the *travel*, or length, of the pipe, is calculated using the lengths of the *advance* and *offset*.

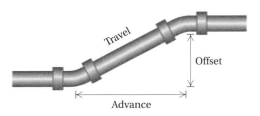

Find the travel if the offset is 17.75 in. and the advance is 10.25 in.

30. *Carpentry.* Dale is laying out the footer of a house. To see if the corner is square, she measures 16 ft from the corner along one wall and 20 ft from the corner along the other wall. How long should the diagonal be between those two points if the corner is a right angle?

Solve. [5.6c]

31. *Commuter Travel.* The speed of the Zionsville Flash commuter train is 14 mph faster than that of the Carmel Crawler. The Flash travels 290 mi in the same time that it takes the Crawler to travel 230 mi. Find the speed of each train.

32. *Marine Travel.* A motor boat travels three times as fast as the current in the Saskatee River. A trip up the river and back takes 10 hr, and the total distance of the trip is 100 mi. Find the speed of the current.

Solve. [4.7a], [5.5a]

33. $2x^2 + 11x - 21 = 0$

34. $x^2 + 24 = 11x$

35. $\dfrac{x + 2}{x + 3} = \dfrac{x - 4}{x - 5}$

36. $3x^2 - 12 = 0$

37. $\dfrac{x - 5}{x - 7} = \dfrac{4}{3}$

38. $\dfrac{x - 1}{x - 3} = \dfrac{6}{x - 3}$

Synthesis

39. ◆ Write a problem for a classmate to solve in which the solution is "The height of the tepee is $5\sqrt{3}$ yd."

40. ◆ Write a problem for a classmate to solve in which the solution is "The height of the window is $15\sqrt{3}$ yd."

41. *Roofing.* Kit's cottage, which is 24 ft wide and 32 ft long, needs a new roof. By counting clapboards that are 4 in. apart, Kit determines that the peak of the roof is 6 ft higher than the sides. If one packet of shingles covers 100 square feet, how many packets will the job require?

42. *Painting.* (Refer to Exercise 41.) A gallon of paint covers about 275 square feet. If Kit's first floor is 10 ft high, how many gallons of paint should be bought to paint the house? What assumption(s) is made in your answer?

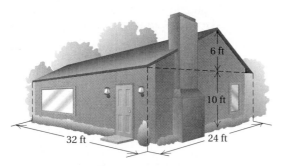

43. A cube measures 5 cm on each side. How long is the diagonal that connects two opposite corners of the cube? Give an exact answer.

44. *Wind Chill Temperature.* We can use square roots to consider an application involving the effect of wind on the feeling of cold in the winter. Because wind enhances the loss of heat from the skin, we feel colder when there is wind than when there is not. The *wind chill temperature* is what the temperature would have to be with no wind in order to give the same chilling effect. A formula for finding the wind chill temperature, T_w, is

$$T_w = 91.4 - \frac{(10.45 + 6.68\sqrt{v} - 0.447v)(457 - 5T)}{110},$$

where T is the actual temperature given by a thermometer, in degrees Fahrenheit, and v is the wind speed, in miles per hour. Use a calculator to find the wind chill temperature in each case. Round to the nearest degree.

a) $T = 40°F$, $v = 25$ mph

b) $T = 20°F$, $v = 25$ mph

c) $T = 10°F$, $v = 20$ mph

d) $T = 10°F$, $v = 40$ mph

e) $T = -5°F$, $v = 35$ mph

f) $T = -16°F$, $v = 40$ mph

Collaborative Learning Manual

Develop a formula for the swing time of a pendulum.

6.8 The Complex Numbers

a | Imaginary and Complex Numbers

Negative numbers do not have square roots in the real-number system. However, mathematicians have invented a larger number system that contains the real-number system, but is such that negative numbers have square roots. That system is called the **complex-number system.** We begin by creating a number that is a square root of -1. We call this new number i.

> We define the number i to be $\sqrt{-1}$. That is, $i = \sqrt{-1}$ and $i^2 = -1$.

To express roots of negative numbers in terms of i, we can use the fact that in the complex numbers, $\sqrt{-p} = \sqrt{-1 \cdot p} = \sqrt{-1}\sqrt{p}$ when p is a positive real number.

Examples Express in terms of i.

1. $\sqrt{-7} = \sqrt{-1 \cdot 7} = \sqrt{-1} \cdot \sqrt{7} = i\sqrt{7}$, or $\sqrt{7}i$ i is *not* under the radical.

2. $\sqrt{-16} = \sqrt{-1 \cdot 16} = \sqrt{-1} \cdot \sqrt{16} = i \cdot 4 = 4i$

3. $-\sqrt{-13} = -\sqrt{-1 \cdot 13} = -\sqrt{-1} \cdot \sqrt{13} = -i\sqrt{13}$, or $-\sqrt{13}\,i$

4. $-\sqrt{-64} = -\sqrt{-1 \cdot 64} = -\sqrt{-1} \cdot \sqrt{64} = -i \cdot 8 = -8i$

5. $\sqrt{-48} = \sqrt{-1 \cdot 48} = \sqrt{-1} \cdot \sqrt{48} = i\sqrt{48} = i \cdot 4\sqrt{3} = 4\sqrt{3}\,i$, or $4i\sqrt{3}$

Do Exercises 1–5.

> An **imaginary*** **number** is a number that can be named
>
> bi,
>
> where b is some real number and $b \neq 0$.

To form the system of **complex numbers,** we take the imaginary numbers and the real numbers and all possible sums of real and imaginary numbers. These are complex numbers:

     $7 - 4i$, $-\pi + 19i$, 37, $i\sqrt{8}$.

> A **complex number** is any number that can be named
>
> $a + bi$,
>
> where a and b are any real numbers. (Note that either a or b or both can be 0.)

---

*Don't let the name "imaginary" fool you. The imaginary numbers are very important in such fields as engineering and the physical sciences.

Express in terms of i.

1. $\sqrt{-5}$

2. $\sqrt{-25}$

3. $-\sqrt{-11}$

4. $-\sqrt{-36}$

5. $\sqrt{-54}$

Answers on page A-37

6. $(7 + 4i) + (8 - 7i)$

7. $(-5 - 6i) + (-7 + 12i)$

8. $(8 + 3i) - (5 + 8i)$

9. $(5 - 4i) - (-7 + 3i)$

Since $0 + bi = bi$, every imaginary number is a complex number. Similarly, $a + 0i = a$, so every real number is a complex number. The relationships among various real and complex numbers are shown in the following diagram.

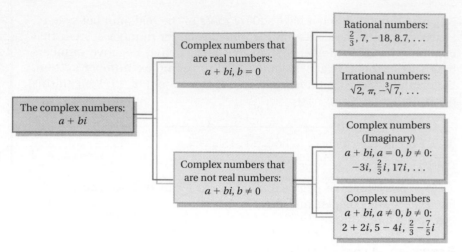

It is important to keep in mind some comparisons between numbers that have real-number roots and those that have complex-number roots that are not real. For example, $\sqrt{-48}$ is a complex number that is not a real number because we are taking the square root of a negative number. But, $\sqrt[3]{-125}$ is a real number because we are taking the cube root of a negative number and *any* real number has a cube root that is a real number.

b | Addition and Subtraction

The complex numbers follow the commutative and associative laws of addition. Thus we can add and subtract them as we do binomials with real-number coefficients, that is, we collect like terms.

Examples Add or subtract.

6. $(8 + 6i) + (3 + 2i) = (8 + 3) + (6 + 2)i = 11 + 8i$

7. $(3 + 2i) - (5 - 2i) = (3 - 5) + [2 - (-2)]i = -2 + 4i$

Do Exercises 6–9.

c | Multiplication

The complex numbers obey the commutative, associative, and distributive laws. But although the property $\sqrt{a}\sqrt{b} = \sqrt{ab}$ does *not* hold for complex numbers in general, it does hold when $a = -1$ and b is a positive real number. To multiply square roots of negative real numbers, we first express them in terms of i. For example,

$$\sqrt{-2} \cdot \sqrt{-5} = \sqrt{-1} \cdot \sqrt{2} \cdot \sqrt{-1} \cdot \sqrt{5} = i\sqrt{2} \cdot i\sqrt{5}$$
$$= i^2\sqrt{10} = -\sqrt{10} \quad \text{is correct!}$$

But $\sqrt{-2} \cdot \sqrt{-5} = \sqrt{(-2)(-5)} = \sqrt{10}$ is wrong!

Keeping this and the fact that $i^2 = -1$ in mind, we multiply in much the same way that we do with real numbers.

Examples Multiply.

8. $\sqrt{-49} \cdot \sqrt{-16} = \sqrt{-1} \cdot \sqrt{49} \cdot \sqrt{-1} \cdot \sqrt{16}$
$= i \cdot 7 \cdot i \cdot 4$
$= i^2(28)$
$= (-1)(28) \qquad i^2 = -1$
$= -28$

9. $\sqrt{-3} \cdot \sqrt{-7} = \sqrt{-1} \cdot \sqrt{3} \cdot \sqrt{-1} \cdot \sqrt{7}$
$= i \cdot \sqrt{3} \cdot i \cdot \sqrt{7}$
$= i^2(\sqrt{21})$
$= (-1)\sqrt{21} \qquad i^2 = -1$
$= -\sqrt{21}$

10. $-2i \cdot 5i = -10 \cdot i^2$
$= (-10)(-1) \qquad i^2 = -1$
$= 10$

11. $(-4i)(3 - 5i) = (-4i) \cdot 3 - (-4i)(5i)$ Using a distributive law
$= -12i + 20i^2$
$= -12i + 20(-1) \qquad i^2 = -1$
$= -12i - 20$
$= -20 - 12i$

12. $(1 + 2i)(1 + 3i) = 1 + 3i + 2i + 6i^2$ Multiplying each term of one number by every term of the other (FOIL)
$= 1 + 3i + 2i - 6 \qquad i^2 = -1$
$= -5 + 5i$ Collecting like terms

13. $(3 - 2i)^2 = 3^2 - 2(3)(2i) + (2i)^2$ Squaring the binomial
$= 9 - 12i + 4i^2$
$= 9 - 12i - 4$
$= 5 - 12i$

Do Exercises 10–17.

---

d   **Powers of *i***

We now want to simplify certain expressions involving powers of i. To do so, we first see how to simplify powers of i. Simplifying powers of i can be done by using the fact that $i^2 = -1$ and expressing the given power of i in terms of even powers, and then in terms of powers of i^2. Consider the following:

$i,$

$i^2 = -1,$

$i^3 = i^2 \cdot i = (-1)i = -i,$

$i^4 = (i^2)^2 = (-1)^2 = 1,$

$i^5 = i^4 \cdot i = (i^2)^2 \cdot i = (-1)^2 \cdot i = i,$

$i^6 = (i^2)^3 = (-1)^3 = -1.$

Note that the powers of i cycle themselves through the values $i, -1, -i,$ and 1.

Multiply.

10. $\sqrt{-25} \cdot \sqrt{-4}$

11. $\sqrt{-2} \cdot \sqrt{-17}$

12. $-6i \cdot 7i$

13. $-3i(4 - 3i)$

14. $5i(-5 + 7i)$

15. $(1 + 3i)(1 + 5i)$

16. $(3 - 2i)(1 + 4i)$

17. $(3 + 2i)^2$

Answers on page A-37

Simplify.

18. i^{47}

19. i^{68}

20. i^{85}

21. i^{90}

Simplify.

22. $8 - i^5$

23. $7 + 4i^2$

24. $6i^{11} + 7i^{14}$

25. $i^{34} - i^{55}$

Find the conjugate.

26. $6 + 3i$

27. $-9 - 5i$

28. $\pi - \dfrac{1}{4}i$

Answers on page A-37

Examples   Simplify.

14. $i^{37} = i^{36} \cdot i = (i^2)^{18} \cdot i = (-1)^{18} \cdot i = 1 \cdot i = i$
15. $i^{58} = (i^2)^{29} = (-1)^{29} = -1$
16. $i^{75} = i^{74} \cdot i = (i^2)^{37} \cdot i = (-1)^{37} \cdot i = -1 \cdot i = -i$
17. $i^{80} = (i^2)^{40} = (-1)^{40} = 1$

Do Exercises 18–21.

Now let's simplify other expressions.

Examples   Simplify to the form $a + bi$.

18. $8 - i^2 = 8 - (-1) = 8 + 1 = 9$
19. $17 + 6i^3 = 17 + 6 \cdot i^2 \cdot i = 17 + 6(-1)i = 17 - 6i$
20. $i^{22} - 67i^2 = (i^2)^{11} - 67(-1) = (-1)^{11} + 67 = -1 + 67 = 66$
21. $i^{23} + i^{48} = (i^{22}) \cdot i + (i^2)^{24} = (i^2)^{11} \cdot i + (-1)^{24} = (-1)^{11} \cdot i + (-1)^{24}$
$\qquad\qquad = -i + 1 = 1 - i$

Do Exercises 22–25.

e   Conjugates and Division

Conjugates of complex numbers are defined as follows.

> The **conjugate** of a complex number $a + bi$ is $a - bi$, and the **conjugate** of $a - bi$ is $a + bi$.

Examples   Find the conjugate.

22. $5 + 7i$ The conjugate is $5 - 7i$.
23. $14 - 3i$ The conjugate is $14 + 3i$.
24. $-3 - 9i$ The conjugate is $-3 + 9i$.
25. $4i$ The conjugate is $-4i$.

Do Exercises 26–28.

When we multiply a complex number by its conjugate, we get a real number.

Examples   Multiply.

26. $(5 + 7i)(5 - 7i) = 5^2 - (7i)^2$ Using $(A + B)(A - B) = A^2 - B^2$
$\qquad\qquad\qquad\quad = 25 - 49i^2$ $i^2 = -1$
$\qquad\qquad\qquad\quad = 25 - 49(-1)$
$\qquad\qquad\qquad\quad = 25 + 49$
$\qquad\qquad\qquad\quad = 74$

27. $(2 - 3i)(2 + 3i) = 2^2 - (3i)^2$
$\qquad\qquad\qquad\quad = 4 - 9i^2$
$\qquad\qquad\qquad\quad = 4 - 9(-1)$ $i^2 = -1$
$\qquad\qquad\qquad\quad = 4 + 9$
$\qquad\qquad\qquad\quad = 13$

Do Exercises 29 and 30.

We use conjugates in dividing complex numbers.

Example 28 Divide and simplify to the form $a + bi$: $\dfrac{-5 + 9i}{1 - 2i}$.

$$\frac{-5 + 9i}{1 - 2i} \cdot \frac{1 + 2i}{1 + 2i} = \frac{(-5 + 9i)(1 + 2i)}{(1 - 2i)(1 + 2i)} \quad \begin{array}{l}\textbf{Multiplying by 1 using the}\\ \textbf{conjugate of the denominator}\\ \textbf{in the symbol for 1}\end{array}$$

$$= \frac{-5 - 10i + 9i + 18i^2}{1^2 - 4i^2}$$

$$= \frac{-5 - i + 18(-1)}{1 - 4(-1)} \qquad i^2 = -1$$

$$= \frac{-5 - i - 18}{1 + 4}$$

$$= \frac{-23 - i}{5} = -\frac{23}{5} - \frac{1}{5}i$$

Note the similarity between the preceding example and rationalizing denominators. In both cases, we used the conjugate of the denominator to write another name for 1. In Example 28, the symbol for the number 1 was chosen using the conjugate of the divisor, $1 - 2i$.

Example 29 What symbol for 1 would you use to divide?

Division to be done	*Symbol for 1*
$\dfrac{3 + 5i}{4 + 3i}$	$\dfrac{4 - 3i}{4 - 3i}$

Example 30 Divide and simplify to the form $a + bi$: $\dfrac{3 + 5i}{4 + 3i}$.

$$\frac{3 + 5i}{4 + 3i} \cdot \frac{4 - 3i}{4 - 3i} = \frac{(3 + 5i)(4 - 3i)}{(4 + 3i)(4 - 3i)} \quad \textbf{Multiplying by 1}$$

$$= \frac{12 - 9i + 20i - 15i^2}{4^2 - 9i^2}$$

$$= \frac{12 + 11i - 15(-1)}{16 - 9(-1)} \qquad i^2 = -1$$

$$= \frac{27 + 11i}{25} = \frac{27}{25} + \frac{11}{25}i$$

Do Exercises 31 and 32.

Multiply.

29. $(7 - 2i)(7 + 2i)$

30. $(-3 - i)(-3 + i)$

Divide and simplify to the form $a + bi$.

31. $\dfrac{6 + 2i}{1 - 3i}$

32. $\dfrac{2 + 3i}{-1 + 4i}$

33. Determine whether $-i$ is a solution of $x^2 + 1 = 0$.

34. Determine whether $1 - i$ is a solution of $x^2 - 2x + 2 = 0$.

Answers on page A-37

f Solutions of Equations

The equation $x^2 + 1 = 0$ has no real-number solution, but it has two non-real complex solutions.

Example 31 Determine whether i is a solution of the equation $x^2 + 1 = 0$.

We substitute i for x in the equation.

$$
\begin{array}{c|c}
x^2 + 1 = 0 & \\
\hline
i^2 + 1 \ ? \ 0 & \\
-1 + 1 & \\
0 & \text{TRUE}
\end{array}
$$

The number i is a solution.

Do Exercise 33 on the preceding page.

Any equation consisting of a polynomial in one variable on one side and 0 on the other has complex-number solutions (some may be real). It is not always easy to find the solutions, but they always exist.

Example 32 Determine whether $1 + i$ is a solution of the equation $x^2 - 2x + 2 = 0$.

We substitute $1 + i$ for x in the equation.

$$
\begin{array}{c|c}
x^2 - 2x + 2 = 0 & \\
\hline
(1 + i)^2 - 2(1 + i) + 2 \ ? \ 0 & \\
1 + 2i + i^2 - 2 - 2i + 2 & \\
1 + 2i - 1 - 2 - 2i + 2 & \\
(1 - 1 - 2 + 2) + (2 - 2)i & \\
0 + 0i & \\
0 & \text{TRUE}
\end{array}
$$

The number $1 + i$ is a solution.

Example 33 Determine whether $2i$ is a solution of $x^2 + 3x - 4 = 0$.

$$
\begin{array}{c|c}
x^2 + 3x - 4 = 0 & \\
\hline
(2i)^2 + 3(2i) - 4 \ ? \ 0 & \\
4i^2 + 6i - 4 & \\
-4 + 6i - 4 & \\
-8 + 6i & \text{FALSE}
\end{array}
$$

The number $2i$ is not a solution.

Do Exercise 34 on the preceding page.

Exercise Set 6.8

a Express in terms of i.

1. $\sqrt{-35}$ **2.** $\sqrt{-21}$ **3.** $\sqrt{-16}$ **4.** $\sqrt{-36}$

5. $-\sqrt{-12}$ **6.** $-\sqrt{-20}$ **7.** $\sqrt{-3}$ **8.** $\sqrt{-4}$

9. $\sqrt{-81}$ **10.** $\sqrt{-27}$ **11.** $\sqrt{-98}$ **12.** $-\sqrt{-18}$

13. $-\sqrt{-49}$ **14.** $-\sqrt{-125}$ **15.** $4 - \sqrt{-60}$ **16.** $6 - \sqrt{-84}$

17. $\sqrt{-4} + \sqrt{-12}$ **18.** $-\sqrt{-76} + \sqrt{-125}$

b Add or subtract and simplify.

19. $(7 + 2i) + (5 - 6i)$ **20.** $(-4 + 5i) + (7 + 3i)$ **21.** $(4 - 3i) + (5 - 2i)$

22. $(-2 - 5i) + (1 - 3i)$ **23.** $(9 - i) + (-2 + 5i)$ **24.** $(6 + 4i) + (2 - 3i)$

25. $(6 - i) - (10 + 3i)$ **26.** $(-4 + 3i) - (7 + 4i)$ **27.** $(4 - 2i) - (5 - 3i)$

28. $(-2 - 3i) - (1 - 5i)$ **29.** $(9 + 5i) - (-2 - i)$ **30.** $(6 - 3i) - (2 + 4i)$

$\boxed{\text{c}}$ Multiply.

31. $\sqrt{-36} \cdot \sqrt{-9}$

32. $\sqrt{-16} \cdot \sqrt{-64}$

33. $\sqrt{-7} \cdot \sqrt{-2}$

34. $\sqrt{-11} \cdot \sqrt{-3}$

35. $-3i \cdot 7i$

36. $8i \cdot 5i$

37. $-3i(-8 - 2i)$

38. $4i(5 - 7i)$

39. $(3 + 2i)(1 + i)$

40. $(4 + 3i)(2 + 5i)$

41. $(2 + 3i)(6 - 2i)$

42. $(5 + 6i)(2 - i)$

43. $(6 - 5i)(3 + 4i)$

44. $(5 - 6i)(2 + 5i)$

45. $(7 - 2i)(2 - 6i)$

46. $(-4 + 5i)(3 - 4i)$

47. $(3 - 2i)^2$

48. $(5 - 2i)^2$

49. $(1 + 5i)^2$

50. $(6 + 2i)^2$

51. $(-2 + 3i)^2$

52. $(-5 - 2i)^2$

$\boxed{\text{d}}$ Simplify.

53. i^7

54. i^{11}

55. i^{24}

56. i^{35}

57. i^{42}

58. i^{64}

59. i^9

60. $(-i)^{71}$

61. i^6

62. $(-i)^4$

63. $(5i)^3$

64. $(-3i)^5$

Simplify to the form $a + bi$.

65. $7 + i^4$

66. $-18 + i^3$

67. $i^{28} - 23i$

68. $i^{29} + 33i$

69. $i^2 + i^4$

70. $5i^5 + 4i^3$

71. $i^5 + i^7$

72. $i^{84} - i^{100}$

73. $1 + i + i^2 + i^3 + i^4$

74. $i - i^2 + i^3 - i^4 + i^5$

75. $5 - \sqrt{-64}$

76. $\sqrt{-12} + 36i$

77. $\dfrac{8 - \sqrt{-24}}{4}$

78. $\dfrac{9 + \sqrt{-9}}{3}$

e Divide and simplify to the form $a + bi$.

79. $\dfrac{4 + 3i}{3 - i}$

80. $\dfrac{5 + 2i}{2 + i}$

81. $\dfrac{3 - 2i}{2 + 3i}$

82. $\dfrac{6 - 2i}{7 + 3i}$

83. $\dfrac{8 - 3i}{7i}$

84. $\dfrac{3 + 8i}{5i}$

85. $\dfrac{4}{3 + i}$

86. $\dfrac{6}{2 - i}$

87. $\dfrac{2i}{5 - 4i}$

88. $\dfrac{8i}{6 + 3i}$

89. $\dfrac{4}{3i}$

90. $\dfrac{5}{6i}$

91. $\dfrac{2 - 4i}{8i}$

92. $\dfrac{5 + 3i}{i}$

93. $\dfrac{6 + 3i}{6 - 3i}$

94. $\dfrac{4 - 5i}{4 + 5i}$

f Determine whether the complex number is a solution of the equation.

95. $1 - 2i$;
$x^2 - 2x + 5 = 0$

96. $1 + 2i$;
$x^2 - 2x + 5 = 0$

97. $2 + i$;
$x^2 - 4x - 5 = 0$

98. $1 - i$;
$x^2 + 2x + 2 = 0$

Skill Maintenance

Solve.   [5.5a]

99. $\dfrac{196}{x^2 - 7x + 49} - \dfrac{2x}{x + 7} = \dfrac{2058}{x^3 + 343}$

100. $\dfrac{5}{t} - \dfrac{3}{2} = \dfrac{4}{7}$

Solve.   [1.6c, d, e]

101. $|3x + 7| = 22$

102. $|3x + 7| < 22$

103. $|3x + 7| \geq 22$

104. $|3x + 7| = |2x - 5|$

Synthesis

105. ◆ How are conjugates of complex numbers similar to the conjugates used in Section 6.5?

106. ◆ Is every real number a complex number? Why or why not?

107. A complex function g is given by

$g(z) = \dfrac{z^4 - z^2}{z - 1}.$

Find $g(2i)$, $g(1 + i)$, and $g(-1 + 2i)$.

108. Evaluate $\dfrac{1}{w - w^2}$ for $w = \dfrac{1 - i}{10}$.

Express in terms of i.

109. $\dfrac{1}{8}(-24 - \sqrt{-1024})$

110. $12\sqrt{-\dfrac{1}{32}}$

111. $7\sqrt{-64} - 9\sqrt{-256}$

Simplify.

112. $\dfrac{i^5 + i^6 + i^7 + i^8}{(1 - i)^4}$

113. $(1 - i)^3(1 + i)^3$

114. $\dfrac{5 - \sqrt{5}\,i}{\sqrt{5}\,i}$

115. $\dfrac{6}{1 + \dfrac{3}{i}}$

116. $\left(\dfrac{1}{2} - \dfrac{1}{3}i\right)^2 - \left(\dfrac{1}{2} + \dfrac{1}{3}i\right)^2$

117. $\dfrac{i - i^{38}}{1 + i}$

118. Find all numbers a for which the opposite of a is the same as the reciprocal of a.

Summary and Review Exercises: Chapter 6

Important Properties and Formulas

$\sqrt{a^2} = |a|$; $\sqrt[k]{a^k} = |a|$, when k is even; $\sqrt[k]{a^k} = a$, when k is odd; $i = \sqrt{-1}$, $i^2 = -1$, $i^3 = -i$, $i^4 = 1$

$\sqrt[k]{ab} = \sqrt[k]{a} \cdot \sqrt[k]{b}$; $\sqrt[k]{\dfrac{a}{b}} = \dfrac{\sqrt[k]{a}}{\sqrt[k]{b}}$; $\sqrt[k]{a^m} = (\sqrt[k]{a})^m$;

Imaginary Numbers: bi, $i^2 = -1$, $b \neq 0$

Complex Numbers: $a + bi$, $i^2 = -1$

$a^{1/n} = \sqrt[n]{a}$; $a^{m/n} = \sqrt[n]{a^m} = (\sqrt[n]{a})^m$; $a^{-m/n} = \dfrac{1}{a^{m/n}}$

Conjugates: $a + bi$, $a - bi$

Principle of Powers: If $a = b$ is true, then $a^n = b^n$ is true.

Pythagorean Theorem: $a^2 + b^2 = c^2$, in a right triangle.

The objectives to be tested in addition to the material in this chapter are [4.7a], [5.1d, e], [5.5a], and [5.6a, b, c].

Use a calculator to approximate to three decimal places.   [6.1a]

1. $\sqrt{778}$

2. $\sqrt{\dfrac{963.2}{23.68}}$

3. For the given function, find the indicated function values.   [6.1a]

$f(x) = \sqrt{3x - 16}$; $f(0), f(-1), f(1)$, and $f\left(\dfrac{41}{3}\right)$

4. Find the domain of the function f in Exercise 3.   [6.1a]

Simplify. Assume that letters represent *any* real number.   [6.1b]

5. $\sqrt{81a^2}$

6. $\sqrt{(-7z)^2}$

7. $\sqrt{(c-3)^2}$

8. $\sqrt{x^2 - 6x + 9}$

Simplify.   [6.1c]

9. $\sqrt[3]{-1000}$

10. $\sqrt[3]{-\dfrac{1}{27}}$

11. For the given function, find the indicated function values.   [6.1c]

$f(x) = \sqrt[3]{x + 2}$; $f(6), f(-10)$, and $f(25)$

Simplify. Assume that letters represent *any* real number.   [6.1d]

12. $\sqrt[10]{x^{10}}$

13. $-\sqrt[13]{(-3)^{13}}$

Rewrite without rational exponents, and simplify, if possible.   [6.2a]

14. $a^{1/5}$

15. $64^{3/2}$

Rewrite with rational exponents.   [6.2a]

16. $\sqrt{31}$

17. $\sqrt[5]{a^2 b^3}$

Rewrite with positive exponents, and simplify, if possible.   [6.2b]

18. $49^{-1/2}$

19. $(8xy)^{-2/3}$

20. $5a^{-3/4} b^{1/2} c^{-2/3}$

21. $\dfrac{3a}{\sqrt[4]{t}}$

Use the laws of exponents to simplify. Write answers with positive exponents.   [6.2c]

22. $(x^{-2/3})^{3/5}$

23. $\dfrac{7^{-1/3}}{7^{-1/2}}$

Use rational exponents to simplify. Write the answer in radical notation if appropriate.   [6.2d]

24. $\sqrt[3]{x^{21}}$

25. $\sqrt[3]{27x^6}$

Use rational exponents to write a single radical expression.   [6.2d]

26. $x^{1/3} y^{1/4}$

27. $\sqrt[4]{x}\,\sqrt[3]{x}$

Simplify by factoring. Assume that all expressions under radicals represent nonnegative numbers.   [6.3a]

28. $\sqrt{245}$

29. $\sqrt[3]{-108}$

30. $\sqrt[3]{250a^2 b^6}$

Simplify. Assume that all expressions under radicals represent positive numbers.   [6.3b]

31. $\sqrt{\dfrac{49}{36}}$

32. $\sqrt[3]{\dfrac{64x^6}{27}}$

33. $\sqrt[4]{\dfrac{16x^8}{81y^{12}}}$

Perform the indicated operations and simplify. Assume that all expressions under radicals represent positive numbers. [6.3a, b]

34. $\sqrt{5x}\sqrt{3y}$

35. $\sqrt[3]{a^5b}\sqrt[3]{27b}$

36. $\sqrt[3]{a}\sqrt[5]{b^3}$

37. $\dfrac{\sqrt[3]{60xy^3}}{\sqrt[3]{10x}}$

38. $\dfrac{\sqrt{75x}}{2\sqrt{3}}$

39. $\dfrac{\sqrt[3]{x^2}}{\sqrt[4]{x}}$

Add or subtract. Assume that all expressions under radicals represent nonnegative numbers. [6.4a]

40. $5\sqrt[3]{x} + 2\sqrt[3]{x}$

41. $2\sqrt{75} - 7\sqrt{3}$

42. $\sqrt[3]{8x^4} + \sqrt[3]{xy^6}$

43. $\sqrt{50} + 2\sqrt{18} + \sqrt{32}$

Multiply. [6.4b]

44. $(\sqrt{5} - 3\sqrt{8})(\sqrt{5} + 2\sqrt{8})$

45. $(1 - \sqrt{7})^2$

46. $(\sqrt[3]{27} - \sqrt[3]{2})(\sqrt[3]{27} + \sqrt[3]{2})$

Rationalize the denominator. [6.5a, b]

47. $\sqrt{\dfrac{8}{3}}$

48. $\dfrac{2}{\sqrt{a} + \sqrt{b}}$

Solve. [6.6a, b]

49. $\sqrt[4]{x + 3} = 2$

50. $1 + \sqrt{x} = \sqrt{3x - 3}$

51. The diagonal of a square has length $9\sqrt{2}$ cm. Find the length of a side of the square. [6.7a]

52. A bookcase is 5 ft tall and has a 7-ft diagonal brace, as shown. How wide is the bookcase? [6.7a]

In a right triangle, find the length of the side not given. Give an exact answer and an answer to three decimal places. [6.7a]

53. $a = 7$, $b = 24$

54. $a = 2$, $c = 5\sqrt{2}$

55. Express in terms of i: $\sqrt{-25} + \sqrt{-8}$. [6.8a]

Add or subtract. [6.8b]

56. $(-4 + 3i) + (2 - 12i)$

57. $(4 - 7i) - (3 - 8i)$

Multiply. [6.8c, d]

58. $(2 + 5i)(2 - 5i)$

59. i^{13}

60. $(6 - 3i)(2 - i)$

Divide. [6.8e]

61. $\dfrac{-3 + 2i}{5i}$

62. $\dfrac{6 - 3i}{2 - i}$

63. Determine whether $1 + i$ is a solution of $x^2 + x + 2 = 0$. [6.8f]

64. Graph: $f(x) = \sqrt{x}$. [6.1a]

Skill Maintenance

Solve.

65. $\dfrac{7}{x + 2} + \dfrac{5}{x^2 - 2x + 4} = \dfrac{84}{x^3 + 8}$ [5.5a]

66. $3x^2 - 5x - 12 = 0$ [4.7a]

67. Multiply and simplify: [5.1d]
$$\dfrac{x^2 + 3x}{x^2 - y^2} \cdot \dfrac{x^2 - xy + 2x - 2y}{x^2 - 9}.$$

68. A motorcycle and a sport utility vehicle (SUV) are driven out of a car dealership at the same time. The SUV travels 8 mph faster than the motorcycle. The SUV travels 105 mi in the same time that the motorcycle travels 93 mi. Find the speed of each vehicle. [5.6c]

Synthesis

69. ◆ We learned a new method of equation solving in this chapter. Explain how this procedure differs from others we have used. [6.6a, b]

70. Simplify: i, i^2, i^3, i^{99}, i^{100}. [6.8c, d]

71. Solve: $\sqrt{11x + \sqrt{6 + x}} = 6$. [6.6a]

Test: Chapter 6

1. Use a calculator to approximate $\sqrt{148}$ to three decimal places.

2. For the given function, find the indicated function values.

$f(x) = \sqrt{8 - 4x};\quad f(1) \text{ and } f(3)$

3. Find the domain of the function f in Exercise 2.

Simplify. Assume that letters represent *any* real number.

4. $\sqrt{(-3q)^2}$

5. $\sqrt{x^2 + 10x + 25}$

6. $\sqrt[3]{-\dfrac{1}{1000}}$

7. $\sqrt[5]{x^5}$

8. $\sqrt[10]{(-4)^{10}}$

Rewrite without rational exponents, and simplify, if possible.

9. $a^{2/3}$

10. $32^{3/5}$

Rewrite with rational exponents.

11. $\sqrt{37}$

12. $\left(\sqrt{5xy^2}\right)^5$

Rewrite with positive exponents, and simplify, if possible.

13. $1000^{-1/3}$

14. $8a^{3/4}b^{-3/2}c^{-2/5}$

Use the laws of exponents to simplify. Write answers with positive exponents.

15. $(x^{2/3}y^{-3/4})^{12/5}$

16. $\dfrac{2.9^{-5/8}}{2.9^{2/3}}$

Use rational exponents to simplify. Write the answer in radical notation if appropriate.

17. $\sqrt[8]{x^2}$

18. $\sqrt[4]{16x^6}$

Use rational exponents to write a single radical expression.

19. $a^{2/5}b^{1/3}$

20. $\sqrt[4]{2y}\,\sqrt[3]{y}$

Simplify by factoring. Assume that all expressions under radicals represent nonnegative numbers.

21. $\sqrt{148}$

22. $\sqrt[4]{80}$

23. $\left(\sqrt[3]{16a^2b}\right)^2$

Simplify. Assume that all expressions under radicals represent positive numbers.

24. $\sqrt[3]{-\dfrac{8}{x^6}}$

25. $\sqrt{\dfrac{25x^2}{36y^4}}$

Answers

1. _____

2. _____

3. _____

4. _____

5. _____

6. _____

7. _____

8. _____

9. _____

10. _____

11. _____

12. _____

13. _____

14. _____

15. _____

16. _____

17. _____

18. _____

19. _____

20. _____

21. _____

22. _____

23. _____

24. _____

25. _____

Perform the indicated operations and simplify. Assume that all expressions under radicals represent positive numbers.

26. $\sqrt[3]{2x}\,\sqrt[3]{5y^2}$ **27.** $\sqrt[4]{x^3y^2}\,\sqrt{xy}$ **28.** $\dfrac{\sqrt[5]{x^3y^4}}{\sqrt[5]{xy^2}}$ **29.** $\dfrac{\sqrt{300a}}{5\sqrt{3}}$

30. Add: $3\sqrt{128} + 2\sqrt{18} + 2\sqrt{32}$.

Multiply.

31. $(\sqrt{20} + 2\sqrt{5})(\sqrt{20} - 3\sqrt{5})$ **32.** $(3 + \sqrt{x})^2$

33. Rationalize the denominator: $\dfrac{1 + \sqrt{2}}{3 - 5\sqrt{2}}$.

Solve.

34. $\sqrt[5]{x - 3} = 2$ **35.** $\sqrt{x - 6} = \sqrt{x + 9} - 3$

36. The diagonal of a square has length $7\sqrt{2}$ ft. Find the length of a side of the square.

In a right triangle, find the length of the side not given. Give an exact answer and an answer to three decimal places.

37. $a = 7, \quad b = 7$ **38.** $a = 1, \quad c = \sqrt{5}$

39. Express in terms of i: $\sqrt{-9} + \sqrt{-64}$. **40.** Subtract: $(5 + 8i) - (-2 + 3i)$.

Multiply.

41. $(3 - 4i)(3 + 7i)$ **42.** i^{95}

43. Divide: $\dfrac{-7 + 14i}{6 - 8i}$. **44.** Determine whether $1 + 2i$ is a solution of $x^2 + 2x + 5 = 0$.

45. Graph: $f(x) = 4 - \sqrt{x}$.

Skill Maintenance

Solve.

46. $\dfrac{11x}{x + 3} + \dfrac{33}{x} + 12 = \dfrac{99}{x^2 + 3x}$ **47.** $6x^2 = 13x + 5$

48. Divide and simplify:

$$\dfrac{x^3 - 27}{x^2 - 16} \div \dfrac{x^2 + 3x + 9}{x + 4}.$$

49. *Mowing Time.* Fran and Juan do all the mowing for the local community college. It takes Juan 9 hr more than Fran to do the mowing. Working together, they can complete the job in 20 hr. How long would it take each, working alone, to do the mowing?

Synthesis

50. Simplify: $\dfrac{1 - 4i}{4i(1 + 4i)^{-1}}$. **51.** Solve:

$$\sqrt{2x - 2} + \sqrt{7x + 4} = \sqrt{13x + 10}.$$

Cumulative Review: Chapters R–6

Simplify.

1. $-\dfrac{3}{5}\left(-\dfrac{7}{10}\right)$

2. $36.2 - 73.4$

3. $7c - [4 - 3(6c - 9)]$

4. $6^2 - 3^3 \cdot 2 + 3 \cdot 8^2$

5. $\left(\dfrac{4x^4 y^{-5}}{3x^{-3}}\right)^3$

6. $(2x^2 - 3x + 1) + (6x - 3x^3 + 7x^2 - 4)$

7. $(2x^2 - y)^2$

8. $(5x^2 - 2x + 1)(3x^2 + x - 2)$

9. $\dfrac{x^3 + 64}{x^2 - 49} \cdot \dfrac{x^2 - 14x + 49}{x^2 - 4x + 16}$

10. $\dfrac{x}{x + 2} + \dfrac{1}{x - 3} - \dfrac{x^2 - 2}{x^2 - x - 6}$

11. $\dfrac{\dfrac{y^2 - 5y - 6}{y^2 - 7y - 18}}{\dfrac{y^2 + 3y + 2}{y^2 + 4y + 4}}$

12. $(y^3 + 3y^2 - 5) \div (y + 2)$

13. $\sqrt[3]{-8x^3}$

14. $\sqrt{16x^2 - 32x + 16}$

15. $9\sqrt{75} + 6\sqrt{12}$

16. $\sqrt{2xy^2} \cdot \sqrt{8xy^3}$

17. $\dfrac{3\sqrt{5}}{\sqrt{6} - \sqrt{3}}$

18. $\sqrt[6]{\dfrac{m^{12} n^{24}}{64}}$

19. $6^{2/9} \cdot 6^{2/3}$

20. $(6 + i) - (3 - 4i)$

21. $\dfrac{2 - i}{6 + 5i}$

Solve.

22. $\dfrac{1}{5} + \dfrac{3}{10}x = \dfrac{4}{5}$

23. $M = \dfrac{1}{8}(c - 3)$, for c

24. $3a - 4 < 10 + 5a$

25. $-8 < x + 2 < 15$

26. $|3x - 6| = 2$

27. $3x + 5y = 30,$
$5x + 3y = 34$

28. $3x + 2y - z = -7,$
$-x + y + 2z = 9,$
$5x + 5y + z = -1$

29. $625 = 49y^2$

30. $\dfrac{6x}{x - 5} - \dfrac{300}{x^2 + 5x + 25} = \dfrac{2250}{x^3 - 125}$

31. $\dfrac{3x^2}{x + 2} + \dfrac{5x - 22}{x - 2} = \dfrac{-48}{x^2 - 4}$

32. $I = \dfrac{nE}{R + nr}$, for R

33. $\sqrt{4x + 1} - 2 = 3$

34. $2\sqrt{1 - x} = \sqrt{5}$

Graph.

35. $f(x) = -\dfrac{2}{3}x + 2$

36. $4x - 2y = 8$

37. $4x \geq 5y + 20$

38. $y \geq -3,$
$y \leq 2x + 3$

39. $g(x) = x^2 - x - 2$

40. $f(x) = |x + 4|$

41. $g(x) = \dfrac{4}{x - 3}$

42. $f(x) = 2 - \sqrt{x}$

Find the domain and the range of the function.

43.

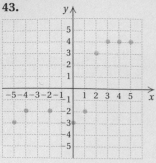

44.

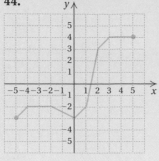

45.

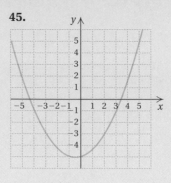

46.

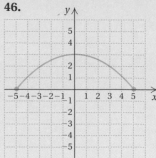

Factor.

47. $12x^2y^2 - 30xy^3$

48. $3x^2 - 17x - 28$

49. $y^2 - y - 132$

50. $27y^3 + 8$

51. $4x^2 - 625$

52. *Movie Costs.* The average cost of a movie ticket in Finland is $1.54 more than the average cost of a movie ticket in Sweden. The sum of the average costs is $14.44. How much does a movie ticket cost in each country?

53. Find the slope and the y-intercept of the line $3x - 2y = 8$.

54. Find an equation for the line perpendicular to the line $3x - y = 5$ and passing through $(1, 4)$.

55. *Heart Transplants.* In a recent year, the University of Pittsburgh Medical Center and the Medical College of Virginia had together performed 669 heart transplants. The University of Pittsburgh had performed 33 more than twice the number of transplants at the Medical College of Virginia. How many transplants had each hospital performed?

56. *Plowing Time.* One tractor can plow a field in 3 hr. Another tractor can plow the same field in 1.5 hr. How long should it take them to plow the field together?

57. The height h of triangles of fixed area varies inversely as the base b. Suppose the height is 100 ft when the base is 20 ft. Find the height when the base is 16 ft. What is the fixed area?

Synthesis

58. *The Pythagorean Formula in Three Dimensions.* The length d of a diagonal of a rectangular box is given by $d = \sqrt{a^2 + b^2 + c^2}$, where a, b, and c are the lengths of the sides. Find the length of a diagonal of a box whose sides have lengths 2 ft, 4 ft, and 5 ft.

59. Graph: $|x + y| \le 1$.

60. Solve: $\dfrac{x + \sqrt{x + 1}}{x - \sqrt{x + 1}} = \dfrac{5}{11}$.

61. Factor: $x^2 - 3x + \dfrac{5}{4}$.

7

Quadratic Equations and Functions

Introduction

We began our study of quadratic equations in Chapter 4 in connection with polynomials. Here we extend our equation-solving skills to those quadratic equations that do not lend themselves to being solved easily by factoring. We consider both real-number and complex-number solutions found using the quadratic formula.

We also learn to graph quadratic functions in various forms and to solve related applied problems.

An Application

Typically rivers are deepest in the middle, with the depth decreasing to 0 at the edges. A hydrologist measures the depths D, in feet, of a river at distances x, in feet, from one bank. Some of the data points found by the hydrologist are (0, 0), (50, 20), and (100, 0). Find a quadratic function that fits the data points and graph the function.

This problem appears as Example 7 in Section 7.7.

The Mathematics

The quadratic function is

$$D(x) = -0.008x^2 + 0.8x.$$

The graph is as shown below.

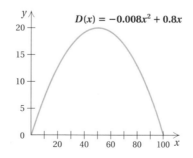

Pretest: Chapter 7

Solve.

1. $5x^2 + 15x = 0$

2. $y^2 + 4y + 8 = 0$

3. $x^2 - 10x + 25 = 0$

4. $3x^4 - 7x^2 + 2 = 0$

5. $\dfrac{2x}{2x + 1} + \dfrac{x + 1}{2x - 1} = \dfrac{6}{4x^2 - 1}$

6. $x + 3\sqrt{x} - 4 = 0$

7. Solve. Give the exact solution and approximate the solutions to three decimal places.
$$2x^2 + 4x - 1 = 0$$

8. Solve for T:
$$W = \sqrt{\dfrac{1}{RT}}.$$

9. Write a quadratic equation having solutions $\frac{2}{3}$ and -2.

10. For $f(x) = 2x^2 - 12x + 16$:
  a) Find the vertex.
  b) Find the line of symmetry.
  c) Graph the function.

11. Find the x-intercepts: $f(x) = x^2 - 6x + 4$.

12. Find the quadratic function that fits the data points $(0, 1)$, $(1, 0)$, and $(2, 7)$.

13. Find three consecutive even integers such that twice the product of the first two is equal to the square of the third plus five times the third.

14. *Car Travel.* During the first part of a trip, a car travels 100 mi at a certain speed. It travels 120 mi on the second part of the trip at a speed that is 8 mph faster. The total time for the trip is 5 hr. Find the speed of the car during each part of the trip.

15. *Corral Design.* What is the area of the largest rectangular horse corral that a rancher can enclose with 300 ft of fencing?

Solve.

16. $x(x - 3)(x + 5) < 0$

17. $\dfrac{x - 2}{x + 3} \geq 0$

Objectives for Retesting

The objectives to be tested in addition to the material in this chapter are as follows.

[2.7b]   Using a set of data, choose two representative points, find a linear function using the two points, and make predictions from the function.

[5.2b]   Add and subtract rational expressions.

[6.3a]   Multiply and simplify radical expressions.

[6.6a, b]   Solve radical equations.

7.1 The Basics of Solving Quadratic Equations

Objectives

a	Solve quadratic equations using the principle of square roots and find the *x*-intercepts of the graph of a related function.
b	Solve quadratic equations by completing the square.
c	Solve applied problems using quadratic equations.

For Extra Help

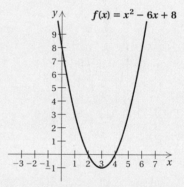

TAPE 15   TAPE 13B   MAC   CD-ROM
                          WIN

 Algebraic–Graphical Connection

Let's reexamine the graphical connections to the algebraic equation-solving concepts we have studied before.

In Chapter 2, we introduced the graph of a quadratic function:

$$f(x) = ax^2 + bx + c, \quad a \neq 0.$$

For example, the graph of the function $f(x) = x^2 + 6x + 8$ and its *x*-intercepts are shown below.

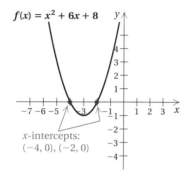

$f(x) = x^2 + 6x + 8$

x-intercepts:
$(-4, 0), (-2, 0)$

The *x*-intercepts are $(-4, 0)$ and $(-2, 0)$. These pairs are also the points of intersection of the graphs of $f(x) = x^2 + 6x + 8$ and $g(x) = 0$ (the *x*-axis). We will analyze the graphs of quadratic functions in greater detail in Sections 7.5–7.7.

In Chapter 4, we solved quadratic equations like $x^2 + 6x + 8 = 0$ using factoring, as here:

$$x^2 + 6x + 8 = 0$$
$$(x + 4)(x + 2) = 0 \qquad \text{Factoring}$$
$$x + 4 = 0 \quad or \quad x + 2 = 0 \qquad \text{Using the principle of zero products}$$
$$x = -4 \quad or \qquad x = -2.$$

We see that the solutions of $0 = x^2 + 6x + 8$, -4 and -2, are the first coordinates of the *x*-intercepts, $(-4, 0)$ and $(-2, 0)$, of the graph of $f(x) = x^2 + 6x + 8$.

Do Exercise 1.

We now enhance our ability to solve quadratic equations.

a   The Principle of Square Roots

The quadratic equation

$$5x^2 + 8x - 2 = 0$$

is said to be in **standard form.** The quadratic equation

$$5x^2 = 2 - 8x$$

is equivalent to the preceding, but it is *not* in standard form.

1. Consider solving the equation

$$x^2 - 6x + 8 = 0.$$

Below is the graph of

$$f(x) = x^2 - 6x + 8.$$

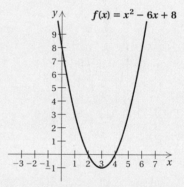

$f(x) = x^2 - 6x + 8$

a) What are the *x*-intercepts of the graph?

b) What are the solutions of $x^2 - 6x + 8 = 0$?

c) What relationship exists between the answers to parts (a) and (b)?

Answers on page A-39

> An equation of the type $ax^2 + bx + c = 0$, where a, b, and c are real-number constants and $a > 0$, is called the **standard form of a quadratic equation.**

For example, the equation $-5x^2 + 4x - 7 = 0$ is not in standard form. However, we can find an equivalent equation that is in standard form by multiplying by -1 on both sides:

$$-1(-5x^2 + 4x - 7) = -1(0)$$
$$5x^2 - 4x + 7 = 0.$$

In Section 4.7, we studied the use of factoring and the principle of zero products to solve certain quadratic equations. Let's review that procedure and introduce a new one.

2. a) Solve: $5x^2 = 8x - 3$.

Example 1

a) Solve: $3x^2 = 2 - x$.

b) Find the x-intercepts of $f(x) = 3x^2 + x - 2$.

a) We first find standard form. Then we factor and use the principle of zero products.

$$3x^2 = 2 - x$$

$3x^2 + x - 2 = 0$ Adding x and subtracting 2 to get the standard form

$(3x - 2)(x + 1) = 0$ Factoring

$3x - 2 = 0 \quad or \quad x + 1 = 0$ Using the principle of zero products

$\quad 3x = 2 \quad or \qquad x = -1$

$\quad\quad x = \frac{2}{3} \quad or \qquad x = -1$

b) Find the x-intercepts of $f(x) = 5x^2 - 8x + 3$.

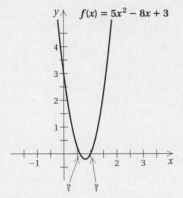

CHECK: 　For $\frac{2}{3}$:

$$3x^2 = 2 - x$$

$$3\left(\frac{2}{3}\right)^2 \;?\; 2 - \left(\frac{2}{3}\right)$$
$$3 \cdot \frac{4}{9} \;\bigg|\; \frac{6}{3} - \frac{2}{3}$$
$$\frac{4}{3} \;\bigg|\; \frac{4}{3} \qquad \text{TRUE}$$

For -1:

$$3x^2 = 2 - x$$

$$3(-1)^2 \;?\; 2 - (-1)$$
$$3 \cdot 1 \;\bigg|\; 2 + 1$$
$$3 \;\bigg|\; 3 \qquad \text{TRUE}$$

The solutions are -1 and $\frac{2}{3}$.

b) The x-intercepts of $f(x) = 3x^2 + x - 2$ are $(-1, 0)$ and $\left(\frac{2}{3}, 0\right)$. The solutions of the equation $3x^2 = 2 - x$ are the first coordinates of the x-intercepts of the graph of $f(x) = 3x^2 + x - 2$.

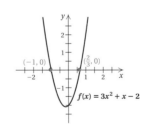

Do Exercise 2.

Answers on page A-39

Example 2

a) Solve: $x^2 = 25$.

b) Find the x-intercepts of $f(x) = x^2 - 25$.

a) We first find standard form and then factor:

$$x^2 - 25 = 0 \qquad \text{Subtracting 25}$$
$$(x - 5)(x + 5) = 0 \qquad \text{Factoring}$$
$$x - 5 = 0 \quad \text{or} \quad x + 5 = 0 \qquad \text{Using the principle of zero products}$$
$$x = 5 \quad \text{or} \qquad x = -5.$$

The solutions are 5 and -5.

b) The x-intercepts of $f(x) = x^2 - 25$ are $(-5, 0)$ and $(5, 0)$.

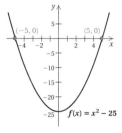

Example 3 Solve: $6x^2 - 15x = 0$.

We factor and use the principle of zero products:

$$6x^2 - 15x = 0$$
$$3x(2x - 5) = 0$$
$$3x = 0 \quad \text{or} \quad 2x - 5 = 0$$
$$x = 0 \quad \text{or} \qquad x = \tfrac{5}{2}.$$

The solutions are 0 and $\frac{5}{2}$. We leave the check to the student.

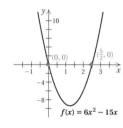

Do Exercises 3 and 4.

Solving Equations of the Type $x^2 = d$

Consider the equation $x^2 = 25$ again. We know from Chapter 6 that the number 25 has two real-number square roots, namely, 5 and -5. Note that these are the solutions of the equation in Example 2. This exemplifies the principle of square roots, which provides a quick method for solving equations of the type $x^2 = d$.

> **THE PRINCIPLE OF SQUARE ROOTS**
>
> The equation $x^2 = d$ has two real-number solutions when $d > 0$. The solutions are $\sqrt{d}$ and $-\sqrt{d}$.
>
> The equation $x^2 = 0$ has 0 as its only solution.
>
> The equation $x^2 = d$ has two imaginary-number solutions when $d < 0$.

Solve.

3. $x^2 = 16$

4. $4x^2 + 14x = 0$

Answers on page A-39

5. Solve: $5x^2 = 15$. Give the exact solution and approximate the solutions to three decimal places.

6. Solve: $-3x^2 + 8 = 0$. Give the exact solution and approximate the solutions to three decimal places.

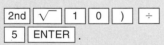
Answers on page A-39

Example 4 Solve: $3x^2 = 6$. Give the exact solution and approximate the solutions to three decimal places.

We have

$$3x^2 = 6$$
$$x^2 = 2$$
$$x = \sqrt{2} \quad or \quad x = -\sqrt{2}.$$

We often use the symbol $\pm\sqrt{2}$ to represent both of the solutions.

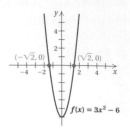

CHECK:

For $\sqrt{2}$:

$$\begin{array}{c} 3x^2 = 6 \\ \hline 3(\sqrt{2})^2 \ ? \ 6 \\ 3 \cdot 2 \ \Big| \\ 6 \ \Big| \quad \text{TRUE} \end{array}$$

For $-\sqrt{2}$:

$$\begin{array}{c} 3x^2 = 6 \\ \hline 3(-\sqrt{2})^2 \ ? \ 6 \\ 3 \cdot 2 \ \Big| \\ 6 \ \Big| \quad \text{TRUE} \end{array}$$

The solutions are $\sqrt{2}$ and $-\sqrt{2}$, or $\pm\sqrt{2}$, which are about 1.414 and -1.414 when written to three decimal places.

Do Exercise 5.

Sometimes we rationalize denominators to simplify answers.

Example 5 Solve: $-5x^2 + 2 = 0$. Give the exact solution and approximate the solutions to three decimal places.

$$-5x^2 + 2 = 0$$
$$x^2 = \frac{2}{5} \qquad \text{Subtracting 2 and dividing by } -5$$
$$x = \sqrt{\frac{2}{5}} \quad or \quad x = -\sqrt{\frac{2}{5}} \qquad \text{Using the principle of square roots}$$
$$x = \sqrt{\frac{2}{5} \cdot \frac{5}{5}} \quad or \quad x = -\sqrt{\frac{2}{5} \cdot \frac{5}{5}} \qquad \text{Rationalizing the denominators}$$
$$x = \frac{\sqrt{10}}{5} \quad or \quad x = -\frac{\sqrt{10}}{5}$$

CHECK: We check both numbers at once, since there is no x-term in the equation.

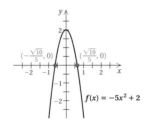

$$\begin{array}{c} -5x^2 + 2 = 0 \\ \hline -5\left(\pm\dfrac{\sqrt{10}}{5}\right)^2 + 2 \ ? \ 0 \\ -5\left(\dfrac{10}{25}\right) + 2 \\ -2 + 2 \\ 0 \ \Big| \quad \text{TRUE} \end{array}$$

The solutions are $\dfrac{\sqrt{10}}{5}$ and $-\dfrac{\sqrt{10}}{5}$, or $\pm\dfrac{\sqrt{10}}{5}$. We can use a calculator for an approximation:

$$\pm\frac{\sqrt{10}}{5} \approx \pm 0.632.$$

Do Exercise 6 on the preceding page.

Sometimes we get solutions that are imaginary numbers.

Example 6   Solve: $4x^2 + 9 = 0$.

$$4x^2 + 9 = 0$$

$$x^2 = -\frac{9}{4} \qquad \text{Subtracting 9 and dividing by 4}$$

$$x = \sqrt{-\frac{9}{4}} \quad \text{or} \quad x = -\sqrt{-\frac{9}{4}} \qquad \text{Using the principle of square roots}$$

$$x = \frac{3}{2}i \qquad \text{or} \quad x = -\frac{3}{2}i \qquad \text{Simplifying}$$

CHECK:

$$\frac{4x^2 + 9 = 0}{4\left(\pm\frac{3}{2}i\right)^2 + 9 \; ? \; 0}$$

$$4\left(-\frac{9}{4}\right) + 9$$

$$-9 + 9$$

$$0 \qquad \text{TRUE}$$

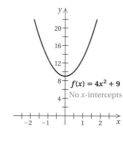

$f(x) = 4x^2 + 9$

No x-intercepts

The solutions are $\frac{3}{2}i$ and $-\frac{3}{2}i$, or $\pm\frac{3}{2}i$.

We see that the graph of $f(x) = 4x^2 + 9$ does not cross the x-axis. This is true because the equation $4x^2 + 9 = 0$ has imaginary solutions.

Do Exercise 7.

Solving Equations of the Type $(x + c)^2 = d$

The equation $(x - 2)^2 = 7$ can also be solved using the principle of square roots.

Example 7

a) Solve: $(x - 2)^2 = 7$.

b) Find the x-intercepts of $f(x) = (x - 2)^2 - 7$.

a) We have

$$(x - 2)^2 = 7$$

$$x - 2 = \sqrt{7} \qquad \text{or} \quad x - 2 = -\sqrt{7} \qquad \text{Using the principle of square roots}$$

$$x = 2 + \sqrt{7} \quad \text{or} \qquad x = 2 - \sqrt{7}.$$

The solutions are $2 + \sqrt{7}$ and $2 - \sqrt{7}$, or $2 \pm \sqrt{7}$.

b) The x-intercepts of $f(x) = (x - 2)^2 - 7$ are $(2 - \sqrt{7}, 0)$ and $(2 + \sqrt{7}, 0)$.

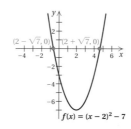

$(2 - \sqrt{7}, 0)$ $(2 + \sqrt{7}, 0)$

$f(x) = (x - 2)^2 - 7$

Do Exercise 8.

7. Solve: $2x^2 + 1 = 0$.

Calculator Spotlight

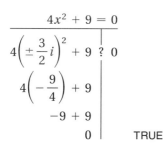 What happens when we use the SOLVE or INTERSECT feature to solve the equations in Example 6 and Margin Exercise 7?

8. a) Solve: $(x - 1)^2 = 5$.

b) Find the x-intercepts of $f(x) = (x - 1)^2 - 5$.

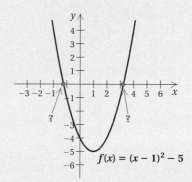

$f(x) = (x - 1)^2 - 5$

Answers on page A-39

9. Solve: $x^2 + 16x + 64 = 11$.

If we can express the left side of an equation as the square of a binomial, we can proceed as we did in Example 7.

Example 8 Solve: $x^2 + 6x + 9 = 2$.

We have

$$x^2 + 6x + 9 = 2 \qquad \text{The left side is the square of a binomial.}$$
$$(x + 3)^2 = 2$$
$$x + 3 = \sqrt{2} \qquad or \quad x + 3 = -\sqrt{2} \qquad \text{Using the principle of square roots}$$
$$x = -3 + \sqrt{2} \quad or \qquad x = -3 - \sqrt{2}.$$

The solutions are $-3 + \sqrt{2}$ and $-3 - \sqrt{2}$, or $-3 \pm \sqrt{2}$.

Do Exercise 9.

b | Completing the Square

We can solve quadratic equations like $3x^2 = 6$ and $(x - 2)^2 = 7$ by using the principle of square roots. We can also solve an equation such as $x^2 + 6x + 9 = 2$ in like manner because the expression on the left side is the square of a binomial, $(x + 3)^2$. This second procedure is the basis for a method called **completing the square.** *It can be used to solve any quadratic equation.*

Suppose we have the following quadratic equation:

$$x^2 + 14x = 4.$$

If we could add on both sides of the equation a constant that would make the expression on the left the square of a binomial, we could then solve the equation using the principle of square roots.

How can we determine what to add to $x^2 + 14x$ to construct the square of a binomial? We want to find a number a such that the following equation is satisfied:

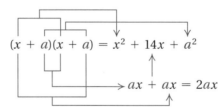

$$(x + a)(x + a) = x^2 + 14x + a^2$$
$$ax + ax = 2ax$$

Thus a is such that $2ax = 14x$. Solving for a, we get

$$a = \frac{14x}{2x} = \frac{14}{2} = 7.$$

That is, a is half of the coefficient of x in $x^2 + 14x$. Since $a^2 = \left(\frac{14}{2}\right)^2 = 7^2 = 49$, we add 49 to our original expression:

$$x^2 + 14x + 49 \text{ is the square of } x + 7;$$

that is,

$$x^2 + 14x + 49 = (x + 7)^2.$$

> When solving an equation, to **complete the square** of an expression like $x^2 + bx$, we take half the x-coefficient, which is $b/2$, and square. Then we add that number, $(b/2)^2$, on both sides.

Answer on page A-39

Returning to solving our original equation, we first add 49 on *both* sides to *complete the square.* Then we solve:

$$x^2 + 14x \qquad = 4 \qquad \text{Original equation}$$
$$x^2 + 14x + 49 = 4 + 49 \qquad \text{Adding 49: } \left(\tfrac{14}{2}\right)^2 = 7^2 = 49$$
$$(x + 7)^2 = 53$$
$$x + 7 = \sqrt{53} \qquad or \quad x + 7 = -\sqrt{53} \qquad \text{Using the principle of square roots}$$
$$x = -7 + \sqrt{53} \quad or \qquad x = -7 - \sqrt{53}.$$

The solutions are $-7 \pm \sqrt{53}$.

We have seen that a quadratic equation $(x + c)^2 = d$ can be solved using the principle of square roots. Any equation, such as $x^2 - 6x + 8 = 0$, can be put in this form by completing the square. Then we can solve as before.

Example 9 Solve: $x^2 - 6x + 8 = 0$.

We have

$$x^2 - 6x + 8 = 0$$
$$x^2 - 6x \qquad = -8. \qquad \text{Subtracting 8}$$

We take half of -6 and square it, to get 9. Then we add 9 on *both* sides of the equation. This makes the left side the square of a binomial, $x - 3$. We have now *completed the square.*

$$x^2 - 6x + 9 = -8 + 9 \qquad \text{Adding 9}$$
$$(x - 3)^2 = 1$$
$$x - 3 = 1 \quad or \quad x - 3 = -1 \qquad \text{Using the principle of square roots}$$
$$x = 4 \quad or \qquad x = 2$$

The solutions are 2 and 4.

Do Exercises 10 and 11.

Example 10 Solve $x^2 + 4x - 7 = 0$ by completing the square.

$$x^2 + 4x - 7 = 0$$
$$x^2 + 4x \qquad = 7 \qquad \text{Adding 7}$$
$$x^2 + 4x + 4 = 7 + 4 \qquad \text{Adding 4: } \left(\tfrac{4}{2}\right)^2 = (2)^2 = 4$$
$$(x + 2)^2 = 11$$
$$x + 2 = \sqrt{11} \qquad or \quad x + 2 = -\sqrt{11} \qquad \text{Using the principle of square roots}$$
$$x = -2 + \sqrt{11} \quad or \qquad x = -2 - \sqrt{11}$$

The solutions are $-2 \pm \sqrt{11}$.

Do Exercise 12.

Solve.

10. $x^2 + 6x + 8 = 0$

11. $x^2 - 8x - 20 = 0$

12. Solve by completing the square:

$$x^2 + 6x - 1 = 0.$$

Answers on page A-39

13. Solve by completing the square:

$$3x^2 - 2x = 7.$$

When the coefficient of x^2 is not 1, we can make it 1, as shown in the following example.

Example 11 Solve $2x^2 = 3x - 7$ by completing the square.

$$2x^2 = 3x - 7$$

$$2x^2 - 3x = -7$$

$$\frac{1}{2}(2x^2 - 3x) = \frac{1}{2} \cdot (-7) \qquad \text{Multiplying by } \tfrac{1}{2} \text{ to make the } x^2\text{-coefficient 1}$$

$$x^2 - \frac{3}{2}x = -\frac{7}{2} \qquad \text{Multiplying and simplifying}$$

$$x^2 - \frac{3}{2}x + \frac{9}{16} = -\frac{7}{2} + \frac{9}{16} \qquad \text{Adding } \tfrac{9}{16}: \left[\tfrac{1}{2}\left(-\tfrac{3}{2}\right)\right]^2 = \left[-\tfrac{3}{4}\right]^2 = \tfrac{9}{16}$$

$$\left(x - \frac{3}{4}\right)^2 = -\frac{56}{16} + \frac{9}{16} \qquad \text{Finding a common denominator}$$

$$\left(x - \frac{3}{4}\right)^2 = -\frac{47}{16}$$

$$x - \frac{3}{4} = \sqrt{-\frac{47}{16}} \quad or \quad x - \frac{3}{4} = -\sqrt{-\frac{47}{16}} \qquad \text{Using the principle of square roots}$$

$$x - \frac{3}{4} = i\sqrt{\frac{47}{16}} \quad or \quad x - \frac{3}{4} = -i\sqrt{\frac{47}{16}} \qquad i\sqrt{-1} = i$$

$$x = \frac{3}{4} + \frac{i\sqrt{47}}{4} \quad or \quad x = \frac{3}{4} - \frac{i\sqrt{47}}{4}$$

The solutions are $\dfrac{3}{4} \pm i\dfrac{\sqrt{47}}{4}$.

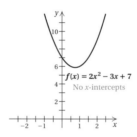

$f(x) = 2x^2 - 3x + 7$

No x-intercepts

SOLVING BY COMPLETING THE SQUARE

To solve an equation $ax^2 + bx + c = 0$ by completing the square:

1. If $a \neq 1$, multiply by $1/a$ so that the x^2-coefficient is 1.

2. If the x^2-coefficient is 1, add or subtract so that the equation is in the form

$$x^2 + bx = -c, \quad \text{or} \quad x^2 + \frac{b}{a}x = -\frac{c}{a} \quad \text{if step (1) has been applied.}$$

3. Take half of the x-coefficient and square it. Add the result on both sides of the equation.

4. Express the side with the variables as the square of a binomial.

5. Use the principle of square roots and complete the solution.

Do Exercise 13.

Answer on page A-39

c | Applications and Problem Solving

If you put money in a savings account, the bank will pay you interest. If interest is **compounded annually,** the bank, at the end of a year, will start paying interest on both the original amount and the interest that has been earned.

> **THE COMPOUND-INTEREST FORMULA**
>
> If an amount of money P is invested at interest rate r, compounded annually, then in t years, it will grow to the amount A given by
>
> $$A = P(1 + r)^t.$$

We can use quadratic equations to solve certain interest problems.

Example 12 *Compound Interest.* $1000 invested at 8.4%, compounded annually, for 2 yr will grow to what amount?

$$A = P(1 + r)^t$$
$$= 1000(1 + 0.084)^2 \quad \text{Substituting into the formula; 8.4\% = 0.0084}$$
$$= 1000(1.084)^2$$
$$= 1000(1.175056)$$
$$\approx 1175.06. \quad \text{Computing and rounding}$$

The amount is $1175.06.

Example 13 *Compound Interest.* $4000 is invested at interest rate r, compounded annually. In 2 yr, it grows to $4410. What is the interest rate?

We know that $4000 is originally invested. Thus P is $4000. That amount grows to $4410 in 2 yr. Thus when t is 2, A is $4410. We substitute 4000 for P, 4410 for A, and 2 for t in the formula, and solve the resulting equation for r:

$$A = P(1 + r)^t$$
$$4410 = 4000(1 + r)^2$$
$$\frac{4410}{4000} = (1 + r)^2$$
$$\frac{441}{400} = (1 + r)^2.$$

We then have

$$\sqrt{\tfrac{441}{400}} = 1 + r \quad or \quad -\sqrt{\tfrac{441}{400}} = 1 + r \quad \text{Using the principle of square roots}$$
$$\tfrac{21}{20} = 1 + r \quad or \quad -\tfrac{21}{20} = 1 + r \quad \text{Simplifying}$$
$$-\tfrac{20}{20} + \tfrac{21}{20} = r \quad or \quad -\tfrac{20}{20} - \tfrac{21}{20} = r$$
$$\tfrac{1}{20} = r \quad or \quad -\tfrac{41}{20} = r.$$

Since the interest rate cannot be negative, we have

$$\tfrac{1}{20} = r$$
$$r = 0.05, \text{ or } 5\%.$$

The interest rate must be 5%.

Do Exercises 14 and 15.

14. *Compound Interest.* Suppose that $3000 is invested at 9.6%, compounded annually, for 2 yr. To what amount will the investment grow?

15. *Compound Interest.* Suppose that $2500 is invested at interest rate r, compounded annually. In 2 yr, the investment grows to $3600. What is the interest rate?

Answers on page A-39

16. *Hang Time.* Anfernee Hardaway, of the Orlando Magic, has a hang time of about 0.866 sec. What is his vertical leap?

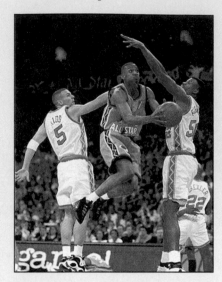

17. *Hang Time.* The record for the greatest vertical leap in the NBA is held by Darryl Griffith of the Utah Jazz. It was 48 in. What was his hang time?

Example 14   *Hang Time.* One of the most exciting plays in basketball is the dunk shot. The amount of time T that passes from the moment a player leaves the ground, goes up, makes the shot, and arrives back on the ground is called the *hang time*. A function relating an athlete's vertical leap V, in inches, to hang time T, in seconds, is given by

$$V(t) = 48T^2.$$

a) Michael Jordan, of the Chicago Bulls, has a hang time of about 0.889 sec. What is his vertical leap?

b) Although his height is only 5 ft 7 in., Spud Webb, formerly of the Sacramento Kings, had a vertical leap of about 44 in. What is his hang time? (***Source:*** Peter Brancazio, "The Mechanics of a Slam Dunk," *Popular Mechanics*, November 1991. Courtesy of Professor Peter Brancazio, Brooklyn College.)

a) To find Jordan's vertical leap, we substitute 0.889 for T in the function and compute V:

$$V(0.889) = 48(0.889)^2 \approx 37.9 \text{ in.}$$

Jordan's vertical leap is about 37.9 in. Surprisingly, Jordan does not have the vertical leap most fans would expect.

b) To find Webb's hang time, we substitute 44 for V and solve for T:

$$44 = 48T^2 \qquad \text{Substituting 44 for } V$$

$$\frac{44}{48} = T^2 \qquad \text{Solving for } T^2$$

$$0.91\overline{6} = T^2$$

$$\sqrt{0.91\overline{6}} = T \qquad \text{Hang time is positive.}$$

$$0.957 \approx T. \qquad \text{Using a calculator}$$

Webb's hang time is 0.957 sec. Note that his hang time is greater than Jordan's.

Do Exercises 16 and 17.

Answers on page A-39

Exercise Set 7.1

a

1. a) Solve:
   $6x^2 = 30$.
   b) Find the
   x-intercepts of
   $f(x) = 6x^2 - 30$.

2. a) Solve:
   $5x^2 = 35$.
   b) Find the
   x-intercepts of
   $f(x) = 5x^2 - 35$.

3. a) Solve:
   $9x^2 + 25 = 0$.
   b) Find the
   x-intercepts of
   $f(x) = 9x^2 + 25$.

4. a) Solve:
   $36x^2 + 49 = 0$.
   b) Find the
   x-intercepts of
   $f(x) = 36x^2 + 49$.

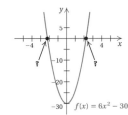
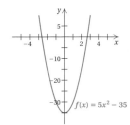
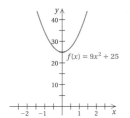
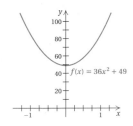

Solve. Give the exact solution and approximate solutions to three decimal places, when appropriate.

5. $2x^2 - 3 = 0$

6. $3x^2 - 7 = 0$

7. $(x + 2)^2 = 49$

8. $(x - 1)^2 = 6$

9. $(x - 4)^2 = 16$

10. $(x + 3)^2 = 9$

11. $(x - 11)^2 = 7$

12. $(x - 9)^2 = 34$

13. $(x - 7)^2 = -4$

14. $(x + 1)^2 = -9$

15. $(x - 9)^2 = 81$

16. $(t - 2)^2 = 25$

17. $\left(x - \frac{3}{2}\right)^2 = \frac{7}{2}$

18. $\left(y + \frac{3}{4}\right)^2 = \frac{17}{16}$

19. $x^2 + 6x + 9 = 64$

20. $x^2 + 10x + 25 = 100$

21. $y^2 - 14y + 49 = 4$

22. $p^2 - 8p + 16 = 1$

b Solve by completing the square. Show your work.

23. $x^2 + 4x = 2$

24. $x^2 + 2x = 5$

25. $x^2 - 22x = 11$

26. $x^2 - 18x = 10$

27. $x^2 + x = 1$

28. $x^2 - x = 3$

29. $t^2 - 5t = 7$

30. $y^2 + 9y = 8$

31. $x^2 + \frac{3}{2}x = 3$

32. $x^2 - \frac{4}{3}x = \frac{2}{3}$

33. $m^2 - \frac{9}{2}m = \frac{3}{2}$

34. $r^2 + \frac{2}{5}r = \frac{4}{5}$

35. $x^2 + 6x - 16 = 0$

36. $x^2 - 8x + 15 = 0$

37. $x^2 + 22x + 102 = 0$ **38.** $x^2 + 18x + 74 = 0$ **39.** $x^2 - 10x - 4 = 0$ **40.** $x^2 + 10x - 4 = 0$

41. a) Solve:
$x^2 + 7x - 2 = 0$.
b) Find the
x-intercepts of
$f(x) = x^2 + 7x - 2$.

42. a) Solve:
$x^2 - 7x - 2 = 0$.
b) Find the
x-intercepts of
$f(x) = x^2 - 7x - 2$.

43. a) Solve:
$2x^2 - 5x + 8 = 0$.
b) Find the
x-intercepts of
$f(x) = 2x^2 - 5x + 8$.

44. a) Solve:
$2x^2 - 3x + 9 = 0$.
b) Find the
x-intercepts of
$f(x) = 2x^2 - 3x + 9$.

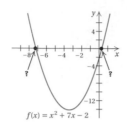
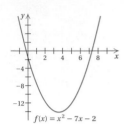
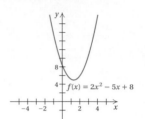
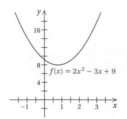

Solve by completing the square. Show your work.

45. $x^2 - \frac{3}{2}x - \frac{1}{2} = 0$ **46.** $x^2 + \frac{3}{2}x - 2 = 0$ **47.** $2x^2 - 3x - 17 = 0$ **48.** $2x^2 + 3x - 1 = 0$

49. $3x^2 - 4x - 1 = 0$ **50.** $3x^2 + 4x - 3 = 0$ **51.** $x^2 + x + 2 = 0$ **52.** $x^2 - x + 1 = 0$

53. $x^2 - 4x + 13 = 0$ **54.** $x^2 - 6x + 13 = 0$

c *Compound Interest.* For Exercises 55–60, use the compound-interest formula $A = P(1 + r)^t$ to determine the interest rate.

55. $1000 grows to $1210 in 2 yr

56. $2560 grows to $2890 in 2 yr

57. $8000 grows to $10,125 in 2 yr

58. $10,000 grows to $10,404 in 2 yr

59. $6250 grows to $6760 in 2 yr

60. $6250 grows to $7290 in 2 yr

Hang Time. For Exercises 61 and 62, use the hang-time function $V(T) = 48T^2$, relating vertical leap to hang time.

61. A basketball player has a vertical leap of 36 in. What is his hang time?

62. Tracy McGrady, an NBA player, has a vertical leap of 40 in. What is his hang time?

Free-Falling Objects. The function $s(t) = 16t^2$ is used to approximate the distance s, in feet, that an object falls freely from rest in t seconds. Use the formula for Exercises 63–66.

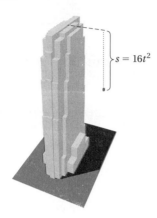

$s = 16t^2$

63. The RCA Building in New York City is 850 ft tall. How long will it take an object to fall from the top?

64. The CN Tower in Toronto, at 1815 ft, is the world's tallest self-supporting tower (no guy wires) (***Source***: *The Guinness Book of Records*). How long would it take an object to fall freely from the top?

65. Reaching 745 ft above the water, the towers of California's Golden Gate Bridge are the world's tallest bridge towers (***Source***: *The Guinness Book of Records*). How long would it take an object to fall freely from the top?

66. The Gateway Arch in St. Louis is 640 ft high. How long would it take an object to fall freely from the top?

Skill Maintenance

Solve.   [2.7b]

67. *Computer Usage in Schools.* Schools are spending more and more on technology. The following graph shows the ratio of number of students to number of computers for recent years.

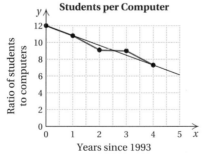

Source: Market Data Retrieval, Inc.

a) Use the two points (0, 12) and (4, 7.3) to find a linear function that fits the data.
b) Graph the function.
c) Use the function to predict the ratio of number of students to number of computers in 1998 and in 2000.

68. *Computers in Schools.* The inventory of computers in schools has soared in recent years. The following graph shows the number of computers, in millions, in schools for recent years.

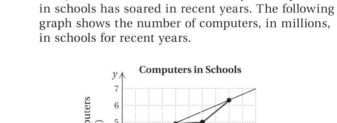

Source: Market Data Retrieval, Inc.

a) Use the two points (0, 3.6) and (4, 6.3) to find a linear function that fits the data.
b) Graph the function.
c) Use the function to predict the number of computers in schools in 1998 and in 2000.

Graph. [2.2c], [2.5a]

69. $f(x) = 5 - 2x$

70. $f(x) = 5 - 2x^2$

71. $f(x) = |5 - 2x|$

72. $2x - 5y = 10$

73. Simplify: $\sqrt{88}$. [6.3a]

74. Rationalize the denominator: $\sqrt{\frac{2}{5}}$. [6.5a]

Synthesis

75. ◆ Explain in your own words a sequence of steps that you might follow to solve any quadratic equation.

76. ◆ Example 13 can be solved with a grapher by graphing each side of
$$4410 = 4000(1 + r)^2$$
and using the INTERSECT feature. How could you determine, from a reading of the problem, a suitable viewing window? What might that window be?

77. Problems such as those in Exercises 17, 21, and 25 can be solved without first finding standard form by using the INTERSECT feature of a grapher. We let y_1 = the left side of the equation and y_2 = the right side. Use a grapher to solve Exercises 17, 21, and 25 in this manner.

78. Use a grapher to solve each of the following equations.

a) $25.55x^2 - 1635.2 = 0$
b) $-0.0644x^2 + 0.0936x + 4.56 = 0$
c) $2.101x + 3.121 = 0.97x^2$

Find b such that the trinomial is a square.

79. $x^2 + bx + 64$

80. $x^2 + bx + 75$

Solve.

81. $x(2x^2 + 9x - 56)(3x + 10) = 0$

82. $\left(x - \frac{1}{3}\right)\left(x - \frac{1}{3}\right) + \left(x - \frac{1}{3}\right)\left(x + \frac{2}{9}\right) = 0$

83. *Boating.* A barge and a fishing boat leave a dock at the same time, traveling at right angles to each other. The barge travels 7 km/h slower than the fishing boat. After 4 hr, the boats are 68 km apart. Find the speed of each vessel.

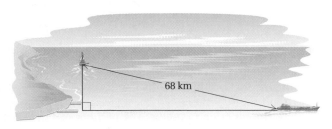

68 km

84. Find three consecutive integers such that the square of the first plus the product of the other two is 67.

Discover the rule for completing the square.

Collaborative Learning Manual

7.2 The Quadratic Formula

There are at least two reasons for learning to complete the square. One is to enhance your ability to graph certain equations that are needed to solve problems later in this chapter. The other is to prove a general formula for solving quadratic equations.

a | Solving Using the Quadratic Formula

Each time you solve by completing the square, the procedure is the same. When we do the same kind of procedure many times, we look for a formula to speed up our work. Consider

$$ax^2 + bx + c = 0, \quad a > 0.$$

Let's solve by *completing the square*. As we carry out the steps, compare them with Example 11 in the preceding section.

$$x^2 + \frac{b}{a}x + \frac{c}{a} = 0 \qquad \text{Multiplying by } \frac{1}{a}$$

$$x^2 + \frac{b}{a}x \phantom{+ \frac{c}{a}} = -\frac{c}{a} \qquad \text{Subtracting } \frac{c}{a}$$

Half of $\frac{b}{a}$ is $\frac{b}{2a}$. The square is $\frac{b^2}{4a^2}$. We add $\frac{b^2}{4a^2}$ on both sides:

$$x^2 + \frac{b}{a}x + \frac{b^2}{4a^2} = -\frac{c}{a} + \frac{b^2}{4a^2} \qquad \text{Adding } \frac{b^2}{4a^2}$$

$$\left(x + \frac{b}{2a}\right)^2 = -\frac{4ac}{4a^2} + \frac{b^2}{4a^2} \qquad \begin{array}{l}\text{Factoring the left side and finding a} \\ \text{common denominator on the right}\end{array}$$

$$\left(x + \frac{b}{2a}\right)^2 = \frac{b^2 - 4ac}{4a^2}$$

$$x + \frac{b}{2a} = \sqrt{\frac{b^2 - 4ac}{4a^2}} \quad \text{or} \quad x + \frac{b}{2a} = -\sqrt{\frac{b^2 - 4ac}{4a^2}} \qquad \begin{array}{l}\text{Using the principle} \\ \text{of square roots}\end{array}$$

Since $a > 0$, $\sqrt{4a^2} = 2a$, so we can simplify as follows:

$$x + \frac{b}{2a} = \frac{\sqrt{b^2 - 4ac}}{2a} \quad \text{or} \quad x + \frac{b}{2a} = -\frac{\sqrt{b^2 - 4ac}}{2a}.$$

Thus,

$$x = -\frac{b}{2a} \pm \frac{\sqrt{b^2 - 4ac}}{2a}, \quad \text{or} \quad x = \frac{-b \pm \sqrt{b^2 - 4ac}}{2a}.$$

Note that the formula also holds when $a < 0$. A similar proof would show this, but we will not consider it here.

AG Algebraic–Graphical Connection

The Quadratic Formula (Algebraic).
The solutions of $ax^2 + bx + c = 0, a \neq 0,$ are given by

$$x = \frac{-b \pm \sqrt{b^2 - 4ac}}{2a}.$$

The Quadratic Formula (Graphical).
The x-intercepts of the graph of the function $f(x) = ax^2 + bx + c, a \neq 0,$ are given by

$$\left(\frac{-b \pm \sqrt{b^2 - 4ac}}{2a}, 0\right).$$

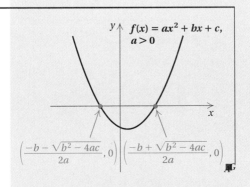

$f(x) = ax^2 + bx + c,$ $a > 0$

$\left(\dfrac{-b - \sqrt{b^2 - 4ac}}{2a}, 0\right)$ $\left(\dfrac{-b + \sqrt{b^2 - 4ac}}{2a}, 0\right)$

Objective

a Solve quadratic equations using the quadratic formula, and approximate solutions using a calculator.

For Extra Help

TAPE 15   TAPE 13B   MAC   CD-ROM
WIN

1. Consider the equation

$$2x^2 = 4 + 7x.$$

a) Solve using the quadratic formula.

Example 1 Solve $5x^2 + 8x = -3$ using the quadratic formula.

We first find standard form and determine a, b, and c:

$$5x^2 + 8x + 3 = 0;$$
$$a = 5, \quad b = 8, \quad c = 3.$$

We then use the quadratic formula:

$$x = \frac{-b \pm \sqrt{b^2 - 4ac}}{2a}$$

$$x = \frac{-8 \pm \sqrt{8^2 - 4 \cdot 5 \cdot 3}}{2 \cdot 5} \qquad \text{Substituting}$$

$$x = \frac{-8 \pm \sqrt{64 - 60}}{10}$$

| Be sure to write the fraction bar all the way across. |

$$x = \frac{-8 \pm \sqrt{4}}{10}$$

$$x = \frac{-8 \pm 2}{10}$$

$$x = \frac{-8 + 2}{10} \quad or \quad x = \frac{-8 - 2}{10}$$

$$x = \frac{-6}{10} \quad or \quad x = \frac{-10}{10}$$

$$x = -\frac{3}{5} \quad or \quad x = -1.$$

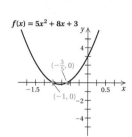

$f(x) = 5x^2 + 8x + 3$

$\left(-\frac{3}{5}, 0\right)$

$(-1, 0)$

The solutions are $-\frac{3}{5}$ and -1.

b) Solve by factoring.

It turns out that we could have solved the equation in Example 1 more easily by factoring as follows:

$$5x^2 + 8x + 3 = 0$$
$$(5x + 3)(x + 1) = 0$$
$$5x + 3 = 0 \quad or \quad x + 1 = 0$$
$$5x = -3 \quad or \qquad x = -1$$
$$x = -\tfrac{3}{5} \quad or \qquad x = -1.$$

We will see in Example 2 that we cannot always rely on factoring.

To solve a quadratic equation:

1. Check for the form $x^2 = d$ or $(x + c)^2 = d$. If it is in this form, use the principle of square roots as in Section 7.1.

2. If it is not in the form of step (1), write it in standard form $ax^2 + bx + c = 0$ with a and b nonzero.

3. Then try factoring.

4. If it is not possible to factor or if factoring seems difficult, use the quadratic formula.

The solutions of a quadratic equation cannot always be found by factoring. They can always be found using the quadratic formula.

Do Exercise 1.

Answers on page A-40

Example 2 Solve: $5x^2 - 8x = 3$. Give the exact solution and approximate the solutions to three decimal places.

We first find standard form and determine a, b, and c:

$$5x^2 - 8x - 3 = 0;$$
$$a = 5, \quad b = -8, \quad c = -3.$$

We then use the quadratic formula:

$$x = \frac{-(-8) \pm \sqrt{(-8)^2 - 4 \cdot 5 \cdot (-3)}}{2 \cdot 5} \quad \text{Substituting}$$

$$= \frac{8 \pm \sqrt{64 + 60}}{10} = \frac{8 \pm \sqrt{124}}{10} = \frac{8 \pm \sqrt{4 \cdot 31}}{10}$$

$$= \frac{8 \pm 2\sqrt{31}}{10} = \frac{2(4 \pm \sqrt{31})}{2 \cdot 5} = \frac{4 \pm \sqrt{31}}{5}.$$

CAUTION! To avoid a common error in simplifying, remember to *factor the numerator and the denominator* and then remove a factor of 1.

We can use a calculator to approximate the solutions:

$$\frac{4 + \sqrt{31}}{5} \approx 1.914; \qquad \frac{4 - \sqrt{31}}{5} \approx -0.314.$$

CHECK: Checking the exact solutions $(4 \pm \sqrt{31})/5$ can be quite cumbersome. It could be done on a calculator or by using the approximations. Here we check 1.914; the check for -0.314 is left to the student.

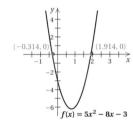

For 1.914:

$$\begin{array}{c}
5x^2 - 8x = 3 \\
\hline
5(1.914)^2 - 8(1.914) \; ? \; 3 \\
5(3.663396) - 15.312 \\
18.31698 - 15.312 \\
3.00498 \quad \text{Approximately TRUE}
\end{array}$$

We do not have a perfect check due to the rounding error. But our check seems to confirm the solutions.

Do Exercise 2.

Some quadratic equations have solutions that are nonreal complex numbers.

Example 3 Solve: $x^2 + x + 1 = 0$.

We have $a = 1$, $b = 1$, $c = 1$. We use the quadratic formula:

$$x = \frac{-1 \pm \sqrt{1^2 - 4 \cdot 1 \cdot 1}}{2 \cdot 1}$$

$$= \frac{-1 \pm \sqrt{1 - 4}}{2}$$

$$= \frac{-1 \pm \sqrt{-3}}{2}$$

$$= \frac{-1 \pm i\sqrt{3}}{2}.$$

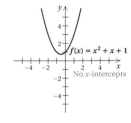

2. Solve using the quadratic formula:

$$3x^2 + 2x = 7.$$

Give the exact solution and approximate solutions to three decimal places.

Answers on page A-40

3. Solve: $x^2 - x + 2 = 0$.

The solutions are

$$\frac{-1 + i\sqrt{3}}{2} \quad \text{and} \quad \frac{-1 - i\sqrt{3}}{2}.$$

The solutions can also be expressed in the form

$$-\frac{1}{2} + i\frac{\sqrt{3}}{2} \quad \text{and} \quad -\frac{1}{2} - i\frac{\sqrt{3}}{2}.$$

Do Exercise 3.

Example 4 Solve: $2 + \dfrac{7}{x} = \dfrac{5}{x^2}$. Give the exact solution and approximate solutions to three decimal places.

We first find standard form:

$$x^2\left(2 + \frac{7}{x}\right) = x^2 \cdot \frac{5}{x^2} \qquad \begin{array}{l}\text{Multiplying by } x^2 \text{ to clear} \\ \text{fractions, noting that } x \neq 0\end{array}$$

$$2x^2 + 7x = 5$$

$$2x^2 + 7x - 5 = 0 \qquad \text{Subtracting 5}$$

$$a = 2, \quad b = 7, \quad c = -5$$

$$x = \frac{-7 \pm \sqrt{7^2 - 4 \cdot 2 \cdot (-5)}}{2 \cdot 2}$$

$$x = \frac{-7 \pm \sqrt{49 + 40}}{4} = \frac{-7 \pm \sqrt{89}}{4}$$

$$x = \frac{-7 + \sqrt{89}}{4} \quad \text{or} \quad x = \frac{-7 - \sqrt{89}}{4}.$$

4. Solve: $3 = \dfrac{5}{x} + \dfrac{4}{x^2}$.

Give the exact solution and approximate solutions to three decimal places.

The quadratic formula always gives correct results when we begin with the standard form. In such cases, we need check only to detect errors in substitution and computation. In this case, since we began with a rational equation, which was *not* in standard form, we *do* need to check. We cleared the fractions before obtaining standard form, and this step could introduce numbers that do not check in the original equation. At the very least, we need to show that neither of the numbers makes a denominator 0. Since neither of them does, the solutions are

$$\frac{-7 + \sqrt{89}}{4} \quad \text{and} \quad \frac{-7 - \sqrt{89}}{4}.$$

We can use a calculator to approximate the solutions:

$$\frac{-7 + \sqrt{89}}{4} \approx 0.608;$$

$$\frac{-7 - \sqrt{89}}{4} \approx -4.108.$$

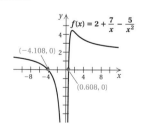

Do Exercise 4.

Answers on page A-40

Exercise Set 7.2

a Solve.

1. $x^2 + 6x + 4 = 0$ **2.** $x^2 - 6x - 4 = 0$ **3.** $3p^2 = -8p - 1$ **4.** $3u^2 = 18u - 6$

5. $x^2 - x + 1 = 0$ **6.** $x^2 + x + 2 = 0$ **7.** $x^2 + 13 = 4x$ **8.** $x^2 + 13 = 6x$

9. $r^2 + 3r = 8$ **10.** $h^2 + 4 = 6h$ **11.** $1 + \dfrac{2}{x} + \dfrac{5}{x^2} = 0$ **12.** $1 + \dfrac{5}{x^2} = \dfrac{2}{x}$

13. a) Solve: $3x + x(x - 2) = 0$.
    b) Find the x-intercepts of
         $f(x) = 3x + x(x - 2)$.

14. a) Solve: $4x + x(x - 3) = 0$.
    b) Find the x-intercepts of
         $f(x) = 4x + x(x - 3)$.

15. a) Solve: $11x^2 - 3x - 5 = 0$.
    b) Find the x-intercepts of
         $f(x) = 11x^2 - 3x - 5$.

16. a) Solve: $7x^2 + 8x = -2$.
    b) Find the x-intercepts of
         $f(x) = 7x^2 + 8x + 2$.

17. a) Solve: $25x^2 = 20x - 4$.
    b) Find the x-intercepts of
         $f(x) = 25x^2 - 20x + 4$.

18. a) Solve: $49x^2 - 14x + 1 = 0$.
    b) Find the x-intercepts of
         $f(x) = 49x^2 - 14x + 1$.

Solve.

19. $4x(x - 2) - 5x(x - 1) = 2$ **20.** $3x(x + 1) - 7x(x + 2) = 6$

21. $14(x - 4) - (x + 2) = (x + 2)(x - 4)$ **22.** $11(x - 2) + (x - 5) = (x + 2)(x - 6)$

23. $5x^2 = 17x - 2$ **24.** $15x = 2x^2 + 16$ **25.** $x^2 + 5 = 4x$ **26.** $x^2 + 5 = 2x$

27. $x + \dfrac{1}{x} = \dfrac{13}{6}$ **28.** $\dfrac{3}{x} + \dfrac{x}{3} = \dfrac{5}{2}$ **29.** $\dfrac{1}{y} + \dfrac{1}{y + 2} = \dfrac{1}{3}$ **30.** $\dfrac{1}{x} + \dfrac{1}{x + 4} = \dfrac{1}{7}$

31. $(2t - 3)^2 + 17t = 15$

32. $2y^2 - (y + 2)(y - 3) = 12$

33. $(x - 2)^2 + (x + 1)^2 = 0$

34. $(x + 3)^2 + (x - 1)^2 = 0$

35. $x^3 - 1 = 0$
(*Hint*: Factor the difference of cubes. Then use the quadratic formula.)

36. $x^3 + 27 = 0$

Solve. Give the exact solution and approximate solutions to three decimal places.

37. $x^2 + 6x + 4 = 0$

38. $x^2 + 4x - 7 = 0$

39. $x^2 - 6x + 4 = 0$

40. $x^2 - 4x + 1 = 0$

41. $2x^2 - 3x - 7 = 0$

42. $3x^2 - 3x - 2 = 0$

43. $5x^2 = 3 + 8x$

44. $2y^2 + 2y - 3 = 0$

Skill Maintenance

Solve. [6.6a, b]

45. $x = \sqrt{x + 2}$

46. $x = \sqrt{15 - 2x}$

47. $\sqrt{x + 2} = \sqrt{2x - 8}$

48. $\sqrt{x + 1} + 2 = \sqrt{3x + 1}$

49. $\sqrt{x + 5} = -7$

50. $\sqrt{2x - 6} + 11 = 2$

51. $\sqrt[3]{4x - 7} = 2$

52. $\sqrt[4]{3x - 1} = 2$

Synthesis

53. ◈ The list of steps on p. 538 does not mention completing the square as a method of solving quadratic equations. Why not?

54. ◈ Given the solutions of a quadratic equation, is it possible to reconstruct the original equation? Why or why not?

55. 〰 Use the SOLVE or INTERSECT feature to solve the equations in Exercises 3, 16, 17, and 37. Then solve $2.2x^2 + 0.5x - 1 = 0$.

56. 〰 Use a grapher to solve the equations in Exercises 9, 27, and 30 using the INTERSECT feature, letting y_1 = the left side and y_2 = the right side. Then solve $5.33x^2 = 8.23x + 3.24$.

Solve.

57. $2x^2 - x - \sqrt{5} = 0$

58. $\dfrac{5}{x} + \dfrac{x}{4} = \dfrac{11}{7}$

59. $ix^2 - x - 1 = 0$

60. $\sqrt{3}\,x^2 + 6x + \sqrt{3} = 0$

61. $\dfrac{x}{x + 1} = 4 + \dfrac{1}{3x^2 - 3}$

62. $(1 + \sqrt{3}\,)x^2 - (3 + 2\sqrt{3}\,)x + 3 = 0$

63. Let $f(x) = (x - 3)^2$. Find all inputs x such that $f(x) = 13$.

64. Let $f(x) = x^2 + 14x + 49$. Find all inputs x such that $f(x) = 36$.

7.3 Applications Involving Quadratic Equations

a | Applications and Problem Solving

Sometimes when we translate a problem to mathematical language, the result is a quadratic equation.

Example 1 *Landscaping.* A rectangular garden is 60 ft by 80 ft. Part of the garden is torn up to install a sidewalk of uniform width around it. The area of the new garden is $\frac{1}{2}$ of the old area. How wide is the sidewalk?

1. **Familiarize.** We first make a drawing and label it with the known information. We don't know how wide the sidewalk is, so we have called its width x.

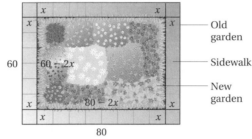

2. **Translate.** Remember, the area of a rectangle is lw (length times width). Then:

> Area of old garden = $60 \cdot 80$;
> Area of new garden = $(60 - 2x)(80 - 2x)$.

Since the area of the new garden is $\frac{1}{2}$ of the area of the old, we have

$$(60 - 2x)(80 - 2x) = \tfrac{1}{2} \cdot 60 \cdot 80.$$

3. **Solve.** We solve the equation:

$$4800 - 120x - 160x + 4x^2 = 2400 \quad \text{Using FOIL on the left}$$
$$4x^2 - 280x + 2400 = 0 \quad \text{Collecting like terms}$$
$$x^2 - 70x + 600 = 0 \quad \text{Dividing by 4}$$
$$(x - 10)(x - 60) = 0 \quad \text{Factoring}$$
$$x = 10 \quad or \quad x = 60. \quad \text{Using the principle of zero products}$$

4. **Check.** We check in the original problem. We see that 60 is not a solution because when $x = 60$, $60 - 2x = -60$, and the width of the garden cannot be negative.

   If the sidewalk is 10 ft wide, then the garden itself will have length $80 - 2 \cdot 10$, or 60 ft. The width will be $60 - 2 \cdot 10$, or 40 ft. The new area is thus $60 \cdot 40$, or 2400 ft². The old area was $60 \cdot 80$, or 4800 ft². The new area of 2400 ft² is $\frac{1}{2}$ of 4800 ft², so the number 10 checks.

5. **State.** The sidewalk is 10 ft wide.

Do Exercise 1.

Objectives

a Solve applied problems involving quadratic equations.

b Solve a formula for a given letter.

For Extra Help

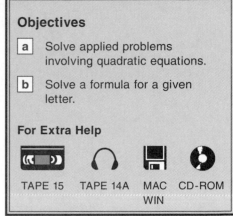

TAPE 15   TAPE 14A   MAC   CD-ROM
                          WIN

1. *Box Construction.* An open box is to be made from a 10-ft by 20-ft rectangular piece of cardboard by cutting a square from each corner. The area of the bottom of the box is to be 96 ft². What is the length of the sides of the squares that are cut from the corners?

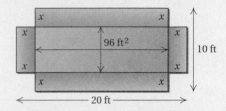

Answer on page A-41

2. *Town Planning.* Three towns A, B, and C are situated as shown. The roads at A form a right angle. The distance from A to B is 2 mi less than the distance from A to C. The distance from B to C is 10 mi. Find the distance from A to B and the distance from A to C.

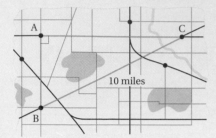

Example 2 *Ladder Location.* A ladder leans against a building, as shown below. The ladder is 20 ft long. The distance to the top of the ladder is 4 ft greater than the distance d from the building. Find the distance d and the distance to the top of the ladder.

1. **Familiarize.** We first make a drawing and label it. We want to find d and $d + 4$.

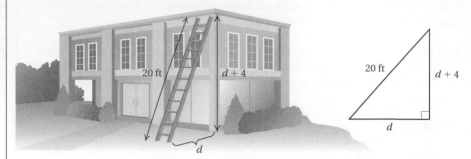

2. **Translate.** As we look at the figure, we see that a right triangle is formed. We can use the Pythagorean equation, which we studied in Chapter 6:

$$c^2 = a^2 + b^2.$$

In this problem then, we have

$$20^2 = d^2 + (d + 4)^2.$$

3. **Solve.** We solve the equation:

$$400 = d^2 + d^2 + 8d + 16 \qquad \text{Squaring}$$
$$2d^2 + 8d - 384 = 0 \qquad \text{Finding standard form}$$
$$d^2 + 4d - 192 = 0 \qquad \text{Dividing by 2}$$
$$(d + 16)(d - 12) = 0 \qquad \text{Factoring}$$
$$d + 16 = 0 \quad or \quad d - 12 = 0 \qquad \text{Using the principle of zero products}$$
$$d = -16 \quad or \quad d = 12.$$

4. **Check.** We know that -16 is not an answer because distances are not negative. The number 12 checks (we leave the check to the student).

5. **State.** The distance d is 12 ft, and the distance to the top of the ladder is $12 + 4$, or 16 ft.

Do Exercise 2.

Example 3 *Ladder Location.* Suppose that the ladder in Example 2 has length 10 ft. Find the distance d and the distance $d + 4$.

Using the same reasoning that we did in Example 2, we translate the problem to the equation

$$10^2 = d^2 + (d + 4)^2.$$

Answers on page A-41

We solve as follows. Note that the quadratic equation we get is not easily factored, so we use the quadratic formula:

$$100 = d^2 + d^2 + 8d + 16 \qquad \text{Squaring}$$

$$2d^2 + 8d - 84 = 0 \qquad \text{Finding standard form}$$

$$d^2 + 4d - 42 = 0 \qquad \text{Multiplying by } \tfrac{1}{2}, \text{ or dividing by 2}$$

$$d = \frac{-b \pm \sqrt{b^2 - 4ac}}{2a}$$

$$= \frac{-4 \pm \sqrt{4^2 - 4(1)(-42)}}{2(1)}$$

$$= \frac{-4 \pm \sqrt{16 + 168}}{2} = \frac{-4 \pm \sqrt{184}}{2}$$

$$= \frac{-4 \pm \sqrt{4(46)}}{2} = \frac{-4 \pm 2\sqrt{46}}{2}$$

$$= -2 \pm \sqrt{46}.$$

Since $-2 - \sqrt{46} < 0$ and $\sqrt{46} > 2$, it follows that d is given by $d = -2 + \sqrt{46}$. Using a calculator, we find that $d \approx -2 + 6.782 \approx 4.782$ ft, and that $d + 4 \approx 4.782 + 4$, or 8.782 ft.

Do Exercise 3.

Some problems translate to rational equations. The solution of such rational equations can involve quadratic equations as well.

Example 4 *Motorcycle Travel.* Karin's motorcycle traveled 300 mi at a certain speed. Had she gone 10 mph faster, she could have made the trip in 1 hr less time. Find her speed.

1. **Familiarize.** We first make a drawing, labeling it with known and unknown information. We can also organize the information, in a table, as we did in Section 5.6. We let $r = $ the speed, in miles per hour, and $t = $ the time, in hours.

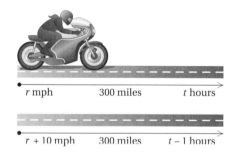

Distance	Speed	Time
300	r	t	$\rightarrow r = \dfrac{300}{t}$
300	$r + 10$	$t - 1$	$\rightarrow r + 10 = \dfrac{300}{t - 1}$

Recalling the motion formula $d = rt$ and solving for r, we get $r = d/t$. From the rows of the table, we obtain

$$r = \frac{300}{t} \quad \text{and} \quad r + 10 = \frac{300}{t - 1}.$$

2. **Translate.** We substitute for r from the first equation into the second and get a translation:

$$\frac{300}{t} + 10 = \frac{300}{t - 1}.$$

3. *Town Planning.* Three towns A, B, and C are situated as shown. The roads at A form a right angle. The distance from A to B is 2 mi less than the distance from A to C. The distance from B to C is 8 mi. Find the distance from A to B and the distance from A to C. Find exact and approximate answers to the nearest hundredth of a mile.

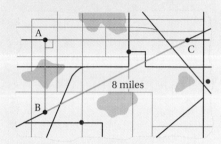

Answers on page A-41

4. *Marine Travel.* Two ships make the same voyage to a destination of 3000 nautical miles. The faster ship travels 10 knots faster than the slower one (a *knot* is 1 nautical mile per hour). The faster ship makes the voyage in 50 hr less time than the slower one. Find the speeds of the two ships.

Complete this table to help with the familiarization.

	Distance	Speed	Time
Faster Ship	3000		$t - 50$
Slower Ship	3000		t

5. Solve $A = \sqrt{\dfrac{w_1}{w_2}}$ for w_2.

3. Solve. We solve as follows:

$$t(t - 1)\left[\frac{300}{t} + 10\right] = t(t - 1) \cdot \frac{300}{t - 1} \qquad \text{Multiplying by the LCM}$$

$$t(t - 1) \cdot \frac{300}{t} + t(t - 1) \cdot 10 = t(t - 1) \cdot \frac{300}{t - 1}$$

$$300(t - 1) + 10(t^2 - t) = 300t$$

$$10t^2 - 10t - 300 = 0 \qquad \text{Standard form}$$

$$t^2 - t - 30 = 0 \qquad \text{Dividing by 10}$$

$$(t - 6)(t + 5) = 0 \qquad \text{Factoring}$$

$$t = 6 \quad or \quad t = -5 \qquad \text{Using the principle of zero products}$$

4. Check. Since negative time has no meaning in this problem, we try 6 hr. Remembering that $r = d/t$, we get $r = 300/6 = 50$ mph.

To check, we take the speed 10 mph faster, which is 60 mph, and see how long the trip would have taken at that speed:

$$t = \frac{d}{r} = \frac{300}{60} = 5 \text{ hr.}$$

This is 1 hr less than the trip actually took, so we have an answer.

5. State. Karin's speed was 50 mph.

Do Exercise 4.

b Solving Formulas

Recall that to solve a formula for a certain letter, we use the principles for solving equations to get that letter alone on one side.

Example 5 *Period of a Pendulum.* The time T required for a pendulum of length l to swing back and forth (complete one period) is given by the formula $T = 2\pi\sqrt{L/g}$, where g is the gravitational constant. Solve for L.

We have

$$T = 2\pi\sqrt{\frac{L}{g}} \qquad \text{This is a radical equation (see Section 6.6).}$$

$$T^2 = \left(2\pi\sqrt{\frac{L}{g}}\right)^2 \qquad \text{Principle of powers (squaring)}$$

$$T^2 = 2^2\pi^2\frac{L}{g}$$

$$gT^2 = 4\pi^2 L \qquad \text{Clearing fractions}$$

$$\frac{gT^2}{4\pi^2} = L. \qquad \text{Multiplying by } \frac{1}{4\pi^2}$$

We now have L alone on one side and L does not appear on the other side, so the formula is solved for L.

Do Exercise 5.

In most formulas, variables represent nonnegative numbers, so we do not need to use absolute-value signs when taking square roots.

Example 6 *Hang Time.* An athlete's *hang time* is the amount of time that the athlete can remain airborne when jumping. A formula relating an athlete's vertical leap V, in inches, to hang time T, in seconds, is $V = 48T^2$. (See Example 14 in Section 7.1.) Solve for T.

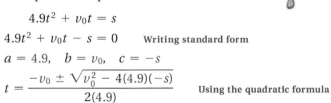

We have

$$48T^2 = V$$

$$T^2 = \frac{V}{48} \qquad \text{Multiplying by } \tfrac{1}{48} \text{ to get } T^2 \text{ alone}$$

$$T = \sqrt{\frac{V}{48}} \qquad \text{Using the principle of square roots; note that } T \geq 0.$$

$$T = \sqrt{\frac{V}{16 \cdot 3} \cdot \frac{3}{3}} \qquad \text{Factoring and multiplying by 1 to rationalize the denominator. Note that 16 is a perfect square.}$$

$$T = \sqrt{\frac{3V}{144}} = \frac{\sqrt{3V}}{12}.$$

Do Exercise 6.

Example 7 *Falling Distance.* An object tossed downward with an initial speed (velocity) of v_0 will travel a distance of s meters, where $s = 4.9t^2 + v_0t$ and t is measured in seconds. Solve for t.

Since t is squared in one term and raised to the first power in the other term, the equation is quadratic in t.

$$4.9t^2 + v_0t = s$$

$$4.9t^2 + v_0t - s = 0 \qquad \text{Writing standard form}$$

$$a = 4.9, \quad b = v_0, \quad c = -s$$

$$t = \frac{-v_0 \pm \sqrt{v_0^2 - 4(4.9)(-s)}}{2(4.9)} \qquad \text{Using the quadratic formula}$$

Since the negative square root would yield a negative value for t, we use only the positive root:

$$t = \frac{-v_0 + \sqrt{v_0^2 + 19.6s}}{9.8}.$$

6. Solve $V = \pi r^2 h$ for r.

(Volume of a right circular cylinder)

Answer on page A-41

7. Solve $s = gt + 16t^2$ for t.

The following list of steps should help you when solving formulas for a given letter. Try to remember that when solving a formula, you do the same things you would do to solve any equation.

> To solve a formula for a letter, say, b:
>
> **1.** Clear the fractions and use the principle of powers, as needed, until b does not appear in any radicand or denominator. (In some cases, you may clear the fractions first, and in some cases you may use the principle of powers first.)
>
> **2.** Collect all terms with b^2 in them. Also collect all terms with b in them.
>
> **3.** If b^2 does not appear, you can finish by using just the addition and multiplication principles.
>
> **4.** If b^2 appears but b does not, solve the equation for b^2. Then take square roots on both sides.
>
> **5.** If there are terms containing both b and b^2, write the equation in standard form and use the quadratic formula.

Do Exercise 7.

Example 8 Solve $q = \dfrac{a}{\sqrt{a^2 + b^2}}$ for a.

In this case, we could either clear the fractions first or use the principle of powers first. Let's clear the fractions. We then have

$$q\sqrt{a^2 + b^2} = a.$$

Now we square both sides and then continue:

$$(q\sqrt{a^2 + b^2})^2 = a^2 \qquad \text{Squaring}$$

> Don't forget to square both q and $\sqrt{a^2 + b^2}$.

$$q^2(a^2 + b^2) = a^2$$
$$q^2a^2 + q^2b^2 = a^2$$
$$q^2b^2 = a^2 - q^2a^2 \qquad \text{Getting all } a^2\text{-terms together}$$
$$q^2b^2 = a^2(1 - q^2) \qquad \text{Factoring out } a^2$$
$$\frac{q^2b^2}{1 - q^2} = a^2 \qquad \text{Dividing by } 1 - q^2$$
$$\sqrt{\frac{q^2b^2}{1 - q^2}} = a \qquad \text{Taking the square root}$$
$$\frac{qb}{\sqrt{1 - q^2}} = a. \qquad \text{Simplifying}$$

You need not rationalize denominators in situations such as this.

Do Exercise 8.

8. Solve $\dfrac{b}{\sqrt{a^2 - b^2}} = p$ for b.

Answers on page A-41

Exercise Set 7.3

a Solve.

1. *Flower Bed.* The width of a rectangular flower bed is 7 ft less than the length. The area is 18 ft². Find the length and the width.

2. *Feed Lot.* The width of a rectangular feed lot is 8 m less than the length. The area is 20 m². Find the length and the width.

3. *Parking Lot.* The length of a rectangular parking lot is twice the width. The area is 162 yd². Find the length and the width.

4. *Computer Part.* The length of a rectangular computer part is twice the width. The area is 242 cm². Find the length and the width.

5. *Picture Framing.* The outside of a picture frame measures 12 cm by 20 cm; 84 cm² of picture shows. Find the width of the frame.

6. *Picture Framing.* The outside of a picture frame measures 14 in. by 20 in.; 160 in² of picture shows. Find the width of the frame.

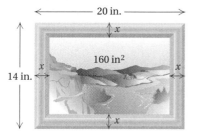

7. *Landscaping.* A landscaper is designing a flower garden in the shape of a right triangle. She wants 10 ft of a perennial border to form the hypotenuse of the triangle, and one leg is to be 2 ft longer than the other. Find the lengths of the legs.

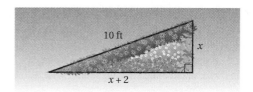

8. The hypotenuse of a right triangle is 25 m long. The length of one leg is 17 m less than the other. Find the lengths of the legs.

9. *Page Numbers.* A student opens a literature book to two facing pages. The product of the page numbers is 812. Find the page numbers.

10. *Page Numbers.* A student opens a mathematics book to two facing pages. The product of the page numbers is 1980. Find the page numbers.

Solve. Find exact and approximate answers rounded to three decimal places.

11. The width of a rectangle is 4 ft less than the length. The area is 10 ft^2. Find the length and the width.

12. The length of a rectangle is twice the width. The area is 328 cm^2. Find the length and the width.

13. *Page Dimensions.* The outside of an oversized book page measures 14 in. by 20 in.; 100 in^2 of printed text shows. Find the width of the margin.

14. *Picture Framing.* The outside of a picture frame measures 12 cm by 20 cm; 80 cm^2 of picture shows. Find the width of the frame.

15. The hypotenuse of a right triangle is 24 ft long. The length of one leg is 14 ft more than the other. Find the lengths of the legs.

16. The hypotenuse of a right triangle is 22 m long. The length of one leg is 10 m less than the other. Find the lengths of the legs.

17. *Car Trips.* During the first part of a trip, Meira's Honda traveled 120 mi at a certain speed. Meira then drove another 100 mi at a speed that was 10 mph slower. If Meira's total trip time was 4 hr, what was her speed on each part of the trip?

18. *Canoeing.* During the first part of a canoe trip, Tim covered 60 km at a certain speed. He then traveled 24 km at a speed that was 4 km/h slower. If the total time for the trip was 8 hr, what was the speed on each part of the trip?

19. *Car Trips.* Petra's Plymouth travels 200 mi at a certain speed. If the car had gone 10 mph faster, the trip would have taken 1 hr less. Find Petra's speed.

20. *Car Trips.* Sandi's Subaru travels 280 mi at a certain speed. If the car had gone 5 mph faster, the trip would have taken 1 hr less. Find Sandi's speed.

21. *Air Travel.* A Cessna flies 600 mi at a certain speed. A Beechcraft flies 1000 mi at a speed that is 50 mph faster, but takes 1 hr longer. Find the speed of each plane.

22. *Air Travel.* A turbo-jet flies 50 mph faster than a super-prop plane. If a turbo-jet goes 2000 mi in 3 hr less time than it takes the super-prop to go 2800 mi, find the speed of each plane.

23. *Bicycling.* Naoki bikes the 40 mi to Hillsboro at a certain speed. The return trip is made at a speed that is 6 mph slower. Total time for the round trip is 14 hr. Find Naoki's speed on each part of the trip.

24. *Car Speed.* On a sales trip, Gail drives the 600 mi to Richmond at a certain speed. The return trip is made at a speed that is 10 mph slower. Total time for the round trip was 22 hr. How fast did Gail travel on each part of the trip?

25. *Navigation.* The current in a typical Mississippi River shipping route flows at a rate of 4 mph. In order for a barge to travel 24 mi upriver and then return in a total of 5 hr, approximately how fast must the barge be able to travel in still water?

26. *Navigation.* The Hudson River flows at a rate of 3 mph. A patrol boat travels 60 mi upriver and returns in a total time of 9 hr. What is the speed of the boat in still water?

b Solve the formula for the given letter. Assume that all variables represent nonnegative numbers.

27. $A = 6s^2$, for s
(Surface area of a cube)

28. $A = 4\pi r^2$, for r
(Surface area of a sphere)

29. $F = \dfrac{Gm_1 m_2}{r^2}$, for r
(Newton's law of gravity)

30. $N = \dfrac{kQ_1 Q_2}{s^2}$, for s
(Number of phone calls between two cities)

31. $E = mc^2$, for c
(Einstein's energy–mass relationship)

32. $V = \frac{1}{3}s^2 h$, for s
(Volume of a pyramid)

33. $a^2 + b^2 = c^2$, for b
(Pythagorean formula in two dimensions)

34. $a^2 + b^2 + c^2 = d^2$, for c
(Pythagorean formula in three dimensions)

35. $N = \dfrac{k^2 - 3k}{2}$, for k
(Number of diagonals of a polygon of k sides)

36. $s = v_0 t + \dfrac{gt^2}{2}$, for t
(A motion formula)

37. $A = 2\pi r^2 + 2\pi rh$, for r
(Surface area of a cylinder)

38. $A = \pi r^2 + \pi rs$, for r
(Surface area of a cone)

39. $T = 2\pi\sqrt{\dfrac{L}{g}}$, for g
(A pendulum formula)

40. $W = \sqrt{\dfrac{1}{LC}}$, for L
(An electricity formula)

41. $I = \dfrac{700W}{H^2}$, for H
(Body mass index)

42. $N + p = \dfrac{6.2A^2}{pR^2}$, for R

43. $m = \dfrac{m_0}{\sqrt{1 - \dfrac{v^2}{c^2}}}$, for v

(A relativity formula)

44. Solve the formula given in Exercise 43 for c.

Skill Maintenance

Add or subtract. [5.2b, c]

45. $\dfrac{1}{x - 1} + \dfrac{1}{x^2 - 3x + 2}$

46. $\dfrac{x + 1}{x - 1} - \dfrac{x + 1}{x^2 + x + 1}$

47. $\dfrac{2}{x + 3} - \dfrac{x}{x - 1} + \dfrac{x^2 + 2}{x^2 + 2x - 3}$

48. Multiply and simplify: $\sqrt{3x^2}\,\sqrt{3x^3}$. [6.3a]

49. Express in terms of i: $\sqrt{-20}$. [6.8a]

Simplify. [5.4a]

50. $\dfrac{\dfrac{3}{x - 1}}{\dfrac{1}{x + 1} + \dfrac{2}{x - 1}}$

51. $\dfrac{\dfrac{4}{a^2 b}}{\dfrac{3}{a} - \dfrac{4}{b^2}}$

Synthesis

52. ◆ ☐ Explain how Exercises 1–26 can be solved without using factoring, completing the square, or the quadratic formula.

53. ◆ Explain how the quadratic formula can be used to factor a quadratic polynomial into two binomials. Use it to factor $5x^2 + 8x - 3$.

54. Solve: $\dfrac{4}{2x + i} - \dfrac{1}{x - i} = \dfrac{2}{x + i}$.

55. Find a when the reciprocal of $a - 1$ is $a + 1$.

56. *Bungee Jumping.* Jesse is tied to one end of a 40-m elasticized (bungee) cord. The other end of the cord is tied to the middle of a train trestle. If Jesse jumps off the bridge, for how long will he fall before the cord begins to stretch? (See Example 7 and let $v_0 = 0$.)

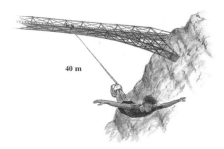

40 m

57. *Surface Area.* A sphere is inscribed in a cube as shown in the figure below. Express the surface area of the sphere as a function of the surface area S of the cube.

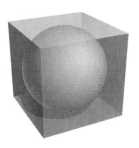

58. *Pizza Crusts.* At Pizza Perfect, Ron can make 100 large pizza crusts in 1.2 hr less than Chad. Together they can do the job in 1.8 hr. How long does it take each to do the job alone?

59. *The Golden Rectangle.* For over 2000 yr, the proportions of a "golden" rectangle have been considered visually appealing. A rectangle of width w and length l is considered "golden" if

$$\frac{w}{l} = \frac{l}{w + l}.$$

Solve for l.

7.4 More on Quadratic Equations

a The Discriminant

From the quadratic formula, we know that the solutions x_1 and x_2 of a quadratic equation are given by

$$x_1 = \frac{-b + \sqrt{b^2 - 4ac}}{2a} \quad \text{and} \quad x_2 = \frac{-b - \sqrt{b^2 - 4ac}}{2a}.$$

The expression $b^2 - 4ac$ is called the **discriminant**. When using the quadratic formula, it is helpful to compute the discriminant first. If it is 0, there will be just one real solution. If it is positive, there will be two real solutions. If it is negative, we will be taking the square root of a negative number; hence there will be two nonreal complex-number solutions, and they will be complex conjugates.

Discriminant $b^2 - 4ac$	Nature of Solutions	x-intercepts
0	Only one solution; it is a real number	Only one
Positive	Two different real-number solutions	Two different
Negative	Two different nonreal complex-number solutions (complex conjugates)	None

If the discriminant is a perfect square, we can solve the equation by factoring, not needing the quadratic formula.

Example 1 Determine the nature of the solutions of $9x^2 - 12x + 4 = 0$.

We have

$$a = 9, \quad b = -12, \quad c = 4.$$

We compute the discriminant:

$$\begin{aligned} b^2 - 4ac &= (-12)^2 - 4 \cdot 9 \cdot 4 \\ &= 144 - 144 \\ &= 0. \end{aligned}$$

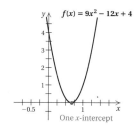

$f(x) = 9x^2 - 12x + 4$

One x-intercept

There is just one solution, and it is a real number. Since 0 is a perfect square, the equation can be solved by factoring.

Example 2 Determine the nature of the solutions of $x^2 + 5x + 8 = 0$.

We have

$$a = 1, \quad b = 5, \quad c = 8.$$

We compute the discriminant:

$$\begin{aligned} b^2 - 4ac &= 5^2 - 4 \cdot 1 \cdot 8 \\ &= 25 - 32 \\ &= -7. \end{aligned}$$

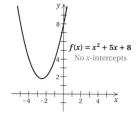

$f(x) = x^2 + 5x + 8$

No x-intercepts

Since the discriminant is negative, there are two nonreal complex-number solutions. The equation cannot be solved by factoring because -7 is not a perfect square.

Determine the nature of the solutions without solving.

1. $x^2 + 5x - 3 = 0$

2. $9x^2 - 6x + 1 = 0$

3. $3x^2 - 2x + 1 = 0$

Example 3 Determine the nature of the solutions of $x^2 + 5x + 6 = 0$.

We have

$$a = 1, \quad b = 5, \quad c = 6;$$
$$b^2 - 4ac = 5^2 - 4 \cdot 1 \cdot 6 = 1.$$

Since the discriminant is positive, there are two solutions, and they are real numbers. The equation can be solved by factoring since the discriminant is a perfect square.

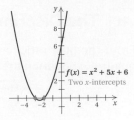

The discriminant, $b^2 - 4ac$, tells us how many real-number solutions the equation $0 = ax^2 + bx + c$ has, so it also indicates how many x-intercepts the graph of $f(x) = ax^2 + bx + c$ has. Compare the following.

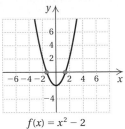

$f(x) = x^2 - 2$
$b^2 - 4ac = 8 > 0$
Two real solutions
Two x-intercepts

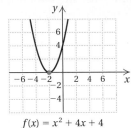

$f(x) = x^2 + 4x + 4$
$b^2 - 4ac = 0$
One real solution
One x-intercept

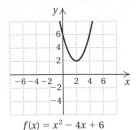

$f(x) = x^2 - 4x + 6$
$b^2 - 4ac = -8 < 0$
No real solutions
No x-intercepts

Do Exercises 1–3.

b Writing Equations from Solutions

We know by the principle of zero products that $(x - 2)(x + 3) = 0$ has solutions 2 and -3. If we know the solutions of an equation, we can write the equation, using this principle in reverse.

Example 4 Find a quadratic equation whose solutions are 3 and $-\frac{2}{5}$.

We have

$$x = 3 \quad or \quad x = -\frac{2}{5}$$
$$x - 3 = 0 \quad or \quad x + \frac{2}{5} = 0 \qquad \text{Getting the 0's on one side}$$
$$x - 3 = 0 \quad or \quad 5x + 2 = 0 \qquad \text{Clearing the fraction}$$
$$(x - 3)(5x + 2) = 0 \qquad \text{Using the principle of zero products in reverse}$$
$$5x^2 - 13x - 6 = 0. \qquad \text{Using FOIL}$$

Answers on page A-41

Example 5 Write a quadratic equation whose solutions are $2i$ and $-2i$.

We have

$$x = 2i \quad or \quad x = -2i$$

$$x - 2i = 0 \quad or \quad x + 2i = 0 \qquad \text{Getting the 0's on one side}$$

$$(x - 2i)(x + 2i) = 0 \qquad \text{Using the principle of zero products in reverse}$$

$$x^2 - (2i)^2 = 0 \qquad \text{Using } (A - B)(A + B) = A^2 - B^2$$

$$x^2 - 4i^2 = 0$$

$$x^2 - 4(-1) = 0$$

$$x^2 + 4 = 0.$$

Example 6 Write a quadratic equation whose solutions are $\sqrt{3}$ and $-2\sqrt{3}$.

We have

$$x = \sqrt{3} \quad or \quad x = -2\sqrt{3}$$

$$x - \sqrt{3} = 0 \quad or \quad x + 2\sqrt{3} = 0 \qquad \text{Getting the 0's on one side}$$

$$(x - \sqrt{3})(x + 2\sqrt{3}) = 0 \qquad \text{Using the principle of zero products}$$

$$x^2 + 2\sqrt{3}x - \sqrt{3}x - 2(\sqrt{3})^2 = 0 \qquad \text{Using FOIL}$$

$$x^2 + \sqrt{3}x - 6 = 0. \qquad \text{Collecting like terms}$$

Do Exercises 4–7.

c | Equations Quadratic in Form

Certain equations that are not really quadratic can still be solved as quadratic. Consider this fourth-degree equation.

$$x^4 \quad - 9x^2 \quad + 8 = 0$$

$$\downarrow \qquad \downarrow \qquad \downarrow \quad \downarrow$$

$$(x^2)^2 - 9(x^2) + 8 = 0 \qquad \text{Thinking of } x^4 \text{ as } (x^2)^2$$

$$\downarrow \qquad \downarrow \qquad \downarrow \quad \downarrow$$

$$u^2 \quad - 9u \quad + 8 = 0 \qquad \text{To make this clearer, write } u \text{ instead of } x^2.$$

The equation $u^2 - 9u + 8 = 0$ can be solved by factoring or by the quadratic formula. After that, we can find x by remembering that $x^2 = u$. Equations that can be solved like this are said to be **quadratic in form,** or **reducible to quadratic.**

Example 7 Solve: $x^4 - 9x^2 + 8 = 0$.

Let $u = x^2$. Then we solve the equation found by substituting u for x^2:

$$u^2 - 9u + 8 = 0$$

$$(u - 8)(u - 1) = 0 \qquad \text{Factoring}$$

$$u - 8 = 0 \quad or \quad u - 1 = 0 \qquad \text{Using the principle of zero products}$$

$$u = 8 \quad or \qquad u = 1.$$

Next, we substitute x^2 for u and solve these equations:

$$x^2 = 8 \qquad or \quad x^2 = 1$$

$$x = \pm\sqrt{8} \quad or \quad x = \pm 1$$

$$x = \pm 2\sqrt{2} \quad or \quad x = \pm 1.$$

Find a quadratic equation having the following solutions.

4. 7 and -2

5. -4 and $\dfrac{5}{3}$

6. $5i$ and $-5i$

7. $-2\sqrt{2}$ and $\sqrt{2}$

Answers on page A-41

8. Solve: $x^4 - 10x^2 + 9 = 0$.

To check, first note that when $x = 2\sqrt{2}$, $x^2 = 8$ and $x^4 = 64$. Also, when $x = -2\sqrt{2}$, $x^2 = 8$ and $x^4 = 64$. Similarly, when $x = 1$, $x^2 = 1$ and $x^4 = 1$, and when $x = -1$, $x^2 = 1$ and $x^4 = 1$. Thus, instead of making four checks, we need make only two.

CHECK:

For $\pm 2\sqrt{2}$:

$$x^4 - 9x^2 + 8 = 0$$

$$\begin{array}{c|c} (\pm 2\sqrt{2})^4 - 9(\pm 2\sqrt{2})^2 + 8 & 0 \\ 64 - 9 \cdot 8 + 8 & \\ 0 & \text{TRUE} \end{array}$$

For ± 1:

$$x^4 - 9x^2 + 8 = 0$$

$$\begin{array}{c|c} (\pm 1)^4 - 9(\pm 1)^2 + 8 & 0 \\ 1 - 9 + 8 & \\ 0 & \text{TRUE} \end{array}$$

The solutions are 1, -1, $2\sqrt{2}$, and $-2\sqrt{2}$.

CAUTION! A common error is to solve for u and then forget to solve for x. Remember that you *must* find values for the *original* variable!

Solving equations quadratic in form can sometimes introduce numbers that are not solutions of the original equation. Thus a check by substituting is necessary.

Do Exercise 8.

Example 8 Solve: $x - 3\sqrt{x} - 4 = 0$.

Let $u = \sqrt{x}$. Then we solve the equation found by substituting u for $\sqrt{x}$ (and, of course, u^2 for x):

$$u^2 - 3u - 4 = 0$$
$$(u - 4)(u + 1) = 0$$
$$u = 4 \quad or \quad u = -1.$$

9. Solve $x + 3\sqrt{x} - 10 = 0$. Be sure to check.

Next, we substitute $\sqrt{x}$ for u and solve these equations:

$$\sqrt{x} = 4 \quad or \quad \sqrt{x} = -1.$$

Squaring the first equation, we get $x = 16$, and 16 checks. Squaring the second equation, we get $x = 1$. But the number 1 does not check. The number 16 is the solution.

Do Exercise 9.

Answers on page A-41

Example 9 Find the x-intercepts of the graph of $f(x) = (x^2 - 1)^2 - (x^2 - 1) - 2$.

The x-intercepts occur where $f(x) = 0$ so we must have

$$(x^2 - 1)^2 - (x^2 - 1) - 2 = 0.$$

Let $u = x^2 - 1$. Then we solve the equation found by substituting u for $x^2 - 1$:

$$u^2 - u - 2 = 0$$
$$(u - 2)(u + 1) = 0$$
$$u = 2 \quad or \quad u = -1.$$

Next, we substitute $x^2 - 1$ for u and solve these equations:

$$x^2 - 1 = 2 \qquad or \quad x^2 - 1 = -1$$
$$x^2 = 3 \qquad or \qquad x^2 = 0$$
$$x = \pm\sqrt{3} \quad or \qquad x = 0.$$

The numbers $\sqrt{3}$, $-\sqrt{3}$, and 0 check. They are the solutions of $(x^2 - 1)^2 - (x^2 - 1) - 2 = 0$. Thus the x-intercepts of the graph of $f(x)$ are $(-\sqrt{3}, 0)$, $(0, 0)$, and $(\sqrt{3}, 0)$.

Do Exercise 10.

Example 10 Solve: $y^{-2} - y^{-1} - 2 = 0$.

Let $u = y^{-1}$. Then we solve the equation found by substituting u for y^{-1} and u^2 for y^{-2}:

$$u^2 - u - 2 = 0$$
$$(u - 2)(u + 1) = 0$$
$$u = 2 \quad or \quad u = -1.$$

Next, we substitute y^{-1} or $1/y$ for u and solve these equations:

$$\frac{1}{y} = 2 \quad or \quad \frac{1}{y} = -1.$$

Solving, we get

$$y = \frac{1}{2} \quad or \quad y = \frac{1}{(-1)} = -1.$$

The numbers $\frac{1}{2}$ and -1 both check. They are the solutions.

Do Exercise 11.

---

10. Find the x-intercepts of

$$f(x) = (x^2 - x)^2 - 14(x^2 - x) + 24.$$

11. Solve: $x^{-2} + x^{-1} - 6 = 0$.

Answers on page A-41

Improving Your Math Study Skills

How Many Women Have Won the Ultimate Math Contest?

Although this Study Skill feature does not contain specific tips on studying mathematics, we hope that you will find this article both challenging and encouraging.

Every year on college campuses across the United States and Canada, the most brilliant math students face the ultimate challenge. For six hours, they struggle with problems from the merely intractable to the seemingly impossible.

Every spring, five are chosen winners of the William Lowell Putnam Mathematical Competition, the Olympics of college mathematics. Every year for 56 years, all have been men.

Until this year.

This spring, Ioana Dumitriu (pronounced yo-AHN-na doo-mee-TREE-oo), 20, a New York University sophomore from Romania, became the first woman to win the award.

Ms. Dumitriu, the daughter of two electrical engineering professors in Romania, who as a girl solved math puzzles for fun, was identified as a math talent early in her schooling in Bucharest. At 11, Ms. Dumitriu was steered into years of math training camps as preparation for the Romanian entry in the International Mathematics Olympiad.

It was this training, and a handsome young coach, that led her to New York City. He was several years older. They fell in love. He chose N.Y.U. for its graduate school in mathematics, and at 19 she joined him in New York.

The test Ms. Dumitriu won is dauntingly difficult, even for math majors. About half of the 2,407 test-takers scored 2 or less of a possible 120, and a third scored 0. Some students simply walk out after staring at the questions for a while.

Ms. Dumitriu said that in the six hours allotted, she had time to do 8 of the 12 problems, each worth a maximum of 10 points. The last one she did in 10 minutes. This year, Ms. Dumitriu and her five co-winners (there was a tie for fifth place) scored between 76 and 98. She does not know her exact score or rank because the organizers do not announce them.

"I didn't ever tell myself that I was unlikely to win, that no woman before had ever won and therefore I couldn't," she said. "It is not that I forget that I'm a woman. It's just that I don't see it as an obstacle or a ——."

Her English is near-perfect, but she paused because she could not find the right word. "The mathematics community is made up of persons, and that is what I am primarily."

Prof. Joel Spencer, who was a Putnam winner himself, said her work for his class in problem solving last year was remarkable. "What really got me was her fearlessness," he said. "To be good at math, you have to go right at it and start playing around with it, and she had that from the start."

In the graduate lounge in the Courant Institute of Mathematical Sciences at N.Y.U., Ms. Dumitriu, a tall, striking redhead, stands out. Instead of jeans and T-shirts, she wears gray pin-striped slacks and a rust-colored turtleneck and vest.

"There is a social perception of women and math, a stereotype," Ms. Dumitriu said during an interview. "What's happening right now is that the stereotype is defied. It starts breaking."

Still, even as women began to flock to sciences, math has remained largely a male bastion.

"Math remains the bottom line of sex differences for many," said Sheila Tobias, author of "Overcoming Math Anxiety" (W.W. Norton & Company, 1994). "It's one thing for women to write books, negotiate bills through Congress, litigate, fire missiles; quite another for them to do math."

Besides collecting the $1,000 awarded to each Putnam fellow, Ms. Dumitriu also won the $500 Elizabeth Lowell Putnam prize for the top woman finisher for the second year in a row, a prize created five years ago to encourage women to take the test. This year 414 did.

In her view, there are never too many problems, never too much practice.

Besides, each new problem holds its own allure: "When you have all the pieces and you put them together and you see the puzzle, that moment always amazes me."

Exercise Set 7.4

a Determine the nature of the solutions of the equation.

1. $x^2 - 8x + 16 = 0$ **2.** $x^2 + 12x + 36 = 0$ **3.** $x^2 + 1 = 0$ **4.** $x^2 + 6 = 0$

5. $x^2 - 6 = 0$ **6.** $x^2 - 3 = 0$ **7.** $4x^2 - 12x + 9 = 0$ **8.** $4x^2 + 8x - 5 = 0$

9. $x^2 - 2x + 4 = 0$ **10.** $x^2 + 3x + 4 = 0$ **11.** $9t^2 - 3t = 0$ **12.** $4m^2 + 7m = 0$

13. $y^2 = \dfrac{1}{2}y + \dfrac{3}{5}$ **14.** $y^2 + \dfrac{9}{4} = 4y$

15. $4x^2 - 4\sqrt{3}\,x + 3 = 0$ **16.** $6y^2 - 2\sqrt{3}\,y - 1 = 0$

b Write a quadratic equation having the given numbers as solutions.

17. -4 and 4 **18.** -11 and 9 **19.** -2 and -7

20. 3 and 10 **21.** 8, only solution
[*Hint*: It must be a double solution,
that is, $(x - 8)(x - 8) = 0$.] **22.** -3, only solution

23. $-\dfrac{2}{5}$ and $\dfrac{6}{5}$ **24.** $-\dfrac{1}{4}$ and $-\dfrac{1}{2}$ **25.** $\dfrac{k}{3}$ and $\dfrac{m}{4}$

26. $\dfrac{c}{2}$ and $\dfrac{d}{2}$

27. $-\sqrt{3}$ and $2\sqrt{3}$

28. $\sqrt{2}$ and $3\sqrt{2}$

$\boxed{\text{c}}$ Solve.

29. $x^4 - 6x^2 + 9 = 0$

30. $x^4 - 7x^2 + 12 = 0$

31. $x - 10\sqrt{x} + 9 = 0$

32. $2x - 9\sqrt{x} + 4 = 0$

33. $(x^2 - 6x)^2 - 2(x^2 - 6x) - 35 = 0$

34. $(x^2 + 5x)^2 + 2(x^2 + 5x) - 24 = 0$

35. $x^{-2} - 5x^{-1} - 36 = 0$

36. $3x^{-2} - x^{-1} - 14 = 0$

37. $(1 + \sqrt{x})^2 + (1 + \sqrt{x}) - 6 = 0$

38. $(2 + \sqrt{x})^2 - 3(2 + \sqrt{x}) - 10 = 0$

39. $(y^2 - 5y)^2 - 2(y^2 - 5y) - 24 = 0$

40. $(2t^2 + t)^2 - 4(2t^2 + t) + 3 = 0$

41. $t^4 - 6t^2 - 4 = 0$

42. $w^4 - 7w^2 + 7 = 0$

43. $2x^{-2} + x^{-1} - 1 = 0$

44. $m^{-2} + 9m^{-1} - 10 = 0$

45. $6x^4 - 19x^2 + 15 = 0$

46. $6x^4 - 17x^2 + 5 = 0$

47. $x^{2/3} - 4x^{1/3} - 5 = 0$

48. $x^{2/3} + 2x^{1/3} - 8 = 0$

49. $\left(\dfrac{x-4}{x+1}\right)^2 - 2\left(\dfrac{x-4}{x+1}\right) - 35 = 0$

50. $\left(\dfrac{x+3}{x-3}\right)^2 - \left(\dfrac{x+3}{x-3}\right) - 6 = 0$

51. $9\left(\dfrac{x+2}{x+3}\right)^2 - 6\left(\dfrac{x+2}{x+3}\right) + 1 = 0$

52. $16\left(\dfrac{x-1}{x-8}\right)^2 + 8\left(\dfrac{x-1}{x-8}\right) + 1 = 0$

53. $\left(\dfrac{x^2-2}{x}\right)^2 - 7\left(\dfrac{x^2-2}{x}\right) - 18 = 0$

54. $\left(\dfrac{y^2-1}{y}\right)^2 - 4\left(\dfrac{y^2-1}{y}\right) - 12 = 0$

Find the x-intercepts of the function.

55. $f(x) = 5x + 13\sqrt{x} - 6$

56. $f(x) = 3x + 10\sqrt{x} - 8$

57. $f(x) = (x^2 - 3x)^2 - 10(x^2 - 3x) + 24$

58. $f(x) = x^{2/5} + x^{1/5} - 6$

Skill Maintenance

Solve. [3.3a]

59. *Coffee Beans.* Twin Cities Roasters has Kenyan coffee worth $6.75 per pound and Peruvian coffee worth $11.25 per pound. How many pounds of each kind should be mixed in order to obtain a 50-lb mixture that is worth $8.55 per pound?

60. *Solution Mixtures.* Solution A is 18% alcohol and solution B is 45% alcohol. How many liters of each should be mixed in order to get 12 L of a solution that is 36% alcohol?

Multiply and simplify. [6.3a]

61. $\sqrt{8x}\ \sqrt{2x}$

62. $\sqrt[3]{x^2}\ \sqrt[3]{27x^4}$

63. $\sqrt[4]{9a^2}\ \sqrt[4]{18a^3}$

64. $\sqrt[5]{16}\ \sqrt[5]{64}$

Graph. [2.2c], [2.5a, c]

65. $f(x) = -\frac{3}{5}x + 4$

66. $5x - 2y = 8$

67. $y = 4$

68. $f(x) = -x - 3$

Synthesis

69. ◈ Describe a procedure that could be used to write an equation having the first seven natural numbers as solutions.

70. ◈ Describe a procedure that could be used to write an equation that is quadratic in $3x^2 + 1$ and has real-number solutions.

71. 📊 Use a grapher to check your answers to Exercises 30, 32, 34, and 37.

72. 📊 Use a grapher to solve each of the following equations.

a) $6.75x - 35\sqrt{x} - 5.26 = 0$
b) $\pi x^4 - \pi^2 x^2 = \sqrt{99.3}$
c) $x^4 - x^3 - 13x^2 + x + 12 = 0$

For each equation under the given condition, **(a)** find k and **(b)** find the other solution.

73. $kx^2 - 2x + k = 0$; one solution is -3.

74. $kx^2 - 17x + 33 = 0$; one solution is 3.

75. Find a quadratic equation for which the sum of the solutions is $\sqrt{3}$ and the product is 8.

76. Find k given that $kx^2 - 4x + (2k - 1) = 0$ and the product of the solutions is 3.

77. The graph of a function of the form

$$f(x) = ax^2 + bx + c$$

is a curve similar to the one shown below. Determine a, b, and c from the information given.

78. ◈ 📊 While solving a quadratic equation of the form $ax^2 + bx + c = 0$ with a grapher, Shawn-Marie gets the following screen.

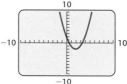

How could the discriminant help her check the graph?

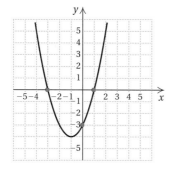

Solve.

79. $\dfrac{x}{x-1} - 6\sqrt{\dfrac{x}{x-1}} - 40 = 0$

80. $\left(\sqrt{\dfrac{x}{x-3}}\right)^2 - 24 = 10\sqrt{\dfrac{x}{x-3}}$

81. $\sqrt{x-3} - \sqrt[4]{x-3} = 12$

82. $a^3 - 26a^{3/2} - 27 = 0$

83. $r^6 - 28r^3 + 27 = 0$

84. $r^6 + 7r^3 - 8 = 0$

7.5 Graphs of Quadratic Functions of the Type $f(x) = a(x - h)^2 + k$

In this section and the next, we develop techniques for graphing quadratic functions.

a Graphs of $f(x) = ax^2$

The most basic quadratic function is $f(x) = x^2$.

Example 1 Graph: $f(x) = x^2$.

We choose some values for x and compute $f(x)$ for each. Then we plot the ordered pairs and connect them with a smooth curve.

x	$f(x) = x^2$	$(x, f(x))$
-3	9	$(-3, 9)$
-2	4	$(-2, 4)$
-1	1	$(-1, 1)$
0	0	$(0, 0)$
1	1	$(1, 1)$
2	4	$(2, 4)$
3	9	$(3, 9)$

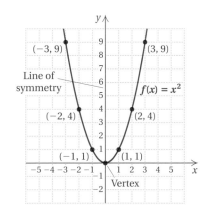

All quadratic functions have graphs similar to the one in Example 1. Such curves are called **parabolas**. They are cup-shaped curves that are symmetric with respect to a vertical line known as the parabola's **line of symmetry,** or **axis of symmetry.** In the graph of $f(x) = x^2$, shown above, the y-axis (or the line $x = 0$) is the line of symmetry. If the paper were to be folded on this line, the two halves of the curve would match. The point $(0, 0)$ is the **vertex** of this parabola.

Let's compare the graphs of $g(x) = \frac{1}{2}x^2$ and $h(x) = 2x^2$ with the graph of $f(x) = x^2$. We choose x-values and plot points for both functions.

x	$g(x) = \frac{1}{2}x^2$
-3	$\frac{9}{2}$
-2	2
-1	$\frac{1}{2}$
0	0
1	$\frac{1}{2}$
2	2
3	$\frac{9}{2}$

x	$h(x) = 2x^2$
-3	18
-2	8
-1	2
0	0
1	2
2	8
3	18

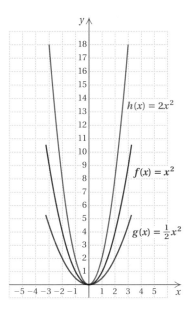

Note that the graph of $g(x) = \frac{1}{2}x^2$ is a wider parabola than the graph of $f(x) = x^2$, and the graph of $h(x) = 2x^2$ is narrower. The vertex and the line of symmetry, however, remain $(0, 0)$ and $x = 0$, respectively.

1. Graph: $f(x) = -\frac{1}{3}x^2$.

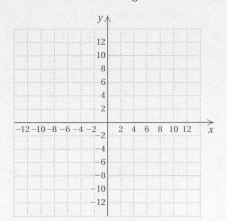

Answer on page A-42

Calculator Spotlight

Exercises

1. a) Graph each of the following equations using the viewing window $[-5, 5, -10, 10]$:

$$y_1 = x^2; \quad y_2 = 3x^2; \quad y_3 = \tfrac{1}{3}x^2.$$

b) See if you can determine a rule that describes the effect of multiplying x^2 by a, in the graph of $y = ax^2$, when $a > 1$ and $0 < a < 1$. Try some other graphs to test your assertion.

2. a) Graph each of the following equations using the viewing window $[-5, 5, -10, 10]$:

$$y_1 = -x^2; \quad y_2 = -4x^2; \quad y_3 = -\tfrac{2}{3}x^2.$$

b) See if you can determine a rule that describes the effect of multiplying x^2 by a, in the graph of $y = ax^2$, when $a < -1$ and $-1 < a < 0$. Try some other graphs to test your assertion.

Graph.

2. $f(x) = 3x^2$

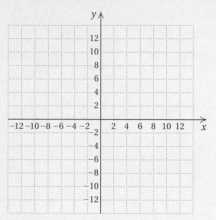

When we consider the graph of $k(x) = -\tfrac{1}{2}x^2$, we see that the parabola opens down and is the same shape as the graph of $g(x) = \tfrac{1}{2}x^2$.

x	$k(x) = -\tfrac{1}{2}x^2$
-3	$-\tfrac{9}{2}$
-2	-2
-1	$-\tfrac{1}{2}$
0	0
1	$-\tfrac{1}{2}$
2	-2
3	$-\tfrac{9}{2}$

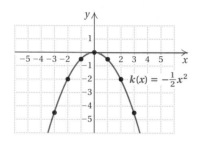

> The graph of $f(x) = ax^2$, or $y = ax^2$, is a parabola with $x = 0$ as its line of symmetry; its vertex is the origin.
>
> For $a > 0$, the parabola opens up; for $a < 0$, the parabola opens down.
>
> If $|a|$ is greater than 1, the parabola is narrower than $y = x^2$.
>
> If $|a|$ is between 0 and 1, the parabola is wider than $y = x^2$.

Do Exercises 1–3. (Exercise 1 is on the preceding page.)

3. $f(x) = -2x^2$

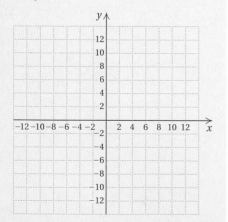

b | Graphs of $f(x) = a(x - h)^2$

It would seem logical now to consider functions of the type

$$f(x) = ax^2 + bx + c.$$

We are heading in that direction, but it is convenient to first consider graphs of $f(x) = a(x - h)^2$ and then $f(x) = a(x - h)^2 + k$, where a, h, and k are constants.

Answers on page A-42

Example 2 Graph: $g(x) = (x - 3)^2$.

We choose some values for x and compute $g(x)$. Then we plot the points and draw the curve.

x	$g(x) = (x - 3)^2$
-1	16
0	9
1	4
2	1
3	0
4	1
5	4
6	9

← Vertex

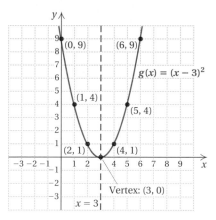

Note that $g(x) = 16$ when $x = -1$ and $g(x)$ gets larger as x gets more negative. Thus we use positive values to fill out the table. Note that the line $x = 3$ is the line of symmetry and the point $(3, 0)$ is the vertex. Had we known earlier that $x = 3$ is the line of symmetry, we could have computed some values on one side, such as $(4, 1)$, $(5, 4)$, and $(6, 9)$, and then used symmetry to get their mirror images $(2, 1)$, $(1, 4)$, and $(0, 9)$ without further computation.

The graph of $g(x) = (x - 3)^2$ in Example 2 looks just like the graph of $f(x) = x^2$ in Example 1, except that it is moved, or translated, 3 units to the right. Comparing the pairs for $g(x)$ with those for $f(x)$, we see that when an input for $g(x)$ is 3 more than an input for $f(x)$, the outputs match.

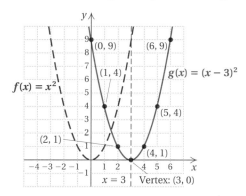

> The graph of $f(x) = a(x - h)^2$ has the same shape as the graph of $y = ax^2$.
> If h is positive, the graph of $y = ax^2$ is shifted h units to the right.
> If h is negative, the graph of $y = ax^2$ is shifted $|h|$ units to the left.
> The vertex is $(h, 0)$ and the axis of symmetry is $x = h$.

Example 3 Graph: $f(x) = -2(x + 3)^2$.

We first rewrite the equation as $f(x) = -2[x - (-3)]^2$. In this case, $a = -2$ and $h = -3$, so the graph looks like that of $g(x) = 2x^2$ translated 3 units to the left and, since $-2 < 0$, the graph opens down. The vertex is $(-3, 0)$, and the line of symmetry is $x = -3$. Plotting points as needed, we obtain the graph shown below.

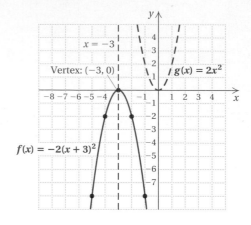

Do Exercises 4 and 5 on the following page.

Do Exercises 4 and 5 on the following page.

c | Graphs of $f(x) = a(x - h)^2 + k$

Given a graph of $f(x) = a(x - h)^2$, what happens if we add a constant k? Suppose that we add 2. This increases each function value $f(x)$ by 2, so the curve is moved up. If k is negative, the curve is moved down. The line of symmetry for the parabola remains $x = h$, but the vertex will be at (h, k), or, equivalently, $(h, f(h))$.

Note that if a parabola opens up $(a > 0)$, the function value, or y-value, at the vertex is a least, or **minimum**, value. That is, it is less than the y-value at any other point on the graph. If the parabola opens down $(a < 0)$, the function value at the vertex is a greatest, or **maximum**, value.

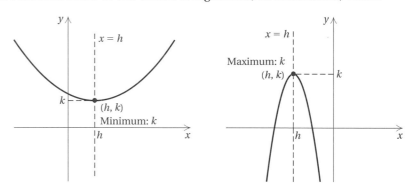

> The graph of $f(x) = a(x - h)^2 + k$ has the same shape as the graph of $y = a(x - h)^2$.
>
> If k is positive, the graph of $y = a(x - h)^2$ is shifted k units up.
>
> If k is negative, the graph of $y = a(x - h)^2$ is shifted $|k|$ units down.
>
> The vertex is (h, k), and the line of symmetry is $x = h$.
>
> For $a > 0$, k is the minimum function value. For $a < 0$, k is the maximum function value.

Calculator Spotlight

〰 Exercises

1. **a)** Graph each of the following equations using the viewing window $[-5, 5, -5, 5]$:

   $y_1 = 7(x - 1)^2$;
   $y_2 = 7(x - 1)^2 + 2$.

   Use the TABLE feature with TblStart $= -2$ and ΔTbl $= 1$ to compare y-values.

   b) See if you can determine a rule that describes the effect of adding k, in the graph of $y = a(x - h)^2 + k$. Try some other graphs to test your assertion.

2. **a)** Graph each of the following equations using the viewing window $[-5, 5, -5, 5]$:

   $y_1 = 7(x - 1)^2$;
   $y_3 = 7(x - 1)^2 - 4$.

   b) See if you can determine a rule that describes the effect of adding k, where k is negative, in the graph of $y = a(x - h)^2 + k$. Try some other graphs to test your assertion.

Example 4 Graph $f(x) = (x - 3)^2 - 5$, and find the minimum function value.

The graph will look like that of $g(x) = (x - 3)^2$ (see Example 2) but translated 5 units down. You can confirm this by plotting some points. For instance, $f(4) = (4 - 3)^2 - 5 = -4$, whereas in Example 2, $g(4) = (4 - 3)^2 = 1$. Note here that $h = 3$, so we calculate points on both sides of $x = 3$.

The vertex is now $(3, -5)$, and the minimum function value is -5.

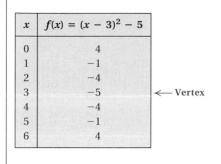

x	$f(x) = (x - 3)^2 - 5$	
0	4	
1	-1	
2	-4	
3	-5	← Vertex
4	-4	
5	-1	
6	4	

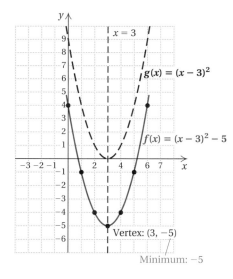

Example 5 Graph $h(x) = \frac{1}{2}(x - 3)^2 + 5$, and find the minimum function value.

The graph looks just like that of $f(x) = \frac{1}{2}x^2$ but moved 3 units to the right and 5 units up. The vertex is $(3, 5)$, and the line of symmetry is $x = 3$. We draw $f(x) = \frac{1}{2}x^2$ and then shift the curve over and up. The minimum function value is 5. By plotting some points, we have a check.

x	$h(x) = \frac{1}{2}(x - 3)^2 + 5$	
0	$9\frac{1}{2}$	
1	7	
3	5	← Vertex
5	7	
6	$9\frac{1}{2}$	

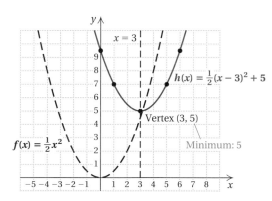

Graph. Find and label the vertex and the line of symmetry.

4. $f(x) = \dfrac{1}{2}(x - 4)^2$

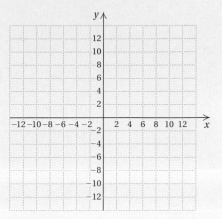

5. $f(x) = -\dfrac{1}{2}(x - 4)^2$

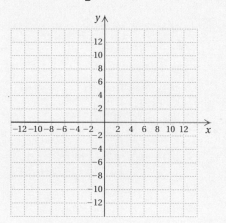

Answers on page A-42

Graph. Find the vertex, the line of symmetry, and the maximum or minimum y-value.

6. $f(x) = \dfrac{1}{2}(x + 2)^2 - 4$

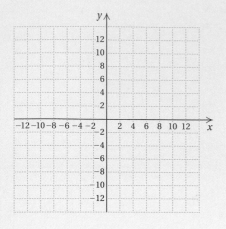

7. $f(x) = -2(x - 5)^2 + 3$

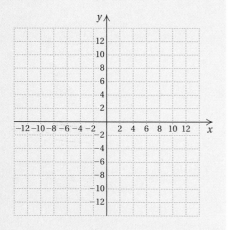

Example 6 Graph $f(x) = -2(x + 3)^2 + 5$. Find the vertex, the line of symmetry, and the maximum or minimum value.

We first express the equation in the equivalent form

$$f(x) = -2[x - (-3)]^2 + 5.$$

The graph looks like that of $g(x) = -2x^2$ translated 3 units to the left and 5 units up. The vertex is $(-3, 5)$, and the line of symmetry is $x = -3$. Since $-2 < 0$, we know that the graph opens down so 5, the second coordinate of the vertex, is the maximum y-value.

We compute a few points as needed. The graph is shown here.

x	$f(x) = -2(x + 3)^2 + 5$	
-4	3	
-3	5	$\leftarrow$ Vertex
-2	3	

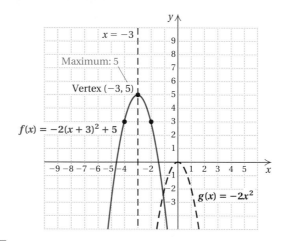

Do Exercises 6 and 7.

Answers on page A-42

Exercise Set 7.5

[a], [b] Graph. Find and label the vertex and the line of symmetry.

1. $f(x) = 4x^2$

x	f(x)
0	
1	
−1	
2	
−2	

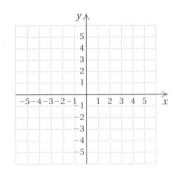

2. $f(x) = 5x^2$

x	f(x)
0	
1	
−1	
2	
−2	

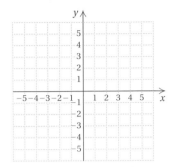

3. $f(x) = \frac{1}{3}x^2$

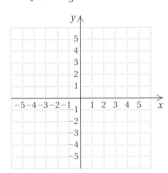

4. $f(x) = \frac{1}{4}x^2$

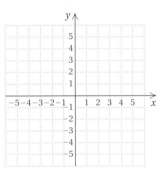

5. $f(x) = -\frac{1}{2}x^2$

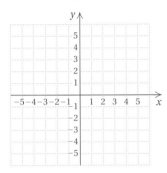

6. $f(x) = -\frac{1}{4}x^2$

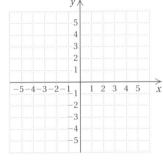

7. $f(x) = -4x^2$

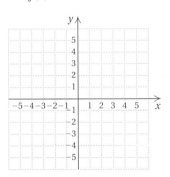

8. $f(x) = -3x^2$

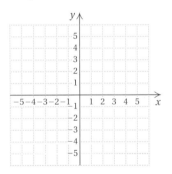

9. $f(x) = (x + 3)^2$

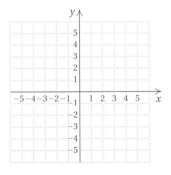

10. $f(x) = (x + 1)^2$

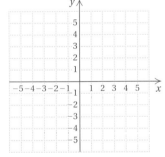

11. $f(x) = 2(x - 4)^2$

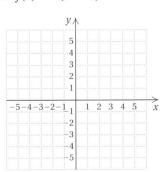

12. $f(x) = 4(x - 1)^2$

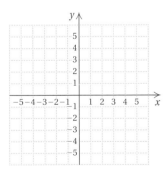

13. $f(x) = -2(x + 2)^2$

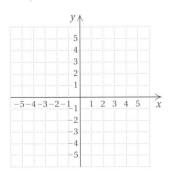

14. $f(x) = -2(x + 4)^2$

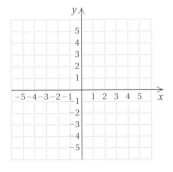

15. $f(x) = 3(x - 1)^2$

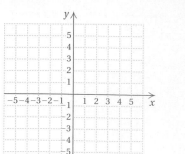

16. $f(x) = 4(x - 2)^2$

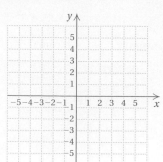

17. $f(x) = -\frac{3}{2}(x + 2)^2$

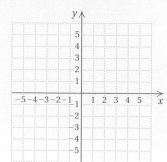

18. $f(x) = -\frac{5}{2}(x + 3)^2$

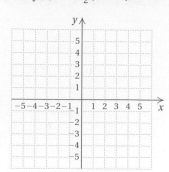

c Graph. Find and label the vertex and the line of symmetry. Find the maximum or minimum value.

19. $f(x) = (x - 3)^2 + 1$

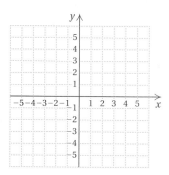

20. $f(x) = (x + 2)^2 - 3$

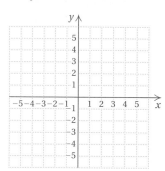

21. $f(x) = -3(x + 4)^2 + 1$

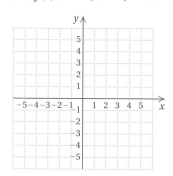

22. $f(x) = \frac{1}{2}(x - 1)^2 - 3$

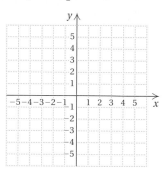

23. $f(x) = \frac{1}{2}(x + 1)^2 + 4$

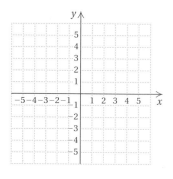

24. $f(x) = -2(x - 5)^2 - 3$

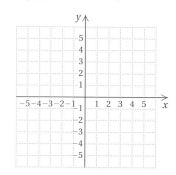

25. $f(x) = -(x + 1)^2 - 2$

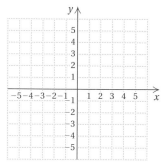

26. $f(x) = 3(x - 4)^2 + 2$

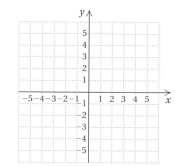

Skill Maintenance

Solve. [6.6a, b]

27. $x - 5 = \sqrt{x + 7}$

28. $\sqrt{2x + 7} = \sqrt{5x - 4}$

29. $\sqrt{x + 4} = -11$

30. $x = 7 + 2\sqrt{x + 1}$

Synthesis

31. ◆ Explain, without plotting points, why the graph of $f(x) = (x + 3)^2$ looks like the graph of $f(x) = x^2$ translated 3 units to the left.

32. ◆ Explain, without plotting points, why the graph of $f(x) = (x + 3)^2 - 4$ looks like the graph of $f(x) = x^2$ translated 3 units to the left and 4 units down.

33. 🖩 Use the TRACE and/or TABLE features on a grapher to confirm the maximum or minimum values given in Exercises 23, 24, and 26. Also, confirm the results using the MINIMUM or MAXIMUM feature under the CALC key.

Collaborative Learning Manual

Practice graphing and identifying the graphs of quadratic functions.

7.6 Graphs of Quadratic Functions of the Type $f(x) = ax^2 + bx + c$

a Graphing and Analyzing $f(x) = ax^2 + bx + c$

By *completing the square*, we can begin with any quadratic polynomial $ax^2 + bx + c$ and find an equivalent expression $a(x - h)^2 + k$. This allows us to combine the skills of Sections 7.1 and 7.5 to graph and analyze any quadratic function $f(x) = ax^2 + bx + c$.

Example 1 For $f(x) = x^2 - 6x + 4$, find the vertex, the line of symmetry, and the minimum value. Then graph.

We first find the vertex and the line of symmetry. To do so, we find the equivalent form $a(x - h)^2 + k$ by completing the square, beginning as follows:

$$f(x) = x^2 - 6x + 4 = (x^2 - 6x \quad) + 4.$$

We complete the square inside the parentheses, but in a different manner than we did before. We take half the x-coefficient, $-6/2 = -3$, and square it: $(-3)^2 = 9$. Then we add 0, or $9 - 9$, inside the parentheses:

$$\begin{aligned}
f(x) &= (x^2 - 6x + 0) + 4 & &\text{Adding 0} \\
&= (x^2 - 6x + 9 - 9) + 4 & &\text{Substituting } 9 - 9 \text{ for 0} \\
&= (x^2 - 6x + 9) + (-9 + 4) & &\text{Using the associative law} \\
& & &\text{of addition to regroup} \\
&= (x - 3)^2 - 5. & &\text{Factoring and simplifying}
\end{aligned}$$

(This equation was graphed in Example 4 of Section 7.5.) The vertex is $(3, -5)$, and the line of symmetry is $x = 3$. The coefficient of x^2 is 1, which is positive, so the graph opens up. This tells us that -5 is a minimum. We plot the vertex and draw the line of symmetry. We choose some x-values on both sides of the vertex and graph the parabola. Suppose we compute the pair $(5, -1)$:

$$f(5) = 5^2 - 6(5) + 4 = 25 - 30 + 4 = -1.$$

We note that it is 2 units to the right of the line of symmetry. There will also be a pair with the same y-coordinate on the graph 2 units to the *left* of the line of symmetry. Thus we get a second point, $(1, -1)$, without making another calculation.

x	$f(x)$	
3	-5	←Vertex
4	-4	
2	-4	
5	-1	
1	-1	
6	4	
0	4	←y-intercept

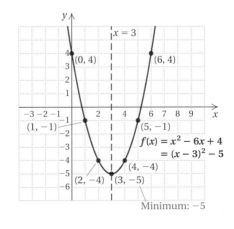

$$f(x) = x^2 - 6x + 4$$
$$= (x - 3)^2 - 5$$

Minimum: -5

1. For $f(x) = x^2 - 4x + 7$, find the vertex, the line of symmetry, and the minimum value. Then graph.

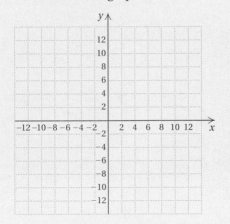

Vertex: _____

Line of symmetry: _____

Minimum value: _____

Do Exercise 1.

Answers on page A-43

2. For $f(x) = 3x^2 - 24x + 43$, find the vertex, the line of symmetry, and the minimum value. Then graph.

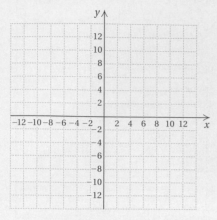

Vertex: _____

Line of symmetry: _____

Minimum value: _____

Example 2 For $f(x) = 3x^2 + 12x + 13$, find the vertex, the line of symmetry, and the minimum value. Then graph.

Since the coefficient of x^2 is not 1, we factor out 3 from only the *first two* terms of the expression. Remember that we want to get to the form $f(x) = a(x - h)^2 + k$:

$$f(x) = 3x^2 + 12x + 13$$
$$= 3(x^2 + 4x) + 13. \quad \text{Factoring 3 out of the first two terms}$$

Next, we complete the square inside the parentheses. We take half the x-coefficient, $\frac{1}{2} \cdot 4 = 2$, and square it: $2^2 = 4$. Then we add 0, or $4 - 4$, inside the parentheses:

$$f(x) = 3(x^2 + 4x + 0) + 13 \quad \text{Adding 0}$$
$$= 3(\underbrace{x^2 + 4x + 4} - 4) + 13 \quad \text{Substituting } 4 - 4 \text{ for 0}$$
$$= 3(x^2 + 4x + 4) + 3(-4) + 13 \quad \begin{array}{l}\text{Using the distributive law to} \\ \text{separate } -4 \text{ from the} \\ \text{trinomial}\end{array}$$
$$= 3(x + 2)^2 + 1. \quad \text{Factoring and simplifying}$$

The vertex is $(-2, 1)$, and the line of symmetry is $x = -2$. The coefficient of x^2 is 3, so the graph is narrow and opens up. This tells us that 1 is a minimum. We choose a few x-values on one side of the line of symmetry, compute y-values, and use the resulting coordinates to find more points on the other side of the line of symmetry. We plot points and graph the parabola.

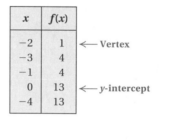

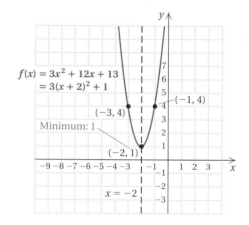

Do Exercise 2.

Example 3 For $f(x) = -2x^2 + 10x - 7$, find the vertex, the line of symmetry, and the maximum value. Then graph.

Again, the coefficient x^2 is not 1. We factor out -2 from only the *first two* terms of the expression. This makes the coefficient of x^2 inside the parentheses 1:

$$f(x) = -2x^2 + 10x - 7$$
$$= -2(x^2 - 5x) - 7.$$

Answer on page A-43

Next, we complete the square as before. We take half the x-coefficient, $\frac{1}{2}(-5) = -\frac{5}{2}$, and square it: $\left(-\frac{5}{2}\right)^2 = \frac{25}{4}$. Then we add 0, or $\frac{25}{4} - \frac{25}{4}$, inside the parentheses:

$$f(x) = -2\left(x^2 - 5x + \frac{25}{4} - \frac{25}{4}\right) - 7$$

$$= -2\left(x^2 - 5x + \frac{25}{4}\right) + (-2)\left(-\frac{25}{4}\right) - 7 \qquad \text{Using the distributive law to separate the } -\frac{25}{4} \text{ from the trinomial}$$

$$= -2\left(x^2 - 5x + \frac{25}{4}\right) + \frac{25}{2} - 7$$

$$= -2\left(x - \frac{5}{2}\right)^2 + \frac{11}{2}.$$

The vertex is $\left(\frac{5}{2}, \frac{11}{2}\right)$, and the line of symmetry is $x = \frac{5}{2}$. The coefficient of x^2 is -2, so the graph is narrow and opens down. This tells us that $\frac{11}{2}$ is a maximum. We choose a few x-values on one side of the line of symmetry, compute y-values, and use the resulting coordinates to find more points on the other side of the line of symmetry. We plot points and graph the parabola.

x	$f(x)$	
$\frac{5}{2}$	$\frac{11}{2}$, or $5\frac{1}{2}$	← Vertex
3	5	
2	5	
4	1	
1	1	
5	-7	
0	-7	← y-intercept

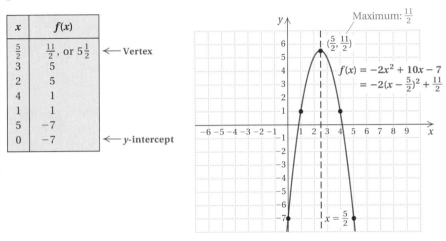

Do Exercise 3.

The method used in Examples 1–3 can be generalized to find a formula for locating the vertex. We complete the square as follows:

$$f(x) = ax^2 + bx + c$$

$$= a\left(x^2 + \frac{b}{a}x\right) + c. \qquad \text{Factoring } a \text{ out of the first two terms. Check by multiplying.}$$

Half of the x-coefficient, $\frac{b}{a}$, is $\frac{b}{2a}$. We square it to get $\frac{b^2}{4a^2}$ and add $\frac{b^2}{4a^2} - \frac{b^2}{4a^2}$ inside the parentheses. Then we distribute the a:

$$f(x) = a\left(x^2 + \frac{b}{a}x + \frac{b^2}{4a^2} - \frac{b^2}{4a^2}\right) + c$$

$$= a\left(x^2 + \frac{b}{a}x + \frac{b^2}{4a^2}\right) + a\left(-\frac{b^2}{4a^2}\right) + c \qquad \text{Using the distributive law}$$

$$= a\left(x + \frac{b}{2a}\right)^2 + \frac{-b^2}{4a} + \frac{4ac}{4a} \qquad \text{Factoring and finding a common denominator}$$

$$= a\left[x - \left(-\frac{b}{2a}\right)\right]^2 + \frac{4ac - b^2}{4a}.$$

3. For $f(x) = -4x^2 + 12x - 5$, find the vertex, the line of symmetry, and the maximum value. Then graph.

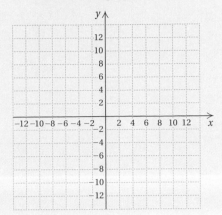

Vertex: _____

Line of symmetry: _____

Maximum value: _____

Answer on page A-43

Find the vertex of the parabola using the formula.

4. $f(x) = x^2 - 6x + 4$

5. $f(x) = 3x^2 - 24x + 43$

6. $f(x) = -4x^2 + 12x - 5$

Find the intercepts.

7. $f(x) = x^2 + 2x - 3$

8. $f(x) = x^2 + 8x + 16$

9. $f(x) = x^2 - 4x + 1$

Answers on page A-43

Thus we have the following.

> The **vertex** of the parabola given by $f(x) = ax^2 + bx + c$ is
> $$\left(-\frac{b}{2a}, \frac{4ac - b^2}{4a}\right), \quad \text{or} \quad \left(-\frac{b}{2a}, f\left(-\frac{b}{2a}\right)\right).$$
> The x-coordinate of the vertex is $-b/(2a)$. The **line of symmetry** is $x = -b/(2a)$. The second coordinate of the vertex is easiest to find by computing $f\left(-\frac{b}{2a}\right)$.

Let's reexamine Example 3 to see how we could have found the vertex directly. From the formula above,

$$\text{the } x\text{-coordinate of the vertex is } -\frac{b}{2a} = -\frac{10}{2(-2)} = \frac{5}{2}.$$

Substituting $\frac{5}{2}$ into $f(x) = -2x^2 + 10x - 7$, we find the second coordinate of the vertex:

$$\begin{aligned} f\left(\tfrac{5}{2}\right) &= -2\left(\tfrac{5}{2}\right)^2 + 10\left(\tfrac{5}{2}\right) - 7 \\ &= -2\left(\tfrac{25}{4}\right) + 25 - 7 \\ &= -\tfrac{25}{2} + 18 = -\tfrac{25}{2} + \tfrac{36}{2} = \tfrac{11}{2}. \end{aligned}$$

The vertex is $\left(\frac{5}{2}, \frac{11}{2}\right)$. The line of symmetry is $x = \frac{5}{2}$.

We have developed two methods for finding the vertex. One is by completing the square and the other is by using a formula. You should check with your instructor about which method to use.

Do Exercises 4–6.

b | Finding the Intercepts of a Quadratic Function

The points at which a graph crosses an axis are called **intercepts**. We determine the y-intercept by finding $f(0)$. For $f(x) = ax^2 + bx + c$, the y-intercept is $(0, c)$.

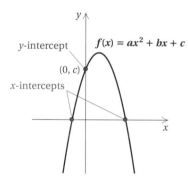

To find the x-intercepts, we look for values of x for which $f(x) = 0$. For $f(x) = ax^2 + bx + c$, we solve

$$0 = ax^2 + bx + c.$$

Example 4   Find the intercepts of $f(x) = x^2 - 2x - 2$.

The y-intercept is $(0, f(0))$. Since $f(0) = 0^2 - 2 \cdot 0 - 2 = -2$, the y-intercept is $(0, -2)$. To find the x-intercepts, we solve

$$0 = x^2 - 2x - 2.$$

Using the quadratic formula gives us $x = 1 \pm \sqrt{3}$. Thus the x-intercepts are $(1 - \sqrt{3}, 0)$ and $(1 + \sqrt{3}, 0)$, or, approximately, $(-0.732, 0)$ and $(2.732, 0)$.

Do Exercises 7–9.

Exercise Set 7.6

a For each quadratic function, find **(a)** the vertex, **(b)** the line of symmetry, and **(c)** the maximum or minimum value. Then **(d)** graph the function.

1. $f(x) = x^2 - 2x - 3$

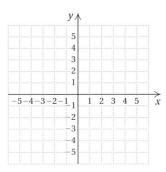

2. $f(x) = x^2 + 2x - 5$

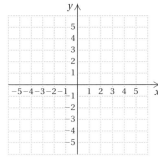

3. $f(x) = -x^2 - 4x - 2$

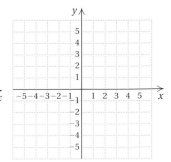

4. $f(x) = -x^2 + 4x + 1$

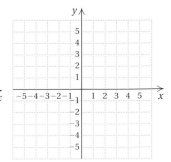

5. $f(x) = 3x^2 - 24x + 50$

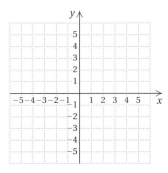

6. $f(x) = 4x^2 + 8x + 1$

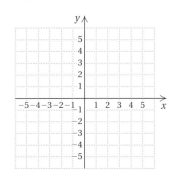

7. $f(x) = -2x^2 - 2x + 3$

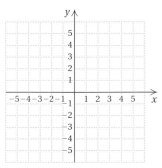

8. $f(x) = -2x^2 + 2x + 1$

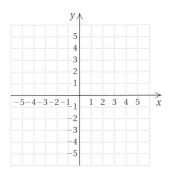

9. $f(x) = 5 - x^2$

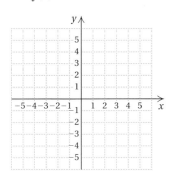

10. $f(x) = x^2 - 3x$

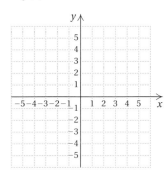

11. $f(x) = 2x^2 + 5x - 2$

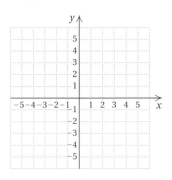

12. $f(x) = -4x^2 - 7x + 2$

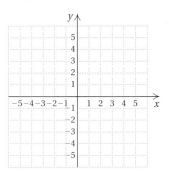

b Find the x- and y-intercepts.

13. $f(x) = x^2 - 6x + 1$

14. $f(x) = x^2 + 2x + 12$

15. $f(x) = -x^2 + x + 20$

16. $f(x) = -x^2 + 5x + 24$

17. $f(x) = 4x^2 + 12x + 9$ **18.** $f(x) = 3x^2 - 6x + 1$ **19.** $f(x) = 4x^2 - x + 8$ **20.** $f(x) = 2x^2 + 4x - 1$

Skill Maintenance

Solve.   [5.8a, b]

21. *Determining Medication Dosage.* A child's dosage D, in milligrams, of a medication varies directly as the child's weight w, in kilograms. To control a fever, a doctor suggests that a child who weighs 28 kg be given 420 mg of Tylenol.

a) Find an equation of variation.
b) How much Tylenol would be recommended for a child who weighs 42 kg?

22. *Calories Burned.* The number C of calories burned while exercising varies directly as the time t, in minutes, spent exercising. Harold exercises for 24 min on a stairmaster and burns 356 calories.

a) Find an equation of variation.
b) How many calories would he burn if he were to exercise for 48 min?

Find the variation constant and an equation of variation in which y varies inversely as x and the following are true. [5.8c]

23. $y = 125$ when $x = 2$

24. $y = 2$ when $x = 125$

Find the variation constant and an equation of variation in which y varies directly as x and the following are true. [5.8a]

25. $y = 125$ when $x = 2$

26. $y = 2$ when $x = 125$

Synthesis

27. ◆ Does the graph of every quadratic function have a y-intercept? Why or why not?

28. ◆ Is it possible for the graph of a quadratic function to have only one x-intercept if the vertex is off the x-axis? Why or why not?

29. 📈 Use the TRACE and/or TABLE features of a grapher to estimate the maximum or minimum values of the following functions. Confirm the results using the MINIMUM or MAXIMUM feature under the CALC key.

a) $f(x) = 2.31x^2 - 3.135x - 5.89$
b) $f(x) = -18.8x^2 + 7.92x + 6.18$

30. 📈 Use the TRACE and/or TABLE features of a grapher to confirm the maximum or minimum values given in Exercises 8, 11, and 12. Confirm the results using the MINIMUM or MAXIMUM feature under the CALC key.

Graph.

31. $f(x) = |x^2 - 1|$ **32.** $f(x) = |x^2 + 6x + 4|$ **33.** $f(x) = |x^2 - 3x - 4|$ **34.** $f(x) = |2(x - 3)^2 - 5|$

35. A quadratic function has $(-1, 0)$ as one of its intercepts and $(3, -5)$ as its vertex. Find an equation for the function.

36. A quadratic function has $(4, 0)$ as one of its intercepts and $(-1, 7)$ as its vertex. Find an equation for the function.

37. Consider

$$f(x) = \frac{x^2}{8} + \frac{x}{4} - \frac{3}{8}.$$

Find **(a)** the vertex, **(b)** the line of symmetry, and **(c)** the maximum or minimum value. Then **(d)** draw the graph.

38. Use only the graph in Exercise 37 to approximate the solutions of each of the following equations.

a) $\dfrac{x^2}{8} + \dfrac{x}{4} - \dfrac{3}{8} = 0$ b) $\dfrac{x^2}{8} + \dfrac{x}{4} - \dfrac{3}{8} = 1$

c) $\dfrac{x^2}{8} + \dfrac{x}{4} - \dfrac{3}{8} = 2$

📈 Use the INTERSECT feature of a grapher to find the points of intersection of the graphs of each pair of functions.

39. $f(x) = x^2 - 4x + 2$, $g(x) = 2 + x$

40. $f(x) = x^2 + 2x + 1$, $g(x) = -2x^2 - 4x + 1$

7.7 Mathematical Modeling with Quadratic Functions

We now consider some of the many situations in which quadratic functions can serve as mathematical models.

a Maximum–Minimum Problems

We have seen that for any quadratic function $f(x) = ax^2 + bx + c$, the value of $f(x)$ at the vertex is either a maximum or a minimum, meaning that either all outputs are smaller than that value for a maximum or larger than that value for a minimum.

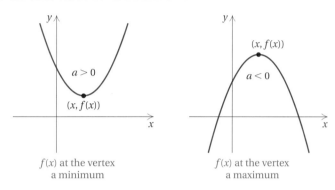

$a > 0$

$(x, f(x))$

$f(x)$ at the vertex
a minimum

$(x, f(x))$

$a < 0$

$f(x)$ at the vertex
a maximum

There are many types of applied problems in which we want to find a maximum or minimum value of a quantity. If a quadratic function can be used as a model, we can find such maximums or minimums by finding coordinates of the vertex.

Example 1 *Fenced-In Land.* A farmer has 64 yd of fencing. What are the dimensions of the largest rectangular pen that the farmer can enclose?

1. **Familiarize.** We first make a drawing and label it. We let l = the length of the pen and w = the width. Recall the following formulas:

   Perimeter: $2l + 2w$;

   Area: $l \cdot w$.

To become familiar with the problem, let's choose some dimensions for which $2l + 2w = 64$ and then calculate the corresponding areas.

l	w	A
22	10	220
20	12	240
18	14	252
18.5	13.5	249.75
12.4	19.6	243.04
15	17	255

What choice of l and w will maximize A?

1. *Fenced-In Land.* A farmer has 100 yd of fencing. What are the dimensions of the largest rectangular pen that the farmer can enclose?

To familiarize yourself with the problem, complete the following table.

l	*w*	*A*
12	38	456
15	35	
24	26	
25	25	
26.2	23.8	

2. Translate. We have two equations, one for perimeter and one for area:

$$2l + 2w = 64,$$
$$A = l \cdot w.$$

Let's use them to express A as a function of l or w, but not both. To express A in terms of w, for example, we solve for l in the first equation:

$$2l + 2w = 64$$
$$2l = 64 - 2w$$
$$l = \frac{64 - 2w}{2}$$
$$= 32 - w.$$

Substituting $32 - w$ for l, we get a quadratic function $A(w)$, or just A:

$$A = lw = (32 - w)w = 32w - w^2 = -w^2 + 32w.$$

3. Carry out. Note here that we are altering the third step of our five-step problem-solving strategy to "carry out" some kind of mathematical manipulation, because we are going to find the vertex rather than solve an equation. To do so, we complete the square as in Section 7.6:

$$A = -w^2 + 32w \qquad \text{This is a parabola opening down, so a maximum exists.}$$

$$= -1(w^2 - 32w) \qquad \text{Factoring out } -1$$

$$= -1(w^2 - 32w + 256 - 256) \qquad \tfrac{1}{2}(-32) = -16; (-16)^2 = 256$$

$$= -1(w^2 - 32w + 256) + (-1)(-256)$$

$$= -(w - 16)^2 + 256.$$

The vertex is (16, 256). Thus the maximum value is 256. It occurs when $w = 16$ and $l = 32 - w = 32 - 16 = 16$.

4. Check. We note that 256 is larger than any of the values found in the *Familiarize* step. To be more certain, we could make more calculations. We leave this to the student. We can also use the graph of the function to check the maximum value.

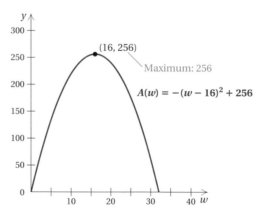

5. State. The largest rectangular pen that can be enclosed is 16 yd by 16 yd; that is, a square.

Do Exercise 1.

Calculator Spotlight

Use the TRACE and/or TABLE features to confirm the maximum or minimum values given in Example 1 and Margin Exercise 1. Also, confirm the results using the MAXIMUM feature under the CALC key.

Answer on page A-44

b Fitting Quadratic Functions to Data

As we move through our study of mathematics, we develop a library of functions. These functions can serve as models for many applications. Some of them are graphed below. We have not considered the cubic or quartic functions in detail other than in the Calculator Spotlights (we leave that discussion to a later course), but we show them here for reference.

Linear function:
$f(x) = mx + b$

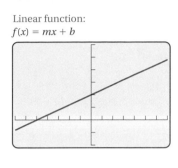

Quadratic function:
$f(x) = ax^2 + bx + c,\ a > 0$

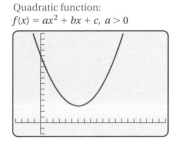

Quadratic function:
$f(x) = ax^2 + bx + c,\ a < 0$

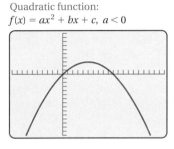

Absolute-value function:
$f(x) = |x|$

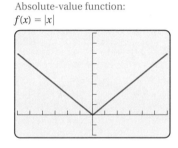

Cubic function:
$f(x) = ax^3 + bx^2 + cx + d,\ a > 0$

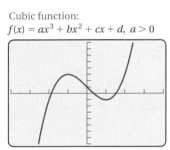

Quartic function:
$f(x) = ax^4 + bx^3 + cx^2 + dx + e,\ a > 0$

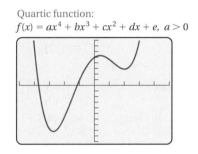

Now let's consider some real-world data. How can we decide which type of function might fit the data of a particular application? One simple way is to graph the data and look for a pattern resembling one of the graphs above. For example, data might be modeled by a linear function if the graph resembles a straight line. The data might be modeled by a quadratic function if the graph rises and then falls, or falls and then rises, in a curved manner resembling a parabola. For a quadratic, it might also just rise or fall in a curved manner as if following only one part of the parabola.

Choosing Models. For the scatterplots and graphs in Margin Exercises 2–5, determine which, if any, of the following functions might be used as a model for the data.

Linear, $f(x) = mx + b$;

Quadratic, $f(x) = ax^2 + bx + c$, $a > 0$;

Quadratic, $f(x) = ax^2 + bx + c$, $a < 0$;

Polynomial, neither quadratic nor linear

2.

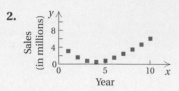

3.

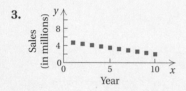

4.

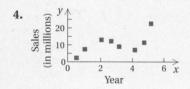

5.

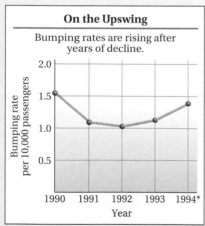

On the Upswing

Bumping rates are rising after years of decline.

* January–June figure
Source: Department of Transportation

Answers on page A-44

Let's now use our library of functions to see which, if any, might fit certain data situations.

Examples *Choosing Models.* For the scatterplots and graphs below, determine which, if any, of the following functions might be used as a model for the data.

Linear, $f(x) = mx + b$;

Quadratic, $f(x) = ax^2 + bx + c$, $a > 0$;

Quadratic, $f(x) = ax^2 + bx + c$, $a < 0$;

Polynomial, neither quadratic nor linear

2.

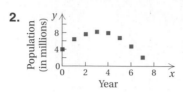

The data rise and then fall in a curved manner fitting a quadratic function $f(x) = ax^2 + bx + c$, $a < 0$.

3.

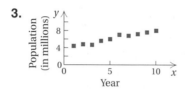

The data seem to fit a linear function $f(x) = mx + b$.

4.

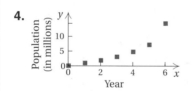

The data rise in a manner fitting the right side of a quadratic function $f(x) = ax^2 + bx + c$, $a > 0$.

5. **Driver Fatalities by Age**

Number of licensed drivers per 100,000 who died in motor vehicle accidents in 1990. The fatality rates for both the 70–79 group and the 80+ age group were lower than for the 15- to 24-year-olds.

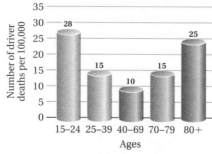

Source: National Highway Traffic Administration

The data fall and then rise in a curved manner fitting a quadratic function $f(x) = ax^2 + bx + c$, $a > 0$.

6.

Use of Rock Salt in Winter

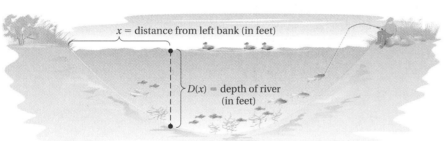

Source: Salt Institute

The data fall, then rise, then fall again, so they do not fit a linear or quadratic function but might fit a polynomial function that is neither quadratic nor linear.

Do Exercises 2–5 on the preceding page.

Whenever a quadratic function seems to fit a data situation, that function can be determined if at least three inputs and their outputs are known.

Example 7   *River Depth.* The drawing below shows the cross section of a river. Typically rivers are deepest in the middle, with the depth decreasing to 0 at the edges. A hydrologist measures the depths D, in feet, of a river at distances x, in feet, from one bank. The results are listed in the table below.

x = distance from left bank (in feet)

$D(x)$ = depth of river (in feet)

Distance, x, from the Riverbank (in feet)	Depth, D, of the River (in feet)
0	0
15	10.2
25	17
50	20
90	7.2
100	0

a) Make a scatterplot of the data.

b) Decide whether the data seem to fit a quadratic function.

c) Use the data points (0, 0), (50, 20), and (100, 0) to find a quadratic function that fits the data.

d) Use the function to estimate the depth of the river at 75 ft.

6. *Ticket Profits.* Valley Community College is presenting a play. The profit P, in dollars, after x days is given in the following table. (Profit can be negative when costs exceed revenue. See Section 3.7.)

Days, x	Profit, P
0	$\$-100$
90	560
180	872
270	870
360	548
450	-100

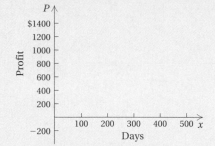

a) Make a scatterplot of the data.
b) Decide whether the data can be modeled by a quadratic function.
c) Use the data points $(0, -100)$, $(180, 872)$, and $(360, 548)$ to find a quadratic function that fits the data.
d) Use the function to estimate the profits after 225 days.

Answers on page A-44

a) The scatterplot is as follows.

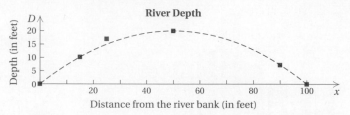

River Depth

b) The data seem to rise and fall in a manner similar to a quadratic function. The dashed black line in the graph represents a sample quadratic function of fit. Note that it may not necessarily go through each point.

c) We are looking for a quadratic function

$$D(x) = ax^2 + bx + c.$$

We need to determine the constants a, b, and c. We use the three data points $(0, 0)$, $(50, 20)$, and $(100, 0)$ and substitute as follows:

$$0 = a \cdot 0^2 + b \cdot 0 + c,$$
$$20 = a \cdot 50^2 + b \cdot 50 + c,$$
$$0 = a \cdot 100^2 + b \cdot 100 + c.$$

After simplifying, we see that we need to solve the system

$$0 = c,$$
$$20 = 2{,}500a + 50b + c,$$
$$0 = 10{,}000a + 100b + c.$$

Since $c = 0$, the system reduces to a system of two equations in two variables:

$$20 = 2{,}500a + 50b, \qquad \textbf{(1)}$$
$$0 = 10{,}000a + 100b. \qquad \textbf{(2)}$$

We multiply equation (1) by -2, add, and solve for a (see Section 3.2):

$$\begin{aligned} -40 &= -5{,}000a - 100b, \\ 0 &= \underline{10{,}000a + 100b} \\ -40 &= 5000a \end{aligned}$$
  Adding

$$\frac{-40}{5000} = a$$
  Solving for a

$$-0.008 = a.$$

Next, we substitute -0.008 for a in equation (2) and solve for b:

$$0 = 10{,}000(-0.008) + 100b$$
$$0 = -80 + 100b$$
$$80 = 100b$$
$$0.8 = b.$$

This gives us the quadratic function:

$$D(x) = -0.008x^2 + 0.8x.$$

d) To find the depth 75 ft from the riverbank, we substitute:

$$D(75) = -0.008(75)^2 + 0.8(75) = 15.$$

At a distance of 75 ft from the riverbank, the depth of the river is 15 ft.

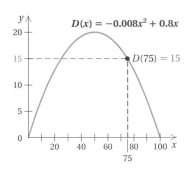

Do Exercise 6 on the preceding page.

Calculator Spotlight

Mathematical Modeling Using Regression: Fitting Quadratic and Other Polynomial Functions to Data. In the Calculator Spotlight of Section 2.7, we considered a method of fitting a linear function to a set of data, linear regression. Regression can be extended to quadratic, cubic, and quartic polynomial functions. The grapher gives us the powerful capability to find these polynomial models, select which seems best, and make predictions.

Example *Live Births to Women of Age x.* The following chart relates the number of live births to women of a particular age.

Age, x	Average Number of Live Births per 1000 Women
16	34
18.5	86.5
22	111.1
27	113.9
32	84.5
37	35.4
42	6.8

Source: Centers for Disease Control and Prevention

a) Fit a quadratic function to the data using the REGRESSION feature on a grapher.

b) Make a scatterplot of the data. Then graph the quadratic function with the scatterplot.

c) Fit a cubic function to the data using the REGRESSION feature on a grapher.

d) Make a scatterplot of the data. Then graph the cubic function with the scatterplot.

e) Decide which function seems to fit the data better.

f) Use the function from part (e) to estimate the average number of live births by women of ages 20 and 30.

We proceed as follows.

a) We fit a quadratic function to the data using the RE-GRESSION feature. The procedure is similar to what is outlined in the Calculator Spotlight of Section 2.7. We just choose QuadReg instead of LinReg(ax + b). (Consult your manual for further details.) We obtain the following function:

$$y_1 = f(x) = -0.49x^2 + 25.95x - 238.49.$$

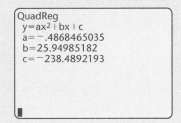

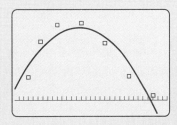

b) The quadratic function is graphed with the scatterplot above.

(continued)

c) We fit a cubic function to the data using the REGRESSION feature. We choose $\boxed{\text{CubicReg}}$ and obtain the following function:

$$y_2 = f(x) = 0.03x^3 - 3.22x^2 + 101.18x - 886.93.$$

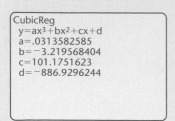

```
CubicReg
y=ax³+bx²+cx+d
 a=.0313582585
 b=-3.219568404
 c=101.1751623
 d=-886.9296244
```

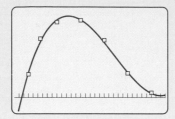

d) The cubic function is graphed with the scatterplot above.

e) The graph of the cubic function seems to fit closer to the data points. Thus we choose it as a model.

f) Using the grapher to do the calculation, we get $Y_2(20) \approx 99.6$ and $Y_2(30) \approx 97.4$ as shown.

```
Y₂(20)
              99.61232795
Y₂(30)
              97.38665968
■
```

Thus the average number of live births is 99.6 per 1000 by women age 20 and 97.4 per 1000 by women age 30.

Exercises

1. a) Use the REGRESSION feature to fit a quartic equation to the live-birth data. Make a scatterplot of the data. Then graph the quartic function with the scatterplot. Decide whether the quartic function gives a better fit than either the quadratic or the cubic function.

b) Explain why the domain of the cubic live-birth function should probably be restricted to women whose ages are in the interval [15, 45].

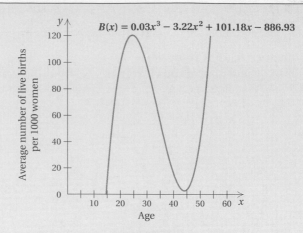

$$B(x) = 0.03x^3 - 3.22x^2 + 101.18x - 886.93$$

2. MEDIAN HOUSEHOLD INCOME BY AGE

Age, x	Median Income in 1996
19.5	$21,438
29.5	35,888
39.5	44,420
49.5	50,472
59.5	39,815
65	19,448

Source: U.S. Bureau of the Census; The Conference Board: Simmons Bureau of Labor Statistics

a) Fit a quadratic function to the data using a REGRESSION feature on a grapher.

b) Make a scatterplot of the data. Then graph the quadratic function with the scatterplot.

c) Fit a cubic function to the data using a REGRESSION feature on a grapher.

d) Make a scatterplot of the data. Then graph the cubic function with the scatterplot.

e) Fit a quartic function to the data using a REGRESSION feature on a grapher.

f) Make a scatterplot of the data. Then graph the quartic function with the scatterplot.

g) Decide which of the quadratic, cubic, or quartic functions seems to fit the data best.

h) Use the function from part (e) to estimate the median household income of people ages 25 and 45.

3. *River Depth.* Use your grapher to rework Example 3.

Exercise Set 7.7

a Solve.

1. *Architecture.* An architect is designing the floors for a hotel. Each floor is to be rectangular and is allotted 720 ft of security piping around walls outside the rooms. What dimensions of the floors will allow an atrium at the bottom to have maximum area?

2. *Stained-Glass Window Design.* An artist is designing a rectangular stained-glass window with a perimeter of 84 in. What dimensions will yield the maximum area?

3. What is the maximum product of two numbers whose sum is 22? What numbers yield this product?

4. What is the maximum product of two numbers whose sum is 45? What numbers yield this product?

5. What is the minimum product of two numbers whose difference is 4? What are the numbers?

6. What is the minimum product of two numbers whose difference is 6? What are the numbers?

7. What is the maximum product of two numbers that add to -12? What numbers yield this product?

8. What is the minimum product of two numbers that differ by 9? What are the numbers?

9. *Garden Design.* A farmer decides to enclose a rectangular garden, using the side of a barn as one side of the rectangle. What is the maximum area that the farmer can enclose with 40 ft of fence? What should the dimensions of the garden be in order to yield this area?

10. *Patio Design.* A stone mason has enough stones to enclose a rectangular patio with 60 ft of perimeter, assuming that the attached house forms one side of the rectangle. What is the maximum area that the mason can enclose? What should the dimensions of the patio be in order to yield this area?

11. *Molding Plastics.* Economite Plastics plans to produce a one-compartment vertical file by bending the long side of an 8-in. by 14-in. sheet of plastic along two lines to form a U shape. How tall should the file be in order to maximize the volume that the file can hold?

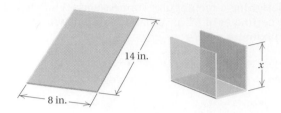

12. *Composting.* A rectangular compost container is to be formed in a corner of a fenced yard, with 8 ft of chicken wire completing the other two sides of the rectangle. If the chicken wire is 3 ft high, what dimensions of the base will maximize the volume of the container?

13. *Minimizing Cost.* Aki's Bicycle Designs has determined that when x hundred bicycles are built, the average cost per bicycle is given by

$$C(x) = 0.1x^2 - 0.7x + 2.425,$$

where $C(x)$ is in hundreds of dollars. How many bicycles should the shop build in order to minimize the average cost per bicycle?

14. *Corral Design.* A rancher needs to enclose two adjacent rectangular corrals, one for sheep and one for cattle. If a river forms one side of the corrals and 180 yd of fencing is available, what is the largest total area that can be enclosed?

Maximizing Profit. Recall (Section 3.7) that total profit P is the difference between total revenue R and total cost C. Given the following total-revenue and total-cost functions, find the total profit, the maximum value of the total profit, and the value of x at which it occurs.

15. $R(x) = 1000x - x^2$,
$\quad C(x) = 3000 + 20x$

16. $R(x) = 200x - x^2$,
$\quad C(x) = 5000 + 8x$

b *Choosing Models.* For the scatterplots and graphs in Exercises 17–24, determine which, if any, of the following functions might be used as a model for the data: Linear, $f(x) = mx + b$; quadratic, $f(x) = ax^2 + bx + c$, $a > 0$; quadratic, $f(x) = ax^2 + bx + c$, $a < 0$; polynomial, neither quadratic nor linear.

17.

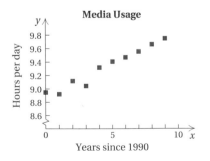

18.

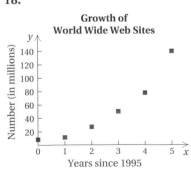

19.

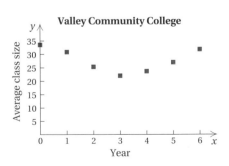

20.

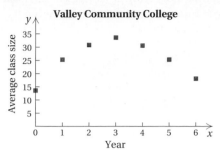

21.

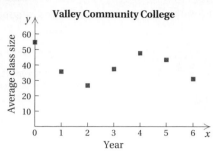

22.

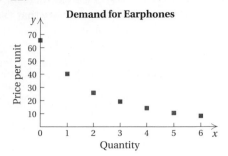

23.

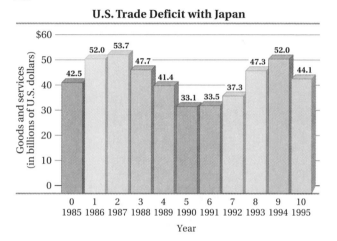

24.

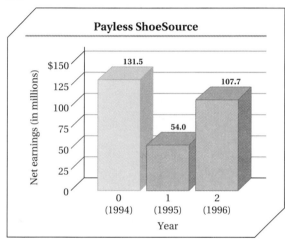

Source: Payless 1996 Annual Report

Find a quadratic function that fits the set of data points.

25. $(1, 4)$, $(-1, -2)$, $(2, 13)$

26. $(1, 4)$, $(-1, 6)$, $(-2, 16)$

27. $(2, 0)$, $(4, 3)$, $(12, -5)$

28. $(-3, -30)$, $(3, 0)$, $(6, 6)$

29. *Nighttime Accidents.*

a) Find a quadratic function that fits the following data.

Travel Speed (in kilometers per hour)	Number of Nighttime Accidents (for every 200 million kilometers driven)
60	400
80	250
100	250

b) Use the function to estimate the number of nighttime accidents that occur at 50 km/h.

30. *Daytime Accidents.*

a) Find a quadratic function that fits the following data.

Travel Speed (in kilometers per hour)	Number of Daytime Accidents (for every 200 million kilometers driven)
60	100
80	130
100	200

b) Use the function to estimate the number of daytime accidents that occur at 50 km/h.

31. *Payless Shoe Source Net Earnings.* Use the data from Exercise 24.

a) Find a quadratic function that fits the data.

b) Use the quadratic function to predict net earnings in 1999 and in 2000.

32. *Rocket Height.* The data in the following table give the height H, in feet, of a rocket t seconds after blastoff.

Time, t (in seconds)	Height, H (in feet)
2	2304
6	2048
10	1280
12	704

a) Use the data points (2, 2304), (6, 2048), and (12, 704) to find a quadratic function that fits the data.

b) Use the function to estimate the height of the rocket after 1 sec, 8 sec, and 11 sec.

c) Determine the maximum height and when it is attained.

d) Determine when the rocket reaches the ground.

Skill Maintenance

Multiply and simplify.   [6.3a]

33. $\sqrt[4]{5x^3y^5} \ \sqrt[4]{125x^2y^3}$

34. $\sqrt{9a^3} \ \sqrt{16ab^4}$

Solve.   [6.6a, b]

35. $\sqrt{4x - 4} = \sqrt{x + 4} + 1$ **36.** $\sqrt{5x - 4} + \sqrt{13 - x} = 7$ **37.** $-35 = \sqrt{2x + 5}$ **38.** $\sqrt{7x - 5} = \sqrt{4x + 7}$

Synthesis

39. ◈ Explain the restrictions that should be placed on the domains of the quadratic functions found in Exercises 15 and 29 and why such restrictions are needed.

40. ◈ Explain how the leading coefficient of a quadratic function can be used to determine whether a maximum or minimum function value exists.

41. 〰 *Japanese Trade Deficit.* Use the REGRESSION feature of a grapher to fit a quartic function to the data in Exercise 23.

42. The sum of the base and the height of a triangle is 38 cm. Find the dimensions for which the area is a maximum, and find the maximum area.

Fit a quadratic function to a set of data.

7.8 Polynomial and Rational Inequalities

a | Quadratic and Other Polynomial Inequalities

Inequalities like the following are called **quadratic inequalities:**

$$x^2 + 3x - 10 < 0, \qquad 5x^2 - 3x + 2 \geq 0.$$

In each case, we have a polynomial of degree 2 on the left. We will solve such inequalities in two ways. The first method provides understanding and the second yields the more efficient method.

The first method for solving a quadratic inequality, such as $ax^2 + bx + c > 0$, is by considering the graph of a related function, $f(x) = ax^2 + bx + c$.

Example 1 Solve: $x^2 + 3x - 10 > 0$.

Consider the function $f(x) = x^2 + 3x - 10$ and its graph. The graph opens up since the leading coefficient ($a = 1$) is positive. We find the intercepts by setting the polynomial equal to 0 and solving:

$$x^2 + 3x - 10 = 0$$
$$(x + 5)(x - 2) = 0$$
$$x + 5 = 0 \quad or \quad x - 2 = 0$$
$$x = -5 \quad or \qquad x = 2.$$

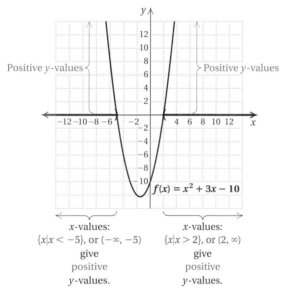

x-values:
$\{x | x < -5\}$, or $(-\infty, -5)$
give
positive
y-values.

x-values:
$\{x | x > 2\}$, or $(2, \infty)$
give
positive
y-values.

Values of y will be positive to the left and right of the intercepts, as shown. Thus the solution set of the inequality is

$$\{x \,|\, x < -5 \ or \ x > 2\}, \quad or \quad (-\infty, -5) \cup (2, \infty).$$

Do Exercise 1.

We can solve any inequality by considering the graph of a related function and finding intercepts as in Example 1. In some cases, we may need to use the quadratic formula to find the intercepts.

Objectives

a Solve quadratic and other polynomial inequalities.

b Solve rational inequalities.

For Extra Help

TAPE 16   TAPE 15A   MAC WIN   CD-ROM

1. Solve by graphing:

$$x^2 + 2x - 3 > 0.$$

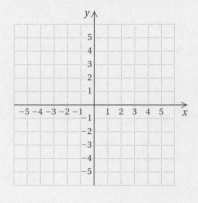

Answer on page A-44

2. Solve by graphing:

$$x^2 + 2x - 3 < 0.$$

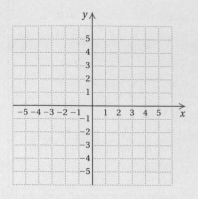

3. Solve by graphing:

$$x^2 + 2x - 3 \le 0.$$

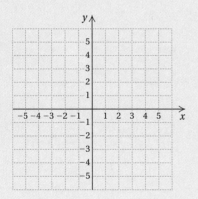

Answers on page A-44

Example 2 Solve: $x^2 + 3x - 10 < 0$.

Looking again at the graph of $f(x) = x^2 + 3x - 10$ or at least visualizing it tells us that y-values are negative for those x-values between -5 and 2.

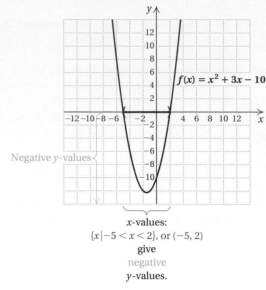

x-values:
$\{x \mid -5 < x < 2\}$, or $(-5, 2)$
give
negative
y-values.

That is, the solution set is $\{x \mid -5 < x < 2\}$, or $(-5, 2)$.

Do Exercise 2.

When an inequality contains $\le$ or $\ge$, the x-values of the intercepts must be included. Thus the solution set of the inequality $x^2 + 3x - 10 \le 0$ is

$$\{x \mid -5 \le x \le 2\}, \quad \text{or} \quad [-5, 2].$$

Do Exercise 3.

We now consider a more efficient method for solving polynomial inequalities. The preceding discussion provides the understanding for this method. In Examples 1 and 2, we see that the intercepts divide the number line into intervals.

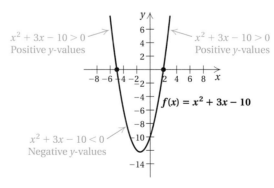

If a function has a positive output for one number in an interval, it will be positive for all the numbers in the interval. The same is true for negative outputs. Thus we can merely make a test substitution in each interval to solve the inequality. This is very similar to our method of using test points to graph a linear inequality in a plane.

Example 3 Solve: $x^2 + 3x - 10 < 0$.

We set the polynomial equal to 0 and solve. The solutions of $x^2 + 3x - 10 = 0$, or $(x + 5)(x - 2) = 0$, are -5 and 2. We then locate them on a number line as follows. Note that the numbers divide the number line into three intervals, which we will call A, B, and C.

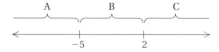

We choose a test number in interval A, say -7, and substitute -7 for x in the function $f(x) = x^2 + 3x - 10$:

$$f(-7) = (-7)^2 + 3(-7) - 10$$
$$= 49 - 21 - 10 = 18. \qquad \text{Thus, } f(-7) > 0.$$

Note that $18 > 0$, so the function values will be positive for any number in interval A.

Next, we try a test number in interval B, say 1, and find the corresponding function value:

$$f(1) = 1^2 + 3(1) - 10$$
$$= 1 + 3 - 10 = -6. \qquad \text{Thus, } f(1) < 0.$$

Note that $-6 < 0$, so the function values will be negative for any number in interval B.

Next, we try a test number in interval C, say 4, and find the corresponding function value:

$$f(4) = 4^2 + 3(4) - 10$$
$$= 16 + 12 - 10 = 18. \qquad \text{Thus, } f(4) > 0.$$

Note that $18 > 0$, so the function values will be positive for any number in interval C.

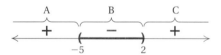

We are looking for numbers x for which $f(x) = x^2 + 3x - 10 < 0$. Thus any number x in interval B is a solution. If the inequality had been $\leq$, it would have been necessary to include the intercepts -5 and 2 in the solution set as well. The solution set is $\{x \mid -5 < x < 2\}$, or the interval $(-5, 2)$.

To solve a polynomial inequality:

1. Get 0 on one side, set the expression on the other side equal to 0, and solve to find the intercepts.

2. Use the numbers found in step (1) to divide the number line into intervals.

3. Substitute a number from each interval into the related function. If the function value is positive, then the expression will be positive for all numbers in the interval. If the function value is negative, then the expression will be negative for all numbers in the interval.

4. Select the intervals for which the inequality is satisfied and write set-builder or interval notation for the solution set.

Do Exercises 4 and 5.

Solve using the method of Example 3.

4. $x^2 + 3x > 4$

5. $x^2 + 3x \leq 4$

Answers on page A-44

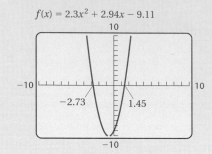
6. Solve: $6x(x + 1)(x - 1) < 0$.

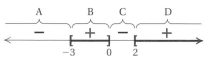

$f(x) = 5x(x + 3)(x - 2)$

Answer on page A-44

Example 4   Solve: $5x(x + 3)(x - 2) \geq 0$.

The solutions of $f(x) = 0$, or $5x(x + 3)(x - 2) = 0$, are -3, 0, and 2. They divide the real-number line into four intervals, as shown below.

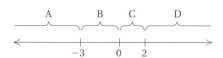

We try test numbers in each interval:

A: Test -5, $f(-5) = 5(-5)(-5 + 3)(-5 - 2) = -350 < 0$.
B: Test -2, $f(-2) = 5(-2)(-2 + 3)(-2 - 2) = 40 > 0$.
C: Test 1, $f(1) = 5(1)(1 + 3)(1 - 2) = -20 < 0$.
D: Test 3, $f(3) = 5(3)(3 + 3)(3 - 2) = 90 > 0$.

The expression is positive for values of x in intervals B and D. Since the inequality symbol is $\geq$, we need to include the intercepts. The solution set of the inequality is

$$\{x \mid -3 \leq x \leq 0 \ or \ 2 \leq x\}, \quad \text{or} \quad [-3, 0] \cup [2, \infty).$$

We visualize this with the graph at left.

Do Exercise 6.

b | Rational Inequalities

We adapt the preceding method when an inequality involves rational expressions. We call these **rational inequalities.**

Example 5 Solve: $\dfrac{x - 3}{x + 4} \geq 2$.

We write a related equation by changing the $\geq$ symbol to $=$:

$$\frac{x - 3}{x + 4} = 2.$$

Then we solve this related equation. First, we multiply on both sides of the equation by the LCM, which is $x + 4$:

$$(x + 4) \cdot \frac{x - 3}{x + 4} = (x + 4) \cdot 2$$
$$x - 3 = 2x + 8$$
$$-11 = x.$$

With rational inequalities, we also need to determine those numbers for which the rational expression is not defined—that is, those numbers that make the denominator 0. We set the denominator equal to 0 and solve: $x + 4 = 0$, or $x = -4$. Next, we use the numbers -11 and -4 to divide the number line into intervals, as shown below.

We try test numbers in each interval to see if each satisfies the original inequality.

A: Test -15,
$$\frac{x - 3}{x + 4} \geq 2$$
$$\frac{-15 - 3}{-15 + 4} \; ? \; 2$$
$$\frac{18}{11} \qquad \text{FALSE}$$

Since the inequality is false for $x = -15$, the number -15 is not a solution of the inequality. Interval A is *not* part of the solution set.

B: Test -8,
$$\frac{x - 3}{x + 4} \geq 2$$
$$\frac{-8 - 3}{-8 + 4} \; ? \; 2$$
$$\frac{11}{4} \qquad \text{TRUE}$$

Since the inequality is true for $x = -8$, the number -8 is a solution of the inequality. Interval B *is* part of the solution set.

C: Test 1,
$$\frac{x - 3}{x + 4} \geq 2$$
$$\frac{1 - 3}{1 + 4} \; ? \; 2$$
$$-\frac{2}{5} \qquad \text{FALSE}$$

Since the inequality is false for $x = 1$, the number 1 is not a solution of the inequality. Interval C is *not* part of the solution set.

Solve.

7. $\dfrac{x+1}{x-2} \geq 3$

The solution set includes the interval B. The number -11 is also included since the inequality symbol is $\geq$ and -11 is a solution of the related equation. The number -4 is not included; it is not an allowable replacement because it results in division by 0. Thus the solution set of the original inequality is

$$\{x \mid -11 \leq x < -4\}, \quad \text{or} \quad [-11, -4).$$

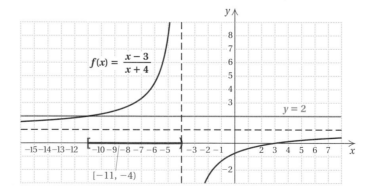

> To solve a rational inequality:
>
> 1. Change the inequality symbol to an equals sign and solve the related equation.
> 2. Find the numbers for which any rational expression in the inequality is not defined.
> 3. Use the numbers found in steps (1) and (2) to divide the number line into intervals.
> 4. Substitute a number from each interval into the inequality. If the number is a solution, then the interval to which it belongs is part of the solution set.
> 5. Select the intervals for which the inequality is satisfied and write set-builder or interval notation for the solution set.

8. $\dfrac{x}{x-5} < 2$

Do Exercises 7 and 8.

AG Algebraic–Graphical Connection

Let's compare the algebraic solution of Example 5 to a graphical solution. Although it is not a goal of this text to instruct you on graphing rational functions, let's look at the graph of the function

$$f(x) = \frac{x-3}{x+4}.$$

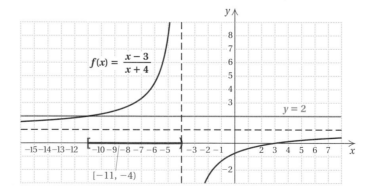

We see that the solutions of the inequality

$$\frac{x-3}{x+4} \geq 2$$

can be found by drawing the line $y = 2$ and determining all x-values for which $f(x) \geq 2$.

Answers on page A-44

Exercise Set 7.8

a Solve algebraically and verify results from the graph.

1. $(x - 6)(x + 2) > 0$

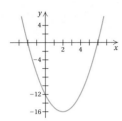

2. $(x - 5)(x + 1) > 0$

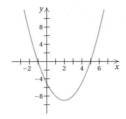

3. $4 - x^2 \geq 0$

4. $9 - x^2 \leq 0$

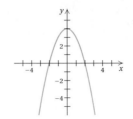

Solve.

5. $3(x + 1)(x - 4) \leq 0$

6. $(x - 7)(x + 3) \leq 0$

7. $x^2 - x - 2 < 0$

8. $x^2 + x - 2 < 0$

9. $x^2 - 2x + 1 \geq 0$

10. $x^2 + 6x + 9 < 0$

11. $x^2 + 8 < 6x$

12. $x^2 - 12 > 4x$

13. $3x(x + 2)(x - 2) < 0$

14. $5x(x + 1)(x - 1) > 0$

15. $(x + 9)(x - 4)(x + 1) > 0$

16. $(x - 1)(x + 8)(x - 2) < 0$

17. $(x + 3)(x + 2)(x - 1) < 0$

18. $(x - 2)(x - 3)(x + 1) < 0$

b Solve.

19. $\dfrac{1}{x - 6} < 0$

20. $\dfrac{1}{x + 4} > 0$

21. $\dfrac{x + 1}{x - 3} > 0$

22. $\dfrac{x - 2}{x + 5} < 0$

23. $\dfrac{3x + 2}{x - 3} \leq 0$

24. $\dfrac{5 - 2x}{4x + 3} \leq 0$

25. $\dfrac{x - 1}{x - 2} > 3$

26. $\dfrac{x + 1}{2x - 3} < 1$

27. $\dfrac{(x-2)(x+1)}{x-5} < 0$ **28.** $\dfrac{(x+4)(x-1)}{x+3} > 0$ **29.** $\dfrac{x+3}{x} \le 0$ **30.** $\dfrac{x}{x-2} \ge 0$

31. $\dfrac{x}{x-1} > 2$ **32.** $\dfrac{x-5}{x} < 1$ **33.** $\dfrac{x-1}{(x-3)(x+4)} < 0$ **34.** $\dfrac{x+2}{(x-2)(x+7)} > 0$

35. $3 < \dfrac{1}{x}$ **36.** $\dfrac{1}{x} \le 2$ **37.** $\dfrac{(x-1)(x+2)}{(x+3)(x-4)} > 0$ **38.** $\dfrac{x^2-11x+30}{x^2-8x-9} \ge 0$

Skill Maintenance

Simplify. [6.3b]

39. $\sqrt[3]{\dfrac{125}{27}}$ **40.** $\sqrt{\dfrac{25}{4a^2}}$ **41.** $\sqrt{\dfrac{16a^3}{b^4}}$ **42.** $\sqrt[3]{\dfrac{27c^5}{343d^3}}$

Add or subtract. [6.4a]

43. $3\sqrt{8} - 5\sqrt{2}$ **44.** $7\sqrt{45} - 2\sqrt{20}$

45. $5\sqrt[3]{16a^4} + 7\sqrt[3]{2a}$ **46.** $3\sqrt{10} + 8\sqrt{20} - 5\sqrt{80}$

Synthesis

47. ◈ Describe a method that could be used to create quadratic inequalities that have no solution.

48. ◈ Describe a method that could be used to create quadratic inequalities that have all real numbers as solutions.

49. ⟋⟍ Use a grapher to solve Exercises 11, 22, and 25 by graphing two curves, one for each side of an inequality.

50. ⟋⟍ Use a grapher to solve each of the following.

    a) $x + \dfrac{1}{x} < 0$ **b)** $x - \sqrt{x} \ge 0$

    c) $\tfrac{1}{3}x^3 - x + \tfrac{2}{3} \le 0$

Solve.

51. $x^2 - 2x \le 2$ **52.** $x^2 + 2x > 4$ **53.** $x^4 + 2x^2 > 0$

54. $x^4 + 3x^2 \le 0$ **55.** $\left| \dfrac{x+2}{x-1} \right| < 3$

56. *Total Profit.* A company determines that its total profit on the production and sale of x units of a product is given by

$$P(x) = -x^2 + 812x - 9600.$$

a) A company makes a profit for those nonnegative values of x for which $P(x) > 0$. Find the values of x for which the company makes a profit.

b) A company loses money for those nonnegative values of x for which $P(x) < 0$. Find the values of x for which the company loses money.

57. *Height of a Thrown Object.* The function

$$H(t) = -16t^2 + 32t + 1920$$

gives the height S of an object thrown from a cliff 1920 ft high, after time t seconds.

a) For what times is the height greater than 1920 ft?

b) For what times is the height less than 640 ft?

Summary and Review Exercises: Chapter 7

The objectives to be tested in addition to the material in this chapter are [2.7b], [5.2b], [6.3a], and [6.6a, b].

1. a) Solve: $2x^2 - 7 = 0$. [7.1a]
   b) Find the x-intercepts of $f(x) = 2x^2 - 7$.

Solve. [7.2a]

2. $14x^2 + 5x = 0$ **3.** $x^2 - 12x + 27 = 0$

4. $4x^2 + 3x + 1 = 0$ **5.** $x^2 - 7x + 13 = 0$

6. $4x(x - 1) + 15 = x(3x + 4)$

7. $x^2 + 4x + 1 = 0$. Give exact solutions and approximate solutions to three decimal places.

8. $\dfrac{x}{x - 2} + \dfrac{4}{x - 6} = 0$ **9.** $\dfrac{x}{4} - \dfrac{4}{x} = 2$

10. $15 = \dfrac{8}{x + 2} - \dfrac{6}{x - 2}$

11. Solve $x^2 + 4x + 1 = 0$ by completing the square. Show your work. [7.1b]

12. *Hang Time.* Use the function $V(T) = 48T^2$. A basketball player has a vertical leap of 39 in. What is his hang time? [7.1c]

13. *VCR Screen.* The width of a rectangular screen on a portable VCR is 5 cm less than the length. The area is 126 cm². Find the length and the width. [7.3a]

14. *Picture Matting.* A picture mat measures 12 in. by 16 in.; 140 in² of picture shows. Find the width of the mat. [7.3a]

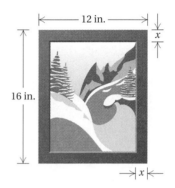

15. *Motorcycle Travel.* During the first part of a trip, a motorcyclist travels 50 mi at a certain speed. The rider travels 80 mi on the second part of the trip at a speed that is 10 mph slower. The total time for the trip is 3 hr. What is the speed on each part of the trip? [7.3a]

Determine the nature of the solutions of the equation. [7.4a]

16. $x^2 + 3x - 6 = 0$ **17.** $x^2 + 2x + 5 = 0$

Write a quadratic equation having the given solutions. [7.4b]

18. $\frac{1}{5}$, $-\frac{3}{5}$ **19.** -4, only solution

Solve for the indicated letter. [7.3b]

20. $N = 3\pi\sqrt{\dfrac{1}{p}}$, for p **21.** $2A = \dfrac{3B}{T^2}$, for T

Solve. [7.4c]

22. $x^4 - 13x^2 + 36 = 0$ **23.** $15x^{-2} - 2x^{-1} - 1 = 0$

24. $(x^2 - 4)^2 - (x^2 - 4) - 6 = 0$

25. $x - 13\sqrt{x} + 36 = 0$

For each quadratic function, find and label (a) the vertex, (b) the line of symmetry, and (c) the maximum or minimum value. Then (d) graph the function. [7.5c], [7.6a]

26. $f(x) = -\frac{1}{2}(x - 1)^2 + 3$

27. $f(x) = x^2 - x + 6$

28. $f(x) = -3x^2 - 12x - 8$

29. Find the x- and y-intercepts: [7.6b]
$$f(x) = x^2 - 9x + 14.$$

30. What is the minimum product of two numbers whose difference is 22? What numbers yield this product? [7.7a]

31. Find the quadratic function that fits the data points $(0, -2)$, $(1, 3)$, and $(3, 7)$. [7.7b]

32. *Hotel Occupancy.* Hotel occupancy has risen and fallen in recent years. Percentages of rooms occupied in various recent years are shown in the graph below. [7.7b]

Hotel Occupancy

Source: Coopers & Lybrand Lodging Research Network

a) Use the data points $(1, 61.8)$, $(5, 65.1)$, and $(8, 64)$ to fit a quadratic function to the data.
b) Use the quadratic function to predict the percentage of rooms occupied in 2000 and in 2004.

Solve. [7.8a, b]

33. $(x + 2)(x - 1)(x - 2) > 0$

34. $\dfrac{(x + 4)(x - 1)}{(x + 2)} < 0$

35. *SAT Verbal Scores.* The following table lists SAT verbal scores for recent years. [2.7b]

Years Since 1990, x	SAT Verbal Score
1	499
2	500
3	500
4	499
5	504
6	505

Source: The College Board

a) Use the data points $(1, 499)$ and $(6, 505)$ to fit a linear function to the data.
b) Use the linear function to predict SAT verbal scores in 2000 and in 2004.

36. Add and simplify: [5.2b]
$$\frac{x}{x^2 - 3x + 2} + \frac{2}{x^2 - 5x + 6}.$$

37. Multiply and simplify: [6.3a]
$$\sqrt[3]{9t^6}\ \sqrt[3]{3s^4t^9}.$$

38. Solve: $\sqrt{5x - 1} + \sqrt{2x} = 5$. [6.6b]

Synthesis

39. ◈ Explain as many characteristics as you can of the graph of the quadratic function $f(x) = ax^2 + bx + c$. [7.5a, b, c], [7.6a, b]

40. ◈ Explain how the x-intercepts of a quadratic function can be used to help find the vertex of the function. What piece of information would still be missing? [7.6a, b]

41. A quadratic function has x-intercepts $(-3, 0)$ and $(5, 0)$ and y-intercept $(0, -7)$. Find an equation for the function. What is its maximum or minimum value? [7.6a, b]

42. Find h and k such that $3x^2 - hx + 4k = 0$, the sum of the solutions is 20, and the product of the solutions is 80. [7.2a], [7.7b]

43. The average of two numbers is 171. One of the numbers is the square root of the other. Find the numbers. [7.3a]

Test: Chapter 7

1. **a)** Solve: $3x^2 - 4 = 0$.
   b) Find the x-intercepts of $f(x) = 3x^2 - 4$.

Solve.

2. $x^2 + x + 1 = 0$

3. $x - 8\sqrt{x} + 7 = 0$

4. $4x(x - 2) - 3x(x + 1) = -18$

5. $x^4 - 5x^2 + 5 = 0$

6. $x^2 + 4x = 2$. Give exact solutions and approximate solutions to three decimal places.

7. $\dfrac{1}{4 - x} + \dfrac{1}{2 + x} = \dfrac{3}{4}$

8. Solve $x^2 - 4x + 1 = 0$ by completing the square. Show your work.

9. *Free-Falling Objects.* The Peachtree Plaza in Atlanta, Georgia, is 723 ft tall. Use the function $s(t) = 16t^2$ to approximate how long it would take an object to fall from the top.

10. *Marine Travel.* The Columbia River flows at a rate of 2 mph for the length of a popular boating route. In order for a motorized dinghy to travel 3 mi upriver and then return in a total of 4 hr, how fast must the boat be able to travel in still water?

11. *Memory Board.* A computer-parts company wants to make a rectangular memory board that has a perimeter of 28 cm. What dimensions will allow the board to have a maximum area?

12. *Hang Time.* Use the function $V(T) = 48T^2$. A basketball player has a vertical leap of 35 in. What is his hang time?

13. Determine the nature of the solutions of the equation $x^2 + 5x + 17 = 0$.

14. Write a quadratic equation having the solutions $\sqrt{3}$ and $3\sqrt{3}$.

15. Solve $V = 48T^2$ for T.

For each quadratic function, find and label **(a)** the vertex, **(b)** the line of symmetry, and **(c)** the maximum or minimum value. Then **(d)** graph the function.

16. $f(x) = -x^2 - 2x$

17. $f(x) = 4x^2 - 24x + 41$

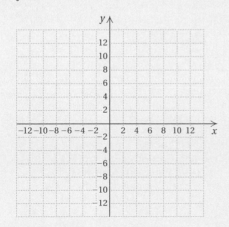

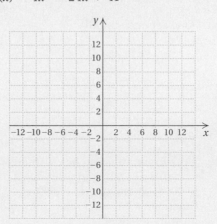

Answers

1. a) _____

b) _____

2. _____

3. _____

4. _____

5. _____

6. _____

7. _____

8. _____

9. _____

10. _____

11. _____

12. _____

13. _____

14. _____

15. _____

16. a) _____

b) _____

c) _____

d) _____

17. a) _____

b) _____

c) _____

d) _____

18. _____

19. _____

20. _____

21. a) _____

b) _____

22. _____

23. _____

24. a) _____

b) _____

25. _____

26. _____

27. _____

28. _____

29. _____

30. _____

18. Find the x- and y-intercepts:
$$f(x) = -x^2 + 4x - 1.$$

19. What is the maximum value of two numbers whose difference is 8? What numbers yield this product?

20. Find the quadratic function that fits the data points $(0, 0)$, $(3, 0)$, and $(5, 2)$.

21. _Hotel Construction._ Hotel-room construction has risen and fallen in recent years. The numbers of rooms constructed in various recent years are shown in the graph at right.

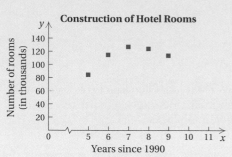

Construction of Hotel Rooms

Source: Coopers & Lybrand Lodging Research Network

a) Use the data points $(5, 84.8)$, $(7, 127.5)$, and $(9, 114.2)$ to fit a quadratic function to the data.

b) Use the quadratic function to predict the number of rooms constructed in 2000 and in 2001.

Solve.

22. $x^2 < 6x + 7$

23. $\dfrac{x - 5}{x + 3} < 0$

Skill Maintenance

24. _SAT Math Scores._ The following table lists SAT math scores for recent years.

Years Since 1990, x	SAT Math Score
1	500
2	501
3	503
4	504
5	506
6	508

Source: The College Board

a) Use the data points $(1, 500)$ and $(6, 508)$ to fit a linear function to the data.

b) Use the linear function to predict SAT math scores in 2000 and in 2004.

25. Subtract and simplify:
$$\frac{x}{x^2 + 15x + 56} - \frac{7}{x^2 + 13x + 42}.$$

26. Multiply and simplify:
$$\sqrt[4]{2a^2b^3}\ \sqrt[4]{a^4b}.$$

27. Solve: $\sqrt{x + 3} = x - 3$.

Synthesis

28. A quadratic function has x-intercepts $(-2, 0)$ and $(7, 0)$ and y-intercept $(0, 8)$. Find an equation for the function. What is its maximum or minimum value?

29. One solution of $kx^2 + 3x - k = 0$ is -2. Find the other solution.

30. Solve: $x^8 - 20x^4 + 64 = 0$.

Cumulative Review: Chapters R–7

Simplify.

1. $\dfrac{-9.1}{-13}$

2. $-3(x - 1) + 4(2x + 5)$

3. $(-4a^2b^{-3})^{-3}$

4. $\dfrac{3.2 \times 10^{-7}}{8.0 \times 10^8}$

5. $(4 + 8x^2 - 5x) - (-2x^2 + 3x - 2)$

6. $(2x^2 - x + 3)(x - 4)$

7. $\dfrac{a^2 - 16}{5a - 15} \cdot \dfrac{2a - 6}{a + 4}$

8. $\dfrac{y}{y^2 - y - 42} \div \dfrac{y^2}{y - 7}$

9. $\dfrac{2}{m + 1} + \dfrac{3}{m - 5} - \dfrac{m^2 - 1}{m^2 - 4m - 5}$

10. $\dfrac{\dfrac{1}{x} - \dfrac{1}{y}}{x + y}$

11. $(9x^3 + 5x^2 + 2) \div (x + 2)$

12. $\sqrt{0.36}$

13. $\sqrt{9x^2 - 36x + 36}$

14. $6\sqrt{45} - 3\sqrt{20}$

15. $\dfrac{2\sqrt{3} - 4\sqrt{2}}{\sqrt{2} - 3\sqrt{6}}$

16. $(8^{2/3})^4$

17. $(3 + 2i)(5 - i)$

18. $\dfrac{6 - 2i}{3i}$

Solve.

19. $3(4x - 5) + 6 = 3 - (x + 1)$

20. $F = \dfrac{mv^2}{r}$, for r

21. $5 - 3(2x + 1) \le 8x - 3$

22. $3x - 2 < -6 \text{ or } x + 3 > 9$

23. $|4x - 1| \le 14$

24. $\begin{aligned} 5x + 10y &= -10, \\ -2x - 3y &= 5 \end{aligned}$

25. $\begin{aligned} 2x + y - z &= 9, \\ 4x - 2y + z &= -9, \\ 2x - y + 2z &= -12 \end{aligned}$

26. $10x^2 + 28x - 6 = 0$

27. $\dfrac{2}{n} - \dfrac{7}{n} = 3$

28. $\dfrac{1}{2x - 1} = \dfrac{3}{5x}$

29. $A = \dfrac{mh}{m + a}$, for m

30. $\sqrt{2x - 1} = 6$

31. $\sqrt{x - 2} + 1 = \sqrt{2x - 6}$

32. $16(t - 1) = t(t + 8)$

33. $x^2 - 3x + 16 = 0$

34. $\dfrac{18}{x + 1} - \dfrac{12}{x} = \dfrac{1}{3}$

35. $P = \sqrt{a^2 - b^2}$, for a

36. $\dfrac{(x + 3)(x + 2)}{(x - 1)(x + 1)} < 0$

Factor.

37. $2t^2 - 7t - 30$

38. $a^2 + 3a - 54$

39. $24a^3 + 18a^2 - 20a - 15$

40. $-3a^3 + 12a^2$

41. $64a^2 - 9b^2$

42. $3a^2 - 36a + 108$

43. $\frac{1}{27}a^3 - 1$

44. $(x + 1)(x - 1) + (x + 1)(x + 2)$

Graph.

45. $2x - y = 11$

46. $x + y = 2$

47. $y \geq 6x - 5$

48. $x < -3$

49. $3x - y > 6,$
$4x + y \leq 3$

50. $f(x) = x^2 - 1$

51. $f(x) = -2x^2 + 3$

52. Solve: $4x^2 - 25 > 0$.

53. Find an equation of the line with slope $\frac{1}{2}$ and through the point $(-4, 2)$.

54. Find an equation of the line parallel to the line $3x + y = 4$ and through the point $(0, 1)$.

55. *Toys "R" Us.* Toys "R" Us, Inc., an international company specializing in toys, has experienced tremendous growth in recent years. Its net sales S, in billions of dollars, are shown in the graph at right.

a) Use the data points $(1, 5.5)$ and $(7, 9.9)$ to fit a linear function to the data.
b) Use the linear function to predict net sales in 2000 and in 2004.

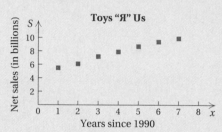

Source: Toys "Я" Us Annual Report, 1997

56. *Marine Travel.* The Connecticut River flows at a rate of 4 km/h for the length of a popular scenic route. In order for a cruiser to travel 60 km upriver and then return in a total of 8 hr, how fast must the boat be able to travel in still water?

57. *Architecture.* An architect is designing a rectangular family room with a perimeter of 56 ft. What dimensions will yield the maximum area? What is the maximum area?

58. The perimeter of a hexagon with all six sides congruent is the same as the perimeter of a square. One side of the hexagon is 3 less than the side of the square. Find the perimeter of each polygon.

59. Two pipes can fill a tank in $1\frac{1}{2}$ hr. One pipe requires 4 hr longer running alone to fill the tank than the other. How long would it take the faster pipe, working alone, to fill the tank?

Synthesis

60. Solve $ax^2 - bx = 0$ for x.

61. Simplify: $\left[\frac{1}{(-3)^{-2}} - (-3)^1 \right] \cdot [(-3)^2 + (-3)^{-2}]$.

62. Solve: $\frac{2x + 1}{x} = 3 + 7\sqrt{\frac{2x + 1}{x}}$.

63. Factor: $\frac{a^3}{8} + \frac{8b^3}{729}$.

64. Pam can do a certain job in a hours working alone. Elaine can do the same job in b hours working alone. Working together, it takes them t hours to do the job.

a) Find a formula for t.
b) Solve the formula for a.
c) Solve the formula for b.

8

Exponential and Logarithmic Functions

Introduction

The functions that we consider in this chapter are important for their rich applications to many fields. We will look at such applications as compound interest, population growth, and radioactive decay, but there are many others.

Exponents are the basis of the theory in this chapter. We define some functions having variable exponents, called exponential functions. The logarithmic functions follow from the exponential functions.

An Application	**The Mathematics**

The world is experiencing an exponential demand for lumber. The amount of timber N, in billions of cubic feet, consumed t years after 1997, can be approximated by

$$N(t) = 62(1.018)^t,$$

where $t = 0$ corresponds to 1997 (**Source**: U. N. Food and Agricultural Organization; American Forest and Paper Association). Estimate the amount of timber to be consumed in 2000.

This problem appears as Exercise 27 in Exercise Set 8.1.

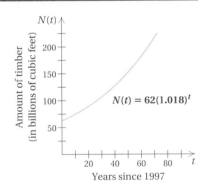

Since the year 2000 is 3 yr after 1997, we substitute 3 for t:

$$N(3) = 62(1.018)^3 \approx 65.4.$$

We estimate that in 2000, 65.4 billion cubic feet of timber will be consumed.

World Wide Web For more information, visit us at www.mathmax.com

Pretest: Chapter 8

Graph.

1. $f(x) = 2^x$

2. $f(x) = \log_2 x$

3. Convert to a logarithmic equation:
$a^3 = 1000.$

4. Convert to an exponential equation:
$\log_2 32 = t.$

5. Express as a single logarithm:
$$2 \log_a M + \log_a N - \frac{1}{3} \log_a Q.$$

6. Express in terms of logarithms of x, y, and z:
$$\log_a \sqrt[4]{\frac{x^3 y}{z^2}}.$$

Solve.

7. $\log_x \dfrac{1}{4} = -2$

8. $\log_5 x = 2$

9. $3^x = 8.6$

10. $16^{x-3} = 4$

11. $\log (x^2 - 4) - \log (x - 2) = 1$

12. $\ln x = 0$

Find each of the following using a calculator.

13. $\log 714$

14. $\ln 0.0008464$

15. $\ln 1$

16. $e^{4.2}$

17. $e^{-2.13}$

18. $\log (0.0234)$

19. *Consumer Price Index.* The consumer price index compares the cost of goods and services over various years using \$100 worth of goods and services in 1967 as a base. It is approximated by the function

$$P(t) = \$100 e^{0.06t},$$

where t is the number of years since 1967. Goods and services that cost \$100 in 1967 will cost how much in 2000?

20. *Carbon Dating.* How old is an animal bone that has lost 76% of its carbon-14?

Objectives for Retesting

The objectives to be tested in addition to the material in this chapter are as follows.

[6.8b, c, d, e] Add, subtract, multiply, and divide complex numbers. Write expressions involving powers of i in the form $a + bi$.

[7.4c] Solve equations that are quadratic in form.

[7.6a] For a quadratic function, find the vertex, the line of symmetry, and the maximum or minimum value, and graph the function.

[7.7b] Fit a quadratic function to a set of data to form a mathematical model, and solve related applied problems.

8.1 Exponential Functions

The rapidly rising graph shown below approximates the graph of an *exponential function*. We will consider such functions and some of their applications.

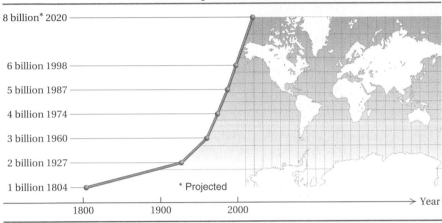

World Population Growth

8 billion* 2020
6 billion 1998
5 billion 1987
4 billion 1974
3 billion 1960
2 billion 1927
1 billion 1804

* Projected

1800   1900   2000   → Year

Source: U.S. Bureau of the Census

Objectives

a Graph exponential equations and functions.

b Graph exponential equations in which *x* and *y* have been interchanged.

c Solve applied problems involving applications of exponential functions and their graphs.

For Extra Help

TAPE 17   TAPE 15B   MAC   CD-ROM
                        WIN

a | Graphing Exponential Functions

We now develop the meaning of exponential expressions with irrational exponents. In Chapter 6, we gave meaning to exponential expressions with rational-number exponents such as

$$8^{1/4}, \quad 3^{-3/4}, \quad 7^{2.34}, \quad 5^{1.73}.$$

For example, $5^{1.73}$, or $5^{173/100}$, or $\sqrt[100]{5^{173}}$, means to raise 5 to the 173rd power and then take the 100th root. Examples of expressions with irrational exponents are

$$5^{\sqrt{3}}, \quad 7^{\pi}, \quad 9^{-\sqrt{2}}.$$

Since we can approximate irrational numbers with decimal approximations, we can also approximate expressions with irrational exponents. For example, consider $5^{\sqrt{3}}$. We know that $5^{\sqrt{3}} \approx 5^{1.73} = \sqrt[100]{5^{173}}$. As rational values of r get close to $\sqrt{3}$, 5^r gets close to some real number. Note the following:

r closes in on $\sqrt{3}$.	5^r closes in on some real number p.
r	5^r
$1 < \sqrt{3} < 2$	$5 = 5^1 < p < 5^2 = 25$
$1.7 < \sqrt{3} < 1.8$	$15.426 = 5^{1.7} < p < 5^{1.8} = 18.119$
$1.73 < \sqrt{3} < 1.74$	$16.189 = 5^{1.73} < p < 5^{1.74} = 16.452$
$1.732 < \sqrt{3} < 1.733$	$16.241 = 5^{1.732} < p < 5^{1.733} = 16.267$

As r closes in on $\sqrt{3}$, 5^r closes in on some real number p. We define $5^{\sqrt{3}}$ to be that number p. To seven decimal places, we have

$$5^{\sqrt{3}} \approx 16.2424508.$$

Any positive irrational exponent can be defined in a similar way. Negative irrational exponents are then defined in the same way as negative integer exponents. Then the expression a^x has meaning for any real number x. The general laws of exponents still hold, but we will not prove that here.

Calculator Spotlight

We can approximate $5^{\sqrt{3}}$ using the scientific keys.

Exercises

Approximate.

1. $5^{\sqrt{3}}$ **2.** 7^{π} **3.** $9^{-\sqrt{2}}$

1. Graph: $f(x) = 3^x$.

a) Complete this table of solutions.

x	$f(x)$
0	
1	
2	
3	
−1	
−2	
−3	

b) Plot the points from the table and connect them with a smooth curve.

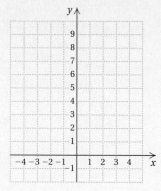

We now define exponential functions.

> **EXPONENTIAL FUNCTION**
>
> The function $f(x) = a^x$, where a is a positive constant different from 1, is called the **exponential function,** base a.

We restrict the base a to be positive to avoid the possibility of taking even roots of negative numbers such as the square root of -1, $(-1)^{1/2}$, which is not a real number. We restrict the base from being 1 because for $a = 1$, $f(x) = 1^x = 1$, which is a constant.

The following are examples of exponential functions:

$$f(x) = 2^x, \qquad f(x) = \left(\tfrac{1}{2}\right)^x, \qquad f(x) = (0.4)^x.$$

Note that in contrast to polynomial functions like $f(x) = x^2$ and $f(x) = x^3$, the variable is *in the exponent*. Let's consider graphs of exponential functions.

Example 1 Graph the exponential function $f(x) = 2^x$.

We compute some function values and list the results in a table. It is a good idea to begin by letting $x = 0$.

$f(0) = 2^0 = 1;$

$f(1) = 2^1 = 2;$

$f(2) = 2^2 = 4;$

$f(3) = 2^3 = 8;$

$f(-1) = 2^{-1} = \dfrac{1}{2^1} = \dfrac{1}{2};$

$f(-2) = 2^{-2} = \dfrac{1}{2^2} = \dfrac{1}{4};$

$f(-3) = 2^{-3} = \dfrac{1}{2^3} = \dfrac{1}{8}.$

x	$f(x)$
0	1
1	2
2	4
3	8
−1	$\frac{1}{2}$
−2	$\frac{1}{4}$
−3	$\frac{1}{8}$

Next, we plot these points and connect them with a smooth curve.

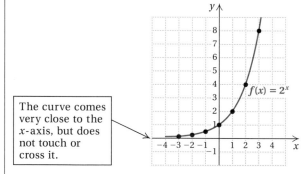

The curve comes very close to the x-axis, but does not touch or cross it.

$f(x) = 2^x$

In graphing, be sure to plot enough points to determine how steeply the curve rises.

Note that as x increases, the function values increase indefinitely. As x decreases, the function values decrease, getting very close to 0. The x-axis, or the line $y = 0$, is an *asymptote*, meaning here that as x gets very small, the curve comes very close to but never touches the axis.

Do Exercise 1.

Answers on page A-47

Example 2 Graph the exponential function $f(x) = \left(\frac{1}{2}\right)^x$.

We compute some function values and list the results in a table. Before we do so, note that

$$f(x) = \left(\tfrac{1}{2}\right)^x = (2^{-1})^x = 2^{-x}.$$

Then we have

$f(0) = 2^{-0} = 1;$

$f(1) = 2^{-1} = \dfrac{1}{2^1} = \dfrac{1}{2};$

$f(2) = 2^{-2} = \dfrac{1}{2^2} = \dfrac{1}{4};$

$f(3) = 2^{-3} = \dfrac{1}{2^3} = \dfrac{1}{8};$

$f(-1) = 2^{-(-1)} = 2^1 = 2;$

$f(-2) = 2^{-(-2)} = 2^2 = 4;$

$f(-3) = 2^{-(-3)} = 2^3 = 8.$

x	f(x)
0	1
1	$\frac{1}{2}$
2	$\frac{1}{4}$
3	$\frac{1}{8}$
-1	2
-2	4
-3	8

Next, we plot these points and draw the curve. Note that this graph is a reflection across the y-axis of the graph in Example 1. The line $y = 0$ is again an asymptote.

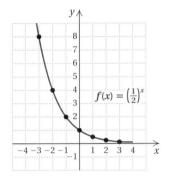

$f(x) = \left(\tfrac{1}{2}\right)^x$

Do Exercise 2.

The preceding examples illustrate exponential functions with various bases. Let's list some of their characteristics. Keep in mind that the definition of an exponential function, $f(x) = a^x$, requires that the base be positive and different from 1.

When $a > 1$, the function $f(x) = a^x$ increases from left to right. The greater the value of a, the steeper the curve. As x gets smaller and smaller, the curve gets closer to the line $y = 0$: It is an asymptote.

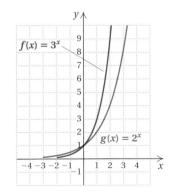

$f(x) = 3^x$

$g(x) = 2^x$

2. Graph: $f(x) = \left(\dfrac{1}{3}\right)^x$.

a) Complete this table of solutions.

x	f(x)
0	
1	
2	
3	
-1	
-2	
-3	

b) Plot the points from the table and connect them with a smooth curve.

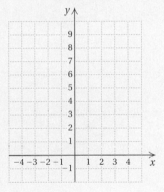

Answers on page A-47

Graph.

3. $f(x) = 4^x$

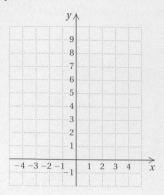

When $0 < a < 1$, the function $f(x) = a^x$ decreases from left to right. As a approaches 1, the curve becomes less steep. As x gets larger and larger, the curve gets closer to the line $y = 0$: It is an asymptote.

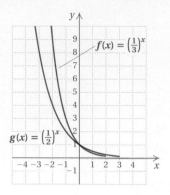

> Note that all functions $f(x) = a^x$ go through the point (0, 1). That is, the y-intercept is (0, 1).

Do Exercises 3 and 4.

4. $f(x) = \left(\dfrac{1}{4}\right)^x$

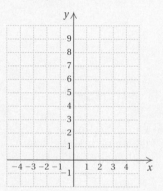

Example 3 Graph: $f(x) = 2^{x-2}$.

We construct a table of values. Then we plot the points and connect them with a smooth curve. Be sure to note that $x - 2$ is the *exponent*.

$$f(0) = 2^{0-2} = 2^{-2} = \frac{1}{2^2} = \frac{1}{4};$$

$$f(1) = 2^{1-2} = 2^{-1} = \frac{1}{2^1} = \frac{1}{2};$$

$$f(2) = 2^{2-2} = 2^0 = 1;$$

$$f(3) = 2^{3-2} = 2^1 = 2;$$

$$f(4) = 2^{4-2} = 2^2 = 4;$$

$$f(-1) = 2^{-1-2} = 2^{-3} = \frac{1}{2^3} = \frac{1}{8};$$

$$f(-2) = 2^{-2-2} = 2^{-4} = \frac{1}{2^4} = \frac{1}{16}.$$

x	$f(x)$
0	$\frac{1}{4}$
1	$\frac{1}{2}$
2	1
3	2
4	4
-1	$\frac{1}{8}$
-2	$\frac{1}{16}$

5. Graph: $f(x) = 2^{x+2}$.

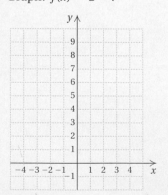

The graph looks just like the graph of $g(x) = 2^x$, but it is translated 2 units to the right. The y-intercept of $g(x) = 2^x$ is (0, 1). The y-intercept of $f(x) = 2^{x-2}$ is $\left(0, \frac{1}{4}\right)$. The line $y = 0$ is still an asymptote.

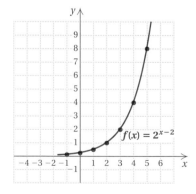

Do Exercise 5.

Answers on page A-47

Example 4   Graph: $f(x) = 2^x - 3$.

We construct a table of values. Then we plot the points and connect them with a smooth curve. Note that the only expression in the exponent is x.

$$f(0) = 2^0 - 3 = 1 - 3 = -2;$$
$$f(1) = 2^1 - 3 = 2 - 3 = -1;$$
$$f(2) = 2^2 - 3 = 4 - 3 = 1;$$
$$f(3) = 2^3 - 3 = 8 - 3 = 5;$$
$$f(4) = 2^4 - 3 = 16 - 3 = 13;$$
$$f(-1) = 2^{-1} - 3 = \frac{1}{2} - 3 = -\frac{5}{2};$$
$$f(-2) = 2^{-2} - 3 = \frac{1}{4} - 3 = -\frac{11}{4}.$$

x	$f(x)$
0	-2
1	-1
2	1
3	5
4	13
-1	$-\frac{5}{2}$
-2	$-\frac{11}{4}$

The graph looks just like the graph of $g(x) = 2^x$, but it is translated down 3 units. The y-intercept is $(0, -2)$. The line $y = -3$ is an asymptote. The curve gets closer to this line as x gets smaller and smaller.

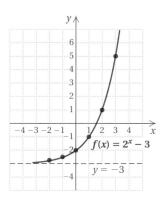

Do Exercise 6.

6. Graph: $f(x) = 2^x - 4$.

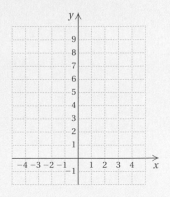

Answer on page A-47

Calculator Spotlight

Graphers are especially helpful when we are graphing exponential functions because bases may not be whole numbers and function values can quickly become very large or very small. To graph

$$f(x) = 5000(1.075)^x,$$

we enter $y_1 = 5000(1.075) \wedge x$. Because y-values are *always* positive and will become quite large, we use the viewing window $[-10, 10, 0, 15000]$, with Yscl = 1000.

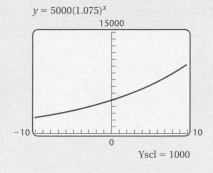

$y = 5000(1.075)^x$

Yscl = 1000

Exercises

Graph the given pair of exponential functions. Adjust the viewing window and scale as needed. Then compare each pair of graphs.

1. $y_1 = \left(\frac{5}{2}\right)^x$, $y_2 = \left(\frac{2}{5}\right)^x$
2. $y_1 = 3.2^x$, $y_2 = 3.2^{-x}$
3. $y_1 = 9.34^x$, $y_2 = 9.34^{-x}$
4. $y_1 = \left(\frac{3}{7}\right)^x$, $y_2 = \left(\frac{7}{3}\right)^x$
5. $y_1 = 5000(1.08)^x$, $y_2 = 5000(1.08)^{x-3}$
6. $y_1 = 2000(1.09)^x$, $y_2 = 2000(1.09)^{x+3}$

7. Graph: $x = 3^y$.

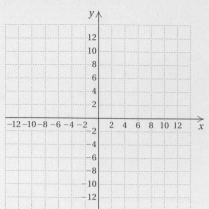

b   Equations with *x* and *y* Interchanged

It will be helpful in later work to be able to graph an equation in which the
x and the y in $y = a^x$ are interchanged.

Example 5   Graph: $x = 2^y$.

Note that x is alone on one side of the equation. We can find ordered
pairs that are solutions more easily by choosing values for y and then
computing the x-values.

For $y = 0$, $x = 2^0 = 1$.
For $y = 1$, $x = 2^1 = 2$.
For $y = 2$, $x = 2^2 = 4$.
For $y = 3$, $x = 2^3 = 8$.
For $y = -1$, $x = 2^{-1} = \dfrac{1}{2^1} = \dfrac{1}{2}$.
For $y = -2$, $x = 2^{-2} = \dfrac{1}{2^2} = \dfrac{1}{4}$.
For $y = -3$, $x = 2^{-3} = \dfrac{1}{2^3} = \dfrac{1}{8}$.

x	y
1	0
2	1
4	2
8	3
$\frac{1}{2}$	-1
$\frac{1}{4}$	-2
$\frac{1}{8}$	-3

(1) Choose values for y.
(2) Compute values for x.

We plot the points and connect
them with a smooth curve.

This curve does not touch
or cross the y-axis.

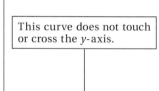

Note that this curve looks just like the graph of $y = 2^x$, except that it is
reflected, or flipped, across the line $y = x$, as shown below.

$y = 2^x$	
x	y
0	1
1	2
2	4
3	8
-1	$\frac{1}{2}$
-2	$\frac{1}{4}$
-3	$\frac{1}{8}$

$x = 2^y$	
x	y
1	0
2	1
4	2
8	3
$\frac{1}{2}$	-1
$\frac{1}{4}$	-2
$\frac{1}{8}$	-3

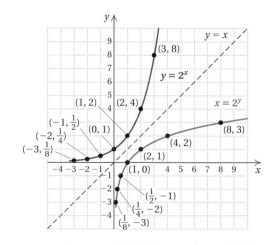

Do Exercise 7.

c | Applications of Exponential Functions

Example 6 *Interest Compounded Annually.* The amount of money A that a principal P will grow to after t years at interest rate r, compounded annually, is given by the formula

$$A = P(1 + r)^t.$$

Suppose that $100,000 is invested at 8% interest, compounded annually.

a) Find a function for the amount in the account after t years.

b) Find the amount of money in the account at $t = 0$, $t = 4$, $t = 8$, and $t = 10$.

c) Graph the function.

We solve as follows:

a) If $P = \$100,000$ and $r = 8\% = 0.08$, we can substitute these values and form the following function:

$$A(t) = \$100,000(1 + 0.08)^t = \$100,000(1.08)^t.$$

b) To find the function values, you might find a calculator with a power key helpful.

$$A(0) = \$100,000(1.08)^0 = \$100,000(1) = \$100,000;$$
$$A(4) = \$100,000(1.08)^4 = \$100,000(1.36048896) \approx \$136,048.90;$$
$$A(8) = \$100,000(1.08)^8 = \$100,000(1.85093021) \approx \$185,093.02;$$
$$A(10) = \$100,000(1.08)^{10} = \$100,000(2.158924997) \approx \$215,892.50$$

c) We use the function values computed in (b) with others, if we wish, to draw the graph as follows. Note that the axes are scaled differently because of the large numbers.

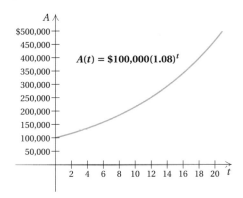

Do Exercise 8.

8. *Interest Compounded Annually.* Suppose that $40,000 is invested at 5% interest, compounded annually.

a) Find a function for the amount in the account after t years.

b) Find the amount of money in the account at $t = 0$, $t = 4$, $t = 8$, and $t = 10$.

c) Graph the function.

Calculator Spotlight

 Graph

$$y_1 = 100{,}000(1.08)^x.$$

Then use the TABLE feature or the $Y_1(\ )$ notation to check Example 6(b).

Answers on page A-48

Calculator Spotlight

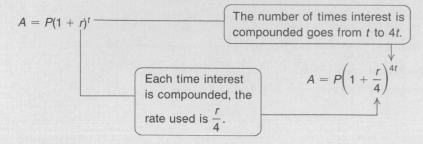 The Compound-Interest Formula. When interest is compounded quarterly, we can find a formula like the one considered in this section as follows:

$A = P(1 + r)^t$ ————————— The number of times interest is compounded goes from t to $4t$.

Each time interest is compounded, the rate used is $\frac{r}{4}$.

$$A = P\left(1 + \frac{r}{4}\right)^{4t}$$

In general, the following formula for compound interest can be used. The amount of money A in an account is given by

$$A = P\left(1 + \frac{r}{n}\right)^{nt},$$

where P is the principal, r is the annual interest rate, n is the number of times per year that interest is compounded, and t is the time, in years.

Example Suppose that $1000 is invested at 8%, compounded quarterly. How much is in the account at the end of 2 yr?

We use the equation above, substituting 1000 for P, 0.08 for r, 4 for n (compounding quarterly), and 2 for t. Then we get

$$A = P\left(1 + \frac{r}{n}\right)^{nt} = 1000\left(1 + \frac{0.08}{4}\right)^{4 \cdot 2} = 1000(1.02)^8 \approx \$1171.66.$$

Exercises

1. Suppose that $1000 is invested at 6%, compounded semiannually. How much is in the account at the end of 2 yr?
2. Suppose that $1000 is invested at 6%, compounded monthly. How much is in the account at the end of 2 yr?
3. Suppose that $20,000 is invested at 4.2%, compounded quarterly. How much is in the account at the end of 10 yr?
4. Suppose that $420,000 is invested at 4.2%, compounded daily. How much is in the account at the end of 30 yr?
5. Suppose that $10,000 is invested at 5.4%. How much is in the account at the end of 1 yr, if interest is compounded
   (a) annually? (b) semiannually? (c) quarterly? (d) daily? (e) hourly?
6. Suppose that $10,000 is invested at 3%. How much is in the account at the end of 1 yr, if interest is compounded
   (a) annually? (b) semiannually? (c) quarterly? (d) daily? (e) hourly?

Exercise Set 8.1

a Graph.

1. $f(x) = 2^x$

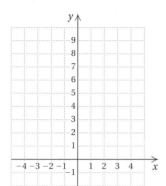

2. $f(x) = 3^x$

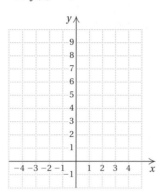

3. $f(x) = 5^x$

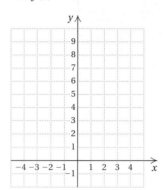

4. $f(x) = 6^x$

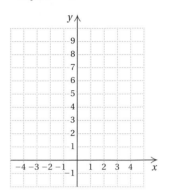

5. $f(x) = 2^{x+1}$

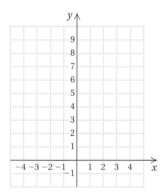

6. $f(x) = 2^{x-1}$

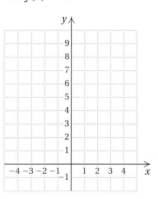

7. $f(x) = 3^{x-2}$

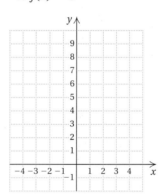

8. $f(x) = 3^{x+2}$

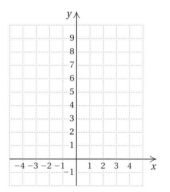

9. $f(x) = 2^x - 3$

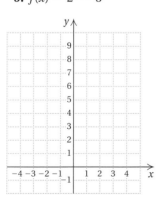

10. $f(x) = 2^x + 1$

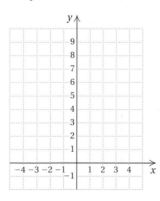

11. $f(x) = 5^{x+3}$

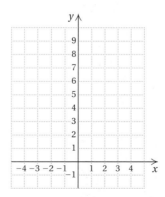

12. $f(x) = 6^{x-4}$

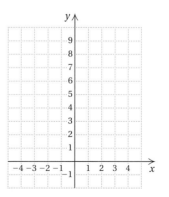

13. $f(x) = \left(\dfrac{1}{2}\right)^x$

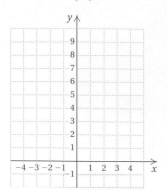

14. $f(x) = \left(\dfrac{1}{3}\right)^x$

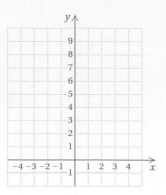

15. $f(x) = \left(\dfrac{1}{5}\right)^x$

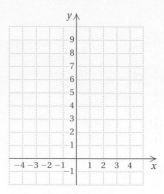

16. $f(x) = \left(\dfrac{1}{4}\right)^x$

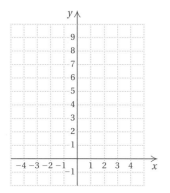

17. $f(x) = 2^{2x-1}$

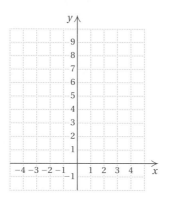

18. $f(x) = 3^{3-x}$

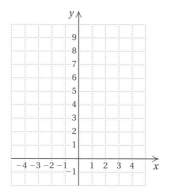

b Graph.

19. $x = 2^y$

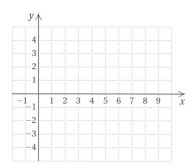

20. $x = 6^y$

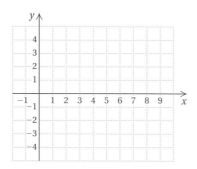

21. $x = \left(\dfrac{1}{2}\right)^y$

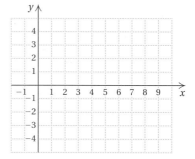

22. $x = \left(\dfrac{1}{3}\right)^y$

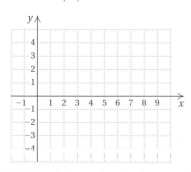

23. $x = 5^y$

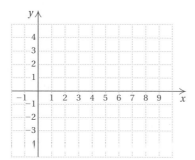

24. $x = \left(\dfrac{2}{3}\right)^y$

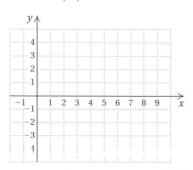

Chapter 8   Exponential and Logarithmic Functions

Graph both equations using the same set of axes.

25. $y = 2^x$, $x = 2^y$

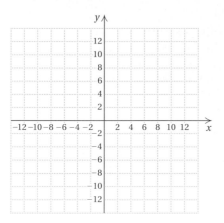

26. $y = \left(\dfrac{1}{2}\right)^x$, $x = \left(\dfrac{1}{2}\right)^y$

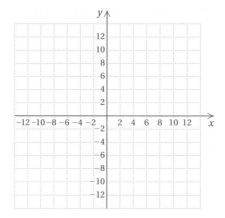

c Solve.

27. *World Demand for Lumber.* The world is experiencing an exponential demand for lumber. The amount of timber N, in billions of cubic feet, consumed t years after 1997, can be approximated by

$$N(t) = 62(1.018)^t,$$

where $t = 0$ corresponds to 1997 (***Source:*** U. N. Food and Agricultural Organization; American Forest and Paper Association).

a) How much timber was consumed in 1997? in 1998?
b) Estimate the amount of timber to be consumed in 2000 and in 2010.
c) Graph the function.

28. *Interest Compounded Annually.* Suppose that $50,000 is invested at 8% interest, compounded annually.

a) Find a function for the amount in the account after t years.
b) Find the amount of money in the account at $t = 0$, $t = 4$, $t = 8$, and $t = 10$.
c) Graph the function.

29. *Growth of Bacteria.* Bladder infections are often caused when the bacteria *Escherichia coli* reach the human bladder. Suppose that 3000 of the bacteria are present at time $t = 0$. Then t min later, the number of bacteria present will be

$$N(t) = 3000(2)^{t/20}$$

(***Source:*** Chris Hayes, "Detecting a Human Health Risk: E. coli," *Laboratory Medicine* 29, no. 6, June 1998: 347–355).

a) How many bacteria will be present after 10 min? 20 min? 30 min? 40 min? 60 min?
b) Graph the function.

30. *Salvage Value.* An office machine is purchased for $5200. Its value each year is about 80% of the value the preceding year. Its value after t years is given by the exponential function

$$V(t) = \$5200(0.8)^t.$$

a) Find the value of the machine after 0 yr, 1 yr, 2 yr, 5 yr, and 10 yr.
b) Graph the function.

31. *Spread of Zebra Mussels.* Beginning in 1988, infestations of zebra mussels started spreading throughout North American waters. These mussels spread with such speed that water treatment facilities, power plants, and entire ecosystems can become threatened. The function

$$A(t) = 10 \cdot 34^t$$

can be used to estimate the number of square centimeters of lake bottom that will be covered with mussels t years after an infestation covering 10 cm^2 first occurs. (**Source:** Many thanks to Dr. Gerald Mackie of the Department of Zoology at the University of Guelph in Ontario for the background information for this exercise.)

a) How many square centimeters of lake bottom will be covered with mussels 5 yr after an infestation covering 10 cm^2 first appears? 7 yr after the infestation first appears?

b) Graph the function.

32. *Cases of AIDS.* The total number of Americans who have contracted AIDS, in thousands, can be approximated by the exponential function

$$N(t) = 1476(1.4)^t,$$

where $t = 0$ corresponds to 1997.

a) According to the function, how many Americans had been infected as of 1998?

b) Estimate the total number of Americans who will have been infected as of 2010.

c) Graph the function.

Skill Maintenance

33. Multiply and simplify: $x^{-5} \cdot x^3$. [R.7a]

34. Simplify: $(x^{-3})^4$. [R.7b]

Simplify. [R.3a]

35. 9^0

36. $\left(\frac{2}{3}\right)^0$

37. $\left(\frac{2}{3}\right)^1$

38. 2.7^1

Divide and simplify. [R.7a]

39. $\dfrac{x^{-3}}{x^4}$

40. $\dfrac{x}{x^{11}}$

41. $\dfrac{x}{x^0}$

42. $\dfrac{x^{-3}}{x^{-4}}$

Synthesis

43. ◆ Why was it necessary to discuss irrational exponents before graphing exponential functions?

44. ◆ Suppose that $1000 is invested for 5 yr at 6% interest, compounded annually. In what year will the greatest amount of interest be earned? Why?

45. Simplify: $(5^{\sqrt{2}})^{2\sqrt{2}}$.

46. Which is larger: $\pi^{\sqrt{2}}$ or $(\sqrt{2})^{\pi}$?

Graph.

47. $y = 2^x + 2^{-x}$

48. $y = |2^x - 2|$

49. $y = \left|\left(\frac{1}{2}\right)^x - 1\right|$

50. $y = 2^{-x^2}$

Graph both equations using the same set of axes.

51. $y = 3^{-(x-1)}$, $x = 3^{-(y-1)}$

52. $y = 1^x$, $x = 1^y$

53. 📈 Use a grapher to graph each of the equations in Exercises 47–50.

8.2 Inverse and Composite Functions

When we go from an output of a function back to its input or inputs, we get what is called an *inverse relation*. When that relation is a function, we have what is called an *inverse function*. We now study such inverse functions and how to find formulas when the original function has a formula. We do so to understand the relationships among the special functions that we study in this chapter.

a Inverses

A set of ordered pairs is called a **relation**. When we consider the graph of a function, we are thinking of a set of ordered pairs. Thus a function can be thought of as a special kind of relation, in which to each first coordinate there corresponds one and only one second coordinate.

Consider the relation h given as follows:

$$h = \{(-7, 4), (3, -1), (-6, 5), (0, 2)\}.$$

Suppose we *interchange* the first and second coordinates. The relation we obtain is called the **inverse** of the relation h and is given as follows:

$$\text{Inverse of } h = \{(4, -7), (-1, 3), (5, -6), (2, 0)\}.$$

> Interchanging the ordered pairs in a relation produces the **inverse relation.**

Example 1 Consider the relation g given by

$$g = \{(2, 4), (-1, 3), (-2, 0)\}.$$

In the figure below, the relation g is shown in red. The inverse of the relation is

$$\{(4, 2), (3, -1), (0, -2)\}$$

and is shown in blue.

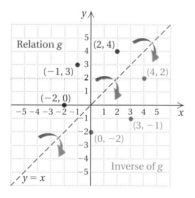

Do Exercise 1.

Objectives

a Find the inverse of a relation if it is described as a set of ordered pairs or as an equation.

b Given a function, determine whether it is one-to-one and has an inverse that is a function.

c Find a formula for the inverse of a function, if it exists.

d Graph inverse relations and functions.

e Find the composition of functions and express certain functions as a composition of functions.

f Determine whether a function is an inverse by checking its composition with the original function.

For Extra Help

TAPE 17   TAPE 15B   MAC WIN   CD-ROM

1. Consider the relation g given by

$$g = \{(2, 5), (-1, 4), (-2, 1)\}.$$

The graph of the relation is shown below in red. Find the inverse and draw its graph in blue.

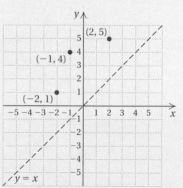

Answer on page A-49

2. Find an equation of the inverse relation. Then graph both the original relation and its inverse.

Relation:
$y = 6 - 2x$

x	y
0	6
2	2
3	0
5	-4

Inverse:

_____

x	y
6	
2	
0	
-4	

Complete the table and draw the graphs using the same set of axes.

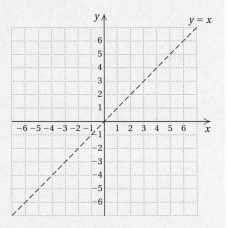

> If a relation is defined by an equation, interchanging the variables produces an equation of the **inverse relation.**

Example 2 Find an equation of the inverse of the relation

$$y = 3x - 4.$$

Then graph both the relation and its inverse.

We interchange x and y and obtain an equation of the inverse:

$$x = 3y - 4.$$

Relation: $y = 3x - 4$  *Inverse:* $x = 3y - 4$

x	y
0	-4
1	-1
2	2
3	5

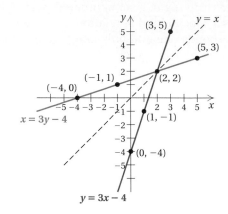

x	y
-4	0
-1	1
2	2
5	3

Example 3 Find an equation of the inverse of the relation

$$y = 6x - x^2.$$

Then graph both the original relation and its inverse.

We interchange x and y and obtain an equation of the inverse:

$$x = 6y - y^2.$$

Relation: $y = 6x - x^2$ *Inverse:* $x = 6y - y^2$

x	y
-1	-7
0	0
1	5
3	9
5	5

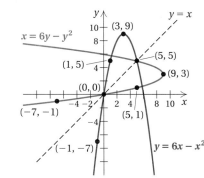

x	y
-7	-1
0	0
5	1
9	3
5	5

Do Exercises 2 and 3. (Exercise 3 is on the following page.)

Answers on page A-49

b | Inverses and One-To-One Functions

Let's consider the following two functions.

Number (Domain)	Cube (Range)
−3 ⟶	−27
−2 ⟶	−8
−1 ⟶	−1
0 ⟶	0
1 ⟶	1
2 ⟶	8
3 ⟶	27

Year (Domain)	First-class Postage Cost, in Cents (Range)
1978 ⟶	15
1983 ⟶	20
1984 ⟶	
1989 ⟶	25
1991 ⟶	29
1995 ⟶	32

Source: U.S. Postal Service

Suppose we reverse the arrows. Are these inverse relations functions?

Cube Root (Range)	Number (Domain)
−3 ⟵	−27
−2 ⟵	−8
−1 ⟵	−1
0 ⟵	0
1 ⟵	1
2 ⟵	8
3 ⟵	27

Year (Range)	First-class Postage Cost, in Cents (Domain)
1978 ⟵	15
1983 ⟵	20
1984 ⟵	
1989 ⟵	25
1991 ⟵	29
1995 ⟵	32

We see that the inverse of the cubing function is a function. The inverse of the postage function is not a function, however, because the input 20 has *two* outputs, 1983 and 1984. Recall that for a function, each input has exactly one output. However, it can happen that the same output comes from two or more different inputs. If this is the case, the inverse cannot be a function. When this possibility is excluded, the inverse is also a function.

In the cubing function, different inputs have different outputs. Thus its inverse is also a function. The cubing function is what is called a **one-to-one function.** If the inverse of a function f is also a function, it is named f^{-1} (read "f-inverse").

CAUTION! The −1 in f^{-1} is *not* an exponent and f^{-1} does not represent a reciprocal!

> **ONE-TO-ONE FUNCTION AND INVERSES**
>
> A function f is **one-to-one** if different inputs have different outputs — that is,
>
> if $a \neq b$, then $f(a) \neq f(b)$. Or,
>
> A function f is **one-to-one** if when the outputs are the same, the inputs are the same — that is,
>
> if $f(a) = f(b)$, then $a = b$.
>
> If a function is one-to-one, then its inverse is a function.
>
> The domain of a one-to-one function f is the range of the inverse f^{-1}.
>
> The range of a one-to-one function f is the domain of the inverse f^{-1}.

3. Find an equation of the inverse relation. Then graph both the original relation and its inverse.

Relation:
$y = x^2 - 4x + 7$

x	y
0	7
1	4
2	3
3	4
4	7

Inverse:

_____

x	y
7	
4	
3	
4	
7	

Complete the table and draw the graphs using the same set of axes.

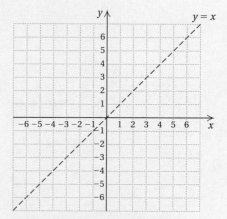

4. Determine whether the function is one-to-one and thus has an inverse that is also a function.

$$f(x) = 4 - x$$

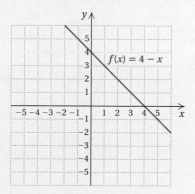

Answers on page A-49

Determine whether the function is one-to-one and thus has an inverse that is also a function.

5. $f(x) = x^2 - 1$

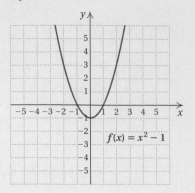

6. $f(x) = 4^x$
(Sketch this graph yourself.)

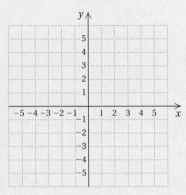

7. $f(x) = |x| - 3$
(Sketch this graph yourself.)

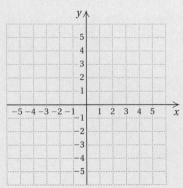

Answers on page A-49

How can we tell graphically whether a function is one-to-one and thus has an inverse that is a function?

Example 4   Here is the graph of the exponential function $f(x) = 2^x$. Determine whether the function is one-to-one and thus has an inverse that is a function.

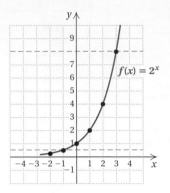

A function is one-to-one if different inputs have different outputs. In other words, no two x-values will have the same y-value. For this function, we cannot find two x-values that have the same y-value. Note also that no horizontal line can be drawn that will cross the graph more than once. The function is thus one-to-one and its inverse is a function.

> **THE HORIZONTAL-LINE TEST**
>
> If it is possible for a horizontal line to intersect the graph of a function more than once, then the function is not one-to-one and therefore its inverse is not a function.

Recall that a graph is that of a function if no vertical line crosses the graph more than once. A function has an inverse that is also a function if no horizontal line crosses the graph more than once.

Example 5   Determine whether the function $f(x) = x^2$ is one-to-one and has an inverse that is also a function.

The graph is shown below. There are many horizontal lines that cross the graph more than once, as shown, so this function is not one-to-one and does not have an inverse that is a function.

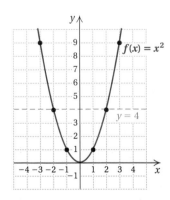

Do Exercises 4–7. (Exercise 4 is on the preceding page.)

c Finding a Formula for an Inverse

Suppose that a function is described by a formula. If it has an inverse that is a function, how do we find a formula for its inverse? If for any equation with two variables such as x and y we interchange the variables, we obtain an equation of the inverse relation. We proceed as follows to find a formula for f^{-1}.

If a function f is one-to-one, a formula for its inverse f^{-1} can be found as follows:

1. Replace $f(x)$ with y.
2. Interchange x and y. (This gives the inverse relation.)
3. Solve for y.
4. Replace y with $f^{-1}(x)$.

Example 6   Given $f(x) = x + 1$:

a) Determine whether the function is one-to-one.

b) If it is one-to-one, find a formula for $f^{-1}(x)$.

We solve as follows.

a) The graph of $f(x) = x + 1$ is shown below. It passes the horizontal-line test, so it is one-to-one. Thus its inverse is a function.

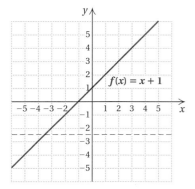

b) **1.** Replace $f(x)$ with y: $y = x + 1$.
   2. Interchange x and y: $x = y + 1$.   This gives the inverse relation.
   3. Solve for y: $x - 1 - y$.
   4. Replace y with $f^{-1}(x)$: $f^{-1}(x) = x - 1$.

Given each function:

a) Determine whether it is one-to-one.

b) If it is one-to-one, find a formula for the inverse.

8. $f(x) = 3 - x$

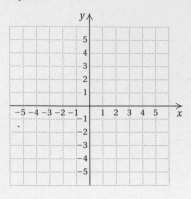

9. $g(x) = 3x - 2$

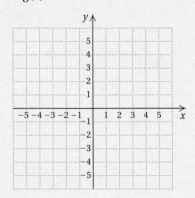

Example 7 Given $f(x) = 2x - 3$:

a) Determine whether the function is one-to-one.

b) If it is one-to-one, find a formula for $f^{-1}(x)$.

We solve as follows.

a) The graph of $f(x) = 2x - 3$ is shown below. It passes the horizontal-line test and is one-to-one.

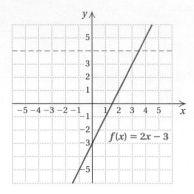

b) **1.** Replace $f(x)$ with y: $y = 2x - 3$.

    2. Interchange x and y: $x = 2y - 3$.

    3. Solve for y: $x + 3 = 2y$

$$\frac{x + 3}{2} = y.$$

    4. Replace y with $f^{-1}(x)$: $f^{-1}(x) = \dfrac{x + 3}{2}.$

Do Exercises 8 and 9.

Let's now consider inverses of functions in terms of a function machine. Suppose that a one-to-one function f is programmed into a machine. If the machine has a reverse switch, when the switch is thrown, the machine performs the inverse function f^{-1}. Inputs then enter at the opposite end, and the entire process is reversed.

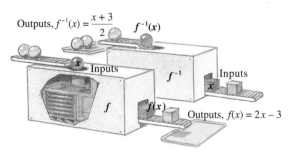

Answers on page A-49

Consider $f(x) = 2x - 3$ and $f^{-1}(x) = (x + 3)/2$ from Example 7. For the input 5,

$$f(5) = 2 \cdot 5 - 3 = 10 - 3 = 7.$$

The output is 7. Now we use 7 for the input in the inverse:

$$f^{-1}(7) = \frac{7 + 3}{2} = \frac{10}{2} = 5.$$

The function f takes 5 to 7. The inverse function f^{-1} takes the number 7 back to 5.

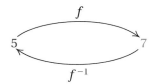

d | Graphing Functions and Their Inverses

How do the graphs of a function and its inverse compare?

Example 8   Graph $f(x) = 2x - 3$ and $f^{-1}(x) = (x + 3)/2$ using the same set of axes. Then compare.

The graph of each function follows. Note that the graph of f^{-1} can be drawn by reflecting the graph of f across the line $y = x$. That is, if we graph $f(x) = 2x - 3$ in wet ink and fold the paper along the line $y = x$, the graph of $f^{-1}(x) = (x + 3)/2$ will appear as the impression made by f.

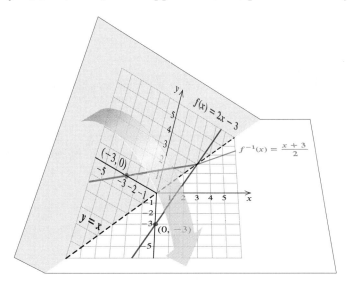

When x and y are interchanged to find a formula for the inverse, we are, in effect, flipping the graph of $f(x) = 2x - 3$ over the line $y = x$. For example, when the coordinates of the y-intercept of the graph of f, $(0, -3)$, are reversed, we get the x-intercept of the graph of f^{-1}, $(-3, 0)$.

10. Graph $g(x) = 3x - 2$ and $g^{-1}(x) = (x + 2)/3$ using the same set of axes.

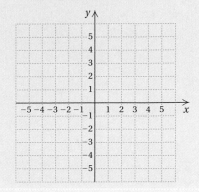

Answer on page A-49

11. Given $f(x) = x^3 + 1$:

a) Determine whether the function is one-to-one.

b) If it is one-to-one, find a formula for its inverse.

c) Graph the function and its inverse using the same set of axes.

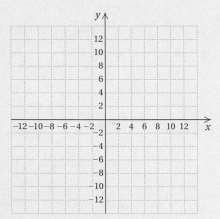

The graph of f^{-1} is a reflection of the graph of f across the line $y = x$.

Do Exercise 10 on the preceding page.

Example 9 Consider $g(x) = x^3 + 2$.

a) Determine whether the function is one-to-one.

b) If it is one-to-one, find a formula for its inverse.

c) Graph the inverse, if it exists.

We solve as follows.

a) The graph of $g(x) = x^3 + 2$ is shown below. It passes the horizontal-line test and thus is one-to-one.

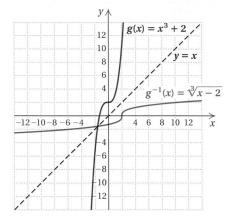

b) 1. Replace $g(x)$ with y: $\qquad y = x^3 + 2$.

  2. Interchange x and y: $\qquad x = y^3 + 2$.

  3. Solve for y: $\qquad x - 2 = y^3$

  $\qquad\qquad\qquad\qquad \sqrt[3]{x - 2} = y.$ **Since a number has only one cube root, we can solve for y.**

  4. Replace y with $g^{-1}(x)$: $\qquad g^{-1}(x) = \sqrt[3]{x - 2}$.

c) To find the graph, we reflect the graph of $g(x) = x^3 + 2$ across the line $y = x$, as we did in Example 8. It can also be found by substituting into $g^{-1}(x) = \sqrt[3]{x - 2}$ and plotting points. The graphs of g and g^{-1} are shown together above.

Do Exercise 11.

Answers on page A-50

We can now see why we exclude 1 as a base for an exponential function. Consider

$$f(x) = a^x = 1^x = 1.$$

The graph of f is the horizontal line $y = 1$. The graph is not one-to-one. The function does not have an inverse that is a function. All other positive bases yield exponential functions that are one-to-one.

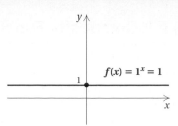

$f(x) = 1^x = 1$

Calculator Spotlight

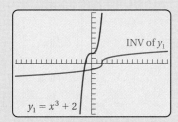 Graphing Inverses. The TI-82 and TI-83 graphers have the capability to automatically draw the graph of the inverse of a function. Consider the function $g(x) = x^3 + 2$ of Example 9. We enter $y_1 = x^3 + 2$. Then we press 2nd | DRAW | 8 to select the DRAWINV feature. On the TI-83, y_1 is selected by pressing VARS | ▷ | 1 | 1 . Next, we press ENTER . The graph of the function and its inverse follows.

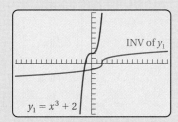

INV of y_1

$y_1 = x^3 + 2$

If you press the y= key, you will note that the DRAW-INV feature does not give you a formula for the inverse. Suppose we solved for the inverse and incorrectly got the function $h(x) = \sqrt[3]{x} + 2$. Then we could graph $h(x)$ and use the DRAWINV feature again to graph the inverse of $g(x) = x^3 + 2$. We would see all three graphs and could tell that a mistake had been made.

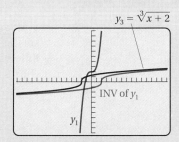

$y_3 = \sqrt[3]{x} + 2$

INV of y_1

y_1

Exercises

1. Use your grapher to check the results of Examples 7–9 and Margin Exercises 8–11.

Use your grapher to determine whether the given function g is the inverse of the given function f.

2. $f(x) = \dfrac{x^2 - 10}{3}, \quad g(x) = \sqrt{3x + 12}$

3. $f(x) = \dfrac{2}{3}x, \quad g(x) = \dfrac{3}{2}x$

4. $f(x) = 3x - 8, \quad g(x) = \dfrac{x + 8}{3}$

5. $f(x) = x^3 - 5, \quad g(x) = \sqrt[3]{x + 5}$

e | Composite Functions

In the real world, functions frequently occur in which some quantity depends on a variable that, in turn, depends on another variable. For instance, the number of employees hired by a firm may depend on the firm's profits, which may in turn depend on the number of items the firm produces. Functions like this are called **composite functions.**

For example, the function g that gives a correspondence between women's shoe sizes in the United States and those in Italy is given by $g(x) = 2x + 24$, where x is the U.S. size and $g(x)$ is the Italian size. Thus a U.S. size 4 corresponds to a shoe size of $g(4) = 2 \cdot 4 + 24$, or 32, in Italy.

There is also a function that gives a correspondence between women's shoe sizes in Italy and those in Britain. The function is given by $f(x) = \frac{1}{2}x - 14$, where x is the Italian size and $f(x)$ is the corresponding British size. Thus an Italian size 32 corresponds to a British size $f(32) = \frac{1}{2}(32) - 14$, or 2.

It seems reasonable to conclude that a shoe size of 4 in the United States corresponds to a size of 2 in Britain and that some function h describes this correspondence. Can we find a formula for h? If we look at the following tables, we might guess that such a formula is $h(x) = x - 2$, and that is indeed correct. But, for more complicated formulas, we would need to use algebra.

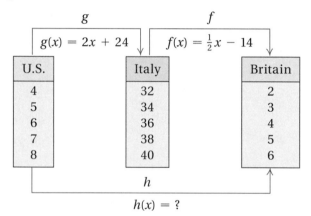

A shoe size x in the United States corresponds to a shoe size $g(x)$ in Italy, where

$$g(x) = 2x + 24.$$

Now $2x + 24$ is a shoe size in Italy. If we replace x in $f(x)$ with $2x + 24$, we can find the corresponding shoe size in Britain:

$$f(g(x)) = \frac{1}{2}[2x + 24] - 14$$
$$= x + 12 - 14 = x - 2.$$

This gives a formula for h: $h(x) = x - 2$. Thus a shoe size of 4 in the United States corresponds to a shoe size of $h(4) = 4 - 2$, or 2 in Britain. The function h is the **composition** of f and g, symbolized by $f \circ g$. To find $f \circ g(x)$, we substitute $g(x)$ for x in $f(x)$.

> The **composite function** $f \circ g$, the **composition** of f and g, is defined as
> $$f \circ g(x) = f(g(x)), \quad \text{or} \quad (f \circ g)(x) = f[g(x)].$$

Example 10   Given $f(x) = 3x$ and $g(x) = 1 + x^2$:

a) Find $f \circ g(5)$ and $g \circ f(5)$.

b) Find $f \circ g(x)$ and $g \circ f(x)$.

We consider each function separately:

$$f(x) = 3x \qquad \text{This function multiplies each input by 3.}$$

and $\quad g(x) = 1 + x^2.\qquad$ This function adds 1 to the square of each input.

a) $f \circ g(5) = f(g(5)) = f(1 + 5^2) = f(26) = 3(26) = 78;$

$\quad g \circ f(5) = g(f(5)) = g(3 \cdot 5) = g(15) = 1 + 15^2 = 1 + 225 = 226$

b) $f \circ g(x) = f(g(x))$

$$\begin{aligned} &= f(1 + x^2) \qquad \text{Substituting } 1 + x^2 \text{ for } x\\ &= 3(1 + x^2)\\ &= 3 + 3x^2; \end{aligned}$$

$\quad g \circ f(x) = g(f(x))$

$$\begin{aligned} &= g(3x) \qquad \text{Substituting } 3x \text{ for } x\\ &= 1 + (3x)^2\\ &= 1 + 9x^2 \end{aligned}$$

As a check, note that $g \circ f(5) = 1 + 9 \cdot 5^2 = 1 + 9 \cdot 25 = 226$, as expected from part (a) above.

Example 10 shows that $f \circ g(5) \neq g \circ f(5)$ and, in general, $f \circ g(x) \neq g \circ f(x)$.

Do Exercise 12.

Example 11   Given $f(x) = \sqrt{x}$ and $g(x) = x - 1$, find $f \circ g(x)$ and $g \circ f(x)$.

$$f \circ g(x) = f(g(x)) = f(x - 1) = \sqrt{x - 1};$$
$$g \circ f(x) = g(f(x)) = g(\sqrt{x}) = \sqrt{x} - 1$$

Do Exercise 13.

It is important to be able to recognize how a function can be expressed as a composition. Such a situation can occur in a study of calculus.

Example 12   Find $f(x)$ and $g(x)$ such that $h(x) = f \circ g(x)$:

$$h(x) = (7x + 3)^2.$$

This is $7x + 3$ to the 2nd power. Two functions that can be used for the composition are $f(x) = x^2$ and $g(x) = 7x + 3$. We can check by forming the composition:

$$h(x) = f \circ g(x) = f(g(x)) = f(7x + 3) = (7x + 3)^2.$$

This is the most "obvious" answer to the question. There can be other less obvious answers. For example, if

$$f(x) = (x - 1)^2 \quad \text{and} \quad g(x) = 7x + 4,$$

then

$$h(x) = f \circ g(x) = f(g(x)) = f(7x + 4) = (7x + 4 - 1)^2 = (7x + 3)^2.$$

Do Exercise 14.

12. Given $f(x) = x + 5$ and $g(x) = x^2 - 1$, find $f \circ g(x)$ and $g \circ f(x)$.

13. Given $f(x) = 4x + 5$ and $g(x) = \sqrt[3]{x}$, find $f \circ g(x)$ and $g \circ f(x)$.

14. Find $f(x)$ and $g(x)$ such that $h(x) = f \circ g(x)$. Answers may vary.

a) $h(x) = \sqrt[3]{x^2 + 1}$

b) $h(x) = \dfrac{1}{(x + 5)^4}$

Answers on page A-50

15. Let $f(x) = \frac{2}{3}x - 4$.

Use composition to show that

$$f^{-1}(x) = \frac{3x + 12}{2}.$$

Answer on page A-50

Suppose that we used some input x for the function f and found its output, $f(x)$. The function f^{-1} would then take that output back to x. Similarly, if we began with an input x for the function f^{-1} and found its output, $f^{-1}(x)$, the original function f would then take that output back to x.

> If a function f is one-to-one, then f^{-1} is the unique function for which
> $$f^{-1} \circ f(x) = x \quad \text{and} \quad f \circ f^{-1}(x) = x.$$

Example 13 Let $f(x) = 2x - 3$. Use composition to show that

$$f^{-1}(x) = \frac{x + 3}{2}. \quad \text{(See Example 7.)}$$

We find $f^{-1} \circ f(x)$ and $f \circ f^{-1}(x)$ and check to see that each is x.

$$
\begin{aligned}
f^{-1} \circ f(x) &= f^{-1}(f(x)) \\
&= f^{-1}(2x - 3) \\
&= \frac{(2x - 3) + 3}{2} \\
&= \frac{2x}{2} \\
&= x;
\end{aligned}
\qquad
\begin{aligned}
f \circ f^{-1}(x) &= f(f^{-1}(x)) \\
&= f\left(\frac{x + 3}{2}\right) \\
&= 2 \cdot \frac{x + 3}{2} - 3 \\
&= x + 3 - 3 \\
&= x
\end{aligned}
$$

Do Exercise 15.

Calculator Spotlight

 Checking Inverses Using Composites. Consider $f(x) = 2x + 6$ and $g(x) = \frac{1}{2}x - 3$. Suppose that we want to check that g is the inverse of f. We enter

$$y_1 = 2x + 6, \quad y_2 = \frac{1}{2}x - 3,$$

and use the ⏐VARS⏐ key to enter the composition of y_1 and y_2 as $y_3 = y_1(y_2)$ and then press ⏐GRAPH⏐.

X	Y2	Y3
0	−3	0
1	−2.5	1
2	−2	2
3	−1.5	3
4	−1	4
5	−.5	5
6	0	6

Y3 = 0

$y_1 = 2x + 6, \quad y_2 = \frac{1}{2}x - 3, \quad y_3 = x$

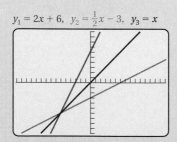

If g is the inverse of f, then the graph of $y_3 = y_1(y_2)$ should be $y_3 = x$. We can check this by using the TRACE feature along y_3 or by comparing the values of x and y_3 with the TABLE feature. The x-values and y_3-values should agree.

Exercises

1. Use your grapher to check the results of Example 13 and Margin Exercise 15.

Use composites to determine whether the given function g is the inverse of the given function f.

2. $f(x) = \frac{x^2 - 12}{3}, \quad g(x) = \sqrt{3x + 12}$

3. $f(x) = \frac{2}{3}x, \quad g(x) = -\frac{2}{3}x$

4. $f(x) = 3x - 7, \quad g(x) = \frac{x + 7}{3}$

5. $f(x) = x^3 - 5, \quad g(x) = \sqrt[3]{x - 5}$

Exercise Set 8.2

a Find the inverse of the relation. Graph the original relation and then graph the inverse relation in blue.

1. {(1, 2), (6, −3), (−3, −5)}

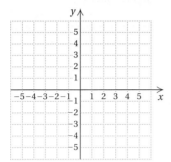

2. {(3, −1), (5, 2), (5, −3), (2, 0)}

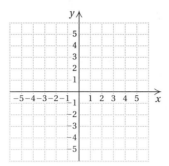

Find an equation of the inverse of the relation. Then complete the second table and graph both the original relation and its inverse.

3. $y = 2x + 6$

x	y
−1	4
0	6
1	8
2	10
3	12

x	y
4	
6	
8	
10	
12	

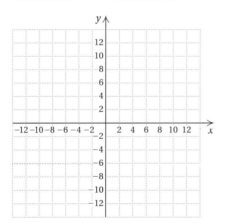

4. $y = \frac{1}{2}x^2 - 8$

x	y
−4	0
−2	−6
0	−8
2	−6
4	0

x	y
0	
−6	
−8	
−6	
0	

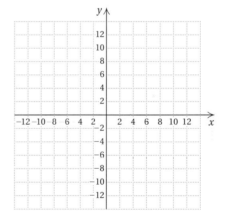

b Determine whether the function is one-to-one.

5. $f(x) = x - 5$

6. $f(x) = 3 - 6x$

7. $f(x) = x^2 - 2$

8. $f(x) = 4 - x^2$

9. $f(x) = |x| - 3$

10. $f(x) = |x - 2|$

11. $f(x) = 3^x$

12. $f(x) = \left(\frac{1}{2}\right)^x$

c Determine whether the function is one-to-one. If it is, find a formula for its inverse.

13. $f(x) = 5x - 2$

14. $f(x) = 4 + 7x$

15. $f(x) = \dfrac{-2}{x}$

16. $f(x) = \dfrac{1}{x}$

17. $f(x) = \frac{4}{3}x + 7$

18. $f(x) = -\frac{7}{8}x + 2$

19. $f(x) = \dfrac{2}{x + 5}$

20. $f(x) = \dfrac{1}{x - 8}$

21. $f(x) = 5$

22. $f(x) = -2$

23. $f(x) = \dfrac{2x + 1}{5x + 3}$

24. $f(x) = \dfrac{2x - 1}{5x + 3}$

25. $f(x) = x^3 - 1$

26. $f(x) = x^3 + 5$

27. $f(x) = \sqrt[3]{x}$

28. $f(x) = \sqrt[3]{x - 4}$

d Graph the function and its inverse using the same set of axes.

29. $f(x) = \frac{1}{2}x - 3$

30. $g(x) = x + 4$

31. $f(x) = x^3$

32. $f(x) = x^3 - 1$

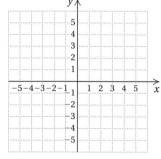
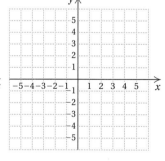
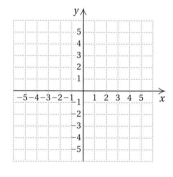
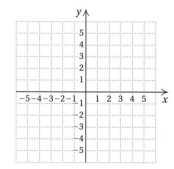

e Find $f \circ g(x)$ and $g \circ f(x)$.

33. $f(x) = 2x - 3,$
$g(x) = 6 - 4x$

34. $f(x) = 9 - 6x,$
$g(x) = 0.37x + 4$

35. $f(x) = 3x^2 + 2,$
$g(x) = 2x - 1$

36. $f(x) = 4x + 3,$
$g(x) = 2x^2 - 5$

37. $f(x) = 4x^2 - 1$,

$\quad g(x) = \dfrac{2}{x}$

38. $f(x) = \dfrac{3}{x}$,

$\quad g(x) = 2x^2 + 3$

39. $f(x) = x^2 + 5$,

$\quad g(x) = x^2 - 5$

40. $f(x) = \dfrac{1}{x^2}$,

$\quad g(x) = x - 1$

Find $f(x)$ and $g(x)$ such that $h(x) = f \circ g(x)$. Answers may vary.

41. $h(x) = (5 - 3x)^2$

42. $h(x) = 4(3x - 1)^2 + 9$

43. $h(x) = \sqrt{5x + 2}$

44. $h(x) = (3x^2 - 7)^5$

45. $h(x) = \dfrac{1}{x - 1}$

46. $h(x) = \dfrac{3}{x} + 4$

47. $h(x) = \dfrac{1}{\sqrt{7x + 2}}$

48. $h(x) = \sqrt{x - 7} - 3$

49. $h(x) = (\sqrt{x} + 5)^4$

50. $h(x) = \dfrac{x^3 + 1}{x^3 - 1}$

---

f For each function, use composition to show that the inverse is correct.

51. $f(x) = \frac{4}{5}x$,

$\quad f^{-1}(x) = \frac{5}{4}x$

52. $f(x) = \dfrac{x + 7}{2}$,

$\quad f^{-1}(x) = 2x - 7$

53. $f(x) = \dfrac{1 - x}{x}$,

$\quad f^{-1}(x) = \dfrac{1}{x + 1}$

54. $f(x) = x^3 - 5$,

$\quad f^{-1}(x) = \sqrt[3]{x + 5}$

Find the inverse of the given function by thinking about the operations of the function and then reversing, or undoing, them. Then use composition to show whether the inverse is correct.

Function	Inverse

55. $f(x) = 3x$ $\qquad\qquad f^{-1}(x) = $ _____

56. $f(x) = \frac{1}{4}x + 7$ $\qquad f^{-1}(x) = $ _____

57. $f(x) = -x$ $\qquad\qquad f^{-1}(x) = $ _____

58. $f(x) = \sqrt[3]{x} - 5$ $\qquad f^{-1}(x) = $ _____

59. $f(x) = \sqrt[3]{x - 5}$ $\qquad f^{-1}(x) = $ _____

60. $f(x) = x^{-1}$ $\qquad\qquad f^{-1}(x) = $ _____

61. *Dress Sizes in the United States and France.* A size-6 dress in the United States is size 38 in France. A function that converts dress sizes in the United States to those in France is

$$f(x) = x + 32.$$

a) Find the dress sizes in France that correspond to sizes of 8, 10, 14, and 18 in the United States.

b) Determine whether this function has an inverse that is a function. If so, find a formula for the inverse.

c) Use the inverse function to find dress sizes in the United States that correspond to sizes of 40, 42, 46, and 50 in France.

62. *Dress Sizes in the United States and Italy.* A size-6 dress in the United States is size 36 in Italy. A function that converts dress sizes in the United States to those in Italy is

$$f(x) = 2(x + 12).$$

a) Find the dress sizes in Italy that correspond to sizes of 8, 10, 14, and 18 in the United States.

b) Determine whether this function has an inverse that is a function. If so, find a formula for the inverse.

c) Use the inverse function to find dress sizes in the United States that correspond to sizes of 40, 44, 52, and 60 in Italy.

Skill Maintenance

Use rational exponents to simplify. [6.2d]

63. $\sqrt[6]{a^2}$

64. $\sqrt[6]{x^4}$

65. $\sqrt{a^4 b^6}$

66. $\sqrt[3]{8t^6}$

67. $\sqrt[8]{81}$

68. $\sqrt[4]{32}$

69. $\sqrt[12]{64x^6 y^6}$

70. $\sqrt[8]{p^4 t^2}$

71. $\sqrt[5]{32a^{15} b^{40}}$

72. $\sqrt[3]{1000x^9 y^{18}}$

73. $\sqrt[4]{81a^8 b^8}$

74. $\sqrt[3]{27p^3 q^9}$

Synthesis

75. ◆ ⏞ How can a grapher be used to determine whether a function is one-to-one?

76. ◆ The function $V(t) = 750(1.2)^t$ is used to predict the value V of a certain rare stamp t years from 1999. Do not calculate $V^{-1}(t)$ but explain how V^{-1} could be used.

⏞ In Exercises 77–80, use a grapher to help determine whether or not the given functions are inverses of each other.

77. $f(x) = 0.75x^2 + 2; \quad g(x) = \sqrt{\dfrac{4(x - 2)}{3}}$

78. $f(x) = 1.4x^3 + 3.2; \quad g(x) = \sqrt[3]{\dfrac{x - 3.2}{1.4}}$

79. $f(x) = \sqrt{2.5x + 9.25}; \quad g(x) = 0.4x^2 - 3.7, \ x \geq 0$

80. $f(x) = 0.8x^{1/2} + 5.23; \quad g(x) = 1.25(x^2 - 5.23), \ x \geq 0$

81. ⏞ Use a grapher to help match each function in Column A with its inverse from Column B.

Column A	*Column B*
(1) $y = 5x^3 + 10$	**A.** $y = \dfrac{\sqrt[3]{x} - 10}{5}$
(2) $y = (5x + 10)^3$	**B.** $y = \sqrt[3]{\dfrac{x}{5}} - 10$
(3) $y = 5(x + 10)^3$	**C.** $y = \sqrt[3]{\dfrac{x - 10}{5}}$
(4) $y = (5x)^3 + 10$	**D.** $y = \dfrac{\sqrt[3]{x - 10}}{5}$

Create composite functions and deduce the original functions.

Collaborative
Learning Manual

8.3 Logarithmic Functions

We are now ready to study inverses of exponential functions. These functions have many applications and are referred to as *logarithm*, or *logarithmic, functions*.

a | Graphing Logarithmic Functions

Consider the exponential function $f(x) = 2^x$. Like all exponential functions, f is one-to-one. Can a formula for f^{-1} be found? To answer this, we use the method of Section 8.2:

1. Replace $f(x)$ with y: $y = 2^x$.

2. Interchange x and y: $x = 2^y$.

3. Solve for y: $y =$ the power to which we raise 2 to get x.

4. Replace y with $f^{-1}(x)$: $f^{-1}(x) =$ the power to which we raise 2 to get x.

We now define a new symbol to replace the words "the power to which we raise 2 to get x".

> $\log_2 x$, read "the logarithm, base 2, of x", or "log, base 2, of x," means "the power to which we raise 2 to get x."

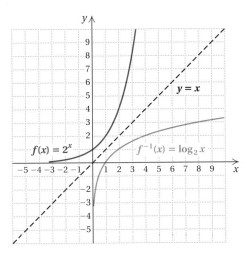

Thus if $f(x) = 2^x$, then $f^{-1}(x) = \log_2 x$. Note that $f^{-1}(8) = \log_2 8 = 3$, because 3 *is the power to which we raise* 2 *to get* 8; that is, $2^3 = 8$.

Although expressions like $\log_2 13$ can only be approximated, remember that $\log_2 13$ represents *the power to which we raise* 2 *to get* 13. That is, $2^{\log_2 13} = 13$.

Do Exercise 1.

Objectives

a   Graph logarithmic functions.

b   Convert from exponential equations to logarithmic equations and from logarithmic equations to exponential equations.

c   Solve logarithmic equations.

d   Find common logarithms on a calculator.

For Extra Help

TAPE 17   TAPE 16A   MAC WIN   CD-ROM

1. Write the meaning of $\log_2 64$. Then find $\log_2 64$.

Answer on page A-50

2. Graph: $y = f(x) = \log_3 x$.

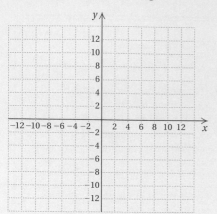

For any exponential function $f(x) = a^x$, the inverse is called a **logarithmic function, base a.** The graph of the inverse can, of course, be drawn by reflecting the graph of $f(x) = a^x$ across the line $y = x$. It will be helpful to remember that the inverse of $f(x) = a^x$ is given by $f^{-1}(x) = \log_a x$. Normally, we use a number a that is greater than 1 for the logarithm base.

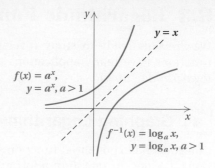

> **DEFINITION OF LOGARITHMS**
>
> The inverse of $f(x) = a^x$ is given by
>
> $$f^{-1}(x) = \log_a x.$$
>
> We read "$\log_a x$" as "the logarithm, base a, of x." We define $y = \log_a x$ as that number y such that $a^y = x$, where $x > 0$ and a is a positive constant other than 1.

It is helpful in dealing with logarithmic functions to remember that the logarithm of a number is an **exponent**. It is the exponent y in $x = a^y$. Keep thinking, "The logarithm, base a, of a number x is the power to which a must be raised in order to get x."

> A logarithm is an exponent.

Exponential Function	Logarithmic Function
$y = a^x$	$x = a^y$
$f(x) = a^x$	$f^{-1}(x) = \log_a x$
$a > 0, a \neq 1$	$a > 0, a \neq 1$
Domain = The set of real numbers	Range = The set of real numbers
Range = The set of positive numbers	Domain = The set of positive numbers

Why do we exclude 1 from being a logarithm base? If we did include it, we would be considering $x = 1^y = 1$. The graph of this equation is a vertical line, which is not a function. It does not pass the vertical-line test.

Example 1 Graph: $y = f(x) = \log_5 x$.

The equation $y = \log_5 x$ is equivalent to $5^y = x$. We can find ordered pairs that are solutions by choosing values for y and computing the corresponding x-values.

For $y = 0$, $x = 5^0 = 1$.

For $y = 1$, $x = 5^1 = 5$.

For $y = 2$, $x = 5^2 = 25$.

For $y = 3$, $x = 5^3 = 125$.

For $y = -1$, $x = 5^{-1} = \dfrac{1}{5}$.

For $y = -2$, $x = 5^{-2} = \dfrac{1}{25}$.

x, or 5^y	y
1	0
5	1
25	2
125	3
$\frac{1}{5}$	-1
$\frac{1}{25}$	-2

(1) Select y.
(2) Compute x.

Answer on page A-51

We plot the ordered pairs and connect them with a smooth curve. The graph of $y = 5^x$ has been shown only for reference.

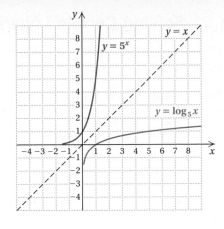

Do Exercise 2 on the preceding page.

b   Converting Between Exponential Equations and Logarithmic Equations

We use the definition of logarithms to convert from exponential equations to logarithmic equations.

> $y = \log_a x \implies a^y = x; \qquad a^y = x \implies y = \log_a x$
>
> Be sure to memorize this relationship! It is probably the most important definition in the chapter. Many times this definition will be a justification for a proof or a procedure that we are considering.

Examples   Convert to a logarithmic equation.

2. $8 = 2^x \implies x = \log_2 8$ The exponent is the logarithm.
         The base remains the base.

3. $y^{-1} = 4 \implies -1 = \log_y 4$

4. $a^b = c \implies b = \log_a c$

Do Exercises 3–6.

We also use the definition of logarithms to convert from logarithmic equations to exponential equations.

Examples   Convert to an exponential equation.

5. $y = \log_3 5 \implies 3^y = 5$ The logarithm is the exponent.
         The base does not change.

6. $-2 = \log_a 7 \implies a^{-2} = 7$

7. $a = \log_b d \implies b^a = d$

Do Exercises 7–10.

Convert to a logarithmic equation.

3. $6^0 = 1$

4. $10^{-3} = 0.001$

5. $16^{0.25} = 2$

6. $m^T = P$

Convert to an exponential equation.

7. $\log_2 32 = 5$

8. $\log_{10} 1000 = 3$

9. $\log_a Q = 7$

10. $\log_t M = x$

Answers on page A-51

Solve.

11. $\log_{10} x = 4$

12. $\log_x 81 = 4$

13. $\log_2 x = -2$

Answers on page A-51

c   Solving Certain Logarithmic Equations

Certain equations involving logarithms can be solved by first converting to exponential equations. We will solve more complicated equations later.

Example 8   Solve: $\log_2 x = -3$.

$$\log_2 x = -3$$

$2^{-3} = x$ Converting to an exponential equation

$$\frac{1}{2^3} = x$$

$$\frac{1}{8} = x$$

CHECK: $\log_2 \frac{1}{8}$ is the exponent to which we raise 2 to get $\frac{1}{8}$. Since $2^{-3} = \frac{1}{8}$, we know that $\frac{1}{8}$ checks and is the solution.

Example 9   Solve: $\log_x 16 = 2$.

$$\log_x 16 = 2$$

$x^2 = 16$ Converting to an exponential equation

$x = 4$ *or* $x = -4$ Using the principle of square roots

CHECK: $\log_4 16 = 2$ because $4^2 = 16$. Thus, 4 is a solution. Since all logarithm bases must be positive, $\log_{-4} 16$ is not defined. Therefore, -4 is not a solution.

Do Exercises 11–13.

To think of finding logarithms as solving equations may help in some cases.

Example 10   Find $\log_{10} 1000$.

METHOD 1.   Let $\log_{10} 1000 = x$. Then

$10^x = 1000$ Converting to an exponential equation

$10^x = 10^3$

$x = 3$.   The exponents are the same.

Therefore, $\log_{10} 1000 = 3$.

METHOD 2.   Think of the meaning of $\log_{10} 1000$. It is the exponent to which we raise 10 to get 1000. That exponent is 3. Therefore, $\log_{10} 1000 = 3$.

Example 11   Find $\log_{10} 0.01$.

METHOD 1.   Let $\log_{10} 0.01 = x$. Then

$10^x = 0.01$ Converting to an exponential equation

$$10^x = \frac{1}{100}$$

$10^x = 10^{-2}$

$x = -2$.   The exponents are the same.

Therefore, $\log_{10} 0.01 = -2$.

METHOD 2. $\log_{10} 0.01$ is the exponent to which we raise 10 to get 0.01. Noting that

$$0.01 = \frac{1}{100} = \frac{1}{10^2} = 10^{-2},$$

we see that the exponent is -2. Therefore, $\log_{10} 0.01 = -2$.

Example 12 Find $\log_5 1$.

METHOD 1. Let $\log_5 1 = x$. Then

$\quad 5^x = 1$ Converting to an exponential equation

$\quad 5^x = 5^0$

$\quad\ x = 0.$ The exponents are the same.

Therefore, $\log_5 1 = 0$.

METHOD 2. $\log_5 1$ is the exponent to which we raise 5 to get 1. That exponent is 0. Therefore, $\log_5 1 = 0$.

Do Exercises 14–16.

Example 12 illustrates an important property of logarithms.

For any base a,

$\quad \log_a 1 = 0.$

The logarithm, base a, of 1 is always 0.

The proof follows from the fact that $a^0 = 1$. This is equivalent to the logarithmic equation $\log_a 1 = 0$.

Another property follows similarly. We know that $a^1 = a$ for any real number a. In particular, it holds for any positive number a. This is equivalent to the logarithmic equation $\log_a a = 1$.

For any base a,

$\quad \log_a a = 1.$

Do Exercises 17–20.

d | Finding Common Logarithms on a Calculator

Base-10 logarithms are called **common logarithms.** Before calculators became so widely available, common logarithms were used extensively to do complicated calculations. In fact, that is why logarithms were invented. The abbreviation **log**, with no base written, is used for the common logarithm, base-10. Thus,

$\quad \log 29$ means $\log_{10} 29$.

We can approximate $\log 29$. Note the following:

$\quad \log 100 = \log_{10} 100 = 2;$

$\quad\quad\quad \log 29 = ?;$ It seems reasonable that log 29 is between 1 and 2.

$\quad \log 10 = \log_{10} 10 = 1.$

Find each of the following.

14. $\log_{10} 10{,}000$

15. $\log_{10} 0.0001$

16. $\log_7 1$

Simplify.

17. $\log_3 1$

18. $\log_3 3$

19. $\log_c c$

20. $\log_c 1$

Answers on page A-51

Find the common logarithm, to four decimal places, on a scientific calculator or a grapher.

21. log 78,235.4

22. log 0.0000309

23. log (−3)

24. Find

log 1000   and   log 10,000

without using a calculator. Between what two whole numbers is log 9874? Then approximate log 9874 on a calculator rounded to four decimal places.

25. Find $10^{4.8934}$ using a calculator. (Compare your computation to that of Margin Exercise 21.)

Answers on page A-51

On a scientific calculator or grapher, the scientific key for common logarithms is generally marked $\boxed{\text{LOG}}$. We find that

log 29 ≈ 1.462397998 ≈ 1.4624,

rounded to four decimal places. This also tells us that $10^{1.4624} \approx 29$.

On some scientific calculators, the keystrokes for doing such a calculation might be

$\boxed{2}$ $\boxed{9}$ $\boxed{\text{LOG}}$ $\boxed{=}$.   The display would then read 1.462398.

Using the scientific keys on a grapher, the keystrokes might be

$\boxed{\text{LOG}}$ $\boxed{2}$ $\boxed{9}$ $\boxed{\text{ENTER}}$.   The display would then read 1.462397998.

Examples Find each common logarithm, to four decimal places, on a scientific calculator or grapher.

Function Value	Readout	Rounded
13. log 287,523	5.458672591	5.4587
14. log 0.000486	−3.313363731	−3.3134
15. log (−5)	NONREAL ANS	NONREAL ANS

In Example 15, log (−5) does not exist as a real number because there is no real-number power to which we can raise 10 to get −5. The number 10 raised to any power is nonnegative. The logarithm of a negative number does not exist.

Do Exercises 21–24.

The inverse of a logarithmic function is an exponential function. Thus, if $f(x) = \log x$, then $f^{-1}(x) = 10^x$. Because of this, on many calculators, the $\boxed{\text{LOG}}$ key doubles as the $\boxed{10^x}$ key after a $\boxed{\text{2nd}}$ or $\boxed{\text{SHIFT}}$ key has been pressed. To find $10^{5.4587}$ on a scientific calculator, we might enter 5.4587 and press $\boxed{10^x}$. On a grapher, we press $\boxed{\text{2nd}}$ $\boxed{10^x}$, followed by 5.4587. In either case, we get the approximation

$10^{5.4587} \approx 287{,}541.1465.$

Compare this computation to Example 13. Note that, apart from the rounding error, $10^{5.4587}$ takes us back to about 287,523.

Do Exercise 25.

Using the scientific keys on a calculator or grapher would allow us to construct a graph of $f(x) = \log_{10} x = \log x$ by finding function values directly, rather than converting to exponential form as we did in Example 1.

x	$f(x)$
0.5	−0.3010
1	0
2	0.3010
3	0.4771
5	0.6990
9	0.9542
10	1

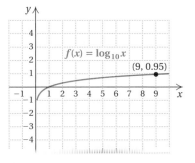

Calculator Spotlight

To graph $f(x) = \log_{10} x = \log x$, we press $\boxed{y=}$ $\boxed{\text{LOG}}$ $\boxed{X}$. You will see $y_1 = \log (x$ on the display. Note that there is no right parenthesis. It does not need to be included unless you are entering a function like

$g(x) = 2 \cdot \log (x - 3) + 1.$

Exercises

Graph.

1. $f(x) = \log_{10} x = \log x$

2. $f(x) = 2 \cdot \log x$

3. $g(x) = 2 \cdot \log (x - 3) + 1$

Exercise Set 8.3

a Graph.

1. $f(x) = \log_2 x$

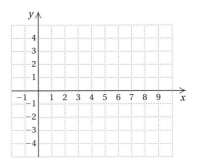

2. $f(x) = \log_{10} x$

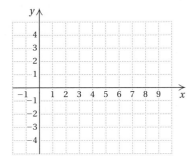

3. $f(x) = \log_{1/3} x$

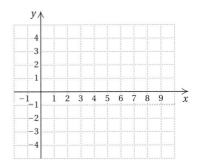

4. $f(x) = \log_{1/2} x$

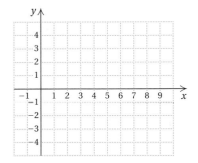

Graph both functions using the same set of axes.

5. $f(x) = 3^x, \quad f^{-1}(x) = \log_3 x$

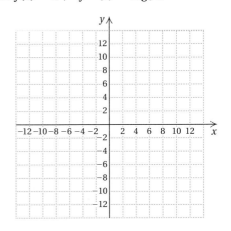

6. $f(x) = 4^x, \quad f^{-1}(x) = \log_4 x$

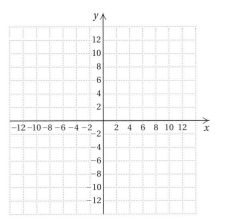

b Convert to a logarithmic equation.

7. $10^3 = 1000$

8. $10^2 = 100$

9. $5^{-3} = \dfrac{1}{125}$

10. $4^{-5} = \dfrac{1}{1024}$

11. $8^{1/3} = 2$

12. $16^{1/4} = 2$

13. $10^{0.3010} = 2$

14. $10^{0.4771} = 3$

15. $e^2 = t$

16. $p^k = 3$

17. $Q^t = x$

18. $P^m = V$

19. $e^2 = 7.3891$

20. $e^3 = 20.0855$

21. $e^{-2} = 0.1353$

22. $e^{-4} = 0.0183$

Convert to an exponential equation.

23. $w = \log_4 10$

24. $t = \log_5 9$

25. $\log_6 36 = 2$

26. $\log_7 7 = 1$

27. $\log_{10} 0.01 = -2$

28. $\log_{10} 0.001 = -3$

29. $\log_{10} 8 = 0.9031$

30. $\log_{10} 2 = 0.3010$

31. $\log_e 100 = 4.6052$

32. $\log_e 10 = 2.3026$

33. $\log_t Q = k$

34. $\log_m P = a$

c Solve.

35. $\log_3 x = 2$

36. $\log_4 x = 3$

37. $\log_x 16 = 2$

38. $\log_x 64 = 3$

39. $\log_2 16 = x$

40. $\log_5 25 = x$

41. $\log_3 27 = x$

42. $\log_4 16 = x$

43. $\log_x 25 = 1$

44. $\log_x 9 = 1$

45. $\log_3 x = 0$

46. $\log_2 x = 0$

47. $\log_2 x = -1$

48. $\log_3 x = -2$

49. $\log_8 x = \dfrac{1}{3}$

50. $\log_{32} x = \dfrac{1}{5}$

Find each of the following.

51. $\log_{10} 100$

52. $\log_{10} 100{,}000$

53. $\log_{10} 0.1$

54. $\log_{10} 0.001$

55. $\log_{10} 1$

56. $\log_{10} 10$

57. $\log_5 625$

58. $\log_2 64$

59. $\log_7 49$

60. $\log_5 125$

61. $\log_2 8$

62. $\log_8 64$

63. $\log_9 \dfrac{1}{81}$

64. $\log_5 \dfrac{1}{125}$

65. $\log_8 1$

66. $\log_6 6$

67. $\log_e e$

68. $\log_e 1$

69. $\log_{27} 9$

70. $\log_8 2$

$\boxed{\text{d}}$ Find the common logarithm, to four decimal places, on a calculator.

71. log 78,889.2   **72.** log 9,043,788   **73.** log 0.67   **74.** log 0.0067

75. log (−97)   **76.** log 0   **77.** $\log\left(\dfrac{289}{32.7}\right)$ **78.** $\log\left(\dfrac{23}{86.2}\right)$

Skill Maintenance

For each quadratic function, find and label (**a**) the vertex, (**b**) the line of symmetry, and (**c**) the maximum or minimum value. Then (**d**) graph the function.   [7.6a]

79. $f(x) = 4 - x^2$ **80.** $f(x) = (x + 3)^2$ **81.** $f(x) = -2(x - 1)^2 - 3$ **82.** $f(x) = x^2 - 2x - 5$

Solve for the given letter.   [7.3b]

83. $E = mc^2$, for c **84.** $B = 3a^2 - 4a$, for a **85.** $A = \sqrt{3ab}$, for b **86.** $T = 2\pi\sqrt{\dfrac{L}{g}}$, for L

87. *Automobile Accidents by Age.* The following table lists the percentages of drivers of age a involved in automobile accidents in a recent year.   [7.7b]

Age, *a*	Percentage Involved in an Accident
20	31
24	34
34	22
44	18

Source: National Safety Council

a) Use the data points (20, 31), (24, 34), and (34, 22) to fit a quadratic function to the data.
b) Use the quadratic function to predict the percentage of drivers involved in accidents at age 30 and at age 37.

Synthesis

88. ◆ Explain in your own words what is meant by $\log_a b = c$.

89. ◆ John Napier (1550–1617) of Scotland is credited by mathematicians as the inventor of logarithms. Make a report on Napier and his work.

Graph.

90. $f(x) = \log_2 (x - 1)$ **91.** $f(x) = \log_3 |x + 1|$

Solve.

92. $|\log_3 x| = 3$ **93.** $\log_{125} x = \frac{2}{3}$ **94.** $\log_4 (3x - 2) = 2$

95. $\log_8 (2x + 1) = -1$ **96.** $\log_{10} (x^2 + 21x) = 2$

Simplify.

97. $\log_{1/4} \frac{1}{11}$ **98.** $\log_{81} 3 \cdot \log_3 81$ **99.** $\log_{10} (\log_4 (\log_9 81))$

100. $\log_2 (\log_2 (\log_4 256))$ **101.** $\log_{1/5} 25$

Practice graphing logarithmic functions and their inverses.

Collaborative Learning Manual

8.4 Properties of Logarithmic Functions

The ability to manipulate logarithmic expressions is important in many applications and in more advanced mathematics. We now establish some basic properties that are useful in manipulating logarithmic expressions.

a | Logarithms of Products

> **PROPERTY 1: THE PRODUCT RULE**
>
> For any positive numbers M and N,
>
> $$\log_a (M \cdot N) = \log_a M + \log_a N.$$
>
> (The logarithm of a product is the sum of the logarithms of the factors. The number a can be any logarithm base.)

Example 1   Express as a sum of logarithms: $\log_2 (4 \cdot 16)$.

$$\log_2 (4 \cdot 16) = \log_2 4 + \log_2 16 \qquad \text{By Property 1}$$

Example 2   Express as a single logarithm: $\log_{10} 0.01 + \log_{10} 1000$.

$$\log_{10} 0.01 + \log_{10} 1000 = \log_{10} (0.01 \times 1000) \qquad \text{By Property 1}$$
$$= \log_{10} 10$$

Do Exercises 1–4.

A PROOF OF PROPERTY 1 (*OPTIONAL*).   Let $\log_a M = x$ and $\log_a N = y$. Converting to exponential equations, we have $a^x = M$ and $a^y = N$. Then we multiply to obtain

$$M \cdot N = a^x \cdot a^y = a^{x+y}.$$

Converting $M \cdot N = a^{x+y}$ back to a logarithmic equation, we get

$$\log_a M \cdot N = x + y.$$

Remembering what x and y represent, we get

$$\log_a M \cdot N = \log_a M + \log_a N.$$

b | Logarithms of Powers

> **PROPERTY 2: THE POWER RULE**
>
> For any positive number M and any real number k,
>
> $$\log_a M^k = k \cdot \log_a M.$$
>
> (The logarithm of a power of M is the exponent times the logarithm of M. The number a can be any logarithm base.)

Objectives

 a   Express the logarithm of a product as a sum of logarithms, and conversely.

 b   Express the logarithm of a power as a product, and conversely.

 c   Express the logarithm of a quotient as a difference of logarithms, and conversely.

 d   Convert from logarithms of products, quotients, and powers to expressions in terms of individual logarithms, and conversely.

e   Simplify expressions of the type $\log_a a^k$.

For Extra Help

TAPE 17	TAPE 16A	MAC WIN	CD-ROM

Express as a sum of logarithms.

1. $\log_5 25 \cdot 5$

2. $\log_b PQ$

Express as a single logarithm.

3. $\log_3 7 + \log_3 5$

4. $\log_a J + \log_a A + \log_a M$

Calculator Spotlight

 Exercises

Use the TABLE and GRAPH features to determine whether each of the following is correct.

1. $\log (5x) = \log 5 + \log x$
2. $\log x^2 = 2 \log x$
3. $\log \left(\dfrac{x}{3} \right) = \log x - \log 3$
4. $\log (x + 2) = \log x + \log 2$

Answers on page A-51

Express as a product.

5. $\log_7 4^5$

6. $\log_a \sqrt{5}$

7. Express as a difference of logarithms:

$$\log_b \frac{P}{Q}.$$

8. Express as a single logarithm:

$$\log_2 125 - \log_2 25.$$

Answers on page A-51

Examples Express as a product.

3. $\log_a 9^{-5} = -5 \log_a 9$ **By Property 2**

4. $\log_a \sqrt[4]{5} = \log_a 5^{1/4}$ Writing exponential notation

$= \frac{1}{4} \log_a 5$ **By Property 2**

Do Exercises 5 and 6.

A PROOF OF PROPERTY 2 (*OPTIONAL*). Let $x = \log_a M$. Then we convert to an exponential equation to get $a^x = M$. Raising both sides to the kth power, we obtain

$$(a^x)^k = M^k, \quad \text{or} \quad a^{xk} = M^k.$$

Converting back to a logarithmic equation with base a, we get $\log_a M^k = xk$. But $x = \log_a M$, so

$$\log_a M^k = (\log_a M)k = k \cdot \log_a M.$$

c   Logarithms of Quotients

> **PROPERTY 3: THE QUOTIENT RULE**
>
> For any positive numbers M and N,
>
> $$\log_a \frac{M}{N} = \log_a M - \log_a N.$$
>
> (The logarithm of a quotient is the logarithm of the numerator minus the logarithm of the denominator. The number a can be any logarithm base.)

Example 5 Express as a difference of logarithms: $\log_t \frac{6}{U}$.

$$\log_t \frac{6}{U} = \log_t 6 - \log_t U \quad \text{By Property 3}$$

Example 6 Express as a single logarithm: $\log_b 17 - \log_b 27$.

$$\log_b 17 - \log_b 27 = \log_b \frac{17}{27} \quad \text{By Property 3}$$

Example 7 Express as a single logarithm: $\log_{10} 10{,}000 - \log_{10} 100$.

$$\log_{10} 10{,}000 - \log_{10} 100 = \log_{10} \frac{10{,}000}{100} = \log_{10} 100$$

Do Exercises 7 and 8.

A PROOF OF PROPERTY 3 (*OPTIONAL*). The proof makes use of Property 1 and Property 2.

$$\log_a \frac{M}{N} = \log_a MN^{-1}$$

$$= \log_a M + \log_a N^{-1} \quad \text{By Property 1}$$

$$= \log_a M + (-1) \log_a N \quad \text{By Property 2}$$

$$= \log_a M - \log_a N$$

d Using the Properties Together

Examples Express in terms of logarithms.

8. $\log_a \dfrac{x^2 y^3}{z^4} = \log_a (x^2 y^3) - \log_a z^4$ Using Property 3

$\qquad\qquad = \log_a x^2 + \log_a y^3 - \log_a z^4$ Using Property 1

$\qquad\qquad = 2 \log_a x + 3 \log_a y - 4 \log_a z$ Using Property 2

9. $\log_a \sqrt[4]{\dfrac{xy}{z^3}} = \log_a \left(\dfrac{xy}{z^3}\right)^{1/4}$ Writing exponential notation

$\qquad\qquad = \tfrac{1}{4} \log_a \dfrac{xy}{z^3}$ Using Property 2

$\qquad\qquad = \tfrac{1}{4} (\log_a xy - \log_a z^3)$ Using Property 3 (note the parentheses)

$\qquad\qquad = \tfrac{1}{4} (\log_a x + \log_a y - 3 \log_a z)$ Using Properties 1 and 2

$\qquad\qquad = \tfrac{1}{4} \log_a x + \tfrac{1}{4} \log_a y - \tfrac{3}{4} \log_a z$ Distributive law

10. $\log_b \dfrac{xy}{m^3 n^4} = \log_b xy - \log_b m^3 n^4$ Using Property 3

$\qquad\qquad = (\log_b x + \log_b y) - (\log_b m^3 + \log_b n^4)$ Using Property 1

$\qquad\qquad = \log_b x + \log_b y - \log_b m^3 - \log_b n^4$ Removing parentheses

$\qquad\qquad = \log_b x + \log_b y - 3 \log_b m - 4 \log_b n$ Using Property 2

Do Exercises 9–11.

Examples Express as a single logarithm.

11. $\dfrac{1}{2} \log_a x - 7 \log_a y + \log_a z$

$\qquad = \log_a x^{1/2} - \log_a y^7 + \log_a z$ Using Property 2

$\qquad = \log_a \dfrac{\sqrt{x}}{y^7} + \log_a z$ Using Property 3

$\qquad = \log_a \dfrac{z\sqrt{x}}{y^7}$ Using Property 1

12. $\log_a \dfrac{b}{\sqrt{x}} + \log_a \sqrt{bx}$

$\qquad = \log_a b - \log_a \sqrt{x} + \log_a \sqrt{bx}$ Using Property 3

$\qquad = \log_a b - \tfrac{1}{2} \log_a x + \tfrac{1}{2} \log_a (bx)$ Using Property 2

$\qquad = \log_a b - \tfrac{1}{2} \log_a x + \tfrac{1}{2} (\log_a b + \log_a x)$ Using Property 1

$\qquad = \log_a b - \tfrac{1}{2} \log_a x + \tfrac{1}{2} \log_a b + \tfrac{1}{2} \log_a x$

$\qquad = \tfrac{3}{2} \log_a b$ Collecting like terms

$\qquad = \log_a b^{3/2}$ Using Property 2

Example 12 could also be done as follows:

$\log_a \dfrac{b}{\sqrt{x}} + \log_a \sqrt{bx} = \log_a \dfrac{b}{\sqrt{x}} \sqrt{bx}$ Using Property 1

$\qquad\qquad = \log_a \dfrac{b}{\sqrt{x}} \cdot \sqrt{b} \cdot \sqrt{x}$

$\qquad\qquad = \log_a b\sqrt{b}, \text{ or } \log_a b^{3/2}.$

Do Exercises 12 and 13.

Express in terms of logarithms of $x, y, z,$ and w.

9. $\log_a \sqrt{\dfrac{z^3}{xy}}$

10. $\log_a \dfrac{x^2}{y^3 z}$

11. $\log_a \dfrac{x^3 y^4}{z^5 w^9}$

Express as a single logarithm.

12. $5 \log_a x - \log_a y + \dfrac{1}{4} \log_a z$

13. $\log_a \dfrac{\sqrt{x}}{b} - \log_a \sqrt{bx}$

Answers on page A-51

8.4 Properties of Logarithmic Functions

645

Given

$$\log_a 2 = 0.301,$$
$$\log_a 5 = 0.699,$$

find each of the following.

14. $\log_a 4$ **15.** $\log_a 10$

16. $\log_a \dfrac{2}{5}$ **17.** $\log_a \dfrac{5}{2}$

18. $\log_a \dfrac{1}{5}$ **19.** $\log_a \sqrt{a^3}$

20. $\log_a 5a$ **21.** $\log_a 16$

Simplify.

22. $\log_2 2^6$

23. $\log_{10} 10^{3.2}$

24. $\log_e e^{12}$

CAUTION! Keep in mind that, in general,

$$\log_a (M + N) \neq \log_a M + \log_a N,$$
$$\log_a (M - N) \neq \log_a M - \log_a N,$$
$$\log_a (MN) \neq (\log_a M)(\log_a N),$$

and

$$\log_a (M/N) \neq (\log_a M) \div (\log_a N).$$

Answers on page A-51

Examples Given $\log_a 2 = 0.301$ and $\log_a 3 = 0.477$, find each of the following.

13. $\log_a 6 = \log_a (2 \cdot 3) = \log_a 2 + \log_a 3$ Property 1
$$= 0.301 + 0.477 = 0.778$$

14. $\log_a \dfrac{2}{3} = \log_a 2 - \log_a 3$ Property 3
$$= 0.301 - 0.477 = -0.176$$

15. $\log_a 81 = \log_a 3^4 = 4 \log_a 3$ Property 2
$$= 4(0.477) = 1.908$$

16. $\log_a \dfrac{1}{3} = \log_a 1 - \log_a 3$ Property 3
$$= 0 - 0.477 = -0.477$$

17. $\log_a \sqrt{a} = \log_a a^{1/2} = \dfrac{1}{2} \log_a a = \dfrac{1}{2} \cdot 1 = \dfrac{1}{2}$ Property 2

18. $\log_a 2a = \log_a 2 + \log_a a$ Property 1
$$= 0.301 + 1 = 1.301$$

19. $\log_a 5$ No way to find using these properties.
($\log_a 5 \neq \log_a 2 + \log_a 3$)

20. $\dfrac{\log_a 3}{\log_a 2} = \dfrac{0.477}{0.301} \approx 1.58$ We simply divide the logarithms, not using any property.

Do Exercises 14–21.

e   The Logarithm of the Base to a Power

▶ **PROPERTY 4**

For any base a,
$$\log_a a^k = k.$$
(The logarithm, base a, of a to a power is the power.)

A PROOF OF PROPERTY 4 (*OPTIONAL*). The proof involves Property 2 and the fact that $\log_a a = 1$:

$$\log_a a^k = k(\log_a a)$$ Using Property 2
$$= k \cdot 1$$ Using $\log_a a = 1$
$$= k.$$

Examples Simplify.

21. $\log_3 3^7 = 7$ **22.** $\log_{10} 10^{5.6} = 5.6$
23. $\log_e e^{-t} = -t$

Do Exercises 22–24.

Exercise Set 8.4

a Express as a sum of logarithms.

1. $\log_2 (32 \cdot 8)$

2. $\log_3 (27 \cdot 81)$

3. $\log_4 (64 \cdot 16)$

4. $\log_5 (25 \cdot 125)$

5. $\log_a Qx$

6. $\log_r 8Z$

Express as a single logarithm.

7. $\log_b 3 + \log_b 84$

8. $\log_a 75 + \log_a 5$

9. $\log_c K + \log_c y$

10. $\log_t H + \log_t M$

b Express as a product.

11. $\log_c y^4$

12. $\log_a x^3$

13. $\log_b t^6$

14. $\log_{10} y^7$

15. $\log_b C^{-3}$

16. $\log_c M^{-5}$

c Express as a difference of logarithms.

17. $\log_a \dfrac{67}{5}$

18. $\log_t \dfrac{T}{7}$

19. $\log_b \dfrac{2}{5}$

20. $\log_a \dfrac{z}{y}$

Express as a single logarithm.

21. $\log_c 22 - \log_c 3$

22. $\log_d 54 - \log_d 9$

d Express in terms of logarithms.

23. $\log_a x^2 y^3 z$

24. $\log_a 5xy^4 z^3$

25. $\log_b \dfrac{xy^2}{z^3}$

26. $\log_b \dfrac{p^2 q^5}{m^4 n^7}$

27. $\log_c \sqrt[3]{\dfrac{x^4}{y^3 z^2}}$

28. $\log_a \sqrt{\dfrac{x^6}{p^5 q^8}}$

29. $\log_a \sqrt[4]{\dfrac{m^8 n^{12}}{a^3 b^5}}$

30. $\log_a \sqrt{\dfrac{a^6 b^8}{a^2 b^5}}$

Express as a single logarithm and simplify if possible.

31. $\dfrac{2}{3} \log_a x - \dfrac{1}{2} \log_a y$

32. $\dfrac{1}{2} \log_a x + 3 \log_a y - 2 \log_a x$

33. $\log_a 2x + 3(\log_a x - \log_a y)$

34. $\log_a x^2 - 2 \log_a \sqrt{x}$

35. $\log_a \dfrac{a}{\sqrt{x}} - \log_a \sqrt{ax}$

36. $\log_a (x^2 - 4) - \log_a (x - 2)$

Given $\log_b 3 = 1.099$ and $\log_b 5 = 1.609$, find each of the following.

37. $\log_b 15$

38. $\log_b \dfrac{3}{5}$

39. $\log_b \dfrac{5}{3}$

40. $\log_b \dfrac{1}{3}$

41. $\log_b \dfrac{1}{5}$

42. $\log_b \sqrt{b}$

43. $\log_b \sqrt{b^3}$

44. $\log_b 3b$

45. $\log_b 5b$

46. $\log_b 9$

$\boxed{\text{e}}$ Simplify.

47. $\log_e e^t$

48. $\log_w w^8$

49. $\log_p p^5$

50. $\log_Y Y^{-4}$

Solve for x.

51. $\log_2 2^7 = x$

52. $\log_9 9^4 = x$

53. $\log_e e^x = -7$

54. $\log_a a^x = 2.7$

Skill Maintenance

Compute and simplify. Express answers in the form $a + bi$, where $i^2 = -1$. [6.8b, c, d, e]

55. i^{29}

56. i^{34}

57. $(2 + i)(2 - i)$

58. $\dfrac{2 + i}{2 - i}$

59. $(7 - 8i) - (-16 + 10i)$

60. $2i^2 \cdot 5i^3$

61. $(8 + 3i)(-5 - 2i)$

62. $(2 - i)^2$

Synthesis

63. ◈ Find a way to express $\log_a (x/5)$ as a difference of logarithms without using the quotient rule. Explain your work.

64. ◈ A student incorrectly reasons that

$$\log_a \frac{1}{x} = \log_a \frac{x}{x \cdot x}$$
$$= \log_a x - \log_a x + \log_a x$$
$$= \log_a x.$$

What mistake has the student made? Explain what the answer should be.

65. ⟡ Use the TABLE and GRAPH features to show that $\log x^2 \neq (\log x)(\log x)$.

66. ⟡ Use the TABLE and GRAPH features to show that $\dfrac{\log x}{\log 4} \neq \log x - \log 4$.

Express as a single logarithm and, if possible, simplify.

67. $\log_a (x^8 - y^8) - \log_a (x^2 + y^2)$

68. $\log_a (x + y) + \log_a (x^2 - xy + y^2)$

Express as a sum or a difference of logarithms.

69. $\log_a \sqrt{1 - s^2}$

70. $\log_a \dfrac{c - d}{\sqrt{c^2 - d^2}}$

Determine whether each is true or false.

71. $\dfrac{\log_a P}{\log_a Q} = \log_a \dfrac{P}{Q}$

72. $\dfrac{\log_a P}{\log_a Q} = \log_a P - \log_a Q$

73. $\log_a 3x = \log_a 3 + \log_a x$

74. $\log_a 3x = 3 \log_a x$

75. $\log_a (P + Q) = \log_a P + \log_a Q$

76. $\log_a x^2 = 2 \log_a x$

8.5 Natural Logarithmic Functions

Any positive number other than 1 can serve as the base of a logarithmic function. Common, or base-10, logarithms, which were introduced in Section 8.3, are useful because they have the same base as our "commonly" used decimal system of naming numbers.

Today, another base is widely used. It is an irrational number named e. We now consider e and base-e, or **natural, logarithms.**

Objectives

a Find logarithms or powers, base e, using a calculator.

b Use the change-of-base formula to find logarithms to bases other than e or 10.

c Graph exponential and logarithmic functions, base e.

For Extra Help

TAPE 18   TAPE 16B   MAC WIN   CD-ROM

a   The Base *e* and Natural Logarithms

When interest is computed n times per year, the compound-interest formula is

$$A = P\left(1 + \frac{r}{n}\right)^{nt},$$

where A is the amount that an initial investment P will grow to after t years at interest rate r. Suppose that \$1 could be invested at 100% interest for 1 year. (In reality, no financial institution would pay such an interest rate.) The preceding formula becomes a function A defined in terms of the number of compounding periods n:

$$A(n) = \left(1 + \frac{1}{n}\right)^{n}.$$

Let's find some function values, using a calculator and rounding to six decimal places. The numbers in this table approach a very important number called e. It is an irrational number, so its decimal representation neither terminates nor repeats.

n	$A(n) = \left(1 + \dfrac{1}{n}\right)^{n}$
1 (compounded annually)	\$2.00
2 (compounded semiannually)	\$2.25
3	\$2.370370
4 (compounded quarterly)	\$2.441406
5	\$2.488320
100	\$2.704814
365 (compounded daily)	\$2.714567
8760 (compounded hourly)	\$2.718127

> The number e: $e \approx 2.7182818284\ldots$

Logarithms, base e, are called **natural logarithms,** or **Naperian logarithms,** in honor of John Napier (1550–1617), a Scotsman who invented logarithms.

The abbreviation **ln** is commonly used with natural logarithms. Thus,

    ln 29   means $\log_e 29$.

We usually read "ln 29" as "the natural log of 29," or simply "el en of 29."

On a scientific calculator or grapher, the scientific key for natural logarithms is generally marked | LN | . Using that key, we find that

    $\ln 29 \approx 3.36729583 \approx 3.3673$,

rounded to four decimal places. This also tells us that $e^{3.3673} \approx 29$.

On some scientific calculators, the keystrokes for doing such a calculation might be

    | 2 | | 9 | | LN | | = | .

The display would then read 3.3672958 .

If we were to use the scientific keys on a grapher, the keystrokes might be

    | LN | | 2 | | 9 | | ENTER |

The display would then read 3.36729583 .

Find the natural logarithm, to four decimal places, on a scientific calculator or a grapher.

1. ln 78,235.4

2. ln 0.0000309

3. ln (−3)

4. ln 0

5. ln 10

6. Find $e^{11.2675}$ using a calculator. (Compare this computation to that of Margin Exercise 1.)

7. Find e^{-2} using a calculator.

Answers on page A-52

Examples Find each natural logarithm, to four decimal places, on a scientific calculator or grapher.

Function Value	Readout	Rounded
1. ln 287,523	12.56905814	12.5691
2. ln 0.000486	−7.629301934	−7.6293
3. ln (−5)	NONREAL ANS	NONREAL ANS
4. ln (e)	1	1
5. ln 1	0	0

Do Exercises 1–5.

The inverse of a logarithmic function is an exponential function. Thus, if $f(x) = \ln x$, then $f^{-1}(x) = e^x$. Because of this, on many calculators, the LN key doubles as the e^x key after a 2nd or SHIFT key has been pressed.

Example 6 Find $e^{12.5691}$ using a calculator.

On a scientific calculator, we might enter 12.5691 and press e^x. On a grapher, we might press 2nd e^x, followed by 12.5691. In either case, we get the approximation

$$e^{12.5691} \approx 287,535.0371.$$

Compare this computation to Example 1. Note that, apart from the rounding error, $e^{12.5691}$ takes us back to about 287,523.

Example 7 Find $e^{-1.524}$ using a calculator.

On a scientific calculator, we might enter −1.524 and press e^x. On a grapher, we might press 2nd e^x, followed by −1.524. In either case, we get the approximation

$$e^{-1.524} \approx 0.2178.$$

Do Exercises 6 and 7.

b Changing Logarithm Bases

Most calculators give the values of both common logarithms and natural logarithms. To find a logarithm with some other base, we can use the following conversion formula.

> **THE CHANGE-OF-BASE FORMULA**
>
> For any logarithm bases a and b and any positive number M,
>
> $$\log_b M = \frac{\log_a M}{\log_a b}.$$

A PROOF OF THE CHANGE-OF-BASE FORMULA (*OPTIONAL*). Let $x = \log_b M$. Then, writing an equivalent exponential equation, we have $b^x = M$. Next, we take the logarithm base a on both sides. This gives us

$$\log_a b^x = \log_a M.$$

By Property 2,

$$x \log_a b = \log_a M,$$

and solving for x, we obtain

$$x = \frac{\log_a M}{\log_a b}.$$

But $x = \log_b M$, so we have

$$\log_b M = \frac{\log_a M}{\log_a b},$$

which is the change-of-base formula.

Example 8   Find $\log_4 7$ using common logarithms.

Let $a = 10$, $b = 4$, and $M = 7$. Then we substitute into the change-of-base formula:

$$\log_b M = \frac{\log_a M}{\log_a b}$$

$$\log_4 7 = \frac{\log_{10} 7}{\log_{10} 4} \qquad \text{Substituting 10 for } a, \\ \text{4 for } b, \text{ and 7 for } M$$

$$\approx 1.4036.$$

To check, we use a calculator with a power key $\boxed{y^x}$ to verify that

$$4^{1.4036} \approx 7.$$

We can also use base e for a conversion.

Example 9   Find $\log_5 29$ using natural logarithms.

Substituting e for a, 5 for b, and 29 for M, we have

$$\log_5 29 = \frac{\log_e 29}{\log_e 5} \qquad \text{Using the change-of-base formula}$$

$$= \frac{\ln 29}{\ln 5} \approx 2.0923.$$

Do Exercises 8 and 9.

c   Graphs of Exponential and Logarithmic Functions, Base e

Example 10   Graph $f(x) = e^x$ and $g(x) = e^{-x}$.

We use a calculator with an $\boxed{e^x}$ key to find approximate values of e^x and e^{-x}. Using these values, we can graph the functions.

x	e^x	e^{-x}
0	1	1
1	2.7	0.4
2	7.4	0.1
−1	0.4	2.7
−2	0.1	7.4

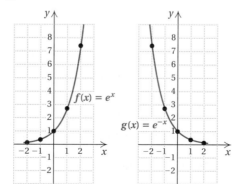

$f(x) = e^x$

$g(x) = e^{-x}$

8. Find $\log_6 7$ using common logarithms.

9. Find $\log_2 46$ using natural logarithms.

Graph.

10. $f(x) = e^{2x}$

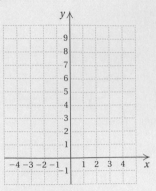

11. $g(x) = \frac{1}{2}e^{-x}$

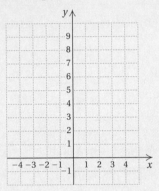

Graph.

12. $f(x) = 2 \ln x$

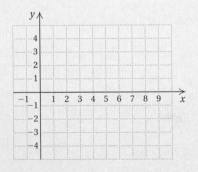

13. $g(x) = \ln (x - 2)$

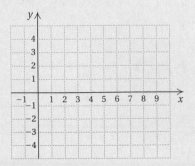

Answers on page A-52

Example 11 Graph: $f(x) = e^{-0.5x}$.

We find some solutions with a calculator, plot them, and then draw the graph. For example,

$$f(2) = e^{-0.5(2)} = e^{-1} \approx 0.4.$$

x	$e^{-0.5x}$
0	1
1	0.6
2	0.4
3	0.2
-1	1.6
-2	2.7
-3	4.5

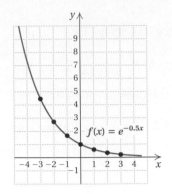

Do Exercises 10 and 11.

Example 12 Graph: $g(x) = \ln x$.

We find some solutions with a calculator and then draw the graph. As expected, the graph is a reflection across the line $y = x$ of the graph of $y = e^x$.

x	$\ln x$
1	0
4	1.4
7	1.9
0.5	-0.7

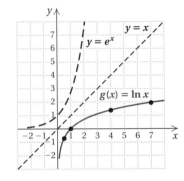

Example 13 Graph: $f(x) = \ln (x + 3)$.

We find some solutions with a calculator, plot them, and then draw the graph.

x	$\ln (x + 3)$
0	1.1
1	1.4
2	1.6
3	1.8
4	1.9
-1	0.7
-2	0
-2.5	-0.7

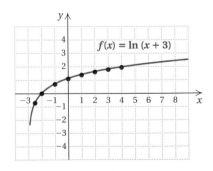

The graph of $y = \ln (x + 3)$ is the graph of $y = \ln x$ translated 3 units to the left.

Do Exercises 12 and 13.

Exercise Set 8.5

a Find each of the following logarithms or powers, base e, using a calculator. Round answers to four decimal places.

1. $\ln 2$ **2.** $\ln 5$ **3.** $\ln 62$ **4.** $\ln 30$ **5.** $\ln 4365$ **6.** $\ln 901.2$

7. $\ln 0.0062$ **8.** $\ln 0.00073$ **9.** $\ln 0.2$ **10.** $\ln 0.04$ **11.** $\ln 0$ **12.** $\ln(-4)$

13. $\ln\left(\dfrac{97.4}{558}\right)$ **14.** $\ln\left(\dfrac{786.2}{77.2}\right)$ **15.** $\ln e$ **16.** $\ln e^2$ **17.** $e^{2.71}$ **18.** $e^{3.06}$

19. $e^{-3.49}$ **20.** $e^{-2.64}$ **21.** $e^{4.7}$ **22.** $e^{1.23}$ **23.** $\ln e^5$ **24.** $e^{\ln 7}$

b Find each of the following logarithms using the change-of-base formula.

25. $\log_6 100$ **26.** $\log_3 100$ **27.** $\log_2 100$ **28.** $\log_7 100$ **29.** $\log_7 65$ **30.** $\log_5 42$

31. $\log_{0.5} 5$ **32.** $\log_{0.1} 3$ **33.** $\log_2 0.2$ **34.** $\log_2 0.08$ **35.** $\log_\pi 200$ **36.** $\log_\pi \pi$

c Graph.

37. $f(x) = e^x$ **38.** $f(x) = e^{0.5x}$ **39.** $f(x) = e^{-0.5x}$ **40.** $f(x) = e^{-x}$

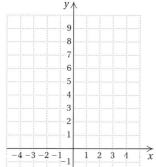 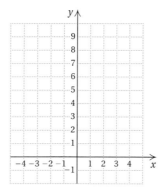 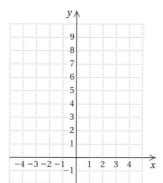 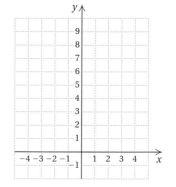

41. $f(x) = e^{x-1}$ **42.** $f(x) = e^{-x} + 3$ **43.** $f(x) = e^{x+2}$ **44.** $f(x) = e^{x-2}$

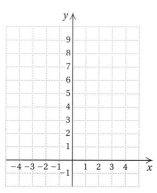 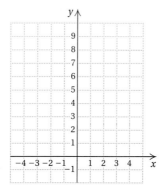 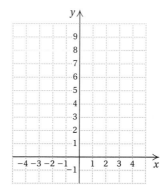 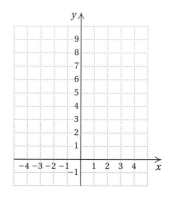

45. $f(x) = e^x - 1$

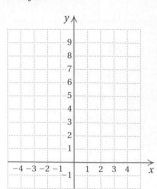

46. $f(x) = 2e^{0.5x}$

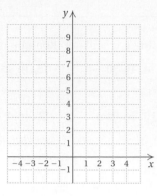

47. $f(x) = \ln(x + 2)$

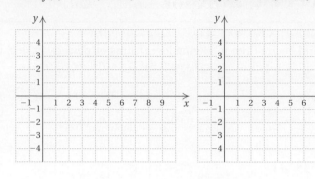

48. $f(x) = \ln(x + 1)$

49. $f(x) = \ln(x - 3)$

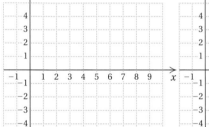

50. $f(x) = 2\ln(x - 2)$

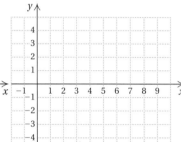

51. $f(x) = 2\ln x$

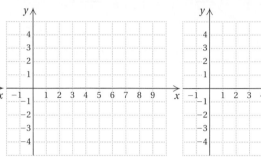

52. $f(x) = \ln x - 3$

53. $f(x) = \frac{1}{2}\ln x + 1$

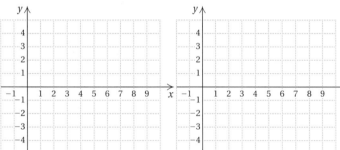

54. $f(x) = \ln x^2$

55. $f(x) = |\ln x|$

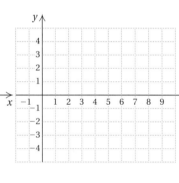

56. $f(x) = \ln|x|$

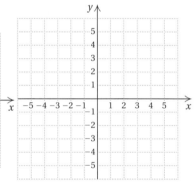

Skill Maintenance

Solve.   [7.4c]

57. $x^{1/2} - 6x^{1/4} + 8 = 0$

58. $2y - 7\sqrt{y} + 3 = 0$

59. $x - 18\sqrt{x} + 77 = 0$

60. $x^4 - 25x^2 + 144 = 0$

Synthesis

61. ◈ Explain how the graph of $f(x) = e^x$ could be used to obtain the graph of $g(x) = 1 + \ln x$.

62. Find a formula for converting common logarithms to natural logarithms. Then find a formula for converting natural logarithms to common logarithms.

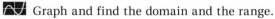 Graph and find the domain and the range.

63. $f(x) = 10x^2e^{-x}$

64. $f(x) = 7.4e^x \ln x$

65. $f(x) = 100(1 - e^{-0.3x})$

Find the domain.

66. $f(x) = \log_3 x^2$

67. $f(x) = \log(2x - 5)$

8.6 Solving Exponential and Logarithmic Equations

a Solving Exponential Equations

Equations with variables in exponents, such as $5^x = 12$ and $2^{7x} = 64$, are called **exponential equations.** Sometimes, as is the case with $2^{7x} = 64$, we can write each side as a power of the same number:

$$2^{7x} = 2^6.$$

Since the base is the same, 2, the exponents are the same. We can set them equal and solve:

$$7x = 6$$
$$x = \tfrac{6}{7}.$$

We use the following property, which is true because exponential functions are one-to-one.

For any $a > 0$, $a \neq 1$,
$$a^x = a^y \implies x = y.$$

Example 1 Solve: $2^{3x-5} = 16$.

Note that $16 = 2^4$. Thus we can write each side as a power of the same number:

$$2^{3x-5} = 2^4.$$

Since the base is the same, 2, the exponents must be the same. Thus,

$3x - 5 = 4$	**CHECK:**	$2^{3x-5} = 16$
$3x = 9$		$2^{3 \cdot 3 - 5}$? 16
$x = 3.$		2^{9-5}
		2^4
		16 | TRUE

The solution is 3.

Do Exercises 1 and 2.

A⫶G Algebraic–Graphical Connection

The solution, 3, of the equation $2^{3x-5} = 16$ in Example 1 is the x-coordinate of the point of intersection of the graphs of $y = 2^{3x-5}$ and $y = 16$, as we see in the graph on the left below.

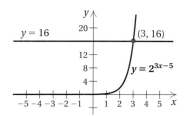
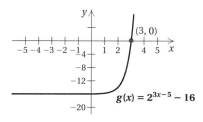

If we subtract 16 on both sides of $2^{3x-5} = 16$, we get $2^{3x-5} - 16 = 0$. The solution, 3, is then the x-coordinate of the x-intercept of the function $g(x) = 2^{3x-5} - 16$, as we see in the graph on the right above.

Solve.

1. $3^{2x} = 9$

2. $4^{2x-3} = 64$

Answers on page A-53

3. Solve: $7^x = 20$.

When it does not seem possible to write each side as a power of the same base, we can take the common or natural logarithm on each side and then use Property 2.

Example 2 Solve: $5^x = 12$.

$$5^x = 12$$
$$\log 5^x = \log 12 \qquad \text{Taking the common logarithm on both sides}$$
$$x \log 5 = \log 12 \qquad \text{Property 2}$$
$$x = \frac{\log 12}{\log 5} \leftarrow \boxed{\textit{CAUTION!} \quad \text{This is not } \log \frac{12}{5}, \text{ or } \log 12 - \log 5!}$$

This is an exact answer. We cannot simplify further, but we can approximate using a calculator:

$$x = \frac{\log 12}{\log 5} \approx 1.5439.$$

We can also partially check this answer by finding $5^{1.5439}$ using a calculator.

CHECK:

$$\frac{5^x = 12}{5^{1.5439} \; ? \; 12}$$
$$11.99885457 \mid \qquad \text{TRUE}$$

We get an answer close to 12, due to the rounding. This checks.

Do Exercise 3.

4. Solve: $e^{0.3t} = 80$.

If the base is e, we can make our work easier by taking the logarithm, base e, on both sides.

Example 3 Solve: $e^{0.06t} = 1500$.

We take the natural logarithm on both sides:

$$e^{0.06t} = 1500$$
$$\ln e^{0.06t} = \ln 1500 \qquad \text{Taking ln on both sides}$$
$$\log_e e^{0.06t} = \ln 1500 \qquad \text{Definition of natural logarithms}$$
$$0.06t = \ln 1500 \qquad \text{Here we use Property 4: } \log_a a^k = k.$$
$$t = \frac{\ln 1500}{0.06}.$$

We can approximate using a calculator:

$$t = \frac{\ln 1500}{0.06} \approx \frac{7.3132}{0.06} \approx 121.89.$$

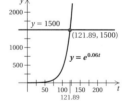

We can also partially check this answer using a calculator.

CHECK:

$$\frac{e^{0.06t} = 1500}{e^{0.06(121.89)} \; ? \; 1500}$$
$$e^{7.3134} \mid$$
$$1500.269444 \mid \qquad \text{TRUE}$$

The solution is about 121.89.

Answers on page A-53

Do Exercise 4.

Calculator Spotlight

Use the SOLVE or INTERSECT feature to check the results of Examples 1–3 and Margin Exercises 1–4.

b | Solving Logarithmic Equations

Equations containing logarithmic expressions are called **logarithmic equations.** We solved some logarithmic equations in Section 8.3 by converting to equivalent exponential equations.

Example 4 Solve: $\log_2 x = 3$.

We obtain an equivalent exponential equation:

$$x = 2^3$$
$$x = 8.$$

The solution is 8.

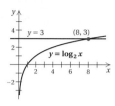

Do Exercise 5.

> To solve a logarithmic equation, first try to obtain a single logarithmic expression on one side and then write an equivalent exponential equation.

Example 5 Solve: $\log_4 (8x - 6) = 3$.

We already have a single logarithmic expression, so we write an equivalent exponential equation:

$$8x - 6 = 4^3 \quad \text{Writing an equivalent exponential equation}$$
$$8x - 6 = 64$$
$$8x = 70$$
$$x = \frac{70}{8}, \text{ or } \frac{35}{4}$$

CHECK:
$$\frac{\log_4 (8x - 6) = 3}{\log_4 \left(8 \cdot \frac{35}{4} - 6\right) \ ? \ 3}$$
$$\log_4 (70 - 6)$$
$$\log_4 64$$
$$3 \quad \Big| \quad \text{TRUE}$$

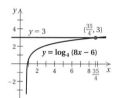

The solution is $\frac{35}{4}$.

Do Exercise 6.

Example 6 Solve: $\log x + \log (x - 3) = 1$.

Here we have common logarithms. It will help us follow the solution to first write in the 10's and obtain a single logarithmic expression on the left.

$$\log_{10} x + \log_{10} (x - 3) = 1$$
$$\log_{10} [x(x - 3)] = 1 \quad \text{Using Property 1 to obtain a single logarithm}$$
$$x(x - 3) = 10^1 \quad \text{Writing an equivalent exponential expression}$$
$$x^2 - 3x = 10$$
$$x^2 - 3x - 10 = 0$$
$$(x + 2)(x - 5) = 0 \quad \text{Factoring}$$
$$x + 2 = 0 \quad or \quad x - 5 = 0 \quad \text{Using the principle of zero products}$$
$$x = -2 \quad or \quad x = 5$$

Answers on page A-53

7. Solve: $\log x + \log (x + 3) = 1$.

CHECK: For -2:

$$\frac{\log x + \log (x - 3) = 1}{\log (-2) + \log (-2 - 3) \; \overset{?}{|} \; 1}$$

The number -2 does *not* check because negative numbers do not have logarithms.

For 5:

$$\frac{\log x + \log (x - 3) = 1}{\log 5 + \log (5 - 3) \; \overset{?}{|} \; 1}$$
$$\log 5 + \log 2 \;|$$
$$\log (5 \cdot 2) \;|$$
$$\log 10 \;|$$
$$1 \;| \quad \text{TRUE}$$

The solution is 5.

Do Exercise 7.

Example 7 Solve: $\log_2 (x + 7) - \log_2 (x - 7) = 3$.

$\log_2 (x + 7) - \log_2 (x - 7) = 3$

$$\log_2 \frac{x + 7}{x - 7} = 3 \qquad \textbf{Using Property 3 to obtain a single logarithm}$$

$$\frac{x + 7}{x - 7} = 2^3 \qquad \textbf{Writing an equivalent exponential expression}$$

$$\frac{x + 7}{x - 7} = 8$$

$$x + 7 = 8(x - 7) \qquad \textbf{Multiplying by the LCM, } x - 7$$

$$x + 7 = 8x - 56 \qquad \textbf{Using a distributive law}$$

$$63 = 7x$$

$$\frac{63}{7} = x$$

$$9 = x$$

8. Solve:

$\log_3 (2x - 1) - \log_3 (x - 4) = 2$.

CHECK:

$$\frac{\log_2 (x + 7) - \log_2 (x - 7) = 3}{\log_2 (9 + 7) - \log_2 (9 - 7) \; \overset{?}{|} \; 3}$$
$$\log_2 16 - \log_2 2 \;|$$
$$\log_2 \tfrac{16}{2} \;|$$
$$\log_2 8 \;|$$
$$3 \;| \quad \text{TRUE}$$

The solution is 9.

Do Exercise 8.

Answers on page A-53

Exercise Set 8.6

a Solve.

1. $2^x = 8$

2. $3^x = 81$

3. $4^x = 256$

4. $5^x = 125$

5. $2^{2x} = 32$

6. $4^{3x} = 64$

7. $3^{5x} = 27$

8. $5^{7x} = 625$

9. $2^x = 11$

10. $2^x = 20$

11. $2^x = 43$

12. $2^x = 55$

13. $5^{4x-7} = 125$

14. $4^{3x+5} = 16$

15. $3^{x^2} \cdot 3^{4x} = \dfrac{1}{27}$

16. $3^{5x} \cdot 9^{x^2} = 27$

17. $4^x = 8$

18. $6^x = 10$

19. $e^t = 100$

20. $e^t = 1000$

21. $e^{-t} = 0.1$

22. $e^{-t} = 0.01$

23. $e^{-0.02t} = 0.06$

24. $e^{0.07t} = 2$

25. $2^x = 3^{x-1}$

26. $3^{x+2} = 5^{x-1}$

27. $(3.6)^x = 62$

28. $(5.2)^x = 70$

b Solve.

29. $\log_4 x = 4$

30. $\log_7 x = 3$

31. $\log_2 x = -5$

32. $\log_9 x = \dfrac{1}{2}$

33. $\log x = 1$

34. $\log x = 3$

35. $\log x = -2$

36. $\log x = -3$

37. $\ln x = 2$

38. $\ln x = 1$

39. $\ln x = -1$

40. $\ln x = -3$

41. $\log_3 (2x + 1) = 5$

42. $\log_2 (8 - 2x) = 6$

43. $\log x + \log (x - 9) = 1$

44. $\log x + \log (x + 9) = 1$

45. $\log x - \log (x + 3) = -1$

46. $\log (x + 9) - \log x = 1$

47. $\log_2 (x + 1) + \log_2 (x - 1) = 3$

48. $\log_2 x + \log_2 (x - 2) = 3$

49. $\log_4 (x + 6) - \log_4 x = 2$

50. $\log_4 (x + 3) - \log_4 (x - 5) = 2$

51. $\log_4 (x + 3) + \log_4 (x - 3) = 2$

52. $\log_5 (x + 4) + \log_5 (x - 4) = 2$

53. $\log_3 (2x - 6) - \log_3 (x + 4) = 2$

54. $\log_4 (2 + x) - \log_4 (3 - 5x) = 3$

Skill Maintenance

Solve. [7.4c]

55. $x^4 + 400 = 104x^2$

56. $x^{2/3} + 2x^{1/3} = 8$

57. $(x^2 + 5x)^2 + 2(x^2 + 5x) = 24$

58. $10 = x^{-2} + 9x^{-1}$

59. Simplify: $(125x^3y^{-2}z^6)^{-2/3}$. [6.2c]

60. Simplify: i^{79}. [6.8d]

Synthesis

61. ◈ Explain how Exercises 37–40 could be solved using only the graph of $f(x) = \ln x$.

62. ◈ Christina first determines that the solution of $\log_3 (x + 4) = 1$ is -1, but then rejects it. What mistake do you think she might be making?

63. 〰 Find the value of x for which the natural logarithm is the same as the common logarithm.

64. 〰 Use a grapher to check your answers to Exercises 4, 20, 36, and 54.

65. 〰 Use a grapher to solve each of the following equations.

  a) $e^{7x} = 14$ b) $8e^{0.5x} = 3$
  c) $xe^{3x-1} = 5$ d) $4 \ln (x + 3.4) = 2.5$

66. 〰 Use the INTERSECT feature of a grapher to find the points of intersection of the graphs of each pair of functions.

  a) $f(x) = e^{0.5x-7}$, $g(x) = 2x + 6$
  b) $f(x) = \ln 3x$, $g(x) = 3x - 8$
  c) $f(x) = \ln x^2$, $g(x) = -x^2$

Solve.

67. $2^{2x} + 128 = 24 \cdot 2^x$

68. $27^x = 81^{2x-3}$

69. $8^x = 16^{3x+9}$

70. $\log_x (\log_3 27) = 3$

71. $\log_6 (\log_2 x) = 0$

72. $x \log \frac{1}{8} = \log 8$

73. $\log_5 \sqrt{x^2 - 9} = 1$

74. $2^{x^2+4x} = \frac{1}{8}$

75. $\log (\log x) = 5$

76. $\log_5 |x| = 4$

77. $\log x^2 = (\log x)^2$

78. $\log_3 |5x - 7| = 2$

79. $\log_a a^{x^2+4x} = 21$

80. $\sqrt{r} \cdot \sqrt[3]{r} \cdot \sqrt[4]{r} \cdot \sqrt[5]{r} = 146$

81. $3^{2x} - 8 \cdot 3^x + 15 = 0$

82. If $x = (\log_{125} 5)^{\log_5 125}$, what is the value of $\log_3 x$?

8.7 Mathematical Modeling with Exponential and Logarithmic Functions

Exponential and logarithmic functions can now be added to our library of functions that can serve as models for many kinds of applications. Let's review some of their graphs.

Objectives

a Solve applied problems involving logarithmic functions.

b Solve applied problems involving exponential functions.

For Extra Help

TAPE 18   TAPE 17A   MAC WIN   CD-ROM

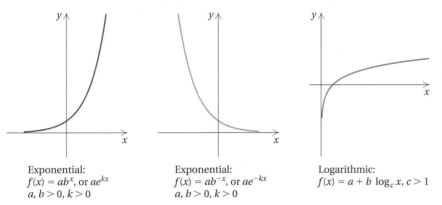

Exponential:
$f(x) = ab^x$, or ae^{kx}
$a, b > 0, k > 0$

Exponential:
$f(x) = ab^{-x}$, or ae^{-kx}
$a, b > 0, k > 0$

Logarithmic:
$f(x) = a + b \log_c x, c > 1$

a   Applications of Logarithmic Functions

Example 1   *Sound Levels.* To measure the "loudness" of any particular sound, the decibel scale is used. The loudness L, in decibels (dB), of a sound is given by

$$L = 10 \cdot \log \frac{I}{I_0},$$

where I is the intensity of the sound, in watts per square meter (W/m²), and $I_0 = 10^{-12}$ W/m². (I_0 is approximately the intensity of the softest sound that can be heard.)

a) It is common for the intensity of sound at live performances of rock music to reach 10^{-1} W/m² (even higher close to the stage). How loud, in decibels, is this sound level?

b) Audiologists and physicians recommend that earplugs be worn when one is exposed to sounds in excess of 90 db. What is the intensity of such sounds?

1. *Acoustics.* The intensity of sound in normal conversation is about 3.2×10^{-6} W/m^2. How high is this sound level in decibels?

a) To find the loudness, in decibels, we use the above formula:

$$L = 10 \cdot \log \frac{I}{I_0}$$

$$= 10 \cdot \log \frac{10^{-1}}{10^{-12}} \qquad \text{Substituting}$$

$$= 10 \cdot \log 10^{11} \qquad \text{Subtracting exponents}$$

$$= 10 \cdot 11 \qquad \log 10^a = a$$

$$= 110.$$

The volume of the music is 110 decibels.

b) We substitute and solve for I:

$$L = 10 \cdot \log \frac{I}{I_0}$$

$$90 = 10 \cdot \log \frac{I}{10^{-12}} \qquad \text{Substituting}$$

$$9 = \log \frac{I}{10^{-12}} \qquad \text{Dividing by 10}$$

$$9 = \log I - \log 10^{-12} \qquad \text{Using Property 3}$$

$$9 = \log I - (-12) \qquad \log 10^a = a$$

$$-3 = \log I \qquad \text{Adding } -12$$

$$10^{-3} = I. \qquad \text{Converting to an exponential equation}$$

Earplugs are recommended for sounds with intensities exceeding 10^{-3} W/m^2.

Do Exercises 1 and 2.

2. *Audiology.* Overexposure to excessive sound levels can diminish one's hearing to the point where the softest sound that is audible is 28 dB. What is the intensity of such a sound?

Example 2 *Chemistry: pH of Liquids.* In chemistry the pH of a liquid is defined as:

$$pH = -\log [H^+],$$

where $[H^+]$ is the hydrogen ion concentration in moles per liter.

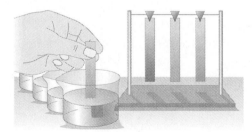

a) The hydrogen ion concentration of human blood is normally about 3.98×10^{-8} moles per liter. Find the pH.

b) The pH of seawater is about 8.3. Find the hydrogen ion concentration.

a) To find the pH of human blood, we use the above formula:

$$pH = -\log [H^+]$$

$$= -\log [3.98 \times 10^{-8}]$$

$$\approx -(-7.400117) \qquad \text{Using a calculator}$$

$$\approx 7.4$$

The pH of human blood is normally about 7.4.

Answers on page A-53

b) We substitute and solve for $[H^+]$:

$$8.3 = -\log [H^+] \qquad \text{Using pH} = -\log [H^+]$$

$$-8.3 = \log [H^+] \qquad \text{Dividing by } -1$$

$$10^{-8.3} = [H^+] \qquad \text{Converting to an exponential equation}$$

$$5.01 \times 10^{-9} \approx [H^+]. \qquad \text{Using a calculator; writing scientific notation}$$

The hydrogen ion concentration of seawater is about 5.01×10^{-9} moles per liter.

Do Exercises 3 and 4.

b | Applications of Exponential Functions

Example 3 *Interest Compounded Annually.* Suppose that $30,000 is invested at 8% interest, compounded annually. In t years, it will grow to the amount A given by the function

$$A(t) = 30,000(1.08)^t.$$

(See Example 6 in Section 8.1.)

a) How long will it take to accumulate $150,000 in the account?

b) Let T = the amount of time it takes for the $30,000 to double itself; T is called the **doubling time.** Find the doubling time.

a) We set $A(t) = 150,000$ and solve for t:

$$150,000 = 30,000(1.08)^t$$

$$\frac{150,000}{30,000} = (1.08)^t \qquad \text{Dividing by 30,000}$$

$$5 = (1.08)^t$$

$$\log 5 = \log (1.08)^t \qquad \text{Taking the common logarithm on both sides}$$

$$\log 5 = t \log 1.08 \qquad \text{Using Property 2}$$

$$\frac{\log 5}{\log 1.08} = t \qquad \text{Dividing by log 1.08}$$

$$20.9 \approx t. \qquad \text{Using a calculator}$$

It will take about 20.9 yr for the $30,000 to grow to $150,000.

b) To find the *doubling time T,* we replace $A(t)$ with 60,000 and t with T and solve for T:

$$60,000 = 30,000(1.08)^T$$

$$2 = (1.08)^T \qquad \text{Dividing by 30,000}$$

$$\log 2 = \log (1.08)^T \qquad \text{Taking the common logarithm on both sides}$$

$$\log 2 = T \log 1.08 \qquad \text{Using Property 2}$$

$$T = \frac{\log 2}{\log 1.08} \approx 9.0. \qquad \text{Using a calculator}$$

The doubling time is about 9 yr.

Do Exercise 5.

---

3. *Coffee.* The hydrogen ion concentration of freshly brewed coffee is about 1.3×10^{-5} moles per liter. Find the pH.

4. *Acidosis.* When the pH of a patient's blood drops below 7.4, a condition called *acidosis* sets in. Acidosis can be fatal at a pH level of 7.0. What would the hydrogen ion concentration of the patient's blood be at that point?

5. *Interest Compounded Annually.* Suppose that $40,000 is invested at 7% interest, compounded annually.

a) After what amount of time will there be $250,000 in the account?

b) Find the doubling time.

Answers on page A-53

6. *Population Growth of the United States.* What will the population of the United States be in 2000? in 2005?

The function in Example 3 illustrates exponential growth. Populations often grow exponentially according to the following model.

> An **exponential growth model** is a function of the form
>
> $$P(t) = P_0 e^{kt}, \quad k > 0,$$
>
> where P_0 is the population at time 0, $P(t)$ is the population at time t, and k is the **exponential growth rate** for the situation. The **doubling time** is the amount of time necessary for the population to double in size.

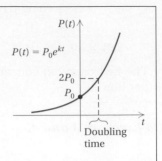

The exponential growth rate is the rate of growth of a population at any *instant* in time. Since the population is continually growing, the percent of total growth after one year will exceed the exponential growth rate.

Example 4 *Population Growth of the United States.* In 1998, the population of the United States was 267 million, and the exponential growth rate was 0.8% per year.

a) Find the exponential growth function.

b) What will the population be in 2010?

a) We are trying to find a model. The given information allows us to create one. At $t = 0$ (1998), the population was 267 million. We substitute 267 for P_0 and 0.8%, or 0.008, for k to obtain the exponential growth function:

$$P(t) = P_0 e^{kt}$$
$$P(t) = 267 e^{0.008t}.$$

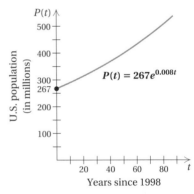

b) In 2010, we have $t = 12$. That is, 12 yr have passed since 1998. To find the population in 2010, we substitute 12 for t:

$$P(12) = 267 e^{0.008(12)} \quad \text{Substituting 12 for } t$$
$$\approx 294 \text{ million.} \quad \text{Using a calculator}$$

The population of the United States will be about 294 million in 2010.

Do Exercise 6.

Example 5 *Interest Compounded Continuously.* Suppose that an amount of money P_0 is invested in a savings account at interest rate k, compounded continuously. That is, suppose that interest is computed every "instant" and added to the amount in the account. The balance $P(t)$, after t years, is given by the exponential growth model

$$P(t) = P_0 e^{kt}.$$

a) Suppose that $30,000 is invested and grows to $44,754.75 in 5 yr. Find the interest rate and then the exponential growth function.

b) What is the balance after 10 yr?

c) What is the doubling time?

Answer on page A-53

a) We have $P(0) = 30{,}000$. Thus the exponential growth function is

$$P(t) = 30{,}000e^{kt}, \quad \text{where } k \text{ must still be determined.}$$

We know that at $t = 5$, $P(5) = 44{,}754.75$. We substitute and solve for k:

$$44{,}754.75 = 30{,}000e^{k(5)} = 30{,}000e^{5k}$$

$$\frac{44{,}754.75}{30{,}000} = e^{5k} \qquad \text{Dividing by 30,000}$$

$$1.491825 = e^{5k}$$

$$\ln 1.491825 = \ln e^{5k} \qquad \text{Taking the natural logarithm on both sides}$$

$$0.4 \approx 5k \qquad \text{Finding ln 1.491825 on a calculator and simplifying } \ln e^{5k}$$

$$\frac{0.4}{5} = 0.08 \approx k.$$

The interest rate is about 0.08, or 8%, compounded continuously. Note that since interest is being compounded continuously, the interest earned each year is more than 8%. The exponential growth function is

$$P(t) = 30{,}000e^{0.08t}.$$

b) We substitute 10 for t:

$$P(10) = 30{,}000e^{0.08(10)} = 66{,}766.23.$$

The balance in the account after 10 yr will be $66,766.23.

c) To find the doubling time T, we replace $P(t)$ with 60,000 and solve for T:

$$60{,}000 = 30{,}000e^{0.08T}$$

$$2 = e^{0.08T} \qquad \text{Dividing by 30,000}$$

$$\ln 2 = \ln e^{0.08T} \qquad \text{Taking the natural logarithm on both sides}$$

$$\ln 2 = 0.08T$$

$$\frac{\ln 2}{0.08} = T \qquad \text{Dividing}$$

$$8.7 \approx T.$$

Thus the original investment of $30,000 will double in about 8.7 yr, as shown in the following graph of the growth function.

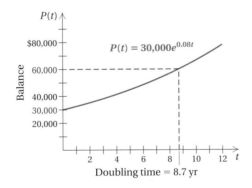

Doubling time = 8.7 yr

Do Exercise 7.

Compare the results of Examples 3(b) and 5(c). Note that continuous compounding gives a higher yield and shorter doubling time. Greater understanding about continuous compounding belongs to a course in business calculus.

---

7. *Interest Compounded Continuously.*

a) Suppose that $5000 is invested and grows to $6356.25 in 4 yr. Find the interest rate and then the exponential growth function.

b) What is the balance after 1 yr? 2 yr? 10 yr?

c) What is the doubling time?

Answers on page A-53

8. *Bagged Lettuce.* The convenience of bagged lettuce has resulted in huge increases in the sales of lettuce in recent years, as shown in the graph below (***Source:*** International Fresh-Cut Produce Association, Information Resources).

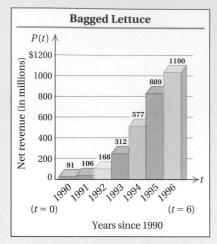

Bagged Lettuce

Net revenue (in millions) vs. Years since 1990

Data values: 1990 (t=0): 91, 1991: 106, 1992: 168, 1993: 312, 1994: 577, 1995: 889, 1996 (t=6): 1100

a) Let $t = 0$ correspond to 1990 and $t = 6$ correspond to 1996. Then t is the number of years since 1990. Use the data points (0, 91) and (6, 1100) to find the exponential growth rate and fit an exponential growth function $P(t) = P_0 e^{kt}$ to the data, where $P(t)$ is the revenue, in millions, t years after 1990.

b) Use the function found in part (a) to predict the net revenue in 2004.

c) Assuming exponential growth continues, predict the year in which revenue from the sale of lettuce will reach $8000 million.

Take a look at the graph that follows in Example 6. Note that the data tend to be rising steeply in a manner that might fit an exponential function. To fit an exponential function to the data, we choose a pair of data points and use them to determine P_0 and k.

Example 6 *Intel Corporation.* The Intel Corporation manufactures computer chips such as the Pentium II microprocessor. Its net revenues have skyrocketed in recent years, as shown in the graph.

a) If we let $t = 0$ correspond to 1992 and $t = 4$ to 1996, then t is the numbers of years since 1992. Use the data points (0, 5844) and (4, 20,847) to find the exponential growth rate and fit an exponential growth function to the data.

b) Use the function found in part (a) to predict the net revenue in 2000.

c) Assuming that exponential growth continues, predict the year in which the net revenue of Intel will reach $100,000 million.

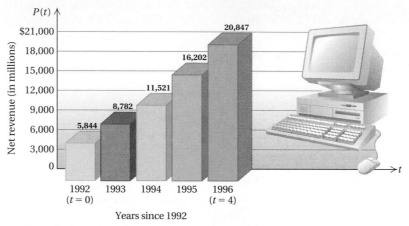

Net revenue (in millions) vs. Years since 1992

Data values: 1992 (t=0): 5,844; 1993: 8,782; 1994: 11,521; 1995: 16,202; 1996 (t=4): 20,847

Source: Intel Corporation, 1996 Annual Report

a) We use the equation $P(t) = P_0 e^{kt}$, where $P(t)$ is the net revenue of Intel, in millions of dollars, t years after 1992. In 1992, at $t = 0$, the net revenue was $5844 million. Thus we substitute 5844 for P_0:

$$P(t) = 5844e^{kt}.$$

To find the exponential growth rate, k, we note that 4 yr later, at the end of 1996, the net revenue was $20,847 million. We substitute and solve for k:

$$\left.\begin{array}{l} P(4) = 5844e^{k(4)} \\ 20{,}847 = 5844e^{4k} \end{array}\right\} \quad \textbf{Substituting}$$

$$3.5672 \approx e^{4k} \qquad \textbf{Dividing by 5844}$$

$$\ln 3.5672 = \ln e^{4k} \qquad \textbf{Taking the natural logarithm on both sides}$$

$$1.2718 = 4k \qquad \textbf{ln } e^a = a$$

$$0.3180 \approx k. \qquad \textbf{Dividing by 4}$$

The exponential growth rate was 0.318, or 31.8%, and the exponential growth function is $P(t) = 5844e^{0.318t}$.

b) Since 2000 is 8 yr from 1992, we substitute 8 for t:

$$P(8) = 5844e^{0.318(8)} \approx 74{,}397.$$

The net revenue in 2000 will be about $74,397 million, or $74.397 billion.

Answers on page A-53

c) To predict when net revenues will reach $100,000 million, we substitute 100,000 for $P(t)$ and solve for t:

$$100,000 = 5844e^{0.318t}$$

$$17.1116 \approx e^{0.318t} \qquad \text{Dividing by 5844}$$

$$\ln 17.1116 \approx \ln e^{0.318t} \qquad \text{Taking the natural logarithm on both sides}$$

$$2.8398 \approx 0.318t \qquad \ln e^a = a$$

$$8.93 \approx t. \qquad \text{Dividing by 0.318}$$

Rounding to 9 yr, we see that, according to this model, by the end of 1992 + 9, or 2001, the net revenue of Intel will reach $100,000 million, or $100 billion.

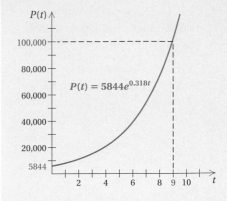

Do Exercise 8 on the preceding page.

In some real-life situations, a quantity or population is *decreasing* or *decaying* exponentially.

▶ An **exponential decay model** is a function of the form

$$P(t) = P_0e^{-kt}, \quad k > 0,$$

where P_0 is the quantity present at time 0, $P(t)$ is the amount present at time t, and k is the **decay rate.** The **half-life** is the amount of time necessary for half of the quantity to decay.

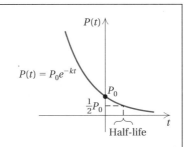

Example 7   *Carbon Dating.* The radioactive element carbon-14 has a half-life of 5750 yr. The percentage of carbon-14 present in the remains of organic matter can be used to determine the age of that organic matter. Recently, while digging in Chaco Canyon, New Mexico, archaeologists found corn pollen that had lost 38.1% of its carbon-14. The age of this corn pollen was evidence that Indians had been cultivating crops in the Southwest centuries earlier than scientists had thought. What was the age of the pollen? (**Source:** *American Anthropologist*)

We first find k. To do so, we use the concept of half-life. When $t = 5750$ (the half-life), $P(t)$ will be half of P_0. Then

$$0.5P_0 = P_0e^{-k(5750)}$$

$$0.5 = e^{-5750k} \qquad \text{Dividing by } P_0$$

$$\ln 0.5 = \ln e^{-5750k} \qquad \text{Taking the natural logarithm on both sides}$$

$$\ln 0.5 = -5750k$$

$$\frac{\ln 0.5}{-5750} = k$$

$$0.00012 \approx k.$$

Now we have a function for the decay of carbon-14:

$$P(t) = P_0e^{-0.00012t}. \qquad \text{This completes the first part of our solution.}$$

(*Note*: This equation can be used for any subsequent carbon-dating problem.)

9. *Carbon Dating.* How old is an animal bone that has lost 30% of its carbon-14?

If the corn pollen has lost 38.1% of its carbon-14 from an initial amount P_0, then $100\% - 38.1\%$, or 61.9%, of P_0 is still present. To find the age t of the pollen, we solve the following equation for t:

$$61.9\%P_0 = P_0e^{-0.00012t} \qquad \text{We want to find } t \text{ for which } P(t) = 0.619P_0.$$

$$0.619 = e^{-0.00012t} \qquad \text{Dividing by } P_0$$

$$\ln 0.619 = \ln e^{-0.00012t} \qquad \text{Taking the natural logarithm on both sides}$$

$$-0.4797 = -0.00012t \qquad \ln e^a = a$$

$$\frac{-0.4797}{-0.00012} \approx t \qquad \text{Dividing by } -0.00012$$

$$4000 \approx t. \qquad \text{Rounding}$$

The pollen is about 4000 yr old.

Do Exercise 9.

Calculator Spotlight

Mathematical Modeling Using Regression: Fitting Exponential Functions to Data.   In the Calculator Spotlight of Section 2.7, we considered a method to fit a linear function to a set of data, *linear regression.* In the Calculator Spotlight of Section 7.7, we extended that regression to quadratic, cubic, and quartic polynomial functions. The grapher also gives us the powerful capability to find exponential models and make predictions. One of the advantages of using regression is that it takes into account all the data, whereas the method we used in Example 6 used only two data points.

Example   *Intel Corporation.* The following data relate the net revenue of the Intel Corporation to the number of years t since 1992. They are illustrated in the graph of Example 6.

Years, t, Since 1992	Net Revenue, $P(t)$ (in millions)
0	$ 5,844
1	8,782
2	11,521
3	16,202
4	20,847

Source: Intel Corporation, 1996 Annual Report

a) Fit an exponential function to the data using the REGRESSION feature on a grapher.

b) Make a scatterplot of the data. Then graph the exponential function with the scatterplot.

c) Use the function from part (a) to estimate the net revenue of Intel in 2000. Compare your answer to that found in Example 6(c).

a) We fit an exponential function to the data using the REGRESSION feature on a TI-83 grapher. The procedure is similar to what is outlined in the Calculator Spotlight of Section 2.7. The only difference is that we choose ExpReg instead of LinReg($ax + b$) . (Consult your manual for further details.) We obtain the following function.

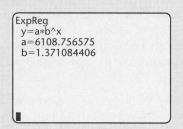

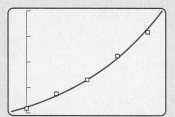

(continued)

Answer on page A-53

On some graphers, there may be a REGRESSION feature that yields an exponential function, base e. If not, and you wish to find such a function, a conversion can be done using $b^x = e^{x(\ln b)}$. For the equation given here, we would have

$$y = 6108.76(1.371)^x$$
$$= 6108.76(e^{x(\ln 1.371)})$$
$$= 6108.76e^{0.316x}.$$

b) The exponential function is graphed with the scatterplot shown on the preceding page.

c) The year 2000 is 8 yr from 1992. We can use the grapher to do the calculation. We get Y1(8) $\approx$ 76,290, as shown. Thus the net revenue of Intel in 2000 will be about $76,290 million, or $76.29 billion. In Example 6(b), we predict a net of $74,906 million.

```
Y1(8)
                    76289.77784
```

Exercises

1. *Bagged Lettuce.* The following data relate the net revenue from bagged lettuce to the number of years t since 1990. They are illustrated in the graph of Margin Exercise 8.

Years, t, Since 1990	Net Revenue, $P(t)$ (in millions)
0	$ 91
1	106
2	168
3	312
4	577
5	889
6	1100

Source: International Fresh-Cut Produce Association, Information Resources

a) Use the REGRESSION feature to fit an exponential function to the data.

b) Make a scatterplot of the data. Then graph the exponential function with the scatterplot.

c) Use the function from part (a) to estimate the net revenue from bagged lettuce in 2004. Compare your answer to that found in Margin Exercise 8(b).

2. *Average Price of a 30-Second Super Bowl Commercial.* The following data relate the cost of a Super Bowl commercial to the number of years t since 1990.

Years, t, Since 1990	Cost
1	$ 800,000
3	850,000
4	900,000
5	1,000,000
8	1,300,000

Source: National Football League

a) Use the REGRESSION feature to fit an exponential function to the Super Bowl data.

b) Make a scatterplot of the data. Then graph the exponential function with the scatterplot.

c) Use the function from part (a) to estimate the cost of a 30-second commercial in 2000 and in 2010.

d) Predict the year in which the cost of a Super Bowl commercial will be $5 million.

Improving Your Math Study Skills

Preparing for a Final Exam

Best Scenario: Two Weeks of Study Time

The best scenario for preparing for a final exam is to do so over a period of at least two weeks. Work in a diligent, disciplined manner, doing some final-exam preparation *each* day. Here is a detailed plan that many find useful.

1. **Begin by browsing through each chapter, reviewing the highlighted or boxed information regarding important formulas in both the text and the Summary and Review.** There may be some formulas that you will need to memorize.

2. **Retake each chapter test that you took in class, assuming your instructor has returned it. Otherwise, use the chapter test in the book.** Restudy the objectives in the text that correspond to each question you missed.

3. **Then work the Cumulative Review that covers all chapters up to that point.** Be careful to avoid any questions corresponding to objectives not covered. Again, restudy the objectives in the text that correspond to each question you missed.

4. **If you are still missing questions, use supplements for extra review.** For example, you might check out the video- or audiotapes, the *Student's Solutions Manual*, or the Interact Math Tutorial Software.

5. **For remaining difficulties, see your instructor, go to a tutoring session, or participate in a study group.**

6. **Check for former final exams that may be on file in the math department or a study center, or with students who have already taken the course.** Use them for practice, being alert to trouble spots.

7. **Take the Final Examination in the text during the last couple of days before the final.** Set aside the same amount of time that you will have for the final. See how much of the final exam you can complete under test-like conditions.

Moderate Scenario: Three Days to Two Weeks of Study Time

1. **Begin by browsing through each chapter, reviewing the highlighted or boxed information regarding important formulas in both the text and the Summary and Review.** There may be some formulas that you will need to memorize.

2. **Retake each chapter test that you took in class, assuming your instructor has returned it. Otherwise, use the chapter test in the book.** Restudy the objectives in the text that correspond to each question you missed.

3. **Then work the last Cumulative Review in the text.** Be careful to avoid any questions corresponding to objectives not covered. Again, restudy the objectives in the text that correspond to each question you missed.

4. **For remaining difficulties, see your instructor, go to a tutoring session, or participate in a study group.**

5. **Take the Final Examination in the text during the last couple of days before the final.** Set aside the same amount of time that you will have for the final. See how much of the final exam you can complete under test-like conditions.

Worst Scenario: One or Two Days of Study Time

1. **Begin by browsing through each chapter, reviewing the highlighted or boxed information regarding important formulas in both the text and the Summary and Review.** There may be some formulas that you will need to memorize.

2. **Then work the last Cumulative Review in the text.** Be careful to avoid any questions corresponding to objectives not covered. Restudy the objectives in the text that correspond to each question you missed.

3. **Attend a final-exam review session if one is available.**

4. **Take the Final Examination in the text during the last couple of days before the final.** Set aside the same amount of time that you will have for the final. See how much of the final exam you can complete under test-like conditions.

Promise yourself that next semester you will allow a more appropriate amount of time for final exam preparation.

Other "Improving Your Math Study Skills" concerning test preparation appear in Sections 2.3 and 3.2.

Exercise Set 8.7

a Solve.

Sound Levels. Use the decibel formula from Example 1 for Exercises 1–4.

1. *Acoustics.* The intensity of sound of a riveter at work is about 3.2×10^{-3} W/m². How high is this sound level in decibels?

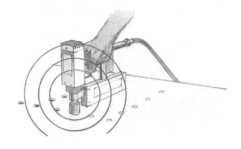

2. *Acoustics.* The intensity of sound of a dishwasher is about 2.5×10^{-6} W/m². How high is this sound level in decibels?

3. *Music.* At a recent performance of the rock group Phish, sound measurements of 105 dB were recorded. What is the intensity of such sounds?

4. *Jet Plane at Takeoff.* A jet plane at takeoff can generate sound measurements of 130 dB. What is the intensity of such sounds?

pH. Use the pH formula from Example 2 for Exercises 5–8.

5. *Milk.* The hydrogen ion concentration of milk is about 1.6×10^{-7} moles per liter. Find the pH.

6. *Mouthwash.* The hydrogen ion concentration of mouthwash is about 6.3×10^{-7} moles per liter. Find the pH.

7. *Alkalosis.* When the pH of a patient's blood rises above 7.4, a condition called *alkalosis* sets in. Alkalosis can be fatal at a pH level above 7.8. What would the hydrogen ion concentration of the patient's blood be at that point?

8. *Orange Juice.* The pH of orange juice is 3.2. What is its hydrogen ion concentration?

Walking Speed. In a study by psychologists Bornstein and Bornstein, it was found that the average walking speed w, in feet per second, of a person living in a city of population P, in thousands, is given by the function

$$w(P) = 0.37 \ln P + 0.05$$

(**Source:** *International Journal of Psychology*). In Exercises 9–12, various cities and their populations are given. Find the walking speed of people in each city.

9. Phoenix, Arizona: 2,892,000

10. Salt Lake City, Utah: 1,178,000

11. Savannah, Georgia: 139,000

12. Plano, Texas: 142,000

b Solve.

13. *Spread of Rumor.* The number of people who hear a rumor increases exponentially. If 20 people start a rumor and if each person who hears the rumor repeats it to two people a day, the number of people N who have heard the rumor after t days is given by the function

$$N(t) = 20(3)^t.$$

a) How many people have heard the rumor after 5 days?

b) After what amount of time will 1000 people have heard the rumor?

c) What is the doubling time for the number of people who have heard the rumor?

14. *Salvage Value.* A color photocopier is purchased for $5200. Its value each year is about 80% of its value in the preceding year. Its value in dollars after t years is given by the exponential function

$$V(t) = 5200(0.8)^t.$$

a) Find the salvage value of the copier after 3 yr.

b) After what amount of time will the salvage value be $1200?

c) After what amount of time will the salvage value be half the original value?

15. *Credit-Card Spending.* The amount S, in billions, charged on credit cards between Thanksgiving and Christmas each year has been increasing exponentially according to the function

$$S(t) = 52.4(1.16)^t,$$

where t is the number of years since 1990 (***Source:*** RAM Research Group, National Credit Counseling Services).

a) Estimate the amount spent between Thanksgiving and Christmas in 2000.

b) In what year will the spending be $1 trillion?

c) What is the doubling time?

16. *Projected College Costs.* In the new millennium, the cost of tuition, books, room, and board at a state university is projected to follow the exponential function

$$C(t) = 11,054(1.06)^t,$$

where C is the cost, in dollars, and t is the number of years after 2000 (***Source:*** College Board, Senate Labor Committee).

a) Find the college costs in 2005.

b) In what year will the cost be $21,000?

c) What is the doubling time of the costs?

Growth. Use the exponential growth model $P(t) = P_0 e^{kt}$ for Exercises 17–24.

17. *World Population Growth.* In 1998, the population of the world reached 6 billion, and the exponential growth rate was 1.5% per year.

a) Find the exponential growth function.

b) What will the population be in 2010?

c) In what year will the population be 10 billion?

d) What is the doubling time?

18. *Anchorage Population Growth.* In 1998, the population of Anchorage, Alaska, reached 253,750, and the exponential growth rate was 2.9% per year.

a) Find the exponential growth function.

b) What will the population be in 2010?

c) In what year will the population be 600,000?

d) What is the doubling time?

19. *Interest Compounded Continuously.* Suppose that P_0 is invested in a savings account in which interest is compounded continuously at 6% per year.

a) Express $P(t)$ in terms of P_0 and 0.06.

b) Suppose that $5000 is invested. What is the balance after 1 yr? 2 yr? 10 yr?

c) When will the investment of $5000 double itself?

20. *Interest Compounded Continuously.* Suppose that P_0 is invested in a savings account in which interest is compounded continuously at 5.4% per year.

a) Express $P(t)$ in terms of P_0 and 0.054.

b) Suppose that $10,000 is invested. What is the balance after 1 yr? 2 yr? 10 yr?

c) When will the investment of $10,000 double itself?

21. *Interest Compounded Continuously.* Suppose that P_0 is invested in a savings account in which interest is compounded continuously.

   a) Suppose that $40,000 is invested and grows to $56,762.70 in 5 yr. Find the interest rate and then the exponential growth function.

   b) What is the balance after 1 yr? 2 yr? 10 yr?

   c) What is the doubling time?

22. *Interest Compounded Continuously.* Suppose that P_0 is invested in a savings account in which interest is compounded continuously.

   a) Suppose that $100,000 is invested and grows to $164,213.96 in 8 yr. Find the interest rate and then the exponential growth function.

   b) What is the balance after 1 yr? 3 yr? 12 yr?

   c) What is the doubling time?

23. *First-Class Postage.* First-class postage (for the first ounce) was 20¢ in 1981 and 32¢ in 1995 (**Source:** U.S. Postal Service). Assume that the cost increases according to an exponential growth function.

   a) Let $t = 0$ correspond to 1981 and $t = 14$ correspond to 1995. Then t is the number of years since 1981. Use the data points $(0, 20)$ and $(14, 32)$ to find the exponential growth rate and fit an exponential growth function $P(t) = P_0 e^{kt}$ to the data, where $P(t)$ is the cost of first-class postage, in cents, t years after 1995.

   b) Use the function found in part (a) to predict the cost of first-class postage in 2000.

   c) When will the cost of first-class postage be $1.00 or 100¢?

24. *Books on Tape.* Sales of books on audiotape totaled $1.6 billion in 1996, increasing rapidly from $0.25 billion in 1989. Assume that sales increase according to an exponential growth function (**Source:** Audio Book Club).

   a) Let $t = 0$ correspond to 1989 and $t = 7$ correspond to 1996. Then $t =$ the number of years since 1989. Use the data points $(0, 0.25)$ and $(7, 1.6)$ to find the exponential growth rate and fit an exponential growth function $P(t) = P_0 e^{kt}$ to the data, where $P(t)$ is the sales, in billions of dollars, t years after 1989.

   b) Use the function found in part (a) to predict the sales of books on audiotape in 2000.

   c) When will sales reach $4.0 billion?

Carbon Dating. Use the carbon-14 decay function $P(t) = P_0 e^{-0.00012t}$ for Exercises 25 and 26.

25. *Carbon Dating.* When archaeologists found the Dead Sea scrolls, they determined that the linen wrapping had lost 22.3% of its carbon-14. How old was the linen wrapping?

26. *Carbon Dating.* In 1998, researchers found an ivory tusk that had lost 18% of its carbon-14. How old was the tusk?

Decay. Use the exponential decay function $P(t) = P_0 e^{-kt}$ for Exercises 27 and 28.

27. *Chemistry.* The decay rate of iodine-131 is 9.6% per day. What is the half-life?

28. *Chemistry.* The decay rate of krypton-85 is 6.3% per day . What is the half-life?

29. *Home Construction.* The chemical urea formaldehyde was found in some insulation used in houses built during the mid to late 1960s. Unknown at the time was the fact that urea formaldehyde emitted toxic fumes as it decayed. The half-life of urea formaldehyde is 1 yr. What is its decay rate?

30. *Plumbing.* Lead pipes and solder are often found in older buildings. Unfortunately, as lead decays, toxic chemicals can get in the water resting in the pipes. The half-life of lead is 22 yr. What is its decay rate?

31. *Value of a Sports Card.* Because he objected to smoking, and because his first baseball card was issued in cigarette packs, the great shortstop Honus Wagner halted production of his card before many were produced. One of these cards was sold in 1991 for $451,000 and again in 1996 for $640,500. For the following questions, assume that the card's value increases exponentially.

WAGNER, PITTSBURG

a) Find an exponential function $V(t)$, where t is the number of years after 1991, if $V_0 = 451,000$.
b) Predict the card's value in 2005.
c) What is the doubling time for the value of the card?
d) In what year will the value of the card first exceed $1,000,000?

32. *Value of a Van Gogh Painting.* The Van Gogh painting *Irises*, shown below, sold for $84,000 in 1947 and was sold again for $53,900,000 in 1987. Assume that the growth in the value V of the painting is exponential.

Van Gogh's *Irises*, a 28-in. by 32-in. oil on canvas.

a) Find the exponential growth rate k, and determine the exponential growth function, where t is the number of years after 1947, assuming $V_0 = 84,000$.
b) Estimate the value of the painting in 2001.
c) What is the doubling time for the value of the painting?
d) How long after 1947 will the value of the painting be $1 billion?

Skill Maintenance

Compute and simplify. Express answers in the form $a + bi$, where $i^2 = -1$.   [6.8c, d, e]

33. i^{46}

34. i^{53}

35. $i^{14} + i^{15}$

36. $i^{18} - i^{16}$

37. $\dfrac{8 - i}{8 + i}$

38. $\dfrac{2 + 3i}{5 - 4i}$

39. $(5 - 4i)(5 + 4i)$

40. $(-10 - 3i)^2$

Synthesis

41. ◈ Write a problem for a classmate to solve in which data that seem to fit an exponential growth function are provided. Try to find data in a newspaper to make the problem as realistic as possible.

42. ◈ Do some research or consult with a chemist to determine how carbon dating is carried out. Write a report.

Use a grapher to solve each of the following equations.

43. $2^x = x^{10}$

44. $(\ln 2)x = 10 \ln x$

45. $x^2 = 2^x$

Fit an exponential function to a set of data.

Collaborative Learning Manual

Summary and Review Exercises: Chapter 8

The objectives to be tested in addition to the material in this chapter are [6.8b, c, d, e], [7.4c], [7.6a], and [7.7b].

Graph.

1. $f(x) = 3^{x-1}$ [8.1a]   **2.** $f(x) = \log_3 x$ [8.3a]

3. $f(x) = e^{x+1}$ [8.5c]

4. $f(x) = \ln (x - 1)$ [8.5c]

5. Find the inverse of the relation   [8.2a]

$\{(-4, 2), (5, -7), (-1, -2), (10, 11)\}.$

Determine whether the function is one-to-one. If it is, find a formula for its inverse.   [8.2b, c]

6. $f(x) = 4 - x^2$ **7.** $g(x) = \dfrac{2x - 3}{7}$

8. $f(x) = 8x^3$ **9.** $f(x) = \dfrac{4}{3 - 2x}$

10. Graph the function $f(x) = x^3 + 1$ and its inverse using the same set of axes.   [8.2d]

11. Find $f \circ g(x)$ and $g \circ f(x)$ if $f(x) = x^2$ and $g(x) = 3x - 5$.   [8.2e]

12. If $h(x) = \sqrt{4 - 7x}$, find $f(x)$ and $g(x)$ such that $h(x) = f \circ g(x)$.   [8.2e]

Convert to a logarithmic equation.   [8.3b]

13. $10^4 = 10,000$ **14.** $25^{1/2} = 5$

Convert to an exponential equation.   [8.3b]

15. $\log_4 16 = x$ **16.** $\log_{1/2} 8 = -3$

Find each of the following.   [8.3c]

17. $\log_3 9$ **18.** $\log_{10} \frac{1}{10}$

19. $\log_m m$ **20.** $\log_m 1$

Find the common logarithm, to four decimal places, using a calculator.   [8.3e]

21. $\log \left(\dfrac{78}{43,112}\right)$ **22.** $\log (-4)$

Express in terms of logarithms of x, y, and z.   [8.4d]

23. $\log_a x^4 y^2 z^3$ **24.** $\log \sqrt[4]{\dfrac{z^2}{x^3 y}}$

Express as a single logarithm.   [8.4d]

25. $\log_a 8 + \log_a 15$

26. $\frac{1}{2} \log a - \log b - 2 \log c$

Simplify.   [8.4e]

27. $\log_m m^{17}$ **28.** $\log_m m^{-7}$

Given $\log_a 2 = 1.8301$ and $\log_a 7 = 5.0999$, find each of the following. [8.4d]

29. $\log_a 28$

30. $\log_a 3.5$

31. $\log_a \sqrt{7}$

32. $\log_a \frac{1}{4}$

Find each of the following, to four decimal places, using a calculator. [8.5a]

33. $\ln 0.06774$

34. $e^{-0.98}$

35. $e^{2.91}$

36. $\ln 1$

37. $\ln 0$

38. $\ln e$

Find the logarithm using the change-of-base formula. [8.5b]

39. $\log_5 2$

40. $\log_{12} 70$

Solve. Where appropriate, give approximations to four decimal places. [8.6a, b]

41. $\log_3 x = -2$

42. $\log_x 32 = 5$

43. $\log x = -4$

44. $3 \ln x = -6$

45. $4^{2x-5} = 16$

46. $2^{x^2} \cdot 2^{4x} = 32$

47. $4^x = 8.3$

48. $e^{-0.1t} = 0.03$

49. $\log_4 16 = x$

50. $\log_4 x + \log_4 (x - 6) = 2$

51. $\log x + \log (x - 15) = 2$

52. $\log_3 (x - 4) = 3 - \log_3 (x + 4)$

Solve. [8.7a, b]

53. *Acoustics.* The intensity of the sound at the foot of Niagara Falls is about 10^{-3} W/m^2. How high is this sound level in decibels? [Use $L = 10 \cdot \log (I/I_0)$.]

54. *Stock Index Funds.* Standard & Poor's 500 Stock Index is an indicator of the stock-price performance of 500 selected stocks. Many people purchase mutual funds that include these stocks. In recent years, the amount A, in billions of dollars, invested in such funds has grown exponentially according to the function

$$A(t) = 4.55(1.54)^t,$$

where t is the number of years since 1990 (*Source:* Lipper Analytical Services).

a) Find the amount invested in these funds in 2000.
b) In what year did the amount invested reach $80 billion?
a) Find the doubling time.

55. *Aeronautics.* There were 6 commercial space launches in 1994 and 12 in 1996. Assume that the number N is growing exponentially according to the function

$$N(t) = N_0 e^{kt}.$$

(*Source:* Federal Aviation Administration, U.S. Department of Transportation)

a) Find k and write the exponential growth function.
b) Predict the number of launches in 2003.
c) When will there be 192 annual scheduled launches?

56. The population of Riverton doubled in 16 yr. Find the exponential growth rate.

57. How long will it take $7600 to double itself if it is invested at 8.4%, compounded continuously?

58. How old is a skeleton that has lost 34% of its carbon-14? (Use $P(t) = P_0 e^{-0.00012t}$.)

Skill Maintenance

59. *Pizza Prices.* Pizza Unlimited has the following prices for pizza.

Diameter	Price
8 in.	$ 6.00
12 in.	8.50
16 in.	11.50

a) Use the data points (8, 6), (12, 8.5), and (16, 11.5) to fit a quadratic function to the data.
b) Use the function to estimate the price of a 24-in. pizza. [7.7b]

60. Solve: $x^4 - 11x^2 - 80 = 0$. [7.4c]

61. For $f(x) = x^2 + 2x + 3$, find and label **(a)** the vertex, **(b)** the line of symmetry, and **(c)** the maximum or minimum value. Then **(d)** graph the function. [7.6a]

Compute and simplify. Express answers in the form $a + bi$, where $i^2 = -1$. [6.8b, c, d, e]

62. $(2 - 3i)(4 + i)$

63. i^{20}

64. $\dfrac{4 - 5i}{1 + 3i}$

65. $(4 - 5i) + (1 + 3i)$

Synthesis

66. ◈ Explain why you cannot take the logarithm of a negative number. [8.3a]

67. ◈ Explain why $\log_a 1 = 0$. [8.3a]

Solve. [8.6a, b]

68. $\ln (\ln x) = 3$

69. $5^{x+y} = 25,\ 2^{2x-y} = 64$

Test: Chapter 8

Graph.

1. $f(x) = 2^{x+1}$

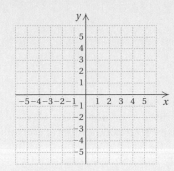

2. $y = \log_2 x$

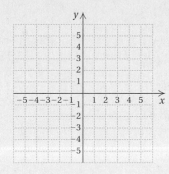

3. $f(x) = e^{x-2}$

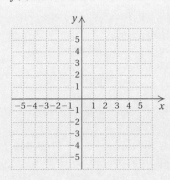

4. $f(x) = \ln (x - 4)$

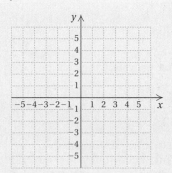

5. Find the inverse of the relation $\{(-4, 3), (5, -8), (-1, -3), (10, 12)\}$.

Determine whether the function is one-to-one. If it is, find a formula for its inverse.

6. $f(x) = 4x - 3$

7. $f(x) = (x + 1)^3$

8. $f(x) = 2 - |x|$

9. Find $f \circ g(x)$ and $g \circ f(x)$ if $f(x) = x + x^2$ and $g(x) = 5x - 2$.

10. Convert to a logarithmic equation: $256^{1/2} = 16$.

11. Convert to an exponential equation: $m = \log_7 49$.

Find each of the following.

12. $\log_5 125$

13. $\log_t t^{23}$

14. $\log_p 1$

Find the common logarithm, to four decimal places, using a calculator.

15. $\log 0.0123$

16. $\log (-5)$

17. Express in terms of logarithms of x, y, and z:

$$\log \frac{a^3 b^{1/2}}{c^2}.$$

18. Express as a single logarithm: $\frac{1}{3} \log_a x - 3 \log_a y + 2 \log_a z$.

Given $\log_a 2 = 0.301$, $\log_a 6 = 0.778$, and $\log_a 7 = 0.845$, find each of the following.

19. $\log_a \frac{2}{7}$

20. $\log_a 12$

Find each of the following, to four decimal places, using a calculator.

21. $\ln 807.39$

22. $e^{4.68}$

23. $\ln 1$

24. Find $\log_{18} 31$ using the change-of-base formula.

Answers

1. _____

2. _____

3. _____

4. _____

5. _____

6. _____

7. _____

8. _____

9. _____

10. _____

11. _____

12. _____

13. _____

14. _____

15. _____

16. _____

17. _____

18. _____

19. _____

20. _____

21. _____

22. _____

23. _____

24. _____

25. _____

26. _____

27. _____

28. _____

29. _____

30. _____

31. _____

32. _____

33. a) _____

b) _____

c) _____

34. a) _____

b) _____

c) _____

d) _____

35. _____

36. _____

37. a) _____

b) _____

38. _____

39. _____

40. _____

41. _____

42. _____

43. _____

Solve. Where appropriate, give approximations to four decimal places.

25. $\log_x 25 = 2$

26. $\log_4 x = \frac{1}{2}$

27. $\log x = 4$

28. $\ln x = \frac{1}{4}$

29. $7^x = 1.2$

30. $\log (x^2 - 1) - \log (x - 1) = 1$

31. $\log_5 x + \log_5 (x + 4) = 1$

32. _Tomatoes._ What is the pH of tomatoes if the hydrogen ion concentration is 6.3×10^{-5} moles per liter? (Use pH $= -\log [H^+]$.)

33. _Projected College Costs._ In the new millennium, the cost of tuition, books, room, and board at a private university is projected to follow the exponential function

$$C(t) = 24{,}534(1.06)^t,$$

where C is the cost, in dollars, and t is the number of years after 2000 (**_Source_**: College Board, Senate Labor Committee).

a) Find the college costs in 2008.
b) In what year will the cost be $43,937?
c) What is the doubling time of the costs?

34. _Population Growth of Canada._ The population of Canada was 30 million in 1996 and is expected to be 30.753 million in 2000 (**_Source_**: Census Bureau of Canada). Assume that the number N is growing exponentially according to the equation

$$N(t) = N_0 e^{kt}.$$

a) Find k and write the exponential growth function.
b) What will the population be in 2003? in 2010?
c) When will the population be 50 million?
d) What is the doubling time?

35. An investment with interest compounded continuously doubled in 15 yr. What is the interest rate?

36. How old is an animal bone that has lost 43% of its carbon-14? (Use $P(t) = P_0 e^{-0.00012t}$.)

Skill Maintenance

37. _Audio Equipment._ The owner's manual for a top-rated cassette deck includes a table relating the time that a tape has to run to the counter reading.

a) Use the data points (0, 0), (400, 25), and (700, 50) to fit a quadratic function to the data.
b) Use the function to estimate how long a tape has run when the counter reading is 200.

Counter Reading	Time Tape Has Run (in minutes)
000	0
400	25
700	50

38. Solve: $y - 9\sqrt{y} + 8 = 0$.

39. For $f(x) = x^2 - 4x - 2$, find and label **(a)** the vertex, **(b)** the line of symmetry, and **(c)** the maximum or minimum value. Then **(d)** graph the function.

Compute and simplify. Express answers in the form $a + bi$, where $i^2 = -1$.

40. i^{29}

41. $\dfrac{4 - i}{1 + i}$

Synthesis

42. Solve: $\log_3 |2x - 7| = 4$.

43. If $\log_a x = 2$, $\log_a y = 3$, and $\log_a z = 4$, find

$$\log_a \frac{\sqrt[3]{x^2 z}}{\sqrt[3]{y^2 z^{-1}}}.$$

Cumulative Review: Chapters R–8

1. Evaluate $\dfrac{x^0 + y}{-z}$ for $x = 6$, $y = 9$, and $z = -5$.

Simplify.

2. $\left| -\dfrac{5}{2} + \left(-\dfrac{7}{2} \right) \right|$

3. $(-5x^4y^{-3}z^2)(-4x^2y^2)$

4. $2x - 3 - 2[5 - 3(2 - x)]$

5. $3^3 + 2^2 - (32 \div 4 - 16 \div 8)$

Solve.

6. $8(2x - 3) = 6 - 4(2 - 3x)$

7. $x(x - 3) = 10$

8. $\begin{aligned} x + y - 3z &= -1, \\ 2x - y + z &= 4, \\ -x - y + z &= 1 \end{aligned}$

9. $\begin{aligned} 4x - 3y &= 15, \\ 3x + 5y &= 4 \end{aligned}$

10. $\dfrac{7}{x^2 - 5x} - \dfrac{2}{x - 5} = \dfrac{4}{x}$

11. $\sqrt{x - 1} = \sqrt{x + 4} - 1$

12. $x - 8\sqrt{x} + 15 = 0$

13. $x^4 - 13x^2 + 36 = 0$

14. $\log_8 x = 1$

15. $3^{5x} = 7$

16. $\log x - \log(x - 8) = 1$

17. $x^2 + 4x > 5$

18. $|2x - 3| \geq 9$

19. If $f(x) = x^2 + 6x$, find a such that $f(a) = 11$.

20. Solve $D = \dfrac{ab}{b + a}$ for a.

21. Solve $\dfrac{1}{p} + \dfrac{1}{q} = \dfrac{1}{f}$ for q.

22. Find the domain of the function f given by

$$f(x) = \dfrac{-4}{3x^2 - 5x - 2}.$$

23. *Urban Farmland.* Lowell King, a farmer in the greater Indianapolis metropolitan area, is losing farmland to urban development. He farmed 1400 acres in 1994 and 950 in 1996 (**Source:** *Indianapolis Star*, 7/14/96). Let A represent the number of acres farmed and t the number of years since 1994.

a) Find a linear function $A(t)$ that fits the data.
b) Use the function of part (a) to predict the number of acres farmed in 2000.

24. *Work.* Jenny can build a shed from a lumber kit in 10 hr. Phil can build the same shed in 12 hr. How long would it take Phil and Jenny, working together, to build the shed?

25. *Acid Mixtures.* Swim Clean is 30% muriatic acid. Pure Swim is 80% muriatic acid. How many liters of each should be mixed together in order to get 100 L of a solution that is 50% muriatic acid?

26. *Marine Travel.* A fishing boat with a trolling motor can move at a speed of 5 km/h in still water. The boat travels 42 km downstream in the same time that it takes to travel 12 km upstream. What is the speed of the stream?

27. *Forgetting.* Students in a biology class took a final examination. A forgetting formula for determining what the average exam score would be on a retest t months later is

$$S(t) = 78 - 15 \log(t + 1).$$

a) The average score when the students first took the test occurs when $t = 0$. Find the students' average score on the final exam.
b) What would the average score be on a retest after 4 months?

28. *Population Growth of Ukraine.* The population of Ukraine was 52 million in 1996 and 52.31 million in 1999.

a) Find k and write the exponential growth function.
b) What will the population be in 2002? in 2010?
c) When will the population be 100 million?
d) What is the doubling time?

29. *Landscaping.* A rectangular lawn measures 60 ft by 80 ft. Part of the lawn is torn up to install a sidewalk of uniform width around it. The area of the new lawn is 2400 ft^2. How wide is the sidewalk?

30. Given that y varies directly as the square of x and inversely as z, and $y = 2$ when $x = 5$ and $z = 100$. What is y when $x = 3$ and $z = 4$?

Perform the indicated operations and simplify.

31. $(11x^2 - 6x - 3) - (3x^2 + 5x - 2)$

32. $(3x^2 - 2y)^2$

33. $(5a + 3b)(2a - 3b)$

34. $\dfrac{x^2 + 8x + 16}{2x + 6} \div \dfrac{x^2 + 3x - 4}{x^2 - 9}$

35. $\dfrac{1 + \dfrac{3}{x}}{x - 1 - \dfrac{12}{x}}$

36. $\dfrac{3}{x + 6} - \dfrac{2}{x^2 - 36} + \dfrac{4}{x - 6}$

Factor.

37. $1 - 125x^3$

38. $6x^2 + 8xy - 8y^2$

39. $x^4 - 4x^3 + 7x - 28$

40. $2m^2 + 12mn + 18n^2$

41. $x^4 - 16y^4$

42. For the function described by
$$h(x) = -3x^2 + 4x + 8,$$
find $h(-2)$.

43. Divide: $(x^4 - 5x^3 + 2x^2 - 6) \div (x - 3)$.

44. Multiply $(5.2 \times 10^4)(3.5 \times 10^{-6})$. Write scientific notation for the answer.

For the radical expressions that follow, assume that all variables represent positive numbers.

45. Divide and simplify: $\dfrac{\sqrt[3]{40xy^8}}{\sqrt[3]{5xy}}$.

46. Multiply and simplify: $\sqrt{7xy^3} \cdot \sqrt{28x^2y}$.

47. Rationalize the denominator: $\dfrac{3 - \sqrt{y}}{2 - \sqrt{y}}$.

48. Multiply these complex numbers:
$$(1 + i\sqrt{3})(6 - 2i\sqrt{3}).$$

49. Find the inverse of f if $f(x) = 7 - 2x$.

50. Find an equation of the line containing the point $(-3, 5)$ and perpendicular to the line whose equation is $2x + y = 6$.

Graph.

51. $5x = 15 + 3y$

52. $f(x) = 2x^2 - 4x - 1$

53. $y = \log_3 x$

54. $y = 3^x$

55. $-2x - 3y \le 6$

56. Express as a single logarithm:
$$3 \log x - \tfrac{1}{2} \log y - 2 \log z.$$

57. Convert to an exponential equation:
$$\log_a 5 = x.$$

Find each of the following using a calculator. Round answers to four decimal places.

58. $\log 0.05566$

59. $10^{2.89}$

60. $\ln 12.78$

61. $e^{-1.4}$

Synthesis

Solve.

62. $\dfrac{5}{3x - 3} + \dfrac{10}{3x + 6} = \dfrac{5x}{x^2 + x - 2}$

63. $\log \sqrt{3x} = \sqrt{\log 3x}$

64. A train travels 280 mi at a certain speed. If the speed had been increased by 5 mph, the trip could have been made in 1 hr less time. Find the actual speed.

Cumulative Review: Chapters R–8

Conic Sections

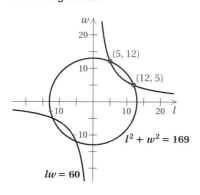

Introduction

We now consider graphs of equations in two variables in which one or both variables are raised to the second power. It is an interesting fact that we can obtain these graphs by cutting a cone with a plane. Thus they are called *conic sections*.

We also consider systems of equations in which at least one of the equations is of second degree. In many cases, solutions of these systems are intersections of conic sections.

9.1 Conic Sections: Parabolas and Circles

9.2 Conic Sections: Ellipses

9.3 Conic Sections: Hyperbolas

9.4 Nonlinear Systems of Equations

An Application

The cargo area of a delivery van must be 60 ft², and the length of a diagonal must accommodate a 13-ft board. Find the dimensions of the cargo area.

The Mathematics

We let l = the length of the van and w = the width of the van. We can then translate the problem to the system of equations

$l^2 + w^2 = 13^2$,   (The graph is a circle.)

$lw = 60$.   (The graph is a hyperbola.)

Algebraically, we find the solution to be that the length is 12 ft and the width is 5 ft. We can check this on the graph at left.

This problem appears as Exercise 33 in Exercise Set 9.4.

World Wide Web For more information, visit us at www.mathmax.com

Pretest: Chapter 9

1. Find the distance between the points $(-6, 2)$ and $(2, -3)$. Give an exact answer and an approximation to three decimal places.

2. Find the midpoint of the segment with endpoints $(-6, 2)$ and $(2, -3)$.

3. Find the center and the radius of the circle $(x - 2)^2 + (y + 7)^2 = 16$.

4. Find the center and the radius of the circle $x^2 + y^2 - 8x + 6y - 11 = 0$.

5. Find an equation of the circle having center $(2, -9)$ and radius 3.

Graph.

6. $x^2 - y^2 = 4$

7. $x^2 + y^2 = 4$

8. $\dfrac{x^2}{4} - \dfrac{y^2}{25} = 1$

9. $\dfrac{x^2}{9} + \dfrac{y^2}{16} = 1$

10. $y = x^2 - 4x + 2$

11. $x = y^2 + 4y$

12. $\dfrac{(x - 1)^2}{4} + \dfrac{(y + 5)^2}{9} = 1$

13. $xy = -8$

Solve.

14. $2x^2 - y^2 = 2,$
    $3x + 2y = -1$

15. $2x^2 - y^2 = 7,$
    $xy = 20$

16. *Tile Design.* The New World tile company wants to make a new rectangular tile that has a perimeter of 6 in. and a diagonal of length $\sqrt{5}$ in. What should the dimensions of the tile be?

17. The perimeter of a square is 12 ft more than the perimeter of another square. The area of the larger square exceeds the area of the smaller square by 81 ft^2. Find the perimeter of each square.

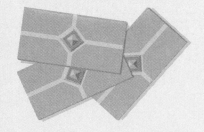
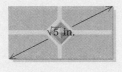

$\sqrt{5}$ in.

Objectives for Retesting

The objectives to be tested in addition to the material in this chapter are as follows.

[4.5b] Factor differences of squares.
[7.2a] Solve quadratic equations using the quadratic formula.
[8.2c] Find a formula for the inverse of a function, if it exists.
[8.3b] Convert between exponential and logarithmic equations.

9.1 Conic Sections: Parabolas and Circles

This section and the next two examine curves formed by cross sections of cones. These curves are graphs of second-degree equations in two variables. Some are shown below:

CONIC SECTIONS IN THREE DIMENSIONS

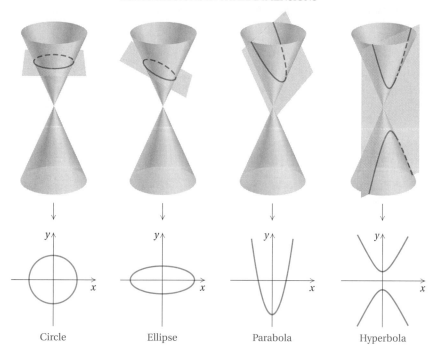

| Circle | Ellipse | Parabola | Hyperbola |

CONIC SECTIONS GRAPHED IN A PLANE

Objectives

a Graph parabolas.

b Use the distance formula to find the distance between two points whose coordinates are known.

c Use the midpoint formula to find the midpoint of a segment when the coordinates of its endpoints are known.

d Given an equation of a circle, find its center and radius and graph it; and given the center and radius of a circle, write an equation of the circle and graph the circle.

For Extra Help

TAPE 19   TAPE 17A   MAC WIN   CD-ROM

a   Parabolas

When a cone is cut by a plane parallel to a side of the cone, as shown in the third figure above, the conic section formed is a **parabola**. General equations of parabolas are quadratic. Parabolas have many applications in electricity, mechanics, and optics. A cross section of a satellite dish is a parabola, and arches that support certain bridges are parabolas. (Free-hanging cables have a different shape, called a *catenary*.)

> **EQUATIONS OF PARABOLAS**
>
> $y = ax^2 + bx + c$ (Line of symmetry is parallel to the y-axis.);
> $x = ay^2 + by + c$ (Line of symmetry is parallel to the x-axis.).

Recall from Chapter 7 that the graph of $f(x) = ax^2 + bx + c$ (with $a \neq 0$) is a parabola.

1. Graph: $y = x^2 + 4x + 7$.

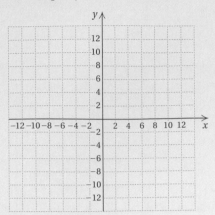

Example 1 Graph: $y = x^2 - 4x + 9$.

First, we must locate the vertex. To do so, we can use either of two approaches. One way is to complete the square:

$$y = (x^2 - 4x) + 9$$
$$= (x^2 - 4x + 0) + 9 \qquad \text{Adding 0}$$
$$= (x^2 - 4x + 4 - 4) + 9 \qquad \tfrac{1}{2}(-4) = -2; (-2)^2 = 4; \text{ substituting } 4 - 4$$
$$= (x^2 - 4x + 4) + (-4 + 9) \qquad \text{Regrouping}$$
$$= (x - 2)^2 + 5. \qquad \text{Factoring and simplifying}$$

The vertex is (2, 5).

A second way to find the vertex is to recall that the x-coordinate of the vertex of the parabola given by $y = ax^2 + bx + c$ is $-b/(2a)$:

$$x = -\frac{b}{2a} = -\frac{-4}{2(1)} = 2.$$

To find the y-coordinate of the vertex, we substitute 2 for x:

$$y = x^2 - 4x + 9 = 2^2 - 4(2) + 9 = 5.$$

Either way, the vertex is (2, 5). Next, we calculate and plot some points on each side of the vertex. Since the x^2-coefficient, 1, is positive, the graph opens up.

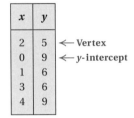

x	y	
2	5	← Vertex
0	9	← y-intercept
1	6	
3	6	
4	9	

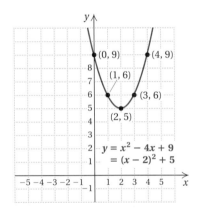

To graph an equation of the type $y = ax^2 + bx + c$ (see Section 7.6):

1. Find the vertex (h, k) either by completing the square to find an equivalent equation

   $$y = a(x - h)^2 + k,$$

   or by using $-b/(2a)$ to find the x-coordinate and substituting to find the y-coordinate.

2. Choose other values for x on each side of the vertex, and compute the corresponding y-values.

3. The graph opens up for $a > 0$ and down for $a < 0$.

Do Exercise 1.

Equations of the form $x = ay^2 + by + c$ represent horizontal parabolas. These parabolas open to the right for $a > 0$ and open to the left for $a < 0$ and have lines of symmetry parallel to the x-axis.

Answer on page A-56

Example 2 Graph: $x = y^2 - 4y + 9$.

This equation is like that in Example 1 except that x and y are interchanged. That is, the equations are inverses of each other (see Section 8.2). The vertex is $(5, 2)$ instead of $(2, 5)$. To find ordered pairs, we choose values for y on each side of the vertex. Then we compute values for x. Note that the x- and y-values of the table in Example 1 are interchanged. The graph in Example 2 is the reflection of the graph in Example 1 across the line $y = x$. You should confirm that, by completing the square, we get $x = (y - 2)^2 + 5$.

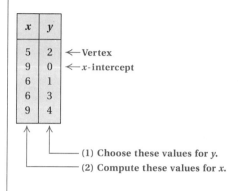

x	y	
5	2	←Vertex
9	0	←x-intercept
6	1	
6	3	
9	4	

(1) Choose these values for y.
(2) Compute these values for x.

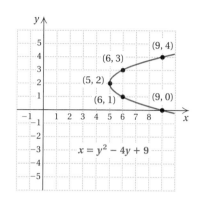

2. Graph: $x = y^2 + 4y + 7$.

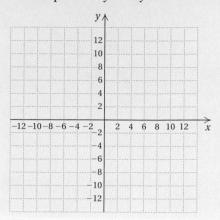

To graph an equation of the type $x = ay^2 + by + c$ (see Section 7.6):

1. Find the vertex (h, k) either by completing the square to find an equivalent equation

$$x = a(y - k)^2 + h,$$

or by using $-b/(2a)$ to find the y-coordinate and substituting to find the x-coordinate.

2. Choose other values for y that are above and below the vertex, and compute the corresponding x-values.

3. The graph opens to the right if $a > 0$ and to the left if $a < 0$.

Do Exercise 2.

Example 3 Graph: $x = -2y^2 + 10y - 7$.

We use the method of completing the square to find the vertex:

$$x = -2y^2 + 10y - 7$$
$$= -2(y^2 - 5y) - 7$$
$$= -2(y^2 - 5y + 0) - 7 \qquad \text{Adding 0}$$
$$= -2\left(y^2 - 5y + \tfrac{25}{4} - \tfrac{25}{4}\right) - 7 \qquad \tfrac{1}{2}(-5) = -\tfrac{5}{2}; \left(-\tfrac{5}{2}\right)^2 = \tfrac{25}{4}; \text{ substituting } \tfrac{25}{4} - \tfrac{25}{4}$$
$$= -2\left(y^2 - 5y + \tfrac{25}{4}\right) + (-2)\left(-\tfrac{25}{4}\right) - 7 \qquad \text{Using the distributive law}$$
$$= -2\left(y^2 - 5y + \tfrac{25}{4}\right) + \tfrac{25}{2} - 7$$
$$= -2\left(y - \tfrac{5}{2}\right)^2 + \tfrac{11}{2}. \qquad \text{Factoring and simplifying}$$

The vertex is $\left(\tfrac{11}{2}, \tfrac{5}{2}\right)$.

For practice, we also find the vertex by first computing its y-coordinate, $-b/(2a)$, and then substituting to find the x-coordinate:

$$y = -\frac{b}{2a} = -\frac{10}{2(-2)} = \frac{5}{2}$$
$$x = -2y^2 + 10y - 7 = -2\left(\tfrac{5}{2}\right)^2 + 10\left(\tfrac{5}{2}\right) - 7$$
$$= \tfrac{11}{2}.$$

Answer on page A-56

3. Graph: $x = 4y^2 - 12y + 5$.

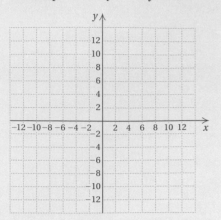

To find ordered pairs, we first choose values for y and then compute values for x. A table is shown below, together with the graph. The graph opens to the left because the y^2-coefficient, -2, is negative.

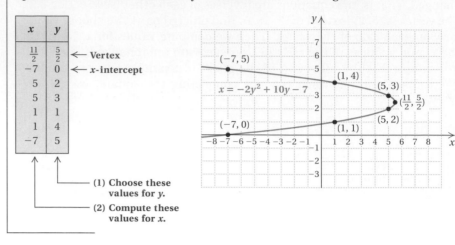

x	y	
$\frac{11}{2}$	$\frac{5}{2}$	← Vertex
-7	0	← x-intercept
5	2	
5	3	
1	1	
1	4	
-7	5	

(1) Choose these values for y.

(2) Compute these values for x.

Do Exercise 3.

b | The Distance Formula

Suppose that two points are on a horizontal line, and thus have the same second coordinate. We can find the distance between them by subtracting their first coordinates. This difference may be negative, depending on the order in which we subtract. So, to make sure we get a positive number, we take the absolute value of this difference. The distance between two points on a horizontal line (x_1, y_1) and (x_2, y_1) is thus $|x_2 - x_1|$. Similarly, the distance between two points on a vertical line (x_2, y_1) and (x_2, y_2) is $|y_2 - y_1|$.

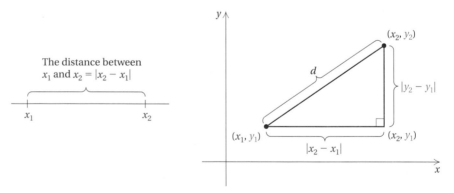

Now consider *any* two points (x_1, y_1) and (x_2, y_2). If $x_1 \neq x_2$ and $y_1 \neq y_2$, these points are vertices of a right triangle, as shown. The other vertex is then (x_2, y_1). The lengths of the legs are $|x_2 - x_1|$ and $|y_2 - y_1|$. We find d, the length of the hypotenuse, by using the Pythagorean theorem:

$$d^2 = |x_2 - x_1|^2 + |y_2 - y_1|^2.$$

Since the square of a number is the same as the square of its opposite, we don't really need these absolute-value signs. Thus,

$$d^2 = (x_2 - x_1)^2 + (y_2 - y_1)^2.$$

Taking the principal square root, we obtain the distance between two points.

Answer on page A-56

> **THE DISTANCE FORMULA**
>
> The distance between any two points (x_1, y_1) and (x_2, y_2) is given by
> $$d = \sqrt{(x_2 - x_1)^2 + (y_2 - y_1)^2}.$$

This formula holds even when the two points *are* on a vertical or a horizontal line.

Example 4 Find the distance between $(4, -3)$ and $(-5, 4)$. Give an exact answer and an approximation to three decimal places.

We substitute into the distance formula:

$$d = \sqrt{(-5 - 4)^2 + [4 - (-3)]^2}$$ **Substituting**
$$= \sqrt{(-9)^2 + 7^2}$$
$$= \sqrt{130} \approx 11.402.$$ **Using a calculator**

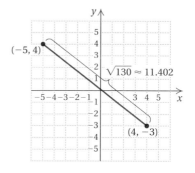

Do Exercises 4 and 5.

Find the distance between the pair of points. Where appropriate, give an approximation to three decimal places.

4. $(2, 6)$ and $(-4, -2)$

5. $(-2, 1)$ and $(4, 2)$

c | Midpoints of Segments

The distance formula can be used to derive a formula for finding the midpoint of a segment when the coordinates of the endpoints are known.

> **THE MIDPOINT FORMULA**
>
> If the endpoints of a segment are (x_1, y_1) and (x_2, y_2), then the coordinates of the midpoint are
> $$\left(\frac{x_1 + x_2}{2}, \frac{y_1 + y_2}{2} \right).$$
>
> (To locate the midpoint, determine the average of the x-coordinates and the average of the y-coordinates.)

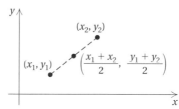

Example 5 Find the midpoint of the segment with endpoints $(-2, 3)$ and $(4, -6)$.

Using the midpoint formula, we obtain

$$\left(\frac{-2 + 4}{2}, \frac{3 + (-6)}{2} \right), \quad \text{or} \quad \left(\frac{2}{2}, \frac{-3}{2} \right), \quad \text{or} \quad \left(1, -\frac{3}{2} \right).$$

Do Exercises 6 and 7.

Find the midpoint of the segment with the given endpoints.

6. $(-3, 1)$ and $(6, -7)$

7. $(10, -7)$ and $(8, -3)$

Answers on page A-56

8. Find the center and the radius of the circle

$$(x - 5)^2 + \left(y + \tfrac{1}{2}\right)^2 = 9.$$

Then graph the circle.

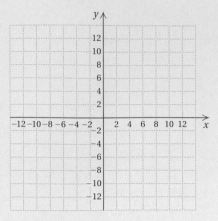

9. Find the center and the radius of the circle $x^2 + y^2 = 64$.

Answers on page A-56

d | Circles

Another conic section, or curve, shown in the figure at the beginning of this section is a *circle*. A **circle** is defined as the set of all points in a plane that are a fixed distance from a point in that plane.

Let's find an equation for a circle. We call the center (h, k) and let the radius have length r. Suppose that (x, y) is any point on the circle. By the distance formula, we have

$$\sqrt{(x - h)^2 + (y - k)^2} = r.$$

Squaring both sides gives an equation of the circle in standard form: $(x - h)^2 + (y - k)^2 = r^2$. When $h = 0$ and $k = 0$, the circle is centered at the origin. Otherwise, we can think of that circle being translated $|h|$ units horizontally and $|k|$ units vertically.

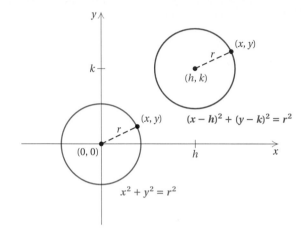

> **EQUATIONS OF CIRCLES**
>
> A circle centered at the origin with radius r has equation
> $$x^2 + y^2 = r^2.$$
> A circle with center (h, k) and radius r has equation
> $$(x - h)^2 + (y - k)^2 = r^2. \qquad \text{(Standard form)}$$

Example 6   Find the center and the radius and graph this circle:

$$(x + 2)^2 + (y - 3)^2 = 16.$$

First, we find an equivalent equation in standard form:

$$[x - (-2)]^2 + (y - 3)^2 = 4^2.$$

Thus the center is $(-2, 3)$ and the radius is 4. We draw the graph, shown at right, by locating the center and then using a compass, setting its radius at 4, to draw the circle.

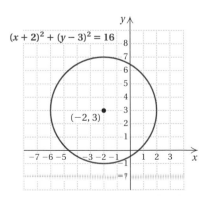

Do Exercises 8 and 9.

Example 7 Write an equation of a circle with center $(9, -5)$ and radius $\sqrt{2}$.

We use standard form $(x - h)^2 + (y - k)^2 = r^2$ and substitute:

$(x - 9)^2 + [y - (-5)]^2 = (\sqrt{2})^2$ **Substituting**

$(x - 9)^2 + (y + 5)^2 = 2.$ **Simplifying**

Do Exercise 10.

With certain equations not in standard form, we can complete the square to show that the equations are equations of circles. We proceed in much the same manner as we did in Section 7.6.

Example 8 Find the center and the radius and graph this circle:

$$x^2 + y^2 + 8x - 2y + 15 = 0.$$

First, we regroup the terms and then complete the square twice, once with $x^2 + 8x$ and once with $y^2 - 2y$:

$x^2 + y^2 + 8x - 2y + 15 = 0$

$(x^2 + 8x) + (y^2 - 2y) + 15 = 0$

$(x^2 + 8x + 0) + (y^2 - 2y + 0) + 15 = 0$ **Adding 0**

$(x^2 + 8x + 16 - 16) + (y^2 - 2y + 1 - 1) + 15 = 0$ $\left(\frac{8}{2}\right)^2 = 4^2 = 16; \left(\frac{-2}{2}\right)^2 = 1;$ **substituting 16 − 16 and 1 − 1**

$(x^2 + 8x + 16) + (y^2 - 2y + 1) - 16 - 1 + 15 = 0$ **Regrouping**

$(x + 4)^2 + (y - 1)^2 - 2 = 0$ **Factoring and simplifying**

$(x + 4)^2 + (y - 1)^2 = 2$ **Adding 2**

$[x - (-4)]^2 + (y - 1)^2 = (\sqrt{2})^2.$ **Writing standard form**

The center is $(-4, 1)$ and the radius is $\sqrt{2}$.

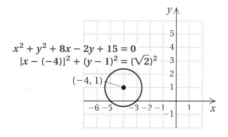

$x^2 + y^2 + 8x - 2y + 15 = 0$
$[x - (-4)]^2 + (y - 1)^2 = (\sqrt{2})^2$
$(-4, 1)$

Do Exercise 11.

10. Find an equation of a circle with center $(-3, 1)$ and radius 6.

11. Find the center and the radius of the circle

$$x^2 + 2x + y^2 - 4y + 2 = 0.$$

Then graph the circle.

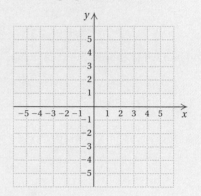

Answers on page A-56

Calculator Spotlight

Graphing Circles. Equations of circles are not functions, so they cannot be entered directly on a grapher. They can be graphed, however, using either of two methods. Suppose we want to graph the circle

$$(x - 3)^2 + (y + 1)^2 = 16.$$

One way to create a graph is to use the DRAW feature. We press $\boxed{\text{2nd}}$ $\boxed{\text{DRAW}}$ $\boxed{9}$ and enter h, k, and r; that is, 3, -1, and 4, respectively.

Another way to graph a circle is to first solve the equation for y. This gives us an expression with a $\pm$ sign in front of a radical. Then we graph two functions, one with the $+$ sign and one with the $-$ sign, using the same set of axes. To graph $(x - 3)^2 + (y + 1)^2 = 16$, we first solve for y:

$$(y + 1)^2 = 16 - (x - 3)^2$$
$$y + 1 = \pm\sqrt{16 - (x - 3)^2}$$
$$y = -1 \pm\sqrt{16 - (x - 3)^2}.$$

Then we graph the two functions

$$y_1 = -1 + \sqrt{16 - (x - 3)^2}$$

and

$$y_2 = -1 - \sqrt{16 - (x - 3)^2}$$

in a squared viewing window to eliminate distortion. This gives us the following graph.

$$y_1 = -1 + \sqrt{16 - (x - 3)^2},$$
$$y_2 = -1 - \sqrt{16 - (x - 3)^2}$$

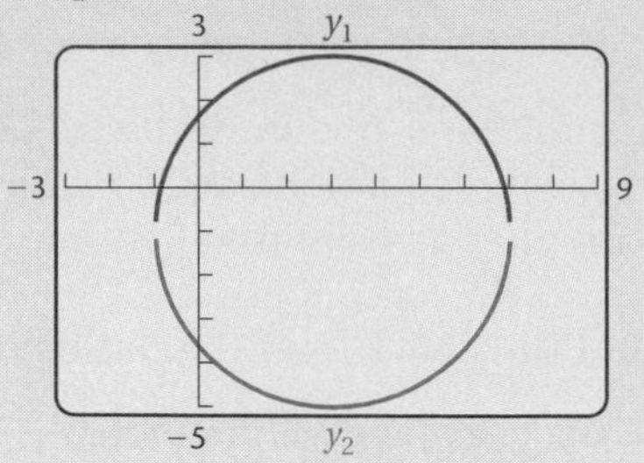

Some graphers will not connect the two pieces of the graph (as shown here) because of approximations made near the endpoints.

Exercises

Graph the circle. Try to use both methods.

1. $x^2 + y^2 - 16 = 0$
2. $4x^2 + 4y^2 = 100$
3. $x^2 + y^2 - 10x - 11 = 0$
4. $x^2 + y^2 + 14x - 16y + 54 = 0$

Exercise Set 9.1

a Graph the equation.

1. $y = x^2$

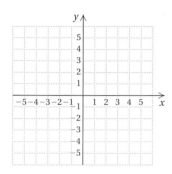

2. $x = y^2$

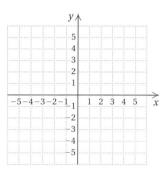

3. $x = y^2 + 4y + 1$

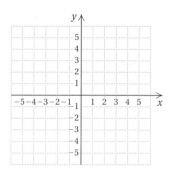

4. $y = x^2 - 2x + 3$

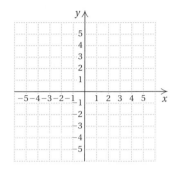

5. $y = -x^2 + 4x - 5$

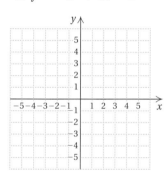

6. $x = 4 - 3y - y^2$

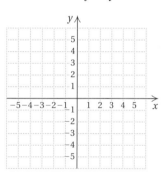

7. $x = -3y^2 - 6y - 1$

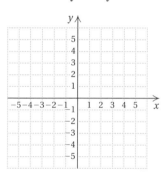

8. $y = -5 - 8x - 2x^2$

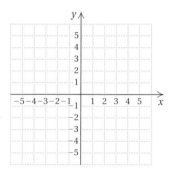

b Find the distance between the pair of points. Where appropriate, give an approximation to three decimal places.

9. $(6, -4)$ and $(2, -7)$

10. $(1, 2)$ and $(-4, 14)$

11. $(0, -4)$ and $(5, -6)$

12. $(8, 3)$ and $(8, -3)$

13. $(9, 9)$ and $(-9, -9)$

14. $(2, 22)$ and $(-8, 1)$

15. $(2.8, -3.5)$ and $(-4.3, -3.5)$

16. $(6.1, 2)$ and $(5.6, -4.4)$

17. $\left(\dfrac{5}{7}, \dfrac{1}{14}\right)$ and $\left(\dfrac{1}{7}, \dfrac{11}{14}\right)$

18. $(0, \sqrt{7})$ and $(\sqrt{6}, 0)$

19. $(-23, 10)$ and $(56, -17)$

20. $(34, -18)$ and $(-46, -38)$

21. (a, b) and $(0, 0)$

22. $(0, 0)$ and (p, q)

23. $(\sqrt{2}, -\sqrt{3})$ and $(-\sqrt{7}, \sqrt{5})$

24. $(\sqrt{8}, \sqrt{3})$ and $(-\sqrt{5}, -\sqrt{6})$

25. $(1000, -240)$ and $(-2000, 580)$

26. $(-3000, 560)$ and $(-430, -640)$

c Find the midpoint of the segment with the given endpoints.

27. $(-1, 9)$ and $(4, -2)$

28. $(5, 10)$ and $(2, -4)$

29. $(3, 5)$ and $(-3, 6)$

30. $(7, -3)$ and $(4, 11)$

31. $(-10, -13)$ and $(8, -4)$

32. $(6, -2)$ and $(-5, 12)$

33. $(-3.4, 8.1)$ and $(2.9, -8.7)$

34. $(4.1, 6.9)$ and $(5.2, -6.9)$

35. $\left(\dfrac{1}{6}, -\dfrac{3}{4}\right)$ and $\left(-\dfrac{1}{3}, \dfrac{5}{6}\right)$

36. $\left(-\dfrac{4}{5}, -\dfrac{2}{3}\right)$ and $\left(\dfrac{1}{8}, \dfrac{3}{4}\right)$

37. $(\sqrt{2}, -1)$ and $(\sqrt{3}, 4)$

38. $(9, 2\sqrt{3})$ and $(-4, 5\sqrt{3})$

d Find the center and the radius of the circle. Then graph the circle.

39. $(x + 1)^2 + (y + 3)^2 = 4$

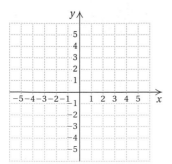

40. $(x - 2)^2 + (y + 3)^2 = 1$

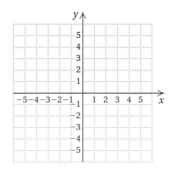

41. $(x - 3)^2 + y^2 = 2$

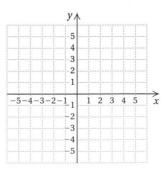

42. $x^2 + (y - 1)^2 = 3$

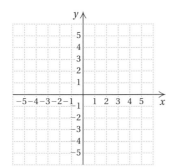

43. $x^2 + y^2 = 25$

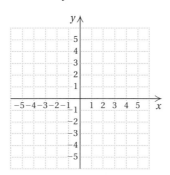

44. $x^2 + y^2 = 9$

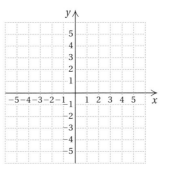

Find an equation of the circle having the given center and radius.

45. Center (0, 0), radius 7

46. Center (0, 0), radius 4

47. Center (−5, 3), radius $\sqrt{7}$

48. Center (4, 1), radius $3\sqrt{2}$

Find the center and the radius of the circle.

49. $x^2 + y^2 + 8x - 6y - 15 = 0$

50. $x^2 + y^2 + 6x - 4y - 15 = 0$

51. $x^2 + y^2 - 8x + 2y + 13 = 0$

52. $x^2 + y^2 + 6x + 4y + 12 = 0$

53. $x^2 + y^2 - 4x = 0$

54. $x^2 + y^2 + 10y - 75 = 0$

Solve.   [3.2a, b]

55. $x + y = 8,$
$\quad x - y = -24$

56. $y = 3x - 2,$
$\quad 2x - 4y = 50$

57. $2x + 3y = 8,$
$\quad x - 2y = -3$

58. $-4x + 12y = -9,$
$\quad x - 3y = 2$

Factor.   [4.5b]

59. $x^2 - 16$

60. $a^2 - 9b^2$

61. $64p^2 - 81q^2$

62. $400c^2d^2 - 225$

Synthesis

63. ◆ Is the center of a circle part of the circle? Why or why not?

64. ◆ 〰 How could a grapher be used to graph an equation of the form $x = ay^2 + by + c$?

Find an equation of a circle satisfying the given conditions.

65. Center $(0, 0)$, passing through $(1/4, \sqrt{31}/4)$

66. Center $(-4, 1)$, passing through $(2, -5)$

67. Center $(-3, -2)$ and tangent to the y-axis

68. The endpoints of a diameter are $(7, 3)$ and $(-1, -3)$.

Find the distance between the given points.

69. $(-1, 3k)$ and $(6, 2k)$

70. (a, b) and $(-a, -b)$

71. $(6m, -7n)$ and $(-2m, n)$

72. $(-3\sqrt{3}, 1 - \sqrt{6})$ and $(\sqrt{3}, 1 + \sqrt{6})$

If the sides of a triangle have lengths a, b, and c and $a^2 + b^2 = c^2$, then the triangle is a right triangle. Determine whether the given points are vertices of a right triangle.

73. $(-8, -5)$, $(6, 1)$, and $(-4, 5)$

74. $(9, 6)$, $(-1, 2)$, and $(1, -3)$

75. Find the midpoint of the segment with the endpoints $(2 - \sqrt{3}, 5\sqrt{2})$ and $(2 + \sqrt{3}, 3\sqrt{2})$.

76. Find the point on the y-axis that is equidistant from $(2, 10)$ and $(6, 2)$.

77. *Snowboarding.* Each side edge of a Burton® Twin snowboard is an arc of a circle. The Twin 53 has a "running length" of 1150 mm and a "sidecut depth" of 19.5 mm (see the figure below).

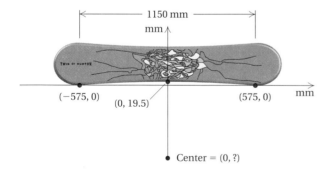

a) Using the coordinates shown, locate the center of the circle. (*Hint*: Equate distances.)
b) What radius is used for the edge of the board?

78. *Doorway Construction.* Ace Carpentry is to cut an arch for the top of an entranceway. The arch needs to be 8 ft wide and 2 ft high. To draw the arch, the carpenters will use a stretched string with chalk attached at an end as a compass.

a) Using a coordinate system, locate the center of the circle.
b) What radius should the carpenters use to draw the arch?

9.2 Conic Sections: Ellipses

a | Ellipses

When a cone is cut at an angle, as shown below, the conic section formed is an *ellipse*.

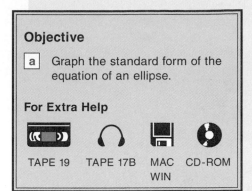
AN ELLIPSE IN
THREE DIMENSIONS

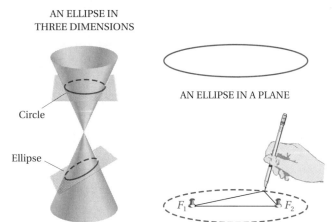

Circle

Ellipse

AN ELLIPSE IN A PLANE

You can draw an ellipse by securing two tacks in a piece of cardboard, tying a string around them, placing a pencil as shown, and drawing with the string kept taut. The formal mathematical definition is related to this method of drawing.

An **ellipse** is defined as the set of all points in a plane such that the *sum* of the distances from two fixed points F_1 and F_2 (called the **foci**; singular, **focus**) is constant. In the preceding drawing, the tacks are at the foci. Ellipses have equations as follows.

> **EQUATION OF AN ELLIPSE**
>
> An ellipse with its center at the origin has equation
>
> $$\frac{x^2}{a^2} + \frac{y^2}{b^2} = 1, \quad a, b > 0, \quad a \neq b. \qquad \text{(Standard form)}$$

We can think of a circle as a special kind of ellipse. A circle is formed when $a = b$ and the cutting plane is perpendicular to the axis of the cone. It is also formed when the foci, F_1 and F_2, are the same point. An ellipse with its foci close together is very nearly a circle.

When graphing ellipses, it helps to first find the intercepts. If we replace x with 0 in the standard form of the equation, we can find the y-intercepts:

$$\frac{0^2}{a^2} + \frac{y^2}{b^2} = 1$$

$$\frac{y^2}{b^2} = 1$$

$$y^2 = b^2$$

$$y = \pm b.$$

Thus the y-intercepts are $(0, b)$ and $(0, -b)$. Similarly, the x-intercepts are $(a, 0)$ and $(-a, 0)$. If $a > b$, the ellipse is horizontal and $(-a, 0)$ and $(a, 0)$ are the **vertices** (singular, **vertex**). If $b > a$, the ellipse is vertical and $(0, -b)$ and $(0, b)$ are the vertices.

Graph the ellipse.

1. $\dfrac{x^2}{9} + \dfrac{y^2}{4} = 1$

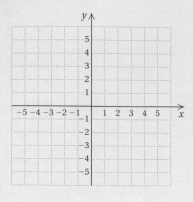

2. $\dfrac{x^2}{9} + \dfrac{y^2}{25} = 1$

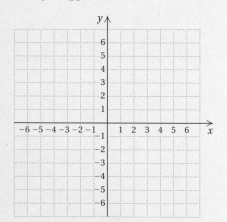

Answers on page A-56

Plotting these points and filling in an oval-shaped curve, we get a graph of the ellipse. If a more precise graph is desired, we can plot more points.

> For the ellipse
> $$\dfrac{x^2}{a^2} + \dfrac{y^2}{b^2} = 1,$$
> the **x-intercepts** are $(-a, 0)$ and $(a, 0)$, and the **y-intercepts** are $(0, -b)$ and $(0, b)$.

Example 1   Graph: $\dfrac{x^2}{4} + \dfrac{y^2}{9} = 1$.

Note that

$$\dfrac{x^2}{4} + \dfrac{y^2}{9} = \dfrac{x^2}{2^2} + \dfrac{y^2}{3^2}.$$

Thus the x-intercepts are $(-2, 0)$ and $(2, 0)$, and the y-intercepts are $(0, -3)$ and $(0, 3)$. We plot these points and connect them with an oval-shaped curve. To be accurate, we might find some other points on the curve. We let $x = 1$ and solve for y:

$$\dfrac{1^2}{4} + \dfrac{y^2}{9} = 1$$

$$36\left(\dfrac{1}{4} + \dfrac{y^2}{9}\right) = 36 \cdot 1$$

$$36 \cdot \dfrac{1}{4} + 36 \cdot \dfrac{y^2}{9} = 36$$

$$9 + 4y^2 = 36$$

$$4y^2 = 27$$

$$y^2 = \dfrac{27}{4}$$

$$y = \pm\sqrt{\dfrac{27}{4}} \approx \pm 2.6.$$

Thus, $(1, 2.6)$ and $(1, -2.6)$ can also be plotted and used to draw the graph. Similarly, the points $(-1, -2.6)$ and $(-1, 2.6)$ can also be computed and plotted.

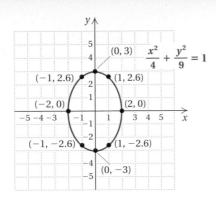

$$\frac{x^2}{4} + \frac{y^2}{9} = 1$$

3. Graph: $16x^2 + 9y^2 = 144$.

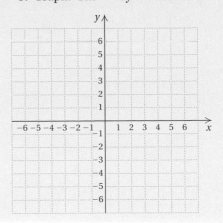

Do Exercises 1 and 2 on the preceding page.

Example 2 Graph: $4x^2 + 25y^2 = 100$.

To write the equation in standard form, we multiply by $\frac{1}{100}$ on both sides:

$$\frac{1}{100}(4x^2 + 25y^2) = \frac{1}{100}(100) \qquad \begin{array}{l}\text{Multiplying by } \frac{1}{100} \\ \text{to get 1 on the right side}\end{array}$$

$$\left.\begin{array}{c}\dfrac{1}{100}(4x^2) + \dfrac{1}{100}(25y^2) = 1 \\[2mm] \dfrac{x^2}{25} + \dfrac{y^2}{4} = 1\end{array}\right\} \qquad \text{Simplifying}$$

$$\frac{x^2}{5^2} + \frac{y^2}{2^2} = 1. \qquad a = 5, b = 2$$

The x-intercepts are $(-5, 0)$ and $(5, 0)$, and the y-intercepts are $(0, -2)$ and $(0, 2)$. We plot the intercepts and connect them with an oval-shaped curve. Other points can also be computed and plotted.

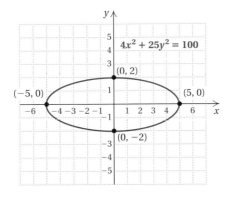

Do Exercise 3.

Horizontal and vertical translations, similar to those used in Chapter 7, can be used to graph ellipses that are not centered at the origin.

The standard form of a horizontal or vertical ellipse centered at (h, k) is

$$\frac{(x - h)^2}{a^2} + \frac{(y - k)^2}{b^2} = 1.$$

The vertices are $(h + a, k)$ and $(h - a, k)$ if horizontal; $(h, k + b)$ and $(h, k - b)$ if vertical.

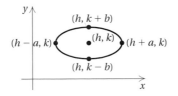

Answer on page A-57

4. Graph:

$$\frac{(x + 2)^2}{16} + \frac{(y - 3)^2}{9} = 1.$$

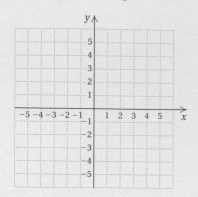

Example 3 Graph: $\frac{(x - 1)^2}{4} + \frac{(y + 5)^2}{9} = 1$.

Note that

$$\frac{(x - 1)^2}{4} + \frac{(y + 5)^2}{9} = \frac{(x - 1)^2}{2^2} + \frac{(y + 5)^2}{3^2}.$$

Thus, $a = 2$ and $b = 3$. To determine the center of the ellipse, (h, k), note that

$$\frac{(x - 1)^2}{2^2} + \frac{(y + 5)^2}{3^2} = \frac{(x - 1)^2}{2^2} + \frac{(y - (-5))^2}{3^2}.$$

Thus the center is $(1, -5)$. We locate $(1, -5)$ and then plot $(1 + 2, -5)$, $(1 - 2, -5)$, $(1, -5 + 3)$, and $(1, -5 - 3)$. These are the points $(3, -5)$, $(-1, -5)$, $(1, -2)$, and $(1, -8)$.

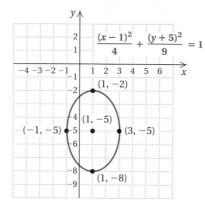

Note that this ellipse is the same as the ellipse in Example 1 but translated 1 unit to the right and 5 units down.

Do Exercise 4.

Calculator Spotlight

Graphing Ellipses. Graphing an ellipse is much like graphing a circle: We graph it in two pieces after solving for *y*. Let's do this for the equation in Example 2:

$$4x^2 + 25y^2 = 100$$

$$25y^2 = 100 - 4x^2$$

$$y^2 = \frac{100 - 4x^2}{25}$$

$$y = \pm\sqrt{\frac{100 - 4x^2}{25}}, \text{ or } \pm\sqrt{4 - \frac{4}{25}x^2}.$$

We let

$$y_1 = \sqrt{4 - \frac{4}{25}x^2} \quad \text{and} \quad y_2 = -\sqrt{4 - \frac{4}{25}x^2}$$

and graph them in a squared window.

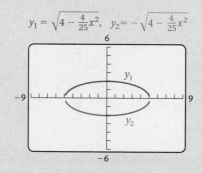

Exercises

Graph the ellipse.

1. $\frac{x^2}{9} + \frac{y^2}{4} = 1$ **2.** $16x^2 + 9y^2 = 1$

3. $\frac{(x - 1)^2}{4} + \frac{(y + 5)^2}{9} = 1$ **4.** $\frac{(x + 2)^2}{16} + \frac{(y - 3)^2}{9} = 1$

Answer on page A-57

Ellipses have many applications. The orbits of planets and some comets around the sun are ellipses. The sun is located at one focus. Whispering galleries are also ellipses. A person standing at one focus will be able to hear the whisper of a person standing at the other focus. One example of a whispering gallery is found in the rotunda of the capital building in Washington, D.C.

Planetary orbit

Whispering gallery

Wind-driven forest fires can be roughly approximated as the union of "half"-ellipses. Shown below is an illustration of such a forest fire. The smaller half-ellipse on the left moves into the wind, and the elongated half-ellipse on the right moves out in the direction of the wind. The wind tends to spread the fire to the right, but part of the fire will still spread into the wind.

Wind Direction

Source: "Predicting Wind-Driven Wild Land Fire Size and Shape," Hal E. Anderson, Research Paper INT-305, U.S. Department of Agriculture, Forest Service, February 1983.

Improving Your Math Study Skills

Study Tips for Trouble Spots

By now you have probably encountered certain topics that gave you more difficulty than others. It is important to know that this happens to every person who studies mathematics. Unfortunately, frustration is often part of the learning process and it is important not to give up when difficulty arises.

One source of frustration for many students is not being able to set aside sufficient time for studying. Family commitments, work schedules, and athletics are just a few of the time demands that many students face. Couple these demands with a math lesson that seems to require a greater than usual amount of study time, and it is no wonder that many students often feel frustrated. Below are some study tips that might be useful if troubles arise.

- **Realize that everyone—even your instructor—has been stymied at times when studying math.** You are not the first person, nor will you be the last, to encounter a "roadblock."

- **Whether working alone or with a classmate, try to allow enough study time so that you won't need to constantly glance at a clock.** Difficult material is best mastered when your mind is completely focused on the subject matter. Thus, if you are tired, it is usually best to study early the next morning or to take a ten-minute "power-nap" in order to make the most productive use of your time.

- **Talk about your trouble spot with a classmate.** It is possible that she or he is also having difficulty with the same material. If that is the case, perhaps the majority of your class is confused and the class can ask the instructor to go over the material again. If your classmate *does* understand the topic that is troubling you, patiently allow him or her to explain it to you. By verbalizing the math in question, your classmate may help clarify the material for both of you. Perhaps you will be able to return the favor for your classmate when he or she is struggling with a topic that you understand.

- **Try to study in a "controlled" environment.** What we mean by this is to put yourself in a setting that will enable you to maximize your powers of concentration. For example, whereas some students may succeed in studying at home or in a dorm room, for many these settings are filled with distractions. Consider a trip to a library, classroom building, or perhaps the attic or basement if such a setting is more conducive to studying. If you plan on working with a classmate, try to find a location in which conversation will not be bothersome to others.

- **When working on difficult material, it is often helpful to first "back up" and review the most recent material that *did* make sense.** This can build your confidence and create a momentum that can often carry you through the roadblock. Sometimes a small piece of information that appeared in a previous section is all that is needed for your problem spot to disappear. When the difficult material is finally mastered, try to make use of what is fresh in your mind by taking a "sneak preview" of what your next topic for study will be.

Exercise Set 9.2

a Graph the ellipse.

1. $\dfrac{x^2}{9} + \dfrac{y^2}{36} = 1$

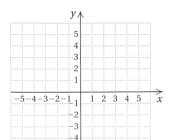

2. $\dfrac{x^2}{16} + \dfrac{y^2}{25} = 1$

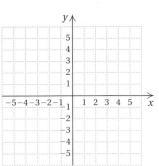

3. $\dfrac{x^2}{1} + \dfrac{y^2}{4} = 1$

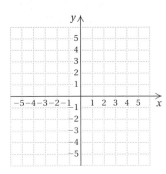

4. $\dfrac{x^2}{4} + \dfrac{y^2}{1} = 1$

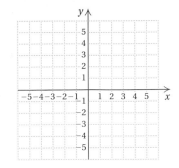

5. $4x^2 + 9y^2 = 36$
(*Hint*: Divide by 36.)

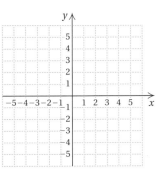

6. $9x^2 + 4y^2 = 36$

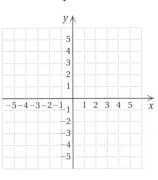

7. $x^2 + 4y^2 = 4$

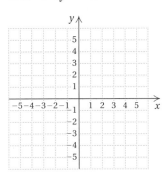

8. $9x^2 + 16y^2 = 144$

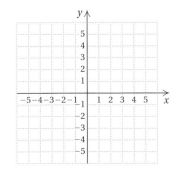

9. $2x^2 + 3y^2 = 6$

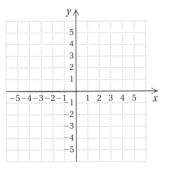

10. $5x^2 + 7y^2 = 35$

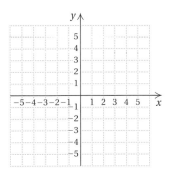

11. $12x^2 + 5y^2 - 120 = 0$

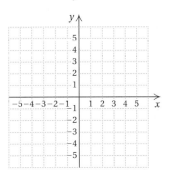

12. $3x^2 + 7y^2 - 63 = 0$

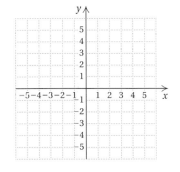

13. $\dfrac{(x-2)^2}{9} + \dfrac{(y-1)^2}{25} = 1$

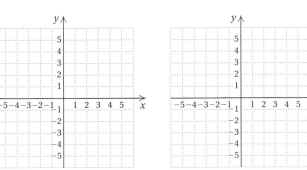

14. $\dfrac{(x-3)^2}{4} + \dfrac{(y-4)^2}{9} = 1$

15. $\dfrac{(x+1)^2}{16} + \dfrac{(y+2)^2}{25} = 1$

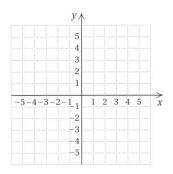

16. $\dfrac{(x+3)^2}{4} + \dfrac{(y-2)^2}{36} = 1$

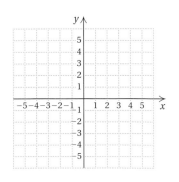

17. $12(x - 1)^2 + 3(y + 2)^2 = 48$

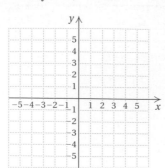

18. $4(x - 2)^2 + 9(y + 2)^2 = 36$

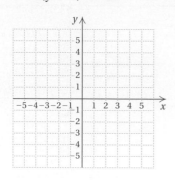

19. $(x + 3)^2 + 4(y + 1)^2 - 10 = 6$

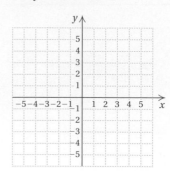

20. $8(x + 1)^2 + (y + 1)^2 - 12 = 4$

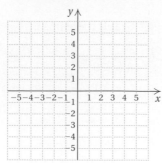

Skill Maintenance

Solve. Give exact solutions.   [7.2a]

21. $3x^2 - 2x + 7 = 0$

22. $3x^2 - 12x + 7 = 0$

23. $x^2 + 2x - 20 = 0$

24. $x^2 + 2x = 10$

Solve. Give both exact and approximate solutions to the nearest tenth.   [7.2a]

25. $x^2 - 2x = 10$

26. $x^2 + 2x - 17 = 0$

27. $3x^2 - 11x + 7 = x$

28. $3x^2 - 12x + 7 = 10 - x^2 + 5x$

Synthesis

29. ◆ An eccentric person builds a pool table in the shape of an ellipse with a hole at one focus and a tiny dot at the other. Guests are amazed at how many bank shots the owner of the pool table makes. Explain how this can happen.

30. ◆ Explain how a circle can be thought of as a special kind of ellipse.

Find an equation of an ellipse that contains the following points.

31. $(-9, 0)$, $(9, 0)$, $(0, -11)$, and $(0, 11)$

32. $(-7, 0)$, $(7, 0)$, $(0, -5)$, and $(0, 5)$

33. $(-2, -1)$, $(6, -1)$, $(2, -4)$, and $(2, 2)$

34. *Planetary Motion.* The maximum distance of the planet Mars from the sun is 2.48×10^8 mi. The minimum distance is 3.46×10^7 mi. The sun is at one focus of the elliptical orbit. Find the distance from the sun to the other focus.

For each of the following equations, complete the square as needed and find an equivalent equation in standard form.

35. $x^2 - 4x + 4y^2 + 8y - 8 = 0$

36. $4x^2 + 24x + y^2 - 2y - 63 = 0$

37. 〰 Use a grapher to check your answers to Exercises 7, 12, and 20.

38. ◆ 〰 *Wind-Driven Forest Fires*

a) Graph the wind-driven fire formed as the union of the following two curves:

$$\frac{x^2}{10.3^7} + \frac{y^2}{4.8^7} = 1, \quad x \geq 0; \qquad \frac{x^2}{3.6^9} + \frac{y^2}{4.8^9} = 1, \quad x \leq 0,$$

b) What other factors do you think affect the shape of forest fires?

Collaborative
Learning Manual

Investigate the effect on the graph of an ellipse of varying a and b.

9.3 Conic Sections: Hyperbolas

a | Hyperbolas

A **hyperbola** looks like a pair of parabolas, but the actual shapes are different. A hyperbola has two **vertices** and the line through the vertices is known as an **axis**. The point halfway between the vertices is called the **center**.

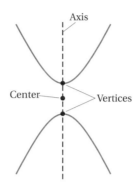

Parabola   Hyperbola in three dimensions   Hyperbola in a plane

> ▶ **EQUATIONS OF HYPERBOLAS**
>
> Hyperbolas with their centers at the origin have equations as follows:
>
> $$\frac{x^2}{a^2} - \frac{y^2}{b^2} = 1; \qquad \text{(Axis horizontal)}$$
>
> $$\frac{y^2}{b^2} - \frac{x^2}{a^2} = 1. \qquad \text{(Axis vertical)}$$

To graph a hyperbola, it helps to begin by graphing two lines called **asymptotes**. Although the asymptotes themselves are not part of the graph, they serve as guidelines for an accurate sketch.

> ▶ **ASYMPTOTES OF A HYPERBOLA**
>
> For hyperbolas with equations as given above, the **asymptotes** are the lines
>
> $$y = \frac{b}{a}x \quad \text{and} \quad y = -\frac{b}{a}x.$$

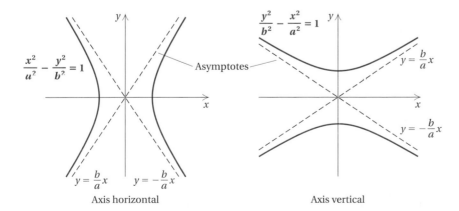

Axis horizontal   Axis vertical

Objectives

a Graph the standard form of the equation of a hyperbola.

b Graph equations (nonstandard form) of hyperbolas.

For Extra Help

TAPE 19   TAPE 17B   MAC WIN   CD-ROM

1. Graph: $\dfrac{x^2}{16} - \dfrac{y^2}{25} = 1$.

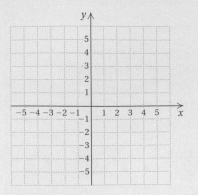

As a hyperbola gets further away from the origin, it gets closer and closer to its asymptotes. The larger $|x|$ gets, the closer the graph gets to an asymptote. The asymptotes act to "constrain" the graph of a hyperbola. Parabolas are *not* constrained by any asymptotes.

The next thing to do after sketching asymptotes is to plot vertices. Then it is easy to sketch the curve.

Example 1   Graph: $\dfrac{x^2}{4} - \dfrac{y^2}{9} = 1$.

Note that

$$\frac{x^2}{4} - \frac{y^2}{9} = \frac{x^2}{2^2} - \frac{y^2}{3^2}, \qquad \textbf{Identifying } \textit{a} \textbf{ and } \textit{b}$$

so $a = 2$ and $b = 3$. The asymptotes are thus

$$y = \frac{3}{2}x \quad \text{and} \quad y = -\frac{3}{2}x.$$

We sketch them, as shown in the graph on the left below.

For horizontal or vertical hyperbolas centered at the origin, the vertices also serve as intercepts. Since this hyperbola is horizontal, we replace y with 0 and solve for x. We see that $x^2/2^2 = 1$ when $x = \pm 2$. The intercepts are $(2, 0)$ and $(-2, 0)$. You can check that no y-intercepts exist.

Finally, we plot the intercepts and sketch the graph. Through each intercept, we draw a smooth curve that approaches the asymptotes closely, as shown.

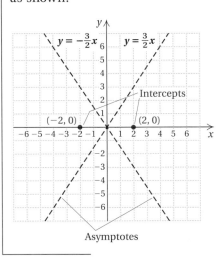

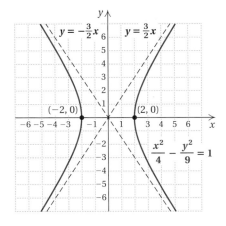

Do Exercise 1.

Example 2   Graph: $\dfrac{y^2}{36} - \dfrac{x^2}{4} = 1$.

Note that

$$\frac{y^2}{36} - \frac{x^2}{4} = \frac{y^2}{6^2} - \frac{x^2}{2^2} = 1.$$

> The intercept distance is found in the term without the minus sign. Here there is a y in this term, so the intercepts are on the y-axis.

The asymptotes are thus $y = \frac{6}{2}x$ and $y = -\frac{6}{2}x$, or $y = 3x$ and $y = -3x$.

Answer on page A-57

The numbers 6 and 2 can be used to sketch a rectangle that helps with graphing. Using ± 2 as x-coordinates and ± 6 as y-coordinates, we form all possible ordered pairs: $(2, 6)$, $(2, -6)$, $(-2, 6)$, and $(-2, -6)$. We plot these pairs and lightly sketch a rectangle through them. The asymptotes pass through the corners (see the figure on the left below). Since the hyperbola is vertical, we plot its y-intercepts, $(0, 6)$ and $(0, -6)$. Finally, we draw curves through the intercepts toward the asymptotes, as shown below.

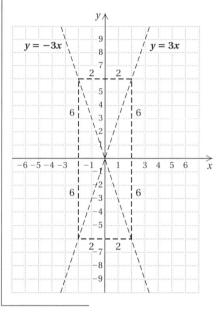
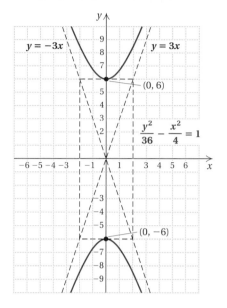

Although we will not consider these equations here, hyperbolas with center at (h, k) are given by

$$\frac{(x - h)^2}{a^2} - \frac{(y - k)^2}{b^2} = 1 \quad \text{or} \quad \frac{(y - k)^2}{b^2} - \frac{(x - h)^2}{a^2} = 1.$$

Do Exercise 2.

b Hyperbolas (Nonstandard Form)

The equations for hyperbolas just examined are the standard ones, but there are other hyperbolas. We consider some of them.

> Hyperbolas having the x- and y-axes as asymptotes have equations as follows:
>
> $$xy = c, \quad \text{where } c \text{ is a nonzero constant.}$$

Example 3 Graph: $xy = -8$.

We first solve for y:

$$y = -\frac{8}{x}. \qquad \textbf{Dividing by } x \textbf{ on both sides. Note that } x \neq 0.$$

Next, we find some solutions, keeping the results in a table. Note that x cannot be 0 and that for large values of $|x|$, y will be close to 0. Thus the x- and y-axes serve as asymptotes. We plot the points and draw the hyperbola.

2. Graph.

a) $\dfrac{y^2}{9} - \dfrac{x^2}{49} = 1$

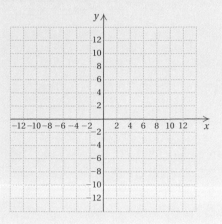

b) $\dfrac{x^2}{49} - \dfrac{y^2}{9} = 1$

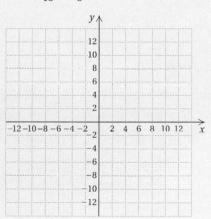

Answers on page A-57

3. Graph: $xy = 8$.

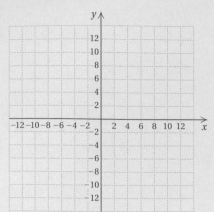

x	y
2	−4
−2	4
4	−2
−4	2
1	−8
−1	8
8	−1
−8	1

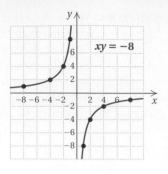

Do Exercise 3.

Graphs of $xy = k$, for $x \geq 0$, $y \geq 0$, and $k > 0$, describe inverse variation (see Section 5.8).

Hyperbolas have many applications. A jet breaking the sound barrier creates a sonic boom with a wave front the shape of a cone. The intersection of the cone with the ground is one branch of a hyperbola. Some comets travel in hyperbolic orbits, and a cross section of certain lenses may be hyperbolic in shape.

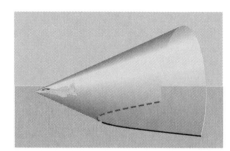
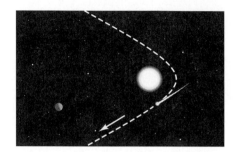

Calculator Spotlight

Graphing Hyperbolas. Graphing hyperbolas is similar to graphing circles and ellipses. First, we solve for y and then graph the two parts of the graph. Let's graph

$$\frac{x^2}{25} - \frac{y^2}{49} = 1.$$

Solving for y, we get

$$y_1 = \frac{7}{5} \sqrt{x^2 - 25}$$

and

$$y_2 = -\frac{7}{5} \sqrt{x^2 - 25}.$$

Now graph the two pieces using a squared viewing window. This gives us the following graph. Note again that the grapher has not connected the pieces at the points where the graph is nearly vertical.

$$y_1 = \frac{7}{5}\sqrt{x^2 - 25}, \ y_2 = -\frac{7}{5}\sqrt{x^2 - 25}$$

Exercises

Graph the hyperbola.

1. $\dfrac{x^2}{16} - \dfrac{y^2}{60} = 1$ **2.** $\dfrac{y^2}{20} - \dfrac{x^2}{64} = 1$

3. $16x^2 - 3y^2 = 48$ **4.** $45y^2 - 9x^2 = 405$

Answer on page A-57

Exercise Set 9.3

a Graph the hyperbola.

1. $\dfrac{y^2}{9} - \dfrac{x^2}{9} = 1$

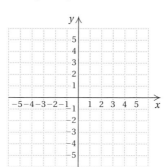

2. $\dfrac{x^2}{16} - \dfrac{y^2}{16} = 1$

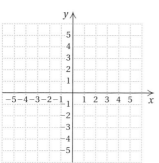

3. $\dfrac{x^2}{4} - \dfrac{y^2}{25} = 1$

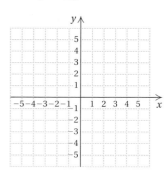

4. $\dfrac{y^2}{16} - \dfrac{x^2}{9} = 1$

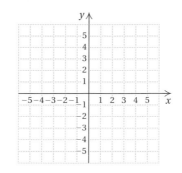

5. $\dfrac{y^2}{36} - \dfrac{x^2}{9} = 1$

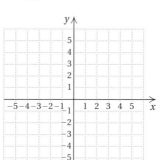

6. $\dfrac{x^2}{25} - \dfrac{y^2}{36} = 1$

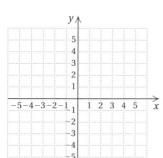

7. $y^2 - x^2 = 25$

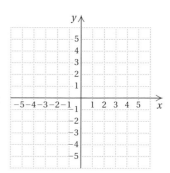

8. $x^2 - y^2 = 4$

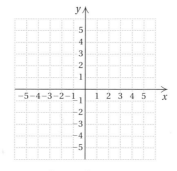

9. $x^2 = 1 + y^2$

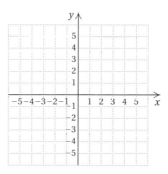

10. $9y^2 = 36 + 4x^2$

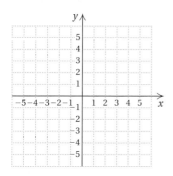

11. $25x^2 - 16y^2 = 400$

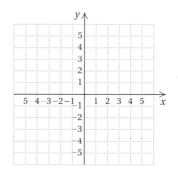

12. $4y^2 - 9x^2 = 36$

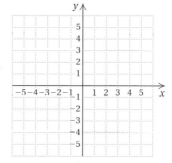

b Graph the hyperbola.

13. $xy = -4$

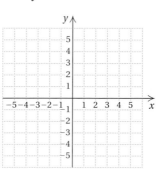

14. $xy = 6$

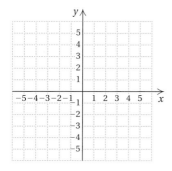

15. $xy = 3$

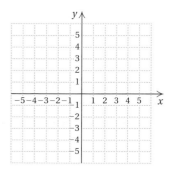

16. $xy = -9$

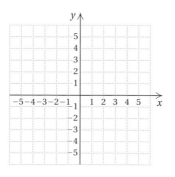

17. $xy = -2$

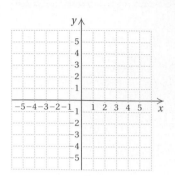

18. $xy = -1$

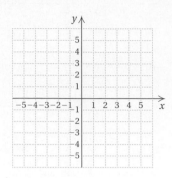

19. $xy = \dfrac{1}{2}$

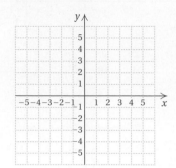

20. $xy = \dfrac{3}{4}$

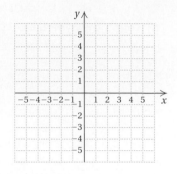

Skill Maintenance

Convert to a logarithmic equation.   [8.3b]

21. $5^2 = 25$

22. $8^a = 17$

23. $a^{-t} = b$

24. $e^{3k} = Q$

25. $10^{2.3} = 199.53$

26. $e^3 = 20.086$

27. $a^{23.4} = 200$

28. $e^{-4} = 0.0183$

Convert to an exponential equation.   [8.3b]

29. $\ln 24 = 3.1781$

30. $\log 1000 = 3$

31. $\log_a N = t$

32. $p = \log_e W$

33. $\log M = b$

34. $\ln P = kt$

35. $\log 456 = 2.659$

36. $\log_a Y = t$

Synthesis

37. ◈ Consider the standard equations of a circle, a parabola, an ellipse, and a hyperbola. Which, if any, are functions? Explain.

38. ◈ If, in
$$\frac{x^2}{a^2} - \frac{y^2}{b^2} = 1,$$
$a = b$, what are the asymptotes of the graph? Explain.

39. 〰 Use a grapher to check your answers to Exercises 1, 8, 12, and 20.

40. Graph: $\dfrac{(x-2)^2}{16} - \dfrac{(y-2)^2}{9} = 1$.

Classify the graph of each of the following equations as a circle, an ellipse, a parabola, or a hyperbola.

41. $x^2 + y^2 - 10x + 8y - 40 = 0$

42. $y + 1 = 2x^2$

43. $1 - 3y = 2y^2 - x$

44. $9x^2 - 4y^2 - 36x + 24y - 36 = 0$

45. $4x^2 + 25y^2 - 8x - 100y + 4 = 0$

46. $\dfrac{x^2}{7} + \dfrac{y^2}{7} = 1$

9.4 Nonlinear Systems of Equations

All the systems of equations we studied in Chapter 3 were linear. We now consider systems of two equations in two variables in which at least one equation is nonlinear.

a | Algebraic Solutions

We first consider systems of one first-degree and one second-degree equation. For example, the graphs may be a circle and a line. If so, there are three possibilities for solutions.

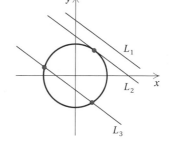

For L_1 there is no point of intersection of the line and the circle, hence no solution of the system in the set of real numbers. For L_2 there is one point of intersection, hence one real-number solution. For L_3 there are two points of intersection, hence two real-number solutions.

These systems can be solved graphically by finding the points of intersection. In solving algebraically, we use the substitution method.

Example 1   Solve this system:

$$x^2 + y^2 = 25, \quad \textbf{(1)} \qquad \text{(The graph is a circle.)}$$
$$3x - 4y = 0. \quad \textbf{(2)} \qquad \text{(The graph is a line.)}$$

We first solve the linear equation (2) for x:

$$x = \tfrac{4}{3}y. \quad \textbf{(3)}$$

We then substitute $\tfrac{4}{3}y$ for x in equation (1) and solve for y:

$$\left(\tfrac{4}{3}y\right)^2 + y^2 = 25$$
$$\tfrac{16}{9}y^2 + y^2 = 25$$
$$\tfrac{25}{9}y^2 = 25$$
$$y^2 = 9$$
$$y = \pm 3.$$

Now we substitute these numbers for y in equation (3) and solve for x:

$$x = \tfrac{4}{3}(3) = 4; \qquad x = \tfrac{4}{3}(-3) = -4.$$

CHECK: For (4, 3):

$x^2 + y^2 = 25$		$3x - 4y = 0$	
$4^2 + 3^2$	25	$3(4) - 4(3)$	0
$16 + 9$		$12 - 12$	
25	TRUE	0	TRUE

For (−4, −3):

$x^2 + y^2 = 25$		$3x - 4y = 0$	
$(-4)^2 + (-3)^2$	25	$3(-4) - 4(-3)$	0
$16 + 9$		$-12 + 12$	
25	TRUE	0	TRUE

Objectives

a Solve systems of equations where at least one equation is nonlinear.

b Solve applied problems involving nonlinear systems.

For Extra Help

TAPE 19   TAPE 18A   MAC   CD-ROM
                          WIN

Solve. Sketch the graphs to confirm the solutions.

1. $x^2 + y^2 = 25,$
   $y - x = -1$

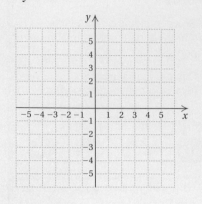

2. $y = x^2 - 2x - 1,$
   $y = x + 3$

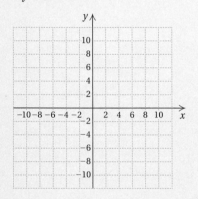

Answers on page A-58

The pairs $(4, 3)$ and $(-4, -3)$ check, so they are solutions. We can see the solutions in the graph. The graph of equation (1) is a circle, and the graph of equation (2) is a line. The graphs intersect at the points $(4, 3)$ and $(-4, -3)$.

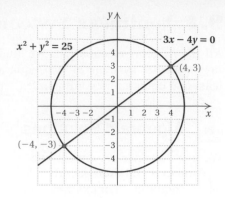

Do Exercises 1 and 2.

Example 2   Solve this system:

$$y + 3 = 2x, \qquad \textbf{(1)}$$
$$x^2 + 2xy = -1. \qquad \textbf{(2)}$$

We first solve the linear equation (1) for y:

$$y = 2x - 3. \qquad \textbf{(3)}$$

We then substitute $2x - 3$ for y in equation (2) and solve for x:

$$x^2 + 2x(2x - 3) = -1$$
$$x^2 + 4x^2 - 6x = -1$$
$$5x^2 - 6x + 1 = 0$$
$$(5x - 1)(x - 1) = 0 \qquad \text{Factoring}$$
$$5x - 1 = 0 \quad or \quad x - 1 = 0 \qquad \text{Using the principle of zero products}$$
$$x = \tfrac{1}{5} \quad or \qquad x = 1.$$

Now we substitute these numbers for x in equation (3) and solve for y:

$$y = 2\left(\tfrac{1}{5}\right) - 3 = -\tfrac{13}{5}; \qquad y = 2(1) - 3 = -1.$$

The check is left to the student. The pairs $\left(\tfrac{1}{5}, -\tfrac{13}{5}\right)$ and $(1, -1)$ are solutions.

Do Exercise 3 on the following page.

Calculator Spotlight

Solving Nonlinear Systems.   Because the algebra is sometimes complicated, a grapher can be useful in solving nonlinear systems. Consider the system in Example 2. We first solve for y:

$$y_1 = 2x - 3 \quad \text{and} \quad y_2 = \frac{-1 - x^2}{2x}.$$

We then graph both equations. Using the INTERSECT feature, we find the solutions to be $(0.2, -2.6)$ and $(1, -1)$.

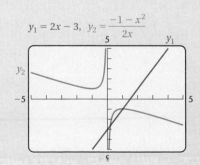

Exercises

Solve the system using a grapher.

1. $4xy - 7 = 0, \quad x - 3y - 2 = 0$

2. $x^2 + y^2 = 14, \quad 16x + 7y^2 = 0$

Example 3 Solve this system:

$$x + y = 5, \quad \text{(The graph is a line.)}$$
$$y = 3 - x^2. \quad \text{(The graph is a parabola.)}$$

We substitute $3 - x^2$ for y in the first equation:

$$x + 3 - x^2 = 5$$
$$-x^2 + x - 2 = 0$$
$$x^2 - x + 2 = 0.$$

To solve this equation, we need the quadratic formula:

$$x = \frac{-b \pm \sqrt{b^2 - 4ac}}{2a} = \frac{-(-1) \pm \sqrt{(-1)^2 - 4(1)(2)}}{2(1)}$$

$$= \frac{1 \pm \sqrt{1 - 8}}{2} = \frac{1 \pm \sqrt{-7}}{2} = \frac{1}{2} \pm \frac{\sqrt{7}}{2}i.$$

Then solving the first equation for y, we obtain $y = 5 - x$. Substituting values for x gives us

$$y = 5 - \left(\frac{1}{2} + \frac{\sqrt{7}}{2}i \right) = \frac{9}{2} - \frac{\sqrt{7}}{2}i$$

and

$$y = 5 - \left(\frac{1}{2} - \frac{\sqrt{7}}{2}i \right) = \frac{9}{2} + \frac{\sqrt{7}}{2}i.$$

The solutions are

$$\left(\frac{1}{2} + \frac{\sqrt{7}}{2}i, \frac{9}{2} - \frac{\sqrt{7}}{2}i \right)$$

and

$$\left(\frac{1}{2} - \frac{\sqrt{7}}{2}i, \frac{9}{2} + \frac{\sqrt{7}}{2}i \right).$$

There are no real-number solutions. Note in the figure at right that the graphs do not intersect. Getting only nonreal complex-number solutions tells us that the graphs do not intersect.

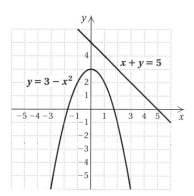

Do Exercise 4.

Two second-degree equations can have common solutions in various ways. If the graphs happen to be a circle and a hyperbola, for example, there are six possibilities, as shown below.

4 real solutions   3 real solutions   2 real solutions

2 real solutions   1 real solution   0 real solutions

3. Solve:

$$y + 3x = 1,$$
$$x^2 - 2xy = 5.$$

4. Solve:

$$9x^2 - 4y^2 = 36,$$
$$5x + 2y = 0.$$

Answers on page A-58

5. Solve:

$$2y^2 - 3x^2 = 6,$$
$$5y^2 + 2x^2 = 53.$$

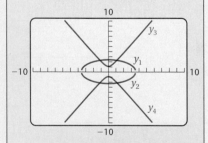
Answer on page A-58

To solve systems of two second-degree equations, we can use either the substitution method or the elimination method. The elimination method is generally used when each equation is of the form $Ax^2 + By^2 = C$. Then we can eliminate an x^2- or a y^2-term in a manner similar to the procedure that we used for systems of linear equations in Chapter 3.

Example 4 Solve:

$$2x^2 + 5y^2 = 22, \qquad \textbf{(1)}$$
$$3x^2 - y^2 = -1. \qquad \textbf{(2)}$$

In this case, we use the elimination method:

$$
\begin{array}{ll}
2x^2 + 5y^2 = 22 & \\
\underline{15x^2 - 5y^2 = -5} & \text{Multiplying by 5 on both sides of equation (2)} \\
17x^2 \qquad\;\; = 17 & \text{Adding} \\
\qquad x^2 = 1 & \\
\qquad x = \pm 1. & \\
\end{array}
$$

If $x = 1$, $x^2 = 1$, and if $x = -1$, $x^2 = 1$, so substituting either 1 or -1 for x in equation (2) gives us

$$
\begin{array}{ll}
3x^2 - y^2 = -1 & \\
3 \cdot 1 - y^2 = -1 & \text{Substituting 1 for } x^2 \\
3 - y^2 = -1 & \\
-y^2 = -4 & \\
y^2 = 4 & \\
y = \pm 2. & \\
\end{array}
$$

Thus if $x = 1$, $y = 2$ or $y = -2$, yielding the pairs $(1, 2)$ and $(1, -2)$. If $x = -1$, $y = 2$ or $y = -2$, yielding the pairs $(-1, 2)$ and $(-1, -2)$.

CHECK: Since $(2)^2 = 4$, $(-2)^2 = 4$, $(1)^2 = 1$, and $(-1)^2 = 1$, we can check all four pairs at one time.

$$
\begin{array}{c|c}
2x^2 + 5y^2 = 22 & \\
\hline
2(\pm 1)^2 + 5(\pm 2)^2 & 22 \\
2 + 20 & \\
22 & \text{TRUE} \\
\end{array}
\qquad
\begin{array}{c|c}
3x^2 - y^2 = -1 & \\
\hline
3(\pm 1)^2 - (\pm 2)^2 & -1 \\
3 - 4 & \\
-1 & \text{TRUE} \\
\end{array}
$$

The solutions are $(1, 2)$, $(1, -2)$, $(-1, 2)$, and $(-1, -2)$.

Do Exercise 5.

When one equation contains a product of variables and the other equation is of the form $Ax^2 + By^2 = C$, we often solve for one of the variables in the equation with the product and then substitute in the other.

Example 5 Solve:

$$x^2 + 4y^2 = 20, \qquad \textbf{(1)}$$
$$xy = 4. \qquad \textbf{(2)}$$

Here we use the substitution method. First, we solve equation (2) for *y*:

$$y = \frac{4}{x}.$$

Then we substitute $4/x$ for y in equation (1) and solve for x:

$$x^2 + 4\left(\frac{4}{x}\right)^2 = 20$$

$$x^2 + \frac{64}{x^2} = 20$$

$$x^4 + 64 = 20x^2 \qquad \text{Multiplying by } x^2$$

$$x^4 - 20x^2 + 64 = 0 \qquad \text{Obtaining standard form. This equation is quadratic in form.}$$

$$u^2 - 20u + 64 = 0 \qquad \text{Letting } u = x^2$$

$$(u - 16)(u - 4) = 0 \qquad \text{Factoring}$$

$$u = 16 \quad or \quad u = 4. \qquad \text{Using the principle of zero products}$$

Next, we substitute x^2 for u and solve these equations:

$$x^2 = 16 \quad or \quad x^2 = 4$$

$$x = \pm 4 \quad or \quad x = \pm 2.$$

Then $x = 4$ or $x = -4$ or $x = 2$ or $x = -2$. Since $y = 4/x$, if $x = 4$, $y = 1$; if $x = -4$, $y = -1$; if $x = 2$, $y = 2$; and if $x = -2$, $y = -2$. The ordered pairs $(4, 1)$, $(-4, -1)$, $(2, 2)$, and $(-2, -2)$ check. They are the solutions.

Do Exercise 6.

b | Solving Applied Problems

We now consider applications in which the translation is to a system of equations in which at least one equation is nonlinear.

Example 6 *Architecture.* For a building at a community college, an architect wants to lay out a rectangular piece of ground that has a perimeter of 204 m and an area of 2565 m². Find the dimensions of the piece of ground.

1. **Familiarize.** We make a drawing of the area, labeling it using l for the length and w for the width.

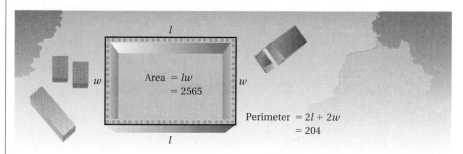

2. **Translate.** We then have the following translation:

Perimeter: $2l + 2w = 204$;

Area: $lw = 2565$.

3. **Solve.** We solve the system

$$2l + 2w = 204, \qquad \text{(The graph is a line.)}$$

$$lw = 2565. \qquad \text{(The graph is a hyperbola.)}$$

6. Solve:

$$x^2 + xy + y^2 = 19,$$

$$xy = 6.$$

Answer on page A-58

7. The perimeter of a rectangular field is 34 m, and the length of a diagonal is 13 m. Find the dimensions of the field.

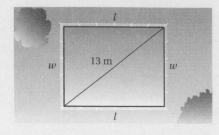

8. *Standard-Sized TV Set.* For any standard-sized television, the ratio of the width to the height of the screen is 4 to 3. A standard-sized TV has a 30-in. diagonal screen. Find the width and the height of a standard-sized 30-in. screen.

30-inch screen

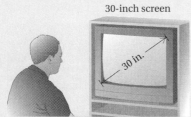

4-to-3 ratio, width to height

Answers on page A-58

We solve the second equation for l and get $l = 2565/w$. Then we substitute $2565/w$ for l in the first equation and solve for w:

$$2\left(\frac{2565}{w}\right) + 2w = 204$$

$$2(2565) + 2w^2 = 204w \qquad \text{Multiplying by } w$$

$$2w^2 - 204w + 2(2565) = 0 \qquad \text{Standard form}$$

$$w^2 - 102w + 2565 = 0 \qquad \text{Dividing by 2}$$

$$w = \frac{-(-102) \pm \sqrt{(-102)^2 - 4 \cdot 1 \cdot 2565}}{2 \cdot 1}$$

Quadratic formula. Factoring could also be used, but the numbers are quite large.

$$w = \frac{102 \pm \sqrt{144}}{2} = \frac{102 \pm 12}{2}$$

$$w = 57 \quad or \quad w = 45.$$

If $w = 57$, then $l = 2565/w = 2565/57 = 45$. If $w = 45$, then $l = 2565/w = 2565/45 = 57$. Since length is generally considered to be greater than width, we have the solution $l = 57$ and $w = 45$, or (57, 45).

4. Check. If $l = 57$ and $w = 45$, the perimeter is $2 \cdot 57 + 2 \cdot 45$, or 204. The area is $57 \cdot 45$, or 2565. The numbers check.

5. State. The length is 57 m and the width is 45 m.

Do Exercise 7.

Example 7 *HDTV Dimensions.* In the not-too-distant future, a new kind of high-definition television (HDTV) with a larger screen and greater clarity will be available. The ratio of the width to the height of the screen of an HDTV is 16 to 9. Suppose a large-screen HDTV has a $\sqrt{4901}$-in. (about 70 in.) diagonal screen. Find the width and the height of the screen.

1. Familiarize. We first make a drawing and label it. Note that there is a right triangle in the figure. We let h = the height and w = the width.

70-inch screen

16-to-9 ratio, width to height

2. Translate. Next, we translate to a system of equations:

$$w^2 + h^2 = (\sqrt{4901})^2, \text{ or } 4901, \qquad \textbf{(1)}$$

$$\frac{w}{h} = \frac{16}{9}. \qquad \textbf{(2)}$$

3. Solve. We solve the system and get $(w, h) \approx (61, 34)$ and $(-61, -34)$.

4. Check. Widths cannot be negative, so we need check only (61, 34). In the right triangle, $34^2 + 61^2 = 1156 + 3721 = 4877 \approx 4901 = (\sqrt{4901})^2$.

5. State. The width is about 61 in. and the height is about 34 in.

Do Exercise 8.

Exercise Set 9.4

Solve.

1. $x^2 + y^2 = 100,$
$y - x = 2$

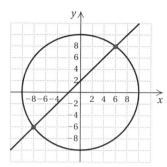

2. $x^2 + y^2 = 25,$
$y - x = 1$

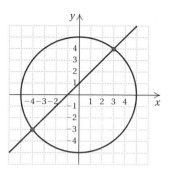

3. $9x^2 + 4y^2 = 36,$
$3x + 2y = 6$

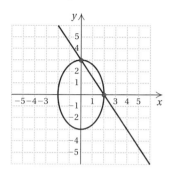

4. $4x^2 + 9y^2 = 36,$
$3y + 2x = 6$

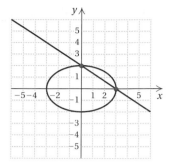

5. $y^2 = x + 3,$
$2y = x + 4$

6. $y = x^2,$
$3x = y + 2$

7. $x^2 - xy + 3y^2 = 27,$
$x - y = 2$

8. $2y^2 + xy + x^2 = 7,$
$x - 2y = 5$

9. $x^2 - xy + 3y^2 = 5,$
$x - y = 2$

10. $a^2 + 3b^2 = 10,$
$a - b = 2$

11. $a + b = -6,$
$ab = -7$

12. $2y^2 + xy = 5,$
$4y + x = 7$

13. $2a + b = 1,$
$\quad b = 4 - a^2$

14. $4x^2 + 9y^2 = 36,$
$\quad x + 3y = 3$

15. $x^2 + y^2 = 5,$
$\quad x - y = 8$

16. $4x^2 + 9y^2 = 36,$
$\quad y - x = 8$

17. $x^2 + y^2 = 25,$
$\quad y^2 = x + 5$

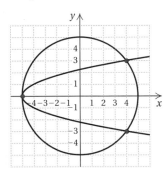

18. $y = x^2,$
$\quad x = y^2$

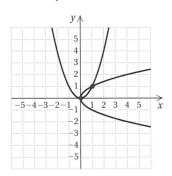

19. $x^2 + y^2 = 9,$
$\quad x^2 - y^2 = 9$

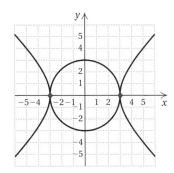

20. $y^2 - 4x^2 = 4,$
$\quad 4x^2 + y^2 = 4$

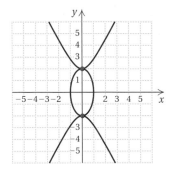

21. $x^2 + y^2 = 20,$
$\quad xy = 8$

22. $x^2 + y^2 = 5,$
$\quad xy = 2$

23. $x^2 + y^2 = 13,$
$\quad xy = 6$

24. $x^2 + y^2 + 6y + 5 = 0,$
$\quad x^2 + y^2 - 2x - 8 = 0$

25. $2xy + 3y^2 = 7,$
$3xy - 2y^2 = 4$

26. $xy - y^2 = 2,$
$2xy - 3y^2 = 0$

27. $4a^2 - 25b^2 = 0,$
$2a^2 - 10b^2 = 3b + 4$

28. $m^2 - 3mn + n^2 + 1 = 0,$
$3m^2 - mn + 3n^2 = 13$

29. $ab - b^2 = -4,$
$ab - 2b^2 = -6$

30. $a^2 + b^2 = 14,$
$ab = 3\sqrt{5}$

31. $x^2 + y^2 = 25,$
$9x^2 + 4y^2 = 36$

32. $x^2 + y^2 = 1,$
$9x^2 - 16y^2 = 144$

b Solve.

33. *Design of a Van.* The cargo area of a delivery van must be 60 ft^2, and the length of a diagonal of the van must accommodate a 13-ft board. Find the dimensions of the cargo area.

34. *Computer Parts.* Dataport Electronics needs a rectangular memory board that has a perimeter of 28 cm and a diagonal of length 10 cm. What should the dimensions of the board be?

35. A rectangle has an area of 14 in^2 and a perimeter of 18 in. Find its dimensions.

36. A rectangle has an area of 40 yd^2 and a perimeter of 26 yd. Find its dimensions.

37. The diagonal of a rectangle is 1 ft longer than the length of the rectangle and 3 ft longer than twice the width. Find the dimensions of the rectangle.

38. It will take 210 yd of fencing to enclose a rectangular field. The area of the field is 2250 yd^2. What are the dimensions of the field?

39. The product of the lengths of the legs of a right triangle is 156. The hypotenuse has length $\sqrt{313}$. Find the lengths of the legs.

40. The product of two numbers is 96. The sum of their squares is 208. Find the numbers.

41. *Garden Design.* A garden contains two square peanut beds. Find the length of each bed if the sum of their areas is 832 ft^2 and the difference of their areas is 320 ft^2.

42. *Investments.* An amount of money invested for 1 yr at a certain interest rate yielded $225 in interest. If $750 more had been invested and the rate had been 1% less, the interest would have been the same. Find the principal and the rate.

43. The area of a rectangle is $\sqrt{2}$ m^2, and the length of a diagonal is $\sqrt{3}$ m. Find the dimensions.

44. The area of a rectangle is $\sqrt{3}$ m^2, and the length of a diagonal is 2 m. Find the dimensions.

Skill Maintenance

Find a formula for the inverse of the function, if it exists. [8.2c]

45. $f(x) = 2x - 5$

46. $f(x) = \dfrac{3}{2x - 7}$

47. $f(x) = \dfrac{x - 2}{x + 3}$

48. $f(x) = \dfrac{3x + 8}{5x - 4}$

49. $f(x) = |x|$

50. $f(x) = 4 - x^2$

51. $f(x) = 10^x$

52. $f(x) = e^x$

53. $f(x) = x^3 - 4$

54. $f(x) = \sqrt[3]{x + 2}$

55. $f(x) = \ln x$

56. $f(x) = \log x$

Synthesis

57. ◈ Write a problem for a classmate to solve that translates to a system of two equations in which at least one equation is nonlinear.

58. ◈ Write a problem for a classmate to solve that translates to a system of two equations in which both equations are nonlinear.

59. ◪ Use a grapher to check your answers to Exercises 1, 8, and 12.

60. Find the equation of an ellipse centered at the origin that passes through the points $(2, -3)$ and $(1, \sqrt{13})$.

61. A piece of wire 100 cm long is to be cut into two pieces and those pieces are each to be bent to make a square. The area of one square is to be 144 cm^2 greater than that of the other. How should the wire be cut?

62. Find the equation of a circle that passes through $(-2, 3)$ and $(-4, 1)$ and whose center is on the line $5x + 8y = -2$.

63. *Railing Sales.* Fireside Castings finds that the total revenue R from the sale of x units of railing is given by the function

$$R(x) = 100x + x^2.$$

Fireside also finds that the total cost C of producing x units of the same product is given by the function

$$C(x) = 80x + 1500.$$

A break-even point is a value of x for which total revenue is the same as total cost; that is, $R(x) = C(x)$. How many units must be sold to break even? (See Section 3.7a.)

Solve.

64. $p^2 + q^2 = 13,$

$\dfrac{1}{pq} = -\dfrac{1}{6}$

65. $a + b = \dfrac{5}{6},$

$\dfrac{a}{b} + \dfrac{b}{a} = \dfrac{13}{6}$

Summary and Review Exercises: Chapter 9

Important Properties and Formulas

Distance Formula: $d = \sqrt{(x_2 - x_1)^2 + (y_2 - y_1)^2}$

Circle: $(x - h)^2 + (y - k)^2 = r^2$

Ellipse: $\dfrac{(x - h)^2}{a^2} + \dfrac{(y - k)^2}{b^2} = 1$

Parabola

$y = ax^2 + bx + c, a > 0$
  $= a(x - h)^2 + k$, opens up;
$x = ay^2 + by + c, a > 0$
  $= a(y - k)^2 + h$, opens to the right;

$y = ax^2 + bx + c, a < 0$
  $= a(x - h)^2 + k$, opens down;
$x = ay^2 + by + c, a < 0$
  $= a(y - k)^2 + h$, opens to the left

Hyperbola

$\dfrac{x^2}{a^2} - \dfrac{y^2}{b^2} = 1$, axis is horizontal; $\dfrac{y^2}{b^2} - \dfrac{x^2}{a^2} = 1$, axis is vertical

The objectives to be tested in addition to the material in this chapter are [4.5b], [7.2a], [8.2c], and [8.3b].

Find the distance between each pair of points. Where appropriate, give an approximation to three decimal places.   [9.1b]

1. (2, 6) and (6, 6)

2. (−1, 1) and (−5, 4)

3. (1.4, 3.6) and (4.7, −5.3)

4. (2, 3a) and (−1, a)

Find the midpoint of the segment with the given endpoints.   [9.1c]

5. (1, 6) and (7, 6)

6. (−1, 1), and (−5, 4)

7. (1, $\sqrt{3}$) and $\left(\frac{1}{2}, -\sqrt{2}\right)$

8. (2, 3a) and (−1, a)

Find the center and the radius of the circle.   [9.1d]

9. $(x + 2)^2 + (y - 3)^2 = 2$

10. $(x - 5)^2 + y^2 = 49$

11. $x^2 + y^2 - 6x - 2y + 1 = 0$

12. $x^2 + y^2 + 8x - 6y - 10 = 0$

13. Find an equation of the circle with center (−4, 3) and radius $4\sqrt{3}$.   [9.1d]

14. Find an equation of the circle with center (7, −2) and radius $2\sqrt{5}$.   [9.1d]

Graph.

15. $\dfrac{x^2}{16} + \dfrac{y^2}{4} = 1$ [9.2a]

16. $\dfrac{y^2}{9} - \dfrac{x^2}{4} = 1$ [9.3a]

17. $x^2 + y^2 = 16$ [9.1d]

18. $x = y^2 + 2y - 2$ [9.1a]

19. $y = -2x^2 - 2x + 3$ [9.1a]

20. $x^2 + y^2 + 6x - 8y - 39 = 0$ [9.1d]

21. $\dfrac{(x - 3)^2}{9} + \dfrac{(y + 4)^2}{4} = 1$ [9.2a]

22. $xy = 9$ [9.3b]

Solve. [9.4a]

23. $x^2 - y^2 = 33,$
      $x + y = 11$

24. $x^2 - 2x + 2y^2 = 8,$
      $2x + y = 6$

25. $x^2 - y = 3,$
      $2x - y = 3$

26. $x^2 + y^2 = 25,$
      $x^2 - y^2 = 7$

27. $x^2 - y^2 = 3,$
      $y = x^2 - 3$

28. $x^2 + y^2 = 18,$
      $2x + y = 3$

29. $x^2 + y^2 = 100,$
      $2x^2 - 3y^2 = -120$

30. $x^2 + 2y^2 = 12,$
      $xy = 4$

31. *Athletic Field.* A rectangular athletic field has a perimeter of 38 m and an area of 84 m². What are the dimensions of the field? [9.4b]

32. Find two positive integers whose sum is 12 and the sum of whose reciprocals is $\frac{3}{8}$. [9.4b]

33. The perimeter of a square is 12 cm more than the perimeter of another square. Its area exceeds the area of the other by 39 cm². Find the perimeter of each square. [9.4b]

34. *Flower Beds.* The sum of the areas of two circular flower beds is 130π ft². The difference of the circumferences is 16π ft. Find the radius of each flower bed. [9.4b]

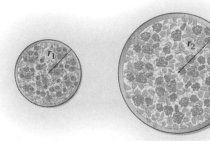

Factor. [4.5b]

35. $a^2 - 49b^2$

36. $12p^2 - 192q^2$

Solve. [7.2a]

37. $x^2 + 2x + 5 = 0$

38. $3x^2 - 12x + 3 = 0$

Find a formula for the inverse, if it exists. [8.2c]

39. $f(x) = 8 - 3x$

40. $f(x) = \sqrt[3]{2x - 5}$

Convert to a logarithmic equation. [8.3b]

41. $7^x = y$

42. $e^{-1.8} = 0.1653$

Convert to an exponential equation. [8.3c]

43. $\ln 250 = 5.522$

44. $\log_a M = N$

Synthesis

45. ◆ We have studied techniques for solving systems of equations in this chapter. How do the equations differ from those systems that we studied earlier in the text? [9.4a]

46. ◆ How does the graph of a hyperbola differ from the graph of a parabola? [9.1a], [9.3a]

47. Solve: [9.4a]
      $4x^2 - x - 3y^2 = 9,$
      $-x^2 + x + y^2 = 2.$

48. Find an equation of the circle that passes through $(-2, -4)$, $(5, -5)$, and $(6, 2)$. [9.1d]

49. Find an equation of the ellipse with the intercepts $(-7, 0)$, $(7, 0)$, $(0, -3)$, and $(0, 3)$. [9.2a]

50. Find the point on the x-axis that is equidistant from $(-3, 4)$ and $(5, 6)$. [9.1b]

Test: Chapter 9

Find the distance between each pair of points. Where appropriate, give an approximation to three decimal places.

1. $(-6, 2)$ and $(6, 8)$

2. $(3, -a)$ and $(-3, a)$

Find the midpoint of the segment with the given endpoints.

3. $(-6, 2)$ and $(6, 8)$

4. $(3, -a)$ and $(-3, a)$

Find the center and the radius of the circle.

5. $(x + 2)^2 + (y - 3)^2 = 64$

6. $x^2 + y^2 + 4x - 6y + 4 = 0$

7. Find an equation of the circle with center $(-2, -5)$ and radius $3\sqrt{2}$.

Graph.

8. $y = x^2 - 4x - 1$

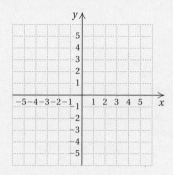

9. $x^2 + y^2 = 36$

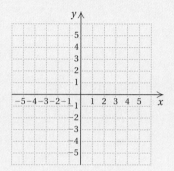

10. $\dfrac{x^2}{9} - \dfrac{y^2}{4} = 1$

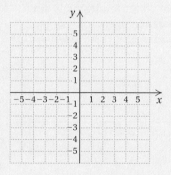

11. $\dfrac{(x + 2)^2}{16} + \dfrac{(y - 3)^2}{9} = 1$

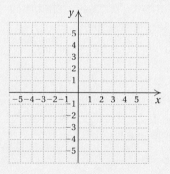

12. $x^2 + y^2 - 4x + 6y + 4 = 0$

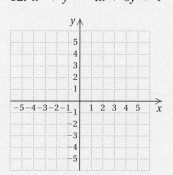

13. $9x^2 + y^2 = 36$

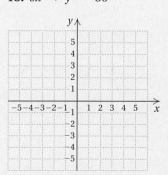

Answers

1. _____

2. _____

3. _____

4. _____

5. _____

6. _____

7. _____

8. _____

9. _____

10. _____

11. _____

12. _____

13. _____

14. _____

15. _____

16. _____

17. _____

18. _____

19. _____

20. _____

21. _____

22. _____

23. _____

24. _____

25. _____

26. _____

27. _____

28. _____

29. _____

30. _____

14. $xy = 4$

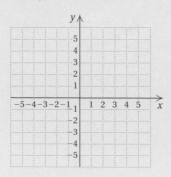

15. $x = -y^2 + 4y$

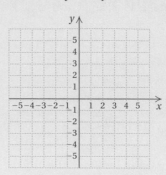

Solve.

16. $\dfrac{x^2}{16} + \dfrac{y^2}{9} = 1$,

  $3x + 4y = 12$

17. $x^2 + y^2 = 16$,

  $\dfrac{x^2}{16} - \dfrac{y^2}{9} = 1$

18. *Home Office.* A rectangular home office has a diagonal of 20 ft and a perimeter of 56 ft. What are the dimensions of the office?

19. In a rational expression, the sum of the values of the numerator and the denominator is 23. The product of their values is 120. Find the values of the numerator and the denominator.

20. A rectangle with a diagonal of length $5\sqrt{5}$ yd has an area of 22 yd². Find the dimensions of the rectangle.

21. *Water Fountains.* The sum of the areas of two square water fountains is 8 m² and the difference of their areas is 2 m². Find the length of a side of each square.

Skill Maintenance

22. Factor: $48a^3 - 75ab^2$.

23. Solve: $x^2 + 2x = 5$.

24. Find a formula for the inverse, if it exists, of $f(x) = 8 - x^3$.

25. Convert to a logarithmic equation: $e^4 = 54.5981$.

26. Convert to an exponential equation: $\log_2 8 = 3$.

Synthesis

27. Find an equation of the ellipse passing through (6, 0) and (6, 6) with vertices at (1, 3) and (11, 3).

28. Find the points whose distance from (8, 0) is 10.

29. The sum of two numbers is 36, and the product is 4. Find the sum of the reciprocals of the numbers.

30. Find the point on the y-axis that is equidistant from $(-3, -5)$ and $(4, -7)$.

Cumulative Review: Chapters R–9

Simplify.

1. $\left| \frac{2}{3} - \frac{4}{5} \right|$

2. $\dfrac{63x^2y^3}{-7x^{-4}y}$

3. $1000 \div 10^2 \cdot 25 \div 4$

4. $5x - 3[4(x - 2) - 2(x + 1)]$

Solve.

5. $\frac{1}{3}x - \frac{1}{5} \geq \frac{1}{5}x - \frac{1}{3}$

6. $|x| > 6.4$

7. $3 \leq 4x + 7 < 31$

8. $3x + y = 4,$
$-6x - y = -3$

9. $x - y + 2z = 3,$
$-x \quad + z = 4,$
$2x + y - z = -3$

10. $2x^2 = x + 3$

11. $3x - \dfrac{6}{x} = 7$

12. $\sqrt{x + 5} = x - 1$

13. $x(x + 10) = -21$

14. $2x^2 + x + 1 = 0$

15. $x^4 - 13x^2 + 36 = 0$

16. $\dfrac{3}{x - 3} - \dfrac{x + 2}{x^2 + 2x - 15} = \dfrac{1}{x + 5}$

17. $-x^2 + 2y^2 = 7,$
$x^2 + \quad y^2 = 5$

18. $\log_2 x + \log_2 (x + 7) = 3$

19. $7^x = 30$

20. $\log_3 x = 2$

21. $x^2 - 1 \geq 0$

22. $\dfrac{x + 1}{x - 2} > 0$

23. $P = \dfrac{3}{4}(M + 2N),$ for N

24. $\dfrac{1}{p} + \dfrac{1}{q} = \dfrac{1}{f},$ for p

Simplify.

25. $(2x + 3)(x^2 - 2x - 1)$

26. $(3x^2 + x^3 - 1) - (2x^3 + x + 5)$

27. $\dfrac{2m^2 + 11m - 6}{m^3 + 1} \cdot \dfrac{m^2 - m + 1}{m + 6}$

28. $\dfrac{x}{x - 1} + \dfrac{2}{x + 1} - \dfrac{2x}{x^2 - 1}$

29. $\dfrac{1 - \dfrac{5}{x}}{x - 4 - \dfrac{5}{x}}$

30. $(x^4 + 3x^3 - x + 4) \div (x + 1)$

31. $\dfrac{\sqrt{75x^5y^2}}{\sqrt{3xy}}$

32. $4\sqrt{50} - 3\sqrt{18}$

33. $(16^{3/2})^{1/2}$

34. $(2 - i\sqrt{2})(5 + 3i\sqrt{2})$

35. $\dfrac{5 + i}{2 - 4i}$

Graph.

36. $4y - 3x = 12$

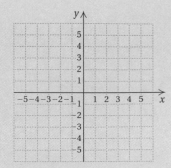

37. $y < -2$

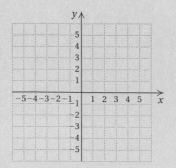

38. $x + y \leq 0,$
$\qquad x \geq -4,$
$\qquad y \geq -1$

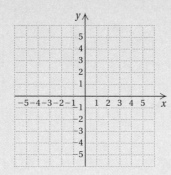

39. $f(x) = 2x^2 - 8x + 9$

40. $(x - 1)^2 + (y + 1)^2 = 9$

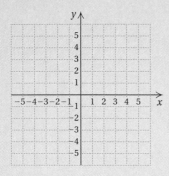

41. $x = y^2 + 1$

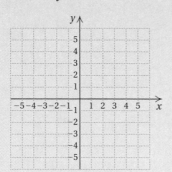

42. $f(x) = e^{-x}$

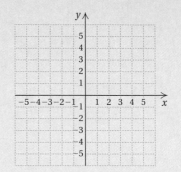

43. $f(x) = \log_2 x$

Factor.

44. $2x^4 - 12x^3 + x - 6$

45. $3a^2 - 12ab - 135b^2$

46. $x^2 - 17x + 72$

47. $81m^4 - n^4$

48. $16x^2 - 16x + 4$

49. $81a^3 - 24$

50. $10x^2 + 66x - 28$

51. $6x^3 + 27x^2 - 15x$

52. Find an equation of the line containing the points $(1, 4)$ and $(-1, 0)$.

53. Find an equation of the line containing the point $(1, 2)$ and perpendicular to the line whose equation is $2x - y = 3$.

54. Find the center and the radius of the circle
$$x^2 - 16x + y^2 + 6y + 68 = 0.$$

55. Find $f^{-1}(x)$ when $f(x) = 2x - 3$.

56. z varies directly as x and inversely as the cube of y, and $z = 5$ when $x = 4$ and $y = 2$. What is z when $x = 10$ and $y = 5$?

57. Given the function f described by $f(x) = x^3 - 2$, find $f(-2)$.

58. Find the distance between the points $(2, 1)$ and $(8, 9)$.

59. Find the midpoint of the segment with endpoints $(-1, -3)$ and $(3, 0)$.

60. Rationalize the denominator: $\dfrac{5 + \sqrt{a}}{3 - \sqrt{a}}$.

61. Find the domain: $f(x) = \dfrac{4x - 3}{3x^2 + x}$.

62. Given that $f(x) = 3x^2 + x$, find a such that $f(a) = 2$.

Solve.

63. The square of a number plus the number is 156. Find the number.

64. A rectangle has an area of 20 in^2 and a perimeter of 18 in. Find the dimensions of the rectangle.

65. *Work.* Jane can finish a quilted pillow sham in 3 hr. Laura can finish the same pillow sham in $1\frac{1}{2}$ hr. How long would it take to finish the pillow sham if they work together?

66. The sum of the squares of three consecutive even integers is equal to 8 more than three times the square of the second number. Find the integers.

67. *America Online.* America Online is an internet, e-mail, and information service. In addition to monthly service fees, it makes money through on-line sales of various products and through advertising. In 1996, such sales were $102 million. In 1997, sales were $256 million. (**Source:** America Online)

a) Find an exponential growth function
   $P(t) = P_0 e^{kt}$.
b) Predict the sales in the year 2000.
c) What is the doubling time?

68. *Train Travel.* A passenger train travels at twice the speed of a freight train. The freight train leaves a station at 2 A.M. and travels north at 34 mph. The passenger train leaves the station at 11 A.M. traveling north on a parallel track. How far from the station will the passenger train overtake the freight train?

69. *Carbon Dating.* Use the function $P(t) = P_0 e^{-0.00012t}$ to find the age of a bone that has lost 25% of its carbon-14.

70. *Beam Load.* The weight W that a horizontal beam can support varies inversely as the length L of the beam. If a 14-m beam can support 1440 kg, what weight can a 6-m beam support?

71. Fit a linear function to the data points $(2, -3)$ and $(5, -4)$.

72. Fit a quadratic function to the data points $(-2, 4)$, $(-5, -6)$, and $(1, -3)$.

73. Convert to a logarithmic equation: $10^6 = r$.

74. Convert to an exponential equation: $\log_3 Q = x$.

75. Express as a single logarithm:
$$\tfrac{1}{5}(7 \log_b x - \log_b y - 8 \log_b z).$$

76. Express in terms of logarithms of x, y, and z:
$$\log_b \left(\frac{xy^5}{z} \right)^{-6}.$$

77. What is the maximum product of two numbers whose sum is 26?

78. Determine whether the function $f(x) = 4 - x^2$ is one-to-one.

79. For the graph of function f shown here, determine **(a)** $f(2)$; **(b)** the domain; **(c)** all x-values such that $f(x) = -5$; and **(d)** the range.

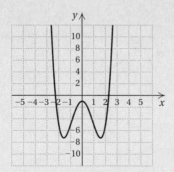

Synthesis

80. Solve:
$$\frac{x^3 + 8}{x + 2} = x^2 - 2x + 4.$$

81. Describe the graph of
$$\frac{x^2}{a^2} + \frac{y^2}{b^2} = 1$$
when $a^2 = b^2$.

82. The square of a certain number exceeds twice the square of another number by $\frac{1}{8}$. The sum of their squares is $\frac{5}{16}$. Find the numbers.

83. Solve:
$$x^2 + y^2 = 208,$$
$$xy = 96.$$

Final Examination

Simplify.

1. $(-9x^2y^3)(5x^4y^{-7})$

2. $|-3.5 + 9.8|$

3. $2y - [3 - 4(5 - 2y) - 3y]$

4. $(10 \cdot 8 - 9 \cdot 7)^2 - 54 \div 9 - 3$

5. Evaluate $\dfrac{ab - ac}{bc}$ for $a = -2$, $b = 3$, and $c = -4$.

Perform the indicated operations and simplify.

6. $(5a^2 - 3ab - 7b^2) - (2a^2 + 5ab + 8b^2)$

7. $(-3x^2 + 4x^3 - 5x - 1) + (9x^3 - 4x^2 + 7 - x)$

8. $(2a - 1)(3a + 5)$

9. $(3a^2 - 5y)^2$

10. $\dfrac{1}{x - 2} - \dfrac{4}{x^2 - 4} + \dfrac{3}{x + 2}$

11. $\dfrac{x^2 - 6x + 8}{3x + 9} \cdot \dfrac{x + 3}{x^2 - 4}$

12. $\dfrac{3x + 3y}{5x - 5y} \div \dfrac{3x^2 + 3y^2}{5x^3 - 5y^3}$

13. $\dfrac{x - \dfrac{a^2}{x}}{1 + \dfrac{a}{x}}$

Factor.

14. $4x^2 - 12x + 9$

15. $27a^3 - 8$

16. $a^3 + 3a^2 - ab - 3b$

17. $15y^4 + 33y^2 - 36$

18. For the function described by $f(x) = 3x^2 - 4x$, find $f(-2)$.

19. Divide:

$(7x^4 - 5x^3 + x^2 - 4) \div (x - 2)$.

Solve.

20. $9(x - 1) - 3(x - 2) = 1$

21. $x^2 - 2x = 48$

22. $\dfrac{6}{x} + \dfrac{6}{x + 2} = \dfrac{5}{2}$

23. $\dfrac{7x}{x - 3} - \dfrac{21}{x} + 11 = \dfrac{63}{x^2 - 3x}$

24. $5x + 3y = 2,$
$3x + 5y = -2$

25. $x + y - z = 0,$
$3x + y + z = 6,$
$x - y + 2z = 5$

26. $\sqrt{x - 5} = 5 - \sqrt{x}$

27. $x^4 - 29x^2 + 100 = 0$

28. $x^2 + y^2 = 8,$
$x^2 - y^2 = 2$

29. $5^x = 8$

Answers

1.

2.

3.

4.

5.

6.

7.

8.

9.

10.

11.

12.

13.

14.

15.

16.

17.

18.

19.

20.

21.

22.

23.

24.

25.

26.

27.

28.

29.

30. $\log (x^2 - 25) - \log (x + 5) = 3$

31. $\log_4 x = -2$

32. $7^{2x+3} = 49$

33. $|2x - 1| \le 5$

34. $7x^2 + 14 = 0$

35. $x^2 + 4x = 3$

36. $|2y + 3| > 7$

37. Given that $f(x) = x^2 - 2x$, find a such that $f(a) = 48$.

Solve.

38. *Music Club.* A music club offers two types of membership. Limited members pay a fee of $10 a year and can buy CDs for $10 each. Preferred members pay $20 a year and can buy CDs for $7.50 each. For what numbers of annual CD purchases would it be less expensive to be a preferred member?

39. Find three consecutive integers whose sum is 198.

40. A pentagon with all five sides the same size has a perimeter equal to that of an octagon in which all eight sides are the same size. One side of the pentagon is 2 less than three times one side of the octagon. What is the perimeter of each figure?

41. *Ammonia Solutions.* A chemist has two solutions of ammonia and water. Solution A is 6% ammonia and solution B is 2% ammonia. How many liters of each solution are needed in order to obtain 80 L of a solution that is 3.2% ammonia?

42. *Air Travel.* An airplane can fly 190 mi with the wind in the same time that it takes to fly 160 mi against the wind. The speed of the wind is 30 mph. How fast can the plane fly in still air?

43. *Work.* Bianca can do a certain job in 21 min. Dahlia can do the same job in 14 min. How long would it take to do the job if the two worked together?

44. *Centripetal Force.* The centripetal force F of an object moving in a circle varies directly as the square of the velocity v and inversely as the radius r of the circle. If $F = 8$ when $v = 1$ and $r = 10$, what is F when $v = 2$ and $r = 16$?

45. The perimeter of a rectangle is 34 ft. The length of a diagonal is 13 ft. Find the dimensions of the rectangle.

46. *Dimensions of a Rug.* The diagonal of a Persian rug is 25 ft. The area of the rug is 300 ft². Find the length and the width of the rug.

47. *Maximizing Area.* A farmer wants to fence in a rectangular area next to a river. (Note that no fence will be needed along the river.) What is the area of the largest region that can be fenced in with 100 ft of fencing?

Graph.

48. $3x - y = 6$

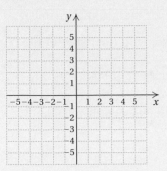

49. $\dfrac{x^2}{25} + \dfrac{y^2}{4} = 1$

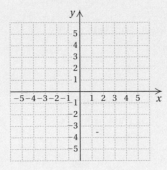

50. $f(x) = \log_2 x$

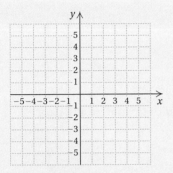

51. $2x - 3y < -6$

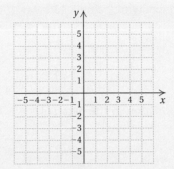

52. Graph: $f(x) = -2(x - 3)^2 + 1$.
   a) Label the vertex.
   b) Draw the line of symmetry.
   c) Find the maximum or minimum value.

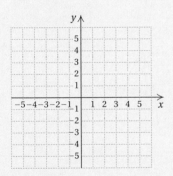

53. Graph. Find the vertices.
$$y - x \geq 1,$$
$$y - x \leq 3,$$
$$2 \leq x \leq 5$$

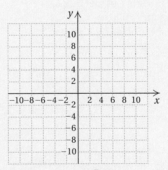

54. Solve $A = P + Prt$ for r.

55. Solve $I = \dfrac{R}{R + r}$ for R.

56. Find an equation of the line containing the point $(-1, 4)$ and perpendicular to the line whose equation is $x - y = 6$.

Find the domain of the function.

57. $f(x) = \sqrt{5 - 3x}$

58. $g(x) = \dfrac{x - 4}{x^2 - 2x + 1}$

59. Multiply $(8.9 \times 10^{-17})(7.6 \times 10^4)$. Write scientific notation for the answer.

60. Multiply and simplify: $\sqrt{8x}\sqrt{8x^3y}$.

61. Divide and simplify: $\dfrac{\sqrt[3]{15x}}{\sqrt[3]{3y^2}}$.

62. Rationalize the denominator:
$$\dfrac{1 - \sqrt{x}}{1 + \sqrt{x}}.$$

48. _____

49. _____

50. _____

51. _____

52. _____

53. _____

54. _____

55. _____

56. _____

57. _____

58. _____

59. _____

60. _____

61. _____

62. _____

63. _____

64. _____

65. _____

66. _____

67. _____

68. _____

69. _____

70. _____

71. _____

72. _____

73. _____

74. _____

75. _____

76. a) _____

b) _____

c) _____

77. a) _____

b) _____

c) _____

d) _____

78. _____

79. _____

80. _____

81. _____

82. _____

83. _____

63. Write a single radical expression:
$$\frac{\sqrt[3]{(x + 1)^5}}{\sqrt{(x + 1)^3}}.$$

64. Multiply these complex numbers:
$$(3 + 2i)(4 - 7i).$$

65. Write a quadratic equation whose solutions are $5\sqrt{2}$ and $-5\sqrt{2}$.

66. Find the center and the radius of the circle
$$x^2 + y^2 - 4x + 6y - 23 = 0.$$

67. Express as a single logarithm:
$$\tfrac{2}{3} \log_a x - \tfrac{1}{2} \log_a y + 5 \log_a z.$$

68. Convert to an exponential equation:
$$\log_a c = 5.$$

69. Find a formula for the inverse of $f(x) = x^3 - 8$.

Find each of the following using a calculator.

70. log 5677.2

71. $10^{-3.587}$

72. ln 5677.2

73. $e^{-3.587}$

74. Fit a linear function to the data points $(-2, 3)$ and $(-5, -4)$.

75. Fit a quadratic function to the data points $(-2, 3)$, $(-5, -4)$, and $(1, -2)$.

76. _Portable Telephones._ There were 1000 portable telephones sold in 1984 and 5,560,000 sold in 1994 (**Source:** CTIA, Herschel Shoestack Associates).
a) Find an exponential growth function $P(t) = P_0 e^{kt}$.
b) Predict the sales of portable telephones in 2000.
c) What is the doubling time?

77. For the graph of function f shown here, determine (a) $f(2)$; (b) the domain; (c) all x-values such that $f(x) = -8$; and (d) the range.

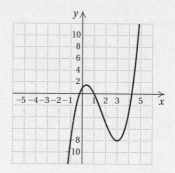

78. Simplify: $(25x^{4/3}y^{1/2})^{3/2}$.

Synthesis

Solve.

79. $\dfrac{9}{x} - \dfrac{9}{x + 12} = \dfrac{108}{x^2 + 12x}$

80. $\log_2 (\log_3 x) = 2$

81. y varies directly as the cube of x and x is multiplied by 0.5. What is the effect on y?

82. Divide:
$$\frac{2\sqrt{6} + 4\sqrt{5}\,i}{2\sqrt{6} - 4\sqrt{5}\,i}.$$

83. Diaphantos, a famous mathematician, spent $\tfrac{1}{6}$ of his life as a child, $\tfrac{1}{12}$ as a young man, and $\tfrac{1}{7}$ as a bachelor. Five years after he was married, he had a son who died 4 yr before his father at half his father's final age. How long did Diaphantos live?

Appendix A Handling Dimension Symbols

a Calculating with Dimension Symbols

Objectives

a Perform calculations with dimension symbols.

b Make unit changes.

In many applications, we add, subtract, multiply, and divide quantities having units, or dimensions, such as ft, km, sec, and hr. For example, to find average speed, we divide total distance by total time. What results is notation very much like a rational expression.

Example 1 A car travels 150 km in 2 hr. What is its average speed?

$$\text{Speed} = \frac{150 \text{ km}}{2 \text{ hr}}, \text{ or } 75 \frac{\text{km}}{\text{hr}}$$

(The standard abbreviation for km/hr is km/h, but it does not suit our present discussion well.)

The symbol km/hr makes it look as if we are dividing kilometers by hours. It can be argued that we can divide only numbers. Nevertheless, we treat dimension symbols, such as km, ft, and hr, as if they were numerals or variables, obtaining correct results mechanically.

Do Exercise 1.

Examples Compare the following.

2. $\frac{150x}{2y} = \frac{150}{2} \cdot \frac{x}{y} = 75\frac{x}{y}$ with $\frac{150 \text{ km}}{2 \text{ hr}} = \frac{150}{2} \frac{\text{km}}{\text{hr}} = 75\frac{\text{km}}{\text{hr}}$

3. $3x + 2x = (3 + 2)x = 5x$ with $3 \text{ ft} + 2 \text{ ft} = (3 + 2) \text{ ft} = 5 \text{ ft}$

4. $5x \cdot 3x = 15x^2$ with $5 \text{ ft} \cdot 3 \text{ ft} = 15 \text{ ft}^2$ (square feet)

Do Exercises 2–4.

If 5 men work 8 hours, the total amount of labor is 40 man-hours.

Example 5 Compare

$5x \cdot 8y = 40xy$ with $5 \text{ men} \cdot 8 \text{ hours} = 40 \text{ man-hours}$.

Do Exercise 5.

1. A truck travels 210 mi in 3 hr. What is its average speed?

Perform these calculations.

2. $\frac{100 \text{ m}}{4 \text{ sec}}$

3. 7 yd + 9 yd

4. 24 in. · 3 in.

5. Calculate: 6 men · 11 hours.

Answers on page A-62

6. Calculate:

$$\frac{200 \text{ kW} \cdot 140 \text{ hr}}{35 \text{ da}}.$$

7. Convert 7 ft to inches.

8. Convert 90 mi/hr to ft/sec.

Answers on page A-62

Example 6 Compare

$$\frac{300x \cdot 240y}{15t} = 4800\frac{xy}{t} \quad \text{with} \quad \frac{300 \text{ kW} \cdot 240 \text{ hr}}{15 \text{ da}} = 4800\frac{\text{kW-hr}}{\text{da}}.$$

If an electrical device uses 300 kW (kilowatts) for 240 hr over a period of 15 days, its rate of usage of energy is 4800 kilowatt-hours per day. The standard abbreviation for kilowatt-hours is kWh.

Do Exercise 6.

b | Making Unit Changes

We can treat dimension symbols much like numerals or variables, because we obtain correct results that way. We can change units by substituting or by multiplying by 1, as shown below.

Example 7 Convert 3 ft to inches.

METHOD 1. We have 3 ft. We know that 1 ft = 12 in., so we substitute 12 in. for ft:

$$3 \text{ ft} = 3 \cdot 12 \text{ in.} = 36 \text{ in.}$$

METHOD 2. We want to convert from "ft" to "in." We multiply by 1 using a symbol for 1 with "ft" on the bottom since we are converting from "ft," and with "in." on the top since we are converting to "in."

$$3 \text{ ft} = 3 \text{ ft} \cdot \frac{12 \text{ in.}}{1 \text{ ft}}$$

$$= \frac{3 \cdot 12}{1} \cdot \frac{\text{ft}}{\text{ft}} \cdot \text{in.} = 36 \text{ in.}$$

Do Exercise 7.

We can multiply by 1 several times to make successive conversions. In the following example, we convert mi/hr to ft/sec by converting successively from mi/hr to ft/hr to ff/min to ft/sec.

Example 8 Convert 60 mi/hr to ft/sec.

$$60\frac{\text{mi}}{\text{hr}} = 60\frac{\text{mi}}{\text{hr}} \cdot \frac{5280 \text{ ft}}{1 \text{ mi}} \cdot \frac{1 \text{ hr}}{60 \text{ min}} \cdot \frac{1 \text{ min}}{60 \text{ sec}}$$

$$= \frac{60 \cdot 5280}{60 \cdot 60} \cdot \frac{\text{mi}}{\text{mi}} \cdot \frac{\text{hr}}{\text{hr}} \cdot \frac{\text{min}}{\text{min}} \cdot \frac{\text{ft}}{\text{sec}} = 88\frac{\text{ft}}{\text{sec}}.$$

Do Exercise 8.

Exercise Set A

a Add these measures.

1. 45 ft + 23 ft

2. 55 km/hr + 27 km/hr

3. 17 g + 28 g

4. 3.4 lb + 5.2 lb

Find the average speeds, given total distance and total time.

5. 90 mi, 6 hr

6. 640 km, 20 hr

7. 9.9 m, 3 sec

8. 76 ft, 4 min

Perform these calculations.

9. $\dfrac{3 \text{ in.} \cdot 8 \text{ lb}}{6 \text{ sec}}$

10. $\dfrac{60 \text{ men} \cdot 8 \text{ hr}}{20 \text{ da}}$

11. $36 \text{ ft} \cdot \dfrac{1 \text{ yd}}{3 \text{ ft}}$

12. $55\dfrac{\text{mi}}{\text{hr}} \cdot 4 \text{ hr}$

13. $5 \text{ ft}^3 + 11 \text{ ft}^3$

14. $\dfrac{3 \text{ lb}}{14 \text{ ft}} \cdot \dfrac{7 \text{ lb}}{6 \text{ ft}}$

15. Divide $4850 by 5 days.

16. Divide $25.60 by 8 hr.

b Make these unit changes.

17. Change 3.2 lb to oz (16 oz = 1 lb).

18. Change 6.2 km to m.

19. Change 35 mi/hr to ft/min.

20. Change $375 per day to dollars per minute.

21. Change 8 ft to in.

22. Change 25 yd to ft.

23. How many years ago is 1 million sec ago? Let 365 days = 1 yr.

24. How many years ago is 1 billion sec ago?

25. How many years ago is 1 trillion sec ago?

26. Change 20 lb to oz.

27. Change $60\dfrac{\text{lb}}{\text{ft}}$ to $\dfrac{\text{oz}}{\text{in.}}$.

28. Change $44\dfrac{\text{ft}}{\text{sec}}$ to $\dfrac{\text{mi}}{\text{hr}}$.

29. Change 2 days to seconds.

30. Change 128 hr to days.

31. Change 216 in^2 to ft^2.

32. Change 1440 man-hours to man-days.

33. Change $80\dfrac{\text{lb}}{\text{ft}^3}$ to $\dfrac{\text{ton}}{\text{yd}^3}$.

34. Change the speed of light, 186,000 mi/sec, to mi/yr.

Appendix B Determinants and Cramer's Rule

In Chapter 3, you probably noticed that the elimination method concerns itself primarily with the coefficients and constants of the equations. Here we learn a method for solving a system of equations using just the coefficients and constants. This method involves *determinants*.

Objectives

a Evaluate second-order determinants.

b Evaluate third-order determinants.

c Solve systems of equations using Cramer's Rule.

a Evaluating Determinants

The following symbolism represents a **second-order determinant**:

$$\begin{vmatrix} a_1 & b_1 \\ a_2 & b_2 \end{vmatrix}.$$

To evaluate a determinant, we do two multiplications and subtract.

Example 1 Evaluate:

$$\begin{vmatrix} 2 & -5 \\ 6 & 7 \end{vmatrix}.$$

We multiply and subtract as follows:

$$\begin{vmatrix} 2 & -5 \\ 6 & 7 \end{vmatrix} = 2 \cdot 7 - 6 \cdot (-5) = 14 + 30 = 44.$$

Determinants are defined according to the pattern shown in Example 1.

> The determinant $\begin{vmatrix} a_1 & b_1 \\ a_2 & b_2 \end{vmatrix}$ is defined to mean $a_1 b_2 - a_2 b_1$.

The value of a determinant is a *number*. In Example 1, the value is 44.

Do Exercises 1 and 2.

Evaluate.

1. $\begin{vmatrix} 3 & 2 \\ 4 & 1 \end{vmatrix}$

2. $\begin{vmatrix} 5 & -2 \\ -1 & -1 \end{vmatrix}$

b Third-Order Determinants

A **third-order determinant** is defined as follows.

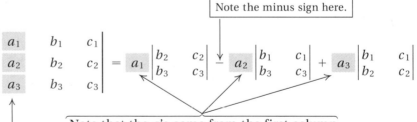

Note the minus sign here.

Note that the a's come from the first column.

$$\begin{vmatrix} a_1 & b_1 & c_1 \\ a_2 & b_2 & c_2 \\ a_3 & b_3 & c_3 \end{vmatrix} = a_1 \begin{vmatrix} b_2 & c_2 \\ b_3 & c_3 \end{vmatrix} - a_2 \begin{vmatrix} b_1 & c_1 \\ b_3 & c_3 \end{vmatrix} + a_3 \begin{vmatrix} b_1 & c_1 \\ b_2 & c_2 \end{vmatrix}$$

Answers on page A-62

Note that the second-order determinants on the right can be obtained by crossing out the row and the column in which each a occurs.

For a_1:
$$\begin{vmatrix} a_1 & b_1 & c_1 \\ a_2 & b_2 & c_2 \\ a_3 & b_3 & c_3 \end{vmatrix}$$

For a_2:
$$\begin{vmatrix} a_1 & b_1 & c_1 \\ a_2 & b_2 & c_2 \\ a_3 & b_3 & c_3 \end{vmatrix}$$

For a_3:
$$\begin{vmatrix} a_1 & b_1 & c_1 \\ a_2 & b_2 & c_2 \\ a_3 & b_3 & c_3 \end{vmatrix}$$

Example 2 Evaluate this third-order determinant:

$$\begin{vmatrix} -1 & 0 & 1 \\ -5 & 1 & -1 \\ 4 & 8 & 1 \end{vmatrix} = -1 \begin{vmatrix} 1 & -1 \\ 8 & 1 \end{vmatrix} - (-5) \begin{vmatrix} 0 & 1 \\ 8 & 1 \end{vmatrix} + 4 \begin{vmatrix} 0 & 1 \\ 1 & -1 \end{vmatrix}.$$

We calculate as follows:

$$-1 \begin{vmatrix} 1 & 1 \\ 8 & 1 \end{vmatrix} - (-5) \begin{vmatrix} 0 & 1 \\ 8 & 1 \end{vmatrix} + 4 \begin{vmatrix} 0 & 1 \\ 1 & -1 \end{vmatrix}$$

$$= -1[1 \cdot 1 - 8(-1)] + 5(0 \cdot 1 - 8 \cdot 1) + 4[0 \cdot (-1) - 1 \cdot 1]$$
$$= -1(9) + 5(-8) + 4(-1)$$
$$= -9 - 40 - 4$$
$$= -53.$$

Do Exercises 3 and 4.

c | Solving Systems by Determinants

Here is a system of two equations in two variables:

$$a_1 x + b_1 y = c_1,$$
$$a_2 x + b_2 y = c_2.$$

We form three determinants, which we call D, D_x, and D_y.

$$D = \begin{vmatrix} a_1 & b_1 \\ a_2 & b_2 \end{vmatrix}$$
In D, we have the coefficients of x and y.

$$D_x = \begin{vmatrix} c_1 & b_1 \\ c_2 & b_2 \end{vmatrix}$$
To form D_x, we replace the x-coefficients in D with the constants on the right side of the equations.

$$D_y = \begin{vmatrix} a_1 & c_1 \\ a_2 & c_2 \end{vmatrix}$$
To form D_y, we replace the y-coefficients in D with the constants on the right.

It is important that the replacement be done *without changing the order of the columns.* Then the solution of the system can be found as follows. This is known as **Cramer's rule.**

Evaluate.

3.
$$\begin{vmatrix} 2 & -1 & 1 \\ 1 & 2 & -1 \\ 3 & 4 & -3 \end{vmatrix}$$

4.
$$\begin{vmatrix} 3 & 2 & 2 \\ -2 & 1 & 4 \\ 4 & -3 & 3 \end{vmatrix}$$

Answers on page A-62

5. Solve using Cramer's rule.

$$4x - 3y = 15,$$
$$x + 3y = 0$$

Answer on page A-62

> **CRAMER'S RULE**
>
> $$x = \frac{D_x}{D}, \qquad y = \frac{D_y}{D}$$

Example 3   Solve using Cramer's rule:

$$3x - 2y = 7,$$
$$3x + 2y = 9.$$

We compute D, D_x, and D_y:

$$D = \begin{vmatrix} 3 & -2 \\ 3 & 2 \end{vmatrix} = 3 \cdot 2 - 3 \cdot (-2) = 6 + 6 = 12;$$

$$D_x = \begin{vmatrix} 7 & -2 \\ 9 & 2 \end{vmatrix} = 7 \cdot 2 - 9(-2) = 14 + 18 = 32;$$

$$D_y = \begin{vmatrix} 3 & 7 \\ 3 & 9 \end{vmatrix} = 3 \cdot 9 - 3 \cdot 7 = 27 - 21 = 6.$$

Then

$$x = \frac{D_x}{D} = \frac{32}{12}, \text{ or } \frac{8}{3} \quad \text{and} \quad y = \frac{D_y}{D} = \frac{6}{12} = \frac{1}{2}.$$

The solution is $\left(\frac{8}{3}, \frac{1}{2}\right)$.

Do Exercise 5.

Cramer's rule for three equations is very similar to that for two.

$$a_1 x + b_1 y + c_1 z = d_1$$
$$a_2 x + b_2 y + c_2 z = d_2$$
$$a_3 x + b_3 y + c_3 z = d_3$$

$$D = \begin{vmatrix} a_1 & b_1 & c_1 \\ a_2 & b_2 & c_2 \\ a_3 & b_3 & c_3 \end{vmatrix} \qquad D_x = \begin{vmatrix} d_1 & b_1 & c_1 \\ d_2 & b_2 & c_2 \\ d_3 & b_3 & c_3 \end{vmatrix}$$

$$D_y = \begin{vmatrix} a_1 & d_1 & c_1 \\ a_2 & d_2 & c_2 \\ a_3 & d_3 & c_3 \end{vmatrix}$$

D is again the determinant of the coefficients of x, y, and z. This time we have one more determinant, D_z. We get it by replacing the z-coefficients in D with the constants on the right:

$$D_z = \begin{vmatrix} a_1 & b_1 & d_1 \\ a_2 & b_2 & d_2 \\ a_3 & b_3 & d_3 \end{vmatrix}.$$

The solution of the system is given by

> **CRAMER'S RULE**
>
> $$x = \frac{D_x}{D}, \qquad y = \frac{D_y}{D}, \qquad z = \frac{D_z}{D}$$

Example 4 Solve using Cramer's rule:

$$x - 3y + 7z = 13,$$
$$x + y + z = 1,$$
$$x - 2y + 3z = 4.$$

We compute D, D_x, D_y, and D_z:

$$D = \begin{vmatrix} 1 & -3 & 7 \\ 1 & 1 & 1 \\ 1 & -2 & 3 \end{vmatrix} = -10; \qquad D_x = \begin{vmatrix} 13 & -3 & 7 \\ 1 & 1 & 1 \\ 4 & -2 & 3 \end{vmatrix} = 20;$$

$$D_y = \begin{vmatrix} 1 & 13 & 7 \\ 1 & 1 & 1 \\ 1 & 4 & 3 \end{vmatrix} = -6; \qquad D_z = \begin{vmatrix} 1 & -3 & 13 \\ 1 & 1 & 1 \\ 1 & -2 & 4 \end{vmatrix} = -24.$$

Then

$$x = \frac{D_x}{D} = \frac{20}{-10} = -2;$$

$$y = \frac{D_y}{D} = \frac{-6}{-10} = \frac{3}{5};$$

$$z = \frac{D_z}{D} = \frac{-24}{-10} = \frac{12}{5}.$$

The solution is $\left(-2, \frac{3}{5}, \frac{12}{5}\right)$.

In Example 4, we would not have needed to evaluate D_z. Once we found x and y, we could have substituted them into one of the equations to find z. In practice, it is faster to use determinants to find only two of the numbers; then we find the third by substitution into an equation.

Do Exercise 6.

In using Cramer's rule, we divide by D. If D were 0, we could not do so.

> If $D = 0$ and at least one of the other determinants is not 0, then the system does not have a solution, and we say that it is *inconsistent*.
>
> If $D = 0$ and all the other determinants are also 0, then there is an infinite set of solutions. In that case, we say that the system is *dependent*.

6. Solve using Cramer's rule:

$$x - 3y - 7z = 6,$$
$$2x + 3y + z = 9,$$
$$4x + y = 7.$$

Answer on page A-62

Exercise Set B

a Evaluate.

1. $\begin{vmatrix} 3 & 7 \\ 2 & 8 \end{vmatrix}$

2. $\begin{vmatrix} 5 & 4 \\ 4 & -5 \end{vmatrix}$

3. $\begin{vmatrix} -3 & -6 \\ -5 & -10 \end{vmatrix}$

4. $\begin{vmatrix} 4 & 5 \\ -7 & 9 \end{vmatrix}$

5. $\begin{vmatrix} 8 & 2 \\ 12 & -3 \end{vmatrix}$

6. $\begin{vmatrix} 1 & 1 \\ 9 & 8 \end{vmatrix}$

7. $\begin{vmatrix} 2 & -7 \\ 0 & 0 \end{vmatrix}$

8. $\begin{vmatrix} 0 & -4 \\ 0 & -6 \end{vmatrix}$

b Evaluate.

9. $\begin{vmatrix} 0 & 2 & 0 \\ 3 & -1 & 1 \\ 1 & -2 & 2 \end{vmatrix}$

10. $\begin{vmatrix} 3 & 0 & -2 \\ 5 & 1 & 2 \\ 2 & 0 & -1 \end{vmatrix}$

11. $\begin{vmatrix} -1 & -2 & -3 \\ 3 & 4 & 2 \\ 0 & 1 & 2 \end{vmatrix}$

12. $\begin{vmatrix} 1 & 2 & 2 \\ 2 & 1 & 0 \\ 3 & 3 & 1 \end{vmatrix}$

13. $\begin{vmatrix} 3 & 2 & -2 \\ -2 & 1 & 4 \\ -4 & -3 & 3 \end{vmatrix}$

14. $\begin{vmatrix} 2 & -1 & 1 \\ 1 & 2 & -1 \\ 3 & 4 & -3 \end{vmatrix}$

15. $\begin{vmatrix} 3 & 2 & 4 \\ 1 & 1 & 1 \\ 1 & 1 & 1 \end{vmatrix}$

16. $\begin{vmatrix} -1 & 6 & -5 \\ 2 & 4 & 4 \\ 5 & 3 & 10 \end{vmatrix}$

c Solve using Cramer's rule.

17. $3x - 4y = 6,$
$5x + 9y = 10$

18. $5x + 8y = 1,$
$3x + 7y = 5$

19. $-2x + 4y = 3,$
$3x - 7y = 1$

20. $5x - 4y = -3,$
$7x + 2y = 6$

21. $4x + 2y = 11,$
$3x - y = 2$

22. $3x - 3y = 11,$
$9x - 2y = 5$

23. $x + 4y = 8,$
$3x + 5y = 3$

24. $x + 4y = 5,$
$-3x + 2y = 13$

25. $2x - 3y + 5z = 27,$
$x + 2y - z = -4,$
$5x - y + 4z = 27$

26. $x - y + 2z = -3,$
$x + 2y + 3z = 4,$
$2x + y + z = -3$

27. $r - 2s + 3t = 6,$
$2r - s - t = -3,$
$r + s + t = 6$

28. $a - 3c = 6,$
$b + 2c = 2,$
$7a - 3b - 5c = 14$

29. $4x - y - 3z = 1,$
$8x + y - z = 5,$
$2x + y + 2z = 5$

30. $3x + 2y + 2z = 3,$
$x + 2y - z = 5,$
$2x - 4y + z = 0$

31. $p + q + r = 1,$
$p - 2q - 3r = 3,$
$4p + 5q + 6r = 4$

32. $x + 2y - 3z = 9,$
$2x - y + 2z = -8,$
$3x - y - 4z = 3$

Appendix C Elimination Using Matrices

The elimination method concerns itself primarily with the coefficients and constants of the equations. In what follows, we learn a method for solving systems using just the coefficients and the constants. This procedure involves what are called *matrices*.

Objective

a Solve systems of two or three equations using matrices.

a In solving systems of equations, we perform computations with the constants. The variables play no important role until the end. Thus we can simplify writing a system by omitting the variables. For example, the system

$$\begin{matrix} 3x + 4y = 5, \\ x - 2y = 1 \end{matrix} \quad \text{simplifies to} \quad \begin{matrix} 3 & 4 & 5 \\ 1 & -2 & 1 \end{matrix}$$

if we omit the variables, the operation of addition, and the equals signs, the result is a rectangular array of numbers. Such an array is called a **matrix** (plural, **matrices**). We ordinarily write brackets around matrices. The following are matrices:

$$\begin{bmatrix} 4 & 1 & 3 & 5 \\ 1 & 0 & 1 & 2 \\ 6 & 3 & -2 & 0 \end{bmatrix}, \quad \begin{bmatrix} 6 & 2 & 1 & 4 & 7 \\ 1 & 2 & 1 & 3 & 1 \\ 4 & 0 & -2 & 0 & -3 \end{bmatrix}, \quad \begin{bmatrix} 1 & 2 \\ 145 & 0 \\ -7 & 9 \\ 8 & 1 \\ 0 & 0 \end{bmatrix}.$$

The **rows** of a matrix are horizontal, and the **columns** are vertical.

$$\begin{bmatrix} 5 & -2 & 2 \\ 1 & 0 & 1 \\ 0 & 1 & 2 \end{bmatrix} \begin{matrix} \longleftarrow & \text{row 1} \\ \longleftarrow & \text{row 2} \\ \longleftarrow & \text{row 3} \end{matrix}$$

column 1   column 2   column 3

Let's now use matrices to solve systems of linear equations.

Example 1 Solve the system

$$\begin{matrix} 5x - 4y = -1, \\ -2x + 3y = 2. \end{matrix}$$

We write a matrix using only the coefficients and the constants, keeping in mind that x corresponds to the first column and y to the second. A dashed line separates the coefficients from the constants at the end of each equation:

$$\begin{bmatrix} 5 & -4 & | & -1 \\ -2 & 3 & | & 2 \end{bmatrix}. \quad \begin{matrix} \textbf{The individual numbers are} \\ \textbf{called } \textit{elements} \textbf{ or } \textit{entries.} \end{matrix}$$

Our goal is to transform this matrix into one of the form

$$\begin{bmatrix} a & b & | & c \\ 0 & d & | & e \end{bmatrix}.$$

The variables can then be reinserted to form equations from which we can complete the solution.

1. Solve using matrices:

$$5x - 2y = -44,$$
$$2x + 5y = -6.$$

We do calculations that are similar to those that we would do if we wrote the entire equations. The first step, if possible, is to multiply and/or interchange the rows so that each number in the first column below the first number is a multiple of that number. In this case, we do so by multiplying Row 2 by 5. This corresponds to multiplying the second equation by 5.

$$\begin{bmatrix} 5 & -4 & \vdots & -1 \\ -10 & 15 & \vdots & 10 \end{bmatrix}$$ **New Row 2 = 5(Row 2)**

Next, we multiply the first row by 2 and add the result to the second row. This corresponds to multiplying the first equation by 2 and adding the result to the second equation. Although we write these equations out here, we generally try to do them mentally:

$$2 \cdot 5 + (-10) = 0; \qquad 2(-4) + 15 = 7; \qquad 2(-1) + 10 = 8.$$

$$\begin{bmatrix} 5 & -4 & \vdots & -1 \\ 0 & 7 & \vdots & 8 \end{bmatrix}$$ **New Row 2 = 2(Row 1) + (Row 2)**

If we now reinsert the variables, we have

$$5x - 4y = -1, \qquad \textbf{(1)}$$
$$7y = 8. \qquad \textbf{(2)}$$

We can now proceed as before, solving equation (2) for y:

$$7y = 8 \qquad \textbf{(2)}$$
$$y = \tfrac{8}{7}.$$

Next, we substitute $\tfrac{8}{7}$ for y back in equation (1). This procedure is called *back-substitution.*

$$5x - 4y = -1 \qquad \textbf{(1)}$$
$$5x - 4 \cdot \tfrac{8}{7} = -1 \qquad \text{Substituting } \tfrac{8}{7} \text{ for } y \text{ in equation (1)}$$
$$x = \tfrac{5}{7} \qquad \text{Solving for } x$$

The solution is $\left(\tfrac{5}{7}, \tfrac{8}{7} \right)$.

Do Exercise 1.

Example 2   Solve the system

$$2x - y + 4z = -3,$$
$$x \qquad - 4z = 5,$$
$$6x - y + 2z = 10.$$

We first write a matrix, using only the coefficients and the constants. Where there are missing terms, we must write 0's:

$$\begin{bmatrix} 2 & -1 & 4 & \vdots & -3 \\ 1 & 0 & -4 & \vdots & 5 \\ 6 & -1 & 2 & \vdots & 10 \end{bmatrix}$$ **(P1)**
(P2)
(P3)

(P1), (P2), and (P3) designate the equations that are in the first, second, and third position, respectively.

Our goal is to find an equivalent matrix of the form

$$\begin{bmatrix} a & b & c & \vdots & d \\ 0 & e & f & \vdots & g \\ 0 & 0 & h & \vdots & i \end{bmatrix}.$$

A matrix of this form can be rewritten as a system of equations from which a solution can be found easily.

Answer on page A-62

The first step, if possible, is to interchange the rows so that each number in the first column below the first number is a multiple of that number. In this case, we do so by interchanging Rows 1 and 2:

$$\begin{bmatrix} 1 & 0 & -4 & \vdots & 5 \\ 2 & -1 & 4 & \vdots & -3 \\ 6 & -1 & 2 & \vdots & 10 \end{bmatrix}$$ **This corresponds to interchanging the first two equations.**

Next, we multiply the first row by -2 and add it to the second row:

$$\begin{bmatrix} 1 & 0 & -4 & \vdots & 5 \\ 0 & -1 & 12 & \vdots & -13 \\ 6 & -1 & 2 & \vdots & 10 \end{bmatrix}.$$ **This corresponds to multiplying new equation (P1) by -2 and adding it to new equation (P2). The result replaces the former (P2). We perform the calculations mentally.**

Now we multiply the first row by -6 and add it to the third row:

$$\begin{bmatrix} 1 & 0 & -4 & \vdots & 5 \\ 0 & -1 & 12 & \vdots & -13 \\ 0 & -1 & 26 & \vdots & -20 \end{bmatrix}.$$ **This corresponds to multiplying equation (P1) by -6 and adding it to equation (P3).**

Next, we multiply Row 2 by -1 and add it to the third row:

$$\begin{bmatrix} 1 & 0 & -4 & \vdots & 5 \\ 0 & -1 & 12 & \vdots & -13 \\ 0 & 0 & 14 & \vdots & -7 \end{bmatrix}.$$ **This corresponds to multiplying equation (P2) by -1 and adding it to equation (P3).**

Reinserting the variables gives us

$$x \quad\quad - 4z = 5, \quad\quad \textbf{(P1)}$$
$$-y + 12z = -13, \quad\quad \textbf{(P2)}$$
$$14z = -7. \quad\quad \textbf{(P3)}$$

We now solve (P3) for z:

$$14z = -7 \quad\quad \textbf{(P3)}$$
$$z = -\tfrac{7}{14} \quad\quad \text{Solving for } z$$
$$z = -\tfrac{1}{2}.$$

Next, we back-substitute $-\tfrac{1}{2}$ for z in (P2) and solve for y:

$$-y + 12z = -13 \quad\quad \textbf{(P2)}$$
$$-y + 12\left(-\tfrac{1}{2}\right) = -13 \quad\quad \text{Substituting } -\tfrac{1}{2} \text{ for } z \text{ in equation (P2)}$$
$$-y - 6 = -13$$
$$-y = -7$$
$$y = 7. \quad\quad \text{Solving for } y$$

Since there is no y-term in (P1), we need only substitute $-\tfrac{1}{2}$ for z in (P1) and solve for x:

$$x - 4z = 5 \quad\quad \textbf{(P1)}$$
$$x - 4\left(-\tfrac{1}{2}\right) = 5 \quad\quad \text{Substituting } -\tfrac{1}{2} \text{ for } z \text{ in equation (P1)}$$
$$x + 2 = 5$$
$$x = 3. \quad\quad \text{Solving for } x$$

The solution is $\left(3, 7, -\tfrac{1}{2}\right)$.

Do Exercise 2.

---

2. Solve using matrices:

$$x - 2y + 3z = 4,$$
$$2x - y + z = -1,$$
$$4x + y + z = 1.$$

Answer on page A-62

All the operations used in the preceding example correspond to operations with the equations and produce equivalent systems of equations. We call the matrices **row-equivalent** and the operations that produce them **row-equivalent operations.**

ROW-EQUIVALENT OPERATIONS

Each of the following row-equivalent operations produces an equivalent matrix:

a) Interchanging any two rows.

b) Multiplying each element of a row by the same nonzero number.

c) Multiplying each element of a row by a nonzero number and adding the result to another row.

The best overall method for solving systems of equations is by row-equivalent matrices; graphers and computers are programmed to use them. Matrices are part of a branch of mathematics known as linear algebra. They are also studied in more detail in many courses in finite mathematics.

Exercise Set C

Solve using matrices.

1. $4x + 2y = 11,$
   $3x - y = 2$

2. $3x - 3y = 11,$
   $9x - 2y = 5$

3. $x + 4y = 8,$
   $3x + 5y = 3$

4. $x + 4y = 5,$
   $-3x + 2y = 13$

5. $5x - 3y = -2,$
   $4x + 2y = 5$

6. $3x + 4y = 7,$
   $-5x + 2y = 10$

7. $2x - 3y = 50,$
   $5x + y = 40$

8. $4x + 5y = -8,$
   $7x + 9y = 11$

9. $4x - y - 3z = 1,$
   $8x + y - z = 5,$
   $2x + y + 2z = 5$

10. $3x + 2y + 2z = 3,$
    $x + 2y - z = 5,$
    $2x - 4y + z = 0$

11. $p + q + r = 1,$
    $p - 2q - 3r = 3,$
    $4p + 5q + 6r = 4$

12. $x + 2y - 3z = 9,$
    $2x - y + 2z = -8,$
    $3x - y - 4z = 3$

13. $x - y + 2z = 0,$
    $x - 2y + 3z = -1,$
    $2x - 2y + z = -3$

14. $4a + 9b = 8,$
    $8a + 6c = -1,$
    $6b + 6c = -1$

15. $3p + 2r = 11,$
    $q - 7r = 4,$
    $p - 6q = 1$

16. $m + n + t = 6,$
    $m - n - t = 0,$
    $m + 2n + t = 5$

17. $2x + 2y - 2z - 2w = -10,$
    $x + y + z + w = -5,$
    $x - y + 4z + 3w = -2,$
    $3x - 2y + 2z + w = -6$

18. $2x - 3y + z - w = -8,$
    $x + y - z - w = -4,$
    $x + y + z + w = 22,$
    $x - y - z - w = -14$

Appendix D The Algebra of Functions

Objective

a Given two functions f and g, find their sum, difference, product, and quotient.

a   The Sum, Difference, Product, and Quotient of Functions

Suppose that a is in the domain of two functions, f and g. The input a is paired with $f(a)$ by f and with $g(a)$ by g. The outputs can then be added to get $f(a) + g(a)$.

Example 1   Let $f(x) = x + 4$ and $g(x) = x^2 + 1$. Find $f(2) + g(2)$.

We visualize two function machines. Because 2 is in the domain of each function, we can compute $f(2)$ and $g(2)$.

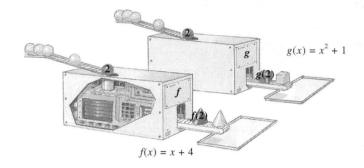

Since

$$f(2) = 2 + 4 = 6 \quad \text{and} \quad g(2) = 2^2 + 1 = 5,$$

we have

$$f(2) + g(2) = 6 + 5 = 11.$$

In Example 1, suppose that we were to write $f(x) + g(x)$ as $(x + 4) + (x^2 + 1)$, or $f(x) + g(x) = x^2 + x + 5$. This could then be regarded as a "new" function: $(f + g)(x) = x^2 + x + 5$. We can alternatively find $f(2) + g(2)$ with $(f + g)(x)$:

$$(f + g)(x) = x^2 + x + 5$$
$$(f + g)(2) = 2^2 + 2 + 5 \qquad \text{Substituting 2 for } x$$
$$= 4 + 2 + 5$$
$$= 11.$$

Similar notations exist for subtraction, multiplication, and division of functions.

> From any functions f and g, we can form new functions defined as:
>
> **1.** The **sum $f + g$:** $(f + g)(x) = f(x) + g(x)$;
> **2.** The **difference $f - g$:** $(f - g)(x) = f(x) - g(x)$;
> **3.** The **product fg:** $(f \cdot g)(x) = f(x) \cdot g(x)$;
> **4.** The **quotient f/g:** $(f/g)(x) = f(x)/g(x)$, where $g(x) \neq 0$.

Example 2   Given f and g described by $f(x) = x^2 - 5$ and $g(x) = x + 7$, find $(f + g)(x)$, $(f - g)(x)$, $(f \cdot g)(x)$, $(f/g)(x)$, and $(g \cdot g)(x)$.

$$(f + g)(x) = f(x) + g(x) = (x^2 - 5) + (x + 7) = x^2 + x + 2;$$
$$(f - g)(x) = f(x) - g(x) = (x^2 - 5) - (x + 7) = x^2 - x - 12;$$
$$(f \cdot g)(x) = f(x) \cdot g(x) = (x^2 - 5)(x + 7) = x^3 + 7x^2 - 5x - 35;$$
$$(f/g)(x) = f(x)/g(x) = \frac{x^2 - 5}{x + 7};$$
$$(g \cdot g)(x) = g(x) \cdot g(x) = (x + 7)(x + 7) = x^2 + 14x + 49$$

Note that the sum, difference, and product of polynomials are also polynomial functions, but the quotient may not be.

Do Exercise 1.

Example 3   For $f(x) = x^2 - x$ and $g(x) = x + 2$, find $(f + g)(3)$, $(f - g)(-1)$, $(f \cdot g)(5)$, and $(f/g)(-4)$.

We first find $(f + g)(x)$, $(f - g)(x)$, $(f \cdot g)(x)$, and $(f/g)(x)$.

$$(f + g)(x) = f(x) + g(x) = x^2 - x + x + 2$$
$$= x^2 + 2;$$

$$(f - g)(x) = f(x) - g(x) = x^2 - x - (x + 2)$$
$$= x^2 - x - x - 2$$
$$= x^2 - 2x - 2;$$

$$(f \cdot g)(x) = f(x) \cdot g(x) = (x^2 - x)(x + 2)$$
$$= x^3 + 2x^2 - x^2 - 2x$$
$$= x^3 + x^2 - 2x;$$

$$(f/g)(x) = \frac{f(x)}{g(x)} = \frac{x^2 - x}{x + 2}.$$

Then we substitute.

$$(f + g)(3) = 3^2 + 2 \qquad \text{Using } (f + g)(x) = x^2 + 2$$
$$= 9 + 2 = 11;$$

$$(f - g)(-1) = (-1)^2 - 2(-1) - 2 \qquad \text{Using } (f - g)(x) = x^2 - 2x - 2$$
$$= 1 + 2 - 2 = 1;$$

$$(f \cdot g)(5) = 5^3 + 5^2 - 2 \cdot 5 \qquad \text{Using } (f \cdot g)(x) = x^3 + x^2 - 2x$$
$$= 125 + 25 - 10 = 140;$$

$$(f/g)(-4) = \frac{(-4)^2 - (-4)}{-4 + 2} \qquad \text{Using } (f/g)(x) = (x^2 - x)/(x + 2)$$
$$= \frac{16 + 4}{-2} = \frac{20}{-2} = -10$$

Do Exercise 2.

1. Given $f(x) = x^2 + 3$ and $g(x) = x^2 - 3$, find each of the following.
   a) $(f + g)(x)$

   b) $(f - g)(x)$

   c) $(f \cdot g)(x)$

   d) $(f/g)(x)$

   e) $(f \cdot f)(x)$

2. Given $f(x) = x^2 + x$ and $g(x) = 2x - 3$, find each of the following.
   a) $(f + g)(-2)$

   b) $(f - g)(4)$

   c) $(f \cdot g)(-3)$

   d) $(f/g)(2)$

Answers on page A-62

Appendix D   The Algebra of Functions

745

Exercise Set D

Let $f(x) = -3x + 1$ and $g(x) = x^2 + 2$. Find the following.

1. $f(2) + g(2)$

2. $f(-1) + g(-1)$

3. $f(5) - g(5)$

4. $f(4) - g(4)$

5. $f(-1) \cdot g(-1)$

6. $f(-2) \cdot g(-2)$

7. $f(-4)/g(-4)$

8. $f(3)/g(3)$

9. $g(1) - f(1)$

10. $g(2)/f(2)$

11. $g(0)/f(0)$

12. $g(6) - f(6)$

Let $f(x) = x^2 - 3$ and $g(x) = 4 - x$. find the following.

13. $(f + g)(x)$

14. $(f - g)(x)$

15. $(f + g)(-4)$

16. $(f + g)(-5)$

17. $(f - g)(3)$

18. $(f - g)(2)$

19. $(f \cdot g)(x)$

20. $(f/g)(x)$

21. $(f \cdot g)(-3)$

22. $(f \cdot g)(-4)$

23. $(f/g)(0)$

24. $(f/g)(1)$

25. $(f/g)(-2)$

26. $(f/g)(-1)$

For each pair of functions f and g, find $(f + g)(x)$, $(f - g)(x)$, $(f \cdot g)(x)$, and $(f/g)(x)$.

27. $f(x) = x^2$,
$g(x) = 3x - 4$

28. $f(x) = 5x - 1$,
$g(x) = 2x^2$

29. $f(x) = \dfrac{1}{x - 2}$,
$g(x) = 4x^3$

30. $f(x) = 3x^2$,
$g(x) = \dfrac{1}{x - 4}$

31. $f(x) = \dfrac{3}{x - 2}$,
$g(x) = \dfrac{5}{4 - x}$

32. $f(x) = \dfrac{5}{x - 3}$,
$g(x) = \dfrac{1}{x - 2}$

Answers

Diagnostic Pretest, p. xxi

1. [R.2c] 3.57   **2.** [R.2d] $-\frac{1}{9}$ **3.** [R.6b] $6x + 8$

4. [R.7b] $\dfrac{4y^{14}}{9x^6}$ **5.** [1.1d] $\frac{1}{4}$

6. [1.5a] $\{x \mid -3 \le x \le 2\}$, or $[-3, 2]$

7. [1.6e] $\{x \mid x < -2 \ or \ x > -1\}$, or $(-\infty, -2) \cup (-1, \infty)$

8. [1.2a] $P = \dfrac{A}{1 + rt}$

9. [1.3a] Length = 94 ft; width = 50 ft

10. [2.5a]

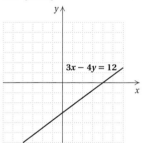

11. [2.5c]

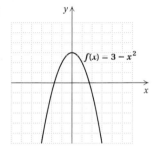

12. [2.2c]

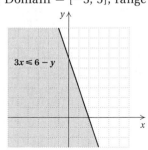

13. [2.6c] $y = -x + 4$ **14.** [2.6d] $y = -\frac{1}{3}x + 3$

15. [2.2b] $f(-1) = 2;\ f(0) = 3;\ f(2) = -1$

16. [2.3a] Domain = $[-3, 5]$; range = $[1, 4]$

17. [3.6b]

18. [3.2b] $\left(\frac{13}{7}, \frac{4}{7}\right)$ **19.** [3.4a] $(3, 0, -2)$

20. [3.3b] 21 mph

21. [3.6c]

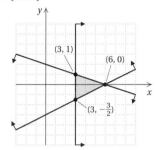

22. [4.2c] $9x^2 - 30xy + 25y^2$

23. [4.5b] $(x^2 + 1)(x + 1)(x - 1)$ **24.** [4.7a] $-2, 9$

25. [4.7b] $-\frac{1}{3}, 2$ **26.** [4.1b] **(a)** 247 thousand;

(b) 65 thousand   **27.** [5.1e] $\dfrac{2x + 6}{x^2 - 2x - 3}$

28. [5.4a] $x - 1$ **29.** [5.1a] $\{x \mid x \ne 3 \ and \ x \ne -3\}$

30. [5.5a] 1   **31.** [5.6a] $1\frac{5}{7}$ hr

32. [6.4b] $6 - 4y\sqrt{3} + 4x\sqrt{3x} - 8xy\sqrt{x}$

33. [6.4a] $24\sqrt{3}$ **34.** [6.5b] $\dfrac{2x + \sqrt{xy} - y}{x - y}$

35. [6.6a] -3 **36.** [6.1a] $\{x \mid x \le 2\}$, or $(-\infty, 2]$

37. [7.2a] $-1 \pm 2i$ **38.** [7.4c] $-1, 6$ **39.** [7.4c] $-2, 1$

40. [7.6a]

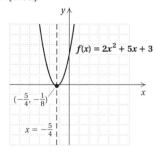

41. [8.2c] $f^{-1}(x) = \dfrac{x + 5}{2}$

42. [8.1a]

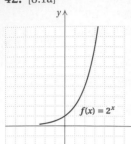

43. [8.3a]

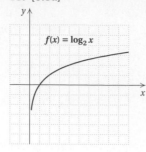

44. [8.4d] $\log_3 \dfrac{\sqrt{x}}{y}$ **45.** [8.3c] $2^y = x$ **46.** [8.6b] 0

47. [8.6b] $\sqrt{3}$

48. [9.2a]

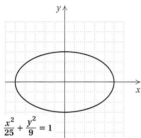

$\dfrac{x^2}{25} + \dfrac{y^2}{9} = 1$

49. [9.1d]

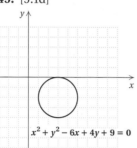

$x^2 + y^2 - 6x + 4y + 9 = 0$

50. [9.4a] $(2\sqrt{2}, \sqrt{2}), (-2\sqrt{2}, -\sqrt{2}), (\sqrt{2}\,i, -2\sqrt{2}\,i),$
$(-\sqrt{2}\,i, 2\sqrt{2}\,i)$

Chapter R

Pretest: Chapter R, p. 2

1. [R.1a] $-43, -\frac{2}{3}, 2.3\overline{76}$ **2.** [R.1b] $<$ **3.** [R.1b] $2 < x$
4. [R.1b] False   **5.** [R.1d] 3.6   **6.** [R.2a] 16
7. [R.2c] -11.6 **8.** [R.2d] $-\frac{6}{7}$ **9.** [R.2e] 8

10. [R.3a] q^3 **11.** [R.3a] 64^5 **12.** [R.3b] $\dfrac{1}{x^3}$

13. [R.3c] 16   **14.** [R.3c] $-\frac{16}{25}$ **15.** [R.4a] $47\%x$, or

$0.47x$ **16.** [R.4b] 17   **17.** [R.5b] $\dfrac{35x}{28x}$ **18.** [R.5b] $-\frac{3}{2}$

19. [R.5d] $-4x + 12y$ **20.** [R.5d] $6(x - 3xy + 4)$
21. [R.6a] $-x + 6y + 11$ **22.** [R.6b] $2x - 5y + 24$

23. [R.6b] $-4x - 56$ **24.** [R.7a] $-\dfrac{20}{x^3}$

25. [R.7a] $-\dfrac{4}{x^2y^2}$ **26.** [R.7b] $\dfrac{x^{12}}{25y^{10}}$

27. [R.7c] 7.86×10^{-8} **28.** [R.7c] 4.5789×10^{11}
29. [R.7c] 78,900,000,000,000   **30.** [R.7c] 0.00000789

Margin Exercises, Section R.1, pp. 3–9

1. -9 **2.** 6   **3.** 0   **4.** {1, 3, 5, 7, 9, 11, 13}; {$x \mid x$ is
one of the first seven odd whole numbers}
5. 0.6875; terminating   **6.** $1.0\overline{769230}$; repeating
7. (a) 20; (b) 20, 0; (c) 20, -10, 0; (d) $\sqrt{7}, -\sqrt{2}$,
9.34334333433334...; (e) 20, -10, -5.34, 18.999, $\frac{11}{45}$, 0,

$-\frac{2}{3}$; (f) 20, -10, -5.34, 18.999, $\frac{11}{45}$, $\sqrt{7}$, $-\sqrt{2}$, 0, $-\frac{2}{3}$,
9.34334333433334...   **8.** $<$ **9.** $>$ **10.** $>$ **11.** $<$
12. $>$ **13.** $>$ **14.** $<$ **15.** $>$ **16.** $>$ **17.** $6 < x$
18. $7 > -4$ **19.** True   **20.** True   **21.** False
22.

23.

24. $\frac{1}{4}$ **25.** 2   **26.** $\frac{3}{2}$ **27.** 2.3

Calculator Spotlight, p. 5

1. 9.327379053   **2.** 21.37755833   **3.** 145.4647724
4. 37.8556546   **5.** 3.141592654   **6.** 103.6725576
7. 178.4424627   **8.** 138.8663978

Exercise Set R.1, p. 11

1. 1, 12, $\sqrt{25}$ **3.** $-6, 0, 1, -\frac{1}{2}, -4, \frac{7}{9}, 12, -\frac{6}{5}, 3.45, 5\frac{1}{2}$,
$\sqrt{25}, -\frac{12}{3}$ **5.** $-6, 0, 1, -\frac{1}{2}, -4, \frac{7}{9}, 12, -\frac{6}{5}, 3.45, 5\frac{1}{2}$,
$\sqrt{3}, \sqrt{25}, -\frac{12}{3}$, 0.131331333133331...   **7.** 12, 0
9. $-11, 12, 0$ **11.** $-\sqrt{5}, \pi, -3.565665666566665...$
13. {m, a, t, h}   **15.** {1, 2, 3, 4, 5, 6, 7, 8, 9, 10, 11, 12}
17. {2, 4, 6, 8, ...}
19. {$x \mid x$ is a whole number less than or equal to 5}
21. $\left\{ \dfrac{a}{b} \,\middle|\, a \text{ and } b \text{ are integers and } b \neq 0 \right\}$
23. {$x \mid x > -3$}   **25.** $>$ **27.** $<$ **29.** $<$ **31.** $<$
33. $>$ **35.** $<$ **37.** $>$ **39.** $<$ **41.** $x < -8$
43. $y \geq -12.7$ **45.** False   **47.** True
49.

51.

53.

55.

57. 6   **59.** 28   **61.** 35   **63.** $\frac{2}{3}$ **65.** 0   **67.** ◆
69. $\leq$ **71.** $\leq$ **73.** $\frac{1}{8}\%, 0.3\%, 0.009, 1\%, 1.1\%, \frac{9}{100}, \frac{1}{11}$,
$\frac{99}{1000}, 0.11, \frac{1}{8}, \frac{2}{7}, 0.286$

Margin Exercises, Section R.2, pp. 13–18

1. 2   **2.** 4   **3.** -4 **4.** 0   **5.** -18 **6.** -18.6
7. $-\frac{29}{5}$ **8.** $-\frac{7}{10}$ **9.** 0   **10.** -7.4 **11.** -3
12. -3.3 **13.** $-\frac{1}{4}$ **14.** $\frac{1}{10}$ **15.** 14   **16.** $-\frac{2}{3}$
17. 0   **18.** -9 **19.** $\frac{3}{5}$ **20.** -5.9 **21.** $\frac{2}{3}$
22. (a) -11; (b) 17; (c) 0; (d) $-x$; (e) x **23.** 17
24. -16 **25.** -3 **26.** -29.6 **27.** 3   **28.** 1
29. $\frac{3}{2}$ **30.** (a) -6; (b) -40; (c) 6   **31.** 10, 5, 0, -5;
$-10, -15, -20, -25, -30$ **32.** -24 **33.** -28.35
34. -8 **35.** $-10, -5, 0; 5, 10, 15, 20, 25$ **36.** 72
37. $\frac{8}{15}$ **38.** 42.77   **39.** 2   **40.** -25 **41.** -3
42. 0.2   **43.** Undefined   **44.** 0   **45.** Undefined
46. Undefined   **47.** $\frac{8}{3}$ **48.** $-\frac{5}{4}$ **49.** $\frac{1}{18}$ **50.** $-\frac{1}{43}$
or $-\frac{10}{43}$ **51.** $\frac{1}{0.5}$, or 2   **52.** $-\frac{4}{5}, \frac{5}{4}; \frac{3}{4}, -\frac{4}{3}; -0.25, 4;$
$-8, \frac{1}{8}; 5, -\frac{1}{5}; 0$, does not exist   **53.** $-\frac{6}{7}$ **54.** $\frac{36}{7}$

Exercise Set R.2, p. 19

1. -28 **3.** 5 **5.** -16 **7.** -4 **9.** -10 **11.** -26
13. 1.2 **15.** -8.86 **17.** $-\frac{1}{3}$ **19.** $-\frac{4}{3}$ **21.** $\frac{1}{10}$
23. $\frac{7}{20}$ **25.** 4 **27.** -3.7 **29.** -10 **31.** 0 **33.** -4
35. -14 **37.** 0 **39.** -46 **41.** 5 **43.** 15
45. -11.6 **47.** -29.25 **49.** $-\frac{7}{2}$ **51.** $-\frac{1}{4}$ **53.** $-\frac{19}{12}$
55. $-\frac{7}{15}$ **57.** -21 **59.** -8 **61.** 24 **63.** -112
65. 34.2 **67.** $-\frac{12}{35}$ **69.** 2 **71.** 60 **73.** 26.46
75. 1 **77.** $-\frac{8}{27}$ **79.** -2 **81.** -7 **83.** 7 **85.** 0.3
87. Undefined **89.** 0 **91.** Undefined **93.** $\frac{4}{3}$
95. $-\frac{8}{7}$ **97.** $\frac{1}{25}$ **99.** 5 **101.** $-\frac{b}{a}$ **103.** $-\frac{6}{77}$
105. 25 **107.** -6 **109.** 5 **111.** -120 **113.** $-\frac{9}{8}$
115. $\frac{5}{3}$ **117.** $\frac{3}{2}$ **119.** $\frac{9}{64}$ **121.** -2 **123.** $0.\overline{923076}$
125. -1.62 **127.** Undefined **129.** $-\frac{2}{3}, \frac{3}{2}; \frac{5}{4}, -\frac{4}{5};$
0, does not exist; -1, 1; 4.5, $-\frac{1}{4.5}; -x, \frac{1}{x}$
130. 26, 0 **131.** 26 **132.** -13, 26, 0
133. $\sqrt{3}$, π, 4.57557555755557... **134.** -12.47, -13, 26,
0, $-\frac{23}{32}, \frac{7}{11}$ **135.** $\sqrt{3}$, -12.47, -13, 26, π, 0, $-\frac{23}{32}, \frac{7}{11}$,
4.57557555755557... **136.** $<$ **137.** $>$ **138.** $<$
139. $>$ **141.** ◆ **143.** $\frac{1}{4}$

Margin Exercises, Section R.3, pp. 23–26

1. 8^4 **2.** m^6 **3.** $\left(\frac{7}{8}\right)^3$ **4.** 81 **5.** $\frac{1}{16}$ **6.** 1,000,000
7. 0.008 **8.** 1131.6496 **9.** 8 **10.** -31 **11.** 1
12. 1 **13.** 1 **14.** $\frac{1}{m^4}$ **15.** $-\frac{1}{64}$ **16.** x^3 **17.** q^{-3}
18. $(-5)^{-4}$ **19.** -4117 **20.** 551 **21.** 9; 33
22. -25 **23.** 6 **24.** $\frac{54}{241}$ **25.** 34

Calculator Spotlight, p. 27

1. [(−)] [3] [ENTER] **2.** [(−)] [5] [0] [8] [ENTER]
3. [(−)] [.] [1] [7] [ENTER]
4. [−] [5] [÷] [8] [ENTER] **5.** 1024 **6.** 40,353,607
7. 361 **8.** 32.695524 **9.** 10.4976 **10.** 12,230.59046
11. 0.0423150361 **12.** 0.0033704702 **13.** 117,649
14. $-1,419,857$ **15.** $-1,124,864$ **16.** $-117,649$
17. $-1,419,857$ **18.** $-1,124,864$ **19.** 38 **20.** 81
21. 72 **22.** -11 **23.** 9 **24.** 114 **25.** 5932
26. -774.989 **27.** 743.027 **28.** 450
29. 14,321,949.1 **30.** 4 **31.** 783 **32.** 228,112.96
33. -2 **34.** 2.000011025

Exercise Set R.3, p. 29

1. 4^5 **3.** 5^6 **5.** m^3 **7.** $\left(\frac{7}{12}\right)^4$ **9.** $(123.7)^2$ **11.** 128
13. -32 **15.** $\frac{1}{81}$ **17.** -64 **19.** 31.36 **21.** 5
23. 1 **25.** 1 **27.** $\frac{7}{8}$ **29.** $\frac{1}{y^5}$ **31.** a^2 **33.** $-\frac{1}{11}$
35. 3^{-4} **37.** b^{-3} **39.** $(-16)^{-2}$ **41.** -4 **43.** -117
45. 2 **47.** 8 **49.** -358 **51.** 144; 74 **53.** -576
55. 2599 **57.** 36 **59.** 5619.712 **61.** $-200,167,769$

63. 3 **65.** 3 **67.** 16 **69.** -310 **71.** 2 **73.** 1875
75. 7804.48 **77.** 12 **79.** 8 **81.** 16 **83.** -86
85. 37 **87.** -1 **89.** 22 **91.** -39 **93.** 12
95. -549 **97.** -144 **99.** 2 **101.** $-\frac{31}{76}$ **103.** $\frac{61}{13}$
105. $\frac{9}{7}$ **106.** 2.3 **107.** 0 **108.** 900 **109.** -33
110. -79 **111.** 33 **112.** -79 **113.** -23 **114.** 23
115. -23 **116.** $\frac{5}{8}$ **117.** ◆ **119.** $25\frac{1}{4}$
121. $9 \cdot 5 + 2 - (8 \cdot 3 + 1) = 22$ **123.** 3125

Margin Exercises, Section R.4, pp. 34–36

1. $42.8 + x = 45.9$; 3.1 million **2.** $y - 16$ **3.** $47 + y$,
or $y + 47$ **4.** $16 - x$ **5.** $\frac{1}{4}t$ **6.** $8x + 6$, or $6 + 8x$
7. $99\% \cdot \frac{a}{b} - 8$, or $(0.99) \cdot \frac{a}{b} - 8$ **8.** 84 **9.** -126
10. -50 **11.** 120 **12.** 42 **13.** -146 **14.** 96 ft^2
15. 64 **16.** 64 **17.** 16

Exercise Set R.4, p. 37

1. $x + 7$, or $7 + x$ **3.** $12b$ **5.** $65\%a$, or $0.65a$
7. $2y - 9$ **9.** $10\%w + 8$, or $0.1w + 8$ **11.** $x - y - 1$
13. $\frac{90}{g}$ **15.** $a + b$, or $b + a$ **17.** $x^2 - y$ **19.** $75t$
21. -92 **23.** 3 **25.** 4 **27.** $\frac{45}{2}$, or 22.5 **29.** 16
31. 19 **33.** 57 **35.** $440.70 **37.** $A = 113.04$ cm^2;
$C = 37.68$ cm **39.** 243 **40.** -243 **41.** 100
42. 10,000 **43.** 28.09 **44.** $\frac{9}{25}$ **45.** 1 **46.** 4.5
47. $3x$ **48.** 1 **49.** ◆ **51.** $d = r \cdot t$ **53.** 9

Margin Exercises, Section R.5, pp. 39–44

1. -10, -10; 40, 40; 0, 0; yes **2.** 1, 13; 64, 34; 60.84,
32.04; no **3.** $\frac{2y}{7y}$ **4.** $\frac{6x}{10x}$ **5.** $\frac{2}{3}$ **6.** $-\frac{5}{3}$ **7.** 2; 2
8. -14; -14 **9.** 21; 21 **10.** 440; 440 **11.** $5 + y$; ba;
$mn + 8$, or $nm + 8$, or $8 + nm$ **12.** $2 \cdot (x \cdot y)$;
$(2 \cdot y) \cdot x$; $(y \cdot 2) \cdot x$; answers may vary **13.** 180; 180
14. 27; 27 **15.** 10; 10 **16.** 24; 24 **17.** $8y - 80$
18. $ax + ay - az$ **19.** $40x - 60y + 5z$ **20.** $-5x$,
$-7y$, $67t$, $-\frac{4}{5}$ **21.** $9(x + y)$ **22.** $a(c - y)$
23. $6(x - 2)$ **24.** $5(7x - 5y + 3w + 1)$
25. $b(s + t - w)$

Exercise Set R.5, p. 45

1. -10, -10; 25, 25; 0, 0; yes **3.** -12, -16; 38.4, 51.2;
0, 0; no **5.** $\frac{7x}{8x}$ **7.** $\frac{6a}{8a}$ **9.** $\frac{5}{3}$ **11.** -4 **13.** $3 + w$
15. tr **17.** $cd + 4$, $dc + 4$, or $4 + dc$ **19.** $x + yz$,
$x + zy$, or $zy + x$ **21.** $(m + n) + 2$ **23.** $7 \cdot (x \cdot y)$
25. $a + (8 + b)$, $(a + 8) + b$, $b + (a + 8)$; others are
possible **27.** $(7 \cdot b) \cdot a$, $b \cdot (a \cdot 7)$, $(b \cdot a) \cdot 7$; others are
possible **29.** $4a + 4$ **31.** $8x - 8y$ **33.** $-10a - 15b$
35. $2ab - 2ac + 2ad$ **37.** $2\pi rh + 2\pi r$
39. $\frac{1}{2}ha + \frac{1}{2}hb$ **41.** $4a$, $-5b$, 6 **43.** $2x$, $-3y$, $-2z$
45. $24(x + y)$ **47.** $7(p - 1)$ **49.** $7(x - 3)$
51. $x(y + 1)$ **53.** $2(x - y + z)$ **55.** $3(x + 2y - 1)$

57. $a(b + c - d)$ **59.** $\frac{1}{4}\pi r(r + s)$ **61.** $(x + y)^2$
62. $x^2 + y^2$ **63.** -102 **64.** $-\frac{1}{2}$ **65.** $\frac{1}{8}$ **66.** -46
67. ◆   **69.** No   **71.** Yes

Margin Exercises, Section R.6, pp. 47–50

1. $20x$ **2.** $-7x$ **3.** $6x$ **4.** $-6x$ **5.** $23.4x + 3.9y$
6. $\frac{22}{15}x - \frac{19}{12}y + 23$ **7.** -24 **8.** 0 **9.** 10 **10.** $-9x$
11. $24t$ **12.** $y - 7$ **13.** $y - x$ **14.** $-9x - 6y - 11$
15. $-23x + 7y + 2$ **16.** $3x + 2y + 1$
17. $2x + 5z - 24$ **18.** $-3x + 2y$
19. $-\frac{1}{4}t - 41w + 5d + 23$ **20.** $3x - 8$ **21.** $4y + 1$
22. $-2x - 9y - 6$ **23.** $15x + 9y - 6$ **24.** $-x - 2y$
25. $23x - 10y$ **26.** $3a - 3b + \frac{23}{4}$ **27.** $23x + 52$
28. $12a + 12$

Exercise Set R.6, p. 51

1. $12x$ **3.** $-3b$ **5.** $15y$ **7.** $11a$ **9.** $-8t$ **11.** $10x$
13. $11x - 5y$ **15.** $-4c + 12d$ **17.** $22x + 18$
19. $1.19x + 0.93y$ **21.** $-\frac{2}{15}a - \frac{1}{3}b - 27$
23. $P = 2(l + w)$ **25.** $2c$ **27.** $-b - 4$ **29.** $-b + 3$,
or $3 - b$ **31.** $-t + y$, or $y - t$ **33.** $-x - y - z$
35. $-8x + 6y - 13$ **37.** $2c - 5d + 3e - 4f$
39. $1.2x - 56.7y + 34z + \frac{1}{4}$ **41.** $3a + 5$ **43.** $m + 1$
45. $9d - 16$ **47.** $-7x + 14$ **49.** $-9x + 17$
51. $17x + 3y - 18$ **53.** $10x - 19$ **55.** $22a - 15$
57. -190 **59.** $12x + 30$ **61.** $3x + 30$ **63.** $9x - 18$
65. $-4x + 808$ **67.** $-14y - 186$ **69.** -37 **70.** -71
71. -23.1 **72.** $\frac{5}{24}$ **73.** -16 **74.** 16 **75.** -16
76. $-\frac{1}{6}$ **77.** $8a - 8b$ **78.** $-16a + 24b - 32$
79. $6ax - 6bx + 12cx$ **80.** $16x - 8y + 10$
81. $24(a - 1)$ **82.** $8(3a - 2b)$ **83.** $a(b - c + 1)$
84. $5(3p + 9q - 2)$ **85.** ◆   **87.** $(3 - 8)^2 + 9 = 34$
89. $5 \cdot 2^3 \div (3 - 4)^4 = 40$ **91.** $23a - 18b + 184$
93. $-9z + 5x$ **95.** $-x + 19$

Margin Exercises, Section R.7, pp. 55–60

1. 8^4 **2.** y^5 **3.** $-18x^3$ **4.** $-\dfrac{75}{x^{14}}$ **5.** $-42x^{8n}$
6. $-\dfrac{10y^2}{x^{12}}$ **7.** $60y$ **8.** 4^3 **9.** 5^6 **10.** $\dfrac{1}{10^6}$
11. $-5x^{2n}$ **12.** $-\dfrac{2y^{10}}{x^4}$ **13.** $\frac{3}{2}a^3b^2$ **14.** 3^{42} **15.** z^{20}
16. $\dfrac{1}{t^{14m}}$ **17.** $8x^3y^3$ **18.** $\dfrac{16y^{14}}{x^4}$ **19.** $-32x^{20}y^{10}$
20. $\dfrac{1000y^{21}}{x^{12}z^6}$ **21.** x^9y^{12} **22.** $\dfrac{9x^4}{y^{16}}$ **23.** $-\dfrac{8a^9}{27b^{21}}$
24. 5.88×10^{12} mi   **25.** 2.57×10^{-10}
26. 0.0000000000004567 **27.** $93,000,000$ mi
28. 7.462×10^{-13} **29.** 5.6×10^{-15} **30.** 2.0×10^3
31. 5.5×10^2 **32.** 5×10^2 sec   **33.** 1.90164×10^{27} kg

Calculator Spotlight, p. 50

1. 2.6×10^8; [2] [.] [6] [2nd] [EE] [8]

2. 6.709×10^{-11};
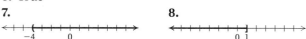

Exercise Set R.7, p. 61

1. 3^9 **3.** $\dfrac{1}{6^4}$ **5.** $\dfrac{1}{8^6}$ **7.** $\dfrac{1}{b^3}$ **9.** a^3 **11.** $72x^5$
13. $-28m^5n^5$ **15.** $-\dfrac{14}{x^{11}}$ **17.** $\dfrac{105}{x^{2t}}$ **19.** 8^7 **21.** 6^5
23. $\dfrac{1}{10^9}$ **25.** 9^2 **27.** $\dfrac{1}{x^{10n}}$ **29.** $\dfrac{1}{w^{5q}}$ **31.** a^5
33. $-3x^5z^4$ **35.** $-\dfrac{4x^9}{3y^2}$ **37.** $\dfrac{3x^3}{2y^2}$ **39.** 4^6 **41.** $\dfrac{1}{8^{12}}$
43. 6^{12} **45.** $125a^6b^6$ **47.** $\dfrac{y^{12}}{9x^6}$ **49.** $\dfrac{a^4}{36b^6c^2}$
51. $\dfrac{1}{4^9 \cdot 3^{12}}$ **53.** $\dfrac{8x^9y^3}{27}$ **55.** $\dfrac{a^{10}b^5}{5^{10}}$ **57.** $\dfrac{6^{30}2^{12}y^{36}}{z^{48}}$
59. $\dfrac{64}{x^{24}y^{12}}$ **61.** $\dfrac{5^7b^{28}}{3^7a^{35}}$ **63.** 10^a **65.** $3a^{-x-4}$
67. $\dfrac{-5x^{a+1}}{y}$ **69.** 8^{4xy} **71.** 12^{6b-2ab}
73. $5^{2c}x^{2ac-2c}y^{2bc+2c}$, or $25^cx^{2ac-2c}y^{2bc+2c}$
75. $2x^{a+2}y^{b-2}$ **77.** 4.7×10^{10} **79.** 2.5×10^8
81. 1.6×10^{-8} **83.** $673,000,000$ **85.** 0.000066 cm
87. $1,000,000,000$ **89.** 9.66×10^{-5}
91. 1.3338×10^{-11} **93.** 2.5×10^3 **95.** 5.0×10^{-4}
97. 9.922×10^{-3} lb   **99.** 4.08 light-years
101. 6.3072×10^{10} sec   **103.** 1.4675×10^{12} mi
105. 3.33×10^{-2} **107.** $19x + 4y - 20$
108. $-23t + 21$ **109.** -11 **110.** -231 **111.** -8
112. 8 **113.** ◆   **115.** 2^{21} **117.** $\dfrac{1}{a^{14}b^{27}}$
119. $4x^{2a}y^{2b}$

Summary and Review: Chapter R, p. 65

1. $2, -\frac{2}{3}, 0.45\overline{45}, -23.788$ **2.** $\{x \mid x$ is a real number
less than or equal to 46\}   **3.** $<$ **4.** $x < 19$ **5.** False
6. True
7.   **8.**

<— + + (+ + + + + + + —>   <— + + + + + + (+ + + + + —>
    -4 0 $0\ 1$

9. 7.23 **10.** 0 **11.** -2 **12.** -7.9 **13.** $-\frac{31}{28}$
14. -5 **15.** -26.7 **16.** $\frac{19}{4}$ **17.** 10.26 **18.** $-\frac{3}{7}$
19. 168 **20.** -4 **21.** 21 **22.** -7 **23.** $-\frac{7}{12}$
24. $\frac{8}{3}$ **25.** Undefined   **26.** -24 **27.** 7
28. -2.3 **29.** 0 **30.** a^5 **31.** $\left(-\frac{7}{8}\right)^3$ **32.** $\dfrac{1}{a^4}$
33. x^{-8} **34.** 59 **35.** -116 **36.** $5x$ **37.** $28\%y$,
or $0.28y$ **38.** $t - 9$ **39.** $\dfrac{a}{b} - 8$ **40.** -17 **41.** -8
42. 84 ft^2 **43.** $-4, 36; 95, 25; -5, 25;$ no
44. $-16, -16; 6, 6; -14, -14;$ yes   **45.** $\dfrac{21x}{9x}$ **46.** -12
47. $a + 11$ **48.** $y \cdot 8$ **49.** $9 + (a + b)$ **50.** $(8x)y$
51. $-6x + 3y$ **52.** $8abc + 4ab$ **53.** $5(x + 2y - z)$
54. $pt(r + s)$ **55.** $-3x + 5y$ **56.** $12c - 4$

57. $9c - 4d + 3$ **58.** $x + 3$ **59.** $6x + 15$

60. $22x - 14$ **61.** $-17m - 12$ **62.** $-\dfrac{10x^7}{y^5}$

63. $-\dfrac{3y^3}{2x^4}$ **64.** $\dfrac{a^8}{9b^2c^6}$ **65.** $\dfrac{81y^{40}}{16x^{24}}$ **66.** 6.875×10^9

67. 1.312×10^{-1} **68.** 1.422×10^{-2} m^3

69. ◈ The commutative laws deal with order and tell us that we can change the order when adding or when multiplying without changing the result. The associative laws deal with grouping and tell us that numbers can be grouped in any manner when adding or multiplying without affecting the result. The distributive laws deal with multiplication together with addition or subtraction. They tell us that adding or subtracting two numbers b and c and then multiplying the sum or difference by a number a is equivalent to first multiplying b and c by a and then adding or subtracting those products.

70. ◈ The expressions $2x + 3x$ and $5x$ are equivalent. (We can show this using the distributive law.) The expressions $(x + 1)^2$ and $x^2 + 1$ are not equivalent, because when $x = 3$, for example, we have $(x + 1)^2 = (3 + 1)^2 = 16$ but $x^2 + 1 = 3^2 + 1 = 10$.

71. x^{12y} **72.** 32 **73.** a, i; d, f; h, j

Test: Chapter R, p. 67

1. [R.1a] $\sqrt{7}$, π **2.** [R.1a] $\{x \mid x$ is a real number greater than 20$\}$ **3.** [R.1b] $>$ **4.** [R.1b] $5 \geq a$

5. [R.1b] True   **6.** [R.1b] True

7. [R.1c]

8. [R.1d] 0 **9.** [R.1d] $\frac{7}{8}$ **10.** [R.2a] -2

11. [R.2a] -13.1 **12.** [R.2a] -6 **13.** [R.2c] -1

14. [R.2c] -29.7 **15.** [R.2c] $\frac{25}{4}$ **16.** [R.2d] -33.62

17. [R.2d] $\frac{3}{4}$ **18.** [R.2d] -528 **19.** [R.2e] 15

20. [R.2e] -5 **21.** [R.2e] $\frac{8}{3}$ **22.** [R.2e] -82

23. [R.2e] Undefined   **24.** [R.2b] 13 **25.** [R.2b] 0

26. [R.3a] q^4 **27.** [R.3b] a^{-9} **28.** [R.3c] 0

29. [R.3c] $-\frac{16}{7}$ **30.** [R.4a] $t + 9$, or $9 + t$

31. [R.4a] $\dfrac{x}{y} - 12$ **32.** [R.4b] 18 **33.** [R.4b] 3.75 cm^2

34. [R.5a] 4, 4; 70, 70; 0, 0; yes   **35.** [R.5a] 2, 8; 530, 800; 0, 0; no   **36.** [R.5b] $\dfrac{27x}{36x}$ **37.** [R.5b] $\frac{3}{2}$

38. [R.5c] qp **39.** [R.5c] $4 + t$

40. [R.5c] $(3 + t) + w$ **41.** [R.5c] $4(ab)$

42. [R.5d] $-6a + 8b$ **43.** [R.5d] $3\pi rs + 3\pi r$

44. [R.5d] $a(b - c + 2d)$ **45.** [R.5d] $h(2a + 1)$

46. [R.6a] $10y - 5x$ **47.** [R.6a] $21a + 14$

48. [R.6b] $9x - 7y + 22$ **49.** [R.6b] $-7x + 14$

50. [R.6b] $10x - 21$ **51.** [R.6b] $-a + 38$

52. [R.7a] $-\dfrac{3y^2}{2x^4}$ **53.** [R.7a] $-\dfrac{6a^9}{b^5}$ **54.** [R.7a] $-50a^{9n}$

55. [R.7a] $-\dfrac{5}{x^{4t}}$ **56.** [R.7b] $\dfrac{a^{12}}{81b^8c^4}$ **57.** [R.7b] $\dfrac{16a^{48}}{b^{48}}$

58. [R.7c] 4.37×10^{-5} **59.** [R.7c] 3.741×10^7

60. [R.7c] 1.875×10^{-6} **61.** [R.7c] 1.196×10^{22} kg

62. [R.7c] 4.8×10^9 **63.** b, e; d, f, h; i, j

Chapter 1

Pretest: Chapter 1, p. 70

1. [1.1c] -7 **2.** [1.1b] 21.3 **3.** [1.1d] -1 **4.** [1.1d] $\frac{14}{5}$

5. [1.4c] $\{x \mid x > -3\}$ **6.** [1.4c] $\{x \mid x \geq -3\}$

7. [1.5a] $\{x \mid -1 \leq x \leq \frac{2}{3}\}$ **8.** [1.6c] -1, 8

9. [1.5b] $\{a \mid a < -2 \text{ or } a > -\frac{1}{2}\}$

10. [1.6e] $\{a \mid a < -2 \text{ or } a > -\frac{4}{3}\}$ **11.** [1.6d] $-\frac{11}{2}$, $\frac{3}{4}$

12. [1.4c] $\{x \mid x \leq \frac{19}{14}\}$ **13.** [1.2a] $r = \dfrac{P}{3q}$ **14.** [1.6b] 2.4

15. [1.3a] Length = 60 ft; width = 30 ft

16. [1.4d] $\{t \mid t > 3.23\}$, 3.23 yr after 1988

17. [1.3a] \$178,000   **18.** [1.5a] $\{-3, 4\}$

19. [1.5b] $\{-1, 0, 1\}$ **20.** [1.6a] $\dfrac{5}{2}\left|\dfrac{y}{x}\right|$

Margin Exercises, Section 1.1, pp. 71–78

1. (a), (f)   **2.** (b), (c)   **3.** (d), (e)   **4.** Yes   **5.** No

6. Yes   **7.** Yes   **8.** No   **9.** -7 **10.** $-\frac{17}{20}$ **11.** 38

12. 140.3 **13.** $\frac{5}{4}$ **14.** -49 **15.** $\frac{3}{14}$ **16.** -3

17. -136 **18.** 5 **19.** $\frac{4}{3}$ **20.** $4 - 5y = 2$; $\frac{2}{5}$

21. $63x - 91 = 30x$; $\frac{91}{33}$ **22.** 3 **23.** $-\frac{29}{5}$ **24.** $\frac{17}{2}$

25. No solution   **26.** All real numbers are solutions.

27. -2 **28.** $-\frac{19}{8}$

Calculator Spotlight, p. 78

1. 9 on each side. The solution is -2.   **2.** -11.125 on each side. The solution is $-\frac{19}{8}$.

Exercise Set 1.1, p. 79

1. Yes   **3.** No   **5.** No   **7.** No   **9.** Yes   **11.** No

13. 7 **15.** -8 **17.** 27 **19.** -39 **21.** 86.86 **23.** $\frac{1}{6}$

25. 6 **27.** -4 **29.** -147 **31.** 32 **33.** -6

35. $-\frac{1}{6}$ **37.** 10 **39.** 11 **41.** -12 **43.** 8 **45.** 2

47. 21 **49.** -12 **51.** 2 **53.** -1 **55.** $\frac{18}{5}$ **57.** 0

59. 1 **61.** All real numbers   **63.** No solution   **65.** 7

67. 2 **69.** 7 **71.** 5 **73.** $-\frac{3}{2}$ **75.** $\frac{79}{32}$ **77.** All real numbers   **79.** 5 **81.** $\frac{23}{66}$ **83.** $\frac{5}{32}$ **85.** a^{14}

86. $\dfrac{1}{a^{32}}$ **87.** $-\dfrac{18x^2}{y^{11}}$ **88.** $-2x^8y^3$ **89.** $12 - 20x$

90. $-5 + 6x$ **91.** $-12x + 8y - 4z$

92. $-10x + 35y - 20$ **93.** $2(x - 3y)$ **94.** $-4(x + 6y)$

95. $2(2x - 5y + 1)$ **96.** $-5(2x - 7y + 4)$

97. $\{1, 2, 3, 4, 5, 6, 7, 8, 9\}$; $\{x \mid x$ is a positive integer less than 10$\}$ **98.** $\{-8, -7, -6, -5, -4, -3, -2, -1\}$; $\{x \mid x$ is a negative integer greater than $-9\}$ **99.** ◈

101. Approximately -4.176 **103.** $\frac{3}{2}$ **105.** 8

Margin Exercises, Section 1.2, pp. 83–86

1. The interval time is about $17\frac{1}{4}$ min.　**2.** $m = \dfrac{rF}{v^2}$

3. $c = \dfrac{5}{3}P - 10$, or $\dfrac{5P - 30}{3}$　**4.** $m = \dfrac{H - 2r}{3}$

5. $b = \dfrac{2A - ha}{h}$, or $\dfrac{2A}{h} - a$　**6.** $Q = \dfrac{T}{1 + iy}$

7. (a) About 1251; **(b)** $W = \dfrac{NR - Nr + 400L}{400}$,

or $W = L + \dfrac{NR - Nr}{400}$

Exercise Set 1.2, p. 87

1. $r = \dfrac{d}{t}$　**3.** $h = \dfrac{A}{b}$　**5.** $w = \dfrac{P - 2l}{2}$, or $\dfrac{P}{2} - l$

7. $b = \dfrac{2A}{h}$　**9.** $a = 2A - b$　**11.** $m = \dfrac{F}{a}$　**13.** $t = \dfrac{I}{Pr}$

15. $c^2 = \dfrac{E}{m}$　**17.** $p = 2Q + q$　**19.** $y = \dfrac{c - Ax}{B}$

21. $N = \dfrac{1.08T}{I}$　**23. (a)** 1614 g; **(b)** $a = \dfrac{P + 299}{9.337d}$

25. (a) 50 mg; **(b)** $d = \dfrac{c(a + 12)}{a}$, or $c + \dfrac{12c}{a}$

27. (a) 2715 calories; **(b)** $a = \dfrac{19.18w + 7h + 92.4 - K}{9.52}$;

$h = \dfrac{K - 19.18w + 9.52a - 92.4}{7}$;

$w = \dfrac{K - 7h + 9.52a - 92.4}{19.18}$　**29.** -5　**30.** 250

31. -2　**32.** -25　**33.** $\frac{4}{5}$　**34.** 6　**35.** -6　**36.** -5

37. ◈　**39.** $s = \dfrac{A - \pi r^2}{\pi r}$　**41.** $V_1 = \dfrac{P_2 V_2 T_1}{P_1 T_2}$;

$P_2 = \dfrac{P_1 V_1 T_2}{T_1 V_2}$　**43.** 0.8 yr　**45. (a)** Approximately

120.5 horsepower; **(b)** approximately 94.1 horsepower

Margin Exercises, Section 1.3, pp. 92–99

1. 22 in., 33 in., 51 in.　**2.** 30　**3.** $124　**4.** $260,000
5. (a) $1.4532 billion; $1.7842 billion; **(b)** 2015
6. 48 in. and 72 in.　**7.** 31°, 93°, 56°　**8.** 421, 422, 423
9. $2\frac{7}{9}$ hr; $1\frac{2}{3}$ hr

Exercise Set 1.3, p. 101

1. $109\frac{1}{3}$ km　**3.** 2　**5.** 22　**7.** 23　**9.** $1120
11. 26°, 130°, 24°　**13.** Length: 94 ft; width: 50 ft
15. $66\frac{2}{3}$ ft and $33\frac{1}{3}$ ft　**17.** $265,000　**19.** 9, 11, 13
21. 348 and 349　**23. (a)** 9.9704 sec; **(b)** 2008
25. 843 sq ft　**27.** $18,950　**29.** Upstream: $\frac{2}{3}$ hr;
downstream: $\frac{18}{73}$ hr　**31.** 6 min　**33.** -2442　**34.** 208
35. 49　**36.** -119　**37.** 49　**38.** -119　**39.** $\frac{78}{1010}$
40. $\frac{17}{10}$　**41.** -1　**42.** -256　**43.** 32　**44.** 2,097,152
45. ◈　**47.** ◈　**49.** 130　**51.** 25% increase

---

53. $110,000　**55.** 62,208,000 sec　**57.** $m\angle 2 = 120°$; $m\angle 1 = 60°$

Margin Exercises, Section 1.4, pp. 107–114

1. Yes　**2.** No　**3.** Yes　**4.** $[-4, 5)$　**5.** $(-\infty, -2]$
6. $[10, \infty)$　**7.** $[-30, 30]$
8. $\{x \mid x > 3\}$, or $(3, \infty)$;　**9.** $\{x \mid x \le 3\}$, or $(-\infty, 3]$;

10. $\{x \mid x \le -2\}$, or $(-\infty, -2]$;

11. $\left\{y \mid y \le \frac{3}{10}\right\}$, or $\left(-\infty, \frac{3}{10}\right]$;

12. $\{y \mid y < -5\}$, or $(-\infty, -5)$

13. $\{x \mid x \ge 12\}$, or $[12, \infty)$

14. $\left\{y \mid y \le -\frac{1}{5}\right\}$, or $\left(-\infty, -\frac{1}{5}\right]$
15. $\left\{x \mid x < \frac{1}{2}\right\}$, or $\left(-\infty, \frac{1}{2}\right)$　**16.** $\left\{y \mid y \ge \frac{17}{9}\right\}$, or $\left[\frac{17}{9}, \infty\right)$
17. $\{t \mid t > 4.29\}$, or approximately after 2000
18. For $\{n \mid n < 100 \text{ hr}\}$, plan A is better.

Exercise Set 1.4, p. 115

1. No, no, no, yes　**3.** No, yes, yes, no, no
5. $(-\infty, 5)$　**7.** $[-3, 3]$　**9.** $(-2, 5)$　**11.** $(-\sqrt{2}, \infty)$

13. $\{x \mid x > -1\}$, or $(-1, \infty)$　**15.** $\{y \mid y < 6\}$, or $(-\infty, 6)$

17. $\{a \mid a \le -22\}$, or $(-\infty, -22]$

19. $\{t \mid t \ge -4\}$, or $[-4, \infty)$

21. $\{y \mid y > -6\}$, or $(-6, \infty)$　**23.** $\{x \mid x \le 9\}$, or $(-\infty, 9]$

25. $\{x \mid x \ge 3\}$, or $[3, \infty)$

27. $\{x \mid x < -60\}$, or $(-\infty, -60)$

29. $\{x \mid x \le 0.9\}$, or $(-\infty, 0.9]$　**31.** $\left\{x \mid x \le \frac{5}{6}\right\}$, or $\left(-\infty, \frac{5}{6}\right]$
33. $\{x \mid x < 6\}$, or $(-\infty, 6)$　**35.** $\{y \mid y \le -3\}$, or $(-\infty, -3]$
37. $\left\{y \mid y > \frac{2}{3}\right\}$, or $\left(\frac{2}{3}, \infty\right)$　**39.** $\left\{x \mid x \ge \frac{45}{4}\right\}$, or $\left[\frac{45}{4}, \infty\right)$
41. $\left\{x \mid x \le \frac{1}{2}\right\}$, or $\left(-\infty, \frac{1}{2}\right]$　**43.** $\left\{y \mid y \le -\frac{75}{2}\right\}$, or $\left(-\infty, -\frac{75}{2}\right]$
45. $\left\{x \mid x > -\frac{2}{17}\right\}$, or $\left(-\frac{2}{17}, \infty\right)$　**47.** $\left\{m \mid m > \frac{1}{3}\right\}$, or $\left(\frac{1}{3}, \infty\right)$
49. $\{r \mid r < -3\}$, or $(-\infty, -3)$　**51.** $\{x \mid x \ge 2\}$, or $[2, \infty)$

53. $\{y \mid y < 5\}$, or $(-\infty, 5)$ 55. $\left\{x \mid x \le \frac{4}{7}\right\}$, or $\left(-\infty, \frac{4}{7}\right]$
57. $\{x \mid x < 8\}$, or $(-\infty, 8)$ 59. $\left\{x \mid x \ge \frac{13}{2}\right\}$, or $\left[\frac{13}{2}, \infty\right)$
61. $\left\{x \mid x < \frac{11}{18}\right\}$, or $\left(-\infty, \frac{11}{18}\right)$
63. $\left\{x \mid x \ge -\frac{51}{31}\right\}$, or $\left[-\frac{51}{31}, \infty\right)$ 65. $\{a \mid a \le 2\}$, or $(-\infty, 2]$
67. (a) 28.11; (b) $\{W \mid W < \text{(approximately) } 190.3 \text{ lb}\}$
69. $\{S \mid S \ge 84\}$ 71. $\{B \mid B \ge \$11{,}500\}$
73. $\{S \mid S > \$7000\}$ 75. $\{n \mid n > 25\}$ 77. $\{p \mid p > 80\}$
79. $\{s \mid s > 8\}$ 81. (a) 8.398 gal, 12.063 gal, 15.728 gal;
(b) years after 1999   83. $-9a + 30b$ 84. $32x - 72y$
85. $-8a + 22b$ 86. $-6a + 17b$ 87. $10(3x - 7y - 4)$
88. $-6a(2 - 5b)$ 89. $-4(2x - 6y + 1)$
90. $5(2n - 9mn + 20m)$ 91. -11.2 92. 6.6
93. -11.2 94. 6.6   95. ◈   97. (a) $\{p \mid p > 10\}$;
(b) $\{p \mid p < 10\}$ 99. True   101. All real numbers
103. All real numbers

Margin Exercises, Section 1.5, pp. 121–127

1. $\{0, 3\}$ 2.

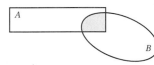

3. ────; $(-1, 4)$

4. $\{x \mid -5 < x \le 10\}$, or $(-5, 10]$;

5. $\{x \mid x < -3\}$;

6. $\varnothing$ 7. $\{x \mid 2 \le x \le 6\}$, or $[2, 6]$
8. $\{0, 1, 3, 4, 7, 9\}$ 9.

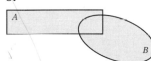

10. ────; $(-\infty, -2] \cup (4, \infty)$

11. $\{x \mid x < 1 \text{ or } x \ge 6\}$, or $(-\infty, 1) \cup [6, \infty)$;

12. $\left\{x \mid x \ge \frac{7}{2} \text{ or } x < -2\right\}$, or $(-\infty, -2) \cup \left[\frac{7}{2}, \infty\right)$;

13. $\{x \mid x > -2 \text{ or } x < -5\}$, or $(-\infty, -5) \cup (-2, \infty)$
14. All real numbers;   15. $\{s \mid 8 < s < 19\}$

Exercise Set 1.5, p. 129

1. $\{9, 11\}$ 3. $\{b\}$ 5. $\{9, 10, 11, 13\}$
7. $\{a, b, c, d, f, g\}$ 9. $\varnothing$ 11. $\{3, 5, 7\}$

13. ────; $(-4, 1]$

15. ────; $(1, 6)$

17. $\{x \mid -4 \le x < 5\}$, or $[-4, 5)$; ────

19. $\{x \mid x \ge 2\}$, or $[2, \infty)$; ────

21. $\varnothing$ 23. $\{x \mid -8 < x < 6\}$, or $(-8, 6)$
25. $\{x \mid -6 < x \le 2\}$, or $(-6, 2]$ 27. $\{y \mid -1 < y \le 5\}$, or $(-1, 5]$ 29. $\left\{x \mid -\frac{5}{3} \le x \le \frac{4}{3}\right\}$, or $\left[-\frac{5}{3}, \frac{4}{3}\right]$
31. $\{x \mid -1 < x \le 6\}$, or $(-1, 6]$
33. $\left\{x \mid -\frac{3}{2} \le x < \frac{9}{2}\right\}$, or $\left[-\frac{3}{2}, \frac{9}{2}\right]$ 35. $\{x \mid 10 < x \le 14\}$, or $(10, 14]$ 37. $\left\{x \mid -\frac{7}{2} < x < \frac{37}{2}\right\}$, or $\left(-\frac{7}{2}, \frac{37}{2}\right)$

39. ────; $(-\infty, -2) \cup (1, \infty)$

41. ────; $(-\infty, -3] \cup (1, \infty)$

43. $\{x \mid x < -5 \text{ or } x > -1\}$, or $(-\infty, -5) \cup (-1, \infty)$;
────

45. $\left\{x \mid x \le \frac{5}{2} \text{ or } x \ge 4\right\}$, or $\left(-\infty, \frac{5}{2}\right] \cup [4, \infty)$;
────

47. $\{x \mid x \ge -3\}$, or $[-3, \infty)$; ────

49. $\left\{x \mid x \le -\frac{5}{4} \text{ or } x > -\frac{1}{2}\right\}$, or $\left(-\infty, -\frac{5}{4}\right] \cup \left(-\frac{1}{2}, \infty\right)$
51. All real numbers, or $(-\infty, \infty)$
53. $\{x \mid x > 2 \text{ or } x < -4\}$, or $(-\infty, -4) \cup (2, \infty)$
55. $\left\{x \mid x < \frac{79}{4} \text{ or } x > \frac{89}{4}\right\}$, or $\left(-\infty, \frac{79}{4}\right) \cup \left(\frac{89}{4}, \infty\right)$
57. $\left\{x \mid x \le -\frac{13}{2} \text{ or } x \ge \frac{29}{2}\right\}$, or $\left(-\infty, -\frac{13}{2}\right] \cup \left[\frac{29}{2}, \infty\right)$
59. $\{d \mid 0 \le d \le 198\}$ 61. $\{t \mid 14 \le t \le 19\}$, or from 2005 to 2010   63. Rounding to the nearest pound, $\{W \mid 152 \text{ lb} < W < 228 \text{ lb}\}$ 65. 3.2   66. 12   67. 2
68. 0   69. $-\dfrac{32y^{30}}{x^{20}}$ 70. $-\dfrac{20b}{a^{7}}$ 71. $-\dfrac{4a^{17}}{5b^{15}}$

72. $25p^{12}q^{22}$ 73. $\dfrac{a^{6}}{8b^{6}}$ 74. $25p^{10}q^{8}$ 75. ◈

77. All real numbers; $\varnothing$ 79. $\{x \mid -4 < x \le 1\}$, or
$(-4, 1]$ 81. $\left\{x \mid \frac{2}{5} \le x \le \frac{4}{5}\right\}$, or $\left[\frac{2}{5}, \frac{4}{5}\right]$
83. $\left\{x \mid -\frac{1}{8} < x < \frac{1}{2}\right\}$, or $\left(-\frac{1}{8}, \frac{1}{2}\right)$
85. $\{x \mid 10 < x \le 18\}$, or $(10, 18]$

Margin Exercises, Section 1.6, pp. 133–138

1. $7|x|$ 2. x^{8} 3. $5a^{2}|b|$ 4. $\dfrac{7|a|}{b^{2}}$ 5. $9|x|$ 6. 29
7. 5   8. $|p|$
9. $\{6, -6\}$;   10. $\varnothing$

11. $\{0\}$ **12.** $\{2, -2\}$ **13.** $\left\{\frac{17}{4}, -\frac{17}{4}\right\}$ **14.** $\{4, -4\}$
15. $\{3, 5\}$ **16.** $\left\{-\frac{13}{3}, 7\right\}$ **17.** $\varnothing$ **18.** $\left\{\frac{7}{4}, -\frac{1}{6}\right\}$
19. $\left\{-\frac{7}{2}\right\}$
20. $\{5, -5\}$;

21. $\{x \mid -5 < x < 5\}$, or $(-5, 5)$;

22. $\{x \mid x \le -5 \text{ or } x \ge 5\}$, or $(-\infty, -5] \cup [5, \infty)$;

23. $\{x \mid -2 < x < 5\}$, or $(-2, 5)$;

24. $\left\{x \mid 1 \le x \le \frac{11}{3}\right\}$, or $\left[1, \frac{11}{3}\right]$
25. $\left\{x \mid x \le -\frac{7}{3} \text{ or } x \ge 1\right\}$, or $\left(-\infty, -\frac{7}{3}\right] \cup [1, \infty)$;

Exercise Set 1.6, p. 139

1. $9|x|$ **3.** $2x^2$ **5.** $2x^2$ **7.** $6|y|$ **9.** $\dfrac{2}{|x|}$ **11.** $\dfrac{x^2}{|y|}$
13. $4|x|$ **15.** 38   **17.** 19   **19.** 6.3   **21.** 5
23. $\{3, -3\}$ **25.** $\varnothing$ **27.** $\{0\}$ **29.** $\{15, -9\}$
31. $\left\{\frac{7}{2}, -\frac{1}{2}\right\}$ **33.** $\left\{\frac{23}{4}, -\frac{5}{4}\right\}$ **35.** $\{11, -11\}$
37. $\{291, -291\}$ **39.** $\{8, -8\}$ **41.** $\{7, -7\}$
43. $\{2, -2\}$ **45.** $\{8, -7\}$ **47.** $\{2, -12\}$ **49.** $\left\{\frac{7}{2}, -\frac{5}{2}\right\}$
51. $\varnothing$ **53.** $\left\{-\frac{13}{54}, -\frac{7}{54}\right\}$ **55.** $\left\{\frac{3}{8}, -\frac{11}{8}\right\}$ **57.** $\left\{\frac{3}{2}\right\}$
59. $\left\{5, -\frac{3}{5}\right\}$ **61.** All real numbers   **63.** $\left\{-\frac{3}{2}\right\}$
65. $\left\{\frac{24}{23}, 0\right\}$ **67.** $\left\{32, \frac{8}{3}\right\}$ **69.** $\{x \mid -3 < x < 3\}$, or $(-3, 3)$
71. $\{x \mid x \le -2 \text{ or } x \ge 2\}$, or $(-\infty, -2] \cup [2, \infty)$
73. $\{x \mid 0 < x < 2\}$, or $(0, 2)$ **75.** $\{x \mid -5 \le x \le -3\}$, or
$[-5, -3]$ **77.** $\left\{x \mid -\frac{1}{2} \le x \le \frac{7}{2}\right\}$, or $\left[-\frac{1}{2}, \frac{7}{2}\right]$
79. $\left\{y \mid y < -\frac{3}{2} \text{ or } y > \frac{17}{2}\right\}$, or $\left(-\infty, -\frac{3}{2}\right) \cup \left(\frac{17}{2}, \infty\right)$
81. $\left\{x \mid x \le -\frac{5}{4} \text{ or } x \ge \frac{23}{4}\right\}$, or $\left(-\infty, -\frac{5}{4}\right] \cup \left[\frac{23}{4}, \infty\right)$
83. $\{y \mid -9 < y < 15\}$, or $(-9, 15)$
85. $\left\{x \mid -\frac{7}{2} \le x \le \frac{1}{2}\right\}$, or $\left[-\frac{7}{2}, \frac{1}{2}\right]$
87. $\left\{y \mid y < -\frac{4}{3} \text{ or } y > 4\right\}$, or $\left(-\infty, -\frac{4}{3}\right) \cup (4, \infty)$
89. $\left\{x \mid x \le -\frac{5}{4} \text{ or } x \ge \frac{23}{4}\right\}$, or $\left(-\infty, -\frac{5}{4}\right] \cup \left[\frac{23}{4}, \infty\right)$
91. $\left\{x \mid -\frac{9}{2} < x < 6\right\}$, or $\left(-\frac{9}{2}, 6\right)$
93. $\left\{x \mid x \le -\frac{25}{6} \text{ or } x \ge \frac{23}{6}\right\}$, or $\left(-\infty, -\frac{25}{6}\right] \cup \left[\frac{23}{6}, \infty\right)$
95. $\{x \mid -5 < x < 19\}$, or $(-5, 19)$
97. $\left\{x \mid x \le -\frac{2}{15} \text{ or } x \ge \frac{14}{15}\right\}$, or $\left(-\infty, -\frac{2}{15}\right] \cup \left[\frac{14}{15}, \infty\right)$
99. $\{m \mid -12 \le m \le 2\}$, or $[-12, 2]$
101. $\left\{x \mid \frac{1}{2} \le x \le \frac{5}{2}\right\}$, or $\left[\frac{1}{2}, \frac{5}{2}\right]$
103. $\{x \mid 0.49705 \le x \le 0.50295\}$, or $[0.49705, 0.50295]$
105. $49\%x$, or $0.49x$ **106.** $4n + 6$, or $6 + 4n$

107. $\dfrac{x}{y} - 50$ **108.** $a + 3b$, or $3b + a$ **109.** -49.3
110. -37.7 **111.** 252.3   **112.** 7.5   **113.** $-\frac{7}{6}$ **114.** -1
115. $-\frac{13}{8}$ **116.** $-\frac{1}{8}$ **117.** $\diamondsuit$
119. $\left\{d \mid 5\frac{1}{2} \text{ ft} \le d \le 6\frac{1}{2} \text{ ft}\right\}$
121. $\{x \mid x \ge -5\}$, or $[-5, \infty)$ **123.** $\left\{1, -\frac{1}{4}\right\}$ **125.** $\varnothing$
127. All real numbers   **129.** All real numbers
131. $|x| < 3$ **133.** $|x| \ge 6$ **135.** $|x + 3| > 5$

Summary and Review: Chapter 1, p. 143

1. 8   **2.** $\frac{3}{7}$ **3.** $\frac{22}{5}$ **4.** $-\frac{1}{13}$ **5.** -0.2 **6.** 5
7. $d = \frac{11}{4}(C - 3)$ **8.** $b = \dfrac{A - 2a}{-3}$, or $\dfrac{2a - A}{3}$
9. 185 and 186   **10.** 15 m, 12 m   **11.** 160,000
12. 40 sec   **13.** $[-8, 9)$ **14.** $(-\infty, 40]$
15. ; $(-\infty, -2]$

16. ; $(1, \infty)$

17. $\{a \mid a \le -21\}$, or $(-\infty, -21]$
18. $\{y \mid y \ge -7\}$, or $[-7, \infty)$ **19.** $\{y \mid y > -4\}$, or $(-4, \infty)$
20. $\{y \mid y > -30\}$, or $(-30, \infty)$
21. $\{x \mid x > -3\}$, or $(-3, \infty)$
22. $\left\{y \mid y \le -\frac{6}{5}\right\}$, or $\left(-\infty, -\frac{6}{5}\right]$
23. $\{x \mid x < -3\}$, or $(-\infty, -3)$ **24.** $\{y \mid y > -10\}$, or
$(-10, \infty)$ **25.** $\left\{x \mid x \le -\frac{5}{2}\right\}$, or $\left(-\infty, -\frac{5}{2}\right]$
26. $\left\{t \mid t > 4\frac{1}{4} \text{ hr}\right\}$ **27.** \$10,000
28. ; $[-2, 5)$

29. ; $(-\infty, -2] \cup (5, \infty)$

30. $\varnothing$ **31.** $\{x \mid -7 < x \le 2\}$, or $(-7, 2]$
32. $\left\{x \mid -\frac{5}{4} < x < \frac{5}{2}\right\}$, or $\left(-\frac{5}{4}, \frac{5}{2}\right)$
33. $\{x \mid x < -3 \text{ or } x > 1\}$, or $(-\infty, -3) \cup (1, \infty)$
34. $\{x \mid x < -11 \text{ or } x \ge -6\}$, or $(-\infty, -11) \cup [-6, \infty)$
35. $\{x \mid x \le -6 \text{ or } x \ge 8\}$, or $(-\infty, -6] \cup [8, \infty)$
36. $\{t \mid 3.23 < t < 7.85\}$, or approximately between
1991 and 1996   **37.** $\dfrac{3}{|x|}$ **38.** $\dfrac{2|x|}{y^2}$ **39.** $\dfrac{4}{|y|}$ **40.** 62
41. $6, -6$ **42.** $\{9, -5\}$ **43.** $\left\{-14, \frac{4}{3}\right\}$ **44.** $\varnothing$
45. $\left\{x \mid -\frac{17}{2} < x < \frac{7}{2}\right\}$, or $\left(-\frac{17}{2}, \frac{7}{2}\right)$
46. $\{x \mid x \le -3.5 \text{ or } x \ge 3.5\}$, or $(-\infty, -3.5] \cup [3.5, \infty)$
47. $\left\{x \mid x \le -\frac{11}{3} \text{ or } x \ge \frac{19}{3}\right\}$, or $\left(-\infty, -\frac{11}{3}\right] \cup \left[\frac{19}{3}, \infty\right)$
48. $\varnothing$ **49.** $\{1, 5, 9\}$ **50.** $\{1, 2, 3, 5, 6, 9\}$ **51.** 33
52. $\frac{1}{6}$ **53.** $\frac{4}{5}$ **54.** 2349   **55.** $20x - 30y + 70$
56. $8(5x - y + 2)$ **57.** $-8x - 32y + 12$
58. $\diamondsuit$ **(1)** $-9(x + 2) = -9x - 18$, not $-9x + 2$.
(2) This would be correct if (1) were correct.   **(3)** If (2)
were correct, the right-hand side would be -5, not 8.

(4) The inequality symbol should be reversed. The correct solution is

$$7 - 9x + 6x < -9(x + 2) + 10x$$
$$7 - 9x + 6x < -9x - 18 + 10x$$
$$7 - 3x < x - 18$$
$$-4x < -25$$
$$x > \frac{25}{4}.$$

59. ◈"Solve" can mean to find all the replacements that make an equation or inequality true. It can also mean to express a formula as an equivalent equation with a given variable alone on one side.
60. $\left\{x \mid -\frac{8}{3} \le x \le -2\right\}$, or $\left[-\frac{8}{3}, -2\right]$

Test: Chapter 1, p. 145

1. [1.1c] $\frac{2}{3}$ **2.** [1.1b] $\frac{19}{15}$ **3.** [1.1d] 2.2 **4.** [1.1d] -2
5. [1.6c] $\{-6, 12\}$ **6.** [1.2a] $B = 3A + C$
7. [1.3a] Length: $14\frac{2}{5}$ ft; width: $9\frac{3}{5}$ ft **8.** [1.3a] 43,333
9. [1.3a] 180,000 **10.** [1.3a] 59°, 60°, 61°
11. [1.3b] $2\frac{2}{5}$ hr; 4 hr **12.** [1.4b] $(-3, 2]$
13. [1.4b] $(-4, \infty)$
14. [1.4c] 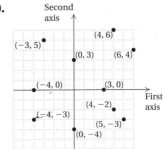 ; $(-\infty, 6]$

15. [1.4c] 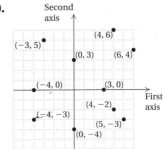 ; $(-\infty, -2]$

16. [1.4c] $\{y \mid y > -50\}$, or $(-50, \infty)$
17. [1.4c] $\left\{a \mid a \le \frac{11}{5}\right\}$, or $\left(-\infty, \frac{11}{5}\right]$ **18.** [1.4c] $\{y \mid y > 1\}$, or $(1, \infty)$ **19.** [1.4c] $\left\{x \mid x > \frac{5}{2}\right\}$, or $\left(\frac{5}{2}, \infty\right)$
20. [1.4c] $\left\{x \mid x \le \frac{7}{4}\right\}$, or $\left(-\infty, \frac{7}{4}\right]$ **21.** [1.4d] $\left\{h \mid h > 2\frac{1}{10}\text{hr}\right\}$
22. [1.4d] $\{d \mid 33 \text{ ft} \le d \le 231 \text{ ft}\}$
23. [1.5a] 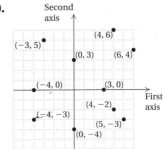 ; $[-3, 4]$

24. [1.5b] 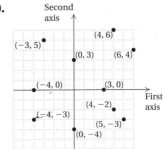 ; $(-\infty, -3) \cup (4, \infty)$

25. [1.5a] $\{x \mid x \ge 4\}$, or $[4, \infty)$
26. [1.5a] $\{x \mid -1 < x < 6\}$, or $(-1, 6)$
27. [1.5a] $\left\{x \mid -\frac{2}{5} < x \le \frac{9}{5}\right\}$, or $\left(-\frac{2}{5}, \frac{9}{5}\right]$
28. [1.5b] $\left\{x \mid x < -4 \text{ or } x > -\frac{5}{2}\right\}$, or $(-\infty, -4) \cup \left(-\frac{5}{2}, \infty\right)$
29. [1.5b] All real numbers, or $(-\infty, \infty)$
30. [1.5b] $\{x \mid x < 3 \text{ or } x > 6\}$, or $(-\infty, 3) \cup (6, \infty)$
31. [1.6a] $\dfrac{7}{|x|}$ **32.** [1.6a] $2|x|$ **33.** [1.6b] 8.4
34. [1.6c] $\{9, -9\}$ **35.** [1.6e] $\{x \mid x < -3 \text{ or } x > 3\}$, or $(-\infty, -3) \cup (3, \infty)$ **36.** [1.6e] $\left\{x \mid -\frac{7}{8} < x < \frac{11}{8}\right\}$, or $\left(-\frac{7}{8}, \frac{11}{8}\right)$ **37.** [1.6e] $\left\{x \mid x \le -\frac{13}{5} \text{ or } x \ge \frac{7}{5}\right\}$, or $\left(-\infty, -\frac{13}{5}\right] \cup \left[\frac{7}{5}, \infty\right)$ **38.** [1.6d] $\{1\}$ **39.** [1.6c] $\varnothing$
40. [1.6e] $\{x \mid -99 \le x \le 111\}$, or $[-99, 111]$
41. [1.5a] $\{3, 5\}$ **42.** [1.5b] $\{1, 3, 5, 7, 9, 11, 13\}$
43. [R.2a] $\frac{1}{8}$ **44.** [R.2c] $\frac{11}{8}$ **45.** [R.2e] $\frac{4}{5}$
46. [R.2d] 2349 **47.** [R.5d] $-16a + 24b$
48. [R.5d] $2(3a - 5b + 6)$
49. [R.6b] $-18x - 27y + 29$ **50.** [1.6e] $\varnothing$
51. [1.5a] $\left\{x \mid \frac{1}{5} < x < \frac{4}{5}\right\}$, or $\left(\frac{1}{5}, \frac{4}{5}\right)$

Chapter 2

Pretest: Chapter 2, p. 148

1. [2.5a]
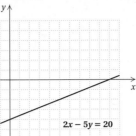

2. [2.5c]
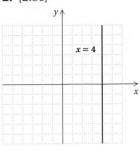

3. [2.1c], [2.5b]
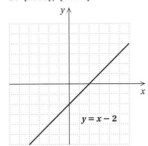

4. [2.5c]
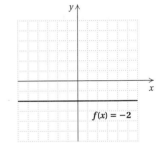

5. [2.2c]
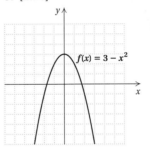

6. [2.4b] 5; $(0, -3)$ **7.** [2.4b] 0 **8.** [2.6c] $y = x + 10$
9. [2.6b] $y = 2x + 7$ **10.** [2.6d] $y = \frac{2}{7}x + \frac{31}{7}$
11. [2.6d] $y = -\frac{7}{2}x + 12$ **12.** [2.5d] Perpendicular
13. [2.5d] Parallel **14.** [2.5a] y-intercept: $(0, -4)$; x-intercept: $(6, 0)$ **15.** [2.2b] -3; -1; 1
16. [2.3a] $\left\{x \mid x \text{ is a real number } and \text{ } x \ne \frac{5}{2}\right\}$, or $\left(-\infty, \frac{5}{2}\right) \cup \left(\frac{5}{2}, \infty\right)$ **17.** [2.2d] **(a)** Yes; **(b)** no
18. [2.7b] **(a)** $m(t) = 3.383t + 37.73$, where t is the number of years since 1991; **(b)** \$105.39

Margin Exercises, Section 2.1, pp. 150–156

1.–10.

11. Both negative **12.** First positive, second negative
13. No **14.** Yes **15.** (−6, 4), (−2, 2); answers may vary

16.

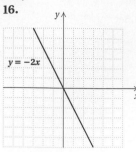

$y = -2x$

17.

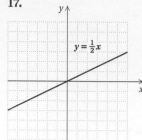

$y = \frac{1}{2}x$

18.

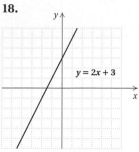

$y = 2x + 3$

19.

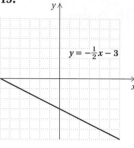

$y = -\frac{1}{2}x - 3$

20.

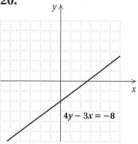

$4y - 3x = -8$

21.

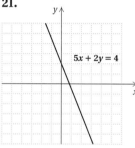

$5x + 2y = 4$

22.

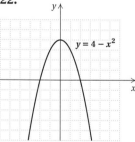

$y = 4 - x^2$

23.

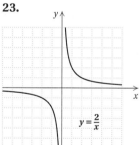

$y = \frac{2}{x}$

24.

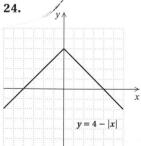

$y = 4 - |x|$

Calculator Spotlight, pp. 157–158

1.–20. Left to the student.

Exercise Set 2.1, p. 159

1.

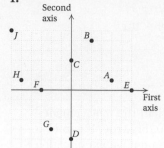

3.

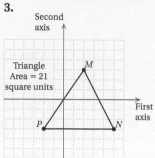

Triangle
Area = 21
square units

5. Yes **7.** Yes **9.** No

11.

$$\begin{array}{c|c} y = 4 - x\ ; \\ \hline 5\ ?\ 4 - (-1) \\ \ \ \ \ 4 + 1 \\ \ \ \ \ 5 \ \ \ \ \ \ \ \ \text{TRUE} \end{array}$$

$$\begin{array}{c|c} y = 4 - x; \\ \hline 1\ \ ?\ \ 4 - 3 \\ 1 \ \ \ \ \ \ \text{TRUE} \end{array}$$

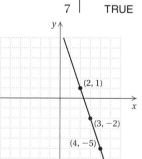

$(-1, 5)$
$(1, 3)$
$(3, 1)$

13.

$$\begin{array}{c|c} 3x + y = 7; \\ \hline 3 \cdot 2 + 1\ ?\ 7 \\ 6 + 1\ \ \big| \\ 7\ \big| \ \ \text{TRUE} \end{array}$$

$$\begin{array}{c|c} 3x + y = 7; \\ \hline 3 \cdot 4 + (-5)\ ?\ 7 \\ 12 - 5\ \ \big| \\ 7\ \big| \ \ \text{TRUE} \end{array}$$

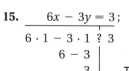

$(2, 1)$
$(3, -2)$
$(4, -5)$

15.

$$\begin{array}{c|c} 6x - 3y = 3\ ; \\ \hline 6 \cdot 1 - 3 \cdot 1\ ?\ 3 \\ 6 - 3\ \ \big| \\ 3\ \big| \ \ \text{TRUE} \end{array}$$

$$\begin{array}{c|c} 6x - 3y = 3\ ; \\ \hline 6(-1) - 3(-3)\ ?\ 3 \\ -6 + 9\ \ \big| \\ 3\ \big| \ \ \text{TRUE} \end{array}$$

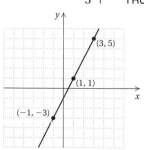

$(3, 5)$
$(1, 1)$
$(-1, -3)$

17.

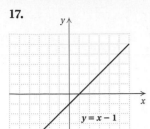

$y = x - 1$

19.

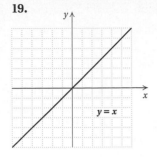

$y = x$

37.

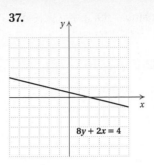

$8y + 2x = 4$

39.

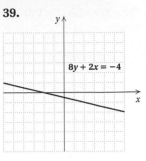

$8y + 2x = -4$

21.

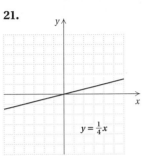

$y = \frac{1}{4}x$

23.

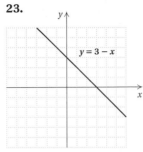

$y = 3 - x$

41. **(a)** \$25.51; \$29.68; \$35.93;
(b) \$28.75; **(c)** 78 lb

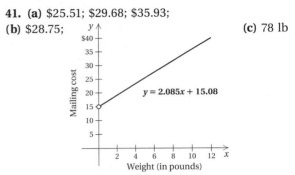

$y = 2.085x + 15.08$

25.

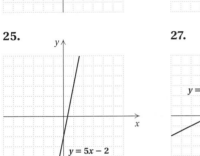

$y = 5x - 2$

27.

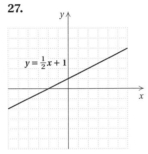

$y = \frac{1}{2}x + 1$

43.

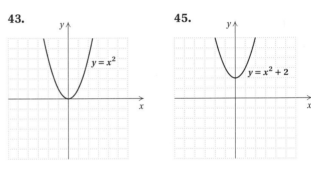

$y = x^2$

45.

$y = x^2 + 2$

29.

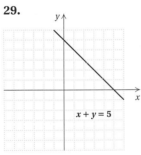

$x + y = 5$

31.

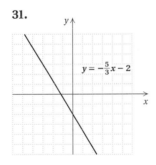

$y = -\frac{5}{3}x - 2$

47.

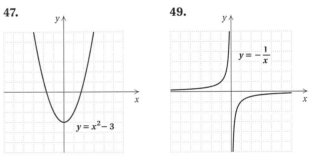

$y = x^2 - 3$

49.

$y = -\frac{1}{x}$

33.

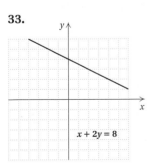

$x + 2y = 8$

35.

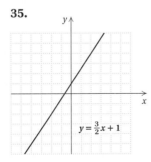

$y = \frac{3}{2}x + 1$

51.

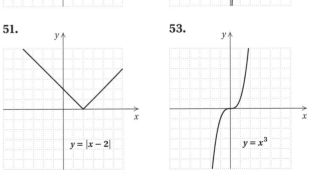

$y = |x - 2|$

53.

$y = x^3$

55. $\left\{x \mid 1 < x \leq \frac{15}{2}\right\}$, or $\left(1, \frac{15}{2}\right]$ **56.** $\{x \mid x > -3\}$, or $(-3, \infty]$ **57.** $\left\{x \mid x \leq -\frac{7}{3} \text{ or } x \geq \frac{17}{3}\right\}$, or $\left(-\infty, -\frac{7}{3}\right] \cup \left[\frac{17}{3}, \infty\right)$ **58.** $\{x \mid -6 < x < 6\}$, or $(-6, 6)$ **59.** 25 ft **60.** 24 **61.** 4 mi **62.** \$330,000

63. ◆   **65.–67.** Left to the student   **69.** $y = -x + 4$
71. $y = |x| - 3$

Margin Exercises, Section 2.2, pp. 165–171

1. Yes   **2.** No   **3.** Yes   **4.** No   **5.** Yes   **6.** Yes
7. (a) -4; (b) 148; (c) 31; (d) $98a^2 + 21a - 4$
8. (a) -33; (b) 2; (c) 7; (d) $5a + 2$
9.

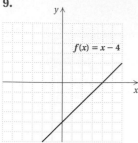

10.

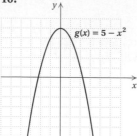

11.

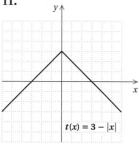

12. Yes   **13.** No   **14.** No   **15.** Yes   **16.** $40 million
17. $7 million

Calculator Spotlight, p. 169

1. 6, 3.99, 150, $-1.\overline{5}$ **2.** -21.3, -18.39, -117.3, $3.2\overline{5}$
3. -75, -65.466, -420.6, $1.6\overline{8}$

Calculator Spotlight, p. 172

1. (0, 1), $(-2.765957, -6.331227)$, $(-1.914894, 3.5529025)$,
$(1.2765957, -3.302515)$, $(2.7659574, 8.3312272)$, answers
may vary   **2.** 951; 42,701   **3.** The table is cleared.
Any number can be entered for x; 21,813

Exercise Set 2.2, p. 173

1. Yes   **3.** Yes   **5.** Yes   **7.** No   **9.** Function
11. A relation but not a function   **13.** Function
15. (a) 9; (b) 12; (c) 2; (d) 5; (e) 7.4; (f) $5\frac{2}{3}$
17. (a) -21; (b) 15; (c) 42; (d) 0; (e) 2; (f) $3a + 3$
19. (a) 7; (b) -17; (c) 24.1; (d) 4; (e) -26; (f) 6
21. (a) 0; (b) 5; (c) 2; (d) 170; (e) 65; (f) $32a^2 - 12a$
23. (a) 1; (b) 3; (c) 3; (d) 4; (e) 11; (f) $|a - 1| + 1$
25. (a) 0; (b) -1; (c) 8; (d) 1000; (e) -125; (f) -1000
27. (a) 159.48 cm; (b) 167.73 cm   **29.** $1\frac{20}{33}$ atm;
$1\frac{10}{11}$ atm; $4\frac{1}{33}$ atm   **31.** 1.792 cm; 2.8 cm; 11.2 cm

33.

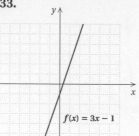

35.

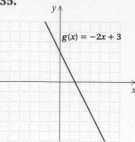

37.

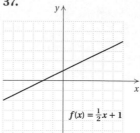

39.

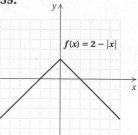

41.

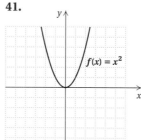

43.

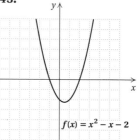

45. Yes   **47.** Yes   **49.** No   **51.** No   **53.** 75
55. 3.7 billion   **57.** 1994   **59.** 18, 20, 22   **60.** 12 ft;
20.8 ft   **61.** $l = \dfrac{S - 2wh}{2h + 2w}$ **62.** $w = \dfrac{S - 2lh}{2l + 2h}$
63. $\{x \mid x \geq -9\}$, or $[-9, \infty)$ **64.** $\left\{x \mid x > \frac{5}{2}\right\}$, or $\left(\frac{5}{2}, \infty\right)$
65. $\{x \mid x > -24\}$, or $(-24, \infty)$ **66.** $\{x \mid x \leq -24\}$, or
$(-\infty, -24]$ **67.** $\left\{-\frac{17}{5}, \frac{3}{5}\right\}$ **68.** $\{x \mid -2 \leq x \leq 3\}$, or
$[-2, 3]$ **69.** $\left\{x \mid x \leq -\frac{17}{3} \ or \ x \geq \frac{31}{3}\right\}$, or
$\left(-\infty, -\frac{17}{3}\right] \cup \left[\frac{31}{3}, \infty\right)$ **70.** $\left\{x \mid x < -\frac{128}{3} \ or \ x > \frac{112}{3}\right\}$, or
$\left(-\infty, -\frac{128}{3}\right) \cup \left(\frac{112}{3}, \infty\right)$ **71.** ◆   **73.** ◆   **75.** 26; 99
77. $g(x) = \frac{15}{4}x - \frac{13}{4}$

Margin Exercises, Section 2.3, pp. 179–181

1. Domain = $\{-3, -2, 0, 2, 5\}$; range = $\{-3, -2, 2, 3\}$
2. (a) -4; (b) all real numbers; (c) -3, 3;
(d) $\{y \mid y \geq -5\}$, or $[-5, \infty)$ **3.** All real numbers
4. $\left\{x \mid x \text{ is a real number } and \ x \neq -\frac{2}{3}\right\}$, or
$\left(-\infty, -\frac{2}{3}\right) \cup \left(-\frac{2}{3}, \infty\right)$

Calculator Spotlight, p. 182

1. Domain = all real numbers; range = $[-7, \infty)$
2. Domain = all real numbers; range = all real
numbers   **3.** Domain = $(-\infty, 0) \cup (0, \infty)$;
range = $(-\infty, 0) \cup (0, \infty)$ **4.** Domain = all real
numbers; range = $[-8, \infty)$

Answers

Exercise Set 2.3, p. 183

1. (a) 3; (b) $\{-4, -3, -2, -1, 0, 1, 2\}$; (c) $-2, 0$;
(d) $\{1, 2, 3, 4\}$ **3.** (a) 2; (b) $\{-6, -4, -2, 0, 1, 3, 4\}$;
(c) 1, 3; (d) $\{-5, -2, 0, 2, 5\}$ **5.** (a) $2\frac{1}{2}$; (b) $[-3, 5]$;
(c) $2\frac{1}{4}$; (d) $[1, 4]$ **7.** (a) $2\frac{1}{4}$; (b) $[-4, 3]$; (c) 0;
(d) $[-5, 4]$ **9.** (a) 2; (b) $[-5, 4]$; (c) $[1, 4]$; (d) $[-3, 2]$
11. (a) -1; (b) $[-6, 5]$; (c) $-4, 0, 3$; (d) $[-2, 2]$
13. $\{x \mid x \text{ is a real number } and \ x \neq -3\}$, or
$(-\infty, -3) \cup (-3, \infty)$ **15.** All real numbers **17.** All
real numbers **19.** $\{x \mid x \text{ is a real number } and \ x \neq \frac{14}{5}\}$,
or $(-\infty, \frac{14}{5}) \cup (\frac{14}{5}, \infty)$ **21.** All real numbers
23. $\{x \mid x \text{ is a real number } and \ x \neq \frac{3}{2}\}$, or $(-\infty, \frac{3}{2}) \cup (\frac{3}{2}, \infty)$
25. $\{x \mid x \text{ is a real number } and \ x \neq 1\}$, or $(-\infty, 1) \cup (1, \infty)$
27. All real numbers **29.** All real numbers
31. $\{x \mid x \text{ is a real number } and \ x \neq \frac{5}{2}\}$, or $(-\infty, \frac{5}{2}) \cup (\frac{5}{2}, \infty)$
33. All real numbers
35. $\{x \mid x \text{ is a real number } and \ x \neq -\frac{5}{4}\}$, or
$(-\infty, -\frac{5}{4}) \cup (-\frac{5}{4}, \infty)$ **37.** $-8; 0; -2$
39. $\{S \mid S > \$42,500\}$ **40.** $\{x \mid x \geq 90\}$ **41.** $\{-8, 8\}$
42. $\{\ \}$, or $\varnothing$ **43.** $\{-4, 18\}$ **44.** $\{-8, 5\}$ **45.** $\{\frac{1}{2}, 3\}$
46. $\{-1, \frac{9}{13}\}$ **47.** $\{\ \}$, or $\varnothing$ **48.** $\{\frac{8}{3}\}$ **49.** ◈
51. $(-\infty, 0) \cup (0, \infty); [2, \infty); [-4, \infty); [0, \infty)$

Margin Exercises, Section 2.4, pp. 185–190

1.

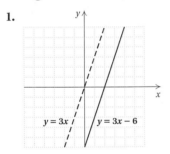

The graph of $y = 3x - 6$ looks just like the graph of
$y = 3x$, but it is moved down 6 units.

2.

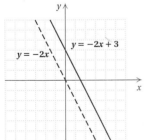

The graph of $y = -2x + 3$ looks just like the graph of
$y = -2x$, but it is moved up 3 units.

3.

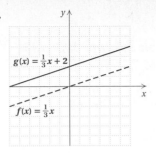

The graph of $g(x)$ looks just like the graph of $f(x)$, but it
is moved up 2 units.
4. $(0, 8)$ **5.** $\left(0, -\frac{2}{3}\right)$
6. $m = -1$ **7.** $m = -\frac{1}{3}$

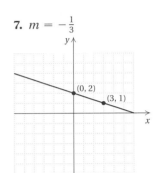

8. $m = -\frac{2}{3}$ **9.** Slope: -8; y-intercept: $(0, 23)$
10. Slope: $\frac{1}{2}$; y-intercept: $\left(0, -\frac{5}{2}\right)$ **11.** The rate of
change is $-10\dfrac{\$ \text{ billion}}{\text{yr}}$.

Calculator Spotlight, p. 186

1. The graph of $y = x - 5$ is the same as the graph of
$y = x$, but it is moved down 5 units. **2.** The values of
y_2 are 3 more than the values of y_1. The values of y_3 are
4 less than the values of y_1.

Calculator Spotlight, p. 187

1. The graph of $y = 247x + 1$ will be steeper than the
others. **2.** The graph of $y = 0.000018x$ will be less
steep than the others. **3.** The graph of $y = -247x$
will be steeper than the others. **4.** The graph of
$y = -0.000043x - 1$ will be less steep than the others.

Exercise Set 2.4, p. 191

1. $m = 4$; y-intercept: $(0, 5)$ **3.** $m = -2$;
y-intercept: $(0, -6)$ **5.** $m = -\frac{3}{8}$; y-intercept: $\left(0, -\frac{1}{5}\right)$
7. $m = 0.5$; y-intercept: $(0, -9)$ **9.** $m = \frac{2}{3}$;
y-intercept: $\left(0, -\frac{8}{3}\right)$ **11.** $m = 3$; y-intercept: $(0, -2)$
13. $m = -8$; y-intercept: $(0, 12)$ **15.** $m = 0$;
y-intercept: $\left(0, \frac{4}{17}\right)$ **17.** $m = -\frac{1}{2}$ **19.** $m = \frac{1}{3}$
21. $m = 2$ **23.** $m = \frac{2}{3}$ **25.** $m = -\frac{1}{3}$ **27.** $\frac{2}{25}$, or 8%
29. 3.5% **31.** $100\dfrac{\$}{\text{yr}}$ **33.** -1323 **34.** $45x + 54$
35. $350x - 60y + 120$ **36.** 25 **37.** Square: 15 yd;
triangle: 20 yd **38.** $\{x \mid x \leq -\frac{24}{5} \text{ or } x \geq 8\}$, or
$\left(-\infty, -\frac{24}{5}\right] \cup [8, \infty)$ **39.** $\{x \mid -\frac{24}{5} < x < 8\}$, or $\left(-\frac{24}{5}, 8\right)$

40. $\left\{-\frac{24}{5}, 8\right\}$ **41.** { }, or $\varnothing$ **43.** ◆

Margin Exercises, Section 2.5, pp. 193–198

1.
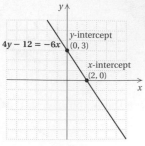

2.
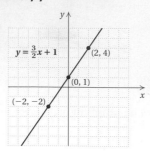

3.
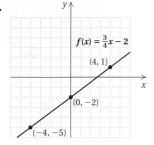

4.
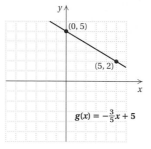

5.
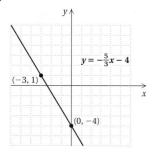

6. $m = 0$
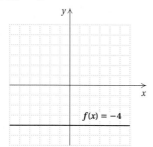

7. $m = 0$
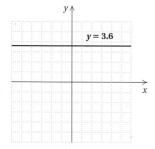

8.
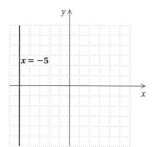

9.
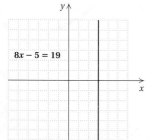

10. (a) Undefined; (b) $m = 0$; (c) $m = 0$; (d) undefined;
(e) $m = 0$; (f) undefined **11.** Yes **12.** No **13.** No
14. Yes **15.** No

Calculator Spotlight, p. 198

1.–2. Left to the student.

Exercise Set 2.5, p. 199

1.
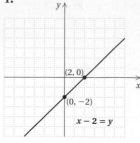

3.
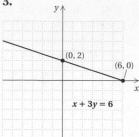

5.
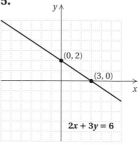

7.
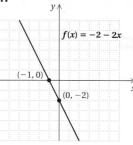

9.
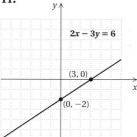

11.
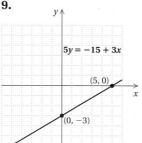

13.
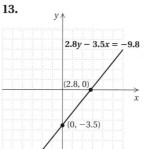

15.
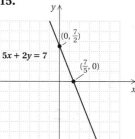

17.
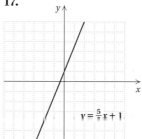

19.
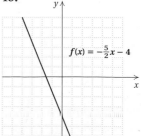

21.

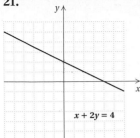

$x + 2y = 4$

23.

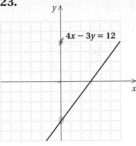

$4x - 3y = 12$

25.

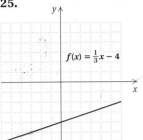

$f(x) = \frac{1}{3}x - 4$

27.

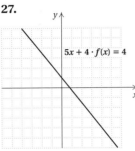

$5x + 4 \cdot f(x) = 4$

29. Undefined

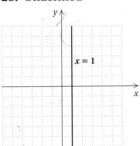

$x = 1$

31. $m = 0$

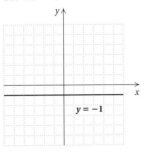

$y = -1$

33. $m = 0$

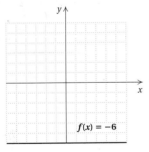

$f(x) = -6$

35. $m = 0$

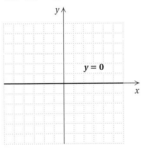

$y = 0$

37. $m = 0$

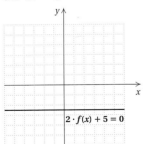

$2 \cdot f(x) + 5 = 0$

39. Undefined

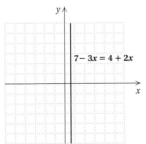

$7 - 3x = 4 + 2x$

41. Yes   **43.** No   **45.** Yes   **47.** Yes   **49.** Yes
51. No   **53.** No   **55.** Yes   **57.** 5.3×10^{10}
58. 4.7×10^{-5} **59.** 1.8×10^{-2} **60.** 9.9902×10^{7}
61. 0.0000213   **62.** 901,000,000   **63.** 20,000

64. 0.085677   **65.** $3(3x - 5y)$ **66.** $3a(4 + 7b)$
67. $7p(3 - q + 2)$ **68.** $64(x - 2y + 4)$ **69.** ◈
71. $y = 3$ **73.** $a = 2$ **75.** $y = \frac{2}{15}x + \frac{2}{5}$ **77.** $y = 0$;
yes   **79.** $m = -\frac{3}{4}$ **81.** (a) II; (b) IV; (c) I; (d) III

Margin Exercises, Section 2.6, pp. 203–206

1. $y = 3.4x - 8$ **2.** $y = -5x - 18$ **3.** $y = 3x - 5$
4. $y = 8x - 19$ **5.** $y = -\frac{2}{3}x + \frac{14}{3}$ **6.** $y = -\frac{5}{3}x + \frac{11}{3}$
7. $y = -17x - 56$ **8.** $y = \frac{8}{7}x - \frac{23}{7}$ **9.** $y = -2x + 14$
10. (a) $C(t) = 20t + 25$;
(b)   (c) $195

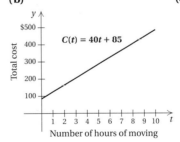

$C(t) = 20t + 25$

Number of months
of service

Exercise Set 2.6, p. 207

1. $y = -8x + 4$ **3.** $y = 2.3x - 1$ **5.** $f(x) = -\frac{7}{3}x - 5$
7. $f(x) = \frac{2}{3}x + \frac{5}{8}$ **9.** $y = 5x - 17$ **11.** $y = -3x + 33$
13. $y = x - 6$ **15.** $y = -2x + 16$ **17.** $y = -7$
19. $y = \frac{2}{3}x - \frac{8}{3}$ **21.** $y = \frac{1}{2}x + \frac{7}{2}$ **23.** $y = x$
25. $y = \frac{7}{4}x + 7$ **27.** $y = \frac{3}{2}x$ **29.** $y = \frac{1}{6}x$
31. $y = 13x - \frac{15}{4}$ **33.** $y = -\frac{1}{2}x + \frac{17}{2}$ **35.** $y = \frac{5}{7}x - \frac{17}{7}$
37. $y = \frac{1}{3}x + 4$ **39.** $y = \frac{1}{2}x + 4$ **41.** $y = \frac{4}{3}x - 6$
43. $y = \frac{5}{2}x + 9$ **45.** (a) $C(t) = 40t + 85$;
(b)   (c) $345

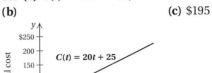

$C(t) = 40t + 85$

Number of hours of moving

47. (a) $V(t) = 750 - 25t$;
(b)   (c) $425

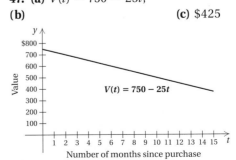

$V(t) = 750 - 25t$

Number of months since purchase

49. $\{x \mid x > 24\}$, or $(24, \infty)$ **50.** $\{-27, 24\}$
51. $\{x \mid x \le 24\}$, or $(-\infty, 24]$ **52.** $\left\{x \mid x \ge \frac{7}{3}\right\}$, or $\left[\frac{7}{3}, \infty\right)$

53. $\{x \mid -8 \leq x \leq 5\}$, or $[-8, 5]$ **54.** $\left\{-7, \frac{1}{3}\right\}$

55. No solution **56.** $\left\{x \mid -\frac{15}{2} \leq x < 24\right\}$, or $\left[-\frac{15}{2}, 24\right)$

57. ◈

Margin Exercises, Section 2.7, pp. 210–211

1. (a)   **(b)** 94%

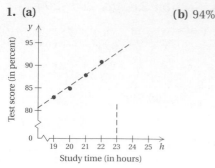

2. (a) $T(x) = 0.08625x + 8.97075$; **(b)** $T(10) = 9.83325$. This prediction differs slightly from the number found in Example 2 and the number 9.85 found in Example 1.

Calculator Spotlight, pp. 212–213

1. (a) $y = 2.7x + 31.4$; **(b)**

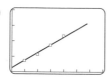

(c) 93.5%; **(d)** This prediction differs by 0.5% from the score of 94% found in Margin Exercise 1.

Exercise Set 2.7, p. 215

Answers may vary in Exercises 1–12.
1. (a) and **(b)**

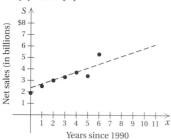

(c) $5.5 billion; $6.2 billion
3. (a) and **(b)**

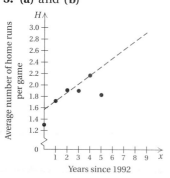

(c) 2.5; 2.8

5. (a) $B(x) = 1879x + 7403$;
(b)

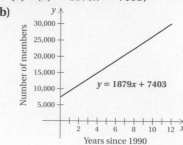

(c) 26,193; 44,983 **7. (a)** $H(x) = 0.15x + 1.57$;
(b) see graph in Exercise 3; **(c)** 2.62; 3.07
9. (a) $S(x) = 0.35x + 2.3$; **(b)** see graph in Exercise 1;
(c) $5.8 billion; $7.55 billion **11. (a)** $y = 0.75x + 71.25$;
(b) 84.75; 90 **13.** $35 **14.** 83 **15.** $y = \dfrac{C - Ax}{B}$

16. $y = \dfrac{3x - 12}{7}$ **17.** $q = \frac{3}{2}(P + y)$ **18.** $b = \dfrac{2A}{h}$

19. $[-9, 2]$ **20.** $[-9, 2)$ **21.** $[-9, \infty)$ **22.** $(-\infty, -9)$
23. ◈ **25. (a)** $y = 0.114x + 1.52$; **(b)** left to the
student; **(c)** 2.66 **27. (a)** $M(c) = -0.0408975991x + 151.6882381$; **(b)** left to the student; **(c)** 20.2; **(d)** 44.5

Summary and Review: Chapter 2, p. 219

1.

$3x - y = 2$		$3x - y = 2$	
$3 \cdot 0 - (-2)$? 2		$3(-1) - (-5)$? 2	
$0 + 2$		$-3 + 5$	
2	TRUE	2	TRUE

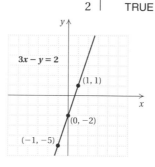

2.

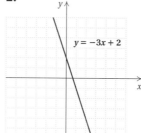

3.

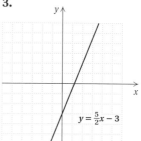

4.

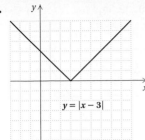

$y = |x - 3|$

5.

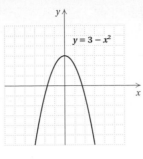

$y = 3 - x^2$

6. No   **7.** Yes   **8.** $g(0) = 5; g(-1) = 7$ **9.** $f(0) = 7;$
$f(-1) = 12$ **10.** $18,185   **11.** Yes   **12.** No
13. (a) $f(2) = 3$; **(b)** $[-2, 4]$; **(c)** -1; **(d)** $[1, 5]$
14. $\{x \mid x$ is a real number *and* $x \neq 4\}$, or
$(-\infty, 4) \cup (4, \infty)$ **15.** All real numbers
16. Slope: -3; y-intercept: $(0, 2)$ **17.** Slope: $-\frac{1}{2}$;
y-intercept: $(0, 2)$ **18.** $m = \frac{11}{3}$
19.

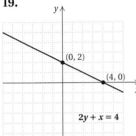

$(0, 2)$ $(4, 0)$
$2y + x = 4$

20.

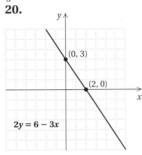

$(0, 3)$
$(2, 0)$
$2y = 6 - 3x$

21.

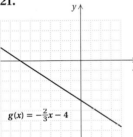

$g(x) = -\frac{2}{3}x - 4$

22.

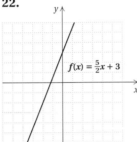

$f(x) = \frac{5}{2}x + 3$

23.

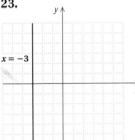

$x = -3$

24.

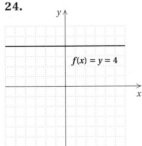

$f(x) = y = 4$

25. Perpendicular   **26.** Parallel   **27.** Parallel
28. Perpendicular   **29.** $f(x) = 4.7x - 23$
30. $y = -3x + 4$ **31.** $y = -\frac{3}{2}x$
32. $y = -\frac{5}{7}x + 9$ **33.** $y = \frac{1}{3}x + \frac{1}{3}$
34. (a) $y = 32.8286x + 175.9$;
(b) $504.2 billion; $668.3 billion   **35.** $0.90
36. $\left\{x \mid x < -\frac{77}{4}\right\}$, or $\left(-\infty, -\frac{77}{4}\right)$
37. $\left\{x \mid x < -\frac{7}{2} \ or \ x > \frac{21}{2}\right\}$, or $\left(-\infty, -\frac{7}{2}\right) \cup \left(\frac{21}{2}, \infty\right)$
38. $\{x \mid -11 \le x \le 18\}$, or $[-11, 18]$ **39.** $\left\{-3, \frac{13}{2}\right\}$

40. $\left\{x \mid -3 \le x \le \frac{13}{2}\right\}$, or $\left[-3, \frac{13}{2}\right]$
41. $\left\{x \mid x \le -3 \ or \ x \ge \frac{13}{2}\right\}$, or $(-\infty, -3] \cup \left[\frac{13}{2}, \infty\right)$
42. ◈ The concept of slope is useful in describing how
a line slants. A line with positive slope slants up from
left to right. A line with negative slope slants down
from left to right. The larger the absolute value of the
slope, the steeper the slant.
43. ◈ The notation $f(x)$ can be read "f of x" or "f at x"
or "the value of f at x." It represents the output of the
function f for the input x. The notation $f(a) = b$
provides a concise way to indicate that for the input a,
the output of the function f is b.
44. (a) $H(x) = 33.6893x + 175.85$; **(b)** left to the student;
(c) $680.5 billion

Test: Chapter 2, p. 221

1. [2.1b] Yes   **2.** [2.1b] No
3. [2.1c]

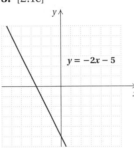

$y = -2x - 5$

4. [2.2c]

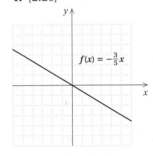

$f(x) = -\frac{3}{5}x$

5. [2.2c]

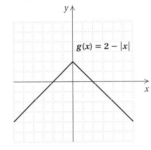

$g(x) = 2 - |x|$

6. [2.1d]

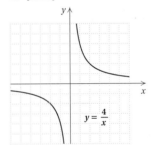

$y = \frac{4}{x}$

7. [2.2b] **(a)** 8.666; **(b)** 1998   **8.** [2.2a] Yes
9. [2.2a] No   **10.** [2.2b] -4; 2   **11.** [2.2b] 7; 8
12. [2.2d] Yes   **13.** [2.2d] No
14. [2.2e] **(a)** $120 million; **(b)** $20 million
15. [2.3a] **(a)** 1.2; **(b)** $[-3, 4]$; **(c)** -3; **(d)** $[-1, 2]$
16. [2.3a] All real numbers
17. [2.3a] $\left\{x \mid x$ is a real number *and* $x \neq -\frac{3}{2}\right\}$, or
$\left(-\infty, -\frac{3}{2}\right) \cup \left(-\frac{3}{2}, \infty\right)$ **18.** [2.4b] Slope: $-\frac{3}{5}$;
y-intercept: $(0, 12)$ **19.** [2.4b] Slope: $-\frac{2}{5}$;
y-intercept: $\left(0, -\frac{7}{5}\right)$ **20.** [2.4b] $m = \frac{5}{8}$
21. [2.4b] $m = 0$ **22.** [2.4c] m (or rate of change) $= \frac{4}{5}$

23. [2.5a]

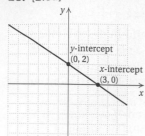
y-intercept (0, 2)
x-intercept (3, 0)

24. [2.5b]

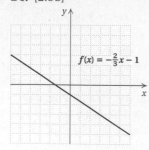
$f(x) = -\frac{2}{3}x - 1$

28. [2.1c]

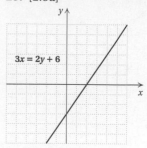
$y = -2x + 3$

29. [2.5a]

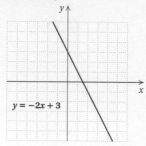
$3x = 2y + 6$

25. [2.5c]

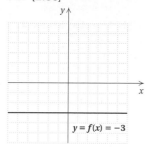
$y = f(x) = -3$

26. [2.5c]

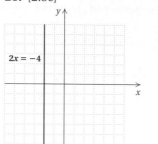
$2x = -4$

30. [2.5c]

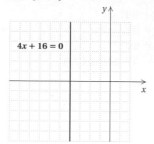
$4x + 16 = 0$

31. [2.5c]

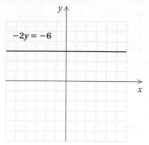
$-2y = -6$

27. [2.5d] Parallel   **28.** [2.5d] Perpendicular
29. [2.6a] $y = -3x + 4.8$ **30.** [2.6a] $f(x) = 5.2x - \frac{5}{8}$
31. [2.6b] $y = -4x + 2$ **32.** [2.6c] $y = -\frac{3}{2}x$
33. [2.6d] $y = \frac{1}{2}x - 3$ **34.** [2.6d] $y = 3x - 1$
35. [2.7b] **(a)** $S(x) = 0.2167x + 0.3$;
(b) $2.0 billion; $2.5 billion   **36.** [1.3a] 170 ft
37. [1.4c] $\{x \mid x \geq \frac{13}{9}\}$, or $\left[\frac{13}{9}, \infty\right)$
38. [1.5b] $\{x \mid x < -2 \; or \; x > \frac{32}{7}\}$, or $(-\infty, -2) \cup \left(\frac{32}{7}, \infty\right)$
39. [1.6e] $\{x \mid -2 < x < \frac{32}{7}\}$, or $\left(-2, \frac{32}{7}\right)$
40. [1.6c] $\left\{-2, \frac{32}{7}\right\}$ **41.** [2.6d] $\frac{24}{5}$ **42.** [2.2b] $f(x) = 3$; answers may vary

Cumulative Review: Chapters R–2, p. 225

1. [R.3a] 31   **2.** [R.1d] 2.2   **3.** [R.1d] $\frac{2}{3}$ **4.** [R.2c] $\frac{1}{2}$
5. [R.2d] 36.48   **6.** [R.6b] $-10x + 32$ **7.** [R.7a] $\dfrac{32x^8}{y}$
8. [R.7a] $-\dfrac{9}{x^2y^2}$ **9.** [R.6b] $23x + 31$ **10.** [R.3c] -328
11. [R.3c] -1 **12.** [1.1b] -22 **13.** [1.1d] $\frac{15}{88}$
14. [1.1c] 20   **15.** [1.1d] $-\frac{21}{4}$ **16.** [1.1d] -5
17. [1.2a] $x = \dfrac{W - By}{A}$ **18.** [1.2a] $A = \dfrac{M}{1 + 4B}$
19. [1.4c] $\{y \mid y \leq 7\}$, or $(-\infty, 7]$ **20.** [1.4c] $\{x \mid x < -\frac{3}{2}\}$, or $\left(-\infty, -\frac{3}{2}\right)$ **21.** [1.4c] $\{x \mid x > -\frac{1}{11}\}$, or $\left(-\frac{1}{11}, \infty\right)$
22. [1.5b] All real numbers
23. [1.5a] $\{x \mid -7 < x \leq 4\}$, or $(-7, 4]$
24. [1.5a] $\{x \mid -2 \leq x \leq \frac{3}{2}\}$, or $\left[-2, \frac{3}{2}\right]$
25. [1.6c] $\{-8, 8\}$
26. [1.6e] $\{y \mid y < -4 \; or \; y > 4\}$, or $(-\infty, -4) \cup (4, \infty)$
27. [1.6e] $\{x \mid -\frac{3}{2} \leq x \leq 2\}$, or $\left[-\frac{3}{2}, 2\right]$

32. [2.2c]

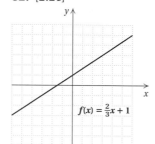
$f(x) = \frac{2}{3}x + 1$

33. [2.2c]

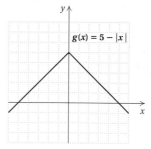
$g(x) = 5 - |x|$

34. [2.6d] $y = -4x - 22$ **35.** [2.6d] $y = \frac{1}{4}x - 5$
36. [2.3a] **(a)** 6; **(b)** [0, 30]; **(c)** 25; **(d)** [0, 15]
37. [2.7b] **(a)** $W(x) = 0.2125x + 6.15$;
(b)

W
7.0
6.8
6.6
6.4
6.2
6.0
Average number of walks per game
1 2 3 4 5 6 7 8 x
Years since 1992

(c) 7.6375; 7.85   **38.** [2.4b] Slope: $\frac{9}{4}$;
y-intercept: (0, −3)   **39.** [2.4b] $m = \frac{4}{3}$
40. [2.6b] $y = -3x - 5$ **41.** [2.6c] $y = -\frac{1}{10}x + \frac{12}{5}$
42. [1.3a] 32.88   **43.** [1.3a] $w = 17$ m, $l = 23$ m
44. [1.3a] 4   **45.** [1.3a] $9000   **46.** [2.7b] $151,000
47. [R.7a] $-12x^{2a}y^{b+y+3}$ **48.** [1.5a] $\{x \mid 6 < x \leq 10\}$, or (6, 10]   **49.** [2.5d] (1), (4)

CHAPTER 3

Pretest: Chapter 3, p. 228

1. [3.1a] (1, 2)   **2.** [3.2a] (1, 2)   **3.** [3.2b] (−3, 2)
4. [3.1a] Consistent   **5.** [3.1a] Independent
6. [3.4a] (1, 2, 3)   **7.** [3.3a] Cashews: 20 lb;
Brazil nuts: 30 lb   **8.** [3.3a] $3300 at 8%; $4200 at 6%
9. [3.3a] A: $64\frac{8}{13}$ lb; B: $55\frac{5}{13}$ lb   **10.** [3.3b] 54 mph
11. [3.6b]   **12.** [3.6c]

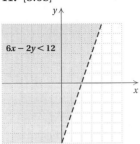

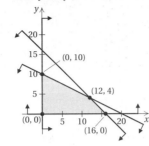

13. [3.7a] **(a)** $P(x) = 11x − 90{,}000$; **(b)** (8182, $212,732)

Margin Exercises, Section 3.1, pp. 229–233

1. (0, 1)   **2.** (2, 1)   **3.** No solution
4. Consistent: 1, 2; inconsistent: 3   **5.** Infinitely many
solutions   **6.** Independent: 1, 2, 3; dependent: 5
7. (a) −3; **(b)**   ; −3

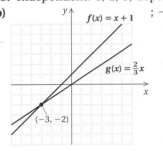

8.   ; −2

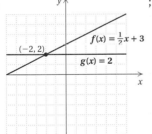

9. (a)   , −3; **(b)** All are −3.

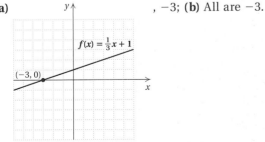

10.   ; −2

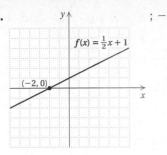

Calculator Spotlight, p. 234

1. −2   **2.** −2   **3.** 2.5455   **4.** 1.3333   **5.** −1.3825,
3.1825   **6.** 2.2790   **7.–12.** See answers for
Exercises 1–6.

Exercise Set 3.1, p. 235

1. (3, 1); consistent; independent   **3.** (1, −2);
consistent; independent   **5.** (4, −2); consistent;
independent   **7.** (2, 1); consistent; independent
9. $\left(\frac{5}{2}, −2\right)$; consistent; independent   **11.** (3, −2);
consistent; independent   **13.** No solution; inconsistent;
independent   **15.** Infinitely many solutions;
consistent; dependent   **17.** (4, −5); consistent;
independent   **19.** (2, −3); consistent; independent
21. −3   **22.** −20   **23.** $\frac{9}{20}$ **24.** −38   **25.** ◈
27. (2.23, 1.14)

Margin Exercises, Section 3.2, pp. 238–245

1. (2, 4)   **2.** (5, 2)   **3.** (−3, 2)   **4.** (13, 16)
5. (1, 4)   **6.** $\left(\frac{1}{3}, \frac{1}{2}\right)$ **7.** (2, 3)   **8.** (−127, 100)
9. (−138, −118)   **10.** No solution   **11.** Infinitely
many solutions   **12.** $2x + 3y = 1$, $3x − y = 7$; (2, −1)
13. $9x + 10y = 5$, $9x − 4y = 3$; $\left(\frac{25}{63}, \frac{1}{7}\right)$
14. 2-pointers: 31; 3-pointers: 9
15. (a) and **(b)** Length: 160 ft; width: 100 ft

Exercise Set 3.2, p. 247

1. (−3, 2)   **3.** (−2, −6)   **5.** (2, −2)   **7.** (−2, 1)
9. (1, 2)   **11.** (−1, 3)   **13.** $\left(\frac{128}{31}, −\frac{17}{31}\right)$ **15.** (6, 2)
17. $\left(\frac{140}{13}, −\frac{50}{13}\right)$ **19.** (4, 6)   **21.** $\left(\frac{110}{19}, −\frac{12}{19}\right)$
23. Infinitely many solutions   **25.** No solution
27. $\left(\frac{1}{2}, −\frac{1}{2}\right)$ **29.** $\left(−\frac{4}{3}, −\frac{19}{3}\right)$ **31.** $\left(\frac{1000}{11}, −\frac{1000}{11}\right)$
33. (140, 60)   **35.** Length: 40 ft; width: 20 ft
37. 48° and 132°   **39.** 5 and −47   **41.** 23 wins;
14 ties   **43.** 1.3   **44.** −15y − 39   **45.** $p = \dfrac{7A}{q}$
46. $\frac{7}{3}$ **47.** −23   **48.** $\frac{29}{22}$ **49.** 1   **50.** 5   **51.** 3
52. 291   **53.** 15   **54.** $12a^2 − 2a + 1$ **55.** 53
56. 8.92   **57.** ◈   **59.** (23.12, −12.04)
61. $\left(\dfrac{a + 2b}{7}, \dfrac{a − 5b}{7}\right)$ **63.** $p = 2$; $q = −\frac{1}{3}$
65. $A = 2$; $B = 4$

1. White: 22; red: 8

White	Red	Totals
w	r	30	$\rightarrow w + r = 30$
18.95	19.50	
$18.95w$	$19.50r$	572.90	$\rightarrow 18.95w + 19.50r = 572.90$

2. Kenyan: 8 lb; Sumatran: 12 lb
3. $1800 at 7%; $1900 at 9%

First Investment	Second Investment	Total
x	y	$3700	$\rightarrow x + y = 3700$
7%	9%	
1 yr	1 yr	
$0.07x$	$0.09y$	$297	$\rightarrow 0.07x + 0.09y = 297$

4. Attack: 15 qt; Blast: 45 qt

Attack	Blast	Mixture
a	b	60	$\rightarrow a + b = 60$
2%	6%	5%
$0.02a$	$0.06b$	0.05×60, or 3	$\rightarrow 0.02a + 0.06b = 3$

5. 280 km

Distance	Rate	Time
d	35 km/h	t	$\rightarrow d = 35t$
d	40 km/h	$t - 1$	$\rightarrow d = 40(t - 1)$

6. 180 mph

Distance	Rate	Time
d	$r + 20$	4 hr	$\rightarrow d = 4(r + 20)$
d	$r - 20$	5 hr	$\rightarrow d = 5(r - 20)$

Exercise Set 3.3, p. 259

1. 32 at $8.50; 13 at $9.75   **3.** Humulin: 42; Novolin: 23   **5.** 30-sec: 4; 60-sec: 8   **7.** 5 lb of each
9. 4 L of 25%; 6 L of 50%   **11.** $7500 at 6%; $4500 at 9%   **13.** Whole milk: $169\frac{3}{13}$ lb; cream: $30\frac{10}{13}$ lb
15. $5 bills: 7; $1 bills: 15   **17.** 375 mi   **19.** 14 km/h
21. 144 mi   **23.** 2 hr   **25.** $1\frac{1}{3}$ hr   **27.** About 1489 mi
29. -7 **30.** -11 **31.** -3 **32.** 33   **33.** -15
34. $8a - 7$ **35.** -23 **36.** 0.2   **37.** -4 **38.** -17
39. $-12h - 7$ **40.** 3993   **41.** ◈   **43.** $4\frac{4}{7}$ L
45. City: 261 mi; highway: 204 mi   **47.** Brown: 0.8 gal; neutral: 0.2 gal

1. $(2, 1, -1)$ **2.** $\left(2, -2, \frac{1}{2}\right)$ **3.** $(20, 30, 50)$

Exercise Set 3.4, p. 267

1. $(1, 2, -1)$ **3.** $(2, 0, 1)$ **5.** $(3, 1, 2)$ **7.** $(-3, -4, 2)$
9. $(2, 4, 1)$ **11.** $(-3, 0, 4)$ **13.** $(2, 2, 4)$
15. $\left(\frac{1}{2}, 4, -6\right)$ **17.** $\left(\frac{1}{2}, \frac{1}{3}, \frac{1}{6}\right)$ **19.** $\left(\frac{1}{2}, \frac{2}{3}, -\frac{5}{6}\right)$
21. $(15, 33, 9)$ **23.** $(4, 1, -2)$ **25.** $a = \dfrac{F}{3b}$
26. $a = \dfrac{Q - 4b}{4}$, or $\dfrac{Q}{4} - b$ **27.** $c = \dfrac{2F + td}{t}$, or $\dfrac{2F}{t} + d$ **28.** $d = \dfrac{tc - 2F}{t}$, or $c - \dfrac{2F}{t}$
29. $y = \dfrac{Ax - c}{B}$ **30.** $y = \dfrac{c - Ax}{B}$ **31.** Slope: $-\frac{2}{3}$; y-intercept: $\left(0, -\frac{5}{4}\right)$ **32.** Slope: -4; y-intercept: $(0, 5)$
33. Slope: $\frac{2}{5}$; y-intercept: $(0, -2)$ **34.** Slope: 1.09375; y-intercept: $(0, -3.125)$ **35.** ◈   **37.** $(1, -2, 4, -1)$

1. $64°, 32°, 84°$ **2.** $4000 at 5%; $5000 at 6%; $16,000 at 7%

Exercise Set 3.5, p. 271

1. 10-oz: 8; 14-oz: 20; 20-oz: 6   **3.** $32°, 96°, 52°$
5. Automatic transmission: $865; power door locks: $520; air conditioning: $375   **7.** A: 1500; B: 1900; C: 2300   **9.** First fund: $45,000; second fund: $10,000; third fund: $25,000   **11.** Asian-American: 385; African-American: 200; Caucasian: 154   **13.** Roast beef: 2; baked potato: 1; broccoli: 2   **15.** Par-3: 6; par-4: 8; par-5: 4   **17.** 4, 2, -1 **19.** 2-point: 32; 3-point: 5; foul shots: 13   **21.** No   **22.** Yes
23. $\{x \,|\, x$ is a real number *and* $x \neq -7\}$, or $(-\infty, -7) \cup (-7, \infty)$ **24.** Domain: all real numbers; range: $\{y \,|\, y \leq 5\}$, or $(-\infty, 5]$ **25.** $y = -\frac{3}{5}x - 7$
26. $\dfrac{a^3}{b}$ **27.** ◈   **29.** $180°$

1. No   **2.** Yes
3.   **4.**

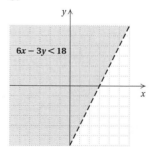

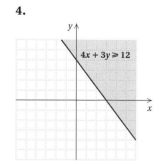

5.

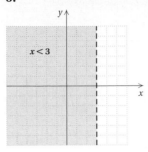

6.

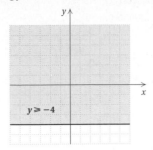

13.

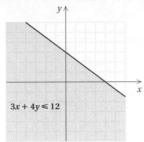

15.

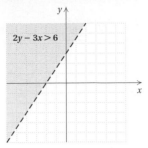

7.

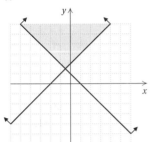

8.

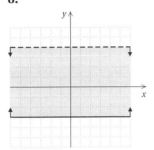

17.

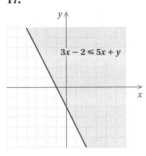

19.

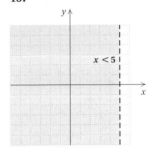

9.

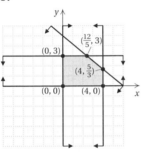

10.

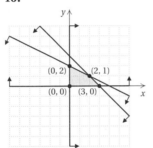

21.

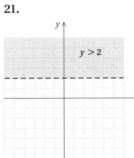

23.

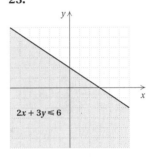

Exercise Set 3.6, p. 283

1. Yes   **3.** Yes

5.

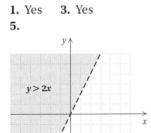

7.

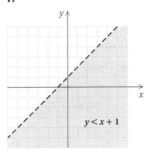

25.

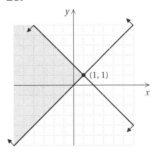

27.

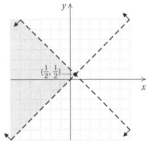

9.

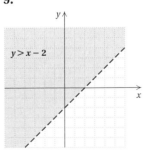

11.

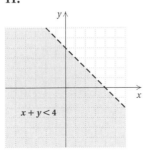

29.

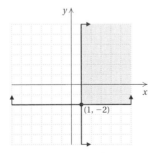

31.

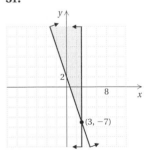

33.

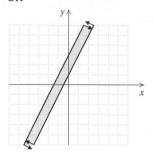

35.

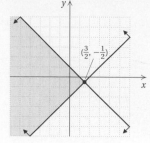

37.

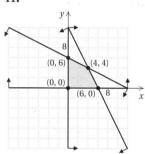

39.

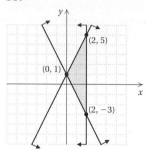

41.

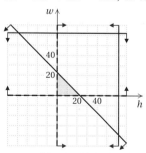

43.

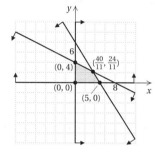

45. $\frac{10}{17}$ **46.** $-\frac{14}{13}$ **47.** -2 **48.** $\frac{29}{11}$ **49.** -12
50. $\frac{333}{245}$ **51.** 2 **52.** 3 **53.** 1 **54.** 8 **55.** 4
56. $|2 - 2a|$, or $2|1 - a|$ **57.** 6 **58.** 0.2 **59.** ◈
61. $0 < w \le 62$,
$0 < h \le 62$,
$62 + 2w + 2h \le 108$, or $w + h \le 23$

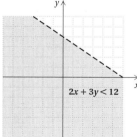

63.–67. Left to the student.

Margin Exercises, Section 3.7, pp. 288–290

1. (a) $C(x) = 80,000 + 20x$; **(b)** $R(x) = 36x$;
(c) $P(x) = 16x - 80,000$; **(d)** a loss of $16,000; a profit of
$176,000

(e)

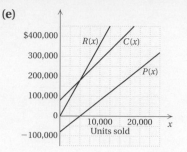

Break-even point: (5000, $180,000)
2. ($14, 356)

Exercise Set 3.7, p. 291

1. (a) $P(x) = 45x - 270,000$; **(b)** (6000, $420,000)
3. (a) $P(x) = 50x - 120,000$; **(b)** (2400, $144,000)
5. (a) $P(x) = 80x - 10,000$; **(b)** (125, $12,500)
7. (a) $P(x) = 18x - 16,000$; **(b)** (889, $35,560)
9. (a) $P(x) = 75x - 195,000$; **(b)** (2600, $325,000)
11. (a) $C(x) = 22,500 + 40x$; **(b)** $R(x) = 85x$;
(c) $P(x) = 45x - 22,500$; **(d)** $112,500 profit; $4500 loss;
(e) (500, $42,500) **13. (a)** $C(x) = 16,404 + 6x$;
(b) $R(x) = 18x$; **(c)** $P(x) = 12x - 16,404$; **(d)** $19,596
profit; $4404 loss; **(e)** (1367, $24,606) **15.** ($70, 300)
17. ($22, 474) **19.** ($50, 6250) **21.** ($10, 1070)
23. Slope: $\frac{3}{5}$; y-intercept: $\left(0, \frac{8}{5}\right)$ **24.** Slope: $-\frac{6}{7}$;
y-intercept: $\left(0, \frac{13}{7}\right)$ **25.** Slope: 1.7; y-intercept: (0, 49)
26. Slope: $-\frac{4}{3}$; y-intercept: (0, 4) **27.** ◈

Summary and Review: Chapter 3, p. 293

1. $(-2, 1)$; consistent; independent
2. Infinitely many solutions; consistent; dependent
3. No solution; inconsistent; independent **4.** $\left(\frac{2}{5}, -\frac{4}{5}\right)$
5. No solution **6.** $\left(-\frac{11}{15}, -\frac{43}{30}\right)$ **7.** $\left(\frac{37}{19}, \frac{53}{19}\right)$
8. $\left(\frac{76}{17}, -\frac{2}{119}\right)$ **9.** (2, 2) **10.** Infinitely many solutions
11. CD: $14; cassette: $9 **12.** 5 L of each **13.** $5\frac{1}{2}$ hr
14. (10, 4, −8) **15.** $\left(-\frac{7}{3}, \frac{125}{27}, \frac{20}{27}\right)$ **16.** (2, 0, 4)
17. $\left(2, \frac{1}{3}, -\frac{2}{3}\right)$ **18.** $90°, 67\frac{1}{2}°, 22\frac{1}{2}°$ **19.** $20 bills: 5;
$5 bills: 15; $1 bills: 19

20.

21.

22.

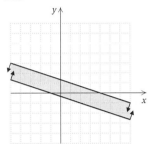

$x + y \geq 1$

23.

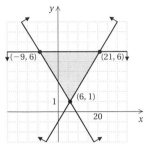

(2, −3)

16. [3.6b]

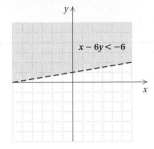

$x - 6y < -6$

17. [3.6c]

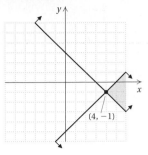

(4, −1)

24.

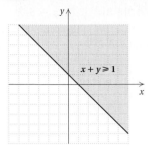

25.

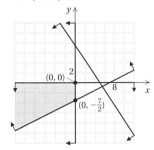

(−9, 6)   (21, 6)
(6, 1)
1
20

18. [3.6c]

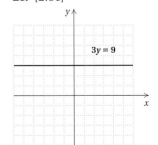

(0, 0) 2
8
(0, −7/2)

26. (a) $C(x) = 35{,}000 + 175x$; (b) $R(x) = 300x$;
(c) $P(x) = 125x - 35{,}000$; (d) $115{,}000 profit; $10{,}000
loss (e) (280, $84{,}000)   **27.** ($3, 81)   **28.** $-\frac{31}{28}$

29. $t = \dfrac{Q}{a - 4}$ **30.** $f(0) = 8$; $f(-2) = 14$ **31.** Slope: $\frac{5}{8}$;

y-intercept: (0, −5)

32. ◆ The comparison is summarized in the table in Section 3.2.

33. ◆ Many problems that deal with more than one unknown quantity are often easier to translate to a system of equations than to a single equation. Problems involving complementary or supplementary angles, the dimensions of a geometric figure, mixtures, and the measures of the angles of a triangle are examples.

34. (0, 2) and (1, 3)   **35.** (a) 165.91 cm; (b) 160.60 cm;
(c) (120.857, 308.521); (d) 120.857 cm

Test: Chapter 3, p. 295

1. [3.1a] (−2, 1); consistent; independent
2. [3.1a] No solution; inconsistent; independent
3. [3.1a] Infinitely many solutions; consistent; dependent   **4.** [3.2a] $\left(3, -\frac{11}{3}\right)$ **5.** [3.2a] $\left(\frac{15}{7}, -\frac{18}{7}\right)$
6. [3.2b] $\left(-\frac{3}{2}, -\frac{3}{2}\right)$ **7.** [3.2b] No solution
8. [3.3a] 34% solution: $48\frac{8}{9}$ mL;
61% solution: $71\frac{1}{9}$ mL   **9.** [3.3a] Buckets: 17;
dinners: 11   **10.** [3.3b] 120 km/h
11. [3.2c] Length: 93 ft; width: 51 ft
12. [3.4a] $\left(2, -\frac{1}{2}, -1\right)$ **13.** [3.7b] ($3, 55)
14. [3.5a] 3.5 hr   **15.** [3.7a] (a) $C(x) = 40{,}000 + 30x$;
(b) $R(x) = 80x$; (c) $P(x) = 50x - 40{,}000$; (d) $20{,}000
profit; $30{,}000 loss; (e) (800, $64{,}000)

19. [1.1d] $-\frac{37}{11}$ **20.** [1.2a] $a = \dfrac{P + 3b}{4}$

21. [2.4b] Slope: $-\frac{7}{2}$; y-intercept: (0, 7)
22. [2.2b] $f(0) = -8$; $f(-3) = 1$ **23.** [3.2b] $m = 7$;
$b = 10$

Cumulative Review: Chapters R–3, p. 297

1. [R.4b] −1   **2.** [R.4b] −14   **3.** [R.1d] 13
4. [R.1d] 0   **5.** [R.2d] 87.78   **6.** [R.2e] $-\frac{5}{21}$

7. [R.6b] $3x + 33$ **8.** [R.3c] 10   **9.** [R.7a] $-\dfrac{2a^{11}}{5b^{33}}$

10. [R.6b] $16b + 1$ **11.** [R.7a] y^{10} **12.** [1.1d] $\frac{10}{9}$

13. [1.1d] 6   **14.** [1.2a] $h = \dfrac{A}{\pi r^2}$

15. [1.2a] $P = \dfrac{3L}{m} - k$, or $\dfrac{3L - km}{m}$

16. [1.4c] $\{x \mid x > -1\}$, or $(-1, \infty)$ **17.** [1.5b] $\{x \mid x \leq 3$ *or*
$x \geq 7\}$, or $(-\infty, 3] \cup [7, \infty)$ **18.** [1.5a] $\left\{x \mid \frac{1}{3} < x \leq \frac{13}{3}\right\}$, or
$\left(\frac{1}{3}, \frac{13}{3}\right]$ **19.** [1.6e] $\left\{y \mid y \leq -\frac{3}{2} \text{ or } y \geq \frac{9}{4}\right\}$, or
$\left(-\infty, -\frac{3}{2}\right] \cup \left[\frac{9}{4}, \infty\right)$ **20.** [1.6c] $\{-5, 3\}$
21. [2.5c]   **22.** [2.5b]

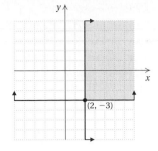

$3y = 9$

$y = -\frac{1}{2}x - 3$

23. [2.1c]

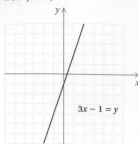

$3x - 1 = y$

24. [3.6b]

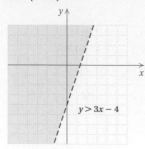

$y > 3x - 4$

25. [2.5a]

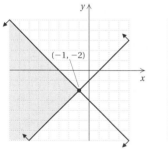

$3x + 5y = 15$

26. [3.6b]

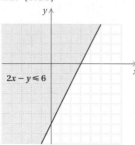

$2x - y \leq 6$

27. [3.1a] $(3, -1)$ **28.** [3.2b] $\left(\frac{8}{5}, -\frac{1}{5}\right)$ **29.** [3.2b] $(1, 1)$
30. [3.2b] $\left(-\frac{23}{13}, \frac{22}{13}\right)$ **31.** [3.4a] $(0, -1, 2)$
32. [3.4a] $(2, 0, -1)$
33. [3.6c]

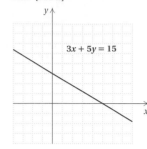

$(-1, -2)$

34. [3.6c]

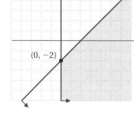

$(0, -2)$

35. [2.2b] $f(-1) = -5; f(0) = -1; f(2) = 7$
36. [2.2b], [2.3a] **(a)** Domain $= \{-5, -3, -1, 1, 3\}$;
(b) range $= \{-3, -2, 1, 4, 5\}$; **(c)** $f(-3) = -2$; **(d)** 3
37. [2.3a] $\left\{x \mid x \text{ is a real number } and \, x \neq \frac{1}{2}\right\}$, or
$\left(-\infty, \frac{1}{2}\right) \cup \left(\frac{1}{2}, \infty\right)$ **38.** [2.2b] $g(-1) = -1; g(0) = 1$;
$g(3) = -17$ **39.** [2.4b] Slope: $\frac{4}{5}$; y-intercept: $(0, 4)$
40. [2.4b] Slope: $-\frac{1}{2}$; y-intercept: $\left(0, \frac{7}{4}\right)$
41. [2.6b] $y = -3x + 17$ **42.** [2.6d] $y = \frac{1}{3}x + 4$
43. [1.3a] 4 m; 6 m **44.** [3.3a] Soakem: $48\frac{8}{9}$ oz;
Rinsem: $71\frac{1}{9}$ oz **45.** [3.3a] Scientific: 18; graphing: 27
46. [3.5a] \$120 **47.** [2.4c] **(a)** 123,000,000; **(b)** $m = 3$;
(c) years after 2009 **48.** [3.5a] 74.5, 68.5, 82
49. [R.7a] $-12x^{2a}y^{b+y+3}$ **50.** [2.7b] \$151,000
51. [3.2b] $m = -\frac{5}{9}; b = -\frac{2}{9}$

Chapter 4

Pretest: Chapter 4, p. 300

1. [4.1b] 6; 1 **2.** [4.1c] $6x^2y + xy$
3. [4.1d] $5m^3 - 6m^2 + 12$

4. [4.1a] $4xy^5 + x^2y^3 - 3x^6y^2 - 2y$
5. [4.2a] $x^4 - 2x^3 + 2x - 1$ **6.** [4.2b] $8y^2 + 18yz - 5z^2$
7. [4.2d] $a^2 - 9b^2$ **8.** [4.2c] $25t^2 - 30tm^2 + 9m^4$
9. [4.4a, b] $(2x - 1)(2x + 3)$ **10.** [4.5a] $2(5m + 2)^2$
11. [4.5d] $4t^3(t + 1)(t^2 - t + 1)$
12. [4.3c] $(a + 2)(a + 4)$ **13.** [4.5b] $(x + 7y)(x - 7y)$
14. [4.3b] $(y^2 + 4)(y + 3)$ **15.** [4.7a] $-\frac{2}{3}, \frac{4}{3}$
16. [4.7b] 100 m by 75 m **17.** [4.1b] 224 ft; 320 ft

Margin Exercises, Section 4.1, pp. 302–307

1. $-92x^5, -8x^4, x^2, 5; -92x^5$ **2.** $5, -4, -2, 1, -1$,
$-5; 5$ **3. (a)** $6x^2, -5x^3, 2x, -7; 2, 3, 1, 0; 3; -5x^3; -5$;
(b) $2y, -4, -5x, 7x^2y^3z^2, 5xy^2; 1, 0, 1, 7, 3; 7; 7x^2y^3z^2; 7$
4. Monomials: (c), (d), (e); binomials: (a), (f), (h);
trinomials: (b), (g) **5. (a)** $5 + 10x - 6x^2 + 7x^3 - x^4$;
(b) $-x^4 + 7x^3 - 6x^2 + 10x + 5$
6. (a) $-5 + x^3 + 5x^4y - 3y^2 + 3x^2y^3$;
(b) $3x^2y^3 - 3y^2 + 5x^4y + x^3 - 5$ **7.** 5; 13; 13
8. (a) $C(3) = 4.326$ mcg/mL; **(b)** $C(9) \approx 2$ mcg/mL
9. $3y - 4x + 4xy^2$ **10.** $8xy^3 + 2x^3y + 3x + 7 - 9x^2y$
11. $-4x^3 + 2x^2 - 4x + 2$ **12.** $10y^5 - 4y^2 + 5$
13. $5p^2q^4 - 8p^2q^2 + 5$ **14.** $-\left(4x^3 - 5x^2 + \frac{1}{4}x - 10\right)$;
$-4x^3 + 5x^2 - \frac{1}{4}x + 10$ **15.** $-\left(8xy^2 - 4x^3y^2 - 9x - \frac{1}{5}\right)$;
$-8xy^2 + 4x^3y^2 + 9x + \frac{1}{5}$
16. $-\left(-9y^5 - 8y^4 + \frac{1}{2}y^3 - y^2 + y - 1\right)$;
$9y^5 + 8y^4 - \frac{1}{2}y^3 + y^2 - y + 1$ **17.** $3x^2 + 5$
18. $14y^3 - 2y + 4$ **19.** $p^2 - 6p - 2$
20. $3y^5 - 3y^4 + 5y^3 - 2y^2 - 3$
21. $9p^4q - 10p^3q^2 + 4p^2q^3 + 9q^4$
22. $y^3 - y^2 + \frac{4}{3}y + 0.1$

Calculator Spotlight, p. 305

All graphs in Exercises 1–7 are left to the student.
Function values are given to the nearest hundredth.
3. $-1; -53; 222.11$ **4.** 8; 0; $-1; 25,691.88$ **5.** -0.33;
$-1; -28.11$ **6.** $0; 49,000; 0$ **7.** $0; 0; -20; 147,798; 18$

Calculator Spotlight, p. 308

1.–5. Left to the student. **6.** The polynomial is in
several variables.

Exercise Set 4.1, p. 309

1. $-9x^4, -x^3, 7x^2, 6x, -8; 4, 3, 2, 1, 0; 4; -9x^4; -9$
3. $t^3, 4t^7, s^2t^4, -2; 3, 7, 6, 0; 7; 4t^7; 4$ **5.** $u^7, 8u^2v^6$,
$3uv, 4u, -1; 7, 8, 2, 1, 0; 8; 8u^2v^6; 8$
7. $-4y^3 - 6y^2 + 7y + 23$ **9.** $-xy^3 + x^2y^2 + x^3y + 1$
11. $-9b^5y^5 - 8b^2y^3 + 2by$ **13.** $5 + 12x - 4x^3 + 8x^5$
15. $3xy^3 + x^2y^2 - 9x^3y + 2x^4$
17. $-7ab + 4ax - 7ax^2 + 4x^6$ **19.** 45; 21; 5
21. $-168; -9; 4; -7\frac{7}{8}$ **23. (a)** 144 ft; **(b)** 1024 ft
25. (a) \$2.3 billion; **(b)** \$1.3 billion **27. (a)** \$18,750;
(b) \$24,000 **29.** $P(x) = -x^2 + 280x - 7000$ **31.** $2x^2$
33. $3x + 4y$ **35.** $7a + 14$ **37.** $-6a^2b - 2b^2$

39. $9x^2 + 2xy + 15y^2$ **41.** $-x^2y + 4y + 9xy^2$
43. $5x^2 + 2y^2 + 5$ **45.** $6a + b + c$
47. $-4a^2 - b^2 + 3c^2$ **49.** $-3x^2 + 2x + xy - 1$
51. $5x^2y - 4xy^2 + 5xy$ **53.** $9r^2 + 9r - 9$
55. $-\frac{2}{15}xy + \frac{19}{12}xy^2 + 1.7x^2y$
57. $-(5x^3 - 7x^2 + 3x - 6); -5x^3 + 7x^2 - 3x + 6$
59. $-(-13y^2 + 6ay^4 - 5by^2); 13y^2 - 6ay^4 + 5by^2$
61. $11x - 7$ **63.** $-4x^2 - 3x + 13$ **65.** $2a - 4b + 3c$
67. $-2x^2 + 6x$ **69.** $-4a^2 + 8ab - 5b^2$
71. $16ab + 8a^2b + 3ab^2$
73. $0.06y^4 + 0.032y^3 - 0.94y^2 + 0.93$ **75.** $x^4 - x^2 - 1$

77.

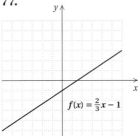

$f(x) = \frac{2}{3}x - 1$

78.

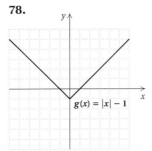

$g(x) = |x| - 1$

79.

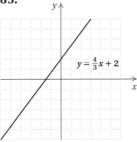

$g(x) = \dfrac{4}{x - 3}$

80.

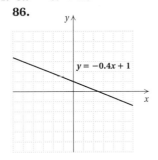

$f(x) = 1 - x^2$

81. $3y - 6$ **82.** $-10x - 20y + 70$
83. $-42p + 28q + 140$ **84.** $8w - 6t + 20$

85.

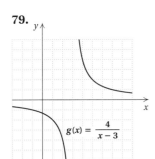

$y = \frac{4}{3}x + 2$

86.

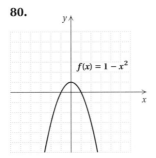

$y = -0.4x + 1$

87.

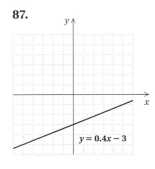

$y = 0.4x - 3$

88.

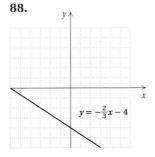

$y = -\frac{2}{3}x - 4$

89. ◆   **91.** 494.55 cm^3
93. $47x^{4a} + 40x^{3a} + 30x^{2a} + x^a + 4$
95. Left to the student.

Margin Exercises, Section 4.2, pp. 313–317

1. $-18y^3$ **2.** $24x^8y^3$ **3.** $-90x^4y^9z^{12}$ **4.** $-6y^2 - 18y$
5. $8xy^3 - 10xy$ **6.** $5x^3 + 15x^2 - 4x - 12$
7. $6y^2 + y - 12$ **8.** $p^4 + p^3 - 12p^2 - 5p + 15$
9. $2x^4 - 8x^3 + 4x^2 - 21x + 20$
10. $8x^5 + 12x^4 - 20x^3 + 4x^2 - 15x + 6$
11. $a^5 - 2a^4b + 3a^3b + a^3b^2 - 7a^2b^2 + 5ab^3 - b^4$
12. $y^2 + 6y - 40$ **13.** $2p^2 + 7pq - 15q^2$
14. $x^3y^3 + x^2y^3 + 2x^2y^2 + 2xy^2$
15. $a^2 - 2ab + b^2$ **16.** $x^2 + 16x + 64$
17. $9x^2 - 42x + 49$ **18.** $m^6 + \frac{1}{2}m^3n + \frac{1}{16}n^2$
19. $x^2 - 64$ **20.** $16y^2 - 49$ **21.** $7.84a^2 - 16.81b^2$
22. $9w^2 - \frac{9}{25}q^4$ **23.** $49x^4y^2 - 4y^2$
24. $4x^2 + 12x + 9 - 25y^2$ **25.** $81x^4 - 16y^4$
26. $f(a + 1) = a^2 + 4a - 4;$
$f(a + h) - f(a) = 2ah + h^2 + 2h$

Exercise Set 4.2, p. 319

1. $24y^3$ **3.** $-20x^3y$ **5.** $-10x^6y^7$ **7.** $14z - 2zx$
9. $6a^2b + 6ab^2$ **11.** $15c^3d^2 - 25c^2d^3$
13. $15x^2 + x - 2$ **15.** $s^2 - 9t^2$ **17.** $x^2 - 2xy + y^2$
19. $x^6 + 3x^3 - 40$ **21.** $a^4 - 5a^2b^2 + 6b^4$
23. $x^3 - 64$ **25.** $x^3 + y^3$
27. $a^4 + 5a^3 - 2a^2 - 9a + 5$
29. $4a^3b^2 + 4a^3b - 10a^2b^2 - 2a^2b + 3ab^3 +$
$7ab^2 - 6b^3$ **31.** $x^2 + \frac{1}{2}x + \frac{1}{16}$ **33.** $\frac{1}{8}x^2 - \frac{2}{9}$
35. $3.25x^2 - 0.9xy - 28y^2$ **37.** $a^2 + 13a + 40$
39. $y^2 + 3y - 28$ **41.** $9a^2 + 3a + \frac{1}{4}$
43. $x^2 - 4xy + 4y^2$ **45.** $b^2 - \frac{5}{6}b + \frac{1}{6}$
47. $2x^2 + 13x + 18$ **49.** $400a^2 - 6.4ab + 0.0256b^2$
51. $4x^2 - 4xy - 3y^2$ **53.** $x^6 + 4x^3 + 4$
55. $4x^4 - 12x^2y^2 + 9y^4$ **57.** $a^6b^4 + 2a^3b^2 + 1$
59. $0.01a^4 - a^2b + 25b^2$ **61.** $A = P + 2Pi + Pi^2$
63. $d^2 - 64$ **65.** $4c^2 - 9$ **67.** $36m^2 - 25n^2$
69. $x^4 - y^2z^2$ **71.** $m^4 - m^2n^2$ **73.** $16p^4 - 9p^2q^2$
75. $\frac{1}{4}p^2 - \frac{4}{9}q^2$ **77.** $x^4 - 1$ **79.** $a^4 - 2a^2b^2 + b^4$
81. $a^2 + 2ab + b^2 - 1$ **83.** $4x^2 + 12xy + 9y^2 - 16$
85. (a) $t^2 + 3t - 4$; (b) $5h + 2ah + h^2$
87. (a) $-3a^2 - 16a - 18$; (b) $-4h - 6ah - 3h^2$
89. 5.5 hr   **90.** 180 mph   **91.** $\left(\frac{4}{3}, -\frac{14}{27}\right)$ **92.** (1, 3)
93. Infinitely many solutions   **94.** $\left(\frac{10}{21}, \frac{11}{14}\right)$ **95.** ◆
97. Left to the student.   **99.** z^{5n^5}
101. $r^8 - 2r^4s^4 + s^8$ **103.** $9x^{10} - \frac{30}{11}x^5 + \frac{25}{121}$
105. $x^{4a} - y^{4b}$ **107.** $x^6 - 1$

Margin Exercises, Section 4.3, pp. 323–328

1. (a) $x - 5$ and $x + 1$; (b) x^2, $-4x$, and -5
2. $3(x^2 - 2)$ **3.** $4x^3(x^2 - 2)$ **4.** $3y^2(3y^2 - 5y + 1)$
5. $3x^2y(2 - 7xy + y^2)$ **6.** $-8(x - 4)$
7. $-3(x^2 + 5x - 3)$ **8.** (a) $h(t) = -16t(t - 6)$;
(b) $h(2) = 128$ in each   **9.** $(p + q)(2x + y + 2)$
10. $(y + 3)(2y - 11)$ **11.** $(y^2 - 2)(5y + 2)$
12. Cannot be factored by grouping
13. $(x + 2)(x + 3)$ **14.** $(y + 2)(y + 5)$

15. $(m - 2)(m - 6)$ **16.** $(t - 3)(t - 8)$, or $(3 - t)(8 - t)$
17. (a) $(x - 5)(x + 4)$; **(b)** The product of each pair is positive. **18.** $x(x - 9)(x + 6)$ **19.** $2x(x - 7)(x + 6)$
20. $x(x + 6)(x - 2)$ **21.** $(y - 6)(y + 2)$
22. $(x + 10)(x - 11)$ **23.** Not factorable
24. $(x - 2y)(x - 3y)$ **25.** $(p - 8q)(p + 2q)$
26. $(x^2 - 2)(x^2 - 7)$ **27.** $(p^3 + 3)(p^3 - 2)$

Exercise Set 4.3, p. 329

1. $3a(2a + 1)$ **3.** $x^2(x + 9)$ **5.** $4x^2(2 - x^2)$
7. $4xy(x - 3y)$ **9.** $3(y^2 - y - 3)$
11. $2a(2b - 3c + 6d)$ **13.** $5(2a^4 + 3a^2 - 5a - 6)$
15. $-5(x + 9)$ **17.** $-6(a + 14)$ **19.** $-2(x^2 - x + 12)$
21. $-3y(y - 8)$ **23.** $-3(y^3 - 4y^2 + 5y - 8)$
25. $N(x) = \frac{1}{6}x(x^2 + 3x + 2)$ **27. (a)** $h(t) = -8t(2t - 9)$;
(b) $h(2) = 80$ **29.** $R(x) = 0.4x(700 - x)$
31. $(a + c)(b - 2)$ **33.** $(x - 2)(2x + 13)$
35. $2a^2(x - y)$ **37.** $(a + b)(c + d)$
39. $(b^2 + 2)(b - 1)$ **41.** $(y^2 + 1)(y - 8)$
43. $12(x^2 + 3)(2x - 3)$ **45.** $a(a^3 - a^2 + a + 1)$
47. $(2y^2 + 5)(y^2 + 3)$ **49.** $(x + 4)(x + 9)$
51. $(t - 5)(t - 3)$ **53.** $(x - 11)(x + 3)$
55. $2(y - 4)(y - 4)$ **57.** $(p + 9)(p - 6)$
59. $(x + 3)(x + 9)$ **61.** $\left(y - \frac{1}{3}\right)\left(y - \frac{1}{3}\right)$
63. $(t - 3)(t - 1)$ **65.** $(x + 7)(x - 2)$
67. $(x + 2)(x + 3)$ **69.** $(8 - x)(7 + x)$
71. $y(8 - y)(4 + y)$ **73.** $(x^2 + 16)(x^2 - 5)$
75. Not factorable **77.** $(x + 9y)(x + 3y)$
79. $(x^2 + 49)(x^2 + 1)$ **81.** $(x^3 + 9)(x^3 + 2)$
83. $(x^4 - 3)(x^4 - 8)$ **85.** Countryside: $9\frac{3}{8}$ lb;

Mystic: $15\frac{5}{8}$ lb **86.** 8 **87.** Yes **88.** No **89.** No
90. Yes **91.** All real numbers **92.** All real numbers
93. $\left\{x \mid x \text{ is a real number } and\ x \neq \frac{7}{4}\right\}$, or $\left(-\infty, \frac{7}{4}\right) \cup \left(\frac{7}{4}, \infty\right)$
94. All real numbers **95.** ◈
97. $76, -76, 28, -28, 20, -20$ **99.** $x - 365$
101. Left to the student.

Margin Exercises, Section 4.4, pp. 334–337

1. $(2x + 3)(2x - 1)$ **2.** $(4x + 1)(x + 9)$
3. $y^2(5y + 4)(2y - 3)$ **4.** $a(36a^2 + 21a + 1)$
5. $(x - 7)(3x + 8)$ **6.** $(3x + 2)(x + 1)$
7. $2(4y - 1)(3y - 5)$ **8.** $2x^3(2x - 3)(5x - 4)$
9. $(3x + 4)(x + 5)$ **10.** $4(2x - 1)(2x + 3)$
11. $(7x - 4y)(3x + y)$ **12.** $3(4a + 9b)(5a - b)$

Exercise Set 4.4, p. 339

1. $(3x + 1)(x - 5)$ **3.** $y(5y - 7)(2y + 3)$
5. $(3c - 8)(c - 4)$ **7.** $(5y + 2)(7y + 4)$
9. $2(5t - 3)(t + 1)$ **11.** $4(2x + 1)(x - 4)$
13. $x(3x - 4)(4x - 5)$ **15.** $x^2(7x + 1)(2x - 3)$
17. $(3a - 4)(a + 1)$ **19.** $(3x + 1)(3x + 4)$
21. $(3 - z)(1 + 12z)$ **23.** $(-2t + 3)(2t + 5)$
25. $x(3x + 1)(x - 2)$ **27.** $(24x + 1)(x - 2)$
29. $(7x + 3)(3x + 4)$ **31.** $4(10x^4 + 4x^2 - 3)$
33. $(4a - 3b)(3a - 2b)$ **35.** $(2x - 3y)(x + 2y)$
37. $2(3x - 4y)(2x - 7y)$ **39.** $(3x - 5y)(3x - 5y)$

41. $(3x^3 - 2)(x^3 + 2)$ **43. (a)** 224 ft; 288 ft; 320 ft;
288 ft; 128 ft; **(b)** $h(t) = -16(t - 7)(t + 2)$
45. $(2, -1, 0)$ **46.** $\left(\frac{3}{2}, -4, 3\right)$ **47.** $(1, -1, 2)$
48. $(2, 4, 1)$ **49.** Parallel **50.** Parallel **51.** Neither
52. Perpendicular **53.** $y = -\frac{1}{7}x - \frac{23}{7}$
54. $y = -\frac{1}{3}x - \frac{7}{3}$ **55.** $y = -\frac{7}{17}x - \frac{19}{17}$
56. $y = -\frac{5}{2}x - \frac{2}{3}$ **57.** ◈ **59.** Left to the student.
61. $(pq + 4)(pq + 3)$ **63.** $\left(x + \frac{4}{5}\right)\left(x - \frac{1}{5}\right)$
65. $(y + 0.5)(y - 0.1)$ **67.** $(7ab + 6)(ab + 1)$
69. $3(x + 15)(x - 11)$ **71.** $6x(x + 9)(x + 4)$
73. $(x^a + 8)(x^a - 3)$

Margin Exercises, Section 4.5, pp. 341–346

1. (a), (b), (d) **2.** $(x + 7)^2$ **3.** $(3y - 5)^2$
4. $(4x + 9y)^2$ **5.** $(4x^2 - 5y^3)^2$ **6.** $-2(2a - 3b)^2$
7. $3(a - 5b)^2$ **8.** $(y + 2)(y - 2)$
9. $(7x^2 + 5y^5)(7x^2 - 5y^5)$ **10.** $\left(m + \frac{1}{3}\right)\left(m - \frac{1}{3}\right)$
11. $(5xy + 2a)(5xy - 2a)$ **12.** $(3x + 4y)(3x - 4y)$
13. $5(2x + y)(2x - y)$ **14.** $y^2(9x^2 + 4)(3x + 2)(3x - 2)$
15. $(a + 4)(a - 4)(a + 1)$ **16.** $(x + 1 + p)(x + 1 - p)$
17. $(y - 4 + 3m)(y - 4 - 3m)$
18. $(x + 4 - 10t)(x + 4 + 10t)$
19. $[8p + (x + 4)][8p - (x + 4)]$, or
$(8p + x + 4)(8p - x - 4)$ **20.** $(x - 2)(x^2 + 2x + 4)$
21. $(4 - y)(16 + 4y + y^2)$
22. $(3x + y)(9x^2 - 3xy + y^2)$
23. $(2y + z)(4y^2 - 2yz + z^2)$
24. $(m + n)(m^2 - mn + n^2)(m - n)(m^2 + mn + n^2)$
25. $2xy(2x^2 + 3y^2)(4x^4 - 6x^2y^2 + 9y^4)$
26. $(3x + 2y)(9x^2 - 6xy + 4y^2)(3x - 2y) \times$
$(9x^2 + 6xy + 4y^2)$ **27.** $(x - 0.3)(x^2 + 0.3x + 0.09)$

Exercise Set 4.5, p. 347

1. $(x - 2)^2$ **3.** $(y + 9)^2$ **5.** $(x + 1)^2$ **7.** $(3y + 2)^2$
9. $y(y - 9)^2$ **11.** $3(2a + 3)^2$ **13.** $2(x - 10)^2$
15. $(1 - 4d)^2$, or $(4d - 1)^2$ **17.** $(y + 2)^2(y - 2)^2$
19. $(0.5x + 0.3)^2$ **21.** $(p - q)^2$ **23.** $(a + 2b)^2$
25. $(5a - 3b)^2$ **27.** $(y^3 + 13)^2$ **29.** $(4x^5 - 1)^2$
31. $(x^2 + y^2)^2$ **33.** $(x + 4)(x - 4)$
35. $(p + 7)(p - 7)$ **37.** $(pq + 5)(pq - 5)$
39. $6(x + y)(x - y)$ **41.** $4x(y^2 + z^2)(y + z)(y - z)$
43. $a(2a + 7)(2a - 7)$
45. $3(x^4 + y^4)(x^2 + y^2)(x + y)(x - y)$
47. $a^2(3a + 5b^2)(3a - 5b^2)$ **49.** $\left(\frac{1}{6} + z\right)\left(\frac{1}{6} - z\right)$
51. $(0.2x + 0.3y)(0.2x - 0.3y)$
53. $(m + 2)(m - 2)(m - 7)$ **55.** $(a + b)(a - b)(a - 2)$
57. $(a + b + 10)(a + b - 10)$
59. $(a + b + 3)(a + b - 3)$
61. $(r - 1 + 2s)(r - 1 - 2s)$
63. $2(m + n + 5b)(m + n - 5b)$
65. $(3 + a + b)(3 - a - b)$ **67.** $(z + 3)(z^2 - 3z + 9)$
69. $(x - 1)(x^2 + x + 1)$ **71.** $(y + 5)(y^2 - 5y + 25)$
73. $(2a + 1)(4a^2 - 2a + 1)$ **75.** $(v - 2)(v^2 + 2v + 4)$
77. $(2 - 3b)(4 + 6b + 9b^2)$
79. $(4y + 1)(16y^2 - 4y + 1)$

81. $(2x + 3)(4x^2 - 6x + 9)$ **83.** $(a - b)(a^2 + ab + b^2)$
85. $\left(a + \frac{1}{2}\right)\left(a^2 - \frac{1}{2}a + \frac{1}{4}\right)$ **87.** $2(y - 4)(y^2 + 4y + 16)$
89. $3(2a + 1)(4a^2 - 2a + 1)$
91. $r(s + 4)(s^2 - 4s + 16)$
93. $5(x - 2z)(x^2 + 2xz + 4z^2)$
95. $(x + 0.1)(x^2 - 0.1x + 0.01)$
97. $8(2x^2 - t^2)(4x^4 + 2x^2t^2 + t^4)$
99. $2y(y - 4)(y^2 + 4y + 16)$
101. $(z - 1)(z^2 + z + 1)(z + 1)(z^2 - z + 1)$
103. $(t^2 + 4y^2)(t^4 - 4t^2y^2 + 16y^4)$ **105.** $\left(-\frac{41}{53}, \frac{148}{53}\right)$
106. $\left(-\frac{26}{7}, -\frac{134}{7}\right)$ **107.** $(1, 13)$ **108.** No solution
109.

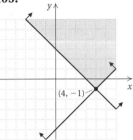

110.

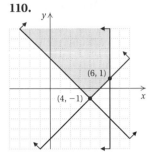

(6, 1)   (4, −1)

111.

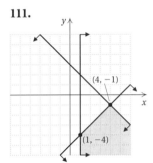

(4, −1)   (1, −4)

112.

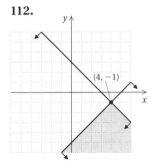

(4, −1)

113. $y = x - 2$; $y = -x - 6$ **114.** $y = \frac{2}{3}x - \frac{23}{3}$;
$y = -\frac{3}{2}x - \frac{11}{2}$ **115.** $y = -\frac{1}{2}x + 7$; $y = 2x - 3$
116. $y = \frac{1}{4}x - \frac{3}{2}$; $y = -4x + 24$ **117.** ◈
119. $h(2a + h)$ **121.** $h[3a^2 + 3ah + h^2]$ **123.** ◈
125. $(xy + 2)^2(xy - 2)^2$
127. $5(c^{50} + 4d^{50})(c^{25} + 2d^{25})(c^{25} - 2d^{25})$
129. $(x^{2a} - y^b)(x^{2a} + y^b)$
131. $(x^{2a} + y^b)(x^{4a} - x^{2a}y^b + y^{2b})$
133. $3(x^a + 2y^b)(x^{2a} - 2x^ay^b + 4y^{2b})$
135. $\frac{1}{3}\left(\frac{1}{2}xy + z\right)\left(\frac{1}{4}x^2y^2 - \frac{1}{2}xyz + z^2\right)$
137. $y(3x^2 + 3xy + y^2)$ **139.** $4(3a^2 + 4)$

Margin Exercises, Section 4.6, pp. 351–352

1. $3y(y + 2x)(y - 2x)$ **2.** $7(a - 1)(a^2 + a + 1)$
3. $(2x + 3y)(4x^2 - 6xy + 9y^2)(2x - 3y) \times$
$(4x^2 + 6xy + 9y^2)$ **4.** $(x - 2)(3 - bx)$ **5.** $5(y^4 + 4x^6)$
6. $3(x - 2)(2x + 3)$ **7.** $(a - b)^2(a + b)$ **8.** $3(x + 3a)^2$
9. $2(x - 5 + 3b)(x - 5 - 3b)$

Exercise Set 4.6, p. 353

1. $(y + 15)(y - 15)$ **3.** $(2x + 3)(x + 4)$
5. $5(x^2 + 2)(x^2 - 2)$ **7.** $(p + 6)^2$ **9.** $2(x - 11)(x + 6)$
11. $(3x + 5y)(3x - 5y)$

13. $(m + 1)(m^2 - m + 1)(m - 1)(m^2 + m + 1)$
15. $(x + 3 + y)(x + 3 - y)$
17. $2(5x - 4y)(25x^2 + 20xy + 16y^2)$
19. $(m^3 + 10)(m^3 - 2)$ **21.** $(c - b)(a + d)$
23. $(5b - a)(10b + a)$ **25.** $(x^2 + 2)(2x - 7)$
27. $2(x + 2)(x - 2)(x + 3)$
29. $2(2x + 3y)(4x^2 - 6xy + 9y^2)$ **31.** $(6y - 5)(6y + 7)$
33. $(a^4 + b^4)(a^2 + b^2)(a + b)(a - b)$
35. $ab(a + 4b)(a - 4b)$ **37.** $\left(\frac{1}{4}x - \frac{1}{3}y^2\right)^2$
39. $5(x - y)^2(x + y)$ **41.** $(9ab + 2)(3ab + 4)$
43. $y(2y - 5)(4y^2 + 10y + 25)$
45. $(a - b - 3)(a + b + 3)$ **47.** Correct: 55;
incorrect: 20   **48.** $\frac{80}{7}$ **49.** ◈
51. $(6y^2 - 5x)(5y^2 - 12x)$ **53.** $5\left(x - \frac{1}{3}\right)\left(x^2 + \frac{1}{3}x + \frac{1}{9}\right)$
55. $x(x - 2p)$ **57.** $y(y - 1)^2(y - 2)$
59. $(2x + y - r + 3s)(2x + y + r - 3s)$ **61.** $c(c^w + 1)^2$
63. $3x(x + 5)$ **65.** $(x - 1)^3(x^2 + 1)(x + 1)$
67. $y(y^4 + 1)(y^2 + 1)(y + 1)(y - 1)$

Margin Exercises, Section 4.7, pp. 356–362

1. 4, 2   **2.** $\frac{1}{2}$, −3   **3.** 0, 2   **4.** −5   **5.** 0, 2, −3
6. $-\frac{3}{2}$, $\frac{1}{5}$ **7.** $\{x \,|\, x$ is a real number *and* $x \neq -4$ *and*
$x \neq 7\}$ **8. (a)** (2, 0) and (4, 0); **(b)** 2, 4; **(c)** The
solutions of $x^2 - 6x + 8 = 0$, 2 and 4, are the first
coordinates of the x-intercepts, (2, 0) and (4, 0), of the
graph of $f(x) = x^2 - 6x + 8$.   **9.** 11 sec   **10.** 12 m
and 13 m   **11. (a)** 240; **(b)** 7

Calculator Spotlight, p. 359

1. −5, 2   **2.** −4, 6   **3.** −2, 1   **4.** −1.414214, 0,
1.414214   **5.** 0, 700   **6.** −2.079356, 0.46295543,
3.1164004   **7.** −3.095574, −0.6460838, 0.64608382,
3.0955736   **8.** −1, 1   **9.** −2, −1.414214, 1.414214, 1
10. −3, −1, 2, 3

Exercise Set 4.7, p. 363

1. −7, 4   **3.** 3   **5.** −10   **7.** −5, −4   **9.** 0, −8
11. −5, 5   **13.** −12, 12   **15.** 7, −9   **17.** −4, 8
19. −2, $-\frac{2}{3}$ **21.** $\frac{1}{2}$, $\frac{3}{4}$ **23.** 0, 6   **25.** $\frac{2}{3}$, $-\frac{3}{4}$
27. −1, 1   **29.** $\frac{2}{3}$, $-\frac{5}{7}$ **31.** 0, $\frac{1}{5}$ **33.** 7, −2
35. 0, −2, 3   **37.** 0, −8, 8   **39.** 5, −5, 1, −1
41. −8, −4   **43.** −4, $\frac{3}{2}$ **45.** −9, −3
47. $\{x \,|\, x$ is a real number *and* $x \neq -1$ *and* $x \neq 5\}$
49. $\{x \,|\, x$ is a real number *and* $x \neq -3$ *and* $x \neq 3\}$
51. $\left\{x \,|\, x \text{ is a real number } and\ x \neq 0 \text{ and } x \neq \frac{1}{2}\right\}$
53. $\{x \,|\, x$ is a real number *and* $x \neq 0$ *and* $x \neq 2$ *and*
$x \neq 5\}$ **55.** 11, −12   **57.** Length: 12 cm; width: 8 cm
59. Height: 6 ft; base: 4 ft   **61.** 8   **63.** Length: 12 m;
width: 9 m   **65.** $d = 12$ ft; $h = 16$ ft   **67.** 16, 18, 20
69. 6 cm   **71.** 6   **73.** 150 ft by 200 ft   **75.** 40 m,
41 m   **77.** 7 sec   **79.** 7   **80.** 1   **81.** 7   **82.** 2.5
83. 1.3   **84.** $\frac{19}{15}$ **85.** 691   **86.** 1023
87. $y = \frac{11}{6}x + \frac{32}{3}$ **88.** $y = -\frac{11}{10}x + \frac{24}{5}$
89. $y = -\frac{3}{10}x + \frac{32}{5}$ **90.** $y = \frac{26}{31}x + \frac{934}{31}$ **91.** ◈

93. $\{-3, 1\}$; $\{x \mid -4 \le x \le 2\}$, or $[-4, 2]$
95. Left to the student. **97. (a)** 1.2522305, 3.1578935;
(b) -0.3027756, 3.3027756; **(c)** 2.1387475, 2.7238657;
(d) -0.7462555, 3.3276509

Summary and Review: Chapter 4, p. 367

1. (a) 7, 11, 3, 2; 11; **(b)** $-7x^8y^3$; -7;
(c) $-3x^2 + 2x^3 + 3x^6y - 7x^8y^3$;
(d) $-7x^8y^3 + 3x^6y + 2x^3 - 3x^2$ **2.** 0; -6 **3.** 4; -31
4. (a) Approximately 8 million; **(b)** approximately
5 million **5.** $-x^2y - 2xy^2$ **6.** $ab + 12ab^2 + 4$
7. $-x^3 + 2x^2 + 5x + 2$ **8.** $x^3 + 6x^2 - x - 4$
9. $13x^2y - 8xy^2 + 4xy$ **10.** $9x - 7$
11. $-2a + 6b + 7c$ **12.** $16p^2 - 8p$
13. $6x^2 - 7xy + y^2$ **14.** $-18x^3y^4$
15. $x^8 - x^6 + 5x^2 - 3$ **16.** $8a^2b^2 + 2abc - 3c^2$
17. $4x^2 - 25y^2$ **18.** $4x^2 - 20xy + 25y^2$
19. $20x^4 - 18x^3 - 47x^2 + 69x - 27$
20. $x^4 + 8x^2y^3 + 16y^6$ **21.** $x^3 - 125$
22. $x^2 - \frac{1}{2}x + \frac{1}{18}$ **23.** $a^2 - 4a - 4$; $2ah + h^2 - 2h$
24. $3y^2(3y^2 - 1)$ **25.** $3x(5x^3 - 6x^2 + 7x - 3)$
26. $(a - 9)(a - 3)$ **27.** $(3m + 2)(m + 4)$
28. $(5x + 2)^2$ **29.** $4(y + 2)(y - 2)$
30. $(x - y)(a + 2b)$ **31.** $4(x^4 + x^2 + 5)$
32. $(3x - 2)(9x^2 + 6x + 4)$
33. $(0.4b - 0.5c)(0.16b^2 + 0.2bc + 0.25c^2)$
34. $y(y^2 + 1)(y + 1)(y - 1)$ **35.** $2z^6(z^2 - 8)$
36. $2y(3x^2 - 1)(9x^4 + 3x^2 + 1)$
37. $(1 + a)(1 - a + a^2)$ **38.** $4(3x - 5)^2$
39. $(3t + p)(2t + 5p)$ **40.** $(x + 3)(x - 3)(x + 2)$
41. $(a - b + 2t)(a - b - 2t)$ **42.** 10 **43.** $\frac{2}{3}, \frac{3}{2}$
44. $0, \frac{7}{4}$ **45.** $-4, 4$ **46.** $-4, 11$
47. $\left\{x \mid x \text{ is a real number } and \ x \ne \frac{2}{3} \text{ and } x \ne -7\right\}$
48. Length: 8 in.; width: 5 in. **49.** $-7, -5, -3$; 3, 5, 7
50. 7 **51.** Conventional: 36; surround-sound: 42
52. $\left(\frac{87}{43}, -\frac{25}{43}\right)$ **53.** No solution **54.** No **55.** Yes
56.   **57.**

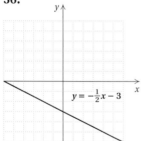

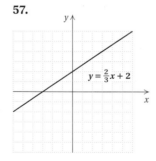

$y = -\frac{1}{2}x - 3$

$y = \frac{2}{3}x + 2$

58. ◆ The discussion could include the following
points:
    a) We can now solve certain quadratic equations.
    b) Whereas most linear equations have exactly one
        solution, quadratic equations can have two
        solutions.
    c) We used factoring and the principle of zero
        products to solve quadratic equations.
59. ◆ **a)** The middle term, $2 \cdot a \cdot 3$, is missing.
    $(a + 3)^2 = a^2 + 6a + 9$

b) The middle term of the trinomial factor should
    be $-ab$.
    $a^3 + b^3 = (a + b)(a^2 - ab + b^2)$
c) The middle term, $-2ab$, is missing and the sign
    preceding b^2 is incorrect.
    $(a - b)(a - b) = a^2 - 2ab + b^2$
d) The product of the outside terms and the product
    of the inside terms are missing.
    $(x + 3)(x - 4) = x^2 - x - 12$
e) There should be a minus sign between the terms
    of the product.
    $(p + 7)(p - 7) = p^2 - 49$
f) The middle term, $-2 \cdot t \cdot 3$, is missing and the
    sign preceding 9 is incorrect.
    $(t - 3)^2 = t^2 - 6t + 9$
60. $2(2x + y)(4x^2 - 2xy + y^2)(2x - y)(4x^2 + 2xy + y^2)$
61. $2(3x^2 + 1)$ **62.** $a^3 - (b - 1)^3$ **63.** $0, \frac{1}{8}, -\frac{1}{8}$

Test: Chapter 4, p. 369

1. [4.1a] **(a)** 4, 3, 9, 5; 9; **(b)** $5x^5y^4$; 5;
(c) $3xy^3 - 4x^2y - 2x^4y + 5x^5y^4$;
(d) $5x^5y^4 + 3xy^3 - 4x^2y - 2x^4y$ **2.** [4.1b] 4; 2
3. [4.1b] **(a)** 247 thousand; **(b)** 65 thousand
4. [4.1c] $3xy + 3xy^2$ **5.** [4.1c] $-3x^3 + 3x^2 - 6y - 7y^2$
6. [4.1c] $7a^3 - 6a^2 + 3a - 3$
7. [4.1c] $7m^3 + 2m^2n + 3mn^2 - 7n^3$
8. [4.1d] $6a - 8b$ **9.** [4.1d] $7x^2 - 7x + 13$
10. [4.1d] $2y^2 + 5y + y^3$ **11.** [4.2a] $64x^3y^3$
12. [4.2b] $12a^2 - 4ab - 5b^2$
13. [4.2a] $x^3 - 2x^2y + y^3$
14. [4.2a] $-3m^4 - 13m^3 + 5m^2 + 26m - 10$
15. [4.2c] $16y^2 - 72y + 81$ **16.** [4.2d] $x^2 - 4y^2$
17. [4.2e] $a^2 + 15a + 50$; $2ah + h^2 - 5h$
18. [4.3a] $x(9x + 7)$ **19.** [4.3a] $8y^2(3y + 2)$
20. [4.5c] $(y + 2)(y - 2)(y + 5)$
21. [4.3c] $(p - 14)(p + 2)$
22. [4.4a, b] $(6m + 1)(2m + 3)$
23. [4.5b] $(3y + 5)(3y - 5)$
24. [4.5d] $3(r - 1)(r^2 + r + 1)$
25. [4.5a] $(3x - 5)^2$ **26.** [4.5b] $(z + 1 + b)(z + 1 - b)$
27. [4.5b] $(x^4 + y^4)(x^2 + y^2)(x + y)(x - y)$
28. [4.5c] $(y + 4 + 10t)(y + 4 - 10t)$
29. [4.5b] $5(2a + b)(2a - b)$
30. [4.4a, b] $2(4x - 1)(3x - 5)$
31. [4.5d] $2ab(2a^2 + 3b^2)(4a^4 - 6a^2b^2 + 9b^4)$
32. [4.7a] $-3, 6$ **33.** [4.7a] $-5, 5$ **34.** [4.7a] $-\frac{3}{2}, -7$
35. [4.7a] $0, 5$ **36.** [4.7a] $\{x \mid x \text{ is a real number } and$
$x \ne -1\}$, or $(-\infty, -1) \cup (-1, \infty)$ **37.** [4.7b] $12, -13$
38. [4.7b] Length: 8 cm; width: 5 cm
39. [4.7b] 24 ft **40.** [4.7b] 5
41. [4.3a] $f(n) = \frac{1}{2}n(n - 1)$
42. [3.3b] 54 mph and 39 mph
43. [3.2b] $\left(\frac{40}{23}, -\frac{38}{69}\right)$ **44.** [2.2d] Yes

45. [2.5b]

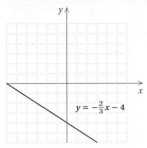

$y = -\frac{2}{3}x - 4$

46. [4.6a] $(3x^n + 4)(2x^n - 5)$ **47.** [4.2c] 19

Cumulative Review: Chapters R–4, p. 371

1. [R.4b] 1   **2.** [R.1d] 0   **3.** [R.2c] −4
4. [R.6b] $-a - 5$ **5.** [R.6b] $5x + 20$
6. [R.7b] $\dfrac{a^{60}}{(-3)^{10}b^{20}}$ **7.** [4.1c] $-2x^2 + x - xy - 1$
8. [4.1d] $-2x^2 + 6x$ **9.** [4.2a] $a^4 + a^3 - 8a^2 - 3a + 9$
10. [4.2b] $x^2 + 13x + 36$ **11.** [1.1d] 2   **12.** [1.1d] 13
13. [1.2a] $b = \dfrac{2A - ha}{h}$, or $\dfrac{2A}{h} - a$
14. [1.4c] $\left\{x \mid x \geq -\frac{7}{9}\right\}$, or $\left[-\frac{7}{9}, \infty\right)$
15. [1.5b] $\left\{x \mid x < \frac{5}{4} \text{ or } x > 4\right\}$, or $\left(-\infty, \frac{5}{4}\right) \cup (4, \infty)$
16. [1.6e] $\{x \mid -2 < x < 5\}$, or $(-2, 5)$
17. [3.4a] $(1, 3, -9)$ **18.** [3.2b] $(4, -2)$
19. [3.2b] $\left(\frac{19}{8}, \frac{1}{8}\right)$ **20.** [3.4a] $(-1, 0, -1)$
21. [3.2b] $\left(\frac{3}{2}, -\frac{1}{3}\right)$ **22.** [3.2b] $\left(\frac{1}{3}, \frac{1}{2}\right)$ **23.** [3.4a] $\left(\frac{1}{2}, \frac{1}{2}, \frac{1}{4}\right)$
24. [4.7a] $-3, -8$ **25.** [4.7a] $\frac{1}{2}, 7$ **26.** [4.7a] $\frac{2}{3}, -2$
27. [4.7a] $\{x \mid x \text{ is a real number } and \ x \neq 5 \ and \ x \neq -3\}$
28. [4.3a] $3x^2(x - 4)$
29. [4.3b] $(2x + 1)(x + 1)(x^2 - x + 1)$
30. [4.3c] $(x - 2)(x + 7)$ **31.** [4.4a, b] $(4a - 3)(5a - 2)$
32. [4.5b] $(2x + 5)(2x - 5)$ **33.** [4.5a] $2(x - 7)^2$
34. [4.5d] $(a + 4)(a^2 - 4a + 16)$
35. [4.5d] $(2x - 1)(4x^2 + 2x + 1)$
36. [4.3c] $(a^3 + 6)(a^3 - 2)$
37. [4.5b] $x^2y^2(2x + y)(2x - y)$
38. [1.5b]

39. [2.1c]

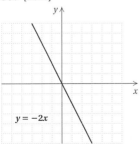

$y = -2x$

40. [2.1c]

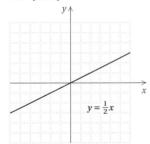

$y = \frac{1}{2}x$

41. [2.5c]

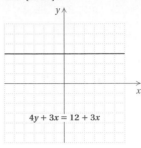

$4y + 3x = 12 + 3x$

42. [2.5c]

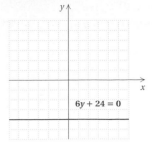

$6y + 24 = 0$

43. [3.6b]

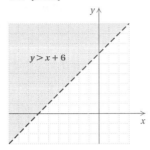

$y > x + 6$

44. [3.6b]

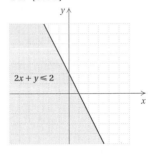

$2x + y \leq 2$

45. [2.2c]

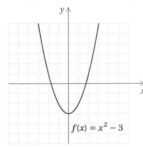

$f(x) = x^2 - 3$

46. [2.2c]

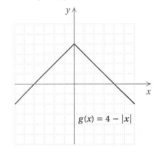

$g(x) = 4 - |x|$

47. [3.6c]

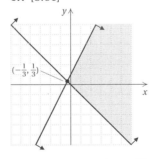

$\left(-\frac{1}{3}, \frac{1}{3}\right)$

48. [3.6c]

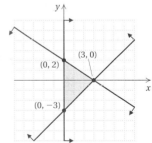

$(3, 0)$ $(0, 2)$ $(0, -3)$

49. [2.6d] $y = -\frac{1}{2}x + \frac{17}{2}$ **50.** [2.6d] $y = \frac{4}{3}x - 6$
51. [2.6c] $x = -1$ **52.** [2.6b] $y = -3x + 7$
53. [3.2c] $1170 for service; $1716 for disposables
54. [1.3a] Yes   **55.** [3.3a] Caracas: $0.14/gal;
Lagos: $0.30/gal   **56.** [3.5a] A: 1500; B: 1900; C: 2300
57. [2.2e] **(a)** 4.8 lb; **(b)** 2030; **(c)** $\{t \mid 34 \leq t \leq 39\}$, or
from 2025 to 2030   **58.** [1.6e] $\{x \mid x \leq 1\}$, or $(-\infty, 1]$
59. [3.2c] Bert: 32 yr; Sally: 14 yr

CHAPTER 5

Pretest: Chapter 5, p. 374

1. [5.2a] $2x^2(x-1)(x-2)$

2. [5.2b] $\dfrac{y+9}{(y+4)^2(y+5)}$

3. [5.2b] $\dfrac{4-a}{a-1}$ 4. [5.2c] $\dfrac{1}{2(y+8)}$ 5. [5.1e] $\dfrac{(y-11)^2}{(y-2)^2}$

6. [5.1d] $\dfrac{2}{3(y+1)}$ 7. [5.4a] $\dfrac{y-1}{y+1}$

8. [5.3b] $x+12$, R 45; or $x+12+\dfrac{45}{x-4}$ 9. [5.5a] $\frac{132}{17}$

10. [5.5a] -25 11. [5.7a] $b=\dfrac{c}{aM-a}$

12. [5.1a] $\{x\,|\,x$ is a real number $and\ x\neq -5\ and$

$x=-2\}$ 13. [5.8e] $y=\dfrac{1}{4}\cdot\dfrac{xz^2}{w}$ 14. [5.8b] 5.625 yd^3

15. [5.6c] A: 250 km/h; B: 200 km/h 16. [5.6a] $1\frac{5}{7}$ hr

Margin Exercises, Section 5.1, pp. 375–382

1. $-\frac{5}{2}$ 2. $\left(-\infty,-\frac{5}{2}\right)\cup\left(-\frac{5}{2},\infty\right)$ 3. 2, 5

4. $(-\infty,2)\cup(2,5)\cup(5,\infty)$ 5. $\dfrac{(3x+2y)x}{(5x+4y)x}$

6. $\dfrac{(2x^2-y)(3x+2)}{(3x+4)(3x+2)}$ 7. $\dfrac{-2a+5}{-a+b}$ 8. $\frac{4}{5}$ 9. $7x$

10. $2a+3$ 11. $\dfrac{3x+2}{2(x+2)}$ 12. $\dfrac{2(y+2)}{y-1}$

13. $\dfrac{5(a+2b)}{4(a-2b)}$ 14. $\dfrac{3(x-y)(x-y)}{x+y}$ 15. $a-b$

16. $\dfrac{x-5}{x+3}$ 17. $\dfrac{1}{x+7}$ 18. y^3-9 19. $\dfrac{x+5}{2(x-5)}$

20. $\dfrac{2ab(a+b)}{a-b}$ 21. a^2-4

Calculator Spotlight, p. 376

1.–8. Left to the student.

Calculator Spotlight, p. 381

1. Correct 2. Correct 3. Incorrect 4. Incorrect
5. Incorrect 6. Correct 7. Correct

Exercise Set 5.1, p. 383

1. $-\frac{17}{3}$ 3. $-7,-5$ 5. $\left(-\infty,-\frac{17}{3}\right)\cup\left(-\frac{17}{3},\infty\right)$

7. $(-\infty,-7)\cup(-7,-5)\cup(-5,\infty)$ 9. $\dfrac{7x(x+2)}{7x(x+8)}$

11. $\dfrac{(q-5)(q+5)}{(q+3)(q+5)}$ 13. $3y$ 15. $\dfrac{2}{3p^4}$ 17. $a-3$

19. $\dfrac{4x-5}{7}$ 21. $\dfrac{y-3}{y+3}$ 23. $\dfrac{t+4}{t-4}$ 25. $\dfrac{x-8}{x+4}$

27. $\dfrac{w^2+wz+z^2}{w+z}$ 29. $\dfrac{1}{3x^3}$ 31. $\dfrac{(x-4)(x+4)}{x(x+3)}$

33. $\dfrac{y+4}{2}$ 35. $\dfrac{(2x+3)(x+5)}{7x}$ 37. $c-2$

---

39. $\dfrac{1}{x+y}$ 41. $\dfrac{3x^5}{2y^3}$ 43. 3 45. $\dfrac{(y-3)(y+2)}{y}$

47. $\dfrac{2a+1}{a+2}$ 49. $\dfrac{(x+4)(x+2)}{3(x-5)}$ 51. $\dfrac{y(y^2+3)}{(y+3)(y-2)}$

53. $\dfrac{x^2+4x+16}{(x+4)(x+4)}$ 55. $\dfrac{2s}{r+2s}$

57. Domain $=\{-4,-2,0,2,4,6\}$;
range $=\{-3,-2,0,1,3,4\}$ 58. Domain $=[-4,5]$;
range $=[-3,2]$ 59. Domain $=[-5,5]$;
range $=[-4,4]$ 60. Domain $=[-4,5]$; range $=[0,2]$
61. $(3a-5b)(2a+5b)$ 62. $(3a-5b)^2$
63. $10(x-7)(x-1)$ 64. $(5x-4)(2x-1)$
65. $(7p+5)(3p-2)$ 66. $2(3m+1)(2m-5)$
67. $2x(x-11)(x+3)$ 68. $10(y+13)(y-5)$

69. ◈ 71. $\frac{13}{19}$; -3; undefined; $\dfrac{2a+2h+3}{4a+4h-1}$

73. $\dfrac{x-3}{(x+1)(x+3)}$ 75. $\dfrac{m-t}{m+t+1}$

Margin Exercises, Section 5.2, pp. 387–392

1. 90 2. 72 3. $\frac{47}{60}$ 4. $\frac{97}{72}$ 5. $5a^3b^2$
6. $(y+3)(y+4)(y+4)$ 7. $2x^2(x-3)(x+3)(x+2)$

8. $2(a+b)(a-b)$, or $2(a+b)(b-a)$ 9. $\dfrac{12+y}{y}$

10. $3x+1$ 11. $\dfrac{a-b}{b+2}$ 12. $\dfrac{y+12}{x^2+y^2}$ 13. $\dfrac{1-b^2}{3}$

14. $\dfrac{2x^2+11}{x-5}$ 15. $\dfrac{3+7x}{4y}$ 16. $\dfrac{11x^2}{2x-y}$ 17. $\dfrac{9x^2+28y}{21x}$

18. $\dfrac{3}{x+y}$ 19. $\dfrac{a+12}{a(a+3)}$ 20. $\dfrac{3y^2+12y+3}{(y-4)(y-3)(y+5)}$

21. $\dfrac{2}{x-1}$

Exercise Set 5.2, p. 393

1. 120 3. 144 5. 210 7. 45 9. $\frac{11}{10}$ 11. $\frac{17}{72}$
13. $\frac{251}{240}$ 15. $21x^2y$ 17. $10(y-10)(y+10)$
19. $30a^3b^2$ 21. $5(y-3)(y-3)$ 23. $(y+5)(y-5)$,
or $(y+5)(5-y)$ 25. $(2r+3)(r-4)(3r-1)(r+4)$
27. $x^3(x-2)(x-2)(x^2+4)$

29. $10x^3(x-1)(x-1)(x+1)(x^2+1)$ 31. $\dfrac{2x+7y}{x+y}$

33. $\dfrac{3y+5}{y-2}$ 35. $a+b$ 37. $\dfrac{13}{y}$ 39. $\dfrac{1}{a+7}$

41. a^2+ab+b^2 43. $\dfrac{2y^2+22}{y^2-y-20}$ 45. $\dfrac{x+y}{x-y}$

47. $\dfrac{3x-4}{x^2-3x+2}$ 49. $\dfrac{8x+1}{x^2-1}$ 51. $\dfrac{2x-14}{15(x+5)}$

53. $\dfrac{-a^2+7ab-b^2}{a^2-b^2}$ 55. $\dfrac{y}{y^2-5y+6}$

57. $\dfrac{3y-10}{y^2-y-20}$ 59. $\dfrac{3y^2-3y-29}{(y+8)(y-3)(y-4)}$

61. $\dfrac{2x^2-13x+7}{(x+3)(x-1)(x-3)}$ 63. 0 65. $\dfrac{3}{x+2}$

67. $\dfrac{-3x^2 - 3x - 4}{x^2 - 1}$ **69.** $\dfrac{-2}{x - y}$, or $\dfrac{2}{y - x}$

71.
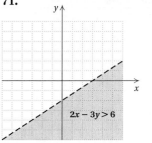

72.
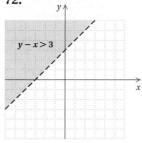

73.
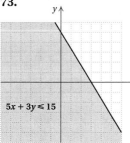

74.
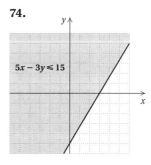

75. $(t - 2)(t^2 + 2t + 4)$ **76.** $(q + 5)(q^2 - 5q + 25)$
77. $23x(x + 1)(x^2 - x + 1)$

78. $(4a - 3b)(16a^2 + 12ab + 9b^2)$ **79.** $\dfrac{3y^6z^7}{7x^5}$

80. $y = \frac{5}{4}x + \frac{11}{2}$ **81.**
83. Domain $= (-\infty, 2) \cup (2, \infty)$;
range $= (-\infty, 0) \cup (0, \infty)$ **85.** 516,600
87. $8a^4,\ 8a^4b,\ 8a^4b^2,\ 8a^4b^3,\ 8a^4b^4,\ 8a^4b^5,\ 8a^4b^6,\ 8a^4b^7$
89. -1 **91.** $\dfrac{-x^3 + x^2y + x^2 - xy^2 + xy + y^3}{(x + y)(x + y)(x - y)(x^2 + y^2)}$

Margin Exercises, Section 5.3, pp. 397–401

1. $\dfrac{x^2}{2} + 8x + 3$ **2.** $5y^3 - 2y^2 + 6y$

3. $\dfrac{x^2y^2}{2} + 5xy + 8$ **4.** $x + 5$

5. $2x^3 - 5x^2 + 19x - 83$, R 341; or
$2x^3 - 5x^2 + 19x - 83 + \dfrac{341}{x + 4}$
6. $3y^3 - 2y^2 + 6y - 4$ **7.** $y^2 - 8y - 24$, R -66; or
$y^2 - 8y - 24 + \dfrac{-66}{y - 3}$ **8.** $x^2 + 10x + 10$, R 5; or
$x^2 + 10x + 10 + \dfrac{5}{x - 1}$ **9.** $y - 11$, R $3y - 27$; or
$y - 11 + \dfrac{3y - 27}{y^2 - 3}$ **10.** $2x^2 + 2x + 14$, R 34; or
$2x^2 + 2x + 14 + \dfrac{34}{x - 3}$ **11.** $x^2 - 4x + 13$, R -30; or
$x^2 - 4x + 13 + \dfrac{-30}{x + 2}$ **12.** $y^2 - y + 1$

Exercise Set 5.3, p. 403

1. $4x^4 + 3x^3 - 6$ **3.** $9y^5 - 4y^2 + 3$
5. $16a^2b^2 + 7ab - 11$ **7.** $x + 7$

9. $a - 12$, R 32; or $a - 12 + \dfrac{32}{a + 4}$ **11.** $x + 2$, R 4; or
$x + 2 + \dfrac{4}{x + 5}$ **13.** $2y^2 - y + 2$, R 6; or
$2y^2 - y + 2 + \dfrac{6}{2y + 4}$ **15.** $2y^2 + 2y - 1$, R 8; or
$2y^2 + 2y - 1 + \dfrac{8}{5y - 2}$ **17.** $2x^2 - x - 9$, R $(3x + 12)$;
or $2x^2 - x - 9 + \dfrac{3x + 12}{x^2 + 2}$ **19.** $2x^3 + 5x^2 + 17x + 51$,
R $152x$; or $2x^3 + 5x^2 + 17x + 51 + \dfrac{152x}{x^2 - 3x}$
21. $x^2 - x + 1$, R -4; or $x^2 - x + 1 + \dfrac{-4}{x - 1}$
23. $a + 7$, R -47; or $a + 7 + \dfrac{-47}{a + 4}$ **25.** $x^2 - 5x - 23$,
R -43; or $x^2 - 5x - 23 + \dfrac{-43}{x - 2}$ **27.** $3x^2 - 2x + 2$,
R -3; or $3x^2 - 2x + 2 + \dfrac{-3}{x + 3}$ **29.** $y^2 + 2y + 1$, R 12;
or $y^2 + 2y + 1 + \dfrac{12}{y - 2}$ **31.** $3x^3 + 9x^2 + 2x + 6$
33. $x^2 + 2x + 4$ **35.** $y^3 + 2y^2 + 4y + 8$
37.
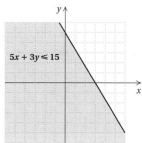

38.
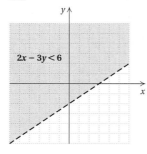

39.
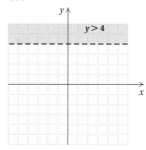

40.
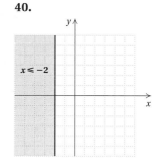

41.
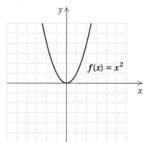

42.
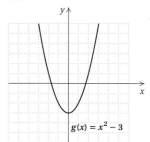

43.

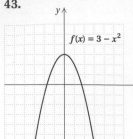

$f(x) = 3 - x^2$

44.

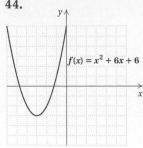

$f(x) = x^2 + 6x + 6$

45. $0, 5$ **46.** $-\frac{8}{5}, \frac{8}{5}$ **47.** $-\frac{1}{4}, \frac{5}{3}$ **48.** $-\frac{1}{4}, -\frac{2}{3}$
49. ◆ **51.** $0; -3, -\frac{5}{2}, \frac{3}{2}$ **53.** $0; -\frac{7}{2}, \frac{5}{3}, 4$

Margin Exercises, Section 5.4, pp. 405–408

1. $\dfrac{14y + 7}{14y - 2}$ **2.** $\dfrac{x}{x + 1}$ **3.** $\dfrac{b + a}{b - a}$ **4.** $\dfrac{a^2b^2}{b^2 + ab + a^2}$

5. $\dfrac{14y + 7}{14y - 2)}$ **6.** $\dfrac{x}{x + 1}$ **7.** $\dfrac{b + a}{b - a}$ **8.** $\dfrac{a^2b^2}{b^2 + ab + a^2}$

Exercise Set 5.4, p. 409

1. $\dfrac{1 + 2a}{1 - a}$ **3.** $\dfrac{x^2 - 1}{x^2 + 1}$ **5.** $\dfrac{3y + 4x}{4y - 3x}$ **7.** $\dfrac{3x + y}{x}$

9. $\dfrac{a^2(b - 3)}{b^2(a - 1)}$ **11.** $\dfrac{1}{a - b}$ **13.** $\dfrac{-1}{x(x + h)}$

15. $\dfrac{(x - 4)(x - 7)}{(x - 5)(x + 6)}$ **17.** $\dfrac{x + 1}{5 - x}$ **19.** $\dfrac{5x - 16}{4x + 1}$

21. $\dfrac{zw(w - z)}{w^2 - wz + z^2}$ **23.** $2x(2x^2 + 10x + 3)$

24. $(y + 2)(y^2 - 2y + 4)$ **25.** $(y - 2)(y^2 + 2y + 4)$
26. $2x(x - 9)(x - 7)$ **27.** $(10x + 1)(100x^2 - 10x + 1)$
28. $(1 - 10a)(1 + 10a + 100a^2)$
29. $(y - 4x)(y^2 + 4xy + 16x^2)$
30. $\left(\frac{1}{2}a - 7\right)\left(\frac{1}{4}a^2 + \frac{7}{2}a + 49\right)$ **31.** $s = 3T - r$

32.

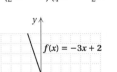

$f(x) = -3x + 2$

33. 22 **34.** $-1, 6$ **35.** ◆ **37.** $\dfrac{-6a - 3h}{a^2(a + h)^2}$

39. $\dfrac{1}{(1 - a - h)(1 - a)}$

41. Left to the student.

43. $\dfrac{x^8}{3^8}$ **45.** a **47.** $\dfrac{x^2}{x^4 + x^3 + x^2 + x + 1}$

Margin Exercises, Section 5.5, pp. 411–415

1. $\frac{2}{3}$ **2.** -31 **3.** No solution **4.** -3 **5.** 7
6. -13 **7.** $4, -3$

Improving Your Math Study Skills, p. 416

1. Rational expression **2.** Solutions **3.** Rational expression **4.** Rational expression **5.** Rational expression **6.** Solutions **7.** Rational expression
8. Solutions **9.** Solutions **10.** Solutions
11. Rational expression **12.** Solutions
13. Rational expression

Exercise Set 5.5, p. 417

1. $\frac{31}{4}$ **3.** $-\frac{12}{7}$ **5.** 144 **7.** $-1, -8$ **9.** 2 **11.** 11
13. 11 **15.** No solution **17.** 2 **19.** 5 **21.** -145
23. $-\frac{10}{3}$ **25.** -3 **27.** $\frac{31}{5}$ **29.** $\frac{85}{12}$ **31.** $-6, 5$
33. No solution **35.** $\frac{17}{4}$ **37.** No solution **39.** $\frac{67}{35}, 4$
41. $-\frac{3}{2}, 2$ **43.** $\frac{17}{4}$ **45.** $\frac{3}{5}$
46. $4(t + 5)(t^2 - 5t + 25)$
47. $(1 - t)(1 + t + t^2)(1 + t)(1 - t + t^2)$
48. $(a + 2b)(a^2 - 2ab + b^2)$
49. $(a - 2b)(a^2 + 2ab + b^2)$
51. ◆ **53.** Left to the student. **55.** ◆

Margin Exercises, Section 5.6, pp. 420–426

1. $2\frac{2}{5}$ hr **2.** Pipe A: 32 hr; pipe B: 96 hr **3.** 525 mg
4. 1620 **5.** Jaime: 23 km/h; Mara: 15 km/h
6. 35.5 mph

Exercise Set 5.6, p. 427

1. $3\frac{3}{14}$ hr **3.** $8\frac{4}{7}$ hr **5.** $2\frac{19}{40}$ hr **7.** $-1, -5$ **9.** $\frac{72}{17}$
11. Juan: 10 days; Ariel: 40 days **13.** Skyler: 12 hr;
Jake: 6 hr **15.** 34 in. or less **17.** (a) 81; (b) yes;
(c) It is 11 more than the final count of 70. **19.** 1160
21. 287 **23.** (a) 4.8 T; (b) 48 lb **25.** 1 **27.** 12 mph
29. 2 mph **31.** 5.2 ft/sec **33.** Simone: $5\frac{1}{3}$ mph;
Rosanna: $3\frac{1}{3}$ mph **35.** Domain: $[-5, 5]$; range: $[-4, 3]$
36. Domain: $\{-4, -2, 0, 1, 2, 4\}$; range: $\{-2, 0, 2, 4, 5\}$
37. Domain: $[-5, 5]$; range: $[-5, 3]$
38. Domain: $[-5, 5]$; range: $[-5, 0]$

39.

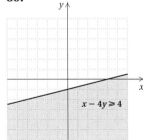

$x - 4y \geq 4$

40.

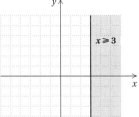

$x \geq 3$

41.

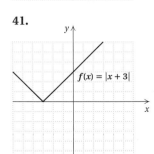

$f(x) = |x + 3|$

42.

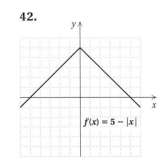

$f(x) = 5 - |x|$

43. ◈ **45.** 30 mi **47.** City: 261 mi; highway: 204 mi
49. $21\frac{9}{11}$ min after 4:00

Margin Exercises, Section 5.7, pp. 431–432

1. $T = \dfrac{PV}{k}$ **2.** $n = \dfrac{PT - IM}{Ip}$ **3.** $s = \dfrac{Fv}{g - F}$

4. $t = \dfrac{ab}{b + a}$

Exercise Set 5.7, p. 433

1. $W_2 = \dfrac{d_2 W_1}{d_1}$ **3.** $r_2 = \dfrac{Rr_1}{r_1 - R}$ **5.** $t = \dfrac{2s}{v_1 + v_2}$

7. $s = \dfrac{Rg}{g - R}$ **9.** $p = \dfrac{qf}{q - f}$ **11.** $a = \dfrac{bt}{b - t}$

13. $E = \dfrac{Inr}{n - I}$ **15.** $H = Sm(t_1 - t_2)$ **17.** $r = \dfrac{eR}{E - e}$

19. $R = \dfrac{3V + \pi h^3}{3\pi h^2}$ **21.** $r = \dfrac{A}{p} - 1$, or $\dfrac{A - P}{P}$

23. $h = \dfrac{2gR^2}{V^2} - R$, or $\dfrac{2gR^2 - RV^2}{V^2}$ **25.** $Q = \dfrac{2Tt - 2AT}{A - q}$

26. Dimes: 2; nickels: 5; quarters: 5 **27.** 30-min: 4;
60-min: 8 **28.** -6 **29.** 6 **30.** 0 **31.** $8a^3 - 2a$
32. $-\frac{4}{5}$ **33.** $y = -\frac{4}{5}x + \frac{17}{5}$ **35.** ◈

Margin Exercises, Section 5.8, pp. 435–440

1. $\frac{2}{5}$; $y = \frac{2}{5}x$ **2.** 0.7; $y = 0.7x$ **3.** 50 volts

4. 1,575,000 tons **5.** 0.6; $y = \dfrac{0.6}{x}$ **6.** $7\frac{1}{2}$ hr

7. $y = 7x^2$ **8.** $y = \dfrac{9}{x^2}$ **9.** $y = \frac{1}{2}xz$ **10.** $y = \dfrac{5xz^2}{w}$

11. 490 m **12.** 1 ohm

Exercise Set 5.8, p. 441

1. 5; $y = 5x$ **3.** $\frac{2}{15}$; $y = \frac{2}{15}x$ **5.** $\frac{9}{4}$; $y = \frac{9}{4}x$

7. 241,920,000 **9.** $66\frac{2}{3}$ cm **11.** 90 g **13.** 40 kg

15. 98; $y = \dfrac{98}{x}$ **17.** 36; $y = \dfrac{36}{x}$ **19.** 0.05; $y = \dfrac{0.05}{x}$

21. 3.5 hr **23.** $\frac{2}{9}$ ampere **25.** 1.92 ft **27.** 160 cm³

29. $y = 15x^2$ **31.** $y = \dfrac{0.0015}{x^2}$ **33.** $y = xz$

35. $y = \frac{3}{10}xz^2$ **37.** $y = \dfrac{xz}{5wp}$ **39.** 36 mph **41.** 2.5 m

43. About 106 **45.** 729 gal **47.** $-4, 3$ **48.** 3
49. $\frac{1}{4}, \frac{2}{3}$ **50.** $-7, 7$ **51.** $y = -\frac{2}{3}x - 5$
52. $y = -\frac{2}{7}x + \frac{48}{7}$ **53.** (a) $y = 97.8x + 1115.2$;

(b) 1800; 1898 **55.** ◈ **57.** $\dfrac{\pi}{4}$

59. Q varies directly as the square of p and inversely as
the cube of q.

Summary and Review: Chapter 5, p. 445

1. $-3, 3$ **2.** $\{x \mid x$ is a real number $and\ x \neq -3\ and$

$x \neq 3\}$, or $(-\infty, -3) \cup (-3, 3) \cup (3, \infty)$ **3.** $\dfrac{x - 2}{3x + 2}$

4. $\dfrac{1}{a - 2}$ **5.** $48x^3$ **6.** $(x - 7)(x + 7)(3x + 1)$

7. $(x + 5)(x - 4)(x - 2)$ **8.** $\dfrac{y - 8}{2}$ **9.** $\dfrac{(x - 2)(x + 5)}{x - 5}$

10. $\dfrac{3a - 1}{a - 3}$ **11.** $\dfrac{(x^2 + 4x + 16)(x - 6)}{(x + 4)(x + 2)}$

12. $\dfrac{x - 3}{(x + 1)(x + 3)}$ **13.** $\dfrac{2x^3 + 2x^2y + 2xy^2 - 2y^3}{(x - y)(x + y)}$

14. $\dfrac{-y}{(y + 4)(y - 1)}$ **15.** $4b^2c - \frac{5}{2}bc^2 + 3abc$

16. $y - 14$, R -20; or $y - 14 + \dfrac{-20}{y - 6}$ **17.** $6x^2 - 9$,

R $(5x + 22)$; or $6x^2 - 9 + \dfrac{5x + 22}{x^2 + 2}$ **18.** $x^2 + 9x + 40$,

R 153; or $x^2 + 9x + 40 + \dfrac{153}{x - 4}$

19. $3x^3 - 8x^2 + 8x - 6$, R -1; or

$3x^3 - 8x^2 + 8x - 6 + \dfrac{-1}{x + 1}$ **20.** $\frac{3}{4}$

21. $\dfrac{a^2b^2}{2(a^2 - ab + b^2)}$ **22.** $\dfrac{(x - 9)(x - 6)}{(x - 3)(x + 6)}$

23. $\dfrac{4x^2 - 14x + 2}{3x^2 + 7x - 11}$ **24.** $\frac{28}{11}$ **25.** 6 **26.** No solution

27. 3 **28.** $-\frac{11}{3}$ **29.** 2 **30.** $5\frac{1}{7}$ hr **31.** 24 mph

32. 4000 mi **33.** $s = \dfrac{Rg}{g - R}$; $g = \dfrac{Rs}{s - R}$

34. $b = \dfrac{ta}{Sa - p}$; $t = \dfrac{Sab - pb}{a}$ **35.** $y = 4x$

36. $y = \dfrac{2500}{x}$ **37.** 20 min **38.** About 77.7

39. 500 watts **40.** Domain: $[-3, 5]$; range: $[-2, 2]$
41. **42.**

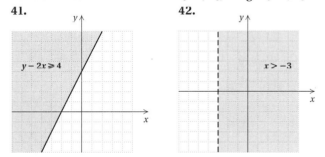

43. $(5x - 2y)(25x^2 + 10xy + 4y^2)$ **44.** $(x + 6)(6x - 7)$
45. ◈ When adding or subtracting rational
expressions, we use the LCM of the denominators (the
LCD). When solving a rational equation or when solving
a formula for a given letter, we multiply by the LCM of
all the denominators to clear fractions. When
simplifying a complex rational expression, we can use
the LCM in either of two ways. We can multiply by a/a,
where a is the LCM of all the denominators occurring
in the expressions. Or we can use the LCM to add or
subtract as necessary in the numerator and in the
denominator. **46.** ◈ Rational equations differ from

those previously studied because they contain variables in denominators. Because of this, possible solutions must be checked in the original equation to avoid division by 0.   **47.** All real numbers except 0 and 13
48. $a^2 + ab + b^2$

Test: Chapter 5, p. 447

1. [5.1a] 1, 2   **2.** [5.1a] $\{x \mid x$ is a real number *and* $x \neq 1$ *and* $x \neq 2\}$, or $(-\infty, 1) \cup (1, 2) \cup (2, \infty)$
3. [5.1c] $\dfrac{3x + 2}{x - 2}$ **4.** [5.1c] $\dfrac{p^2 - p + 1}{p - 2}$
5. [5.2a] $(x + 3)(x - 2)(x + 5)$ **6.** [5.1d] $\dfrac{2(x + 5)}{x - 2}$
7. [5.2b] $\dfrac{x - 6}{(x + 4)(x + 6)}$ **8.** [5.1e] $\dfrac{y + 4}{2}$
9. [5.2b] $x + y$ **10.** [5.2c] $\dfrac{3x}{(x - 1)(x + 1)}$
11. [5.2c] $\dfrac{a^3 + a^2b + ab^2 + ab - b^2 - 2}{a^3 - b^3}$
12. [5.3a] $4s^2 + 3s - 2rs^2$ **13.** [5.3b] $y^2 - 5y + 25$
14. [5.3b] $4x^2 + 3x - 4$, R $(-8x + 2)$; or
$4x^2 + 3x - 4 + \dfrac{-8x + 2}{x^2 + 1}$ **15.** [5.3c] $x^2 + 6x + 20$;
R 54; or $x^2 + 6x + 20 + \dfrac{54}{x - 3}$
16. [5.3c] $4x^2 - 26x + 130$, R -659; or
$4x^2 - 26x + 130 + \dfrac{-659}{x + 5}$ **17.** [5.4a] $\dfrac{x + 1}{x}$
18. [5.4a] $\dfrac{b^2 - ab + a^2}{a^2b^2}$ **19.** [5.5a] 9
20. [5.5a] No solution   **21.** [5.5a] No solution
22. [5.5a] $\frac{17}{8}$ **23.** [5.5a] -1, 4   **24.** [5.6a] 2 hr
25. [5.6c] $3\frac{3}{11}$ mph   **26.** [5.6b] $14\frac{2}{17}$ gal
27. [5.7a] $a = \dfrac{Tb}{T - b}$; $b = \dfrac{Ta}{a + T}$ **28.** [5.7a] $a = \dfrac{2b}{Qb + t}$
29. [5.8e] $Q = \frac{5}{2}xy$ **30.** [5.8c] $y = \dfrac{250}{x}$
31. [5.8b] \$495   **32.** [5.8f] 615.44 cm^2
33. [2.3a] Domain: $[-2, 2]$; range: $[-1, 3]$
34. [3.6b]

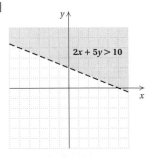

35. [4.4a, b] $8(2t + 3)(t - 3)$
36. [4.5d] $(4a + 1)(16a^2 - 4a + 1)$
37. [5.5a] All real numbers except 0 and 15
38. [5.2a] $(1 - t^6)(1 + t^6)$
39. [5.5a] x-intercept: $(11, 0)$; y-intercept: $\left(0, -\frac{33}{5}\right)$

Cumulative Review: Chapters R–5, p. 449

1. [R.4b] $\frac{4}{3}$ **2.** [R.1d] 12   **3.** [R.2e] $-\frac{8}{9}$
4. [R.6b] $47b - 51$ **5.** [R.3c] 224   **6.** [R.7b] $\dfrac{x^6}{4y^8}$
7. [4.1d] $16p^2 - 8p$ **8.** [4.2c] $36m^2 - 12mn + n^2$
9. [4.2b] $15a^2 - 14ab - 8b^2$ **10.** [5.1d] $\dfrac{y - 2}{3}$
11. [5.1e] $\dfrac{3x - 5}{x + 4}$ **12.** [5.2b] $\dfrac{6x + 13}{20(x - 3)}$
13. [5.2c] $\dfrac{4x + 1}{(x + 2)(x - 2)}$ **14.** [5.4a] $\dfrac{y + 2x}{3y - x}$
15. [5.4a] $\dfrac{y^3 - 2y}{y^3 - 1}$ **16.** [5.3b] $2x^2 - 11x + 23 + \dfrac{-49}{x + 2}$
17. [1.1d] $\frac{15}{2}$ **18.** [1.2a] $C = \frac{5}{9}(F - 32)$
19. [1.5a] $\left\{x \mid -3 < x < -\frac{3}{2}\right\}$, or $\left(-3, -\frac{3}{2}\right)$
20. [1.6e] $\{x \mid x \leq -2.1$ *or* $x \geq 2.1\}$, or
$(-\infty, -2.1] \cup [2.1, \infty)$ **21.** [3.2b] Infinite number of
solutions   **22.** [3.4a] $(3, 2, -1)$ **23.** [4.7a] $\frac{1}{4}$
24. [4.7a] $-2, \frac{7}{2}$ **25.** [5.5a] No solution   **26.** [5.5a] -1
27. [5.7a] $a = -\dfrac{bP}{P - 3}$, or $\dfrac{bP}{3 - P}$ **28.** [3.2a] $(-2, 1)$
29. [3.4a] $\left(\frac{5}{8}, \frac{1}{16}, -\frac{3}{4}\right)$ **30.** [4.3a] $2x^2(2x + 9)$
31. [4.3b] $(2a - 1)(4a^2 - 3)$ **32.** [4.3c] $(x - 6)(x + 14)$
33. [4.4a, b] $(2x + 5)(3x - 2)$
34. [4.5b] $(4y + 9)(4y - 9)$ **35.** [4.5a] $(t - 8)^2$
36. [4.5d] $8(2x + 1)(4x^2 - 2x + 1)$
37. [4.5d] $(0.3b - 0.2c)(0.09b^2 + 0.06bc + 0.04c^2)$
38. [4.6a] $x^2(x^2 + 1)(x + 1)(x - 1)$
39. [4.4a, b] $(4x - 1)(5x + 3)$
40. [2.1c]   **41.** [2.5c]

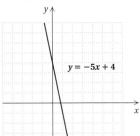

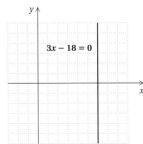

42. [3.6b]   **43.** [3.6c]

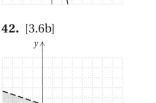

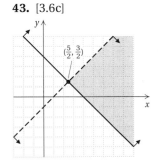

44. [2.6b] $y = -\frac{1}{2}x - 1$ **45.** [2.6d] $y = \frac{1}{2}x - \frac{5}{2}$
46. [1.3a] \$45,333.33   **47.** [1.3a] 20.3%
48. [3.3a] Detroit: \$4.09; Chicago: \$1.66
49. [3.5a] Win: 38; lose: 30; tie: 13   **50.** [5.6a] $6\frac{2}{3}$ hr

51. [5.8b] 60 yr old   **52.** [2.3a] $\{x \mid x$ is a real number *and $x \neq -5$ and $x \neq 5$*$\}$, or $(-\infty, -5) \cup (-5, 5) \cup (5, \infty)$
53. [2.3a] Domain: $[-5, 5]$; range: $[-2, 4]$
54. [3.5a] $a = 1$, $b = -5$, $c = 6$ **55.** [5.5a] All real numbers except 9 and -5 **56.** [4.7a] $0, \frac{1}{4}, -\frac{1}{4}$

Chapter 6

Pretest: Chapter 6, p. 452

1. [6.1b] $|t|$ **2.** [6.1c] $3x$ **3.** [6.1d] $|y|$
4. [6.2d] $3a\sqrt[3]{3ab^2}$ **5.** [6.4a] $4\sqrt{5}$ **6.** [6.3a] $12x^2\sqrt{x}$
7. [6.4b] $25 - 4\sqrt{6}$ **8.** [6.3b] $2\sqrt{a}$
9. [6.5b] $-6 + 3\sqrt{5}$ **10.** [6.6a] 13   **11.** [6.6a] $-\frac{7}{6}$
12. [6.6b] 0, 8   **13.** [6.7a] $\sqrt{39}$; 6.245
14. [6.7a] $\sqrt{32}$ ft; 5.657 ft   **15.** [6.2d] $\sqrt[6]{x^5}$
16. [6.8f] No   **17.** [6.8b] $-5 + 13i$ **18.** [6.8c] $24 - 7i$
19. [6.8e] $-\frac{14}{13} - \frac{5}{13}i$ **20.** [6.8d] $-i$

Margin Exercises, Section 6.1, pp. 453–458

1. 3, -3 **2.** 6, -6 **3.** 11, -11 **4.** 1   **5.** 6   **6.** $\frac{9}{10}$
7. 0.08   **8. (a)** 4; **(b)** -4; **(c)** does not exist as a real number   **9. (a)** 7; **(b)** -7; **(c)** does not exist as a real number   **10. (a)** 12; **(b)** -12; **(c)** does not exist as a real number   **11.** 4.123   **12.** 6.325   **13.** 33.734
14. -29.455 **15.** 0.793   **16.** -5.569 **17.** $28 + x$
18. $\dfrac{y}{y + 3}$ **19.** 2; $\sqrt{22} \approx 4.690$; does not exist as a real number   **20.** -2; $-\sqrt{7}$; does not exist as a real number   **21.** Domain $= \{x \mid x \geq 5\} = [5, \infty)$
22. Domain $= \left\{x \mid x \geq -\frac{3}{2}\right\} = \left[-\frac{3}{2}, \infty\right)$
23. 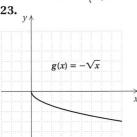   **24.**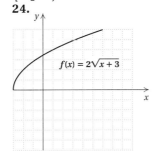

25. $|y|$ **26.** 24   **27.** $5|y|$ **28.** $4|y|$ **29.** $|x + 7|$
30. $2|x - 2|$ **31.** $7|y + 5|$ **32.** $|x - 3|$ **33.** -4
34. $3y$ **35.** $2(x + 2)$ **36.** $-\frac{7}{4}$ **37.** -3; 0; $\sqrt[3]{-5} \approx -1.710$; $\sqrt[3]{7} \approx 1.913$ **38.** 3   **39.** -3 **40.** x
41. y **42.** 0   **43.** $-2x$ **44.** $3x + 2$ **45.** 3
46. -3 **47.** Does not exist as a real number
48. 0   **49.** $2|x - 2|$ **50.** $|x|$ **51.** $|x + 3|$

Calculator Spotlight, p. 457

1. 2.802039331   **2.** -3.50339806 **3.** 3.601131332
4. -12.68265141

Calculator Spotlight, p. 458

1. Domain $= [0, \infty)$; range $= [0, \infty)$
2. Domain $= [-2, \infty)$; range $= [0, \infty)$
3. Domain $= (-\infty, \infty)$; range $= (-\infty, \infty)$
4. Domain $= (-\infty, \infty)$; range $= (-\infty, \infty)$
5. Domain $= [1, \infty)$; range $= [0, \infty)$
6. Domain $= (-\infty, \infty)$; range $= (-\infty, \infty)$
7. Domain $= [-3, \infty)$; range $= (-\infty, 5]$
8. Domain $= (-\infty, \infty)$; range $= (-\infty, \infty)$
9. Incorrect   **10.** Correct

Exercise Set 6.1, p. 459

1. 4, -4 **3.** 12, -12 **5.** 20, -20 **7.** $-\frac{7}{6}$ **9.** 14
11. 0.06   **13.** 18.628   **15.** 1.962   **17.** $y^2 + 16$
19. $\dfrac{x}{y - 1}$ **21.** $\sqrt{20}$; 0; does not exist as a real number; does not exist as a real number   **23.** $\sqrt{11}$; does not exist as a real number; $\sqrt{11}$; 12
25. Domain $= \{x \mid x \geq 2\} = [2, \infty)$
27. About 24.5 mph; about 54.8 mph
29. 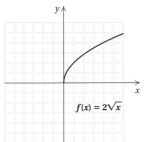   **31.**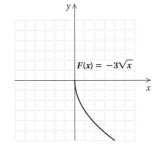

33. 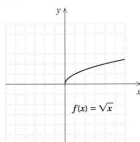   **35.**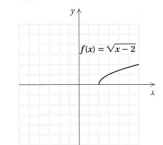

37. 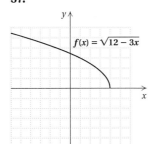   **39.**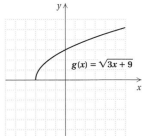

41. $4|x|$ **43.** $12|c|$ **45.** $|p + 3|$ **47.** $|x - 2|$
49. 3 **51.** $-4x$ **53.** -6 **55.** $0.7(x + 1)$ **57.** 2; 3;
-2; -4 **59.** -1; approximately 2.7144; -4; -10
61. 5 **63.** -1 **65.** $-\frac{2}{3}$ **67.** $|x|$ **69.** $5|a|$ **71.** 6
73. $|a + b|$ **75.** y **77.** $x - 2$ **79.** $-2, 1$ **80.** $-1, 0$
81. $-\frac{7}{2}, \frac{7}{2}$ **82.** 4, 9 **83.** $-2, \frac{5}{3}$ **84.** $\frac{5}{2}$ **85.** $0, \frac{5}{2}$
86. 0, 1 **87.** $a^9 b^6 c^{15}$ **88.** $10 a^{10} b^9$ **89.** ◈
91. $[-3, 2)$ **93.** 1.7; 2.2; 3.2
95. **(a)** Domain $= (-\infty, \infty)$, range $= (-\infty, \infty)$;
(b) domain $= (-\infty, \infty)$, range $= (-\infty, \infty)$;
(c) domain $= [-3, \infty)$, range $= (-\infty, 2]$;
(d) domain $= [0, \infty)$, range $= [0, \infty)$; **(e)** domain $= [3, \infty)$,
range $= [0, \infty)$

Margin Exercises, Section 6.2, pp. 463–466

1. $\sqrt[4]{y}$ **2.** $\sqrt{3a}$ **3.** 2 **4.** 5 **5.** $\sqrt[5]{a^3 b^2 c}$
6. $(19ab)^{1/3}$ **7.** $19(ab)^{1/3}$ **8.** $\left(\dfrac{x^2 y}{16}\right)^{1/5}$
9. $7(2ab)^{1/4}$ **10.** $\sqrt{x^3}$ **11.** 4 **12.** 32 **13.** $(7abc)^{4/3}$
14. $6^{7/5}$ **15.** $\frac{1}{2}$ **16.** $\dfrac{1}{(3xy)^{7/8}}$ **17.** $\frac{1}{27}$ **18.** $\dfrac{7p^{3/4}}{q^{6/5}}$
19. $\left(\dfrac{7n}{11m}\right)^{2/3}$ **20.** $7^{14/15}$ **21.** $5^{1/3}$ **22.** $9^{2/5}$
23. $\dfrac{q^{1/8}}{p^{1/3}}$ **24.** $\sqrt{a}$ **25.** x **26.** $\sqrt{2}$ **27.** $\sqrt[4]{xy^2}$
28. $a^2 \sqrt{b}$ **29.** ab^2 **30.** $\sqrt[4]{63}$ **31.** $\sqrt[6]{x^4 y^3 z^5}$
32. $\sqrt[4]{ab}$ **33.** $\sqrt[7]{5m}$ **34.** $\sqrt[6]{m}$ **35.** $a^{20} b^{12} c^4$
36. $\sqrt[10]{x}$

Calculator Spotlight, p. 464

1. 2.924 **2.** 5.278 **3.** 0.283 **4.** 11.053 **5.** 0.00006
6.–11. Left to the student.

Exercise Set 6.2, p. 467

1. $\sqrt[7]{y}$ **3.** 2 **5.** $\sqrt[5]{a^3 b^3}$ **7.** 8 **9.** 343 **11.** $17^{1/2}$
13. $18^{1/3}$ **15.** $(xy^2 z)^{1/5}$ **17.** $(3mn)^{3/2}$ **19.** $(8x^2 y)^{5/7}$
21. $\frac{1}{3}$ **23.** $\frac{1}{1000}$ **25.** $\dfrac{1}{x^{1/4}}$ **27.** $\dfrac{1}{(2rs)^{3/4}}$ **29.** $\dfrac{2a^{3/4} c^{2/3}}{b^{1/2}}$
31. $\left(\dfrac{8yz}{7x}\right)^{3/5}$ **33.** $x^{2/3}$ **35.** $\dfrac{x^4}{2^{1/3} y^{2/7}}$ **37.** $\dfrac{7x}{z^{1/3}}$
39. $\dfrac{5ac^{1/2}}{3}$ **41.** $5^{7/8}$ **43.** $7^{1/4}$ **45.** $4.9^{1/2}$ **47.** $6^{3/28}$
49. $a^{23/12}$ **51.** $a^{8/3} b^{5/2}$ **53.** $\dfrac{1}{x^{2/7}}$ **55.** $\sqrt[3]{a}$ **57.** x^5
59. $\dfrac{1}{x^3}$ **61.** $a^5 b^5$ **63.** $\sqrt[4]{2^5}$ **65.** $\sqrt[3]{2x}$ **67.** $x^2 y^3$
69. $2c^2 d^3$ **71.** $\sqrt[12]{7^4 \cdot 5^3}$ **73.** $\sqrt[20]{5^5 \cdot 7^4}$ **75.** $\sqrt[6]{4x^5}$
77. $a^6 b^{12}$ **79.** $\sqrt[18]{m}$ **81.** $\sqrt[12]{x^4 y^3 z^2}$ **83.** $\sqrt[30]{\dfrac{d^{35}}{c^{99}}}$
85. $a = \dfrac{Ab}{b - A}$ **86.** $s = \dfrac{Qt}{Q - t}$ **87.** $t = \dfrac{Qs}{s + Q}$
88. $b = \dfrac{la}{t - a}$ **89.** ◈ **91.** Left to the student.

Margin Exercises, Section 6.3, pp. 469–472

1. $\sqrt{133}$ **2.** $\sqrt{21pq}$ **3.** $\sqrt[4]{2821}$ **4.** $\sqrt[3]{\dfrac{10}{pq}}$ **5.** $\sqrt[6]{500}$
6. $\sqrt[6]{25x^3 y^2}$ **7.** $4\sqrt{2}$ **8.** $2\sqrt[3]{10}$ **9.** $10\sqrt{3}$ **10.** $6y$
11. $2a\sqrt{3b}$ **12.** $2bc\sqrt{3ab}$ **13.** $2\sqrt{2}$
14. $3xy^2 \sqrt[3]{3xy^2}$ **15.** $3\sqrt{2}$ **16.** $6y\sqrt{7}$ **17.** $3x\sqrt[3]{4y}$
18. $7\sqrt{3ab}$ **19.** 5 **20.** $56\sqrt{xy}$ **21.** $5a$ **22.** $\frac{20}{7}$
23. $\frac{5}{6}$ **24.** $\dfrac{x}{10}$ **25.** $\dfrac{3x\sqrt[3]{2x^2}}{5}$ **26.** $\sqrt[12]{xy^2}$

Exercise Set 6.3, p. 473

1. $2\sqrt{6}$ **3.** $3\sqrt{10}$ **5.** $5\sqrt[3]{2}$ **7.** $6x^2 \sqrt{5}$
9. $3x^2 \sqrt[3]{2x^2}$ **11.** $2t^2 \sqrt[3]{10t^2}$ **13.** $2\sqrt[4]{5}$ **15.** $4a\sqrt{2b}$
17. $3x^2 y^2 \sqrt[4]{3y^2}$ **19.** $2xy^3 \sqrt[5]{3x^2}$ **21.** $5\sqrt{2}$ **23.** $3\sqrt{10}$
25. 2 **27.** $30\sqrt{3}$ **29.** $3x^4 \sqrt{2}$ **31.** $5bc^2 \sqrt{2b}$
33. $a^3 \sqrt{10}$ **35.** $2y^3 \sqrt[3]{2}$ **37.** $4\sqrt[4]{4}$ **39.** $4a^3 b\sqrt{6ab}$
41. $\sqrt[6]{200}$ **43.** $\sqrt[4]{12}$ **45.** $a\sqrt[4]{a}$ **47.** $b\sqrt[10]{b^9}$
49. $xy\sqrt[6]{xy^5}$ **51.** $3\sqrt{2}$ **53.** $\sqrt{5}$ **55.** 3 **57.** $y\sqrt{7y}$
59. $2\sqrt[3]{a^2 b}$ **61.** $4\sqrt{xy}$ **63.** $2x^2 y^2$ **65.** $\dfrac{1}{\sqrt[6]{a}}$
67. $\sqrt[12]{a^5}$ **69.** $\sqrt[12]{x^2 y^5}$ **71.** $\frac{5}{6}$ **73.** $\frac{4}{7}$ **75.** $\frac{5}{3}$ **77.** $\dfrac{7}{y}$
79. $\dfrac{5y\sqrt{y}}{x^2}$ **81.** $\dfrac{3a\sqrt[3]{a}}{2b}$ **83.** $\dfrac{3x}{2}$ **85.** $\dfrac{2x\sqrt[5]{x^3}}{y^2}$
87. $\dfrac{x^2 \sqrt[6]{x}}{yz^2}$ **89.** $-10, 9$ **90.** Height: 4 in.; base: 6 in.
91. 8 **92.** $\frac{15}{2}$ **93.** No solution **94.** No solution
95. ◈ **97.** **(a)** 1.62 sec; **(b)** 1.99 sec; **(c)** 2.20 sec
99. $2yz\sqrt{2z}$

Margin Exercises, Section 6.4, pp. 477–478

1. $13\sqrt{2}$ **2.** $10\sqrt[4]{5x} - \sqrt{7}$ **3.** $19\sqrt{5}$
4. $(3y + 4)\sqrt[3]{y^2} + 2y^2$ **5.** $2\sqrt{x - 1}$ **6.** $5\sqrt{6} + 3\sqrt{14}$
7. $a\sqrt[3]{3} - \sqrt[3]{2a^2}$ **8.** $-4 - 9\sqrt{6}$
9. $3\sqrt{ab} - 4\sqrt{3a} + 6\sqrt{3b} - 24$ **10.** -3 **11.** $p - q$
12. $20 - 4y\sqrt{5} + y^2$ **13.** $58 + 12\sqrt{6}$

Exercise Set 6.4, p. 479

1. $11\sqrt{5}$ **3.** $\sqrt[3]{7}$ **5.** $13\sqrt[3]{y}$ **7.** $-8\sqrt{6}$ **9.** $6\sqrt[3]{3}$
11. $21\sqrt{3}$ **13.** $38\sqrt{5}$ **15.** $122\sqrt{2}$ **17.** $9\sqrt[3]{2}$
19. $29\sqrt{2}$ **21.** $(1 + 6a)\sqrt{5a}$ **23.** $(2 - x)\sqrt[3]{3x}$
25. $15\sqrt[3]{4}$ **27.** $4\sqrt{5} - 10$ **29.** $\sqrt{6} - \sqrt{21}$
31. $2\sqrt{15} - 6\sqrt{3}$ **33.** -6 **35.** $3a\sqrt[3]{2}$ **37.** 1
39. -12 **41.** 44 **43.** 1 **45.** 3 **47.** -19
49. $a - b$ **51.** $1 + \sqrt{5}$ **53.** $7 + 3\sqrt{3}$ **55.** -6
57. $a + \sqrt{3a} + \sqrt{2a} + \sqrt{6}$ **59.** $2\sqrt[3]{9} - 3\sqrt[3]{6} - 2\sqrt[3]{4}$
61. $7 + 4\sqrt{3}$ **63.** $3 - \sqrt[5]{24} - \sqrt[5]{81} + \sqrt[5]{72}$
65. $\dfrac{x(x^2 + 4)}{(x + 4)(x + 3)}$ **66.** $\dfrac{(a + 2)(a + 4)}{a}$ **67.** $a - 2$
68. $\dfrac{(y - 3)(y - 3)}{y + 3}$ **69.** $\dfrac{4(3x - 1)}{3(4x + 1)}$ **70.** $\dfrac{x}{x + 1}$
71. $\dfrac{pq}{q + p}$ **73.** ◈ **75.** Left to the student.
77. $2x - 2\sqrt{x^2 - 4}$ **79.** $\sqrt[3]{y} - y$ **81.** $12 + 6\sqrt{3}$

Margin Exercises, Section 6.5, pp. 483–485

1. $\dfrac{\sqrt{10}}{5}$ **2.** $\dfrac{\sqrt[3]{10}}{2}$ **3.** $\dfrac{2\sqrt{3ab}}{3b}$ **4.** $\dfrac{\sqrt[4]{56}}{2}$

5. $\dfrac{x\sqrt[3]{12x^2y^2}}{2y}$ **6.** $\dfrac{7\sqrt[3]{2x^2y}}{2y^2}$ **7.** $c^2 - b$ **8.** $a - b$

9. $\dfrac{\sqrt{3}+y}{\sqrt{3}+y}$ **10.** $\dfrac{\sqrt{2}-\sqrt{3}}{\sqrt{2}-\sqrt{3}}$ **11.** $-7 - 6\sqrt{2}$

12. $6 - 2\sqrt{2}$

Exercise Set 6.5, p. 487

1. $\dfrac{\sqrt{15}}{3}$ **3.** $\dfrac{\sqrt{22}}{2}$ **5.** $\dfrac{2\sqrt{15}}{35}$ **7.** $\dfrac{2\sqrt[3]{6}}{3}$ **9.** $\dfrac{\sqrt[3]{75ac^2}}{5c}$

11. $\dfrac{y\sqrt[3]{9yx^2}}{3x^2}$ **13.** $\dfrac{\sqrt[4]{s^3t^3}}{st}$ **15.** $\dfrac{\sqrt{15x}}{10}$ **17.** $\dfrac{\sqrt[3]{100xy}}{5x^2y}$

19. $\dfrac{\sqrt[4]{2xy}}{2x^2y}$ **21.** $\dfrac{54 + 9\sqrt{10}}{26}$ **23.** $-2\sqrt{35} - 2\sqrt{21}$

25. $\dfrac{\sqrt{15} + 20 - 6\sqrt{2} - 8\sqrt{30}}{-77}$ **27.** $\dfrac{6 - 5\sqrt{a} + a}{9 - a}$

29. $\dfrac{3\sqrt{6}+4}{2}$ **31.** $\dfrac{x - 2\sqrt{xy} + y}{x - y}$ **33.** 30 **34.** $-\dfrac{19}{5}$

35. 1 **36.** $\dfrac{x - 2}{x + 3}$ **37.** ◆ **39.** Left to the student.

41. $-\dfrac{3\sqrt{a^2 - 3}}{a^2 - 3}$

Margin Exercises, Section 6.6, pp. 489–494

1. 100 **2.** No solution **3.** 1 **4.** 2, 5 **5.** 4 **6.** 17
7. 9 **8.** 27 **9.** 5 **10.** About 276 mi **11.** About
8.9 mi **12.** About 169 ft

Calculator Spotlight, p. 491

1. 9 **2.** 1 **3.** 2, 5 **4.** 15.6 **5.** 9.5 **6.** 2.0, 7.2

Exercise Set 6.6, p. 495

1. $\dfrac{19}{2}$ **3.** $\dfrac{49}{6}$ **5.** 57 **7.** $\dfrac{92}{5}$ **9.** -1 **11.** No solution
13. 3 **15.** 19 **17.** -6 **19.** $\dfrac{1}{64}$ **21.** 5 **23.** 9
25. 7 **27.** $\dfrac{80}{9}$ **29.** 6, 2 **31.** -1 **33.** No solution
35. 3 **37.** About 232 mi **39.** 16,200 ft
41. 211.25 ft; 281.25 ft **43.** 49 **45.** 5 **47.** $4\dfrac{4}{9}$ hr
48. Jeff: $1\dfrac{1}{3}$ hr; Grace: 4 hr **49.** 2808 mi **50.** 84 hr
51. 0, -2.8 **52.** 0, $\dfrac{5}{3}$ **53.** -8, 8 **54.** -3, $\dfrac{7}{2}$
55. $2ah + h^2$ **56.** $2ah + h^2 - h$
57. $4ah + 2h^2 - 3h$ **58.** $4ah + 2h^2 + 3h$ **59.** ◆
61. Left to the student. **63.** 6912 **65.** 0
67. -3, -6 **69.** 2 **71.** 0, $\dfrac{125}{4}$ **73.** 2 **75.** $\dfrac{1}{2}$ **77.** 3

Margin Exercises, Section 6.7, pp. 499–500

1. $\sqrt{41}$; 6.403 **2.** $\sqrt{6}$; 2.449 **3.** $\sqrt{200}$; 14.142
4. $\sqrt{8200}$; 90.554 ft **5.** 7.4 ft

Exercise Set 6.7, p. 501

1. $\sqrt{34}$; 5.831 **3.** $\sqrt{450}$; 21.213 **5.** 5 **7.** $\sqrt{43}$; 6.557

9. $\sqrt{12}$; 3.464 **11.** $\sqrt{n - 1}$ **13.** $\sqrt{10{,}561}$; 102.767 ft
15. $\sqrt{116}$; 10.770 ft **17.** $s + s\sqrt{2}$ **19.** 7.1 ft
21. $\sqrt{181}$; 13.454 cm **23.** 12 in. **25.** (3, 0), (-3, 0)
27. $\sqrt{340} + 8$; 26.439 ft **29.** 20.497 in.
31. Flash: $67\dfrac{2}{3}$ mph; Crawler: $53\dfrac{2}{3}$ mph **32.** $3\dfrac{3}{4}$ mph
33. -7, $\dfrac{3}{2}$ **34.** 3, 8 **35.** 1 **36.** -2, 2 **37.** 13
38. 7 **39.** ◆ **41.** 9 **43.** $\sqrt{75}$ cm

Margin Exercises, Section 6.8, pp. 505–509

1. $i\sqrt{5}$, or $\sqrt{5}\,i$ **2.** $5i$ **3.** $-i\sqrt{11}$, or $-\sqrt{11}\,i$
4. $-6i$ **5.** $3i\sqrt{6}$, or $3\sqrt{6}\,i$ **6.** $15 - 3i$ **7.** $-12 + 6i$
8. $3 - 5i$ **9.** $12 - 7i$ **10.** -10 **11.** $-\sqrt{34}$
12. 42 **13.** $-9 - 12i$ **14.** $-35 - 25i$
15. $-14 + 8i$ **16.** $11 + 10i$ **17.** $5 + 12i$ **18.** $-i$
19. 1 **20.** i **21.** -1 **22.** $8 - i$ **23.** 3
24. $-7 - 6i$ **25.** $-1 + i$ **26.** $6 - 3i$ **27.** $-9 + 5i$
28. $\pi + \dfrac{1}{4}i$ **29.** 53 **30.** 10 **31.** $2i$ **32.** $\dfrac{10}{17} - \dfrac{11}{17}i$
33. Yes **34.** Yes

Calculator Spotlight, p. 510

1. $-2 - 9i$ **2.** $20 + 17i$ **3.** $-47 - 161i$
4. $-0.5207 + 0.2517i$ **5.** $4i$ **6.** -20 **7.** -28.3725
8. $-0.64 - 0.02i$ **9.** 81 **10.** $-i$ **11.** -2
12. $-0.6358 - 0.4227i$ **13.** $-i$

Exercise Set 6.8, p. 511

1. $i\sqrt{35}$, or $\sqrt{35}\,i$ **3.** $4i$ **5.** $-2i\sqrt{3}$, or $-2\sqrt{3}\,i$
7. $i\sqrt{3}$, or $\sqrt{3}\,i$ **9.** $9i$ **11.** $7i\sqrt{2}$, or $7\sqrt{2}\,i$
13. $-7i$ **15.** $4 - 2\sqrt{15}\,i$ **17.** $(2 + 2\sqrt{3})i$
19. $12 - 4i$ **21.** $9 - 5i$ **23.** $7 + 4i$ **25.** $-4 - 4i$
27. $-1 + i$ **29.** $11 + 6i$ **31.** -18 **33.** $-\sqrt{14}$
35. 21 **37.** $-6 + 24i$ **39.** $1 + 5i$ **41.** $18 + 14i$
43. $38 + 9i$ **45.** $2 - 46i$ **47.** $5 - 12i$
49. $-24 + 10i$ **51.** $-5 - 12i$ **53.** $-i$ **55.** 1
57. -1 **59.** i **61.** -1 **63.** $-125i$ **65.** 8
67. $1 - 23i$ **69.** 0 **71.** 0 **73.** 1 **75.** $5 - 8i$
77. $2 - \dfrac{\sqrt{6}}{2}i$ **79.** $\dfrac{9}{10} + \dfrac{13}{10}i$ **81.** $-i$ **83.** $-\dfrac{3}{7} - \dfrac{8}{7}i$
85. $\dfrac{6}{5} - \dfrac{2}{5}i$ **87.** $-\dfrac{8}{41} + \dfrac{10}{41}i$ **89.** $-\dfrac{4}{3}i$ **91.** $-\dfrac{1}{2} - \dfrac{1}{4}i$
93. $\dfrac{3}{5} + \dfrac{4}{5}i$ **95.** Yes **97.** No **99.** 7 **100.** $\dfrac{70}{29}$
101. $-\dfrac{29}{3}$, 5 **102.** $\left\{x \mid -\dfrac{29}{3} < x < 5\right\}$, or $\left(-\dfrac{29}{3}, 5\right)$
103. $\left\{x \mid x \le -\dfrac{29}{3} \text{ or } x \ge 5\right\}$, or $\left(-\infty, -\dfrac{29}{3}\right] \cup [5, \infty)$
104. $-\dfrac{2}{5}$, -12 **105.** ◆ **107.** $-4 - 8i$; $-2 + 4i$;
$8 - 6i$ **109.** $-3 - 4i$ **111.** $-88i$ **113.** 8
115. $\dfrac{3}{5} + \dfrac{9}{5}i$ **117.** 1

Summary and Review: Chapter 6, p. 515

1. 27.893 **2.** 6.378 **3.** $f(0)$, $f(-1)$, and $f(1)$ do not
exist as real numbers; $f\left(\dfrac{41}{3}\right) = 5$
4. Domain $= \left\{x \mid x \ge \dfrac{16}{3}\right\}$, or $\left[\dfrac{16}{3}, \infty\right)$ **5.** $9|a|$ **6.** $7|z|$
7. $|c - 3|$ **8.** $|x - 3|$ **9.** -10 **10.** $-\dfrac{1}{3}$ **11.** 2; -2; 3
12. $|x|$ **13.** 3 **14.** $\sqrt[5]{a}$ **15.** 512 **16.** $31^{1/2}$

17. $(a^2b^3)^{1/5}$ **18.** $\frac{1}{7}$ **19.** $\frac{1}{4x^{2/3}y^{2/3}}$ **20.** $\frac{5b^{1/2}}{a^{3/4}c^{2/3}}$

21. $\frac{3a}{t^{1/4}}$ **22.** $\frac{1}{x^{2/5}}$ **23.** $7^{1/6}$ **24.** x^7 **25.** $3x^2$

26. $\sqrt[12]{x^4y^3}$ **27.** $\sqrt[12]{x^7}$ **28.** $7\sqrt{5}$ **29.** $-3\sqrt[3]{4}$

30. $5b^2\sqrt[3]{2a^2}$ **31.** $\frac{7}{6}$ **32.** $\frac{4}{3}x^2$ **33.** $\frac{2x^2}{3y^3}$ **34.** $\sqrt{15xy}$

35. $3a\sqrt[3]{a^2b^2}$ **36.** $\sqrt[15]{a^5b^9}$ **37.** $y\sqrt[3]{6}$ **38.** $\frac{5}{2}\sqrt{x}$

39. $\sqrt[12]{x^5}$ **40.** $7\sqrt[3]{x}$ **41.** $3\sqrt{3}$ **42.** $(2x + y^2)\sqrt[3]{x}$

43. $15\sqrt{2}$ **44.** $-43 - 2\sqrt{10}$ **45.** $8 - 2\sqrt{7}$

46. $9 - \sqrt[3]{4}$ **47.** $\frac{2\sqrt{6}}{3}$ **48.** $\frac{2\sqrt{a} - 2\sqrt{b}}{a - b}$ **49.** 13

50. 4   **51.** 9 cm   **52.** $\sqrt{24}$; 4.899 ft   **53.** 25

54. $\sqrt{46}$; 6.782   **55.** $(5 + 2\sqrt{2})i$ **56.** $-2 - 9i$

57. $1 + i$ **58.** 29   **59.** i **60.** $9 - 12i$ **61.** $\frac{2}{5} + \frac{3}{5}i$

62. 3   **63.** No   **64.**

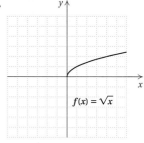

65. $\frac{23}{7}$ **66.** $-\frac{4}{3}, 3$ **67.** $\frac{x(x + 2)}{(x + y)(x - 3)}$

68. Motorcycle: 62 mph; SUV: 70 mph

69. ◆ The procedure for solving radical equations is to isolate one of the radical terms, use the principle of powers, repeat these steps if necessary until all radicals are eliminated, and then check the possible solutions. A check is necessary since the principle of powers does not always yield equivalent equations.   **70.** -1 **71.** 3

Test: Chapter 6, p. 517

1. [6.1a] 12.166   **2.** [6.1a] 2; does not exist as a real number   **3.** [6.1a] Domain = $\{x \mid x \le 2\}$, or $(-\infty, 2]$

4. [6.1b] $3|q|$ **5.** [6.1b] $|x + 5|$ **6.** [6.1c] $-\frac{1}{10}$

7. [6.1d] x **8.** [6.1d] 4   **9.** [6.2a] $\sqrt[3]{a^2}$ **10.** [6.2a] 8

11. [6.2a] $37^{1/2}$ **12.** [6.2a] $(5xy^2)^{5/2}$ **13.** [6.2b] $\frac{1}{10}$

14. [6.2b] $\frac{8a^{3/4}}{b^{3/2}c^{2/5}}$ **15.** [6.2c] $\frac{x^{8/5}}{y^{9/5}}$ **16.** [6.2c] $\frac{1}{2.9^{31/24}}$

17. [6.2d] $\sqrt[4]{x}$ **18.** [6.2d] $2x\sqrt{x}$ **19.** [6.2d] $\sqrt[15]{a^6b^5}$

20. [6.2d] $\sqrt[12]{8y^7}$ **21.** [6.3a] $2\sqrt{37}$ **22.** [6.3a] $2\sqrt[4]{5}$

23. [6.3a] $4a\sqrt[3]{4ab^2}$ **24.** [6.3b] $-\frac{2}{x^2}$ **25.** [6.3b] $\frac{5x}{6y^2}$

26. [6.3a] $\sqrt[3]{10xy^2}$ **27.** [6.3a] $xy\sqrt[4]{x}$ **28.** [6.3b] $\sqrt[5]{x^2y^2}$

29. [6.3b] $2\sqrt{a}$ **30.** [6.4a] $38\sqrt{2}$ **31.** [6.4b] -20

32. [6.4b] $9 + 6\sqrt{x} + x$ **33.** [6.5b] $\frac{13 + 8\sqrt{2}}{-41}$

34. [6.6a] 35   **35.** [6.6b] 7   **36.** [6.7a] 7 ft

37. [6.7a] $\sqrt{98}$; 9.899   **38.** [6.7a] 2   **39.** [6.8a] $11i$

40. [6.8b] $7 + 5i$ **41.** [6.8c] $37 + 9i$ **42.** [6.8d] $-i$

43. [6.8e] $-\frac{77}{50} + \frac{7}{25}i$ **44.** [6.8f] No

45. [6.1a]

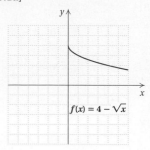

$f(x) = 4 - \sqrt{x}$

46. [5.5a] No solution   **47.** [4.7a] $-\frac{1}{3}, \frac{5}{2}$

48. [5.1e] $\frac{x - 3}{x - 4}$ **49.** [5.6a] Fran: 36 hr; Juan: 45 hr

50. [6.8c, e] $-\frac{17}{4}i$ **51.** [6.6b] 3

Cumulative Review: Chapters R–6, p. 519

1. [R.2d] $\frac{21}{50}$ **2.** [R.2c] -37.2 **3.** [R.3c] $25c - 31$

4. [R.3c] 174   **5.** [R.7b] $\frac{64x^{21}}{27y^{15}}$

6. [4.1c] $-3x^3 + 9x^2 + 3x - 3$

7. [4.2c] $4x^4 - 4x^2y + y^2$

8. [4.2a] $15x^4 - x^3 - 9x^2 + 5x - 2$

9. [5.1d] $\frac{(x + 4)(x - 7)}{x + 7}$ **10.** [5.2c] $\frac{-2x + 4}{(x + 2)(x - 3)}$

11. [5.4a] $\frac{y - 6}{y - 9}$ **12.** [5.3b] $y^2 + y - 2 + \frac{-1}{y + 2}$

13. [6.1c] $-2x$ **14.** [6.3a] $4|x - 1|$ **15.** [6.4a] $57\sqrt{3}$

16. [6.3a] $4xy^2\sqrt{y}$ **17.** [6.5b] $\sqrt{30} + \sqrt{15}$

18. [6.1d] $\frac{m^2n^4}{2}$ **19.** [6.2c] $6^{8/9}$ **20.** [6.8b] $3 + 5i$

21. [6.8e] $\frac{7}{61} - \frac{16}{61}i$ **22.** [1.1d] 2

23. [1.2a] $c = 8M + 3$ **24.** [1.4c] $\{a \mid a > -7\}$, or $(-7, \infty)$ **25.** [1.5a] $\{x \mid -10 < x < 13\}$, or $(-10, 13)$

26. [1.6c] $\frac{4}{3}, \frac{8}{3}$ **27.** [3.2b] $(5, 3)$ **28.** [3.4a] $(-1, 0, 4)$

29. [4.7a] $\frac{25}{7}, -\frac{25}{7}$ **30.** [5.5a] -5 **31.** [5.5a] $\frac{1}{3}$

32. [5.7a] $R = \frac{nE - nrI}{I}$ **33.** [6.6a] 6   **34.** [6.6b] $-\frac{1}{4}$

35. [2.2c]   **36.** [2.5a]

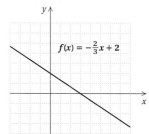

$f(x) = -\frac{2}{3}x + 2$

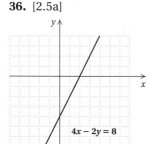

$4x - 2y = 8$

37. [3.6b]

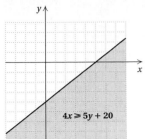

38. [3.6c]

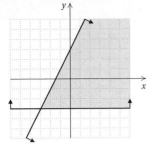

39. [2.2c]

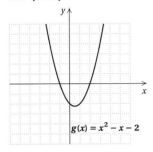

$g(x) = x^2 - x - 2$

40. [2.2c]

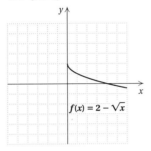

$f(x) = |x + 4|$

41. [2.2c]

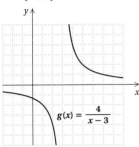

$g(x) = \dfrac{4}{x - 3}$

42. [2.2c]

$f(x) = 2 - \sqrt{x}$

43. [2.3a] Domain = {−5, −4, −2, 0, 1, 2, 3, 4, 5};
range = {−3, −2, 3, 4} **44.** [2.3a] Domain = [−5, 5];
range = [−3, 4] **45.** [2.3a] Domain = (−∞, ∞);
range = [−5, ∞) **46.** [2.3a] Domain = [−5, 5];
range = [0, 3] **47.** [4.3a] $6xy^2(2x - 5y)$
48. [4.4a, b] $(3x + 4)(x - 7)$
49. [4.3c] $(y + 11)(y - 12)$
50. [4.5d] $(3y + 2)(9y^2 - 6y + 4)$
51. [4.5b] $(2x + 25)(2x - 25)$ **52.** [3.3a] Finland:
$7.99; Sweden: $6.45 **53.** [2.4b] $m = \frac{3}{2}$;
y-intercept = (0, −4) **54.** [2.6d] $y = -\frac{1}{3}x + \frac{13}{3}$
55. [3.2c] Pittsburgh: 457; Virginia: 212
56. [5.6a] 1 hr **57.** [5.8d] $h = 125$ ft, $A = 1000$ ft^2
58. [6.7a] $d = \sqrt{45} \approx 6.708$ ft
59. [1.6e], [3.6c]

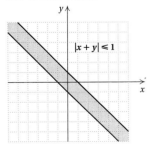

$|x + y| \le 1$

60. [6.6b] $-\frac{8}{9}$ **61.** [4.3c] $\left(x - \frac{5}{2}\right)\left(x - \frac{1}{2}\right)$

Chapter 7

Pretest: Chapter 7, p. 522

1. [7.1b], [7.2a] 0, −3 **2.** [7.1b], [7.2a] −2 + 2i, −2 − 2i
3. [7.1a, b], [7.2a] 5 **4.** [7.4c] $\dfrac{\sqrt{3}}{3}, -\dfrac{\sqrt{3}}{3}, \sqrt{2}, -\sqrt{2}$
5. [7.2a] $-1, \dfrac{5}{6}$ **6.** [7.4c] 1 **7.** [7.2a] $\dfrac{-2 \pm \sqrt{6}}{2}$;

0.225, −2.225 **8.** [7.3b] $T = \dfrac{1}{RW^2}$
9. [7.4b] $3x^2 + 4x - 4 = 0$
10. [7.6a] **(a)** (3, −2); **(b)** $x = 3$;
(c)

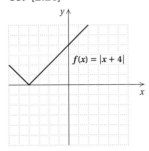

$x = 3$

Vertex: (3, −2)

$f(x) = 2x^2 - 12x + 16$

11. [7.6b] $(3 + \sqrt{5}, 0), (3 - \sqrt{5}, 0)$
12. [7.7b] $f(x) = 4x^2 - 5x + 1$ **13.** [7.3a] 12, 14, 16
14. [7.3a] 40 mph; 48 mph **15.** [7.7a] 5625 ft^2
16. [7.8a] {$x|x < -5$ *or* $0 < x < 3$}, or (−∞, −5) ∪ (0, 3)
17. [7.8b] {$x|x < -3$ *or* $x \ge 2$}, or (−∞, −3) ∪ [2, 0)

Margin Exercises, Section 7.1, pp. 523–532

1. (a) (2, 0), (4, 0); **(b)** 2, 4; **(c)** The solutions of
$x^2 - 6x + 8 = 0$, 2 and 4, are the first coordinates of
the x-intercepts, (2, 0) and (4, 0), of the graph of
$f(x) = x^2 - 6x + 8$. **2. (a)** $\frac{3}{5}$, 1; **(b)** $\left(\frac{3}{5}, 0\right)$, (1, 0)
3. 4 and −4, or +4 **4.** 0, $-\frac{7}{2}$ **5.** $\sqrt{3}$ and $-\sqrt{3}$, or
$\pm\sqrt{3}$; 1.732 and −1.732, or ±1.732 **6.** $\dfrac{2\sqrt{6}}{3}$ and $-\dfrac{2\sqrt{6}}{3}$,
or $\pm\dfrac{2\sqrt{6}}{3}$; 1.633 and −1.633, or ±1.633 **7.** $\dfrac{\sqrt{2}}{2}i$ and
$-\dfrac{\sqrt{2}}{2}i$, or $\pm\dfrac{\sqrt{2}}{2}i$ **8. (a)** $1 \pm \sqrt{5}$; **(b)** $(1 + \sqrt{5}, 0)$,
$(1 - \sqrt{5}, 0)$ **9.** $-8 \pm \sqrt{11}$ **10.** −2, −4 **11.** 10, −2
12. $-3 \pm \sqrt{10}$ **13.** $\dfrac{1 \pm \sqrt{22}}{3}$ **14.** $3603.65
15. 20% **16.** About 36.0 in. **17.** 1 sec

Calculator Spotlight, p. 526

1. 1.633, −1.633 **2.** 0.650, −0.650

The graphs of $y_1 = 4x^2 + 9$ and $y_2 = 0$ in Example 6 do not intersect. The graphs of $y_1 = 2x^2 + 1$ and $y_2 = 0$ in Margin Exercise 7 do not intersect. The calculator indicates an error since there are no real-number solutions.

Exercise Set 7.1, p. 533

1. (a) $\sqrt{5}$, $-\sqrt{5}$, or $\pm\sqrt{5}$; (b) $(\sqrt{5}, 0)$, $(-\sqrt{5}, 0)$
3. (a) $\frac{5}{3}i$, $-\frac{5}{3}i$, or $\pm\frac{5}{3}i$; (b) no x-intercepts
5. $\pm\dfrac{\sqrt{6}}{2}$; ± 1.225 **7.** $5, -9$ **9.** $8, 0$
11. $11 \pm \sqrt{7}$; $13.646, 8.354$ **13.** $7 \pm 2i$ **15.** $18, 0$
17. $\dfrac{3 \pm \sqrt{14}}{2}$; $3.371, -0.371$ **19.** $5, -11$ **21.** $9, 5$
23. $-2 \pm \sqrt{6}$ **25.** $11 \pm 2\sqrt{33}$ **27.** $\dfrac{-1 \pm \sqrt{5}}{2}$
29. $\dfrac{5 \pm \sqrt{53}}{2}$ **31.** $\dfrac{-3 \pm \sqrt{57}}{4}$ **33.** $\dfrac{9 \pm \sqrt{105}}{4}$
35. $2, -8$ **37.** $-11 \pm \sqrt{19}$ **39.** $5 \pm \sqrt{29}$
41. (a) $\dfrac{-7 \pm \sqrt{57}}{2}$; (b) $\left(\dfrac{-7 + \sqrt{57}}{2}, 0\right)$, $\left(\dfrac{-7 - \sqrt{57}}{2}, 0\right)$
43. (a) $\dfrac{5}{4} \pm i\dfrac{\sqrt{39}}{4}$; (b) no x-intercepts **45.** $\dfrac{3 \pm \sqrt{17}}{4}$
47. $\dfrac{3 \pm \sqrt{145}}{4}$ **49.** $\dfrac{2 \pm \sqrt{7}}{3}$ **51.** $-\dfrac{1}{2} \pm i\dfrac{\sqrt{7}}{2}$
53. $2 \pm 3i$ **55.** 10% **57.** 12.5% **59.** 4%
61. 0.866 sec **63.** About 7.3 sec **65.** About 6.8 sec
67. (a) $f(x) = -1.175x + 12$;
(b)

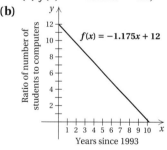

(c) 6.1; 3.8
68. (a) $f(x) = 0.675x + 3.6$;
(b)

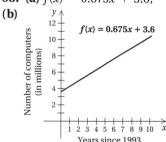

(c) 7.0 million; 8.3 million

69.

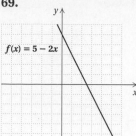

$f(x) = 5 - 2x$

70.

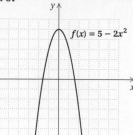

$f(x) = 5 - 2x^2$

71.

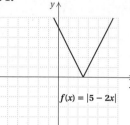

$f(x) = |5 - 2x|$

72.

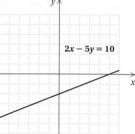

$2x - 5y = 10$

73. $2\sqrt{22}$ **74.** $\dfrac{\sqrt{10}}{5}$ **75.** ◈ **77.** Left to the student. **79.** $16, -16$ **81.** $0, \frac{7}{2}, -8, -\frac{10}{3}$
83. Fishing boat: 15 km/h; barge: 8 km/h

Margin Exercises, Section 7.2, pp. 538–540

1. (a) $-\frac{1}{2}, 4$; (b) $-\frac{1}{2}, 4$ **2.** $\dfrac{-1 \pm \sqrt{22}}{3}$; $1.230, -1.897$
3. $\dfrac{1 \pm i\sqrt{7}}{2}$, or $\dfrac{1}{2} \pm i\dfrac{\sqrt{7}}{2}$ **4.** $\dfrac{5 \pm \sqrt{73}}{6}$; $2.257, -0.591$

Calculator Spotlight, p. 539

1. $-1.897, 1.230$ **2.** $-0.309, 1.946$

Exercise Set 7.2, p. 541

1. $-3 \pm \sqrt{5}$ **3.** $\dfrac{-4 \pm \sqrt{13}}{3}$ **5.** $\dfrac{1}{2} \pm i\dfrac{\sqrt{3}}{2}$ **7.** $2 \pm 3i$
9. $\dfrac{-3 \pm \sqrt{41}}{2}$ **11.** $-1 \pm 2i$ **13.** (a) $0, -1$;
(b) $(0, 0)$, $(-1, 0)$ **15.** (a) $\dfrac{3 \pm \sqrt{229}}{22}$;
(b) $\left(\dfrac{3 + \sqrt{229}}{22}, 0\right)$, $\left(\dfrac{3 - \sqrt{229}}{22}, 0\right)$ **17.** (a) $\frac{2}{5}$;
(b) $\left(\frac{2}{5}, 0\right)$ **19.** $-1, -2$ **21.** $5, 10$ **23.** $\dfrac{17 \pm \sqrt{249}}{10}$
25. $2 \pm i$ **27.** $\frac{2}{3}, \frac{3}{2}$ **29.** $2 \pm \sqrt{10}$ **31.** $\frac{3}{4}, -2$
33. $\dfrac{1}{2} \pm \dfrac{3}{2}i$ **35.** $1, -\dfrac{1}{2} \pm i\dfrac{\sqrt{3}}{2}$ **37.** $-3 \pm \sqrt{5}$; -0.764,
-5.236 **39.** $3 \pm \sqrt{5}$; $5.236, 0.764$ **41.** $\dfrac{3 \pm \sqrt{65}}{4}$; 2.766,
-1.266 **43.** $\dfrac{4 \pm \sqrt{31}}{5}$; $1.914, -0.314$ **45.** 2 **46.** 3
47. 10 **48.** 8 **49.** No solution **50.** No solution

51. $\frac{15}{4}$ **52.** $\frac{17}{3}$ **53.** ◈   **55.** $-0.797, 0.570$
57. $\dfrac{1 \pm \sqrt{1 + 8\sqrt{5}}}{4}$ **59.** $\dfrac{-i \pm i\sqrt{1 + 4i}}{2}$
61. $\dfrac{-1 \pm 3\sqrt{5}}{6}$ **63.** $3 \pm \sqrt{13}$

Margin Exercises, Section 7.3, pp. 543–548

1. 2 ft   **2.** A to B is 6 mi; A to C is 8 mi   **3.** A to B is $\sqrt{31} - 1$, or 4.57 mi; A to C is $1 + \sqrt{31}$, or 6.57 mi

4.

	Distance	Speed	Time
Faster Ship	3000	$r + 10$	$t - 50$
Slower Ship	3000	r	t

20 knots, 30 knots

5. $w_2 = \dfrac{w_1}{A^2}$ **6.** $r = \sqrt{\dfrac{V}{\pi h}}$ **7.** $t = \dfrac{-g + \sqrt{g^2 + 64s}}{32}$
8. $b = \dfrac{pa}{\sqrt{1 + p^2}}$

Exercise Set 7.3, p. 549

1. Length: 9 ft; width: 2 ft   **3.** Length: 18 yd; width: 9 yd   **5.** 3 cm   **7.** 6 ft, 8 ft
9. 28 and 29   **11.** Length: $2 + \sqrt{14} \approx 5.742$ ft; width: $\sqrt{14} - 2 \approx 1.742$ ft   **13.** $\dfrac{17 - \sqrt{109}}{2} \approx 3.280$ in.
15. $7 + \sqrt{239} \approx 22.460$ ft; $\sqrt{239} - 7 \approx 8.460$ ft
17. First part: 60 mph; second part: 50 mph
19. 40 mph   **21.** Cessna: 150 mph, Beechcraft: 200 mph; or Cessna: 200 mph, Beechcraft: 250 mph
23. Hillsboro: 10 mph; return trip: 4 mph
25. About 11 mph   **27.** $s = \sqrt{\dfrac{A}{6}}$ **29.** $r = \sqrt{\dfrac{Gm_1 m_2}{F}}$
31. $c = \sqrt{\dfrac{E}{m}}$ **33.** $b = \sqrt{c^2 - a^2}$
35. $k = \dfrac{3 + \sqrt{9 + 8N}}{2}$ **37.** $r = \dfrac{-\pi h + \sqrt{\pi^2 h^2 + 2\pi A}}{2\pi}$
39. $g = \dfrac{4\pi^2 L}{T^2}$ **41.** $H = \sqrt{\dfrac{700W}{I}}$, or $10\sqrt{\dfrac{7W}{I}}$
43. $v = \dfrac{c\sqrt{m^2 - m_0^2}}{m}$ **45.** $\dfrac{1}{x - 2}$
46. $\dfrac{x^3 + x^2 + 2x + 2}{(x - 1)(x^2 + x + 1)}$ **47.** $\dfrac{-x}{(x + 3)(x - 1)}$
48. $3x^2\sqrt{x}$ **49.** $2i\sqrt{5}$ **50.** $\dfrac{3x + 3}{3x + 1}$ **51.** $\dfrac{4b}{a(3b^2 - 4a)}$
53. ◈   **55.** $\pm\sqrt{2}$ **57.** $A(S) = \dfrac{\pi S}{6}$
59. $l = \dfrac{w + w\sqrt{5}}{2}$

Margin Exercises, Section 7.4, pp. 554–557

1. Two real   **2.** One real   **3.** Two nonreal
4. $x^2 - 5x - 14 = 0$ **5.** $3x^2 + 7x - 20 = 0$
6. $x^2 + 25 = 0$ **7.** $x^2 + \sqrt{2}\,x - 4 = 0$
8. $\pm 3, \pm 1$ **9.** 4   **10.** $(-3, 0), (-1, 0), (2, 0), (4, 0)$
11. $-\frac{1}{3}, \frac{1}{2}$

Exercise Set 7.4, p. 559

1. One real   **3.** Two nonreal   **5.** Two real   **7.** One real   **9.** Two nonreal   **11.** Two real   **13.** Two real
15. One real   **17.** $x^2 - 16 = 0$ **19.** $x^2 + 9x + 14 = 0$
21. $x^2 - 16x + 64 = 0$ **23.** $25x^2 - 20x - 12 = 0$
25. $12x^2 - (4k + 3m)x + km = 0$
27. $x^2 - \sqrt{3}x - 6 = 0$ **29.** $\pm\sqrt{3}$ **31.** 1, 81
33. $-1, 1, 5, 7$ **35.** $-\frac{1}{4}, \frac{1}{9}$ **37.** 1   **39.** $-1, 1, 4, 6$
41. $\pm\sqrt{3 + \sqrt{13}}, \pm\sqrt{3 - \sqrt{13}}$ **43.** $-1, 2$
45. $\pm\dfrac{\sqrt{15}}{3}, \pm\dfrac{\sqrt{6}}{2}$ **47.** $-1, 125$ **49.** $-\frac{11}{6}, -\frac{1}{6}$
51. $-\frac{3}{2}$ **53.** $\dfrac{9 \pm \sqrt{89}}{2}, -1 \pm \sqrt{3}$ **55.** $\left(\frac{4}{25}, 0\right)$
57. $(4, 0), (-1, 0), \left(\dfrac{3 + \sqrt{33}}{2}, 0\right), \left(\dfrac{3 - \sqrt{33}}{2}, 0\right)$
59. Kenyan: 30 lb; Peruvian: 20 lb   **60.** 4 L of A; 8 L of B   **61.** $4x$ **62.** $3x^2$ **63.** $3a^4\sqrt{2a}$ **64.** 4
65.

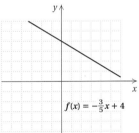

$f(x) = -\frac{3}{5}x + 4$

66.

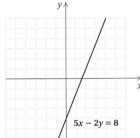

$5x - 2y = 8$

67.

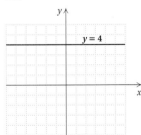

$y = 4$

68.

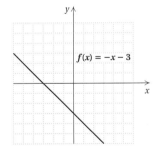

$f(x) = -x - 3$

69. ◈   **71.** Left to the student.   **73.** (a) $-\frac{3}{5}$; (b) $-\frac{1}{3}$
75. $x^2 - \sqrt{3}x + 8 = 0$ **77.** $a = 1, b = 2, c = -3$
79. $\frac{100}{99}$ **81.** 259   **83.** 1, 3

Margin Exercises, Section 7.5, pp. 563–568

1.

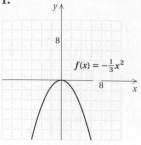

$f(x) = -\frac{1}{3}x^2$

2.

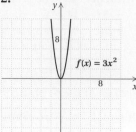

$f(x) = 3x^2$

3.

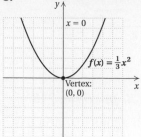

$x = 0$

$f(x) = \frac{1}{3}x^2$

Vertex: (0, 0)

5.

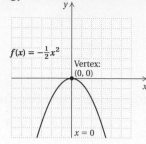

$f(x) = -\frac{1}{2}x^2$

Vertex: (0, 0)

$x = 0$

3.

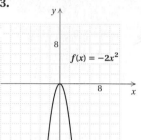

$f(x) = -2x^2$

4.

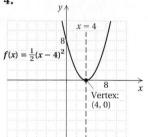

$x = 4$

$f(x) = \frac{1}{2}(x - 4)^2$

Vertex: (4, 0)

7.

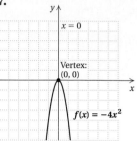

$x = 0$

Vertex: (0, 0)

$f(x) = -4x^2$

9.

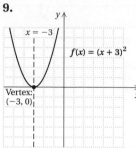

$x = -3$

$f(x) = (x + 3)^2$

Vertex: (−3, 0)

5.

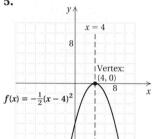

$x = 4$

Vertex: (4, 0)

$f(x) = -\frac{1}{2}(x - 4)^2$

6.

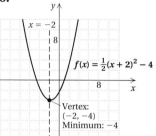

$x = -2$

$f(x) = \frac{1}{2}(x + 2)^2 - 4$

Vertex: (−2, −4)
Minimum: −4

11.

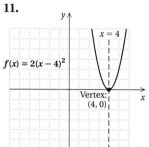

$x = 4$

$f(x) = 2(x - 4)^2$

Vertex: (4, 0)

13.

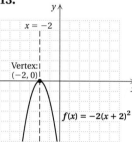

$x = -2$

Vertex: (−2, 0)

$f(x) = -2(x + 2)^2$

7.

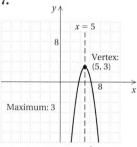

$x = 5$

Vertex: (5, 3)

Maximum: 3

$f(x) = -2(x - 5)^2 + 3$

15.

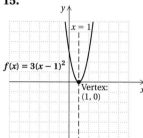

$x = 1$

$f(x) = 3(x - 1)^2$

Vertex: (1, 0)

17.

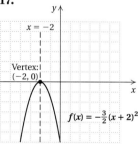

$x = -2$

Vertex: (−2, 0)

$f(x) = -\frac{3}{2}(x + 2)^2$

Exercise Set 7.5, p. 569

1.

x	f(x)
0	0
1	4
−1	4
2	16
−2	16

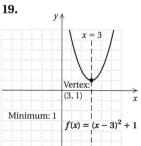

$f(x) = 4x^2$

Vertex: (0, 0)

$x = 0$

19.

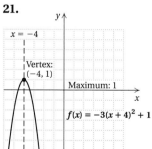

$x = 3$

Vertex: (3, 1)

Minimum: 1

$f(x) = (x - 3)^2 + 1$

21.

$x = -4$

Vertex: (−4, 1)

Maximum: 1

$f(x) = -3(x + 4)^2 + 1$

23.

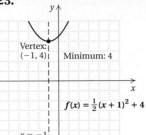

Vertex: $(-1, 4)$ Minimum: 4

$f(x) = \frac{1}{2}(x + 1)^2 + 4$

$x = -1$

25.

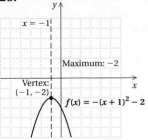

$x = -1$ Maximum: -2

Vertex: $(-1, -2)$ $f(x) = -(x + 1)^2 - 2$

27. 9   **28.** $\frac{11}{3}$ **29.** No solution   **30.** 15

31. ◈

33. Left to the student.

Margin Exercises, Section 7.6, pp. 571–574

1.

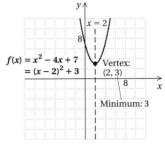

$f(x) = x^2 - 4x + 7$
$= (x - 2)^2 + 3$ $x = 2$

Vertex: $(2, 3)$

Minimum: 3

2.

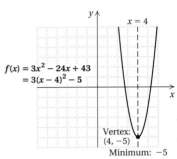

$f(x) = 3x^2 - 24x + 43$
$= 3(x - 4)^2 - 5$ $x = 4$

Vertex: $(4, -5)$ Minimum: -5

3.

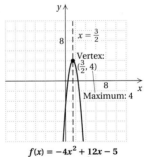

$x = \frac{3}{2}$ Vertex: $\left(\frac{3}{2}, 4\right)$ Maximum: 4

$f(x) = -4x^2 + 12x - 5$
$= -4\left(x - \frac{3}{2}\right)^2 + 4$

4. $(3, -5)$ **5.** $(4, -5)$ **6.** $\left(\frac{3}{2}, 4\right)$

7. y-intercept: $(0, -3)$; x-intercepts: $(-3, 0)$, $(1, 0)$

8. y-intercept: $(0, 16)$; x-intercept: $(-4, 0)$

9. y-intercept: $(0, 1)$; x-intercepts: $(2 + \sqrt{3}, 0)$, $(2 - \sqrt{3}, 0)$, or $(3.732, 0)$, $(0.268, 0)$

Exercise Set 7.6, p. 575

1.

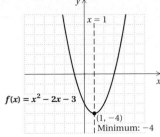

$x = 1$

$f(x) = x^2 - 2x - 3$

$(1, -4)$ Minimum: -4

3.

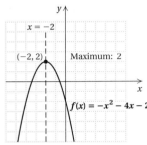

$x = -2$

$(-2, 2)$ Maximum: 2

$f(x) = -x^2 - 4x - 2$

5.

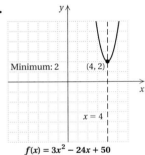

Minimum: 2 $(4, 2)$

$x = 4$

$f(x) = 3x^2 - 24x + 50$

7.

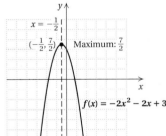

$x = -\frac{1}{2}$

$\left(-\frac{1}{2}, \frac{7}{2}\right)$ Maximum: $\frac{7}{2}$

$f(x) = -2x^2 - 2x + 3$

9.

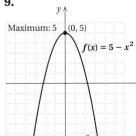

Maximum: 5 $(0, 5)$

$f(x) = 5 - x^2$

$x = 0$

11.

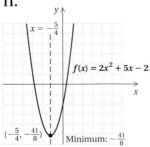

$x = -\frac{5}{4}$

$f(x) = 2x^2 + 5x - 2$

$\left(-\frac{5}{4}, -\frac{41}{8}\right)$ Minimum: $-\frac{41}{8}$

13. y-intercept: (0, 1); x-intercepts: $(3 + 2\sqrt{2}, 0)$, $(3 - 2\sqrt{2}, 0)$ **15.** y-intercept: (0, 20); x-intercepts: (5, 0), (−4, 0) **17.** y-intercept: (0, 9); x-intercept: $\left(-\frac{3}{2}, 0\right)$ **19.** y-intercept: (0, 8); x-intercepts: none **21.** (a) $D = 15w$; (b) 630 mg **22.** (a) $C = \frac{89}{6}t$; (b) 712 calories **23.** 250; $y = \frac{250}{x}$ **24.** 250; $y = \frac{250}{x}$ **25.** $\frac{125}{2}$; $y = \frac{125}{2}x$ **26.** $\frac{2}{125}$; $y = \frac{2}{125}x$ **27.** ◈ **29.** (a) Minimum: −6.954; (b) maximum: 7.014

31.

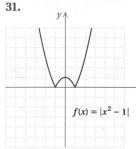

$f(x) = |x^2 - 1|$

33.

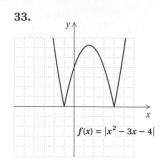

$f(x) = |x^2 - 3x - 4|$

35. $f(x) = \frac{5}{16}x^2 - \frac{15}{8}x - \frac{35}{16}$ **37.** (a) Vertex: $\left(-1, -\frac{1}{2}\right)$; (b) line of symmetry: $x = -1$; (c) minimum: $-\frac{1}{2}$; (d)

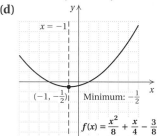

$x = -1$

$\left(-1, -\frac{1}{2}\right)$ Minimum: $-\frac{1}{2}$

$f(x) = \frac{x^2}{8} + \frac{x}{4} - \frac{3}{8}$

39. (0, 2), (5, 7)

Margin Exercises, Section 7.7, pp. 578–582

1. 25 yd by 25 yd **2.** $f(x) = ax^2 + bx + c, a > 0$ **3.** $f(x) = mx + b$ **4.** Polynomial, neither quadratic nor linear **5.** $f(x) = ax^2 + bx + c, a > 0$ **6.** (a)

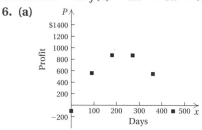

(b) yes; (c) $f(x) = -0.02x^2 + 9x - 100$; (d) $912.50

Calculator Spotlight, p. 583

1. (a)

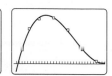

The quartic function is better than the quadratic, but it seems to fit equally as well as the cubic. (b) For the cubic function, domain values less than 15 give negative range values (negative number of live births). As domain values increase beyond 45, range values increase indefinitely (continual increase of live births). Neither of these facts is true for those ages.

2. (a), (b)

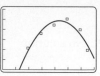

(c), (d)

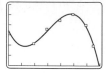

(e), (f)

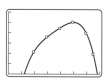

(g) quartic; (h) $30,675; $48,581
3. Left to the student.

Exercise Set 7.7, p. 585

1. 180 ft by 180 ft **3.** 121; 11 and 11 **5.** −4; 2 and −2 **7.** 36; −6 and −6 **9.** 200 ft²; 10 ft by 20 ft **11.** 3.5 in. **13.** 3.5 hundred, or 350 **15.** $P(x) = -x^2 + 980x - 3000$; $237,100 at $x = 490$ **17.** $f(x) = mx + b$ **19.** $f(x) = ax^2 + bx + c, a > 0$ **21.** Polynomial, neither quadratic nor linear **23.** Polynomial, neither quadratic nor linear **25.** $f(x) = 2x^2 + 3x - 1$ **27.** $f(x) = -\frac{1}{4}x^2 + 3x - 5$ **29.** (a) $A(s) = \frac{3}{16}s^2 - \frac{135}{4}s + 1750$; (b) about 531 per 200,000,000 km driven **31.** (a) $f(x) = 65.6x^2 - 143.1x + 131.5$; (b) $1056 million, $1634.5 million **33.** $5xy^2\sqrt[4]{x}$ **34.** $12a^2b^2$ **35.** 5 **36.** 4 **37.** No solution **38.** 4 **39.** ◈ **41.** $f(x) = -0.087x^4 + 1.837x^3 - 12.036x^2 + 23.694x + 41.232$

Margin Exercises, Section 7.8, pp. 589–594

1. $\{x \mid x < -3 \text{ or } x > 1\}$, or $(-\infty, -3) \cup (1, \infty)$
2. $\{x \mid -3 < x < 1\}$, or $(-3, 1)$ **3.** $\{x \mid -3 \le x \le 1\}$, or $[-3, 1]$ **4.** $\{x \mid x < -4 \text{ or } x > 1\}$, or $(-\infty, -4) \cup (1, \infty)$
5. $\{x \mid -4 \le x \le 1\}$, or $[-4, 1]$
6. $\{x \mid x < -1 \text{ or } 0 < x \le 1\}$, or $(-\infty, -1) \cup (0, 1)$
7. $\{x \mid 2 < x \le \frac{7}{2}\}$, or $\left(2, \frac{7}{2}\right]$
8. $\{x \mid x < 5 \text{ or } x > 10\}$, or $(-\infty, 5) \cup (10, \infty)$

1. $\{x \mid x < -5 \ or \ x > 2\}$, or $(-\infty, -5) \cup (2, \infty)$
2. $\{x \mid x \le -5 \ or \ x \ge 2\}$, or $(-\infty, -5] \cup [2, \infty)$
3. $\{x \mid -5 \le x \le 2\}$, or $[-5, 2]$ **4.** $\{x \mid -5 < x < 2\}$, or $(-5, 2)$ **5.** $\{x \mid -0.78 \le x \le 1.59\}$, or $[-0.78, 1.59]$
6. $\{x \mid x \le -0.21 \ or \ x \ge 2.47\}$, or $(-\infty, -0.21] \cup [2.47, \infty)$ **7.** $\{x \mid x < -1.26 \ or \ x > 2.33\}$, or $(-\infty, -1.26) \cup (2.33, \infty)$ **8.** $\{x \mid x > -1.37\}$, or $(-1.37, \infty)$

Exercise Set 7.8, p. 595

1. $\{x \mid x < -2 \ or \ x > 6\}$, or $(-\infty, -2) \cup (6, \infty)$
3. $\{x \mid -2 \le x \le 2\}$, or $[-2, 2]$ **5.** $\{x \mid -1 \le x \le 4\}$, or $[-1, 4]$ **7.** $\{x \mid -1 < x < 2\}$, or $(-1, 2)$ **9.** All real numbers, or $(-\infty, \infty)$ **11.** $\{x \mid 2 < x < 4\}$, or $(2, 4)$
13. $\{x \mid x < -2 \ or \ 0 < x < 2\}$, or $(-\infty, -2) \cup (0, 2)$
15. $\{x \mid -9 < x < -1 \ or \ x > 4\}$, or $(-9, -1) \cup (4, \infty)$
17. $\{x \mid x < -3 \ or \ -2 < x < 1\}$, or $(-\infty, -3) \cup (-2, 1)$
19. $\{x \mid x < 6\}$, or $(-\infty, 6)$ **21.** $\{x \mid x < -1 \ or \ x > 3\}$, or $(-\infty, -1) \cup (3, \infty)$ **23.** $\left\{x \mid -\frac{2}{3} \le x < 3\right\}$, or $\left[-\frac{2}{3}, 3\right)$
25. $\left\{x \mid 2 < x < \frac{5}{2}\right\}$, or $\left(2, \frac{5}{2}\right)$
27. $\{x \mid x < -1 \ or \ 2 < x < 5\}$, or $(-\infty, -1) \cup (2, 5)$
29. $\{x \mid -3 \le x < 0\}$, or $[-3, 0)$ **31.** $\{x \mid 1 < x < 2\}$, or $(1, 2)$ **33.** $\{x \mid x < -4 \ or \ 1 < x < 3\}$, or $(-\infty, -4) \cup (1, 3)$ **35.** $\left\{x \mid 0 < x \le \frac{1}{3}\right\}$, or $\left(0, \frac{1}{3}\right]$
37. $\{x \mid x < -3 \ or \ -2 < x < 1 \ or \ x > 4\}$, or
$(-\infty, -3) \cup (-2, 1) \cup (4, \infty)$ **39.** $\frac{5}{3}$ **40.** $\frac{5}{2a}$
41. $\frac{4a}{b^2}\sqrt{a}$ **42.** $\frac{3c}{7d}\sqrt[3]{c^2}$ **43.** $\sqrt{2}$ **44.** $17\sqrt{5}$
45. $(10a + 7)\sqrt[3]{2a}$ **46.** $3\sqrt{10} - 4\sqrt{5}$ **47.** ◆
49. Left to the student.
51. $\{x \mid 1 - \sqrt{3} \le x \le 1 + \sqrt{3}\}$, or $[1 - \sqrt{3}, 1 + \sqrt{3}]$
53. All real numbers except 0, or $(-\infty, 0) \cup (0, \infty)$
55. $\left\{x \mid x < \frac{1}{4} \ or \ x > \frac{5}{2}\right\}$, or $\left(-\infty, \frac{1}{4}\right) \cup \left(\frac{5}{2}, \infty\right)$
57. (a) $\{t \mid 0 < t < 2\}$, or $(0, 2)$;   **(b)** $\{t \mid t > 10\}$, or $(10, \infty)$

Summary and Review: Chapter 7, p. 597

1. (a) $\pm\frac{\sqrt{14}}{2}$; **(b)** $\left(\frac{\sqrt{14}}{2}, 0\right), \left(-\frac{\sqrt{14}}{2}, 0\right)$ **2.** $0, -\frac{5}{14}$
3. $3, 9$ **4.** $-\frac{3}{8} \pm i\frac{\sqrt{7}}{8}$ **5.** $\frac{7}{2} \pm i\frac{\sqrt{3}}{2}$ **6.** $3, 5$
7. $-2 \pm \sqrt{3}$; $-0.268, -3.732$ **8.** $4, -2$ **9.** $4 \pm 4\sqrt{2}$
10. $\frac{1 \pm \sqrt{481}}{15}$ **11.** $-2 \pm \sqrt{3}$ **12.** 0.90 sec
13. Length: 14 cm; width: 9 cm   **14.** 1 in.
15. First part: 50 mph; second part: 40 mph
16. Two real   **17.** Two nonreal
18. $25x^2 + 10x - 3 = 0$ **19.** $x^2 + 8x + 16 = 0$
20. $p = \frac{9\pi^2}{N^2}$ **21.** $T = \sqrt{\frac{3B}{2A}}$ **22.** $2, -2, 3, -3$
23. $3, -5$ **24.** $\pm\sqrt{7}, \pm\sqrt{2}$ **25.** $81, 16$

26. (a) $(1, 3)$; **(b)** $x = 1$; **(c)** maximum: 3;
(d)

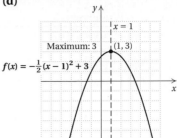

27. (a) $\left(\frac{1}{2}, \frac{23}{4}\right)$; **(b)** $x = \frac{1}{2}$; **(c)** minimum: $\frac{23}{4}$;
(d)

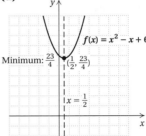

28. (a) $(-2, 4)$; **(b)** $x = -2$; **(c)** maximum: 4;
(d)

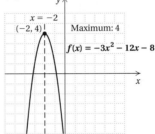

29. y-intercept: $(0, 14)$; x-intercepts: $(2, 0)$, $(7, 0)$
30. -121; 11 and -11 **31.** $f(x) = -x^2 + 6x - 2$
32. (a) $f(x) = -0.170x^2 + 1.846x + 60.124$; **(b)** 61.6%;
52.6%   **33.** $\{x \mid -2 < x < 1 \ or \ x > 2\}$, or
$(-2, 1) \cup (2, \infty)$ **34.** $\{x \mid x < -4 \ or \ -2 < x < 1\}$, or
$(-\infty, -4) \cup (-2, 1)$ **35. (a)** $f(x) = 1.2x + 497.8$;
(b) $f(10) = 509.8$, $f(14) = 514.6$ **36.** $\frac{x + 1}{(x - 3)(x - 1)}$
37. $3t^5s\sqrt[3]{s}$ **38.** 2   **39.** ◆ The graph of
$f(x) = ax^2 + bx + c$ is a parabola with vertex
$\left(-\frac{b}{2a}, \frac{4ac - b^2}{4a}\right)$. The line of symmetry is $x = -\frac{b}{2a}$. If
$a < 0$, then the parabola opens down and $\frac{4ac - b^2}{4a}$ is
the maximum function value. If $a > 0$, the parabola
opens up and $\frac{4ac - b^2}{4a}$ is the minimum function value.
The x-intercepts are $\left(\frac{-b - \sqrt{b^2 - 4ac}}{2a}, 0\right)$ and
$\left(\frac{-b + \sqrt{b^2 - 4ac}}{2a}, 0\right)$, for $b^2 - 4ac > 0$. When

$b^2 - 4ac = 0$, there is just one x-intercept, $\left(-\frac{b}{2a}, 0\right)$.
When $b^2 - 4ac < 0$, there are no x-intercepts.
40. ◆ The x-coordinate of the maximum or minimum point lies halfway between the x-coordinates of the x-intercepts. The function must be evaluated for this value of x in order to determine the maximum or minimum value. **41.** $f(x) = \frac{7}{15}x^2 - \frac{14}{15}x - 7$; minimum: $-\frac{112}{15}$ **42.** $h = 60, k = 60$ **43.** 18 and 324

Test: Chapter 7, p. 599

1. [7.1a] **(a)** $\pm\frac{2\sqrt{3}}{3}$; **(b)** $\left(\frac{2\sqrt{3}}{3}, 0\right), \left(-\frac{2\sqrt{3}}{3}, 0\right)$

2. [7.2a] $-\frac{1}{2} \pm i\frac{\sqrt{3}}{2}$ **3.** [7.4c] 49, 1 **4.** [7.2a] 9, 2

5. [7.4c] $\pm\sqrt{\frac{5+\sqrt{5}}{2}}, \pm\sqrt{\frac{5-\sqrt{5}}{2}}$

6. [7.2a] $-2 \pm \sqrt{6}$; 0.449, -4.449 **7.** [7.2a] 0, 2
8. [7.1b] $2 \pm \sqrt{3}$ **9.** [7.1c] About 6.7 sec
10. [7.3a] About 2.89 mph **11.** [7.3a] 7 cm by 7 cm
12. [7.1c] About 0.85 sec **13.** [7.4a] Two nonreal
14. [7.4b] $x^2 - 4\sqrt{3}x + 9 = 0$
15. [7.3b] $T = \sqrt{\frac{V}{48}}$, or $\frac{1}{4}\sqrt{\frac{V}{3}}$

16. [7.6a] **(a)** $(-1, 1)$; **17.** [7.6a] **(a)** $(3, 5)$;
(b) $x = -1$; **(b)** $x = 3$;
(c) maximum: 1; **(c)** minimum: 5;
(d) **(d)**

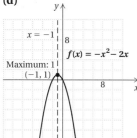

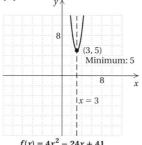

$f(x) = 4x^2 - 24x + 41$

18. [7.6b] y-intercept: $(0, -1)$; x-intercepts: $(2 - \sqrt{3}, 0)$, $(2 + \sqrt{3}, 0)$ **19.** [7.7a] -16; 4 and -4
20. [7.7b] $f(x) = \frac{1}{5}x^2 - \frac{3}{5}x$
21. [7.7b] **(a)** $f(x) = -7x^2 + 105.35x - 266.95$;
(b) 86.550; 44,900 **22.** [7.8a] $\{x\,|\,-1 < x < 7\}$, or $(-1, 7)$ **23.** [7.8b] $\{x\,|\,-3 < x < 5\}$, or $(-3, 5)$
24. [2.7b] **(a)** $f(x) = 1.6x + 498.4$; **(b)** 514.4, 520.8
25. [5.2b] $\frac{x-8}{(x+6)(x+8)}$ **26.** [6.3a] $ab\sqrt[4]{2a^2}$
27. [6.6a] 6 **28.** [7.6a, b], [7.7b]
$f(x) = -\frac{4}{7}x^2 + \frac{20}{7}x + 8$; maximum: $\frac{81}{7}$
29. [7.2a] $\frac{1}{2}$ **30.** [7.4c] $\pm 2, \pm\sqrt[4]{4}$

Cumulative Review: Chapters R–7, p. 601

1. [R.2e] 0.7 **2.** [R.6b] $5x + 23$ **3.** [R.7b] $-\frac{b^9}{64u^6}$

4. [R.7c] 4.0×10^{-16} **5.** [4.1d] $10x^2 - 8x + 6$
6. [4.2a] $2x^3 - 9x^2 + 7x - 12$ **7.** [5.1d] $\frac{2a-8}{5}$
8. [5.1e] $\frac{1}{y^2 + 6y}$ **9.** [5.2c] $\frac{-m^2 + 5m - 6}{(m+1)(m-5)}$
10. [5.4a] $\frac{y-x}{xy(x+y)}$
11. [5.3b] $9x^2 - 13x + 26 + \frac{-50}{x+2}$ **12.** [6.1b] 0.6
13. [6.1b] $3(x-2)$ **14.** [6.4a] $12\sqrt{5}$
15. [6.5b] $\frac{\sqrt{6} + 9\sqrt{2} - 12\sqrt{3} - 4}{-26}$ **16.** [6.2c] 2^8, or 256
17. [6.8c] $17 + 7i$ **18.** [6.8e] $-\frac{2}{3} - 2i$ **19.** [1.1d] $\frac{11}{13}$
20. [1.2a] $r = \frac{mv^2}{F}$ **21.** [1.4c] $\{x\,|\,x \geq \frac{5}{14}\}$, or $[\frac{5}{14}, \infty)$
22. [1.5b] $\{x\,|\,x < -\frac{4}{3} \text{ or } x > 6\}$, or $(-\infty, -\frac{4}{3}) \cup (6, \infty)$
23. [1.6e] $\{x\,|\,-\frac{13}{4} \leq x \leq \frac{15}{4}\}$, or $[-\frac{13}{4}, \frac{15}{4}]$
24. [3.2b] $(-4, 1)$ **25.** [3.4a] $(\frac{1}{2}, 3, -5)$
26. [4.7a] $\frac{1}{5}, -3$ **27.** [5.5a] $-\frac{5}{3}$ **28.** [5.5a] 3
29. [5.7a] $m = \frac{aA}{h-A}$ **30.** [6.6a] $\frac{37}{2}$ **31.** [6.6b] 11
32. [4.7a] 4 **33.** [7.2a] $\frac{3}{2} \pm i\frac{\sqrt{55}}{2}$
34. [7.2a] $\frac{17 \pm \sqrt{145}}{2}$ **35.** [7.3b] $a = \sqrt{P^2 + b^2}$
36. [7.8b] $\{x\,|\,-3 < x < -2 \text{ or } -1 < x < 1\}$, or $(-3, -2) \cup (-1, 1)$ **37.** [4.4a, b] $(2t + 5)(t - 6)$
38. [4.3c] $(a + 9)(a - 6)$ **39.** [4.3b] $(6a^2 - 5)(4a + 3)$
40. [4.3a] $-3a^2(a - 4)$ **41.** [4.5b] $(8a + 3b)(8a - 3b)$
42. [4.5a] $3(a - 6)^2$ **43.** [4.5d] $(\frac{1}{3}a - 1)(\frac{1}{9}a^2 + \frac{1}{3}a + 1)$
44. [4.3a] $(x + 1)(2x + 1)$
45. [2.5a] **46.** [2.5a]

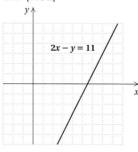

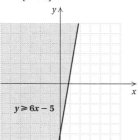

47. [3.6b] **48.** [3.6b]

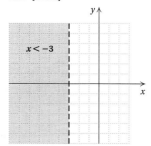

49. [3.6c]

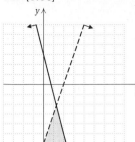

50. [7.6a]

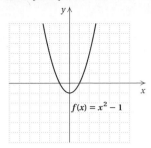

$f(x) = x^2 - 1$

51. [7.6a]

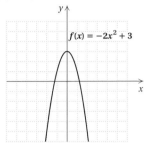

$f(x) = -2x^2 + 3$

52. [7.8a] $\left\{x \mid x < -\frac{5}{2} \text{ or } x > \frac{5}{2}\right\}$, or $\left(-\infty, -\frac{5}{2}\right) \cup \left(\frac{5}{2}, \infty\right)$
53. [2.6b] $y = \frac{1}{2}x + 4$ **54.** [2.6d] $y = -3x + 1$
55. [2.7b] **(a)** $S(x) = \frac{11}{15}x + \frac{143}{30}$;
(b) \$12.1 billion; \$15.0 billion
56. [7.3a] 16 km/h   **57.** [7.7a] 14 ft by 14 ft; 196 ft^2
58. [3.2c] 36   **59.** [5.6a] 2 hr   **60.** [4.7a] $0, \dfrac{b}{a}$
61. [R.7a] $\dfrac{328}{3}$ **62.** [6.6a] $\dfrac{2}{51 + 7\sqrt{61}}$, or $\dfrac{-51 + 7\sqrt{61}}{194}$
63. [4.5d] $\left(\dfrac{a}{2} + \dfrac{2b}{9}\right)\left(\dfrac{a^2}{4} - \dfrac{ab}{9} + \dfrac{4b^2}{81}\right)$
64. [5.6a], [5.7a] **(a)** $t = \dfrac{ab}{a + b}$; **(b)** $a = \dfrac{bt}{b - t}$;
(c) $b = \dfrac{at}{a - t}$

Chapter 8

Pretest: Chapter 8, p. 604

1. [8.1a]

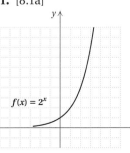

$f(x) = 2^x$

2. [8.3a]

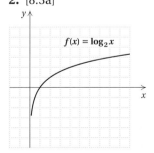

$f(x) = \log_2 x$

3. [8.3b] $\log_a 1000 = 3$ **4.** [8.3b] $2^t = 32$
5. [8.4d] $\log_a \dfrac{M^2N}{\sqrt[3]{Q}}$ **6.** [8.4d] $\frac{3}{4}\log_a x + \frac{1}{4}\log_a y -$

$\frac{1}{2}\log_a z$ **7.** [8.6b] 2   **8.** [8.6b] 25   **9.** [8.6a] $\dfrac{\log 8.6}{\log 3} \approx$
1.9586   **10.** [8.6a] $\frac{7}{2}$ **11.** [8.6b] 8   **12.** [8.6b] 1
13. [8.3d] 2.8537   **14.** [8.5a] -7.0745 **15.** [8.5a] 0
16. [8.5a] 66.6863   **17.** [8.5a] 0.1188
18. [8.3d] -1.6308 **19.** [8.1c] \$724.27
20. [8.7b] About 11,893 yr old

Margin Exercises, Section 8.1, pp. 606–611

1. (a)

x	$f(x)$
0	1
1	3
2	9
3	27
-1	$\frac{1}{3}$
-2	$\frac{1}{9}$
-3	$\frac{1}{27}$

(b)

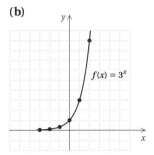

$f(x) = 3^x$

2. (a)

x	$f(x)$
0	1
1	$\frac{1}{3}$
2	$\frac{1}{9}$
3	$\frac{1}{27}$
-1	3
-2	9
-3	27

(b)

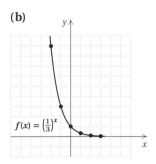

$f(x) = \left(\frac{1}{3}\right)^x$

3.

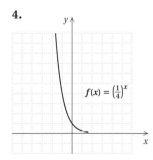

$f(x) = 4^x$

4.

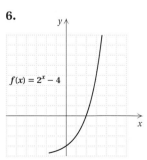

$f(x) = \left(\frac{1}{4}\right)^x$

5.

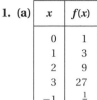

$f(x) = 2^{x+2}$

6.

$f(x) = 2^x - 4$

7.

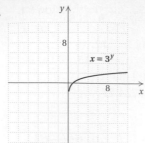

8. (a) $A(t) = \$40,000(1.05)^t$; (b) $40,000, $48,620.25, $59,098.22, $65,155.79;

(c)

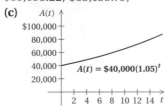

Calculator Spotlight, p. 605

1. 16.24   **2.** 451.81   **3.** 0.045

Calculator Spotlight, p. 612

1. $1125.51   **2.** $1127.16   **3.** $30,372.65
4. $1,480,569.70   **5.** (a) $10,540; (b) $10,547.29;
(c) $10,551.03; (d) $10,554.80; (e) $10,554.84
6. (a) $10,300; (b) $10,302.25; (c) $10,303.39;
(d) $10,304.53; (e) $10,304.54

Exercise Set 8.1, p. 613

1.

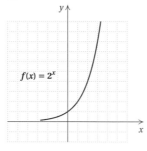

3.

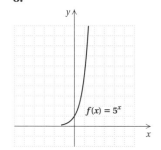

5.

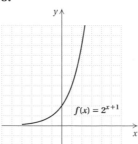

7.

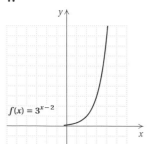

9.

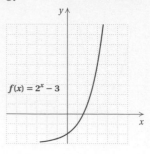

11.

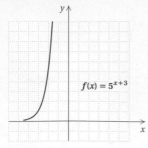

13.

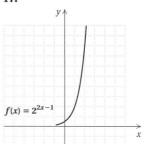

15.

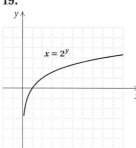

17.

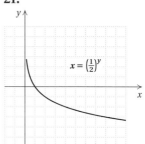

19.

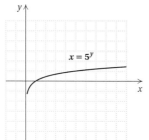

21.

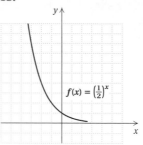

23.

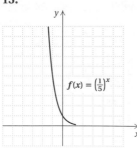

25.

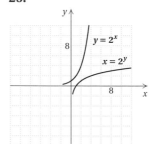

27. (a) 62 billion ft³, 63.116 billion ft³; **(b)** 65.409 billion ft³, 78.183 billion ft³;

(c)

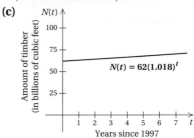

29. (a) 4243, 6000, 8485, 12,000, 24,000

(b)

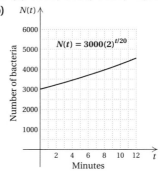

31. (a) 454,354,240 cm², 525,233,501,400 cm²;

(b)

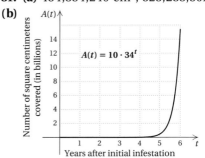

33. $\dfrac{1}{x^2}$　**34.** $\dfrac{1}{x^{12}}$　**35.** 1　**36.** 1　**37.** $\dfrac{2}{3}$　**38.** 2.7

39. $\dfrac{1}{x^7}$　**40.** $\dfrac{1}{x^{10}}$　**41.** x　**42.** x　**43.** ◈

45. 5^4, or 625

47.

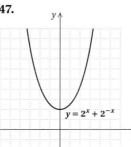

49.

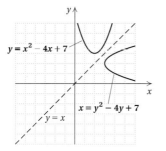

51.

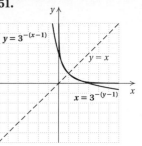

53. Left to the student.

Margin Exercises, Section 8.2, pp. 617–627

1. Inverse of $g = \{(5, 2), (4, -1), (1, -2)\}$

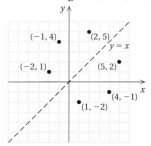

2. Inverse: $x = 6 - 2y$;

x	y
6	0
2	2
0	3
-4	5

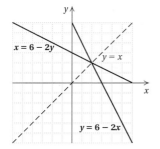

3. Inverse: $x = y^2 - 4y + 7$;

x	y
7	0
4	1
3	2
4	3
7	4

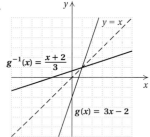

4. Yes　**5.** No　**6.** Yes　**7.** No　**8. (a)** Yes;

(b) $f^{-1}(x) = 3 - x$　**9. (a)** Yes; **(b)** $g^{-1}(x) = \dfrac{x + 2}{3}$

10.

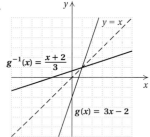

11. (a) Yes; **(b)** $f^{-1}(x) = \sqrt[3]{x-1}$;
(c)

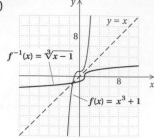

12. $x^2 + 4$; $x^2 + 10x + 24$ **13.** $4\sqrt[3]{x} + 5$; $\sqrt[3]{4x+5}$
14. (a) $f(x) = \sqrt[3]{x}$, $g(x) = x^2 + 1$; **(b)** $f(x) = \dfrac{1}{x^4}$,

$g(x) = x + 5$
15. $f^{-1} \circ f(x) = f^{-1}(f(x)) = f^{-1}\left(\frac{2}{3}x - 4\right)$

$$= \frac{3\left(\frac{2}{3}x - 4\right) + 12}{2} = \frac{2x - 12 + 12}{2}$$

$$= \frac{2x}{2} = x;$$

$$f \circ f^{-1}(x) = f(f^{-1}(x)) = f\left(\frac{3x + 12}{2}\right)$$

$$= \frac{2}{3}\left(\frac{3x + 12}{2}\right) - 4 = \frac{6x + 24}{6} - 4$$

$$= x + 4 - 4 = x$$

Calculator Spotlight, p. 624

1. Left to the student. **2.** No **3.** Yes **4.** Yes
5. Yes

Calculator Spotlight, p. 627

1. Left to the student. **2.** No **3.** No **4.** Yes
5. No

Exercise Set 8.2, p. 629

1. Inverse:
$\{(2, 1), (-3, 6), (-5, -3)\}$

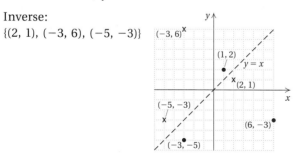

3. Inverse: $x = 2y + 6$

x	y
4	-1
6	0
8	1
10	2
12	3

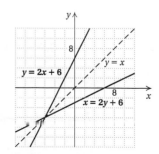

5. Yes **7.** No **9.** No **11.** Yes **13.** $f^{-1}(x) = \dfrac{x+2}{5}$

15. $f^{-1}(x) = \dfrac{-2}{x}$ **17.** $f^{-1}(x) = \frac{3}{4}(x - 7)$

19. $f^{-1}(x) = \dfrac{2}{x} - 5$ **21.** Not one-to-one

23. $f^{-1}(x) = \dfrac{1 - 3x}{5x - 2}$ **25.** $f^{-1}(x) = \sqrt[3]{x + 1}$

27. $f^{-1}(x) = x^3$
29. **31.**

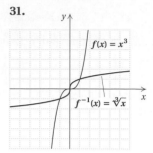

33. $-8x + 9$; $-8x + 18$ **35.** $12x^2 - 12x + 5$; $6x^2 + 3$

37. $\dfrac{16}{x^2} - 1$; $\dfrac{2}{4x^2 - 1}$ **39.** $x^4 - 10x^2 + 30$;

$x^4 + 10x^2 + 20$ **41.** $f(x) = x^2$; $g(x) = 5 - 3x$

43. $f(x) = \sqrt{x}$; $g(x) = 5x + 2$ **45.** $f(x) = \dfrac{1}{x}$;

$g(x) = x - 1$ **47.** $f(x) = \dfrac{1}{\sqrt{x}}$; $g(x) = 7x + 2$

49. $f(x) = x^4$; $g(x) = \sqrt{x} + 5$
51. $f^{-1} \circ f(x) = f^{-1}(f(x)) = f^{-1}\left(\frac{4}{5}x\right)$

$$= \frac{5}{4}\left(\frac{4}{5}x\right) = x;$$

$$f \circ f^{-1}(x) = f(f^{-1}(x)) = f\left(\frac{5}{4}x\right)$$

$$= \frac{4}{5}\left(\frac{5}{4}x\right) = x$$

53. $f^{-1} \circ f(x) = f^{-1}(f(x)) = f^{-1}\left(\dfrac{1 - x}{x}\right)$

$$= \frac{1}{\dfrac{1 - x}{x} + 1} = \frac{1}{\dfrac{1}{x}} = x;$$

$$f \circ f^{-1}(x) = f(f^{-1}(x)) = f\left(\dfrac{1}{x + 1}\right)$$

$$= \frac{1 - \dfrac{1}{x + 1}}{\dfrac{1}{x + 1}} = \frac{\dfrac{x}{x + 1}}{\dfrac{1}{x + 1}} = x$$

55. $f^{-1}(x) = \frac{1}{3}x$ **57.** $f^{-1}(x) = -x$
59. $f^{-1}(x) = x^3 + 5$ **61. (a)** 40, 42, 46, 50;
(b) $f^{-1}(x) = x - 32$; **(c)** 8, 10, 14, 18 **63.** $\sqrt[3]{a}$
64. $\sqrt[3]{x^2}$ **65.** a^2b^3 **66.** $2t^2$ **67.** $\sqrt{3}$ **68.** $2\sqrt[4]{2}$
69. $\sqrt{2xy}$ **70.** $\sqrt[4]{p^2t}$ **71.** $2a^3b^8$ **72.** $10x^3y^6$
73. $3a^2b^2$ **74.** $3pq^3$ **75.** ◆ **77.** No **79.** Yes
81. (1) C; **(2)** A; **(3)** B; **(4)** D

Margin Exercises, Section 8.3, pp. 633–638

1. $\log_2 64$ is the power to which we raise 2 to get 64; 6

2.

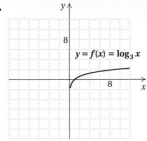

3. $0 = \log_6 1$ **4.** $-3 = \log_{10} 0.001$
5. $0.25 = \log_{16} 2$ **6.** $T = \log_m P$ **7.** $2^5 = 32$
8. $10^3 = 1000$ **9.** $a^7 = Q$ **10.** $t^x = M$ **11.** $10,000$
12. 3   **13.** $\frac{1}{4}$ **14.** 4   **15.** -4 **16.** 0   **17.** 0
18. 1   **19.** 1   **20.** 0   **21.** 4.8934   **22.** -4.5100
23. Does not exist   **24.** $\log 1000 = 3$, $\log 10,000 = 4$;
between 3 and 4; 3.9945   **25.** 78,234.8042

Exercise Set 8.3, p. 639

1.

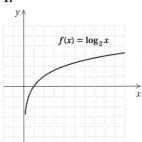

3.

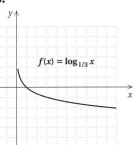

5.

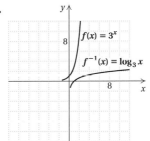

7. $3 = \log_{10} 1000$
9. $-3 = \log_5 \frac{1}{125}$
11. $\frac{1}{3} = \log_8 2$
13. $0.3010 = \log_{10} 2$
15. $2 = \log_e t$
17. $t = \log_Q x$
19. $2 = \log_e 7.3891$
21. $-2 = \log_e 0.1353$

23. $4^w = 10$ **25.** $6^2 = 36$ **27.** $10^{-2} = 0.01$
29. $10^{0.9031} = 8$ **31.** $e^{4.6052} = 100$ **33.** $t^k = Q$ **35.** 9
37. 4   **39.** 4   **41.** 3   **43.** 25   **45.** 1   **47.** $\frac{1}{2}$ **49.** 2
51. 2   **53.** -1 **55.** 0   **57.** 4   **59.** 2   **61.** 3
63. -2 **65.** 0   **67.** 1   **69.** $\frac{2}{3}$ **71.** 4.8970
73. -0.1739 **75.** Does not exist   **77.** 0.9464
79.

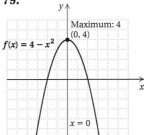

80.

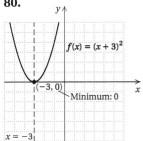

81.

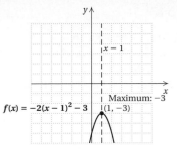

82.

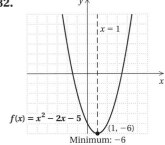

83. $c = \sqrt{\dfrac{E}{m}}$ **84.** $a = \dfrac{2 + \sqrt{4 + 3B}}{3}$ **85.** $b = \dfrac{A^2}{3a}$
86. $L = \dfrac{T^2 g}{4\pi^2}$ **87. (a)** $f(x) = -\dfrac{39}{280}x^2 + \dfrac{963}{140}x - \dfrac{356}{7}$;
(b) 30%, 13%   **89.** ◈
91.

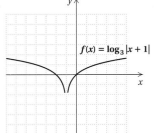

93. 25   **95.** $-\dfrac{7}{16}$ **97.** 3   **99.** 0   **101.** -2

Margin Exercises, Section 8.4, p. 643

1. $\log_5 25 + \log_5 5$ **2.** $\log_b P + \log_b Q$ **3.** $\log_3 35$
4. $\log_a (JAM)$ **5.** $5 \log_7 4$ **6.** $\frac{1}{2} \log_a 5$
7. $\log_b P - \log_b Q$ **8.** $\log_2 5$
9. $\frac{3}{2} \log_a z - \frac{1}{2} \log_a x - \frac{1}{2} \log_a y$
10. $2 \log_a x - 3 \log_a y - \log_a z$
11. $3 \log_a x + 4 \log_a y - 5 \log_a z - 9 \log_a w$
12. $\log_a \dfrac{x^5 z^{1/4}}{y}$ **13.** $\log_a \dfrac{1}{b\sqrt{b}}$, or $\log_a b^{-3/2}$
14. 0.602   **15.** 1   **16.** -0.398 **17.** 0.398
18. -0.699 **19.** $\frac{3}{2}$ **20.** 1.699   **21.** 1.204   **22.** 6
23. 3.2   **24.** 12

Calculator Spotlight, p. 643

1. Correct   **2.** Correct   **3.** Correct   **4.** Not correct

Exercise Set 8.4, p. 647

1. $\log_2 32 + \log_2 8$ **3.** $\log_4 64 + \log_4 16$
5. $\log_a Q + \log_a x$ **7.** $\log_b 252$ **9.** $\log_c Ky$
11. $4 \log_c y$ **13.** $6 \log_b t$ **15.** $-3 \log_b C$
17. $\log_a 67 - \log_a 5$ **19.** $\log_b 2 - \log_b 5$ **21.** $\log_c \frac{22}{3}$
23. $2 \log_a x + 3 \log_a y + \log_a z$
25. $\log_b x + 2 \log_b y - 3 \log_b z$
27. $\frac{4}{3} \log_c x - \log_c y - \frac{2}{3} \log_c z$

29. $2 \log_a m + 3 \log_a n - \frac{3}{4} - \frac{5}{4} \log_a b$ **31.** $\log_a \frac{x^{2/3}}{y^{1/2}}$,

or $\log_a \dfrac{\sqrt[3]{x^2}}{\sqrt{y}}$ **33.** $\log_a \dfrac{2x^4}{y^3}$ **35.** $\log_a \dfrac{\sqrt{a}}{x}$ **37.** 2.708

39. 0.51   **41.** -1.609 **43.** $\frac{3}{2}$ **45.** 2.609   **47.** t
49. 5   **51.** 7   **53.** -7 **55.** i **56.** -1 **57.** 5
58. $\frac{3}{5} + \frac{4}{5}i$ **59.** $23 - 18i$ **60.** $10i$ **61.** $-34 - 31i$
62. $3 - 4i$ **63.** ◈   **65.** Left to the student.
67. $\log_a (x^6 - x^4 y^2 + x^2 y^4 - y^6)$
69. $\frac{1}{2} \log_a (1 - s) + \frac{1}{2} \log_a (1 + s)$ **71.** False
73. True   **75.** False

Margin Exercises, Section 8.5, pp. 650–652

1. 11.2675   **2.** -10.3848 **3.** Does not exist   **4.** Does
not exist   **5.** 2.3026   **6.** 78,237.1596   **7.** 0.1353
8. 1.086   **9.** 5.5236
10.

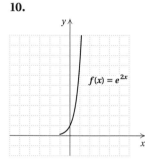

11.

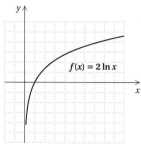

12.

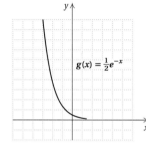

13.

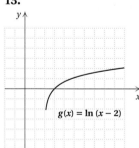

Exercise Set 8.5, p. 653

1. 0.6931   **3.** 4.1271   **5.** 8.3814   **7.** -5.0832
9. -1.6094 **11.** Does not exist   **13.** -1.7455 **15.** 1
17. 15.0293   **19.** 0.0305   **21.** 109.9472   **23.** 5
25. 2.5702   **27.** 6.6439   **29.** 2.1452   **31.** -2.3219
33. -2.3219 **35.** 4.6284

37.

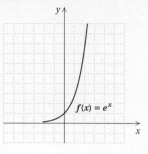

39.

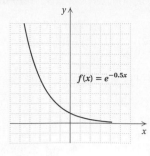

41.

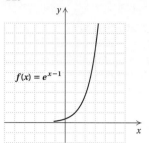

43.

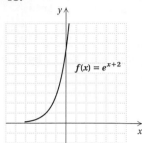

45.

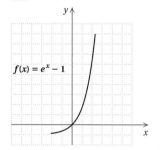

47.

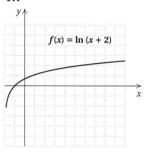

49.

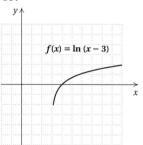

51.

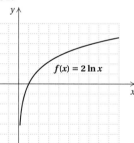

53.

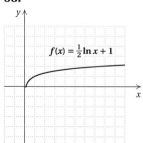

55.

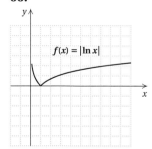

57. 16, 256   **58.** $\frac{1}{4}$, 9   **59.** 49, 121   **60.** ± 3, ± 4
61. ◈   **63.** Domain: $(-\infty, \infty)$; range: $[0, \infty)$
65. Domain: $(-\infty, \infty)$; range: $(-\infty, 100)$
67. Domain: $\left(\frac{5}{2}, \infty\right)$

Margin Exercises, Section 8.6, pp. 655–658

1. 1   **2.** 3   **3.** 1.5395   **4.** 14.6068   **5.** 25   **6.** $\frac{2}{5}$
7. 2   **8.** 5

Exercise Set 8.6, p. 659

1. 3   **3.** 4   **5.** $\frac{5}{2}$ **7.** $\frac{3}{5}$ **9.** 3.4594   **11.** 5.4263
13. $\frac{5}{2}$ **15.** −3, −1   **17.** $\frac{3}{2}$ **19.** 4.6052   **21.** 2.3026
23. 140.6705   **25.** 2.7095   **27.** 3.2220   **29.** 256
31. $\frac{1}{32}$ **33.** 10   **35.** $\frac{1}{100}$ **37.** $e^2 \approx 7.3891$
39. $\dfrac{1}{e} \approx 0.3679$ **41.** 121   **43.** 10   **45.** $\frac{1}{3}$ **47.** 3
49. $\frac{2}{5}$ **51.** 5   **53.** No solution   **55.** ±10, ±2
56. −64, 8   **57.** −2, −3, $\dfrac{-5 \pm \sqrt{41}}{2}$ **58.** $-\frac{1}{10}$, 1
59. $\dfrac{y^{4/3}}{25x^2 z^4}$ **60.** $-i$ **61.** ◈   **63.** 1   **65.** (a) 0.377;
(b) −1.962; **(c)** 0.904; **(d)** −1.532   **67.** 3, 4   **69.** −4
71. 2   **73.** $\pm\sqrt{34}$ **75.** $10^{100,000}$ **77.** 1, 100
79. 3, −7   **81.** 1, 1.465

Margin Exercises, Section 8.7, pp. 662–668

1. About 65 decibels   **2.** $10^{-9.2}$ W/m^2 **3.** About 4.9
4. 10^{-7} moles per liter   **5.** (a) 27.1 yr; (b) 10.2 yr
6. About 271 million; about 282 million
7. (a) $k \approx 6\%$, $P(t) = 5000e^{0.06t}$; (b) $5309.18, $5637.48,
$9110.59; **(c)** about 11.6 yr   **8.** (a) $k \approx 0.415$,
$P(t) = 91e^{0.415t}$; **(b)** $30,359 million, or $30.359 billion;
(c) 2001   **9.** 2972 yr

Calculator Spotlight, p. 668

1. (a) $y = 77.694(1.589)^x$;
(b)
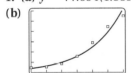
(c) $50,750 million, or $50.75 billion
2. (a) $y = 706,278.268(1.075)^x$;
(b)
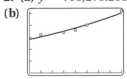
(c) $1,449,669; $2,975,514; **(d)** 2017

Exercise Set 8.7, p. 671

1. About 95 dB   **3.** $10^{-1.5}$, or about 3.2×10^{-2} W/m^2
5. About 6.8   **7.** 1.58×10^{-8} moles per liter
9. 3.00 ft/sec   **11.** 1.88 ft/sec   **13. (a)** 4860;
(b) 3.6 days; **(c)** 0.63 day   **15. (a)** $231 billion; **(b)** 2010;
(c) 4.7 yr   **17. (a)** $P(t) = 6e^{0.015t}$; **(b)** 7.2 billion; **(c)** 2032;
(d) 46.2 yr   **19. (a)** $P(t) = P_0 e^{0.06t}$; **(b)** $5309.18,

$5637.48, $9110.59; **(c)** in 11.6 yr   **21. (a)** 7%,
$P(t) = 40,000e^{0.07t}$; **(b)** $42,900.33, $46,010.95, $80,550.11;
(c) 9.9 yr   **23. (a)** $k \approx 0.034$, $P(t) = 20e^{0.034t}$; **(b)** 38¢;
(c) 2028   **25.** About 2103 yr   **27.** About 7.2 days
29. 69.3% per year   **31. (a)** $V(t) = 451,000e^{0.07t}$, where
t is the number of years since 1991; **(b)** $1,201,670;
(c) 9.9 yr; **(d)** 2002   **33.** −1   **34.** i **35.** $-1 - i$
36. −2   **37.** $\frac{63}{65} - \frac{16}{65}i$ **38.** $-\frac{2}{41} + \frac{23}{41}i$ **39.** 41
40. $91 + 60i$ **41.** ◈   **43.** −0.937, 1.078, 58.770
45. −0.767, 2, 4

Summary and Review: Chapter 8, p. 675

1.
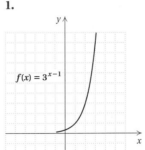

2.
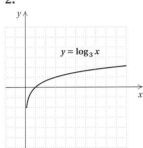

3.
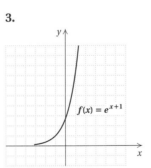

4.
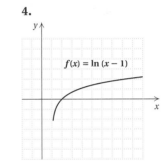

5. {(2, −4), (−7, 5), (−2, −1), (11, 10)}
6. Not one-to-one.   **7.** $g^{-1}(x) = \dfrac{7x + 3}{2}$
8. $f^{-1}(x) = \frac{1}{2}\sqrt[3]{x}$ **9.** $f^{-1}(x) = \dfrac{3x - 4}{2x}$
10.
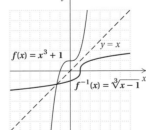

11. $f \circ g(x) = 9x^2 - 30x + 25$; $g \circ f(x) = 3x^2 - 5$
12. $f(x) = \sqrt{x}$, $g(x) = 4 - 7x$; answers may vary
13. $\log 10,000 = 4$ **14.** $\log_{25} 5 = \frac{1}{2}$ **15.** $4^x = 16$
16. $\left(\frac{1}{2}\right)^{-3} = 8$ **17.** 2   **18.** −1   **19.** 1   **20.** 0
21. −2.7425   **22.** Does not exist
23. $4 \log_a x + 2 \log_a y + 3 \log_a z$
24. $\frac{1}{2} \log z - \frac{3}{4} \log x - \frac{1}{4} \log y$ **25.** $\log_a 120$

26. $\log \dfrac{a^{1/2}}{bc^2}$ **27.** 17   **28.** -7 **29.** 8.7601

30. 3.2698   **31.** 2.54995   **32.** -3.6602 **33.** -2.6921
34. 0.3753   **35.** 18.3568   **36.** 0   **37.** Does not exist
38. 1   **39.** 0.4307   **40.** 1.7097   **41.** $\frac{1}{9}$ **42.** 2
43. $\frac{1}{10,000}$ **44.** $e^{-2} \approx 0.1353$ **45.** $\frac{7}{2}$ **46.** 1, -5
47. 1.5266   **48.** 35.0656   **49.** 2   **50.** 8   **51.** 20
52. $\sqrt{43}$ **53.** 90 dB   **54. (a)** \$341.4 billion; **(b)** 1997;
(c) 1.6 yr   **55. (a)** $k \approx 0.347$, $N(t) = 6e^{0.347t}$, where t is
the number of years since 1994; **(b)** 136; **(c)** 2004
56. $k \approx 0.043$, or 4.3%   **57.** About 8.25 yr   **58.** About
3463 yr   **59. (a)** $f(x) = \frac{1}{64}x^2 + \frac{5}{16}x + \frac{5}{2}$; **(b)** \$19
60. ± 4, $\pm i\sqrt{5}$
61.

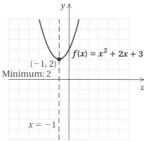

62. $11 - 10i$ **63.** 1   **64.** $-\frac{11}{10} - \frac{17}{10}i$ **65.** $5 - 2i$
66. ◆ You cannot take the logarithm of a negative
number because logarithm bases are positive and there
is no power to which a positive number can be raised
to yield a negative number.   **67.** ◆ $\log_a 1 = 0$ because
$a^0 = 1$.   **68.** e^{e^3} **69.** $\left(\frac{8}{3}, -\frac{2}{3}\right)$

Test: Chapter 8, p. 677

1. [8.1a]

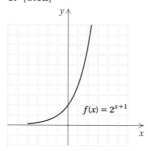

2. [8.3a]

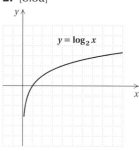

3. [8.5c]

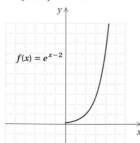

4. [8.5c]

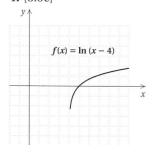

5. [8.2a] $\{(3, -4), (-8, 5), (-3, -1), (12, 10)\}$
6. [8.2b, c] $f^{-1}(x) = \dfrac{x + 3}{4}$
7. [8.2b, c] $f^{-1}(x) = \sqrt[3]{x} - 1$

8. [8.2b, c] Not one-to-one
9. [8.2e] $f \circ g(x) = 25x^2 - 15x + 2$,
$g \circ f(x) = 5x^2 + 5x - 2$
10. [8.3b] $\log_{256} 16 = \frac{1}{2}$ **11.** [8.3b] $7^m = 49$
12. [8.3c] 3   **13.** [8.3c] 23   **14.** [8.3c] 0
15. [8.3e] -1.9101 **16.** [8.3e] Does not exist
17. [8.4d] $3 \log a + \frac{1}{2} \log b - 2 \log c$
18. [8.4d] $\log_a \dfrac{x^{1/3}z^2}{y^3}$ **19.** [8.4d] -0.544
20. [8.4d] 1.079   **21.** [8.5a] 6.6938   **22.** [8.5a] 107.7701
23. [8.5a] 0   **24.** [8.5b] 1.1881   **25.** [8.6b] 5
26. [8.6b] 2   **27.** [8.6b] 10,000
28. [8.6b] $e^{1/4} \approx 1.2840$ **29.** [8.6a] 0.0937
30. [8.6b] 9   **31.** [8.6b] 1   **32.** [8.7a] 4.2
33. [8.7b] **(a)** \$39,103; **(b)** 2010; **(c)** 11.9 yr
34. [8.7b] **(a)** $k \approx 0.006$, $N(t) = 30e^{0.006t}$, where t is the
number of years since 1996 and N is in millions;
(b) 31.287 million, 32.629 million; **(c)** 2081; **(d)** 115.5 yr
35. [8.7b] About 4.6%   **36.** [8.7b] About 4684 yr
37. [7.7b] **(a)** $\frac{1}{33,600}x^2 + \frac{17}{336}x$; **(b)** about 11.3 min
38. [7.4c] 64, 1
39. [7.6a]

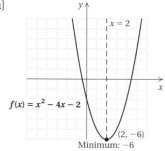

40. [6.8d] i **41.** [6.8e] $\frac{3}{2} - \frac{5}{2}i$ **42.** [8.6b] 44, -37
43. [8.4d] 2

Cumulative Review: Chapters R–8, p. 679

1. [R.4b] 2   **2.** [R.1d], [R.2a] 6   **3.** [R.7a] $\dfrac{20x^6z^2}{y}$
4. [R.6b] $-4x - 1$ **5.** [R.3c] 25   **6.** [1.1d] $\frac{11}{2}$
7. [4.7a] -2, 5   **8.** [3.4a] $(1, -2, 0)$ **9.** [3.2b] $(3, -1)$
10. [5.5a] $\frac{9}{2}$ **11.** [6.6b] 5   **12.** [7.4c] 9, 25
13. [7.4c] ± 3, ± 2 **14.** [8.6b] 8   **15.** [8.6a] 0.3542
16. [8.6b] $\frac{80}{9}$ **17.** [7.8a] $\{x \,|\, x < -5 \text{ or } x > 1\}$, or
$(-\infty, -5) \cup (1, \infty)$ **18.** [1.6e] $\{x \,|\, x \le -3 \text{ or } x \ge 6\}$, or
$(-\infty, -3] \cup [6, \infty)$ **19.** [7.2a] $-3 \pm 2\sqrt{5}$
20. [5.7a] $a = \dfrac{Db}{b - D}$ **21.** [5.7a] $q = \dfrac{pf}{p - f}$
22. [5.1a] $\left(-\infty, -\frac{1}{3}\right) \cup \left(-\frac{1}{3}, 2\right) \cup (2, \infty)$
23. [2.7b] **(a)** $A(t) = -225t + 1400$; **(b)** 50 acres
24. [5.6a] $5\frac{5}{11}$ hr   **25.** [3.3a] 60 L of Swim Clean; 40 L
of Pure Swim   **26.** [5.6c] $2\frac{7}{9}$ km/h   **27.** [8.7a] **(a)** 78;
(b) 67.5   **28.** [8.7b] **(a)** $k \approx 0.002$, $P(t) = 52e^{0.002t}$,
where t is the number of years after 1996 and P is in
millions; **(b)** 52.63 million, 53.48 million; **(c)** in 327 yr,
or 2323; **(d)** 346.6 yr   **29.** [4.7b] 10 ft   **30.** [5.8e] 18
31. [4.1d] $8x^2 - 11x - 1$ **32.** [4.2c] $9x^4 - 12x^2y + 4y^2$

33. [4.2b] $10a^2 - 9ab - 9b^2$ **34.** [5.1e] $\dfrac{(x+4)(x-3)}{2(x-1)}$

35. [5.4a] $\dfrac{1}{x-4}$ **36.** [5.2c] $\dfrac{7x+4}{(x+6)(x-6)}$

37. [4.5d] $(1-5x)(1+5x+25x^2)$

38. [4.3a], [4.4a, b] $2(3x-2y)(x+2y)$

39. [4.3b] $(x^3+7)(x-4)$ **40.** [4.3a, c] $2(m+3n)^2$

41. [4.5b] $(x-2y)(x+2y)(x^2+4y^2)$ **42.** [2.2b] -12

43. [5.3b, c] $x^3 - 2x^2 - 4x - 12 + \dfrac{-42}{x-3}$

44. [R.7c] 1.82×10^{-1} **45.** [6.3b] $2y^2\sqrt[3]{y}$

46. [6.3a] $14xy^2\sqrt{x}$ **47.** [6.5b] $\dfrac{6+\sqrt{y}-y}{4-y}$

48. [6.8c] $12+4\sqrt{3}i$ **49.** [8.2c] $f^{-1}(x) = \dfrac{x-7}{-2}$

50. [2.6d] $y = \frac{1}{2}x + \frac{13}{2}$

51. [2.5a]

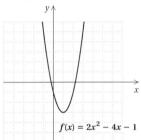

52. [7.5c], [7.6a]

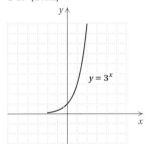

53. [8.3a]

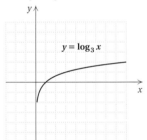

54. [8.1a]

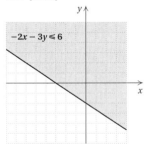

55. [3.6b]

56. [8.4d] $\log\left(\dfrac{x^3}{y^{1/2}z^2}\right)$ **57.** [8.3b] $a^x = 5$

58. [8.3d] -1.2545 **59.** [8.3d] 776.2471

60. [8.5a] 2.5479 **61.** [8.5a] 0.2466

62. [5.5a] All real numbers except 1 and -2

63. [8.6b] $\frac{1}{3}, \frac{10,000}{3}$ **64.** [5.6c] 35 mph

Chapter 9

Pretest: Chapter 9, p. 682

1. [9.1b] $\sqrt{89}$; 9.434   **2.** [9.1c] $\left(-2, -\frac{1}{2}\right)$

3. [9.1d] Center: $(2, -7)$; radius: 4

4. [9.1d] Center: $(4, -3)$; radius: 6

5. [9.1d] $(x-2)^2 + (y+9)^2 = 9$

6. [9.3a]

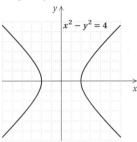

7. [9.1d]

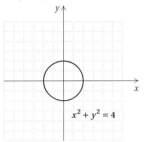

8. [9.3a]

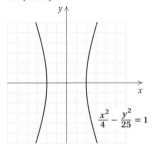

9. [9.2a]

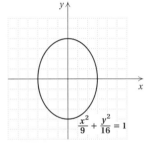

10. [9.1a]

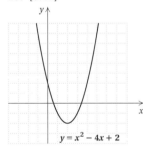

11. [9.1a]

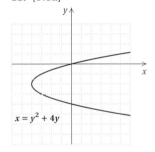

12. [9.2a]

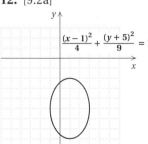

13. [9.3b]

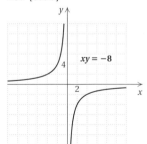

14. [9.4a] $(-3, 4)$ **15.** [9.4a] $(4, 5), (-4, -5),$
$\left(-\dfrac{5i\sqrt{2}}{2}, 4i\sqrt{2}\right), \left(\dfrac{5i\sqrt{2}}{2}, -4i\sqrt{2}\right)$

16. [9.4b] 2 in. by 1 in.   **17.** [9.4b] 60 ft; 48 ft

1.
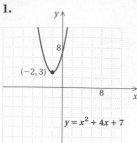
$y = x^2 + 4x + 7$
$(-2, 3)$

2.
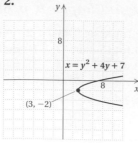
$x = y^2 + 4y + 7$
$(3, -2)$

3.
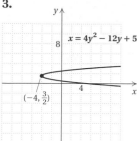
$x = 4y^2 - 12y + 5$
$\left(-4, \frac{3}{2}\right)$

4. 10   **5.** $\sqrt{37} \approx 6.083$
6. $\left(\frac{3}{2}, -3\right)$ **7.** $(9, -5)$

8.
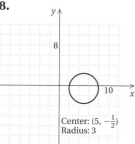
Center: $\left(5, -\frac{1}{2}\right)$
Radius: 3

9. $(0, 0)$; $r = 8$ **10.** $(x + 3)^2 + (y - 1)^2 = 36$
11. Center: $(-1, 2)$; radius: $\sqrt{3}$;
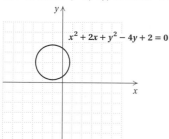
$x^2 + 2x + y^2 - 4y + 2 = 0$

5.
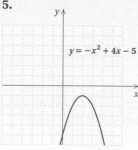
$y = -x^2 + 4x - 5$

7.
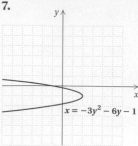
$x = -3y^2 - 6y - 1$

9. 5   **11.** $\sqrt{29} \approx 5.385$ **13.** $\sqrt{648} \approx 25.456$
15. 7.1   **17.** $\dfrac{\sqrt{41}}{7} \approx 0.915$ **19.** $\sqrt{6970} \approx 83.487$
21. $\sqrt{a^2 + b^2}$ **23.** $\sqrt{17 + 2\sqrt{14} + 2\sqrt{15}} \approx 5.677$
25. $\sqrt{9{,}672{,}400} \approx 3110.048$ **27.** $\left(\frac{3}{2}, \frac{7}{2}\right)$ **29.** $\left(0, \frac{11}{2}\right)$
31. $\left(-1, -\frac{17}{2}\right)$ **33.** $(-0.25, -0.3)$ **35.** $\left(-\frac{1}{12}, \frac{1}{24}\right)$
37. $\left(\dfrac{\sqrt{2} + \sqrt{3}}{2}, \dfrac{3}{2}\right)$

39.
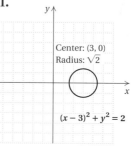
Center: $(-1, -3)$
Radius: 2
$(x + 1)^2 + (y + 3)^2 = 4$

41.
Center: $(3, 0)$
Radius: $\sqrt{2}$
$(x - 3)^2 + y^2 = 2$

43.
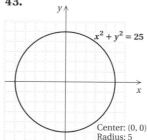
$x^2 + y^2 = 25$
Center: $(0, 0)$
Radius: 5

45. $x^2 + y^2 = 49$
47. $(x + 5)^2 + (y - 3)^2 = 7$
49. $(-4, 3)$; $r = 2\sqrt{10}$
51. $(4, -1)$; $r = 2$
53. $(2, 0)$; $r = 2$
55. $(-8, 16)$
56. $\left(-\frac{21}{5}, -\frac{73}{5}\right)$ **57.** $(1, 2)$
58. No solution
59. $(x - 4)(x + 4)$
60. $(a - 3b)(a + 3b)$
61. $(8p - 9q)(8p + 9q)$ **62.** $25(4cd - 3)(4cd + 3)$
63. ◈   **65.** $x^2 + y^2 = 2$
67. $(x + 3)^2 + (y + 2)^2 = 9$ **69.** $\sqrt{49 + k^2}$
71. $8\sqrt{m^2 + n^2}$ **73.** Yes   **75.** $(2, 4\sqrt{2})$
77. **(a)** $(0, -8467.8)$; **(b)** 8487.3 mm

Exercise Set 9.1, p. 691

1.
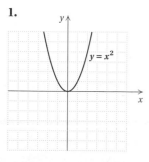
$y = x^2$

3.
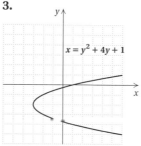
$x = y^2 + 4y + 1$

1.
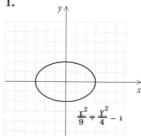
$\dfrac{x^2}{9} + \dfrac{y^2}{4} = 1$

2.
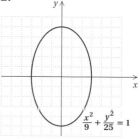
$\dfrac{x^2}{9} + \dfrac{y^2}{25} = 1$

3.

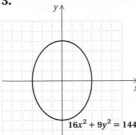

$16x^2 + 9y^2 = 144$

4.

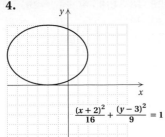

$$\frac{(x+2)^2}{16} + \frac{(y-3)^2}{9} = 1$$

17.

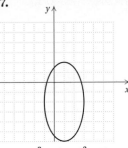

$12(x-1)^2 + 3(y+2)^2 = 48$

19.

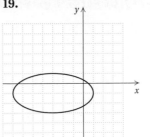

$(x+3)^2 + 4(y+1)^2 - 10 = 6$

Exercise Set 9.2, p. 701

1.

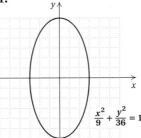

$$\frac{x^2}{9} + \frac{y^2}{36} = 1$$

3.

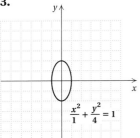

$$\frac{x^2}{1} + \frac{y^2}{4} = 1$$

5.

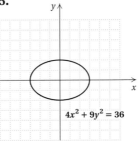

$4x^2 + 9y^2 = 36$

7.

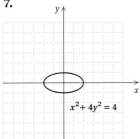

$x^2 + 4y^2 = 4$

9.

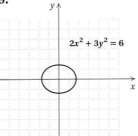

$2x^2 + 3y^2 = 6$

11.

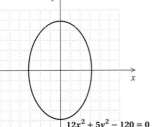

$12x^2 + 5y^2 - 120 = 0$

13.

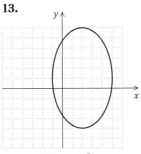

$$\frac{(x-2)^2}{9} + \frac{(y-1)^2}{25} = 1$$

15.

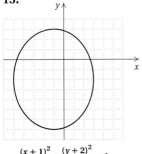

$$\frac{(x+1)^2}{16} + \frac{(y+2)^2}{25} = 1$$

21. $\dfrac{1 \pm 2i\sqrt{5}}{3}$ **22.** $\dfrac{6 \pm \sqrt{15}}{3}$ **23.** $-1 \pm \sqrt{21}$

24. $-1 \pm \sqrt{11}$ **25.** $1 \pm \sqrt{11}$; 4.3, -2.3

26. $-1 \pm 3\sqrt{2}$; 3.2, -5.2 **27.** $\dfrac{6 \pm \sqrt{15}}{3}$; 3.3, 0.7

28. $\dfrac{17 \pm \sqrt{337}}{8}$; 4.4, -0.2 **29.** ◈   **31.** $\dfrac{x^2}{81} + \dfrac{y^2}{121} = 1$

33. $\dfrac{(x-2)^2}{16} + \dfrac{(y+1)^2}{9} = 1$

35. $\dfrac{(x-2)^2}{16} + \dfrac{(y+1)^2}{4} = 1$ **37.** Left to the student.

Margin Exercises, Section 9.3, pp. 704–706

1.

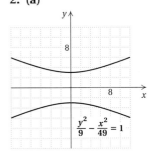

$$\frac{x^2}{16} - \frac{y^2}{25} = 1$$

2. (a)

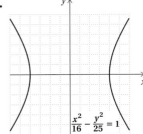

$$\frac{y^2}{9} - \frac{x^2}{49} = 1$$

(b)

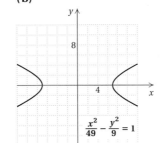

$$\frac{x^2}{49} - \frac{y^2}{9} = 1$$

3.

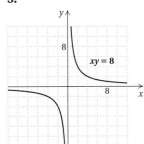

$xy = 8$

Exercise Set 9.3, p. 707

1.

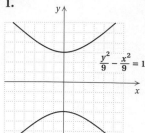
$$\frac{y^2}{9} - \frac{x^2}{9} = 1$$

3.

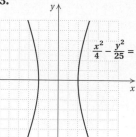
$$\frac{x^2}{4} - \frac{y^2}{25} = 1$$

5.

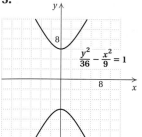
$$\frac{y^2}{36} - \frac{x^2}{9} = 1$$

7.

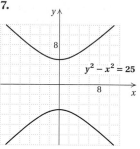
$$y^2 - x^2 = 25$$

9.

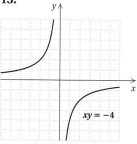
$$x^2 = 1 + y^2$$

11.

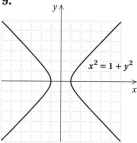
$$25x^2 - 16y^2 = 400$$

13.

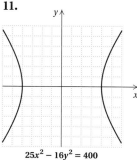
$$xy = -4$$

15.

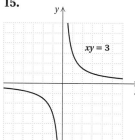
$$xy = 3$$

17.

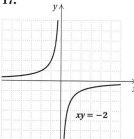
$$xy = -2$$

19.

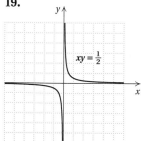
$$xy = \frac{1}{2}$$

21. $\log_5 25 = 2$ **22.** $\log_8 17 = a$ **23.** $\log_a b = -t$
24. $\ln Q = 3k$ **25.** $\log 199.53 = 2.3$
26. $\ln 20.086 = 3$ **27.** $\log_a 200 = 23.4$

28. $\ln 0.0183 = -4$ **29.** $e^{3.1781} = 24$
30. $10^3 = 1000$ **31.** $a^t = N$ **32.** $e^p = W$
33. $10^b = M$ **34.** $e^{kt} = P$ **35.** $10^{2.659} = 456$
36. $a^t = Y$ **37.** ◆   **39.** Left to the student.
41. Circle   **43.** Parabola   **45.** Ellipse

Margin Exercises, Section 9.4, pp. 710–714

1.

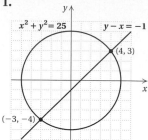

2.

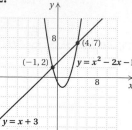

3. $\left(-\frac{5}{7}, \frac{22}{7}\right)$, $(1, -2)$ **4.** $\left(-\frac{3}{2}i, \frac{15}{4}i\right)$, $\left(\frac{3}{2}i, -\frac{15}{4}i\right)$
5. $(2, 3)$, $(2, -3)$, $(-2, 3)$, $(-2, -3)$
6. $(3, 2)$, $(-3, -2)$, $(2, 3)$, $(-2, -3)$
7. 12 m by 5 m   **8.** Width: 24 in.; height: 18 in.

Calculator Spotlight, p. 710

1. $(3.5, 0.5)$, $(-1.5, -1.2)$ **2.** $(-2.769, 2.516)$, $(-2.769, -2.516)$

Exercise Set 9.4, p. 715

1. $(-8, -6)$, $(6, 8)$ **3.** $(2, 0)$, $(0, 3)$ **5.** $(-2, 1)$
7. $\left(\frac{5 + \sqrt{70}}{3}, \frac{-1 + \sqrt{70}}{3}\right)$, $\left(\frac{5 - \sqrt{70}}{3}, \frac{-1 - \sqrt{70}}{3}\right)$
9. $\left(\frac{7}{3}, \frac{1}{3}\right)$, $(1, -1)$ **11.** $(-7, 1)$, $(1, -7)$ **13.** $(3, -5)$, $(-1, 3)$ **15.** $\left(\frac{8 + 3i\sqrt{6}}{2}, \frac{-8 + 3i\sqrt{6}}{2}\right)$, $\left(\frac{8 - 3i\sqrt{6}}{2}, \frac{-8 - 3i\sqrt{6}}{2}\right)$ **17.** $(-5, 0)$, $(4, 3)$, $(4, -3)$
19. $(3, 0)$, $(-3, 0)$ **21.** $(2, 4)$, $(-2, -4)$, $(4, 2)$, $(-4, -2)$
23. $(2, 3)$, $(-2, -3)$, $(3, 2)$, $(-3, -2)$ **25.** $(2, 1)$, $(-2, -1)$ **27.** $(5, 2)$, $(-5, 2)$, $\left(2, -\frac{4}{5}\right)$, $\left(-2, -\frac{4}{5}\right)$
29. $(\sqrt{2}, -\sqrt{2})$, $(-\sqrt{2}, \sqrt{2})$ **31.** $\left(\frac{8i\sqrt{5}}{5}, \frac{3\sqrt{105}}{5}\right)$, $\left(-\frac{8i\sqrt{5}}{5}, \frac{3\sqrt{105}}{5}\right)$, $\left(\frac{8i\sqrt{5}}{5}, -\frac{3\sqrt{105}}{5}\right)$, $\left(-\frac{8i\sqrt{5}}{5}, -\frac{3\sqrt{105}}{5}\right)$ **33.** Length: 12 ft; width: 5 ft
35. Length: 7 in.; width: 2 in.   **37.** Length: 12 ft; width: 5 ft   **39.** 13 and 12   **41.** 24 ft, 16 ft
43. Length: $\sqrt{2}$ m; width: 1 m   **45.** $f^{-1}(x) = \dfrac{x + 5}{2}$
46. $f^{-1}(x) = \dfrac{7x + 3}{2x}$ **47.** $f^{-1}(x) = \dfrac{3x + 2}{1 - x}$
48. $f^{-1}(x) = \dfrac{4x + 8}{5x - 3}$ **49.** Does not exist   **50.** Does not exist   **51.** $f^{-1}(x) = \log x$ **52.** $f^{-1}(x) = \ln x$
53. $f^{-1}(x) = \sqrt[3]{x + 4}$ **54.** $f^{-1}(x) = x^3 - 2$

55. $f^{-1}(x) = e^x$ **56.** $f^{-1}(x) = 10^x$ **57.** ◈
59. Left to the student. **61.** One piece: $38\frac{12}{25}$ cm; the other: $61\frac{13}{25}$ cm **63.** 30 **65.** $\left(\frac{1}{2}, \frac{1}{3}\right), \left(\frac{1}{3}, \frac{1}{2}\right)$

Summary and Review: Chapter 9, p. 719

1. 4 **2.** 5 **3.** $\sqrt{90.1} \approx 9.492$ **4.** $\sqrt{9 + 4a^2}$
5. (4, 6) **6.** $\left(-3, \frac{5}{2}\right)$ **7.** $\left(\frac{3}{4}, \frac{\sqrt{3} - \sqrt{2}}{2}\right)$ **8.** $\left(\frac{1}{2}, 2a\right)$
9. (−2, 3), $\sqrt{2}$ **10.** (5, 0), 7 **11.** (3, 1), 3
12. (−4, 3), $\sqrt{35}$ **13.** $(x + 4)^2 + (y - 3)^2 = 48$
14. $(x - 7)^2 + (y + 2)^2 = 20$
15.

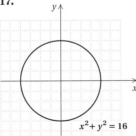

16.

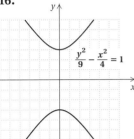

17.

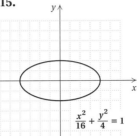

18.

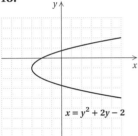

19.

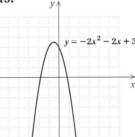

20.

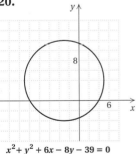

21.

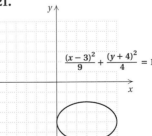

22.

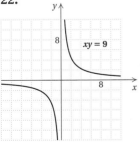

23. (7, 4) **24.** (2, 2), $\left(\frac{32}{9}, -\frac{10}{9}\right)$ **25.** (0, −3), (2, 1)
26. (4, 3), (4, −3), (−4, 3), (−4, −3) **27.** (2, 1), ($\sqrt{3}$, 0), (−2, 1), (−$\sqrt{3}$, 0) **28.** (3, −3), $\left(-\frac{3}{5}, \frac{21}{5}\right)$
29. (6, 8), (6, −8), (−6, 8), (−6, −8) **30.** (2, 2),

(−2, −2), (2$\sqrt{2}$, $\sqrt{2}$), (−2$\sqrt{2}$, −$\sqrt{2}$)
31. Length: 12 m; width: 7 m **32.** 4 and 8
33. 32 cm, 20 cm **34.** 11 ft, 3 ft
35. $(a - 7b)(a + 7b)$ **36.** $12(p - 4q)(p + 4q)$
37. $-1 \pm 2i$ **38.** $2 \pm \sqrt{3}$ **39.** $f^{-1}(x) = \dfrac{8 - x}{3}$
40. $f^{-1}(x) = \dfrac{x^3 + 5}{2}$ **41.** $\log_7 y = x$
42. $\ln 0.1653 = -1.8$ **43.** $e^{5.522} = 250$ **44.** $a^N = M$
45. ◈ Earlier, we studied systems of linear equations. In this chapter, we studied systems of two equations in which at least one equation is of second degree.
46. ◈ The graph of a parabola has one branch whereas the graph of a hyperbola has two branches. A hyperbola has asymptotes, but a parabola does not.
47. $(-5, -4\sqrt{2}), (-5, 4\sqrt{2}), (3, -2\sqrt{2}), (3, 2\sqrt{2})$
48. $(x - 2)^2 + (y + 1)^2 = 25$ **49.** $\dfrac{x^2}{49} + \dfrac{y^2}{9} = 1$ **50.** $\left(\frac{9}{4}, 0\right)$

Test: Chapter 9, p. 721

1. [9.1b] $\sqrt{180} \approx 13.416$ **2.** [9.1b] $\sqrt{36 + 4a^2}$, or $2\sqrt{9 + a^2}$ **3.** [9.1c] (0, 5) **4.** [9.1c] (0, 0)
5. [9.1d] Center: (−2, 3); radius: 8
6. [9.1d] Center: (−2, 3); radius: 3
7. [9.1d] $(x + 2)^2 + (y + 5)^2 = 18$
8. [9.1a]

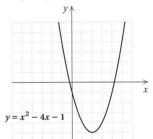

9. [9.1d]

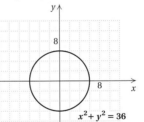

10. [9.3a]

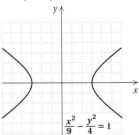

11. [9.2a]

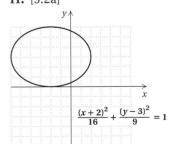

12. [9.1d]

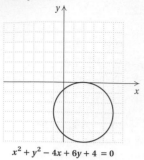

$x^2 + y^2 - 4x + 6y + 4 = 0$

13. [9.2a]

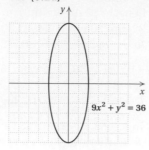

$9x^2 + y^2 = 36$

34. [6.8c] $16 + i\sqrt{2}$ **35.** [6.8e] $\frac{3}{10} + \frac{11}{10}i$

36. [2.5a]

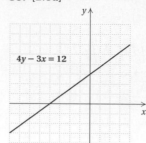

$4y - 3x = 12$

37. [3.6b]

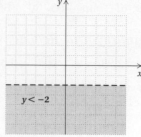

$y < -2$

14. [9.3b]

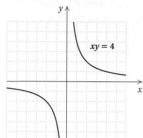

$xy = 4$

15. [9.1a]

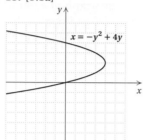

$x = -y^2 + 4y$

38. [3.6c]

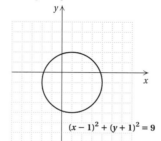

39. [7.6a]

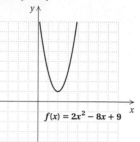

$f(x) = 2x^2 - 8x + 9$

16. [9.4a] (0, 3), (4, 0)   **17.** [9.4a] (4, 0), (−4, 0)
18. [9.4b] 16 ft by 12 ft   **19.** [9.4b] 8 and 15
20. [9.4b] 11 yd by 2 yd   **21.** [9.4b] $\sqrt{5}$ m, $\sqrt{3}$ m
22. [4.5b] $3a(4a - 5b)(4a + 5b)$ **23.** [7.2a] $-1 \pm \sqrt{6}$
24. [8.2c] $f^{-1}(x) = \sqrt[3]{8 - x}$ **25.** [8.3b] $\ln 54.5981 = 4$
26. [8.3b] $2^3 = 8$ **27.** [9.2a] $\dfrac{(x - 6)^2}{25} + \dfrac{(y - 3)^2}{9} = 1$
28. [9.1d] $\{(x, y)\,|\,(x - 8)^2 + y^2 = 100\}$ **29.** [9.4b] 9
30. [9.1b] $\left(0, -\frac{31}{4}\right)$

Cumulative Review: Chapters R–9, p. 723

1. [R.1d] $\frac{2}{15}$ **2.** [R.7a] $-9x^6y^2$ **3.** [R.3c] 62.5
4. [R.6b] $-x + 30$ **5.** [1.4c] $\{x\,|\,x \geq -1\}$, or $[-1, \infty)$
6. [1.6e] $\{x\,|\,x < -6.4 \text{ or } x > 6.4\}$, or
$(-\infty, -6.4) \cup (6.4, \infty)$ **7.** [1.5a] $\{x\,|\,-1 \leq x < 6\}$, or
$[-1, 6)$ **8.** [3.2b] $\left(-\frac{1}{3}, 5\right)$ **9.** [3.4a] $(-1, 2, 3)$
10. [4.7a] $-1, \frac{3}{2}$ **11.** [5.5a] $-\frac{2}{3}, 3$ **12.** [6.6a] 4
13. [4.7a] $-7, -3$ **14.** [7.2a] $-\dfrac{1}{4} \pm i\dfrac{\sqrt{7}}{4}$
15. [7.4c] $-3, -2, 2, 3$ **16.** [5.5a] -16
17. [9.4a] (1, 2), (−1, 2), (1, −2), (−1, −2)   **18.** [8.6b] 1
19. [8.6a] 1.748   **20.** [8.6b] 9
21. [7.8a] $\{x\,|\,x \leq -1 \text{ or } x \geq 1\}$, or $(-\infty, -1] \cup [1, \infty)$
22. [7.8b] $\{x\,|\,x < -1 \text{ or } x > 2\}$, or $(-\infty, -1) \cup (2, \infty)$
23. [1.2a] $N = \dfrac{4P - 3M}{6}$ **24.** [5.7a] $p = \dfrac{qf}{q - f}$
25. [4.2a] $2x^3 - x^2 - 8x - 3$
26. [4.1d] $-x^3 + 3x^2 - x - 6$ **27.** [5.1d] $\dfrac{2m - 1}{m + 1}$
28. [5.2c] $\dfrac{x + 2}{x + 1}$ **29.** [5.4a] $\dfrac{1}{x + 1}$
30. [5.3b], [5.3c] $x^3 + 2x^2 - 2x + 1 + \dfrac{3}{x + 1}$
31. [6.3b] $5x^2\sqrt{y}$ **32.** [6.4a] $11\sqrt{2}$ **33.** [6.2d] 8

40. [9.1d]

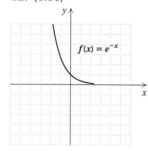

$(x - 1)^2 + (y + 1)^2 = 9$

41. [9.1a]

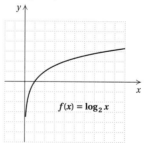

$x = y^2 + 1$

42. [8.5c]

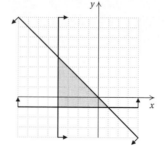

$f(x) = e^{-x}$

43. [8.3a]

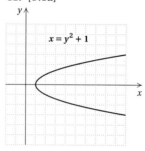

$f(x) = \log_2 x$

44. [4.3b] $(2x^3 + 1)(x - 6)$
45. [4.3c] $3(a - 9b)(a + 5b)$
46. [4.3c] $(x - 8)(x - 9)$
47. [4.5b] $(9m^2 + n^2)(3m + n)(3m - n)$
48. [4.5a] $4(2x - 1)^2$
49. [4.5d] $3(3a - 2)(9a^2 + 6a + 4)$
50. [4.4a] $2(5x - 2)(x + 7)$
51. [4.4a] $3x(2x - 1)(x + 5)$ **52.** [2.6c] $y = 2x + 2$
53. [2.6d] $y = -\frac{1}{2}x + \frac{5}{2}$ **54.** [9.1d] Center: (8, −3);
radius: $\sqrt{5}$ **55.** [8.2c] $f^{-1}(x) = \frac{1}{2}(x + 3)$ **56.** [5.8e] $\frac{4}{5}$
57. [2.2b] -10 **58.** [9.1b] 10   **59.** [9.1c] $\left(1, -\frac{3}{2}\right)$
60. [6.5b] $\dfrac{15 + 8\sqrt{a} + a}{9 - a}$

61. [5.1a] $\left(-\infty, -\frac{1}{3}\right) \cup \left(-\frac{1}{3}, 0\right) \cup (0, \infty)$ **62.** [4.7a] $-1, \frac{2}{3}$
63. [7.3a] $-13, 12$ **64.** [9.4b] 5 in. by 4 in.
65. [5.6a] 1 hr   **66.** [7.3a] True for any three
consecutive even integers
67. [8.7b] **(a)** $P(t) = 102e^{0.92t}$; **(b)** \$4044 million, or
\$4.044 billion; **(c)** 0.75 yr   **68.** [5.6c] 612 mi
69. [8.7b] 2397 yr   **70.** [5.8d] 3360 kg
71. [2.7b] $f(x) = -\frac{1}{3}x - \frac{7}{3}$
72. [7.7b] $f(x) = -\frac{17}{18}x^2 - \frac{59}{18}x + \frac{11}{9}$
73. [8.3b] $\log r = 6$ **74.** [8.3b] $3^x = Q$
75. [8.4d] $\log_b \left(\dfrac{x^7}{yz^8}\right)^{1/5}$
76. [8.4d] $-6 \log_b x - 30 \log_b y + 6 \log_b z$
77. [7.7a] 169   **78.** [8.2b] No   **79.** [2.3a] **(a)** -5;
(b) $(-\infty, \infty)$; **(c)** $-2, -1, 1, 2$; **(d)** $[-7, \infty)$
80. [5.5a] $(-\infty, -2) \cup (-2, \infty)$
81. [9.1d] Circle centered at the origin with radius a
82. [9.4b] $\frac{1}{2}, \frac{1}{4}; \frac{1}{2}, -\frac{1}{4}; -\frac{1}{2}, \frac{1}{4}; -\frac{1}{2}, -\frac{1}{4}$
83. [9.4a] $(-12, -8), (-8, -12), (8, 12), (12, 8)$

Final Examination, p. 727

1. [R.7a] $-\dfrac{45x^6}{y^4}$ **2.** [R.1d] 6.3   **3.** [R.6b] $-3y + 17$
4. [R.3c] 280   **5.** [R.4b] $\frac{7}{6}$ **6.** [4.1d] $3a^2 - 8ab - 15b^2$
7. [4.1c] $13x^3 - 7x^2 - 6x + 6$ **8.** [4.2b] $6a^2 + 7a - 5$
9. [4.2c] $9a^4 - 30a^2y + 25y^2$ **10.** [5.2c] $\dfrac{4}{x+2}$
11. [5.1d] $\dfrac{x-4}{3(x+2)}$ **12.** [5.1e] $\dfrac{(x+y)(x^2+xy+y^2)}{x^2+y^2}$
13. [5.4a] $x - a$ **14.** [4.5a] $(2x - 3)^2$
15. [4.5d] $(3a - 2)(9a^2 + 6a + 4)$
16. [4.3b] $(a^2 - b)(a + 3)$ **17.** [4.4a] $3(y^2 + 3)(5y^2 - 4)$
18. [2.2b] 20
19. [5.3b, c] $7x^3 + 9x^2 + 19x + 38 + \dfrac{72}{x-2}$
20. [1.1d] $\frac{2}{3}$ **21.** [4.7a] $8, -6$ **22.** [5.5a] $-\frac{6}{5}, 4$
23. [5.5a] No solution   **24.** [3.2b] $(1, -1)$
25. [3.4a] $(2, -1, 1)$ **26.** [6.6b] 9   **27.** [7.4c] $\pm 5, \pm 2$
28. [9.4b] $(\sqrt{5}, \sqrt{3}), (\sqrt{5}, -\sqrt{3}), (-\sqrt{5}, \sqrt{3})$,
$(-\sqrt{5}, -\sqrt{3})$ **29.** [8.6a] 1.2920   **30.** [8.6b] 1005
31. [8.6b] $\frac{1}{16}$ **32.** [8.6a] $-\frac{1}{2}$
33. [1.6e] $\{x \mid -2 \le x \le 3\}$, or $[-2, 3]$ **34.** [7.1a] $\pm\sqrt{2}\,i$
35. [7.2a] $-2 \pm \sqrt{7}$ **36.** [1.6e] $\{y \mid y < -5 \text{ or } y > 2\}$, or
$(-\infty, -5) \cup (2, \infty)$ **37.** [7.2a] $-6, 8$ **38.** [3.3a] More
than 4   **39.** [1.3a] 65, 66, 67   **40.** [1.3a] $11\frac{3}{7}$
41. [3.3a] 24 L of A; 56 L of B   **42.** [5.6c] 350 mph
43. [5.6a] $8\frac{2}{5}$ min   **44.** [5.8e] 20   **45.** [9.4b] 5 ft by
12 ft   **46.** [9.4b] Length: 20 ft; width: 15 ft
47. [7.7a] 1250 ft^2

48. [2.5a]
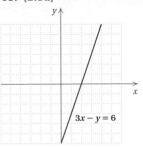

49. [9.2a]
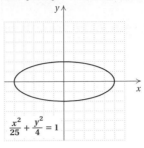

50. [8.3a]
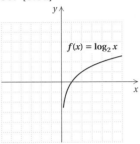

51. [3.6b]
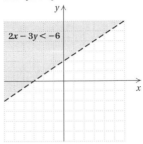

52. [7.6a]
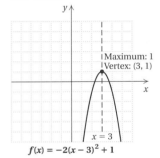

53. [3.6c]
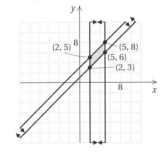

54. [1.2a] $r = \dfrac{A - P}{Pt}$ **55.** [5.7a] $R = \dfrac{Ir}{1 - I}$
56. [2.6d] $y = -x + 3$ **57.** [2.3a] $\left(-\infty, \frac{5}{3}\right]$
58. [5.1a] $(-\infty, 1) \cup (1, \infty)$ **59.** [R.7c] 6.764×10^{-12}
60. [6.3a] $8x^2\sqrt{y}$ **61.** [6.3b] $\dfrac{\sqrt[3]{5xy}}{y}$
62. [6.5b] $\dfrac{1 - 2\sqrt{x} + x}{1 - x}$ **63.** [6.3b] $\sqrt[6]{x+1}$
64. [6.8c] $26 - 13i$ **65.** [7.4b] $x^2 - 50 = 0$
66. [9.1d] Center: $(2, -3)$; radius: 6
67. [8.4d] $\log_a \dfrac{\sqrt[3]{x^2}z^5}{\sqrt{y}}$ **68.** [8.3b] $a^5 = c$
69. [8.2c] $f^{-1}(x) = \sqrt[3]{x+8}$ **70.** [8.3d] 3.7541
71. [8.3d] 2.5882×10^4, or 0.00025882
72. [8.5a] 8.6442   **73.** [8.5a] 0.0277
74. [2.7b] $f(x) = \frac{7}{3}x + \frac{23}{3}$
75. [7.7b] $f(x) = -\frac{2}{3}x^2 - \frac{7}{3}x + 1$
76. [8.7b] **(a)** $P(t) = 1000e^{0.86t}$; **(b)** 946,002,036;
(c) 0.81 yr   **77.** [2.3a] **(a)** -5; **(b)** $(-\infty, \infty)$; **(c)** $-1, 3$;
(d) $(-\infty, \infty)$ **78.** [6.2c] $125x^2y^{3/4}$ **79.** [5.5a] All real
numbers except 0 and -12 **80.** [8.6b] 81
81. [5.8e] y is divided by 8.   **82.** [6.8e] $-\dfrac{7}{13} + \dfrac{2\sqrt{30}}{13}i$
83. [1.3a] 84 yr

Appendixes

Margin Exercises, Appendix A, pp. 731–732

1. $70 \dfrac{\text{mi}}{\text{hr}}$ **2.** $25 \dfrac{\text{m}}{\text{sec}}$ **3.** 16 yd **4.** 72 in^2

5. 66 man-hours **6.** $800 \dfrac{\text{kW-hr}}{\text{da}}$ **7.** 84 in.

8. $132 \dfrac{\text{ft}}{\text{sec}}$

Exercise Set A, p. 733

1. 68 ft **3.** 45 g **5.** $15 \dfrac{\text{mi}}{\text{hr}}$ **7.** $3.3 \dfrac{\text{m}}{\text{sec}}$ **9.** $4 \dfrac{\text{in.-lb}}{\text{sec}}$

11. 12 yd **13.** 16 ft^3 **15.** $\dfrac{\$970}{\text{day}}$ **17.** 51.2 oz

19. 3080 ft/min **21.** 96 in. **23.** Approximately

0.03 yr **25.** Approximately 31,710 yr **27.** $80 \dfrac{\text{oz}}{\text{in.}}$

29. 172,800 sec **31.** $\dfrac{3}{2}$ ft^2 **33.** $1.08 \dfrac{\text{ton}}{\text{yd}^3}$

Margin Exercises, Appendix B, pp. 734–737

1. -5 **2.** -7 **3.** -6 **4.** 93 **5.** $(3, -1)$
6. $(1, 3, -2)$

Exercise Set B, p. 738

1. 10 **3.** 0 **5.** -48 **7.** 0 **9.** -10 **11.** -3
13. 5 **15.** 0 **17.** $(2, 0)$ **19.** $\left(-\frac{25}{2}, -\frac{11}{2}\right)$ **21.** $\left(\frac{3}{2}, \frac{5}{2}\right)$

23. $(-4, 3)$ **25.** $(2, -1, 4)$ **27.** $(1, 2, 3)$
29. $\left(\frac{3}{2}, -4, 3\right)$ **31.** $(2, -2, 1)$

Margin Exercises, Appendix C, pp. 740–742

1. $(-8, 2)$ **2.** $(-1, 2, 3)$

Exercise Set C, p. 743

1. $\left(\frac{3}{2}, \frac{5}{2}\right)$ **3.** $(-4, 3)$ **5.** $\left(\frac{1}{2}, \frac{3}{2}\right)$ **7.** $(10, -10)$
9. $\left(\frac{3}{2}, -4, 3\right)$ **11.** $(2, -2, 1)$ **13.** $(0, 2, 1)$
15. $\left(4, \frac{1}{2}, -\frac{1}{2}\right)$ **17.** $(w, x, y, z) = (1, -3, -2, -1)$

Margin Exercises, Appendix D, p. 745

1. (a) $2x^2$; (b) 6; (c) $x^4 - 9$; (d) $\dfrac{x^2 + 3}{x^2 - 3}$; (e) $x^4 + 6x^2 + 9$

2. (a) -5; (b) 15; (c) -54; (d) 6

Exercise Set D, p. 746

1. 1 **3.** -41 **5.** 12 **7.** $\frac{13}{18}$ **9.** 5 **11.** 2
13. $x^2 - x + 1$ **15.** 21 **17.** 5
19. $-x^3 + 4x^2 + 3x - 12$ **21.** 42 **23.** $-\frac{3}{4}$ **25.** $\frac{1}{6}$
27. $x^2 + 3x - 4$; $x^2 - 3x + 4$; $3x^3 - 4x^2$; $\dfrac{x^2}{3x - 4}$

29. $\dfrac{1}{x - 2} + 4x^3$; $\dfrac{1}{x - 2} - 4x^3$; $\dfrac{4x^3}{x - 2}$; $\dfrac{1}{4x^3(x - 2)}$

31. $\dfrac{3}{x - 2} + \dfrac{5}{4 - x}$; $\dfrac{3}{x - 2} - \dfrac{5}{4 - x}$; $\dfrac{15}{(x - 2)(4 - x)}$;
$\dfrac{3(4 - x)}{5(x - 2)}$

Index

D

Data analysis, 209, 210
Decay model, exponential, 667, 675
Decay rate, 667
Decimal notation. *See also* Decimals.
    converting to scientific notation, 59
    repeating, 4
    terminating, 4
Decimals, clearing, 77. *See also* Decimal notation.
Degree
    of a polynomial, 302
    of a term, 302
Demand, 289
Denominators
    least common, 387
    rationalizing, 483–485
Dependent system of equations, 232, 242
Dependent variable, 156
Descartes, René, 150
Descending order, 303
Determinants, 734
    and solving systems of equations, 735
Diagonals of a polygon, 330
Difference, 15
    of cubes, factoring, 345, 346, 367
    of functions, 744
    of logarithms, 644, 675
    of squares, 316
        factoring, 343, 367
Dimension symbols
    calculating with, 731
    unit changes, 732
Direct variation, 435, 438, 445
Discriminant, 553, 597
Disjoint sets, 123
Disjunction, 125
Distance
    on the number line, 133, 134, 143
        in the plane, 687, 719
            on a vertical or horizontal line, 686
Distance formula, 687, 719
Distance traveled, 87, 98
Distributive laws, 43, 65
Division
    of complex numbers, 509
    definition, 17
    with exponential notation, 56, 65
    of functions, 744
    by a monomial, 397
    of polynomials, 397–402
    quotient, 17
    with radical expressions, 471, 515
    of rational expressions, 381
    of real numbers, 17
    and reciprocals, 18
    remainder, 398
    with scientific notation, 60
    synthetic, 400
    by zero, 17
Divisor, inverting, 381
Domain
    of a function, 165, 179, 375
        and graphs, 179
    on a grapher, 182
    of a relation, 166
Doppler effect, 432
Dot mode on a grapher, 376
Doubling time, 663, 664
Draw feature on a grapher, 625

E

e, 649, 675
Earned run average, 434, 443
Element of a matrix, 739
Elimination method, 239, 243, 263, 264
Ellipse, 683, 695–699, 719
Empty set, 123
Endpoints of an interval, 108
Entries of a matrix, 739
Equation, 33, 71. *See also* Formulas.
    with absolute value, 134, 136
    addition principle, 72, 143
    any real number a solution, 78
    of asymptotes of a hyperbola, 703
    checking the solution, 73, 78
    of a circle, 688, 719
    of an ellipse, 695, 697, 719
    equivalent, 72
    exponential, 655
    false, 71
    fractional, 411
    graphs, *see* Graphs
    of a hyperbola, 703, 705, 719
    linear, 153
    logarithmic, 657
    multiplication principle, 74, 143
    with no solution, 77
    nonlinear, 155
    of a parabola, 683, 719
    containing parentheses, 78
    polynomial, 355
    quadratic, 355, 523
    quadratic in form, 555
    radical, 489
    rational, 411
    reducible to quadratic, 555
    related, 277
    slope–intercept, 188, 219
    solution set, 71
    solutions of, 71, 151
    solving, *see* Solving equations
    systems of, 229, 709. *See also* Systems of equations.
    true, 71
    writing from solutions, 554
Equilibrium point, 290
Equivalent equations, 72
Equivalent expressions, 39
    fractional, 39
    rational, 378
Equivalent inequalities, 109
Escape velocity, 434
Evaluating determinants, 734
Evaluating expressions, 35
Evaluating polynomial functions, 303, 305, 317
Even integers, consecutive, 98
Even root, 457, 458, 515
Exponent, 23. *See also* Exponential notation; Exponents.
Exponential decay model, 667, 675
Exponential equations, 655. *See also* Exponential functions.
    converting to logarithmic equations, 635
    solving, 655
Exponential functions, 606, 675. *See also* Exponential equations.
    graphs of, 606–610, 651
        asymptotes, 606
        y-intercept, 608
    inverse of, 634
Exponential growth model, 664, 675
Exponential notation, 23. *See also* Exponents.
    and division, 56, 65
    and multiplication, 55, 65
    and raising a power to a power, 57, 65
    and raising a product to a power, 57, 65
    and raising a quotient to a power, 58, 65
Exponential regression, 668
Exponents
    addition of, 55, 65
    on a grapher, 27
    integers as, 23, 24
    irrational, 605
    laws of, 464
    and logarithms, 634
    multiplying, 57, 65
    negative, 24, 65
    of one, 24, 65
    power rule, 57, 65
    product rule, 55, 65
    properties of, 65
    quotient rule, 56, 65
    rational, 463, 464, 515
    subtracting, 56, 65
    of zero, 24, 65
Expressions
    algebraic, 33

P

Pairs, ordered, 150
Parabola, 563, 683
  axis of symmetry, 563, 574, 597, 683
  equations of, 683, 719
  graphs of, 563–568, 571–574, 684, 685
  intercepts, 574
  line of symmetry, 563, 574, 597, 683
  vertex, 563, 574, 597, 684, 685
Parallel lines, 196, 204, 219
Parallelogram, area, 38
Parentheses, 25
  in equations, 78
  removing, 49
Pendulum, period, 476
Perfect square, 455
Perfect-square trinomial, 341. *See also* Trinomial square.
Perimeter, rectangle, 51
Period of a pendulum, 476
Perpendicular lines, 197, 205, 219
pH, 662
Pie chart, 149
Pixels, 157
Plane, 151
Plotting points, 150, 151
Points on a plane, 150, 151
Polynomial, 301, 302. *See also* Polynomial function.
  addition of, 306
  additive inverse, 306
  ascending order, 303
  binomial, 303
  coefficients, 302
  combining (or collecting) like terms, 306
  degree of, 302
  degree of a term, 302
  descending order, 303
  division of, 397–400
    synthetic, 400
  factoring
    common factors, 323
    completely, 324
    difference of cubes, 345, 346, 367
    difference of squares, 343, 367
    by FOIL method, 325, 335
    by grouping, 324, 333, 344
    strategy, 351
    sum of cubes, 345, 346, 367
    by trial-and-error, 325
    trinomial squares, 341, 342, 367
    trinomials, 325–328, 333–338, 341, 342, 367
  leading coefficient, 302
  leading term, 302
  monomial, 301
  multiplication, 313–316
  in one variable, 301
  opposite, 306
  prime, 323
  in several variables, 301
  subtraction, 307
  terms, 302
  trinomial, 303
Polynomial equation, 355
Polynomial function, 303. *See also* Polynomial.
  evaluating, 303, 305
Polynomial inequality, 591
Population decay, 667, 675
Population growth, 664, 675
Positive integers, 3
Power raised to a power, 57, 65
Powers
  on a grapher, 27, 464
  of i, 507
  logarithm of, 643, 675
  principle of, 489, 515
Power rule
  for exponents, 57, 65
  for logarithms, 643, 675
Powers, principle of, 489, 515
Prime polynomial, 323
Principal square root, 453, 455
Principle
  absolute-value, 134, 143
  addition
    for equations, 72, 143
    for inequalities, 109, 143
  multiplication
    for equations, 74, 143
    for inequalities, 111, 143
  of powers, 489, 515
  of square roots, 525, 597
  work, 420, 445
  of zero products, 355, 367
Problem solving, five-step strategy, 91
Product. *See also* Multiplication.
  of functions, 744
  logarithm of, 643, 675
  of polynomials, 313–316
  of rational expressions, 377, 380
  raising to a power, 57, 65
  of sums and differences, 316
Product rule
  for exponents, 55, 65
  for logarithms, 643, 675
  for radicals, 469, 515
Profit, 287, 310, 586, 596
Properties of absolute value, 133, 143
Properties of exponents, 65
Properties, identity
  of one, 40, 65
  of zero, 13, 65

Properties of logarithms, 643, 644, 646, 675
Property of −1, 48, 65
Proportion, 422
Proportional, 422
  directly, 435
  inversely, 436
Proportionality, constant of, 435, 436
Pythagorean theorem, 361, 499, 515
  in three dimensions, 520

Q

Quadrants, 151
Quadratic equations, 355. *See also* Quadratic functions.
  approximating solutions, 526, 539
  discriminant, 553, 597
  reducible to quadratic, 555
  solutions, nature of, 553
  solving, 538
    by completing the square, 528–530
    by factoring, 355, 367, 524
    by principle of square roots, 525
    by quadratic formula, 537
  standard form, 355, 523, 524
  writing from solutions, 554
Quadratic in form, equation, 555
Quadratic formula, 537, 597
Quadratic functions
  fitting to data, 579
  graphs of, 563–568, 571–574
  intercepts, 574
  maximum value of, 566
  minimum value of, 566
Quadratic inequalities, 589
Quadratic regression, 583
Quartic regression, 584
Quotient, 17, 398
  of functions, 744
  logarithm of, 644, 675
  raising to a power, 58, 65
  of radical expressions, 471, 515
  of rational expressions, 381
Quotient rule
  for exponents, 56, 65
  for logarithms, 644, 675
  for radicals, 471, 515

R

Radical equations, 489
Radical expressions, 454
  and absolute value, 455, 458, 515
  addition of, 477
  conjugates, 485

ometric Formulas

Plane Geometry

Rectangle
Area: $A = l \cdot w$
Perimeter: $P = 2 \cdot l + 2 \cdot w$

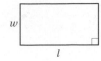

Square
Area: $A = s^2$
Perimeter: $P = 4 \cdot s$

Triangle
Area: $A = \frac{1}{2} \cdot b \cdot h$

Sum of Angle Measures
$A + B + C = 180°$

Right Triangle
Pythagorean Theorem:
$a^2 + b^2 = c^2$

Parallelogram
Area: $A = b \cdot h$

Trapezoid
Area: $A = \frac{1}{2} \cdot h \cdot (a + b)$

Circle
Area: $A = \pi \cdot r^2$
Circumference:
$C = \pi \cdot d = 2 \cdot \pi \cdot r$
$\left(\frac{22}{7} \text{ and } 3.14 \text{ are different approximations for } \pi\right)$

Solid Geometry

Rectangular Solid
Volume: $V = l \cdot w \cdot h$

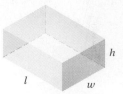

Cube
Volume: $V = s^3$

Right Circular Cylinder
Volume: $V = \pi \cdot r^2 \cdot h$
Surface Area:
$S = 2 \cdot \pi \cdot r \cdot h + 2 \cdot \pi \cdot r^2$

Right Circular Cone
Volume: $V = \frac{1}{3} \cdot \pi \cdot r^2 \cdot h$
Surface Area: $S = \pi \cdot r^2 + \pi \cdot r \cdot s$

Sphere
Volume: $V = \frac{4}{3} \cdot \pi \cdot r^3$
Surface Area: $S = 4 \cdot \pi \cdot r^2$